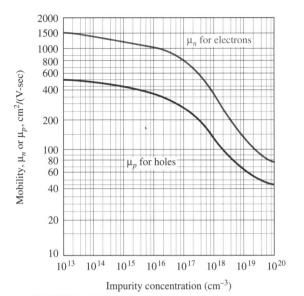
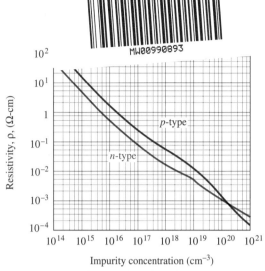

■ **FIGURE 2.18** Properties of doped silicon. (a) Electron and hole mobilities as a function of impurity concentration. (b) Resistivity of n- and p-type silicon[6] at 300 K as a function of impurity concentration.

TABLE 18.2 Key Typical BJT Parameters at 300°K (Room Temperature)

Parameter	npn	pnp (Substrate)	pnp (Lateral)
β_F	200	50	30
β_R	2	4	3
V_A (volts)	150	50	50
ϕ_J (volts)	0.7	0.55	0.55
I_S (amperes)	2×10^{-15}	10^{-14}	2×10^{-15}
r_b (ohms)	200	100	300

TABLE 11.1 Two-Port Feedback Summary

Voltage	Shunt
Sample output voltage	Sum currents in parallel at input
Parallel-input	Parallel-output
Shunt (input)	Shunt (output)
Input impedance decreases	Output impedance decreases

$A = V/I$ (**Transresistance Amplifier**) Use y **Parameters**

Current	Series
Sample output current	Sum voltages in series at input
Series-input	Series-output
Series (input)	Series (output)
Input impedance increases	Output impedance increases

$A = I/V$ (**Transconductance Amplifier**) Use z **Parameters**

Voltage	Series
Sample output voltage	Sum voltages in series at input
Series-input	Parallel-output
Series (input)	Shunt (output)
Input impedance increases	Output impedance decreases

$A = V/V$ (**Voltage Amplifier**) Use h **Parameters**

Current	Shunt
Sample output current	Sum currents in parallel at input
Parallel-input	Series-output
Shunt (input)	Series (output)
Input impedance decreases	Output impedance increases

$A = I/I$ (**Current Amplifier**) Use g **Parameters**

TABLE 3.1 Selected Diode Model Parameters

Name	Parameter	Default Value	Description
IS	Saturation current	10^{-14} A	I_S from Eq. (3.27)
N	Emission coefficient	1	η from Eq. (3.27)
RS	Parasitic resistance	0 Ω	Series lead and bulk resistance
CJO	Zero-bias pn capacitance	0 F	C_0 from Eq. (3.48) or (3.49)
VJ	pn junction potential	1 V	ϕ_J from Eq. (3.13) as in Eq. (3.48)
M	pn grading coefficient	0.5	0.5 = abrupt; 0.33 = graded
FC	Forward-bias-depletion capacitance coefficient	0.5	C_T saturates for $v_R < -\text{FC} \times \text{VJ}$
TT	Transit time	0 sec	τ of Eq. (3.53)
BV	Reverse breakdown voltage	∞ V	Magnitude of BV
IBV	Reverse breakdown current	10^{-10}	Reverse current at BV
EG	Band-gap potential	1.11 eV	0.67 = Schottky; 1.11 = silicon
XTI	I_S temperature coefficient	3	Temperature coefficient for n_i^2
KF	Flicker noise coefficient	0	
AF	Flicker noise exponent	1	

TABLE 4.1 Selected SPICE BJT Model Parameters

Name	Parameter	Default Value	Description
IS	Transport saturation current	10^{-16} A	I_S from Eq. (4.10)
NF	Forward I emission coefficient	1	η if needed in Eq. (4.10)
BF	Ideal maximum forward β_F	100	As in Eq. (4.17)
IKF	Corner for high-current β_F rolloff	∞ A	
VAF	Forward Early voltage	∞ V	V_A
BR	Ideal maximum reverse β_R	1	As in Eq. (4.39)
NR	Reverse I emission coefficient	1	
IKR	Corner for high-current β_R rolloff	∞ A	
RB	Base ohmic resistance	0 Ω	Base spreading resistance r_b
RC	Collector resistance	0 Ω	Collector lead and bulk resistance
RE	Emitter resistance	0 Ω	Emitter lead resistance
TF	Forward transit time	0 s	$\tau_F = 1/2\pi f_T$
TR	Reverse transit time	0 s	
CJE	Zero-bias B-E capacitance	0 F	C_{je0}, transition capacitance
VJE	B-E junction potential	0.75 V	ϕ_J for base-emitter junction
MJE	B-E junction grading coefficient	0.33	0.5 = abrupt; 0.33 = graded
FC	Forward-bias-depletion capacitance coefficient	0.5	C_{je} saturates for $v_{BE} > FC \times VJE$
CJC	Zero-bias B-C capacitance	0 F	$C_{\mu 0}$, as in Eq. (9.1)
VJC	B-C junction potential	0.75 V	ϕ_J for base-collector junction
MJC	B-C junction grading coefficient	0.33	0.5 = abrupt; 0.33 = graded
CJS	Zero-bias collector-substrate C	0 F	C_{CS0}, as in Eq. (9.2)
VJS	C-S junction potential	0.75 V	ϕ_J for collector-substrate junction
MJS	C-S junction grading coefficient	0	0.5 if needed
EG	Band-gap potential	1.11 eV	Silicon
XTI	I_S temperature coefficient	3	Temperature coefficient for I_S
KF	Flicker noise coefficient	0	
AF	Flicker noise exponent	1	

TABLE 5.1 Selected SPICE MOSFET Model Parameters

Name	Parameter	Default Value	Description
LEVEL	Model index	1	Shichman-Hodges
VTO	Zero-bias threshold voltage	0 V	V_t, as in Eq. (5.10)
KP	Transconductance parameter	2×10^{-5} A/V^2	$2k(L/W) = \mu C_{ox}$, as in Eq. (5.10)
LAMBDA	Channel-length modulation	0 V^{-1}	λ, as in Eq. (5.43)
GAMMA	Bulk threshold parameter	0 V$^{0.5}$	γ, as in Eq. (5.27)
RD	Drain ohmic resistance	0 Ω	r_d
RS	Source ohmic resistance	0 Ω	r_s
TOX	Oxide thickness	10^{-5} cm	d, as in Fig. 5.4
NSUB	Substrate doping	0 cm^{-3}	
UO	Surface mobility	600 cm^2/V-sec	

PRINCIPLES OF ELECTRONIC CIRCUITS

SECOND EDITION

Stanley G. Burns
Paul R. Bond
Iowa State University

PWS Publishing Company

I(T)P

An International Thomson Publishing Company

Boston • Albany • Bonn • Cincinnati • Detroit • London • Madrid
Melbourne • Mexico City • New York • Paris • San Francisco
Singapore • Tokyo • Toronto • Washington

PWS PUBLISHING COMPANY
20 Park Plaza, Boston, MA 02116-4324

Copyright © 1997 by PWS Publishing Company, a division of International Thomson Publishing Inc.
All rights reserved. No part of this book may be reproduced, stored in a retrieval system, or transcribed in any form or by any means—electronic, mechanical, photocopying, recording, or otherwise—without the prior written permission of PWS Publishing Company.

I(T)P®
International Thomson Publishing
The trademark ITP is used under license

For more information, contact:

PWS Publishing Co.
20 Park Plaza
Boston, MA 02116

International Thomson Publishing Europe
Berkshire House 168-173
High Holborn
London WC1V 7AA
England

Thomas Nelson Australia
102 Dodds Street
South Melbourne, 3205
Victoria, Australia

Nelson Canada
1120 Birchmount Road
Scarborough, Ontario
Canada M1K 5G4

International Thomson Editores
Campos Eliseos 385, Piso 7
Col. Polanco
11560 Mexico D.F., Mexico

International Thomson Publishing GmbH
Königswinterer Strasse 418
53227 Bonn, Germany

International Thomson Publishing Asia
221 Henderson Road
#05-10 Henderson Building
Singapore 0315

International Thomson Publishing Japan
Hirakawcho Kyowa Building, 31
2-2-1 Hirakawacho
Chiyoda-ku, Tokyo 102
Japan

Acquisitions Editor: Bill Barter
Assistant Editor: Angie Mlinko
Editorial Assistant: Monica Block
Production Editor: Pamela Rockwell
Editorial Coordination: Monotype Editorial Services
Art: Monotype Editorial Services
Proofreading: Monotype Editorial Services
Cover Design: Julia Gecha
Manufacturing Coordinator: Wendy Kilborn
Marketing Manager: Nathan Wilbur
Composition: Monotype Composition Company
Cover Printer: John Pow Company
Text Printer and Binder: Quebecor/Hawkins

Printed and bound in the United States of America.

96 97 98 99 20-10 9 8 7 6 5 4 3 2 1

CONTENTS

Preface	x
Preface for Students	xviii

PART I Semiconductor Devices and Basic Circuits — 1

CHAPTER 1 ELECTRONIC CIRCUIT FUNDAMENTALS — 3

1.1 Historical Perspective	4
1.2 Electronic Signals	8
1.3 Amplification	14
1.4 Ideal Operational Amplifier	16

Summary 20/Survey Questions 21/Problems 21

CHAPTER 2 INTRODUCTION TO SEMICONDUCTORS — 25

2.1 Charged Particles in a Solid: Drift and Mobility	26
2.2 Conductivity	28
2.3 Diffusion	31
2.4 Energy-Band Theory of Solids: An Overview	34
2.5 Semiconductor Materials	39
2.6 Properties of Intrinsic Silicon	40
2.7 Properties of Doped Silicon	44
2.8 Experimental Studies of Drift and Diffusion	49

Summary 52/Survey Questions 53/Problems 53/
References 58/Suggested Readings 58

CHAPTER 3 SEMICONDUCTOR DIODES AND DIODE CIRCUITS — 59

3.1 The *pn* Junction in Equilibrium	61
3.2 The Externally Biased Junction	67
3.3 The Diode Equation	71

3.4	Practical Diode Characteristics	73
3.5	Load Lines and Piecewise-Linear Diode Models	78
3.6	Dynamic Resistance	81
3.7	Rectifier Circuits	83
3.8	Breakdown-Diode Voltage Regulator	90
3.9	Diode Wave-Shaping Circuits	92
3.10	Diode Logic Circuits	96
3.11	Diode Analog Switch	97
3.12	Diode Capacitance and Switching Times	99
3.13	Metal-Semiconductor Junctions	103
3.14	Photonic and Microwave Diodes	104
3.15	Diode Heating	107
3.16	SPICE Model for the Diode	108

Summary 113/Survey Questions 114/Problems 114/References 122/Suggested Readings 122

CHAPTER 4 THE BIPOLAR JUNCTION TRANSISTOR 123

4.1	Basic BJT Operation	124
4.2	Volt-Ampere Equations for the BJT	129
4.3	BJT Regions of Operation	130
4.4	The Common-Base Configuration	133
4.5	The Common-Emitter Configuration	138
4.6	Cutoff in the CE Configuration	146
4.7	Saturation in the CE Configuration	148
4.8	Inverse Mode	154
4.9	CE Current Gain Considerations	156
4.10	Maximum Ratings for the BJT	157
4.11	Switching Times for the BJT	162
4.12	SPICE Model for the BJT	165

Summary 171/Survey Questions 171/Problems 179/References 180/Suggested Readings 180

CHAPTER 5 THE FIELD-EFFECT TRANSISTOR 181

5.1	Notation and Symbols	184
5.2	MOSFET Operation	185
5.3	MOSFET Specifications	194
5.4	MOS Circuits	197
5.5	SPICE Model for the MOSFET	205
5.6	JFET Operation	209
5.7	JFET Specifications	217
5.8	FET Small-Signal Models	221

5.9	JFET Circuits	224
5.10	SPICE Model for the JFET	228
5.11	MESFET Operation	231

Summary 233/Survey Questions 234/Problems 235/References 242/Suggested Readings 243

PART II Linear Circuits   245

CHAPTER 6   BASIC ANALYSIS TECHNIQUES   247

6.1	Operational Amplifier Specifications	248
6.2	General Feedback Concepts	257
6.3	Bode Magnitude and Phase Plots: Standard Forms	261

Summary 279/Survey Questions 280/Problems 281/References 293/Suggested Readings 293

CHAPTER 7   BIASING AND STABILITY   294

7.1	BJT Q-Point Selection	295
7.2	BJT Q-Point Stabilization	299
7.3	FET Bias and Stabilization	304
7.4	BJT Current Sources	310
7.5	MOSFET Current Sources	321

Summary 325/Survey Questions 326/Problems 326/References 335/Suggested Readings 335

CHAPTER 8   SMALL-SIGNAL MIDFREQUENCY AMPLIFIER ANALYSIS   336

8.1	The Hybrid-π Model	338
8.2	The Common-Emitter Amplifier: Impedances and Amplification	340
8.3	The Common-Base Amplifier: Impedances and Amplification	346
8.4	The Emitter-Follower Amplifier: Impedances and Amplification	348
8.5	The FET Model	351
8.6	MOSFET Amplifiers with Active Loads	358
8.7	Discrete Cascaded Amplifier Stages	367
8.8	Direct-Coupled and BiCMOS Amplifiers	370
8.9	Measurement of BJT Hybrid Parameters	373

Summary 377/Survey Questions 377/Problems 378/Reference 386/Suggested Readings 387

CHAPTER 9 FREQUENCY EFFECTS IN SMALL-SIGNAL AMPLIFIERS 388

9.1 High-Frequency Small-Signal BJT and FET Models 389
9.2 The Effect of C_μ and C_{gd}: The Miller Effect 399
9.3 Small-Signal Amplifier Low-Frequency Analysis 412
9.4 Direct-Coupled and BiCMOS Amplifier Examples 424
Summary 429/Survey Questions 430/Problems 430/References 439

CHAPTER 10 OPERATIONAL AMPLIFIER CIRCUITRY 440

10.1 Differential BJT Amplifiers with Resistive Loading 441
10.2 Differential MOS Amplifiers with Resistive Loading 453
10.3 Differential JFET Amplifiers with Resistive Loading 457
10.4 BJT Amplifiers with Active Loading 463
10.5 MOS Amplifiers with Active Loading 473
10.6 Power Output Stages 477
10.7 Effects of Device Mismatch 498
Summary 501/Survey Questions 502/Problems 503/References 515/Suggested Readings 515

CHAPTER 11 FEEDBACK 516

11.1 Overview of General Feedback Concepts 517
11.2 Voltage-Shunt Feedback 519
11.3 Current-Series Feedback 534
11.4 Voltage-Series Feedback 544
11.5 Current-Shunt Feedback 549
11.6 Amplifier Frequency Dependence and Compensation 556
11.7 Oscillators 563
Summary 574/Survey Questions 576/Problems 577/Suggested Readings 585

CHAPTER 12 OPERATIONAL AMPLIFIER EXAMPLES 587

12.1 Design of an Operational Amplifier using a CA3096 *npn/pnp* Transistor Array 589
12.2 Analysis of a μA741 Operational Amplifier 599
12.3 Analysis of a CA3140 Operational Amplifier 617
12.4 The LM111 Comparator 622
12.5 Trends in Operational Amplifier Performance 626
Summary 628/Survey Questions 629/Problems 629/References 635/Suggested Readings 636

CONTENTS

PART III Digital Circuits 637

CHAPTER 13 INTEGRATED-CIRCUIT LOGIC GATES 639

13.1 Digital Operation of Circuits 640
13.2 Basic Gate Terminology 641
13.3 Early Integrated-Circuit Logic Families 647
13.4 Transistor-Transistor Logic 653
13.5 Emitter-Coupled Logic 665
13.6 NMOS Logic 670
13.7 Complementary MOS Logic 678
13.8 Integrated-Injection Logic 688
13.9 Comparison of Logic Families 694
Summary 695/Survey Questions 696/Problems 696/References 700/Suggested Readings 701

CHAPTER 14 SMALL DIGITAL SUBSYSTEMS 702

14.1 Combinational Logic Circuits 704
14.2 Sequential Logic Circuits 715
14.3 Monostable and Astable Timing Circuits 728
14.4 Data Systems Examples 734
Summary 742/Survey Questions 743/Problems 743/References 747/Suggested Readings 747

CHAPTER 15 SEMICONDUCTOR MEMORIES 748

15.1 Overview of Semiconductor Memories 749
15.2 Introduction to Memory Organization 752
15.3 Read-Only Memories 757
15.4 Static Read/Write Random-Access Memories 764
15.5 Dynamic Random-Access Memory 769
15.6 Charge-Coupled Devices 775
15.7 Gate Arrays and Programmable Logic 777
Summary 782/Survey Questions 783/Problems 784/References 786/Suggested Reading 787

CHAPTER 16 ANALOG-TO-DIGITAL AND DIGITAL-TO-ANALOG CONVERSION 800

16.1 Analog-to-Digital Conversion Process 803
16.2 Digital-to-Analog Conversion 806
16.3 Analog-to-Digital Conversion 811
Summary 823/Survey Questions 824/Problems 825/References 826

CHAPTER 17 ADDITIONAL EXAMPLES OF ANALOG INTEGRATED CIRCUITS — 827

17.1	Analog Systems Overview	828
17.2	Series-Pass Voltage Regulator	830
17.3	Switching Regulators	837
17.4	Analog Multipliers	842
17.5	Phase-Locked Loop, System, and Circuit Description	853
17.6	Phase-Locked Loop Applications	860

Summary 866/Survey Questions 867/Problems 867/References 874

PART IV Semiconductor Technology — 875

CHAPTER 18 BASIC FABRICATION TECHNOLOGY AND DEVICE CONSTRAINTS — 877

18.1	Impurity Diffusion	880
18.2	Ion Implantation	886
18.3	Resistive Properties of Doped Layers	889
18.4	Photolithography and Masking	892
18.5	Resistors	897
18.6	Capacitors	904
18.7	*npn* Transistors	907
18.8	*pnp* Transistors	911
18.9	Diodes	914
18.10	Junction Field-Effect Transistors	914
18.11	Metal-Oxide Semiconductor Transistors	916

Summary 924/Survey Questions 924/Problems 924/References 931/Suggested Readings 932

APPENDIX A LIST OF SYMBOLS — 933

APPENDIX B PHYSICAL CONSTANTS AND CONVERSION FACTORS — 938

APPENDIX C NUMBER SYSTEMS AND BOOLEAN ALGEBRA — 940

C.1	Number Systems	941
C.2	Basic Functions of Binary Variables	942

APPENDIX D INTRODUCTION TO SPICE — 947

D.1	Basic Circuit Models for SPICE	948
D.2	Element Statements	949

D.3	Sources	**951**
D.4	Active Device Models for SPICE	**953**

References 958

APPENDIX E TWO-PORT MODEL SUMMARY 959

APPENDIX F ANSWERS TO SELECTED PROBLEMS 961

INDEX 967

PREFACE

Principles of Electronic Circuits, Second Edition, is an introductory text for upper-level sophomores and juniors in electrical or computer engineering. There have been significant changes in organization and content between the first and second editions to reflect recent technological and pedagogical trends and the need to provide the fundamental knowledge in electronic circuits for students practicing their discipline in the 21st century.

New to This Edition

The first edition of this book has been used at over 40 engineering schools. The manuscript leading to this second edition has been class-tested by hundreds of students and their professors. There are numerous changes in the second edition, representing an over 80% revision from the first edition, both in content and pedagogy.

▪ *Organization*   The second edition has been divided into four sections, consisting of a total of 18 chapters, to make it easier to tailor the text to a wide variety of course sequences prevalent in electrical engineering and computer engineering curricula.

▪ *Over 1000 Analysis and Design Problems*   Selected problem answers are given in Appendix F.

▪ *Integration of SPICE*   Of special note is the integration of SPICE throughout the text and the expanded emphasis on MOS transistor circuit analysis and design. The use of SPICE includes the development and justification of SPICE models as well as their incorporation into a number of annotated analysis and design examples. The SPICE solutions are compared to analytically derived solutions.

▪ *The CD and Other Ancillaries*   The second edition includes a package of electronic tools provided on the CD-ROM bound into the back of this book. The CD-ROM contains:

- The Evaluation Version of PSPICE® from MicroSim for Windows®-based computers.*
- Electronic copies of all of the SPICE netlists printed in this book, plus some additional ones that demonstrate important concepts.
- Sample Electronics Workbench circuit models, derived from selected Burns/Bond SPICE files, for use with *The Student Edition of Electronics Workbench®* (available from PWS).
- A slide show (Quicktime®/Video for Windows movies) and demonstration version of the Electronics Workbench software for Windows.

---

*Windows is a registered trademark of Microsoft, Inc., Electronics Workbench is a registered trademark of Interactive Image Technologies, Ltd., Acrobat is a registered trademark of Adobe Systems, Inc., Quicktime is a registered trademark of Apple Computer, Inc., and PSPICE is a trademark of MicroSim Corporation.

- Acrobat® transparencies of selected Burns/Bond text illustrations (provided with the Acrobat reader for Macintosh and Windows computers), which can be used by instructors to display text illustrations during lectures, or as a study aid by students working with computers at home.

In addition, the following ancillaries are available with this book:

- The *Instructor's Solutions Manual,* which has worked solutions to every problem in the book.
- The second edition *Lab Manual,* containing labs written by the authors covering a wide spectrum of key topics studied in the course.

▪ *BJT and MOS Circuit Design*   The student is guided in designing basic BJT and MOS circuits into a variety of digital and analog integrated circuits. There are detailed discussions of widely used diode circuits, BJT and MOS amplifier circuits, current sources and mirrors, multistage amplifiers including the operational amplifier, oscillators, logic families, MSI circuits, memories, A/D and D/A converters, voltage regulators, multipliers, phase detectors, and phase-locked loops.

▪ *CMOS and BiCMOS Circuits*   There is significant increased emphasis on CMOS and BiCMOS circuits.

▪ *Coverage of Semiconductor Fabrication*   Because electronic circuit performance and concomitant limitations ultimately depend upon the device technology employed, we have included a chapter on semiconductor fabrication that could be used in a course sequence or as a stand-alone reference.

▪ *Other Contemporary Topics*   Other topics of contemporary interest, such as the GaAS MESFET and microwave and photonic diodes, are included in this edition.

▪ *Pedagogy*   **Two Colors** have been used effectively to clarify diagrams, of which there are over 900, and guide students in their study of the topic. **Margin notes** are included throughout the text to provide clarification and expansion of key points. In addition to a wide assortment of problems, we have expanded our use of **examples with solutions** and added **Drill** exercises and **Checkup** questions in each section. **More challenging end-of-chapter problems are marked**. **Actual device data sheets** are used to support the technical material and chapter problems. To assist the student, each chapter starts with a list of **Important Concepts** and concludes with a list of **Survey Questions**. Students are also encouraged to read **Preface for Students**, which contains valuable hints on using our pedagogical tools.

The **appendices**, which include an updated list of symbols, physical constants and conversion factors useful throughout the text, number systems and Boolean algebra, an overview of SPICE netlist syntax, a two-port model summary, and answers to selected problems, are designed to support the student's study of electronic circuits.

Features

Contemporary electronic circuit analysis and design uses CAD tools, especially SPICE. SPICE-based solutions are integrated throughout the text and SPICE models are presented for the diode in Chapter 3, the BJT in Chapter 4, and the FET in Chapter 5. SPICE solutions and the approximate "calculator-based" algebraic solutions, which offer insights into electronic circuit behavior, are compared. Virtually all versions of SPICE programs use the same models and syntax, although there are differences especially in output format and user friendliness. The SPICE listings presented in this text are based on PSPICE® without schematic capture. The authors believe it is important for the student to be able to write a netlist for a circuit even though schematic capture is often mechanically easier to use. Appendix D provides a summary of SPICE syntax and formalisms used to write netlists.

In keeping with the general use of mixed CGS, MKS, and English units in the technical and trade literature, we will use these mixed unit conventions, explaining variations as necessary. Key conversions are provided in Appendix B.

A large selection of problems is included at the end of each chapter (over 1000 in total). There is a variety of both analysis and design problems, some using SPICE.

References and Selected Reading are provided at the end of each chapter. A list of symbols is given in Appendix A. Inclusion of selected manufacturer's data sheets throughout the text makes it easier for the teacher to assign design problems.

Overview and Organization

The emphasis and organization of the second edition are tailored to serve a variety of curricula. The text is organized into four sections. Part I, "Semiconductor Devices and Basic Circuits," consists of Chapters 1 through 5. Part II, "Linear Circuits," consists of Chapters 6 through 12. Part III, "Digital Circuits," consists of Chapters 13 through 17. Part IV includes a single chapter, Chapter 18, "Semiconductor Device Fabrication." Figures P.1 through P.4, which we will discuss shortly, illustrate the organization of these chapters into various course sequences as found in most schools.

The text assumes that sophomore engineering and physics students have an elementary calculus and physics background, as well as experience in the solution of ac and dc circuits by means of Kirchhoff's laws. Several of these key circuit concepts are reviewed in Chapter 1, "Electronic Circuit Fundamentals," which is included to serve as a review of basic circuit analysis techniques as well as an overview of the operational amplifier as a system building block. A cursory overview of semiconductor device physics is provided in Chapter 2 to give students insight into the validity of device models that are used throughout the text. Chapter 2 may be used as a reference resource in those curricula having a more formal solid-state physics course. It is assumed that the student has access to and a working knowledge of computers because SPICE is used throughout the text to support electronic circuit analysis and design.

Most curricula include a separate course in digital systems, where number systems, Boolean algebra, sequential logic theory, and logic reduction are studied. For this reason, and because this material is not essential to our purpose, we have elected not to include a chapter on these topics. Chapters 13 through 15 in Part III do summarize key terms as needed for those students with no previous exposure to these topics. Number systems and Boolean algebra are summarized in Appendix C.

The microprocessor has become the central feature in virtually all electronic systems. Indeed, virtually all electronic systems are "smart," that is they include microprocessors and associated logic and memory circuits. We have chosen not to treat microprocessor internal architectures or features in depth, preferring to consider these topics worthy of a separate course. However, we do discuss the microprocessor in system design and in interfacing to system buses in Chapters 15 and 16.

Possible Course Topic Sequences

■ *Two-semester Sequence*   The text can be adapted to several curricula, as shown in Figs. P.1 through P.4. Figure P.1 illustrates the use of this text in a two-semester, three-lecture credits per semester, sequence. Most schools will also support these courses with a hands-on laboratory equivalent to one extra semester credit every semester. The first semester begins with Chapter 1, "Electronic Circuit Fundamentals." The topics in

PREFACE

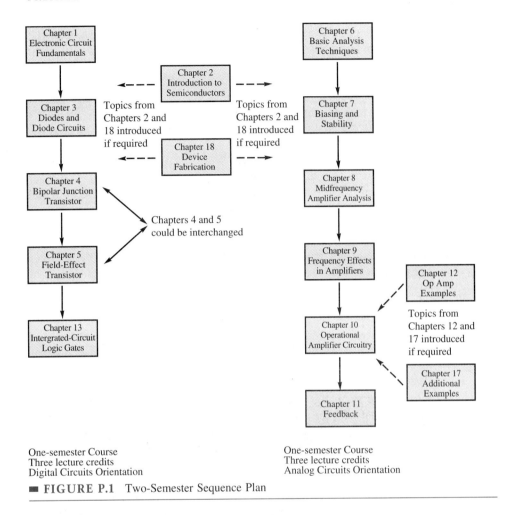

FIGURE P.1 Two-Semester Sequence Plan

Chapter 2, "Introduction to Semiconductors," may be optional depending upon the students' background and interest. Chapter 2 includes a description of drift and diffusion as well as an overview of the basic concepts and definitions important to semiconductor materials and devices. It is likely that most students will study these topics in more depth in physics courses or in an advanced semiconductor course; nevertheless all or portions of Chapter 2 may be assigned as appropriate.

The major emphasis of the course begins with Chapter 3, "Semiconductor Diodes and Diode Circuits." The theory of the *pn* junction, *pn* junction characteristics, and models, followed by a selection of representative diode circuit applications, are presented. Applications include power-supply rectifier circuits and various signal-conditioning and switching circuits likely to be encountered as subsystems in integrated circuits. A SPICE model for the diode is introduced and used in a number of examples. Bipolar junction transistor principles are presented in Chapter 4, "The Bipolar Junction Transistor." Models are developed for cutoff, active, and saturation regions, and representative circuit configurations are discussed. A SPICE model is introduced and applied to a number of examples. Temperature effects, internal junction capacitances, and switching speed are studied here to provide an understanding of the limitations of device structures. Similarly, Chapter 5, "The Field-Effect Transistor," takes up the operation, models, SPICE models and circuit configurations for MOS and junction field-effect transistors. One may delete JFET studies although understanding the operation of the JFET is useful for undertstanding the MESFET. There is an emphasis on describing the transfer characteristic for the CMOS gate. Some instructors may choose to interchange the order of Chapters 4 and 5, and this may be done with no loss of continuity and clarity.

Because this first-semester course of this two-semester sequence is designed to have a digital emphasis, we conclude this first semester with Chapter 13, "Integrated-Circuit Logic Gates," in Part III, "Digital Circuits." Chapter 13 considers and compares various digital integrated-circuit families. There is a focus on CMOS, although CMOS, TTL, and ECL are discussed in some detail to provide experience in active-circuit analysis and to familiarize the student with terminal characteristics and manufacturer's data sheets. The topics of fan-out and noise margin are discussed, as are 3-state and open-collector (drain) configurations. All technologies have performance constraints established by the device fabrication. Any or all of Chapter 18, "Semiconductor Device Fabrication" in Part IV, "Semiconductor Technology," may be used as a reference to provide insight into the design and manufacture of integrated circuits. Often, Chapter 18 topics are included in more advanced theoretical and laboratory courses in electronic materials and devices.

The second semester uses Part II, "Linear Circuits," which focuses on the analysis, design, and application of linear electronic circuits. We start with Chapter 6, "Basic Analysis Techniques." Topics include operational amplifier specifications, a simplified operational amplifier SPICE model, frequency-dependent characterization of networks using Bode plots, and general feedback definitions and concepts. Chapter 7, "Biasing and Stability," takes up the study of biasing and dc stability for both bipolar and field-effect transistor circuits. Current sources, with an emphasis on the simple, Widlar, and Wilson topologies, are introduced as the key biasing building blocks for integrated circuits. In Chapter 8, "Small-Signal Midfrequency Amplifier Analysis," the hybrid-π and other models are used to analyze bipolar, field-effect, and BiMOS frequency-independent transistor amplifiers. Biasing concepts from Chapter 7 are applied. SPICE simulations are used to illustrate many of the examples. Chapter 9, "Frequency Effects in Small-Signal Amplifiers," presents the design and analysis of BJT, MOS, and BiCMOS amplifiers at high and low frequencies. The Bode plot presented in Chapter 6 is used extensively. SPICE is used to support the algebraically complex calculations. Chapter 10, "Operational Amplifier Circuitry," which leads to the internal topology of an operational amplifier whose terminal characteristics were outlined in Chapter 1, presents the analysis and design for the emitter- and source-coupled pair (current sources from Chapter 7 operating as active loads) and classes A, B, and AB power output stages. Although feedback concepts are used in Chapter 1 and defined in Chapter 6, feedback theory and applications to circuits are more formally presented in Chapter 11, "Feedback." This includes negative feedback as applied not only to single-stage and multistage amplifiers but to internal operational amplifier circuits and operational amplifier-based systems. Stability analysis is presented using Bode plot techniques. An overview of oscillators illustrates the application of stability and positive feedback. The instructor may choose to limit the coverage in Chapter 11 and select some topics from Chapter 12, "Operational Amplifier Examples," which incorporates basic building block circuits from earlier chapters in the design of an operational amplifier based upon a CA3096 transistor array. These analyses are then applied to the μA741 BJT and the CA3140 BiMOS operational amplifiers. If time permits, other optional circuits and applications, including the voltage regulator, analog mutiplier, and phase-locked loop are presented in Chapter 17, "Additional Examples of Analog Integrated Circuits," can be included in the course sequence.

▪ *One-semester Course*   This text, as illustrated in Figure P.2, could be used for a four- or five-credit, one-semester electronics course. Again, it is assumed that an additional one-credit laboratory would be included as part of a one-semester comprehensive package. This course includes an in-depth treatment of the topics in Chapters 3 through 8. Chapter 1, as in

PREFACE

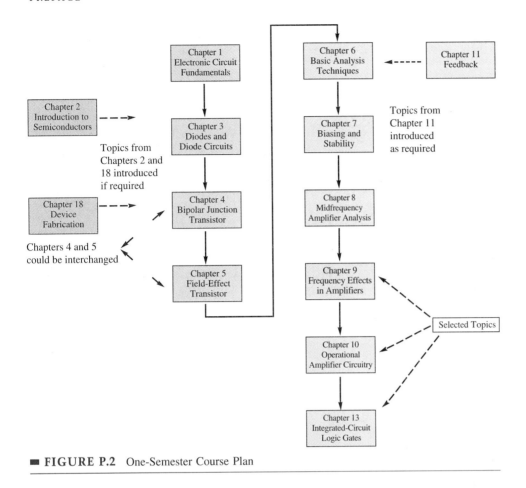

■ **FIGURE P.2** One-Semester Course Plan

the two-semester course sequence, provides a transition and review from earlier circuits courses. Chapters 4 and 5 could be interchanged as with the two-semester course sequence. Topics from Chapters 2 and 18 are selected, by the instructor. For example, one could delete JFETs from Chapter 5, Chapters 9, 10, and 13 could be covered almost completely; however, some of the low-frequency response topics in Chapter 9 and the JFET source-coupled pair from Chapter 10 could be omitted. Similarly, one might focus on the CMOS logic-gate in Chapter 13 and devote less time to other logic-gate families. The feedback discussion offered in Chapter 6 could be augmented as appropriate from the more detailed analyses presented in Chapter 11. The depth and breadth of coverage from Chapters 9, 10, 11, and 13 will depend upon whether four or five credits are allocated to the course.

■ *Three-quarter Course* Figure P.3 illustrates a three-credit-per-quarter, three-quarter packaging of the text topics. One additional credit per quarter for a laboratory is often included. Quarter one would include Chapters 2 through 5 with Chapter 1 providing the transition between basic circuits courses and this course sequence. Again, Chapters 4 and 5 could be interchanged. Quarter two would include Chapters 6 through 9 and 11 with some topic selectivity appropriate in Chapters 9 and 11. Quarter three could include Chapters 10, 12, 13, and 17. As discussed in the two-semester outline, Chapters 12 and 17 focus heavily on linear circuit applications and Chapter 13 has a digital circuit emphasis. For all three-quarter courses, topics from Chapters 2 and 18 are selected by the instructor.

■ *Courses with Digital Emphasis* We have observed that there are some curricula offering an electronic circuits course with a digital emphasis. Figure P.4 outlines such a

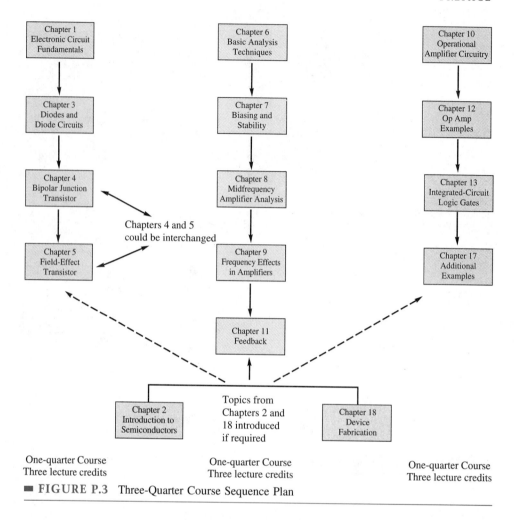

FIGURE P.3 Three-Quarter Course Sequence Plan

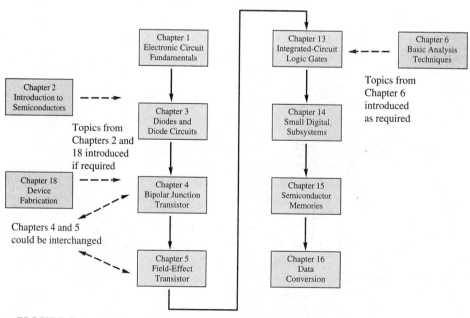

FIGURE P.4 One-Semester, Course with a Digital Emphasis

plan based upon a four- or five-lecture hour one-semester course. In essence, all of Part III, "Digital Circuits," is covered after studying the fundamentals. As with all of the previous course outlines, we propose starting out with Chapters 1 through 5. From there we continue to Chapter 13, "Integrated-Circuit Logic Gates." Chapter 14, "Small Digital Subsystems," presents selected medium-scale integration (MSI) circuit configurations, including both combinational and sequential logic circuits. Application examples include parallel and serial data transmission systems. Semiconductor memories are discussed in Chapter 15 in the context of microprocessor-based applications. Examples of read-only memory (ROM), static random-access memory (RAM), dynamic RAM (DRAM), and electrically erasable programmable read-only memory (EEPROM) structures illustrate current VLSI technology, with an emphasis on MOS. Charge-coupled devices (CCDs) for direct optical image processing are also discussed. It is useful at this point to include topics from Chapter 18. Analog-to-digital (A/D) and digital-to-analog (D/A) data conversion techniques are presented in Chapter 16. Comparator applications are also presented in this chapter.

We believe that all electronics courses should include a strong laboratory component that merges a student's analytical knowledge base, computer modeling and simulation tools, and hands-on laboratory design and measurement experiments.

Acknowledgments

We really appreciate the comments and suggestions we have received from innumerable students and faculty as they used the first edition and class-tested versions of the second edition manuscript. In particular, we want to thank Thomas M. Scott, Professor Emeritus, for his detailed reading of the manuscript, pedagogical feedback, and preparation of the Solutions Manual.

We also wish to thank the following reviewers for their comments in developing this text:

Professor Mary R. Anderson
Arizona State University

Professor Ronald L. Carter
University of Texas at Arlington

Professor Roy H. Cornely
New Jersey Institute of Technology

Professor W. T. Easter
North Carolina State University

Professor Mahmoud El Nokali
University of Pittsburgh

Professor B. J. Farbrother
Rose-Hulman Institute of Technology

Professor Robert D. Hatch
Lawrence Technological University

Professor Ted Higman
University of Minnesota

Professor Marian K. Kazimierczuk
Wright State University

Professor Mohan Krishnan
University of Detroit, Mercy

Professor Satish M. Mahajan
Tennessee Technological University

Professor David H. Navon
University of Massachussetts, Amherst

Dr. V. Rajaravivarma
North Carolina A&T State University

Of course, it goes without saying that our wives, Janice Burns and Donna Bond, were very supportive of our work and most eager for a successful and timely completion of this project.

Stanley G. Burns and Paul R. Bond

PREFACE FOR STUDENTS

The objective of this book is to introduce the electrical and computer engineering student to the broad range of electronic circuits—both discrete and integrated and both analog and digital. Our intent is to develop your ability to analyze and design a variety of electronic circuits systematically through examples ranging from the simple to the complex of classic and contemporary practice.

Our procedure will be to present just enough physical concepts to give an insight into the behavior of the semiconductor junction. We will then proceed to propose models for semiconductor devices such as diodes, bipolar junction transistors, and field-effect transistors. The major emphasis will be in studying the behavior of the device models in the context of practical circuits. Inherent in this approach is an awareness of the accuracy and range of validity of the models that are being used for the various devices.

The text is divided into four parts as follows: Part I, "Semiconductor Devices"; Part II, "Linear Circuits"; Part III, "Digital Circuits"; and Part IV, "Semiconductor Technology." After Part I, the parts may be read in any order.

You need to have taken an introductory linear circuit analysis course and mastered Kirchhoff's current and voltage laws and to have a mathematics background through simple integration. You should also be familiar with a computer circuit simulation program such as SPICE. For Part III, it is helpful if you have had some exposure to logic systems and the binary number system.

We hope to help you develop your skills as a problem solver. Educational researchers have learned that problem-solving skills are not necessarily learned simply by routinely solving more and more problems. The use of proven problem-solving strategies and techniques greatly accelerates the development of these skills, and later in this preface we discuss sound approaches to learning technical material and to problem solving.

You will find that we prefer to make simplifying assumptions and quick approximate calculations when that approach is justified. This develops a "feel" for what is going on in the circuit. You will also see when the more accurate computer simulation is better. Computer simulation of electronic circuits with SPICE is used as a powerful analysis tool in a number of examples. While programs that perform the simulation after "schematic capture" are readily available, we find that much more attention to detail takes place when students first practice developing a netlist. Our examples show typical netlists and model statements.

You will want to use some of the resources provided on the CD-ROM included with this text (see pp. x–xi for a description of the CD-ROM contents).

Structure Your Studying and Your Learning

You might think that your engineering studies are difficult. Well, they are! However, some of this difficulty may be due to improper or ineffective study methods. To develop good study habits, you may need to add structure and discipline to your studying. You may benefit from use of formal study techniques, which can greatly increase your learning efficiency. Techniques that you can try to improve your studying and learning that we address include reading, note-taking, problem solving, and assessment of your learning. However, in the end, you must decide what works for you. If you solve the problem of developing strong study methods, you have taken the first step on the way to becoming an engineer, a problem solver.

The methods we outline below are not all our own. In the following discussion we acknowledge the contributions made by various authors and educators as we adapt their work for this book. Additional insight to effective teaching and learning can be found in the book by Stice.[1]

Reading: The SQ3R Method

The process of reading a textbook for understanding is not the same as that of reading a novel for entertainment. One does not ordinarily skim over a novel to get the high points, jump to the end to find out how it ends, and then go back to fill in the gaps. This approach would defeat the objective of reading a novel. However, that is just what you should do when studying a textbook to maximize your potential for understanding and retention. A formal procedure that has proven to be very effective in improving textbook reading comprehension, *and grades*, is the SQ3R method developed by Francis Robinson.[2] The symbols S-Q-R-R-R stand for Survey, Question, Read, Recite, and Review. The five steps are discussed below.

- *Survey:* Skim over the assignment before you begin to read for details. At this point, read only the article headings, figure captions, and the chapter summary. Also look for the highlighted material and other points of emphasis that indicate key ideas. You should get a global perspective of the assignment so that you know your objectives before you start to read.
- *Question:* Break the assignment into manageable pieces by formulating questions that you will want to ask and answer about each article. Use the article heading, figure captions, etc., as guides to ask yourself questions about the material. Be as specific as possible. Jot down your questions in outline form and leave room for the answers that will follow. This step is intended to create a structure for storing (remembering) the detailed information that you will learn when you read. This book is printed with liberal margins to give you space to take notes and write questions. Use this space to your advantage.
- *Read:* Now, you have prepared yourself to absorb the new information. Next, read for details. Read a small amount of material at a time. If the material is difficult or confusing after the first reading, don't go on. Instead, go back and re-read the article, maybe just a paragraph or two at a time.
- *Recite:* After you have read a section, stop and recite what you have read. Try to answer your questions. Take notes on your outline. If your initial set of questions is

---

1. Stice, J. E. (ed), *Developing Critical Thinking and Problem-Solving Abilities,* New Directions for Teaching and Learning, No 30, San Francisco: Jossey-Bass, June 1987.
2. Robinson, F. P., *Effective Behavior,* New York: Harper & Row, 1941.

incomplete, ask (and answer) more questions. Don't go on to the next section before you have summarized the main ideas in your own words.
- Repeat the Question, Read, and Recite cycle for each section until you have completed the assignment.
- *Review:* After you have read the entire assignment, skim over the assignment again, review your notes, and ask and answer your questions again. Try to apply the material in ways that are not discussed in the text. If possible, work with a companion and quiz each other on the material.

In electronic circuits courses, it will usually take several class periods to cover the material on a particular subject or chapter, so the topic does not have to be completed in a single study session. In fact, it is probably best that you use several shorter study sessions to complete your reading of a single chapter or section. This will give you a chance to think about the material and review it more frequently.

We have organized this text to facilitate use of the SQ3R method. At the beginning of each chapter, introductory remarks provide an overview and a general perspective on the chapter. A list of important concepts is presented, and sample survey questions are posed to guide your reading; however, you are encouraged to perform your own survey and develop your own set of questions as well. The book is liberally illustrated with sketches, circuit diagrams, graphs, and tables that are tightly integrated with the text. Equations and figures are systematically numbered, so that once developed, they can be referenced easily from anywhere in the text.

CHECKUPs are inserted at the end of various sections to remind you to recite. This is an opportunity to answer your questions and a few short questions that we have prepared. This is also a chance to complete your outline. Also, *new terms* are italicized, and **definitions**, as well as **key concepts**, are flagged by notes in the margin. At the end of each chapter, highlights of the chapter are provided for your survey and review exercises.

Notetaking

Your task in taking notes in a lecture is far different from that of a court stenographer. The stenographer is obliged to record every word that is spoken. Your objective is to capture the main ideas from the lecture with sufficient accuracy and organization that you can use your notes later as a study aid. Some simple hints for effective notetaking are :

- *Come to class prepared.* Do the assigned reading before class and bring your SQ3R notes. Ask your questions when the instructor is discussing the topic that confuses you. Jot down his or her answer on your outline.
- *Imagine that you are taking notes for your best friend who could not make it to class.* Your objective is to give your friend an organized set of notes that record the key points of the lecture. Your lecture notes can be a supplement to your SQ3R notes.
- *Listen for key words* that signal main ideas *(The three primary reasons that . . .),* an alternative explanation *(In other words . . .),* a conclusion or summary statement *(Finally we can say that . . .),* or a change of direction *(On the other hand . . .).*
- *Do not assume that the instructor will write everything* that you are expected to know on the board. Read everything that is written and listen to everything that is said; but be selective in what you write down.
- *After class,* but before the next class, *rewrite your* SQ3R *notes and your lecture notes* into a single, organized unit. This is another opportunity to review the material and solidify your understanding.

Develop Problem-Solving Skills

Before we discuss how to become a better problem solver, perhaps we should decide what we mean by problem solving. The following definition was proposed by D. R. Woods,[3] and is appropriate for our use: "Problem solving is the process of obtaining a satisfactory solution to a novel problem, or at least a problem that the problem solver has not seen before." This definition means that problem solving is more than just working homework problems by following examples in the book. Real problem solving requires that we use our knowledge and understanding of facts and principles, our ability to apply those facts and principles, and our perceptiveness to comprehend the new situation. With all this, we can solve novel problems, problems for which we do not see an immediate solution.

An important element of problem solving is an organized strategy to follow. We outline the strategy developed by Polya[4] and modified by Woods[3]. Initially it is important to pay close attention to each of the steps. Then as you become more experienced and successful, you will be able to develop your own personal variation on the strategy. For now, try the following steps when you solve a problem:

- *Define:* Define the actual problem. Often, unnecessary, incorrect, or misleading information is given or collected that might distract you from the real problem.
- *Think About It:* Don't jump into your solution without "getting acquainted" with the problem. Identify the attributes of the problem. Summarize and organize what is given. Draw sketches. What area of knowledge is involved? Do you have enough information to solve the problem? Have you solved similar problems before?
- *Plan:* Flowchart a solution. Where do you start? What sequence of steps will lead you to your solution? Can the problem be broken down into manageable pieces that can be solved one at a time? Think about alternative plans in case you get stuck.
- *Carry Out Plan:* Follow your plan to solve the problem.
- *Look Back:* Did you solve the problem that was posed, or did you solve some other problem? Is your solution reasonable? Did you make any math errors? Check for correct signs, decimal point locations, transposed digits, etc. Did you present your solution, and your answer, in a fashion that is logical and readable?

Throughout the book, we include step-by-step procedures for problem solving in a large number of examples.

Assessment Techniques

Assessment is the process by which we find out what we are learning and how well we are learning it. When we think of *assessment,* we usually think of exams. However, there are several other assessment techniques that we can (and should) use continually to monitor our learning. We have adapted a selection of assessment techniques from Angelo[5] as components of homework problems. The techniques are outlined below. Keep in mind that using the assessment technique itself contributes to enhanced learning.

- *Documented Problem Solutions:* When you solve a problem, it is not sufficient to get the right answer! You must also understand how you solved the problem, and how

---

3. Woods, D. R. et al., "Teaching Problem-Solving Skills," *Engineering Education*, 66(3): 238–243, December 1975.

4. Polya, G., *How to Solve It*, 2nd Ed., Princeton: Princeton University Press, 1971.

5. Angelo, T. A. and Cross, K. P., *Classroom Assessment Techniques: A Handbook for College Teachers*, 2nd Ed., San Francisco: Jossey-Bass, 1993.

the approach can be adapted to other situations. Documented problem solutions can help you achieve this understanding. The technique involves adding to your solution a detailed written explanation of the process that you used to obtain your answer. It is important that you clearly explain each step that you take and why you make certain decisions. Of course, it is valuable to refer to example problems when you are solving your homework problems. However, it is not good simply to compare your problem statement to an example and then blindly follow the procedure used in the example.

- *Directed Paraphrasing:* One of the most difficult tasks that an engineer faces is communicating highly technical and specialized information in everyday language that can be understood by nontechnical people (clients, customers, friends, parents, etc.). Directed paraphrasing is a technique in which you use your own words to paraphrase a concept for a specific audience with a specific need to know. Directed paraphrasing will help you to develop your communications skills. In addition, the exercise of composing, in your own words, an explanation of a concept will help you gain a more complete understanding of the concept.

- *Student-Generated Test Questions:* The process of solving the homework problems at the end of each chapter is basically a passive response. That is, you answer questions that are posed but, generally, you do not expand your solution beyond what is required of you. Ideally, your studying should include more active involvement in developing both questions and answers. The process of generating your own test questions is a valuable technique that allows you to organize the many facts that you have gathered and synthesize them into something of your own creation. Then you can proceed to solve your exam questions as a drill on the topic material. It is also an opportunity for you to discover what aspects of the topic material you do not understand. As you develop your own test questions, you are actually beginning to prepare for the upcoming test. You must identify what concepts are most important, how knowledge of these concepts can be demonstrated through the test question, and what constitutes a complete solution to the question. Generating *good* test questions is a difficult task. If you take these exercises seriously, you are more likely to develop greater understanding and improve your performance on real exams.

Final Words

An old saying goes: "What I hear, I forget; what I see, I may remember; what I experience, I know for life." Students in engineering tend to be visual, sensory learners. For this reason, this book is heavily illustrated with circuit diagrams, transfer characteristics, timing waveforms, and computer-generated plots to help you visualize what is going on in electronic circuits. We encourage you to become personally involved with the subject so that you can *experience* the excitement of electronics. Apply what you learn to your own life and know it for life. We have done all we can to get you off on the right foot. The rest is up to you. Now start your study of electronic circuits with our best wishes for your success.

PART I

SEMICONDUCTOR DEVICES AND BASIC CIRCUITS

Part I, consisting of Chapters 1 through 5, provides you with a basic understanding of key semiconductor devices and their circuit applications.

Chapter 1 is a review of basic circuit theory and concepts. Chapter 2, introduces the concept of drift and diffusion as well as the basics of the physics of semiconductor devices. The material properties of resistivity, conductivity, mobility, and temperature dependence are introduced along with typical values for electronically interesting materials. Although many of you may cover these topics in more depth in physics courses or in an advanced semiconductor course, it is important that key concepts and terminology be introduced early to support the device models and specifications that are required for electronic circuit design and analysis. Although the focus is on the properties of doped and undoped silicon, there are references to III-V materials, metals, and insulators, especially SiO_2, which is a technologically important material. You may elect to use Chapter 2 as a reference, especially if, as mentioned, the Chapter 2 material is covered elsewhere in your curriculum.

In Chapter 3, *pn* junction theory, including *pn* junction characteristics and large- and small-signal diode models, is presented. This is followed by a selection of representative diode circuit applications, including power-supply rectifier circuits and various signal-conditioning and switching circuits you are likely to encounter as subsystems in integrated circuits. The SPICE model for the diode is presented along with examples of its use in designing power-supply circuits.

Bipolar junction transistor prinicples are presented in Chapter 4. Models are developed for cutoff, active, and saturation regions, and representative circuit configurations are discussed. Small-signal models are also introduced. As in Chapter 3, the SPICE model is introduced and applied to some example circuits. Temperature effects, capacitances, and switching speed are studied here to provide an understanding of the limitations of device structures.

Similarly, Chaper 5 takes up the operation, models, and circuit configurations for MOS and junction field-effect transistors. Specifically, the basic operation of the MOS transistor is explained along with key I-V characteristic equations. From these equations, we show the derived SPICE model and apply this model to some circuit examples. Although the focus is on MOS, we also explain the operation of the JFET and its high-frequency GaAs counterpart, the MESFET. The SPICE model for the JFET is introduced and applied to some examples.

After this exposure to basic device operation, constraints, large- and small-signal models, SPICE models, and circuits, you are in a position to apply your background to a study of linear circuits in Part II or digital circuits in Part III.

CHAPTER 1

ELECTRONIC CIRCUIT FUNDAMENTALS

1.1 Historical Perspective
1.2 Electronic Signals
1.3 Amplification
1.4 Ideal Operational Amplifier
  Summary
  Survey Questions
  Problems

IMPORTANT CONCEPTS IN THIS CHAPTER

- Fundamentals of circuit theory including basic definition of signals, difference between ac and dc, and series and parallel circuits
- Standard notation for representing a total, a dc, and an ac voltage or current
- Double-subscript notation
- Linear superposition
- Phasors and phasor notation
- Difference between a digital and an analog signal
- Definition of the two binary states: ONE–ZERO, HIGH–LOW, ON–OFF
- Ohmic and nonlinear I-V characteristics
- Thévenin and Norton source representation, equivalent circuits, and source transformations
- Definitions of power gain and voltage gain
- Decibels and decibel and phase calculations in cascaded linear networks
- Ideal operational amplifier: circuit diagram, terminal labels, characteristics, and model
- Inverting amplifier topology and calculations
- Summing point constraints and application to ideal operational amplifiers
- Noninverting amplifier topology and calculations
- Unity gain buffer
- Difference amplifier topology and calculations

1.1 HISTORICAL PERSPECTIVE

On June 30, 1948, Bell Telephone Laboratories announced the invention of the transistor (see Fig. 1.1). The phenomenal growth in semiconductor technology and application since that time has had an impact on every phase of our lives. The seeds culminating in the invention of the transistor were planted in the nineteenth century. Indeed, from a physics viewpoint, it was the Greeks, millennia ago, who "invented" the word electricity. In addition, they proposed, from a philosophical viewpont, that matter was composed of small particles, or atoms, meaning indivisible bits of matter.

In the late nineteenth and early twentieth centuries, scientists such as Rutherford, Mosely, and Thompson, using X-ray data of scattering experiments, developed the theories that atomic structures have both positive and negative particles as the primary constituent. Millikan, in his famous "oil-drop" experiment, was the first to deduce the discrete nature of electric charge. His experimental work and subsequent studies have shown that the charge, q, of an *electron*, the basic unit of electric charge, has a magnitude of $q \cong 1.6 \times 10^{-19}$ C. In 1900 Planck, in his quantum theory development, provided the foundation for the theory to describe the behavior of electrons in solids. Quantum theory suggested that photons were small wave packets of energy whose energy was inversely proportional to wavelength. Using this theory, Einstein, in 1905, was able to explain the photoelectric effect observed by Hertz in 1887 and by Becquerel. Schrödinger in 1926

■ $q = 1.6 \times 10^{-19}$ C is worth remembering.

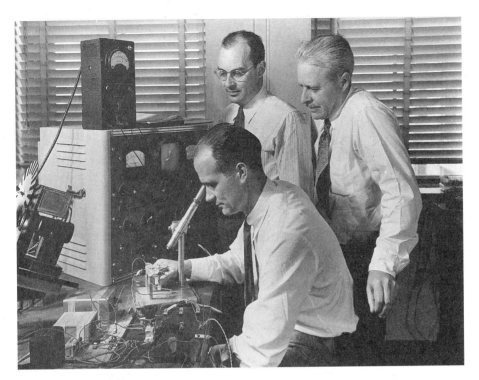

■ **FIGURE 1.1** (a) **Pioneers.** In 1947, physicists William Shockley (seated), John Bardeen (left), and Walter H. Brattain of Bell Telephone Laboratories developed the point-contact transistor—the first solid-state amplifier. The three received the Nobel Prize in physics in 1956. (Courtesy of AT&T Archives.) (b) **Solid amplifier.** The first transistor placed two closely spaced contacts (the emitter and collector) on the surface of a germanium slab (the base). The V-shaped object is a plastic triangle around which gold foil was wrapped and slit at the apex with a razor; it was held on the slab by spring pressure. (Courtesy of AT&T Archives.)

published his famous wave equation, which provided a unifying treatment to many of the solid-state phenomena we will study. The explosion of knowledge in the underlying physics continues to this day.

Even prior to this tremendous expansion of knowledge in physics, many interesting nineteenth-century experiments hinted at the potential of our present utilization of semiconductors. Laboratory experiments often led to and supported the development of scientific theory. Several scientists were working independently in the mid-1800s, studying rectifying and photovoltaic properties of what we now consider semiconducting compounds. In 1833 Faraday, a man whose name is linked with many fundamentals of electrical engineering, discovered that silver sulfide's conductivity increased with temperature as compared to the opposite effect in metals. A. C. Becquerel, in 1839, showed a photovoltaic output when illuminating a junction of an electrolyte and metal. Selenium, used in industrial rectifiers in the 1940s and 1950s and in early photovoltaic devices, was first studied by W. Smith in 1873. Smith observed a conductivity change when he illuminated a sample of selenium. Prior to 1905, a prevacuum-tube detector diode was fabricated from a pressure contact of wire on galena, or lead sulfide. This rectifying property was first observed in 1874 by a German physics professor, Ferdinand Braun. The so-called "cat's whisker" wire contact was a precursor to the metal-semiconductor junction device, the Schottky diode. Interest in semiconductor phenomena then waned until the 1920s and 1930s. (Schottky and others developed the theoretical background for the rectification mechanism in the late 1930s.) It is interesting to note that the pervasive advancements in vacuum-tube technology during the late 1920s and early 1930s led first to thoughts about solid-state amplifiers using field-effect principles that were more closely allied with vacuum-tube ideas as opposed to bipolar principles. Lilienfield (1925), Heil (1935), and Pohl (1938) all proposed various precursors to the field-effect transistor (FET). A limited materials technology and an incomplete understanding of device preparation and physics delayed implementation of the FET until well after the establishment of the bipolar transistor as a viable device.

Bell Telephone Laboratories, which was able to apply expertise in materials, solid-state physics, chemistry, and engineering, started working on the solid-state amplifier problem in the 1930s. The three key people in this effort were Walter H. Brattain, William Shockley, and John Bardeen. Not only did they work with copper oxide, but for the first time, they and others used adequately pure crystalline silicon in rectifying and photovoltaic experiments. By adding impurities, inadvertently at first, they noted changes in the electrical properties. Work on semiconductor devices proceeded rapidly, not only at Bell Labs but elsewhere. Spurred by wartime needs, the point-contact diode received new attention as a detector of microwave radar signals. On December 23, 1947, Brattain and Shockley first fabricated a point-contact germanium transistor. This first transistor used two closely spaced gold wires pressing on a small slab of germanium. A voltage gain of about 100 into the audio frequency range was demonstrated that day. An oscillator was demonstrated the very next day. First called a transfer resistor, its name was subsequently shortened to transistor. In 1956 Bardeen, Brattain, and Shockley received the Nobel Prize in physics; Bardeen and Brattain for their work on the point-contact transistor and Shockley for his subsequent contribution to the development of the junction transistor.

Development of transistor manufacturing capability proceeded behind the scenes for several years, and in the early 1950s a minimally performing point-contact or junction transistor cost more than $10. A key step toward improving repeatability during manufacturing came in 1959 when Fairchild Semiconductor Corporation developed the planar process for diffusing semiconducting layers. After that, developments happened quickly. Fairchild introduced the first monolithic integrated circuit (IC) for commercial use in 1961. It consisted of four bipolar transistors and two diffused resistors connected as a resistor-transistor logic (RTL) flip-flop. Texas Instruments was also contributing to basic IC development during this time frame (see Fig. 1.2). In 1963 RCA demonstrated a metal-oxide semiconductor (MOS) integrated circuit with 16 MOS transistors. By 1964 General Microelectronics was building circuits such as an MOS 20-bit shift register. By 1965 more than 25 companies were manufacturing digital ICs, although the major market share was held by Fairchild, which along with Texas Instruments and a few other companies had initiated manufacturing by early 1962. In the late 1960s Fairchild was the first to introduce commercially available analog ICs. The μA709 was the first operational amplifier available; it cost $75. Subsequently, the μA710, the first comparator, and the still-pervasive μA741, the first internally compensated operational amplifier, were introduced to the marketplace.

■ The μA741 is the subject of detailed analysis in Chapter 12.

Component densities increased rapidly from the late 1960s onward. This was particularly evident in the digital integrated circuit. Whereas a 64-bit MOS memory was an-

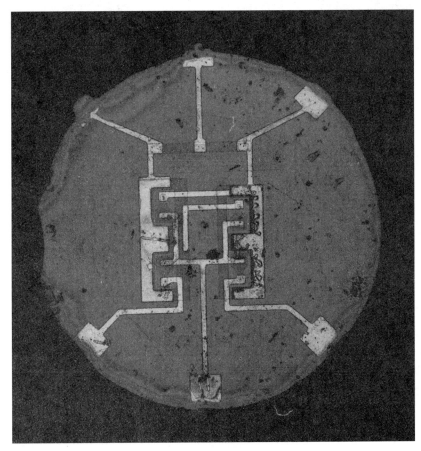

■ **FIGURE 1.2** **(a) Flip-flop.** The industry's first commercially available monolithic integrated circuit was this flip-flop from Fairchild Semiconductor, introduced in 1961. A member of the Micrologic family, the chip used resistor-transistor logic (RTL) and held four bipolar transistors and two resistors. (Reprinted by permission of National Semiconductor.) (*Figure continues on next page*)

■ **FIGURE 1.2** (*Continued*) **(b) First integrated circuit.**
The first integrated circuit, developed at Texas Instruments, proved an entire circuit could be built with silicon. The bars, actually sections from a wafer of mesa transistors, became breadboard in this oscillator using silicon as resistors, capacitors, and transistors. (Reprinted by permission of Texas Instruments.)

nounced by Fairchild in 1967, a 1024-bit (1-kilobit) memory was announced only one year later. By 1994, 16-megabit memories were available, and research into the production of 64-megabit and even 256-megabit memories was proceeding rapidly. Similarly, only a few years after 1970, the industry went from a modest 4-bit microprocessor integrated circuit to 16-, 32-, and even 64-bit microprocessors with considerable on-board ancillary circuitry and memory. To achieve these densities, production feature sizes are now on the order of 0.5 to 0.8 micrometers, and research is being conducted in the best approaches to fabricating ICs with sub-0.5-μm feature sizes.

With these few historical highlights in mind, let us start our study of electronics by examining some of the fundamentals of electronic signals and amplification.

■ Superposition, Thévenin's and Norton's equivalent circuits, and Kirchoff's laws are very important basic circuit theory concepts.

1.2 ELECTRONIC SIGNALS

Our study of electronic circuits will use extensions of concepts presented in basic circuit theory. We will use concepts such as ***linear superposition***, ***Thévenin's*** and ***Norton's*** *equivalent circuits*, *Kirchhoff's circuit laws*, and *ac signal analysis* to assist us in understanding, analyzing, modeling, and designing electronic circuits.

Notation

Throughout the text, we use the following conventions to distinguish between total values, quiescent values, and small signal or ac values. Subscripts also denote a terminal for current or a terminal pair for voltage drops. For example, in Fig. 1.3(a), we have $v_{AB} = V_{AB} + v_{ab}$ and $i_A = I_A + i_a$ where

v_{AB}, i_A = total instantaneous value: lowercase variable, uppercase subscripts

V_{AB}, I_A = dc value (alternatively V_{ABQ}, I_{AQ} for the quiescent value): uppercase variable, uppercase subscripts

v_{ab}, i_a = instantaneous value of the time-varying component (alternatively, $v_{ab}(t)$, $i_a(t)$: lowercase variable and lowercase subscripts

V_{ab}, I_a = magnitude of sinusoidal variation: uppercase variable, lowercase subscripts (see below).

If $v_{ab}(t)$ is a sinusoidal waveform [Fig. 1.3(c)] with an arbitrary phase reference, θ, then

$$v_{ab}(t) = V_{ab} \cos(\omega t + \theta) = \text{real part } [V_{ab} e^{j(\omega t + \theta)}]. \tag{1.1}$$

The quantity $v_{ab}(t)$ is also implied by the phasor notation $V_{ab} \angle \theta$, where the phasor magnitude variable is uppercase with lowercase subscripts.

■ Cosine reference for phasors is now almost universally used.

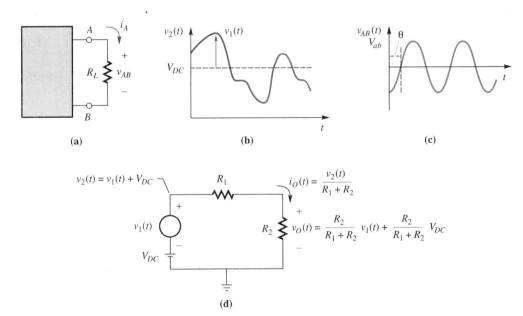

■ **FIGURE 1.3** Fundamental circuit concepts: (a) Double-subscript notation. (b) Arbitrary waveform with a dc and ac component. (c) Sinusoidal signal definition. (d) Linear superposition illustration.

DRILL 1.1

Identify the dc and ac components of the following signals.

SIGNALS
(a) $v_{CD} = 10 + 150\cos(377t)$ V
(b) $V_{cd} = 200\angle\theta$ V
(c) $V_A = 12$ V
(d) $v_{CE} = 10 + 0.01\cos(\omega t)$ V

ANSWERS
(a) dc = 10 volts, ac = 150-V_{peak} value; 60 Hz = 377 rps; phase, 0°
(b) dc = 0 volts, ac = 200 V_{peak} value; phase, θ
(c) dc = 12 V, ac = 0 V
(d) dc = 10 V from point C to E and ac = 10 mV at frequency ω; phase, 0°

Let us initially consider the concept of applying Kirchhoff's voltage law. Figure 1.3(d) shows an ac voltage source, $v_1(t)$, in series with a dc source, V_{DC}, denoted by a battery symbol. We can write that

$$v_2(t) = V_{DC} + v_1(t). \tag{1.2}$$

The resultant $v_2(t)$ is illustrated in Figure 1.3(b) for an arbitrary $v_1(t)$.

As we will observe in the following chapters, electronic devices require a dc source to establish device operation. The dc voltage and current level of a device operating in a circuit are often called the **quiescent point**, **Q-point**, **bias point**, or **operating point**. We then interpret $v_1(t)$ as the signal in which we are interested.

Figure 1.3(d) is used to illustrate linear superposition. Applying Ohm's law, we have

$$i_O(t) = \frac{v_2(t)}{R_1 + R_2}. \tag{1.3}$$

Observe that $v_O(t) = R_2 i_O(t)$, across R_2 is the sum of a contribution from $v_1(t)$ given by $\frac{R_2}{R_1 + R_2} v_1(t)$ and a contribution from V_{DC} given by $\frac{R_2}{R_1 + R_2} V_{DC}$.

Analog and Digital Signal Definition

■ We live in an analog world, but signal manipulation is often easier if the signals are digital.

We can further describe electronic signals by whether they are analog or digital (Fig. 1.4). The **analog** voltage, $v(t)$ in Fig. 1.4(a), changes continuously with time. The electronic signals in **digital** circuits assume only discrete levels. As shown in Fig. 1.4, the voltage takes on only two values. These *binary* states are variously referred to as TRUE–FALSE, HIGH–LOW, ON–OFF, etc. We shall refer to them as the 1 state and the 0 state.

The actual circuit voltages used to represent these two states may vary from system to system. When the voltage used to represent the 1 state is more positive than the voltage that represents the 0 state, the circuit is said to be *positive logic* or *asserted positive*. We will use positive logic throughout this text, in accord with many system designs. Negative logic is used for data signals in the EIA-232 interface convention. The Electronic Industries Association (EIA) sets many standards within the electronics industry as does the Institute of Electrical and Electronics Engineers (IEEE).

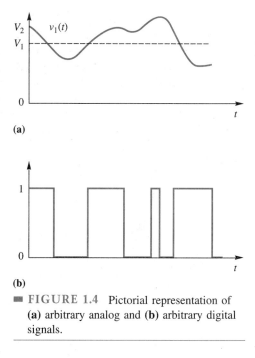

FIGURE 1.4 Pictorial representation of (a) arbitrary analog and (b) arbitrary digital signals.

To accommodate normal variations in component values, temperature, and noise, the voltage levels assigned to each binary state have a range, separated by a region of prohibited voltages. As an example, a certain logic family may define the 1 state as any voltage between 2 and 5 V, whereas the 0 state is defined as any voltage between 0 and 0.8 V. Voltages ranging between 0.8 and 2 V must not be used because the circuit may not be able to define the state unambiguously as a 1 or a 0. This 1.2-V difference between the highest voltage that will always be recognized as a 0 and the lowest voltage that will always be recognized as a 1 is called the *transition region* of the circuit. We will spend considerably more time on these concepts in Chapter 13 where they are defined in the context of digital integrated circuit families.

Small-Signal Discussion

Earlier, we used Ohm's law, $V = I \times R$, which describes a linear relation between voltage and current in a circuit. Refer to Fig. 1.5. Assume the signal generator $v(t)$ provides a linear ramp waveform to the unknown load. The current, $i(t)$, is measured across the

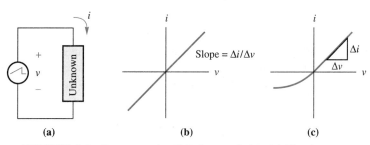

FIGURE 1.5 Representative I-V characteristics. (a) Circuit diagram. (b) Linear I-V, ideal resistor. (c) Nonlinear I-V.

unknown and displayed graphically as an *I-V characteristic*, Figs. 1.5(b) and (c). If the unknown load is a resistor, R_L, such that $v(t) = i(t)R_L$, the I-V characteristic is a straight line [Fig. 1.5(b)] with a slope $\Delta i/\Delta v = 1/R_L$, which does not change with the voltage level. This represents ideal linear behavior, that is, an ideal resistor.

Most electronic devices, however, as we will study in later chapters, have nonlinear I-V characteristics [Fig. 1.5(c)]. Because the slope of the I-V characteristic changes depending on the value of the voltage level, the resistance also changes as a function of the voltage level. We use this concept of resistance, defined as the slope of the I-V characteristic at a given voltage, in later chapters.

Often, this I-V characteristic is described by an exponential equation of the form

$$i(t) = I_X e^{[v(t)/V_{REF}]}, \tag{1.4}$$

where V_{REF} and I_x are constants. This exponential equation can be expanded using a Taylor series to yield

$$i(t) = I_x\{1 + v(t)/V_{REF} + (1/2)[v(t)/V_{REF}]^2 + (1/6)[v(t)/V_{REF}]^3 + \cdots\}. \tag{1.5}$$

■ <10% is an arbitrary level but useful for calculations.

Typically, if the $v(t)$ is sufficiently small such that the second-order and higher terms, $(1/2)[v(t)/V_{REF}]^2 + (1/6)[v(t)/V_{REF}]^3 + \cdots$, are less than 10% of the first-order term, $v(t)/V_{REF}$, $v(t)$ is considered to be a *small signal*. If $v(t) < 20\% \, V_{REF}$, then Eq. (1.5) can be approximated by

$$i(t) \cong I_x[1 + v(t)/V_{REF}] \tag{1.6}$$

DRILL 1.2

Which of the following would represent the I-V characteristics of an ideal resistor?

(a) $I = 10 \times V^2$
(b) $I = 12 \times V$
(c) $I = I_O \exp(V/V_R)$
(d) $I = V/3$

ANSWER Equations (b) and (d) yield linear, resistor transfer functions.

---

because $(1/2)[v(t)/V_{REF}]^2 + (1/6)[v(t)/V_{REF}]^3 + \cdots = 0.02 + 0.00133 + \cdots$ can be neglected compared to $v(t)/V_{REF} = 0.2$. Equation (1.6) describes a linear relationship between $i(t)$ and $v(t)$ added to a dc Q-point value of I_x.

Thévenin and Norton Equivalent Sources

One way to represent a signal source, as we implicitly assumed in Fig. 1.3 is to model it as a one-port equivalent circuit consisting of a variety of voltage and current generators and resistors [Fig. 1.6(a)]. This general one-port circuit can be represented by an equivalent voltage generator, v_T, in series with an equivalent output resistance R_T, as shown in Fig. 1.6(b). We establish the equivalency between Figs. 1.6(a) and (b) by realizing that in both cases the short-circuit current, i_{SC}, obtained when terminals A and B are connected is given by

$$i_{SC} = v_T/R_T. \tag{1.7}$$

Figure 1.6(b) is called the **Thévenin equivalent circuit** where the **Thévenin resistance**, R_T, is given as the ratio of the open-circuit voltage, v_T, and the short-circuit current, i_{SC}.

■ Source transformations are another very important concept from basic circuit theory.

It is often useful to represent the one-port in Fig. 1.6(a) as an equivalent current generator, i_N, in parallel with an equivalent output resistance, R_T, as shown in Fig. 1.6(c). A *source-transformation* between the Thévenin equivalent circuit in Fig. 1.6(b) and the circuit in Fig. 1.6(c) is given by

$$i_N = i_{SC} = v_T/R_T. \tag{1.8}$$

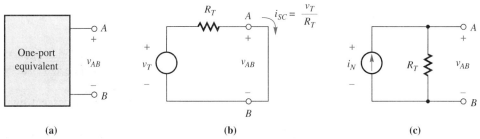

FIGURE 1.6 Thévenin and Norton equivalent sources.
(a) General one-port circuit. (b) Thévenin equivalent circuit.
(c) Norton equivalent circuit.

We also observe that the open circuit voltage from Fig. 1.3(b), v_T, is equivalent to v_{AB} in Fig. 1.6(c). The circuit in Fig. 1.6(c) is called the *Norton equivalent circuit*.

EXAMPLE 1.1

Find the Thévenin equivalent circuit for Fig. 1.7.

SOLUTION Refer to Fig. 1.6(a). The Thévenin open-circuit voltage, V_T, is obtained by using voltage division at the output, where

$$V_T = V_{DC} \frac{(R_3 \parallel R_4 + R_2)}{(R_3 \parallel R_4 + R_2) + R_1}.$$

The symbol $R_3 \parallel R_4$ means that R_3 is in parallel with R_4 and is computed as $(R_3 \times R_4)/(R_3 + R_4)$. The Thévenin resistance is obtained by realizing the V_{DC} independent source resistance is 0 Ω so that $R_T = (R_3 \parallel R_4 + R_2) \parallel R_1$.

EXAMPLE 1.2

Find the Norton equivalent circuit for Fig. 1.7.

SOLUTION Refer to Fig. 1.6(c). The Norton short-circuit current, I_N, is obtained by short-circuiting the output and computing $I_N = V_{DC}/R_1$. From Example 1.1, we have $R_T = (R_3 \parallel R_4 + R_2) \parallel R_1$.

DRILL 1.3

Find the Thévenin equivalent circuit parameters, V_T and R_T, for both the networks shown in Fig. 1.8.

ANSWER For part (a), $V_T = 10$ V, $R_T = 83$ Ω; for part (b) $V_T = -10.1$ V, $R_T = 84$ Ω.

DRILL 1.4

Find the Norton equivalent circuit parameters, i_N and R_T, for both networks shown in Fig. 1.8.

ANSWER For part (a) $I_N = 120$ mA, $R_T = 83$ Ω; for part (b) $I_N = -120$ mA, $R_T = 84$ Ω.

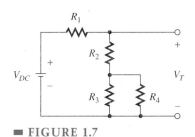

FIGURE 1.7

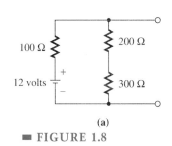 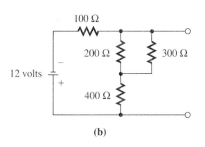

FIGURE 1.8

> **CHECK UP**
>
> 1. **TRUE OR FALSE?** Phasor notation can only be used for sinusoidal signals.
> 2. Write the correct notation for the total voltage of a series-connected battery and ac source.
> 3. Sketch the I-V characteristics for an ideal resistor and an incandescent light bulb.
> 4. **TRUE OR FALSE?** Double-subscript notation is used for both dc and ac signals.
> 5. **TRUE OR FALSE?** A circuit can be modeled either as a Thévenin or Norton equivalent circuit but not both.

1.3 AMPLIFICATION

One of the most important concepts and applications in electronic circuits is amplification. Virtually all electronic analog, digital, or mixed analog and digital systems require amplifiers for scaling signals to a useful level. We will spend a considerable portion of this book on the analysis and design of amplifiers for a large number of applications. Consider the diagram of the two-port network shown in Fig. 1.9. The output signal, V_2, is an enlarged, *amplified*, version of the input signal, V_1. Implicit to this diagram of an amplifier, but usually not shown, is the dc power supply. Essentially, the power from this dc power source is converted to the ac signal, V_2, across R_L. R_1 is the input resistance of the amplifier. Referring to Fig. 1.9, the power gain is defined as follows:

$$A_p = \left| \frac{V_2 I_2}{V_1 I_1} \right| = A_v A_i = \left| \frac{V_2}{V_1} \times \frac{I_2}{I_1} \right|. \tag{1.9}$$

■ There is nothing wrong with a ratio <1 although amplification usually implies a ratio >1.

The voltage gain is defined by $A_v = V_2/V_1$ and the current gain is defined by $A_i = I_2/I_1$.

The Decibel

■ Signal ratios can not only be very large, but also can be very small, depending upon the context and application.

Consider again the two-port network (amplifier) in Fig. 1.9 and Eq. (1.9). We assume that V_1, I_1, V_2, and I_2 in Fig. 1.9 are phasors. In an electronic circuit, the ratios computed using Eq. (1.9) can be large, so large in fact that it is more meaningful to express the ratio logarithmically as

$$A_p \text{ (dB)} = 10 \log_{10} \frac{P_2}{P_1}, \tag{1.10}$$

$$A_p = \frac{P_2}{P_1} = \frac{|V_2 I_2|}{|V_1 I_1|} = A_v A_i$$

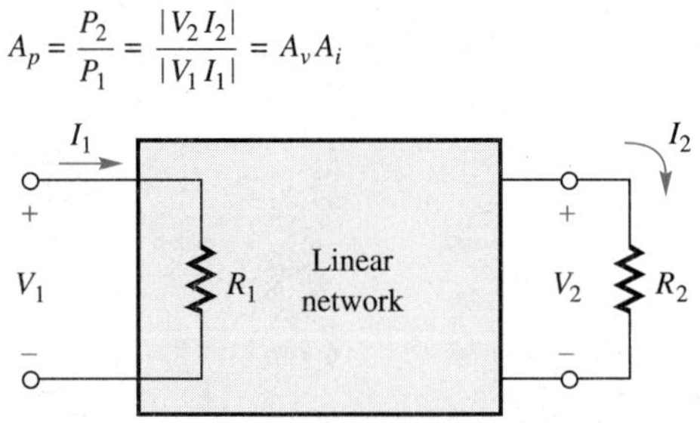

■ **FIGURE 1.9** Two-port network used to define power gain and voltage gain.

where A_p is expressed in decibels (dB). Throughout this text, $\log_{10}$ will be just written as log, never to be confused with $\log_e = \ln$ (natural log). The dB is also used directly for voltage and current ratios. For example,

$$A_p = 10 \log \frac{P_2}{P_1} = 10 \log \frac{V_2^2 R_1}{R_2 V_1^2} = 20 \log \left|\frac{V_2}{V_1}\right| + 10 \log \left|\frac{R_1}{R_2}\right|. \tag{1.11}$$

From Eq. (1.11), the voltage gain in decibels is given by

$$A_v \text{ (dB)} = 20 \log \left|\frac{V_2}{V_1}\right|. \tag{1.12}$$

It is useful to note from Eq. (1.11) that the power gain in decibels is equal to the voltage gain if the input and output resistances are identical.

We can illustrate the power of using the decibel with the aid of Fig. 1.10. The overall voltage gain of this *cascaded* series of two-port circuits, or amplifiers, if you wish, is given by

■ Examples of these cascaded circuits will appear in later chapters.

$$A_v = \frac{V_1}{V_0} \frac{V_2}{V_1} \frac{V_3}{V_2} \cdots \frac{V_n}{V_{n-1}} = A_1 A_2 A_3 \ldots A_n. \tag{1.13}$$

Each A_n consists of a magnitude with a phase angle, θ. Then, in general, Eq. (1.13) can be expanded to

$$A_v = |A_1 A_2 A_3 \ldots A_n| e^{j(\theta_1 + \theta_2 + \theta_3 + \cdots + \theta_n)}. \tag{1.14}$$

If the voltage gain from Eq. (1.14) is converted to decibels, the result is

$$\begin{aligned} 20 \log|A_v| &= 20 \log|A_1 A_2 A_3 \ldots A_n| \\ &= 20 \log|A_1| + 20 \log|A_2| + 20 \log|A_3| + \cdots + 20 \log|A_n|. \end{aligned} \tag{1.15}$$

From Eq. (1.15), we note that decibels are added when considering cascaded networks, a much more convenient technique for handling large numbers and small numbers. In a similar fashion the total phase shift through the network is the sum of all individual phase shifts:

$$\theta_v = \theta_1 + \theta_2 + \theta_3 + \cdots + \theta_n \tag{1.16}$$

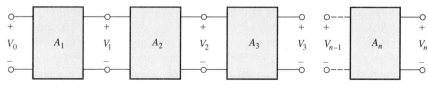

■ **FIGURE 1.10** Cascaded two-port circuits (amplifiers) used to define voltage gain.

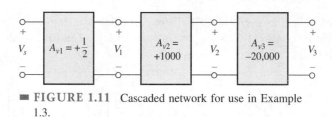

FIGURE 1.11 Cascaded network for use in Example 1.3.

DRILL 1.5

Compute the gain in decibels for each of the following:

(a) $A_v = -10,000$
(b) $A_v = -0.005$
(c) $A_v = 200 \angle \theta$
(d) $A_v = 5000 \times 60$

ANSWERS

(a) 80 dB
(b) −46 dB
(c) 46 dB
(d) 109.5 dB

DRILL 1.6

Compute the voltage gain, A_v, if the voltage gain in decibels is given by the following:

(a) 30 dB
(b) −70 dB
(c) 10 dB

ANSWERS

(a) 31.62
(b) 3.16×10^{-4}
(c) 3.16

EXAMPLE 1.3

Compute the overall voltage gain in decibels for the cascaded network shown in Fig. 1.11.

SOLUTION From Eq. (1.13), we can write

$$\frac{V_3}{V_s} = \frac{V_3}{V_2}\frac{V_2}{V_1}\frac{V_1}{V_s}$$

and individually

$$V_3/V_2 = -20,000, \quad V_2/V_1 = +1000, \quad V_1/V_s = +1/2.$$

Then $V_3/V_s = -10^7$, an unwieldy number to handle. Using the decibel, Eq. (1.15),

$$V_3/V_2 \text{ (dB)} = 86 \text{ dB}, \quad V_2/V_1 \text{ (dB)} = 60 \text{ dB}, \quad V_1/V_s \text{ (dB)} = -6 \text{ dB}$$

or

$$V_3/V_s \text{ (dB)} = 86 + 60 - 6 \text{ dB} = 140 \text{ dB}.$$

The phase is 180° as a result of the negative sign in $A_v = V_3/V_2$.

CHECK UP

1. **TRUE OR FALSE?** Gain in decibels adds arithmetically in a cascaded network.
2. **TRUE OR FALSE?** Phase adds arithmetically in a cascaded network.
3. **TRUE OR FALSE?** The voltage gain and power gain in decibels of a network are the same.
4. **TRUE OR FALSE?** A voltage gain of 1000 is computed to be 20 ln(1000) = 138 dB.
5. **TRUE OR FALSE?** A voltage gain of 1000 is computed to be 20 log(1000) = 60 dB.

1.4 IDEAL OPERATIONAL AMPLIFIER

The operational amplifier consists of a high-gain amplifier with high input impedance and low output impedance. Virtually all operational amplifiers are manufactured with monolithic IC technology. Selected technologies are presented in Chapter 18. Prior to the late 1960s, operational amplifiers designed with discrete components were not cost effective for use in many applications. Now IC-based operational amplifiers are used in virtually all electronic systems. We present a sampling of these applications here and throughout this text. Even digital systems require operational amplifiers for analog signal processing at input and output, display and peripheral interfaces, and for power-supply regulation and conditioning. A major objective of this section is to document the versatility of the operational amplifier. Additional material on the operational amplifier is presented in Chapter 11 when we discuss feedback principles and examples.

The circuit diagram for an operational amplifier connected to a load resistor R_L is shown in Fig. 1.12(a). There are three terminals explicitly shown on the operational amplifier symbol. These are:

1. The inverting input terminal designated by the minus sign $(-)$
2. The noninverting input terminal designated by the plus sign $(+)$
3. The output terminal.

The positive power supply, V^+, and the negative power supply, V^-, connections must be provided to make the operational amplifier work, but are usually not shown. The input voltage is V_i and is applied from the inverting to the noninverting terminals. The output voltage, V_o, is obtained across R_L and is constrained to lie between V^+ and V^-. Almost always, as will be presented in detail in Chapter 11, a circuit element is connected between the output terminal and the inverting input terminal.

We will develop a two-port circuit model [Fig. 1.12(b)] for the operational amplifier by modifying the circuits introduced in Figs. 1.6(b) and 1.9. The voltage source will be replaced by a voltage-controlled voltage generator with a voltage gain, a_v. The controlling voltage is the input voltage, V_i. The input resistance is given by r_i. The Thévenin resistance, R_T, is denoted as an output resistance, r_o.

A classic, often-studied operational amplifier is the μA741. The prefix often can be related to the major manufacturer. For example, μA741 refers to a Fairchild product, CA741 to an RCA (owned by Thompson CSF) product, LM741 to a National Semiconductor product, and so forth. Throughout this text the μA741 designation will be used although most manufacturers have essentially the same specifications for similarly labeled products.

The μA741 has the following key specifications: $a_v = 200{,}000$, $r_i = 2$ MΩ, and $r_o = 75$ Ω. Implicit in this model is that for $V_i = 0$, $V_o = 0$. This holds also for dc levels. The actual dc level present at the output for a 0-dc input is called the ***offset***. Any electronic circuit or system has a limited frequency range or bandwidth. In our initial studies, we will find it useful to remove this limited bandwidth specification. Therefore, we define the following specifications for the ***ideal operational amplifier:***

■ Real operational amplifiers have offsets as low as a few μV.

■ One of the most useful electronic circuit element concepts.

■ These ideal specifications are worth remembering.

1. Infinite a_v
2. Infinite r_i
3. Zero r_o
4. Infinite bandwidth
5. Zero offset.

Thus, the model of Fig. 1.12(b) reduces to that shown in Fig. 1.12(c). Of course, ideally,

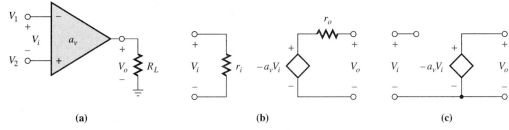

■ **FIGURE 1.12** Operational amplifier models. (a) Circuit symbol. (b) Low-frequency ac model. (c) Ideal model.

all of the preceding specifications are independent of temperature, power supply variations, and manufacturing variability. These ideal assumptions are examined in Chapters 11 and 12. Specifications and examples of other operational amplifiers are also presented in Chapter 12.

Inverting Amplifier

A widely-used electronic circuit.

A fundamental operational amplifier circuit is shown in Fig. 1.13(a). This is called an ***inverting amplifier***. In Chapter 11 we will call R_F a feedback resistor. The amplifier is driven from V_s through R_S. A sample of the output voltage is being summed as a current through R_F at the inverting node.

To compute the voltage gain, $A_v = V_o/V_s$, the ideal operational amplifier model, Fig. 1.12(c) is used to obtain the equivalent circuit—Fig. 1.13(b)—from Fig. 1.13(a). Summing currents are at the inverting node,

$$\frac{V_s - V_i}{R_S} = \frac{V_i - V_o}{R_F}. \tag{1.17}$$

For the voltage amplifier,

$$V_o = -a_v V_i. \tag{1.18}$$

Substitute this into Eq. (1.17) and solve for

$$A_{vs} = \frac{V_o}{V_s} = -\frac{1}{R_S}\left(\frac{1}{\dfrac{1}{a_v R_S} + \dfrac{1}{a_v R_F} + \dfrac{1}{R_F}}\right). \tag{1.19}$$

Eq. 1.20 is a very useful approximation.

If $a_v \to \infty$, then Eq. (1.19) reduces to

$$A_{vs} = \frac{-R_F}{R_S}. \tag{1.20}$$

Summing point constraints are a very practical use of negative feedback (Chapter 11) for operational amplifier circuits.

The analogous quantity for summing point constraints $V_i = 0$ and $I_i = 0$ in control systems is called a zero-level error signal.

Examining Eq. (1.17) and applying the ideal operational amplifier definitions suggest that Eq. (1.20) could have been obtained directly by assuming that $V_i = 0$ and $I_i = 0$. These are called the ***summing point constraints***. The summing point is the inverting node connection of the operational amplifier. We study this concept in more depth in Chapter 11.

DRILL 1.7

Use summing point constraints to compute A_{vs} for the inverting amplifier configuration where (a) $R_F = 100$ kΩ and $R_S = 22$ kΩ and (b) $R_F = R_S = 100$ kΩ.

ANSWERS

(a) $A_{vs} = -4.54$
(b) $A_{vs} = -1.0$

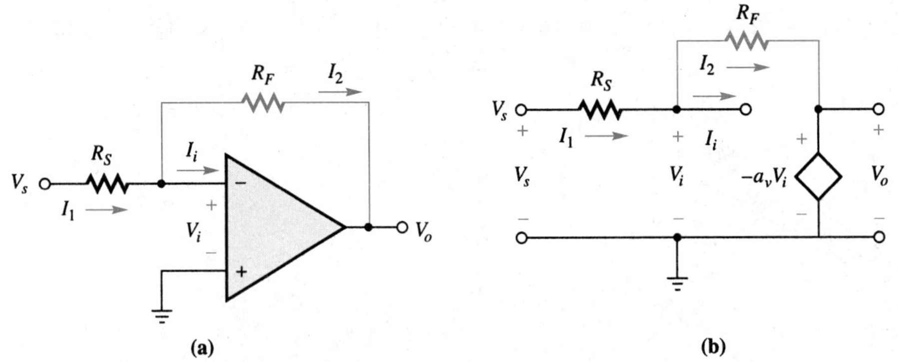

FIGURE 1.13 **(a)** Basic inverting operational amplifier using a simplified operational amplifier model. **(b)** Equivalent circuit.

EXAMPLE 1.4

Let $R_S = 10$ kΩ exactly and $R_F = 100$ kΩ exactly and use μA741 typical gain specifications to obtain A_{vs}. Compare the results by using summing point constraints directly.

SOLUTION Substituting the circuit element values into Eq. (1.19) yields

$$A_{vs} = \frac{V_o}{V_s} = \frac{-1}{10^4 \, \Omega \left[\dfrac{1}{(2 \times 10^5)(10^4 \, \Omega)} + \dfrac{1}{(2 \times 10^5)(10^5)} + \dfrac{1}{10^5} \right]} = -9.99945003.$$

Equation (1.20), derived from Eq. (1.17) by setting $V_i = 0$, yields $A_{vs} = -10$, which is obviously a very good approximation to $A_{vs} = -9.99945003$.

DRILL 1.8

Use summing point constraints, compute A_{vs} for the noninverting amplifier configuration where (a) $R_2 = 100$ kΩ and $R_1 = 22$ kΩ and (b) $R_2 = R_1 = 100$ kΩ.

ANSWERS
(a) $A_{vs} = +5.54$
(b) $A_{vs} = +2.0$

We now study several typical and widely used configurations using summing point constraints. Additional examples are explored in the problems at the end of the chapter.

Noninverting Amplifier

The basic **noninverting amplifier** circuit is shown in Fig. 1.14. Setting $V_i = 0$ and $I_i = 0$,

$$V^+ = V_s = \frac{R_1}{R_1 + R_2} \times V_o. \quad (1.21)$$

Rearrange Eq. (1.21) to solve for the closed-loop source voltage gain:

$$A_{vs} = \frac{V_o}{V_s} = \frac{R_1 + R_2}{R_1} = 1 + \frac{R_2}{R_1}. \quad (1.22)$$

A very useful version of the noninverting amplifier is called the **unity gain buffer**. If $R_1 \to \infty$ and $R_2 = 0$ in Fig. 1.14, the result is as illustrated in Fig. 1.15. By inspection, $V_s = V_o$ since $V_i = 0$. This type of circuit has the very useful characteristics of an extremely high input impedance and an effectively zero output impedance. It is used primarily for circuit isolation and as a means for driving low-impedance loads from a high-impedance source.

■ The unity-gain buffer is a circuit topology used in both analog and digital systems.

Difference Amplifier

The **difference amplifier**, sometimes called the differential amplifier (Fig. 1.16), is used to amplify the difference between two signals, such as those found as output from biomedi-

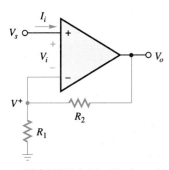

■ FIGURE 1.14 Noninverting amplifier.

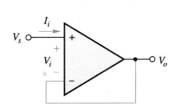

■ FIGURE 1.15 Unity gain buffer.

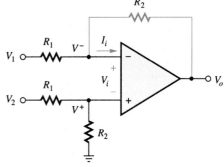

■ FIGURE 1.16 Difference amplifier.

cal and other types of sensors. Again, assume $I_i = 0$ and $V_i = 0$. By summing currents at the inverting-input node,

$$\frac{V_1 - V^-}{R_1} = \frac{V^- - V_o}{R_2}. \tag{1.23}$$

At the noninverting-input node,

$$V^+ = \frac{R_2}{R_1 + R_2} V_2. \tag{1.24}$$

Since $V_i = 0$, $V^- = V^+$ so that substituting Eq. (1.24) into Eq. (1.23) yields

$$V_o = \frac{-R_2}{R_1}(V_1 - V_2) = \frac{R_2}{R_1}(V_2 - V_1). \tag{1.25}$$

CHECK UP

1. List five key properties of an ideal operational amplifier.
2. List the two summing point constraints.
3. **TRUE OR FALSE?** Summing point constraint analysis does not work well using real operational amplifiers.

Use ideal operational amplifiers and summing point constraints to address the following:

4. Design a unity gain buffer.
5. Design an inverting amplifier with a voltage gain $= -10$.
6. Design a noninverting amplifier with a voltage gain $= +10$.

SUMMARY

In this chapter, we have introduced some of the concepts that will be required throughout the rest of this text. These include the basic circuit concepts of superposition, source transformations, and Kirchhoff's circuit laws. We recommend that if these concepts are unclear, you review them again before proceeding into later chapters, especially Chapter 3 and beyond.

The definition of an analog and digital signal was graphically presented. Notation conventions for dc, ac, and total voltage and current quantities as well as cosine reference phasors were given. Voltage and current ratios are often stated using the decibel and its use was illustrated for cascaded two-port networks. The ability to add phase shifts and gain in decibels arithmetically for cascaded two-ports will be used extensively in Part II.

Most importantly, from an applications viewpoint, the basic principles of the operational amplifier along with conventions and a model and the use of the summing point constraints $V_i = 0$ and $I_i = 0$ were introduced. Summing point constraints, as a consequence of employing negative feedback, were then used to derive the voltage gain for inverting and noninverting amplifiers. Much more will be done with the operational amplifier in Chapters 6 and 11.

SURVEY QUESTIONS

1. What are some of the commercial, industrial, and military products made possible by the key technological advances presented in Section 1.1?
2. What notation is used to represent a signal consisting of a dc signal only, an arbitrary ac signal only, a signal including both a dc and arbitary ac signal, and how does it demonstrate superposition and Kirchhoff's laws?
3. What would be the phasor notation for a 110-V_{rms} 60-Hz power line?
4. How do you compute the Thévenin and Norton equivalent circuits for a network?
5. How do you convert a voltage gain of 500 to decibels?
6. How do you compute the voltage gain in dB of two cascaded amplifiers, each with a voltage gain of 500?
7. What are five key characteristics of an ideal operational amplifier?
8. What are the two summing point constraints?
9. How would you use summing point constraints and ideal operational amplifiers to design inverting, noninverting, and difference amplifiers?

PROBLEMS

Problems marked with an asterisk are more challenging.

1.1 Write the phasor notation for the following signals:
   (a) $v(t) = 150 \cos(377t + 30°)$ V
   (b) $v(t) = 150 \sin(377t + 30°)$ V
   (c) $v(t) = 100 \cos(\omega t) + 200 \cos(\omega t + 45°)$ V

1.2 Write the sinusoidal time function for the following functions:
   (a) $V_a = 100 \angle 20°$ V
   (b) $v(t) = \text{Re}[150 e^{j(\omega t + 45°)}]$ V
   (c) $V_a = 200$ V

1.3 Sketch the following *I-V* characteristics:
   (a) $I = 10 \, V^2$
   (b) $I = 12 \, V$
   (c) $I = I_o e^{(V/V_R)}$, V_R a constant

1.4 * Use a linear approximation for $i = I_x/[1 - (v/V_{REF})]$. For small values of v/V_{REF}, sketch the resultant *I-V* characteristics. *Hint:* Rewrite as $i = I_x[1 - (v/V_{REF})]^{-1}$ and use the binomial expansion

$$(x + y)^n = x^n + nx^{n-1}y + \frac{n(n-1)}{2}x^{n-2}y^2 + \ldots$$

1.5 * Use a binomial expansion for $i = I_x/\{[1 - (v/V_{REF})]\}^{1/2}$. Look at the terms. Establish and justify a small-signal criterion for v/V_{REF}. Look at the hint in Problem 1.4.

1.6 Find the slope of $I = I_o e^{(V/V_R)}$ at $V = V_R$.

1.7 Derive the Thévenin equivalent circuit in terms of V_s, R_1, R_2, and R_x for the circuit shown in Fig. 1.17.

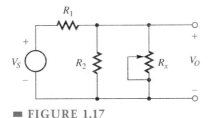

■ **FIGURE 1.17**

1.8 Derive the Norton equivalent circuit in terms of V_s, R_1, R_2, and R_x for the circuit shown in Fig. 1.17.

1.9 Derive the Thévenin equivalent circuit for the circuit shown in Fig. 1.18.

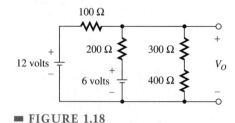

■ FIGURE 1.18

1.10 Derive the Norton equivalent circuit for the circuit shown in Fig. 1.18.

1.11 Remove the 12-V source and 100-Ω resistor from Fig. 1.18 and find the Thévenin equivalent circuit at V_O.

1.12 Remove the 12-V source and 100-Ω resistor from Fig. 1.18 and find the Norton equivalent circuit at V_O.

1.13 Replace the 6-V source in Fig. 1.18 with a 0-V source and find the Thévenin equivalent circuit at V_O.

1.14 A $1.0 \cos(\omega t)$ mV signal is input to an amplifier with a 35-dB gain and a 180° phase shift. Write an expression for the output voltage.

1.15 Repeat Problem 1.14 for two identical amplifiers in a cascade.

1.16 The "half-power point" reference is widely used in electrical engineering. What does half power correspond to in decibels?

1.17 Usually, a good FM receiver can respond to a 1-μV_{rms} signal. How much gain in decibels is required to amplify this signal to a level necessary to deliver 10 watts to an 8-Ω speaker?

1.18 Suppose your 200-W audio system yields an 80-dB signal as measured by the local authorities in response to a noise ordinance violation complaint. If the limit is 65 dB, at what power level should you run your system to avoid being cited?

1.19 Refer to Problem 1.18. Suppose you suggest, politely of course, that the measurement be taken further away. How much more distance, compared to the original distance, would be required to meet the 65-dB requirement?

1.20 Design an inverting amplifier for a voltage gain of 30 dB. Use standard resistor values and an ideal operational amplifier.

1.21 Use two ideal operational amplifiers to obtain an overall gain of $+0.5$.

1.22 Design a difference amplifier to obtain a voltage transfer function given by $V_o = 10(V_2 - V_1)$. Use standard resistor values.

1.23 Use summing point constraints to derive an expression for v_o as a function of v_1, v_2, and v_3 for the circuit shown in Fig. 1.19.

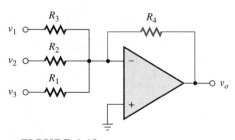

■ FIGURE 1.19

1.24 Use the results of Problem 1.23 to design a circuit to obtain a digitally weighed output voltage, V_o, i.e., the effect of v_1 at the output is twice the effect of v_2 and v_2 has twice the effect of v_3.

1.25* Use summing point constraints to derive an expression for V_o as a function of V_s for the circuit shown in Fig. 1.20. Sketch the transfer function. We will find in Chapter 3 that

$$i_D = \begin{cases} I_S[e^{(qv_D/kT)} - 1] \cong I_S e^{(qv_D/kT)}, & \text{for } v_D > 0 \\ 0 & \text{for } v_D < 0. \end{cases}$$

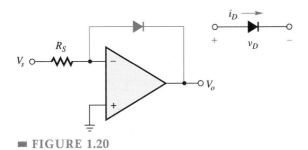

■ FIGURE 1.20

1.26* Use summing point constraints and the results from Problem 1.25 to derive an expression for V_o as a function of V_s for the circuit shown in Fig. 1.21. Sketch the transfer function.

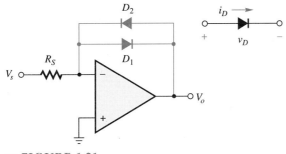

■ FIGURE 1.21

1.27 Use summing point constraints to derive an expression for $A_{vs} = V_o/V_s$ in Fig. 1.22.

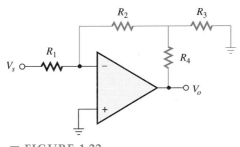

■ FIGURE 1.22

1.28 Show that the operational amplifier circuit given in Fig. 1.23 will function as an integrator, i.e., $v_o \propto \int v_s(t)\, dt$.

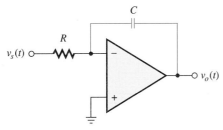

■ FIGURE 1.23

1.29* A transistor is included in the feedback loop of an operational amplifier (Fig. 1.24). Use summing point constraints to derive an expression for V_o/V_s. Sketch the transfer function of V_o versus $V_s > 0$ and $V_s < 0$. As will be shown in Chapter 4 for the transistor,

$$i_C = \begin{cases} I_S[e^{(qv_{BE}/kT)} - 1] \cong I_S e^{(qv_{BE}/kT)} & \text{for } v_{BE} > 0 \\ 0 & \text{for } v_{BE} < 0. \end{cases}$$

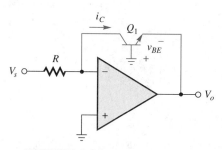

■ FIGURE 1.24

1.30 The multiplier whose transfer function is given by $V_y = KV_x V_o$ is incorporated into the feedback loop of the operational amplifier shown in Fig. 1.25. Derive the transfer function V_o/V_s, and explain why this circuit can be called an analog divider.

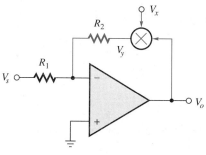

■ FIGURE 1.25

1.31* Using Problem 1.30 as a hint, design a circuit block diagram with ideal operational amplifiers and multipliers to generate the square root of the signal, $v_1(t)$; i.e.,

$$v_o(t) = \left[-\frac{R_2 v_1(t)}{R_1 K} \right]^{1/2}.$$

What restrictions are required on $v_1(t)$?

1.32* Using Problems 1.29 and 1.30 as a hint, design a circuit block diagram with ideal operational amplifiers and multipliers to generate the cube root of a signal, $v_1(t)$; i.e.,

$$v_o(t) \propto [v_1(t)]^{1/3}.$$

CHAPTER 2

INTRODUCTION TO SEMICONDUCTORS

2.1 Charged Particles in a Solid: Drift and Mobility
2.2 Conductivity
2.3 Diffusion
2.4 Energy-Band Theory of Solids: An Overview
2.5 Semiconductor Materials
2.6 Properties of Intrinsic Silicon
2.7 Properties of Doped Silicon
2.8 Experimental Studies of Drift and Diffusion
Summary
Survey Questions
Problems
References
Suggested Readings

Our study of the principles of electronics will include selected topics in device physics, semiconductor materials, applications of semiconductor devices in circuits, and circuit design. This chapter treats some elements of solid-state physics that are helpful in understanding the fabrication and functioning of semiconductor devices. We will see, for example, what is meant when silicon wafers are specified as *n*- or *p*-type with a certain resistivity in Ω-cm and a certain dopant. We will see what is meant by drift mobility, bandgap, and lifetime, which are of interest in, for example, research on new compound semiconductors.

It will become readily apparent that the field of electronics is a keystone in modern electrical engineering. In this introductory text, we intend to provide you with a basic foundation in semiconductor device operation and circuit design. These topics are supported by a brief qualitative treatment of solid-state physics necessary for an understanding of the functioning of devices and circuits.

IMPORTANT CONCEPTS IN THIS CHAPTER

- Definitions of current, drift velocity, and mobility, μ
- Conductivity, $\sigma = nq\mu$, and typical values
- Resistivity, $\rho = 1/\sigma$, and typical values
- Typical values of resistivity and conductivity for good conductors, insulators, and semiconductors
- Diffusion and diffusion current
- Definition of $V_T = kT/q$ and its value at 300 K $\cong$ 26 mV
- Fick's second law, diffusion equation, derivation, and solution interpretation
- Energy quantization given by $E = hf = hc/\lambda$ supported by calculations
- Qualitative descriptions of the energy band structure for conductors, insulators, and semiconductors
- Definition and interpretation of conduction band and valence band energy levels, and of the energy gap in a semiconductor
- Qualitative description of a hole
- Computation of the intrinsic electron concentration in a semiconductor
- Periodic Table entries as related to elements used in semiconductor devices
- Properties of intrinsic silicon-structure, energy-band diagram, carrier concentration, conductivity (resistivity), mobility
- Definition of a dopant and the relationship between doping concentration and carrier concentration
- Minority carrier suppression
- Energy-band diagrams for n-doped and p-doped semiconductors
- Mobility and resistivity properties of doped silicon
- Typical n-type and p-type dopants in silicon
- Computation of resistivity and conductivity as a function of doping density
- Minority carrier lifetime and diffusion length and their experimental verification
- Method and results in the Haynes-Shockley experiment

2.1 CHARGED PARTICLES IN A SOLID: DRIFT AND MOBILITY

Two mechanisms are used to describe the transport of charged particles in a solid. The first mechanism is the ***drift*** associated with a charged particle in an electric field; the second is the ***diffusion*** of charged particles that results from their density variations within a solid. Both processes are three dimensional in nature. However, to avoid unneeded algebraic complexity, we will pursue only a one-dimensional analysis, which will satisfactorily illustrate the major transport phenomena.

■ Diffusion is an important concept, as discussed in Chapter 18.

An outer-shell electron in an atom of a conductor is relatively free to move. The electron has an average velocity of about 10^7 cm/sec at room temperature, 300 K. This random motion is strictly due to a thermal effect. As illustrated in Fig. 2.1(a), with no electric field, the resultant path is random, the direction continuously being changed by "collisions" with lattice ***phonons***, atomic lattice vibrations, ***photons***, and other electrons. After each collision, the new velocity is nearly independent of the velocity before the

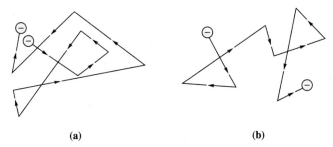

■ **FIGURE 2.1** Electron motion in a solid. **(a)** Electron random motion with no external electric field. **(b)** Electron random motion with a superimposed external electric field.

collision. The path lengths vary, as do the times between collisions, but the net displacement over a period of time is zero. Therefore, by definition,

$$\text{Average current} = \frac{\Delta Q}{\Delta t} = 0, \tag{2.1}$$

where Q is the net charge transfer.

When an electric field, $\overline{\mathbf{E}}$, a vector quantity, is applied as illustrated in Fig. 2.1(b), there will be a net displacement of the charged particle, yielding a net current. This electron drift displacement per unit time is defined by

$$\overline{v}_d = \mu_n \overline{\mathbf{E}}. \tag{2.2}$$

Here $\overline{v}_d$ is the electron **drift velocity**, and μ_n is the **electron mobility**. Drift velocity is typically small compared to thermal velocity. Therefore, the application of an electric field changes the velocities only slightly. The change in the velocity can never be substantial because the electron will suffer a collision with the lattice and the velocity must effectively start from zero. We can estimate this incremental velocity change to obtain an approximation for μ_n. The electron, with mass m_e, will undergo an acceleration between collisions given by

■ Mobility is a very important parameter for semiconductor materials.

$$\frac{\overline{F}}{m_e} = \overline{a} = -\frac{q\overline{\mathbf{E}}}{m_e} = \frac{d\overline{v}}{dt}, \tag{2.3}$$

where $q = 1.6 \times 10^{-19}$ C, the electronic charge. Assuming that the average time between collisions is given by t_c, the maximum velocity reached just before a collision is given by

$$v_{\max} = \frac{t_c q \overline{\mathbf{E}}}{m_e}. \tag{2.4}$$

The quantity t_c is called the **relaxation time**. Because the electron accelerates to $v_{\max}$ linearly in time t_c from zero velocity just after a collision, the average increment in velocity would be

$$\overline{v}_d = \frac{t_c q \overline{\mathbf{E}}}{2m_e}. \tag{2.5}$$

Comparing this to Eq. (2.2) yields an expression for electron mobility:

$$\mu_n = \frac{t_c q}{2m_e}. \tag{2.6}$$

This mobility is a complicated function of the physical structure of the material as well as temperature. For example, the mobility μ_n for copper is 0.0032 m²/(Vsec) and for aluminum 0.0012 m²/(Vsec) at T = 300 K = 27°C. In Section 2.8, we present empirical techniques for the measurement of mobility.

EXAMPLE 2.1

Estimate the relaxation time and drift velocity in a copper conductor supporting an electric field of 1 V/m.

SOLUTION   From Eq. (2.6), we have

$$t_c = \frac{2\mu_n m_e}{q}$$

$$= \frac{(2)[0.0032 \text{ m}^2/(\text{V-sec})](9.1 \times 10^{-31} \text{ kg})}{1.6 \times 10^{-19} \text{ C}}$$

$$= 3.64 \times 10^{-14} \text{ sec}$$

From Eq. (2.5),

$$v_d = \frac{t_c q \mathbf{E}}{2m_e} = \mu_n \mathbf{E} = [0.0032 \text{ m}^2/(\text{V-sec})](1 \text{ V/m})$$

$$= 0.0032 \text{ m/sec} = 0.32 \text{ cm/sec}.$$

Note that this is much less than the thermal velocity of $\approx 10^7$ cm/sec.

Also observe that we will usually be talking about one-dimensional motion so that whereas Eq. (2.5) uses general vector notation, we will suppress this notation when the direction property of a quantity is clear.

CHECK UP

1. What is the relationship between mobility, drift velocity, and the applied electric field?
2. **TRUE OR FALSE?** Current is equal to the amount of charge crossing a given plane in a conductor.
3. **TRUE OR FALSE?** Drift and diffusion are two different mechanisms.

2.2 CONDUCTIVITY

■ Conductivity is a very important, widely-used semiconductor material parameter.

The definition of **conductivity**, σ, which is a bulk property of a material, requires the combination of the mobility with a population density of charge carriers. Assume a homogeneous material as illustrated in Fig. 2.2. Let

$$n = \frac{N}{\Delta L \times A}. \tag{2.7}$$

where n is the charged particle concentration usually given in cm^{-3} and N is the total number of charge carriers within the volume given by $\Delta L \times A$. Then the current is given by

$$I = \frac{\Delta Q}{\Delta t} = \frac{Nq}{\Delta t}. \tag{2.8}$$

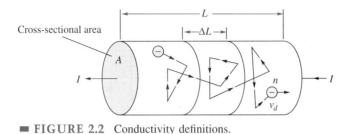

■ **FIGURE 2.2** Conductivity definitions.

The N charge carriers are traveling with the average velocity

$$v_d = \frac{\Delta L}{\Delta t}. \qquad (2.9)$$

Combining Eqs. (2.8) and (2.9) yields

$$I = \frac{Nqv_d}{\Delta L}. \qquad (2.10)$$

Assuming one-dimensional motion, the **current density**, J, is given by

$$J = \frac{I}{A} = \frac{Nqv_d}{\Delta LA} = nqv_d \qquad (2.11)$$

■ Current density (not "current") is used to quantify charge transport in a semiconductor.

Substituting Eq. (2.2) into Eq. (2.11) yields

$$J = nqv_d = nq\mu_n\overline{\mathbf{E}}. \qquad (2.12)$$

Conductivity can now be defined by comparing the coefficients in Eq. (2.12) with the point or differential form of Ohm's law, in one dimension, given by

$$J = \sigma\overline{\mathbf{E}} \qquad (2.13)$$

and, consequently,

$$\sigma = nq\mu_n. \qquad (2.14)$$

■ Eq. 2.14 is a very useful equation. Refer also to Eqs. (2.34) and (2.36).

Therefore, we end up with the very reasonable result that conductivity depends on the number of charge carriers and their ability to move.

The conductivity of a solid ranges over 25 orders of magnitude, one of the largest ranges of any physical property in nature. Figure 2.3 illustrates this variation.[1] The units of conductivity are given in $(\Omega\text{-cm})^{-1}$. The **resistivity** of a material is given by $\rho = 1/\sigma$ where the units are given in $\Omega\text{-cm}$.

■ Resistivity is a very important, widely-used semiconductor material parameter.

The best **conductor** at $T = 300$ K is silver, with $\sigma = 6.3 \times 10^5$ $(\Omega\text{-cm})^{-1}$, although there is usually economic justification for using copper [$\sigma = 5.8 \times 10^5$ $(\Omega\text{-cm})^{-1}$] in the fabrication of electronic circuits. At the *insulator* end, there is quartz, with $\sigma \cong 10^{-16}$ $(\Omega\text{-cm})^{-1}$. The group of materials whose conductivities range between 10^{-6} and 1 $(\Omega\text{-cm})^{-1}$ are called **semiconductors**. The focus of this book is on the utilization of devices fabricated from semiconducting materials.

■ Being able to list typical values and materials in these three categories is important.

Combining Eqs. (2.12) through (2.14), along with Ohm's law, $R = V/I$, and referring to Fig. 2.2, we can obtain the resistance of the sample by using

$$R = \frac{V}{I} = \frac{\rho L}{A} = \frac{L}{\sigma A}. \qquad (2.15)$$

DRILL 2.1

Compute the resistivity ρ for (a) Cu, (b) Ag, and (c) Al.

ANSWERS
(a) 1.72×10^{-6} Ω-cm
(b) 1.58×10^{-6} Ω-cm
(c) 2.86×10^{-6} Ω-cm

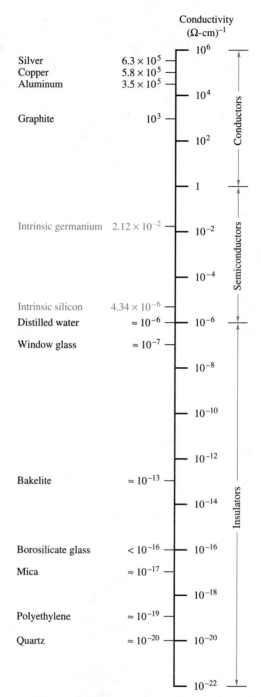

■ **FIGURE 2.3** Conductivity of common substances (from Ref. 1).

EXAMPLE 2.2

Compute the resistance of a 10-m length of American Wire Gauge (AWG) #20 copper wire. This #20 wire has a cross-sectional area of 0.52 mm².

■ AWG is an old standard but almost universally used.

SOLUTION For copper, $\sigma = 5.8 \times 10^7\ (\Omega\text{-m})^{-1}$. Using Eq. (2.15)

$$R = \frac{L}{\sigma A} = \frac{10\ \text{m}}{5.8 \times 10^7\ (\Omega\text{-m})^{-1}(0.52 \times 10^{-6}\ \text{m}^2)} = 0.33\ \Omega.$$

DRILL 2.2

Using the wire specifications from Example 2.2, compare the results if the wire were fabricated from aluminum.

ANSWER
$R = 0.55\ \Omega$

CHECK UP

1. **TRUE OR FALSE?** Current density is given by charge per unit time crossing a unit area.
2. What is the relationship between resistivity and conductivity?
3. What is the relationship between mobility, resistivity, and conductivity?
4. **TRUE OR FALSE?** The better conductors have lower conductivities.
5. What are the three best conductors and what are their conductivities and resistivities?
6. Provide examples of good insulators including typical values for the resistivities and conductivities.

2.3 DIFFUSION

Particles that are free to move in a solid will be in continual random thermal motion. If there is a high concentration of particles in one region compared to another, there is also net particle motion that tends to equalize the concentration after a period of time. This net motion of the charged particles per unit time that equalizes the concentration results in what is called a *diffusion current*. Figure 2.4(a) illustrates a nonuniform concentration of charge carriers. Figure 2.4(b) illustrates the eventual uniform concentration.

■ For diffusion, refer also to Chapter 18.

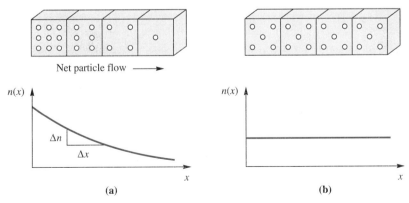

■ **FIGURE 2.4** Diffusion process. (a) Initial concentration of particles. (b) Final concentration of particles.

It seems reasonable that the number of charged particles crossing a unit area in a given time interval will be proportional to some constant of the material and the slope of the charged-particle concentraton itself. Mathematically, this is called **Fick's first law** and in a one-dimensional analysis is given by

$$f = -D\frac{\partial n}{\partial x}. \tag{2.16}$$

where

$f =$ the particle flux, number per unit area per unit time, in the x direction

$D =$ *the **diffusion constant** or **coefficient***, a material-dependent property

$\lim_{\Delta x \to 0} \frac{\Delta n}{\Delta x} = \frac{\partial n}{\partial x}$, the differential slope of the concentration gradient.

The negative sign is used because the concentration gradient as shown results in diffusion to the right. The diffusion coefficient, D, is a function of the material and temperature. The use of the partial derivative implies that n will be a function of position x and time t. By comparing Eq. (2.16) with Eq. (2.12) and realizing that the current density J is a "particle flux," we can see that the diffusion current density for electrons is

$$J_n = qD_n\frac{\partial n}{\partial x}, \tag{2.17}$$

and that the diffusion current density for positively charged particles is given by

$$J_p = -qD_p\frac{\partial p}{\partial x}. \tag{2.18}$$

It would seem reasonable that the diffusion constant and mobility would be related because both terms relate to particle motion and collision phenomena, and indeed they are related; however, the derivation of the relation between D_p and μ_p, and D_n and μ_n, which is called the **Einstein relation**, is beyond the scope of this text.[2] These relationships are given by

$$\frac{D_n}{\mu_n} = \frac{kT}{q}, \tag{2.19}$$

$$\frac{D_p}{\mu_p} = \frac{kT}{q}, \tag{2.20}$$

■ $V_T \approx 26$ mV at 300 K is a useful number to remember.

where $k = 1.38 \times 10^{-23}$ J/K (Boltzmann's constant) and $T =$ temperature in kelvins. (Note that $kT/q \approx 0.026$ V at 300 K.) Although the quantity kT/q is often represented by the symbol V_T (**thermal voltage**), we often prefer to show the temperature dependence explicitly. Clearly the mobility and diffusion cofficients are uniquely related. Again, μ_n and μ_p (D_n and D_p) are measured quantities for a given material.

To categorize completely the current due to diffusion, it is important to know $n(x,y,z,t)$, the charge-carrier concentration as a function of position and time. However, to simplify matters, we will restrict ourselves to one-dimensional variation in charge density. In Fig. 2.5 the net flux entering area 1 is f and the net flux leaving area 2 is given by

$$f + \frac{\partial f}{\partial x}\Delta x.$$

Therefore, by subtraction, the net flux into the volume $A\Delta x$ is

$$f - \left(f + \frac{\partial f}{\partial x}\Delta x\right) = -\frac{\partial f}{\partial x}\Delta x. \tag{2.21}$$

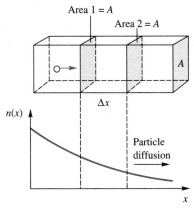

FIGURE 2.5 One-dimensional diffusion equation derivation definitions.

Applying Fick's first law from Eq. (2.16) to Eq. (2.21) yields

$$-\frac{\partial f}{\partial x}\Delta x = -\frac{\partial}{\partial x}\left[-D\frac{\partial n(x,t)}{\partial x}\right]\Delta x = D\frac{\partial^2 n(x,t)}{\partial x^2}\Delta x. \quad (2.22)$$

Because the rate of accumulation of charged particles in the volume $A x$ is also given by

$$\frac{\partial n(x,t)}{\partial t}\Delta x, \quad (2.23)$$

the result, by combining Eqs. (2.22) and (2.23) is

$$\frac{\partial n(x,t)}{\partial t} = D\frac{\partial^2 n(x,t)}{\partial x^2}. \quad (2.24)$$

■ This form of the diffusion equation is found in many applications throughout physics and chemistry.

■ Observe D is assumed to be a constant and this equation is valid for a one-dimensional case

This is known as **Fick's second law**, the **diffusion equation** (in one dimension). Equation (2.24) is also used in Chapter 18 to describe the diffusion of impurities (doping) in semiconductors.

Solutions to Fick's second law appear in a number of advanced differential equation textbooks.[3] Assuming one starts with a fixed quantity of particles, all concentrated at $x = 0$, the one-dimensional solution in rectangular coordinates is given by

$$n(x,t) = \frac{n_o}{2(\pi D t)^{1/2}}e^{-x^2/4Dt} \quad \text{for } -\infty < x < \infty. \quad (2.25)$$

Equation (2.25) is graphed in Fig. 2.6 for three different times, $t = 0$, t_1, and t_2. We assume that we start with a fixed quantity of particles, n_o, all concentrated at the origin, $x = 0$, at time $t = 0$. Equation (2.25) then predicts that the concentration decreases at $x = 0$ for $t > 0$ and that the concentration tends to flatten out as t becomes large. It is important to observe that the total number of particles remains the same for all t; this is essentially an intuitive statement that the area under each of the curves is a constant. The area under each curve, n_o, is called the **dose** and has units of number/cm^2. Often Q is used for the dose instead of n_o.

■ The dose, number per unit area, is widely used. Analogous to describing a surface charge density from electromagnetic field theory.

DRILL 2.3

Compare the concentration of particles at $x = 0$ for times t_1, t_2, and t_3 where $t_3 = 2t_2 = 4t_1$.

ANSWER
$n(0, t_3 = 4t_1)$
$\quad = n(0, t_1)/2,$
$n(0, t_2 = 2t_1)$
$\quad = n(0, t_1)/(2)^{1/2}$

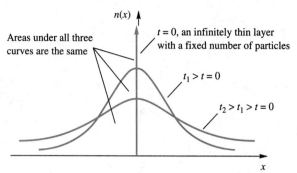

■ **FIGURE 2.6** Graphical representation of the solution [Eq. (2.25)] to Fick's second law [Eq. (2.24)] with a fixed number of particles as the boundary condition. Equation (2.25) illustrates a Gaussian function.

CHECK UP

1. Define diffusion current.
2. **TRUE OR FALSE?** Particles tend to diffuse from a high concentration to a low concentration.
3. **TRUE OR FALSE?** The ratio of diffusivity to mobility is a constant.
4. **TRUE OR FALSE?** The ratio of diffusivity to mobility is proportional to absolute temperature.
5. Show that 26 mV is a good approximation for kT/q at room temperature.
6. Sketch the solution to Fick's second law for time $t = 0$, t intermediate, and t approaching ∞.

2.4 ENERGY-BAND THEORY OF SOLIDS: AN OVERVIEW

As shown in Fig. 2.3, it is useful to divide materials into three categories: insulators, semiconductors, and conductors. This division is based on the density of charge carriers that are able to move under the influence of an electric field and the ease (mobility) of movement. Charge-carrier availability and mobility can be explained by means of energy-band theory. Because this discussion is qualitative in nature, we suggest that you refer to two of the excellent references on this subject given at the end of this chapter[2,4] for more quantitative information.

A single isolated atom contains electrons with certain precisely defined or quantized amounts of energy. This is most simply illustrated for the hydrogen atom. A few of these quantized amounts of energy or energy levels are diagrammed for a hydrogen atom in Fig. 2.7, which is called an *energy-level diagram*. Each level corresponds to a different allowed orbit of an electron around the nucleus.

By definition, the energy scale is in *electron-volts,* where

$$1 \text{ eV} = 1.6 \times 10^{-19} \text{ J}.$$

■ This is one of those useful conversion factors worth remembering.

This is the amount of energy gained by a single electron moving through a potential difference of one volt. Electrons move downward on the diagram so as to occupy the lowest possible energy level. To move an electron from one discrete energy level to

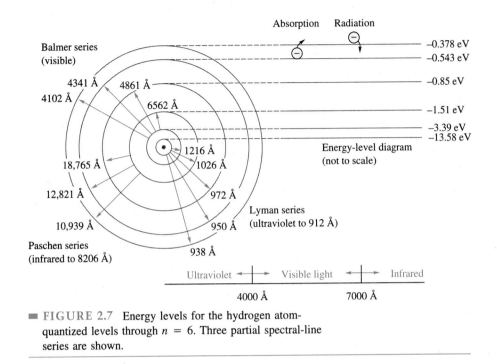

■ **FIGURE 2.7** Energy levels for the hydrogen atom-quantized levels through $n = 6$. Three partial spectral-line series are shown.

another higher level requires a precise amount of energy. An electron losing energy, or falling to a lower level, releases a precise amount of energy. In both cases, this quantized amount of energy is given by

$$E = hf = \frac{hc}{\lambda}, \qquad (2.26)$$

where

$c = f\lambda \approx 2.998 \times 10^8$ m/sec or 299,792,458 m/sec by definition and for practical purposes
$\approx 3 \times 10^8$ m/sec (velocity of light in a vacuum)

h = Planck's constant (6.626×10^{-34} J-sec)

f = Frequency in hertz (Hz).

Equation (2.26) is an important equation used in quantum mechanics. Also observe that energy in electron-volts must be converted to joules to obtain consistent units (1 eV = 1.6×10^{-19} J).

This quantum of energy can be added in the form of **photon** excitation (light impinging on the crystal lattice), **phonon** excitation (crystal-lattice vibrations in the form of heat), or external electrical excitation. Similarly, energy can be released in any of these forms when an electron falls to a lower energy level. This is illustrated in the following example, using the isolated (single) hydogen atom.

■ Photon and phonon are similar concepts.

EXAMPLE 2.3

Show that the Balmer series spectral lines of Fig. 2.7 will emit in the visible spectrum.

SOLUTION   An electron falling from the -1.51-eV level to the -3.39-eV level has a net energy loss of $3.39 - 1.51 = 1.88$ eV. From Eq. (2.26), the frequency of the

> **DRILL 2.4**
>
> Compute the energy-level transition associated with the middle of the visible spectrum defined by $\lambda = 5500$ Å.
>
> **ANSWER**
> $E = 3.614 \times 10^{-19}$ J
> $ = 2.25$ eV

emission is given by

$$f = \frac{(1.88 \text{ eV}) \times \left(\frac{1.6 \times 10^{-19} \text{ J}}{\text{eV}}\right)}{(6.626 \times 10^{-34} \text{ J-sec})} = 4.5 \times 10^{14} \text{ Hz}$$

and

$$\lambda = \frac{c}{f} = \frac{3 \times 10^8 \text{ m/sec}}{4.54 \times 10^{14} \text{ Hz}} = 6.6 \times 10^{-7} \text{ m} = 6600 \text{ Å}$$

The other transitions can be computed similarly. Note 1 Å = 10^{-10} m = 10^{-8} cm, and the velocity of light $c = 3 \times 10^8$ m/sec = 3×10^{10} cm/sec. Similarly, a photon with a wavelength of 6600 Å (6562 Å to be precise) incident on this atom will be absorbed, raising the electron from -3.39 to -1.51 eV. The visible light spectrum runs from about 4000 Å (violet) to 7000 Å (deep red) so that 6600 Å corresponds to a red spectral emission.

As a number of atoms are brought closer together to form a solid, the individual energy levels interact, resulting in a broadening or coupling of the discrete energy levels to form essentially continuous bands of allowed energy levels. This plot of energy levels with respect to a zero-energy reference is called an *energy-band diagram*. Three key features are present in an energy-band diagram:

1. In the completely filled energy bands, called *valence bands*, electrons are not mobile; therefore, they do not contribute to electric current conduction.
2. The first energy band above the valence band which is occupied by an electron is called the *conduction band* and is usually only partially filled. Thus, it is easy to add only a small amount of energy, such as that provided by an external electric field, and initiate electron motion, which is an electric current.
3. There may be gaps between valence bands and the conduction band. These *forbidden-energy band gaps* are a continuum of energy levels that are not allowed for a charge carrier. We can now qualitatively sketch the energy-band diagrams for a typical conductor, insulator, and semiconductor.

■ Bandgaps are a key feature of semiconductor materials. Bandgap engineering—i.e., tailoring materials to obtain a particular bandgap is an important technical area.

The energy-band diagram for an insulator is shown in Fig. 2.8. An insulator such as SiO_2 (glass) has completely filled valance bands. There is a large band gap of about 9 eV between the top of the last valence band and the bottom of the conduction band. However, since 9 eV—a large value as we shall soon observe—of energy is required to excite an electron from the valence band to the conduction band, the conduction band for

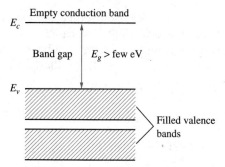

■ **FIGURE 2.8** Insulator energy-band diagram.

CHAPTER 2   INTRODUCTION TO SEMICONDUCTORS

this insulator is virtually empty. We can estimate the number of electrons that can be thermally excited into the conduction band from the formula

$$n \approx \left[2\left(\frac{2\pi m_e kT}{h^2}\right)^{3/2}\right] e^{-E_g/2kT} \qquad (2.27)$$

$$\cong 4.8 \times 10^{15} T^{3/2} e^{-5797 E_g/T} \text{ electrons/cm}^3,$$

where

T = temperature in kelvins
k = Boltzmann's constant, 1.38×10^{-23} J/K or 8.617×10^{-5} eV/K
m_e = mass of an electron, 9.1×10^{-31} kg
h = Planck's constant, 6.626×10^{-34} joule-sec
E_g = energy gap that may be expressed in electron volts where 1 eV = 1.6×10^{-19} joule. The ($-E_g/2kT$) term must be computed with consistent units.

The exponential coefficient 5797 is computed from

$$q/2k \cong (1.6 \times 10^{-19})/2 \times 1.38 \times 10^{-23}.$$

The use of $m_e = 9.1 \times 10^{-31}$ kg is not exact. It can be shown[2,4] using quantum mechanics that m_e, called an **effective mass**, is somewhat less than 9.1×10^{-31} kg. For our qualitative solid-state physics discussion, we will ignore this difference. You will find in more advanced texts that the m_e rigorously should be replaced by $(m_e^* m_h^*)^{1/2}$ where m_e^* and m_h^* are the effective masses of the electrons and holes, respectively. Essentially, when an elecron or hole moves in a solid, the effective mass serves as a correction to the relationships between the particle energy, its potential, and the lattice properties as compared to an electron in free space.

The exponential dependence of carrier density on energy gap guarantees that, for large gaps, we do indeed have a good insulator, as shown in the following example.

EXAMPLE 2.4

Estimate the number of electrons that can be thermally excited into the conduction band at 300 K for an insulator whose band-gap energy, E_g, is 3 eV. Perform the same calculation if we increase the temperature by 50 K.

SOLUTION   From Eq. (2.27),

$$n \approx 2\left(\frac{2\pi m_e kT}{h^2}\right)^{3/2} e^{-E_g/2kT}$$

$$= 2\left[\frac{(2\pi)(9.1 \times 10^{-31} \text{ kg})(1.38 \times 10^{-23} \text{ J/K})(300 \text{ K})}{(6.626 \times 10^{-34} \text{ J-sec})^2}\right]^{3/2}$$

$$\times \exp[-(1.6 \times 10^{-19} \text{ J/eV})(3 \text{ eV})/(2)(1.38 \times 10^{-23} \text{ J/K})(300 \text{ K})]$$

$$\approx 1.7 \text{ electrons/m}^3$$

$$= 1.7 \times 10^{-6} \text{ electrons/cm}^3.$$

Compare this number to the $\sim 10^{22}$ atoms/cm^3 for a solid. Therefore, this is a very good insulator. If we increase the temperature by 50 K, the number of carriers increases dramatically to $n = 8.3 \times 10^3$ electrons/cm^3. This is still a small number compared to 10^{22} atoms/cm^3 but the percentage increase is significant. For this reason, $E_g \approx 3$ eV is considered to be close to the lower limit for material to be considered as an insulator.

DRILL 2.5

Estimate the number of electrons that can be thermally excited into the conduction band at 300 K for a material whose band-gap energy, E_g, is 2.0 eV. How would you classify this material?

ANSWER
$n \approx 410$ electrons/cm^3, which is quite low. This material could be classified as an insulator.

If we have a very large external electric field from an externally applied large voltage, we can cause electrons to jump the gap from the valence band into the conduction band in an insulator. This is the mechanism for dielectric breakdown and is usually irreversibly destructive to the insulating material.

Conductors have virtually no band gaps, so at reasonable temperatures, a large number of electrons are available for conduction in the partially filled conduction band (Fig. 2.9). Essentially, each conductor atom contributes one electron to the conduction band.

Semiconductors have an energy-band diagram very similar to that of an insulator with two very important differences. First, the band gap is relatively small, for example, for silicon about 1.1 eV. Second, the smaller band gap means that thermally excited electrons can jump the gap from the valence band to the conduction band and hence are available for conduction (Fig. 2.10).

■ Depending upon the reference, the bandgap in Si is listed between 1.1 and 1.12 eV. The 0.02 eV does make a difference in carrier concentration.

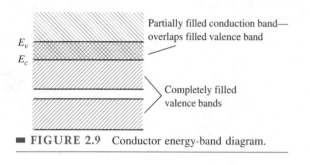

■ **FIGURE 2.9** Conductor energy-band diagram.

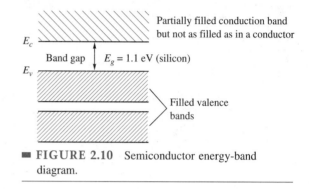

■ **FIGURE 2.10** Semiconductor energy-band diagram.

EXAMPLE 2.5

Silicon, as we will see in Section 2.5, has a band gap of 1.1 eV. Compute the number of thermally excited electrons in the conduction band at 300 K.

SOLUTION From Eq. (2.27),

$n \cong 2.494 \times 10^{19} e^{-E_g/2kT}$ electrons/cm^3

$= 2.494 \times 10^{-19}$

$\quad \times \exp[-(1.6 \times 10^{-19}\,\text{J/eV})(1.1\,\text{eV})/(2)(1.38 \times 10^{-23}\,\text{J/K})(300\,\text{K})]$ electrons/cm^3

$= 1.46 \times 10^{10}$ electrons/cm^3.

■ Numbers from 1.4 to 1.5×10^{10} electrons/cm^3 are found in the literature.

Usually the number 1.5×10^{10} electrons/cm^3 at 300 K is used because approximations have been made in the derivation of Eq. (2.27) and one must consider the effective mass.

We will use the concept of energy-band diagrams when we discuss the operation of diodes and transistors. It is important to remember that an electron loses a quantum of energy if it is displaced downward and gains a quantum of energy if it is excited upward.

CHECK UP

1. Illustrate the application of the formula $E = hf = hc/\lambda$.
2. **TRUE OR FALSE?** The visible spectrum extends from 4000 to 7000 Å.

3. **TRUE OR FALSE?** Conduction bands are never filled.
4. **TRUE OR FALSE?** An energy gap is a principal feature of semiconducting materials.
5. **TRUE OR FALSE?** Carrier concentration is a much stronger function of temperature than of energy band gap.
6. Provide a numerical value for the intrinsic silicon electron and hole carrier concentration. Compare this value to those of a good conductor and a good insulator.
7. **TRUE OR FALSE?** A photon results from an electron losing a quantum of energy falling from the conduction band to the valence band.
8. Provide a numerical value for the energy gap in silicon and compare this value to those of germanium, gallium arsenide, and a good insulator.

DRILL 2.6

Compute the number of thermally excited electrons in the conduction band in silicon at 300 K if the band gap is assumed to be 1.12 eV rather than 1.10 eV.

ANSWER About 1×10^{10} electrons/cm^3.

2.5 SEMICONDUCTOR MATERIALS

Semiconductor devices are typically fabricated from either silicon or to a lesser extent, gallium arsenide, and to a much lesser extent germanium. Silicon and germanium are located in Column IV of the periodic table (see Table 2.1). We are also interested in boron, aluminum, gallium, and indium in Column III and phosphorus, arsenic, and antimony in Column V. Compounds of elements in Columns III and V are used for a wide variety of photonic, microwave, and specialty high-speed devices and circuits.

Although some specialized devices, primarily photonic devices such as light-emitting

TABLE 2.1 Periodic Table of the Elements—Selected Entries of Important Dopants and Column IV Semiconductors

Column III		Column IV		Column V	
Acceptors (*p*-dopants)				Donors (*n*-dopants)	
Boron 5 (atomic number)	B 10.81 (atomic weight)	[Carbon]		[Nitrogen]	
Aluminum 13	Al 26.98	Silicon 14	Si 28.09	Phosphorus 15	P 30.97
Gallium 31	Ga 69.72	Germanium 32	Ge 72.59	Arsenic 33	As 74.92
Indium 49	In 114.82	[Tin]		Antimony 51	Sb 121.75

diodes and lasers, microwave transistors, and very high speed logic, are fabricated from gallium arsenide, a compound semiconductor produced from gallium and arsenic, most devices we are interested in studying in this text are fabricated from silicon. Semiconductor devices are also fabricated from indium and phosphorus or other materials from Columns III and V in the periodic table. There is considerable interest in devices fabricated from III–V materials and we will discuss the special features of gallium arsenide when we consider these other devices. Germanium was used early in the semiconductor industry and there is some current research interest in germanium heterojunction structures, but as silicon-processing technology improved, silicon became the most important material. Consequently, we will restrict our detailed discussion to silicon, although most of what we have to say is applicable to the other semiconductors as well.

■ Because of bandgap differences in III–V materials, many devices are called heterostructures or heterojunctions.

Silicon is one of the most abundant elements in the earth's crust. Indeed it comprises more than 25% of the earth's crust, and only oxygen is more abundant. The most recognizable form of silicon is SiO_2, common sand. Silicon almost always occurs in compounds with other elements. Common minerals that include silicon are asbestos, mica, quartz, granite, clay, and opal. In the semiconductor industry, silicon must be purified to the extent that there is no more than one impurity atom per 10^{10} silicon atoms. Actually, once one has pure silicon, which is called *intrinsic* material, one will want to add impurities in small controlled amounts to tailor the properties in a predictable fashion. We will soon discuss this addition of impurities into intrinsic silicon, a process called *doping*.

■ Truly intrinsic Si is almost impossible to obtain.

■ Doping is a very important semi-conductor fabrication technique, and is discussed further in Chapter 18.

2.6 PROPERTIES OF INTRINSIC SILICON

Intrinsic silicon can exist in *amorphous, polycrystalline*, and *crystalline* forms. For example, the amorphous form of carbon is graphite, and the crystalline form is diamond. This example illustrates a definite difference in properties (and cost!). Metals typically exist as amorphous or polycrystalline materials. That is, there is no long-range ordering to their atomic structure. The atoms are surrounded by a sea of electrons. With such loose (low-energy) bonding for the conduction-band electrons, metals exhibit a large conductivity. Although research interest in amorphous silicon is considerable, especially for low-cost photovoltaic solar cells, we will focus on crystalline silicon and crystalline silicon-based devices. Polycrystalline silicon consists of small crystals with no long-range order and is widely used as a conducting material in integrated cicuits. It is beyond the scope of this text to dwell on crystal structures as such, but a number of excellent references on the subject appear at the end of this chapter.

■ amorphous silicon, polycrystalline silicon (or polysilicon for short) crystalline silicon

■ There is growing interest in specialty semiconductor devices fabricated using amorphous silicon.

Crystalline silicon exhibits a diamond structure, as illustrated in Fig. 2.11. Each sphere represents the nucleus and bound electrons of a single silicon atom. The double line represents the shared two-electron bond between adjacent silicon atoms. This sharing of two electrons is called a *covalent bond*. Conceptually, it is much easier to diagram this crystal structure in two dimensions as illustrated in Fig. 2.12. Each circle with $+4$ represents a silicon atom with a valence of $+4$. The double lines, as before, are the covalent bonds. We are now in a position to explain the properties of intrinsic silicon and, more important, to visualize the results of introducing impurities.

A very legitimate question at this point is "How can intrinsic silicon exhibit a nonzero conductivity if all the valence electrons form covalent bonds?" Indeed, intrinsic silicon would have zero conductivity at $T = 0$ K. Above absolute zero, energy is available in the form of thermally induced crystal-lattice vibrations. These packets of vibrational energy, which are called phonons, are analogous to packets of radiation (light), which are called photons. If a phonon, or photon for that matter, strikes a covalent bond as shown in Fig. 2.13, there is a statistical probability that the bond will be broken. When

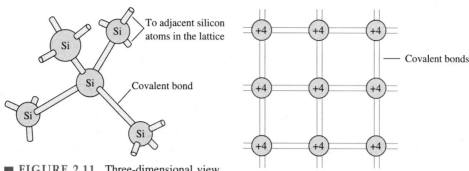

FIGURE 2.11 Three-dimensional view of a diamond (tetrahedral) crystal structure for silicon.

FIGURE 2.12 Two-dimensional representation of the silicon crystal lattice.

a bond is broken, the electron is released, and this free electron can then move under the influence of an external electric field, resulting in an electron current.

Simultaneously, with the breaking of the bond and the electron release, a net positive charge is left behind in the valence structure. This net positive charge is called a ***hole*** and is a feature of a semiconducting material. This process is called *electron-hole pair generation*. The ***generation rate*** $G(T)$, the number of electron-hole pairs generated per second, is dependent on temperature.

■ A hole is a critically important concept.

In intrinsic material, the electron and hole concentrations are equal:

$$n = p = n_i = p_i. \qquad (2.28)$$

As shown in the energy-band diagram in Fig. 2.14, one electron has been moved into the conduction band, and this results in one unfilled state with a positive charge in the valence band, i.e., a hole. It is of vital interest that a valence electron dislodged from an adjacent bond can move to fill the original unfilled state. In doing so, this net positive charge, a *hole,* then appears in the adjacent bond. This occurs repeatedly in the crystal, so that the hole, a positive charge, appears to be moving when we impress an electric field. Because this is, in effect, a moving positively charged particle, we can treat it in the same way as a moving electron. The total semiconductor current has both an electron and a hole component, given by

$$\overline{J} = (nq\mu_n + pq\mu_p)\overline{E} = \sigma\overline{E} \quad [\text{A/cm}^2], \qquad (2.29)$$

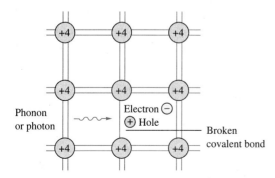

FIGURE 2.13 Electron-hole pair generation.

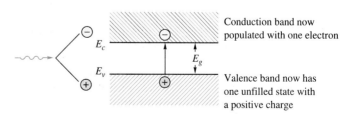

FIGURE 2.14 Electron-hold pair generation, band diagram.

where

n = electron concentration in cm^{-3}
μ_n = electron mobility in cm^2/(V-sec)
p = hole concentration in cm^{-3}
μ_p = hole mobility in cm^2/(V-sec).

In intrinsic silicon at 300 K, one bond out of every 3×10^{12} is broken at any one time; see Table 2.2. Similarly, the reverse process, as diagrammed in Fig. 2.15, is called *recombination*. The *recombination rate* $R(T)$, or the number of electron-hole pairs recombined per second, is proportional to the concentration of holes and electrons with which one starts. In the equilibrium condition, $G = R$, and we can write

$$G(T) = R(T) = Knp \tag{2.30}$$

Qualitatively, the **intrinsic electron** and **hole carrier concentrations**, n_i and p_i, as a function of increasing temperature and as illustrated in the energy-band diagrams in Fig. 2.16, are in equilibrium. More specifically, the intrinsic electron and hole concentrations as a function of absolute temperature are given by Eq. (2.27) and are repeated here for convenience:

$$n_i = p_i \approx \left[2\left(\frac{2\pi m_e kT}{h^2}\right)^{3/2} \right] e^{-E_g/2kT}$$
$$\cong 4.8 \times 10^{15} T^{3/2} e^{-5797 E_g/T} \text{ charge carriers/cm}^3 \tag{2.27}$$

For an intrinsic material in equilibrium and at a given temperature, Eq. (2.30) must hold so that we can write

$$R(T) = G(T) = Knp = Kn_i^2(T) = Kp_i^2(T), \tag{2.31}$$

which reduces to the so-called **law of mass action**[2]

$$np = n_i^2 = p_i^2, \tag{2.32}$$

where $n_i = p_i$ is computed from Eq. (2.27). Equation (2.32) permits us to compute the product of the electron and hole concentrations if we know the intrinsic charge-carrier concentration. This is an important equation used for the *extrinsic* situation, where

TABLE 2.2 Useful Properties of Crystalline Silicon, Germanium, and Gallium Arsenide

Property (at 300 K)	Silicon	Germanium	Gallium Arsenide (GaAs)
Atomic number	14	32	Gallium 31
			Arsenic 33
Atoms/cm^3	5×10^{22}	4.41×10^{22}	4.42×10^{22}
Relative dielectric constant, ϵ_r	11.8	15.8	13.1
Energy gap, E_g (eV)	1.12	0.66	1.42
Electron mobility, μ_n (cm^2/V-sec)	1500	3900	8500
Hole mobility, μ_p (cm^2/V-sec)	480	1900	400
Nominal $n_i = p_i$ (number/cm^3)	1.5×10^{10}	2.4×10^{13}	1.8×10^6
Melting point, °C	1420	936	1238

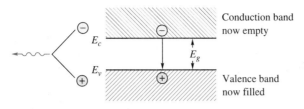

FIGURE 2.15 Electron-hole pair recombination.

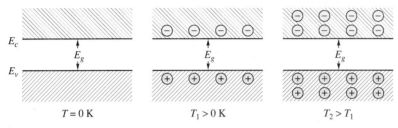

FIGURE 2.16 Intrinsic material electron-hole pair generation as a function of temperature.

$n \neq p$, which we will consider in the next section. If we know n, we can find p, and vice versa.

Table 2.2 summarizes a number of physical properties for intrinsic silicon and germanium.[1,5] Gallium arsenide also appears here because it is expected to become increasingly important as semiconductor devices are required to operate at higher switching speeds and at higher frequencies.

We can observe several key differences among the physical properties. Electron mobility is several times larger than hole mobility. Operationally, this means that devices relying on electron conduction rather than hole conduction will be faster. Expanding on this idea, the electron mobilities in gallium arsenide are more than five times the electron mobilities in silicon so that the former automatically exhibits faster operation. The larger energy gap in gallium arsenide (1.42 eV) compared to that in silicon (1.12 eV) means that the conductivity of undoped gallium arsenide is much lower than that of silicon. Therefore, undoped gallium arsenide is often called *semi-insulating*.

EXAMPLE 2.6

Compute the conductivity of intrinsic silicon at 300 K.

SOLUTION We have already calculated the intrinsic electron and hole concentration, $n_i = p_i = 1.5 \times 10^{10}$ charge carriers/cm^3. In Table 2.2, μ_n and μ_p are given as 1500 cm^2/(V-sec) and 480 cm^2/(V-sec), and according to Eq. (2.29),

$$\sigma = (1.5 \times 10^{10})(1500)(1.6 \times 10^{-19}) + (1.5 \times 10^{10})(480)(1.6 \times 10^{-19})$$
$$= 4.75 \times 10^{-6} \ (\Omega\text{-cm})^{-1}.$$

It is interesting to note that this is 11 orders of magnitude lower than the conductivity for copper, a good conductor.

DRILL 2.7

Compute the resistivity ρ for intrinsic silicon at 300 K.

ANSWER
$\rho = 210$ kΩ-cm.

CHECK UP

1. **TRUE OR FALSE?** Intrinsic silicon is undoped.
2. **TRUE OR FALSE?** One cannot obtain electrons and holes in intrinsic silicon.
3. **TRUE OR FALSE?** The electron and hole concentrations are the same in intrinsic silicon.
4. Demonstrate the computation of the resistivity and conductivity of intrinsic silicon using electron and hole mobility.
5. **TRUE OR FALSE?** Thermal electron and hole generation and recombination are continuous processes.

2.7 PROPERTIES OF DOPED SILICON

The enormous versatility in semiconductor applications is the result of our ability to tailor the conducting properties of intrinsic silicon by adding impurities. We can either enhance the number of electrons or the number of holes available for conduction. Let us consider each type of enhancement in turn.

Donors

The elements phosphorus, arsenic, and antimony are located in Column V of the periodic table shown in Table 2.1, indicating that they have a valence of $+5$. If we add a small amount of one of the elements in Column V as an impurity by substitutionally displacing one of the silicon atoms, we have a crystalline structure as shown in Fig. 2.17. As noted earlier, this addition of an impurity is called *doping*. Four of the five valence electrons

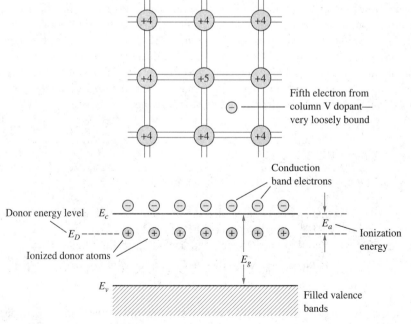

■ **FIGURE 2.17** Representation of *n*-type doped crystal and energy bands (thermally generated carriers are not shown).

form the covalent bonds with surrounding silicon atoms. The fifth electron is so weakly bound to the impurity atom that for all practical purposes, it acts as if it were a free electron in a metal. We represent this fifth electron as a circle with a minus sign inside. We will also include the Column V element on an energy-band diagram. Consider what it means to have the fifth electron so weakly bound that it is usually available for conduction. This means that the electron from the Column V impurity can easily be excited into the conduction band. If this ionization energy is small, we can hypothesize that the impurity ion remaining will be located just below the edge of the conduction band. Indeed, this is the case, as illustrated in Fig. 2.17. The binding energy, or *ionization energy*, E_a, is on the order of 0.045 eV. More specifically, $E_a = 0.044$ eV for phosphorus, 0.049 eV for arsenic, and 0.039 eV for antimony. This energy is sufficiently low that at room temperature (300 K), virtually all of the fifth electrons lie in the conduction band. Atoms of Column V elements that are added as an impurity are called **donor atoms**, or **donors** for short. The donor is said to be ionized when the electron is excited into the conduction band. The donor (now ion) retains a positive charge. The donor ion energy level is labeled E_D.

■ Comes from "donating" an electron to the conduction band.

We will now consider how the inclusion of donors expressed as N_D atoms/cm^3 affects the conductivity. Charge neutrality dictates that the total positive and negative charge in the donor-doped semiconductor must be the same. This means that

$$n = N_D + p \cong N_D. \tag{2.33}$$

■ A very useful approximation along with the results of Eqn. (2.34).

Recall that the ionized donors have a net positive charge and because doping concentrations are typically between 10^{14} and 10^{19} cm^{-3}, we can neglect the remaining p-hole concentration compared to N_D in Eq. (2.33). Thus, for electron conductivity, Eq. (2.14) reduces to

$$\sigma_n \cong N_D q \mu_n. \tag{2.34}$$

The addition of impurities also changes the mobility from the intrinsic value due to lattice distortion. Mobility and resistivity data[5,6] are plotted in Figs. 2.18(a) and (b), respectively. Note that the electron mobilities are at least twice the hole mobilities.

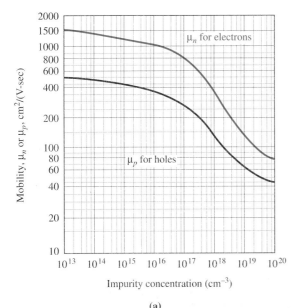
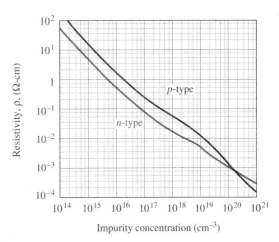

(a)   (b)

■ **FIGURE 2.18** Properties of doped silicon. (a) Electron and hole mobilities as a function of impurity concentration. (b) Resistivity of n- and p-type silicon[6] at 300 K as a function of impurity concentration.

DRILL 2.8

Compute the electron and hole concentrations and conductivities in silicon doped at
(a) $N_D = 10^{15}$ atoms/cm^3 and (b) $N_D = 10^{19}$ atoms/cm^3.

ANSWERS
(a) $n = 10^{15}$ electrons/cm^3,
$p = 2.25 \times 10^5$ holes/cm^3,
$\sigma \approx 0.208$ (Ω-cm)$^{-1}$
(b) $n = 10^{19}$ electrons/cm^3,
$p = 22.5$ holes/cm^3,
$\sigma \approx 176$ (Ω-cm)$^{-1}$

EXAMPLE 2.7

Compute the electron and hole concentration and conductivity in silicon doped at $N_D = 10^{17}$ atoms/cm^3.

SOLUTION The electron concentration n is equal to N_D. From Eq. (2.32), $p = n_i^2/n = n_i^2/N_D = (1.5 \times 10^{10})^2/10^{17} = 2250$ holes/cm^3, which is much less than N_D, thus validating the approximation given in Eq. (2.33). From Eq. (2.34) and using Fig. 2.18 to obtain σ_n, $\cong (10^{17})(1.6 \times 10^{-19})(750) = 12.0$ (Ω-cm)$^{-1}$.

The holes in a semiconductor doped with a donor element are the *minority carriers*. Note that in Example 2.7 the hole concentration in a donor-doped semiconductor is much less than p_i. This is called *minority carrier suppression* and results from excess electrons "filling" up holes, i.e., filling the unfilled valence bonds (energy levels). It is important to remember that in a donor-doped material, that is, an *n*-type material, current conduction is primarily by means of electrons.

DRILL 2.9

Compute the electron concentration, hole concentration, conductivity σ, and resistivity ρ for $N_D = 5 \times 10^{17}$ atoms/cm^3 phosphorus-doped silicon at $T = 300$ K. What value of electron mobility, μ_n, are you using?

ANSWER
$n = 5 \times 10^{17}$ electrons/cm^3
$p = 450$ holes/cm^3
$\sigma = 40$ (Ω-cm)$^{-1}$
$\rho = 0.025$ Ω-cm
$\mu_n = 500$ cm^2/(V-sec).

EXAMPLE 2.8

A silicon sample is doped with one atom of phosphorus for every 10^8 atoms of silicon. Compute the electron and hole concentration, conductivity, and resistivity.

SOLUTION Phosphorus is a donor dopant. Because almost complete ionization occurs at 300 K, $n \approx N_D$. There are 5×10^{22}/cm^3 atoms of silicon so that $N_D = 5 \times 10^{22}/10^8 = 5 \times 10^{14}$/cm^3 atoms of phosphorus. From Eq. (2.32)

$$p = \frac{n_i^2}{N_D} = \frac{p_i^2}{N_D} = \frac{(1.5 \times 10^{10} \text{ cm}^{-3})^2}{5 \times 10^{14} \text{ cm}^{-3}}$$
$$= 4.5 \times 10^5 \text{ holes/cm}^3.$$

This is a reduction of almost 5 orders of magnitude in the hole concentration as compared to intrinsic material and again illustrates nicely minority carrier suppression. We can estimate μ_n from Fig. 2.18(a) as 1250 cm^2/(V-sec) so that by using Eq. (2.34) we get

$$\sigma = (5 \times 10^{14} \text{ cm}^{-3})(1.6 \times 10^{-19} \text{ C})[1250 \text{ cm}^2/(\text{V-sec})^{-1}] = 0.10 \text{ }(\Omega\text{-cm})^{-1}.$$

The conductivity has increased by almost 4 orders of magnitude as compared to the intrinsic material. Resistivity using Fig. 2.18(b) is

$$\rho = \frac{1}{\sigma} = 10.0 \text{ }\Omega\text{-cm}.$$

Acceptors

The elements boron, gallium, and indium are located in Column III of the periodic table shown in Table 2.1, indicating that they have a valence of +3. If we dope an intrinsic silicon sample with a small amount of one of the elements in Column III, we have a crystalline structure as shown in Fig. 2.19. A Column III element has one too few electrons to participate in the silicon covalent bonds of the surrounding atoms. To complete the covalent bond structure, an electron is attracted from the valence band (see Fig. 2.19). Thus, the Column III element becomes a singly charged negative ion.

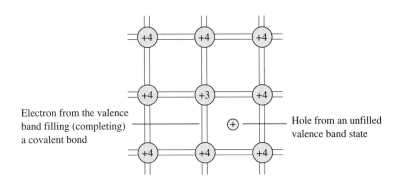

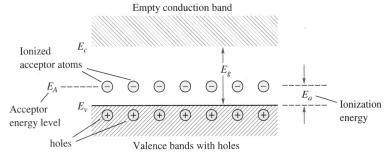

■ **FIGURE 2.19** Representation of *p*-type doped crystal and energy bands (thermally generated carriers are not shown).

Analogous to the giving up of an electron for a donor material, the Column III element is called an *acceptor* for acquiring an electron. When an electron is removed from the valence band, a net positive charge is left in the valence band. Because the valence band is no longer filled, this positive charge can effectively move under the application of an external electric field. As previously indicated, this is called a hole. In other words, when we incorporate an acceptor material, very little energy is required to move an electron from the valence band to complete the covalent bond. Therefore, referring to Fig. 2.19, we can hypothesize that the impurity acceptor ion must be located just above the edge of the valence band. This binding or ionization energy, E_a, is 0.045 eV for boron, 0.065 eV for gallium, and 0.16 eV for indium. Again, as in the case of the donor, this energy is sufficiently low so that at room temperature virtually all the acceptors are ionized. The acceptor ion energy level is labeled as E_A. As an added note, boron is by far the most common acceptor material used in semiconductor technology.

■ Comes from "accepting" an electron from the valence band leaving a hole— vacancy.

We can now consider the inclusion of acceptors expressed as N_A atoms/cm³ and how they affect conductivity. Charge neutrality requires that

$$p = N_A + n \cong N_A. \qquad (2.35)$$

■ A very useful approximation along with the results of Eq. (2.36).

Recall that the ionized acceptors have a net negative charge and because doping concentrations are typically between 10^{14} and 10^{19} atoms/cm³, we can neglect the remaining *n*-electron concentration compared to N_A in Eq. (2.35). Thus, for hole conductivity, Eq. (2.14) becomes

$$\sigma_p \cong N_A q \mu_p. \qquad (2.36)$$

In this case, the electrons are the minority carriers and the holes are the majority carriers. In an acceptor-doped material (a *p*-type material), current conduction is primarily by means of holes.

DRILL 2.10

Compute the electron concentration, hole concentration, conductivity σ, and resistivity ρ for $N_A = 5 \times 10^{17}$ atoms/cm³ boron-doped silicon at $T = 300$ K. What value of hole mobility, μ_p, are you using?

ANSWER
$n = 450$ electrons/cm³
$p = 5 \times 10^{17}$ holes/cm³
$\sigma = 14$ (Ω-cm)$^{-1}$
$\rho = 0.0714$ Ω-cm
$\mu_p = 175$ cm²/(V-sec).

EXAMPLE 2.9

A silicon sample is doped with 2×10^{16}/cm³ boron atoms. Compute the electron and hole concentrations and the conductivity at 300 K.

SOLUTION Boron is an acceptor dopant. Because complete ionization occurs at 300 K,
$$p \approx N_A = 2 \times 10^{16} \text{ holes/cm}^3.$$
From Eq. (2.32),
$$n = \frac{n_i^2}{p} = \frac{p_i^2}{p} = \frac{p_i^2}{N_A} = \frac{(1.5 \times 10^{10} \text{ cm}^{-3})^2}{2 \times 10^{16} \text{ cm}^{-3}} = 11{,}250 \text{ electrons/cm}^3.$$
We can now estimate from Fig. 2.18 that $\mu_p = 350$ cm²/(V-sec), so according to Eq. (2.36),
$$\sigma = (2 \times 10^{16} \text{ cm}^{-3})(1.6 \times 10^{-19} \text{ C})[350 \text{ cm}^2/(\text{V-sec})]$$
$$= 1.12 \, (\Omega\text{-cm})^{-1}.$$

It is interesting to note that adding only one impurity atom to 2.5×10^6 silicon atoms increases the conductivity from 4.75×10^{-6} to 1.12 (Ω-cm)$^{-1}$, an increase of 6 orders of magnitude, but still 7 orders of magnitude below the conductivity of copper. When both donors and acceptors are present,

$$\begin{aligned} n &\approx N_D - N_A \quad \text{for } N_D > N_A, \\ p &\approx N_A - N_D \quad \text{for } N_D < N_A, \end{aligned} \quad (2.37)$$

although the resultant net material doping is such that either Eq. (2.33) or Eq. (2.35) is typically used.

Consider your graphical accuracy as you perform these calculations and compare your answers to those given here.

CHECK UP

1. **TRUE OR FALSE?** Column V elements are donors for silicon.
2. **TRUE OR FALSE?** Column III elements are acceptors for silicon.
3. **TRUE OR FALSE?** Acceptors have five covalently bonded electrons in the outer shell.
4. Estimate the electron and hole concentration in donor-doped silicon.
5. Estimate the electron and hole concentration in acceptor-doped silicon.
6. List two common donor and acceptor elements for silicon.
7. **TRUE OR FALSE?** The conductivity increases with doping.
8. **TRUE OR FALSE?** The mobility increases with doping.
9. **TRUE OR FALSE?** In general, at a given doping density, the mobility of electrons is higher than that of holes in silicon.
10. **TRUE OR FALSE?** The resistivity of doped silicon is higher than that found in intrinsic silicon.
11. Compare electron and hole mobilities in silicon and gallium arsenide.

2.8 EXPERIMENTAL STUDIES OF DRIFT AND DIFFUSION

Stevenson and Keyes[7] presented an early experiment of interest. Assume we have an n-type semiconductor sample in the circuit of Fig. 2.20(a). We will measure the voltage across R_L with an oscilloscope. If there is no external illumination, the voltage will be constant with time, and the majority of the current is due to the electrons. The background hole concentration, p_{no}, is quite low, as one might expect with an n-type material. If we now illuminate the sample with a narrow pulse of light, where each photon has an energy $E = hf$, there will be a step increase in $v(t)$ with a subsequent decay to the quiescent values, as illustrated in Fig. 2.20(b). We can understand why $v(t)$ behaves as it does by studying the hole concentration as a function of time. A key feature of the graph in Fig. 2.20(c) is the exponential decay of the hole concentration after the light generating the electron-hole pairs is switched off. The excess hole concentration decays to $1/e$ of the original excess value in a time τ_p, defined as the ***minority carrier lifetime*** for holes. As we would expect, in the absence of photon excitation, the electron-hole pairs will recombine, with the final hole concentration being determined by Eq. (2.32). If we define the excess hole concentration being equal to $\tau_p G$, where G is a generation rate $(cm^3\text{-sec})^{-1}$, then

$$p_n(t) = p_{no} + \tau_p G e^{-t/\tau_p}. \tag{2.38}$$

The minority carrier lifetime will be a significant term in our consideration of the importance of the base dimension of a biopolar transitor in Chapter 4.

■ Usually τ_p doesn't exceed a few μsec and is often much lower. The more perfection to the crystal, the longer the τ_p.

■ Of course, τ_n would be the minority carrier lifetime for electrons in p-doped material.

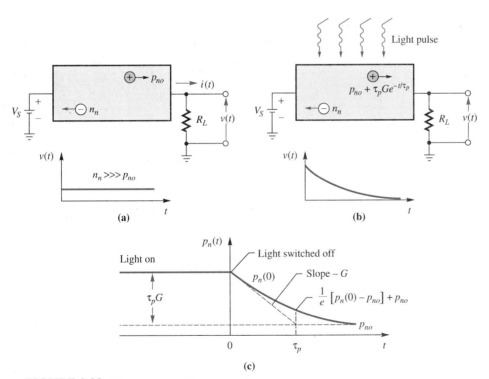

■ **FIGURE 2.20** Stevenson and Keyes experiment.
(a) n-type material current flow, not illuminated.
(b) Hole-electron pair generation by light pulse photons.
(c) Hole concentration.

Suppose we modify the Stevenson-Keyes experiment to the form illustrated in Fig. 2.21(a). We will generate electron-hole pairs continuously by illuminating the end cross section. If we were to sample $p_n(x)$ with a probe, as illustrated, we would observe the hole concentration graphed in Fig. 2.21(b). Because of collisions and recombination, the density decreases exponentially with depth. The distance L_p where the density decreases to $1/e$ times the difference between the surface value, $p_n(0)$, and the initial background value p_{no}, is called the **diffusion length**. Mathematically, Fig. 2.21(b) is described by

$$p_n(x) = p_{no} + [p_n(0) - p_{no}]e^{-x/L_p}. \tag{2.39}$$

The relationship between L_p and τ_p is given by

$$L_p = (D_p \tau_p)^{1/2}. \tag{2.40}$$

Similarly, if we wish to create electron-hole pairs in a p-type sample, we can compute

$$L_n = (D_n \tau_n)^{1/2}. \tag{2.41}$$

From a practical viewpoint, we want L_p and L_n to be as large as possible, since these quantities are related to the degree of perfection of the crystal. Diffusion lengths of up to 1 cm are possible in silicon.

If we combine the experimental methodology illustrated in Figs. 2.20 and 2.21, we obtain the situation illustrated in Fig. 2.22. With this experiment, which was originally performed by Haynes and Shockley,[8] we can determine whether we have n- or p-type material and measure the mobilities. If we generate electron-hole pairs at a small spot on the n-type sample, we will observe a carrier concentration as shown in Fig. 2.22(b) when

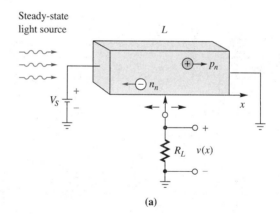

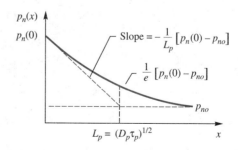

■ **FIGURE 2.21** Modified Stevenson and Keyes experiment. (a) n-type material illuminated at one end. (b) Hole concentration as a function of x when $L \gg L_p$.

$V_S = 0$. Since both diffusion and recombination are working, Eqs. (2.25) and (2.38) can be combined to yield

$$p_n(x, t) = \frac{p_o}{2(\pi D_p t)^{1/2}} \exp\left(\frac{-x^2}{4D_p t} - \frac{t}{\tau_p}\right) + p_{no}. \qquad (2.42)$$

The interesting features of this experiment occur when $V_S > 0$ V. When this condition exists, the holes will move to the right with a velocity v_p. The carrier densities at two different times are shown in Fig. 2.22(c). Mathematically, if we replace x in Eq. (2.42) with

$$x - v_p t = x - \mu_p E t, \qquad (2.43)$$

where

$$\mathbf{E} \cong \frac{V_S}{L}, \qquad (2.44)$$

we will obtain the complete $p_n(x, t)$.

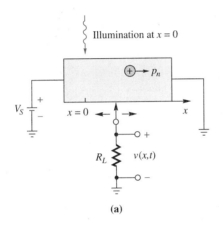

(a)

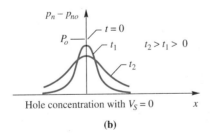

Hole concentration with $V_S = 0$

(b)

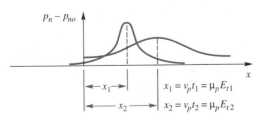

Hole concentration with $V_S = V_S u(t)$, measure with a $t = 0$ reference

(c)

■ **FIGURE 2.22** Haynes-Shockley experiment, n-type sample.

■ This technique is available as a commercial instrument with the illumination provided by a switched LASER.

In a p-type sample, the minority electrons will move to the left with their own drift velocity v_n. The complete $n_p(x,t)$ would be found by making appropriate changes to Eq. (2.42) and using

$$x - v_n t = x - \mu_n \mathbf{E} t. \tag{2.45}$$

We used Eq. (2.2) to derive Eqs. (2.43) and (2.45). Therefore, we can measure mobility from the peak amplitude time-displacement measurement and the type of carrier from the relative pulse scaling and direction of pulse movement.

EXAMPLE 2.10

Assume we have an infinitely long n-type semiconductor bar at 300 K with a $+x$-directed electric field of 1000 V/cm. At $t = 0$, electron-hole pairs are generated at $x = 0$ by a light pulse. At $t = 1$ μs, the charge concentration maximum is measured as a function of x, and it occurs at $x_p = 0.5$ cm. Compute μ_p and D_p.

SOLUTION With the electric field oriented as described, holes will move to the right. The hole concentration maximum has moved 0.5 cm in 1 μs so that from Eq. (2.43),

$$\mu_p = \frac{x_p}{\mathbf{E}t} = \frac{0.5 \text{ cm}}{(1000 \text{ V/cm} \times 1 \text{ μs})} = 500 \text{ cm}^2/(\text{V-sec}).$$

Using the Einstein relation, Eq. (2.20), we have for holes

$$D_p = \frac{\mu_p kT}{q}$$

$$= \frac{[500 \text{ cm}^2/(\text{V-sec})](1.38 \times 10^{-23} \text{ J/K})(300 \text{ K})}{(1.6 \times 10^{-19} \text{ C})}$$

$$= 12.9 \text{ cm}^2/\text{sec}.$$

CHECK UP

1. **TRUE OR FALSE?** The longer the minority carrier lifetime, the longer the diffusion length.
2. **TRUE OR FALSE?** D_n is usually smaller than D_p.
3. **TRUE OR FALSE?** Light can be used to generate electron-hole pairs.
4. Graphically interpret the results you expect to find for $p_n(x)$ and $p_n(t)$ in the Stevenson and Keyes experiment.
5. Explain why an electric field is required to measure mobility in the Haynes-Shockley experiment.

SUMMARY

Drift and diffusion describe the transport of charged particles in a solid. The drift mechanism defines the charged-particle mobility and the resultant conductivity or resistivity of conductors, semiconductors, and insulators. A qualitative description of energy-band theory was presented to obtain the number of electrons and holes contributing to the

CHAPTER 2   INTRODUCTION TO SEMICONDUCTORS

conductivity in a semiconductor. Properties of specific semiconducting materials were discussed with particular emphasis on silicon, the most widely used semiconductor for electronic devices. By adding selected impurities in a controlled fashion, i.e., by doping, either *n*-type or *p*-type material is obtained and the majority carriers are electrons or holes, respectively. This ability to modify semiconductor properties makes possible a large array of electronic devices. The sign and mobility of the charge carriers can be determined experimentally by means of the Stevenson and Keyes experiments. Other measurement techniques, especially the Hall effect technique,[9] are widely used.

SURVEY QUESTIONS

1. What are the basic definitions for current, mobility, drift velocity, conductivity and resistivity?
2. How would you compute the resistance of a material using the resistivity or conductivity and that material's dimensions?
3. What are the ranges for conductivity of common good conductors, semiconductors, and insulators?
4. What is the Einstein relationship and how is it used?
5. How would you explain and interpret graphically the basic solutions to a one-dimensional diffusion equation?
6. What is valence-band energy? Conduction-band energy? Energy gap?
7. How would you compare the energy-band diagrams for a good conductor, semiconductor, and insulator?
8. How do you compute the intrinsic carrier concentration as a function of either temperature or energy gap?
9. What are some of the key features of intrinsic and *n*-type and *p*-type doped silicon with respect to the energy-band diagram, carrier concentration, conductivity (resistivity), and mobility?
10. What is the effect of a donor atom and an acceptor atom in a semiconductor?
11. What are some of the common *n*-type and *p*-type dopants for silicon?
12. How do you compute the resistivity and conductivity of *n*-type and *p*-type silicon?
13. How do you use the graphs of Fig. 2.18 to obtain the mobility, resistivity, and conductivity of *n*-type and *p*-type silicon?
14. How is the equation $np = n_i^2$ used to illustrate the effects of minority carrier suppression?
15. What are the definitions for minority carrier lifetime and diffusion length?
16. How is the Stevenson and Keyes experiment used to obtain minority carrier lifetime and diffusion length?
17. How is the Haynes-Shockley experiment used to obtain mobility?

PROBLEMS

Unless otherwise stated, all calculations assume $T = 300$ K. Problems marked with an asterisk are more challenging.

2.1   Compare the resistances of wires made from Ag, Cu, and Al. Each wire is 1 m long and has a cross section of 1 mm². Also compute the current densities for an applied potential of 1 V.

2.2   How many electrons per second are equivalent to a current of 1 mA?

2.3   If a conductor of a given diameter D and a given length L has a resistance of 1 Ω, what is the resistance for (a) a conductor of length $2L$ and (b) a diameter of $2D$?

2.4   The resistance of AWG #18 wire is 6.385×10^{-3} Ω/ft. The diameter is 0.0403 in. There are about 8×10^{28} atoms/m³. For a 100-mA current, compute v_d, μ_n, and σ_n.

2.5 The resistance of #18 copper wire is given as 6.385 Ω/1000 ft at 20°C. The diameter of #18 copper wire is 0.0403 in. The wire is supporting a current of 500 mA. Are there any changes, from Problem 2.4, in the values of the drift mobility, drift velocity, resistivity, and conductivity of this wire?

2.6 Aluminum rectangular stripes are often used for the interconnection metallization of individual circuit elements in an integrated circuit (IC).

    (a) What is the resistance of a strip of aluminum 5 μm wide, 5000 Å thick, and 100 μm long?
    (b) Gold is also often used for IC metallization. Compute the resistance of a strip of gold with the same dimensions as given in part (a). The σ for Au is 4.1×10^7 $(\Omega\text{-m})^{-1}$.

2.7 Compute D_n and D_p for undoped Si, Ge, and GaAs at 300 K. Use Table 2.2.

2.8 Show that Eq. (2.25) is a solution to Fick's second law, Eq. (2.24).

2.9 Prepare a sketch of Eq. (2.25)

    (a) What does the area under the curve represent?
    (b) Graphically compare $n(x,t)$ at some time $t > 0$ for two different types of particles where the individual mobilities are such that $\mu_1 > \mu_2$. Explain your results.

2.10 Compute the wavelength and type of radiation emitted from energy-level transitions, E, of 100, 10, 5, 1.0, 0.1, and 0.01 eV.

2.11 Show that the energy levels associated with the Paschen series and Lyman series spectral lines illustrated in Fig. 2.7 emit in the infrared and ultraviolet ranges, respectively.

2.12 Show that the energy levels associated with the Balmer series spectral lines are as illustrated in Fig. 2.7.

2.13 Compute the energy-level transition required for emission at each extreme of the visible spectrum.

2.14 What is the maximum wavelength for a photon to excite an electron-hole pair in silicon? Compare your answer with the wavelengths associated with the visible spectrum.

2.15 Repeat Problem 2.14 for germanium.

2.16 Repeat Problem 2.14 for GaAs.

2.17 If you were able to select the appropriate compound semiconductor material, what bandgap energy would you choose to obtain maximum optical emission at 5000 Å (blue) and 7000 Å (red)?

2.18 You are given that $E_g = 1.42$ eV for GaAs (gallium arsenide), $E_g = 2.26$ eV for GaP (gallium phosphide), and $E_g = 3.36$ eV for GaN (gallium nitride). What material would you use to obtain maximum optical emission in the center of the visible spectrum? Justify your answer numerically.

2.19 Compute the wavelengths of radiation required to generate electron-hole pairs in Si, Ge, GaAs, and SiO_2. Where does this radiation lie in the spectrum?

2.20 For a 10°C increase in temperature, compute the relative importance of the exponential term with respect to the $(\ )^{3/2}$ term for the intrinsic carrier concentration and resistivity in Si at 300 K.

2.21 Compute the conductivity σ and resistivity ρ in intrinsic silicon at $T = 500$ K.

2.22 Compute the conductivity σ and resistivity ρ in intrinsic germanium at $T = 300$ K and compare with Figure 2.3 data.

2.23 Compute the ratio of electron to hole current density in intrinsic silicon and germanium at 300 K.

2.24 Plot the intrinsic electron and hole concentration in silicon for temperatures between 150 and 500 K. Use a log scale to plot n_i. Comment on your results and the potential problems of using silicon at very low or very high temperatures. Assume the effective masses are equal to the rest mass of the electron.

2.25 Although the following is not done in practice, suppose we dope silicon with aluminum $N_A = 10^{14}$ atoms/cm^3.

    (a) Explain why the metallurgical properties of the doped silicon are not changed significantly but the electrical properties are.
    (b) Compute the conductivity and compare the result with that of intrinsic silicon and aluminum at $T = 300$ K.

CHAPTER 2 INTRODUCTION TO SEMICONDUCTORS

2.26* A silicon sample is uniformly doped with $1.0 \times 10^{16}/cm^3$ phosphorus atoms. Find the resistivity. Suppose $1.0 \times 10^{18}/cm^3$ boron atoms are now added to the phosphorus-doped sample. Now compute the resistivity.

2.27 Find the concentration of holes and electrons in 0.01 Ω-cm, n-type and 0.01 Ω-cm, p-type silicon at 300 K.

2.28 Compare the intrinsic carrier concentration for silicon at $T = 0°C = 273$ K and $T = 27°C = 300$ K. Numerically show the dominance of the exponential term.

2.29* At what temperature would the intrinsic concentration of electron-hole pairs become comparable to typical doping densities in silicon? Use 10^{17} atoms/cm^3 as a typical doping density. Assume the effective mass is equal to the rest mass of the electron and neglect the decrease in E_g with T.

2.30 SiO_2 (silicon dioxide), a quartz-type glass, is considered a good insulator. Show that this is true using $E_g \approx 9$ eV.

2.31 GaAs (gallium arsenide) is often called semi-insulating. Explain this realizing that $E_g \approx$ 1.424 eV. Assume the effective mass of an electron is 6.55% of the rest mass of the electron and the effective mass of the hole is 52.4% the rest mass of the electron.

2.32 Silicon is doped with $8 \times 10^{17}/cm^3$ arsenic atoms. Assume $T = 300$ K.

(a) What are the electron and hole concentrations?
(b) What is the conductivity of this doped-silicon sample?

2.33 A silicon sample is initially doped with $5 \times 10^{16}/cm^3$ phosphorus atoms. Assume $T = 300$ K. Compute the electron and hole concentrations and provide an estimate of the conductivity and resistivity.

2.34 How do the results in Problem 2.33 change if the temperature is increased to 310 K?

2.35* A silicon sample is initially doped with $1 \times 10^{16}/cm^3$ phosphorus atoms. Assume $T = 300$ K. We now take this $1 \times 10^{16}/cm^3$ phosphorus-doped silicon sample and dope it with $5 \times 10^{16}/cm^3$ boron atoms. Compute the electron and hole concentration and provide an estimate of the conductivity and resistivity. This is called *compensation* in a semiconductor.

2.36* Refer to Figure 2.18(a). Rescale the y axis to obtain values of diffusivity as a function of doping concentration. You need only show representative values for low, midrange, and high doping concentrations.

2.37* Reasonable minority carrier lifetimes are greater than 1 μs. Using results from Problem 2.36, compute diffusion length values for low, midrange, and high doping concentrations.

2.38 From a practical viewpoint, it is difficult to purchase doped silicon with doping densities of $>10^{20}$ atoms/cm^3. What resistivity does this correspond to for both n-doped and p-doped silicon? Compute the respective electron and hole concentrations.

2.39 From a practical viewpoint, it is difficult to purchase doped silicon with doping densities of $<10^{13}$ atoms/cm^3. What resistivity does this correspond to for both n-doped and p-doped silicon? Compute the respective electron and hole concentrations.

2.40 Calculate the resistance between the contacts for the silicon sample shown in Fig. 2.23 under the following conditions:

(a) Intrinsic material
(b) Doped uniformly with boron at 3×10^{17} atoms/cm^3
(c) Doped uniformly with arsenic at 3×10^{17} atoms/cm^3.

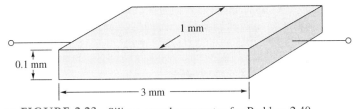

■ **FIGURE 2.23** Silicon sample geometry for Problem 2.40.

2.41 Repeat Problem 2.40 but now assume a germanium sample.

2.42 Refer to Fig. 2.24 and Problem 2.40. If a 10-V source is applied between the two terminals, compute the current density and the number of charge carriers per second crossing a given plane of the sample under the three doping conditions.

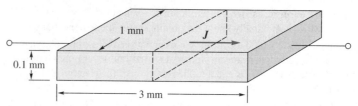

■ **FIGURE 2.24** Silicon sample geometry for Problem 2.42 with reference plane shown.

2.43* Doping in semiconductor devices is often nonuniform as shown in Fig. 2.25. Suppose the boron doping profile were given by

$$N_A(x) = 3 \times 10^{17} e^{-x/1.5 \text{ mm}} \text{ atoms/cm}^3$$

Compute the resistivity of the sample. Use an average value for $\mu_p \approx 225$ cm^2/V-sec.

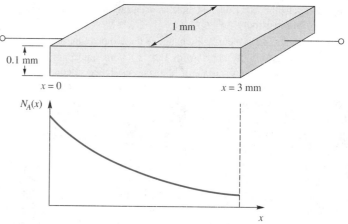

■ **FIGURE 2.25** Silicon sample geometry for Problem 2.43 with x-dependent doping density profile shown.

2.44 Calculate the density of electrons and holes in 2 Ω-cm, n-type silicon at 300 and 400 K.

2.45 Calculate the electron and hole concentration in n-type silicon for $10^{14} < N_D < 10^{18}$ atoms/cm^3. Use the results to plot the conductivity at $T = 300$ K for this doping range.

2.46 Calculate the electron and hole concentration in p-type silicon for $10^{14} < N_A < 10^{18}$ atoms/cm^3. Use the results to plot the conductivity at $T = 300$ K for this doping range. Compare this graph with the one you obtained in Problem 2.45 and explain the differences.

2.47* In silicon, the density of states effective mass is $1.182m_e$ for electrons and $0.810m_e$ for holes where m_e is the rest mass of the electron. If these values were used to calculate n_i, would the results of Problem 2.24 be substantially different?

2.48* Many III–V compounds have E_g between 1.42 and 2.5 eV. Compute n_i. Even though not strictly true for all of these compounds, assume the effective mass of an electron is 6.55% of the rest mass of the electron and the effective mass of the hole is 52.4% of the rest mass of the electron as we did in Problem 2.31.

2.49 Compute the range in diffusivities, D_n and D_p, one can expect at 300 and 350 K for lightly doped and heavily doped silicon.

2.50 Semiconductor devices fabricated from silicon often specify a maximum operating temperature in the 150 to 175°C range. Compute the intrinsic carrier concentrations and compare this value to an equivalent or effective doping density.

2.51 Estimate the electron and hole concentration in a 5 $(\Omega\text{-m})^{-1}$, n-type sample of silicon at $T = 300$ K.

2.52 Use the mobility plot to calculate the density of electrons and holes in an n-type silicon whose resistivity is 1 Ω-cm at 300 K. Repeat the calculation at 400K and compare your results. Assume mobility decreases 30% at 400 K.

2.53* Refer to Fig. 2.26, where $V_S = 0$. An incident light pulse at $t = 0$ results in a 1-V signal at $x = 0$. After 1 μs, the peak amplitude has decreased to 0.5 V. Compute τ_p.

2.54* Assume we have an infinitely long semiconducting bar with an $+x$-directed electric field of 500 V/cm (Fig. 2.27). At $t = 0$, electron-hole pairs are generated at $x = 0$ by a light pulse. After 1 μs, we notice a voltage response at $x = -0.5$ cm. Is the sample n- or p-type? Compute the corresponding μ and D at 300 K.

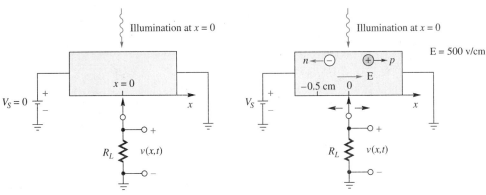

■ **FIGURE 2.26** Experimental setup for Problem 2.53.

■ **FIGURE 2.27** Experimental setup for Problem 2.54.

2.55* Assume we have an infinitely long semiconducting bar. At $t = 0$, electron-hole pairs are generated at $x = 0$ by a light pulse (Fig. 2.28). There is no external electric field. At $t = 3$ μs, the peak amplitude has decreased to 10% of the peak value at $t = 1$ μs. From these data and the results of the Problem 2.54 measurements, compute the minority carrier lifetime τ and diffusion length L.

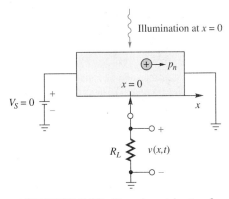

■ **FIGURE 2.28** Experimental setup for Problem 2.55.

REFERENCES

1. *Handbook of Chemistry and Physics,* 75 ed. (Robert C. Weast, Ed.), Cleveland, Ohio: The Chemical Rubber Company, 1995.
2. van der Ziel, A. *Solid State-Physical Electronics.* Englewood Cliffs, N.J.: Prentice-Hall, 1976. An advanced-level textbook with detailed derivations of many of the solid-state physical relationships.
3. Weinberger, H. F. *Partial Differential Equations.* New York: John Wiley & Sons, 1965. One of a large number of differential equation advanced texts that discuss the diffusion question.
4. Omar, M. A. *Elementary Solid-State Physics: Principles and Applications.* Reading, Mass.: Addison-Wesley, 1975. An advanced-level textbook with good physical interpretations of solid-state physical phenomena.
5. Conwell, E. M. "Properties of Silicon and Germanium," *Proceedings of IRE,* June 1958, pp. 1281–1300.
6. Irvin, J. C. "Resistivity of Bulk Silicon and of Diffused Layers in Silicon," *Bell System Technical Journal,* March 1962, pp. 387–410.
7. Stevenson, D. T., and R. J. Keyes. "Measurement of Carrier Lifetime in Germanium and Silicon," *Journal of Applied Physics,* Vol. 26, 1955, p. 190.
8. Haynes, J. R., and W. Shockley. "The Mobility and Life of Injected Holes and Electrons in Germanium," *Physical Review,* Vol. 81, 1951, p. 835.
9. Streetman, B. G., *Solid State Electronic Devices,* 4th ed. Englewood Cliffs, N.J.: Prentice-Hall, 1995. Very good textbook written for the senior or first-year graduate student in electrical engineering.

SUGGESTED READINGS

Electronics, April 7, 1980. The entire issue is devoted to a 50-year historical review of developments in electronic theory and applications through 1980.

Kittel, C. *Introduction to Solid-State Physics,* 6th ed. New York: John Wiley & Sons, 1986. An intermediate to advanced textbook.

Pierret, R. F. *Semiconductor Fundamentals, Modular Series on Solid State Devices,* Vol. 1, 2nd ed. Reading, Mass.: Addison-Wesley, 1989. Several volumes of this set have been combined in an intermediate-level text: Pierret, R. F., *Semiconductor Device Fundamentals.* Reading, Mass.: Addison-Wesley, 1996.

Sproull, R. L., and W. A. Phillips. *Modern Physics: The Quantum Physics of Atoms, Solids, and Nuclei,* 3rd ed. New York: John Wiley & Sons, 1980. A basic physics textbook.

Sze, S. M. *Physics of Semiconductor Devices,* 2nd ed. New York: John Wiley & Sons, 1981. A classic reference written at an intermediate to advanced level.

Wang, F. Y. *Introduction to Solid State Electronics.* Amsterdam: North-Holland Publishing Company, 1980. An intermediate-level text for solid-state and quantum electronics.

Wang, S. *Fundamentals of Semiconductor Theory and Device Physics.* Englewood Cliffs, N.J.: Prentice-Hall, 1989. A textbook written at the intermediate to advanced level.

CHAPTER 3

SEMICONDUCTOR DIODES AND DIODE CIRCUITS

3.1 The *pn* Junction in Equilibrium
3.2 The Externally Biased Junction
3.3 The Diode Equation
3.4 Practical Diode Characteristics
3.5 Load Lines and Piecewise-Linear Diode Models
3.6 Dynamic Resistance
3.7 Rectifier Circuits
3.8 Breakdown-Diode Voltage Regulator
3.9 Diode Wave-Shaping Circuits
3.10 Diode Logic Circuits
3.11 Diode Analog Switch
3.12 Diode Capacitance and Switching Times
3.13 Metal-Semiconductor Junctions
3.14 Photonic and Microwave Diodes
3.15 Diode Heating
3.16 Spice Model for the Diode
Summary
Survey Questions
Problems
References
Suggested Readings

In this chapter we introduce the *pn* junction. The *pn* junction is the major part of a device called a *junction diode*, which is very important in its own right. However, perhaps of even more significance, the *pn* junction is the basis for nearly all other solid-state devices and for integrated circuits. Thus, you need to understand the *pn* junction before proceeding to the electronic devices discussed in subsequent chapters.

Figure 3.1 shows pictures of several commercial diode packages to help you visualize what a diode might look like. First, we present diode structure and physical processes, which develop naturally into the diode equation. Next we discuss diode characteristics. Representative diode data sheets are presented for comparison. Then we develop simple circuit models for the diode characteristic.

We immediately present a number of common diode circuits with the dual purpose of exposing you to a variety of typical diode applications and of providing experience in the analysis of nonlinear circuits. We treat the various rectifier circuits in some detail because of their widespread use in power supplies for electronic equipment. We also consider the application of the breakdown diode to the problem of maintaining a constant voltage but reserve the topics of active regulators, both series and switching, for study in Chapter 17. Then we introduce diode limiter circuits in preparation for application as input protection for the various integrated circuits to be studied in subsequent chapters. In this section we introduce the *transfer characteristic*, in which a circuit's output voltage is plotted as a function of its input voltage. The transfer characteristic is a valuable circuit analysis tool. We also present a variety of other circuits, including the diode clamp (dc restorer), diode logic gates, and a diode analog switch.

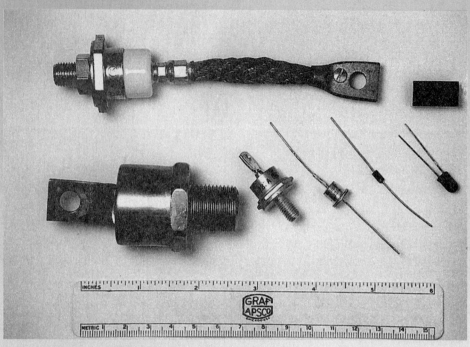

■ **FIGURE 3.1** Several commercial diode packages. Top left, 40A diode (DO-8 package); top right, 7-segment LED display (dual-inline-package); bottom from left to right, 150A diode, 6A diode (DO-9 package), 1A diode (metal package), 1A diode (plastic package DO-41), LED diode (light-emitting-diode).

After some practice on the solution of simple diode circuits, we treat the topics of junction capacitance, diode switching speeds, light-emitting diodes, metal-semiconductor junctions, and diode self-heating.

A computer analysis of a diode circuit is illustrated, using the very common circuit simulation program SPICE. A SPICE model of the diode is presented, with several example programs, including a simulation of the peak rectifier circuit, which demonstrates an appropriate use of the computer where numerical iteration is involved.

As you study this chapter, note that the key terms are marked in the margin, and be on the lookout for the important concepts listed below. Answer each checkup question before proceeding. Having read the material, consider the survey questions, because they will help you gauge your comprehension of the material.

IMPORTANT CONCEPTS IN THIS CHAPTER

- Majority carriers diffuse across a *pn* junction in equilibrium, creating a space charge layer, and a built-in potential barrier.
- Applying a forward-bias voltage to a diode lowers the junction potential barrier, and allows a diffusion current to flow from *p* to *n*.
- Applying a reverse-bias voltage to a diode raises the junction potential barrier, and allows only a very small drift current to flow from *n* to *p*.
- Diode current is an exponential function of applied diode voltage v_D.
- Saturation current is a parameter of the diode, and it doubles for a 5°C rise in temperature near 27°C.
- Light energy falling on a *pn* junction causes the saturation current to increase.
- Reverse-biased *pn* junctions will conduct if the breakdown voltage is reached.
- An *ideal diode* has 0 V drop for $i_D > 0$, and zero current for $v_D < 0$.
- A practical diode is often modeled by an ideal diode in series with an offset V_F.
- The dynamic resistance of a diode is the reciprocal of the slope of the diode equation at the operating point.
- Rectifiers change alternating current to direct current. Half-wave rectifiers allow load current during one-half of the ac waveform, and full-wave rectifiers allow load current during both halves of the ac waveform.
- Peak rectifier ripple voltage is proportional to the load current, and inversely proportional to the filter capacitance and the frequency.
- A breakdown diode has nearly constant breakdown voltage for a wide range of breakdown current, and can be used to improve the *voltage regulation* of a circuit.
- A clipper circuit prevents a circuit voltage from exceeding a reference voltage.
- A clamp circuit fixes one extreme of a waveform to some reference voltage.
- Diode transition capacitance varies inversely with the square root of the reverse-bias voltage for an abrupt junction.
- Diode diffusion capacitance is proportional to diode current and minority carrier lifetime.
- When a diode v_D is reversed, a delay occurs before the diode turns OFF.
- Metal-semiconductor contacts can be made nonrectifying by heavily doping the semiconductor at the interface.
- Schottky metal-semiconductor diodes are very fast and have low offset voltage.
- Forward-biased *pn* diodes can be made to emit light or to operate as a semiconductor laser.
- Junction heating raises junction temperature above ambient.
- The SPICE model for the diode allows the user to control many parameters, providing for accurate simulation of circuits containing diodes.

3.1 THE *pn* JUNCTION IN EQUILIBRIUM

Consider the situation in a semiconductor crystal having two physically isolated regions, one doped with acceptors and one doped with donors. In Fig. 3.2, the *p*-type region is depicted as an array of paired charges: the negative acceptor ions (circled) and the positive holes. Likewise, the *n*-type region contains an array of positive donor ions (circled) paired

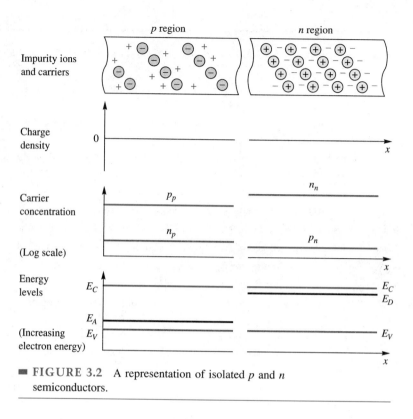

■ **FIGURE 3.2** A representation of isolated p and n semiconductors.

with electrons. This figure assumes a situation in which the concentration of donors on the n side is greater than the concentration of acceptors on the p side. The atoms of the semiconductor itself are not shown, but as you will recall, they are orders of magnitude more numerous than the impurity atoms.

The energy levels in Fig. 3.2 reveal a band of energy states at E_A provided by the acceptors in the p region and a band of energy states at E_D provided by the donors in the n region. At room temperature, most of the donors are ionized; that is, each unbound electron has acquired enough thermal energy to escape from its atom and is free to circulate, leaving the positive ion bound in the crystal structure. A relatively small number of the basic semiconductor atoms are also ionized because it requires much more thermal energy to break a bond. The energy-level diagram shows that the donor electrons only need to acquire a small amount of energy to move from E_D to E_C. Likewise, most of the acceptors have acquired an electron from the valence band, leaving holes free to circulate from bond to bond. In other words, the E_A states fill as valence-band electrons acquire thermal energy. In the absence of external forces, the carrier concentration is uniform. Pair generation and recombination processes are also in equilibrium, so that a relatively small number of minority carriers are present.

EXAMPLE 3.1

Suppose that the regions of Fig. 3.2 are silicon doped with 10^{14} acceptors and 10^{16} donors/cm^3, respectively. Assume $T = 300$ K. Find the majority and minority carrier concentrations of both regions.

SOLUTION From Table 2.2 in Chapter 2 we find $n_i = 1.5 \times 10^{10}$ electrons/cm^3. For the p region, assuming all of the acceptors ionized require $p_p = 10^{14}$ holes/cm^3, and $n_p = n_i^2/p_p = 2.25 \times 10^6$ electrons/cm^3. Similarly, $n_n = 10^{16}$ electrons/cm^3 and $p_n = 2.25 \times 10^4$ holes/cm^3 in the n region.

The basic electronic device known as a **junction diode** consists of a p-type semiconductor in contact with an n-type semiconductor. This *pn junction* is formed within a single-crystal semiconductor by doping one region with donor impurities and an adjacent region with acceptors. We want to emphasize that there must be no disruption in the crystal lattice at the interface. Several procedures for fabricating diodes are discussed in Chapter 18.

Figure 3.3 illustrates some of the many relationships at a junction between p and n regions. Curve (a) shows the n side to be doped with N_D donors/cm^3, and the p side to be doped with a lower concentration, N_A acceptors/cm^3. As before, in part (b), only the impurity ions and free carriers are represented. We assume that no external potentials are applied to the crystal, and carrier densities are functions only of x.

Consider the situation on the n side of the junction. The bulk of the region is electrically neutral, because there are about the same number of electrons and donor ions. However, the electrons that are very near the junction will diffuse away from the region of high electron concentration on the n side to the lower electron concentration on the p side. The electron current density is, from Eq. (2.17),

$$J_n = qD_n \frac{dn}{dx}. \quad (3.1)$$

A similar situation exists on the p side, where holes diffuse toward the n side, and the analogous hole current density is

$$J_p = -qD_p \frac{dp}{dx}. \quad (3.2)$$

Both the electron diffusion current and the hole diffusion current constitute a conventional current flow to the right.

The ionized donors at the edge of the junction, whose charge is no longer compensated by the charge of the electrons that have diffused to the p region, are said to be *uncovered*. Over in the p region, acceptor ions are similarly uncovered by the diffusion of holes to the n side. These thin layers of unneutralized ions are termed the **space-charge layer, space-charge region,** or **depletion region.**

The charge distribution for an abrupt junction is depicted in Fig. 3.3(c). Charge density in the space-charge layer is equal to the impurity concentration times the charge per ion. Charge neutrality requires that the areas of the two rectangles be the same. Because the donors have the higher density in this example, the positive space-charge layer of thickness w_n is thinner than the negative space-charge region, which has thickness w_p.

From the charge distribution, we may obtain the electric field distribution by means of Poisson's equation in one dimension,

$$\frac{d\mathbf{E}}{dx} = \frac{\rho}{\varepsilon}, \quad (3.3)$$

where ρ is the charge density, and $\varepsilon = \varepsilon_r \varepsilon_o$ is permittivity of the semiconductor, with $\varepsilon_o = 8.85 \times 10^{-14}$ F/cm being the permittivity of free space. We assume the electric

---

DRILL 3.1

Given a silicon *pn* junction, doped with 10^{15} acceptors/cm^3 on one side and with 10^{15} donors/cm^3 on the other side, what are the minority carrier concentrations on each side at room temperature?

ANSWER
$p_n = 2.25 \times 10^5$ holes/cm^3,
$n_p = 2.25 \times 10^5$ electrons/cm^3.

■ The *pn* junction is the basic building block of all semiconductor devices and integrated circuits.

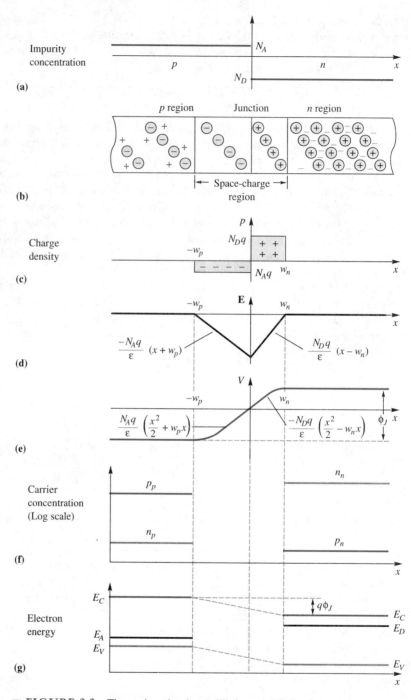

FIGURE 3.3 The *pn* junction in equilibrium. (a) Step impurity profile. (b) Uncovered charges at the junction. (c) Charge density. (d) Electric field. (e) Potential. (f) Carrier concentration. (g) Energy-level diagram.

CHAPTER 3   SEMICONDUCTOR DIODES AND DIODE CIRCUITS

field is zero in the neutral region to the left of $x = -w_p$, where there is no trapped charge, and integrate Eq. (3.3) over the space-charge region with $\rho = -qN_A$:

$$\mathbf{E} = \int_{-w_p}^{x} \frac{\rho}{\varepsilon} dx = \frac{-qN_A}{\varepsilon}(x + w_p) \qquad -w_p < x < 0. \qquad (3.4)$$

On the n side, noting that $\mathbf{E}$ must be zero at $x = w_n$, where a neutral region begins,

$$\mathbf{E} = \int_{w_n}^{x} \frac{qN_D}{\varepsilon} dx = \frac{qN_D}{\varepsilon}(x - w_n) \qquad 0 < x < w_n. \qquad (3.5)$$

The electric field intensity is plotted in Fig. 3.3(d), and we see at once that the maximum field intensity occurs at the metallurgical junction ($x = 0$). Recognizing that the electric field must be continuous at $x = 0$, we find that

$$N_A w_p = N_D w_n, \qquad (3.6)$$

which illustrates the statement made earlier that the space-charge layer extends further into the more lightly doped side. This electric field will sweep the carriers out of the space-charge region, resulting in an electron drift current, from Eq. (2.12), of

$$J_n = qn\mu_n \mathbf{E}. \qquad (3.7)$$

Because we can see from the figure that $\mathbf{E}$ is negative ($\mathbf{E}$ points in the opposite direction from increasing x), the electron drift current is from the p to the n side. Recall that electrons diffuse from the n side to the p side, so these two components of electron current density always oppose each other in a pn junction. Analogous statements can be made for hole currents.

In the equilibrium situation (no external potential applied to the junction, so that no net current flows through the junction), diffusion current to the right will exactly equal drift current to the left. This means that the width of the space-charge layer will stabilize at such a value that every electron diffusing to the p region under the influence of the high concentration gradient for electrons will be matched by an electron that has entered the space-charge region from the p region and been swept by the electric field to the n side. The same argument can be applied to holes.

Figure 3.3(f) reveals uniform distribution of carriers in each bulk semiconductor region. Within the space-charge region, large numbers of carriers are in motion. Carrier distribution will be such that the large diffusion current will just cancel the large drift current created by the "built-in" field at equilibrium. We shall see that this balance of drift and diffusion currents may be disturbed by application of external potential to the junction.

■ A pn junction with no applied voltage at the terminals has no net junction current, and is at equilibrium.

The energy-level diagram in Fig. 3.3(g) shows that the energy states of the conduction band (and valence band) have a higher value in the p material than in the n material. Consider the fact that the average energy of free electrons in the p region is closer to the valence band because of the acceptor states, whereas in the n material their average energy is closer to the conduction band because of the many electrons in donor states. However, for equilibrium to exist when the two regions are in contact, the average electron energy must be uniform, otherwise a transfer of energy will occur. The higher energy levels of the conduction and valence bands on the p side of the junction reflect a **contact potential**, ϕ_J, that exists across the depleted region. In essence, the contact potential represents a potential barrier that must be surpassed by a charge carrier in order for it to diffuse across the junction.

Potential variation across the space-charge region may be found by integration of the electric field, $V = -\int \mathbf{E}\, dx$. Refer to the plot of Fig. 3.3(e). Using the electric field from Eqs. (3.4) and (3.5), we obtain

$$V_J = -\int_{-w_p}^{w_n} \mathbf{E}\, dx = \frac{N_A q}{\varepsilon}\int_{-w_p}^{0}(x + w_p)\, dx + \int_{0}^{w_n} \frac{N_D q}{\varepsilon}(w_n - x)\, dx \qquad (3.8)$$

$$= \frac{q}{2\varepsilon}(N_A w_p^2 + N_D w_n^2).$$

In the equilibrium situation, this is the contact potential ϕ_J.

To determine the value of ϕ_J, we use the fact that drift current is equal and opposite to diffusion current at equilibrium. From Eqs. (3.1) and (3.7), we may equate the electron components:

$$qD_n \frac{dn_o}{dx} = -qn_o \mu_n \mathbf{E}, \qquad (3.9)$$

where n_o is the equilibrium concentration of electrons in the space-charge region. Combining with the Einstein relation [Eq. (2.19)] leads to

$$V_T \frac{dn_o}{n_o} = -\mathbf{E}\, dx. \qquad (3.10)$$

Integrating across the space-charge region involves

$$V_T \int_{n_{po}}^{n_{no}} \frac{dn_o}{n_o} = \int_{-w_p}^{w_n} -\mathbf{E}\, dx, \qquad (3.11)$$

where n_{no} is the equilibrium electron concentration on the n side and n_{po} is the equilibrium electron concentration on the p side. The right-hand integral is the contact potential ϕ_J, so that

$$\phi_J = V_T \ln \frac{n_{no}}{n_{po}}. \qquad (3.12)$$

Because $n_{no} = N_D$ and $n_{po} = n_i^2/N_A$, Eq. (3.12) may be expressed as

$$\phi_J = V_T \ln \frac{N_D N_A}{n_i^2}. \qquad (3.13)$$

We can see that the contact potential is related only to the junction doping levels and to temperature.

In summary, the junction potential is a built-in potential barrier that is necessary to maintain equilibrium conditions. We must not expect to measure this contact potential with a voltmeter, however, because of offsetting contact potentials wherever we make contact with the crystal.

We can find the width of the space-charge layers by combining Eqs. (3.6) and (3.8):

$$w_n = \left[\frac{2\varepsilon V_J}{q\left(N_D + \frac{N_D^2}{N_A}\right)}\right]^{1/2} \quad \text{and} \quad w_p = \left[\frac{2\varepsilon V_J}{q\left(N_A + \frac{N_A^2}{N_D}\right)}\right]^{1/2}. \qquad (3.14)$$

■ Doping profiles are discussed in detail in Chapter 18.

Note that in developing Eqs. (3.3) through (3.14) we have assumed an *abrupt junction*. Similar calculations may be carried out for *graded junctions*, where the doping profile varies continuously from the p side to the n side, as in certain diffused junctions. We shall point out any significant differences as we proceed.

CHAPTER 3 SEMICONDUCTOR DIODES AND DIODE CIRCUITS

EXAMPLE 3.2

Given a silicon *pn* junction with $N_A = 10^{14}$ acceptors/cm^3 and $N_D = 10^{16}$ donors/cm^3 at T = 300 K, find the junction contact potential and the space-charge layer thicknesses w_p and w_n.

SOLUTION $\varepsilon_r = 11.8$, and $n_i = 1.5 \times 10^{10}$ electrons/cm^3. Then

$$\varepsilon = \varepsilon_r \varepsilon_o = (11.8)(8.85 \times 10^{-14}) = 1.04 \times 10^{-12} \text{ F/cm.}$$

At 300 K, $V_T \cong 0.026$ V. Thus, from Eq. (3.13),

$$\phi_J = 0.026 \ln \frac{10^{14}\, 10^{16}}{2.25 \times 10^{20}} = 0.58 \text{ V.}$$

Using this value in Eq. (3.14),

$$w_p = \left[\frac{2(1.04 \times 10^{-12})(0.58)}{(1.6 \times 10^{-19})(10^{14} + 10^{12})}\right]^{1/2} = 2.73 \times 10^{-4} \text{ cm}$$

$$w_n = w_p \frac{N_A}{N_D} = 2.73 \times 10^{-6} \text{ cm}$$

CHECK UP

1. **TRUE OR FALSE?** Drift current and diffusion current are equal and opposite in a *pn* junction at equilibrium.
2. **TRUE OR FALSE?** Electron current and hole current are equal in a *pn* junction.
3. **TRUE OR FALSE?** The space-charge region extends further into the side of the junction that is more highly doped.
4. **TRUE OR FALSE?** Contact potential increases with doping concentrations.

DRILL 3.2

A silicon *pn* junction having 10^{18} donors/cm^3 and 10^{15} acceptors/cm^3 is at 300 K. Find the junction potential of this diode.

ANSWER 0.76 V.

DRILL 3.3

Find the width of the depletion region on each side of the junction of the diode of Drill 3.2.

ANSWER
$w_p = 9.9 \times 10^{-5}$ cm,
$w_n = 9.9 \times 10^{-8}$ cm.

3.2 THE EXTERNALLY BIASED JUNCTION

In electronic circuits, the term **bias** concerns the dc operating constraints placed on a device by the external circuit. Consider the diagram of Fig. 3.4, where the standard symbol

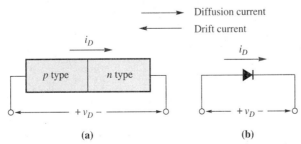

FIGURE 3.4 (a) External voltage v_D applied to *pn* junction. (b) Diode symbol.

for the diode is also shown. The arrow alludes to the direction of conventional current flow in the *forward-biased* diode (from *p* to *n*). The *p* terminal is referred to as the **anode**, the *n* terminal as the **cathode**. An external potential v_D has been applied to the diode. Internally, the potential barrier of the junction is effectively lowered by an amount v_D. Assuming no voltage drop in the neutral regions, the voltage v_D superimposes a field in the space-charge region opposed to the built-in field, effectively weakening it. The junction potential decreases: $v_J = \phi_J - v_D$. This can be seen in the energy-level diagram in Fig. 3.5(d). As the potential barrier is lowered, many more electrons have the energy to diffuse to the *p* side (and holes to the *n* side). Thus, the external forward-bias potential creates a small imbalance between the diffusion current and the drift current components. Diffusion current dominates, and the net current is from *p* to *n* in the forward-biased diode. Figure 3.5(b) also shows that the number of uncovered charges is reduced, and the space-charge region narrows, reflecting the fact that the barrier potential is less.

■ In a forward-biased *pn* junction, the *p* terminal is positive with respect to the *n* terminal, and current flows easily from *p* to *n*. The diode is ON.

We are now ready to consider the plot of carrier concentration in Fig. 3.5(c). Holes that diffuse to the *n* side are said to be minority carriers *injected* into that region. Because of the great number of majority carriers (electrons) in this region, the holes will recombine within a short distance of the junction. Thus, an exponential concentration gradient causes a continuous diffusion away from the space-charge region boundary. We can predict the shape of this curve. Let $x = 0$ at the *n* side of the depletion region, so that $p_n(0)$ is the

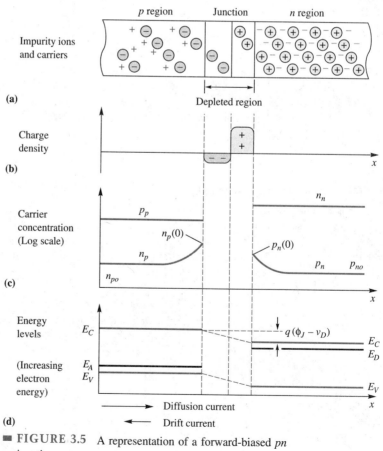

■ **FIGURE 3.5** A representation of a forward-biased *pn* junction.

density of holes at this boundary. We shall see that $p_n(0)$ is determined by the magnitude of forward-bias voltage v_D. Note that p_{no} is the equilibrium hole concentration on the n side. Having defined L_p (the diffusion length for holes) as the average distance that holes will diffuse before recombining, we find from Eq. (2.39) that

$$p_n(x) = [p_n(0) - p_{no}]e^{-x/L_p} + p_{no}. \tag{3.15}$$

This plot assumes that the n region is several diffusion lengths thick (generally a realistic assumption for diodes). At the boundary of the space-charge region, the current consists entirely of the net excess holes diffusing away from the junction. Because these holes are continuously recombining, the current gradually becomes electron current, and the hole density decreases. At a distance of one diffusion length from the boundary, 63% of the excess holes have recombined, and the current is becoming majority carrier current. An analogous situation will exist on the other side of the junction, where electrons are injected into the p region and become hole current there.

At equilibrium we see from Eq. (3.12) that

$$n_{po} = n_{no}e^{-\phi_J/V_T}. \tag{3.16}$$

It does not follow that the equilibrium equations will apply to the forward-biased junction, but it turns out[1] that the situation is very near to that of equilibrium as long as we maintain what is termed **low-level injection.** This means that the injected minority carrier levels remain well below the majority carrier levels [$n_p(0) \ll p_{po}$, and $p_n(0) \ll n_{no}$]. By contrast, **high-level injection** implies that this inequality is not satisfied. This assumption is also required to maintain a negligible electric field in the neutral regions. Thus, we repeat the integration of Eq. (3.11), this time using $n_p(0)$ as the electron concentration at the p boundary:

$$V_T \int_{n_p(0)}^{n_{no}} \frac{dn}{n} = \int_{-w_p}^{w} -\mathbf{E}\, dx. \tag{3.17}$$

The right-hand integral is the junction potential $\phi_J - v_D$, so that

$$\phi_J - v_D = V_T \ln \frac{n_{no}}{n_p(0)}. \tag{3.18}$$

Rearranging, and combining with Eq. (3.16), we obtain

$$n_p(0) = n_{po}e^{v_D/V_T}. \tag{3.19}$$

The same process would yield the injected hole concentration at the n boundary:

$$p_n(0) = p_{no}e^{v_D/V_T}. \tag{3.20}$$

Thus, Eq. (3.15) becomes

$$p_n(x) = p_{no}(e^{v_D/V_T} - 1)e^{-x/L_p} + p_{no}. \tag{3.21}$$

An analogous expression relates to electron concentration in the p region. With $x = 0$ at the p edge of the space-charge layer,

$$n_p(x) = n_{po}(e^{v_D/V_T} - 1)e^{+x/L_n} + n_{po} \qquad x < 0. \tag{3.22}$$

EXAMPLE 3.3

A silicon diode at $T = 300$ K having doping levels of $N_D = 10^{18}$ donors/cm^3 and $N_A = 10^{15}$ acceptors/cm^3 is forward biased with $V_D = 0.5$ V. Sketch minority carrier concentrations, assuming L_p and $L_n \ll$ diode length.

SOLUTION $n_n \approx N_D = 10^{18}$ electrons/cm^3. Then,

$$p_{no} = n_i^2/n_n = 2.25 \times 10^2 \text{ holes/cm}^3$$
$$p_n(0) = p_{no}e^{v_D/V_T} = p_{no}e^{qv_D/kT}$$
$$= 2.25 \times 10^2 \exp[(1.6 \times 10^{-19})(0.5)/(1.38 \times 10^{-23})(300)]$$
$$= 5.5 \times 10^{10} \text{ holes/cm}^3.$$

Similarly,

$p_p = 10^{15}$ cm^{-3}, $n_{po} = 2.25 \times 10^5$ holes/cm^3, $n_p(0) = 5.5 \times 10^{13}$ electrons/cm^3.

Figure 3.6 shows the distribution of carriers. This is low-level injection, since $n_p(0) \ll p_p$.

One might expect that the junction potential barrier could be reduced to zero with the application of sufficient external forward bias, but there are practical mechanisms to prevent this. We have been tacitly neglecting the potential drop in the bulk semiconductor and in the terminal metal-semiconductor contacts. With sufficient forward bias, the diode current increases dramatically, and voltage drops develop in the "neutral" regions, so that we are unable to reduce the potential barrier to zero. More will be said about the metal-semiconductor contacts in Section 3.13.

If the external potential v_D in Fig. 3.4 is made negative, the diode is said to be ***reverse biased***. Refer to Fig. 3.7. As you might except from the foregoing discussion, the internal energy barrier at the junction is effectively raised. The electric field is strengthened, the space-charge layer widens, and the diffusion current is reduced to near zero.

Note again that the density of holes at the n-side edge of the space-charge region, $p_n(0)$, is found to be $p_{no}e^{qv_D/kT}$. This time, however, this represents a concentration lower than the equilibrium concentration for holes. Only a small reverse voltage is required to reduce $p_n(0)$ to virtually zero, so that the diode current consists only of those thermally generated minority carriers that diffuse to the space-charge region and are swept by the field across the junction. There are now relatively few free carriers in the space-charge layer, which is often referred to as the *depleted region* when reverse biased. In the depleted region, drift current exceeds diffusion current. Thus, electrons move from left to right,

■ In a reverse-biased pn junction, the p-terminal is negative with respect to the n terminal, and only a negligible current flows from n to p. The diode is OFF.

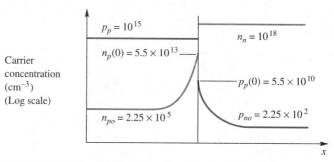

■ **FIGURE 3.6** Distribution of carriers for Example 3.3.

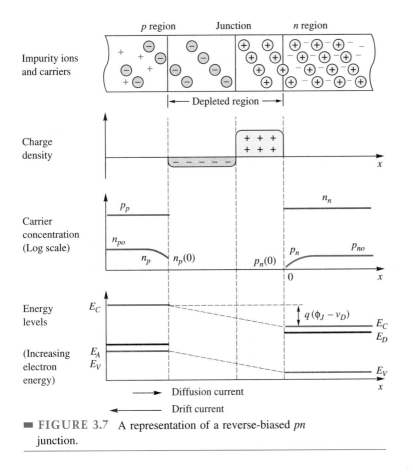

FIGURE 3.7 A representation of a reverse-biased *pn* junction.

and holes from right to left. This very small current, in the opposite direction from the forward-bias current, is known as *reverse current*.

3.3 THE DIODE EQUATION

We are now ready to determine diode current as a function of diode voltage. Figure 3.4(a) applies; n and p regions are assumed to be several diffusion lengths long, and the space-charge layer is so thin on this scale that it is represented by the plane $x = 0$. We shall assume low-level injection and no potential gradient in the neutral regions. At $x = 0^+$, all hole current is assumed to be diffusion current, and at $x = 0^-$, all electron current is assumed to be due to diffusion. This is true because under low-level injection there is no significant field in the neutral regions; charge neutrality is maintained by the high concentration of majority carriers with minimal potential gradient. From Eqs. (2.17) and (2.18), in one dimension, we use the diffusion equation

$$i_D = A_j\left(qD_n\frac{dn}{dx} - qD_p\frac{dp}{dx}\right), \quad (3.23)$$

assuming negligible recombination in the very thin space-charge layer, so that the sum of the two diffusion components will yield the total junction current.

DRILL 3.4

The diode of Example 3.3 is reverse biased with $V_D = -0.5$ V. Estimate the minority carrier concentration at the edge of each depletion region.

ANSWER
$p_n(0) \cong 0$,
$n_p(0) \cong 0$.

Differentiating Eq. (3.22), and letting $x = 0$ at the p edge,

$$\frac{dn}{dx} = \frac{n_{po}}{L_n}(e^{qv_D/kT} - 1) = \frac{n_{po}}{L_n}(e^{v_D/V_T} - 1). \tag{3.24}$$

Similarly, using Eq. (3.21), where $x = 0$ at the n boundary of the space-charge region,

$$\frac{dp}{dx} = -\frac{p_{no}}{L_p}(e^{v_D/V_T} - 1). \tag{3.25}$$

Substituting these concentration gradients into Eq. (3.23),

$$i_D = A_j q \left(\frac{D_n n_{po}}{L_n} + \frac{D_p p_{no}}{L_p}\right)(e^{v_D/V_T} - 1) \tag{3.26}$$

$$= I_S(e^{v_D/V_T} - 1).$$

■ The diode equation is a very important relationship for all *pn* junctions.

This is the **diode equation**, and I_S is the diode **reverse saturation current**. The term *saturation* relates to the idea that this current is reached for small negative v_D and does not increase further for more negative values of v_D. This is because as $n_p(0)$ and $p_n(0)$ reach zero, dp/dx cannot change further as v_D becomes more negative. The diffusion constants and the diffusion lengths in Eq. (3.26) are for minority carriers. Note that at $x \neq 0$, the junction current is not all diffusion current, but the total current must remain constant. Equation (3.26) is more generally written

$$i_D = I_S(e^{v_D/\eta V_T} - 1), \tag{3.27}$$

where $\eta = 1$ for pure diffusion current. However, where recombination in the depleted region is very significant (as in silicon with values of v_D less than 0.5 V), η may need to be increased to 2. It is also true that $\eta = 2$ for high-level injection. At moderate diode currents between these conditions, $1 < \eta < 2$. For most cases in this text, we will assume $\eta = 1$.

■ EXAMPLE 3.4

A silicon *pn* junction has a cross-sectional area of 10^{-2} cm^2 and $N_A = 10^{18}$ acceptors/cm^3 and $N_D = 10^{16}$ donors/cm^3. Excess carrier lifetimes are both 1 μs, T = 300 K, and $\eta = 1$. Assume Fig. 2.18 may be used to find minority carrier mobilities. Find the diode equation for this diode.

SOLUTION From Fig. 2.18, $\mu_n \approx 360$ cm^2/(V-s),

$$\mu_p = 390 \text{ cm}^2/\text{(V-s)}$$

$$p_{no} = \frac{n_i^2}{N_D} = \frac{(1.5 \times 10^{10})^2}{10^{16}} = 2.25 \times 10^4 \text{ holes/cm}^3.$$

Similarly,

$$n_{po} = 2.25 \times 10^2 \text{ electrons/cm}^3.$$

The Einstein relation yields

$$D_n = \frac{kT}{q}\mu_n = 9.3 \text{ cm}^2/\text{sec} \quad \text{and} \quad D_p = 10.1 \text{ cm}^2/\text{sec}.$$

Then

$$L_n = \sqrt{D_n \tau} = 3.1 \times 10^{-3} \text{ cm} \quad \text{and} \quad L_p = 3.2 \times 10^{-3} \text{ cm}.$$

From Eq. (3.26),

$$I_S = (1.6 \times 10^{-19})(10^{-2})\left[\frac{(2.25 \times 10^2)(9.3)}{3.1 \times 10^{-3}} + \frac{(2.25 \times 10^4)(10.1)}{3.2 \times 10^{-3}}\right]$$

$$= 1.1 \times 10^{-13} \text{ A}$$

$$i_D = 1.1 \times 10^{-13}(e^{39v_D} - 1) \text{ A}$$

DRILL 3.5

Assume a silicon diode with $\eta = 1$, $V_T = 26$ mV, $v_D = 0.6$ V, and $i_D = 1$ mA. Find I_S.

ANSWER
9.5×10^{-14} A.

CHECK UP

1. **TRUE OR FALSE?** In a forward-biased *pn* junction drift current exceeds diffusion current.
2. **TRUE OR FALSE?** The reverse current in a reverse-biased *pn* junction is drift current.
3. **TRUE OR FALSE?** The space-charge region is thinner in a forward-biased junction than in a reverse-biased junction.
4. **TRUE OR FALSE?** Diode saturation current is determined by junction cross-sectional area and the doping levels of the *p* and *n* regions.

DRILL 3.6

Suppose the diode of Drill 3.5 has v_D changed to 0.5 V. Find I_D.

ANSWER
2.1×10^{-5} A.

3.4 PRACTICAL DIODE CHARACTERISTICS

Figure 3.8 shows a plot of the diode equation. Because V_T is approximately 26 mV at room temperature (300 K), i_D behaves exponentially for v_D more positive than 50 mV. Also, for v_D more negative than -50 mV, the diode current is "saturated" at $-I_S$. The negative current scale is exaggerated to reveal the very small I_S. The figure also takes note of the fact that the diode equation becomes invalid at v_D sufficiently negative, where current increases quite markedly due to ***voltage breakdown***.

Reverse Diode Current

As shown in Eq. (3.26), the reverse saturation current is a function of junction area, minority carrier diffusion constants, the equilibrium minority carrier concentrations, and the diffusion lengths for minority carriers. These parameters are, in turn, functions of temperature and doping levels. Saturation current I_S may be in the range of microamperes

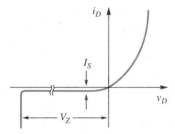

■ **FIGURE 3.8** A plot of the diode characteristic, with a breakdown shown at reverse voltage V_Z. Not to scale, so that I_S may be seen.

for germanium diodes and in the range of nanoamperes for small silicon diodes. Many diodes exhibit an increase in reverse current with increasing reverse voltage not predicted by the diode equation. One cause of increased reverse current is the leakage "around" the junction at the surface of the crystal. Second, in developing the diode equation, we neglected the thermal generation of hole-electron pairs in the space-charge layer. For reverse-biased silicon junctions this is not valid; indeed, this **thermal generation current** dominates the saturation current at room temperature and below. This current is somewhat voltage dependent, because the depleted layer becomes thicker at higher reverse voltages.

Temperature Effects

■ This temperature dependence of I_S has a significant effect on the diode equation. The approximations of Eq. (3.28) will be very useful near room temperature.

We have seen that reverse saturation current varies with minority carrier density, which, in turn, varies with n_i^2, where n_i^2 is a function of temperature, Eq. (2.27). It can be shown that I_S doubles for silicon for a 5°C increase in temperature near 25°C (see Problem 3.7). For constant forward v_D, i_D doubles for a 10°C increase in temperature near 25°C. At constant i_D, v_D decreases approximately 2 mV for each degree Celsius increase in temperature near 25°C, depending on η:

$$\left.\frac{dv_D}{dT}\right|_{i_D} \cong -2 \text{ mV/°C} \qquad \left.\frac{di_D}{dT}\right|_{v_D} = I_D(0.072) \text{ A/°C}. \qquad (3.28)$$

DRILL 3.7

Assume a silicon diode with $\eta = 1$, $T = 300$ K, $I_D = 1$ mA, $V_D = 0.6$ V, and $I_S = 9.5 \times 10^{-14}$ A. Let the temperature increase to 310 K with the same diode voltage. Find the diode current.

ANSWER   2 mA.

Photodiodes

The reverse saturation current just described depends on the generation of hole-electron pairs by the average thermal energy of the crystal. This reverse current may be greatly increased by illuminating the crystal. When the manufacturer makes provision for light to reach the semiconductor, we have what is called a **photodiode**. The photodiode package may include a lens or some type of optical conductor to concentrate the light on the junction region. This type of device has rapidly increasing application in optoelectronics, where light intensity is used to control a current.

The ability of the photodiode to cause light energy to create hole-electron pairs varies with the semiconductor material and the wavelength of the light source. Silicon responds broadly from 4000 to 10,000 Å, with a peak at 8500 Å, and germanium responds even more broadly from 3000 to 20,000 Å, peaking at about 15,000 Å. A metal-semiconductor junction, such as gold-silicon, is more compatible with the visible light spectrum, 3500 to 7000 Å. Other structures allow optimization of device performance. The p-i-n diode, a diode having a thin intrinsic layer between the p and n regions, can be designed to optimize quantum efficiency (photons emitted/photons absorbed) because of the depletion region thickness. *Heterojunctions*, where the p- and n-type semiconductors have different band gaps, such as AlGaAs-GaAs, may be designed to optimize quantum efficiency and speed for a given optical wavelength.[2]

DRILL 3.8

For the diode of Drill 3.7, find I_S at 290 K with $V_D = 0.6$ V.

ANSWER
2.4×10^{-14} A.

DRILL 3.9

For the diode of Drill 3.7, find V_D at 310 K with I_D held at 1 mA.

ANSWER   0.58 V.

The **dark current** is the current that will exist in the absence of illumination. The application of light to the junction results in the transfer of energy from the incident photons to the atomic structure, resulting in an increase in minority carriers, and an increase in reverse current. The number of carriers generated is proportional to the intensity of the incident light, usually measured in lumens/ft² = footcandles (1 fc = 1.6×10^{-12} W/m²).

■ This ability of a photodiode to sense light and essentially "turn ON" a diode switch makes this an important sensor.

Turn-on and turn-off times for these devices can be very small, on the order of a nanosecond. Thus, they are used in conjunction with infrared light sources in high-speed scanning and optical communications. More information on photonic sources appears in Section 3.14.

Reverse Breakdown Mechanisms

Consideration of Eqs. (3.4) and (3.14) reveals that the maximum electric field in the space-charge region occurs at the metallurgical junction and that it increases with doping levels.

When the electric field is strong enough, the electrons of the saturation current are accelerated into collisions with valence electrons in the space-charge region. Bonds are broken, and new carriers are freed. They, in turn, are accelerated into more collisions. The result is an uncontrolled current called *avalanche breakdown*.

Should the doping levels be high enough, the space-charge region becomes so narrow that collisions are less likely, but the even more intense field has the force to break the bonds directly. This mechanism is called *Zener breakdown*, and the result is the same as before without the need for collisions.

The two mechanisms can be distinguished by the magnitude of the breakdown voltage. When breakdown occurs at voltages of less than 6 V (as in very heavily doped junctions) it is Zener breakdown. Breakdown at higher voltages is the result of the avalanche mechanism.

Breakdown Diodes

As we can see in Fig. 3.8, breakdown (whether Zener or avalanche) occurs at a rather specific reverse voltage and is nearly independent of increases in current. The manufacturer is able to control this voltage over a wide range by choice of impurity concentrations and profiles. Sometimes breakdown at a specific voltage is desirable, and the so-called *breakdown diode* or *Zener diode* or *avalanche diode* is available over a range of fractions of a volt to several hundred volts. Technically, these terms are not synonymous, but in common usage the term Zener diode may refer to all breakdown diodes. Applications include circuits in which a voltage reference is desired. The manufacturer must make provision for dissipating the junction power ($V_Z \times I_Z$) and will specify a power rating. The temperature coefficient, or percentage change in V_Z per degree, is another parameter of interest. For avalanche diodes, this coefficient is positive, perhaps 0.1%/°C. The Zener mechanism yields a negative temperature coefficient, and diodes in which V_Z is near 6 V and in which both mechanisms are present can have a nearly zero coefficient.

Figure 3.9 shows a typical manufacturer's data sheet for a 500-mW family of breakdown diodes. Note breakdown voltages ranging from 2.4 to 200 V. Three grades are alluded to: voltage tolerance is 5% for B, 10% for A; and 20% for nonsuffix devices. Note also the voltage temperature coefficient. The power derating curve is discussed in Section 3.15. Further discussion of breakdown diodes appears in Section 3.8.

■ Note that all of the breakdown diodes in the Fig. 3.9 data sheet are referred to as "Zener" diodes according to common usage of that name.

Cut-in

Figure 3.10 shows the forward volt-ampere characteristics of typical silicon and germanium diodes at room temperature. When the current scale is chosen appropriately for the maximum rated current, we see that each diode has a threshold forward-bias voltage below which the current is very small, say, less than 1% of rated current. We call this threshold *cut-in*. Because I_S is very much larger for germanium, we observe that the cut-in voltage for the germanium diode is on the order of 0.2 V, as compared to approximately 0.6 V for the silicon diode. In many common applications, the diode may be considered OFF at voltages less than the cut-in voltage.

Maximum Diode Ratings

Refer to the manufacturer's data sheet given in Fig. 3.11. The 1N4001 through 1N4007 is a family of low-power silicon rectifiers. Each diode in the series is rated at some

ELECTRICAL CHARACTERISTICS

($T_A = 25°C$ unless otherwise noted. Based on dc measurements at thermal equilibrium; lead length = 3/8"; thermal resistance of heat sink = $30°C/W$) $V_F = 1.1$ max @ $I_F = 200$ mA for all types.

JEDEC Type No. (Note 1)	Nominal Zener Voltage V_Z @ I_{ZT} Volts (Note 2)	Test Current I_{ZT} mA	Max Zener Impedance A and B Suffix only		Max Reverse Leakage Current				Max Zener Voltage Temperature Coeff. (A and B Suffix only) θ_{VZ} (%/°C) (Note 3)
			Z_{ZT} @ I_{ZT} Ohms	Z_{ZK} @ I_{ZK} = 0.25 mA Ohms	A and B Suffix only			Non-Suffix	
					I_R µA	@ V_R Volts A	B	I_R @ V_R Used for Suffix A µA	
1N5221	2.4	20	30	1200	100	0.95	1.0	200	−0.085
1N5222	2.5	20	30	1250	100	0.95	1.0	200	−0.085
1N5223	2.7	20	30	1300	75	0.95	1.0	150	−0.080
1N5224	2.8	20	30	1400	75	0.95	1.0	150	−0.080
1N5225	3.0	20	29	1600	50	0.95	1.0	100	−0.075
1N5226	3.3	20	28	1600	25	0.95	1.0	100	−0.070
1N5227	3.6	20	24	1700	15	0.95	1.0	100	−0.065
1N5228	3.9	20	23	1900	10	0.95	1.0	75	−0.060
1N5229	4.3	20	22	2000	5.0	0.95	1.0	50	±0.055
1N5230	4.7	20	19	1900	5.0	1.9	2.0	50	±0.030
1N5231	5.1	20	17	1600	5.0	1.9	2.0	50	±0.030
1N5232	5.6	20	11	1600	5.0	2.9	3.0	50	+0.038
1N5233	6.0	20	7.0	1600	5.0	3.3	3.5	50	+0.038
1N5234	6.2	20	7.0	1000	5.0	3.8	4.0	50	+0.045
1N5235	6.8	20	5.0	750	3.0	4.8	5.0	30	+0.050
1N5236	7.5	20	6.0	500	3.0	5.7	6.0	30	+0.058
1N5237	8.2	20	8.0	500	3.0	6.2	6.5	30	+0.062
1N5238	8.7	20	8.0	600	3.0	6.2	6.5	30	+0.065
1N5239	9.1	20	10	600	3.0	6.7	7.0	30	+0.068
1N5240	10	20	17	600	3.0	7.6	8.0	30	+0.075
1N5241	11	20	22	600	2.0	8.0	8.4	30	+0.076
1N5242	12	20	30	600	1.0	8.7	9.1	10	+0.077
1N5243	13	9.5	13	600	0.5	9.4	9.9	10	+0.079
1N5244	14	9.0	15	600	0.1	9.5	10	10	+0.082
1N5245	15	8.5	16	600	0.1	10.5	11	10	+0.082
1N5246	16	7.8	17	600	0.1	11.4	12	10	+0.083
1N5247	17	7.4	19	600	0.1	12.4	13	10	+0.084
1N5248	18	7.0	21	600	0.1	13.3	14	10	+0.085
1N5249	19	6.6	23	600	0.1	13.3	14	10	+0.086
1N5250	20	6.2	25	600	0.1	14.3	15	10	+0.086
1N5251	22	5.6	29	600	0.1	16.2	17	10	+0.087
1N5252	24	5.2	33	600	0.1	17.1	18	10	+0.088
1N5253	25	5.0	35	600	0.1	18.1	19	10	+0.089
1N5254	27	4.6	41	600	0.1	20	21	10	+0.090
1N5255	28	4.5	44	600	0.1	20	21	10	+0.091
1N5256	30	4.2	49	600	0.1	22	23	10	+0.091
1N5257	33	3.8	58	700	0.1	24	25	10	+0.092
1N5258	36	3.4	70	700	0.1	26	27	10	+0.093
1N5259	39	3.2	80	800	0.1	29	30	10	+0.094
1N5260	43	3.0	93	900	0.1	31	33	10	+0.095
1N5261	47	2.7	105	1000	0.1	34	36	10	+0.095
1N5262	51	2.5	125	1100	0.1	37	39	10	+0.096
1N5263	56	2.2	150	1300	0.1	41	43	10	+0.096
1N5264	60	2.1	170	1400	0.1	44	46	10	+0.097
1N5265	62	2.0	185	1400	0.1	45	47	10	+0.097
1N5266	68	1.8	230	1600	0.1	49	52	10	+0.097
1N5267	75	1.7	270	1700	0.1	53	56	10	+0.098
1N5268	82	1.5	330	2000	0.1	59	62	10	+0.098
1N5269	87	1.4	370	2200	0.1	65	68	10	+0.099
1N5270	91	1.4	400	2300	0.1	66	69	10	+0.099
1N5271	100	1.3	500	2600	0.1	72	76	10	+0.110
1N5272	110	1.1	750	3000	0.1	80	84	10	+0.110

NOTE 1. Tolerance — The JEDEC type numbers shown indicate a tolerance of ±10% with guaranteed limits on only V_Z, I_R and V_F as shown in the electrical characteristics table. Units with guaranteed limits on all six parameters are indicated by suffix "A" for ±10% tolerance and suffix "B" for ±5.0% units.

†For more information on special selections contact your nearest Motorola representative.

NOTE 2. Special Selections† Available Include:
1. Nominal zener voltages between those shown.
2. Two or more units for series connection with specified tolerance on total voltage. Series matched sets make zener voltages in excess of 200 volts possible as well as providing lower temperature coefficients, lower dynamic impedance and greater power handling ability.
3. Nominal voltages at non-standard test currents.

■ **FIGURE 3.9** Partial data sheet 1N5221 through 1N5281 500-mW silicon Zener diodes (courtesy Motorola, Inc.).

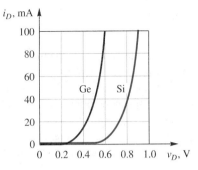

■ **FIGURE 3.10** Typical forward characteristics of a germanium diode and a silicon diode at 25°C. Voltage increases faster than the diode equation predicts because of resistance.

GENERAL-PURPOSE RECTIFIERS

... subminiature size, axial lead mounted rectifiers for general-purpose low-power applications.

LEAD MOUNTED SILICON RECTIFIERS

50-1000 VOLTS
DIFFUSED JUNCTION

*MAXIMUM RATINGS

Rating	Symbol	1N4001	1N4002	1N4003	1N4004	1N4005	1N4006	1N4007	Unit
Peak Repetitive Reverse Voltage, Working Peak Reverse Voltage, DC Blocking Voltage	V_{RRM}, V_{RWM}, V_R	50	100	200	400	600	800	1000	Volts
Non-Repetitive Peak Reverse Voltage (halfwave, single phase, 60 Hz)	V_{RSM}	60	120	240	480	720	1000	1200	Volts
RMS Reverse Voltage	$V_{R(RMS)}$	35	70	140	280	420	560	700	Volts
Average Rectified Forward Current (single phase, resistive load, 60 Hz, see Figure 8, $T_A = 75°C$)	I_O	1.0							Amp
Non-Repetitive Peak Surge Current (surge applied at rated load conditions, see Figure 2)	I_{FSM}	30 (for 1 cycle)							Amp
Operating and Storage Junction Temperature Range	T_J, T_{stg}	−65 to +175							°C

*ELECTRICAL CHARACTERISTICS

Characteristic and Conditions	Symbol	Typ	Max	Unit
Maximum Instantaneous Forward Voltage Drop ($i_F = 1.0$ Amp, $T_J = 25°C$) Figure 1	v_F	0.93	1.1	Volts
Maximum Full-Cycle Average Forward Voltage Drop ($I_O = 1.0$ Amp, $T_L = 75°C$, 1 inch leads)	$V_{F(AV)}$	—	0.8	Volts
Maximum Reverse Current (rated dc voltage) $T_J = 25°C$ $T_J = 100°C$	I_R	0.05 1.0	10 50	μA
Maximum Full-Cycle Average Reverse Current ($I_O = 1.0$ Amp, $T_L = 75°C$, 1 inch leads)	$I_{R(AV)}$	—	30	μA

*Indicates JEDEC Registered Data.

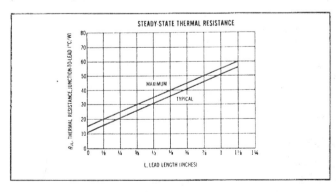

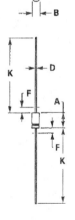

NOTES:
1. ALL RULES AND NOTES ASSOCIATED WITH JEDEC DO-41 OUTLINE SHALL APPLY.
2. POLARITY DENOTED BY CATHODE BAND.
3. LEAD DIAMETER NOT CONTROLLED WITHIN "F" DIMENSION.

DIM	MILLIMETERS		INCHES	
	MIN	MAX	MIN	MAX
A	4.07	5.20	0.160	0.205
B	2.04	2.71	0.080	0.107
D	0.71	0.86	0.028	0.034
F	—	1.27	—	0.050
K	27.94	—	1.100	—

CASE 59-03
DO-41
PLASTIC

■ **FIGURE 3.11** 1N4001 through 1N4007 diode data sheet (courtesy Motorola, Inc.).

DRILL 3.10

Refer to the data sheet of Fig. 3.11. What is the maximum rated average current for the 1N400X diode family?

ANSWER 1 A.

maximum safe reverse voltage, from 50 up to 1000 V. This means that the manufacturer has specified a peak voltage safely below breakdown, where avalanche carrier multiplication is not a serious factor. The average and peak currents are specified to maintain power dissipation within the capabilities of the device. Maximum forward voltage drop is specified at rated current. Maximum reverse current is specified at two temperatures.

This series was designed for low-frequency power applications and thus was not optimized for low junction capacitance or high switching speeds. These topics are discussed in the following sections.

CHECK UP

1. **TRUE OR FALSE?** A diode's saturation current doubles for a 10°C rise in temperature.
2. **TRUE OR FALSE?** For a constant current, a diode's voltage drop decreases with a rise in temperature.
3. **TRUE OR FALSE?** When breakdown occurs in a diode at 100 V of reverse bias, the breakdown mechanism is Zener breakdown.
4. **TRUE OR FALSE?** The knee of the diode characteristic, where forward current becomes appreciable, is called *cut-in*.

3.5 LOAD LINES AND PIECEWISE-LINEAR DIODE MODELS

A simple diode circuit consisting of a voltage source v_S, a resistor, and a diode is shown in Fig. 3.12. If the diode characteristic is known, a graphical method may be used to solve for the current and voltages of the diode. Kirchhoff's voltage law yields

$$v_D = v_S - i_D R. \tag{3.29}$$

The plot of this equation is a straight line through $v_D = v_S$, $i_D = 0$ with a slope of $-1/R$ and is known as a **load line**. The load line relates the operating voltage and current of the load (in this case the diode) to the linear part of the circuit. Note that the static load line for any linear circuit can be obtained by determination of Thévenin's voltage and resistance.

A graphical solution requires plotting the load line and the diode characteristic on the same volt-ampere axes. The intersection of the two graphs represents the simultaneous solution of the voltage source-resistor characteristic (the load line) and the diode characteristic and is called the **operating point** of the circuit.

■ The *operating point* defines the dc condition of a circuit. This condition is also called the *quiescent* point (Q-point).

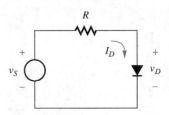

■ **FIGURE 3.12** A basic diode circuit.

EXAMPLE 3.5

Let the circuit shown in Fig. 3.12 use $V_S = 1$ V, and $R = 50\ \Omega$. The diode characteristic is plotted in Fig. 3.13. Find I_D and V_D. Note that this is a dc problem because V_S is constant.

SOLUTION The load line is plotted on the characteristic, extending from the point $V_D = 0$, $I_D = V_S/R = 1\text{ V}/50\ \Omega = 20$ mA to the point $V_D = V_S$, $I_D = 0$. The figure shows that at a current of 6 mA, both the diode voltage and the load line voltage are 0.7 V.

DRILL 3.11

Change the resistor in Example 3.5 to $20\ \Omega$ and repeat.

ANSWER
$I_D = 12$ mA,
$V_D = 0.77$ V.

The graphical technique is a general approach to the solution of a circuit consisting of some nonlinear device in series with a network that can be resolved into a Thévenin's voltage source and resistance.

Piecewise-Linear Diode Models

Let us define a fictitious circuit element called an ***ideal diode***. Such a device would have zero resistance when biased in the forward direction and infinite resistance when biased in the reverse direction. The characteristic would appear as in Fig. 3.14. Observe that the ideal diode is not a good model for the diode whose characteristic is plotted in Fig. 3.13 if the same scale is used, but might be useful to approximate first-order behavior in circuits where the voltage scale is much greater.

Now let us attempt to construct a better model for the diode characteristic using an ideal diode, a resistor, and a voltage source. The diode characteristic has been redrawn in Fig. 3.15. Also sketched in the figure is one possible straight-line approximation to the characteristic. The accuracy of the model may not be impressive, but it is not in error more than 0.1 V at any current over the given range. A more accurate model (one with more segments) would not be justified in many engineering problems.

The circuit shown in Fig. 3.16 is suggested as the realization of the model. Note that no current flows in the model until v_D exceeds an ***offset voltage*** of $V_F = 0.7$ V, and then the characteristic would be $v_D = 0.7 + 3i_D$, and R_F is seen to be

$$\frac{\Delta v_D}{\Delta i_D} = \frac{1.0 - 0.7}{0.1 - 0} = 3\ \Omega.$$

We now demonstrate how to use this model to solve Example 3.5 analytically. Draw

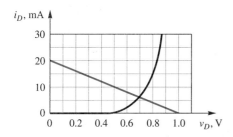

FIGURE 3.13 Graphical solution of the diode circuit of Example 3.5.

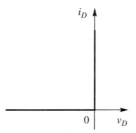

FIGURE 3.14 The ideal diode characteristic.

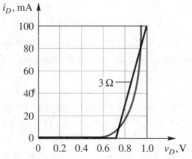

FIGURE 3.15 A typical silicon diode characteristic with a possible piecewise-linear model characteristic.

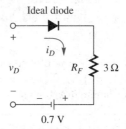

FIGURE 3.16 Piecewise-linear model for the diode characteristic in Fig. 3.15.

the circuit with the model as in Fig. 3.17. Assume the ideal diode to be forward biased (ON) and treat it as zero resistance. The operating point is calculated to be

$$I_D = \frac{1\text{ V} - 0.7\text{ V}}{50\ \Omega + 3\ \Omega} = 5.7\text{ mA}.$$

Because a positive current resulted, the diode is ON as assumed. Then

$$V_D = 0.7\text{ V} + 3\ \Omega(5.7\text{ mA}) = 0.717\text{ V}.$$

These values may be compared with the "exact" graphical answers of 6 mA and 0.7 V. We should point out, however, that the "exact" diode characteristic is not likely to be known, and it is usually adequate to assume a piecewise-linear model. If I_D had been negative in the preceding calculation, then this would have been a contradiction, and we go back and assume the diode to be OFF.

The values of *offset voltage* V_F and **forward resistance** R_F used in the model depend on the type of diode and the desired range of current. The more restricted the operating point range is, the more closely the model can be made to approximate the diode characteristic. For **large signals**, where the input voltage (and hence the *dynamic* operating point) varies over a wide range, the model parameters will be more of a compromise. As a practical matter, for large signals the silicon diode is often represented by an ideal diode in series with the offset voltage $V_F = 0.7$ V, and R_F is neglected.

■ *Dynamic* implies that the circuit is driven by a time-varying input signal, causing the instantaneous operating point to vary around the Q-point.

If the operating point moves across a breakpoint in the diode characteristic because of large swings of the input signal, the solution will require separate models for each segment of the characteristic. It is common in these situations to say that the diode is turned ON and OFF by the signal.

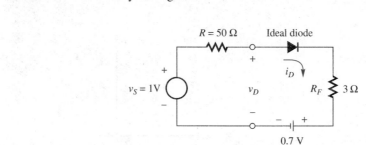

FIGURE 3.17 The simple diode circuit of Example 3.5 with the piecewise-linear model representing the diode.

3.6 DYNAMIC RESISTANCE

When a signal voltage consists of a constant (dc) *bias* component and a much smaller time-varying (ac) component, this ac variation is referred to as **small signal**. As an example of this situation, let us assume that the diode circuit of Fig. 3.12 is driven by a voltage

$$v_S = V_{DC} + V_{ac} \sin \omega t = 1 + 0.05 \sin \omega t \text{ V}.$$

■ *Small signal* has a specific meaning for each circuit, and will be carefully defined.

A diode characteristic and the dc load line using 1 V and 50 Ω are drawn in Fig. 3.18 to show that the solution of the dc operating point, called the *quiescent point* (**Q-point**), is 6 mA and 0.7 V.

We know that the *dynamic* operating point for this circuit will remain in the vicinity of the Q-point because v_S and v_D do not change much in comparison to the dc component. Thus, the best piecewise-linear model for the diode in this region would have a characteristic that passes through the Q-point with the same slope as the diode characteristic. The slope of the diode equation may be found by differentiation. Assuming forward bias,

$$i_D = I_S(e^{v_D/\eta V_T} - 1) \cong I_S e^{v_D/\eta V_T},$$
$$\frac{di_D}{dv_D} = \frac{I_S}{\eta V_T} e^{v_D/\eta V_T} = \frac{i_D}{\eta V_T}. \quad (3.30)$$

The reciprocal of the characteristic slope at any operating point is called the *dynamic resistance* of the diode at that Q-point.

$$r_D = \left.\frac{dv_D}{di_D}\right|_{I_{DQ}} = \frac{\eta V_T}{I_{DQ}}. \quad (3.31)$$

In our example, $I_{DQ} = 6$ mA and, assuming room temperature and $\eta = 1$,

$$r_d = \frac{26 \text{ mV}}{6 \text{ mA}} = 4.33 \text{ }\Omega.$$

■ $\eta = 1$ is a very common approximation.

This is the value of resistance we want for our model, and an offset voltage may be calculated such that the characteristic will pass through the Q-point and be tangent to the diode characteristic:

$$V_F = V_{DQ} - I_{DQ} r_d = 0.7 - 0.006(4.33) = 0.674 \text{ V}.$$

The resulting model characteristic is shown in Fig. 3.18.

Assuming that a linear model is a satisfactory representation of the diode for this small-

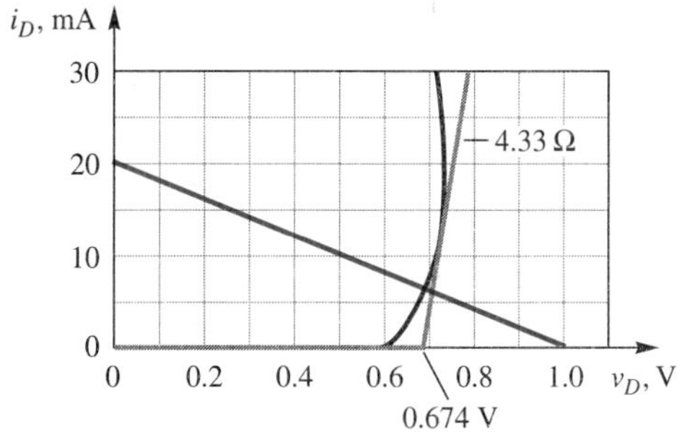

■ **FIGURE 3.18** Piecewise-linear model, using dynamic resistance of the diode, valid near the Q-point.

signal situation, we may find v_D by superposition. We have already found the dc term (you should verify that the new model will yield the same result for I_{DQ} and V_{DQ}). We can find the ac component by properly removing the dc sources, leaving

$$v_d = v_s \frac{r_d}{r_d + R} = 0.05 \sin \omega t \frac{4.33}{54.33} = 0.004 \sin \omega t \text{ V}.$$

By superposition, the composite solution for v_D is

$$v_D = V_{DQ} + v_d = 0.7 + 0.004 \sin \omega t \text{ V}.$$

We now investigate the meaning of *small signal* as it relates to the diode. Rewriting the diode equation, assuming forward bias,

$$i_D = I_S(e^{v_D/\eta V_T} - 1) \cong I_S e^{v_D/\eta V_T}. \tag{3.32}$$

To explicitly show both dc and ac components of diode voltage, let

$$v_D = V_D + v_d. \tag{3.33}$$

Substituting Eq. (3.33) into Eq. (3.32),

$$i_D = I_S e^{V_D/\eta V_T} e^{v_d/\eta V_T} = I_{DQ} e^{v_d/\eta V_T}. \tag{3.34}$$

We recognize that the dc term represents the quiescent operating current. The ac term may be expanded in the Taylor series,

$$i_D = I_{DQ}\left[1 + \frac{v_d}{\eta V_T} + \frac{1}{2}\left(\frac{v_d}{\eta V_T}\right)^2 + \frac{1}{6}\left(\frac{v_d}{\eta V_T}\right)^3 + \cdots\right]. \tag{3.35}$$

The series may be approximated by the constant term plus the linear term provided that

$$v_d \ll \eta V_T \tag{3.36}$$

because the higher order terms will become negligible. Then

$$i_D = I_{DQ} + \frac{I_{DQ}}{\eta V_T} v_d. \tag{3.37}$$

■ *Small signal* for a diode requires v_d to be less than 5 mV (peak).

Again, Eq. (3.30) applies. Thus we see that **small signal** requires that the square term in Eq. (3.35) be much smaller than the linear term. If v_d is less than 5 mV at room temperature, the square term will be less than 10% of the linear term. We note that the 4 mV of our example qualifies as a small signal.

CHECK UP

1. **TRUE OR FALSE?** If you have a plot of the I-V characteristic of a device, its intersection with the dc load line of the Thévenin equivalent for the rest of the circuit is the solution for the operating point.
2. **TRUE OR FALSE?** The piecewise-linear model for the diode is valid in each of two ranges.
3. **TRUE OR FALSE?** The R_F in the diode model can be neglected when it is small compared to the other resistances in the circuit.
4. **TRUE OR FALSE?** The dynamic resistance of a diode has meaning when the diode voltage drop has an ac variation of 50 mV or less.

CHAPTER 3 SEMICONDUCTOR DIODES AND DIODE CIRCUITS

3.7 RECTIFIER CIRCUITS

The process of obtaining a dc voltage from an ac supply is called **rectification**. Rectifiers are used in the power supplies of all types of electronic equipment where the user wishes to obtain power from the ac line rather than from batteries.

The Transformer

Most power supplies that rely on the ac line as a primary energy source employ a power *transformer*. This transformer consists of a primary coil of N_1 turns and a secondary coil of N_2 turns wound around an iron core. The core serves to confine the magnetic flux, which then links the two coils quite tightly. In a power supply, the transformer serves several functions.

First, by choice of the turns ratio (N_2/N_1), the designer may obtain a secondary voltage different from the given primary source voltage. In the case shown in Fig. 3.19, $v_S = N_2/N_1 (120)$ V_{rms}. Typically, the common 110–125 V_{rms} 60-Hz line voltage in the United States, or the 220–250 V_{rms} 50-Hz power line voltage in other parts of the world, is transformed to a lower voltage, in the range of 5 to 20 V_{rms} for transistor or integrated-circuit power supplies.

Second, the transformer provides electrical isolation between the secondary and primary circuits. This is important because it means that the secondary circuit may be connected to any reference voltage, independent of that of the ac primary power. Typically, the secondary circuit may be connected to earth ground, but even if it is not, the danger of hazardous voltages appearing between the secondary circuit and some accidental path to earth ground is reduced.

Finally, transformers allow for developing voltages that are separated by 180° in phase, as in the circuit shown later in Fig. 3.23.

■ Almost all semiconductor electronic circuits use a transformer and a rectifier to obtain 5 to 12 V dc from the ac power line. The alternative is to use batteries, but even then the batteries might be recharged by a transformer and rectifier.

The Half-Wave Rectifier

The circuit shown in Fig. 3.19 is known as a *half-wave rectifier* because the load current flows approximately one-half of the sine-wave period. Figure 3.20 shows that the diode conducts when v_s is more positive than V_F and $v_L = v_s - V_F$. When v_s is less than V_F, the diode is OFF and $v_L = 0$. For silicon diodes, the offset voltage V_F is about 0.7 V.

The plot of output voltage as a function of input voltage for a given device or circuit is known as its voltage *transfer characteristic*. We can readily obtain important information by studying the transfer characteristic of circuits, and in the case of Fig. 3.20, we immediately see how the output voltage is related to the input voltage in the half-wave rectifier.

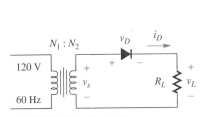

■ FIGURE 3.19 Half-wave rectifier circuit.

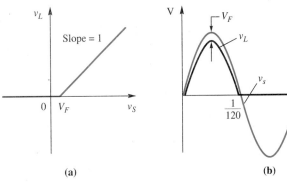

■ FIGURE 3.20 Half-wave rectifier. (a) Transfer characteristic. (b) Waveforms.

We will make extensive use of transfer characteristics in the description of circuits in this text.

EXAMPLE 3.6

Find the average (dc) value of the load voltage V_L for the circuit shown in Fig. 3.19. Assume that the transformer has a secondary voltage of 12.6 V_{rms} and a secondary resistance of 3 Ω. Assume a diode with a forward resistance $R_F = 1\ \Omega$ and an offset voltage $V_F = 0.7$ V. Reverse leakage current is negligible and load resistance $R_L = 50\ \Omega$.

SOLUTION A model for the given circuit is shown in Fig. 3.21(a). The peak value of the source voltage is $12.6\sqrt{2} = 17.82$ V. The diode has been represented by an ideal diode, in series with R_F and V_F. The diode conducts only when v_s is more positive than 0.7 V, so

$$i_L = 0 \quad \text{when } v_s < 0.7 \text{ V}$$

$$i_L = \frac{v_s - V_F}{R_S + R_F + R_L} = \frac{17.82 \sin 377t - 0.7}{3 + 1 + 50} \quad v_s > 0.7 \text{ V}$$

and the load voltage $v_L = i_L R_L$.

Figure 3.21(b) shows a sketch of the load voltage v_L and the source voltage v_s. We can determine the average value of v_L by finding the area under the curve and dividing by the period,

$$V_L = \frac{1}{T} \int_0^T v_L(t)\, dt,$$

and by substituting $\theta = \omega t = 2\pi t/T$:

$$V_L = \frac{1}{2\pi} \int_0^{2\pi} v_L(\theta)\, d\theta.$$

Note that v_L remains zero until 17.82 sin 377t equals 0.7 V, or $\theta = 0.04$ radians (2.25°).

$$V_L = \frac{1}{2\pi} \int_{0.04}^{\pi - 0.04} \frac{50}{54}(17.82 \sin\theta - 0.7)\, d\theta$$

$$= \left(\frac{-16.5 \cos\theta - 0.65\theta}{2\pi}\right)_{0.04}^{3.10} = 4.93 \text{ V}.$$

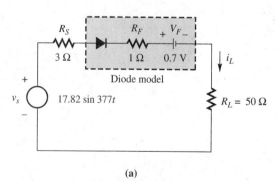
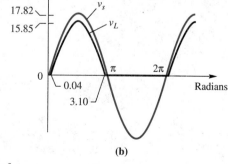

FIGURE 3.21 (a) Model for the half-wave rectifier of Example 3.6. (b) Waveforms for the half-wave rectifier of Example 3.6

Note that the diode must be able to withstand a reverse voltage of $v_{s(peak)}$. Load voltage V_L will approach $v_{s(peak)}/\pi$ as losses in the diode and source are neglected. Often the ac component of the output voltage is reduced by capacitive filtering, as in the peak rectifier circuits that are discussed later in this section.

The Precision Half-Wave Rectifier

The circuit shown in Fig. 3.22(a) uses an op amp in conjunction with the diode in order to eliminate the offset voltage inherent in the practical half-wave diode rectifier circuit; hence the name *precision half-wave rectifier*. Note the absence of the offset in the transfer characteristic of Fig. 3.22(b). When v_{IN} goes positive, the op amp output voltage v_A is fed back to the inverting input through the diode in such a manner that the difference signal $v_O - v_{IN}$ is virtually zero (a few microvolts for good op amps). Also, V_F is essentially reduced by a factor equal to the open-loop gain of the op amp. When v_{IN} goes negative, v_A becomes negative, the diode turns OFF, and v_O is zero.

The transfer characteristic looks ideal, but the circuit does have its problems. When the input v_{IN} becomes negative, the full value of v_{IN} appears across the op amp input, which must be protected from overvoltage. Also, the op amp would likely saturate when v_{IN} is negative, with v_A equal to the negative op amp supply voltage. This would degrade its response to high-frequency signals. Nevertheless, this circuit has application for small signals, where the normal offset voltage of the diode rectifier would overwhelm the signal.

DRILL 3.12

A 60-Hz 115-V_{rms} sine wave is half-wave rectified by a diode having an offset voltage of 0.5 V. Find the average value of the rectified waveform.

ANSWER 51.6 V.

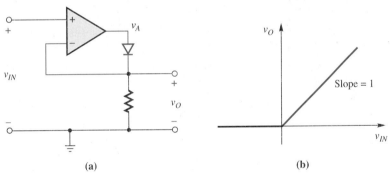

■ **FIGURE 3.22** Precision half-wave rectifier. (a) Circuit. (b) Transfer characteristic.

The Full-Wave Rectifier

The circuit shown in Fig. 3.23 is a *full-wave rectifier* because current flows through the load during both halves of the cycle. This particular arrangement requires two diodes and a center-tapped transformer. Diode D_1 conducts while the top terminal of the transformer is positive with respect to the center tap, and diode D_2 conducts on the next half-cycle, when the bottom terminal is positive.

■ For the same ac input voltage, the full-wave rectifier will have twice the dc output voltage of a half-wave rectifier.

EXAMPLE 3.7

Assume that the transformer shown in Fig. 3.23 has a secondary voltage of 12.6 V_{rms} on each side of the center tap. Suppose the secondary resistance is 1 Ω (each half). Let the load consist of a 12.6-V lead-acid battery with an internal resistance of 0.5 Ω. Find the dc charging current of the circuit. Use $V_F = 0.7$ V and $R_F = 0.1$ Ω for each diode.

SOLUTION The model for the battery charger is shown in Fig. 3.24. The ideal diodes of the model serve to remind us that the top diode only conducts when $17.82 \sin 377t > 0.7 + 12.6$ V, and the bottom diode conducts when $-v_s = -17.82 \sin 377t > 0.7 + 12.6$ V. Figure 3.25 shows a sketch of v_s and v_L. Note that v_L remains 12.6 V and $i_L = 0$, until $17.82 \sin 377t = 12.6 + 0.7$ V, or $\theta = 0.84$ rad. Again, the average current I_L is found by integrating:

$$I_L = \frac{1}{\pi} \int_{0.84}^{\pi - 0.84} \frac{(17.82 \sin \theta - 13.3) \, d\theta}{1.6}$$

$$= \left[\frac{-17.82 \cos \theta - 13.3\theta}{5.03} \right]_{0.84}^{2.30} = 0.86 \text{ A}.$$

In this application, the diode is called on to withstand a peak diode current of

$$\frac{17.82 - 13.3}{1.6} = 2.83 \text{ A}.$$

The average current in each diode is only 0.43 A. The peak battery voltage is $12.6 + (2.83)(0.5) = 14.02$ V, and the peak reverse voltage on the diode is $17.82 + 14.02 = 31.84$ V.

DRILL 3.13

The full-wave rectifier of Fig. 3.23(a) has $v_s = 115$ V$_{rms}$. Assume a diode drop of 0.5 V, and find the average voltage v_L.

ANSWER 103 V.

For the full-wave rectifier with a resistive load, V_L can approach $(2/\pi)v_{S(peak)}$ as losses in the transformer and diodes are neglected.

The circuit shown in Fig. 3.26 is also a full-wave rectifier, known as a ***bridge rectifier***, or full-wave bridge. This arrangement avoids the need for a center-tapped transformer by

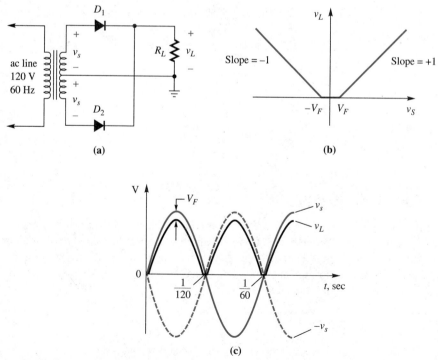

FIGURE 3.23 Full-wave rectifier. **(a)** Circuit. **(b)** Transfer characteristic. **(c)** Waveforms.

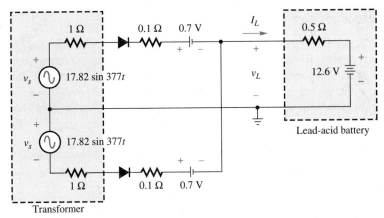

■ **FIGURE 3.24** Model for the full-wave battery charger of Example 3.7.

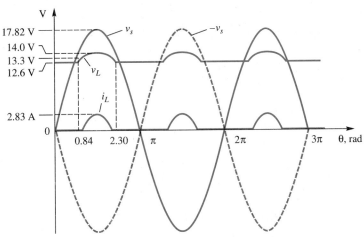

■ **FIGURE 3.25** Voltage and current waveforms for the full-wave battery charger circuit of Example 3.7.

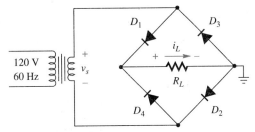

■ **FIGURE 3.26** Full-wave bridge rectifier circuit.

using four diodes. When v_s is positive, diodes D_1 and D_2 conduct, and the current flows through the load as indicated. When v_s is negative, diodes D_3 and D_4 conduct. Again the load current flows from left to right. The output waveform is the same as in Fig. 3.23(c) except that the offset would be $2V_F$, because two conducting diodes are in series.

The Peak Rectifier

The circuit shown in Fig. 3.27 is known as a ***peak rectifier***. This name arises because the capacitor charges to the peak value of v_s (less the diode drop). When R_L is very large, i_L is small, and the capacitor does not discharge as fast as the decrease in v_s after its peak.

The capacitor can only discharge through R_L because the diode is OFF when the source voltage is less than the load voltage (plus the offset voltage of the diode, V_F). The discharge has a time constant of $R_L C$, and thus during discharge the load voltage is

$$v_L = (V_P - V_F)e^{-t/R_L C}. \tag{3.38}$$

Refer to the waveforms shown in Fig. 3.28. The discharge continues until the source voltage again equals $v_L + V_F$ at T_1. Equating $v_s - V_F$ with v_L,

$$V_P \cos \omega_0 T_1 - V_F = (V_P - V_F)e^{-T_1/R_L C}. \tag{3.39}$$

The discharge time T_1 may be found from Eq. (3.39) by iteration. The charge lost during discharge must be restored to the capacitor during that short part of the cycle when the source voltage is rising above the minimum load voltage. This results in a peak diode current that is much in excess of the average current. The amount by which the load voltage sags while the diode is OFF is called the ***ripple voltage***.

When the ripple voltage ΔV is relatively small in comparison with V_P ($\Delta V < 0.1 V_P$), two simplifying assumptions are often made to expedite finding an approximate solution. Load current is assumed constant,

$$I_L = \frac{V_P - V_F}{R_L},$$

and discharge time T_1 is assumed equal to the period of the charging cycle (in this case $2\pi/\omega_o$). Then in the steady state, the charge out of the capacitor during discharge, $Q_D = IT$, must be equal to the charge added during recharge, $Q_C = C\Delta V$. Equating the two expressions allows us to solve for the ripple voltage:

$$\Delta V = \frac{I_L T}{C} = \frac{(V_P - V_F)T}{R_L C}. \tag{3.40}$$

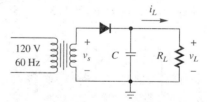

■ **FIGURE 3.27** Peak rectifier.

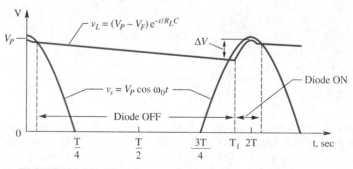

■ **FIGURE 3.28** Waveforms of the half-wave peak rectifier.

Equation (3.40) will predict a ripple voltage slightly greater than the actual case because of the approximations made, but the error is on the pessimistic side because we normally want a small ripple. Often % Ripple = $\Delta V/V_R \times 100\%$ is specified. The reader will realize that the ripple voltage increases with greater load current.

■ The approximate relationship of Eq. (3.40) is very useful. Note that $T_1 \cong 1/f$ for a half-wave peak rectifier, and $T_1 \cong \frac{1}{2f}$ for a full-wave rectifier.

EXAMPLE 3.8

Consider the power supply shown in Fig. 3.29. This circuit is typical of a large class of power supplies using an integrated-circuit (IC) voltage regulator. Suppose that the 5-V regulator requires that its input voltage not drop below 8 V at 1 A of load current. The transformer has a 15-V_{rms} center-tapped secondary. The diodes have a 0.7-V offset, and the transformer plus diode equivalent resistances are 0.2 Ω on each side. Find the minimum capacitor value. The power-line frequency is 60 Hz.

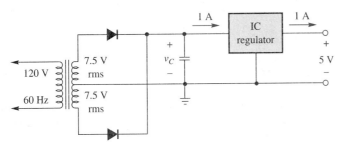

■ **FIGURE 3.29** Full-wave power supply with regulator.

SOLUTION The peak capacitor voltage will be approximately $7.5\sqrt{2} - 0.7 - 0.2 = 9.7$ V. The maximum ripple voltage is $9.7 - 8 = 1.7$ V. The period between full-wave peaks is 1/120 sec. From Eq. (3.40), we have

$$C = \frac{I_L T}{\Delta V} = \frac{(1\text{ A})(1/120 \text{ sec})}{(1.7 \text{ V})} = 4900 \text{ }\mu\text{F}.$$

The approximate solution would lead a designer to use a nominal capacitor value such as 5000 μF. A more careful investigation would yield the additional information that the diode is ON for about 20% of the time. This would require a peak diode current in excess of 5 A during the conduction time (average diode current/% of time conducting). You should expect that a peak current of this magnitude will create an *additional* drop in the equivalent source resistance of about 4 A $\times$ 0.2 Ω = 0.8 V. This will reduce the peak load voltage to about 9 V and require a smaller ΔV. But the losses also increase the conduction angle so that the overall result with the given capacitor value is as shown in Fig. 3.30. This solution is obtained by means of a SPICE computer simulation of this circuit in Example 3.15. In Section 3.16 you will examine the computer analysis of this problem, which also shows the diode current waveform. Computer analysis is particularly well suited to the iterative solution of this nonlinear problem.

DRILL 3.14

Estimate the value of capacitance needed in Example 3.8 to reduce the ripple voltage to 0.5 V peak-to-peak.

ANSWER 0.017 F.

The general implications of the capacitor filter problem are as follows: First, the peak diode current for large filter capacitors is much greater than the average load current. Second, the peak load voltage will be less than the transformer open-circuit voltage by a factor related to this peak diode current. This may put a serious constraint on source resistances.

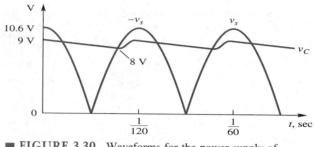

FIGURE 3.30 Waveforms for the power supply of Example 3.8.

The variation in dc output voltage as dc output current changes is called **regulation**. We define power supply regulation as

$$VR = \frac{V_{\text{no load}} - V_{\text{load}}}{V_{\text{no load}}} \times 100\%. \tag{3.41}$$

An ideal supply would have output voltage independent of load current, and thus regulation, $VR = 0\%$.

Some power supplies use a more complex filter circuit than the single capacitor discussed in this section. However, the trend is to accomplish the further reduction in power supply ripple by means of an active regulator circuit such as that suggested in Example 3.8.

CHECK UP

1. **TRUE OR FALSE?** The purpose of a rectifier is to change ac to dc.
2. **TRUE OR FALSE?** A full-wave rectifier output has a period twice that of a half-wave rectifier.
3. **TRUE OR FALSE?** Diode current flows in a half-wave peak rectifier about $\frac{1}{2}$ the cycle.
4. **TRUE OR FALSE?** The peak current in the diodes of a full-wave peak rectifier is slightly greater than the average current in the load.

3.8 BREAKDOWN-DIODE VOLTAGE REGULATOR

The breakdown voltage V_Z of a breakdown diode is relatively constant over a wide current range. The volt-ampere characteristic of such a device is plotted in Fig. 3.31, with the breakdown portion of the curve shown in the third quadrant. As indicated on the circuit symbol for the breakdown diode, the direction of the breakdown current i_Z is reversed from that of a normal diode. At typical values of i_Z, the reverse-breakdown voltage V_Z is quite constant.

■ A piecewise-linear model is one that has a transfer characteristic that is piecewise-linear.

A *piecewise-linear* model for this diode, valid for the normal region of interest, would consist of an ideal diode in series with the voltage source V_Z and a very small resistance, R_Z (see Z_{ZT} in Fig. 3.9). This characteristic is exploited in the voltage regulator circuit in Fig. 3.32. The purpose of the breakdown diode is to maintain a constant voltage across the load, even when the supply voltage v_S varies or when the load current varies. Good regulation requires that the breakdown current be maintained above some minimum current

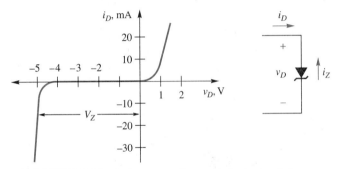

FIGURE 3.31 Volt-ampere characteristic for breakdown diode. Circuit symbol for breakdown diode shows breakdown current reversed from normal direction.

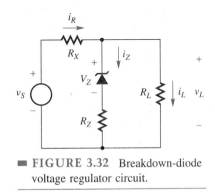

FIGURE 3.32 Breakdown-diode voltage regulator circuit.

where the I-V characteristic is essentially vertical. The circuit design must also protect the diode against exceeding some maximum current.

EXAMPLE 3.9

The load voltage of the regulator shown in Fig. 3.32 is to be maintained at 5 V by use of a 5-V Zener diode. The diode current is not to be less than 5 mA or more than 100 mA. The source voltage v_S varies from 8 to 10 V (this is the capacitor voltage of a peak rectifier, with 2 V of ripple). The load current varies between 0 and 30 mA. Find a suitable resistor, R_X.

SOLUTION Minimum breakdown current will flow when source voltage is minimum and load current is maximum. Let $I_Z = 5$ mA and $I_L = 30$ mA so that $I_R = 35$ mA. Then

$$R_X = \frac{(8-5)\text{ V}}{(30+5)\text{ mA}} = 86\ \Omega.$$

When v_S increases to 10 V,

$$I_R = \frac{(10-5)\text{ V}}{86\ \Omega} = 58\text{ mA}.$$

Then, if i_L becomes zero, i_Z will increase to 58 mA. This is an acceptable value. If the breakdown diode has a dynamic resistance (slope of the characteristic) of 1 Ω, then the

■ Modern voltage regulators use transistors and a breakdown diode in an IC rather than stand-alone breakdown diodes because the efficiency of the diode circuit is low. Refer to Chapter 17 for a discussion of IC voltage regulators.

DRILL 3.15

The regulator circuit of Fig. 3.32 uses a 5-V breakdown diode having a useful i_Z current range between 5 and 50 mA. Let $v_S = 10\ V_{dc}$, and $R_X = 100\ \Omega$. For what range of load current, i_L, will this regulator be effective?

ANSWER 0 to 45 mA.

---

current variation of 53 mA would result in a voltage variation of 53 mV. In practice, the standard resistor size of 82 Ω would be used for R_X, resulting in the reverse diode current range of 6.6 to 61 mA. The peak resistor current would be 61 mA, resulting in a power dissipation of 5 V × 61 mA = 305 mW. The resistor must be capable of dissipating the heat generated by this power. The breakdown diode likewise could dissipate 305 mW at the worst.

3.9 DIODE WAVE-SHAPING CIRCUITS

Two examples of simple diode circuits used to alter the shape of time-varying electrical signals, and that appear in all types of electronic circuit design, follow.

Clipper Circuits

Clipper or *limiter* circuits are so named because their function is to prevent the output voltage from exceeding a given value. The clipper circuit shown in Fig. 3.33 limits the output voltage to no more than V_1. Assume an ideal diode for the moment, although V_1 may be assumed to include the diode offset voltage if desired. For any v_O less than V_1 the diode is OFF and the diode branch is treated as an open circuit; v_O is found from voltage division:

$$v_O = v_S \frac{R_L}{R_1 + R_L}.$$

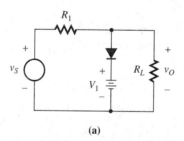

(a)

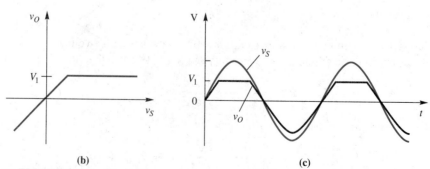

(b)   (c)

■ **FIGURE 3.33** Diode clipper. (**a**) Circuit. (**b**) Transfer characteristic. (**c**) Waveforms.

When v_O reaches V_1, the diode becomes forward biased and any further increase in v_S has no corresponding increase in v_O because the diode is treated as a short circuit. The transfer characteristic in Fig. 3.33(b) illustrates these statements. Figure 3.33(c) shows a sinusoidal v_S applied to this circuit. The output waveform, v_O, is proportional to v_S until it reaches V_1; the excess is "clipped" off.

The clipper circuit in Fig. 3.34 has two branches with diodes. When v_S has a value between the limiting values, the output v_O follows the input, with both diodes acting as open circuits:

$$v_O = \frac{R_L}{R_1 + R_L} v_S. \qquad (3.42)$$

When v_O reaches V_1 (plus the diode offset voltage), D_1 conducts and v_O can become no more positive. As v_O decreases to V_2 (less the diode offset), D_2 conducts and v_O can become no more negative. Note that V_2 carries its own sign, which is negative in this example. The characteristic in Fig. 3.34(b) and the waveforms in Fig. 3.34(c) show an idealized situation in which the clipping levels V_1' and V_2' incorporate the diode offset voltages. In the practical case, diodes also have a small resistance, which would give the output characteristic a small slope in the limiting regions. See Problems 3.37 through 3.42 for variations in clipper circuits.

■ More sophisticated limiter circuits would be used in an audio power amplifier to prevent overdriving the speakers.

Applications include systems in which the circuit driven by v_O must be protected from extremes in voltage and circuits designed to modify the waveform of a signal. In particular, the inputs to the integrated logic circuits presented in Chapter 13 will be protected from damage due to excessive static voltages by limiting diodes.

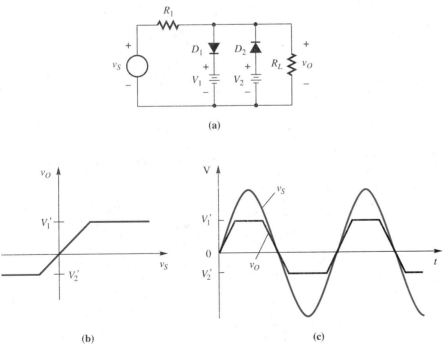

■ **FIGURE 3.34** Double-diode clipper. (a) Circuit. (b) Transfer characteristic. (c) Waveforms.

DRILL 3.16

The clipper circuit of Fig. 3.34(a) has $R_1 = R_L = 1\ k\Omega$, $V_1 = 6\ V$, and $V_2 = -4\ V$. Over what range may v_S vary without clipping? Neglect diode offset voltage.

ANSWER
$-8\ V < v_S < 12\ V$.

EXAMPLE 3.10

Suppose we want to protect a sensitive circuit from voltages exceeding the range from -0.7 to $+5.7\ V$. Assume that the circuit to be protected has an input resistance of $1\ M\Omega$. Given two silicon diodes with $V_F = 0.7\ V$ rated at 10 mA maximum, protect the circuit from hazardous voltages up to $\pm 120\ V$.

SOLUTION We will use the circuit in Fig. 3.34. The statement of the problem indicates that R_L is $1\ M\Omega$. Let $V_1 = +5\ V$ and $V_2 = 0\ V$. Then the clipping levels will be as specified if the diode offset voltages are 0.7 V. When $v_S = -120\ V$ and $V_2 = 0$, the current in D_2 will not exceed 10 mA. $R_1 = 120\ V/10\ mA = 12\ k\Omega$.

Diode Clamping Circuits

■ More specifically, this circuit is sometimes referred to as a *clamped capacitor* circuit.

The circuit shown in Fig. 3.35(a) is known as a *clamp* because its function is to "clamp" or fix one edge of a periodic waveform (in this case the positive peak) to some reference voltage. Sometimes the R is omitted. If R is present, the time constant RC must be long compared to the period of v_S so that the capacitor will not discharge appreciably during each cycle.

Assume that the capacitor is initially uncharged. Should the positive extreme of v_S

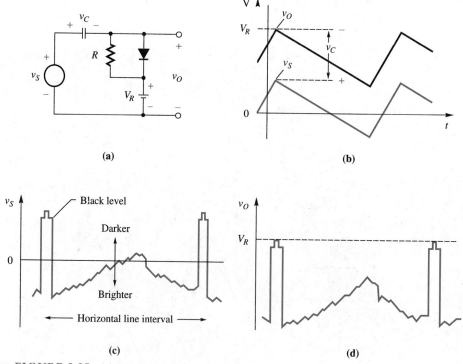

■ **FIGURE 3.35** (a) Diode clamping circuit. (b) Waveforms of clamping circuit. (c) Video amplifier signal of zero average value. (d) Same signal clamped to "fix" the black level.

exceed V_R, the diode will be forward biased and allow the capacitor to charge to the difference between the positive extreme of v_S and the reference voltage V_R. In this case, V_C is positive, and the R is not necessary. Thus, after the initial charge of V_C, $v_O = v_S - V_C$ and the positive peak of v_O appears to be "clamped" at V_R. Representative waveforms are shown in Fig. 3.35(b), assuming an ideal diode. In the practical case, the peak of v_O will be very slightly clipped because the diode must conduct enough to replace the charge that has leaked off (through R) during the remainder of the period. The resistor R serves a useful function only if the positive peak of v_S is less than V_R. In this case, the capacitor charges to the negative value $v_{S(peak)} - V_R$ through the resistor, and the diode does not conduct until v_O attempts to exceed V_R. Again, the positive extreme of v_O is clamped at V_R.

To clamp the negative peak of v_S to reference voltage V_R, it is only necessary to reverse the diode; V_R is assumed to carry its own sign.

Among the applications for the clamping circuit is that of *dc restoration*. Perhaps the signal v_S has passed through amplifier circuits or other circuits where its dc level has been altered. The clamp allows the circuit designer to restore a given dc reference point on the signal waveform. This might be the case where v_S represents picture brightness information, and intensity levels are referenced to one edge of the signal, called the "black" level.

Figure 3.35(c) shows an example in which v_S represents a particular brightness signal for one horizontal line of a television picture. The video amplifier has removed the dc component so that the average value is 0 V. Because the average picture brightness is continually changing, the black level is not a constant voltage for this signal. Figure 3.35(d) portrays v_O clamped at positive peaks by the circuit in Fig. 3.35(a). Now the desired black level has been restored to a desired fixed value (suitable for the desired brightness) independent of the time-varying brightness information of v_S.

Another application of the clamp is illustrated in Fig. 3.36. This circuit, called a *voltage doubler*, consists of a diode clamp followed by a peak rectifier. Diode D_1 and capacitor C_1 constitute the clamp. The first time that the sine wave v_S becomes negative, C_1 charges to the peak value V_P. Waveform v_1 consists of the sine wave with the constant V_P added; thus the negative peaks are clamped at 0 V. Diode D_2 and capacitor C_2 constitute the peak rectifier, yielding $v_2 \approx 2V_P$. If current i_L is drawn from this circuit, the voltage v_2 will sag according to Eq. (3.38).

DRILL 3.17

Let the clamp of Fig. 3.35(a) have $v_S = 3 \sin 377t$, $C = 1\ \mu F$, $R = 1\ M\Omega$, and $V_R = 1$ V. Assume an ideal diode. What is the positive peak voltage and the negative peak voltage of v_O? What is the capacitor voltage V_C?

ANSWER   1 V, −5 V, 2 V.

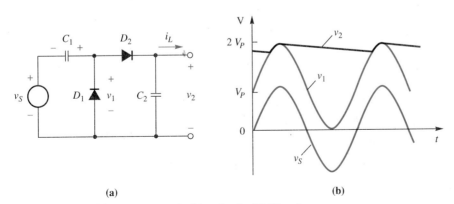

■ **FIGURE 3.36** (a) Voltage doubler circuit. (b) Waveforms of doubler circuit.

CHECK UP

1. **TRUE OR FALSE?** A zener diode in series with the load will stabilize the load voltage.
2. **TRUE OR FALSE?** A diode clipper circuit will protect a sensitive circuit from overvoltage.
3. **TRUE OR FALSE?** A diode clamp circuit is used to fix the average value of a waveform to some predetermined value.
4. **TRUE OR FALSE?** A voltage doubler consists of a diode clamp and a peak rectifier circuit.

3.10 DIODE LOGIC CIRCUITS

Diodes can be used in circuits to perform simple logical operations. As an example, consider Fig. 3.37. If input voltages V_A and V_B are **binary** signals, say, either *LOW* = 0 V or *HIGH* = +3 V, then the output voltage V_C will only be HIGH when input voltages V_A **AND** V_B are both HIGH ($C = A \times B$). The supply voltage V will have to be at least as positive as the HIGH logic state desired at C. When V_A **OR** V_B is LOW, then the corresponding diode conducts and V_C also becomes LOW (one diode drop above the LOW input voltage). A current of V/R flows out of the LOW input so that the impedance of the driving source must be considered. Note that V and R should be chosen such that the diode drop is not an appreciable factor. A more general consideration of this circuit reveals that V_C will be equal to the most negative of the three voltages, V_A, V_B, and V.

EXAMPLE 3.11

Assume the two-input **AND** circuit of Fig. 3.37, with $V = 5$ V, $V_A = 3$ V, $V_B = 1$ V, and $R = 1$ kΩ. Find the voltage at C. Assume that the diode has negligible forward resistance and a forward drop of 0.7 V.

SOLUTION The lower input voltage at B will cause D_B to conduct; V_C will equal

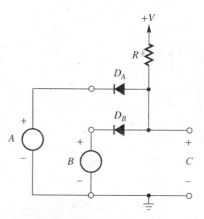

■ **FIGURE 3.37** Diode-logic two-input **AND** circuit.

$V_B + V_D = 1.7$ V; and D_A will be reverse biased with the anode at 1.7 V and the cathode at 3 V. The resistor and D_B will conduct

$$\frac{(5 - 0.7 - 1) \text{ V}}{1 \text{ k}\Omega} = 3.3 \text{ mA}.$$

You should verify that the circuit shown in Fig. 3.38 functions to make C approximately equal to the most positive of the voltages V_A, V_B, and 0. For binary input voltages, this means that C is high when V_A **OR** V_B is high ($C = A + B$). The inputs will require a current of V_A/R or V_B/R from the signal source.

Because of the input current required, and because of the $V_F = 0.7$ V drop in each diode, these diode logic circuits are not readily combined or cascaded because of loading difficulties (see Problem 3.49). Chapter 13 will deal with integrated circuit solutions to the problem of interconnection of logic circuits. Those who are unfamiliar with the logic terms used in this section will find additional information in Appendix C.

■ Diode logic is sometimes incorporated as a part of a more complex IC; the low-power Schottky TTL digital logic family, Ch. 13, provides examples.

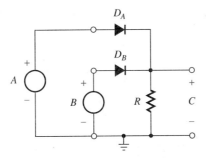

■ **FIGURE 3.38** Diode-logic two-input **OR** circuit.

3.11 DIODE ANALOG SWITCH

Many circuits have been developed to pass an analog signal or to inhibit it, depending on the status of a control voltage.[3] The circuits discussed in Section 3.10 can serve this function (see Problems 3.51 and 3.52). Active (transistor) circuits usually serve this purpose, as we will see in later chapters, but diodes have a speed advantage. A representative diode analog gate is shown in Fig. 3.39.

The basic idea of the diode switch is straightforward. When v_C is sufficiently positive, the diodes are all ON and the signal v_{IN} is transferred because $v_O = v_A$. When v_C is sufficiently negative, the diodes are OFF and there can be no transmission. However, constraints are placed on the magnitude of the signal v_{IN} by the value of v_C.

To understand the nature of these constraints, first consider a case in which v_C is positive and the diodes are all forward biased. Figure 3.40 shows a model of this situation, with each diode replaced by an equal offset voltage V_F. By superposition, we see that each diode has two current components, one caused by the control voltage v_C, and the other caused by the signal v_{IN}. The control current (downward in the figure) is

$$i_C = \frac{2v_C - 2V_F}{2R_C}(1/2), \qquad (3.43)$$

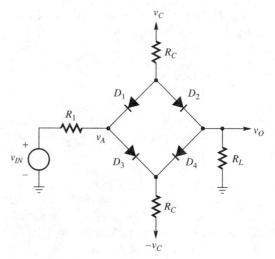

FIGURE 3.39 A four-diode analog gate.

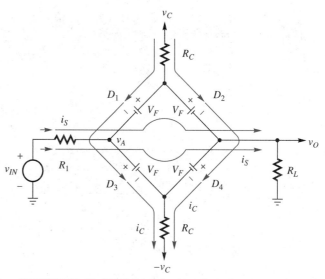

FIGURE 3.40 Model for the four-diode gate, assuming that all diodes are ON, with an equal offset voltage V_F.

and the signal current (to the right) is

$$i_S = \frac{v_{IN}}{R_1 + R_L \| R_C/2}(1/2). \tag{3.44}$$

The terminology $R_L \| R_C/2$ means R_L is in parallel with $R_C/2$. These currents add in D_2 and D_3, but they are opposing in D_1 and D_4. Thus, if i_S should equal or exceed i_C, the current in D_1 and D_4 would not be in the forward direction, the model would be invalid, and the diodes would actually be OFF. The requirement that the diodes remain forward biased imposes the condition $i_S < i_C$; thus,

$$v_{IN} < (v_C - V_F)\left(\frac{R_1}{R_C} + \frac{R_L}{2R_L + R_C}\right). \tag{3.45}$$

CHAPTER 3 SEMICONDUCTOR DIODES AND DIODE CIRCUITS

The assumption that all of the diodes have the same voltage drop is suspect because we have just shown that the diode currents are not equal. Thus, the actual constraint imposed on v_{IN} is slightly more restrictive than we have shown.

For a case in which the control voltage v_C is less than cut-in for D_2 and D_4, they will both be OFF. Here the magnitude of v_{IN} is limited by the breakdown voltage of D_1 and D_3.

CHECK UP

1. **TRUE OR FALSE?** Diode logic gates suffer from severe loading problems when cascaded.
2. **TRUE OR FALSE?** The diode bridge analog gate requires one bipolar control waveform.

3.12 DIODE CAPACITANCE AND SWITCHING TIMES

To this point we have considered the static behavior of the *pn* junction. We now consider some capacitive effects that account for the time-dependent response of the junction to bias changes. These capacitances need to be incorporated into any high-frequency model for the diode.

Transition Capacitance

We have seen that in a reverse-biased *pn* junction there exists a region depleted of carriers, separating two regions of relatively good conductivity. Thus we have in essence, a parallel-plate capacitor, with silicon for the dielectric.

We determined the thickness of the space-charge layer in Section 3.1. For the *n* side, from Eq. (3.14),

$$w_n = \left[\frac{2\varepsilon}{q} \frac{v_J}{N_D^2\left(\frac{1}{N_D} + \frac{1}{N_A}\right)}\right]^{1/2}. \qquad (3.46)$$

The charge in the *n*-depleted region for a cross-sectional junction area of A_j is

$$Q_n = A_j N_D q w_n = A_j \left[\frac{2\varepsilon q(\phi_J + v_R)}{\left(\frac{1}{N_D} + \frac{1}{N_A}\right)}\right]^{1/2}, \qquad (3.47)$$

where the substitution $v_J = \phi_J + v_R$ recognizes the fact that the total junction potential consists of the contact potential plus some external reverse voltage $v_R = -v_D$.

The *transition capacitance* relates the change in charge in the depleted region to the change in bias voltage. Because Q_n is not a linear function of v_R, it is necessary to take the derivative:

$$C_T = \frac{\Delta Q}{\Delta v_R} = \left.\frac{dQ_n}{dv_R}\right|_{V_{RQ}} = A_j\left[\frac{\varepsilon q}{2\left(\dfrac{1}{N_D} + \dfrac{1}{N_A}\right)(\phi_J + v_R)}\right]^{1/2} \quad (3.48)$$

$$= \frac{C_0}{\sqrt{\left(1 + \dfrac{v_R}{\phi_J}\right)}},$$

DRILL 3.18

An abrupt junction with a contact potential of 0.6 V has 8 pF of transition capacitance when $V_R = 5$ V. What is C_T when V_R is increased to 12 V?

ANSWER   5.3 pF.

---

where C_0 is seen to be equal to C_T at $V_R = 0$, and is a constant of the junction, related to area and doping levels.

For other impurity concentration profiles more gradual than the step junction, the transition capacitance may be closer to that for a linear graded junction,

$$C_T = C_0\left(1 + \frac{v_R}{\phi_J}\right)^{-1/3}. \quad (3.49)$$

Varactor Diodes

We have seen that the transition capacitance varies with the reverse-bias voltage. Greater reverse voltage widens the depletion region and reduces C_T. The capability of changing a capacitance by varying a voltage can be exploited in some applications. Diodes designed for these uses are called **varactors** or **varicaps**, depending on the application. Typical values are 10 to 100 pF at reverse biases of 3 to 25 V. Positive biases are avoided, so that usually the shunt conductance and series resistance can be neglected in the device. Silicon is usually used to minimize leakage current. Equation (3.49) applies, with the exponent varying from -0.5 for the step junction, to -0.33 for the graded junction. It is possible to fabricate a *hyperabrupt* junction,[4] where the exponent is -2.

■ Remote operation of TV tuners often is accomplished using this principle.

One application of the varactor is the remote adjustment of the resonant frequency of a tuned circuit. Assume that the frequency of a signal source (or detector) is determined by the resonance of an LC circuit. You will recall that $f_R = 1/(2\pi\sqrt{LC})$ for an LC circuit. Varying the C directly from a distance is impractical because of the effect of parasitic capacitance of the leads, but a varactor can be used for the C, and the dc controlling voltage can be isolated from the sensitive part of the circuit by a large series inductance. Thus, distance presents no difficulty to the dc control circuit. When a hyperabrupt junction is used, the frequency of such a tuned circuit varies linearly with a control voltage.[4]

Diffusion Capacitance

When the junction is *forward* biased, the depletion region narrows and transition capacitance increases. However, the large number of minority carriers injected under forward bias causes a much greater charge *storage* effect, which is called **diffusion capacitance**.

The minority carrier distribution for forward-bias situations is shown in Fig. 3.5. The area between the p_n curve and the p_{no} line represents the density of holes (in excess of the equilibrium value) that has been injected into the n region under the given forward-bias condition. The total number of excess holes is

$$A_j\int_0^\infty p_{no}(e^{v_D/V_T} - 1)e^{-x/L_p}\,dx = A_j p_{no} L_p(e^{v_D/V_T} - 1). \quad (3.50)$$

It also represents an excess charge "stored" in the region

$$Q_p = A_j q p_{no} L_p (e^{v_D/V_T} - 1). \qquad (3.51)$$

Note that this equation is of the same form as the hole-current part of the diode current—see Eq. (3.26)—and that

$$Q_p = \frac{L_p^2}{D_p} i_{Dp}. \qquad (3.52)$$

The term L_p^2/D_p has been shown to be the mean lifetime of holes in the n region.

A similar situation with regard to the injected electrons on the p side means that the total stored charge is nearly linearly related to the total forward diode current:

$$Q_D = \tau i_D, \qquad (3.53)$$

where carrier lifetime τ is a constant of the diode related to doping levels and temperature. The diode current is proportional to the stored charge Q_D of excess minority carriers.

When the bias voltage changes, the stored charge must change also. We define *diffusion capacitance* as

$$C_D = \frac{dQ_D}{dv_D} = \tau \frac{di_D}{dv_D} = \frac{\tau}{r_d} = \tau \frac{I_D}{\eta V_T}. \qquad (3.54)$$

In summary, for reverse bias, C_T may be of the order of pF, with C_D smaller. But for forward bias, C_D may be of the order of nF, with C_T negligible. The manufacturer's data sheet specification of C_T for the 1N4001–14007 series of diodes is given in Fig. 3.41.

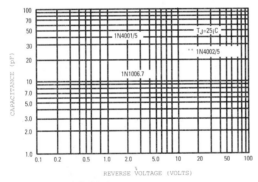

■ **FIGURE 3.41** Manufacturer's data for junction capacitance of the 1N4001–1N4007 silicon diodes (courtesy Motorola, Inc.).

Switching Times for the Junction Diode

When a junction diode's bias voltage is abruptly reversed, the capacitive effects discussed earlier prevent the current from responding instantaneously. This situation is described by the example waveforms shown in Fig. 3.42.

The diode is assumed to have been forward biased with a source voltage V_1 long enough to have reached a steady-state situation with minority carrier densities at the junction determined by the diode forward voltage and with a diode forward current of

$$I_F = \frac{V_1 - v_D(0)}{R}. \qquad (3.55)$$

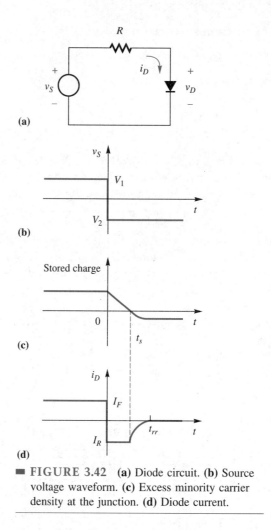

FIGURE 3.42 (a) Diode circuit. (b) Source voltage waveform. (c) Excess minority carrier density at the junction. (d) Diode current.

At time $t = 0$, the source voltage is abruptly changed to a negative voltage V_2. Now the external circuit no longer supports diode current in the forward direction, but the diode voltage v_D cannot suddenly change because of the diffusion capacitance (or the stored charge). The current will jump to

$$I_R = \frac{V_2 - v_D}{R} \tag{3.56}$$

(a negative value), and the excess carrier concentration will begin to decay. The excess carriers are being drawn back across the junction.

After a time t_s, called **storage time**, the excess charge has been removed, the diode voltage has dropped to zero, and the diode current can begin to drop. The minority carriers in the neighborhood of the junction diffuse back to the junction and are removed until the reverse-bias steady-state distribution of Fig. 3.7 is reached. By this time, the diode voltage has reached V_2 (C_T has charged) and the diode current has decreased to the reverse saturation value. The total **reverse recovery time, t_{rr},** is defined as the time required for i_R to decrease to $0.1 I_R$. The time interval $t_{rr} - t_s$ is called the **transition time**, and t_{rr} is a function of I_R/I_F. For an example, see Fig. 3.43, which gives manufacturer's data for the relatively slow 1N400X series.

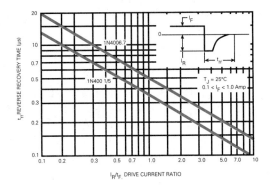

■ **FIGURE 3.43** Reverse recovery time for the 1N4001–1N4007 diodes (courtesy Motorola, Inc.).

When a junction is switched in the other direction, from OFF to ON, a similar process takes place, except this time the charge to be stored is injected in the direction of positive current flow, and the delay is very small.

CHECK UP

1. **TRUE OR FALSE?** Transition capacitance at a *pn* junction is analogous to a parallel-plate capacitance, where two conductors are separated by a dielectric.
2. **TRUE OR FALSE?** Diffusion capacitance is proportional to the amount of charge stored in a nonconducting junction.
3. **TRUE OR FALSE?** Diode storage time is proportional to the diode forward voltage drop.
4. **TRUE OR FALSE?** The capacitance of a varactor is controlled by varying the diode current.

3.13 METAL-SEMICONDUCTOR JUNCTIONS

The contact of metal with semiconductor may create a rectifying junction with properties similar to that of a *pn* junction. This may be desirable or not, depending on the purpose of the contact.

Ohmic Contacts

If the purpose of the metal-semiconductor contact is simply to make connection to the semiconductor from the outside, we desire an ***ohmic*** contact. This means that the contact is not rectifying (sensitive to current direction) because no barrier potential must be overcome. To ensure ohmic contacts, the semiconductor is heavily doped in the neighborhood of the contact. In the case of aluminum to *n*-silicon connections, this heavily doped region is called an n^+ region. Heavy doping causes a high density of electrons at the metal interface so that the semiconductor takes on the characteristics of the metal, and no depleted region exists. In the case of aluminum to *p*-silicon connections, the aluminum

■ The distinction between an ohmic contact and a rectifying contact cannot be overemphasized.

itself (being an acceptor material) can provide p^+ doping to the silicon if the interface is heat-treated.

Schottky Diodes

The *Schottky diode*, a rectifying metal-silicon junction, exhibits a switching speed advantage over the *pn* diode. Usually the semiconductor is *n*-type silicon, and the metal may be aluminum, platinum, molybdenum, chrome, or tungsten. In both the metal and the *n*-type silicon, the electron is the majority carrier. Electrons from the *n*-type silicon flow into the metal, much as they would into acceptor material. The injected electrons have a higher energy level than those in the metal, and are often referred to as "hot carriers." We emphasize that the electrons are still majority carriers in the metal. The flow of electrons into the metal leaves a region in the semiconductor adjacent to the junction that is depleted of carriers (as in a *pn* junction), but there is no corresponding flow of holes from the metal.[5] The additional electrons in the metal establish a surface barrier in the metal at the boundary, which in the equilibrium situation prevents any further current. Schottky diodes are unique in that conduction is entirely by majority carriers. Electrons in the *n*-type silicon face a carrier-free region and a negative potential barrier at the metal interface.

Application of a forward bias, positive on metal, negative on the *n*-type semiconductor, will reduce the negative barrier by attracting electrons away from the boundary. The result is further flow of electrons across the junction, controlled by the external bias. The Schottky barrier is lower than that of a *pn* junction. The result is a lower voltage drop in the forward-biased situation, and also a higher saturation current I_s. The diode equation, Eq. (3.27), applies, with η dependent on fabrication details, but typically near 1. This results from the fact that there is very little recombination in the depleted region. The offset voltage V_F varies with the metal used, but typically is about 0.3 V with aluminum, significantly less than the 0.6 V typical of a silicon *pn* junction.

One of the great advantages of the Schottky diode is its very low storage time. Since it is almost entirely injection from the semiconductor into the metal that accounts for the forward current, there is very little excess minority charge accumulated in the semiconductor. This enables Schottky diodes to switch very rapidly from conduction to the OFF state. Junction capacitance is as predicted by Eq. (3.48). Another feature, the low offset voltage, is often exploited. One example is as a rectifier for low-level signals.

■ The Schottky diode will be encountered again as we study the so-called *Schottky transistor*.

3.14 PHOTONIC AND MICROWAVE DIODES

The objectives of this text do not justify an exhaustive survey of the myriad applications of diodes in their various forms. However, a brief listing will be helpful in demonstrating their ever-increasing importance.

Photonic Devices and Applications

When the injected minority carriers in a forward-biased *pn* junction recombine, energy is released. In silicon and germanium, this is predominantly in the form of heat because recombination consists of multiple transitions,[6] but in the case of gallium arsenide the energy is released in the form of electromagnetic radiation in the infrared (IR).

Infrared Emitters. A typical device is constructed using GaAs, with a thin *p*-layer on an *n*-type substrate. When the junction is forward biased, electrons from the *n*-region will recombine with excess holes of the *p*-layer in a specially designed recombination region located between the *p*- and *n*-regions. Energy is radiated away from the junction in the form of photons during this recombination process. The structure is arranged to minimize the radiant energy reabsorbed by the device, and to emit the infrared perpendicular to the *p*-surface through a window in the package.

When combined in a system with a photodiode, these devices have vast application in optical communications, opto-isolators, and optical sensors. A low-cost, limited-distance communications system might consist of an IR transmitter, an optical fiber, and a photo detector. For greater distances, the semiconductor laser described later would be used for the optical transmitter.

When an IR emitter is packaged with a photodiode in close proximity on the same header, we obtain an **optical coupler** or **optical isolator**. Figure 3.44 shows a pictorial representation of such a device, which is usually fabricated on a single substrate. The opto-isolator is capable of providing signal transfer, and yet complete electrical isolation between input and output circuits. Refer to Problem 3.68 for an example application.

The combination of an IR source and a photo detector, with perhaps some lenses and/or mirrors, is widely used for angular and linear position sensors, and would be incorporated in many types of counters.

Light-Emitting Diodes. As has been mentioned, a *pn* diode fabricated of GaAs emits in the IR region of the spectrum. Other materials, such as gallium-arsenide-phosphide (GaAsP), emit directly in the visible range (red). Also, gallium-phosphide (GaP) can be used by including special impurities in it. This gives rise to impurity levels that act as traps and make radiative recombination possible. Addition of CdO or ZnO gives red light; sulphur or nitrogen shifts the radiation to green; and yellow and orange emission have become available.[7] Such devices, packaged to pass this light to the outside, are called **light-emitting diodes** (LED).

LEDs are widely used as indicators or display devices because of their high efficiency, long life, ruggedness, small size, and compatibility with IC voltages. These are distinct advantages of the LED when compared to the incandescent lamp. Displays of all shapes and sizes are assembled by arranging LEDs in various patterns of "segments." The intensity of the emitted light is controlled by the diode current.

Injection Junction Lasers. It is possible, by using high doping levels and high forward current, to excite a large number of the electrons near the *pn* junction to the conduction band, and a large number of holes to the valence band. Then, when some of the injected electrons start to recombine, the first photons, instead of being reabsorbed or radiating away, stimulate the emission of more photons. The photon, interacting with an excited electron, stimulates a second photon, and the electron drops to a lower energy state. This

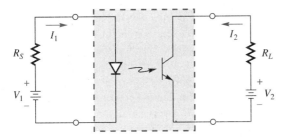

■ **FIGURE 3.44** Opto-isolator circuit with I_1 isolated from I_2.

process of *l*ight *a*mplification by *s*timulated *e*mission of *r*adiation (laser) generates coherent light. The amplification suceeds only when the conditions are met such that the losses are exceeded by the stimulated emission.[8,9]

The **semiconductor laser** is a compact source of highly directional monochromatic light.[10] It is easily modulated by simply varying the diode current. When compared to some other laser structures, the unique advantages of the junction laser are its small size and the fact that its output level is easily modulated. When compared to LEDs, the advantage relates to the efficiency and utility of the collimated monochromatic light beam produced. These lasers are used for light sources in optical fiber communication, remote control units, bar-code readers, optical character readers, compact-disc players, and all sorts of distance ranging systems. Other applications include high-speed printers, burglar alarms, and motion detectors. Sophisticated surgical techniques are being implemented with laser "knives" and laser "welding" or cauterizing.

Photodiodes. *Photodiodes*, either ***p-i-n*** diodes or ***avalanche photodiodes*** (*APDs*), are widely used as the optical detector in the systems just outlined.[10] In addition, they have application in light intensity measurement, and in outside lighting control. These devices were discussed in Section 3.4.

Solar Cells. Semiconductor **solar cells** convert sunlight directly to electricity with reasonable efficiency.[10] They are fabricated either with crystalline silicon, amorphous silicon, or with gallium-arsenide. Most crystalline silicon solar cells operate at ≈12 to 15% efficiency, amorphous silicon solar cells at 6 to 8%, and GaAs to 20%. At 10 to 20% efficiency and no concentration of the solar radiation, silicon solar cells can produce more than 90 to 180 W/m^2. Direct energy conversion is practical in space, and in terrestrial applications where nearly permanent nonpolluting power at low operating cost is required. Many hand-held calculators use small amorphous silicon solar cells instead of batteries.

Microwave Devices

A device in which, for some region of the I-V characteristic, voltage decreases as current increases is said to possess **negative resistance** (NR). A number of semiconductor junction diode devices have been developed that can be made to exhibit NR at frequencies from 1 to 1000 GHz. Without discussing the exploitation of NR in circuit design at this point, this means that these devices can be used for **oscillators** (signal sources) in this range. These devices include the **tunnel diode**, the *impact ionization avalanche transit time (IMPATT) diode*, and the *transferred-electron device (TED or Gunn diode)*. The tunnel diode and the IMPATT diode are fabricated in silicon, and the TED is constructed of gallium-arsenide or indium-phosphide. These devices are described in the literature,[11] and are mentioned here only to suggest to the reader some of the very diverse applications for which semiconductor diodes have been developed.

The first practical semiconductor device was the metal-semiconductor contact realized by a metal whisker pressed against a semiconductor surface (circa 1904). It wasn't until 1938 that Schottky suggested the barrier theory discussed in the previous section. Today *point-contact* diodes are fabricated by pressing a phosphor-bronze spring against an *n*-type substrate, and applying a short high-current pulse. This results in a number of metal atoms passing into the semiconductor, creating a *p*-region in the wafer. The very small area of the *pn* junction results in a very small junction capacitance (typically <1 pF). For this reason, the point-contact diode is used in applications of very high frequency, such as microwave mixers and detectors. The small junction area also means low current ratings, however, and less-than-ideal volt-ampere characteristics.

CHAPTER 3   SEMICONDUCTOR DIODES AND DIODE CIRCUITS

> **CHECK UP**
>
> 1. **TRUE OR FALSE?** An ohmic contact is a metal-semiconductor junction that rectifies.
> 2. **TRUE OR FALSE?** A Schottky diode is a metal-semiconductor junction with a low V_F and a low storage time.
> 3. **TRUE OR FALSE?** A GaAs *pn* junction (LED) is able to radiate light in the visible range when forward-biased.
> 4. **TRUE OR FALSE?** An opto-isolator consists of an LED and a photodiode mounted in the same package.

3.15   DIODE HEATING

Power is dissipated in the junction diode ($v_D \times i_D$), which results in a temperature rise at the junction. We have already discussed temperature effects on the diode characteristics. The temperature must be held well below approximately 300°C to avoid damaging internal metallization and solder alloys. Expansion and contraction create high shear stresses at the interfaces of dissimilar materials. Also, n_i becomes unacceptably high at these temperatures.

■ High temperatures from localized heating allow diffusion of impurity atoms, which will permanently alter the characteristics of the *pn* junction. Localized heating also degrades metallization.

Manufacturers specify the maximum allowable junction temperature. Typical values are 100°C for germanium and 175°C for silicon devices. Manufacturers also provide information concerning the ability of the device package to transfer the heat from the junction to the case. For example, consider the thermal resistance curve given in Fig. 3.11. The ability of the package to transfer the heat away from the junction is shown to be a function of the diode lead length to where it is soldered to the printed circuit board or to some large surface that will conduct the heat away.

Consider the thermal equation

$$T_J = T_L + \theta_{JL} P_J, \qquad (3.57)$$

which implies that the junction temperature is greater than the lead temperature $T_J > T_L$ (both in degrees Celsius) by an amount equal to the product of the junction power dissipation P_J (in watts) and the ***thermal resistance*** between the junction and the case θ_{JL} (°C/W).

The case may pass the heat on to the surrounding air (or other environment) directly by radiation or convection, or for greater dissipation, it may be fastened to a larger surface called a ***heat sink***, which then passes the heat on to the surrounding environment. Verify that the situation represented by the 1-in. lead length in Fig. 3.11 is described by the equation $T_J = 51 P_J + T_L$. We note that the junction must not dissipate any power when it reaches the maximum rated temperature of 175°C. Obviously, the leads must be connected to a surface of lower temperature. Another situation is illustrated in Fig. 3.45(a). The case-to-heat sink, the case-to-ambient, and the heat-sink-to-ambient heat paths all have their own thermal resistance, which is also often provided by the manufacturer of the device or of the heat sink. The thermal system for this figure could be described as follows:

$$T_J = P_J(\theta_{JC} + \theta_{CS} + \theta_{SA}) + T_A. \qquad (3.58)$$

The electrical analog for this thermal equation is shown in Fig. 3.45(b). Thermal dissipation power is analogous to current, and temperature rise is analogous to voltage.

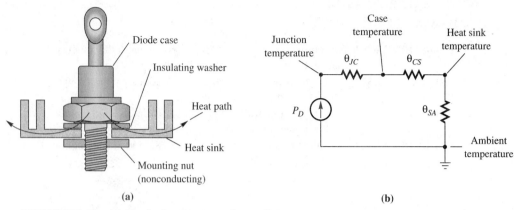

FIGURE 3.45 (a) Physical arrangement for cooling a 12-A diode. (b) Electrical analog of the thermal circuit.

EXAMPLE 3.12

A large 10-V Zener diode has a thermal resistance of $\theta_{JC} = 4°C/W$. The maximum junction temperature is 175°C, and the ambient temperature is 40°C. A heat sink with 6°C/W of thermal resistance from sink to ambient is available, but the heat sink must be electrically insulated from the Zener case. A thin mica washer having a thermal resistance of 2°C/W is used for this. What is the maximum Zener current allowable?

SOLUTION From Eq. (3.55),

$$P_J = \frac{T_J - T_A}{\theta_{JC} + \theta_{CS} + \theta_{SA}} = \frac{175 - 40}{4 + 2 + 6} = 11 \text{ W},$$

$$I_Z = \frac{P_J}{V_2} = \frac{11}{10} = 1.1 \text{ A}.$$

DRILL 3.19

How would Example 3.12 be changed if the ambient temperature were 30°C?

ANSWER $I_Z = 1.2$ A.

CHECK UP

1. **TRUE OR FALSE?** Diode power dissipation is $i_D v_D$, and it causes heating at the junction.
2. **TRUE OR FALSE?** The case temperature of a diode is approximately equal to the junction temperature.
3. **TRUE OR FALSE?** A good heat sink lowers the thermal resistance from case to ambient.

3.16 SPICE MODEL FOR THE DIODE

Another method of solving circuit problems is to simulate the circuit using a computer model. For the reader who may not be familiar with SPICE (*S*imulation *P*rogram, *I*ntegrated *C*ircuit *E*mphasis), reference is made to Appendix D and the information provided there. Although programs are available that will produce a net list after capturing a schematic diagram, we choose to emphasize the details of preparing the net list, especially the device model statements, so that the reader knows what is being done.

The diode, as well as the other semiconductor devices, is described by using an *element* statement and a *model* statement. The model statement is used to describe the device parameters. We will present the commonly used default device parameters.

The diode element model is given in Fig. 3.46. The element statement format is given by

DXXX NI NJ MNAME [AREA] [OFF] [IC=VD]   (3.59)

The associated model statement is

.MODEL MNAME TYPE [PNAME1=PVAL1] PNAME2=PVAL2 . . .]   (3.60)

The anode of the diode is connected to node **NI**; the cathode to **NJ**. **MNAME** is any alphanumeric model designation for the device. Individual diodes with identical parameters can use the same model statement, in this case, **MNAME**. The default value for the cross-sectional **AREA** is 1.0, so it need not be specified unless the parameter values in the model statement need to be scaled. **OFF** allows the device to be set to the nonconducting condition in the dc circuit analysis. **VD** refers to the initial diode voltage for a transient analysis. In the model statement, **TYPE** is **D** for diode, and certain parameters may be assigned values appropriate for the particular device. Selected common default values are provided in Table 3.1. For example, the element statement for a diode **D2**, connected between nodes **NI** = 4 and **NJ** = 7 whose model designation is **HELLO** is

D2 4 7 HELLO   (3.61)

and the model statement for a diode, **TYPE=D**, having the default saturation current of 10^{-14} A, is

.MODEL HELLO D   (3.62)

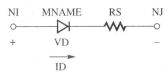

FIGURE 3.46 Diode element model.

TABLE 3.1	Selected Diode Model Parameters		
Name	Parameter	Default Value	Description
IS	Saturation current	10^{-14} A	I_S from Eq. (3.27)
N	Emission coefficient	1	η from Eq. (3.27)
RS	Parasitic resistance	0 Ω	Series lead and bulk resistance
CJO	Zero-bias *pn* capacitance	0 F	C_0 from Eq. (3.48) or (3.49)
VJ	*pn* junction potential	1 V	ϕ_J from Eq. (3.13) as in Eq. (3.48)
M	*pn* grading coefficient	0.5	0.5 = abrupt; 0.33 = graded
FC	Forward-bias-depletion capacitance coefficient	0.5	C_T saturates for $v_R < -$**FC** $\times$ **VJ**
TT	Transit time	0 sec	τ of Eq. (3.53)
BV	Reverse breakdown voltage	∞ V	Magnitude of **BV**
IBV	Reverse breakdown current	10^{-10}	Reverse current at **BV**
EG	Band-gap potential	1.11 eV	0.67 = Schottky; 1.11 = silicon
XTI	I_S temperature coefficient	3	Temperature coefficient for n_i^2
KF	Flicker noise coefficient	0	
AF	Flicker noise exponent	1	

Suppose we decide to connect a diode, **D1**, whose model designation is **ALPHA3** between **N1** = 3 and **NJ** = 0 with a series resistance of 0.5 Ω and a saturation current of 5×10^{-14} A. The element and model statements would be given by

D1 3 0 ALPHA3 (3.63)

and

.MODEL ALPHA3 D RS=0.5 IS=5E-14 (3.64)

It may be that the SPICE default model, which assumes a small silicon IC-type diode with no capacitance or series resistance, and no reverse breakdown, will adequately represent the diode in a particular circuit. Alternatively, the parameters may be obtained from a data sheet, or from actual measurements.

If we have data giving the diode current and forward voltage drop at only a single point, we would assume $\eta = 1$, **RS** = 0 (unless better values are known), and an appropriate temperature; then I_S is found from Eq. (3.27). If current and voltage were known at two points, I_S could be found using Eq. (3.27) at the lower current, and then **RS** could be estimated by the increase of v_D over that predicted by Eq. (3.27) at the higher current. Three points on the diode characteristic would allow independent determination of η.

The parameter **BV** allows the user to specify a reverse breakdown voltage (use a positive value). If it is known that the reverse leakage current as breakdown is approached is in excess of I_S, it may be specified as **IBV** (also positive).

CJO, **VJ**, and **M** allow the user to specify the reverse-biased depletion capacitance according to Eq. (3.48). But Eq. (3.48) fails to model depletion capacitance for a forward-biased junction, so the parameter **FC** specifies a fraction of v_R beyond which forward bias causes only a linear increase in C_T.

SPICE calculates diffusion capacitance according to Eq. (3.54) if $\tau = $ **TT**, transit time, has been specified. **TT** can be determined[12] from a measurement of the storage time, t_s. Referring to Fig. 3.42,

$$TT = \tau = \frac{t_s}{\ln[1 - (I_F/I_R)]}. \tag{3.65}$$

The parameters **EG** and **XTI** affect SPICE's modeling of temperature variation. Experience has shown that the Schottky-barrier diode can best be modeled with **EG** = 0.69 eV and **XTI** = 2.

EXAMPLE 3.13

Write a SPICE program to find the operating point of the circuit of Fig. 3.47. Let $V_s = 1$ V, $R = 50$ Ω, and use the default model for the diode.

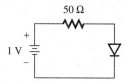

FIGURE 3.47 Circuit for use in Example 3.13.

CHAPTER 3 SEMICONDUCTOR DIODES AND DIODE CIRCUITS

SOLUTION   A satisfactory program is:

```
DIODE CIRCUIT
VS 1 0 DC 1V
R1 1 2 50
D1 2 0 DIODE
.MODEL DIODE D
.OP
.END
```

The `.OP` (operating point) command is assumed in some SPICE programs, but may be required in PSPICE®, for example. The simulation yields $V_D = 0.701$ V, $I_D = 5.97$ mA, and $R_{eq} = 4.33$ Ω.

■ EXAMPLE 3.14

Use SPICE to obtain the forward characteristics ($0 < i_D < 100$ mA) for three silicon diodes having $I_s = 10^{-14}, 10^{-13}$, and 10^{-12} A. Assume $T = 27°C$, $R_s = 1$ Ω, and $\eta = 1$.

■ The answers to these example programs would be found in the output file. In the case of a plot, the graphical interface will vary with the software used.

SOLUTION   A suitable program is shown, with the diodes in parallel with a voltage source. The default temperature is 27°C. As the voltage is swept through the desired range, the current is determined for each diode. The results are printed and plotted in an output file.

```
DIODE COMPARISON
VS 1 0 DC 0
D1 1 2 D1
D2 1 3 D2
D3 1 4 D3
V1 2 0 DC 0
V2 3 0 DC 0
V3 4 0 DC 0
.MODEL D1 D RS=1 IS=1E-14
.MODEL D2 D RS=1 IS=1E-13
.MODEL D3 D RS=1 IS=1E-12
.DC VS 0 1 0.01
.PRINT DC I(V1) I(V2) I(V3)
.PLOT DC I(V1) I(V2) I(V3)
.END
```

Note that the diode currents in the simulation are not limited to 100 mA, but the data for higher current can be discarded. SPICE 2G® requires the dummy (0-V) sources to act as ammeters for the currents desired, while PSPICE®, for example, allows a direct request for the branch currents `I(D1)`, etc.

■ EXAMPLE 3.15

The power supply of Example 3.8 is redrawn in Fig. 3.48. The capacitor is 4900 μF, and is assumed initially uncharged. The transformer and diode resistance totals 0.2 Ω. Write a SPICE program to plot the capacitor voltage and the current through one of the diodes for the first 20 ms after $t = 0$.

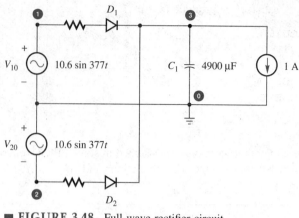

FIGURE 3.48 Full-wave rectifier circuit.

SOLUTION

The nodes are numbered as shown in Fig. 3.48. The SPICE statements follow:

```
PEAK RECTIFIER CIRCUIT
V10 1 0 SIN(0 10.6 60)
V20 0 2 SIN(0 10.6 60)
I30 3 0 1A
D1 1 3 DIODE
D2 2 3 DIODE
C1 3 0 4900UF IC=0
.MODEL DIODE D RS=0.2 IS=1E-13
.TRAN 400US 20MS UIC
.PLOT TRAN V(3) I(V20)
.PRINT TRAN V(3) I(V20)
.END
```

Note that the voltage sources are defined sinusoidal, with offset = 0 V, peak value = $7.5\sqrt{2} = 10.6$ V, and frequency = 60 Hz. These sources are zero for $t < 0$. For convenience, the total diode branch resistance is lumped into the diode model, $RS = 0.2\ \Omega$. Our diode model uses $I_s = 10^{-13}$ A, a value yielding $V_D = 0.7$ V at about 50 mA. The transient analysis will begin at $t = 0$ and proceed at 400-μs intervals for 20 ms. UIC directs the use of initial condition $V_{CAP} = 0$. Capacitor voltage and D_2 current are plotted in Fig. 3.49. Note that the D_2 current is I(V20), flowing from positive to negative through v_{20}. At steady state, v_3 is seen to vary between 8 and 9 V. Peak diode current is 5.41 A, with a pulse width of only about 3 ms per cycle. SPICE will print the first output variable with the plot, but to obtain the exact current through V_{20} we needed the print statement. PSPICE®, with PROBE®, gives much more flexibility in what can be requested for output. Compare these results with the approximate solution of Example 3.8.

DRILL 3.20

Write the model statement for a silicon diode 1Nxx having a saturation current of 10^{-12} A, $\eta = 1$, and a lead resistance of 1 Ω.

ANSWER .MODEL 1Nxx D IS=1e-12 N=1 RS=1 (EG=1.11 is the default)

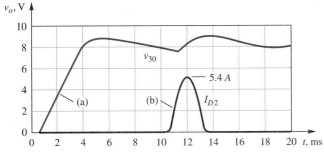

FIGURE 3.49 Full-wave rectifier. (a) Capacitor voltage. (b) D_2 current.

CHECK UP

1. **TRUE OR FALSE?** In the diode element statement, the order of the nodes is not significant.
2. **TRUE OR FALSE?** In the diode model statement, the parameter `IS` is of great importance.
3. **TRUE OR FALSE?** In the diode model statement, specifying a value for `RS` has the same effect on the circuit as using an external series resistance.

SUMMARY

Virtually all semiconductor electronic devices depend on the *pn* junction. In this chapter we presented the charge distribution, electric field, potential distribution, and current density across a diode *pn* junction in equilibrium and with an external bias. The most important result was the derivation of the diode equation, Eq. (3.26), which we used to describe the one-port circuit element model for a diode. The *pn* junction forms the basis for the operation of the bipolar junction transistor, discussed in Chapter 4, and field-effect transistor, discussed in Chapter 5.

Important diode characteristics and specifications including temperature sensitivity, reverse-breakdown voltage, and transition capacitance can be derived and illustrated by means of data sheets. Specialized diode devices include the LED, the breakdown diode, and the varactor diode, whose performance depends on the voltage-dependent transition capacitance. All semiconductor devices generate heat that must be removed from the area of the junction to prevent damage. An electrical analog for a thermal circuit with a power diode as a heat generator can be used to evaluate heat sink performance.

We have demonstrated the graphical load line solution technique for establishing the operating point of a diode circuit with a plot of the diode characteristic. A piecewise-linear model for the diode was used in the analytical solution of a number of key diode-circuit applications. The half-wave rectifier, full-wave rectifier, and full-wave bridge rectifier were treated in some detail because power supply rectifier circuits are used in virtually all electronic systems. The design of breakdown-diode regulators is one way to obtain a power supply with a regulated output voltage.

Diode logic gates, limiter circuits, clamps, and analog switches are among the representative applications that use the switching-type transfer characteristic. We expand on these ideas and applications in Chapter 13, where we will study more complex logic and switching circuits.

The nonlinear characteristic of the diode is often exploited in circuits for obtaining the product of two signals. Such circuits are known as *mixers, modulators,* or *multipliers.* We leave this topic for the communications texts,[13] except to give an example in Chapter 17, as related to the phase-locked-loop circuit. Although advanced photonic and microwave diode applications are beyond the scope of this text, a brief survey of these devices has been included.

We also considered a SPICE model for the diode, to provide examples of computer simulation of electronic circuits containing diodes.

SURVEY QUESTIONS

- What causes a space-charge region at a *pn* junction?
- Is it possible to eliminate the potential barrier at a *pn* junction by forward-biasing the junction? Explain.
- What parameters of a diode must be specified in the design of a full-wave rectifier for a particular power supply application?
- Perhaps the most important element in a peak rectifier power supply is the filter capacitor. How does the designer choose this capacitor?
- Under what circumstances might you want to use a diode clipper circuit in an electronic circuit design? When might you use a diode clamp circuit?
- What measures could be taken to reduce the storage time in a diode switching circuit?

PROBLEMS

Problems marked with an asterisk are more challenging.

3.1 At an abrupt silicon junction at 300 K, $N_A = 10^{17}$ atoms/cm^3 and $N_D = 10^{14}$ atoms/cm^3. Calculate the minority carrier concentrations, and make a sketch of carrier concentrations outside the depletion region versus distance from the junction.

3.2 Repeat Problem 3.1 assuming that the junction is germanium.

3.3 Repeat Problem 3.1 assuming that the diode is forward biased and $V_D = 0.5$ V.

3.4 Repeat Problem 3.1 assuming that the diode is reverse biased and $V_D = -0.1$ V.

3.5 A silicon diode at 25°C has a reverse saturation current of 1 μA. At what applied voltage will the diode current be 100 mA? Assume $\eta = 1$.

3.6 A germanium diode at 25°C has a reverse saturation current of 10 μA. At what applied voltage will the diode current be 100 mA?

3.7* Assuming that I_S varies with n_i^2, use Eq. (2.27) to show that I_S doubles for a 5°C rise at $T = 300$ K.

3.8* What is the reverse saturation current for the diode in Problem 3.5 at 100°C? Does I_S double for a 5°C rise at $T = 100$°C?

3.9 A silicon diode has a current of 100 mA. At 25°C the forward-bias voltage is 0.7 V. Find the forward voltage for this diode at the same current when $T = -55$°C and when $T = 100$°C.

3.10 A silicon diode has a forward bias of 0.7 V. If the diode current is 100 mA at 25°C, what will it be at 100°C? Assume $\eta = 1$.

3.11 Can two similar diodes connected in parallel be expected to share the current equally? Explain your answer.

CHAPTER 3 SEMICONDUCTOR DIODES AND DIODE CIRCUITS

3.12 Why might a typical camera *not* use a silicon photodiode in its light meter?

3.13 Consider the diode whose characteristic is graphed in Fig. 3.50. Assume that $T = 27°C$ and $\eta = 1$. Use v_D at 5 mA to estimate I_S, and then use v_D at 30 mA to estimate the series bulk and lead resistance.

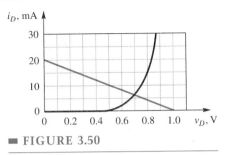

■ FIGURE 3.50

3.14 Distinguish between avalanche breakdown and Zener breakdown.

3.15 An 8-V avalanche diode has a temperature coefficient of $+0.1\%/°C$. In an attempt to give some temperature stabilization, three silicon diodes are placed in series with it (forward biased). What is the temperature coefficient of the combination?

3.16 Using the data from Fig. 3.9, sketch the reverse characteristic of the 1N5234B Zener diode at 25°C. Assume worst-case leakage and resistance.

3.17 Using the data from Fig. 3.9, estimate the breakdown voltage for the 1N5234B at 100°C.

3.18 Two reverse-biased silicon diodes are connected in series across a 100-V source as shown in Fig. 3.51. The saturation currents are 1 and 2 μA. Find the operating voltage and current of each. Let $\eta = 2$.

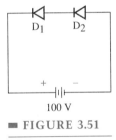

■ FIGURE 3.51

3.19 Repeat Problem 3.18 assuming that each diode has an avalanche breakdown voltage of 80 V.

3.20 Given a silicon diode that has a voltage drop of 0.5 V at 1 mA, estimate the operating point of the diode when it is inserted in series with a 10-V battery and a 1-kΩ resistor as in Fig. 3.52. Assume $\eta = 2$.

3.21 The silicon diode shown in Fig. 3.53 is connected in series with a 1-V battery and a 10-Ω resistor so that it conducts. Use a graphical solution to find the operating point.

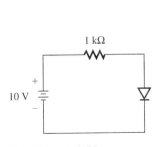

■ FIGURE 3.52

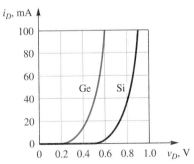

■ FIGURE 3.53 Typical forward characteristics of a germanium diode and a silicon diode at 25°C. Voltage increases faster than the diode equation predicts because of resistance.

3.22 A GaAs LED has $V_D = 1.6$ V. This diode is to be used to indicate the presence of a 5-V power supply in Fig. 3.54. Find the series resistance needed to bias the indicator current at 10 mA.

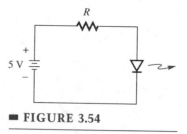

■ **FIGURE 3.54**

3.23 Determine a piecewise-linear model for the germanium diode shown in Fig. 3.53 that is a good approximation over the range of 0 to 50 mA.

3.24 Design a piecewise-linear model for the Zener diode of Fig. 3.31, useful over the range -30 mA $< i_D < +20$ mA. Your model will have two branches, one for each breakpoint. Assume ideal diodes.

3.25 A silicon diode is known to have $I_S = 10^{-14}$ A, $\eta = 1.2$, and a bulk-lead series resistance $R_S = 2\,\Omega$ at 27°C. Find the dynamic resistance of this diode at an operating current of 25 mA.

3.26 Suppose the half-wave rectifier circuit shown in Fig. 3.55 has an input voltage v_S that is a square wave, alternating between $+5$ and -5 V. The diode has an offset voltage $V_F = 0.6$ V and a forward resistance $R_F = 1\,\Omega$; $R_L = 10\,\Omega$. Sketch the load voltage v_L, and find the dc load voltage.

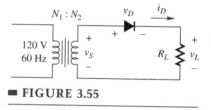

■ **FIGURE 3.55**

3.27 Show that the full-wave rectifier illustrated in Fig. 3.56 and driven by a transformer secondary voltage of $V_P \sin \omega t$ (each side of center top) does yield a dc output voltage $2V_P/\pi$ if losses in the transformer and diode are neglected.

3.28* In the full-wave bridge rectifier shown in Fig. 3.57, $v_S = 12.6 \sin 377t$. The transformer has a secondary resistance of $0.5\,\Omega$, and the diodes have an offset voltage $V_F = 0.7$ V and a forward resistance $R_F = 0.5\,\Omega$; $R_L = 15\,\Omega$. Sketch $i_L(t)$, labeling both axes, and find the dc current I_L.

3.29 In the full-wave bridge rectifier shown in Fig. 3.57, $v_S = 15 \sin 377t$ and $R_L = 500\,\Omega$. Assume ideal diodes.

(a) Sketch load voltage and current of diode D_1.
(b) What peak and average current ratings would you require for the diodes?
(c) Would the answer to part (b) change if the frequency were 400 Hz?

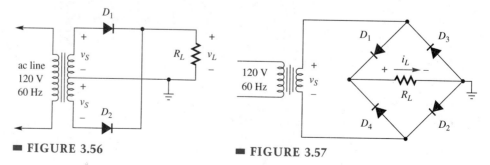

■ **FIGURE 3.56**   ■ **FIGURE 3.57**

3.30 Design a half-wave peak rectifier such as that shown in Fig. 3.58 to be used in a small power supply, for which V_L is 5 V, i_L is 1 A, and the ripple voltage is to be less than 0.1 V_{p-p}. Specify the values of C and v_S that you require.

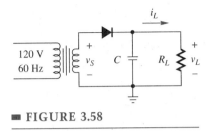

■ FIGURE 3.58

3.31 A full-wave rectifier such as the one shown in Fig. 3.56 has 1000 μF across the load resistor R_L, v_S is 10 V_{peak} at 60 Hz, and the ripple voltage 0.1 V_{p-p}. Find the dc load current.

3.32* Assume the peak rectifier circuit shown in Fig. 3.58. Let $v_S = V_P \sin \omega t$ and neglect all losses in the diode and transformer.

(a) Show that while the diode is conducting the diode current is

$$i_D = V_P \left[\left(\frac{1}{R_L} \right)^2 + (\omega C)^2 \right]^{1/2} \sin[\omega t + \tan^{-1}(\omega R_L C)].$$

(b) Make a sketch showing v_S, v_L with a small ripple, and i_D, all on a common time scale.

3.33 The avalanche diode in Fig. 3.59 has a constant voltage drop of 9 V as long as i_Z is maintained between 10 and 200 mA. Assume the load consists of a fixed 100-Ω resistor.

(a) If $R_X = 100$ Ω, find the limits on v_S such that the diode continues to regulate.
(b) Calculate the maximum power that the diode will dissipate and also the maximum power in R_X.

3.34 In the circuit in Fig. 3.59, let $V_S = 6.3$ V, $V_Z = 5$ V, 5 mA $< i_Z <$ 100 mA, and $R_X = 13$ Ω.

(a) Determine the range of load currents available at 5 V.
(b) Describe the situation if R_X is made 10 Ω.

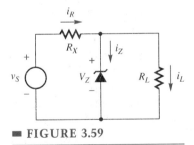

■ FIGURE 3.59

3.35 Design a Zener regulated power supply for the following situation: A given 5-V display circuit has a load current that varies from a low of 5 mA up to a maximum of 100 mA. A 5-V Zener diode is available having a minimum useful Zener current of 5 mA and a maximum power dissipation of 5 W. Find the maximum and minimum values of R_X (Fig. 3.59) such that the circuit maintains regulation with a source voltage of either 6.3 or 12.6 V.

3.36 Design a circuit using a series R_X and a breakdown diode that will deliver 6.0 V from a 12.6-V auto battery. Pick a diode from the chart in Fig. 3.9. Observe maximum power ratings on the device for no-load current. What is the maximum load current that your circuit will deliver without dropping out of regulation?

3.37 Find the voltage regulation at the input to the IC regulator for Example 3.8.

3.38 Given the clipper circuit shown in Fig. 3.60, let $v_S = 10 \sin \omega t$, $V_1 = 1$ V, and $R_1 = R_L = 1$ kΩ. Assume a diode offset voltage of 0.7 V.

(a) Sketch the transfer characteristic v_O versus v_S.
(b) Sketch $v_O(t)$.

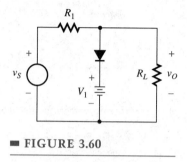

■ FIGURE 3.60

3.39 Repeat Problem 3.38 assuming that the diode has 100-Ω series resistance.

3.40 Given the clipper circuit shown in Fig. 3.61 (see bottom of page), let $v_S = 10 \sin \omega t$, $R_1 = 1$ kΩ, $R_L = 1$ kΩ, $V_1 = +1$ V, and $V_2 = -1$ V. Assume diode offset $V_F = 0.7$ V.

(a) Sketch the transfer characteristic v_O versus v_S.
(b) Sketch $v_O(t)$.

3.41 Design a diode circuit that will protect the input of an audio amplifier from overvoltage. This amplifier has an input resistance of 10 kΩ, one side grounded. Normal input voltages to this amplifier are less than 1 V_{rms}. Your circuit must protect this input from voltages outside the range +2 V without attenuating acceptable signals by more than 25%. You may assume diodes with $V_F = 0.5$ V. Draw a diagram of your circuit, with all component values given.

3.42 The fixed voltage sources of diode clipper circuits may be replaced by breakdown diodes. Sketch the transfer characteristic for Fig. 3.60 if the diode and V_1 are replaced with the Zener diode whose characteristics are given in Fig. 3.31. Let $R_1 = R_L = 1$ kΩ. Expect two breakpoints.

3.43 Sketch the transfer characteristic for Fig. 3.60 if the diode and V_1 are replaced with two breakdown diodes back-to-back. Use the characteristics in Fig. 3.31. Let $R_1 = R_L = 1$ kΩ.

3.44 Consider the clamp circuit shown in Fig. 3.62, with $R = 1$ MΩ, $C = 1$ μF, $V_R = 2$ V, and v_S is a square wave, alternating 10 ms at +5 V, then 10 ms at −5 V.

(a) To what voltage does C charge?
(b) Sketch $v_O(t)$ on the same time scale with v_S.

3.45* (a) Repeat Problem 3.44 with $R = 10$ kΩ.
(b) Make a general statement relating the time constant of the clamp to the period of the waveform for v_S such that the output waveform v_O does not show more than 1% "sag."

3.46 What effect would a source impedance (in series with v_S) have on the results of Problem 3.44?

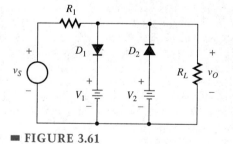

■ FIGURE 3.61

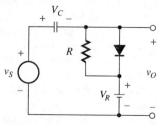

■ FIGURE 3.62

CHAPTER 3 SEMICONDUCTOR DIODES AND DIODE CIRCUITS

3.47 Design a diode circuit that will accept a 1-kHz signal having zero average value, and shift it without distortion so that it is always more negative than −1 V. You may assume that the amplitude of this signal changes very slowly, if at all. You may assume ideal components.

3.48 In the diode logic circuits shown in Fig. 3.63, let $V = 5$ V and $R = 10$ kΩ. The diodes have an offset voltage of $V_F = 0.7$ V but negligible resistance. Find the output voltages at C for each case:

(a) $A = B = 4$ V.
(b) $A = 4$ V, $B = 0$ V.
(c) $A = 0$ V, $B = 4$ V.
(d) $A = B = 0$ V.

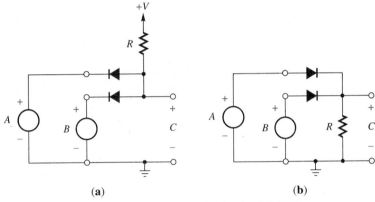

■ FIGURE 3.63 Two-input diode-logic circuits. (a) AND. (b) OR.

3.49 Suppose the output terminal of the AND is connected to an input of the OR as shown in Fig. 3.64. Let the other input of the OR circuit be 0 V. Assume ideal diodes, $V = 5$ V, and $R = 1$ kΩ. Show that if $A = B = 4$ V for the AND circuit, the output voltage at C is not 4 V, as it should be if the circuits were independent. Why is this true?

3.50 Suppose the output terminal of the OR is connected to an input of the AND as in Fig. 3.65. Let the other input of the AND circuit be 4 V. Assume ideal diodes, $V = 5$ V, and $R = 1$ kΩ. Show that when $A = B = 0$ V for the OR, the output voltage C is not 0 V, as it should be if the circuits were independent. Why is this true?

3.51 Consider the AND circuit shown in Fig. 3.63, with $V = 5$ V and $R = 10$ kΩ. Assume ideal diodes. Let input voltage $A = 3 \sin \omega t$.

(a) Sketch the output voltage $C(t)$ when input voltage $B = -5$ V.
(b) Sketch the output voltage $C(t)$ when input voltage $B = +5$ V.
(c) Describe the operation of this circuit as an analog switch.

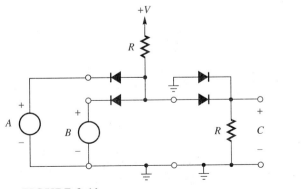

■ FIGURE 3.64

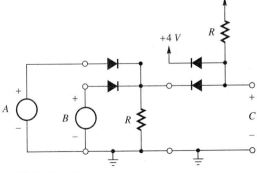

■ FIGURE 3.65

3.52 Consider the **OR** circuit shown in Fig. 3.63, with $R = 10$ kΩ. Assume ideal diodes. Let input voltage $A = 3 \sin \omega t$.

(a) Sketch the output voltage $C(t)$ when input voltage $B = -5$ V.
(b) Sketch the output voltage $C(t)$ when input voltage $B = +5$ V.
(c) Describe the operation of this circuit as an analog switch.

3.53* In the diode gate shown in Fig. 3.66, $R_L = R_C = 10$ kΩ. Let $v_{IN} = 2 \sin \omega t$ and $R_1 = 600$ Ω. Assume the diodes are matched, with $V_F = 0.7$ V.

(a) Find the value of v_C required to enable the gate.
(b) Find the value of v_C required to disable the gate.
(c) What value of breakdown voltage is required of the diodes?

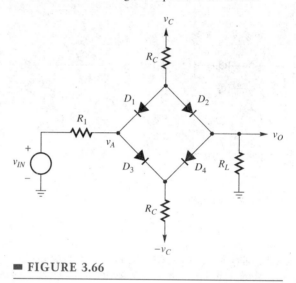

■ **FIGURE 3.66**

3.54 Show that for a step junction,

$$C_T = \frac{\varepsilon A}{w_p + w_n}$$

or the same as for a parallel-plate capacitor with the plate separation equal to the thickness of the depleted region.

3.55 At a silicon step junction, $N_D = 10^{14}$ atoms/cm^3 and $N_A = 10^{16}$ atoms/cm^3. Contact potential is 0.58 V. If the applied reverse voltage is 10 V, find the width of the depleted region and the electric field intensity at the junction.

3.56 At a silicon step junction, $N_D = 10^{14}$ atoms/cm^3 and $N_A = 10^{16}$ atoms/cm^3. Contact potential is 0.58 V and cross-sectional area of the junction is 10^4 μm^2. Compute the $C_j(V_R)$ and graph the result over the range from -12 to 0 V.

3.57 Calculate the following quantities for an abrupt *pn* junction, with cross-sectional area of 5×10^3 μm^2, doped on one side with 5×10^{15} boron atoms/cm^3 and on the other side with 2×10^{17} arsenic atoms/cm^3. The reverse bias is 5 V.

(a) Built-in potential at 300 K.
(b) Depletion-layer dimensions.
(c) A labeled sketch of the electric field profile.
(d) Transition capacitance.
(e) An estimate of the built-in potential at 290 K.

3.58 Given a silicon abrupt junction with $N_A = 10^{19}$ atoms/cm^3, $N_D = 10^{16}$ atoms/cm^3, and $A_j = 10^{-2}$ cm^2, find (a) contact potential, (b) depletion layer thickness, and (c) zero-bias transition capacitance C_0.

3.59 Given the diode described in Problem 3.58, find the transition capacitance C_T for reverse-bias voltages of 1, 2, 5, and 10 V.

3.60 Given a linear **graded** junction, that is, the charge density $\rho = qax$, $-W/2 < x < W/2$, do the following:

(a) Plot charge density $\rho(x)$, electric field $\mathbf{E}(x)$, and potential $V(x)$.
(b) Show that the junction potential is $aqW^3/12\varepsilon$.

3.61 Use the results of Problem 3.60 to derive Eq. (3.49), C_T for the graded junction.

3.62 Consider a silicon diffused junction, where the charge distribution may be considered to have a linear grade $a = 10^{20}$ atoms/cm^4 in the depleted region. The contact potential is 0.63 V, and $A_j = 10^{-2}$ cm^2. Find the zero-bias transition capacitance C_0.

3.63 A silicon step junction has a transition capacitance of 10 pF with no dc bias, and 3.3 pF with a reverse voltage of 4 V. Find the contact potential of the diode.

3.64 A germanium diode has an average carrier lifetime of 10 μs. Estimate the diffusion capacitance at 1 mA of diode current.

3.65* (a) Find the maximum magnitude of the electric field $\mathbf{E}_m$ at a step graded junction.
(b) Assuming breakdown will occur at $\mathbf{E}_m = 2 \times 10^7$ V/m, derive an expression for the breakdown voltage in terms of the doping levels.

3.66* Assume a p^+-n diode and the relationships $Q_p = I_F \tau_p$ for $t < 0$ and

$$i_D = \frac{Q_p(t)}{\tau_p} + \frac{dQ_p(t)}{dt}.$$

At time $t = 0$, the current is switched from I_F to I_R (a negative value). Find the storage time t_s [the time from the switch until $Q_p(t) = 0$]. This is essentially a derivation of Eq. (3.65).

3.67 Assume a p^+-n diode, where the lifetime of the holes injected into the n-region is 1 μs. At $t = 0$, the diode current is switched from I_F to $-I_F$. Find the storage time t_s.

3.68 Refer to the opto-isolator of Fig. 3.67. The ratio of I_2/I_1 is called the *current transfer ratio* (CTR), given in percent. Suppose the opto-isolator has a CTR of 20%, and $R_L = 1$ kΩ. What will the input current I_1 have to be to develop a 1-V output signal?

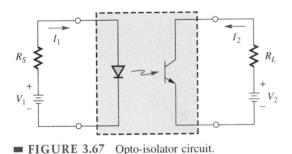

■ **FIGURE 3.67** Opto-isolator circuit.

3.69 A certain power diode is rated at 6 A with a junction voltage of 0.64 V. The case temperature is not to exceed 150°C. Thermal resistance $\theta_{JC} = 4.25$°C/W.

(a) Estimate the junction temperature at maximum case temperature.
(b) If ambient temperature is 30°C, what is the maximum thermal resistance, θ_{CA}, of the heat sink?

3.70 Assume a silicon diode with $I_S = 10^{-14}$ A, $R_S = 0$, and $T = 27$°C. Write a SPICE program to compare the forward characteristics of this diode $0 < i_D < 100$ mA for the three cases $\eta = 1, 1.2,$ and 1.4.

3.71 Assume a silicon diode with $I_S = 10^{-14}$ A, $R_S = 0$, and $\eta = 1$. Write a SPICE program to compare the forward characteristics of this diode $0 < i_D < 100$ mA for the three temperatures $T = -55, 27,$ and 100°C.

3.72 Assume a silicon diode with $I_S = 10^{-14}$ A, $T = 27°C$, and $\eta = 1$. Write a SPICE program to compare the forward characteristics of this diode for $0 < i_D < 100$ mA for the three values of series resistance $R_S = 0$, 1, and 4 Ω.

3.73 Write a SPICE program to solve Problem 3.18.

3.74 Write a SPICE program to solve Problem 3.19.

3.75 Estimate the peak diode current for the rectifier of Problem 3.30. State assumptions. You may want to use SPICE for accurate results.

3.76 Use SPICE to simulate the diode clipper circuit of Problem 3.38. Use the diode default model rather than the given offset voltage.

3.77 Solve the clamp circuit of Problem 3.44 by SPICE simulation. Use the default diode model. Let the initial capacitor voltage be zero, and plot v_O from 0 to 40 ms.

3.78 Simulate Problem 3.52, parts (a) and (b), and plot the output waveform with SPICE.

REFERENCES

1. Allison, J. *Electronic Engineering Semiconductors and Devices.* New York: McGraw-Hill, 1990, pp. 157–164.
2. Sze, S. M. *Semiconductor Devices, Physics and Technology.* New York: John Wiley & Sons, 1985, pp. 281–287.
3. Haznedar, H., *Digital Microelectronics.* Redwood City, Calif.: Benjamin/Cummings, 1991, pp. 394–397.
4. Streetman, B. G. *Solid-State Electronic Devices,* 4th ed. Englewood Cliffs, N.J.: Prentice-Hall, 1995, pp. 207–208.
5. Ref. 2, pp. 160–171.
6. Ref. 4, pp. 222–224.
7. Ref. 2, pp. 258–267.
8. Ref. 4, pp. 387–397.
9. Ref. 2, pp. 267–278.
10. Ref. 2, Chap. 7.
11. Ref. 4, pp. 208–212 and Chap. 12.
12. Ref. 4, p. 173.
13. Couch II, L. W. *Digital and Analog Communication Systems,* 2nd ed. New York: Macmillan Publishing Co., 1987, pp. 231–245.

SUGGESTED READINGS

Banzhaf, W. *Computer-Aided Circuit Analysis Using PSPICE,* 2nd ed. Englewood Cliffs, N.J.: Prentice-Hall, 1992. A complete introduction to PSPICE®; diode models are discussed in Section 8.2.

Neudeck, G. W. *The PN Junction Diode,* Vol. II of *Modular Series on Solid-State Devices,* 2nd ed., Reading, Mass.: Addison Wesley, 1989.

Pulfrey, D. L. *Inrodution to Microelectronic Devices.* Englewood Cliffs, N.J.: Prentice-Hall, 1989.

Streetman, B. G. *Solid-State Electronic Devices,* 4th ed. Englewood Cliffs, N.J.: Prentice-Hall, 1995. Chapter 5 treats junctions and Chapter 6 covers devices.

Sze, S. M. *Semiconductor Devices, Physics and Technology.* New York: John Wiley & Sons, 1985. Chapters 3, 5, and 7 cover *pn* junctions, metal-semiconductor junctions, and photonic devices.

CHAPTER 4

THE BIPOLAR JUNCTION TRANSISTOR

4.1 Basic BJT Operation
4.2 Volt-Ampere Equations for the BJT
4.3 BJT Regions of Operation
4.4 The Common-Base Configuration
4.5 The Common-Emitter Configuration
4.6 Cutoff in the CE Configuration
4.7 Saturation in the CE Configuration
4.8 Inverse Mode
4.9 CE Current Gain Considerations
4.10 Maximum Ratings for the BJT
4.11 Switching Times for the BJT
4.12 SPICE Model for the BJT
Summary
Survey Questions
Problems
References
Suggested Readings

Having considered the operation of the *pn* junction in Chapter 3, we are now ready to turn our attention to a device composed of back-to-back junctions. The **bipolar junction transistor** (BJT) is the realization of a current-controlled current source. This means that we can control the current across one junction by means of a current (voltage) applied to the other junction. Since their invention in 1948, transistors have revolutionized the electronics industry, displacing the vacuum tube in nearly all applications because of their size, power, service life, cost, and reliability.

The *pn* junction theory of Chapter 3 will apply, with suitable modifications, in this chapter. Here we study the operation, terminal characteristics, and simplified models of the transistor. We also present representative circuit applications, both as amplifiers and as switches. SPICE computer simulation of BJT circuits is also demonstrated. As you study this chapter, look for the important concepts listed next.

IMPORTANT CONCEPTS IN THIS CHAPTER

- In an active BJT, majority carriers from the emitter are injected into the base, diffuse across the base, and are collected at the collector.
- The current gain α_F of a BJT is that fraction of emitter current that reaches the collector.
- A BJT has four modes of operation, depending on the bias direction of each of the two junctions.
- The BJT has three possible configurations in a circuit, depending on which of the terminals is used as the reference.
- The common-emitter dc circuit model for an active npn BJT consists of an offset voltage of 0.7 V from base to emitter, and a controlled current source of $\beta_F I_B$ at the collector.
- The effect of collector-base voltage on collector current is characterized by the Early voltage.
- Dynamic base resistance $r_\pi = \beta V_T / I_C$.
- The common-emitter small-signal circuit model for an active BJT consists of a base input resistance r_π and a controlled current source $g_m v_{be} = i_b \beta$ at the collector.
- The BJT is essentially cut off when $|V_{BE}| < 0.5$ V.
- The BJT collector leakage current I_{CEO} is negligible at $T = 300$ K, but doubles with 10°C rise.
- BJT collector current saturates when $|v_{CE}| < 0.3$ V.
- Collector power dissipation, $i_C v_{CE}$, produces a temperature rise of the collector junction.

Most electronic circuits, including integrated circuits (ICs), use transistors as current amplifiers or current switches. Another type of transistor, the field-effect transistor (FET), is the topic of Chapter 5. Detailed analysis of small-signal amplifiers is addressed in Chapters 8 and 9, and switching circuits are the focus of Chapter 13.

4.1 BASIC BJT OPERATION

The device known as a bipolar junction transistor consists of three layers of semiconductor within a monolithic (single-piece) structure. Silicon is the dominant material because it has fabrication, breakdown voltage, and temperature advantages over germanium and other semiconductor materials. The III-V compound gallium arsenide is used for very-high-frequency (microwave) devices and for optical devices. The three layers alternate by type of impurity and are arranged in either a **pnp** or **npn** configuration as shown in Fig. 4.1. The three layers are identified as **emitter** (E), **base** (B), and **collector** (C). This sketch is simplified for the sake of clarity—real transistors do not look like this (see Fig. 4.5 later in this chapter). Realistic geometries, dimensions, and fabrication details are presented in Chapter 18. Recognize that the entire transistor structure is very small (from a few square micrometers for IC BJTs to several square millimeters for discrete power BJTs) and that as a practical circuit component, it is sealed in an airtight metal or plastic package. The package provides mechanical protection and facilitates heat transfer. Figure 4.2 shows several discrete BJT packages.

Figure 4.1 also includes corresponding transistor circuit symbols. In each case, the lead with the arrow identifies the terminal designed to be the emitter, and the direction of the arrow (as in the diode symbol) indicates a transition from p-type to n-type semicon-

CHAPTER 4 THE BIPOLAR JUNCTION TRANSISTOR

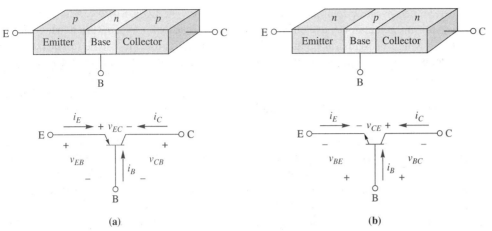

FIGURE 4.1 Two arrangements for a junction transistor with corresponding circuit symbols. (a) *pnp*. (b) *npn*.

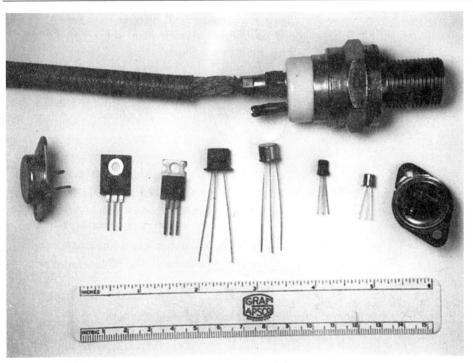

FIGURE 4.2 Some commercial transistor packages. Top, high-power BJT; bottom from left to right, audio power amplifier (TO-3 package), series voltage regulator, 3A power BJT (TO-220 package), very early BJT 250mW, small-signal amplifier 500mW (TO-5 package), small-signal amplifier 500mW (TO-92 plastic package), small-signal amplifier 500mW (TO-18 package), audio power amplifier (TO-3 package cutaway).

■ Often dozens to 10^4+ BJTs are incorporated in a single IC, as we will see in later chapters.

ductor. Thus, when the arrow points away from the base, it signifies an *n*-type emitter, a *p*-type base, and an *npn* transistor. Note the terminal voltage markings on the figure. The order of the double subscripts implies the polarity indicated by the signs. Whether a voltage such as v_{BE} (the voltage drop from base to emitter) is actually positive or negative is determined by the circuit external to the transistor. The choice between these voltage labels is arbitrary, and we have elected to label the transistors in Fig. 4.1 such that positive voltages will result in forward-bias junctions. By standard conventions, all terminal currents are defined into the transistor in both cases, so each current must carry its own sign, determined by the external circuit.

As we shall see, our goal is to have the emitter inject carriers into the base, and have as many of them as possible reach the collector. The transistor gets its name from this "transfer" of current from one junction to the other (*trans*fer res*istor*).

Figure 4.3(a) shows an *npn* transistor, with the base-emitter junction forward-biased and the base-collector junction reverse-biased. This is the "normal" biasing situation, where the collector current is to be controlled by the emitter current. As shown in Fig. 4.3(b), forward-biasing the base-emitter junction lowers the energy barrier for electrons, and they enter the base region from the emitter. As they reach the base-collector junction, they "fall down" the energy "hill" to the collector, giving up energy in the form of heat.

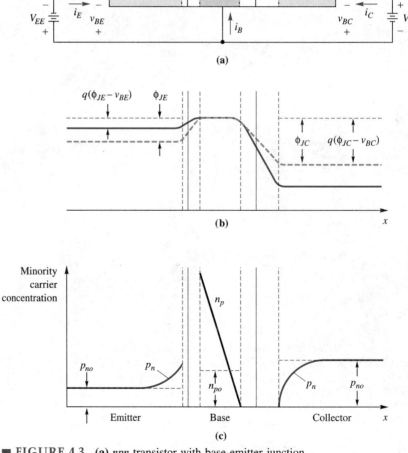

■ **FIGURE 4.3** (a) *npn* transistor with base-emitter junction forward biased, base-collector junction reverse biased. (b) Potential-energy barriers for electrons. (c) Minority carrier concentrations in the various regions of the transistor. (The x dimension is not to scale.)

Note that the horizontal scale is distorted to expand the thin base region. In this case, emitter current is to the left (negative). As shown, the reverse-biased collector junction has a wider space-charge region than the forward-biased emitter junction.

Figure 4.3(c) reveals that large numbers of electrons are injected into the base from the emitter. The sketch assumes that the emitter is most heavily doped, followed by the base, and the collector is least heavily doped. Thus, minority-carrier distributions $p_e < n_b \leq p_c$. The concentration distribution for electrons in the base is steep and nearly linear because the base is relatively thin compared to one diffusion length for electrons, and the concentration must be small at the edge of the collector junction space-charge region, where all electrons are swept by the electric field to the collector.

Fewer holes are injected from the base into the emitter because the equilibrium hole concentration in the base is lower than the electron concentration in the emitter. This is intentional, because this hole current does not contribute to the collector current. Those holes recombine within about one diffusion length as they move into the emitter, and become part of the emitter current.

In the collector region, holes near the junction diffuse to the space-charge region and are swept by the electric field to the base. However, relatively small numbers are involved because the equilibrium hole concentration in the collector is relatively small. This is the normal reverse-biased diode leakage current.

The current flowing across the base-emitter junction is composed of electrons injected into the base and holes injected into the emitter. We define *emitter efficiency*:

$$\gamma = \frac{\text{Electron current injected into the base}}{\text{Total emitter current}}.$$

Emitter efficiency can be made very nearly one (typically 99.6%) by doping the emitter with a much higher impurity concentration than that of the base region. When the base-emitter junction is forward-biased, then the emitter lives up to its name and "emits" or "injects" minority carriers into the base, where they diffuse toward the collector (see Fig. 4.4).

Most of the minority carriers injected into the base (in this case, electrons) diffuse directly across the base and, aided by the electric field in the space-charge region of

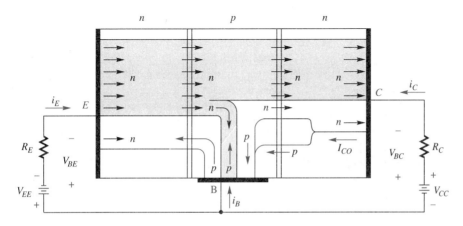

■ **FIGURE 4.4** Current components existing in the transistor when the emitter is forward-biased and the collector is reverse-biased. $p \rightarrow$ represents hole motion, $n \rightarrow$ represents electron motion, and $\rightarrow \leftarrow$ represents recombination.

the base-collector junction, are "collected" at the collector side of the junction. A small fraction of the injected carriers recombines with the majority carriers (holes) in the base region and thus constitutes part of the base current. We define the *base-transport factor*:

$$\alpha_T = \frac{\text{Number of electrons reaching the collector junction}}{\text{Total number of electrons injected into the base region}}.$$

The base-transport factor can also be brought nearly to unity by (1) making the base very narrow so that diffusing minority carriers have less opportunity to recombine, (2) arranging the geometry such that most of the injected electrons can easily "find" the collector (see Fig. 4.5), and (3) making the doping level of the base low to reduce recombination in the base. There are other geometric considerations and constraints; nevertheless, practical base-transport factors are in the range of 99 to 99.8%.

The percentage of total emitter current that actually reaches the collector, then, is the product of emitter efficiency and the base-transport factor. This product, known as the *forward current gain α_F*, is near one for any practical transistor. In Fig. 4.4 i_C is positive because electrons are moving to the right.

■ Typical α_F will be in the range from 0.98 to 0.997.

The only other current shown in Fig. 4.4 is a current, I_{CO}, that flows across the base-collector junction independently of the carriers injected from the emitter. You should recognize this term as the diode reverse-saturation current that will exist because the base-collector junction is a reverse-biased diode. This term is small and can be neglected unless high temperatures are involved.

The terminal currents are now

$$i_C = -\alpha_F i_E + I_{CO} \cong -\alpha_F i_E \tag{4.1}$$

$$i_B = -i_E - i_C = -(1 - \alpha_F)i_E - I_{CO}. \tag{4.2}$$

A *pnp* transistor is shown in Fig. 4.6(b), where the external sources are reversed from those in Fig. 4.6(a) so that again the emitter-base junction is forward-biased and the collector-base junction is reverse-biased. In this circuit, all of the terminal currents will have signs opposite those of the previous case and Eqs. (4.1) and (4.2) again describe the terminal-current relationships. In the *pnp* transistor, of course, the collector current results from holes injected from the emitter into the base, where they diffuse to the collector-base junction.

■ A connection is also made to the substrate of Fig. 4.5(a) for BJT isolation in an IC.

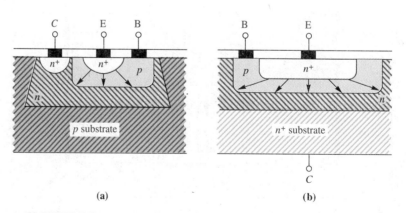

■ **FIGURE 4.5** Geometric configuration for two types of transistors, both illustrating design for good base-transport factor. (a) Planar-diffused IC transistor. (b) Planar-diffused discrete transistor.

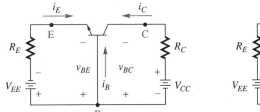 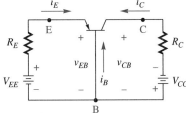

■ **FIGURE 4.6** Transistor circuits with the emitter-base junction forward-biased and the collector-base junction reverse-biased. **(a)** *npn*. **(b)** *pnp*.

4.2 VOLT-AMPERE EQUATIONS FOR THE BJT

In an attempt to characterize the behavior of the BJT for arbitrary bias situations, let us consider the generalized terminal voltages and currents for an *npn* transistor, such as that shown in Fig. 4.7. The sign conventions of the previous section are continued. Lowercase current and voltage symbols indicate possible time dependence.

The collector current consists of two components. One component is the diode current, Eq. (3.26), determined by the base-collector voltage, as in any *pn* junction. As in the case of the diode equation, the thermal voltage V_T should strictly be ηV_T, where η may be greater than one (but less than two) for silicon under certain high- or low-current conditions (see Section 3.3). In most transistors $\eta \cong 1$. The other component consists of the minority carriers (electrons) that may be injected into the base from the emitter and diffuse across the base as discussed in the previous section. The sum of the two components is

$$i_C = -\alpha_F i_E - I_{CO}(e^{v_{BC}/V_T} - 1). \tag{4.3}$$

This equation reduces to Eq. (4.1) when the base-collector junction is sufficiently reverse-biased (at room temperature, $e^{v_{BC}/V_T} \ll 1$ when $v_{BC} < -0.1$ V).

Should the base-collector junction be forward-biased, we find that the emitter current will be a function of minority carriers injected into the base from the collector, plus a diode-like term:

$$i_E = -\alpha_R i_C - I_{EO}(e^{v_{BE}/V_T} - 1). \tag{4.4}$$

The fraction of collector current actually reaching the emitter when the collector is used as an injector of carriers, α_R, is considerably smaller than α_F. As we have already pointed out, the manufacturer has designed the ***forward*** α_F to be a fraction on the order of 99% by careful attention to the relative doping levels in the emitter and base, as well as the geometric configuration of the device. Conflicting design considerations cause the

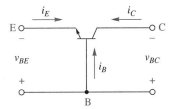

■ **FIGURE 4.7** The *npn* transistor drawn as a two-port network.

reverse α_R to be more like 10 to 50%. The adjectives *forward* and *reverse* indicate a definite preference as to which terminal should be used as the emitter.

Because the two junctions are not symmetric, the collector-junction reverse-saturation current I_{CO} will not be equal to the emitter-junction reverse-saturation current I_{EO}. Actually, we can show by reciprocity[1] that

$$I_{CO}\alpha_R = I_{EO}\alpha_F. \tag{4.5}$$

Equations (4.3) and (4.4) are not exact for all base impurity profiles, but they have very broad application and lead to realistic predictions of transistor action. These relationships are known as the Ebers-Moll equations,[2] named for the men that first proposed them. An equivalent circuit for the transistor that satisfies these equations is shown in Fig. 4.8. You should verify that analogous expressions for *pnp* transistors require sign changes on I_{EO}, I_{CO}, v_{BE}, and v_{BC}.

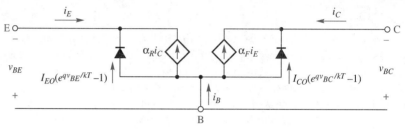

■ **FIGURE 4.8** An Ebers-Moll equivalent circuit for the *npn* transistor using terminal currents and voltages. For a *pnp* transistor, the diodes (and their arrows) would be reversed and the polarity of the voltages interchanged.

Solving Eqs. (4.3) and (4.4) for terminal currents yields

$$i_E = -\frac{I_{EO}}{1-\alpha_F\alpha_R}(e^{v_{BE}/V_T}-1) + \frac{\alpha_R I_{CO}}{1-\alpha_F\alpha_R}(e^{v_{BC}/V_T}-1), \tag{4.6}$$

$$i_C = \frac{\alpha_F I_{EO}}{1-\alpha_F\alpha_R}(e^{v_{BE}/V_T}-1) - \frac{I_{CO}}{1-\alpha_F\alpha_R}(e^{v_{BC}/V_T}-1). \tag{4.7}$$

In terms of terminal currents, junction voltages are

$$v_{BE} = V_T \ln\left[1 - \frac{i_E + \alpha_R i_C}{I_{EO}}\right], \tag{4.8}$$

$$v_{BC} = V_T \ln\left[1 - \frac{i_C + \alpha_F i_E}{I_{CO}}\right]. \tag{4.9}$$

4.3 BJT REGIONS OF OPERATION

When a BJT is used in an electronic circuit, the external circuit will constrain the biasing on the two junctions, according to the preceding equations, to exist in one of four possible states:

1. Base-emitter junction is forward-biased, base-collector junction is reverse-biased. This is usually called the forward-active, or **active** region of operation. We shall find that weak collector-base forward bias ($V_{CB} < 0.2$ V) is permitted (Section 4.7).

CHAPTER 4 THE BIPOLAR JUNCTION TRANSISTOR

2. Base-emitter junction is forward-biased, base-collector junction is forward-biased. This condition is called the **saturation** region.
3. Base-emitter junction is reverse-biased, base-collector junction is reverse-biased. This state is referred to as the **cutoff** region.
4. Base-emitter junction is reverse-biased, base-collector junction is forward-biased. This situation is designated the *reverse-active* or **inverted** mode of operation.

■ External variation of the bias conditions of one or both of the junctions is what controls the action of the BJT.

Active Region

The active region is the normal condition of a BJT when the transistor is being used as an amplifier. The plot in Fig. 4.9 designated n_A shows that the forward-biased emitter-base junction causes the minority-carrier concentration at the base edge of the emitter depleted region to rise above the equilibrium value, specifically to $n_0 e^{v_{BE}/V_T}$, caused by minority carriers injected from the emitter. Because the collector-base junction is reverse-biased, the concentration at the base edge of the collector depleted region decreases to $n_0 e^{v_{BC}/V_T} \approx 0$. The curve itself is nearly a straight line because the base thickness is short compared to one diffusion length for these carriers. The collector current is proportional to the slope of this curve, just as diode current is proportional to the initial slope of the minority-carrier density. The difference is that in the transistor base, recombination is negligible across the base (for a reasonable base-transport factor), and the concentration gradient remains constant across the active base region. The sketched plot of n_A terminates short of the junction to indicate a depleted layer.

Saturated Region

The saturated region usually arises when the forward-biased emitter-base junction causes sufficient collector current to flow such that external circuit constraints cause the collector-base junction to become forward-biased also. The situation, illustrated by the n_S curve in Fig. 4.9, assumes that the collector current is the same as for the active example (n_S has the same slope as n_A). Minority carriers are being injected into the base from both the emitter and collector, resulting in an increased base current for the same collector current.

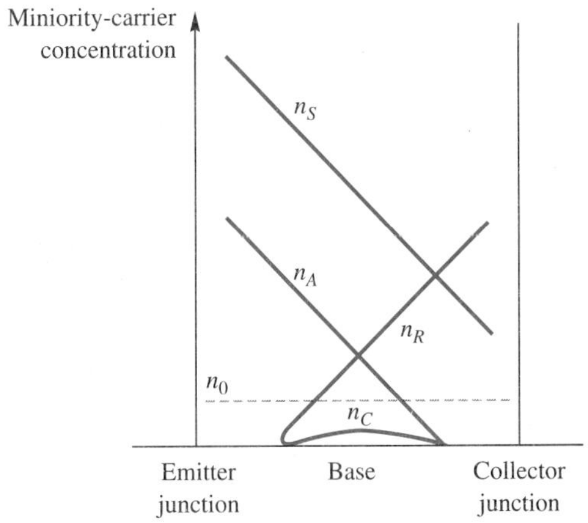

■ **FIGURE 4.9** Minority-carrier concentration in the base region: n_S, saturated condition; n_A, active condition; n_C, cutoff condition; n_R, inverted condition; n_0, equilibrium condition.

Cutoff

When both junctions are reverse-biased, all terminal currents are extremely small and the transistor is said to be OFF. Curve n_C in Fig. 4.9 indicates that there are very few minority carriers in the base region—those being created by thermal generation.

Inverted Mode

The inverted mode is the same situation as the active condition with the collector and emitter interchanged (see curve n_R in Fig. 4.9). Recall that α_R is much smaller than α_F because the injection efficiency of the collector is much lower than that of a highly doped emitter. It is also likely that the transistor geometry yields a lower base-transport factor in this direction.

We next discuss these four situations as they relate to various circuit configurations of the transistor.

■ The inverted mode has limited application in TTL logic circuits. See Chapter 13.

Control of the Collector Current

In the active region, the collector terminal acts as a dependent current source controlled by some other terminal condition. Four models are suggested:

1. The collector current is controlled by base-emitter voltage, as we can see from Eq. (4.7). Since the base-collector junction is reverse-biased, the first term dominates and can be written

$$i_C \approx I_S e^{v_{BE}/V_T}. \tag{4.10}$$

   This model has wide application where the base-emitter junction is voltage driven. We shall use this extensively in the linear amplifier sections of this text.

2. The collector current is controlled by emitter current, as demonstrated in Eq. (4.1). This model is briefly treated in Section 4.4, where emitter current is the independent variable. We will find that signals are usually applied from base to ground, rather than from emitter to ground, but this model is useful in common-base circuits.

3. The collector current is controlled by the base current, as presented in Section 4.5 [see Eq. (4.18)]. This is a very useful model when the base is current driven, as in the dc bias circuits for many amplifiers, and especially in switching circuits. We fully develop this model in this chapter in preparation for BJT-based logic circuits.

4. The collector current is controlled by excess charge in the base region. This concept is reasonable, since the excess charge in the base is a result of the past history of base current. This model is required to predict switching times in the transistor (Section 4.11).

CHECK UP

1. **TRUE OR FALSE?** Almost all of the emitter current consists of carriers that diffuse across the base and reach the collector.
2. **TRUE OR FALSE?** The collector current is a linear function of the base-emitter voltage in an active BJT.
3. **TRUE OR FALSE?** To be in the active region, both junctions of a BJT must be forward-biased.
4. **TRUE OR FALSE?** Saturation in a BJT is the result of external circuit conditions, rather than an inherent property of the BJT itself.

4.4 THE COMMON-BASE CONFIGURATION

Transistors are generally employed in electronic circuits using one terminal pair as an input and another terminal pair as an output. For the three-terminal circuits, one terminal is common to both input and output circuits. When the input is applied between the emitter and base, and the output is taken between collector and base, we speak of the circuit as having a ***common-base*** (CB) configuration. We consider some features of these common-base circuits first, not because they are widely used, but because our introductory sections have already alluded to this configuration. Figures 4.6(a) and (b) are common-base circuits.

Active Region

When the collector is reverse-biased, so that $e^{v_{BC}/V_T} \ll 1$, the general Eq. (4.3) reduces to the linear Eq. (4.1) repeated here:

$$i_C = -\alpha_F i_E + I_{CO} \cong -\alpha_F i_E,$$

and Eq. (4.6) reduces to

$$i_E = -\frac{I_{EO}}{1 - \alpha_F \alpha_R}(e^{v_{BE}/V_T} - 1) - \frac{\alpha_R I_{CO}}{1 - \alpha_F \alpha_R}. \tag{4.11}$$

Eliminating I_{CO} by substituting Eq. (4.5) into Eq. (4.11) yields

$$i_E = -\frac{I_{EO}}{1 - \alpha_F \alpha_R}(e^{v_{BE}/V_T} - 1 + \alpha_F) \approx -\frac{I_{EO}}{1 - \alpha_F \alpha_R} e^{v_{BE}/V_T} \tag{4.12}$$

for positive v_{BE}. Usually α_R is small enough that the denominator is near unity.

CB Collector Characteristics

As an aid in visualizing the current/voltage relationships and transistor operation, a family of static characteristic plots of collector current i_C versus collector-base voltage v_{CB} for several values of emitter current is shown in Fig. 4.10, using measurements of a typical *npn* transistor, the 2N2222A. The first quadrant represents the domain where collector current is positive (base-emitter junction is forward-biased). We have referred to this part of the characteristics as the active region because the collector current responds

■ The curves plotted in Figs. 4.10 and 4.11 can be obtained experimentally. Special oscilloscopes designed to plot these characteristics are called transistor *curve tracers*, or *semiconductor device parameter analyzers*.

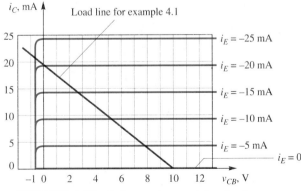

■ **FIGURE 4.10** Collector characteristics for the *npn* transistor 2N2222A in a common-base configuration.

proportionately to changes in the emitter current. Note that Eq. (4.1) describes what we observe in the first quadrant of Fig. 4.10. We have already seen that α_F is a parameter near unity. Note that I_{CO} is negligibly small in this plot. The collector current is nearly flat out to some collector-base breakdown voltage BV_{CBO} (greater than 75 V for the 2N2222A). Breakdown is discussed in Section 4.10.

As we move into the second quadrant, where v_{CB} becomes negative (actually the junction has to reach a cut-in voltage $v_{BC} > 0.5$ V), the collector current abruptly reverses. We are leaving the active region and entering saturation. After that, conventional current increases from base to collector (i_C is negative).

When $i_E = 0$, the only collector current is the base-collector diode reverse-saturation current I_{CO}. This current is relatively small at room temperature and can often be ignored (although we discuss it further in Section 4.6). We have again left the active region and become cut off (both junctions are reverse-biased). Figure 4.10 shows that the active region is bounded by saturation at the left and cutoff below.

CB Emitter Characteristics

A family of curves, emitter current i_E versus base-emitter voltage v_{BE}, is shown in Fig. 4.11. As expected from Eq. (4.12), the curves look exponential, like a diode characteristic, except that they are shifted slightly for different values of v_{CB}, the base-collector reverse-bias voltage. Note also that, as with any *pn* diode, the base-emitter voltage v_{BE} must reach a **cut-in** voltage before the emitter current is significant. Silicon transistors have a cut-in voltage of about 0.5 V at 25°C. The dynamic emitter resistance,

$$r_e = -\left.\frac{\partial v_{BE}}{\partial i_E}\right|_{v_{CE} = V_{CEQ},\, i_E = I_{EQ}} \quad (4.13)$$

■ Discrete *npn* transistors are more common than *pnp* devices for a number of reasons. In integrated circuits, most of the BJTs are *npn*. More will be said on this subject in Chapter 18.

is found to be like the dynamic resistance of the forward-biased diode in Eq. (3.31):

$$r_e = \frac{V_T}{|I_E|}. \quad (4.14)$$

Note that r_e is defined at some dc (quiescent) operating point (Q-point).

Characteristics for *pnp* transistors would have the same appearance as Figs. 4.10 and 4.11, but all currents and voltages would have opposite signs.

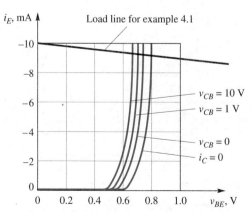

■ **FIGURE 4.11** Emitter characteristic for the *npn* transistor 2N2222A for the common-base configuration.

Base-Width Modulation or Early Effect[3]

As shown in Fig. 4.3(b), the effective base width is reduced by the encroachment of the space-charge regions at the junctions, especially at the reverse-biased collector junction. The depleted region extends further into the base region when the base-collector voltage becomes more negative, as indicated by Eq. (3.14). Thus, the effective base width decreases as the reverse collector voltage increases.

The significance of base narrowing is threefold. First, when the base narrows, the concentration gradient for minority carriers in the base increases for any given injection level. Thus the diffusion current and, hence, i_E will increase for the same v_{BE}. This accounts for the family of i_E curves for various values of v_{CB} in Fig. 4.11. Second, when the base narrows, the base-transport factor increases because there is less chance for injected carriers to recombine in the base. Thus α_F will increase slightly, which accounts for the slight increase of i_C with v_{CB} in Fig. 4.10. We will return to this effect in Section 4.5. Third, if the collector-base voltages were to increase enough, the base could narrow to zero, an effect called **punch-through**. The barrier to electrons from the emitter [Fig. 4.3(b)] would be removed, and excessive emitter current would result, limited only by the external circuit. Collector-base voltage must not exceed the limit imposed by this breakdown phenomenon. Another breakdown phenomenon, avalanche breakdown, is discussed in Section 4.10. Transistors may be destroyed by the heat generated by either type of breakdown unless current is limited by the external circuit.

Circuit Model for the Active Transistor

Equations (4.1) and (4.12) are synthesized by the equivalent circuit shown in Fig. 4.12(a). The diode is presumed to have the characteristic of Eq. (4.12), and the current source constitutes the term(s) of Eq. (4.1). Assuming a piecewise-linear model for the diode with an offset voltage of 0.6 to 0.7 V, we obtain the model for the transistor shown in Fig. 4.12(b). This is a very practical dc model for the *npn* transistor *if it is in the active region*. When attempting an ac model, we might also want to include the dynamic resistance in the emitter branch, using Eq. (4.14) at the dc operating point. The model for a *pnp* transistor would have opposite signs for V_{BE} and I_{CO}. The emitter current, if drawn as indicated in this figure, would be positive for a properly biased *pnp* circuit.

Either the BJT characteristics themselves or the circuit model may be used to find the operating point for the transistor in a circuit. We should point out that although the

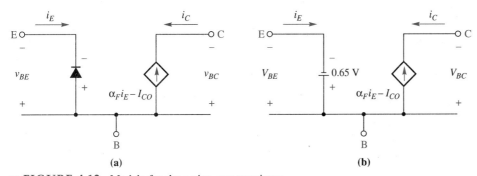

■ **FIGURE 4.12** Models for the active *npn* transistor, common base. For *pnp*, it is necessary to reverse v_{BE} and I_{CO}. (a) An equivalent circuit for Eqs. (4.1) and (4.12). (b) A model in which the diode has been replaced with an offset voltage source.

characteristic curves are very helpful in understanding transistor operation, they are not commonly used in transistor circuit design or analysis. Manufacturer's data sheets provide information for constructing the model and several examples serve to illustrate the procedure.

EXAMPLE 4.1

Consider the circuit of Fig. 4.6(a), with $V_{EE} = V_{CC} = 10$ V, $R_E = 1000$ Ω, and $R_C = 500$ Ω. The transistor is a 2N2222A, whose characteristics are given in Figs. 4.10 and 4.11. Find I_E and V_{CB} (this is the dc Q-point) using a graphical solution.

SOLUTION The 1000-Ω load line in Fig. 4.11 represents the equation $v_{BE} = 10 + 1000 i_E$. Its intersection with the $V_{CB} = 1$ V emitter characteristic gives $V_{BE} = 0.7$ V and $I_E = -9.3$ mA. At this point, we do not know the value of V_{CB} but note that the characteristics for $V_{CB} = 1$ V and $V_{CB} = 10$ V are close, and we try $V_{CB} = 1$ V. The collector load line, shown in Fig. 4.10, expresses $v_{CB} = 10 - 500 i_C$. We estimate a curve representing $I_E = -9.3$ mA between the -5-mA curve and the -10-mA curve. The intersection of this plot with the load line occurs at $I_C \approx 9.3$ mA and $V_{CB} \cong 5.4$ V. This is the operating point of the transistor. Note that our original guess of $V_{CB} = 1$ V is realistic, since V_{CB} is between 1 and 10 V.

EXAMPLE 4.2

Repeat Example 4.1 using a circuit model for the BJT. Let $V_{BE} = 0.7$ V, $I_{CO} = 10$ nA, and $\alpha_F = 0.99$.

SOLUTION Figure 4.13(a) shows the circuit with the transistor model.

$$I_E = \frac{-V_{EE} + V_{BE}}{R_E} = \frac{-10 + 0.7 \text{ V}}{1000 \text{ Ω}} = -9.3 \text{ mA},$$

$$I_C = -I_E \alpha_F + I_{CO} = 9.3(0.99) + 10^{-5} = 9.2 \text{ mA},$$

$$V_{CB} = V_{CC} - I_C R_L = 10 - 9.2(0.5) = 5.4 \text{ V}.$$

Note that I_{CO} is too small to have any effect, which is typical. Before we can accept this result, we must check the model's validity by making sure that $I_E < 0$ and $V_{CB} > 0$. The emitter is forward-biased, and the collector is reverse-biased and is operating in the active region! Using $I_C = 9.2$ mA and $V_{CB} = 5.4$ V, we can determine the collector power dissipation $P_C = (9.2 \text{ mA})(5.4 \text{ V}) = 50$ mW.

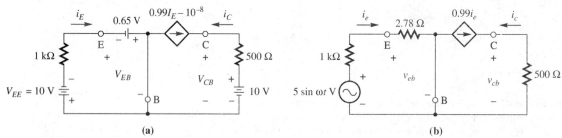

FIGURE 4.13 The circuit for Examples 4.2 through 4.5 using transistor models. (a) dc model. (b) ac model.

EXAMPLE 4.3

Repeat Example 4.2 except that $V_{EE} = 0.5$ V.

SOLUTION   With $V_{EE} \leq 0.5$ V, $I_E \approx 0$ from the characteristics of Fig. 4.11, or from Fig. 4.13(a),

$$I_E = \frac{-0.5 + 0.7}{1000} = +0.2 \text{ mA},$$

a contradiction from the model, because the model assumed $i_E \leq 0$. The transistor is cut off. $I_C = I_{CO} = -10^{-8}$ A, $V_{CB} \approx 10$ V. Note that the active model was revealed to be invalid by the resulting incorrect direction of i_E.

EXAMPLE 4.4

Repeat Example 4.2 except that $V_{EE} = 25$ V.

SOLUTION   Using Figure 4.13(a) with $V_{EE} = 25$ V,

$$I_E = \frac{-25 + 0.7}{1000} = -24.3 \text{ mA},$$
$$I_C = (24.3)(0.99) + 10^{-8} = 24.1 \text{ mA},$$
$$V_{CB} = 10 - 24.1(0.5) = -2.0 \text{ V}!$$

This value of $V_{CB} < -0.5$ V is unacceptable because it would put our operating point in saturation, but we were using a model that assumes we are operating in the active region. Glancing at the collector characteristic (Fig. 4.10), we estimate that $V_{CB} \approx -0.6$ V for $i_E = -24$ mA, and thus I_C would be about

$$\frac{V_{CC} - V_{CB}}{R_C} = \frac{10 - (-0.6)}{500} = 21 \text{ mA}.$$

It appears that any I_E more negative than -21 mA would saturate our transistor (or maybe it is the circuit that saturates?). $V_{CB} \approx -0.6$ V.

EXAMPLE 4.5

Now we try an ac problem. Repeat Example 4.2 except that V_{EE} is 10 V and an ac input signal $v_{in} = 5 \sin \omega t$ V is added in series.

SOLUTION   Our model is linear, so we should be able to use superposition. Example 4.2 has given us the dc solution of $I_E = -9.3$ mA, $I_C = 9.2$ mA, and $V_{CB} = 5.4$ [Fig. 4.13(a)]. The necessary conditions for the active model are met for the operating point, so we proceed with the ac problem. Figure 4.13(b) shows an ac model, with all dc voltage sources replaced by short circuits and the dc current source opened. The emitter resistance is [from Eq. (4.14)]

$$r_e = \frac{V_T}{I_E} = \frac{26 \text{ mV}}{9.3 \text{ mA}} = 2.8 \text{ }\Omega$$

DRILL 4.1

The circuit in Fig. 4.6(a) uses $V_{EE} = 1$ V, $R_E = 60$ Ω, $V_{CC} = 10$ V, and $R_C = 1$ kΩ. Use the transistor characteristics of Figs. 4.10 and 4.11 to solve graphically for I_C and V_{CB}.

ANSWER $I_C = 5$ mA, $V_{CB} = 5$ V.

DRILL 4.2

The circuit of Fig. 4.6(a) uses $V_{EE} = 1$ V, $R_E = 60$ Ω, $V_{CC} = 10$ V, and $R_C = 1$ kΩ. Use the transistor model $\alpha_F = 0.99$ and $V_{BE} = 0.7$ V. Calculate I_E, I_C, and V_{CB}.

ANSWER
$I_E = -5$ mA,
$I_C = 5$ mA, and
$V_{CB} = 5$ V.

DRILL 4.3

The circuit of Fig. 4.6(b) uses $V_{EE} = 5$ V, $R_E = 220$ Ω, $V_{CC} = 10$ V, and $R_C = 330$ Ω. Use the transistor model $\alpha_F = 0.99$ and $V_{EB} = 0.6$ V. Calculate I_E, I_C, and V_{CB}.

ANSWER
$I_E = 20$ mA,
$I_C = -20$ mA, and
$V_{CB} = -3.4$ V.

---

at room temperature. The ac emitter current is

$$i_e = \frac{5.0 \sin \omega t}{1003} = 4.99 \sin \omega t \text{ mA},$$

$$i_c = -\alpha_F i_e = -4.94 \sin \omega t \text{ mA},$$

$$v_{cb} = -R_C i_c = 2.47 \sin \omega t \text{ V},$$

and the complete solution would be

$$i_C = I_C + i_c = 9.2 - 4.94 \sin \omega t \text{ mA},$$

$$v_{CB} = V_{CB} + v_{cb} = 5.4 + 2.47 \sin \omega t \text{ V}.$$

The dynamic operation remains in the active region, so this solution is valid. Note that our model takes no note of parasitic capacitance effects in the transistor, so validity is limited to lower frequencies. We consider more elegant transistor models in Chapters 8 and 9. You may note that the ac current amplification of this circuit is $i_c/i_e = -0.99 = -\alpha_F$. It is always less than one for a CB configuration. The ac voltage amplification is $v_{cb}/v_{in} = 2.47/5.0 = 0.49$. This is a consequence of our choice of R_C and R_E, and it is possible to obtain a voltage amplification magnitude greater than one for the common-base configuration. Try changing the collector resistor shown in Fig. 4.12 to 2 kΩ, and let $V_{CC} = 30$ V. Show that $V_{cb}/v_{in} = 1.98$.

CHECK UP

1. **TRUE OR FALSE?** The model for a forward-biased emitter-base junction is a 0.45-V offset voltage.
2. **TRUE OR FALSE?** The model for the collector-base branch of an active CB BJT is a controlled current source, $\alpha_F I_E$, directed toward the collector terminal.
3. **TRUE OR FALSE?** In order for the active BJT model to be valid, the collector-base junction must be reverse-biased.
4. **TRUE OR FALSE?** For an *npn* BJT, V_{CB} will go no more negative than one diode drop.

4.5 THE COMMON-EMITTER CONFIGURATION

Figure 4.14 illustrates the situation in which the input signal is applied between the base and emitter and the output is taken between the collector and emitter. Lowercase current and voltage symbols (see Section 1.2) indicate a possible time dependence impressed on this circuit at the base-emitter terminals by some external source. We will see that this arrangement offers several advantages over the common-base configuration. **Common-emitter** (CE) amplifier circuits can be constructed with large current and voltage amplifications. CE switches allow large collector currents to be controlled by relatively small base currents.

In the active region, we have seen that the collector current is a linear function of the emitter current. To express collector current as a function of base current, we use the terminal current relationship

$$i_E = -i_C - i_B. \tag{4.15}$$

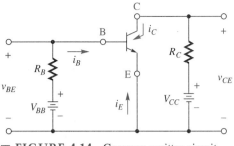

FIGURE 4.14 Common-emitter circuit using an *npn* transistor.

Combining this with Eq. (4.1) yields

$$i_C = \frac{\alpha_F}{1 - \alpha_F} i_B + \frac{I_{CO}}{1 - \alpha_F}. \tag{4.16}$$

It is convenient to define the parameter β_F (also referred to as h_{FE}):

$$\beta_F \equiv \frac{\alpha_F}{1 - \alpha_F}. \tag{4.17}$$

Equation (4.16) becomes

$$i_C = \beta_F i_B + (1 + \beta_F) I_{CO} \equiv \beta_F i_B + I_{CEO}. \tag{4.18}$$

The term β_F designates the *current gain* for the common-emitter configuration. The term I_{CEO}, even though much larger than I_{CO}, is still negligibly small unless the temperature is elevated. Combining Eqs. (4.2), (4.12), and (4.17), we obtain

■ $\beta_F \cong i_C/i_B$ is one of the most important parameters of a BJT.

$$i_B = \frac{I_{EO}}{(1 + \beta_F)(1 - \alpha_R \alpha_F)} e^{v_{BE}/V_T} - I_{CO}. \tag{4.19}$$

CE Collector Characteristics

A family of plots of collector current i_C versus collector-emitter voltage v_{CE} for several values of base current is shown in Fig. 4.15. The data represent measurements of the *npn* 2N2222A. Let us compare Eq. (4.18) with what we observe in the figure. It appears that for $v_{CE} > 0.5$ V, $i_C = 150 i_B$. This means that $\beta_F = i_C/i_B = \Delta i_C/\Delta i_B = 150$. For example, when i_B changes from 50 to 100 μA, i_C changes from 7.5 to 15 mA. In Section 4.9 we discuss how the current gain may vary with operating conditions. The term I_{CEO} is exaggerated very much so that it can be seen on the i_C scale chosen in the figure. The I_{CEO} term is usually negligible at room temperature for silicon transistors.

The regions where v_{CE} is less than 0.5 V represent saturation. The region below $i_C = I_{CEO} \approx 0$ represents cutoff. Maximum v_{CE} is limited by avalanche breakdown, which is discussed in Section 4.10.

Slopes of the plots indicate that i_C and β_F increase with increasing v_{CE}. This is caused by the **Early effect** (base narrowing) and is much more noticeable on the common-emitter characteristics than on the common-base characteristics. This is because in the CB curves I_E was held constant, and the Early effect only caused α to increase with v_{CB}. In the CE characteristic, I_B is held constant, so that when the base narrows due to increased v_{CB}, I_E increases due to the increased minority concentration gradient and α increases due to less recombination, as discussed in the previous section.

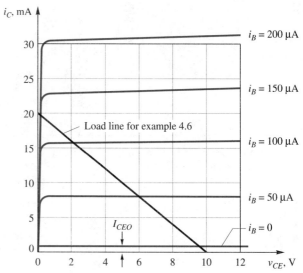

FIGURE 4.15 Common-emitter collector characteristics for a 2N2222A *npn* transistor, where $\beta_F = 150$, and I_{CEO} is negligible. Note that the scale for I_{CEO} is greatly magnified.

In visualizing this effect, it is helpful to think of each member of the collector characteristic family as being extensions of rays originating from a common point on the negative collector-emitter voltage axis (Fig. 4.16). This point, called the **Early voltage, V_A**, has been defined such that

■ The parameter *Early voltage*, though not a physical voltage, describes the effect of changing v_{CB} on i_C for fixed v_{BE}.

$$\begin{aligned} i_C &= I_S e^{v_{BE}/V_T}\left(1 + \frac{V_{CE}}{V_A}\right) \\ &= \beta_F i_B \left(1 + \frac{v_{CE}}{V_A}\right) \end{aligned} \quad (4.20)$$

Typical values of V_A ranges from 50 to 150 V for IC transistors, and 100 to 500 V for discrete BJTs like the 2N2222A. Note that the characteristic slopes increase for higher currents, and that the curves are for constant v_{BE}.

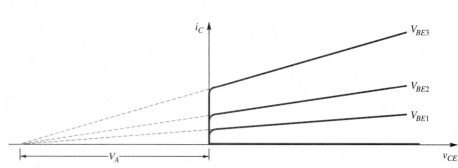

FIGURE 4.16 Projection of collector characteristics to illustrate Early voltage, v_A.

CE Base Characteristics

The curves shown in Fig. 4.17 represent base current i_B as a function of base-emitter voltage v_{BE}. As in the common-base situation (Fig. 4.11), the Early effect causes a slight repositioning of the exponential characteristic for various collector-junction biases. This time, for a fixed V_{BE}, base current should decrease as v_{CE} increases, because recombination is decreasing. Base current only becomes appreciable when base-emitter voltage exceeds a cut-in voltage of 0.5 to 0.6 V. These curves seem consistent with Eq. (4.19). As with any pn junction, v_{BE} decreases ~2.2 mV for each degree Celsius of temperature rise [see Eq. (3.28)].

The dynamic base resistance is the sum of two components called r_b and r_π. The **base-spreading resistance, r_b,** is the average ohmic resistance of the thin base layer in the direction of base current. The other term is found by differentiating Eq. (4.19):

$$r_\pi = \left.\frac{\partial v_{BE}}{\partial i_B}\right|_{Q\text{-point}} = \frac{V_T}{|I_B|} = \frac{V_T \beta_F}{|I_C|} = (\beta_F + 1)r_e. \quad (4.21)$$

These relations are exact only if β_F is constant (see Section 4.9). The r_b term may range from a few ohms to more than 100 Ω; usually the Q-point is chosen so that r_π is much greater than r_b, consequently r_b is often neglected.

Collector and base characteristics of pnp transistors could be expected to have the same shape as Figs. 4.15 and 4.17 except that all currents and voltages would be negative.

■ Thus r_π is seen to be dependent on the dc Q-point, as well as β_F and temperature. See Eq. 4.28.

EXAMPLE 4.6

In the common-emitter circuit shown in Fig. 4.14, $V_{BB} = 2.0$ V, $V_{CC} = 10$ V, $R_B = 27$ kΩ, and $R_C = 500$ Ω. Find V_{CE} and I_C graphically, using the characteristics of Figs. 4.15 and 4.17.

SOLUTION The load line $v_{BE} = V_{BB} - R_B I_B = 2.0 - 27k\, i_B$ is plotted in Fig. 4.17. We try $V_{CE} = 2$ V. The input curve $V_{CE} = 2$ V intersects the load line at $V_{BE} = 0.67$ V and $I_B = 50$ μA. Construct a load line $v_{CE} = V_{CC} - R_C i_C = 10 - 500 i_C$ on Fig. 4.15. The intersection of the load line with the $I_B = 50$ μA curve gives us the operating point for this circuit, $I_C = 7.5$ mA and $V_{CE} = 6.0$ V.

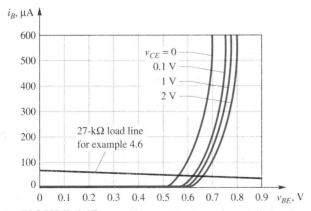

■ **FIGURE 4.17** Common-emitter base characteristics for the 2N2222A npn transistor, where I_{CO} is negligible.

DC Circuit Model for the Active Transistor

As discussed earlier, the volt-ampere characteristics themselves are seldom used in circuit analysis and design because the manufacturer does not normally provide them; they are presented here to aid in visualizing the current/voltage relationships and transistor operation. But the characteristics similar to Figs. 4.15, 4.16, and 4.17 are readily obtained for a BJT experimentally by means of a laboratory instrument called a *curve tracer*. Make sure that you are able to determine β_F for the BJT whose collector characteristic is plotted in Fig. 4.15. Experimentally-obtained characteristics are often used for obtaining parameters for a linear model; also device data sheets (or computer databases) specify a range of parameters and "typical" values for commercial BJTs.

A dc model, valid for active region bias only, is shown in Fig. 4.18. This model systematizes the application of Eq. (4.18) and uses a piecewise-linear model for the base-emitter characteristic that consists of a dc offset voltage V_{BE}. For small silicon transistors at room temperature, 0.6 to 0.7 V is appropriate. Consult manufacturer's data sheets if they are available. The data sheet for the 2N2222A (Fig. 4.32, shown later) does not give a value for $V_{BE(\text{active})}$. We shall discuss finding reasonable values of β_F and I_{CEO} from data sheets presently. Note that Eq. (4.10) represents an equally valid voltage-controlled model.

This simplified model does not recognize dynamic base resistance, the Early effect, or junction capacitive effects. Models that add these refinements for ac small-signal analysis follow, but first we determine the dc operating point of a transistor circuit using the dc model. Note also that I_{CEO} is negligible at $T = 27°C$.

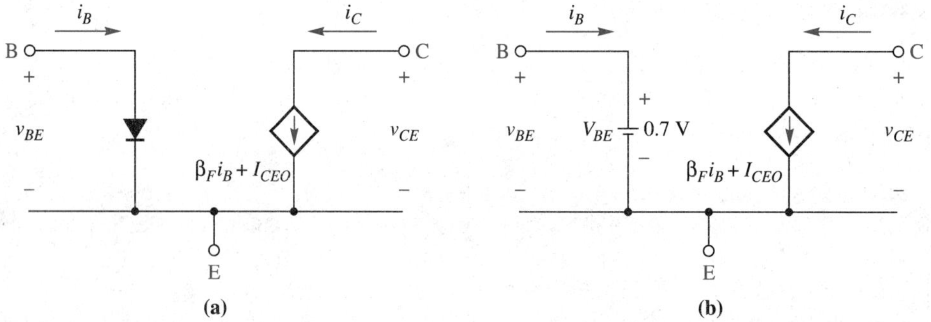

■ You should learn this dc model for the BJT, as it will be used throughout this text to find the operating point of a BJT. The same model is used for a *pnp* BJT except V_{BE} and I_{CEO} are negative.

DRILL 4.4

Suppose that a BJT has $I_C = 1$ mA and $I_S = 10^{-14}$ A at $T = 300$ K.
(a) Find v_{BE} at $T = 300$ K using Eq. (4.10).
(b) Find v_{BE} at $T = 310$ K assuming I_C is constant.

ANSWER 0.66 V at $T = 300$ K and 0.64 V at $T = 310$ K.

■ **FIGURE 4.18** Models for the active *npn* transistor, common-emitter configuration. The *pnp* model would require reversal of polarity on V_{BE} and I_{CEO}. (a) An equivalent circuit for Eqs. (4.18) and (4.19). (b) A simplified model in which the diode has been replaced with an offset voltage source.

EXAMPLE 4.7

Repeat Example 4.6 using a model for the transistor. Let $V_{BE} = 0.65$ V and $\beta_F = 150$. Neglect I_{CEO}.

SOLUTION Figure 4.19(a) shows the circuit of Fig. 4.14 redrawn, and Fig. 4.19(b) shows the circuit with the transistor represented by its model.

$$I_B = \frac{2.0 - 0.65}{27\text{k}} = 50 \ \mu\text{A}$$

and $I_C = 150(50 \ \mu\text{A}) = 7.5$ mA; $V_{CE} = 10 - (7.5 \text{ mA})(500 \ \Omega) = 6.25$ V. These results compare favorably with the answers in Example 4.6.

CHAPTER 4   THE BIPOLAR JUNCTION TRANSISTOR

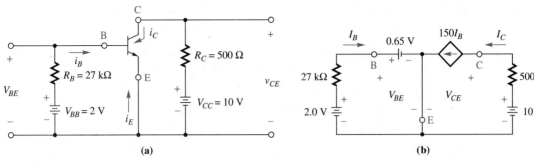

■ **FIGURE 4.19**   The circuit of Examples 4.6 and 4.7:
(a) Schematic. (b) dc model.

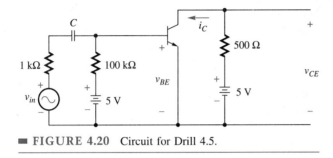

■ **FIGURE 4.20**   Circuit for Drill 4.5.

AC Small-Signal Circuit Model for the Active Transistor

We now consider the effect of time-varying signals on the transistor model. Recall from Eq. (4.10), reported below, that instantaneous collector current is controlled by instantaneous base-emitter voltage as

$$i_C \approx I_S e^{v_{BE}/V_T}. \tag{4.10}$$

Now let the base-emitter voltage consist of a dc term plus a time-varying term:

$$v_{BE} = V_{BE} + v_{be}. \tag{4.22}$$

Then,

$$i_C = I_S e^{(V_{BE}+v_{be})/V_T} = I_S e^{V_{BE}/V_T} e^{v_{be}/V_T} = I_C e^{v_{be}/V_T} \tag{4.23}$$

and we see that the collector current is the product of the dc (Q-point) collector current and an exponential term containing the time-varying part of the base-emitter voltage. The time-varying part can be expanded into the series

$$e^{v_{be}/V_T} = 1 + \frac{v_{be}}{V_T} + \frac{1}{2}\left[\frac{v_{be}}{V_T}\right]^2 + \frac{1}{6}\left[\frac{v_{be}}{V_T}\right]^3 + \cdots, \tag{4.24}$$

which is linear in v_{be} as long as $v_{be} \ll V_T = kT/q$. This inequality is known as the *small-signal condition*. At room temperature, where $V_T \approx 26$ mV, v_{be} would have to be less than about 5 mV at peak. Substituting (4.24) into (4.23), we obtain

$$i_C \cong I_C + \frac{I_C}{V_T} v_{be} = I_C + g_m v_{be} = I_C + i_c, \tag{4.25}$$

DRILL 4.5

Consider the circuit of Fig. 4.20. $V_{BE} = 0.7$ V and $\beta_F = 100$. Find I_C and V_{CE} using a dc model for the circuit. The capacitor is an open circuit to dc.

ANSWER   4.3 mA and 2.85 V.

■ The *small-signal* condition for the BJT is thus the same as it is for *pn* diodes.

■ We assume that $\eta = 1$.

which shows that the total collector current is the superposition of the dc collector current plus a term linearly related to the ac small-signal base-emitter voltage. The term

$$g_m = \frac{I_C}{V_T} \qquad (4.26)$$

is called the **transconductance** of the transistor, an ac parameter that is seen to be dependent on the dc collector current. It should be noted that g_m is the slope of the i_C versus v_{BE} characteristic at I_C:

$$g_m \equiv \left.\frac{\partial i_C}{\partial v_{BE}}\right|_{i_C = I_C} = \left.\frac{i_c}{v_{be}}\right|_{I_C}. \qquad (4.27)$$

Figure 4.21 shows a graph of Eq. (4.10) with a line drawn tangent to the characteristic at the Q-point I_C. The straight line is a good approximation to the characteristic only near the operating current I_C (the small-signal requirement).

We may observe a relationship between g_m and r_π. From Eq. (4.21),

$$r_\pi = \frac{V_T \beta_F}{|I_C|} = \frac{\beta_F}{g_m}. \qquad (4.28)$$

We are now in a position to construct an ac model for the transistor. Figure 4.22 shows a model with small-signal parameters r_π for dynamic base resistance, and $g_m v_{be}$ for control of the collector current.

■ The ac model will be used whenever we analyze the small-signal operation of a BJT circuit.

Note that the Early effect may be modeled in the small-signal model by the **output resistance, r_o**, shown in Fig. 4.22. Realistically, the output characteristic exhibits a nonzero

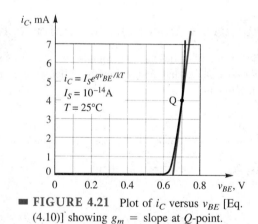

■ **FIGURE 4.21** Plot of i_C versus v_{BE} [Eq. (4.10)] showing g_m = slope at Q-point.

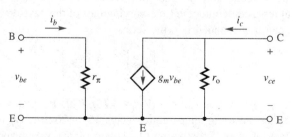

■ **FIGURE 4.22** Small-signal, low-frequency ac model for the BJT.

slope (see Fig. 4.15), causing r_o to have some finite value. We may find r_o if we know the Early voltage and the operating point. As an example, assume $V_A = 100$ V and $I_C = 1$ mA at $V_{CE} = 12$ V. Then, from Fig. 4.16,

$$r_o = \left.\frac{\Delta v_{CE}}{\Delta i_C}\right|_{V_{BE}} = \frac{V_A + v_{CE}}{i_C}$$

$$= \frac{112 \text{ V}}{1 \text{ mA}} = 112 \text{ k}\Omega. \tag{4.29}$$

Problem 4.12 at the end of the chapter considers how Eq. (4.21) could be modified to include the Early effect.

EXAMPLE 4.8

The circuit of Example 4.7 is redrawn in Fig. 4.23(a) with an ac input signal, v_{in}, in series with a capacitor, which blocks the dc so that the Q-point is undisturbed by the signal source. Assume small-signal conditions and an Early voltage of 150 V. Draw an ac model for the circuit and find the circuit voltage amplification v_{ce}/v_{in}.

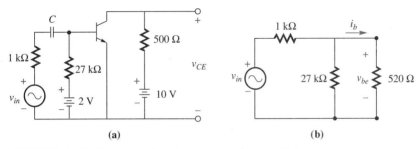

■ **FIGURE 4.23** Example 4.7 with ac voltage source added: (a) Circuit. (b) ac model.

SOLUTION The dc operating point of Example 4.7 was found to be $I_C = 7.5$ mA and $V_{CE} = 6.25$ V. The small-signal model parameters are

$$g_m = \frac{I_C}{V_T} = \frac{7.5 \text{ mA}}{26 \text{ mV}} = 0.29 \text{ S},$$

$$r_\pi = \frac{\beta_F}{g_m} = \frac{150}{0.29 \text{ mS}} = 520 \text{ }\Omega,$$

$$r_o = \frac{V_A + v_{CB}}{i_C} = \frac{156 \text{ V}}{7.5 \text{ mA}} = 20.8 \text{ k}\Omega.$$

The ac model, replacing all dc voltage sources with short circuits, is shown in Fig. 4.23(b); 520∥27k = 510 Ω, 500∥20.8k = 488 Ω. We find that $v_{be} = \dfrac{v_{in}\,510}{1\text{k} + 510} = 0.34 v_{in}$. Note that v_{in} would have to be less than 29 mV at peak for v_{be} to qualify as small-signal (less than 10 mV). Next, $i_c = g_m v_{be} = (0.29)(0.34 v_{in}) = 99 v_{in}$ mA, and $v_{ce} = -i_c(488) = -48.3 v_{in}$ V. We find the voltage amplification $v_{ce}/v_{in} = -48.3$. Note that $i_b = \dfrac{0.34 v_{in}}{520} = 0.65 v_{in}$ mA and $i_c = 99 v_{in}$ mA, so that $i_c \approx 150 i_b$, as it should if β is 150. It is interesting to observe the larger signal situation on the collector characteristic shown in Fig. 4.15. Assume that v_{in} varies from about $+77$ mV to about

DRILL 4.6

Find g_m and r_π for the situation of Drill 4.5.

ANSWER 165 mS and 605 Ω.

−77 mV. The dynamic operating point moves along the load line from $v_{CE} = 2.3$ V at $i_B = I_B + i_b = 50 + 50 = 100$ μA to the other extreme $v_{CE} \cong 10$ V at $i_B = I_B + i_b = 50 − 50 = 0$ μA. Thus, the peak-to-peak collector voltage is 7.7 V, approximately equal to the result found by using the linear model, which would yield (gain = −48.3) × (peak-peak voltage = 154 mV). Note that any increase in v_{in} would result in the transistor's reaching cutoff (v_{CE} cannot exceed 10 V).

DRILL 4.7

Let v_{in} in Fig. 4.20 be 10 sin ωt mV. Find v_o using a small-signal model for the circuit. Assume that the capacitor is a short circuit for this ω.

ANSWER −0.31 sin ωt V.

Amplification is the chief application of BJTs biased in the active region. This is such an important subject that we will treat it in detail in Part II of this text, comparing common-emitter, common-base, and common-collector (*emitter-follower*) amplifier circuits, as well as other configurations. Each configuration offers certain special features. We shall find that common-emitter amplifiers have moderate input resistances and can be designed to have both current gain and voltage gain. Common-base amplifiers possess low input resistance, and only voltage gain is possible (Example 4.5). Emitter-follower circuits may be constructed with high input impedance and low output impedance, but only current gain may be obtained; voltage gain is always less than one.

■ *Emitter follower* is a more descriptive term than common collector, and is often used because the emitter voltage tends to follow the input (base) voltage in this circuit.

CHECK UP

1. **TRUE OR FALSE?** The Early voltage of a BJT (V_A) can be found from the collector characteristics.
2. **TRUE OR FALSE?** The dynamic CE input resistance of a BJT (r_π) can be found if the operating point is known.
3. **TRUE OR FALSE?** The open-base collector leakage current of a BJT (I_{CEO}) is likely to be significant in most circuits.
4. **TRUE OR FALSE?** The transconductance of a BJT (g_m) can be found if the operating point is known.

4.6 CUTOFF IN THE CE CONFIGURATION

In Section 4.3 we defined *cutoff* in the BJT as the condition in which $I_E = 0$, $I_C = I_{CO}$, and thus $I_B = −I_{CO}$. The required value of V_{BE} for this situation can be found by means of Eq. (4.8), substituting $I_E = 0$ and using Eq. (4.5) to eliminate the saturation currents:

$$V_{BE(\text{cutoff})} = V_T \ln(1 - \alpha_F). \qquad (4.30)$$

At room temperature, with $\alpha_F = 0.99$, this would yield about −0.12 V. At very small emitter currents, α_F could be expected to drop significantly in silicon transistors, as we shall see, so that cutoff would approach zero. Consider the situation in which $v_{BE} = 0$ and the collector-base junction is reverse-biased. Equation (4.7) yields

$$i_C = \frac{I_{CO}}{1 - \alpha_F \alpha_R} \approx I_{CO}. \qquad (4.31)$$

This result is again near to cutoff because α_R is almost always less than 0.5, and α_F is much less than unity for silicon transistors at very small emitter current.

Another case of great interest for the common-emitter arrangement is that in which $i_B = 0$. Substituting into Eq. (4.18) yields

$$i_C = (1 + \beta_F)I_{CO} \equiv I_{CEO}. \qquad (4.32)$$

Although this collector current appears to be negligible on the scale of Fig. 4.15, it is β_F times the collector cutoff current I_{CO} and may be significant in circuit operation when I_{CO} increases because of elevated temperature. The value of v_{BE} that corresponds to $i_B = 0$ can be determined from Eqs. (4.6) and (4.7), assuming $v_{BC} \ll 0$. You may verify that

$$v_{BE}\bigg|_{I_B=0} = V_T \ln\left(1 + \frac{\beta_F}{\beta_R}\right). \qquad (4.33)$$

The typical value of open-base base-emitter voltage is about 0.15 V at room temperature for silicon.

Summarizing, we can say that silicon BJTs are cut off when $V_{BE} < 0.1$ V. At cutoff, $i_C = -i_B = I_{CO}$. If I_{CO} can be neglected, the model at cutoff is very simple: $i_C = i_B = 0$.

Collector Cutoff Current I_{CBO}

The actual collector current that flows in the reverse-biased collector-base junction when the emitter current is zero is called I_{CBO}. This current exceeds the theoretical current I_{CO} for several reasons. First, current leaks around the junction because of imperfections in the crystal structure at its surfaces. Second, thermal generation of carriers occurs in the space-charge region. Third, at higher reverse voltages, carrier multiplication due to avalanche collisions in the space-charge layer is possible. Because these three terms are sensitive to the reverse voltage across the junction, they are more noticeable for a large reverse collector-base voltage, but they are less dependent on temperature than I_{CO}. The theoretical I_{CO} used in the foregoing idealized BJT analysis is constant with voltage, but it varies as n_i^2 with temperature. Thus, I_{CO} varies with temperature as does I_S in a pn junction.

For small silicon transistors, I_{CO} might be expressed in picoamperes at room temperature, whereas I_{CBO} might be in the nanoampere range. Thus, I_{CBO} would not appear to increase as rapidly with temperature as would I_{CO} until the temperature had already increased 100°C, so that I_{CO} dominated I_{CBO}.

After saying all this, we find that I_{CBO} approximately doubles for every 10°C rise in temperature for silicon BJTs near room temperature. Refer to the data sheets shown later in Fig. 4.32 for I_{CBO} specifications for the 2N2222A. Verify that the manufacturer guarantees $I_{CBO} < 10^{-8}$ A at 25°C and $I_{CBO} < 10$ μA at 150°C. We find I_{CEO} from I_{CBO} by substituting $I_{CBO} = I_{CO}$ in Eq. (4.32).

■ In most cases I_{CEO} can be neglected.

DRILL 4.8

A certain BJT has $I_{CBO} = 1$ nA at $T = 25°C$. Assume that β_F is 150 at 100°C. Find I_{CBO} and I_{CEO} at 100°C.

ANSWER
0.18 and 27 μA.

EXAMPLE 4.9

The BJT in Fig. 4.24 is biased in the cutoff region by the source $V_{BB} = -5$ V. The silicon transistor has $I_{CBO} = 10$ nA at 25°C. Suppose I_{CBO} doubles for each 10°C rise in temperature. How much can the temperature rise before the transistor will no longer be cut off?

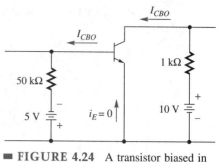

FIGURE 4.24 A transistor biased in the cutoff region provided I_{CBO} does not overcome the base-emitter reverse bias.

SOLUTION Assume that a silicon BJT comes out of cutoff for $v_{BE} \geq 0.1$ V. In the given circuit,

$$v_{BE} \text{ (at } T_2\text{)} = V_{BB} + R_B I_{CBO} \text{ (at } T_2\text{)}$$
$$= V_{BB} + R_B I_{CBO} \text{ (at } 25°\text{)} \, 2^{\Delta T/10}$$
$$0.1 \geq -5 + (50 \text{ k}\Omega)(10^{-8} \text{ A})2^{\Delta T/10}$$
$$2^{\Delta T/10} \leq \frac{4.9}{5 \times 10^{-4}} = 9.8 \times 10^3$$
$$\Delta T/10 \leq \log_2 9.8 \times 10^3 = 13.3$$
$$\Delta T \leq 133°\text{C}.$$

Note that the inverse-breakdown voltage for the emitter-base junction (BV_{EBO}) is likely to be much less than that of the collector-base junction. The manufacturer's data sheet for the 2N2222A, for example, shows $V_{EB(\text{max})} = 6$ V. Thus, V_{BB} in circuits such as the one in Fig. 4.24 must be chosen so that this limit is not exceeded.

4.7 SATURATION IN THE CE CONFIGURATION

We have seen that when a BJT is in the active region, with the emitter-base junction forward-biased and the collector-base junction reverse-biased, the collector current varies linearly with the base current. In the CE configuration in Fig. 4.14, this implies that v_{CE} is larger than v_{BE}. As base current increases, the corresponding increase in collector current causes v_{CE} to decrease because of the voltage drop in R_C. If this continues until the collector voltage v_{CE} becomes less than the base voltage v_{BE}, the base-collector junction becomes forward-biased ($v_{BC} > 0$ in this *npn* case). The collector begins to inject electrons into the base. As shown by curve n_S in Fig. 4.9, increased injection from the emitter is required to maintain the same slope of electron distribution in the base. Thus, base current is increased without a significant increase in collector current. The collector current approaches the limit V_{CC}/R_C and is said to **saturate**. The astute student may notice that it is the limiting effect of the external circuit that causes saturation, but it is the transistor that is said to be saturated.

To study BJT behavior in saturation, we again consider the Ebers-Moll equations, Eqs. (4.6) and (4.7). With both junctions forward-biased, we can neglect the unity terms, which are dominated by the exponential terms. Thus, in Eqs. (4.8) and (4.9), the unity terms are also neglected. Recognizing that

$$v_{CE} = v_{BE} + v_{CB},$$

we subtract Eq. (4.9) from Eq. (4.8) and simplify to find

$$v_{CE(\text{sat})} = V_T \ln \frac{I_{CO}(i_E + \alpha_R i_C)}{I_{EO}(i_C + \alpha_F i_E)}. \tag{4.34}$$

We can substitute $i_E = -i_B - i_C$ and use Eqs. (4.5) and (4.17) to eliminate the reverse-saturation currents. The results can be expressed as follows:

$$v_{CE(\text{sat})} = V_T \ln \frac{i_B \beta_F + i_C \beta_F (1 - \alpha_R)}{i_B \beta_F \alpha_R - i_C \alpha_R}. \tag{4.35}$$

Consider a typical case with $\alpha_R = 0.1$. The collector-emitter voltage becomes

$$v_{CE(\text{sat})} = V_T \ln \frac{10 + 9 i_C/i_B}{1 - i_C/\beta_F i_B}. \tag{4.36}$$

It is instructive to plot this function with the parameter $\sigma \equiv i_C/\beta_F i_B$ as the independent variable. This term is unity in the active region and will become progressively smaller in the saturated region as the base current increases faster than the collector current. Thus, it is a measure of the degree of saturation, or "base overdrive." Figure 4.25 illustrates the way in which the saturated collector-emitter voltage varies with base overdrive at 25°C for several different β_F using Eq. (4.36). The following generalizations can be drawn from this figure. The collector current responds linearly to changes in base current until the collector-emitter voltage $v_{CE} < 0.35$ V. The average value of collector-emitter voltage in the saturated region is somewhat under 0.2 V, so we will use as a reasonable model for the saturated transistor $|V_{CE(\text{sat})}| = 0.2$ V. When the transistor saturates, the base-emitter voltage will be somewhat higher than the active value. We will use $|V_{BE(\text{sat})}| = 0.8$ V in our model, which is appropriate for a range of small silicon transistors. The resulting model is shown in Fig. 4.26. Several examples will illustrate the use of the model.

Because we might not know whether the BJT in a given circuit is active or saturated,

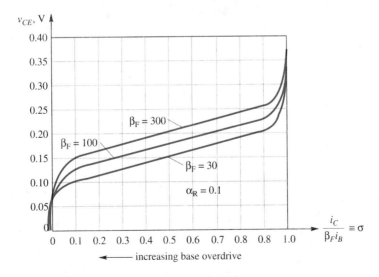

■ **FIGURE 4.25** Saturated collector-emitter voltage as a function of base overdrive at 25°C.

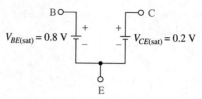

■ **FIGURE 4.26** Model for saturated *npn* transistor. The *pnp* transistors require opposite polarity on voltage sources.

one must assume a model, solve for the operating point, and then verify that the assumed model is consistent with this *Q*-point. One such systematic procedure is as follows:

1. Replace the BJT by a saturated model: $V_{BE} = 0.8$ V and $V_{CE} = 0.2$ V.
2. Redraw the circuit and solve for I_B and I_C.
3. If $I_C < \beta_F I_B$, the BJT is saturated, assumed model is correct, circuit is solved.
4. If $I_C > \beta_F I_B$, the BJT is active, replace the model, and solve with active model.

EXAMPLE 4.10

The transistor in Fig. 4.27(a) is silicon with $\beta_F = 100$. Find i_C and v_{CE}.

(a) Assume a saturated model.
(b) Assume an active model.

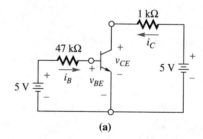

(a)

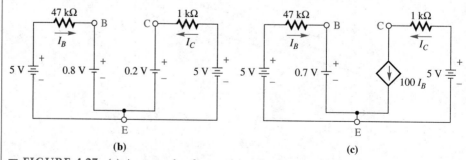

■ **FIGURE 4.27** (a) An example of a transistor circuit. (b) The transistor replaced with a saturated model. (c) The transistor replaced with an active model.

SOLUTION

(a) Assuming that the transistor is saturated, it may be helpful to redraw the circuit as shown in Fig. 4.27(b), using the indicated values for $V_{BE(\text{sat})}$ and $V_{CE(\text{sat})}$. Or you may find it sufficient to label the circuit in Fig. 4.27(a) with 0.8 V from base to emitter and 0.2 V from collector to emitter. We find

$$I_B = \frac{5 - 0.8}{47k} = 0.089 \text{ mA},$$

$$I_C = \frac{5 - 0.2}{1k} = 4.8 \text{ mA}.$$

We consider the ratio $I_C/I_B = 4.8/0.089 = 53.9$. This is less than $\beta_F = 100$, which is what the ratio would have been in the active region; therefore our assumption of saturation is correct. We now may feel comfortable with the assumed value $V_{CE} = 0.2$ V, leading to $I_C = 4.8$ mA. We could obtain a somewhat better approximation for v_{CE} from Fig. 4.25 by using $\sigma = i_C/\beta_F i_B = 0.539$, leading to $v_{CE} = 0.18$.

(b) The circuit is redrawn in Fig. 4.27(c), using an active model for the transistor. Using $V_{BE} = 0.7$ V, we obtain

$$I_B = \frac{5 - 0.7}{47k} = 0.091 \text{ mA}.$$

Then $I_C = \beta_F I_B = 100(0.091) = 9.1$ mA. Now $V_{CE} - I_C R_C = 5 - (9.1)(1) = -4.1$ V! This is a contradiction, because the active model assumes $v_{CE} \geq 0.35$ V. The contradiction is our indication that the active model is incorrect, and we should have used the saturated model. Part (a) is correct.

EXAMPLE 4.11

Find the operating point of the circuit in Fig. 4.28(a), assuming a silicon transistor with $\beta_F = 50$.

SOLUTION If we assume the transistor is saturated, we may draw the model in Fig. 4.28(b). Let V_E be the voltage from emitter to reference. Kirchhoff's current law applied at the emitter node gives

$$\frac{5 - 0.8 - V_E}{47k} + \frac{5 - 0.2 - V_E}{1k} = \frac{V_E}{1k}.$$

Solving, we obtain $V_E = 2.42$ V. Then $V_B = 2.42 + 0.8 = 3.22$ and $V_C = 2.42 + 0.2 = 2.62$. We may now check on the validity of our assumption. We find

$$I_B = \frac{5 - 3.22}{47k} = 0.038 \text{ mA},$$

$$I_C = \frac{5 - 2.62}{1k} = 2.38 \text{ mA}.$$

The ratio $I_C/I_B = 2.38/0.038 = 62.6$! This is greater than $\beta_F = 50$, so the transistor cannot be saturated. We must assume the active model in Fig. 4.28(c). Summing voltage drops around the base loop, $5 = 47k I_B + 0.7 + 1k \times 51 I_B$. Solving for base current, $I_B = 0.044$ mA, and $I_C = 50 I_B = 2.2$ mA. $V_E = 51 I_B \times 1k = 2.24$ V and

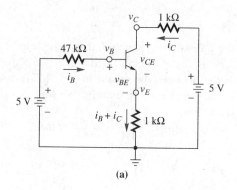

DRILL 4.9

Consider the circuit of Example 4.11, Fig. 4.28(a). What minimum value of β_F will cause the transistor to saturate?

ANSWER 62.

DRILL 4.10

Consider the circuit of Example 4.11, Fig. 4.28(a). What minimum value of the collector resistor will cause the transistor to saturate?

ANSWER 1.2 kΩ.

■ This is a sound procedure; don't waste time trying to guess the circuit situation exactly the first time.

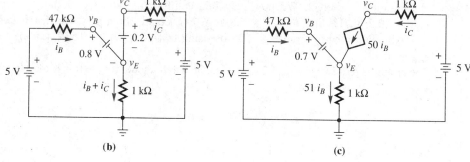

■ **FIGURE 4.28** (a) An example of a transistor circuit. (b) The circuit using a saturated model. (c) The circuit using an active model.

$V_C = 5 - 50 I_B \times 1k = 2.8$ V. We see that $V_{CE} = V_C - V_E = 2.8 - 2.24 = 0.56$ V is greater than 0.35 V where saturation begins. The solution is consistent. The operating point $V_{CE} = 0.56$ V and $I_C = 2.2$ mA. See Example 4.14 for a SPICE computer solution of this problem.

Be sure to observe the proper order of analysis. When we are uncertain as to whether the transistor is saturated or active, we must assume a model, solve the circuit on the basis of the assumed model, and then test the validity of the model. The necessary condition for saturation is $i_C/i_B < \beta_F$. The active model requires $v_{CE} \geq 0.35$ V.

Summary of Common-Emitter Models

The following set of junction voltages is offered as typical of the magnitudes encountered with small silicon transistors. You will do well to assume values such as these for models unless a better value is known.

■ These are good default assumptions to approximate small transistor models.

$$\begin{aligned} V_{BE(\text{cutoff})} &\leq 0, \\ V_{BE(\text{cut-in})} &= 0.5 \text{ V}, \\ V_{BE(\text{active})} &= 0.7 \text{ V}, \\ V_{BE(\text{sat})} &= 0.8 \text{ V}, \\ V_{CE(\text{sat})} &= 0.2 \text{ V}. \end{aligned} \quad (4.37)$$

At temperatures other than 25°C, the $V_{BE(active)}$ and $V_{BE(sat)}$ can be expected to decrease 2.2 mV/°C temperature rise.

As a final example of modeling common-emitter transistor circuits, we consider a problem that spans the three regions of operation: cutoff, active, and saturation.

EXAMPLE 4.12

For the transistor circuit shown in Fig. 4.29(a), plot the output voltage v_o versus the input voltage v_{IN}. Assume $\beta_F = 100$ and neglect I_{CBO}.

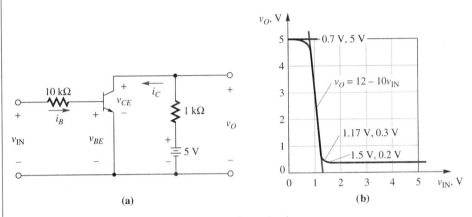

FIGURE 4.29 (a) An example of a transistor circuit.
(b) A plot of v_{CE} versus v_{IN} for the circuit.

SOLUTION Under the assumption that the transistor will be in the active region for certain values of v_{IN}, we use the linear model ($V_{BE} = 0.7$ V, $i_C = 100i_B$). Then,

$$i_B = \frac{v_{IN} - 0.7}{10k} = 0.1v_{IN} - 0.07 \text{ mA},$$

$$i_C = 100i_B = 10v_{IN} - 7 \text{ mA},$$

$$v_o = V_{CC} - (1k)(i_C) = 5 - (10v_{IN} - 7) = 12 - 10v_{IN} \text{ V}.$$

We plot this function on the coordinates as in Fig. 4.29(b). We should recognize that the linear equation will not apply outside of the linear region. One limit for this model occurs at cutoff, when $I_B = 0$, $I_C = 0$, and $V_{CE} = 5$ V. Thus our linear model assumes $v_{IN} > v_{BE}$. Since the model used has a constant $V_{BE} = 0.7$ V, then the edge of the cutoff region must be at $v_{IN} = 0.7$ V. In an actual transistor, cut-in would occur at about $v_{IN} = v_{BE} = 0.5$ V, so that the actual plot would round slightly for 0.5 V $< v_{IN} <$ 0.7 V. This comes about because the actual v_{BE} is a smooth function of base current and does not jump abruptly from 0.5 V at cut-in to 0.7 V (an average value for the active region).

The other limit of the linear region is saturation. As $v_o = v_{CE}$ approaches about 0.3 V (see Fig. 4.25), collector current is no longer 100 times base current. Using $v_o = 0.3$ V as the breakpoint, we find $0.3 = 12 - 10v_{IN}$ and $v_{IN} = 1.17$ V. Now, as

v_{IN} increases above 1.17 V, we move into saturation and the linear model no longer applies. Let us try the saturated model for $V_{IN} = 1.5$ V. Using Eq. (4.37),

$$I_B = \frac{V_{IN} - V_{BE(sat)}}{10k} = \frac{1.5 - 0.8}{10k} = 0.07 \text{ mA},$$

$$I_C = \frac{5 - V_{CE(sat)}}{1k} = \frac{5 - 0.2}{1k} = 4.8 \text{ mA},$$

$$\frac{I_C}{I_B} = \frac{4.8}{0.07} = 68.57 < 100 \quad \text{proves saturation!}$$

We might calculate

$$\sigma = \frac{I_C}{\beta_F I_B} = \frac{4.8}{(100)(0.07)} = 0.686$$

and, from Fig. 4.25, find $V_{CE(sat)} \approx 0.195$ V (close to the assumed 0.2 V!) We can now finish sketching v_o in Fig. 4.29(b) since we know that further increase in v_{IN} only increases i_B, causing $v_{CE(sat)}$ to decrease only slightly. Verify that at $V_{IN} = 3$ V, $I_B = 0.22$ mA, $I_C/\beta_F I_B \approx 0.22$, and $V_{CE(sat)} \approx 0.15$ V. To summarize, we see that for $v_{IN} < 0.5$ V (cut-in), the transistor is OFF and $v_o = 5$ V. As v_{IN} exceeds cut-in, i_B begins to flow. As we will see in Section 4.9, β_F is low for very small base current and i_C is also small. A further increase in v_{IN} leads to further increase in i_B, and soon our active model applies. Output voltage drops linearly as shown. In an actual transistor, v_{BE} might increase from 0.65 to 0.75 V over the active region. This means that our assumed 0.7 is a practical compromise, made in the interest of a reasonably accurate solution without undue complexity for the model. As v_{IN} reaches 1.17 V, the BJT reaches saturation, and a further rise in v_{IN} results in only a slight decrease in v_o, to approximately 0.13 V at $v_{IN} = 5$ V.

This plot of v_o versus v_{IN} is called the *transfer characteristic* of the circuit. This circuit serves as a logical *inverter*; that is, the output is HIGH (5 V) when the input is LOW (<0.5 V), and the output is LOW (<0.3 V) when the input is HIGH (>1.2 V). A SPICE computer-generated solution of this problem is presented in Example 4.15.

CHECK UP

1. **TRUE OR FALSE?** As a practical matter, a silicon BJT is nearly OFF when $v_{BE} < 0.5$ V.
2. **TRUE OR FALSE?** Collector reverse leakage current I_{CBO} is a severe limitation in the design of BJT circuits.
3. **TRUE OR FALSE?** A realistic model for a saturated *npn* BJT is a fixed $V_{BE} = 0.8$-V source, and a fixed $V_{CE} = 0.2$-V source.
4. **TRUE OR FALSE?** In a saturated BJT, I_C is always less than $\beta_F I_B$.

4.8 INVERSE MODE

As we noted in Section 4.3, it is possible for the transistor to be biased such that the collector-base junction is forward-biased and the emitter-base junction is reverse-biased. Then Eq. (4.4) would reduce to

$$i_E = -\alpha_R i_C + I_{EO}, \tag{4.38}$$

and substitution of $i_C = -i_B - i_E$ results in

$$i_E = \frac{\alpha_R}{1 - \alpha_R} i_B + \frac{I_{EO}}{1 - \alpha_R} \equiv \beta_R i_B + (1 + \beta_R) I_{EO}. \quad (4.39)$$

It appears that by interchanging the roles of collector and emitter of a transistor, we achieve an active current gain of β_R, which does not seem desirable because β_R is typically only 0.1 to 0.5. Another disadvantage is that the breakdown voltage BV_{EB} is much lower than BV_{CB}. You may investigate the low saturation voltage $V_{EC(sat)}$ that can be achieved by driving the base with the collector grounded (Problem 4.39).

Let us consider one of the few common applications of the inverse mode. Refer to the circuit shown in Fig. 4.30. This circuit is similar to the input arrangement in certain logic circuits that we will study in Chapter 13. Suppose that the input voltage $v_{IN} = 0$. We will demonstrate that Q_1 is saturated. Then

$$I_{B1} = \frac{5 - V_{BE(sat)} - v_{IN}}{4k} = \frac{5 - 0.8}{4k} = 1.05 \text{ mA}.$$

Since $V_{BE2} = V_{CE(sat)} \approx 0.2$ V, Q_2 is off and $i_{C1} = -i_{B2} \approx 0$. Thus, Q_1 is very saturated! The exact value of i_{C1} is an interesting problem, but it is sufficient to say that it is very small, and Q_2 is cut off.

Next let $v_{IN} = 3$ V. Now current flows through the base-collector junction of Q_1 and the base-emitter junction of Q_2. The base voltage of Q_1 cannot exceed the two diode drops, approximately $0.7 + 0.7 = 1.4$ V.

$$I_{B1} = \frac{5 - 1.4}{4k} = 0.9 \text{ mA}.$$

However, if $v_{B1} = 1.4$ V and $v_{E1} = v_{IN} = 3$ V, then $v_{BE1} = -1.6$ V, and it is reverse-biased. With its base-collector junction forward-biased, and its base-emitter junction reverse-biased, Q_1 is in the inverse mode. Suppose we neglect I_{EO} and assume $\beta_F = 50$ and $\beta_R = 0.1$. Now we can use Eq. (4.39), which results in

$$I_{E1} = \beta_R I_{B1} + (1 + \beta_R) I_{EO} = (0.1)(0.9) = 0.09 \text{ mA}$$

into the emitter. The base current of Q_2 is found to be $I_{B2} = -I_{C1} = I_{B1} + I_{E1} = 0.9 + 0.09 = 0.99$ mA. We can prove Q_2 to be saturated,

$$I_{C2} = \frac{5 - 0.2}{2k} \quad \text{and} \quad \frac{I_{C2}}{I_{B2}} = \frac{2.4}{0.99} < \beta_F.$$

We have seen that Q_1 serves to saturate Q_2 when v_{IN} is HIGH (3 V) by operating in

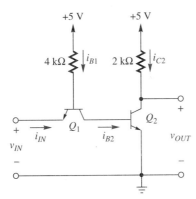

■ **FIGURE 4.30** A circuit that demonstrates the inverse mode.

the inverse mode. We found that the input current i_{IN} was 0.09 mA. We also found that Q_1 saturates when v_{IN} is LOW (0 V), and this turns Q_2 OFF. As we will show later, having Q_1 saturate and accept current from the base of Q_2 will aid in turning this transistor OFF more quickly. We leave it to you to find the transfer function for this circuit (Problem 4.41).

4.9 CE CURRENT GAIN CONSIDERATIONS

Since defining β_F in Eq. (4.17), we have used it for common-emitter current gain as if it were a constant. Recognizing the collector cutoff current I_{CBO}, Eq. (4.18) becomes

$$i_C = \beta_F i_B + (1 + \beta_F)I_{CBO} \tag{4.40}$$

and we see that

$$\beta_F = \frac{i_C - I_{CBO}}{i_B + I_{CBO}}, \tag{4.41}$$

where β_F is the ratio of a collector current change from cutoff to i_C to a base current change from cutoff to i_B. For this reason it is called the ***large-signal*** current gain.

The ***dc current gain*** has been defined as

$$h_{FE} \equiv \frac{I_C}{I_B}. \tag{4.42}$$

We see that for the static case the two definitions are equivalent when the base current is large compared with I_{CBO}, as is almost always the case in active or saturated circuits. In this text, we make no distinction between these terms when using dc models to find circuit operating points or to check for saturation.

Manufacturer's data in Fig. 4.31 shows that h_{FE} varies with collector current, collector-emitter voltage, and temperature. Current gain is relatively constant over a range of I_C, but it drops off for both large currents and small currents. For very small currents (even less than 0.5 mA), recombination within the emitter-base space-charge region, which is usually negligible, begins to be noticeable, reducing emitter efficiency. At high collector currents, the minority-carrier concentration at the emitter edge of the base becomes great enough that majority-carrier concentration increases locally to maintain charge neutrality. This also causes a reduction in emitter efficiency because of increased injection toward

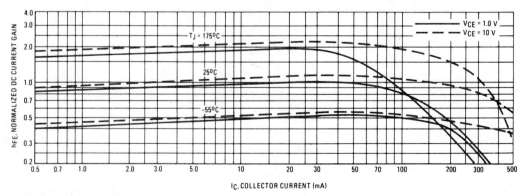

■ **FIGURE 4.31** Plots of normalized dc current gain for the 2N2222A, normalized to unity at $V_{CE} = 1$ V, $I_C = 30$ mA, and $T = 25°C$ (courtesy Motorola, Inc.).

the emitter. The figure shows this drop in h_{FE} above 100 mA for the 2N2222A. Note that a 1% decrease in α_F from 0.99 to 0.98 will cause a 50% decrease in β_F, from 99 to 49 [see Eq. (4.17)].

We have previously noted that increasing collector-base reverse voltage narrows the base region. Thus higher v_{CE} increases the base-transport factor, and thus also h_{FE}, because of the Early effect (Section 4.4). The figure indicates an approximate 10% increase in h_{FE} as v_{CE} increases from 1 to 10 V.

Increasing temperature also increases the base-transport factor because the lifetime of minority carriers is increased. Recombination in the base appears to decrease because of thermal reexcitation from recombination centers. Figure 4.31 indicates that h_{FE} doubles as the temperature increases from 25 to 175°C and halves as the temperature decreases from 25 to −55°C. The data sheet in Fig. 4.32 (see pages 158–159) specifies $100 < h_{FE} < 300$ for the 2N2222A at $I_C = 150$ mA, $V_{CE} = 10$ V, and $T = 25°C$. Locate this entry in the "ON Characteristics" section and compare it with the values given for other conditions.

When i_C and i_B consist of small ac signals superimposed on larger dc components, it is convenient to define an *ac small-signal current gain* as

$$\beta = h_{fe} \equiv \frac{\Delta i_C}{\Delta i_B} = \frac{i_c}{i_b}, \qquad \Delta i_B << I_B. \tag{4.43}$$

This incremental ratio is taken at the dc operating point. This ratio is almost certain to be slightly greater than h_{FE} because the operating point is likely to be in a region where i_C is changing most rapidly with respect to i_B. Note that Fig. 4.32 (see pages 158–159) lists $75 < h_{fe} < 375$ for the 2N2222A at $I_C = 10$ mA, $V_{CE} = 10$ V, and $T = 25°C$. We use this parameter and others in the analysis of amplifier circuits in later chapters.

■ Don't fail to distinguish between the dc $h_{FE} \cong \beta_F$ and the small-signal $h_{fe} = \beta$, even though they often are nearly the same.

DRILL 4.11

Use the manufacturer's data in Figs. 4.31 and 4.32 to find the collector current where β_F (h_{FE}) for the 2N2222A is maximum, and the range of possible values of β_F. Assume $T = 25°C$ and $V_{CE} = 10$ V.

ANSWER 30 to 40 mA; ~100 to 300.

4.10 MAXIMUM RATINGS FOR THE BJT

Typical restrictions that must be observed in the operation of a BJT are shown on the manufacturer's data sheets. As an example, we will refer to the specifications for the 2N2222A given in Fig. 4.32.

Temperature Range

Extremes of temperature are specified for both passive handling (storage) and operation. Very low temperatures (−65°C) endanger the integrity of the device package because of thermal stresses where materials with slightly different expansion coefficients are bonded. Very high temperatures (greater than 200°C) will melt the solder used in connecting leads in addition to creating thermal stress.

Power Dissipation

Dissipation within the device causes internal heating, of course. Usually the average power of the collector junction, $P_D \cong V_{CE}I_C$, is of concern in circuit operation. Limits are specified such that the junction temperature will not exceed some safe maximum value. For example, the 2N2222A data sheet specifies $P_{D(max)} = 1800 - 12(T_C - 25°C)$ mW. This means that P_D may be 1.8 W if case temperature is 25°C or less, but as case temperature rises, dissipation must decrease 12 mW/°C. This implies a maximum junction temperature of 175°C, and thermal resistance $\theta_{JC} = 83.3°C/W$ (see Section 3.15). When the case is in free air (no heat sink), $P_{D(max)} = 500 - 3.3(T_A - 25°C)$ mW. Thermal resistance between case and ambient air $\theta_{CA} = 216.7°C/W$. The ratings are designed to

*MAXIMUM RATINGS

Rating	Symbol	2N2218 2N2219 2N2221 2N2222	2N2218A 2N2219A 2N2221A 2N2222A	2N5581 2N5582	Unit
Collector-Emitter Voltage	V_{CEO}	30	40	40	Vdc
Collector-Base Voltage	V_{CB}	60	75	75	Vdc
Emitter-Base Voltage	V_{EB}	5.0	6.0	6.0	Vdc
Collector Current — Continuous	I_C	800	800	800**	mAdc
		2N2218,A 2N2219,A	2N2221,A 2N2222,A	2N5581 2N5582	
Total Device Dissipation @ $T_A = 25°C$ Derate above 25°C	P_D	0.8 5.33	0.5 3.33	0.5 3.33	Watt mW/°C
Total Device Dissipation @ $T_C = 25°C$ Derate above 25°C	P_D	3.0 20	1.8 12	2.0 11.43	Watts mW/°C
Operating and Storage Junction Temperature Range	T_J, T_{stg}	←	−65 to +200	→	°C

*Indicates JEDEC Registered Data
**Motorola Guarantees this Data in Addition to JEDEC Registered Data.

ELECTRICAL CHARACTERISTICS ($T_A = 25°C$ unless otherwise noted.)

Characteristic		Symbol	Min	Max	Unit
OFF CHARACTERISTICS					
Collector-Emitter Breakdown Voltage ($I_C = 10$ mAdc, $I_B = 0$)	Non-A Suffix A-Suffix, 2N5581, 2N5582	$V_{(BR)CEO}$	30 40	— —	Vdc
Collector-Base Breakdown Voltage ($I_C = 10$ μAdc, $I_E = 0$)	Non-A Suffix A-Suffix, 2N5581, 2N5582	$V_{(BR)CBO}$	60 75	— —	Vdc
Emitter-Base Breakdown Voltage ($I_E = 10$ μAdc, $I_C = 0$)	Non-A Suffix A-Suffix, 2N5581, 2N5582	$V_{(BR)EBO}$	5.0 6.0	— —	Vdc
Collector Cutoff Current ($V_{CE} = 60$ Vdc, $V_{EB(off)} = 3.0$ Vdc)	A-Suffix, 2N5581, 2N5582	I_{CEX}	—	10	nAdc
Collector Cutoff Current ($V_{CB} = 50$ Vdc, $I_E = 0$) ($V_{CB} = 60$ Vdc, $I_E = 0$) ($V_{CB} = 50$ Vdc, $I_E = 0$, $T_A = 150°C$) ($V_{CB} = 60$ Vdc, $I_E = 0$, $T_A = 150°C$)	 Non-A Suffix A-Suffix, 2N5581, 2N5582 Non-A Suffix A-Suffix, 2N5581, 2N5582	I_{CBO}	— — — —	 0.01 0.01 10 10	μAdc
Emitter Cutoff Current ($V_{EB} = 3.0$ Vdc, $I_C = 0$)	A-Suffix, 2N5581, 2N5582	I_{EBO}	—		nAdc
Base Cutoff Current ($V_{CE} = 60$ Vdc, $V_{EB(off)} = 3.0$ Vdc)	A-Suffix	I_{BL}	—	20	nAdc
ON CHARACTERISTICS					
DC Current Gain ($I_C = 0.1$ mAdc, $V_{CE} = 10$ Vdc)	2N2218,A, 2N2221,A, 2N5581(1) 2N2219,A, 2N2222,A, 2N5582(1)	h_{FE}	20 35	— —	—
($I_C = 1.0$ mAdc, $V_{CE} = 10$ Vdc)	2N2218,A, 2N2221,A, 2N5581 2N2219,A, 2N2222,A, 2N5582		25 50	— —	
($I_C = 10$ mAdc, $V_{CE} = 10$ Vdc)	2N2218,A, 2N2221,A, 2N5581(1) 2N2219,A, 2N2222,A, 2N5582(1)		35 75	— —	
($I_C = 10$ mAdc, $V_{CE} = 10$ Vdc, $T_A = -55°C$)	2N2218A, 2N2221A, 2N5581 2N2219A, 2N2222A, 2N5582		15 35	— —	
($I_C = 150$ mAdc, $V_{CE} = 10$ Vdc)(1)	2N2218,A, 2N2221,A, 2N5581 2N2219,A, 2N2222,A, 2N5582		40 100	120 300	
($I_C = 150$ mAdc, $V_{CE} = 1.0$ Vdc)(1)	2N2218,A, 2N2221,A, 2N5581 2N2219,A, 2N2222,A, 2N5582		20 50	— —	
($I_C = 500$ mAdc, $V_{CE} = 10$ Vdc)(1)	2N2218, 2N2221 2N2219, 2N2222 2N2218A, 2N2221A, 2N5581 2N2219A, 2N2222A, 2N5582		20 30 25 40	— — — —	

(1) Pulse Test: Pulse Width ≤ 300 μs, Duty Cycle ≤ 2.0%.

■ **FIGURE 4.32** Electrical characteristics and ratings for the 2N2222A transistor family (courtesy Motorola, Inc.).

ELECTRICAL CHARACTERISTICS (continued) ($T_A = 25°C$ unless otherwise noted.)

Characteristic		Symbol	Min	Max	Unit
Collector-Emitter Saturation Voltage(1) ($I_C = 150$ mAdc, $I_B = 15$ mAdc)	Non-A Suffix	$V_{CE(sat)}$	—	0.4	Vdc
	A-Suffix, 2N5581, 2N5582		—	0.3	
($I_C = 500$ mAdc, $I_B = 50$ mAdc)	Non-A Suffix		—	1.6	
	A-Suffix, 2N5581, 2N5582		—	1.0	
Base-Emitter Saturation Voltage(1) ($I_C = 150$ mAdc, $I_B = 15$ mAdc)	Non-A Suffix	$V_{BE(sat)}$	0.6	1.3	Vdc
	A-Suffix, 2N5581, 2N5582		0.6	1.2	
($I_C = 500$ mAdc, $I_B = 50$ mAdc)	Non-A Suffix		—	2.6	
	A-Suffix, 2N5581, 2N5582		—	2.0	

SMALL-SIGNAL CHARACTERISTICS

Characteristic		Symbol	Min	Max	Unit
Current-Gain — Bandwidth Product(2) ($I_C = 20$ mAdc, $V_{CE} = 20$ Vdc, $f = 100$ MHz)	All Types, Except	f_T	250	—	MHz
	2N2219A, 2N2222A, 2N5582		300	—	
Output Capacitance(3) ($V_{CB} = 10$ Vdc, $I_E = 0$, $f = 100$ kHz)		C_{obo}	—	8.0	pF
Input Capacitance(3) ($V_{EB} = 0.5$ Vdc, $I_C = 0$, $f = 100$ kHz)	Non-A Suffix	C_{ibo}	—	30	pF
	A-Suffix, 2N5581, 2N5582		—	25	
Input Impedance ($I_C = 1.0$ mAdc, $V_{CE} = 10$ Vdc, $f = 1.0$ kHz)	2N2218A, 2N2221A	h_{ie}	1.0	3.5	kohms
	2N2219A, 2N2222A		2.0	8.0	
($I_C = 10$ mAdc, $V_{CE} = 10$ Vdc, $f = 1.0$ kHz)	2N2218A, 2N2221A		0.2	1.0	
	2N2219A, 2N2222A		0.25	1.25	
Voltage Feedback Ratio ($I_C = 1.0$ mAdc, $V_{CE} = 10$ Vdc, $f = 1.0$ kHz)	2N2218A, 2N2221A	h_{re}	—	5.0	$\times 10^{-4}$
	2N2219A, 2N2222A		—	8.0	
($I_C = 10$ mAdc, $V_{CE} = 10$ Vdc, $f = 1.0$ kHz)	2N2218A, 2N2221A		—	2.5	
	2N2219A, 2N2222A		—	4.0	
Small-Signal Current Gain ($I_C = 1.0$ mAdc, $V_{CE} = 10$ Vdc, $f = 1.0$ kHz)	2N2218A, 2N2221A	h_{fe}	30	150	—
	2N2219A, 2N2222A		50	300	
($I_C = 10$ mAdc, $V_{CE} = 10$ Vdc, $f = 1.0$ kHz)	2N2218A, 2N2221A		50	300	
	2N2219A, 2N2222A		75	375	
Output Admittance ($I_C = 1.0$ mAdc, $V_{CE} = 10$ Vdc, $f = 1.0$ kHz)	2N2218A, 2N2221A	h_{oe}	3.0	15	µmhos
	2N2219A, 2N2222A		5.0	35	
($I_C = 10$ mAdc, $V_{CE} = 10$ Vdc, $f = 1.0$ kHz)	2N2218A, 2N2221A		10	100	
	2N2219A, 2N2222A		25	200	
Collector Base Time Constant ($I_E = 20$ mAdc, $V_{CB} = 20$ Vdc, $f = 31.8$ MHz)	A-Suffix	$r_b'C_c$	—	150	ps
Noise Figure ($I_C = 100$ µAdc, $V_{CE} = 10$ Vdc, $R_S = 1.0$ kohm, $f = 1.0$ kHz)	2N2219A, 2N2222A	NF	—	4.0	dB
Real Part of Common-Emitter High Frequency Input Impedance ($I_C = 20$ mAdc, $V_{CE} = 20$ Vdc, $f = 300$ MHz)	2N2218A, 2N2219A 2N2221A, 2N2222A	$Re(h_{ie})$	—	60	Ohms

SWITCHING CHARACTERISTICS

Characteristic		Symbol	Min	Max	Unit
Delay Time	($V_{CC} = 30$ Vdc, $V_{BE(off)} = 0.5$ Vdc, $I_C = 150$ mAdc, $I_{B1} = 15$ mAdc) (Figure 14)	t_d	—	10	ns
Rise Time		t_r	—	25	ns
Storage Time	($V_{CC} = 30$ Vdc, $I_C = 150$ mAdc, $I_{B1} = I_{B2} = 15$ mAdc) (Figure 15)	t_s	—	225	ns
Fall Time		t_f	—	60	ns
Active Region Time Constant ($I_C = 150$ mAdc, $V_{CE} = 30$ Vdc) (See Figure 14 for 2N2218A, 2N2219A, 2N2221A, 2N2222A)		T_A	—	2.5	ns

(1) Pulse Test: Pulse Width ≤ 300 µs, Duty Cycle $\leq 2.0\%$.
(2) f_T is defined as the frequency at which $|h_{fe}|$ extrapolates to unity.
(3) 2N5581 and 2N5582 are Listed C_{cb} and C_{eb} for these conditions and values.

■ **FIGURE 4.32** continued.

keep the operating temperature of the junction lower than 175°C. In addition to the concerns about high temperature expressed earlier, high temperatures cause the collector current to increase as β_F and I_{CEO} increase. Depending on the external circuit, increased I_C may cause increased P_D, leading to increased temperature, followed by increased I_C, etc. This unstable situation is called **thermal runaway**, and it must be prevented by proper design of the Q-point and heat sinking.

EXAMPLE 4.13

(a) Plot the maximum safe power dissipation for the 2N2222A operating without a heat sink with ambient $0°C < T_A < 175°C$.
(b) Plot the maximum safe power dissipation for the 2N2222A operating with the case temperature $0°C < T_C < 175°C$.
(c) Find the case temperature and junction temperature for this device when $V_{CE} = 10$ V and $I_C = 25$ mA at $T_A = 25°C$ and no heat sink is used.

SOLUTION

(a) Data are taken from Fig. 4.32. With no heat sink, $P_{D(\max)} = 500$ mW and decreases 3.33 mW/°C above 25°C. The device may not be operated above 175°C, when $P_D = 0$ [see Fig. 4.33, plot (a)].

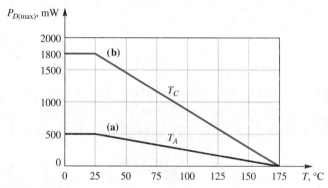

■ **FIGURE 4.33** Maximum power dissipation for the 2N2222A as a function of (a) ambient temperature (no heat sink) and (b) case temperature.

(b) With case temperature controlled, $P_{D(\max)} = 1800$ mW and decreases 12 mW/°C above 25°C [see Fig. 4.33, plot (b)]. Again, the device cannot be used above 175°C.
(c) The slopes of the above plots imply $\theta_{JA} = 300°C/W$ and $\theta_{JC} = 83.3°C/W$. Then $T_J = T_A + \theta_{JA} P_D = 25 + (300)(0.25) = 100°C$, and $T_C = T_J - \theta_{JC} P_D = 100 - 83(0.25) = 79°C$.

DRILL 4.12

A transistor has $I_C = 100$ mA and $V_{CE} = 10$ V.

(a) Find the power dissipation of the transistor.
(b) What would be the maximum thermal resistance permitted a heat sink if the ambient-to-case temperature must not exceed 70°C?

ANSWER (a) 1 W, (b) 70°C/W.

Collector Current

The current-carrying capability of the transistor is limited by the resistance of the wire bonds that connect the chip to the package leads, or perhaps the bulk resistance of the collector material itself. In the case of the 2N2222A, the manufacturer specifies 800 mA as a maximum. Note also that Fig. 4.31 indicates that h_{FE} would be small at this current level due to high-level injection.

Collector-Base Voltage

The maximum allowable collector-base reverse voltage is limited by the lower of two breakdown voltages. We discussed one breakdown phenomenon, punch-through, in Section 4.4. Increasing V_{CB} could extend the collector space-charge region until it contacts the emitter space-charge region, resulting in the loss of any barrier to current flow. Normally the other breakdown mechanism, avalanche, will occur first. We discussed avalanche breakdown in Section 3.4.

Avalanche Multiplication

Avalanche multiplication, as it relates to diodes, involves the acceleration of leakage current carriers in the reverse-biased junction space-charge region, producing ionizing collisions with the crystal atoms. New carriers are generated, and the current increases. This same process occurs with the carriers that constitute normal collector current. The collector current is effectively multiplied by an empirical factor

$$M = \frac{1}{1 - (V_{CB}/BV_{CBO})^n}, \quad (4.44)$$

where BV_{CBO} is the **collector-base breakdown voltage** and n is some constant of the transistor, usually between 2 and 6. Thus Eq. (4.1) becomes

$$i_C = \frac{-\alpha_F i_E + I_{CBO}}{1 - (V_{CB}/BV_{CBO})^n} \quad (4.45)$$

Collector current becomes large at $V_{CB} = BV_{CBO}$. The collector characteristics for the common base appear as in Fig. 4.34(a). One can observe significant increases in i_C at values of V_{CB} of less than BV_{CBO}. The data sheet will specify a maximum V_{CB}, guaranteed to be less than BV_{CBO} (75 V in the case of the 2N2222A). Locate this value in Fig. 4.32.

In the case of the common emitter, we substitute $i_E = -i_C - i_B$ in Eq. (4.45), giving us

$$i_C = \frac{\alpha_F i_B + I_{CBO}}{1 - \alpha_F - (V_{CB}/BV_{CBO})^n}. \quad (4.46)$$

This expression will become large at $V_{CB} = BV_{CBO} \sqrt[n]{1 - \alpha_F} \equiv BV_{CEO}$ (collector-emitter breakdown voltage), which is much lower than BV_{CBO}. Observe that the 2N2222A data sheet indicates that maximum V_{CE} is 40 V. Figure 4.34(b) indicates that considerable nonlinearity occurs well below the breakdown voltage.

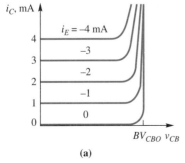
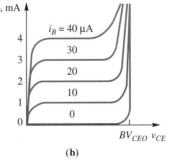

■ **FIGURE 4.34** 2N2222A characteristics showing avalanche breakdown. **(a)** CB collector characteristic. **(b)** CE collector characteristic.

The breakdown voltage of the emitter-base junction is typically much smaller than that of the collector. This results from the much higher doping level in the emitter. Figure 4.32 shows 6 V for the 2N2222A. The emitter-base breakdown voltage is labeled $V_{(BR)EBO}$ in the data sheet.

CHECK UP

1. **TRUE OR FALSE?** The inverse current gain β_R is emitter current divided by base current, with the collector-base junction forward-biased and the emitter-base junction reverse-biased.
2. **TRUE OR FALSE?** The inverse mode is a very common region for BJT operation.
3. **TRUE OR FALSE?** The BJT current gain β_F increases with v_{CE} due to Early effect.
4. **TRUE OR FALSE?** The BJT current gain β_F increases with I_C due to high-level injection.

4.11 SWITCHING TIMES FOR THE BJT

We saw in Section 3.12 that diode currents do not respond instantly to changes in junction voltage. Changes in collector current also lag behind changes in base current. Consider the switching circuit in Fig. 4.35(a). At time $t = 0$, the input voltage $v_{IN} = V_L$, shown negative in waveform in Fig. 4.35(b). The transistor is cut off. The base region is depleted of minority carriers, as shown in Fig. 4.36 by curve n_C. We discussed this minority-carrier distribution in Section 4.3. Four switching times are specified on the data sheet in Fig. 4.32.

Delay Time

At time T_1, v_{IN} changes abruptly to V_H, some positive voltage large enough to cause the BJT to saturate eventually. The base current jumps to the value

$$I_B = \frac{V_H - v_{BE}(T_1)}{R_B}, \tag{4.47}$$

reflecting the fact that electrons are being injected into the base region from the emitter. Minority-carrier concentration begins to build at the emitter edge of the base region, and collector current begins to flow when a concentration gradient is established. After a *delay time* t_d, i_C will reach 10% of the saturation value $I_{CS} \approx V_{CC}/R_C$, and v_{BE} will reach about 0.7 V. Electron distribution in the base is shown by the curve marked $0.1n_A$ in Fig. 4.36. The delay is the time it takes for the base current to charge up the **emitter-base transition capacitance,** C_{ib} on the data sheet, and to provide the charge in the base represented by the difference in the two plots, $0.1n_A$ and n_C, which represents the **base diffusion capacitance**. We discuss transistor junction capacitances further in Chapter 9.

Rise Time

Continued base current causes further buildup of the electron concentration in the base. After an additional delay, called **rise time t_r**, collector current "rises" to $0.9I_{CS}$, and electron concentration is represented by the plot $0.9n_A$. This delay is the time it takes to

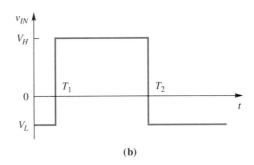
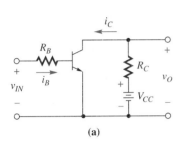
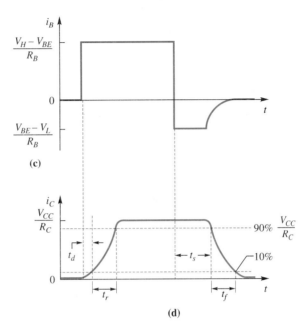

■ **FIGURE 4.35** Transistor switching circuit and waveforms. (a) Transistor switch. (b) Input voltage. (c) Base current. (d) Collector current.

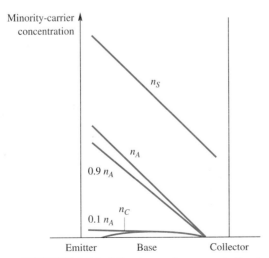

■ **FIGURE 4.36** Minority-carrier concentration in the base region: n_C, cutoff; $0.1n_A$, becoming active, $I_C = 0.1I_{CS}$; $0.9n_A$ active, $I_C = 0.9I_{CS}$; n_A, at edge of saturation; n_S, deeply saturated.

163

■ Note that the rise time is defined when the collector *current* is rising, and the collector voltage would be falling.

charge the ***collector-base transition capacitance,*** C_{ob}, on the data sheet, and the base current to provide the difference in charge between curve $0.1n_A$ and $0.9n_A$. After saturation is reached, electron concentration in the base continues to increase to the excess level shown by curve n_S in Fig. 4.36. This excess charge (between curves n_S and n_A) results from the condition that base current is greater than that required just to saturate the switch ($i_B > I_{CS}/h_{FE}$). Plot n_S is actually the superposition of minority-carrier concentrations resulting from injection from both junctions.

Storage Time

At time T_2, v_{IN} changes abruptly back to V_L. Base current reverses, reflecting the fact the electrons now move back across the emitter junction. However, collector current continues at $i_C = I_{CS}$ until the excess charge stored in the base is depleted. The ***storage time*** t_s is the delay as carrier concentration adjusts from plot n_S down to plot $0.9n_A$. Note that the greater the degree of saturation (the more i_B exceeds I_{CS}/h_{FE}), the greater the storage time.

Fall Time

After entering the active region, but before the transistor is cut off, the collector transition capacitance must discharge, and the base carrier distribution must change from $0.9n_A$ to n_C. The time it takes the collector current to "fall" to 10% I_{CS} is called the ***fall time*** t_f. During this time, the base voltage decreases to V_L and base current decays to zero.

Charge-Control Analysis

Quantitative estimates of BJT switching times are facilitated by a charge-control model.[4,5] The currents i_B and i_C can be related to the changes represented by the minority-carrier concentrations in the base region (Fig. 4.36) and the time rate of change of these charges. The relationship of base current and this stored charge is

$$i_B(t) = \frac{Q_b(t)}{\tau} + \frac{dQ_b(t)}{dt}, \tag{4.48}$$

where τ is the lifetime of minority carriers in the base. With step changes in i_B, each switch results in a charge readjustment characterized by the time constant τ.

A complete simulation of the switching time of a BJT inverter is presented in Example 4.16 using a SPICE transient analysis. Representative values of switching times and time constants for the 2N2222A are shown on the data sheet in Fig. 4.32.

Schottky Transistor

Storage time is regarded as a serious defect for fast switches. One technique in treating this problem is to prevent the transistor from saturating. A BJT fabricated with a Schottky diode connected between the base and the collector is depicted in Fig. 4.37(a). Because the diode has a drop of only 0.4 V, the collector is clamped to go no lower than $v_{CE} = v_{BE} - v_{SD} = 0.7 - 0.4 = 0.3$ V. Further, as noted in Section 3.13, the Schottky diode has no storage time problem of its own. The dc situation for the circuit shown in Fig. 4.37(a) is investigated in Problem 4.48. The standard circuit symbol for the Schottky transistor is shown in Fig. 4.37(b).

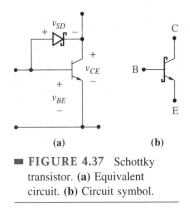

FIGURE 4.37 Schottky transistor. (a) Equivalent circuit. (b) Circuit symbol.

4.12 SPICE MODEL FOR THE BJT

BJT circuit problems may be solved by simulation of the circuit using a computer model. For the reader who may not be familiar with SPICE (**S**imulation **P**rogram, **I**ntegrated **C**ircuit **E**mphasis), reference is made to Appendix D and the information provided there. The BJT requires both an element statement and a model statement. The *npn* and *pnp* transistor element models are given in Fig. 4.38. The element statement format is

■ Often, some versions of SPICE require a substrate connection for IC BJTs.

QXXX NC NB NE MNAME [AREA] [OFF] [IC=VBE, VCE] (4.49)

NC, **NB**, and **NE** are the collector, base, and emitter nodes, respectively. The default value for the base cross-sectional **AREA** is 1.0. **OFF** refers to the device being set to the nonconducting condition in the dc circuit analysis. The circuit will be analyzed using the computer Q-point for the BJT. This is the default condition. Initial conditions other than the Q-point value can be specified optionally for a transient analysis.

The BJT model statement is

.MODEL MNAME TYPE [PNAME1=PVAL1 PNAME2=PVAL2 . .] (4.50)

where **TYPE** must be specified as either **NPN** or **PNP**. The transistor model, depending on the complexity of the parameter selection, is adapted from the Gummel and Poon integral charge control model or from the Ebers-Moll model. Recall that we have been using Ebers-Moll-derived models. Selected common default parameters are provided in Table 4.1.

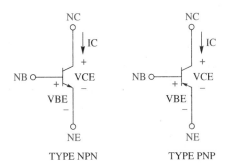

FIGURE 4.38 **MNAME BJT** element model.

TABLE 4.1 Selected SPICE BJT Model Parameters

Name	Parameter	Default Value	Description
IS	Transport saturation current	10^{-16} A	I_S from Eq. (4.10)
NF	Forward I emission coefficient	1	η if needed in Eq. (4.10)
BF	Ideal maximum forward β_F	100	As in Eq. (4.17)
IKF	Corner for high-current β_F rolloff	∞ A	
VAF	Forward Early voltage	∞ V	V_A
BR	Ideal maximum reverse β_R	1	As in Eq. (4.39)
NR	Reverse I emission coefficient	1	
IKR	Corner for high-current β_R rolloff	∞ A	
RB	Base ohmic resistance	0 Ω	Base spreading resistance r_b
RC	Collector resistance	0 Ω	Collector lead and bulk resistance
RE	Emitter resistance	0 Ω	Emitter lead resistance
TF	Forward transit time	0 s	$\tau_F = 1/2\pi f_T$
TR	Reverse transit time	0 s	
CJE	Zero-bias B-E capacitance	0 F	C_{je0}, transition capacitance
VJE	B-E junction potential	0.75 V	ϕ_J for base-emitter junction
MJE	B-E junction grading coefficient	0.33	0.5 = abrupt; 0.33 = graded
FC	Forward-bias-depletion capacitance coefficient	0.5	C_{je} saturates for $v_{BE} > FC \times VJE$
CJC	Zero-bias B-C capacitance	0 F	$C_{\mu 0}$, as in Eq. (9.1)
VJC	B-C junction potential	0.75 V	ϕ_J for base-collector junction
MJC	B-C junction grading coefficient	0.33	0.5 = abrupt; 0.33 = graded
CJS	Zero-bias collector-substrate C	0 F	C_{CS0}, as in Eq. (9.2)
VJS	C-S junction potential	0.75 V	ϕ_J for collector-substrate junction
MJS	C-S junction grading coefficient	0	0.5 if needed
EG	Band-gap potential	1.11 eV	Silicon
XTI	I_S temperature coefficient	3	Temperature coefficient for I_S
KF	Flicker noise coefficient	0	
AF	Flicker noise exponent	1	

■ Additional parameters that may be specified for the BJT are given in Reference 6.

For example, the element and model statements for a default *npn* transistor ($\beta_F = 100$ and $I_S = 10^{-16}$ A), **T15**, whose collector is connected to node 5, emitter to node 0, and base to node 2, are given by

```
Q15  5  2  0  T15
```
(4.51)

and

```
.MODEL T15 NPN.
```
(4.52)

The dc model is defined by **IS**, **NF**, **BF**, and **IKF**, which determine the forward current gain characteristics, **BR**, **NR**, and **IKR**, which determine the reverse current gain characteristics, and **VAF**, which determines the output conductance for the forward region. The three ohmic resistances may be needed for transistors (see Chapter 18 for fabrication details). Base charge storage is modeled by forward and reverse transit times, **TF** and **TR**. Temperature dependence of the saturation current, **IS**, is determined by the energy gap, **EG**, and the temperature exponent, **XTI**. The nonlinear depletion layer capacitances are determined by **CJE**, **VJE**, and **MJE** for the base-emitter junction, and **CJC**, **VJC**, and **MJC** for the base-collector junction.

It may be that the SPICE default model, which assumes a small silicon IC-type BJT, with no capacitance or series resistance, constant β_F, and infinite Early voltage, will adequately represent the BJT in a particular circuit. Alternatively, the parameters may be obtained from a data sheet or from actual measurements. To give an example of customizing the model statement, consider a 2N2222A:

- **IS** $= I_S = 3 \times 10^{-14}$ is calculated using Eq. (4.10) with $V_T = 26$ mV, $I_C = 100$ mA, and $V_{BE} = 0.75$ V.
- **BF** $= \beta_F = 200$ is determined from the data sheet of Fig. 4.32, which gives $100 < h_{FE} < 300$ at 150 mA.
- **IKF** $= 0.3$ from Fig. 4.31, where it appears that β_F is rapidly falling off at $I_C > 0.3$ A.
- **VAF** $= V_A = 115$ V is found by observing the output conductance as given in the data sheet: At $V_{CE} = 10$ V and $I_C = 10$ mA, $25 \ \mu S < g_o < 200 \ \mu S$. Choosing $g_o = 80 \ \mu S$ leads to $(V_A + 10) = 10$ mA$/80 \ \mu S$ and $V_A = 115$ V.
- **TF** $= \tau_F = 530$ ps, since $\tau_F = 1/2\pi f_T$, and the data sheet gives $f_T = 300$ MHz.

■ SPICE computes β from $\beta = \beta_F\left(1 + \dfrac{V_{CE}}{V_A}\right)$

Capacitance parameters are discussed in Chapter 9. A detailed discussion of these and other SPICE BJT parameters is given in Ref. 6.

EXAMPLE 4.14

The BJT in the circuit of Fig. 4.39 has $\beta_F = 50$ and a saturation current $I_S = 10^{-14}$ A.

(a) Write a SPICE program to determine the operating point of the BJT.
(b) Use the sensitivity statement to find the effect on V_{30} of changes in resistance values.

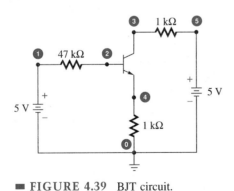

■ **FIGURE 4.39** BJT circuit.

SOLUTION

(a) Using the node numbers of Fig. 4.39, the SPICE statements are as follows:

```
BJT OPERATING POINT
VCC 5 0 DC 5
VBB 1 0 DC 5
RB 1 2 47K
```

```
RC 5 3 1K
RE 4 0 1K
Q1 3 2 4 BJT
.MODEL BJT NPN BF=50 IS=1E-14
.OP
.END
```

Note that nodes 1 and 5 need not be separate, but were separated here for generality. The SPICE **BJT** transistor model has default values of $\beta_F = 100$ and $I_S = 10^{-16}$ A, but the given values illustrate how the model may be specified. The higher saturation current is more typical of larger transistors. The **.OP** command causes SPICE to print all dc node voltages and also the dc terminal voltages and currents of all transistors. The small-signal models will also be determined. The dc results are $I_B = 44.1$ µA, $I_C = 2.21$ mA, and $V_{CE} = 0.543$ V, which agree with the approximate solution of Example 4.11. We are also given that $g_m = 85$ mS and $r_\pi = 586$ Ω.

(b) Adding the program statement

```
.SENS V(3)
```

between the model statement and the **END** statement causes SPICE to generate a sensitivity table, which indicates how V_{30} is affected by variation of any circuit parameter:

```
DC SENSITIVITIES OF OUTPUT V(3)
```

ELEMENT NAME	ELEMENT VALUE	SENSITIVITY VOLTS/UNIT	SENSITIVITY VOLTS/%
RB	47000Ω	0.000022	0.011
RC	1000Ω	−0.0022	−0.022
RE	1000Ω	0.0011	0.011
VBB	5V	−0.51	−0.025
VCC	5V	1.0	0.05
BF	50	−0.021	−0.011
IS	10^{-14} A	-1.3×10^{12}	−0.00013

■ This chart shows at a glance how the Q-point changes with a change in any circuit element.

EXAMPLE 4.15

Write a SPICE program to plot the transfer characteristic (v_{30} versus v_{IN}) of the inverter of Fig. 4.40 for temperatures of −55, 27, and 125°C. Use $\beta_F = 100$ and $I_S = 10^{-14}$ A at 27°C.

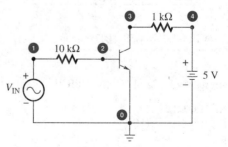

■ **FIGURE 4.40** Inverter circuit.

SOLUTION   Consider the following program:

```
INVERTER TRANSFER CHARACTERISTIC
VIN 1 0 DC
VCC 4 0 DC 5
RB 1 2 10K
RC 4 3 1K
Q1 3 2 0 T1
.MODEL T1 NPN BF=100 IS=1E-14 XTB=1.6
.TEMP -55 27 125
.DC VIN 0 2 0.2
.PLOT DC V(3)
.END
```

SPICE assumes the model parameters are given at 27°C, and recalculates them for the other temperatures. I_S is presumed to vary as in Eq. (2.27), junction voltages vary as in the diode equation, and β_F is changed according to absolute temperature to the **XTB = 1.6** power $(T_2/T_1)^{1.6}$. This constant was chosen to correspond with the data of Fig. 4.31. Calculations of v_{30} will be made as v_{IN} ranges from 0 to 2 V in 0.2-V steps. Note that SPICE assumes that $v_3 = v_{30}$. The resulting plots appear in Fig. 4.41. Compare these results with the approximate solution of Example 4.12.

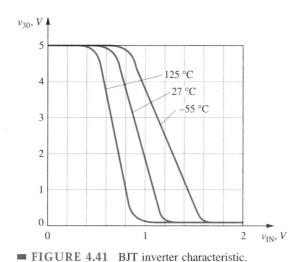

■ FIGURE 4.41 BJT inverter characteristic.

EXAMPLE 4.16

Write a SPICE program to plot the transient response of the inverter of Fig. 4.40 to a pulse input. The input waveform is plotted as curve (a) in Fig. 4.42. Use a transistor model with $\beta_F = 100$, transport saturation current $I_S = 10^{-14}$ A, forward transit time $\tau_F = 0.1$ ns, reverse transit time $\tau_R = 10$ ns, a collector junction capacitance of 2 pF, and an emitter junction capacitance of 2 pF. (These constraints will simulate a discrete transistor equivalent to the 2N2222.)

■ The transient response of a circuit tells a great deal about delays and parasitic capacitances of the BJT(s).

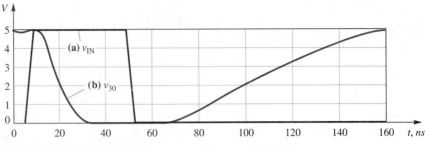

■ **FIGURE 4.42** Transient response of inverter: **(a)** Input voltage **(b)** Collector voltage.

SOLUTION

```
TRANSIENT RESPONSE OF INVERTER
V10 1 0 PULSE (0 5 4NS 4NS 4NS 40NS 160NS)
VCC 4 0 DC 5
RB 1 2 10K
RC 4 3 1K
Q1 3 2 0 T1
.MODEL T1 NPN BF=100 TF=.1NS TR=10NS CJC=2PF
+CJE=2PF IS=1E-14
.TRAN 4NS 160NS
.PLOT TRAN V(3)
.END
```

■ + in first column continues statement to the next line.

The meaning of the pulse definition statement just given is, respectively: initial value = 0 V, pulse height = 5 V, delay time = 4 ns, rise time = 4 ns, fall time = 4 ns, pulse width = 40 ns, and period = 160 ns. The output voltage v_{30} will be found each 4 ns for 160 ns. The result is plotted as curve (b) in Fig. 4.42. You should note the various delay times as defined in Fig. 4.35.

CHECK UP

1. **TRUE OR FALSE?** After a step of base current is applied to a BJT that has been off, the delay in the response of the collector current is called the rise time.
2. **TRUE OR FALSE?** At turn-OFF, a saturated BJT switch experiences a storage time delay, after which the collector-emitter voltage increases from $0.1V_{CC}$ to $0.9V_{CC}$. This later time is called the rise time.
3. **TRUE OR FALSE?** A Schottky transistor has no storage time because it does not saturate.
4. **TRUE OR FALSE?** The default BJT SPICE model values are $\beta_F = 100$, $\beta_R = 1$, and $V_A = \infty$.

SUMMARY

Expanding on the *pn* junction theory presented in Chapter 3, in this chapter we developed the theory of operation, terminal characteristics, and circuit models for the bipolar junction transistor (BJT). Circuit models for the active, saturation, and cutoff regions and for the inverted mode of operation were presented. Each model was demonstrated by means of typical circuit applications including an overview of the design and analysis of amplifiers and switches. We have found that when unsure if the BJT is active, we should assume a model and attempt to verify it. For example, assume $v_{BE} = 0.7$ V and $I_C = \beta_F I_B$, and prove that $v_{CE} > 0.3$ V. When unable to verify the active model, it is incorrect, and we must assume a saturated model.

Key transistor parameters needed for the circuit models including those important for both amplifier and switching operation were illustrated with data sheets. These data sheets will be used in subsequent chapters to develop design and analysis examples. Detailed small-signal amplifier analysis is presented in Chapter 8 and detailed operation as a switch and logic gate element appears in Chapter 13.

Simulation of BJT circuits by means of SPICE was discussed, after introduction to some of the parameters that SPICE uses in its BJT model.

SURVEY QUESTIONS

- How does a collector differ from an emitter in structure?
- What fabrication details contribute to production of higher current gain BJTs?
- What is the major application of a BJT biased in the active region?
- What regions of operation would be useful in designing a BJT switch?
- Why must the dc operating point be known before predicting small-signal behavior in an amplifier circuit?
- What are the effects of operating a BJT at elevated temperatures?

PROBLEMS

*Problems marked with * are more challenging.*

4.1 Equations (4.6) and (4.7) were developed for an *npn* transistor. Write analogous expressions for the *pnp* transistor of Fig. 4.1(a).

4.2 Specify the region of operation of a silicon *npn* transistor for each condition given:

(a) $V_{BE} = 0.7$ V and $V_{CE} = 10$ V.
(b) $V_{BE} = -10$ V and $V_{CB} = -0.7$ V.
(c) $V_{BE} = 0.7$ V and $V_{CE} = 0.2$ V.
(d) $V_{BE} = -0.7$ V and $V_{CE} = 10$ V

4.3 Specify the region of operation of a silicon *pnp* transistor for each condition given:

(a) $V_{BE} = 0.7$ V and $V_{CE} = -10$ V.
(b) $V_{BE} = -0.7$ V and $V_{CE} = -10$ V.
(c) $V_{BE} = -0.7$ V and $V_{CE} = -0.2$ V

4.4 Measurements on a silicon *npn* BJT give $V_{BE} = 0.7$ V when $I_C = 0.5$ mA and $V_{CE} = 10$ V. Find I_S.

4.5 A silicon *npn* BJT has $V_{BE} = 0.7$ V when $I_C = 0.5$ mA. What V_{BE} do you expect for $I_C = 25$ mA? For $I_C = 5$ μA? Assume *active* operation; r_b, I_{CO} are negligible.

4.6 The transistor in Fig. 4.43 has the characteristics shown in Figs. 4.10 and 4.11. Let $V_{EE} = 2$ V, $V_{CC} = 12$ V, $R_E = 270$ Ω, and $R_C = 1200$ Ω.

(a) Find I_E and V_{BE}.
(b) Find I_C and V_{CB}.

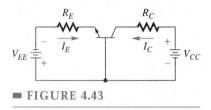

■ FIGURE 4.43

4.7 The transistor in Fig. 4.43 has the characteristics shown in Figs. 4.10 and 4.11. Let $V_{EE} = 1.2$ V, $V_{CC} = 10$ V, and $R_C = 1000$ Ω. Find the value of R_E that will cause $V_{CB} = 5$ V.

4.8 For the transistor in Fig. 4.43, $V_{BE} = 0.65$, $\alpha_F = 0.99$, and $I_{CBO} = 10$ nA. Let $V_{EE} = 5$ V, $V_{CC} = 10$ V, $R_E = 500$ Ω, and $R_C = 1$ kΩ.

(a) Find I_C and V_{CB}.
(b) Change R_C to 2 kΩ and repeat part (a).

4.9 The transistor in Fig. 4.44 is silicon: $V_{EB} = 0.7$ V, $\alpha_F = 0.99$, and $I_{CBO} = 10$ nA.

(a) Find R_B such that the ammeter current is 1 mA.
(b) Write an expression for I as a function of R_B.

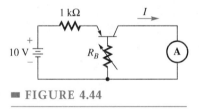

■ FIGURE 4.44

4.10 Sketch the collector characteristics (I_C versus V_{CE}) for an *npn* BJT having $\beta_F = 100$ and $I_{CEO} = 0.5$ mA. Let $i_B = 10, 20, 30$, and 40 μA. Ignore the Early effect.

4.11 Sketch the collector characteristics I_C versus V_{CE} for an *npn* BJT using Eq. (4.10) with $I_S = 10^{-14}$ A and $T = 300$ K. Let $V_{BE} = 0.65, 0.7, 0.71$, and 0.72 V. Ignore the Early effect.

4.12 Using the geometric relationships of Fig. 4.16, show how Eq. (4.10) can be modified to incorporate the Early effect, given V_{CE} and V_A.

4.13 Suppose a silicon *npn* BJT is tested with V_{BE} constant at 0.65 V, and it is found that $I_C = 0.50$ mA at $V_{CE} = 2$ V, and $I_C = 0.55$ mA at $V_{CE} = 10$ V. Find r_o and V_A.

4.14 The transistor whose characteristics are plotted in Figs. 4.15 and 4.17 is incorporated in the circuit of Fig. 4.45. Let $V_{CC} = 12$ V, $R_C = 400$ Ω, $V_{BB} = 5$ V, and $R_B = 43$ kΩ.

(a) Find the quiescent values of i_B, i_C, and v_{CE}.
(b) Suppose an ac base current $i_b = 50 \sin \omega t$ μA is superimposed on the dc value of part (a). Sketch i_C and v_{CE} as time functions.
(c) Suppose the ac base current is increased to $100 \sin \omega t$ μA. Again sketch the waveshape of i_C.

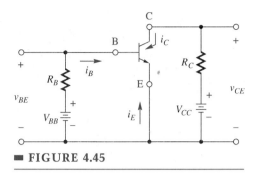

FIGURE 4.45

4.15 Derive Eq. (4.21), assuming Eq. (4.19).

4.16 Measurements on a given BJT yield $I_B = 25\ \mu\text{A}$, $I_C = 2$ mA, and $V_{CE} = 3$ V. Find β_F and α_F.

4.17 Draw a dc model for a silicon *npn* BJT having $I_C = 10$ mA, when $I_B = 25\ \mu\text{A}$ and $V_{BE} = 0.7$ V.

(a) Neglect I_{CO}.
(b) Let $I_{CO} = 0.1\ \mu\text{A}$.

4.18 Draw a dc model for a silicon *pnp* BJT that has $I_C = -20$ mA, when $I_B = -100\ \mu\text{A}$ and $V_{BE} = -0.65$ V. Neglect I_{CO}.

4.19 Design the circuit of Fig. 4.45 using an *npn* transistor having a $\beta_F = 100$ and a $V_{BE} = 0.7$ V. Let $V_{CC} = V_{BB} = 6$ V, and choose values of R_B and R_C such that $V_{CE} = 3$ V and $I_C = 10$ mA.

4.20 The *pnp* transistor of Fig. 4.46 has $\beta_F = 100$ and $V_{BE} = -0.65$ V. Find the quiescent I_B, I_C, and V_{CE}.

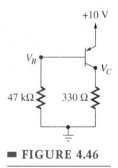

FIGURE 4.46

4.21 The circuit of Fig. 4.46 is found to have $V_B = 9.35$ V and $V_C = 5$ V. What is the β_F of the transistor if you can neglect I_{CO}? What do you expect to happen if the temperature is increased? Discuss qualitatively.

4.22* Carry out the derivation of Eq. (4.33).

4.23* Beginning with Eqs. (4.8) and (4.9), derive Eq. (4.35).

4.24 The transistor in Fig. 4.47 is silicon and $\beta_F = 50$. Apply Thévenin's theorem to the circuit to the left of the base. Assume an active model for the transistor and verify the validity of the model. If it is not valid, then assume a saturated model.

(a) Find the quiescent i_B, i_C, and v_{CE}.
(b) Interchange the 22- and 33-kΩ resistors, and repeat part (a).

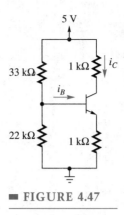

■ FIGURE 4.47

4.25 The transistor in Fig. 4.47 is silicon and $\beta_F = 100$. Find the quiescent I_B, I_C, and V_{CE}.

4.26 The transistor in Fig. 4.47 is silicon and $\beta_F = 50$. Change the 22-kΩ resistor to 33 kΩ and find the quiescent I_B, I_C, and V_{CE}.

4.27 Redesign the circuit of Figure 4.47 by changing the 22-kΩ resistor such that V_{CE} is 2 V for a transistor whose $\beta_F = 250$. Let $V_{BE} = 0.7$ V, and neglect I_{CBO} and the Early effects.

4.28 The transistor in Fig. 4.48 is silicon and $\beta_F = 100$. Assume a model for the transistor and verify assumptions. Find V_{CE} and I_C. Is the transistor active, saturated, or cut off?

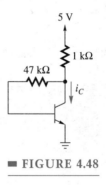

■ FIGURE 4.48

4.29 The circuit of Fig. 4.48 has a silicon transistor with $\beta_F = 300$. Find V_{CE} and I_C. Why is it not possible to saturate any transistor in this circuit arrangement?

4.30 Redraw the ac small-signal model of Fig. 4.22 as a current-controlled model, recognizing that $v_{be} = i_b r_\pi$. Also recall that $\beta = r_\pi g_m$. Are these two models equivalent?

4.31 An *npn* transistor is operating with $I_C = 5$ mA and $V_{CE} = 10$ V. The Early voltage, V_A, is 100 V, and $\beta = 250$. Find g_m, r_π, and r_o for the small-signal model.

4.32 Is there any difference between the small-signal ac model for an *npn* transistor and that for a *pnp* transistor? Explain.

4.33 Consider the power-series expansion of $I_S\, e^{v_{be}/V_T}$ in Eq. (4.24). The *small-signal* approximation neglects the $(v_{be})^2$ term. How large is this term when $v_{be} = 10$ mV and $V_T = 26$ mV (compared to the v_{be} term)?

4.34 Consider the amplifier circuit of Fig. 4.49 for which $\beta_F = 75$ and $V_A = 200$ V. Assume that the capacitors are open circuits to dc, and short circuits to the ac frequency of interest.

(a) Draw a dc model and find the operating point.
(b) Find the small-signal ac model parameters, and draw an ac small-signal model for the circuit. Neglect r_b.
(c) Find the voltage amplification v_{out}/v_{in}.

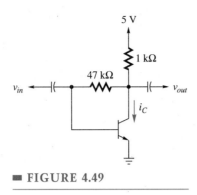

FIGURE 4.49

4.35 Consider the amplifier circuit of Fig. 4.50 for which $\beta_F = 75$ and $V_A = 200$ V. Assume that the capacitors are open circuits to dc, and short circuits to the ac frequency of interest.

(a) Draw a dc model and find the operating point.
(b) Find the small-signal ac model parameters, and draw an ac small-signal model for the circuit.
(c) Find the voltage amplification v_{out}/v_{in}.

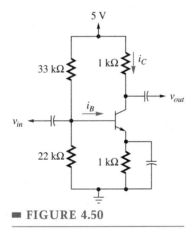

FIGURE 4.50

4.36 The transistor used in the circuit of Fig. 4.51 is silicon; $V_{BE} = 0.7$ V, $\beta_F = 100$, $\alpha_R = 0.1$, and $I_{CBO} = 10$ nA. Let $V_{CC} = V_{BB} = 5$ V and $R_C = 1$ kΩ.

(a) Find R_B for $V_{CE} = 2.5$ V.
(b) Find R_B for $V_{CE} = 0.3$ V.
(c) Using Fig. 4.25, find R_B such that $V_{CE} = 0.1$ V.
(d) Repeat part (a), assuming that the temperature increases 100°C. Consider the effect on I_{CBO} and on V_{BE}. Assume β_F increases 50%.

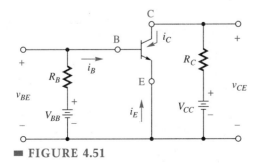

FIGURE 4.51

4.37 The transistors in Fig. 4.52 are silicon and $\beta_F = 50$.

(a) Assume Q_1 is saturated. Is Q_2 active, saturated, or cut off?
(b) Verify that Q_1 is indeed saturated.

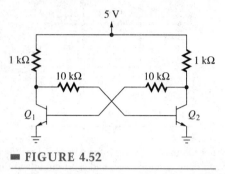

■ FIGURE 4.52

4.38 Repeat Problem 4.37 after changing both 10-kΩ resistors to 47 kΩ. Is Q_1 really saturated?

4.39* In a manner analogous to Eqs. (4.34) and (4.35), derive an expression for v_{EC} as a function of i_B and i_E. Show that the saturation voltage for the inverted transistor in Fig. 4.53 can be made smaller than that of the normal mode, but that i_B must be very large.

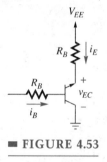

■ FIGURE 4.53

4.40 A transistor in the circuit of Fig. 4.53 is found to have $i_E = 0.5$ mA and $i_B = 1$ mA. Find β_R and α_R.

4.41* Sketch the output voltage v_o versus the input voltage v_{IN} for the circuit in Fig. 4.54. Assume $\beta_F = 50$ and $\beta_R = 0.1$. Two points, $v_{IN} = 0$ and $v_{IN} = 3$ V, were determined in Section 4.8. Here are two hints: First, one breakpoint in the plot will occur at $v_o = 0.3$ V. This is the edge of saturation, and you may use an active model for Q_2. Find i_{C2}, $i_{B2} = -i_{C1}$, i_{B1}, and then v_{CE1}, from Fig. 4.25. Second, another breakpoint is at cut-in for Q_2, when $v_{BE2} \approx 0.5$ V and $i_{B2} = -i_{C2} \approx 0$.

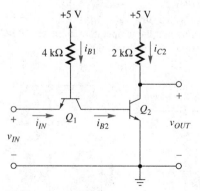

■ FIGURE 4.54 A circuit that demonstrates the inverse mode.

CHAPTER 4 THE BIPOLAR JUNCTION TRANSISTOR

4.42 Consider the circuit in Fig. 4.55. The transistors are silicon and $\beta_F = 100$.

(a) Let $v_{IN} = 0$, so that Q_1 is off. Is Q_2 active or saturated? Find v_o and v_E.
(b) If v_{IN} is increased to v_E [from part (a)] + $v_{BE1(cut-in)}$, Q_1 will saturate and Q_2 will cut off. Verify that this will happen. Find this v_{IN} and the resulting v_o.
(c) Now, to switch back to the original state, v_{IN} must be reduced until Q_1 is active, with $v_{CE1} = V_{BE2(cut-in)}$. Solve for this v_{IN}, where $V_{CE1} = 0.5$ V.
(d) Sketch v_o versus v_{IN} (all the breakpoints are identified above). This circuit is called a *Schmitt trigger*. It serves to convert a continuously varying input voltage into a binary output.

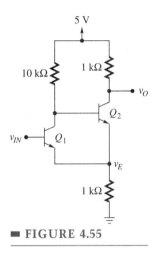

■ **FIGURE 4.55**

4.43 Work Problem 4.42 with $V_{CC} = 12$ V rather than 5 V.

4.44 The transistor used in the circuit in Fig. 4.51 is a 2N2222A with $\beta_F = 100$, $V_{BE(active)} = 0.7$ V, and $I_{CBO} = 10$ nA at 25°C. Let $V_{CC} = V_{BB} = 5$ V, $R_C = 1$ kΩ, and $R_B = 180$ kΩ. Consider the temperature effects of I_{CBO}, V_{BE}, and β_F (refer to Fig. 4.31).

(a) Find I_C and V_{CE} at 25°C.
(b) Find I_C and V_{CE} at 125°C.
(c) Find I_C and V_{CE} at -55°C.

4.45 A given transistor has $\alpha_F = 0.99$ and avalanche breakdown voltage $BV_{CBO} = 75$ V. Collector current in the common-emitter configuration becomes very large at $V_{CB} = 35$ V.

(a) Find the multiplication constant n, as used in Eq. (4.46).
(b) Suppose $I_C = 1$ mA at $V_{CB} = 1$ V. At what value of v_{CB} will avalanche multiplication cause a 20% increase in collector current for this transistor if i_B is held constant?

4.46 The circuit shown in Fig. 4.56 uses silicon transistors with $\beta_F = 100$.

(a) In the steady state, the capacitor current is zero. Is Q_2 saturated? Is Q_1 off? Find i_{B2}, i_{C2}, v_{B2}, v_{C2}, v_{B1}, v_{C1}, and v_{CAP}.
(b) Suppose a narrow pulse of current into the base of Q_1 saturates it at $t = 0$. To what value will v_{C1} drop?
(c) To what value will v_{B2} drop? Note that Q_2 will turn off, and then current through the 1-kΩ + 10-kΩ resistors will hold Q_1 saturated.
(d) The capacitor will charge through the 15-kΩ resistor. Show that $v_{B2}(t) = 5 - 9\,e^{-t/150\,\mu s}$ V until the base of Q_2 reaches cut-in, $V_{BE} \sim 0.5$ V.
(e) What happens in this circuit when v_{B2} reaches cut-in? This circuit has applications in generating a pulse (v_{C2}) of a given length [$T = RC\,\ln(9.0/4.5)$] when Q_1 is triggered.

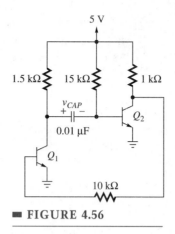

FIGURE 4.56

4.47 The transistors in the circuit shown in Fig. 4.57 are silicon and $\beta_F = 50$.

(a) Find V_{CE2} if both $I_{CBO} = 0$.
(b) Find V_{CE2} if both $I_{CBO} = 10$ nA.

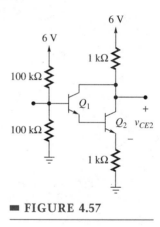

FIGURE 4.57

4.48 The transistor in Fig. 4.58 is silicon with $\beta_F = 100$. The Schottky diode has a forward voltage drop of 0.4 V. Find the currents I_B, I_C, and I_D. Is the transistor active or saturated?

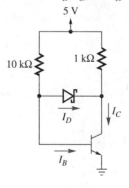

FIGURE 4.58

4.49 The input voltage v_{IN} for the silicon transistor in Fig. 4.59 alternates between 0 and 5 V. Assume $V_{BE(ON)} = 0.7$ V. When the v_{IN} switches to 0 from 5 V, there is a delay before

i_C decreases (Fig. 4.35). A small capacitor (called a *speed-up capacitor*) across the 10-kΩ resistor reduces this turnoff delay.

(a) Explain why this is so.
(b) Suppose it is found experimentally that a speed-up capacitor of 20 pF just reduces this delay time to zero. Estimate the stored charge in the saturated transistor.

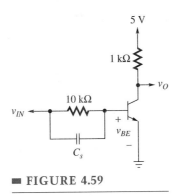

■ **FIGURE 4.59**

4.50 The quiescent point of a 2N2222A amplifier circuit is $I_C = 30$ mA and $V_{CE} = 5$ V. Assume no ac signal power. The device is operating in an ambient temperature of 25°C with no heat sink. Estimate the junction temperature and the case temperature. Use the data sheet (Fig. 4.32).

4.51 A small heat sink is added to the transistor of Problem 4.50, and the case temperature stabilizes at 40°C. Find the thermal resistance of this heat sink and the junction temperature.

4.52 Write a SPICE program to plot the curves of Fig. 4.25. Let $\beta_R = \alpha_R/(1 - \alpha_R) = 0.11$. *Hint:* Set the base current to 1 mA and let SPICE step the collector current.

4.53 Use SPICE to simulate the Schmitt trigger circuit of Fig. 4.55. Let v_{IN} increase from 0 to 4 V in 0.1-V steps, and plot v_o. In order for SPICE to converge on a high-gain circuit like this, the BJTs must exhibit realistic delays; use a model statement like

```
.MODEL BJT NPN IS=1E-15 TF=2N CJC=1P CJE=5P
```

You could also let v_{IN} be a piecewise-linear function in the time domain if you want to get a sweep from 0 to 4 V and back down to 0 in one plot.

4.54 Obtain the transfer characteristic for the BJT inverter of Fig. 4.30 by means of SPICE. Let v_{IN} increase from 0 to 4 V in 0.1-V steps, and plot v_{OUT}.

4.55 Write a SPICE program to simulate the monostable circuit of Fig. 4.56. Use a model statement for the BJTs as given in Problem 4.53, and to trigger the circuit, use a current pulse into the base of Q_1 something like

```
ITRIG 0 N PULSE(0 0.1M 2U 1U 1U 3U)
```

which will give a 0.1-mA current pulse starting from 0 mA, delayed 2 μs, rise time 1 μs, fall time 1 μs, and pulse width 3 μs. Your transient analysis must persist at least 150 μs to include most of the response. Plot the voltages on both collectors and both bases.

REFERENCES

1. Shockley, W., M. Sparks, and G. K. Teal. "The *pn* Junction Transistors," *Physical Review*, July 1951, pp. 151–162.
2. Ebers, J. J., and J. L. Moll, "Large-Signal Behavior of Junction Transistors," *Proceedings of the IRE*, December 1954, pp. 1761–1772.

3. Early, J. M. "Effects of Space-Charge Layer Widening in Junction Transistors," *Proceedings of the IRE,* November 1952, pp. 1401–1406.
4. Beaufoy, R., and J. Sparkes. "A Study of Charge-Controlled Parameters," *Proceedings of the IRE,* October 1960, pp. 1696–1705.
5. Hodges, D. A., and H. G. Jackson. *Analysis and Design of Digital Integrated Circuits,* 2nd ed. New York: McGraw-Hill, 1988, pp. 194–205.
6. Antognetti, P., and G. Massobrio. *Semiconductor Device Modeling with SPICE.* New York: McGraw-Hill, 1988, Chap. 3.

SUGGESTED READINGS

Hodges, D. A., and H. G. Jackson. *Analysis and Design of Digital Integrated Circuits,* 2nd ed. New York: McGraw-Hill, 1988. Chapter 6 contains a complete analysis of the switching time of a BJT inverter.

Streetman, B. G. *Solid-State Electronic Devices,* 4th ed. Englewood Cliffs, N.J.: Prentice-Hall, 1995. Chapter 7 contains a thorough treatment of transistor action.

Sze, S. M. *Physics of Semiconductor Devices,* 2nd ed. New York: John Wiley & Sons, 1981. Chapter 3 deals with all aspects of transistor action and various models.

Taub, H., and D. Schilling. *Digital Integrated Electronics.* New York: McGraw-Hill, 1977. Chapter 1 contains an excellent treatment of saturation.

CHAPTER 5

THE FIELD-EFFECT TRANSISTOR

5.1 Notation and Symbols
5.2 MOSFET Operation
5.3 MOSFET Specifications
5.4 MOS Circuits
5.5 SPICE Model for the MOSFET
5.6 JFET Operation
5.7 JFET Specifications
5.8 FET Small-Signal Models
5.9 JFET Circuits
5.10 SPICE Model for the JFET
5.11 MESFET Operation
Summary
Survey Questions
Problems
References
Suggested Readings

As we mentioned in Chapter 2, the principle of using an electric field to control current flow was introduced well before the development of the bipolar junction transistor. However, practical devices based on field-effect principles were not commercially available until the mid-1960s. These devices are called *field-effect transistors* (FETs) and operate as voltage-controlled amplifiers. FETs are referred to as unipolar devices; that is, the conduction process uses either electrons or holes, not both as in a BJT.

The FET is used in both analog and digital circuits (Fig. 5.1). We will show that the FET has a very high input impedance that enhances its function as a good voltage amplifier. The FET is also used extensively as the basic memory or logic element in digital integrated circuits (ICs). In general, FET devices are used as amplifiers and switches and in nonlinear circuits such as mixers. FETs fabricated from gallium arsenide, in several different morphologies, are used extensively in radio-frequency (rf) amplifiers and in microwave circuits. This wide range of applications does not imply that the FET will replace the BJT. As we will soon see, it really means that the characteristics are sufficiently different to provide the designer with an option not easily available with a BJT. Indeed, a growing number of ICs include both FETs and BJTs in the same circuit (BiMOS or BiCMOS technology) to take advantage of the key features of each.

A number of different types of FETs are available. Their names reflect distinctive features of their fabrication processes or geometries. Figure 5.2 shows a "family tree" of such devices as fabricated in silicon and in GaAs for microwave applications and for high-speed digital circuits. The first commercially available FET was the *junction field-effect transistor* (JFET), which was first analyzed by Shockley[1-3] in 1952 as basically a voltage-controlled resistor. There are two types of JFETs: *n channel* and *p channel*. Because the mobility for electrons in silicon is about three times that of holes in silicon (refer back to Table 2.2), *n*-channel devices are about three times faster than

■ **FIGURE 5.1** Examples of packaged integrated circuits incorporating FETs. (1) Tektronix Application-Specific Integrated Circuit (ASIC) with JFETs and BJTs. 28 pin ceramic (Dual-In-Line Package) DIP. (2) 4000 series CMOS IC in a 14-pin plastic DIP. (3) Discrete JFET in a TO-5 package. (4) Hybrid IC incorporating MOS, BJT, and passive element die within a single flat ceramic package. Top cover has been removed. (5) 4000 series CMOS IC in a TO-5 package with the cover removed. (6) 4000 series CMOS IC in a 14-pin plastic DIP showing the internal die placement and before the lead frame has been trimmed to obtain a completed IC as shown in Item 2. (7) IBM 4 Megabit MOS memory die before packaging.

p-channel devices and have other advantages as we will soon observe. The JFET, as its name implies, uses a *pn* junction to define its operation regions. If, instead of a *pn* junction, we use a metal-semiconductor junction, the result is a ***metal-semiconductor field-effect transistor*** (MESFET). Currently, the MESFET, fabricated in GaAs, is used almost exclusively in microwave applications and very high-speed digital processing. This particular technology was developed in the mid-1960s, but was not commercially exploited until the mid-1970s.

Another key FET structure is the ***metal-oxide semiconductor field-effect transistor*** (MOSFET), somtimes referred to as an ***insulated-gate FET*** (IGFET). The oxide referred to here is usually an extremely thin layer of silicon dioxide, SiO_2. The MOSFET fabrication procedures lend themselves to relatively high-yield, high-density ICs suitable for the large-scale integration (LSI) and very large-scale integration (VLSI) of memory and related microprocessor circuits. It is the dominant device for these types of integrated circuits.

The MOSFET category is divided into a number of subdivisions that reflect key fabrication features. These are:

- PMOS (*p*-channel MOS)
- NMOS (*n*-channel MOS)
- CMOS (complementary MOS)

■ The MOSFET, along with the BJT, is one of the most important semiconductor devices.

- BiMOS or BiCMOS (mixed-mode, bipolar, and MOS)
- HMOS (high-performance MOS)
- DMOS/DIMOS (double-diffused MOS)
- VMOS (vertical MOS)
- SOS/SOI (silicon-on-saphire/silicon-on-insulating substrate)
- HEMT (High-electron-mobility transistor)

Each of the technologies has certain technical and economic advantages. For example, CMOS devices are widely used for medium-speed digital high-density applications and memories. SOS/SOI devices offer very high-speed performance in conjunction with minimal performance degradation under exposure to ionizing radiation. BiMOS and BiCMOS, so called mixed-mode, is a process used to exploit the advantages of BJT and MOS technologies on the same integrated circuit. JFETs offer low-noise performance, and thus are often used as input amplifier stages in high-gain operational amplifiers. Capitalizing on the five times increase in electron mobility in GaAs as compared to Si, GaAs MESFET-based digital ICs are finding their way into very high-speed systems for primarily military and space applications.

After a preliminary discussion about notation and symbols, we will consider the MOSFET. Although JFET operation is a bit easier to explain, MOSFETs, by far, are predominant in the industry; consequently, we focus on the MOSFET. JFET theory will also serve as the basis for a qualitative discussion of the GaAs MESFET. Several basic amplifier and switch applications are introduced with additional applications presented in subsequent chapters.

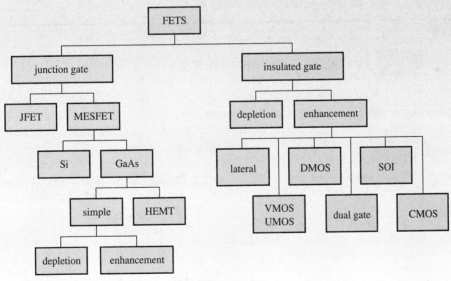

■ FIGURE 5.2 Family tree of field-effect transistors.

IMPORTANT CONCEPTS IN THIS CHAPTER

- Topology and technology-related acronyms and symbols
- Structure of a basic MOSFET, associated biasing conventions
- Regions of MOSFET operation—ohmic, subthreshold, cutoff, saturation
- Difference in biasing for enhancement- and depletion-mode MOSFETs
- Definition of $C_{ox} = (\varepsilon_o \varepsilon_r)/d$ as related to MOSFET topology
- Derivation of the MOS ohmic and saturation region i_D-v_{DS} characteristics
- Definition and interpretation of key terms on a MOSFET data sheet and the derivation and definition of MOSFET transconductance, $g_m = 2k(V_{GS} - V_t)$
- Difference between PMOS and NMOS operation, biasing, and characteristic curves
- Analysis and design of a basic MOS inverter with a passive (resistive) load, an actively-loaded MOS inverter, a CMOS inverter, and MOS-based analog switches
- SPICE MOSFET model and syntax for MODEL and ELEMENT statements
- Qualitative description of JFET operation and regions of JFET operation—ohmic, cutoff, saturation
- Derivation of the JFET ohmic and saturation region i_D-v_{DS} characteristics
- Definition and interpretation of key terms on a JFET data sheet and the derivation and definition of JFET transconductance
- Interpretation of the λ term on the i_D-v_{DS} characteristics
- Drain resistance definition and its measurement from the i_D-v_{DS} characteristics
- JFET-based common-source, common-gate, and common-drain amplifiers
- SPICE JFET model-syntax for MODEL and ELEMENT statements
- Qualitative topology, description, and model of the MESFET

5.1 NOTATION AND SYMBOLS

The three external connections to any type of discrete FET are the **source** (*S*), **gate** (*G*), and **drain** (*D*) (see Fig. 5.3). If we consider MOSFETs as a component in an integrated circuit, we also show a connection for the **substrate**, also often called the **body** or **bulk** (*B*). Often the substrate and source are connected together but not always. The entire FET or multiple-device IC structure is hermetically sealed in an epoxy, ceramic, or metal package, as shown earlier in Fig. 5.1. As in the BJT, the package is designed to provide mechanical protection and appropriate heat transfer and to exclude contaminants such as moisture.

■ Although there are a few applications using discretely packaged FETs, most FETs are integral to VLSI or ULSI circuits. These often include over 100,000 to 10^6+ FETs on a single IC die.

The terminal voltages and currents and their polarity using double subscripting are also illustrated in Fig. 5.3. The order of the double subscripts implies the polarity indicated by the signs. Whether the actual voltage or current is positive or negative is determined by the circuit external to the FET. Terminal currents are defined into the FET terminals in both cases, so each current must carry its own sign. In this respect, the double-subscript voltage and current conventions are identical to those of the BJT.

The symbols used for the FET are somewhat more complex than for the BJT because the conventions developed over the years have tended to distinguish the type of FET by its symbol and circuit application. Figure 5.3 presents a summary of some of the different

CHAPTER 5 THE FIELD-EFFECT TRANSISTOR

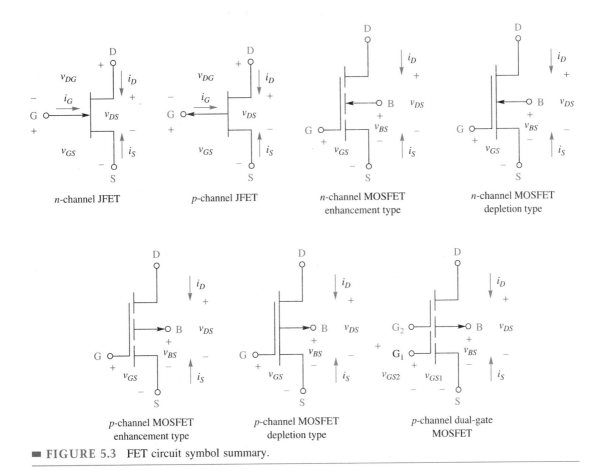

■ **FIGURE 5.3** FET circuit symbol summary.

types of symbols. Considerable variations occur with MOS devices, especially in multiple-device ICs. As we come across a variation, we will note it. In JFET symbols, the gate connection is sometimes placed closer to the source end, although the drain and source are, in most cases, interchangeable.

> ### CHECK UP
>
> 1. **TRUE OR FALSE?** A MOSFET is most accurately represented with three terminals.
> 2. **TRUE OR FALSE?** There are several accepted MOSFET schematic symbols.
> 3. Label the terminal voltages and currents for an NMOS, PMOS, and n-channel JFET.

5.2 MOSFET OPERATION

The key idea in any FET is to control the drain current, i_D, using the gate-source voltage, v_{GS}. In the MOSFET, the voltage applied to the gate, v_G, which is separated from the semiconductor by a thin silicon dioxide (SiO_2) layer, controls charge movement, i_D, within the semiconductor because of an induced electric field.

■ Refer to Chapter 18 to more thoroughly understand how fabrication technology impacts MOSFET topologies.

Enhancement Mode

An idealized *n*-channel MOSFET (NMOS) is illustrated in Fig. 5.4, along with its preferred schematic symbol. As shown in the figure, the source and drain consist of two *n*-type regions in a *p*-type substrate. These regions are separated by a distance L, which we call the *gate length*. A thin layer of thickness d (less than 500 Å and as thin as 50 to 150 Å) of SiO_2, a good electrical insulator, is formed over the intervening region. Originally, in the 1970s, a thin layer of aluminum was deposited on top of this oxide layer, forming a capacitor, and this electrode forms the gate contact. Current technology uses heavily doped polycrystalline silicon as the gate "metal" conductor although the "M" in MOSFET remains the device designation of choice. Additional fabrication details, especially with respect to the gate-electrode formation are presented in Chapter 18. The schematic symbol includes the (1) *source*, (2) *gate*, (3) *drain*, and (4) *substrate* connections.

Our initial goal is to develop a qualitative step-by-step picture of MOSFET operation. We then use this intuitive picture to obtain a small-signal model, along with the i_D-v_{DS} terminal characteristics. A small sample of typical MOSFET circuit topologies is also presented.

As illustrated in Fig. 5.5(a), we will apply a voltage between the drain and source, V_{DS}. Similarly, we will apply a voltage between the gate and the source, V_{GS}. A series of sketches can now be developed to illustrate the formation of the conducting channel and the resultant i_D-v_{DS} terminal characteristic.

If $V_{GS} = 0$ and we adjust $V_{DS} > 0$, the drain and source are effectively isolated from each other because the diode junction formed is reverse-biased and there are essentially no charge carriers (electrons) in the intervening *p*-type material [see Fig. 5.5(a)]. This NMOS FET is now in the OFF state, $I_D = 0$. The result is characteristic curve A in Fig. 5.6.

Now suppose $V_{DS} = 0$ and $V_{GS} > 0$ [Fig. 5.5(b)]. The positive potential will attract electrons that accumulate just under the oxide. At first a depletion layer forms; when V_{GS} is increased further, the free-electron concentration becomes larger than the hole concentration. This occurs in a very thin layer of silicon, just under the oxide, and is called **inversion**. Inversion creates a thin conducting *n* region called the **channel**, between

■ Although it may be confusing to the student new to the topic, the symbol

is sometimes used where the drain, source, gate, and assumed substrate connection are to be inferred from the application schematic diagram.

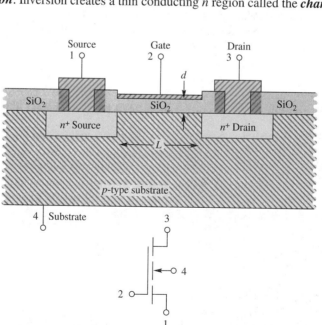

■ **FIGURE 5.4** NMOS (*n*-channel MOSFET) configuration.

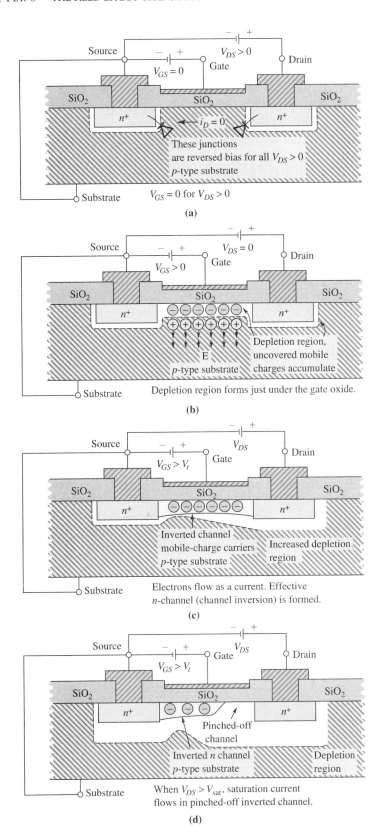

■ **FIGURE 5.5** NMOS (*n*-channel MOSFET) configuration.

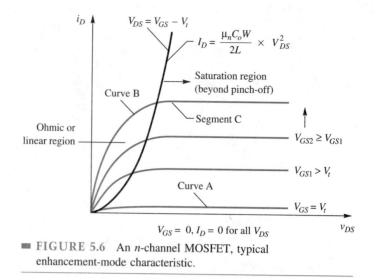

FIGURE 5.6 An *n*-channel MOSFET, typical enhancement-mode characteristic.

■ Do not confuse V_t, the threshold voltage for a MOSFET, with the $V_T = kT/q$ notation used for diodes and BJTs!

the *n*-type source and *n*-type drain. The gate potential V_{GS} at which inversion produces an *n* concentration equal to the unbiased *p* concentration is called the *threshold* voltage, V_t, [Fig. 5.5(c)]. In order for inversion to occur at a reasonable value of V_t, the substrate must be lightly doped.

When V_{DS} is increased with $V_{GS} > V_t$, the depletion layer at the drain will expand, and if V_{DS} is large enough the depletion layer will cut off the channel. This condition is called **pinch-off** [Fig. 5.5(d)]. Conduction from the source to drain still occurs, however, with current passing through the depletion layer just as it does in the reverse-biased collector-base junction of a bipolar transistor. When V_{DS} is small (below pinch-off) the drain current increases with V_{DS}. This represents the **ohmic** or nonsaturation region of operation (curve B in Fig. 5.6). Above pinch-off, the additional voltage is dropped across the depletion layer and the drain current **saturates**; that is, the current becomes a constant.

■ Operating in *saturation* for a FET is distinctly different than operating in *saturation* for a BJT!

There is another intuitive way to consider the pinch-off condition. At first guess it would seem reasonable that at pinch-off, i_D would approach zero, but actually i_D just levels off at some constant value when v_{DS} is above the pinch-off voltage. If $i_D = 0$, then the voltage along any segment of the channel would be zero, and $v_{DS} = 0$. This contradicts the original assumption. If one were to visualize a complete closure of the channel, carriers are essentially being swept through the depletion region in much the same way that carriers are swept through the collector-base depletion region in a BJT.

Because the drain current is increased as the magnitude of V_{GS} increases beyond V_t for $V_t > 0$, this type of MOSFET is called the **enhancement type**. Another term sometimes used in reference to a MOSFET device is **crossed-field** device since the electric field between the drain and source is at right angles to the gate-induced electric field.

There are some second-order effects whose detailed study and application are beyond the scope of this text. For example, the channel becomes very weakly *n*-type when V_{GS} is slightly less than V_t and a drain current in the nanoampere to tens of nanoampere range can be supported. Normally, this effect is of no interest and is ignored, but this so-called **subthreshold** operation is exploited for low-voltage battery-operated circuits, such as digital watches, which are required to operate at very low currents and voltages. A degradation in performance, predominantly switching speed, results, but the performance is adequate for these low-power applications.

CHAPTER 5   THE FIELD-EFFECT TRANSISTOR

Another second-order effect is that the mobility of the charge carriers in the channel of a MOSFET is usually less than that found in bulk silicon because the charge carriers travel in a very thin layer just under the oxide, i.e., at the oxide-silicon interface. This interface is not perfect, although the semiconductor industry spends large sums of money to keep the interface smooth with minimal defects. However, charge-carrier collisons with interface defects reduce the mobility in addition to the bulk silicon effects of charge-carrier collisions with the lattice atoms.

Still another second-order effect occurs when the channel is very short, such as those found in high-speed MOS or rf power devices. Typically, a short channel means less than 0.75 to 1 μm. For long channels, the drift velocity is proportional to the electric field, Eq. (2.2); thus, the current increases linearly with applied voltage. The proportionality between drift velocity and electric field is called the *mobility*, and so far, we have considered it to be constant.

At the higher electric fields present in a short channel, the carrier velocity *saturates*, which appears as a decrease in mobility. Therefore, Eq. (2.2) becomes $v = \mu(\mathbf{E})\mathbf{E}$, where $\mu(\mathbf{E})$ is illustrated as the continually changing slope of the graph in Fig. 5.7. When the carrier velocity saturation becomes significant, the square law model to be presented becomes more nearly linear, and both the saturation current and resultant output resistance decreases. This velocity saturation and resultant mobility change have been intensively studied.[4,5] The entire field of channel behavior in FETs, especially the operation of very short channels resulting from advances in microelectronics technology, is an active research topic.

Drain Characteristics. Using basic electrostatic concepts, we outline a simplified derivation yielding an approximate expression for the MOSFET i_D versus v_{GS} characteristic. We will use Fig. 5.8, which represents a rescaled and relabeled version of the n-channel MOSFET shown in Fig. 5.4, to support this derivation.

Suppose $V_{DS} = 0$, and $V_{GS} > V_t$ is applied. An electric charge $Q(x)$ per unit length (coulombs per meter) is induced along the channel and, for the simple geometry in Fig. 5.8, is given by

$$Q(x) = -C_{ox}W(v_{GS} - V_t), \qquad (5.1)$$

where C_{ox} is the capacitance per unit area for a parallel-plate capacitor given by

$$C_{ox} = \frac{\varepsilon_o \varepsilon_r}{d}, \qquad (5.2)$$

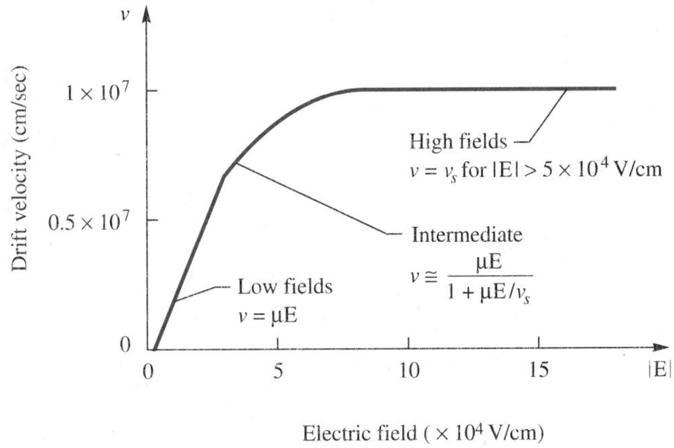

■ **FIGURE 5.7** Velocity saturation.

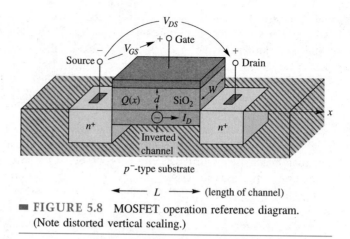

FIGURE 5.8 MOSFET operation reference diagram. (Note distorted vertical scaling.)

where d is the oxide thickness and $\varepsilon_r = 3.9$ is the relative dielectric current for SiO_2. The total gate-to-channel capacitance is $C_{ox} \times L \times W$. Recall that the permittivity of free space is given by $\varepsilon_o = 8.854 \times 10^{-14}$ F/cm.

DRILL 5.1

Oxide thicknesses as low as 50—150 Å are used in contemporary technologies. Compute the capacitance for a $d = 150$-Å-thick oxide with $L = 5$ μm and $W = 100$ μm.

ANSWER 1.15 pF

EXAMPLE 5.1

Compute C_{ox} for a 500-Å SiO_2 dielectric. Then compute the total capacitance for an $L = 5$-μm-long gate of width 50 μm.

SOLUTION From Eq. (5.2), we have

$$C_o = \frac{\varepsilon_o \varepsilon_r}{d} = [(8.854 \times 10^{-14} \text{ F/cm}) \times 3.9]/500 \times 10^{-8} \text{ cm} = 6.9 \times 10^{-8} \text{ F/cm}^2.$$

Then $C = C_o \times L \times W = (6.9 \times 10^{-8} \text{ F/cm}^2) \times (5 \times 10^{-4} \text{ cm}) \times (50 \times 10^{-4} \text{ cm}) = 0.17$ pF.

Implicit in Eq. (5.2) is the fact that in order to induce a capacitance, the resultant electric field must terminate on negative charges within the silicon, just below the SiO_2 layer. Thus, there is an effective n-layer inversion from the lightly p-doped substrate. Effectively, we have formed a conducting channel for electrons between heavily n-doped, n^+ drain and source contacts.

The parameter V_t, which we previously designated as a threshold voltage or turn-on voltage, is also dependent on the interface-trapped electric charge between the silicon and SiO_2 as well as substrate doping density. The actual value of V_t is significantly affected by fabrication technology.

As we might expect, if we now apply a $V_{DS} > 0$, the voltage will increase along the length of the gate. This voltage at any point along the gate is given as $V(x)$. Then Eq. (5.1) becomes

$$Q(x) = -C_{ox}W[v_{GS} - V_t - V(x)]. \tag{5.3}$$

Recall from Eq. (2.14) that the conductivity is given by $\sigma = nq\mu$. Applying Eq. (5.3) and realizing that

$$nq = \frac{-Q(x)}{A} \tag{5.4}$$

yields

$$\sigma = \frac{\mu C_{ox} W}{A}[v_{GS} - V_t - V(x)], \quad (5.5)$$

where μ is an electron mobility and A is the inverted-channel cross-sectional area. By implication, this result is valid for a channel that is not yet pinched off.

Combining Eqs. (2.13) and (5.5), we have

$$-i_D = JA = \sigma EA = \mu C_{ox} W[v_{GS} - V_t - V(x)]E. \quad (5.6)$$

The electric field along the channel is given by

$$E_x = -\frac{dV(x)}{dx}. \quad (5.7)$$

Substituting Eq. (5.7) into Eq. (5.6) yields

$$i_D = +\mu C_{ox} W[v_{GS} - V_t - V(x)]\frac{dV(x)}{dx}. \quad (5.8)$$

We integrate Eq. (5.8) to obtain

$$\int_0^L i_D \, dx = i_D L = \int_0^{v_{DS}} \mu C_{ox} W[v_{GS} - V_t - V(x)] \, dv(x). \quad (5.9)$$

Performing the integration yields the **ohmic** region equation

$$i_D = \frac{\mu C_{ox} W}{2L}[2(v_{GS} - V_t)v_{DS} - v_{DS}^2] \quad v_{DS} \le v_{GS} - V_t. \quad (5.10)$$

Because the inversion layer is present when $v_{GS} - v_{DS} \ge V_t$, the boundary between the ohmic and saturation regions is given by setting $v_{GS} - V_t = v_{DS}$ in Eq. (5.10) to obtain

$$i_D = \frac{\mu C_{ox} W}{2L} v_{DS}^2. \quad (5.11)$$

This function plots as a quadratic as illustrated in Fig. 5.6. In the **saturation** region, Eq. (5.10) becomes

$$i_D = \frac{\mu C_{ox} W}{2L}(v_{GS} - V_T)^2 = k(v_{GS} - V_t)^2 \quad v_{DS} > v_{GS} - V_t, \quad (5.12)$$

when $v_{GS} - V_t$ is substituted for v_{DS} and $k = \left(\frac{\mu C_{ox}}{2}\right)\left(\frac{W}{L}\right)$.

■ We will see how Eqn. (5.12) defines a Level 1 SPICE model.

The symbol k is a device-specific constant. Equation (5.12) is very important because it is used as the fundamental equation to model the MOSFET in the saturation region, which is analogous to the linear-active region in the BJT.

Even though Eq. (5.12) comes from a simplified physical representation for the MOSFET, it works reasonably well for many calculations.[6] More complex models are presented in the literature.[7,8] Variations of these more complex models are used in the SPICE program. Indeed, work on MOSFET device theory and secondary fabrication-induced effects is an active research and development area, especially for devices with submicron geometries.

As you might infer, the operation of the *p*-channel MOSFET (PMOS) can be similarly explained by reversing the dopant classification and the polarity signs on the voltages and currents.

Historically, the PMOS transistor was developed first with large-scale production after 1967.[9,10] It turns out, however, that the *n*-channel MOSFET is more widely used because the mobility for electrons is between 2 and 3 times that for holes. That means that the

ON resistance, R_{ON}, for a p-channel silicon MOSFET will be greater than twice the ON resistance for an n-channel silicon MOSFET. The economic implication is that to achieve a given ON resistance, the n-channel MOSFET occupies only half the area of the p-channel MOSFET. There are some alternative technologies to reduce R_{ON}, which we mention in Chapter 18. Along with this reduction in size and increased packing density, one achieves lower capacitance, which yields higher switching speeds in digital circuits. Another key feature is that n-channel MOSFET operating voltages are compatible with bipolar TTL 0-V to +5-V logic levels (see Chapter 13).

■ A more recent trend has a 3.3 volt standard.

EXAMPLE 5.2

Suppose we measure an n-channel enhancement MOSFET and find that $V_t = 2$ V and that with $V_{DS} = V_{GS} = 5$ V, $I_D = 3$ mA. Find the value of V_{DS} where saturation occurs, and calculate I_D when $V_{DS} = 1$ V.

SOLUTION From Eq. (5.12), 3 mA $= \mu C_{ox}(W/2L)(5 - 2)^2$ if $v_{DS} > 5$ V $- 2$ V. Thus the FET is saturated at $V_{DS} > 3$ V, and $\mu C_{ox}(W/2L) = 3/9 = 0.33$ mA/V^2. At $V_{DS} = 1$ V, Eq. (5.10) yields $I_D = 0.33 [2(5 - 2)1 - 1^2] = 1.67$ mA.

DRILL 5.2

On Fig. 5.9, sketch the I_D versus V_{DS} characteristics for Example 5.2 for $V_{GS} = 2, 3, 4,$ and 5 volts.

ANSWER

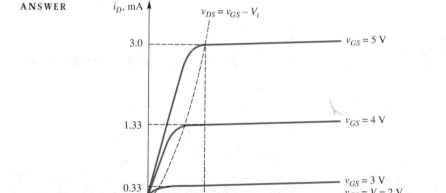

■ FIGURE 5.9

Depletion Mode

In the previous section, we discussed the enhancement-mode MOSFET in which, for an n-channel device, an increase in $V_{GS} > V_t$ resulted in an increase in I_D according to Eq. (5.10) or Eq. (5.12). By reworking the MOSFET fabrication topology, it is also possible to obtain a conducting channel for $V_{GS} = 0$. Therefore, for $V_{GS} < 0$, I_D decreases. This is called **depletion-mode operation** in a MOSFET. The modified topology is illustrated in Fig. 5.10. Let us examine the structure and operation of an

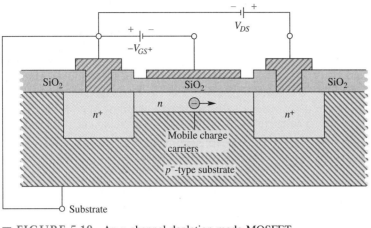

FIGURE 5.10 An *n*-channel depletion-mode MOSFET.

n-channel depletion-mode MOSFET. As illustrated, the source and drain consist of two *n*-type regions in a *p*-type substrate. A thin *n*-type layer connects between the two *n*-type regions. A thin <500-Å SiO$_2$ layer is formed over this intervening region. A thin layer of heavily doped polysilicon is deposited on top of this oxide layer, and this electrode forms the gate contact.

If $v_{GS} = 0$ and $v_{DS} > 0$, an electron current will flow in the thin *n* channel. This is represented as curve A shown in Fig. 5.11. If v_{GS} is made negative, holes are attracted into the *n* channel, effectively reducing the conductivity of the channel. Because the number of majority carriers, electrons in this case, is reduced by recombination, I_D is reduced. This is curve B in Fig. 5.11. Thus, this device is known as a ***depletion-mode MOSFET***. Conversely, in the same manner of our description of NMOS enhancement-mode operation, we can operate this device in the enhancement mode by making $V_{GS} > 0$. The *n* channel becomes more conductive by the introduction of electrons as a reverse-bias leakage current. This feature is illustrated as curve C in Fig. 5.11. The drain characteristics used to describe the depletion-mode MOSFET are given by Eqs. (5.10) and (5.12), shifting V_t to the negative voltage required to obtain $I_D = 0$.

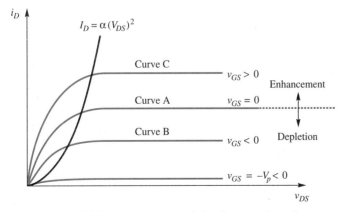

FIGURE 5.11 Typical characteristics for an *n*-channel depletion-mode MOSFET.

CHECK UP

1. Sketch and label cross sections of an NMOS and PMOS transistor.
2. **TRUE OR FALSE?** Operating in the saturation region of a MOSFET is the same as operating in the forward-active region of a BJT.
3. **TRUE OR FALSE?** V_{GS} must exceed a certain threshold value before significant drain current will flow in a MOSFET.
4. Explain qualitatively what happens to the depletion region in a MOSFET when V_{DS} is increased for a given $V_{GS} > V_t$.
5. Compute C_{ox} for typical SiO_2 thicknesses.
6. Explain the difference between a depletion-mode and enhancement-mode MOSFET using typical characteristic curves.
7. **TRUE OR FALSE?** I_D is a linear function of V_{GS} in the saturation region.

5.3 MOSFET SPECIFICATIONS

An engineer uses manufacturer's data sheets when creating a design. These data sheets, whether available as printed documents, CD ROMS, diskettes, or as part of the data bank and design rules in a computer-aided design (CAD) system, provide considerable useful information. These data are often buried in the device models that are called on in a large circuit or system design consisting of thousands of individual devices. The data sheets also illustrate the differences in notation used throughout the industry.

■ The list of MOS specifications are usually resident in a technology-specific computer design package.

Figure 5.12 shows an abbreviated data sheet for the Motorola 3N169-170-171 *n*-channel enhancement-mode MOSFET. Its recommended use is as a low-power switch, but it also exhibits good small-signal amplifier characteristics.

1. As expected, the data sheet starts off with a number of maximum ratings that are similar to those found when we studied BJTs in Chapter 4. However, one added precaution must be observed here. Because of the very thin SiO_2 layer under the gate electrode, it is very susceptible to destructive breakdown from the buildup of electrostatic charge during handling.
2. The **gate-leakage current**, I_{GSS}, represents the current through the SiO_2 and is very small. Observe that we are talking about picoamperes (1 pA = 10^{-12} A) instead of nA to μA as in BJTs. Consequently, we are often able to ignore I_G in the MOSFET.
3. The term I_{DSS} is the **OFF state current** with $V_{GS} = 0$. Note the strong temperature characteristics.
4. The **gate-source threshold voltage**, $V_{GS(Th)}$, is what we have been calling V_t.
5. The term $I_{D(ON)}$ represents the **ON drain current** and is the current specified at some reasonable but arbitrary value of V_{GS}. A more useful elaboration of this parameter is illustrated on the I_D versus V_{GS} transfer characteristics.
6. Directly related to $I_{D(ON)}$, is $V_{DS(ON)}$, the **drain-source ON voltage**. This is the condition for ensuring operation in the saturation region; that is, $v_{DS} = v_{GS} - V_t$. More information is presented graphically in the ON drain-source resistance graph.
7. The y_{fs} term is defined as the transconductance often given as g_m and defined as

■ See also the *y*-parameter model in Fig. 5.35.

$$g_m = \left.\frac{\partial i_D}{\partial v_{GS}}\right|_{Q\text{-point}}. \qquad (5.13)$$

CHAPTER 5　THE FIELD-EFFECT TRANSISTOR

MAXIMUM RATINGS

Rating	Symbol	Value	Unit
Drain-Source Voltage	V_{DS}	25	Vdc
Drain-Gate Voltage	V_{DG}	= 35	Vdc
Gate-Source Voltage	V_{GS}	= 35	Vdc
Drain Current	I_D	30	mAdc
Total Device Dissipation @ $T_A = 25°C$ Derate above 25°C	P_D	300 1.7	mW mW/°C
Total Device Dissipation @ $T_C = 25°C$ Derate above 25°C	P_D	800 4.56	mW mW/°C
Junction Temperature Range	T_J	175	°C
Storage Temperature Range	T_{stg}	−65 to +175	°C

ELECTRICAL CHARACTERISTICS ($T_A = 25°C$ unless otherwise noted.)

Characteristic	Symbol	Min	Max	Unit
OFF CHARACTERISTICS				
Drain-Source Breakdown Voltage ($I_D = 10$ μAdc, $V_{GS} = 0$)	$V_{(BR)DSX}$	25	—	Vdc
Zero-Gate-Voltage Drain Current ($V_{DS} = 10$ Vdc, $V_{GS} = 0$) ($V_{DS} = 10$ Vdc, $V_{GS} = 0$, $T_A = 125°C$)	I_{DSS}	— —	10 1.0	nAdc μAdc
Gate Reverse Current ($V_{GS} = -35$ Vdc, $V_{DS} = 0$) ($V_{GS} = -35$ Vdc, $V_{DS} = 0$, $T_A = 125°C$)	I_{GSS}	— —	10 100	pAdc
ON CHARACTERISTICS				
Gate Threshold Voltage ($V_{DS} = 10$ Vdc, $I_D = 10$ μAdc)　　3N169 　　　　　　　　　　　　　　　　　3N170 　　　　　　　　　　　　　　　　　3N171	$V_{GS(Th)}$	0.5 1.0 1.5	1.5 2.0 3.0	Vdc
Drain-Source On-Voltage ($I_D = 10$ mAdc, $V_{GS} = 10$ Vdc)	$V_{DS(on)}$	—	2.0	Vdc
On-State Drain Current ($V_{GS} = 10$ Vdc, $V_{DS} = 10$ Vdc)	$I_{D(on)}$	10	—	mAdc
SMALL-SIGNAL CHARACTERISTICS				
Drain-Source Resistance ($V_{GS} = 10$ Vdc, $I_D = 0$, $f = 1.0$ kHz)	$r_{ds(on)}$	—	200	Ohms
Forward Transfer Admittance ($V_{DS} = 10$ Vdc, $I_D = 20$ mAdc, $f = 1.0$ kHz)	$|y_{fs}|$	1000	—	μmhos
Input Capacitance ($V_{DS} = 10$ Vdc, $V_{GS} = 0$, $f = 1.0$ MHz)	C_{iss}	—	5.0	pF
Reverse Transfer Capacitance ($V_{DS} = 0$, $V_{GS} = 0$, $f = 1.0$ MHz)	C_{rss}	—	1.3	pF
Drain-Substrate Capacitance ($V_{D(SUB)} = 10$ Vdc, $f = 1.0$ MHz)	$C_{d(sub)}$	—	5.0	pF
SWITCHING CHARACTERISTICS				
Turn-On Delay Time　　　($V_{DD} = 10$ Vdc, $I_{D(on)} = 10$ mAdc,	$t_{d(on)}$	—	3.0	ns
Rise Time　　　　　　　　$V_{GS(on)} = 10$ Vdc, $V_{GS(off)} = 0$,	t_r	—	10	ns
Turn-Off Delay Time　　　$R_G = 50$ Ohms)	$t_{d(off)}$	—	3.0	ns
Fall Time　　　　　　　　See Figure 1	t_f	—	15	ns

■ **FIGURE 5.12** 3N169-171 *n*-channel enhancement-mode MOSFET data sheet (courtesy Motorola, Inc.).

■ Recall, mho has been replaced by Siemen.

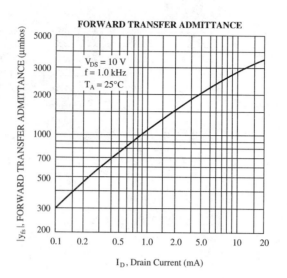

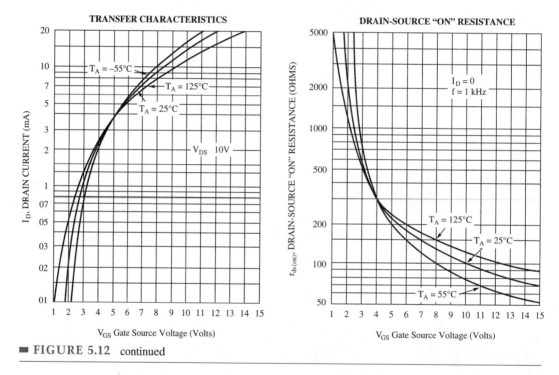

■ **FIGURE 5.12** continued

■ Compare the MOSFET g_m equation with the BJT g_m equation as in Chapter 4.

If we apply Eq. (5.12) to Eq. (5.13), we obtain the transconductance for MOSFET as

$$g_m = \left.\frac{\partial[k(v_{GS} - V_t)^2]}{\partial v_{GS}}\right|_{Q\text{-point}} = 2k(V_{GS} - V_t). \quad (5.14)$$

More detailed information is provided in the forward transfer admittance curve.

8. The switching characteristics and definitions are equivalent to those given in Chapter 4 for the BJT. Again, the tabular information is augmented by additional graphical information.

CHAPTER 5 THE FIELD-EFFECT TRANSISTOR

EXAMPLE 5.3

Compute a value for k for the 3N170 MOSFET using the value for y_{fs} at a $V_{DS} = 10$ V at 25° C.

SOLUTION From Fig. 5.12, observe the minimum value of $V_{GS(th)} = V_t = 1.0$ V. From the forward transfer admittance curve we can pick a reasonable value of I_D. In this case note that at $I_D = 2$ mA, $g_m = y_{fs} = 1500$ μmhos $= 1.5 \times 10^{-3}$ S. Using the transfer characteristics curve, for $I_D = 2$ mA, $V_{GS} = 4.2$ V. Then using Eq. (5.14), $k = g_m/2(v_{GS} - V_t) = 1.5 \times 10^{-3}/2(4.2 - 1.0) = 2.3 \times 10^{-4}$ A/V^2.

DRILL 5.3

Verify that the value of k computed in Example 5.3 can be used to generate values for $y_{fs} = g_m$ on the forward transfer admittance curve for $V_{GS} = 2$, 3, and 5 V.

ANSWER $g_m(V_{GS} = 2$ V$) = 0.46$ mS and $I_D = 0.23$ mA, $g_m(V_{GS} = 3$ V$) = 0.92$ mS and $I_D = 0.92$ mA, $g_m(V_{GS} = 5$ V$) = 1.8$ mS and $I_D = 3.7$ mA.

These values are verified by comparing the forward transfer admittance and transfer characteristics in Fig. 5.12.

CHECK UP

1. **TRUE OR FALSE?** Transconductance increases linearly with V_{GS}.
2. Define V_t, I_{GSS}, $I_{D(ON)}$, and $y_{fs} = g_m$.
3. **TRUE OR FALSE?** Absolutely no drain current flows when $V_{GS} < V_t$.
4. **TRUE OR FALSE?** Most FET parameters are temperature independent.

5.4 MOS CIRCUITS

One of the principal uses of MOS devices is as logic switches in digital circuits. MOS logic gates are often preferred because of their lower power dissipation and smaller size as compared to the BJT. Both of these factors yield denser circuits, which is important from an economic viewpoint. To illustrate these applications, we analyze the MOS device in its use as an inverter with passive and active loads and as an analog switch.

Differences in the symbolic notation exist in the literature. For instance, Fig. 5.13 illustrates the different symbols for an n-channel MOSFET. Figure 5.13(a) is an enhancement-mode device in which the substrate is connected to the source internally. Depletion-mode operation is represented by Fig. 5.13(b). An alternative symbol places an arrow in the direction of positive current flow in the source lead, [see Fig. 5.13(c)]. Because the source and substrate are almost always connected, the symbol shown in Fig. 5.13(d) can be used. These symbols have been drawn with the gate lead asymmetrically located toward the source end. Often the gate lead is symmetrically positioned, as in Fig. 5.13(e), and if there is no confusion in the circuit diagram, the source arrow is omitted, as in Fig. 5.13(f). In the following circuit analyses, we use the symbols shown in Fig. 5.13(a) and (b), which are commonly used because they provide the most information about the FET. Of course, for a p-channel device, the arrow directions are reversed.

■ The INTEL Pentium IC (Microcomputer chip) has over 3.1×10^6 transistors. It uses 0.8 μm minimum feature sizes. The newer still, INTEL P6 has 5.5×10^6 transistors using 0.6 μm minimum feature sizes.

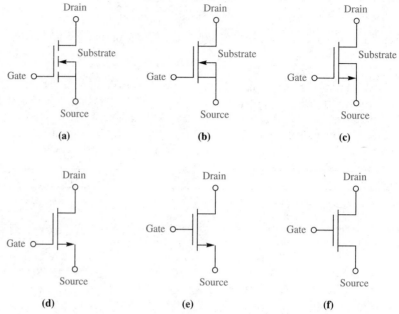

FIGURE 5.13 Alternative schematic diagram symbols for an *n*-channel MOSFET. Enhancement mode (a) and depletion mode (b) are most common.

MOS Inverter with a Passive Load

Analogous to the BJT inverter studied in Chapter 4, the MOSFET can also be used as an inverter. The inverter is the basic building block for all logic circuits. In particular, Fig. 5.14(a) shows a schematic diagram of such a circuit utilizing a passive load (resistor). The load line

$$V_{DD} = i_D R_D + v_{DS} = i_D R_D + v_O \tag{5.15}$$

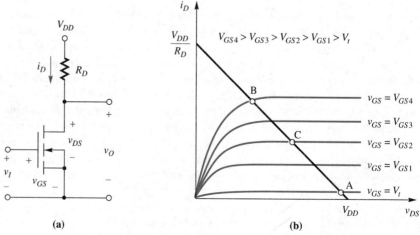

FIGURE 5.14 MOS inverter with passive load (resistor). (a) Inverter circuit. (b) Load-line analysis.

is sketched on a set of i_D versus v_{DS} characteristic curves, as shown in Fig. 5.14(b). If $v_I = v_{GS} = 0$, then $v_O \approx V_{DD}$, point A. If $v_I = v_{GS} = V_{GS4}$, then v_O is small, point B. For amplifier operation, the Q-point would be somewhat centered along the load line, point C. This circuit works; however, the R_D drain resistor requires several tens of times the chip area of the MOSFET itself, and because a complex logic circuit requires hundreds, if not thousands, of these types of circuits, the die area (and associated I^2R power losses) would be economically and technically prohibitive. This subject is discussed further in Chapter 18. Note that at point B, the v_{DS} is small but nonzero and $I_D \approx V_{DD}/R_D$ so that the static power dissipation $P_D = V_{DS} \times I_D$, if multiplied by a large number of devices on a die, can become cumulatively unacceptably large.

EXAMPLE 5.4

The MOSFET from Example 5.2 is incorporated in the circuit shown in Fig. 5.14(a) with $V_{DD} = 12$ V and $R_D = 1$ kΩ. Sketch and label the load line on a set of $i_D - v_{DS}$ characteristics and identify the ohmic and saturation regions of operation.

SOLUTION Referring to Fig. 5.14(b), generate the characteristic shown in Figure 5.15. The y-axis intercept, $I_D = V_{DD}/R_D = 12$ V/ 1 kΩ = 12 mA. The x-axis intercept is $V_{DD} = 12$ V.

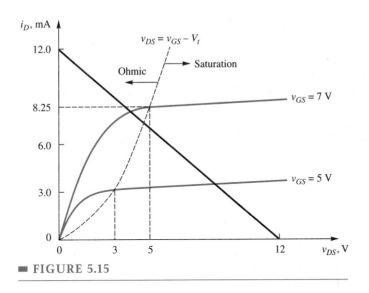

■ FIGURE 5.15

DRILL 5.4

Repeat Example 5.4 but
(a) change R_D to 2 kΩ with $V_{DD} = 12$ V and
(b) change R_D to 2 kΩ with $V_{DD} = 24$ V.

ANSWER See Figure 5.16 on page 200.

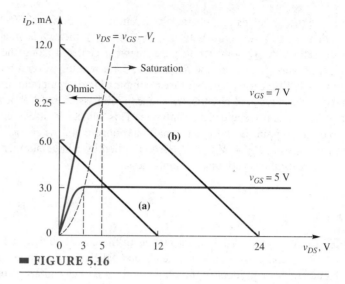
FIGURE 5.16

A Class of circuits called *active loads*, which use active devices to source or sink a given current, will permit the replacement of R_D with an active device and thereby significantly reduce the size and power requirements for the inverter.

MOS Inverter with an Active Load

The MOSFET depletion-mode device, as illustrated in Fig. 5.17(a), will operate as an active load. By arranging the external circuit so that v_{DS1} is greater than a few volts, the drain current will essentially be a constant.

Two n-channel devices are connected as shown in Fig. 5.17(a): M_1 is operating as an enhancement-mode device; M_2 is operating as a depletion-mode device. The characteristic curve for M_1 is given in Fig. 5.17(b), and the characteristic curve for M_2 is given in Fig. 5.17(c). The active-load curve [dashed curve in Fig. 5.17(b)] is obtained by observing that

$$v_O = V_{DS1} = V_{DD} - v_{DS2} \tag{5.16}$$

and using $v_{GS2} = 0$ and any value $I_{D2} = I_{D1}$. Equation (5.16) is plotted as the load curve in Fig. 5.17(b). We illustrate the switching characteristic by plotting the transfer function of v_O versus v_{IN} in Fig. 5.17(d). Refer to Fig. 5.17(b). For $v_{IN} = v_{GS1} = 0$, since $v_{IN} < V_t$ for M_1, the enhancement-mode FET, $I_{D1} \approx 0$, and this characteristic intersects the load curve at point A. As v_{IN} increases, there is no drain current until v_{IN} reaches $V_t = 2$ V. The operating point remains at A.

When $v_{IN} = V_{GS1} = 3$ V, I_{D1} becomes 1 mA at saturation, and this characteristic intersects the load curve at point B. When $v_{IN} = V_{GS1} = 4$ V, the corresponding characteristic intersects the load curve at point C.

If we continue increasing $v_{IN} = V_{GS1}$ to 5 V, the characteristic now intersects the load at point D, much lower in output voltage $V_{DS1} = v_O$.

The output voltage corresponding to the operating points in Fig. 5.17(b) is plotted as a function of input voltage in Fig. 5.17(d). Thus, the circuit functions as an inverter. A SPICE simulation is presented later in this chapter.

Even though there is a "linear" range, this inverter circuit is used exclusively for digital switching. There are two disadvantages with this circuit. First, neglecting power dissipation during switching, this circuit does dissipate more power than necessary in the input HIGH,

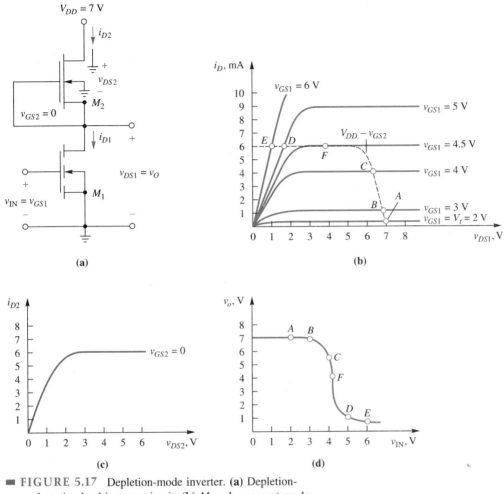

FIGURE 5.17 Depletion-mode inverter. (a) Depletion-mode active-load inverter circuit. (b) M_1 enhancement-mode MOSFET specifications. (c) M_2 depletion-mode MOSFET specifications. (d) Transfer characteristic.

$V_{IN} = 6$ V state. One can observe that the product of $V_{DS1} \times I_{D1} \neq 0$, Q-point (E). Fortunately, this can be made reasonably small. Second, this design requires that both the enhancement-mode (M_1) and depletion-mode (M_2) devices be fabricated on a single die. This and other MOS inverters are treated in depth in Section 13.5. MOS amplifiers with active loads are treated in Chapter 10.

CMOS Inverter

A low-power alternative to the depletion-load inverter circuit is the CMOS inverter. The term **CMOS** stands for Complementary Metal-Oxide Semiconductor. The term **complementary** refers to the fabrication of n- and p-channel MOSFETS on the same chip. An idealized cross section of this construction is illustrated in Fig. 5.18. We say more about this in Chapter 18. The inverter circuit diagram appears symmetrical with respect to the

■ CMOS is one of the most widely-used logic families, as we will see in Chapter 13.

input and output terminals. Figures 5.18 and 5.19 illustrate the interconnections for an inverter. Both gates, points 1 and 2, are connected and are then externally available for connection to v_I. Both drains, points 5 and 6, are connected together and to the output, v_O. A positive voltage V_{DD}, point 3 is connected to the source of the *p*-channel device (PMOS), point 4. A more negative voltage, V_{SS}, point 8 (could be circuit ground), is connected to the source of the *n*-channel device (NMOS), point 7.

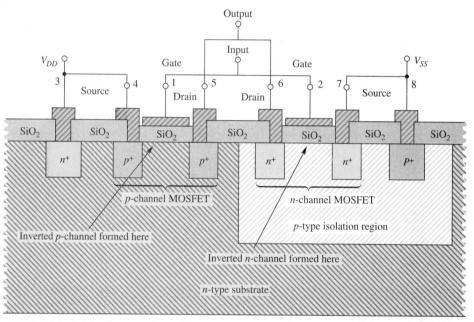

FIGURE 5.18 CMOS cross section.

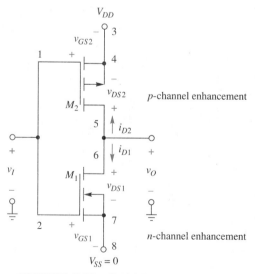

FIGURE 5.19 CMOS inverter.

Both M_1 and M_2 are enhancement devices. A set of i_D versus v_{DS} characteristic curves is sketched in Figs. 5.20(a) and (b), respectively, so we can illustrate the inverter circuit function. Let $V_{GS} = 0$, then $I_{D1} = 0$. Neglecting capacitive loading, $i_{D2} = -i_{D1} = 0$ because both M_1 and M_2 are in series. Also, $V_{GS2} = -V_{DD}$ when $V_I = 0$. Because it is a p-channel enhancement-mode device, M_2 has its Q-point located at A since this is the only point along the $V_{GS2} = -V_{DD}$ curve that also satisfies $I_{D2} = 0$. At Q-point A, $V_{DS2} = 0$, which means that $v_O = V_{DD} - 0 = V_{DD}$. It is important to observe that the static power dissipation is quite small in this state since $V_{DS1} \times I_{D1} \approx 0$ and $V_{DS2} \times I_{D2} \approx 0$. In practice, this is on the order of nanowatts.

If $v_I = V_{DD}$, $V_{GS1} = V_{DD}$. Simultaneously, $V_{GS2} = 0$, and M_2 is switched OFF, yielding $I_{D2} = 0$. However, $I_{D1} = -I_{D2} = 0$. Therefore, the M_1 Q-point switches to point B in Fig. 5.20(a), and $V_O = V_{DS1} = 0$, which means $V_{DS2} = V_{DD}$, point B in Fig. 5.20(b). Inverter action has been demonstrated. The static power dissipation is close to zero since $V_{DS1} \times I_{D2} \approx 0$ and $V_{DS2} \times I_{D2} \approx 0$ again.

Power is dissipated during switching. As v_I makes a transition between 0 V (V_{SS}) and V_{DD}, the output voltage follows a path represented by the dashed line on the M_1 and M_2 characteristics. Since, realistically, this transition does not occur instantaneously, there is a nonzero $v_{DS1} \times i_{D1}$ and $v_{DS2} \times i_{D2}$ product during the transition time. For this reason, the power dissipation for a CMOS inverter is given as a function of frequency. The CMOS inverter, including capacitive loading, is treated in Section 13.6.

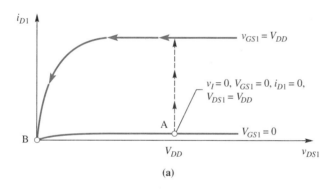

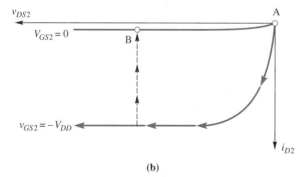

■ **FIGURE 5.20** CMOS inverter switching; arrow indicates V_I switching from 0 to $+V_{DD}$. (a) M_1 n-channel enhancement characteristic. (b) M_2 p-channel enhancement characteristic.

FET Analog Switches

Applications for analog switches used for the digitally gated transmission of analog signals occur in multiplexing, sampling, and analog-to-digital and digital-to-analog conversion. These topics are discussed in Chapters 13 and 16. FET switches may be realized with JFETs or MOSFETs, *p*-channel or *n*-channel, and enhancement-mode or depletion-mode devices. Typically, MOSFETs are used, with the *p*-channel/*n*-channel and enhancement/depletion mode choice depending on the signal magnitudes and polarity of the control voltage. Usually, FET analog switches are available as arrays in either a single integrated circuit or as part of a more complex signal-processing integrated circuit.

■ Of course, an *n*-substrate connects to the most positive node voltage in the circuit.

Figure 5.21 shows three examples with a control voltage, v_C, switching between levels of 0 and 5 V. The voltage v_I is assumed $\ll 5$ V and the *p*-substrate is connected to the most negative voltage node in the circuit. Referring to Fig. 5.21(a), an *n*-channel enhancement-mode FET switch, if $v_C = 0$ V, the FET is OFF, and $v_O = 0$. If $v_C = 5$ V, the FET is ON and

$$v_O = v_I \frac{R_L}{R_i + R_{ON} + R_L}, \tag{5.17}$$

where R_{ON}, the ON resistance of the MOSFET, can be computed by differentiating Eq. (5.10) to obtain

$$g_o = \left.\frac{di_D}{dv_{DS}}\right|_{Q\text{-point}} = 2k(V_{GS} - V_t - V_{DS}), \tag{5.18}$$

Since $V_{DS} \approx 0$ when the FET is ON,

$$R_{ON} \approx [2k(V_{GS} - V_t)]^{-1}. \tag{5.19}$$

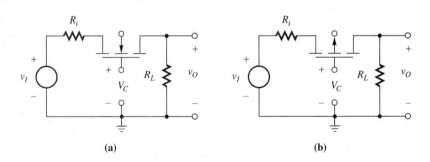

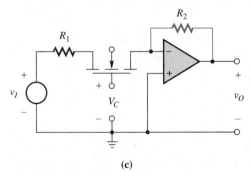

■ **FIGURE 5.21** FET analog switch cicuits. **(a)** An *n*-channel enhancement-mode switch. **(b)** A *p*-channel depletion-mode switch. **(c)** An FET switch with an op-amp.

EXAMPLE 5.5

Find R_{ON} for an n-channel enhancement MOSFET with $k = 2$ mA/V^2 and $V_t = 2$ V. Assume $R_{ON} \ll R_i + R_L$.

SOLUTION From Eq. (5.19), $R_{ON} \approx [2k(V_{GS} - v_t)]^{-1} = 250/V_{GS} - 2)$ Ω, where V_{GS} must be greater than $+2$ V for the FET to be ON. Consider the circuits shown in Figs. 5.21(a) and (b). Note that when the FET is ON, its source voltage is approximately equal to its drain voltage if $R_{ON} \ll R_i + R_L$. In addition, $R_i \ll R_L$ so that the source voltage $v_S \approx v_I$ when the FET is ON.

Suppose the FET in Fig. 5.21(a) has a $V_t = 3$ V. In order for the FET to be ON, $V_{GS} > V_t$, so that $v_C - v_I > 3$ V. Let $v_C = +5$ V. Then $5 - v_I > 3$ V requires that $v_I < 2$ V. Thus the constraint for this switch to be ON at $v_C = +5$ V is that the most positive v_I be less than $+2$ V. When this n-channel FET is OFF, its source voltage, v_S, is the more negative of v_I or $v_O = 0$. To ensure that the FET is OFF, $v_{GS} < v_t$. Thus if $v_C = 0$, we require $v_G - v_S = 0 - v_I < 3$ or $v_I > -3$ V. The most negative value of v_I must be above -3 V. Summarizing, for this analog switch circuit to operate with $v_C = 5$ V (ON) and $v_C = 0$ V (OFF), the input signal must be in the range of $-3 < v_I < 2$.

Similarly, consider the FET in Fig. 5.21(b) to become depleted at $V_t = +3$ V. When $v_C = 0$, the channel is ON when $v_I > -3$ V because $v_{GS} = v_C - v_S = 0 - v_I < 3$ V. When the p-channel FET is OFF, note that the voltage v_S is the more positive value of v_I or $v_O = 0$. For $v_C = +5$ V, the FET is OFF when $v_i < 2$ V because $v_{GS} = v_c - v_s = 5 - v_I > 3$ V.

The circuit shown in Fig. 5.21(c) is often used, because placing the FET at the virtual ground of the operational amplifier circuit causes the FET to be ON for any v_I whenever $v_C > V_t$.

DRILL 5.5

The FET switch shown in Fig. 5.21(a) has $v_I = 2 \sin \omega t$ V. The FET $V_t = 1.0$ V. Assume $R_i \ll R_L$ and $R_{ON} \ll R_i + R_L$.

(a) Find the v_C necessary for the FET to be ON.
(b) Find the v_C necessary for the FET to be OFF.

ANSWERS
(a) $v_C > 3$ V
(b) $v_C < -1$ V

CHECK UP

1. **TRUE OR FALSE?** MOS inverters with active loads occupy less chip area than MOS inverters with a resistive load.
2. Sketch a schematic diagram for an MOS inverter using a depletion-mode active load.
3. **TRUE OR FALSE?** The principal power dissipation mechanism for an MOS resistively loaded inverter is due to the gate current.
4. Sketch a schematic diagram for a CMOS inverter.
5. Explain why the power dissipation in a CMOS inverter is essentially zero except during switching.
6. Sketch a schematic diagram for an analog switch.

5.5 SPICE MODEL FOR THE MOSFET

SPICE computer simulations can be used for the analysis and design of MOSFET circuits. The MOSFET requires both an element statement and a model statement. The n-channel and p-channel MOSFET element models are given in Fig. 5.22. The element statement format is

$$\text{MXXX ND NG NS NB MNAME} \quad (5.20)$$

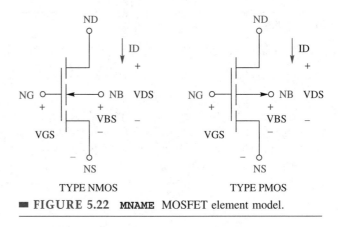

FIGURE 5.22 **MNAME** MOSFET element model.

where **ND, ND, NS, and NB** are the drain, gate, source, and substrate nodes, respectively. Gate width and length can be included in the element statement, as well as other parameters beyond the scope of this text.

There are three or more different MOSFET model levels used in SPICE depending on the particular version of SPICE being used. Each offers different levels of complexity. Each additional model level includes additional subtleties of the technology of fabrication and corrections resulting from empirical data. It is well beyond the scope of this discussion to use anything other than **LEVEL 1**, the default model which is derived from Shichman and Hodges.[5] This model is consistent with the MOSFET model derived in Section 5.2 although SPICE includes, depending on default parameter selection, many of the model complexities we ignored in the manual computations. Even for a **LEVEL 1** model, we omit listing most of the parameters over which the circuit designer may have little knowledge or control. Discussion of these additional model parameters[11] and their associated device physics is beyond the scope of this text. Selected common default parameters for a **LEVEL 1** MOSFET model are provided in Table 5.1.

Model parameters that influence frequency response are introduced, as needed, in later

TABLE 5.1 Selected SPICE MOSFET Model Parameters

Name	Parameter	Default Value	Description
LEVEL	Model index	1	Shichman-Hodges
VTO	Zero-bias threshold voltage	0 V	V_t, as in Eq. (5.10)
KP	Transconductance parameter	2×10^{-5} A/V^2	$2k(L/W) = \mu C_{ox}$, as in Eq. (5.10)
LAMBDA	Channel-length modulation	0 V^{-1}	λ, as in Eq. (5.43)
GAMMA	Bulk threshold parameter	0 V$^{0.5}$	γ, as in Eq. (5.27)
RD	Drain ohmic resistance	0 Ω	r_d
RS	Source ohmic resistance	0 Ω	r_s
TOX	Oxide thickness	10^{-5} cm	d, as in Fig. 5.4
NSUB	Substrate doping	0 cm^{-3}	
UO	Surface mobility	600 cm^2/V-sec	

CHAPTER 5   THE FIELD-EFFECT TRANSISTOR

chapters. It is also important to note that the default is $W/L = 100~\mu\text{m}/100~\mu\text{m} = 1$. Other W and L values may be entered in the .MODEL statement.

The relationship between k and KP is obtained from Eq. (5.12) as

$$i_D = \frac{\mu C_{OX} W}{2L}(v_{GS} - V_t)^2 = k(v_{GS} - V_t)^2$$

$$= \frac{KP}{2} \times \frac{W}{L}(v_{GS} - V_t)^2 \qquad v_{DS} > v_{GS} - V_t, \quad (5.21)$$

where $KP = \mu C_{OX} = 2k\dfrac{L}{W}$.

The model statement format is given by Eq. (5.22). **TYPE** must be specified as either **NMOS** or **PMOS**.

$$\text{.MODEL MNAME TYPE [PNAME1 = PVAL1] PNAME2 = PVAL2...]} \qquad (5.22)$$

For example, we can represent an n-channel MOSFET, **MOS1**, with $V_t = 1$ V, $KP = 1 \times 10^{-5}$ A/V^2, and whose drain is connected to node 5, gate to node 3, and source and substrate to node 1 by

$$\text{MOS1 5 3 1 1 FET} \qquad (5.23)$$

and

$$\text{.MODEL FET NMOS VTO = 1 KP = 1E-5} \qquad (5.24)$$

■ Assumes $\dfrac{W}{L} = 1$.

Alternatively, the **MODEL** statement can use process parameters instead of device parameters. For example, if **UO**, **NSUB**, and **TOX** (see Table 5.1) are given, SPICE will calculate V_t. However, if process parameters and device parameters are both given, SPICE will use the device parameters. The device parameters override the process parameters.

We also observe that even though MOSFETS are essentially three-teminal devices with a source, drain, and gate, they are fabricated within a semiconductor substrate and the potential between the substrate and source influences i_D and V_t. The source-substrate forms a parasitic FET with essentially a "second" gate control on i_D. In Fig. 5.8, we show that the source and substrate are at the same potential, i.e., connected together. In general, the IC substrate is not connected to the source but to the most negative point in the circuit such that the voltage between the body or substrate and source, $V_{BS} < 0$. For example, in Fig. 5.17, we observe that the substrate is at ground potential but the M_2 source is at v_O. Therefore, analogous to Eq. (5.13), the output current is a function of v_{BS} defined by

$$g_{mb} = \frac{\partial i_D}{\partial v_{BS}} = i_d/v_{bs} \qquad (5.25)$$

and we may define

$$\chi = g_{mb}/g_m. \qquad (5.26)$$

Usually χ ranges from 0.1 to 0.3 so its effect in a circuit could be significant. In SPICE, this substrate effect is modeled by the term **GAMMA** where for our purposes,

$$\textbf{GAMMA} = \gamma = \frac{\sqrt{2\varepsilon_s q N_A}}{C_{ox}} = 2\chi\sqrt{2\phi_P - V_{BS}}. \qquad (5.27)$$

This is discussed in Section 8.6. We incorporate **GAMMA** in Example 5.6.

EXAMPLE 5.6

Write a SPICE program to plot the transfer characteristic of the NMOS inverter of Fig. 5.23. The depletion load transistor has $V_t = -2.5$ V and $k_L = 0.02$ mA/V^2, and the enhancement driver transistor has $V_t = 1$ V and $k_D = 0.1$ mA/V^2.

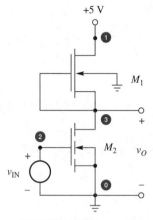

FIGURE 5.23 NMOS inverter.

SOLUTION   Consider the program statements

```
NMOS INVERTER TRANSFER CHARACTERISTIC
V10 1 0 DC 5
VIN 2 0 DC
M1 1 3 3 0 M1
M2 3 2 0 0 M2
.MODEL M1 NMOS VTO=-2.5 KP=4E-5 GAMMA=.3
.MODEL M2 NMOS VTO=1 KP=2E-4 GAMMA=.3
.DC VIN 0 4 .05
.PLOT DC V(3)
.END
```

Note that the MOSFET nodes are, in order, drain, gate, source, and substrate. We are able to show a common substrate for the two FETs. **GAMMA** is the *bulk threshold parameter* for the FET. The dc statement steps the input voltage from 0 to 4 V in 0.05-V increments. The computer plot of the transfer function is shown in Fig. 5.24. Note that KP is double the value of k for the default $W/L = 1$.

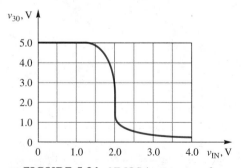

FIGURE 5.24 NMOS inverter transfer characteristic.

CHAPTER 5 THE FIELD-EFFECT TRANSISTOR

CHECK UP

1. Verify the Example 5.6 SPICE results using the SPICE version and syntax available to you. If you have schematic capture capability, compare the resultant netlist with a manually generated program listing.
2. **TRUE OR FALSE?** The model statements shown in Example 5.6 are the most complete ones that you would ever use.
3. **TRUE OR FALSE?** SPICE requires that you specify a substrate connection in addition to the drain, gate, and source.

5.6 JFET OPERATION

An idealized *n*-channel JFET (Junction Field-Effect Transistor) is illustrated in Fig. 5.25. A conductive channel (*n*-type in this case) has an ohmic contact on each end. One contact acts as a source, the other as a drain. The third electrode is called the gate and consists of two *p* regions on either side of the *n* channel. The *p* and *n* region forms junctions, and the behavior of these junctions is the key to understanding the device. Let us first examine the operation qualitatively. We will then be in a position to obtain the small-signal model and transconductance.

■ Except for analog IC input stages, JFETs are rarely used in ICs.

As illustrated in Fig. 5.25, we connect a voltage between the drain and source, V_{DS}. Similarly, we connect a voltage between the gate and source, V_{GS}. The sketches in Fig. 5.26 illustrate the formation of a conducting channel and how changing V_{DS} and V_{GS} is reflected in obtaining the I-V characteristics. From this we can deduce a small-signal model and explain large-signal switching properties.

Suppose the *p*-region doping is much larger than the *n*-region doping. Then with $V_{GS} = 0$ and $V_{DS} = 0$, a uniform depletion region of width y_0 will exist along both *pn* junctions, with most of the depletion region extending into a lightly doped *n* region [see Fig. 5.26(a)]. The JFET is symmetric about $y = T/2$. The reverse junction voltage V_R

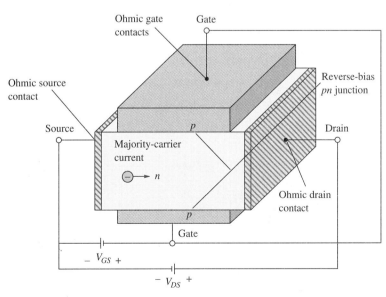

■ **FIGURE 5.25** An *n*-channel JFET model.

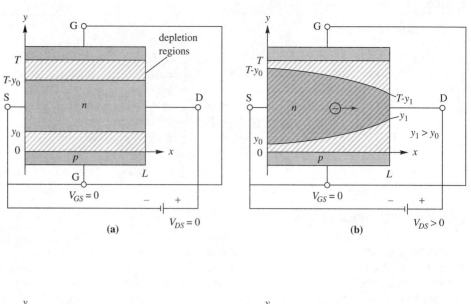
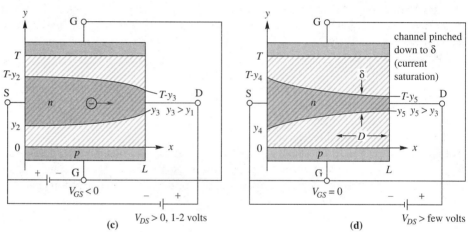
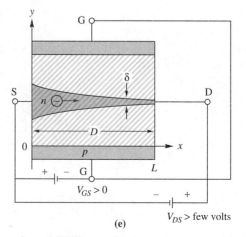

FIGURE 5.26 An n-channel JFET operation.

CHAPTER 5 THE FIELD-EFFECT TRANSISTOR

(from channel to gate) is 0 everywhere. From Eq. (3.14) we can obtain y_0 for the abrupt junction at equilibrium,

$$w(x) = y_0 \approx \sqrt{\frac{2\varepsilon\phi_J}{qN_D}}, \quad (5.28)$$

by assuming $N_A \gg N_D$. ϕ_J is the built-in junction potential, nominally 0.5 to 0.7 V in silicon. Note that since V_R is constant, $w(x)$ is constant along the device in the x direction.

EXAMPLE 5.7

Compute y_0 for an $N_D = 10^{17}$ atoms/cm^3 doped channel. Assume $\phi_J = 0.7$ V.

SOLUTION Using Eq. 5.28,

$$y_0 = \sqrt{\frac{2 \times 8.854 \times 10^{-14} \text{ F/cm} \times 11.8 \times 0.7 \text{ V}}{1.6 \times 10^{-19} \text{ C} \times 10^{17} \text{ cm}^{-3}}} = 9.56 \times 10^{-6} \text{ cm}$$

DRILL 5.6

Compute y_o in micrometers for an $N_D = 5 \times 10^{15}$ atoms/cm^3 doped channel. Assume $\phi_J = 0.7$ V.

ANSWER
$y_o = 0.43$ μm

Let's now change the biasing so that with $V_{GS} = 0$ and V_{DS} is slightly positive by a volt or two. We then have a situation within the device as illustrated in Fig. 5.26(b). The potential at any point in the channel is referred to the source and is $V(x)$ at a distance x from the source. The reverse-junction voltage at any point x along the device is thus $V_R(x) = V(x) - V_{GS}$. Since $V_{GS} = 0$, V_R will increase from $V_R(0) = 0$ at the source to $V_R(L) = V_{DS}$ at the drain. Thus the width of the depletion region varies along the x dimension. By modifying Eq. (5.28), the width is given by

$$w(x) \approx \sqrt{\frac{2\varepsilon[V(x) + \phi_J]}{qN_D}}. \quad (5.29)$$

The result of this is a channel whose dimensions vary from a thickness of $T - 2y_0$ at the source end to a thickness of $T - 2y_1$ at the drain end.

Assuming a cross-sectional dimension of d (into the page), we can compute the bulk ohmic resistance of this channel composed of n-doped material. The nonuniformity of the thickness does not permit us to use $R = \rho L/A$ directly unless we made some approximations. No current flows in the depletion region. Using Ohm's law and an approximate average cross-sectional area, $\bar{A} \approx d(T - y_0 - y_1)$, we can obtain an expression for the approximate channel resistance and drain current

$$R = \frac{L}{N_D q \mu_n \bar{A}}, \quad (5.30)$$

$$I_D = \frac{V_{DS}}{R} = N_D q \mu_n \bar{A} \frac{V_{DS}}{L} = N_D q \mu_n d(T - y_0 - y_1) \frac{V_{DS}}{L}. \quad (5.31)$$

Equation (5.31) describes, for small V_{DS}, a portion of the i_D versus v_{DS} characteristic, shown in Fig. 5.27, which behaves as an ohmic resistance. The slope of the $v_{GS} = 0$ characteristic at the origin is given by Eq. (5.32) and is called the *ON drain resistance*, $r_{DS(ON)}$.

$$r_{DS(ON)} \cong \frac{L}{N_D q \mu_n d(T - y_0 - y_1)}. \quad (5.32)$$

Typically, this is on the order of several ohms to several hundred ohms for an n-channel

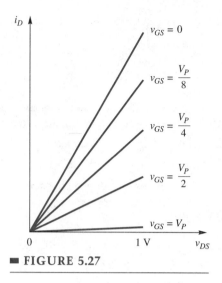

■ FIGURE 5.27

device. Because the mobility for holes is less than that for electrons, p-channel JFETs typically have a higher $r_{DS(ON)}$.

EXAMPLE 5.8

Compute $r_{DS(ON)}$ for an n-channel JFET with $L = 50$ μm, $N_D = 10^{18}$ atoms/cm^3, $d = 20$ μm, and an average channel thickness of 1 μm. Compare the results to a similarly doped p-channel JFET.

SOLUTION From Fig. 2.18, $\mu_n \approx 300$ cm^2/V-sec. Substituting into Eq. (5.32) yields

$$r_{DS(ON)} \cong \frac{50 \ \mu m}{(10^{18} \ \text{atoms/cm}^3)(1.6 \times 10^{-19} \ \text{C})(300 \ \text{cm}^2/\text{V-sec})(20 \ \mu m)(1 \ \mu m)}$$
$$= 520 \ \Omega.$$

Recall that 1 μm $= 10^{-4}$ cm when we perform this calculation. For a p-channel JFET, $\mu_p = 120$ cm^2/V-sec and $r_{DS(ON)} = 1.3$ kΩ.

Figure 5.26(c) represents the condition if $V_{GS} < 0$ while V_{DS} is slightly positive by a volt or two. The reverse-junction voltage is given by $V_R(x) = V(x) - V_{GS}$. Basically, the depletion region widens from y_0 to y_2 at the source end and from y_1 to y_3 at the drain end. The thickness of this depletion region can be computed at any point by modifying Eq. (5.29) to obtain

$$w(x) \approx \sqrt{\frac{2\varepsilon|V(x) - v_{GS} + \phi_J|}{qN_D}}. \tag{5.33}$$

This results in an overall narrowing of the channel, resulting in a higher resistance. Thus for a given V_{DS}, the I_D is less than that for the case of $V_{GS} = 0$. This phenomenon is illustrated in Fig. 5.27 by the curves below the $V_{GS} = 0$ curve. Note the gradual decrease in slope as V_{GS} becomes more negative, corresponding to increasing channel resistance. Again, for small V_{DS}, the behavior is essentially ohmic. At some point, where r_{DS} has become very large and I_D is essentially zero, we say that the JFET is at pinch-off. The gate-source voltage for this condition, $V_{GS} = V_P$, is called the **pinch-off voltage**.

CHAPTER 5 THE FIELD-EFFECT TRANSISTOR

EXAMPLE 5.9

Compute $|V_P|$ for $w = 2\ \mu m$ in an abrupt junction n-channel JFET when $N_D = 10^{-15}$ atoms/cm^3 and $V_{DS} = 0$. Neglect the built-in potential.

SOLUTION From Eq. (5.33), setting $V(x) = 0$,

$$|V_P| \cong \frac{qN_D w^2}{2\varepsilon}, \qquad \varepsilon = \varepsilon_o \varepsilon_r \text{ where } \varepsilon_r = 11.8 \text{ for silicon}$$

$$= \frac{(1.6 \times 10^{-19}\ \text{C})(10^{15}\ \text{atoms/cm}^3)(2 \times 10^{-4}\ \text{cm})^2}{(2)(11.8)(8.854 \times 10^{-14}\ \text{F/cm})}$$

$$= 3.06\ \text{V}.$$

DRILL 5.7

(a) Compute $|V_p|$ for $N_D = 3 \times 10^{15}$ cm^{-3} and $W = 2\ \mu m$.
(b) We will observe that for a 2N5459, $|V_P|$ could range from 2 to 8 V. Assuming $W = 2\ \mu m$, what is the range in channel doping density that yields these V_P values?

ANSWERS
(a) $|V_P| = 9.2$ V
(b) 6.3×10^{14} cm$^{-3} < N_D < 2.6 \times 10^{15}$ cm^{-3}

Let us return to the case where $V_{GS} = 0$ and start to increase V_{DS} [see Fig. 5.26(d)]. Clearly, as we increase V_{DS}, the depletion region thickness increases and the narrowest channel dimension, at $x = L$, given by $= T - 2y_5$, decreases. This limiting value of I_D when $V_{GS} = 0$ is called I_{DSS}, the **drain-source saturation current**. Saturation occurs when $v_{GD} < V_p$, or in this case, $v_{DS} > -V_P$. ($v_{GD} = -v_{DS}$ when $V_{GS} = 0$.)

As we increase V_{GS} more negatively, the channel constriction extends over more of the length of the device [see Fig. 5.26(e)]. Referring to Fig. 5.28, we see that when this channel constriction occurs, the current saturation is at a lower level. Saturation occurs when $v_{GD} < V_P$, or $v_{DS} > v_{GS} - V_P$. The region to the right of $v_{DS} = V_{GS} - V_P$ is called the *saturation region*, or the region beyond pinch-off.

To produce pinch-off at the drain and ensure operation in the saturation region, the total (dc + ac) drain-source voltage must satisfy

$$v_{GD} < V_P, \qquad \text{or}$$
$$v_{DS} \geq v_{GS} - V_P, \tag{5.34}$$

for $v_{DS} = v_{DS(\text{sat})}$ when the equality holds in Eq. (5.34). Equation (5.34) is plotted in Fig. 5.28(a), thus defining the ohmic and saturation regions.

Drain Characteristic in Saturation

Whereas for a BJT the relationship between an output parameter, i_C, and an input parameter, i_B, is given by a constant β_F, the relationship in a JFET between an output parameter, i_D, and an input parameter, v_{GS}, is more complex. In the saturation region, the commonly used square-law transfer relationship is

$$i_D = I_{DSS}\left(1 - \frac{v_{GS}}{V_P}\right)^2 \qquad |v_{DS}| \geq |v_{GS} - V_P|. \tag{5.35}$$

■ Compare Eqn. (5.35) for a JFET with Eqn. (5.12) for a MOSFET.

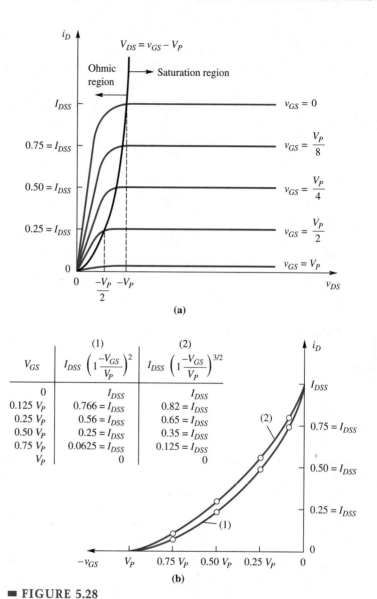

■ FIGURE 5.28

Note that ϕ_J is usually neglected when considering V_{GS}. This equation is used for operating conditions dictated by Eq. (5.34).

Using some of our approximate equations for the ohmic region, it is interesting to outline the development leading to this square-law relationship. Neglecting the effects of ϕ_J, Eq. (5.29) evaluated at the $V_{DS(\text{sat})}$ becomes

$$W(x = L) \cong \frac{T}{2} \cong \left(\frac{2\varepsilon v_{DS(\text{sat})}}{qN_D}\right)^{1/2} = \left[\frac{2\varepsilon(v_{GS} - V_P)}{qN_D}\right]. \tag{5.36}$$

Similarly at pinch-off, referring to Fig. 5.26(b) where $y_1 \cong T/2$ and $y_0 \cong 0$ the drain current I_D given by Eq. (5.31) is rewritten in the form

$$i_D \approx N_D q \mu_n d \left(\frac{T}{2}\right) \frac{v_{DS(\text{sat})}}{L} = \frac{N_D q \mu_n d}{L} \left(\frac{T}{2}\right)(v_{GS} - V_P). \tag{5.37}$$

Substituting $T/2$ from Eq. (5.36) into Eq. (5.37) yields, after some algebraic manipulation,

$$i_D = \frac{N_D q \mu_n d}{L}\left[\frac{2\varepsilon(-V_P)^3}{qN_D}\right]^{1/2}\left(1 - \frac{v_{GS}}{V_P}\right)^{3/2} = I_{DSS}\left(1 - \frac{v_{GS}}{V_P}\right)^{3/2}. \quad (5.38)$$

For $V_{GS} = 0$, we see that $i_D = I_{DSS}$. Equation (5.38) is plotted in Fig. 5.28(b) along with Eq. (5.35). The curves agree quite closely. Furthermore, abrupt junctions are assumed in the derivation of Eq. (5.38). A more detailed development[6,7] taking into account the graded junctions yields the result that Eq. (5.35) best represents the JFET i_D versus v_{GS} characteristic in the saturation region. Thus, the characteristic curves of Eq. (5.35) are entirely specified by knowing I_{DSS} and V_P for any given device when the condition of Eq. (5.34) is met. Equation (5.35) is equally applicable to p-channel JFETs since I_{DSS} and V_P carry their own signs.

Note that the depletion-mode MOSFET has an i_D versus v_{DS} characteristic very similar to that of a JFET with one very important difference. In any MOSFET, the oxide layer, being such a good insulator, ensures that the gate current will always be negligible. Although the JFET can operate marginally at weak forward bias of the gate junction, the JFET then loses one of the most useful properties of the FET—its negligible gate current.

EXAMPLE 5.10

An n-channel JFET is specified by $I_{DSS} = 5$ mA and $V_P = -3$ V. Sketch the i_D versus v_{DS} and i_D versus v_{GS} characteristics.

SOLUTION Equation (5.35) completely satisfies these characteristics in the saturation region, defined by Eq. (5.34). Neglecting the ohmic region, we can plot a set of idealized curves using Eq. (5.35),

$$i_D = 5\left(1 - \frac{v_{GS}}{-3}\right)^2 \text{ mA}$$

when $v_{DS} \geq v_{GS} - V_P$. See Fig. 5.29.

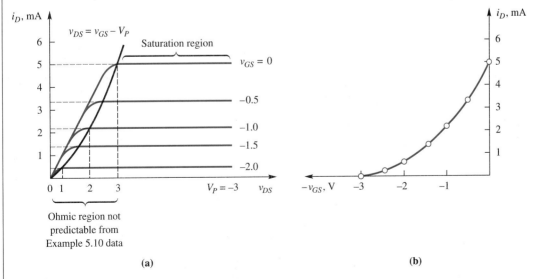

■ **FIGURE 5.29** Idealized characteristics (a) i_D versus v_{DS} and (b) i_D versus v_{GS}.

The following example illustrates the application of Eqs. (5.34) and (5.35) as a JFET model to find the operating point for a practical JFET circuit. Note that I_G is assumed to be zero in practical JFET circuits because the gate-channel junction is reverse-biased.

EXAMPLE 5.11

The n-channel JFET in Example 5.10 is used in the circuit in Fig. 5.30. Find I_D, V_{GS}, and V_{DS}. Check Eq. (5.34) to see if Eq. (5.35) is valid.

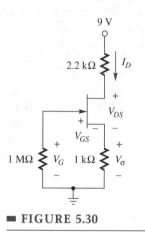

FIGURE 5.30

SOLUTION The gate-source voltage is $V_{GS} = V_G - V_\sigma$. Since there is no gate current, $V_G = 0$, $V_\sigma = I_D(1\ \text{k}\Omega)$. Then $V_{GS} = -1000I_D$. From Eq. (5.35),

$$I_D = 0.005\left(1 - \frac{-1000I_D}{-3}\right)^2.$$

The roots are $I_D = 1.41, 6.39$ mA. Then $V_{GS} = -1000I_D = -1.41, -6.39$ V. Note that the smaller (in magnitude) root must be used, since the quadratic Eq. (5.35) solution is not valid for $|V_{GS}| > |V_P|$. $V_{DS} = 9 - (2.2\ \text{k}\Omega + 1\ \text{k}\Omega)I_D = 4.49$ V. Using Eq. (5.34), we find that $4.49 > -1.41 + 3 = 1.59$ V. Our solution is valid because the FET is operating in the saturation region. Virtually identical results are obtained by means of a SPICE computer modeling of this circuit.

DRILL 5.8

Refer to Examples 5.10 and 5.11. What is the smallest V_{DD} possible such that the Q-point remains in the saturation region? Provide values for the Q-point.

ANSWERS

$I_D = 1.41$ mA.
$V_{GS} = -1.41$ V.
$V_{DS} = 1.59$ V.
$V_{DD} = 6.1$ V.

DRILL 5.9

Again refer to Examples 5.10 and 5.11. If $R_D = 3$ kΩ and $R_\sigma = 1.5$ kΩ, find I_D, V_{GS}, and V_{DS} and verify that the Q-point is in the saturation region.

ANSWERS

$I_D = 1.073$ mA.
$V_{GS} = -1.61$ V.
$V_{DS} = 4.17$ V.
$V_{DS} \geq V_{GS} - V_P$ and
$4.17 > -1.61 - (-3)$
$= 1.39$ V.

CHECK UP

1. Sketch and label cross sections for an n-channel and p-channel JFET.
2. **TRUE OR FALSE?** Operating in the saturation region of a JFET is the same as operating in the forward-active region of a BJT.
3. **TRUE OR FALSE?** The V_{GS} must be less than the pinch-off voltage before significant drain current will flow in a JFET.
4. Explain qualitatively what happens to the depletion region in a JFET when V_{DS} is increased for a given $V_{GS} > V_P$.
5. **TRUE OR FALSE?** I_D is a square-law function of V_{GS} in the saturation region.
6. **TRUE OR FALSE?** JFETs are almost always used as depletion-mode devices.
7. **TRUE OR FALSE?** MOSFETs are far more widely used than JFETs.

CHAPTER 5   THE FIELD-EFFECT TRANSISTOR

5.7   JFET SPECIFICATIONS

Figures 5.31 and 5.32 are data sheets for the Siliconix 2N5457-59 family of *n*-channel JFET depletion-mode devices. These JFETs are designed for use in general-purpose audio and switching applications. Our approach is to explain each entry as we did for the MOSFET in Section 5.3.

1. The maximum voltage ratings for V_{DS}, V_{DG}, and $V_{GS(\text{reverse})}$ are established by the maximum electric field in the device that would result in avalanche breakdown (Chapter 3). The term $V_{DS(\text{max})}$ refers to avalanche breakdown in the channel, whereas $V_{DG(\text{max})}$ and $V_{SG(\text{reverse max})}$ refer to an avalanche breakdown of the reverse-bias junctions defining the channel. Generally, good design practice dictates that operating voltages not exceed 50 to 75% of any of the maximum ratings. Indeed, for critical applications where device replacement is impossible or very expensive, design practice dictates that no operating voltage or current exceeds 33% of a maximum rating.

*ABSOLUTE MAXIMUM RATINGS (25°C)

Drain-Source Voltage	25 V
Drain-Gate Voltage	25 V
Source-Gate Voltage	25 V
Total Device Dissipation at 25°C	310 mW
Derate above 25°C	2.82 mW/°C
Operating Junction Temperature	135°C
Storage Temperature Range	−65 to +150°C

TO-92
See Section 7

*ELECTRICAL CHARACTERISTICS (25°C unless otherwise noted)

		Characteristic	2N5457			2N5458			2N5459			Unit	Test Conditions	
			Min	Typ	Max	Min	Typ	Max	Min	Typ	Max			
1	S T A T I C	I_{GSS}	Gate Reverse Current		−.01	−1.0		−.01	−1.0		−.01	−1.0	nA	$V_{GS} = -15$ V, $V_{DS} = 0$	
2						−200			−200			−200			$T_A = 100°C$
3		BV_{GSS}	Gate-Source Breakdown Voltage	−25	−60		−25	−60		−25	−60		V	$I_G = -10$ μA, $V_{DS} = 0$	
4		$V_{GS(off)}$	Gate-Source Cutoff Voltage	−0.5		−6.0	−1.0		−7.0	−2.0		−8.0		$V_{DS} = 15$ V, $I_D = 10$ nA	
5		I_{DSS}	Saturation Drain Current	1.0		5.0	2.0		9.0	4.0		16	mA	$V_{DS} = 15$ V, $V_{GS} = 0$ (Note 1)	
6	D Y N A M I C	g_{fs}	Common-Source Forward Transconductance	1,000		5,000	1,500		5,500	2,000		6,000	μmho	$V_{DS} = 15$ V, $V_{GS} = 0$	$f = 1$ kHz
7		g_{os}	Common-Source-Output Conductance		10	50		15	50		20	50			
8		C_{iss}	Common-Source Input Capacitance		4.5	7.0		4.5	7.0		4.5	7.0	pF		$f = 1$ MHz
9		C_{rss}	Common-Source Reverse Transfer Capacitance		1.0	3.0		1.0	3.0		1.0	3.0			
10		NF	Noise Figure		.04	3.0		.04	3.0		.04	3.0	dB	$V_{DS} = 15$ V, $V_{GS} = 0$, $R_G = 1$ MΩ, NBW = 1 Hz	$f = 1$ kHz

*JEDEC registered data
NOTE:
1. Pulse test pulsewidth = 2 ms.

NRL

■ **FIGURE 5.31** 2N5457-59 *n*-channel JFET data sheet (courtesy of Siliconix, Inc.).

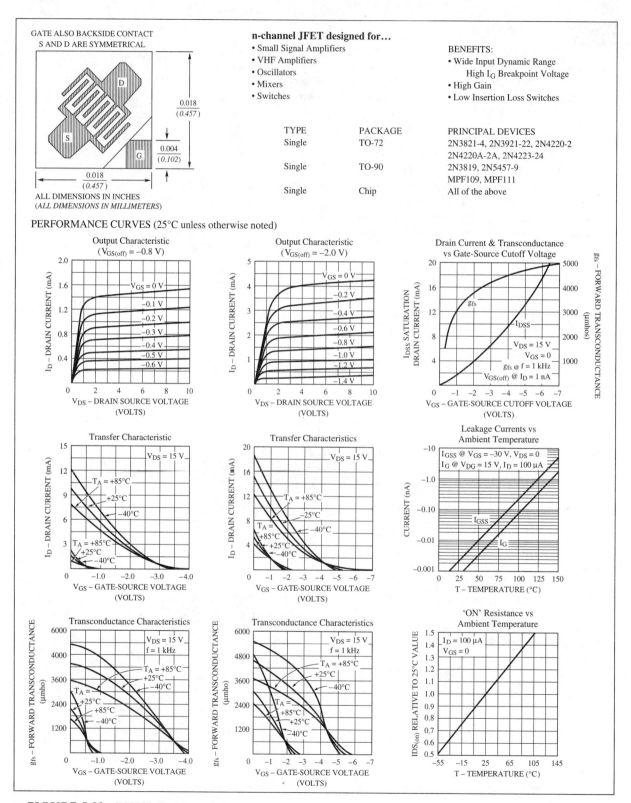

FIGURE 5.32 2N5457-59 characteristics (courtesy of Siliconix, Inc.).

2. Normally, a JFET is operated below I_{DSS} such that I_G is nominally zero. Should operating conditions in the circuit result in forward-bias operation of the junctions, I_G increases rapidly. An I_G maximum is similar to a maximum forward-current rating for a diode.

    The P_D maximum is computed by using

    $$P_{D(\max)} = V_{DS} \times I_D. \tag{5.39}$$

3. As illustrated in Fig. 5.33, all operation must be restricted to the shaded area to avoid exceeding $P_{D(\max)}$, I_{DSS}, and $V_{DS(\max)}$.

    The derating factor is graphically presented in Fig. 5.34. This essentially says that to keep the operating junction temperature below 135°C, the junction power dissipation, P_D, must be kept below the derating curve. It is interesting to observe that recent packaging improvements, which permit enhanced heat flow from the device to ambient, have resulted in an increase of $P_{D(\max)}$ to 1.0 W with a derating of -9.0 mW/°C. Unless otherwise stated, it is assumed that all ratings are applicable at $T = 25°C$ (nominally, room temperature).

4. Usually a wide storage temperature range is permitted. The limitations are established by the device design and the manufacturing capability in matching thermal stresses within a package.

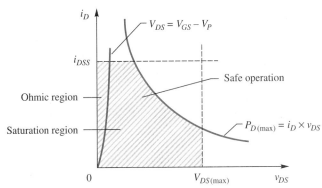

■ **FIGURE 5.33** Operating parameter maxima defining region of "safe" operation.

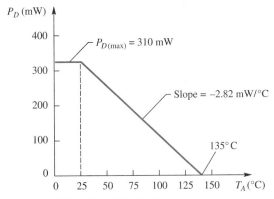

■ **FIGURE 5.34** 2N5457-59 power derating curve.

5. The term BV_{GSS} refers to the **gate-source breakdown voltage**. It is essentially a restatement of $V_{G(\text{reverse-bias max})}$. More specifically, it is given at an I_G that corresponds to the avalanche current knee in an avalanche-breakdown diode characteristic.
6. Similarly, the maximum value of I_{GSS}, the gate reverse current is given. Of more interest, it is given at two different temperatures. A temperature rise of 75°C results in a 200-fold increase in I_{GSS}. This should not be surprising since there is a strong temperature dependence in a diode reverse-bias current, as discussed in Chapter 3. From a practical viewpoint, as long as the operating voltages are reasonably below BV_{GSS}, I_{GSS} can be ignored in most circuits.
7. The **gate-source cutoff voltage**, $V_{GS(\text{off})}$, is another name for V_P, the *pinch-off* voltage. Observe that it is specified at a low but nonzero value for I_D, in this case 10 nA. There is a wide range for $V_{GS(\text{off})}$, reflecting variability in the semiconductor processing and device fabrication. Furthermore, whether a device is labeled as a 2N5457, -8, or -9 depends on the measured parameters during the manufacturer's final testing. Device parameters realized by the manufacturing process usually have a Gaussian distribution. Measured values often govern its final type label.
8. The term V_{GS} at a given I_D is closely related to V_P. It represents an arbitrary point at which the device channel is beginning to conduct. Again, note the large differences for the three different devices.
9. Minimum, and maximum values for I_{DSS}, the zero-gate-voltage drain current, are given.
10. The **forward transfer admittance**, $|g_{fs}|$, is obtained as follows. From Eq. (5.35), we define

■ Compare the JFET g_m with both the MOSFET g_m and the BJT g_m.

$$g_m \equiv \left.\frac{\partial i_D}{\partial v_{GS}}\right|_{Q\text{-point}} = \left.\frac{\partial}{\partial v_{GS}} I_{DSS}\left(1 - \frac{v_{GS}}{V_P}\right)^2\right|_{Q\text{-point}} = \frac{-2I_{DSS}}{V_P}\left(1 - \frac{V_{GS}}{V_P}\right), \quad (5.40)$$

where $g_{m0} = -2I_{DSS}/V_P = |g_{fs}|$ from the data sheet. Refer to Fig. 5.35. This is the maximum useful value for g_m, occurring at $V_{GS} = 0$.

Transconductance was introduced in Chapter 4 and for the MOSFET in Section 5.3. Again, we observe that for a given device, the range in a specification is typically a factor of 2 or 3, if not more. In later chapters, design techniques will have to take into account this wide range of parameters. Although the standard unit for conductance is the **siemen**, a large number of people still use **mho**, and this situation is reflected on the data sheet in Fig. 5.31.

11. The output admittance, g_{os}, is given as the slope of the I_{DSS} characteristic. Often, $r_d = 1/g_{os}$ is preferred. Refer to Fig. 5.35.
12. The effects of C_{iss} and C_{rss} are covered in later chapters. For now, we assume that our frequency of operation will ensure that the capacitive reactance of these two elements presents essentially an open circuit. Note that while the measurements of g_{fs} and g_{os} specify a frequency of 1 kHz, C_{iss} and C_{rss} are specified at 1 MHz.

DRILL 5.10

Compute g_{m0} and g_m for the circuits presented in Example 5.11 and Drill 5.9.

ANSWERS
$g_{m0} = 3.33$ mS for both.
$g_m = 1.77$ mS.
$g_m = 1.54$ mS.

Another very common technqiue is to present device information by means of sets of performance curves. First, observe that the curves in Fig. 5.32 are applicable to a large number of devices. Indeed, there are two sets of characteristics, reflecting a minimum and maximum condition. For example, we see that I_{DSS} ranges between 1.5 and 4.3 mA using the first pair of output characteristics. From the second set of curves (transfer characteristics), V_P ranges between -1.8 and -5 V for the high I_{DSS} device. Temperature variations are also illustrated. The transconductance, g_m, which in this case is called g_{fs}, the forward transconductance, is plotted on the last set of curves. At $V_{GS} = 0$ V, the y-axis intercept can be used to find g_{m0}.

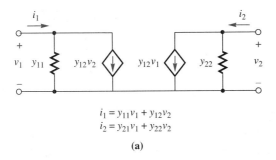

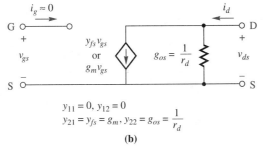

■ **FIGURE 5.35** Comparison of the (a) y-parameter two port with the (b) MOSFET and JFET model.

CHECK UP

1. **TRUE OR FALSE?** The transconductance increases linearly with V_{GS}.
2. Define V_P, I_{GSS}, and $g_{fs} = g_m$.
3. **TRUE OR FALSE?** Absolutely no drain current flows when $V_{GS} = 0$.
4. **TRUE OR FALSE?** Most FET parameters are temperature dependent.
5. Explain why establishing a Q-point to obtain the maximum g_m (g_{m0}) is a poor design practice.
6. **TRUE OR FALSE?** Gate leakage current is greater in a MOSFET than a JFET.

5.8 FET SMALL-SIGNAL MODELS

The saturation region behavior of the JFET and MOSFET exhibit similar properties that are directly based on their theory of operation. This allows us to develop a model (Fig. 5.36) in the same way we developed a model for the diode and BJT in earlier chapters, which can be used for circuit analysis and design.

1. Since the drain current is relatively independent of the drain-source voltage in the saturation region, as demonstrated in Eq. (5.12) for the MOSFET and Eq. (5.38) for the JFET, its effect can be modeled by a voltage-current generator, $i_d = g_m v_{gs}$.

   The transconductance for a MOSFET, given by Eq. (5.14), is repeated here for convenience as

$$g_m = \left.\frac{\partial [k(v_{GS} - V_t)^2]}{\partial v_{GS}}\right|_{Q\text{-point}} = 2k(V_{GS} - V_t). \quad (5.41)$$

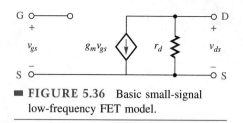

FIGURE 5.36 Basic small-signal low-frequency FET model.

The transconductance for a JFET given by Eq. (5.40) is repeated here for convenience as

$$g_m \equiv \left.\frac{\partial i_D}{\partial v_{GS}}\right|_{Q\text{-point}} = \left.\frac{\partial}{\partial v_{GS}} I_{DSS}\left(1 - \frac{v_{GS}}{V_P}\right)^2\right|_{Q\text{-point}}$$

$$= \frac{-2I_{DSS}}{V_P}\left(1 - \frac{V_{GS}}{V_P}\right) = \frac{-2\sqrt{I_{DSS}I_D}}{V_P}, \qquad (5.42)$$

where $g_{m0} = -2 I_{DSS}/V_P$.

2. Although not explicitly included in Eqs. (5.12) and (5.38), the nonzero slope can be modeled by a resistor, r_d, in parallel with this generator. Typically r_d ranges from tens to hundreds of kilo-ohms. See Fig. 5.35.
3. For a JFET, the reverse-bias junction between the gate and source, when operating in the depletion mode, effectively yields a very high input resistance—typically in the >100-MΩ range, so that the input circuit is modeled as an open circuit. For a MOSFET, the SiO_2 gate is an insulator, consequently, the input impedance is also very high, and this is also modeled as an open circuit.

Drain Resistance

As V_{DS} increases, the gate drain depletion layer widens, reducing the length of the conducting channel. Since the current varies inversely with the channel length L, the I_D increases as the effective L decreases. The small slope of the characteristic, when included in the transfer function for the JFET, changes Eq. (5.35) to Eq. (5.43a), and for the MOSFET changes Eq. (5.12) to Eq. (5.43b) as follows:

$$i_D = I_{DSS}\left(1 - \frac{v_{GS}}{V_P}\right)^2 (1 + \lambda v_{DS}), \qquad (5.43a)$$

$$i_D = \frac{\mu C_{ox} W}{2L}(v_{GS} - V_t)^2 (1 + \lambda v_{DS})$$

$$= k(v_{GS} - V_t)^2 (1 + \lambda v_{DS}) \qquad v_{DS} > v_{GS} - V_t, \qquad (5.43b)$$

where the extrapolated intercept on the negative v_{DS} axis is at $-(1/\lambda)$ V in both cases. Recall that for a BJT, an incremental change in i_C as v_{BC} changes is due to the base-width modulation. Thus λ^{-1} is analogous to V_A, the Early voltage that is used to describe the slope and extrapolated intercept of BJT characteristics. Of course, a second-quadrant characteristic and voltage axis intercept really do not exist, but this extrapolation is a useful concept. For a p-channel FET, λ^{-1} would be negative. The dynamic drain resistance, r_d, is defined at the Q-point

$$r_d = \left.\frac{\partial v_{DS}}{\partial i_D}\right|_{V_{GS}} \cong \left.\frac{\Delta v_{DS}}{\Delta i_D}\right|_{V_{GS}} = \frac{\lambda^{-1}}{I_{DSS}[1 - (V_{GS}/V_P)]^2} = \frac{1 + \lambda V_{DS}}{\lambda I_D} \approx \frac{1}{\lambda I_D}. \qquad (5.44)$$

SPICE models λ by means of the term **LAMBDA** in the **MODEL** statement.

CHAPTER 5 THE FIELD-EFFECT TRANSISTOR

EXAMPLE 5.12

(a) Using the data from Example 5.10, plot g_m versus v_{GS}.
(b) Assuming $\lambda^{-1} = 100$ V, compute r_d at $V_{GS} = 0$ and $V_{DS} = 5$ V.

SOLUTION

(a) From Eq. (5.42) and using $V_P = -3$ V,

$$I_{DSS} = 5 \text{ mA}, \quad g_{mO} = \frac{-2I_{DSS}}{V_P} = 3.33 \text{ mS}.$$

Similarly, we find that

$$g_m = 3.33\left(1 + \frac{V_{GS}}{3}\right) \text{ mS}.$$

See Fig. 5.37.

(b) $\quad r_d = 105 \text{ V}/5 \text{ mA} = 21 \text{ k}\Omega$ at $V_{GS} = 0$.

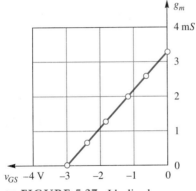

FIGURE 5.37 Idealized characteristic g_m versus v_{GS}.

DRILL 5.11

Compare the value of g_m for a MOSFET ($V_t = 1.5$ V, $k = 1$ mA/V^2) biased at $V_{GS} = 4$ V with the Example 5.12 JFET g_{mO}.

ANSWER 12.5 mS

DRILL 5.12

In Example 5.11, g_{m0} was computed to be 3.33 mS for $I_D = I_{DSS} = 5$ mA. Suppose a BJT is biased to obtain $I_C = 5$ mA. Compare the g_m for this typical JFET with a similarly biased BJT.

ANSWER
$g_m(\text{BJT}) = 192$ mS $\gg$ $g_{m0}(\text{JFET}) = 3.33$ mS

CHECK UP

1. **TRUE OR FALSE?** The small-signal models for a MOSFET and JFET are essentially the same.
2. **TRUE OR FALSE?** The physical mechanism yielding the λ^{-1} intercept in a FET is the same mechanism found for V_A in a BJT.
3. Explain how the drain resistance is computed from the I_D-V_{DS} characteristics.
4. **TRUE OR FALSE?** Transconductance has linear dependence with V_{GS} in both JFETs and MOSFETs.
5. **TRUE OR FALSE?** Transconductance values in FETs are considerably smaller than those found in BJTs assuming the same output terminal current levels in both devices.

5.9 JFET CIRCUITS

- Observe the similarity with the BJT CE, CB, and CC amplifier topologies.
- Small-signal usually implies <10% of a Q-point value of $|V_{GS} - V_P|$ for a JFET or $|V_{GS} - V_t|$ for a MOSFET.

Just as in Chapter 4 where we studied the amplifier application of the BJT, we can use the JFET as an amplifier in three basic configurations. In turn, we will study the operation of the JFET in common-source, common-gate, and common-drain configurations. It is important to remember that since g_m is highly dependent on V_{GS}, what we compute depends on our use of relatively small signals in order that a linear model will best predict actual operating results.

A more detailed analysis and design for the Q-point bias will be presented in Chapter 7. In the meantime, we will represent the bias circuit by the series combination of V_{GG} and R_G.

Common Source

The basic common-source (CS) circuit is illustrated in Fig. 5.38(a), where C_S is a large capacitor whose effective reactance is assumed to be zero at the frequency of interest and presents an open-circuit to dc, the so-called dc blocking capacitor. We will present a more detailed frequency response analysis in Chapter 9. Using a hypothetical set of characteristic curves, we can locate the Q-point $V_{GS} = V_{GG}$ since $i_G = 0$ for the depletion-mode JFET.

- MOSFET amplifier circuits have similar small-signal models to these JFET amplifier circuits, and the analysis is the same.

In the output circuit,

$$V_{DD} = i_D R_D + v_{DS} \quad \text{or}$$

$$i_D = \frac{-v_{DS}}{R_D} + \frac{V_{DD}}{R_D}. \tag{5.45}$$

This is of the form

$$y = mx + b \tag{5.46}$$

- Depletion-mode MOSFETs have the same type of biasing as a JFET.

and can be plotted with the y-axis intercept, $b = V_{DD}/R_D$, and a slope, $m = -(1/R_D)$ [see Fig. 5.38(c)]. The Q-point is located at the intersection of Eq. (5.45) and the $v_{GS} = V_{GG}$ characteristic curve. The transconductance is computed by means of Eq. (5.40).

Referring to the small-signal model in Fig. 5.38(b), we compute the voltage gain with respect to the source defined as $A_{vs} = v_o/v_s$. Expanding this definition yields

$$A_{vs} = \frac{v_o}{v_i} \times \frac{v_i}{v_s}. \tag{5.47}$$

Using voltage division directly since $I_G = 0$,

$$\frac{v_i}{v_s} = \frac{R_G}{R_G + R_S}. \tag{5.48}$$

In the output circuit, since $g_d = 0$,

$$v_o = -g_m v_{gs} R_D. \tag{5.49}$$

Combining Eqs. (5.48) and (5.49) yields

$$A_{vs} = (-g_m R_D)\left(\frac{R_G}{R_G + R_S}\right). \tag{5.50}$$

This voltage-gain mechanism can be illustrated graphically. A small-signal sinusoid $v_i(t) = V \sin \omega t$ is sketched around the Q-point value for $v_{GS} = V_{GG}$ in Fig. 5.38(c). The resultant larger sinusoidal variation in v_{DS} and i_D can then be observed using the load line. With the appropriate selection of R_D, the ratio of $v_{ds} = v_o$ to v_{gs} can be made reasonably large.

The input resistance $R_i = v_i/i_i = R_G$ by inspection; again $i_G = 0$ is correctly assumed.

CHAPTER 5 THE FIELD-EFFECT TRANSISTOR

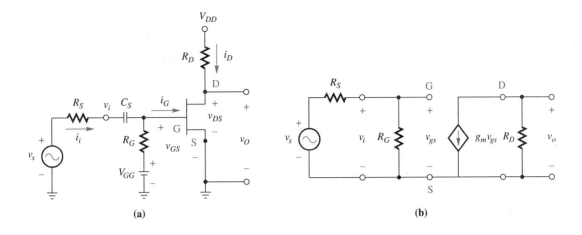

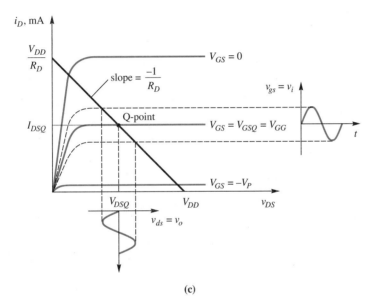

■ **FIGURE 5.38** Common-source amplifier. (a) Common-source amplifier circuit. (b) Small-signal model. (c) Load-line analysis i_D versus v_{DS} and v_{GS}.

EXAMPLE 5.13

For the common-source amplifier, let $R_D = 2$ kΩ, $R_G = 100$ kΩ, $V_{DD} = 15$ V, $V_{GG} = -1.0$ V. Compute $A_{vi} = v_o/v_i$. Use the JFET curves illustrated in Examples 5.10 and 5.12.

SOLUTION For $V_{GSQ} = -1.0$ V, $I_{DQ} = 2.22$ mA; then $V_{DSQ} = V_{DD} - I_D R_D = 15 - (2.22 \text{ mA})(2 \text{ k}\Omega) = 10.56$ V; $g_m = 2.22$ mS is computed using Example 5.12.

$$A_{vi} = -g_m R_D = -(2.22 \text{ mS})(2 \text{ k}\Omega) = -4.44.$$

DRILL 5.13

What is the maximum A_{vi} possible for the circuit described in Example 5.13 and what are the Q-point values for V_{GS}, I_D, and V_{DS} for obtaining this maximum A_{vi}? Is this a good Q-point location? Why or why not?

ANSWERS

$g_m = g_{m0}$
  $= 3.33$ mS when
$V_{GS} = 0$ V,
$A_{vi} = -6.66$,
$I_D = I_{DSS} = 5$ mA,
$V_{DS} = 5$ V.

No, because signal excursions more positive than $V_{GS} = 0$ will forward-bias the gate-source junction.

Common Gate

The common-gate (CG) source-voltage gain is computed in the same way as for the common-source amplifier. From Fig. 5.39(a), the circuit diagram, and Fig. 5.39(b), the small-signal model,

$$v_o = -g_m v_{gs} R_D \tag{5.51}$$

and $v_{gs} = -v_i$, so that

$$\frac{v_o}{v_i} = +g_m R_D. \tag{5.52}$$

Note that there is 0° phase shift in the CG between v_o and v_i as compared to the 180° phase shift in the CS. Using voltage division in the input circuit,

$$A_{vs} = (+g_m R_D)\left(\frac{R_i}{R_i + R_S}\right). \tag{5.53}$$

To compute the input resistance $R_i = v_i/i_i$, we can sum currents at the source node to obtain

$$v_i = i_i R_\sigma + g_m v_{gs} R_\sigma = i_i R_\sigma - g_m R_\sigma v_i \tag{5.54}$$

(observe that $v_{gs} = -v_i$) or

$$R_i = \frac{v_i}{i_i} = \frac{R_\sigma}{1 + g_m R_\sigma} = R_\sigma \left\| \frac{1}{g_m} \right. \tag{5.55}$$

DRILL 5.14

What is the maximum A_{vi} possible for the circuit described in Example 5.14 and what are the Q-point values for V_{GS}, I_D, and V_{DS} for obtaining this maximum A_{vi}? Compute R_i.

ANSWERS

$g_m = g_{m0}$
  $= 3.33$ mS when
$V_{GS} = 0$ V,
$R_i = 231\ \Omega$,
$A_{vi} = +5.47$.
$I_D = I_{DSS}$
  $= 5$ mA,
$V_{DS} = 10$ V.

As with Drill 5.13, signal excursions more positive than $V_{GS} = 0$ will start to forward bias the drain-source junction. Consequently, this is not a good Q-point.

EXAMPLE 5.14

For the common-gate amplifier, let $V_{GG} = 1.22$ V so that $V_{GSQ} = -1$ V as before, $V_{DD} = 15$ V, $R_D = 2$ kΩ, $R_\sigma = 1$ kΩ, and $R_S = 50\ \Omega$. Compute $A_{vs} = v_O/v_S$.

SOLUTION As with the common-source amplifier, $V_{GSQ} = -1.0$ V and $I_{DQ} = 2.22$ mA. Then $g_m = 2.22$ mS.

$$V_{DSQ} = V_{DD} - I_D(R_D + R_\sigma) = 15 - (2.22\ \text{mA})(3\ \text{k}\Omega) = 8.34\ \text{V},$$

$$R_i = 1\ \text{k}\Omega \left\| \frac{1}{2.22\ \text{mS}} \right. = 1000\|450\ \Omega = 310\ \Omega.$$

From Eq. (5.53),

$$A_{vs} = (+g_m R_D)\left(\frac{R_i}{R_i + R_S}\right) = (+2.22\ \text{mS})(2\ \text{k}\Omega)\left(\frac{310}{360}\right) = +3.8.$$

Common Drain

The source-voltage gain for the common-gain (CD) amplifier [Fig. 5.40(a)] is computed by means of the small-signal model [Fig. 5.40(b)] as

$$v_o = g_m v_{gs} R_\sigma \tag{5.56}$$

CHAPTER 5 THE FIELD-EFFECT TRANSISTOR

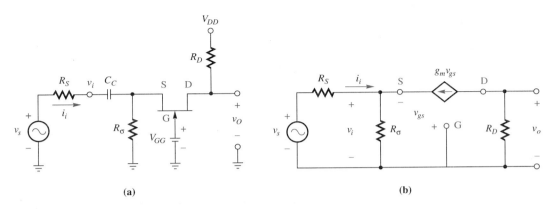

FIGURE 5.39 Common-gate amplifier. (a) Common-gate amplifier circuit. (b) Small-signal model.

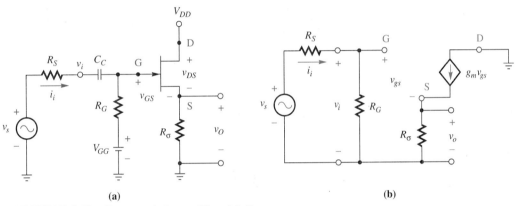

FIGURE 5.40 Common-drain amplifier. (a) Common-drain (source follower) amplifier circuit. (b) Small-signal model.

and

$$v_i = v_{gs} + v_o. \tag{5.57}$$

Rearranging, Eq. (5.56) substituted into Eq. (5.57) yields

$$v_o = g_m(v_i - v_o)R_\sigma \tag{5.58}$$

$$\frac{v_o}{v_i} = \frac{g_m R_\sigma}{1 + g_m R_\sigma} \tag{5.59}$$

If $g_m R_\sigma \gg 1$, $A_{vi} \approx 1$. Then

$$A_{vs} \approx \frac{R_G}{R_G + R_s} \times A_{vi} \approx 1 \quad \text{if } R_G \gg R_s. \tag{5.60}$$

The common-drain configuration is often called a ***source follower***.

■ Recall, a common collector BJT amplifier was called an emitter-follower.

DRILL 5.15

Adjust V_{GG} in the source-follower circuit of Fig. 5.40 to operate at $V_{GS} = 0$. Use JFET and remaining circuit specifications from Example 5.15. Compute A_{vi} and the Q-point values of I_D and V_{DS}.

ANSWERS

$I_D = I_{DSS} = 5$ mA.
$A_{vi} = +0.77$.
$V_{DS} = 10$ V.

EXAMPLE 5.15

Let $R_\sigma = 1$ kΩ and $V_{DD} = 15$ V in a CD circuit using the same JFET parameters from the earlier examples. $V_{GG} = 1.22$ V yields $V_{GS} = -1.0$ V. How close will this amplifier operate to $A_{vi} = 1$? Derive and compute the input resistance defined as $R_i = v_i/i_i$.

SOLUTION $V_{GS} = -1.0$ V with $I_{DQ} = 2.22$ mA and $V_{DSQ} = V_{DD} - I_{DQ}R_\sigma = 15 - (2.22 \text{ mA})(1 \text{ k}\Omega) = 12.78$ V. $g_m = 2.22$ mS. From Eq. (5.59),

$$A_{vi} = \frac{g_m R_\sigma}{1 + g_m R_\sigma} = \frac{(2.22 \times 10^{-3} \text{ s})(1000 \text{ }\Omega)}{1 + (2.22 \times 10^{-3})(1000 \text{ }\Omega)} = +0.69.$$

We can observe that no portion of i_i flows into the gate so $R_i \approx R_G$.

CHECK UP

1. Compare the voltage gain characteristics for all three amplifier topogies.
2. **TRUE OR FALSE?** An amplifier with a voltage gain of ≤ 1 is useless.
3. **TRUE OR FALSE?** The amplifier topologies shown for JFETs could have been just as easily applied using MOSFETs and obtaining similar results.
4. Which amplifiers exhibit 0° phase shift between input and output?
5. Which amplifier topology exhibits a finite input resistance?

5.10 SPICE MODEL FOR THE JFET

SPICE computer simulations can be used for the analysis and design of JFET circuits. The JFET requires both an element statement and a model statement. The *n*-channel and *p*-channel JFET element model are given in Fig. 5.41. The element statement format is

JXXX ND NG NS MNAME [AREA][OFF][IC = VDS, VGS] (5.61)

where **ND**, **NG**, and **NS** are the drain, gate, and source nodes, respectively. The default

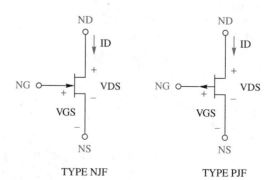

FIGURE 5.41 JFET element model.

value for the cross-sectional **AREA** is 1.0. **OFF** refers to the JFET being set to the nonconducting condition in the dc circuit analysis. The circuit will be analyzed using the computer Q-point for the JFET. This is the default condition. Initial conditions other than the Q-point values of the drain-source voltage, **VDS**, and gate-source voltage, **VGS**, can be specified optionally for a transient analysis. Selected common default parameters are provided in Table 5.2.

The model statement format is given by Eq. (5.22) except that, in this case, **TYPE** must be specified as either **NJF** or **PJF**. Observe that **VTO** is negative for **NJF** and all other parameters are positive. An abrupt junction model is used so that all depletion-layer capacitances are proportional to $V^{-1/2}$. This model is consistent with the JFET model used in Section 5.6, although SPICE includes, depending on default parameter selection, many of the model complexities we ignored in the manual computations.

For example, the element and model statements for an n-channel JFET, **J1**, whose drain is connected to node 1, gate to node 2, and source to node 0 and with $V_P = -4$ V, **BETA** = 1 mA/V^2, and $\lambda^{-1} = 100$ V are given by

$$\text{J1} \quad 1 \quad 2 \quad 0 \quad \text{TEST} \tag{5.62}$$

and

$$\text{.MODEL TEST NJF VTO} = -4 \text{ BETA} = 1\text{M LAMBDA} = 1\text{E}-2 \tag{5.63}$$

Note that **BETA** = I_{DSS}/V_P^2. We use the capacitance model parameters when we consider frequency response in later chapters.

■ Do not confuse BETA for a JFT with β for a BJT!

■ Note that the use of BETA when writing the JFET equation, Eqn. (5.35) becomes $I_D = \text{BETA} (V_P - V_{GS})^2$ and looks similar to the MOSFET equation, Eqn. (5.12).

TABLE 5.2	SPICE JFET Model Parameters		
Name	Parameter	Default Value	Description
VTO	Threshold (pinch-off) voltage	-2.0	Pinch-off voltage, V_P, or $V_{GS(OFF)}$ from data
BETA	Transconductance parameter	10^{-4} A/V^2	Defined using Eq. (5.38) as I_{DSS}/V_P^2
LAMBDA	Channel-length modulation parameter	0 V^{-1}	λ from Eq. (5.43)
RD	Drain ohmic resistance	0 Ω	Typically ignored in most calculations
RS	Source ohmic resistance	0 Ω	Typically ignored in most calculations
CGS	Zero-bias, gate-source capacitance	0 F	C_{iss} from Fig. 5.31 at $V_{GS} = 0$
CGD	Zero-bias, gate-drain capacitance	0 F	C_{rss} from Fig. 5.31 at $V_{GD} = 0$
PB	Gate junction potential	1 V	Junction potential often set at 0.7 V
FC	Coefficient for depletion capacitance formula	0.5	Eq. (3.48) saturates for $V > $ **FC**

EXAMPLE 5.16

The n-channel JFET in Fig. 5.42 has a pinch-off voltage $V_P = -3$ V, a transconductance constant of **BETA** = 0.56 mA/V^2, and channel-length modulation factor of $\lambda = 10^{-4}$ V^{-1}. Write a SPICE program to determine the quiescent point of the circuit, and the FET small-signal parameters.

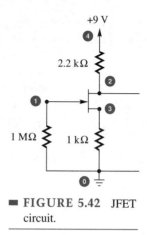

FIGURE 5.42 JFET circuit.

SOLUTION The program may be written as follows:

```
JFT OPERATING POINT
VDD 4 0 DC 9
RG 1 0 1MEG
RD 4 2 2.2K
RS 3 0 1K
J1 2 1 3 J1
.MODEL J1 NJF VTO=-3 BETA=.56M LAMBDA=1E-4
.OP
.END
```

J is the SPICE symbol for JFET, and **NJF** designates the n-channel. The Q-point parameters are requested as output using the **.OP** statement. As always, SPICE finds all dc node voltages, the supply current, and the total power, as well as the small-signal parameters, $I_D = 1.4$ mA, $V_{GS} = -1.4$ V, $V_{DS} = 4.48$ V, $g_m = 1.78$ mS, and $g_d = 1.4 \times 10^{-7}$ S. This compares very favorably with the results of Example 5.11.

CHECK UP

1. Verify the Example 5.16 SPICE results using the SPICE version and syntax available to you. If you have schematic capture capability, compare the resultant netlist with a manually generated program listing.
2. TRUE OR FALSE? I_{DSS} is directly included in the JFET **MODEL** statement.
3. TRUE OR FALSE? **BETA** for a JFET is different than the β for a BJT.
4. Comment on the relative complexity of the MOS and JFET SPICE MODEL statement specification.

5.11 MESFET OPERATION

The basic JFET used the depletion region of a reversed-bias junction to obtain a pinched-off conducting channel (Section 5.6). The junction was synthesized from *p*-doped and *n*-doped silicon. There is an entire class of JFET devices that uses a metal-semiconductor junction to obtain a depletion region when operating in reverse bias. We briefly discussed this metal semiconductor junction, the Schottky diode, in Section 3.13. This class of devices is called a *MESFET* (metal-semiconductor field-effect transistor). A simplified structure of the MESFET is shown in Fig. 5.43. Virtually all MESFETs use *n*-type GaAs as the channel material because *n*-type GaAs has an electron mobility, μ_n, five times that found in *n*-type Si (Table 2.2). The MESFET is far easier to implement in GaAs than the MOSFET-type structure. A thin layer, <1 μm, of *n*-type GaAs is epitaxially grown on an undoped GaAs substrate. As shown in the Chapter 2 problems, undoped GaAs has a very high resistivity, consequently, this substrate material is often called *semi-insulating*. This means that isolation between adjacent devices is often achieved directly because the resistance through the substrate between adjacent devices is quite high. The source and drain regions are heavily doped, n^+-GaAs, to facilitate achieving an ohmic contact. The source and drain ohmic contacts are usually made with a AuGeNi, Au-Sn, or In-Au alloy. The Schottky-barrier gate metal is often Al with Mo, W, and Ti also used.

To lower parasitic resistance in the channel, the gate is usually recessed into the *n*-channel GaAs and often the n^+-GaAs extends to the edge of the gate region to lower series resistance in the channel. This is called a ***recessed gate*** and is widely used in GaAs technology, as discussed in Chapter 18. This technique also effectively lowers the pinch-off voltage because the depletion region does not have to extend as far to obtain pinch-off. It is important to observe that the basic MESFET is a depletion-mode, *n*-channel device. The hole mobility, μ_p, of *p*-type GaAs is quite low, even below that of *p*-type Si (Table 2.2), so that there is no speed advantage to fabricating a *p*-channel GaAs MESFET. We also require that the *n*-type GaAs below the gate electrode be lightly doped to ensure a large depletion region, as illustrated in Fig. 5.43, extending into the channel. The effective gate length, L, is a technology-defined minimum dimension, often as low as $L = 0.5$ μm, with research into the fabrication and properties of short-channel devices leading to $L \approx 0.25$ μm. As with the MOSFET gate width W, W/L is usually >10 and defined by the circuit application.

■ Refer to Chapter 18 for further fabrication technology background.

Basic Model

For a first approximation, we use the JFET equations, Eq. (5.35) and (5.43), to describe the static or dc (low-frequency) operation of the MESFET. For a depletion-mode

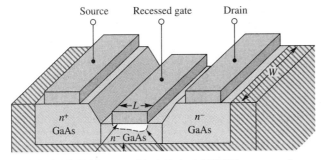

■ **FIGURE 5.43** Simplified GaAs MESFET cross section showing recessed gate and Schottky barrier depletion region.

n-channel MESFET, we have $I_D = 0$ for $v_{GS} < V_P$. In the saturation region, the i_{DS}-v_{DS} characteristics can be described by

$$i_D = I_{DSS}\left(1 - \frac{v_{GS}}{V_P}\right)^2 (1 + \lambda v_{DS}) = BETA(v_{GS} - V_P)^2(1 + \lambda v_{DS}), \quad (5.64)$$

where $v_{DS} > v_{GS} - V_P$ and $BETA = I_{DSS}/V_P^2$. Often in the JFET case, we assumed $\lambda = 0$; however, for the MESFET, this is not a good approximation. The i_{DS}-v_{DS} characteristics have a significant nonzero slope as shown in Fig. 5.44. For the ohmic region, these characteristics are described by

$$i_D = BETA[2(v_{GS} - V_P)v_{DS} - v_{DS}^2](1 + \lambda v_{DS}), \quad (5.65)$$

where $v_{DS} < v_{GS} - V_P$ and, of course, $v_{GS} > V_P$. Reasonable values for a state-of-the-art MESFET include $V_P = -1$ V, $BETA$ on the order of 10^{-4} A/V^2, and $\lambda^{-1} = 10$ to 20 V.

The basic model, which is very similar to that of the JFET is shown in Fig. 5.45. Figure 5.46 illustrates the physical basis of the model parameters. Observe that the device capacitances have been included in this model because all applications for MESFETs depend on their high-speed, high-frequency performance. The capacitances C_{CS} and C_{DG} are the lumped value of the reverse-bias Schottky diode junction transition capacitances as described in Section 3.12. Similarly, C_{DS} is a junction-type parasitic capacitance between the drain and substrate. Resistances r_G, r_D, and r_S represent the ohmic parasitic resistance values of the GaAs material and ohmic contacts. We discuss the ramifications of device

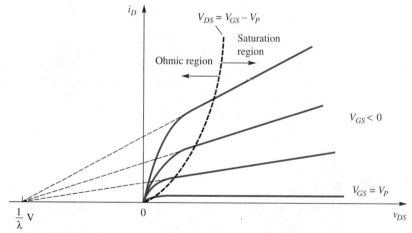

FIGURE 5.44 MESFET i_D-v_{DS} characteristics showing ohmic and saturation regions and the effect of low λ^{-1}.

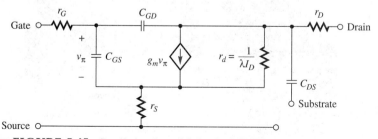

FIGURE 5.45 Small-signal model for the MESFET.

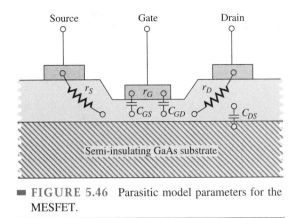

FIGURE 5.46 Parasitic model parameters for the MESFET.

capacitances and geometries on the frequency response of a circuit or system in later chapters. Typically, commercially available MESFETs operate from 1 GHz to in excess of 10 GHz in analog circuits and are used for subnanosecond, special-purpose, digital circuits.

A typical digital circuit using this type of GaAs MESFET and the basic model is presented in Chapter 17.

CHECK UP

1. Discuss the qualitative reasons MESFETs are usually fabricated in GaAs.
2. **TRUE OR FALSE?** The JFET model works well for modeling the MOSFET.
3. Sketch, label, and explain the topological cross sections of a JFET and MESFET.
4. **TRUE OR FALSE?** *p*-channel MESFETs offer a significant speed advantage compared to silicon-based devices.
5. Why aren't MOS structures implemented in GaAs?

SUMMARY

On the basis of a solid understanding of the *pn* junction and BJT, we developed the theory of operation, terminal characteristics, and circuit models for both the metal-oxide semiconductor transistor (MOSFET) and the junction field-effect transistor (JFET). Key relationships are summarized in Table 5.3.

Even though the MOSFET is more widely used, JFET theory was covered in some detail because its operation follows most directly from the basic *pn* junction theory. We then demonstrated that the simplified small-signal models for both MOSFETs and JFETs were identical, although the g_m computation is significantly different. Using the simplified FET small-signal model, we also demonstrated common-gate, common-source, and common-drain amplifier operation with the JFET. There is increasing utilization of the MOSFET for amplifier application, and these topologies, which make use of active loads and the controllable *W/L* ratios during fabrication, are presented in Chapter 8. Basically, only the biasing arrangements are changed when considering a MOSFET as opposed to a JFET, since the simplified small-signal models are the same.

The MOSFET finds its greatest application in logic circuits. In this chapter we illustrated selected applications including the basic switch, active load, and inverter. In Chapter 13

TABLE 5.3 MOSFET and JFET Summary

MOSFET

Parameter	NMOS	PMOS
V_t (enhancement)	>0	<0
V_t (depletion)	<0	>0
V_{DS} and I_D	>0	<0
Quadrant	First	Third
Transfer function (saturation region)	$i_D = k(v_{GS} - V_t)^2(1 + \lambda v_{DS})$, where $k = (KP/2)(W/L)$ and $|v_{DS}| > |v_{GS} - V_t|$	
KP	$\mu_n C_{ox}$	$\mu_p C_{ox}$
Transconductance	$g_m = 2k(V_{GS} - V_t)$	
Output resistance	$r_d = 1/\lambda I_D$	
Transfer function (ohmic region)	$i_D = k[2(v_{GS} - V_t)v_{DS} - v_{DS}^2]$ and $|v_{DS}| < |v_{GS} - V_t|$	

JFET

Parameter	n-Channel	p-Channel
V_P	<0	>0
I_{DSS}	>0	<0
V_{DS} and I_D	>0	<0
Quadrant	First	Third
Transfer function (saturation region)	$i_D = I_{DSS}(1 - v_{GS}/V_P)^2(1 + \lambda v_{DS})$ $i_D = BETA(V_P - v_{GS})^2(1 + \lambda v_{DS})$, where $BETA = I_{DSS}/V_P^2$ and $|v_{DS}| > |v_{GS} - V_P|$	
Transconductance	$g_m = (-2 I_{DSS}/V_P)(1 - V_{GS}/V_P)$	
Output resistance	$r_d = 1/\lambda I_D$	
Transfer function (ohmic region)	$i_D = BETA[2(v_{GS} - V_P)v_{DS} - v_{DS}^2]$ and $|v_{DS}| < |v_{DS} - V_P|$	

we consider MOSFET logic circuits, and in Chapter 15 we provide more depth in the use of MOSFETs as semiconductor memory elements.

The JFET and MOSFET data sheets included in this chapter to illustrate specifications and circuit applications are used in subsequent chapters to develop design and analysis examples. Fabrication of a variety of FET devices is discussed in Chapter 18. A detailed discussion of the use of GaAs MESFETs in microwave and very high-speed digital circuits is beyond the scope of this text.

SURVEY QUESTIONS

1. What do the acronyms in Fig. 5.2 stand for?
2. What circuit diagram symbols are used for NMOS, PMOS, enhancement-mode, and depletion-mode MOSFETs and n- and p-channel JFETs?
3. How would you qualitatively describe the operation of a MOSFET?
4. Can you sketch the I_D-V_{DS} characteristics for both an NMOS and PMOS labeling the regions of operation?

5. What are the definitions and interpretations for key MOSFET parameters as listed in Fig. 5.12?
6. What are the basic equations for the MOSFET saturation region i_D-v_{DS} transfer function and transconductance?
7. Draw the schematic diagrams for MOS inverters with active and passive loads.
8. What are the key design equations describing the operation of MOS inverters with active and passive loads?
9. Draw the schematic diagram and describe the operation of a CMOS inverter and a MOS analog switch.
10. Write a SPICE netlist, using a LEVEL 1 MOS model, for the circuits outlined in the three previous questions.
11. Qualitatively describe the operation of a JFET.
12. Can you sketch the I_D-V_{DS} characteristics for both an *n*-channel and *p*-channel JFET labeling the regions of operation?
13. What are the definitions and interpretations for key JFET parameters as listed in Figs. 5.31 and 5.32?
14. What are the basic equations for the JFET saturation region I_D-V_{DS} transfer function and transconductance?
15. How would you draw the schematic diagrams and desribe the operation of a common-source, common-gate, and common-drain JFET amplifier?
16. How would you write a SPICE netlist for the circuits outlined in the previous question?
17. How does the MESFET relate to the JFET in terms of topology, fabrication technology, and electrical characteristics?

PROBLEMS

Problems marked with an asterisk are more challenging.

5.1 Evaluate *k* for an *n*-channel MOSFET whose SiO_2 thickness is 400 Å and $W/L = 10$. Assume a channel mobility of 700 cm^2/V-sec.

5.2 Evaluate *k* for *p*-channel MOSFET whose SiO_2 thickness is 400 Å and $W/L = 10$. Assume a channel mobility of 250 cm^2/V-sec.

5.3 Compare the transconductances for the NMOS and PMOS devices in Problems 5.1 and 5.2. Assume the drain currents are the same.

5.4 Sketch and label the i_D-v_{DS} characteristics for $V_{GS} = 1, 2, 3,$ and 4 V for an *n*-channel MOSFET whose SiO_2 thickness is 400 Å and $W/L = 10$. Assume a channel mobility of 700 cm^2/V-sec, and $V_t = 1.0$ V and $\lambda = 0$. Extend your results to $V_{DS} = 20$ V.

5.5 * Generate the i_D-v_{DS} characteristics for the NMOS device in Problem 5.4 using SPICE and compare the SPICE graphical output with the sketch prepared in Problem 5.4.

5.6 * Rework Problems 5.4 and 5.5 but assume $\lambda^{-1} = 50$ V.

5.7 Sketch and label the i_D-v_{DS} characteristics for $V_{GS} = -1, -2, -3,$ and -4 V for a *p*-channel MOSFET whose SiO_2 thickness is 400 Å and $W/L = 10$. Assume a channel mobility of 250 cm^2/V-sec, $V_t = -1.0$ V, and $\lambda = 0$. Extend your results to $V_{DS} = -20$ V.

5.8 * Generate the i_D-v_{DS} characteristics for the PMOS device in Problem 5.2 using SPICE and compare the SPICE graphical output with the sketch prepared in Problem 5.7.

5.9 * Rework Problems 5.7 and 5.8 but assume $\lambda^{-1} = -50$ V.

5.10 Plot transconductance g_m as a function of V_{GS} and I_D for the NMOS whose specifications are given in Problem 5.4.

  (a) Compare your plot to the forward transfer admittance, y_{fs}, plot in Fig. 5.12.
  (b) Also compare these g_m values from your calculation and from Fig. 5.12 to that found for an *npn* BJT operating at the same current levels.

5.11 Plot transconductance g_m as a function of V_{GS} and I_D for the PMOS whose specifications are given in Problem 5.7. Compare your plot to the forward transfer admittance, y_{fs}, plot in Fig. 5.47.

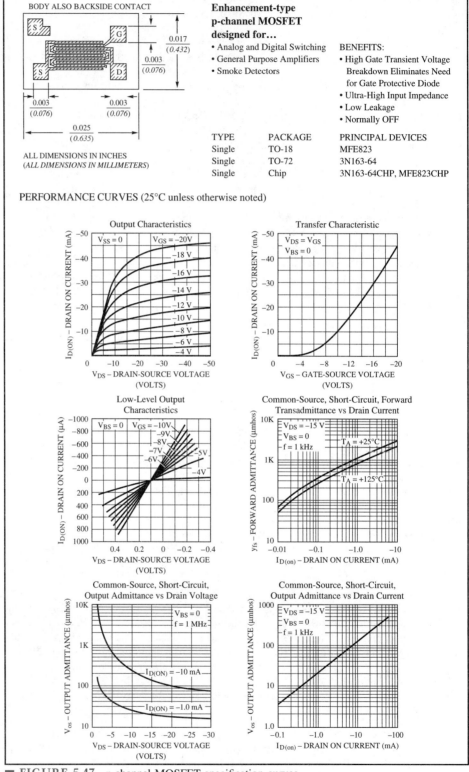

■ **FIGURE 5.47** *p*-channel MOSFET specification curves (courtesy of Siliconix, Inc.).

5.12 A p-channel enhancement-mode MOSFET, whose characteristics are given in Fig. 5.47, is to be used as a current source. Using the i_D versus v_{DS} characteristics shown, a single 20-V battery, and resistors of your choice, design a circuit that will provide a load current of 20 mA for some range of R_L. Label all component values and the range for R_L.

5.13 Using the circuit shown in Fig. 5.48 and the i_D versus v_{DS} characteristics of Fig. 5.47, complete the following:

(a) Draw and label the dc load line.
(b) Plot a bias curve (i_D versus v_{GS} for this circuit) on the transfer characteristic. Determine the Q-point values for V_{GSQ}, and I_{DQ}.
(c) Determine the value for g_m at the Q-point.
(d) Graphically illustrate the maximum v_O signal swing possible for operation in the saturation region.

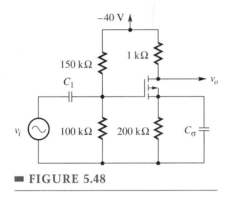

■ **FIGURE 5.48**

5.14 Develop a SPICE model for the Fig. 5.47 PMOS device using the i_D-v_{DS} characteristics.

5.15 Develop a SPICE frequency-independent (i.e., no capacitors need be considered) model for the Fig. 5.12 NMOS device using the data in the Max column.

5.16* Apply your SPICE model from Problem 5.15 to Fig. 5.48 with an NMOSFET. Use $V_{DD} = 40$ V. Using a dc and transient analysis, determine a value for g_m at the Q-point and graphically illustrate the maximum v_O signal swing possible for operation in the saturation region. Compare your results with those obtained in Problem 5.13.

5.17 For the MOSFET in Fig. 5.49, $k = 3$ mA/V^2 and $V_t = 2$ V. Let $R_1 + R_2 = 100$ kΩ.

(a) Compute I_D when $R_1 = R_2$.
(b) What are the values of R_1 and R_2 when the MOSFET is operating just at the edge of saturation?
(c) What are the values of R_1 and R_2 such that the power dissipated by the MOSFET is a maximum? Realize that constant power hyperbolas are obtained using $P_D = V_{DS} \times I_D$.

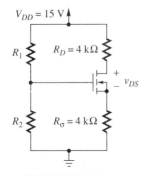

■ **FIGURE 5.49**

5.18 Use SPICE to check your answers to Problem 5.17.

5.19 For the MOSFET in Fig. 5.50, $k = 2.5$ mA/V^2 and $V_t = 2$ V. Let $V_{DD} = 10$ V and $R_D = 10$ kΩ. Sketch a set of characteristic curves, and on these curves plot a load line. What values of v_{IN} are required to obtain operation in the saturation region?

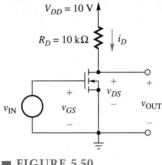

■ **FIGURE 5.50**

5.20 Use SPICE to draw a set of I-V characteristics for the FET in Problem 5.19. Compare your results with Problem 5.19.

5.21 For both MOSFETs in Fig. 5.51, $k = 2.5$ mA/V^2 and $V_t = 2$ V. Let $V_{DD} = 10$ V and $R_D = 5$ kΩ. Find v_{OUT} and i_D for the following:

(a) $V_1 = V_2 = 0$ V.
(b) $V_1 = V_2 = 10$ V.
(c) $V_1 = 0$ V, $V_2 = 10$ V.
(d) $V_1 = 0$ V, $V_2 = 5$ V.
(e) $V_1 = 3$ V, $V_2 = 5$ V.

5.22 Two MOSFETs are connected as shown in Fig. 5.52. Derive expressions for $v_O(t) = f[v_I(t)]$. If $k_1 = 10$ mA/V^2, $k_2 = 1.11$ mA/V^2, $V_{t1} = 1$ V, $V_{t2} = -3$ V, and $V_{DD} = 10$ V, sketch the transfer function, v_O versus v_I, for $0 < v_I < 10$ V.

5.23★ Repeat Problem 5.21 using SPICE.

5.24★ Figure 5.53 shows a circuit that exploits the square-law drain characteristic of Eq. (5.12).

(a) Let $v_1(t)$ and $v_2(t)$ be two independent input voltages, and show that the MOSFET can be used to multiply the two input signals because i_D contains a term that is the product of the input signals.
(b) Assume $v_1(t) = A \sin \omega_1 t$, and $v_2(t) = B \sin \omega_2 t$, and $\omega_1 > \omega_2$. Show that $i_D(t)$ contains frequencies at $\omega_1 \pm \omega_2$.

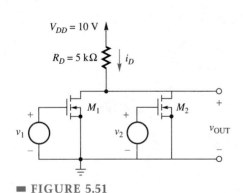

■ **FIGURE 5.51**

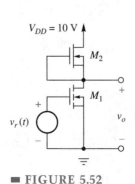

■ **FIGURE 5.52**

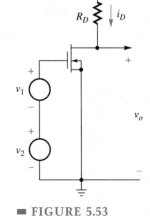

■ **FIGURE 5.53**

5.25 Figure 5.54 shows a FET switch controlling the input to an ideal operational amplifier with the inverting input operaing at virtual ground. For the MOSFET, $k = 1$ mA/V^2 and $V_t = 2$ V. If $v_1(t) = 5 \sin \omega t$ V:

(a) What value of V_C will ensure the FET is ON? Assume $R_{ON} \leq 250\ \Omega$.
(b) What value of V_C will ensure the FET is OFF?

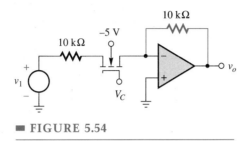

■ FIGURE 5.54

5.26 An enhancement-mode MOSFET has $k = 1$ mA/V^2 and $V_t = 2$ V. Find R_{ON} when $V_{GS} = 2, 3,$ and 4 V.

5.27 Figure 5.55 shows a FET switch using a depletion-mode MOSFET. Let $k = 1$ mA/V^2, $V_t = -2$ V, and $v_1(t) = 5 \sin \omega t$ V.

(a) What value of V_C will ensure that the FET is ON? Assume $V_{GS} - V_t \geq 1$ volt. Assume $R_{ON} \leq 200\ \Omega$?
(b) What value of V_C will ensure that the FET is OFF?

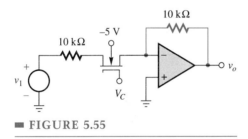

■ FIGURE 5.55

5.28 Assume an n-channel JFET is doped at $N_D = 10^{16}$/cm^3. Plot the width of the depletion region, w, for $-10 < V_R < 0$ V. Assume $\phi_J = 0.8$ V.

5.29 Explain why the doping density for the channel region in a JFET must be less than the doping density for the gate.

5.30 Assume an n-channel JFET is doped at $N_D = 10^{16}$/cm^3. Assume $V_{DS} = 2$ V, $V_{GS} = 0$, $L = 50\ \mu$m, and $\phi_J = 0.8$ V. Plot $w(x)$.

5.31 Compute the voltage required to pinch off an n-channel JFET with $N_D = 5 \times 10^{16}$/cm^3 and a channel thickness of $T = 0.5\ \mu$m.

5.32* Calculate $|V_P|$ as a function of doping density (10^{15} cm$^{-3} < N_D < 10^{18}$ cm^{-3}) in an n-channel JFET with a channel thickness of $T = 0.5\ \mu$m. Do you have any recommendations on doping densities if $|V_P|$ values of several volts are required?

5.33 If an n-channel JFET has $V_P = -5$ V, compute the V_{GS} to yield pinch-off operation when

(a) $V_{DS} = 4$ V.
(b) $V_{GS} = 2$ V.

5.34 Compute $r_{DS(ON)}$ for a p-channel JFET with $L = 3\ \mu$m, $d = 10\ \mu$m, $N_A = 10^{17}$/cm^3, and an average channel thickness of $0.5\ \mu$m. Compare this result to the $r_{DS(ON)}$ of an n-channel JFET with an identical geometry and $N_D = 10^{17}$/cm^3.

5.35 Which of the JFETs in Problem 5.34 would be a good switch? Explain. What parameters would you change in the device design to further improve the switch performance?

5.36 For a *p*-channel JFET, $I_D = -1.0$ mA, $V_{DS} = -10$ V, $V_P = 3$ V, $\lambda^{-1} = -100$ V, and $I_{DSS} = -5$ mA. Compute V_{GS} and r_d and g_m at the *Q*-point.

5.37 Sketch several of the i_D versus v_{DS} characteristics for a *p*-channel JFET with $I_{DSS} = -5$ mA, $V_P = 3$ V, and $\lambda^{-1} = -100$ V. Label all key features.

5.38 Generate the i_D-v_{DS} characteristics for the *p*-channel JFET in Problem 5.37 using SPICE and compare the SPICE graphical output with the sketch prepared in Problem 5.36.

5.39 In the *n*-channel JFET in Fig. 5.56, $I_{DSS} = 2$ mA, $V_P = -3$ V, $V_{DD} = 10$ V, and $R_D = 5$ kΩ. Sketch a set of characteristic curves and on these curves plot a load line. What values of v_{IN} are required to obtain operation in the saturation region?

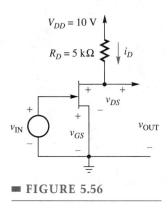

■ FIGURE 5.56

5.40 Use SPICE to plot v_{DS} versus v_{IN} (the transfer characteristic) for Problem 5.39.

5.41 The circuit shown in Fig. 5.57 uses a JFET with $V_P = -4$ V and $I_{DSS} = 4$ mA. In addition, $R_1 = \infty$, $R_2 = 1$ MΩ, $R_D = 2$ kΩ, and $V_{DD} = 10$ V. Assuming saturation, specify a value for R_σ and find the resultant values for I_D, V_{GS}, and V_{DS}. Then prove the assumption of saturation was valid.

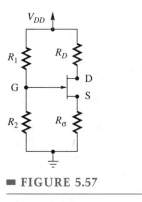

■ FIGURE 5.57

5.42 Use SPICE to verify the *Q*-point computed in Problem 5.41.

5.43 The circuit in Fig. 5.57 uses a JFET with $V_P = -4$ V and $I_{DSS} = 16$ mA. In addition, $R_1 = R_2 = 1$ MΩ, $R_D = 0$ Ω, $R_\sigma = 1$ kΩ, and $V_{DD} = 16$ V. Find I_D, V_{GS}, and V_{DS}.

5.44 Use SPICE to verify the *Q*-point computed in Problem 5.43.

5.45 Design a bias circuit of the type illustrated in Fig. 5.57. Assume an *n*-channel JFET with $V_P = -5$ V and $I_{DSS} = 10$ mA. Let $V_{DD} = 20$ V, and design for a *Q*-point of $i_D = 5$ mA and $V_{DS} = 10$ V.

5.46 The circuit of Fig. 5.49 is to be used as a common-source amplifier, with the input voltage $v_i = V_s \sin 377t$ V applied to the gate node through a capacitor C_1, with $R_S = 0$ Ω and

CHAPTER 5   THE FIELD-EFFECT TRANSISTOR

the output voltage, v_o, taken from the drain node. The resistance of the voltage source can be neglected because of the high input resistance of this amplifier. The enhancement mode NMOS has $k = 1.25$ mA/V^2, $V_t = 1.5$ V, and $\lambda = 0$. Assume $R_1 = R_2 = 100$ kΩ.

(a) Obtain the Q-point and calculate g_m.
(b) Draw a small-signal model for the amplifier circuit, and compute $A_{vi} = v_o/v_i$.
(c) Repeat part (b) for the situation where R_σ is bypassed with a large capacitor, i.e., $X_C \approx 0$.

5.47* Apply your SPICE model to the Problem 5.46 circuit. Using a dc and transient analysis, determine the operating point information and plot the output signal for the case where R_σ is bypassed. Choose the values of C_1 and C_σ such that the X_C is small at the frequency of interest. Try values of V_S to determine the maximum v_o signal amplitude to be obtained without serious distortion.

5.48 The circuit of Fig. 5.49 is to be used as a common-source amplifier, with the input voltage $v_i = V_s \sin 377t$ V applied to the gate node through a capacitor C_1, and the output voltage, v_o, taken from the drain node. The resistance of the voltage source can be neglected because of the high input resistance of this amplifier. The enhancement-node NMOS has $k = 1.25$ mA/V^2, $V_t = 1.5$ V, and $\lambda = 0$. Change R_σ to 1 kΩ, and bypass it with a large capacitor, i.e., assume $X_C \approx 0$.

(a) Design this circuit to obtain $I_D = 2$ mA. You must specify R_1 and R_2 such that $R_G > 100$ kΩ.
(b) Find g_m and draw a small-signal model for this amplifier circuit.
(c) Compute a value for $A_{vi} = v_o/v_i$.
(d) What values of C_1 and C_σ should be used?

5.49 Verify your Problem 5.48 design results using SPICE.

5.50* Use the bias circuit illustrated in Fig. 5.57. For a 2N3684, 2 V $< |V_P| <$ 5 V and 1.6 $< I_{DSS} <$ 7 mA. Let $V_{DD} = 20$ V. Assume that larger values of V_P are associated with larger values of I_{DSS}. Find values for R_1, R_2, and R_σ so that 0.8 mA $< I_D <$ 1.2 mA. What is the largest R_D that can be used such that the transistor is still saturated?

5.51 For your design of Problem 5.50, compute the range in $A_{vi} = v_o/v_i$. Do this calculation using both a large bypass capacitor across R_σ and no bypass capacitor. Use your value of R_D obtained in Problem 5.50.

5.52* Verify your Problem 5.50 design using SPICE. Compare your A_{vi} calculations with that obtained using a SPICE ac analysis.

5.53 Consider the circuit in Fig. 5.57 with the JFET of Probelm 5.50. Suppose R_1 is open circuited. What is the range of values for I_D if $R_D = 1$ kΩ and $R_\sigma = 3$ kΩ? Let $V_{DD} = 12$ V.

5.54 A 2N5457 JFET is used in the circuit shown in Fig. 5.38. Let $V_{DD} = 10$ V, $V_{GG} = -1$ V, and $R_D = 1$ kΩ.

(a) Using the typical I_{DSS} and V_P parameters from Fig. 5.31, find I_{DQ}, V_{GSQ}, V_{DSQ}, and A_{vi}. Assume typical values for I_{DSS} and V_P ($V_{GS(\text{off})}$) are midway between the minimum and maximum values.
(b) Repeat part (a) assuming a 2N5457 with both a minimum or maximum set of electrical parameters.

5.55 Verify your Q-point values for Problem 5.54 using SPICE.

5.56 Using the small-signal FET model with a finite r_d, derive the output resistance, R_o, for the common-source amplifier with a partially bypassed R_σ.

5.57 Using the small-signal FET model, derive the output resistance, R_o, for the common-gate amplifier in Fig. 5.39. The R_o is to include R_D but neglect r_d.

5.58 Using the JFET model that neglects r_d, show that the output resistance in Fig. 5.40 is given by $R_o = R_\sigma \| 1/g_m$.

5.59* Design a circuit to measure g_{m0} and r_d for the FET. Explain your step-by-step procedure.

5.60 The n-channel JFET in Fig. 5.58 is used as a current source, and $I_{DSS} = 2$ mA and $V_P = -4$ V. For what range of resistance values, R, and V_{DD} is the current through R a constant at 2 mA?

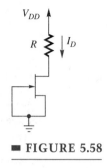

■ **FIGURE 5.58**

5.61 For what range of R_D will the circuit shown in Fig. 5.58 operate as a constant current source? Assume $V_{DD} = 10$ V, $I_{DSS} = 5$ mA, and $V_P = -3$ V.

REFERENCES

1. Shockley, W. "A Unipolar Field-Effect Transistor," *Proceedings of the IRE*, November 1952, p. 1365. The first analysis of a JFET. The original paper.
2. Dacey, G. C., and I. M. Ross. "The Field-Effect Transistor," *Proceedings of the IRE*, Vol. 41, 1953, p. 973. A description of the first working JFET based on Ref. 1.
3. Hansen, J. R. "Junction Field-Effect Transistors," Chap. 3 in *Fundamentals of Silicon Integrated Device Technology*, Vol. 2 (R. M. Burger and R. P. Donovan, Eds.). Englewood Cliffs, N.J.: Prentice-Hall, 1968. Early tutorial summarizing basic device characteristics. Used as a reference for later work.
4. Jacoboni, C., C. Canali, G. Ottaviani, and A. Quaranta. "A Review of Some Change Transport Properties of Silicon," *Solid-State Elecronics*, Vol. 20, 1977, p. 77.
5. Smith, P., M. Inoue, and J. Frey. "Electron Velocity in Silicon and GaAs at Very High Electron Fields," *Applied Physics Letters*, Vol. 37, 1980, p. 797. References 4 and 5 are both advanced-level technical papers that discuss nonohmic effects in silicon, effects that are to a large degree empirical. Nevertheless, the data are used to predict device functioning and performance.
6. Sevin, L. J., Jr., Texas Instruments Staff. *Field-Effect Transistors*. New York: McGraw-Hill, 1965. An early reference on FETs. Texas Instruments, among other large manufacturers, prepares a number of tutorials and application notes that the student should be aware of. These publications, whether available from the semiconductor house or from a commercial publisher, are often the best source of current device information and application.
7. Sze, S. M. *Physics of Semiconductor Devices*, 2d ed. New York: Wiley-Interscience, 1981. An advanced reference covering a broad spectrum of solid-state devices. In particular, Chapter 6 on the JFET and Chapter 8 on the MOSFET are very good in conveying an in-depth understanding of the device physics and developments in current research areas.
8. van der Ziel, A. *Solid-State Physical Electronics*, 3d ed. Englewood Cliffs, N.J.: Prentice-Hall, 1976. An advanced treatment of solid-state physics with an emphasis on detailed theoretical discussion on how a device functions.
9. Zahng, D., and M. Atalla. "Silicon-Silicon Dioxide Field-Induced Surface Devices," *IRE Solid-State Circuits Research Conference*, Carnegie Institute of Technology, Pittsburgh, 1960. An early theoretical presentation on the MOS device.
10. Kahng, D. "A Historical Perspective on the Development of MOS Transistors and Related Devices," *IEEE Transactions on Electron Devices*, Vol. ED-23, 1976, p. 655. A very readable review.
11. Banzhof, W., *Computer-Aided Circuit Analysis Using PSPICE*™, 2nd ed. Englewood Cliffs, N.J.: Prentice-Hall, 1992, especially pp. 117–121.

SUGGESTED READINGS

Divekar, Dileep A. *FET Modeling for Circuit Simulation*, New York: Kluwer Academic Publishers, 1988.

Pierret, Robert F. *Field-Effect Devices*, 2nd ed. Reading, Mass.: Addison Wesley. This is Volume 4 of a 10-volume set, each covering a different aspect of semiconductor device physics and semiconductor devices. Also Volume VII, *Advanced MOS Devices* by D. K. Schroder.

Tsividis, Yannis P. *Operation and Modeling of the MOS Transistor*, New York: McGraw-Hill, 1987.

It is impossible to list the many articles devoted to FET operation and circuits. The interested student is urged to at least scan current copies of *IEEE Spectrum, Electronics*, and *Electronic Design* for tutorial and application information. Theoretical in-depth material can be found in *IEEE Proceedings*, various IEEE transactions, and the *Journal of Applied Physics*. Of course, manufacturers continually publish device specifications and applications, and these too present the latest information.

PART II

LINEAR CIRCUITS

INTRODUCTION

In Chapters 4 and 5, we briefly introduced a variety of circuits with Q-points located in the linear-active region to obtain amplifier operation. This discussion was used to support our discussion of device operation and models. In Part II, we focus on the analysis and design of linear circuits, especially linear circuits that comprise the key building blocks of integrated circuits.

In order to proceed, we introduce some key analysis tools in Chapter 6. These tools include an introduction to Bode plots used to look at frequency response. Many of you are already familiar with the concept of frequency response. For example, whenever you refer to the 3-kHz bandwidth of a telephone channel, the 6-MHz bandwidth of a standard television channel, the greater than 10 MHz required for high-definition TV (HDTV), the hundreds of megahertz found in photonic optical fiber systems, or the very modest 20-Hz to 20-kHz bandwidth requirements of a good stereo sound system, you are explicitly specifying the frequency-dependent behavior of a circuit or system. Another tool we introduce in Chapter 6 is negative feedback, already implicitly invoked in Chapter 1 when we discussed the ideal operational amplifier. Virtually every circuit, integrated circuit, and system employs negative feedback to tailor operating characteristics to given specifications despite variations in device and circuit parameters. Indeed feedback is present as a consequence of the internal topologies in semiconductor devices. You will also observe in your other electrical engineering courses that feedback concepts are employed in control systems and power systems—basically all complex electronic and mechanical systems with which you will be dealing. Additionally in Chapter 6, these key tools are applied to nonidealities of operational amplifiers. Our application vehicle will be data sheets for two classical operational amplifiers, the μA741 and the CA3140. Basic SPICE models are developed from these data sheets and used to illustrate the general validity of the ideal operational amplifier model from Chapter 1 and its use in negative feedback circuits.

Chapter 7 includes details on establishing the Q-point in the linear-active region of BJT, MOS, and JFET circuits. We proceed rapidly from biasing single devices, as we briefly addressed in Chapters 4 and 5, to biasing in integrated circuits (ICs) using current sources. In later chapters, particularly Chapter 10, we will use these current sources as a necessary substitute for resistive loads in ICs because current sources effectively synthesize high resistance values, which are unavailable in IC technologies (Chapter 18).

Detailed frequency-independent and frequency-dependent small-signal hybrid-π models for BJTs, MOSFETs, and JFETs based on device physics and topologies are presented in Chapters 8 and 9, respectively. These models are employed to study the properties of basic, single-transistor circuits. Depending on the level of algebraic complexity, we will use SPICE to support our analysis and design efforts. The Bode plot is our principal vehicle for presenting frequency-dependent behavior. Multistage amplifiers are also studied.

The differential amplifier, employing either emitter or source-coupled pairs, found in virtually all ICs, is presented in Chapter 10. Models and circuit results from Chapter 8 are combined with Chapter 7 current sources to obtain circuits using a minimum number of resistors and no coupling or bypass capacitors. The device matching and specification tracking inherent with basic semiconductor technology is important to meet the required high-performance specifications. It is also interesting to note that many of the circuits in Chapter 10 are used in digital IC families. This aspect is explored in Part III.

Although we introduced basic feedback concepts in Chapter 6, we did not apply them to individual circuits. In Chapter 11, we apply feedback to amplifiers. A key task will be to identify which of the four basic topologies best describes a particular configuration. Key circuit performance characteristics are dependent on which of the four topologies are employed. When we consider the frequency response of multistage amplifiers, we must be concerned about stability. An amplifier might be very stable at low frequencies, where the phase of the feedback signal is negative, but become unstable at higher frequencies where the phase of the feedback signal changes. Indeed, we may deliberately design a circuit to have *positive* feedback at some frequency in order to obtain an oscillator, i.e., become a signal generator whose output is no longer controlled by an input signal. Virtually all electronic systems require an oscillator, which serves as a signal source. For example, a figure of merit for computers is the clock rate, i.e., the clock is an oscillator. We discuss several typical oscillator designs in Chapter 11.

The concepts presented in all of the preceding chapters, supported by SPICE, are used in Chapter 12 to perform a detailed circuit design of an operational amplifier using a BJT array followed by a detailed analysis for the μA741 and CA3140. Other common linear ICs including voltage regulators and analog multipliers are also discussed, as are trends in linear IC performance.

CHAPTER 6

BASIC ANALYSIS TECHNIQUES

6.1 Operational Amplifier Specifications
6.2 General Feedback Concepts
6.3 Bode Magnitude and Phase Plots: Standard Forms
Summary

Survey Questions
Problems
References
Suggested Readings

Two basic tools are discussed in this chapter: techniques for analyzing frequency-dependent circuits and systems, and negative feedback. Expanding on the ideal operational amplifier concepts presented in Chapter 1, we apply both of these tools to nonideal operational amplifiers. It is important to understand that even though we will focus on operational amplifier circuitry and applications, operational amplifiers usually comprise only a small, albeit important, portion of any electronic system. However, the circuits, combinational topologies, and analysis and design techniques are sufficiently universal that a detailed study of the operational amplifier, in this chapter and in Chapter 12, will allow us to extend these concepts to many other analog and digital integrated circuits (ICs) and systems.

Furthermore, we will use the ubiquitous μA741 as our tutorial analysis vehicle. Early versions of the μA741 operational amplifier were introduced in 1966, and in the ensuing decade it became one of the most widely used ICs. Since then, literally thousands of operational amplifiers have been introduced as either stand-alone ICs or as part of a larger IC. Many of these stand-alone operational amplifiers are designed with special features; but it is interesting to observe that many of the newer designs advertise their pin-for-pin compatibility with a μA741. The μA741 is a BJT-based integrated circuit. Our tutorial vehicle for operational amplifiers with an MOS input stage is the CA3140. As with the μA741, the CA3140 represents a mature technology and design whose performance has been superseded.

IMPORTANT CONCEPTS IN THIS CHAPTER

- Ideal operational amplifier criteria
- Reading and interpretating an operational amplifier data sheet
- Definitions of input offset voltage and current, input bias current, input resistance, large-signal voltage gain, output resistance, and output short-circuit current
- Frequency-independent and frequency-dependent SPICE models for the operational amplifier
- Summing point constraint applications
- Basic feedback system block diagram, Figure 6.9
- Negative feedback
- Definitions of loop-gain, open-loop gain, and closed-loop gain
- Use of negative feedback for gain stabilization and distortion reduction
- Definition and numerical manipulation of the decibel
- Construction of the Bode magnitude and phase plots
- Computation and identification of the -3-dB response point
- Bode magnitude and phase plots for a single-pole low-pass filter and for a single-pole high-pass filter
- SPICE ac analysis of the single-section low-pass and high-pass filters
- Bode plot magnitude and phase errors compared to the actual transfer function plots
- Bode magnitude and phase plots for a band-pass filter
- Use of negative feedback for broadbanding
- Definition of the gain-bandwidth product
- Approximations for systems having a multicorner frequency response

6.1 OPERATIONAL AMPLIFIER SPECIFICATIONS

In general, the internal circuitry of an operational amplifier, introduced in Chapter 1, can be divided into three sections (see Fig. 6.1). The *input section* is used to obtain a differential input (Chapter 10) with high input resistance, appropriate level shifting for dc coupling to the following cicuitry, and sometimes differential-to-single-ended signal conversion, all the while providing a significant portion of the overall voltage gain. Circuits appropriate

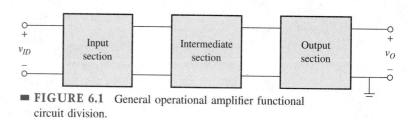

FIGURE 6.1 General operational amplifier functional circuit division.

CHAPTER 6 BASIC ANALYSIS TECHNIQUES

to realizing these specifications in BJT, MOSFET, and JFET technologies will be presented in detail in Chapter 7 through 10.

The *intermediate section* is designed to provide the rest of the voltage gain while not significantly loading the input circuitry, and to provide the appropriate direct-coupled dc level to the output circuitry. As we will observe when we discuss feedback in Chapter 11, the overall frequency response is usually implemented in this section.

The *output section* usually consists of a class B or AB amplifier with unity voltage gain of sufficient power capability to drive relatively low resistance loads. This means the output resistance of this stage should be low. These class B and AB basic circuit topologies are presented in Chapter 10.

Recall from Chapter 1 that the ideal amplifier criteria are given by

1. Infinite a_v
2. Infinite r_i
3. Zero r_o
4. Infinite bandwidth
5. Zero offset.

■ Input stage, often a differential amplifier, as discussed in Chapter 10. The Intermediate section provides a large voltage and/or current gain. The output section provides impedance matching to drive a load.

We present data sheets for the μA741 and CA3140 to operation amplifiers to illustrate how these operational amplifiers specifications compare with the ideal specifications. The bandwidth comparison is delayed until Section 6.3 where we introduce techniques for displaying bandwidth specifications.

■ These data sheets and models are resident in many CAD programs including more advanced versions of PSPICE. Also many manufacturers offer data sheets on computer diskettes and CD ROMs as well as on the World-Wide Web (Internet).

Data Sheet Overview for a μA741 Operational Amplifier

An abbreviated data sheet[1] for a μA741 is given in Fig. 6.2. First observe that the first set of tabulated electrical characteristics is given for $V_{CC} = V_{EE} = 15$ V and ambient temperature of $T_A = 25°$C, $R_S \leq 10$ kΩ, and $R_L \geq 2$ kΩ. The second set of tabulated electrical characteristics extends the temperature performance from $-55°$C $< T_A < +125°$C. We now discuss the following parameters:

Input Offset Voltage. Ideally $V_O = 0$ if $V_I = 0$ but actually, due to component mismatches, this does not happen. The input offset voltage, V_{IO} or V_{OS}, is the dc voltage that, when applied between the input terminals, forces the output to zero [Fig. 6.3(a)]. Typical values are on the order of 1 mV, small enough to be ignored in many applications. Voltage V_{IO} is actually a random variable with an average close to zero, so the typical value from a data sheet is only the magnitude. Most packaged op amps provide terminals for external adjustment of the dc offset in critical applications.

Input Bias Current. If an input terminal is the base of a BJT, the external circuit must carry the dc base current, I_B; otherwise, the BJT would be cut off. If the dc resistance of the input circuit is large (100 kΩ, say), I_B produces a significant voltage drop, which is then amplified at the output—another unwanted dc offset. The effect of I_B can be reduced by adding a comparable resistance in the other input lead (Problem 6.4), sometimes bypassed to prevent any effect on signal gain.

When the input stage uses MOSFETs or JFETs, the bias current is greatly reduced,

μA741 and μA741C
Electrical Characteristics $V_S = \pm 15$ V, $T_A = 25°C$ unless otherwise specified

Characteristic	Condition	μA741 Min	μA741 Typ	μA741 Max	μA741C Min	μA741C Typ	μA741C Max	Unit
Input Offset Voltage	$R_S \leq 10$ kΩ		1.0	5.0		2.0	6.0	mV
Input Offset Current			20	200		20	200	nA
Input Bias Current			80	500		80	500	nA
Power Supply Rejection Ratio	$V_S = +10, -20$ $V_S = +20, -10$ V, $R_S = 50$ Ω		30	150		30	150	μV/V
Input Resistance		.3	2.0		.3	2.0		MΩ
Input Capacitance			1.4			1.4		pF
Offset Voltage Adjustment Range			± 15			± 15		mV
Input Voltage Range					± 12	± 13		V
Common Mode Rejection Ratio	$R_S \leq 10$ kΩ				70	90		dB
Output Short Circuit Current			25			25		mA
Large Signal Voltage Gain	$R_L \geq 2$ kΩ, $V_{OUT} = 10$ V	50k	200k		20k	200k		
Output Resistance			75			75		Ω
Output Voltage Swing	$R_L \geq 10$ kΩ				± 12	± 14		V
	$R_L \geq 2$ kΩ				± 10	± 13		V
Supply Current			1.7	2.8		1.7	2.8	mA
Power Consumption			50	85		50	85	mW
Transient Response (Unity Gain) Rise Time	$V_{IN} = 20$ mV, $R_L = 2$ kΩ, $C_L \leq 100$ pF		.3			.3		μS
Transient Response (Unity Gain) Overshoot			5.0			5.0		%
Bandwidth			1.0			1.0		MHz
Stew Rate	$R_L \geq 2$ kΩ		.5			.5		V/μS

The following specifications apply over the range of $-55°C \leq T_A \leq 125°C$ for μA741, $0°C \leq T_A \leq 70°C$ for μA741C

Characteristic	Condition	μA741 Min	μA741 Typ	μA741 Max	μA741C Min	μA741C Typ	μA741C Max	Unit
Input Offset Voltage							7.5	mV
	$R_S \leq 10$ kΩ		1.0	6.0				mV
Input Offset Current								nA
	$T_A = +125°C$		7.0	200				nA
	$T_A = -55°C$		85	500				nA
Input Bias Current								nA
	$T_A = +125°C$		.03	.5				μA
	$T_A = -55°C$		.3	1.5				μA
Input Voltage Range		± 12	± 13					V
Common Mode Rejection Ratio	$R_S \leq 10$ kΩ	70	90					dB
Adjustment for Input Offset Voltage			± 15			± 15		mV
Supply Voltage Rejection Ratio	$V_S = +10, -20$; $V_S = +20, -10$ V, $R_S = 50$Ω		30	150				μV/V
Output Voltage Swing	$R_L \geq 10$ kΩ	± 12	± 14					V
	$R_L \geq 2$ kΩ	± 10	± 13		± 10	± 13		V
Large Signal Voltage Gain	$R_L = 2$ kΩ, $V_{OUT} = \pm 10$ V	25k			15k			
Supply Current	$T_A = +125°C$		1.5	2.5				mA
	$T_A = -55°C$		2.0	3.3				mA
Power Consumption	$T_A = +125°C$		45	75				mW
	$T_A = -55°C$		60	100				mW

■ **FIGURE 6.2** μA741 op amp electrical characteristics and performance curves.

DATA SHEET OVERVIEW FOR A µA741 OPERATIONAL AMPLIFIER

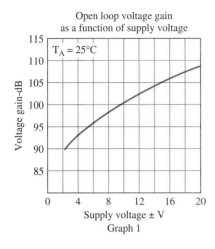

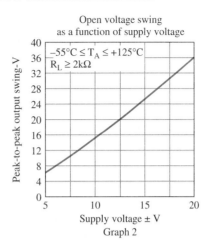

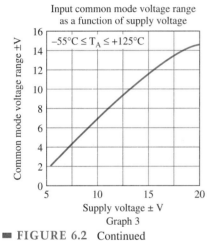

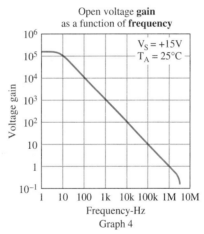

FIGURE 6.2 Continued

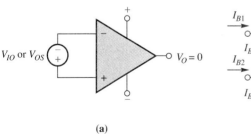

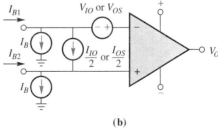

FIGURE 6.3 (a) Definition of input offset voltage.
(b) Equivalent circuit showing all three offset components.

allowing much larger resistors to be used, but the circuit must still provide a dc path for each input.

Input Offset Current. Some random mismatch occurs in the input transistors, so data sheets actually give the average I_B value.

$$I_B = \frac{I_{B1} + I_{B2}}{2}. \tag{6.1}$$

The difference is called the *input offset current*, I_{IO}.

$$I_{IO} = I_{B1} - I_{B2} \tag{6.2}$$

Like V_{IO}, I_{IO} is a random variable with a near-zero mean and data sheets give only the magnitude.

Figure 6.3(b) shows an equivalent circuit, which represents the three offsets as dc sources followed by an ideal op amp. I_B will be positive for an *npn* input stage.

Input resistance. The typical value is 2 MΩ. The ideal R_{id} is infinite. Observe that the minimum value is 15% of the typical value, illustrating the wide variations possible in the IC manufacturing process.

Large-signal voltage gain. Sometimes this is called the *open-loop voltage gain*. The typical value is 200,000. Observe the wide range of values possible. The ideal voltage gain would be infinite.

Output resistance. The typical value is 75 Ω. The ideal value would be $R_o = 0$ Ω. An $R_o = 0$ Ω implies an ideal voltage generator in a Thévenin equivalent circuit.

Output short-circuit current. The typical value is 25 mA. This corresponds to a minimum load resistance value of $R_L = V_o/25$ mA $= 600$ Ω for $V_o = 15$ V. Observe, however, that $R_L > 2$ kΩ is recommended for normal operation. The power supply current and resultant power consumption is sufficiently small that little, if any, heat sinking would be required.

In addition to the tabulated data, several of the key parameters are given graphically in Fig. 6.2 as a function of some of the operating conditions:

- *Graph 1:* Clearly, because all of the internal circuit operation is a function of V_{CC} and V_{EE}, we should expect a_v to be a strong function of V_{CC} and V_{EE}. This is illustrated in this graph.
- *Graph 2:* The key observation here is that for $V_{CC} = V_{EE} = 15$ V, the maximum output voltage swing is ±12.5 V. The output voltage can only come to within about 2 V of the power supply voltage.
- *Graph 3:* Similarly, the input dc voltage level on either of the differential inputs must be at least 3 V less than the power supply voltage.
- *Graph 4:* This graph illustrates the unity gain frequency, f_T, which is discussed in Section 6.3 as part of our discussion on graphically representing frequency response.

Often many more graphs are given, but this sampling does illustrate the type of information available.

■ PSPICE has a number of internally available libraries. For the time-being, it is useful to just use simplified models.

EXAMPLE 6.1

Use a SPICE simulation to obtain the voltage gain $A_v = v_o/v_s$ for the inverting amplifier shown in Fig. 6.4, where $R_S = 10$ kΩ and $R_F = 100$ kΩ, and the typical values for a μA741 operational amplifier are used. Use models as developed in Chapter 1 instead of PSPICE library models. Compare your results with Example 1.5.

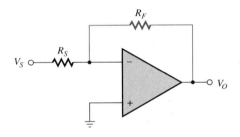

■ **FIGURE 6.4** Inverting amplifier.

■ Refer to Problem 6.59 for a somewhat more accurate two-port SPICE model.

SOLUTION A SPICE listing and equivalent circuit for the circuit in Fig. 6.4 is given in Fig. 6.5. The typical μA741 specifications are $R_{IN} = 2$ MΩ, $R_O = 75$ Ω, and $a_v = 200,000$.

■ We will show feedback elements in blue throughout the text.

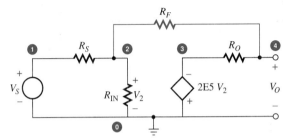

■ **FIGURE 6.5** Inverting amplifier SPICE circuit simulation setup.

```
INVERTING AMPLIFIER
VS 1 0 AC 1 0
RS 1 2 10K
RIN 2 0 2MEG
RF 2 4 100K
E1 0 3 2 0 200E3
RO 3 4 75
.AC DEC 1 1 1E7
.PRINT AC VM(4) VP(4) VM(2) VP(2)

   FREQ        VM(4)       VP(4)       VM(2)       VP(2)
1.000E+00   9.999E+00   1.800E+02   5.003E-05   0.000E+00
1.000E+01   9.999E+00   1.800E+02   5.003E-05   0.000E+00
1.000E+02   9.999E+00   1.800E+02   5.003E-05   0.000E+00
1.000E+03   9.999E+00   1.800E+02   5.003E-05   0.000E+00
1.000E+04   9.999E+00   1.800E+02   5.003E-05   0.000E+00
1.000E+05   9.999E+00   1.800E+02   5.003E-05   0.000E+00
1.000E+06   9.999E+00   1.800E+02   5.003E-05   0.000E+00
1.000E+07   9.999E+00   1.800E+02   5.003E-05   0.000E+00
```

Observe that the **AC ANALYSIS** was conducted at frequencies from 1 Hz to 10 MHz and because there are no frequency-dependent circuit elements the results for all node voltages were frequency independent. The phase shift is 180° as expected for an inverting amplifier. Appendix D has a discussion on SPICE syntax. Observe that by combining Eqs. (1.19) and (1.20) we have,

$$A_{vs} = \frac{V_o}{V_s} = -\frac{1}{R_S}\left(\frac{1}{\frac{1}{a_v R_S} + \frac{1}{a_v R_F} + \frac{1}{R_F}}\right) \cong -\frac{R_F}{R_S} \quad (6.3)$$

and this approximation is extremely good when we observe that $A_v = V_4/V_S = -9.999$, as obtained in Example 1.5 and which compares with $A_v = -R_F/R_S = -10.0$ if one assumes an ideal operational amplifier.

We generated an elementary, frequency-independent μA741 model for Example 6.1. Most commercial versions of SPICE, for example, PSPICE™ by Microsim,[2] offer more complex and accurate models for a variety of popular discrete devices and ICs. Shortly, we will present a frequency-dependent μA741 model and, in Chapter 12, we will model the complete μA741 from the schematic diagram.

DRILL 6.1

Use the worst case values for R_{IN} and a_v and rework Example 6.1

ANSWER With $R_{IN} = 300$ kΩ and $a_v = 50,000$, the SPICE simulation becomes

```
   FREQ        VM(4)       VP(4)       VM(2)       VP(2)
1.000E+00   9.998E+00   1.800E+02   2.001E-04   0.000E+00
```

$$A_v = v_4/V_S = -9.998.$$

Data Sheet Overview for a CA3140 Operational Amplifier[3]

In the mid-1970s, technology advancements permitted the incorporation of MOS and bipolar technology on the same IC die. This is called BiMOS or BiCMOS technology. To illustrate some of the unique features of an operational amplifier incorporating this technology, we present an abbreviated data sheet for the CA3140. This operational amplifier, along with many others, capitalizes on physical and electrical compatibility with the μA741, although the architecture and internal design permits improved performance in several areas. As with the μA741, the CA3140 has been superceded by hundreds, if not thousands, of other BiMOS ICs; nevertheless, the CA3140 serves as a good pedagogical analysis platform.

Key specifications of the CA3140 are given in Fig. 6.6. The CA series was developed by RCA and now marketed by Harris Corp.[3] A block diagram of the form shown in Fig. 6.1 is included in the data sheet. The first-stage voltage gain is 10 (20 dB), the second-stage voltage gain is 10,000 (80 dB), and the output stage is 1 (0 dB) for a total $a_v = 100,000$ (100 dB). This value is also provided in the data sheet typical value column as A_{OL}.

Electrical Characteristics for Equipment Design
At V⁺ = 15 V, V⁻ = 15 V, T_A = 25°C Unless Otherwise Specified

Characteristic		Test Conditions V⁺ = 15 V V⁻ = 15 V T_A = 25°C (Unless Specified Otherwise)	CA3140B			CA3140A			CA3140			Units
			Min	Typ	Max	Min	Typ	Max	Min	Typ	Max	
Input Offset Voltage, V_{IO}			—	0.8	2	—	2	5	—	8	15	mV
Input Offset Current, I_{IO}			—	0.5	10	—	0.5	20	—	0.5	30	pA
Input Current, I_I			—	10	30	—	10	40	—	10	50	pA
Large Signal Voltage Gain, A_{OL}		$V_O = 26\ V_{p-p}$ +12V, −14V $R_L = 2\ k\Omega$	50k	100k	—	20k	100k	—	20k	100k	—	V/V
			94	100	—	86	100	—	86	100	—	dB
Common-Mode Rejection Ratio, CMRR			—	20	50	—	32	320	—	32	320	μV/V
			86	94	—	70	90	—	70	90	—	dB
Common-Mode Input Voltage Range, V_{ICR}			−15	−15.5 to 12.5	1.2	−15	−15.5 to 12.5	12	−15	−15.5 to 12.5	11	V
Power-Supply Rejection Ratio, $\Delta V_{IO}/V^+$			—	32	100	—	100	150	—	100	320	μV/V
			80	90	—	76	80	—	76	80	—	dB
Maximum Output Voltage	V_{OM}^+	$R_L = 2\ k\Omega$	+12	+12.5	—	+12	+12.5	—	+12	+12.5	—	V
	V_{OM}^-		−14	−14.4	—	−14	−14.4	—	−14	−14.4	—	
Input Current, I^+			—	4	5.5	—	4	5.5	—	4	5.5	mA
Device Dissipation, P_D			—	120	165	—	120	165	—	120	165	mW
Input Current, I_I		$T_A = -55$ to +125°C V± = ±15 V	—	10	30	—	10	—	—	10	—	nA
Input Offset Voltage, V_{IO}			—	1.3	3	—	3	—	—	10	—	mV
Large-Signal Voltage Gain, A_{OL}		$V_O = 26\ V_{p-p}$ $R_L = 2\ k\Omega$	20k	100k	—	—	100k	—	—	100k	—	V/V
			86	100	—	—	100	—	—	100	—	dB
Maximum Output Voltage	V_{OM}^+	V± = ±22 V	+19	19.5	—	—	—	—	—	—	—	V
	V_{OM}^-		−21	−21.4	—	—	—	—	—	—	—	
Large-Signal Voltage Gain, A_{OL}		$R_L = 2\ k\Omega$ V_O = +19 V −21 V	20k	50k	—	—	—	—	—	—	—	V/V
			86	94	—	—	—	—	—	—	—	dB

Features:

- MOS/FET Input Stage provides:
a) Very high input impedance
  (Z_{IN}) − 1.5 TΩ typ.
b) Very low input current
  (I_I) − 10 pA typ. at ± 15 V
c) Low input offset voltage
  (V_{IO}) − to 2 mV max.
d) Wide common-mode input voltage range
  (V_{ICR}) − can be swung 0.5 volt below negative rail
e) output swing complements input common mode
- Directly replaces industry type 741 in most applications
- Operation from 4-to-44 volts single or dual supplies
- Internally compensated
- Characterized for +5 volts TTL supply systems with operation down to 4 volts
- Wide bandwidth −4.5 MHz unity gain at ±15 V or 30 V; 3.7 MHz at +5 V
- High slew rate −9 Volts/μs
- Fast settling time −1.4 μs typ. to 10 mV with a 10 V p-p signal
- Output swings to within 0.2 volt of negative supply
- Strobable output stage

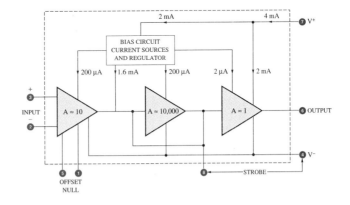

■ **FIGURE 6.6** CA3140 op amp data sheet.

EXAMPLE 6.2

Use a SPICE simulation to obtain the voltage gain v_o/v_s for the noninverting amplifier shown in Fig. 6.7, where $R_1 = 10 \text{ k}\Omega$ and $R_2 = 100 \text{ k}\Omega$, and the typical values for a CA3140 operational amplifier are used. Compare your results with Eq. (1.22).

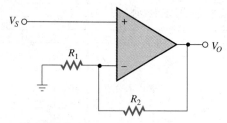

FIGURE 6.7 Noninverting amplifier.

SOLUTION A SPICE listing and equivalent circuit for the circuit in Fig. 6.7 is given in Fig. 6.8. The typical CA3140 specifications are $R_{IN} = 1.5 \text{ T}\Omega$, $R_O = 60 \text{ }\Omega$, and $a_v = 100{,}000$.

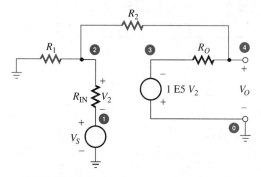

FIGURE 6.8 Noninverting amplifier SPICE circuit simulation setup.

```
VS 1 0 AC 1 0
R1 2 0 10K
RIN 2 1 1.5E12
R2 4 2 100K
E1 0 3 2 1 1E5
RO 3 4 60
.AC DEC 1 1 1E7
.PRINT AC VM(4) VP(4)
.END
```

FREQ	VM(4)	VP(4)
1.000E+00	1.100E+01	0.000E+00
1.000E+01	1.100E+01	0.000E+00
1.000E+02	1.100E+01	0.000E+00
1.000E+03	1.100E+01	0.000E+00
1.000E+04	1.100E+01	0.000E+00
1.000E+05	1.100E+01	0.000E+00
1.000E+06	1.100E+01	0.000E+00
1.000E+07	1.100E+01	0.000E+00

Observe that $V_o/V_s = VM(4)/VS = 1 + (R_2/R_1) = 11$ as expected from Eq. (1.22). Note that for a noninverting amplifier, the phase between the input and output is 0°. As in the case of Example 6.1, there are no frequency-dependent circuit elements; consequently, the results for all node voltages are frequency independent.

We repeatedly refer to these μA741 and CA3140 specifications as we proceed with our analog circuitry studies.

DRILL 6.2

Compare typical values of a_v, I_{OS}, V_{OS}, and I_{IO} for the μA741 and CA3140 and explain why the input currents are orders of magnitude different.

ANSWER

	a_v	I_{os}	V_{os}	I_I
μA741	200,000	20 nA	1 mV	80 nA
CA3140	100,000	0.5 pA	2 mV	10 pA

The FET input stage gate current is essentially zero compared to the μA741 BJT base current. Recall from Chapter 5 that the input resistance of a MOSFET is essentially infinite, which is characteristic of the resistivity of the gate oxide. This mechanism is to be compared to the significantly lower forward-biased base-emitter junction input resistance found in a μA741.

CHECK UP

1. Draw a model for an ideal operational amplifier.
2. **TRUE OR FALSE?** The μA741 meets the ideal amplifier input resistance criteria better than a CA3140.
3. **TRUE OR FALSE?** For all practical purposes, and not considering bandwidth, both the μA741 and CA3140 can be considered close to ideal in designing an inverting amplifier with a voltage gain of -10.
4. Draw and label equivalent circuits for the μA741 and CA3140 that could be used for a SPICE simulation.
5. Write SPICE netlists for the preceding question's equivalent circuits.
6. **TRUE OR FALSE?** FET operational amplifier input stages have lower input bias currents than BJT operational amplifier input stages.

6.2 GENERAL FEEDBACK CONCEPTS

As we will observe in Chapters 7 and 8, the properties of an amplifier are subject to the inherent fluctuations in active device parameters whether due to manufacturing tolerances, power supply voltage fluctuations, temperature changes, or aging. In addition, the transfer characteristic of a BJT or FET is nonlinear. Although we can study the performance of a circuit using a linear small-signal model, the model is, at best, an approximation. In this section, we introduce the theoretical basis for using feedback to improve these performance characteristics. Detailed circuit applications are presented in Chapter 11.

Very broadly speaking, *feedback*, specifically in this section *negative feedback*, is the incorporation of a sample of the output signal at the input to modify and stabilize the performance of a circuit. Indeed, in Chapter 1, we used feedback in the form of an R_F interconnection with an ideal operational amplifier. We also discuss the use of *positive feedback* in the context of designing for amplifier stability and oscillators in Chapter 11.

Any circuit or system incorporating a single feedback path can be represented by the block diagram shown in Fig. 6.9. The use of S for signal means that it can be either a voltage or current. The a block is drawn as an amplifier to emphasize that is always incorporates active elements. This is called the forward amplifier, and the relationship between the output and input is given by

$$S_o = aS_\varepsilon. \tag{6.4}$$

Because S can be either a voltage of current, a can assume the form of a voltage gain, current gain, transresistance gain, or transconductance gain. This is a broader scope definition than previously encountered. The feedback network, represented by the f block, is usually composed of passive components, although it need not be. The f block transfer function is defined by

$$S_{fb} = fS_o. \tag{6.5}$$

We should also note that the symbol G is sometimes used instead of a, especially in control systems applications. The symbol β is sometimes used instead of f to represent the gain of the feedback network. Control system applications often use H. We will not use β because it is too easy to confuse a feedback β with the current gain $\beta = h_{fe}$ of a transistor. Similarly, we will not confuse lowercase f, which is frequency with the feedback ratio written as f.

It is assumed that the feedback network does not load, that is, does not affect the gain of the forward amplifier a. The combining of the output and input signals at the input summing node (subtractor) yields

$$S_\varepsilon = S_i - S_{fb} = S_i - fS_o. \tag{6.6}$$

The difference signal, S_ε, is often called the *error signal*. Substituting Eq. (6.6) into Eq. (6.4) and solving for S_o.

$$S_o = aS_\varepsilon = a(S_i - fS_o); \tag{6.7}$$

and then

$$\frac{S_o}{S_i} = \frac{a}{1 + af} = A, \tag{6.8}$$

where Eq. (6.8) is the classic expression for the closed-loop gain A, or the gain with

■ This is one diagram EE students should never forget.

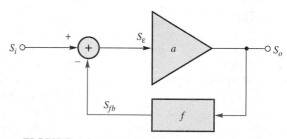

■ **FIGURE 6.9** Basic feedback system.

feedback. A dimensional analysis of Eq. (6.8) shows that A must have the same units as the *open-loop gain* a, and the quantity af must be dimensionless. The quantity af is called the *loop gain* and is often denoted by T. Typically, a is a function of frequency, $a(j\omega)$, as expected for an amplifier. Section 6.3 presents a technique for frequency response analysis. Typically, f is not a function of frequency in simple feedback circuits, although that situation is not precluded by any of what follows.

Immediately, we can observe the gain stabilization property of negative feedback; if $T = af \gg 1$, then

$$\lim_{T=af\to\infty} A = \lim \frac{\dfrac{a}{af}}{\dfrac{1}{af}+1} = \lim_{af\to\infty} \frac{\dfrac{1}{f}}{\dfrac{1}{af}+1} = \frac{1}{f}. \qquad (6.9)$$

That is, A is essentially only dependent on f. Since f is usually composed of relatively precise, stable passive elements, A is now virutally independent of variations in a if a is large enough so that $af \gg 1$. Qualitatively, if a were to increase, S_o would try to increase, but a portion of this increase would then subtract from, S_i, resulting in a smaller S_ε that when amplified by the larger a, would result in an essentially unchanged S_o. The phase relationship in a negative feedback circuit is such as to reduce the S_ε error signal. For $a \gg 1$, S_ε approaches zero; thus the feedback signal S_{fb} tends to track the input signal S_i.

■ It is often useful to find the minimum value of af such that $A \approx \dfrac{1}{f}$ within some arbitrary percentage, e.g., 1%.

Distortion Analysis

Realistically, the transfer function for the forward amplifier is given by the nonlinear relation shown in Fig. 6.10(a). Superimposed on this transfer function is a piecewise-linear curve fit. The transfer function is modeled by slopes of a_1 and a_2, where $a_1 > a_2 \gg 1$; and the saturation region by $a_3 = 0$. By applying Eq. (6.9), we observe that $A_1 \cong A_2 \cong 1/f$, a constant. This feature is illustrated in Fig. 6.10(b). Because $a_3 = 0$, $A_3 = 0$. This means, for instance, that the saturation portion of the transfer function cannot be linearized. Nevertheless, for the region where $a \gg 1$, $A \cong 1/f$. It is important to observe that the x-axis is scaled differently, Fig. 6.10(a) in terms of S_ε and Fig. 6.10(b)

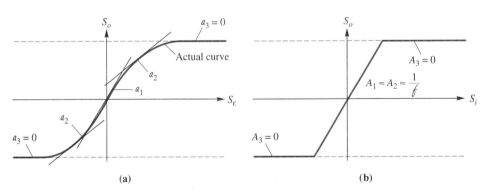

■ **FIGURE 6.10** Linearization of a transfer function to minimize distortion. (*Note:* $A \ll a_1$ or a_2. **(a)** Piecewise-linear fit to the nonlinear transfer function. $a = S_o/S_\varepsilon$. $a_1 > a_2 \gg 1$. $a_3 = 0$. **(b)** Transfer function with feedback. $A = S_o/S_i \approx 1/f$ for a_1 and $a_2 \gg 1$, $A_3 = 0$ since $a_3 = 0$.

■ Observe that $a_1 \gg A_1$ as a consequence of $S_i \gg S_\varepsilon$

in terms of S_i, so that $A \ll a$, which is almost always true in negative feedback circuits. As one observes by the linearization of the closed-loop transfer function, there is considerable improvement in the distortion in the circuit.

Gain Stabilization Sensitivity

Equation (6.9) represents a limit on A for $af \to \infty$. Clearly, a more probable situation would be $af \gg 1$ but not infinite. The gain, as previously mentioned, is a function of semiconductor processing variations, temperature, power supply voltages, and aging. The effect of a changing or poorly defined open-loop gain a on a closed-loop gain A can be determined by differentiating Eq. (6.8) with respect to a:

$$\frac{dA}{da} = \frac{d}{da}\left(\frac{a}{1+af}\right) = \frac{(1+af)-(af)}{(1+af)^2} = \frac{1}{(1+af)^2}. \qquad (6.10)$$

By letting $dA \to \Delta A$ and $da \to \Delta a$, Eq. (6.10) can be written as

$$\Delta A = \frac{1}{(1+af)^2}\Delta a. \qquad (6.11)$$

To determine the percentage change in A given percentage change in a, combine Eq. (6.11) with Eq. (6.8) to obtain

$$\frac{\Delta A}{A} \times 100\% = \frac{\frac{\Delta a}{(1+af)^2}}{\frac{a}{1+af}} \times 100\% = \frac{\frac{\Delta a}{a}}{(1+af)} \times 100\%. \qquad (6.12)$$

Thus, variations in a are effectively reduced by the quantity $(1+af) \gg 1$.

EXAMPLE 6.3

Find values for A and $\Delta A/A$ assuming $a = 1000$, $\Delta a/a = 10\%$, and $f = 0.05$.

SOLUTION From Eq. (6.8), $A = 19.6$ or approximating from Eqn. (6.9), $A \approx 1/f = 20$. Using Eq. (6.12), we have $\Delta A/A = 0.196\%$ compared with $\Delta a/a = 10\%$. Therefore, the gain has been stabilized using feedback.

■ Sensitivity analysis, on a component by component basis, is built into SPICE.

CHECK UP

1. Sketch and label a block diagram of a feedback system.
2. Write the equation of the transfer function for your feedback system block diagram.
3. **TRUE OR FALSE?** There are no circuits employing positive feedback.
4. **TRUE OR FALSE?** Negative feedback can be used to reduce distortion and stabilize gain in a system.
5. Explain the relationship required for a and f to obtain a system gain of $A = 1/f$.
6. **TRUE OR FALSE?** It is highly desirable to have f dependent only on passive components.

6.3 BODE MAGNITUDE AND PHASE PLOTS: STANDARD FORMS

As observed in Chapter 1, we can add decibels and phase shifts directly when computing overall linear system gain. That is, for the system shown in Fig. 6.11, we can write for
$A_v = \dfrac{V_2}{V_0} = |A_v|e^{j\theta}$,

■ Note, we will now use phasor notation.

$$A_v|_{dB} = 20 \log \left|\dfrac{V_2}{V_0}\right| = 20 \log \left|\dfrac{V_2}{V_1}\right| + 20 \log \left|\dfrac{V_1}{V_0}\right| \quad (6.13)$$

$$\theta = \theta_1 + \theta_2. \quad (6.14)$$

In general, both A_v and θ are functions of frequency, ω. Consequently, we need to obtain $A_v(j\omega) = |A_v(j\omega)|e^{j[\theta(j\omega)]}$. Also recall that $\omega = 2\pi f$. In general, these transfer functions are algebraically complex; however, we will develop a straightforward procedure to sketch the asymptotic frequency response of a complicated transfer function by recognizing the functional response of basic circuit topologies. The asymptotic response characteristics are called Bode plots named after H. W. Bode from Bell Laboratories.[4] We then compare our asymptotic approach to SPICE-obtained solutions. Often, combining the exact SPICE frequency response solutions with the Bode plots allows us to interpret and understand circuit behavior as functions of one or more of the key circuit elements.

Define a transfer function in the frequency domain for the system block in Fig. 6.12 as

$$H(j\omega) = \dfrac{V_2}{V_1}(j\omega). \quad (6.15)$$

By studying both the actual and asymptotic responses of commonly occurring $H(j\omega)$, we will compile a glossary of useful standard graphical forms.

From Eqs. (6.13) and (6.14), we observe that plotting an $H(j\omega)$ requires preparing both a magnitude plot and a phase plot. The y-axis will be scaled in decibels, i.e., $20 \log|H(j\omega)|$ for the magnitude plot, and $\theta(j\omega)$ in degrees for the phase plot. For both plots, the x-axis is scaled as $\log \omega$; however, from a practical engineering viewpoint, more

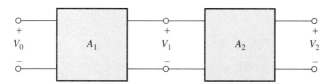

■ FIGURE 6.11 Two-stage linear system to define cascade gain calculations.

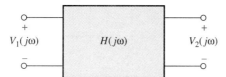

■ FIGURE 6.12 Frequency domain transfer function defined by $(V_2/V_1)(j\omega) = H(j\omega)$.

often log f is used and everyone is careful to recognize the relationship, $\omega = 2\pi f$. It is more convenient from a mathematical viewpoint to proceed with derivations using ω. After this introductory section, we use the frequency scaling, variable f or ω, whichever is most appropriate.

Consider the following standard forms:

$$\textbf{\textit{Real constant}} \quad H(j\omega) = \pm A. \tag{6.16}$$

Equation (6.16) can be written as $H(j\omega) = Ae^{\pm j\pi}$ for $-A$ and $H(j\omega) = Ae^{j0}$ for $+A$. The Bode magnitude plot is given in Fig. 6.13(a) and the Bode phase plot is given in (b). Also observe that in the phase plot, the phase of $-A$ can be sketched at either $+180°$ or $-180°$ corresponding to $e^{\pm j\pi}$, respectively. By convention, the $-180°$ reference is used. For $H(j\omega) = +A$, the actual response and asymptotic response are the same.

$$\textbf{\textit{Imaginary function, case I}} \quad H(j\omega) = \pm j\frac{\omega}{a}. \tag{6.17}$$

Equation (6.17) can be written as

$$H(j\omega) = \frac{\omega}{a} e^{\pm j\pi/2}. \tag{6.18}$$

The Bode magnitude plot is illustrated in Fig. 6.14(a). The y-axis is a plot of $20\log|\omega/a|$ in decibels. For $\omega = a$, $20\log 1 = 0$ dB. For $\omega = 10a$, $20\log 10 = 20$ dB. Therefore, the slope is given by $+20$ dB/decade. An alternative expression is $+6$ dB/octave where an *octave* is defined to be a doubling of the frequency, i.e., from ω to 2ω. Problem 6.21 at the end of the chapter asks you to show the equivalence of

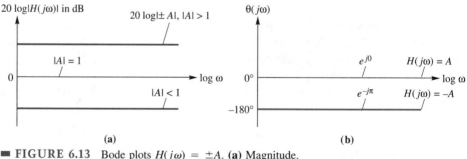

■ **FIGURE 6.13** Bode plots $H(j\omega) = \pm A$. **(a)** Magnitude. **(b)** Phase.

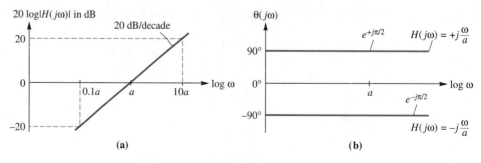

■ **FIGURE 6.14** Bode plots $H(j\omega) = j\frac{\omega}{a}$: **(a)** Magnitude. **(b)** Phase.

CHAPTER 6 BASIC ANALYSIS TECHNIQUES

6 dB/octave and 20 dB/decade. The phase plot, which is shown in Fig. 6.14(b), is either $+90°$ or $-90°$ corresponding to

$$\theta(j\omega) = e^{+j\pi/2} = e^{+j90°} \quad \text{for} \quad \left[H(j\omega) = +j\frac{\omega}{a}\right]$$

and (6.19)

$$\theta(j\omega) = e^{-j\pi/2} = e^{-j90°} \quad \text{for} \quad \left[H(j\omega) = -j\frac{\omega}{a}\right].$$

The actual and asymptotic responses are the same.

Imaginary function, case II $H(j\omega) = \pm j\dfrac{a}{\omega}.$ (6.20)

Equation (6.20) can be written as

$$H(j\omega) = \frac{a}{\omega} e^{\pm j\pi/2}. \quad (6.21)$$

In a manner similar to the previous example, we plot $20 \log|a/\omega|$ in Fig. 6.15(a). For $\omega = a$, $20 \log|H(j\omega)| = 0$ dB. For $\omega = 10a$, $20 \log|H(j\omega)| = -20$ dB. Therefore, the slope is -20 dB/decade as illustrated in Fig. 6.13(a). The phase plot is the same [Fig. 6.15(b)] for imaginary function, case *I*.

Low-pass filter $H(j\omega) = \dfrac{1}{\left(1 + j\dfrac{\omega}{\omega_c}\right)}.$ (6.22)

This transfer function can be realized by the *RC* circuit shown in Fig. 6.16(a) where, using standard voltage divider techniques,

$$V_2 = \frac{\dfrac{1}{j\omega C} V_1}{R + \dfrac{1}{j\omega C}} = \frac{V_1}{1 + j\omega RC}. \quad (6.23)$$

It is advantageous to simplify Eq. (6.23) by writing it as

$$H(j\omega) = \frac{V_2}{V_1}(j\omega) = \frac{1}{1 + j\dfrac{\omega}{\omega_c}}, \quad (6.24)$$

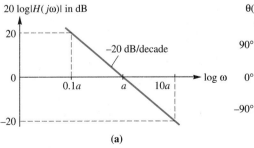
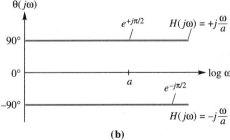

(a) (b)

■ **FIGURE 6.15** Bode plots $H(j\omega) = \pm j\dfrac{a}{\omega}$:

(a) Magnitude. (b) Phase.

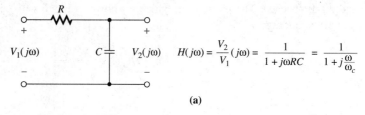

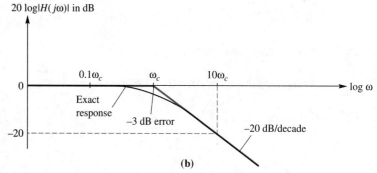

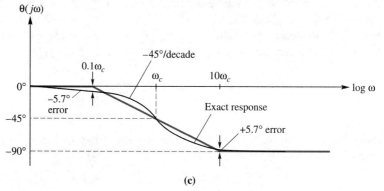

FIGURE 6.16 Bode plots for $H(j\omega) = 1/(1 + j\omega/\omega_c)$ single-section low-pass filter. (a) Filter. (b) Magnitude. (c) Phase.

where $\omega_c = 1/RC$ is defined as the *corner frequency*. We now sketch the low- and high-frequency asymptotes for the low-pass filter whose transfer function is in the form given in Eq. (6.24). The magnitude of $H(j\omega)$ is given by

$$|H(j\omega)| = \frac{1}{\sqrt{1 + \left(\dfrac{\omega}{\omega_c}\right)^2}} \quad (6.25)$$

For $\omega \ll \omega_c$, $|H(j\omega)| = 1$, or equivalently, 0 dB. For our purposes, $\ll$ implies a factor of 10 or more. By convention, we will extend the low-frequency asymptote to ω_c, although the actual response will exhibit its largest error at this point [see Fig. 6.16(b)]. For $\omega \gg \omega_c$, Eq. (6.25) is approximated by

$$|H(j\omega)| \approx \frac{\omega_c}{\omega}. \quad (6.26)$$

From the analysis used to obtain the magnitude plot shown in Fig. 6.15(a), we see that this high-frequency asymptote plots with a slope of -20 dB/decade. This will meet the

CHAPTER 6 BASIC ANALYSIS TECHNIQUES

low-frequency asymptote at ω_c. The point where the low- and high-frequency asymptotes meet is called the *corner frequency*, or the *3-dB point* since the exact value there is 3 dB lower than the low-frequency asymptote. See Table 6.1, which includes an exact computation of Eq. (6.25). Thus we have the classic low-pass single-pole filter response characteristic.

The phase plot of the low-pass filter is

$$\theta_H(j\omega) = -\tan^{-1}\left(\frac{\omega}{\omega_c}\right), \tag{6.27}$$

where for clarity, the notation $\theta_H(j\omega)$ for phase of $H(j\omega)$ is used. The low-frequency asymptote is obtained by letting $\omega \ll \omega_c$ in Eq. (6.27). This yields $\theta_H(j\omega) = 0°$. For $\omega \gg \omega_c$, $\theta_H(j\omega) = -90°$. If $\omega = \omega_c$, then $\theta_H(j\omega) = -45°$ exactly. Using the factor of 10 for $\ll$ and $\gg$, these results are shown in Fig. 6.16(c). The slope is described as $-45°$/decade for this transfer function.

■ If you have not yet been introduced to the concept of *pole* in your basic circuits courses, you can substitute the term *section* for *pole* in the usage in this chapter as it relates to filters.

So far, we have shown only the asymptotes. The entries in Table 6.1 were prepared by calculating the exact response using Eqs. (6.25) and (6.27). Note that at the corner frequency ω_c, the Bode plot phase shift of $-45°$ is exact. The magnitude errors at the $10\omega_c$ and $0.1\omega_c$ point are only 0.0432 dB, which is quite acceptable for most engineering calculations. The phase error at the $10\omega_c$ and $0.1\omega_c$ points is $-5.7°$, which is an acceptable approximation in many situations.

TABLE 6.1 Exact Response for $H(j\omega) = 1 / \left(1 + j\frac{\omega}{\omega_c}\right)$

ω	$20 \log\left[1 + \left(\frac{\omega}{\omega_c}\right)^2\right]^{-1/2}$ in dB	$\theta(j\omega) = -\tan^{-1}\frac{\omega}{\omega_c}$ in degrees
0	0	0
$0.1\omega_c$	-0.0432	-5.7
$0.5\omega_c$	-0.969	-26.6
ω_c	-3.01	-45.0
$5\omega_c$	-14.15	-78.7
$10\omega_c$	-20.04	-84.3
∞	$-\infty$	-90.0

EXAMPLE 6.4

Design a low-pass filter with a 3-dB corner at 3 kHz. Verify your design using SPICE.

SOLUTION This is an open-ended design problem. We want to design a circuit of the form illustrated in Fig. 6.16(a) that will produce the amplitude-frequency characteristic illustrated in Fig. 6.16(b) and the frequency-phase characteristic illustrated in Fig. 6.16(c). The 3-dB corner frequency is given by $f_c = 1/2\pi RC = 3$ kHz. Obviously, with one equation and two unknowns, R and C, we need to provide additional information. In lieu of other constraints, we assume a standard capacitance value of $C = 0.01\ \mu F$. Then,

$$R = \frac{1}{(2\pi)(0.01\ \mu F)(3000\ Hz)} = 5305\ \Omega.$$

> ■ Electronic Industries Association standards ±5% resistor values are commonly available. Of course, one could pay more for ±1% resistor values. $ is a design trade-off. Any catalog listing electronic components has tables of these values.

Practically speaking, we would select a standard E.I.A 5% resistor value of either 5100 Ω ($f_c = 3120$ Hz) or 5600 Ω ($f_c = 2842$ Hz).

The SPICE listing, for the case where $R = 5305$ Ω, is given by

```
LOW PASS FILTER
VS 1 0 AC 1 0
R  1 2 5305OHMS
C  2 0 0.01UF
.AC DEC 10 10 100E3
.PROBE
.END
```

The SPICE-computed frequency responses along with the asymptotic responses are plotted in Figs. 6.17(a) and (b).

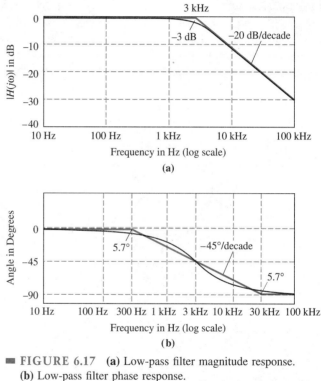

■ **FIGURE 6.17** (a) Low-pass filter magnitude response. (b) Low-pass filter phase response.

High-pass filter $$H(j\omega) = \frac{1}{1 - j\frac{\omega_c}{\omega}}. \tag{6.28}$$

The basic form for a single-pole high-pass filter is illustrated in Fig. 6.18(a). Using voltage division in much the same way as for the low-pass filter, the transfer function in a usable form is given by

$$H(j\omega) = \frac{V_2}{V_1}(j\omega) = \frac{R}{R + \frac{1}{j\omega C}} = \frac{1}{1 + \frac{1}{j\omega RC}} = \frac{1}{1 - \frac{j}{\omega RC}} = \frac{1}{1 - j\frac{\omega_c}{\omega}}, \tag{6.29}$$

CHAPTER 6 BASIC ANALYSIS TECHNIQUES

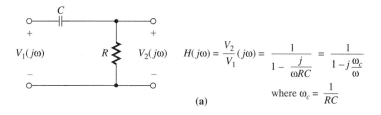

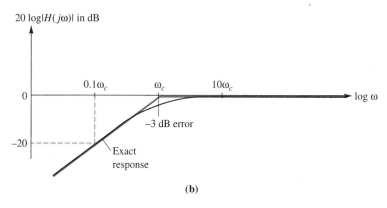

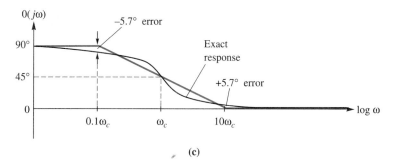

FIGURE 6.18 Bode plots for $H(j\omega) = 1/(1 - j\omega_c/\omega)$ single-pole high-pass filter. **(a)** Filter. **(b)** Magnitude. **(c)** Phase.

where we have again made use of the corner frequency, $\omega_c = 1/RC$. The equations, obtained from Eq. (6.29), for plotting the asymptotic and exact values for the transfer function are given by

$$\left|\frac{V_2}{V_1}(j\omega)\right| = \frac{1}{\sqrt{1 + \left(\frac{\omega_c}{\omega}\right)^2}} \quad (6.30)$$

and

$$\theta_H(j\omega) = +\tan^{-1}\left(\frac{\omega_c}{\omega}\right). \quad (6.31)$$

Using the same techniques we have just studied for the low-pass filter, we can easily identify the corner frequency and asymptotes. So far, we have shown only the asymptotes. The entries in Table 6.2 were prepared by calculating the exact response using Eqs. (6.30) and (6.31). Note that at the corner frequency ω_c, the Bode plot phase shift of $+45°$ is exact. The magnitude errors at the $10\omega_c$ and $0.1\omega_c$ points are only 0.0432 dB, which is

TABLE 6.2 Exact Response for $H(j\omega) = 1/\left(1 - j\dfrac{\omega_c}{\omega}\right)$

ω	$20\log\left[1 + \left(\dfrac{\omega_c}{\omega}\right)^2\right]^{-1/2}$ in dB	$\theta(j\omega) = +\tan^{-1}\left(\dfrac{\omega_c}{\omega}\right)$ in degrees
0	$-\infty$	90.0
$0.1\omega_c$	-20.04	84.3
$0.5\omega_c$	-6.99	63.4
ω_c	-3.01	45.0
$5\omega_c$	-0.17	11.3
$10\omega_c$	-0.0432	5.7
∞	0	0

quite acceptable for most engineering calculations. The phase error at the $10\omega_c$ and $0.1\omega_c$ points is $\pm 5.7°$, which is an acceptable approximation in many situations. The slope is $+20$ dB/decade for the magnitude plot and $45°$/decade for the phase plot. These magnitudes are the same as the low-pass filter case. Both the approximate and exact functions for the magnitude and phase are illustrated in Figs. 6.18(b) and (c), respectively.

EXAMPLE 6.5

Design a high-pass filter with a 3-dB corner at 3 kHz. Use $C = 0.01$ μF. Verify your design using SPICE.

SOLUTION The required circuit is of the form illustrated in Fig. 6.18(a) because it will yield the amplitude-frequency characteristic illustrated in Fig. 6.18(b) and the phase-frequency characteristic illustrated in Fig. 6.18(c). The 3-dB corner is given by $f_c = 1/2\pi RC = 3$ kHz. For $C = 0.01$ μF,

$$R = \frac{1}{(2\pi)(0.01\ \mu\text{F})(3000\ \text{Hz})} = 5305\ \Omega.$$

The SPICE listing is given by

```
HIGH PASS FILTER
VS 1 0 AC 1 0
R  2 0 5305OHMS
C  1 2 0.01UF
.AC DEC 10 10 100E3
.PROBE
.END
```

The SPICE-computed frequency responses along with the asymptotic responses are plotted in Figs. 6.19(a) and (b) (see top of page 269).

Multiple Corners

It is important to note that if the transfer function can be algebraically manipulated so that corner frequencies can be identified, virtually any complicated transfer function's

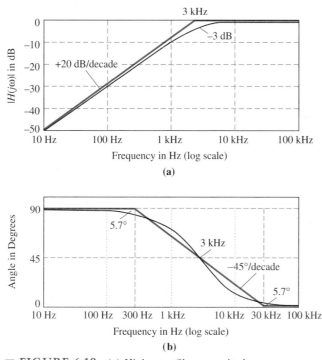

FIGURE 6.19 (a) High-pass filter magnitude response. (b) High-pass filter phase response.

amplitude and phase characteristic can be quickly sketched. For example, consider the amplifier response shown in Fig. 6.20 whose transfer function is given by

$$a(j\omega) = \frac{a_o}{\left(1 + j\dfrac{\omega}{\omega_H}\right)\left(1 - j\dfrac{\omega_L}{\omega}\right)}. \tag{6.32}$$

This type of response characteristic could represent the frequency response for a music system where the low-frequency corner is at 20 Hz (or lower if you like room-shaking bass response) and the high-frequency response is 20 kHz (although most of us cannot hear much beyond 15 to 17 kHz).

The following example illustrates this procedure for a typical amplifier by plotting the amplitude and phase characteristics when both a low- and high-frequency corner are expected.

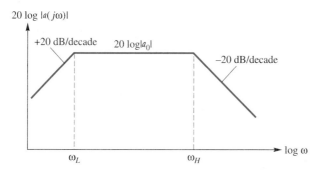

FIGURE 6.20 Band-pass amplifier response. Recall that $\omega = 2\pi f$.

EXAMPLE 6.6

Sketch and label the Bode magnitude and phase plot for

$$H(jf) = \frac{-50}{\left(1 + j\dfrac{f}{10 \text{ kHz}}\right)\left(1 - j\dfrac{500 \text{ Hz}}{f}\right)}.$$

SOLUTION Immediately we can identify the high-frequency corner at 10 kHz since the term

$$\left(\frac{1}{1 + j\dfrac{f}{10 \text{ kHz}}}\right)$$

is of the low-pass filter form illustrated in Eq. (6.24). Similarly, the low-frequency corner

$$\left(\frac{1}{1 - j\dfrac{500 \text{ Hz}}{f}}\right)$$

is of the high-pass filter form illustrated in Eq. (6.29).

The amplitude and phase characteristics for the individual single-pole low-pass and high-pass responses are shown in Fig. 6.21 (see page 271). To this we add a constant term of $20 \log 50 = 34$ dB. If all three individual asymptotic curves are added, the result is as shown. Knowledge of the key errors then permits us to superimpose the estimated actual response on both plots. It is important to observe that if corners are two decades or more apart, their interaction with respect to the errors can be neglected. From a practical viewpoint, a $0.0432 \times 2 = 0.0864$ dB error (at one decade from either corner) can be neglected in most applications.

For subsequent work, we will reduce complicated transfer functions to a form amenable to this technique.

Broadbanding

We are now in a position to combine feedback concepts with a frequency response analysis. Assume that the forward gain of the amplifier is given by Eq. (6.32) where ω_H and ω_L are the single high-frequency and low-frequency corners (3-dB points) as previously shown in the Bode magnitude plot of Fig. 6.20. It is convenient to assume that $\omega_H \gg \omega_L$ so that the effect feedback has on each corner can be considered independently. Applying Eq. (6.8) to the high-frequency portion of the gain expression in Eq. (6.32) yields

$$A(j\omega)|_{\omega_H \text{ only}} = \frac{\dfrac{a_o}{\left(1 + j\dfrac{\omega}{\omega_H}\right)}}{1 + \dfrac{a_o f}{\left(1 + j\dfrac{\omega}{\omega_H}\right)}}. \tag{6.33}$$

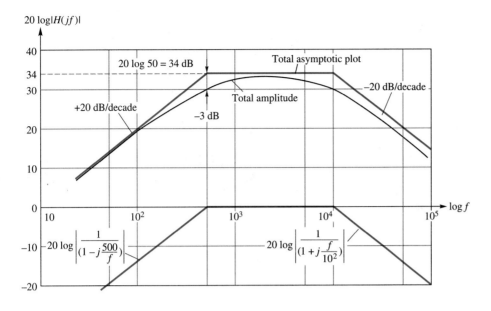

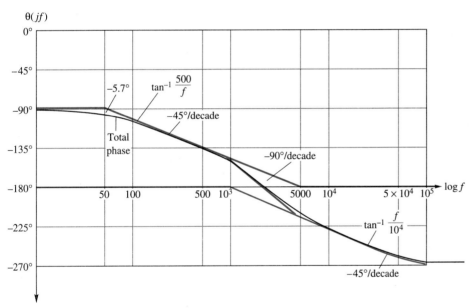

FIGURE 6.21 Bode amplitude and phase plot for Example 6.6.

Clearing fractions and rearranging in a form where the new corner is recognizable yields

$$A(j\omega)|_{\omega_H \text{ only}} = \frac{\left(\dfrac{a_o}{1 + a_o f}\right)}{\left[1 + j\dfrac{\omega}{\omega_H(1 + a_o f)}\right]}. \tag{6.34}$$

We revise Fig. 6.20 to show this result (see Fig. 6.22). The midband gain has decreased

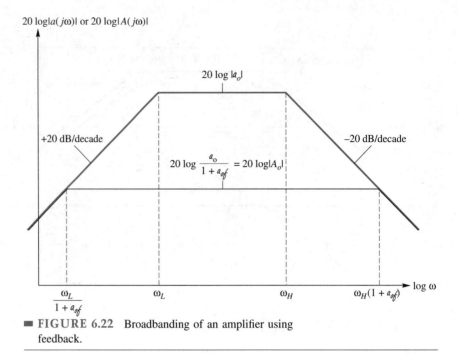

FIGURE 6.22 Broadbanding of an amplifier using feedback.

from a_o to $A_o = a_o/(1 + a_o f)$ and the high-frequency corner has increased by the same $(1 + a_o f)$ factor.

Similarly, applying Eq. (6.8) to the low-frequency portion of Eq. (6.32) yields

$$A(j\omega)|_{\omega_L \text{ only}} = \frac{\dfrac{a_o}{\left(1 - j\dfrac{\omega_L}{\omega}\right)}}{1 + \dfrac{a_o f}{\left(1 - j\dfrac{\omega_L}{\omega}\right)}}. \qquad (6.35)$$

Clearing fractions and rearranging in a form where the corner can be identified, we obtain

$$A(j\omega)|_{\omega_L \text{ only}} = \frac{\left(\dfrac{a_o}{1 + a_o f}\right)}{\left(1 - j\dfrac{\omega_L}{(1 + a_o f)\omega}\right)}. \qquad (6.36)$$

The low-frequency corner, originally ω_L, has decreased by the same $(1 + a_o f)$ factor. This is also illustrated in Fig. 6.22. Since $\omega_H \gg \omega_L$, the overall open-loop bandwidth $\omega = \omega_H - \omega_L \approx \omega_H$. The *gain-bandwidth product* (GBW) is $a_o \omega_H$.

■ Observe that to compute GBW, express a_o in magnitude, not dB.

With feedback,

$$\text{GBW} = \left(\frac{a_o}{1 + a_o f}\right) \times \omega_H (1 + a_o f) = a_o \omega_H. \qquad (6.37)$$

The GBW remains unchanged. Of course, because $\omega = 2\pi f$, we could easily compute our results in f(Hz) as well as ω (rad/sec). As we observe, the use of feedback has allowed us to increase the bandwidth f, however, with a concomitant reduction in gain.

Frequency Response of the µA741

We are now able to discuss the frequency response of a μA741 as represented in Graph 4 of Fig. 6.2. Observe that the low-frequency asymptote is at $a = 200{,}000$ (106 dB) and the corner is at a surprisingly low 5 Hz! Using Eq. (6.37), we compute the gain-bandwidth product to be $200{,}000 \times 5 = 1$ MHz. The high-frequency asymptote has a -20 dB/decade slope illustrative of a single-pole low-pass filter response. The other key point to study on Graph 4 of Fig. 6.2 is the point $a(jf) = 1$ (0 dB).

Although more complex and accurate models will be introduced, Graph 4 can be synthesized by modifying the basic μA741 operational amplifier as shown in Fig. 6.5 by incorporating a low-pass filter to obtain a corner at 5 Hz. As shown in Fig. 6.23, and using Eq. (6.24) along with Fig. 6.16, $C = 425$ µF.

■ A more accurate frequency-dependent model, less subject to loading, is shown in Problem 6.59.

The SPICE listing for Fig. 6.23 is given by

```
741 FREQUENCY-DEPENDENT MODEL
VS 1 0 AC 1 0
RS 1 2 10K
RIN 2 0 2MEG
E1 0 3 2 0 200E3
RO 3 4 75
CO 4 0 425UF
.AC DEC 10 0.1 1E6
.PROBE
```

where the results are computed using 10 steps every decade between 0.1 Hz and 1 MHz. As shown in Fig. 6.24, the simulation agrees nicely with the Graph 4 specifications of Fig. 6.2.

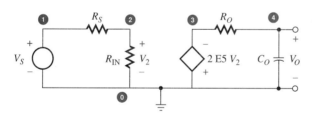

■ **FIGURE 6.23** Model for the μA741.

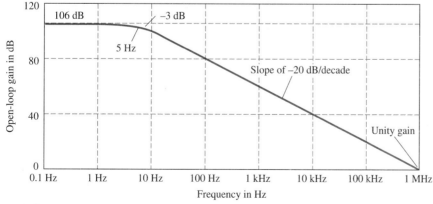

■ **FIGURE 6.24** SPICE simulation of the μA741 frequency-dependent open-loop gain 1-MHz GBW.

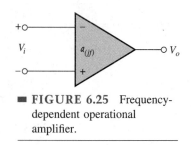

FIGURE 6.25 Frequency-dependent operational amplifier.

■ f_T is an important parameter for operational amplifiers. We address f_T again for BJT and FET circuits in Chapter 9.

The response of the op amp in Figure 6.25 can be modeled as

$$a(jf) = \frac{V_o}{V_i}(jf) = \frac{a_o}{1 + j(f/f_c)}$$
$$= \frac{a_o f_c}{f_c + jf} = \frac{f_T}{f_c + jf} \cong \frac{f_T}{jf} \quad (6.38)$$

Notice that the frequency of unity gain, f_T, occurs at $f = a_o f_c$ provided $f_T \gg f_c$.

The approximate form predicts infinite dc gain rather than a_o but that difference rarely matters. The approximate gain model is what we need to analyze the inverting amplifier of Fig. 6.26. Letting V_i stand for the input voltage to the op amp, we have

$$V_o = -aV_i = -\frac{f_T}{jf}V_i \quad (6.39)$$

or

$$V_i = \frac{-jf}{f_T}V_o. \quad (6.40)$$

At the inverting input node we have

$$\frac{V_i - V_s}{R_S} + \frac{V_i - V_o}{R_F} = 0 \quad (6.41)$$

or

$$V_i\left(\frac{1}{R_S} + \frac{1}{R_F}\right) - \frac{V_o}{R_F} = \frac{V_s}{R_S}. \quad (6.42)$$

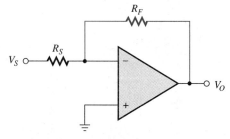

FIGURE 6.26 Inverting amplifier.

Substituting Eq. (6.40) into Eq. (6.42) and simplifying gives

$$\frac{V_o}{V_s} = \frac{-R_F/R_S}{1 + j\frac{f}{f_T}\left(1 + \frac{R_F}{R_S}\right)} = \frac{A_o}{1 + j\frac{f}{f_H}} \qquad (6.43)$$

where

$$A_o = -\frac{R_F}{R_S}, \qquad f_H = \frac{f_T}{1 + R_F/F_S}. \qquad (6.44)$$

EXAMPLE 6.7

Using the frequency-dependent $\mu A741$ model, compute the -3 dB frequency for the inverting amplifier of Fig. 6.26 for $R_F = 100$ kΩ and $R_S = 10$ kΩ. Verify the result using SPICE.

SOLUTION The low-frequency voltage gain is $A_v = -R_F/R_S = -10$. Using $f_T = 1$ MHz, Eq. (6.44) gives

$$f_H = \frac{1 \text{ MHz}}{11} = 91 \text{ kHz}.$$

Notice that designing for $A_v = -1{,}000$ would give a corner frequency of about 1 kHz for a $\mu A741$.
The SPICE-labeled circuit is shown in Fig. 6.27.

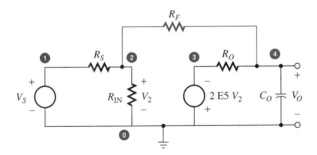

FIGURE 6.27 SPICE-labeled circuit.

The SPICE listing for this circuit model is given by

```
741 FREQUENCY-DEPENDENT MODEL WITH FEEDBACK
VS 1 0 AC 1 0
RS 1 2 10K
RF 2 4 100K
RIN 2 0 2MEG
E1 0 3 2 0 200E3
RO 3 4 75
CO 4 0 425UF
.AC DEC 10 1E3 1E7
.PROBE
```

The frequency is swept from 1 kHz to 10 MHz. As shown in the resultant output (Fig. 6.28), the -3-dB response is at 91 kHz and the closed-loop low-frequency aymptotic gain, A_o, is 20 dB.

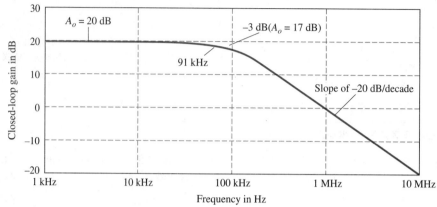

FIGURE 6.28 Closed-loop frequency response for the -10 (20-dB) inverting amplifier.

Multicorner Frequency Response

■ SPICE really shows its computational worth in analysis and design of multi-corner circuits.

It should be obvious that in a multistage circuit, the frequency response characteristic will be complicated. However, usually what we really need to know for a given design is the location of the overall 3-dB down response points. As illustrated in Fig. 6.29, the circuit behavior above ω_{H1} and below ω_{L3} is of relatively minor importance. In general, it is possible to represent this characteristic in the form

$$A_{vs}(j\omega) = \frac{A_{vsm}}{\left(1 - j\frac{\omega_{L1}}{\omega}\right) \cdots \left(1 - j\frac{\omega_{Lq}}{\omega}\right)\left(1 + j\frac{\omega}{\omega_{H1}}\right) \cdots \left(1 + j\frac{\omega}{\omega_{Hr}}\right)}, \quad (6.45)$$

where q and r are integers. From this equation we need to determine the overall 3-dB down response points.

Consider the high-frequency end first. Equation (6.45) reduces to

$$A_{vs}(j\omega) \approx \frac{A_{vsm}}{\left(1 + j\frac{\omega}{\omega_{H1}}\right)\left(1 + j\frac{\omega}{\omega_{H2}}\right) \cdots \left(1 + j\frac{\omega}{\omega_{Hr}}\right)}. \quad (6.46)$$

Theoretically, the 3-dB response can be found by computing the magnitude of Eq. (6.46) and solving Eq. (6.46) for the $\omega = \omega_H$ where

$$|A_{vs}(j\omega = j\omega_H)/A_{vsm}| = \frac{1}{\sqrt{2}}. \quad (6.47)$$

CHAPTER 6 BASIC ANALYSIS TECHNIQUES

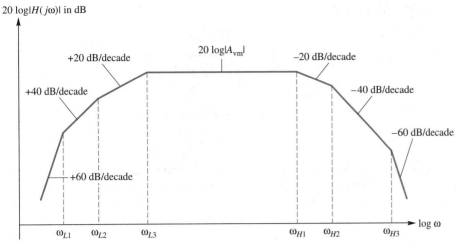

■ **FIGURE 6.29** General amplifier-type response with each corner frequency identified.

An approximation to this calculation is obtained as follows. Expanding Eq. (6.46) by multiplying the terms in the denominator yields

$$A_{vs}(j\omega) = \frac{A_{vsm}}{1 + j\omega\left(\dfrac{1}{\omega_{H1}} + \dfrac{1}{\omega_{H2}} + \cdots + \dfrac{1}{\omega_{Hr}}\right) + \text{Higher order terms}}, \quad (6.48)$$

■ Strictly speaking, higher order terms should not be neglected if poles are within one decade.

where these higher order terms can be neglected when solving for ω_H.

Even though the response is multicornered, Eq. (6.48) suggests that an equivalent 3-dB response is located approximately at the frequency

$$\frac{1}{\omega_H} \cong \frac{1}{\omega_{H1}} + \frac{1}{\omega_{H2}} + \cdots + \frac{1}{\omega_{Hr}}. \quad (6.49)$$

EXAMPLE 6.8

Suppose a circuit has two identical high-frequency corners; estimate the 3-dB response and compare this result with the exact solution.

SOLUTION The circuit response is of the form

$$A_{vs}(j\omega) = \frac{A_{vsm}}{\left(1 + j\dfrac{\omega}{\omega_{H1}}\right)^2},$$

so that from Eq. (6.49),

$$\frac{1}{\omega_H} = \frac{1}{\omega_{H1}} + \frac{1}{\omega_{H1}} = \frac{2}{\omega_{H1}}, \qquad \omega_H = 0.5\omega_{H1}.$$

The exact solution is obtained by solving

$$\left|\frac{A_{vs}(j\omega)}{A_{vs}(\text{mid})}\right| = \frac{1}{\sqrt{2}} = \left|\frac{1}{\left(1 + j\frac{\omega}{\omega_{H1}}\right)\left(1 + j\frac{\omega}{\omega_{H1}}\right)}\right| \quad (6.50)$$

$$= \frac{1}{\sqrt{\left[1 - \left(\frac{\omega}{\omega_{H1}}\right)^2\right]^2 + \left(\frac{2\omega}{\omega_{H1}}\right)^2}}$$

for $\omega_H = 0.64\omega_{H1}$. The linearized version of the solution $\omega_H = 0.5\omega_{H1}$ was actually pessimistic, so even though the approximate and exact solutions differ by about 20%, the error is one favoring better amplifier performance. The approximation becomes much better if the multiple corners are separated as illustrated in the problems.

Similarly, at the low-frequency end, Eq. (6.46) reduces to

$$A_{vs}(j\omega) = \frac{A_{vsm}}{\left(1 - j\frac{\omega_{L1}}{\omega}\right)\left(1 - j\frac{\omega_{L2}}{\omega}\right) \cdots \left(1 - j\frac{\omega_{Lq}}{\omega}\right)}. \quad (6.51)$$

Applying the same linearizing criteria, Eq. (6.51) can be written as

$$A_{vs}(j\omega) = \frac{A_{vsm}}{1 - j\left(\frac{\omega_{L1} + \omega_{L2} + \cdots \omega_{Lq}}{\omega}\right) + \text{Higher order terms}}, \quad (6.52)$$

where an approximate 3-dB response is given by

$$A_{vs}(j\omega) \approx \frac{A_{vsm}}{\left(1 - j\frac{\omega_L}{\omega}\right)}, \quad (6.53)$$

where the higher order terms are neglected and

$$\omega_L = \omega_{L1} + \omega_{L2} + \cdots + \omega_{Lq}. \quad (6.54)$$

EXAMPLE 6.9

Following in the context of the previous example, suppose a circuit has a double low frequency corner; estimate the 3-dB response and compare this result with the exact solution.

SOLUTION The circuit response is of the form

$$A_{vs}(j\omega) = \frac{A_{vsm}}{\left(1 + j\frac{\omega_{L1}}{\omega}\right)^2},$$

so that from Eq. (6.53), $\omega_L = 2\omega_{L1}$. The exact solution is obtained by solving

$$\left|\frac{A_{vs}(j\omega)}{A_{vs}(\text{mid})}\right| = \frac{1}{\sqrt{2}} = \left|\frac{1}{\left(1 - 1j\frac{\omega_{L1}}{\omega}\right)\left(1 - j\frac{\omega_{L1}}{\omega}\right)}\right| \quad (6.55)$$

$$= \frac{1}{\sqrt{\left[1 - \left(\frac{\omega_{L1}}{\omega}\right)^2\right]^2 + \left(\frac{2\omega_{L1}}{\omega}\right)^2}}$$

so that $\omega_L = 1.554\omega_{L1}$. The linearized version of the solution $\omega_L = 2\omega_{L1}$ was actually pessimistic, so even though the approximate and exact solutions differ by about 25%, the error is one favoring better amplifier performance. Again, the approximation becomes much better if the multiple corners are separated.

It is interesting to observe from the previous two examples that multiple corners occurring at the same frequency will result in the largest errors when the linearized and exact solutions are compared.

CHECK UP

1. **TRUE OR FALSE?** -100 dB of voltage gain corresponds to $a = 100{,}000$.
2. Sketch and label Bode magnitude and phase plots for $H(j\omega) = +20$, -20, $(\omega/20)e^{-j\pi/8}$, and $(20/\omega)e^{-j\pi/8}$.
3. **TRUE OR FALSE?** The -3-dB point occurs where $\omega = \omega_C$ in a low-pass or high-pass filter.
4. Sketch and label Bode magnitude and phase plots for a single-pole low-pass filter with a -3-dB point at 20 kHz. Show errors by comparing to the actual responses.
5. Sketch and label Bode magnitude and phase plots for a single-pole high-pass filter with a -3-dB point at 100 Hz. Show errors by comparing to the actual responses.
6. **TRUE OR FALSE?** Negative feedback can be used to broadband a system.

SUMMARY

Key analog integrated circuit and system design and analysis tools were presented. We first focused on the data sheets for a bipolar and a BiMOS operational amplifier. You should be able to read and interpret data sheets. Using these data and the basic operational amplifier model, a simplified two-port SPICE model was synthesized and demonstrated. Currently, many of the larger IC manufacturers have provided detailed circuit-based SPICE models for their products. We study these circuit designs in Chapter 12. As mentioned previously, virtually all analog ICs, whether considering their internal circuit design or their use in small or large systems, employ negative feedback. In this chapter we derived and exploited the key equation

$$A = \frac{a}{1 + af} \cong \frac{1}{f} \quad \text{for } af \gg 1$$

for use when using operational amplifiers for a variety of applications. Applying summing point constraints of $v_i = 0$ and $i_i = 0$ allowed us to design with resistor ratios, thereby ignoring the large active device-based parameter variations found in IC circuits. We also observed that negative feedback is useful to linearize the IC transfer function. Additional feedback concepts are applied to circuit designs in Chapter 11. Virtually all systems have, by design or otherwise, frequency-dependent behavior. A common approach to show this behavior is by use of Bode magnitude and phase plots. Indeed, the key output representations from a SPICE ac analysis are magnitude and phase plots, which are directly related to the asymptotic Bode plots. We presented the derivation of two key functional forms, the single-pole low-pass and high-pass filters. Although SPICE can provide frequency-response information for complex circuits and systems, it is important for the reader to understand and interpret the results using transfer functions and equivalent circuits. The frequency-dependent behavior, along with feedback, was included in the analysis of ideal and non-ideal operational amplifier-based circuits. Indeed, the focus of many of the chapter problems combines operational amplifiers with external frequency-dependent networks in a negative feedback loop. The analytical designs are to be supported by SPICE. The GBW was also introduced as a metric to compare system performance.

Now that we have studied the terminal characteristics of operational amplifier ICs, we will proceed to study the individual internal circuits and their design in analog ICs including frequency response. We then return to use feedback with these circuits in Chapter 11 and combine the results in Chapter 12 by analyzing the internal circuitry of several analog ICs as well as by designing our own basic operational amplifier. Table 6.3 summarizes some properties of common operational amplifiers.

TABLE 6.3 Selected Properties of Some Common Operational Amplifiers

Type Designations	Input Stage	f_T MHz	I_B nA	r_o Ω
µA741	BJT	1	80	75
LF411	JFET	4	0.05	40
CA3140	MOS	4.5	0.01	60
LM318	BJT	15	150	50
HA2841	BJT	54	8000	8.5

SURVEY QUESTIONS

1. What are five criteria for an ideal operational amplifier?
2. What are the basic definitions for input offset voltage, input offset current, input bias current, input resistance, large-signal voltage gain, output resistance, and output short-circuit current?
3. What are values for the quantities listed in the previous question for the µA741 and CA3140?
4. Draw and label a frequency-independent and frequency-dependent SPICE model for a µA741 and CA3140.

5 How do you apply summing point constraints to the inverting and noninverting amplifier configurations?
6 Draw and specifically label a block diagram of a feedback system.
7 What are the definitions for open-loop gain, loop gain, and closed-loop gain?
8 What are the system requirements to obtain gain stabilization such that $A \approx 1/f$?
9 What are the qualitative and quantitative system requirements required to use negative feedback for distortion reduction and gain stabilization?
10 How do you compute the voltage gain and power gain in decibels for a multistage system?
11 How do you compute the total phase shift in a multistage system from the phase shifts of each individual stage?
12 What are the Bode magnitude and phase plots for the following functions:

$$H(j\omega) = \pm A, = \frac{\omega}{a} e^{\pm j\pi/2}, \text{ and } \frac{a}{\omega} e^{\pm j\pi/2}?$$

13 What are the analytical and graphical definitions for the -3-dB response point?
14 What are the Bode magnitude and phase plots for the single-section low-pass filter? For the single-section high-pass filter and a band-pass filter?
15 How do you graphically combine multiple transfer functions on Bode plots?
16 What are the analytical and graphical definitions for the gain-bandwidth product?
17 How do you use negative feedback to broad band a system?
18 What are the Bode magnitude and phase plots for a system with two identical corner frequencies?

PROBLEMS

Problems marked with an asterisk are more difficult.

6.1 Let $V^+ = V^- = 5$ V. Find a numerical value for the a_v for a μA741. Refer to Fig. 6.2.
6.2 For the circuit shown in Fig 6.30, a dc output voltage of 150 mV is observed. What is V_{OS}?
6.3 It is observed that $I_{B1} = 100$ nA and $I_{B2} = 90$ nA. Compute values for I_B and I_{OS}. Are these values most reasonable for a μA741 or a CA3140?
6.4 Derive the value of R_2 in the circuit shown in Fig. 6.31 that will minimize the effect of I_B.
6.5 Why does the μA741 bias current increase with a decrease in temperature? Assume I_C is constant and consider material in Chapter 4.
6.6 Use the CA3140 frequency-independent SPICE model and repeat Example 6.1.

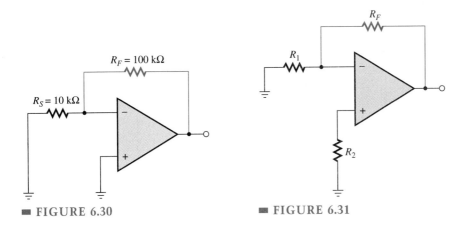

■ FIGURE 6.30

■ FIGURE 6.31

6.7 Using the operational amplifier model, which includes a finite a_v and R_{IN}, and non zero R_O, derive an expression for $A_v\ V_o/V_s$ for the noninverting amplifier shown in Fig. 6.32. Compare your answer to the ideal operational calculation given by $A_v = V_o/V_s = 1 + (R_2/R_1)$.

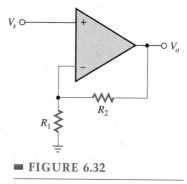

■ FIGURE 6.32

6.8 Use SPICE and the µA741 model given in Fig. 6.5 to compute the frequency-independent $A_v = V_o/V_s$ for the noninverting amplifier shown in Fig. 6.32 where $R_2 = 33$ kΩ and $R_1 = 10$ kΩ. Compare your results to the ideal operational amplifier calculation given by $A_v = V_o/V_s = 1 + (R_2/R_1)$.

6.9 Use summing point constraints to derive an experssion for V_o as a function of V_1, V_2 and V_3. Refer to Fig. 6.33. Assume an ideal operational amplifier.

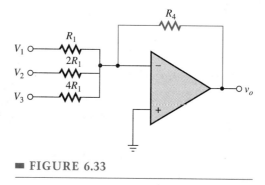

■ FIGURE 6.33

6.10* Show that the negative feedback circuit shown in Figure 6.34(a) can be used to synthesize the circuit shown in Fig. 6.34(b) where $C' = C(R_1/R_2)$. Observe that this circuit can be used as a *capacitance multiplier*. Assume an ideal operational amplifier.

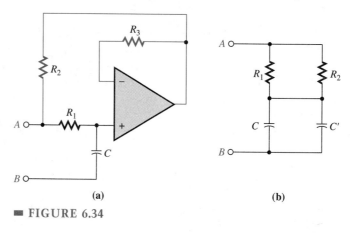

■ FIGURE 6.34

6.11* Show that the negative feedback circuit of Fig. 6.35(a) can be used to synthesize the circuit shown in Fig. 6.35(b) where $L' = CR_1R_2$. Observe that this circuit, sometimes called a *gyrator*, can be used to synthesize an effective inductive reactance. Assume an ideal operational amplifier.

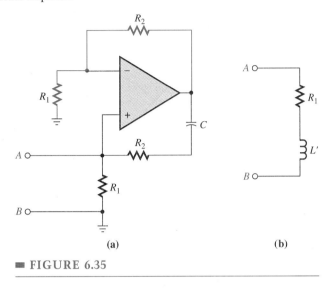

(a)   (b)

■ FIGURE 6.35

6.12 An npn transistor is included in the feedback loop of an operational amplifier, as shown in Fig. 6.36.

(a) Using summing point constraints, derive an expression for V_o/V_s. Sketch the transfer function of V_o versus V_s for $V_s > 0$ and $V_s < 0$. Assume

$$i_c = I_s e^{V_{BE}q/kT}.$$

(b) How would you change Figure 6.36 if you used a pnp transistor? Now sketch the transfer function.

6.13 The multiplier whose transfer function is given by $V_y = KV_xV_o$ is incorporated into the feedback loop of the operational amplifier shown in Fig. 6.37. Derive the relationship between V_0, V_1, and V_2 and explain why this circuit can be called an analog divider.

6.14 The total range for the forward gain of a μA741 is 50,000 to 200,000. Compute the feedback factor needed such that the closed-loop gain would not change by more than 1% over this range.

6.15 The total range for the forward gain of a CA3140A is 20,000 to 100,000. Compute the feedback factor needed such that the closed-loop gain would not change by more than 1% over this range.

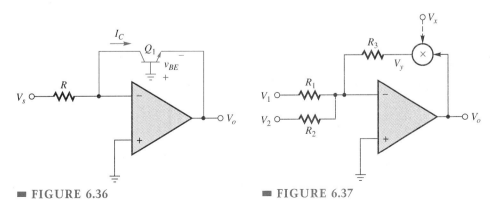

■ FIGURE 6.36   ■ FIGURE 6.37

6.16 The transfer function for $a = S_o/S_\varepsilon$ is defined in Fig. 6.38.
(a) If $f = 0.05$, sketch and label the transfer function $A = S_o/S_i$
(b) Compute a value for f such that A does not vary by more than 1%.

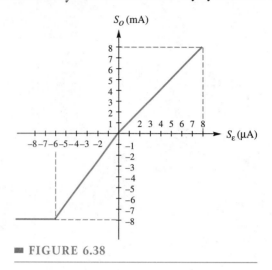

■ FIGURE 6.38

6.17 Derive the transfer function S_o/S_i for the feedback system shown in Fig. 6.39. This type of circuit illustrates a system that includes two layers of feedback.

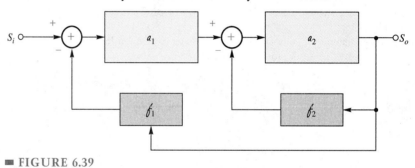

■ FIGURE 6.39

6.18 For the circuit shown in Fig. 6.40, compute the following:
(a) The open-loop voltage gain, $a_v = v_o/v_1$. Assume the switch is open.
(b) The closed-loop voltage gain, $A_V = v_o/v_1$ in terms of R_1 and R_2. Assume the switch is closed.

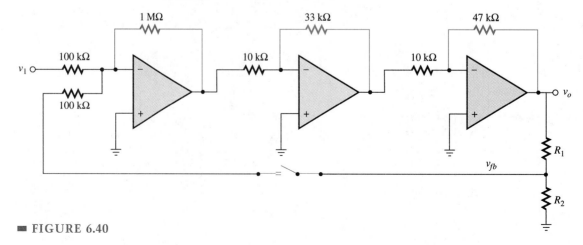

■ FIGURE 6.40

6.19 For $R_1 = R_2 = 1$ MΩ, use SPICE to obtain the closed-loop $A_V = v_o/v_1$ for the circuit shown in Fig. 6.40. Note that the switch is closed for this closed-loop case. Use a frequency-independent SPICE model.

6.20 Use summing point constraints to derive an expression for $V_o(j\omega)$ as a function of V_1, V_2, and V_3. Refer to Fig. 6.41. Assume an ideal operational amplifier.

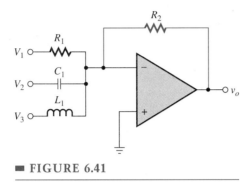

■ FIGURE 6.41

6.21 Show that a slope of 20 dB/decade is equivalent to 6 dB/octave.

6.22 At what other two frequencies between 0.1 ω_c and 10ω_c, besides $\omega = \omega_c$, does $\theta(j\omega)$ for

$$H(j\omega) = \frac{1}{1 + j\dfrac{\omega}{\omega_c}}$$

equal the asymptotic approximation? Refer to Fig. 6.16(c).

6.23 Sketch and label the Bode magnitude and phase plots for the high-pass filter given in Fig. 6.42. Also sketch the actual magnitude and phase response by illustrating the error at key points on the Bode plots.

6.24 Design a single-pole low-pass filter with $f_C = 100$ kHz. Use a 1-kΩ resistor.

6.25 Verify your Problem 6.24 solution using a SPICE simulation.

6.26 Design a single-pole low-pass filter with $f_C = 20$ Hz. Use a 0.1-μF capacitor.

6.27 Verify your Problem 6.26 solution using a SPICE simulation.

6.28 Derive an expression for $A_V(j\omega) = V_o/V_s$ for the circuit shown in Fig. 6.43. Sketch and label the Bode magnitude and phase plots.

6.29 Refer to your Problem 6.28 derivation.

(a) Assuming $C = 0.01$ μF, design a low-pass filter with a -3-dB corner of 3 kHz and a midband voltage gain of $|5|$. Sketch and label the resultant Bode plots.

(b) Verify your analytical solution in part (a) using a SPICE simulation.

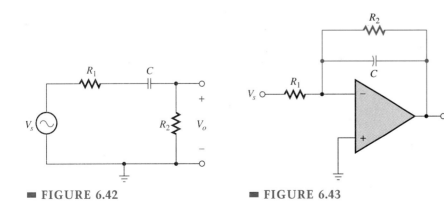

■ FIGURE 6.42   ■ FIGURE 6.43

6.30 Derive an expression for $A_V(j\omega) = V_o/V_s$ for the circuit shown in Fig. 6.44. Sketch and label the Bode magnitude and phase plots. What is ω_c?

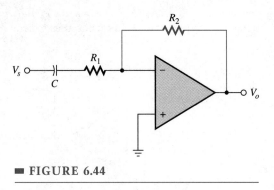

■ FIGURE 6.44

6.31 Refer to your Problem 6.30 derivation.

(a) Assuming $C = 0.001$ μF, design a high-pass filter with a 3-dB corner of 50 Hz and a midband voltage gain of $|5|$. Sketch and label the resultant Bode plots.
(b) Verify your analytical solution in part (a) using a SPICE simulation.

6.32 Sketch and label the Bode magnitude and phase plots for the low-pass filter given in Fig. 6.45. Assume two cases: $R_2 \gg R_1$ and $R_1 \ll R_2$. Sketch the actual magnitude and phase response by illustrating the error at key points on the Bode plots.

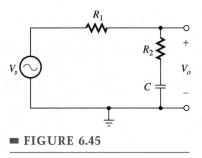

■ FIGURE 6.45

6.33 Sketch and label the Bode magnitude and phase plots for the filter given in Fig. 6.46. Also sketch the actual magnitude and phase response by illustrating the error at key points on the Bode plots. Compare your result with the basic high-pass filter response shown in Fig. 6.18.

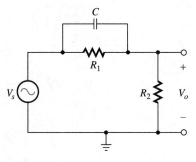

■ FIGURE 6.46

6.34 Sketch and label the Bode magnitude and phase plots for the filter given in Fig. 6.47. Also sketch the actual magnitude and phase response by illustrating the error at key points on the Bode plots. Why should this low-pass filter be a more realistic realization of an actual circuit than that shown in Fig. 6.16?

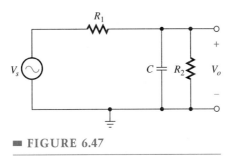

■ FIGURE 6.47

6.35* Consider the circuit shown in Fig. 6.48.

(a) Assume the corner frequencies are widely spaced, so that C_2 has no effect at low frequencies and C_1 acts as a short circuit at high frequencies. Draw a low-frequency approximate model without C_2 and use it to find V_o/V_s, ω_{C1}, and maximum gain A_o. Then draw a high-frequency approximate model where C_2 is present but C_1 is shorted and use it to find ω_{C2} and A_o. Use these results to construct Bode plots (uncorrected) of magnitude and phase for the entire circuit.

(b) Suppose the corner frequencies are not widely spaced. Show that the response is bandpass in form:

$$\frac{V_o}{V_s} = \frac{A_o}{1 + jQ\left(\dfrac{\omega}{\omega_o} - \dfrac{\omega_o}{\omega}\right)}.$$

Find the center frequency and maximum gain (ω_o and A_o).

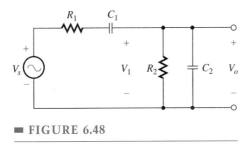

■ FIGURE 6.48

6.36 A Bode magnitude plot for a two-port network is given in Fig. 6.49 (see page 288).

(a) Derive a transfer function $H(j\omega)$ that yields this result.
(b) Sketch and label the resultant Bode phase plot.

6.37 a Bode magnitude plot for a two-port network is given in Fig. 6.50 (see page 288).

(a) Derive a transfer function $H(jf)$ that yields this result.
(b) Sketch and label the resultant Bode phase plot.

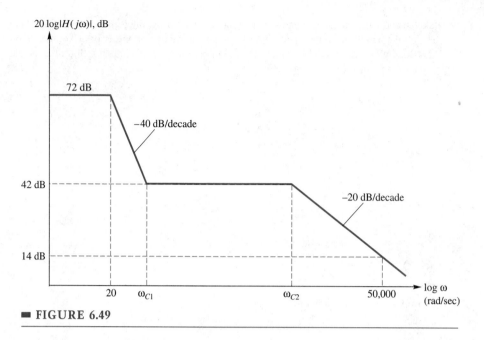

FIGURE 6.49

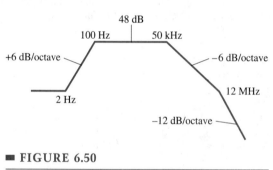

FIGURE 6.50

6.38 Consider the transfer function given by

$$H(j\omega) = \frac{\left(1 + j\dfrac{\omega}{\omega_{c1}}\right)}{\left(1 + j\dfrac{\omega}{\omega_{c2}}\right)\left(1 + j\dfrac{\omega}{\omega_{c3}}\right)}.$$

Sketch and label the Bode magnitude and phase plots (and sketch the estimated actual response) for the following cases:

(a) $\omega_{c2} = \omega_{c3} = 10\omega_{c1}$.
(b) $\omega_{c3} = 10\omega_{c2} = 10\omega_{c1}$.
(c) $\omega_{c1} = 10\omega_{c2} = 10\omega_{c3}$.

6.39 For the transfer function,

$$H(jf) = \frac{-200}{[(1 + jf/10^5)(1 - j50/f)]}$$

draw the Bode magnitude and phase plots. Label all key points and slopes. Sketch the actual magnitude and phase responses showing errors at key points between the Bode and actual plots.

6.40 The circuit in Fig. 6.51 exhibits a 20-Hz high-pass response and a 20-kHz low-pass response.

(a) Write a transfer function $H(jf) = V_3/V_1$ for this circuit.
(b) Compute values for C_1 and C_2.
(c) Sketch and label the Bode magnitude and phase plots.

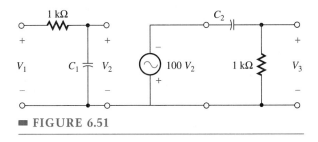

■ **FIGURE 6.51**

6.41 Verify your results from Problem 6.40 using a SPICE simulation.

6.42 Sketch and label the $A_V(j\omega) = V_o/V_s$ Bode magnitude and phase plots for the circuit shown in Fig. 6.52. Assume both operational amplifiers are ideal.

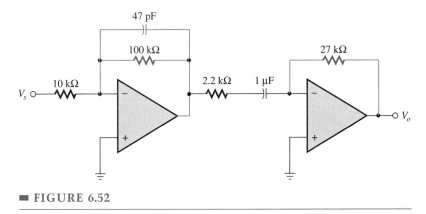

■ **FIGURE 6.52**

6.43 Repeat your Problem 6.42 analysis with a SPICE simulation using both a CA3140 and μA741 operational amplifier model.

6.44 Complete the following tasks related to the Bode magnitude plot shown in Fig. 6.53. The plot is not to scale.

(a) Write a transfer function for $H(jf)$. Your transfer function must include a numerical value for f_{C3}.
(b) For the transfer function obtained in part (a), sketch and label the Bode phase plot. You need not have the drawing to scale but you must label all key points and levels. You may assume that there is 0° of phase shift in the midband frequency region.

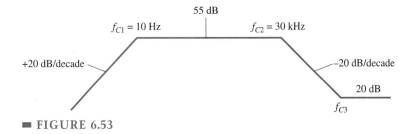

■ **FIGURE 6.53**

6.45* Use a SPICE simulation with typical μA741 a_v, R_{IN}, and R_O specifications to obtain the magnitude and phase frequency response for the circuit shown in Figure 6.52.

6.46 For the circuit shown in Fig. 6.54, complete the following tasks:

(a) Derive the transfer function of this circuit using summing point constraints. Assume an ideal operational amplifier.
(b) Sketch the Bode magnitude and phase plots.
(c) Why won't this circuit work as intended?
(d) How would you redesign the circuit to preserve the original functionality over a reasonable frequency range?

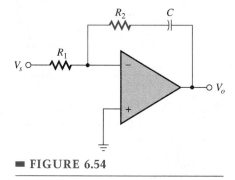

■ FIGURE 6.54

6.47 Why won't the circuit shown in Fig. 6.55 produce a gain of $-\left(1 + \dfrac{R_2}{R_1}\right)$? Draw a circuit which produces this gain.

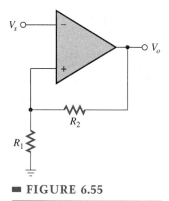

■ FIGURE 6.55

6.48 Using two ideal operational amplifiers, design a headphone audio amplifier and filter. Let $A_O = 20$ dB, $f_L = 20$ Hz, and $f_H = 20$ kHz.

6.49 Verify your Problem 6.48 design using SPICE. Assume typical μA741 a_v, R_{IN}, and R_O specifications.

6.50 Why won't the circuit shown in Fig. 6.56 work as a practical integrator?

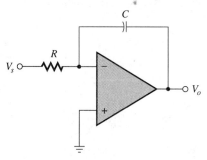

■ FIGURE 6.56

6.51 Sketch and label the Bode magnitude and phase plots for the circuit shown in Fig. 6.57. Assume that $R_1 = 1.5$ kΩ, $R_2 = 15$ kΩ, $C_1 = 1.0$ μF, and $C_2 = 0.5$ nF.

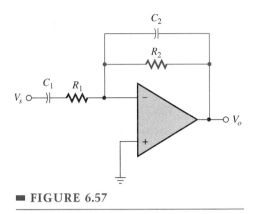

FIGURE 6.57

6.52 Verify your results for Problem 6.51 using a SPICE simulation.
6.53 Reverse the positions of the RC networks in Fig. 6.57. Sketch and label the Bode magnitude and phase plots.
6.54 Verify your results for Problem 6.53 using a SPICE simulation.
6.55 Three identical, noninteracting amplifier stages are connected in cascade. The voltage gain of each stage is given by

$$A_v(j\omega) = \frac{-A_O}{\left(1 + j\frac{f}{f_1}\right)\left(1 - j\frac{f_2}{f}\right)} \quad \text{for } f_1 = 100 \text{ kHz and } f_2 = 20 \text{ Hz.}$$

(a) What are the approximate high- and low-frequency 3-dB points?
(b) Compare your results in part (a) with the exact frequency response at these points.

6.56 Design a circuit to synthesize the transfer function in Problem 6.55. Let $A_O = 35$ dB and assume ideal operational amplifiers.
6.57 Verify your Problem 6.56 design using SPICE.
6.58 An expression analogous to Eq. (6.42) for the high-frequency 3-dB point is

$$\frac{1}{f_H} \cong \sqrt{\frac{1}{f_{H1}^2} + \frac{1}{f_{H2}^2} + \cdots + \frac{1}{f_{Hn}^2}}.$$

Repeat Example 6.8 using this equation.
6.59 Use the model shown in Figure 6.58 to demonstrate that a SPICE simulation will match Graph 4 in Figure 6.24. Why would this model be preferable to that shown in Figure 6.23? Answer this quesion considering a cacaded network with capacitive loading. Refer to Figure 6.23 for additional information.

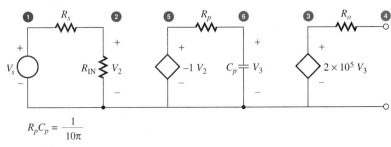

FIGURE 6.58

6.60 Using summing point constraints, show that the transfer function in Figure 6.16 can be synthesized using the circuit in Figure 6.59. Specify relative values for R_1, R_2, R_3, and R_4.

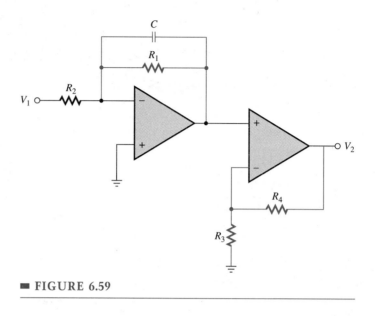

■ **FIGURE 6.59**

6.61 Using summing point constraints, show that the transfer function in Figure 6.18 can be synthesized using the circuit in Figure 6.60. Specify relative values for C_1 and C_2 and R_2 and R_3.

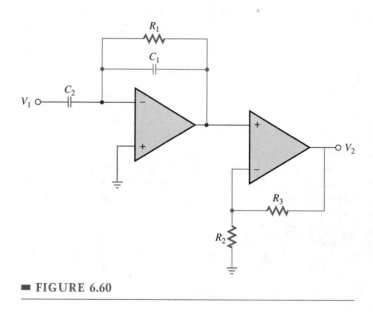

■ **FIGURE 6.60**

6.62 Show that the circuit shown in Figure 6.61 (see page 293) could be used to synthesize the transfer function, $a(j\omega)$, given by Eqn. (6.32). Specify relative component values to obtain a_o as a midband voltage gain. Use summing point constraints.

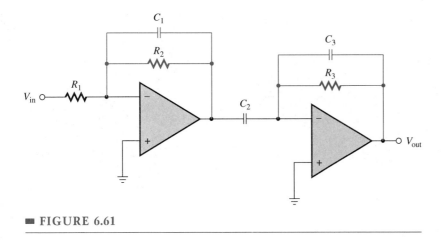

■ **FIGURE 6.61**

REFERENCES

1. *Linear Division Products Data Book 1990*, Fairchild Semiconductor Corp. (A Schlumberger Company), 333 Western Avenue, South Portland Maine. Fairchild prints a series of data books, each covering a different product line. They are updated periodically, often annually.
2. PSPICE™, MicroSim Corp., 20 Fairbanks, Irvine, California 92718. PSPICE™ is one of the widely used commercial versions of SPICE. They make a special student version available at no cost.
3. Harris Corp. Melbourne, Florida Data Books. They print a series of data books covering their product lines. These are updated periodically, often annually.
4. Bode, H.W., *Network Analysis and Feedback Amplifier Design*, New York: Van Nostrand Company, 1945. This is the original reference that forms the basis for the analysis approach and the implementation in SPICE.

SUGGESTED READINGS

Burr-Brown Product Data Book. Tucson, Ariz.: Burr-Brown Corp. 1993. A well-documented data and applications manual. There is also a five-volume series produced by Burr-Brown describing applications, design, and analysis of operational amplifier-based systems. Authors include J. G. Graeme, G. E. Tobey, L. P. Huelsman, Y. J. Wong, and W. E. Ott. Published by McGraw-Hill.

1991 Data Book. Comlinear Corp., Fort Collins, Colorado. They are updated periodically, often annually.

Virtually all manufacturers and vendors of operational amplifier ICs provide a wealth of application, design, and analysis literature for the user. You should avail yourself of this information as well as the information provided in the numerous trade magazines available to the engineering community. Also, many of the IC manufacturers offer printed data books as well as software versions. Many of the manufacturers also prepare device SPICE model netlists for their customers in the engineering community. More recently, many manufacturers offer information through their Internet Home Pages.

CHAPTER 7

BIASING AND STABILITY

7.1 BJT Q-Point Selection
7.2 BJT Q-Point Stabilization
7.3 FET Bias and Stabilization
7.4 BJT Current Sources
7.5 MOSFET Current Sources

Summary
Survey Questions
Problems
References
Suggested Readings

As we continue the study of transistor amplifier circuits, it is necessary to consider the dc bias circuitry that is used to establish the Q-point. Of course, we expect the operation of a linear amplifier to be in the active region, as explained in Chapters 4 and 5. Since selection of the Q-point affects the ac parameters of the transistor, it is not an arbitrary decision in amplifier design. Furthermore, this operating point must be maintained, within limits, as the dc parameters of the transistors vary, either with temperature or because manufacturing variations within the same device type make the exact values uncertain.

We present design criteria for Q-point selection and stabilization, both for bipolar junction transistors (BJTs) and for field-effect transistors (FETs). Resistive bias circuits are investigated first because they are easily understood and are used to construct test circuits and simple low-volume analog circuits. Aspects of bias stabilization such as the benefit of an emitter resistor carry over into integrated-circuit (IC) designs.

Integrated circuits must be biased without large resistors or large capacitors, which cannot be integrated economically, if at all. For ICs, a transistor is usually cheaper than a resistor and much cheaper than a large capacitor. As a consequence, active current sources are used instead of resistors. Current sources are presented as the second topic in bias; their use is described in Chapters 8 through 13.

Power amplifiers have a third type of bias arrangement that is used to avoid wasting power. This topic is discussed in Chapter 10.

IMPORTANT CONCEPTS IN THIS CHAPTER

- BJT and FET Q-points must be selected such that device maximum rated values are not exceeded
- The Q-point for each active device of an amplifier must be chosen such that the dynamic Q-point remains in its linear region at the desired signal level
- The bias network of an amplifier must not load the signal source excessively
- Q-points are selected to obtain desired small-signal parameters
- An effective way to stabilize BJT and FET Q-points for changes in parameters is by the use of an emitter resistor or a source resistor, respectively
- Two or more active devices can be accurately matched in parameters on a die with IC technology, and temperatures also tend to equalize on a die
- Two active devices and a reference current can produce a stable current source equal to the reference current, with a high output resistance
- A Widlar current source produces a current smaller than the reference current, and has somewhat higher output resistance than a simple current source
- A Wilson current source has a very high output resistance
- MOS current sources use another FET to provide the reference current

7.1 BJT Q-POINT SELECTION

We use the typical common-emitter (CE) amplifier circuit shown in Fig. 7.1 to illustrate some of the many constraints to be considered in the design of a bias circuit, including choice of the Q-point. Figure 7.2 shows the dc model of the circuit, which allows analytic solution of the operating conditions if we know the model parameters. Note that the

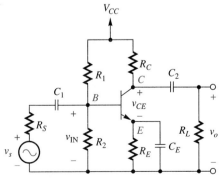

FIGURE 7.1 Typical common-emitter amplifier.

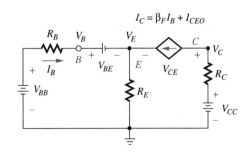

FIGURE 7.2 A dc model for the common-emitter amplifier shown in Fig. 7.1. Note that $I_{CEO} \equiv I_{CBO}(1 + \beta_F)$.

capacitors are treated as open circuits for dc. The base resistor network has been replaced by its Thévenin equivalents:

$$V_{BB} = V_{CC} \frac{R_2}{R_1 + R_2}, \qquad (7.1)$$

$$R_B = R_1 \parallel R_2 = \frac{R_1 R_2}{R_1 + R_2}. \qquad (7.2)$$

Summing around the left loop

$$V_{BB} - I_B R_B - V_{BE} - (1 + \beta_F) I_B R_E - I_{CBO}(\beta_F + 1) R_E = 0.$$

Solving for I_B,

$$I_B = \frac{V_{BB} - V_{BE} - I_{CBO}(\beta_F + 1) R_E}{R_B + (\beta_F + 1) R_E}. \qquad (7.3)$$

The voltage equation around the right-hand loop yields

$$V_{CE} = V_{CC} - I_C R_C - (I_C + I_B) R_E$$

$$= V_{CC} - I_C \left[R_C + \frac{(\beta_F + 1)}{\beta_F} R_E \right] + I_{CBO}\left(\frac{\beta_F + 1}{\beta_F}\right) R_E. \qquad (7.4)$$

■ The I_{CBO} term is often neglected. We will demonstrate the validity of this assumption.

EXAMPLE 7.1

Find the Q-point for the amplifier shown in Fig. 7.1, using $R_1 = R_2 = 6.8$ kΩ, $R_E = R_C = 100$ Ω, and $V_{CC} = 12$ V. Let $V_{BE} = 0.6$ V, $I_{CBO} = 0.01$ μA, and $\beta_F = 100$ at $T = 25°$C. Neglect any change in temperature.

SOLUTION Equation (7.3) gives

$$I_B = \frac{6.0 - 0.6 - 10^{-8}(101)100}{3400 + (101)100} = 0.4 \text{ mA}.$$

Then $I_C = \beta_F I_B + (\beta_F + 1) I_{CBO} = 40$ mA. Note that I_{CBO} did not affect this calculation, which is often true at $T = 25°$C. Using Eq. (7.4), we find that $V_{CE} = 12 - 0.04(100 + 101) = 3.96$ V. The operating point is indicated as Q_1 in Fig. 7.4.

DRILL 7.1

Repeat the problem of Example 7.1 using $\beta_F = 300$.

ANSWER $I_C = 48$ mA, $V_{CE} = 2.32$ V

AC Considerations

Figure 7.3 presents a simplified ac model for the amplifier, using the active small-signal model for the BJT from Section 4.5, which is valid for frequencies sufficiently high that capacitors C_1, C_2, and C_E can be assumed to have negligible reactance, but still low enough that parasitic capacitances of the transistor can be treated as open circuits. Such a model is an example of a **midfrequency** or **midband model**. We study more accurate "small-signal" ac models for the transistor itself in Chapter 8. Capacitors C_1 and C_2 are known as **coupling** capacitors because C_1 serves to couple the input signal to the amplifier and C_2 couples the amplifier output to the output load resistor, without disturbing the dc bias. Capacitor C_E is a **bypass** capacitor because its function is to prevent ac (signal) currents from developing an ac voltage across R_E. The ac input resistance of the CE transistor consists of r_π [Eq. (4.28)].

■ Although many audio amplifiers could be modeled as *midband* amplifiers, the reason we begin with midband models is the convenience of analyzing circuits with no capacitors that would complicate our calculations.

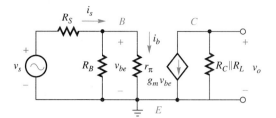

FIGURE 7.3 An ac model for the common-emitter amplifier shown in Fig. 7.1.

The dc and ac Load Lines

Figure 7.4 shows a plot of the collector characteristics of the 2N2222A transistor. We shall make reference to some of the transistor ratings given in Fig. 4.32. The dc load line is a plot of the v_{CE} versus i_C relationship of the circuit external to the transistor, Eq. (7.4) in this case. The intersection of the dc load line with the collector characteristic for $I_B = 0.4$ mA is the dc solution, or Q-point, marked Q_1 in the figure. The ac load line is drawn through this point. Suppose $R_L = R_C = 100\ \Omega$. Then continuing with the amplifier in Example 7.1, the ac load is R_C in parallel with R_L ($R_C \| R_L$), or 50 Ω. Referring to the 50-Ω load line drawn through point Q_1 in Fig. 7.4, we can observe something of the operation of this amplifier. Should the base current be perturbed away from the bias point of 0.4 mA by an input signal, we see at a glance how the collector circuit responds. A base-current decrease to 0 mA would cause the collector current to be 0 mA (cutoff), while a base-current increase to 1 mA would result in a collector current of 100 mA. This Q-point is realistic for amplifier operation because it meets all of the following conditions:

1. There is opportunity for the dynamic operating point to vary up and down along the ac load line without immediately encountering cutoff or saturation. In this example,

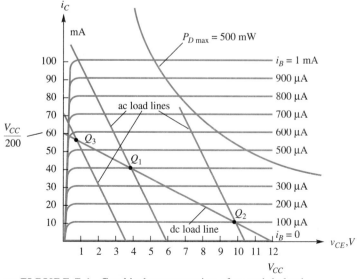

FIGURE 7.4 Graphical representation of ac and dc load lines on the collector characteristic in Example 7.1.

I_C could change 40 mA from the Q-point to cutoff and about 65 mA to saturation. The current-transfer characteristic, i_C/i_B, for this operating point is plotted in Fig. 7.5(a).
2. The operating points remain well below the locus of maximum power dissipation, $P_D = 500$ mW at $T_A = 25°C$ (also plotted in Fig. 7.4).
3. The maximum dynamic operating voltage (6 V at cutoff) remains well below the maximum rated V_{CEO} of 30 V.
4. The maximum dynamic operating current (105 mA at saturation) is well below the maximum rated I_C of 800 mA.
5. At the indicated Q-point, $P_D = (40 \text{ mA})(4.0 \text{ V}) = 160$ mW. In Section 4.10, we showed that the thermal resistance of the 2N2222A is 300°C/W from junction to ambient with no heat sink ($\theta_{JA} = \theta_{JC} + \theta_{CA}$). Thus, we can expect the junction temperature to rise $\Delta T = P_D \theta_{JA} = (0.3)(160) = 48°C$. For an ambient temperature of 25°C, this leads to a junction temperature of 73°C, much less than the maximum rated junction temperature of 175°C. We shall see that this temperature rise will affect the Q-point, however.

You may show that Example 7.1 with R_1 replaced with the larger resistance of 33 kΩ results in $I_B \cong 0.1$ mA. The resulting transistor Q-point is marked Q_2 in Fig. 7.4. Again the ac load line is sketched on the characteristic. This bias situation is much less satisfactory for an amplifier because the instantaneous operating point cannot move very far downward along the ac load line before reaching cutoff. Thus, a negative-going input signal could only move 0.1 mA from the dc bias value of 0.1 mA before the dynamic collector current would reach 0, and no further amplification could occur. Figure 7.5(b) shows graphically that the Q-point is near the cutoff end of the linear region. Note also that an instantaneous peak of signal current greater than 0.7 mA would drive the transistor into a region of excessive power dissipation. This is unacceptable because the transistor will destroy itself if this situation persists more than a few milliseconds, corresponding to the thermal time constant of the junction.

Using Example 7.1 with R_2 increased to 22 kΩ, we find that $I_B \cong 0.56$ mA. This operating point is labeled Q_3 in Fig. 7.4. Again we can sketch in the ac load line and observe the locus of dynamic operation for such an amplifier bias. The corresponding transfer characteristic is shown in Fig. 7.5(c), and you will observe from either of the figures that now the circuit is biased too close to saturation for maximum input signal and linear amplification.

We have considered constraints on the Q-point imposed by maximum ratings of the transistor. It is also important to consider the effect of the bias point on the ac parameters of the transistor. For example, all of the resistances in the examples just considered could

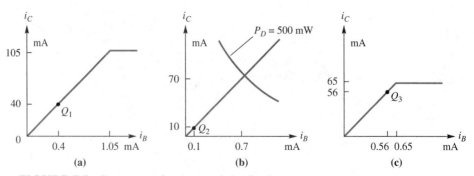

■ **FIGURE 7.5** Current-transfer characteristics for the amplifier in Example 7.1. **(a)** Q-point in linear region. **(b)** Q-point near cutoff. **(c)** Q-point near saturation.

be multiplied by a factor of 10, and the graphs would all be approximately correct if all dc currents were scaled by a factor of 0.1 (you should verify this). However, the parameters of the ac model would be drastically altered. We find that r_π would increase by a factor of almost 10 [from Eq. (4.28)]. Indeed, all of the parameters change; even $\beta(h_{fe})$ changes somewhat. Thus, if we are free to choose I_B, we may choose it to obtain a desired ac transistor model parameter.

A final consideration to be mentioned here is the magnitude of R_B, the base bias resistor network, $(R_1 \| R_2)$. We can observe in the model in Fig. 7.3 that i_b is a fraction of i_{in}, depending on how R_B is related to r_π. Normally, R_B must be much larger than r_π so that most of the input current will flow into the base; otherwise the input current will not be effective in controlling the output current.

Usually the value of R_L will be determined by the situation at hand.

■ The purpose of the amplifier circuit is to develop an appropriate voltage across (or current through) some particular load resistance. Thus the application usually determines R_L.

EXAMPLE 7.2

Consider the amplifier shown in Fig. 7.1 again. This time let $R_E = R_C = R_L = 1\ \text{k}\Omega$, $V_{CC} = 12$ V, $V_{BE} = 0.6$ V, and $\beta_F = 100$. Find values of R_1 and R_2 that will bias this circuit at a Q-point of $I_C = 4$ mA and $V_{CE} = 4$ V. Assume that $r_\pi = 650\ \Omega$ and that the base bias resistance R_B should be 30 times that of r_π so as not to divert significant signal current from the base. Neglect I_{CBO}.

■ $r_\pi = V_T \beta_F / I_C$, so $r_\pi = 650\ \Omega$ can be calculated.

SOLUTION I_B is $I_C/\beta_F = 4/100 = 0.04$ mA. Using Eq. (7.3) with $R_B = 20$ kΩ, $V_{BB} = V_{BE} + I_B[R_B + (\beta_F + 1)R_E] = 0.6 + 0.04[20\ \text{k} + (101)1\ \text{k}] = 5.44$ V. Then, using Eqs. (7.1) and (7.2), we find that $R_1 = 44.1$ kΩ and $R_2 = 36.6$ kΩ. Standard values of 47 kΩ and 39 kΩ would likely be used in practice.

DRILL 7.2

Modify the problem of Example 7.2 by changing V_{CC} to 10 V and I_C to 3 mA. This will increase r_π, so let $R_B = 25$ kΩ. Find values of R_1 and R_2.

ANSWER $R_1 = 57.1$ kΩ and $R_2 = 44.5$ kΩ

CHECK UP

1. **TRUE OR FALSE?** The Q-point of an active circuit can be found with the aid of the dc model only.
2. **TRUE OR FALSE?** The design of the best Q-point for an active circuit does not require an ac model.
3. **TRUE OR FALSE?** The bias resistors of a CE amplifier must not be too small for only one reason: because of the load on the power supply.
4. **TRUE OR FALSE?** The initial choice of the operating current for active devices is influenced by the ac parameters desired for the devices.

7.2 BJT Q-POINT STABILIZATION

We have demonstrated the importance of designing an amplifier with an appropriate Q-point. It is equally important that this Q-point be maintained in the presence of inevitable differences in transistor parameters. Such differences can be caused by variations in the devices used on a production line (compared to those assumed by the circuit designer). The differences can also be caused by temperature changes, either because of power dissipation in circuit components or because of ambient temperature changes. We might also want to stabilize the Q-point should the dc power supply voltage vary from the design value. Recall from Section 6.2 that stabilization is a consequence of negative feedback, a concept that is covered in depth in Chapter 11.

The circuit shown in Fig. 7.1 is typical of CE amplifiers that have some degree of stabilization. We will now investigate how this stabilization can be controlled by the circuit designer. The dc part of Eq. (4.18) when $I_{CO} \cong I_{CBO}$ is

$$I_C = \beta_F I_B + (\beta_F + 1)I_{CBO}. \tag{7.5}$$

Substituting in Eq. (7.3) and rearranging terms gives

$$I_C = \frac{\beta_F(V_{BB} - V_{BE}) + I_{CBO}(\beta_F + 1)(R_B + R_E)}{R_B + (\beta_F + 1)R_E}. \tag{7.6}$$

It appears that the operating current is determined by β_F, V_{BE}, and I_{CBO}, all parameters of the transistor that may vary from device to device or with temperature. In addition, V_{BB} may vary if V_{CC} is not constant. We assume R_B and R_E to be constants under the designer's control.

The significance of the effects of these variables on I_C is much better understood by first neglecting the I_{CBO} term. Referring to Eq. (7.6), we see that this is justified if

$$I_{CBO}(R_B + R_E) << V_{BB} - V_{BE}, \tag{7.7}$$

which is almost always the case. To demonstrate this, we insert the parameters of Example 7.1 into Eq. (7.7) and are rewarded by the very defensible result of 0.01 μA × 3500 Ω = 35 μV << 5.4 V. Neglecting I_{CBO} reduces Eq. (7.6) to

$$I_C = \frac{\beta_F(V_{BB} - V_{BE})}{R_B + (\beta_F + 1)R_E}. \tag{7.8}$$

The effect of changes in β_F and V_{BE} on I_C is considered separately. Because I_C is a function of both variables, treating them independently is not exact, but if each effect is small, the error will be very small.

Effect of Changing β_F on I_C

When V_{BE} is held constant, we find $\Delta I_C = I_{C2} - I_{C1}$ by finding an I_{C2} using β_2 in Eq. (7.8), and I_{C1} using β_1, then subtracting. We then divide by I_{C1} to get the percentage change:

$$\frac{\Delta I_C}{I_{C1}} = \frac{(\beta_2 - \beta_1)(V_{BB} - V_{BE})(R_B + R_E)}{\beta_1(V_{BB} - V_{BE})(R_B + R_E + \beta_2 R_E)} = \frac{\Delta \beta}{\beta_1} \cdot \frac{R_B + R_E}{R_B + R_E + \beta_2 R_E}, \tag{7.9}$$

■ Remember, this $\Delta \beta$ may represent an uncertainty as to what the β really is as compared to what was assumed in the design. Manufacturers specify a tolerance on β such as $\beta_F = 200 \pm 100$.

where $\Delta \beta = \beta_2 - \beta_1$. Examination of Eq. (7.9) reveals that if $R_E = 0$, there will be no stabilization of I_C against changes in β_F. That is, a given percentage change in β_F results in the same percentage change in I_C. Conversely, the larger R_E is, compared to R_B, the smaller the change in I_C for a given change in β_F. Indeed, this is exactly the reason for having R_E in the circuit. Let us refer to our example again.

■ EXAMPLE 7.3

Using the amplifier shown in Fig. 7.1 with the values given in Example 7.1, find the percentage change in the Q-point when β_F changes from 100 to 150. Neglect any other changes.

CHAPTER 7   BIASING AND STABILITY

SOLUTION   Recall that $I_C = 40$ mA for $\beta_F = 100$ (Example 7.1). By the same method [Eq. (7.3)], $I_C = 43.8$ mA can be found for $\beta_F = 150$. We conclude that I_C increased from 40 to 43.8 mA, or 9.5%, as β_F increased from 100 to 150, or 50%. We are now in a position to compare these results with the alternative method using Eq. (7.9):

$$\frac{\Delta I_C}{I_{C1}} = \frac{(150 - 100)(3500)}{100[3500 + (150)(100)]} = 9.5\%.$$

The results agree.

The result of this example indicates good stabilization, where a 50% change in β_F causes only a 9.5% change in I_C. Reference to Fig. 7.4 confirms that any Q-point between 40 and 44 mA would be satisfactory. The general conclusion that we may draw from Eq. (7.9) is that to get significant stabilization of the Q-point with variations of β_F,

$$\beta_F R_E > R_B + R_E. \tag{7.10}$$

Effect of Changing V_{BE} on I_C

When β_F is held constant, we find $\Delta I_C = I_{C2} - I_{C1}$ by finding an I_{C2} using V_{BE2} in Eq. (7.8), and I_{C1} using V_{BE1}, then subtracting. We then divide by I_{C1} to get the percentage change:

$$\frac{\Delta I_C}{I_{C1}} = \frac{-\Delta V_{BE}}{(V_{BB} - V_{BE1})}, \tag{7.11}$$

where $\Delta V_{BE} = V_{BE2} - V_{BE1}$. Immediately we see that the percentage of change in I_C is proportional to the (negative) fractional change in V_{BE} but is small when

$$(V_{BB} - V_{BE1}) \gg V_{BE1}. \tag{7.12}$$

Of course, if ΔV_{BE} is due to a temperature change, β_F is likely to change also, so that Eq. (7.11) will be somewhat in error.

EXAMPLE 7.4

Consider the amplifier shown in Fig. 7.1 with the values given in Example 7.3, including $V_{BE} = 0.6$ V at 25°C. Find the change in I_C that occurs due to ΔV_{BE} when the actual junction temperature, T_j, increases to 125°C because of increased ambient temperature and self-heating. Recall that V_{BE} decreases 2 to 2.5 mV/°C as the temperature rises (Section 4.7).

SOLUTION   A 100°C temperature rise causes V_{BE} to decrease 100×2.5 mV, and $\Delta V_{BE} = -250$ mV. Then, from Eq. (7.11),

$$\frac{\Delta I_C}{I_{C1}} \cong \frac{-(-0.25)}{0.6} \frac{0.6}{6.0 - 0.6} = 4.6\%.$$

DRILL 7.3

In a circuit of the type of Fig. 7.1, I_C increases 10% while β_F increases from 100 to 200. What does this tell you about the ratio of R_B to R_E?

ANSWER $R_B/R_E = 21$

■ This decrease is, of course, a rather rough approximation for this large a temperature change.

DRILL 7.4

In a circuit of the type of Fig. 7.1, V_{BE} decreases from 0.5 to 0.4 V. To hold I_C to a 2% increase, what must V_{BB} be?

ANSWER   5.5 V

Effect of Changing I_{CBO} on I_C

If we hold β_F and V_{BE} constant, we find $\Delta I_C = I_{C2} - I_{C1}$ by finding an I_{C2} using I_{CBO2} in Eq. (7.6), and I_{C1} using I_{CBO1}, then subtracting. We then divide by I_{C1} to get the percentage change:

$$\frac{\Delta I_C}{I_{C1}} = \frac{(I_{CBO2} - I_{CBO1})(R_B + R_E)(1 + \beta_F)}{\beta_F(V_{BB} - V_{BE}) + I_{CBO1}(R_B + R_E)(1 + \beta_F)}. \tag{7.13}$$

If we assume $\beta_F \gg 1$, and when the second term in the denominator is neglected [Eq. (7.7)], the expression reduces to

$$\frac{\Delta I_C}{I_{C1}} = \frac{\Delta I_{CBO}(R_B + R_E)}{(V_{BB} - V_{BE})}. \tag{7.14}$$

The percentage change in I_C is seen to be proportional to the change in I_{CBO}, but this percentage is smaller when the factor

$$\frac{(R_B + R_E)}{(V_{BB} - V_{BE})}$$

is smaller. We can easily show that to make this factor smaller but to maintain the same I_B, R_E must be increased with respect to R_B.

EXAMPLE 7.5

Consider the amplifier shown in Fig. 7.1 with the values given in Example 7.3. Let $I_{CBO} = 0.01\ \mu A$ at 25°C. Find the change in I_C that results as T_j increases to 125°C. Recall that I_{CBO} doubles for every 10° rise. Neglect the other changes.

SOLUTION A 100°C rise corresponds to a $2^{10} = 1024$-fold increase in I_{CBO}. Then $\Delta I_{CBO} = 10.24 - 0.01 = 10.23\ \mu A$. From Eq. (7.14),

$$\frac{\Delta I_C}{I_{C1}} = \frac{0.01023\ \text{mA}\ (3400 + 100)\ \Omega}{(6.0 - 0.6)\ \text{V}} = \frac{0.0358}{5.4} = 0.0066 = 0.66\%.$$

■ This is one more indication that the term I_{CBO} can usually be neglected.

In Example 7.5, the fractional change in I_C due to I_{CBO} is very small, 0.66%. When each of the changes is small ($< 10\%$), the total combined effect of all of these changes on I_C is approximately the sum of the individual effects.

$$\frac{\Delta I_C}{I_{C1}} = \frac{\Delta \beta}{\beta_1} \cdot \frac{R_B + R_E}{R_B + R_E + \beta_2 R_E} - \frac{\Delta V_{BE}}{(V_{BB} - V_{BE1})} + \frac{\Delta I_{CBO}(R_B + R_E)}{(V_{BB} - V_{BE1})}. \tag{7.15}$$

EXAMPLE 7.6

Examples 7.3, 7.4, and 7.5 have considered the effects on I_C of changes in β_F, V_{BE}, and I_{CBO} separately. Calculate the combined effects of all three changes on this amplifier as T_j increases from 25 to 125°C.

SOLUTION Reference to Fig. 4.31 reveals that β_F for the 2N2222A at 40 mA should increase about 40 to 50% as the temperature increases from 25 to 125°C. Then the result of Example 7.3 applies, as do the results of Examples 7.4 and 7.5. The three terms are summed in order: ΔI_C = 40 mA (9.5% + 4.6% + 0.66%) = 5.9 mA. A very careful solution of this problem, not neglecting the interdependence of these terms, yields ΔI_C = 5.8 mA, which is not significantly different (see Drill 7.5).

DRILL 7.5

Solve the problem of Example 7.6 by the exact method of using Eq. (7.6) at both temperatures.

ANSWER 5.8 mA

Summary of BJT Q-point Stabilization

We have shown that a BJT amplifier Q-point can be stabilized against parameter variations by the simple addition of an emitter resistor of appropriate value. Indeed, transistor amplifiers are seldom designed without some bias stabilization. In general, the two conditions, Eqs. (7.10) and (7.12), will be met if $R_E I_E \approx V_{CC}/5$ and $R_B < \beta_F R_E/3$, as in our examples. Often $R_B < \beta_F R_E/10$ is used. Under these conditions, I_C will vary linearly with V_{CC}.

Collector-Base Stabilization

Another bias arrangement that is used to provide Q-point stabilization is shown in Fig. 7.6. The same sort of analysis that was carried out for the circuit shown in Fig. 7.1 will indicate what the constraints are. You can show that all of the I_C expressions are the same for this configuration as for the one we examined, if we substitute R_C for R_E, and V_{CC} for V_{BB} (Problem 7.15 at the end of the chapter). For example, Eq. (7.15) becomes

$$\frac{\Delta I_C}{I_{C1}} = \frac{\Delta \beta}{\beta_1} \cdot \frac{R_B + R_C}{R_B + R_C + \beta_2 R_C} - \frac{\Delta V_{BE}}{(V_{CC} - V_{BE1})} + \frac{\Delta I_{CBO}(R_B + R_C)}{(V_{CC} - V_{BE1})}. \quad (7.16)$$

The ac conditions may be quite different, however. The input resistance of the amplifier is affected by the circuit arrangement, as we will see in Chapter 8. To prevent ac feedback, R_B could be partitioned into two parts, with a bypass capacitor to ground at the midpoint. To examine this idea, see Problem 7.17.

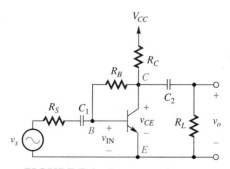

■ **FIGURE 7.6** Common-emitter amplifier with collector-base bias.

> **CHECK UP**
>
> 1. **TRUE OR FALSE?** For a BJT, β_F, V_{BE} and I_{CEO} all increase with a temperature increase.
> 2. **TRUE OR FALSE?** In all cases, increasing R_E with respect to R_B increases BJT stability.
> 3. **TRUE OR FALSE?** In all cases, increasing R_E with respect to R_B increases BJT stability if I_B is maintained constant.
> 4. **TRUE OR FALSE?** Increasing R_E with respect to R_B but maintaining I_B constant will require an increase in V_{BB}.

7.3 FET BIAS AND STABILIZATION

The principles of FET amplifier bias design are similar to those of the BJT; only the models are different. Consider first the circuit shown in Fig. 7.7(a), which is drawn with a depletion-type MOSFET. We could just as well have used a JFET as shown in Fig. 7.7(b) or an enhancement-mode MOSFET, only the dc circuit parameters would be different. You should notice the analogy between this circuit and the self-biased BJT amplifier shown in Fig. 7.1. In general, the feature that all of these circuits have in common is that the operating-point current is sampled by an emitter (source) resistor, and the voltage thus developed is of such a polarity that the dc bias is adjusted in a manner to oppose a change in the Q-point current caused by device parameter variations. In system design, this concept is known as **negative feedback**. The principle would work equally well with *pnp* BJTs and *p*-channel FETs, with the understanding that the supply voltages would be negative. In Chapter 11 we discuss negative feedback as it applies to stabilization of a large number of parameters.

Details of the analysis of the FET bias situation differ from those of the BJT because of the different model used to represent the active device. JFET operation in the saturation region ($V_{DS} > V_{GS} - V_P$) was described in Chapter 5 by Eq. (5.35); repeated here, it is

$$i_D = \begin{cases} I_{DSS}\left(1 - \dfrac{v_{GS}}{V_P}\right)^2 & \text{for} \quad v_{GS} > V_P \\ 0 & \quad\quad\quad v_{GS} < V_P \end{cases}. \quad (7.17)$$

Equation (5.12), often used for MOSFETs, is

$$i_D = k(v_{GS} - V_t)^2 \quad \text{for} \quad v_{DS} > v_{GS} - V_t. \quad (7.18)$$

Note that these two models are equivalent upon substitution of constants. For amplifiers, these **saturation region** models must hold for all values of the instantaneous input signal, which means that the Q-points must be chosen such that Eq. (7.17) or (7.18) will apply over a range of v_{DS}. We consider the two cases for *n*-channel FETs, and the situation is identical for *p*-channel devices except that the signs of all voltages and currents are reversed.

JFETs or Depletion-Type MOSFETs

For the *n*-channel JFETs, V_P would be a negative pinch-off voltage, and Eq. (7.17) would be valid only for v_{GS} less negative than V_P. Also, v_{GS} must not be more positive than 0 V, or it will forward-bias the gate-substrate junction. For depletion-type MOSFETs,

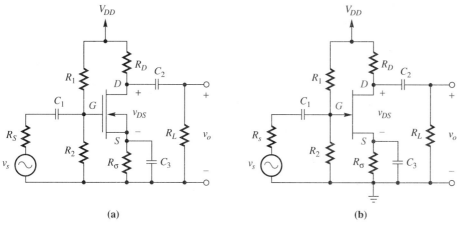

FIGURE 7.7 (a) MOSFET amplifier. (b) JFET amplifier.

V_t is a negative threshold voltage, and Eq. (7.18) would only be valid for v_{GS} less negative than V_t (see left curve in Fig. 7.8). For the depletion MOSFET, a positive v_{GS} would enhance the channel.

Enhancement-Mode MOSFETs

For enhancement-mode MOSFETs, however, V_t is a positive threshold voltage, and Eq. (7.18) would only be valid for v_{GS} more positive than V_t (see right curve in Fig. 7.8).

The DC and AC Models. From Fig. 7.7, we construct the dc model shown in Fig. 7.9 and the ac model shown in Fig. 7.10. Gate current is neglected. Again we ignore the reactances of the coupling and bypass capacitors for ac and do not show the parasitic capacitances that inevitably would be part of a high-frequency model for the FET. Recall from Eq. (5.40) that for JFETs

$$g_m = \left.\frac{\partial i_D}{\partial v_{GS}}\right|_{Q\text{-point}} = \frac{-2I_{DSS}}{V_P}\left(1 - \frac{V_{GS}}{V_P}\right)$$

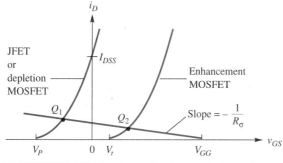

FIGURE 7.8 FET transfer characteristics and bias line.

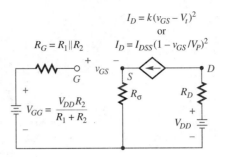

■ **FIGURE 7.9** A dc model for the circuit shown in Fig. 7.7.

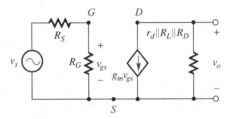

■ **FIGURE 7.10** An ac model for the circuit shown in Fig. 7.7.

and from Eq. (5.14) for MOSFETs

$$g_m = \left.\frac{\partial i_D}{\partial v_{GS}}\right|_{Q\text{-point}} = 2k(V_{GS} - V_t). \tag{7.19}$$

Figure 7.9 reveals that

$$V_{GS} = V_G - V_S = V_{GG} - I_D R_\sigma. \tag{7.20}$$

Substitution of Eq. (7.20) into (7.17) yields the quadratic equation

$$I_D = I_{DSS}\left(1 - \frac{V_{GG}}{V_P} + \frac{I_D R_\sigma}{V_P}\right)^2$$

or

$$I_D^2 + \left[\frac{2(V_P - V_{GG})}{R_\sigma} - \frac{V_P^2}{I_{DSS} R_\sigma^2}\right]I_D + \frac{(V_P - V_{GG})^2}{R_\sigma^2} = 0. \tag{7.21}$$

Figure 7.8 shows the graphical solution of this equation for two different FET transfer characteristics. The left one might represent a JFET or a depletion-type MOSFET. The right characteristic is that of an enhancement-mode MOSFET. Note that by making V_{GG} and R_σ large, we can control the operating I_D nearly independently of the transistor. The practical limit to this is that $R_D I_D$ must not become so large that $v_{DS} = V_{DD} - R_D i_D - R_\sigma i_D$ is too small to allow a reasonable ac variation in v_{DS} without violating the constraints of the saturated model [Eq. (7.17) or (7.18)].

We now undertake an example of bias design to maintain nearly constant I_D.

EXAMPLE 7.7

Design an amplifier of the configuration shown in Fig. 7.7 using the 2N5459 JFET (see Fig. 5.31). The value of I_D is to be within 15% of 2 mA for all FETs that meet the manufacturer's specifications at 25°C. In addition, $V_{DD} = 25$ V, R_G should be 100 kΩ, and R_L is 2.2 kΩ.

SOLUTION The 2N5459 data sheet shows 4 mA $< I_{DSS} <$ 16 mA, and -8 V $< V_P < -2$ V. We will therefore assume that the limiting transfer characteristics are $I_{D1} = 16[1 - (v_{GS}/-8)]^2$ mA and $I_{D2} = 4[1 - (v_{GS}/-2)]^2$ mA. Refer to Fig. 7.11 for a graphical representation of the worst-case situations. The manufacturer guarantees all 2N5459 characteristics to be within these limits, and those with highest I_{DSS} will also have the higher V_P.

1. Let $I_{D1} = 2 + 15\%(2) = 2.3$ mA, and use Eq. (7.17) to find $V_{GS1} = -4.97$ V, labeled Q_1 in the figure.
2. With $I_{D2} = 2 - 15\%(2) = 1.7$ mA, $V_{GS2} = -0.70$ V at point Q_2.
3. The slope of the bias load line through these points must be $R_\sigma = \Delta V_{GS}/\Delta I_D =$ 4.27 V/0.6 mA = 7.12 kΩ. (Use standard values 6800 + 330 Ω.)
4. We now find V_{GG} with Eq. (7.20).

$$V_{GG} = I_D R_\sigma + V_{GS} = (1.7 \text{ mA})(7.13 \text{ k}\Omega) - 0.70 \text{ V} = 11.4 \text{ V}.$$

5. R_G and V_{GG} are the Thévenin equivalent for the resistor network $R_1 = 219$ kΩ and $R_2 = 184$ kΩ (use standard values 220 kΩ and 180 kΩ).

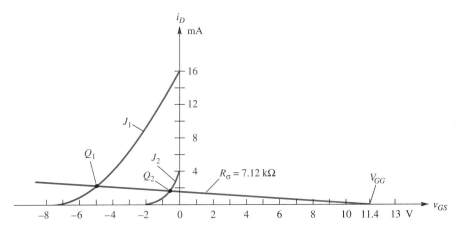

FIGURE 7.11 FET characteristics and bias load line in Example 7.7.

It is noteworthy that this bias design will stabilize I_D quite well over the considerable range in I_{DSS} and V_P, but g_m ranges from 1.52 mS at Q_1 to 2.6 mS at Q_2. $V_{DS} = V_{DD} - I_D(R_\sigma + R_D)$, so we use the median $I_D = 2$ mA, and $R_\sigma = 7.13$ kΩ to find $V_{DS} = 25 - 14.3 - 0.002R_D$. Figure 7.12 shows a plot of $V_{DS} = 10.7 - 0.002R_D$ lines for several values of R_D. The limiting values are $V_{DS} = 10.7$ V for $R_D = 0$, and $V_{DS} = 0$ when $R_D = 5.35$ kΩ at $I_D = 2$ mA. There can be no amplification if $R_D = 0$ ($V_D = V_{DD}$) or if $V_{DS} = 0$ (the FET is inoperative, analogous to a saturated BJT). We attempt a compromise at $R_D = 3.3$ kΩ. Figure 7.12 shows the result: $V_{DS} = 4.1$ V at point Q. The ac load line is $R_D \| R_L = 3.3 \| 2.2 = 1.32$ kΩ. The ac load line

- With a BJT, stabilizing the Q-point also stabilizes the g_m. This is not true for the FET, as shown in this example.

- Actually, V_{DS} must not be less than $V_{GS} - V_P$ or the saturated model is incorrect.

■ Note that the maximum power dissipation in the range of Q-points shown in Fig. 7.12 is only 20 mW, much less than the max $P_D = 310$ mW given for this device in the data sheet.

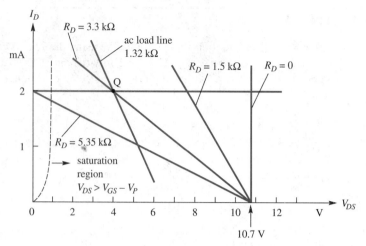

■ **FIGURE 7.12** Plot of V_{DS} versus I_D for several values of R_D in Example 7.7.

is drawn on the figure through Q, indicating the locus of instantaneous ac output voltage. The drain characteristic shown assumes a g_m of 2 mS, which is only a rough estimate because this parameter varies in this example.

The FET dissipates $V_{DS} \times I_D = (4.1)(0.002) = 8.2$ mW, so I_D, V_{DS}, and P_D are all conservative compared to the maximum values specified by the manufacturer in Fig. 5.31. The data sheet implies $\theta_{JA} = 1/2.82 = 0.35°C/mW$, so the temperature of our FET will rise $(0.35)(8.2) = 2.9°C$ above ambient!

In many circumstances it would be preferable to design the bias circuit to minimize g_m variations in the FET. Consider the following example.

EXAMPLE 7.8

Given the same range of FET characteristics as in Example 7.7, design a bias arrangement that will minimize variations in g_m. The circuit is Fig. 7.7(b).

SOLUTION   As before, we begin by finding the extreme transfer characteristics, using Eq. (7.17). These are plotted in Fig. 7.13. The design proceeds as follows:

1. Select $I_{DQ} < I_{DSS}$ for FET A to provide for a reasonable variation about the Q-point with ac signal. In the present case we have chosen $I_{DQA} = 2$ mA, in the center of the range 0 to 4. This point is labeled Q_A in the figure.
2. From Eq. (7.19), we find the corresponding $V_{GSQA} = -0.6$ V.
3. Equation (5.40) allows us to find $g_{mA} = 2.8$ mS. Indeed, we have used Eq. (5.40) to find the equation for g_m as a function of V_{GS} for the two limiting cases: $g_{mA} = 4 + 2V_{GSA}$ mS, and $g_{mB} = 4 + 0.5V_{GSB}$ mS. These relationships are plotted in Fig. 7.13 to show how g_m changes with Q-point V_{GS} for the limiting FETs.
4. We now find $V_{GSB} = -2.4$ V at the point where g_{mB} is also 2.8 mS. We can see this from the plot of g_{mB} or from Eq. (5.40).
5. For $V_{GSB} = -2.4$ V, the corresponding $I_{DBQ} = 8$ mA. See point Q_B in the figure. This can be done graphically, but recognize that this is just an application of Eq. (7.17).

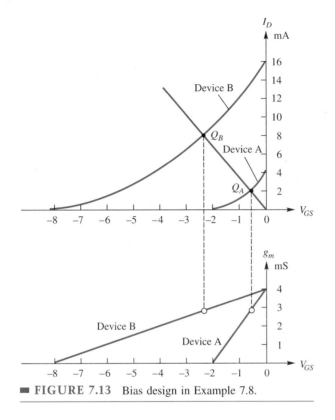

FIGURE 7.13 Bias design in Example 7.8.

6. The bias load line, Eq. (7.20), must pass through Q_A and Q_B. The result is

$$R_\sigma = \frac{\Delta V_{GSQ}}{\Delta I_{DQ}} = \frac{1.8 \text{ V}}{6 \text{ mA}} = 300 \text{ }\Omega, \text{ and } V_{GG} = 0 \text{ V}.$$

It just happens that this example requires no dc voltage V_{GG} and that $R_\sigma = 300$ Ω is a 5% standard resistor value. Should we encounter a 2N5459 with, say, $I_{DSS} = 5$ mA and $V_P = -3$ V (this is still within specifications), g_m would not be exactly the same as for our limiting assumptions, but it would be close.

Enhancement-Mode MOSFET Alternative Bias Design. Figure 7.14 presents an alternative bias design, the MOSFET version of Fig. 7.6. This arrangement is not appropriate for depletion-mode FETs because the V_{GS} polarity is wrong. Because the gate current is negligible, $V_{GS} = V_{DS}$. Observe that this ensures operation in the saturation region for any positive V_t. This circuit does not have the "wasted" voltage drop across R_σ of Fig. 7.7. To eliminate ac feedback from the drain to the gate, R_G could be divided into two parts with the interior point bypassed to ground with a C_3 having small reactance at the lowest frequency of interest.

Temperature Effects on FETs

The data of Fig. 5.32 would indicate that I_{DSS} and g_m of the JFET decrease somewhat with temperature rise. This is true for FETs in general since the mobility of the channel decreases. The magnitude of V_P is shown to increase slightly with temperature rise for the JFET, but this effect is usually very small in MOSFETs. Stabilization of I_D against

DRILL 7.6

Given an enhancement MOSFET with $k = 4$ mA/V^2 and $V_t = 2$ V, let $V_{DD} = 5$ V, and find values of R_D and R_G to obtain $I_D = 1$ mA for the amplifier of Fig. 7.14.

ANSWER
$R_D = 2.5$ kΩ and R_G must be large (like 1 MΩ) so as not to load v_{in}.

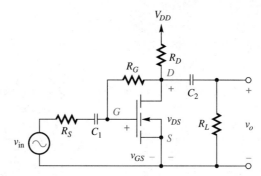

FIGURE 7.14 Preferred bias arrangement for the enhancement-mode MOSFET amplifier.

■ The MOSFET will have essentially no gate current at any temperature, assuming maximum gate voltage is not exceeded, SiO_2 is an excellent insulator.

temperature changes is handled in exactly the same way as suggested by Example 7.7, and stabilization of g_m against changes in temperature is accomplished by the technique of Example 7.8. The 2N5459 data sheet indicates a reverse-biased gate current at 25°C of a negligible 1 nA, but at 100°C it predictably increases to 200 nA. This will limit the size of R_G in high-temperature applications for JFETs, but not for MOSFETs.

CHECK UP

1. **TRUE OR FALSE?** Depletion-mode MOSFETs and JFETs are often operated with $V_{GG} = 0$, and since $I_G = 0$, $V_G = 0$ even when R_G is very large.
2. **TRUE OR FALSE?** Only depletion-mode FETs can be stabilized by means of a source resistor with $V_{GG} = 0$.
3. **TRUE OR FALSE?** Enhancement-mode FETs can be stabilized by connecting R_G to the drain node.
4. **TRUE OR FALSE?** The MOSFET parameter k increases with temperature because mobility increases with temperature.

7.4 BJT CURRENT SOURCES

We saw in Section 7.2 that the transistor can be biased in such a manner that I_C is nearly constant with changes in its parameters, but at certain costs. The configuration shown in Fig. 7.1 "wastes" a significant fraction of the power supply voltage in R_E and requires a bypass capacitor C_E to prevent reduction of ac amplification. The circuit shown in Fig. 7.6 also requires a bypass arrangement in the R_B circuit if ac gain is not to be sacrificed. Coupling capacitors, such as C_1 and C_2, are required to allow dc level shifts while not impeding the ac signal.

When we consider fabrication of such a circuit with monolithic IC techniques, certain difficulties come into view. The rules of thumb in monolithic IC design are (1) use transistors whenever possible, (2) use a minimum number of the smallest possible resistors, (3) avoid capacitors, if possible, because of the area they require on the chip, and (4) do not attempt to fabricate semiconductor inductors until frequencies approach a few tens of megahertz.

CHAPTER 7 BIASING AND STABILITY

To illustrate a typical difficulty, recall the discrete design of Example 7.2, where R_1 was 44 kΩ for $I_C = 4$ mA, and it certainly would increase if we needed smaller currents, which is usually the case. This would lead to uneconomically large resistor die areas in monolithic construction.

■ Many IC designers attempt to limit diffused resistor values to less than 30 kΩ.

Monolithic structures have positive features, however, that we can exploit. First, although the absolute transistor parameters cannot be controlled with any greater precision in ICs than in discrete devices, they can be *matched* very closely. This means that two transistors on the same die are fabricated under identical conditions of doping levels, temperature, and timing (see Chapter 18). This allows the designer to achieve accurate matching of all parameters. Second, since the transistors are in close proximity and thermally coupled through the substrate, they will always be at approximately the same temperature. Let us consider some design arrangements that exploit these features.

Simple Current Source or Simple Current Mirror

Consider the circuit shown in Fig. 7.6 with $R_B = 0$, as drawn in Fig. 7.15(a). Such a circuit would not be a useful amplifier, but we will show that it has the feature that I_C is independent of β_F and I_{CBO} and almost independent of V_{BE}. From Eq. (7.16),

$$\frac{\Delta I_C}{I_{C1}} = -\frac{\Delta V_{BE}}{(V_{CC} - V_{BE})}. \tag{7.22}$$

The transistor is barely in the active region.

We now add a second transistor, as shown in Fig. 7.15(b). We assume the two devices are matched ($I_{S1} = I_{S2}$) and temporarily neglect the output resistance r_o. Recall Eq. (4.10), $I_C = I_S e^{V_{BE}/V_T}$. We take the logarithm of both sides,

$$V_{BE1} = V_T \ln \frac{I_{C1}}{I_{S1}} \qquad V_{BE2} = V_T \ln \frac{I_{C2}}{I_{S2}}, \tag{7.23}$$

and since $V_{BE1} = V_{BE2}$, for identical devices, $I_{C1} = I_{C2} = I_C$ and $\beta_1 = \beta_2 = \beta$. At the collector of Q_1,

$$I_{REF} = I_{C1} + \frac{I_{C1}}{\beta_1} + \frac{I_{C2}}{\beta_2} = I_C\left(1 + \frac{2}{\beta}\right). \tag{7.24}$$

If $\beta \gg 1$,

$$I_{REF} \cong I_C. \tag{7.25}$$

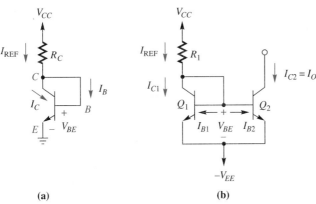

■ **FIGURE 7.15** (a) Common-emitter amplifier with $R_B = 0$. (b) Addition of Q_2 forms simple current source.

We see from Fig. 7.15 that

$$I_{REF} = \frac{V_{CC} - (-V_{EE}) - V_{BE}}{R_1} = \frac{V_{CC} + V_{EE} - V_T \ln \frac{I_C}{I_S}}{R_1}. \quad (7.26)$$

Since $I_C \cong I_{REF}$, this equation is transcendental. Because $V_{CC} + V_{EE}$ is usually an order of magnitude greater than V_{BE}, it is reasonable to approximate $V_{BE(ON)}$ by a constant voltage, say, 0.7 V and to use

$$I_O = I_C = I_{REF} = \frac{V_{CC} - (-V_{EE}) - V_{BE(ON)}}{R_1} = \frac{V_{CC} + V_{EE} - 0.7 \text{ V}}{R_1} \quad (7.27)$$

for a design equation. The output impedance of Q_2, r_o, was assumed infinite, and the *ideal* current source behavior is illustrated as curve (a) in Fig. 7.16.

Since I_O is a copy of I_{REF}, the circuit is often referred to as a **current mirror**. Also, it should be apparent that the current I_O is *into* the "current source," so this circuit might be properly called a *current sink*. To realize a current *out* of the current source collector, one would substitute *pnp* transistors and reverse the polarities of the dc supplies shown in Fig. 7.15(b).

■ In Fig. 7.15(b), transistors Q_1 and Q_2 constitute the current *mirror*, and R_1 determines the reference current.

■ Important: $V_O > V_{CE(Sat)}$, which means Q_2 must be active.

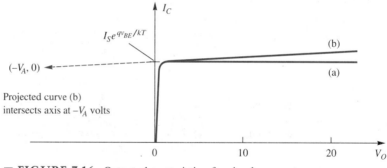

■ **FIGURE 7.16** Output characteristic of a simple current source.

DRILL 7.7

Suppose that I_S for the transistors of Example 7.9 is 10^{-14} A. Find V_{BE} for the transistors at the design current. Does this change R_1 significantly?

ANSWER
$V_{BE} = 0.64$ V. No, $R_1 = 58.7$ kΩ.

EXAMPLE 7.9

Design a simple 500-μA current source using matched *npn* BJTs and a ±15-V power supply. Include the schematic diagram. Neglect the BJT output resistance r_o.

SOLUTION The circuit will have the form shown in Fig. 7.15(b). Assuming $\beta_F > 50$, we justify $I_{REF} \cong I_O$, and from Eq. (7.27),

$$R_1 = [15 - (-15) - 0.7] \text{ V}/500 \text{ μA} = 58.6 \text{ kΩ}.$$

Note that this resistance is somewhat large to be appropriate for monolithic construction. We consider alternatives in the following pages.

Realistically, the output characteristic exhibits a nonzero slope, curve (b) in Fig. 7.16. The Early effect causes r_o to have some finite value, as discussed in Section 4.5. If the characteristic could be projected as a straight line into the second quadrant, it would intersect the voltage axis at the so-called *Early voltage*, V_A, typically 100 V negative. As an example, if I_C (at $V_{CE} = 1$ V) $= 50 \ \mu A$,

$$r_o = \frac{\Delta V_{CE}}{\Delta I_C} = \frac{101 \text{ V}}{50 \ \mu A} \cong 2 \text{ M}\Omega,$$

a respectable value for an actual current-source output impedance. Equation (7.24) can be corrected to reflect a finite r_o. We write

$$I_C = I_S e^{V_{BE}/V_T}\left(1 + \frac{V_{CE}}{V_A}\right), \qquad (7.28)$$

where $I_S e^{V_{BE}/V_T}$ is the current axis (y-axis) intercept of the extrapolated characteristic. Then

$$\frac{I_{C2}}{I_{C1}} = \frac{I_{S2} e^{V_{BE2}/V_T}\left(1 + \dfrac{V_{CE2}}{V_{A2}}\right)}{I_{S1} e^{V_{BE1}/V_T}\left(1 + \dfrac{V_{CE1}}{V_{A1}}\right)} \qquad (7.29)$$

for identical transistors

$$I_{C2} = I_{C1} \frac{\left(1 + \dfrac{V_{CE2}}{V_A}\right)}{1 + \left(\dfrac{0.7}{V_A}\right)} \approx I_{C1}\left(1 + \frac{V_{CE2}}{V_A}\right). \qquad (7.30)$$

■ For a "constant" current source output current to remain relatively constant for varying load voltage, R_O must be large. This is certainly demonstrated in Fig. 7.16, in Eq. (7.30), and in Example 7.10. Subsequent examples will reinforce this observation. Ideally, $R_o \to \infty$.

■ **EXAMPLE 7.10**

Repeat Example 7.9, including the effect of r_o. Let $V_A = 100$ V.

SOLUTION For $I_C = 500 \ \mu A$, $r_o = 100 \text{ V}/500 \ \mu A = 200 \text{ k}\Omega$. This is the current-source output resistance, which is certainly an improvement over just the 58.6-kΩ resistor in series with a 30-V battery. Using Eq. (7.30), $I_O = I_{C2} \cong 500 \ \mu A \ (1 + V_{CE2}/100)$. This means that if V_{CE2} increases to 10 V, and I_O increases to 550 μA. The 10% change is probably quite acceptable. On the other hand, should we have an Early voltage as low as 50 V, $r_o = 50 \text{ V}/500 \ \mu A = 100 \text{ k}\Omega$, and $I_O = 500 \ \mu A \ (1 + V_{CE2}/50)$. This means that for $V_{CE2} = 10$ V, I_O becomes $\cong 600 \ \mu A$, a 20% change, which may not be acceptable. It now appears that we are faced with the following design choices: (1) Design for typical transistor characteristics, and then test either the device or circuit, rejecting unsuitable devices. The IC foundry will experience additional costs for testing and for replacement of rejects. (2) Specify transistors that meet higher standards (in this case on V_A). This would increase the cost of the transistors. (3) Reject this design as too demanding in the requirement on V_A. We find alternatives in the next section.

DRILL 7.8

Design a simple 100-μA current source using a +10-V voltage source, two *pnp*(substrate) transistors as specified in Table 18.2, and a resistor. Find r_o.

ANSWER
$R_1 = 94.5$ kΩ (too large for economical IC technology),
$r_o = 500$ kΩ.

The key stability consideration is the sensitivity of the output current to changes in the supply voltages. Assuming a single supply, V_{CC}, and using Eq. (7.27),

$$\frac{\Delta I_{C2}}{I_{C2}} = \frac{V_{CC2} - V_{CC1}}{V_{CC2} - V_{BE(ON)}} = \frac{\Delta V_{CC}}{V_{CC} - 0.7} \cong \frac{\Delta V_{CC}}{V_{CC}}, \qquad (7.31)$$

which indicates that where $V_{CC} \gg 0.7$ V, the voltage stability factor is 1, or the current changes by the same percentage as the supply voltage. This is no better than could be obtained by a battery and resistor in series. If simple current sources are used in a circuit design, this sensitivity to voltage must be kept in mind.

Typically, more than one bias current is required in a circuit. To accomplish this, a number of sources can be driven from the same reference transistor as shown in Fig. 7.17. Often the reference transistor is replaced with a diode symbol to simplify the schematic diagram. This multiple-source configuration is often referred to as a *current mirror* or **current repeater**. Nonequal junction areas and the errors introduced by neglecting the base currents are explored in the problems for this chapter.

■ If the areas of the BJT junctions are made different, the resulting currents will be different in the same ratio, since I_S is proportional to area.

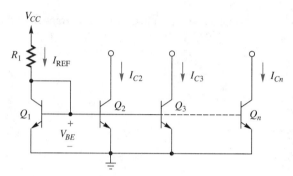

■ **FIGURE 7.17** Multiple simple current sources.

Widlar Current Source

To obtain smaller currents using reasonable resistor values and to obtain a lower power supply stability factor, **Widlar**[1] developed the circuit illustrated in Fig. 7.18. Note that R_2 is usually $\ll R_1$, and for the following analysis, we assume that Q_1 and Q_2 are identical. In addition, we assume that r_o (and V_A) approach infinity and that β_F is large. We can now sum voltages around the base loop, obtaining

$$V_{BE1} \cong V_{BE2} + I_{C2}R_2. \tag{7.32}$$

Since V_{BE1} and V_{BE2} are no longer equal, we use Eq. (7.23) to expand Eq. (7.32), yielding

$$V_T \ln \frac{I_{C1}}{I_{S1}} = V_T \ln \frac{I_{C2}}{I_{S2}} + I_{C2}R_2, \tag{7.33}$$

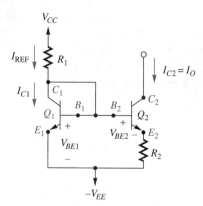

■ **FIGURE 7.18** Widlar current source.

CHAPTER 7 BIASING AND STABILITY

and then combine terms to obtain

$$V_T \ln \frac{I_{C1} I_{S2}}{I_{S1} I_{C2}} = I_{C2} R_2. \qquad (7.34)$$

Since I_{S1} and I_{S2} are matched and $I_{C1} = I_{REF}$ for large β_F,

$$I_{C2} R_2 = V_T \ln \frac{I_{C1}}{I_{C2}} = V_T \ln \frac{I_{REF}}{I_{C2}}. \qquad (7.35)$$

As before, Eq. (7.26) applies for I_{REF},

$$I_{REF} = \frac{V_{CC} - (-V_{EE}) - V_{BE1}}{R_1} = \frac{V_{CC} + V_{EE} - V_T \ln (I_{C1}/I_{S1})}{R_1}.$$

Here a V_{BE1} error of a few tens of millivolts is insignificant when compared to $[V_{CC} - (-V_{EE})]$, and V_{BE1} is approximated by 0.7 V. Substituting Eq. (7.26) into (7.35) results in

$$I_O = I_{C2} = \frac{V_T}{R_2} \ln \left(\frac{V_{CC} + V_{EE} - 0.7}{R_1 I_{C2}} \right). \qquad (7.36)$$

Equation (7.36), however, does not yield to further simplification. The solution of Eq. (7.36) for I_{C2} requires iteration, but is a fundamental step in the general analysis of Widlar current sources. For design, solution for R_2 is straightforward.

EXAMPLE 7.11

Design a Widlar current source using matched 2N2222A transistors (such as the MPQ2222 chip) to deliver 500 μA from a ± 15-V power supply. Select resistor values compatible with monolithic IC technology. Assume $T = 27°C$.

SOLUTION The circuit will be that shown in Fig. 7.18. We must use a large but realistic R_1 for monolithic fabrication. Suppose we select $R_1 = 10$ kΩ. Then from Eq. (7.27), $I_{REF} = 29.3$ V/10 k$\Omega = 2.9$ mA. This is probably on the high side for a reference current but not unreasonably so. Then from Eq. (7.35),

$$R_2 = \frac{26 \text{ mV}}{500 \mu\text{A}} \ln \left(\frac{2.9 \text{ mA}}{500 \mu\text{A}} \right) = 92 \ \Omega.$$

It is important to observe that the die area of $R_1 + R_2 = 10.1$ kΩ is only 20% of the resistor die area used in the simple current-source design in Example 7.9 (58 kΩ).

DRILL 7.9

Rework Example 7.11 after choosing $R_1 = 20$ kΩ.

ANSWER $R_2 = 56 \ \Omega$

It is possible to obtain currents on the order of a few microamperes or less with modest-sized resistors. Problem 7.22 is used to investigate the condition for the minimization of total die area when designing for a given output current.

It is instructive to derive an expression for the output impedance of this circuit. Figure 7.18 is redrawn in Fig. 7.19(a) with the simplified small-signal ac models of Fig. 4.22 inserted for the transistors. Figure 7.19(b) shows the ac model with all dc supplies properly removed. The controlled source $g_{m1} v_{be1}$ is controlled by its own terminal voltage. Any branch with voltage V and current $I = V \times G$ must have a constant conductance G. Thus, this current source may be thought of as a passive conductance g_{m1}. The resistances $R_1, r_{o1}, r_{\pi 1}$, and $1/g_{m1}$ are all in parallel and can be combined into resistance R' as shown in Fig. 7.19(c). As a matter of fact, $r_{\pi 1}$ is always smaller than $r_{\pi 2}$ (because V_{BE1} is always

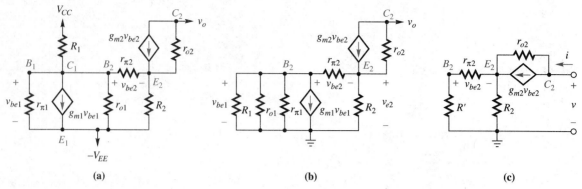

■ **FIGURE 7.19** Finding the output impedance in Fig. 7.18.
(a) Transistor models substituted into Fig. 7.18. (b) ac model
(dc sources removed). (c) $R' = R_1 \| r_{o1} \| r_{\pi 1} \| 1/g_{m1}$.

larger than V_{BE2}), and g_{m1} is $\beta_1/r_{\pi 1}$; thus R' is always small compared to $r_{\pi 2}$ and will be neglected. This makes the voltage across $R_2 = -v_{be2}$. Summing drops from the output terminal, we write

$$v = (i - g_{m2}v_{be2})r_{o2} - v_{be2}. \tag{7.37}$$

Let $r_{\pi 2} \| R_2 = R_2$ since $r_{\pi 2} \gg R_2$. Then $-v_{be2} = iR_2$. Substituting into Eq. (7.37), we find that

$$v = [i - g_{m2}(-iR_2)]r_{o2} + iR_2. \tag{7.38}$$

The output impedance is then

$$R_O = \frac{v}{i} = r_{o2}(1 + g_{m2}R_2) + R_2 \cong r_{o2}(1 + g_{m2}R_2). \tag{7.39}$$

The result is a modest improvement in the output resistance over that of a simple current source.

DRILL 7.10

The major assumptions made in the derivation of Eq. (7.39) are that $r_{\pi 2} \gg R'$, and that $r_{\pi 2} \gg R_2$. Demonstrate that these assumptions are valid in Example 7.12.

ANSWER
$r_{\pi 2} = 5.2$ kΩ,
$R' = 8.9$ Ω, and
$R_2 = 92$ Ω

EXAMPLE 7.12

Include the effects of r_o in the design of Example 7.11. Let $V_A = 100$ V, $\beta_F = 100$ and use $r_\pi = \beta/g_m$.

SOLUTION We see that $g_{m2} = 500$ μA/26 mV $= 19$ mS, and $R_2 = 92$ Ω. From Eq. (7.39), using the typical value of $r_{o2} = 200$ kΩ,

$$R_o = 200 \text{ k}\Omega \, [1 + (19 \text{ mS})(92 \, \Omega)] = 550 \text{ k}\Omega.$$

The 550-kΩ source resistance would produce a current change of $\Delta I_C = \Delta V_{CE}/r_o = 10$ V/550 k$\Omega = 18$ μA for a 10-V change in V_{CE}. This means our 500-μA source would increase to 518 μA. This is a 3.6% change, a significant improvement over the 10% change of the simple current source of Example 7.10.

CHAPTER 7   BIASING AND STABILITY

The Widlar power supply stability factor calculation requires a bit more algebraic manipulation than did the simple current source. We begin by rearranging Eq. (7.36),

$$I_{C2} = \frac{V_T}{R_2}[\ln(V_{CC} + V_{EE} - 0.7) - \ln R_1 I_{C2}], \quad (7.40)$$

and then differentiating with respect to V_{CC}:

$$\frac{dI_{C2}}{dV_{CC}} = \frac{V_T}{R_2}\left[\frac{1}{V_{CC} + V_{EE} - 0.7} - \left(\frac{R_1}{R_1 I_{C2}}\right)\left(\frac{dI_{C2}}{dV_{CC}}\right)\right]. \quad (7.41)$$

Factoring,

$$\frac{dI_{C2}}{dV_{CC}}\left(1 + \frac{V_T}{R_2 I_{C2}}\right) = \frac{V_T}{R_2(V_{CC} + V_{EE} - 0.7)}, \quad (7.42)$$

the derivative becomes

$$\frac{dI_{C2}}{dV_{CC}} = \frac{I_{C2} V_T}{(V_{CC} + V_{EE} - 0.7)(R_2 I_{C2} + V_T)}. \quad (7.43)$$

For small changes,

$$\frac{\Delta I_{C2}}{I_{C2}} = \frac{\Delta V_{CC}}{(V_{CC} + V_{EE} - 0.7)} \cdot \frac{V_T}{(R_2 I_{C2} + V_T)}. \quad (7.44)$$

The last term is the desired stability factor, showing the fractional change of I_{C2} as a function of the fractional change in supply voltage. Note that this factor is always less than unity, which means that the Widlar current source is less sensitive to power supply variations than the simple current source (as R_2 approaches 0, the factor goes to 1).

EXAMPLE 7.13

Compute the stability factor of the Widlar source designed in Example 7.11, and compare the result to that of the simple current source of Example 7.9.

SOLUTION   From Eq. (7.44), the percentage in I_{C2} as a function of the percentage change in V_{CC} (at 25°C) is

$$100\% \cdot \frac{V_T}{(R_2 I_{C2} + V_T)} = 100\% \cdot \frac{26 \text{ mV}}{(92\Omega)(500 \text{ }\mu\text{A}) + 26 \text{ mV}} = 36\%$$

Therefore, this circuit is much less sensitive to voltage changes than is the simple current source, whose stability factor is $\cong 1$ (100%). Whereas a 10% change in V_{CC} would cause the simple current source I_O to increase from 500 to 550 μA, the Widlar source would increase to 518 μA, a significant improvement.

As with the simple current source, the Widlar source can be configured for multiple currents as shown in Fig. 7.20. Each source current can be computed independently and, again, nonequal junction areas and errors introduced by not including base currents are explored in the problems.

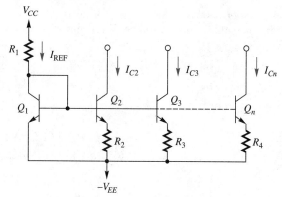

■ **FIGURE 7.20** Multiple Widlar sources.

Wilson Current Source

A circuit called the ***Wilson current source***[2] is shown in Fig. 7.21. This circuit is designed to exhibit a very high output impedance, and to compensate for the finite base current of the transistors. Figure 7.22(a) shows a small-signal model of the circuit used to calculate the output impedance. Since the operating points of all transistors are approximately the same, g_m, r_π, and r_o are equal for identical IC devices. Now refer to Fig. 7.22(b).

- $i_{C1} = i_{C2} = v_1 g_m$, and $i_x \cong 2v_1 g_m$, so that $v_1 g_m \cong i_x/2$.
- $i_{C2} = v_1 g_m$, and Q_2's effective collector resistance is $v_1/i_{C2} = 1/g_m$ Ω.
- The parallel combination $1/g_{m2} \parallel r_{\pi 1} \parallel r_{\pi 2} \parallel r_{o2}$ is approximately $1/g_m$ Ω.
- i_{C1} flows through $r_{\pi 3}$ because $(R_{REF} \parallel r_{o1}) \gg r_{\pi 3}$, and $v_\pi = -v_1 g_m r_\pi = -r_\pi i_x/2$.
- At the output node, $i_x = g_m(-r_\pi i_x/2) + (v_x - v_1)/r_o = -\beta i_x/2 + v_x/r_o - i_x/2g_m r_o$.

$$R_O = \frac{v_x}{i_x} \cong \left(1 + \frac{\beta}{2}\right) r_o. \quad (7.45)$$

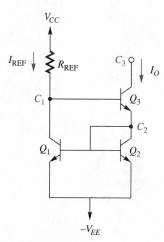

■ **FIGURE 7.21** Wilson current source.

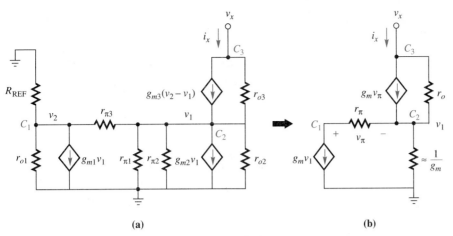

FIGURE 7.22 Small-signal model used to find output impedance of Wilson current source.

This is a significant increase in output impedance over simple and Widlar current sources. The resistance r_μ, usually neglected, would have a small effect here.

Temperature Stabilization

A large number of design variations are available for achieving power supply and thermal stability. The circuit shown in Fig. 7.23 combines some of the more widely used techniques. Again, assume all transistors are matched and the current gains are large. The breakdown diode (recall the material in Chapter 3) is obtained by reverse-biasing the base-emitter junction of an *npn* transistor as illustrated. Resistor R_1 must be small enough to provide bias current for Q_1 and current for the breakdown diode for all possible values of V_{CC}. Then

$$V_Z + V_{BE4} + V_{BE5} = V_{R2} + V_{BE1} + V_{BE2}. \tag{7.46}$$

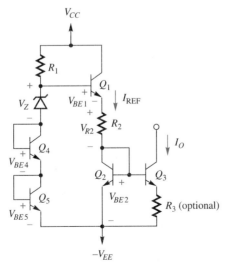

FIGURE 7.23 Stabilized current source.

■ The assumption of equal temperatures is dependent on thermal resistances of the die and power dissipation, so this assumption is approximate.

The key point is that since all the transistors are on a common die, they are all at the same temperature. Essentially, all of the base-emitter drops are identical, so that

$$V_Z = V_{R2} = I_{REF}R_2, \tag{7.47}$$

and if $R_3 = 0$, $I_{REF} = I_O$, so that

$$I_O \cong \frac{V_Z}{R_2}. \tag{7.48}$$

The avalanche diode has a very small temperature coefficient (less than a few millivolts/°C), so Eq. (7.48) describes a relatively temperature-insensitive circuit. There is an economic trade-off in that the circuit is more complex and requires additional die area. We should note that a third resistor can be inserted into the emitter of Q_3 to obtain a much-improved stabilized Widlar current source. When R_3 is included,

$$I_O = \frac{V_T}{R_3} \ln\left(\frac{V_Z/R_2}{I_O}\right). \tag{7.49}$$

■ EXAMPLE 7.14

Use the circuit shown in Fig. 7.21 to stabilize the simple current source designed in Example 7.9. Assume $V_Z = 6.8$ V, $I_{Z(min)} = 20$ μA, and $P_{Z(max)} = 100$ mW. Specify R_1 and R_2. Assume matched 2N2222A transistors, as in the MPQ2222, and $V_{CC} = 15$ V and $V_{EE} = 15$ V.

SOLUTION Since $\beta_F \gg 1$, $I_{REF} \cong I_O = 500$ μA, and Eq. (7.45) gives $R_2 = 6.8$ V/500 μA $= 13.6$ kΩ. To obtain a value for R_1, we sum currents at the base of Q_1, $I_{R1} = I_Z + I_{B1} = I_Z + I_{C1}/\beta_F$. Then $I_{R1(min)} = 20$ μA $+ 500$ μA/100 $= 25$ μA. But this gives

$$R_1 = (V_{CC} + V_{EE} - V_{BE4} - V_{BE5} - V_Z)/25 \text{ μA}$$
$$= (15 + 15 - 0.7 - 0.7 - 6.8)/25 \text{ μA} = 872 \text{ kΩ}.$$

This is absurdly high for monolithic construction! However, $I_{Z(max)} = P_{Z(max)}/V_Z = 100$ mW/6.8 V $= 14$ mA. So we could let I_Z be much greater than 20 μA. Let's allow I_Z to be 1 mA; then $I_{R1} \cong 1$ mA, and R_1 becomes 22 kΩ, a much more realistic value. Note that V_{CC} and V_{EE} can change over a considerable range while satisfying 20 μA $< I_Z < 14$ mA. Verify this.

CHECK UP

1. **TRUE OR FALSE?** Accurate BJT current sources require transistors with matched I_S.
2. **TRUE OR FALSE?** Accurate BJT current sources require transistors to be at the same temperature (thermally coupled, such as on the same substrate).
3. **TRUE OR FALSE?** A Widlar current source has a lower output resistance than a simple current source of the same output current.
4. **TRUE OR FALSE?** A BJT current source requires the output transistor to be saturated.

7.5 MOSFET CURRENT SOURCES

FETs connected as shown in Fig. 7.24 are often used as current sources. This circuit, using a depletion-mode MOSFET has been seen once before, in Section 5.4, Fig. 5.17. The drain current is given by Eq. (5.12), repeated here:

$$i_D = k(v_{GS} - V_t)^2 \qquad v_{DS} > v_{GS} - V_t. \tag{5.12}$$

Because v_{GS} is always zero, I_D is the constant current

$$I_D = k(-V_t)^2 = kV_t^2 \qquad v_{DS} > |V_t|. \tag{7.50}$$

Of course, practical FETs have a drain resistance as discussed in Section 5.6. This can be treated in a manner analogous to the output resistance in the BJT. The FET parameter λ (lambda) has the reciprocal effect as Early voltage in a BJT. Thus Eq. (7.50) can be refined by multiplying by the factor $(1 + \lambda V_{DS})$:

$$I_D = kV_t^2(1 + \lambda V_{DS}) \qquad v_{DS} > |V_t|. \tag{7.51}$$

As in Chapter 5, we define r_d as follows:

$$r_d \cong \left. \frac{\Delta v_{DS}}{\Delta i_D} \right|_{Q\text{-point}} = \frac{\lambda^{-1}}{kV_t^2} \cong \frac{1}{\lambda I_D}. \tag{7.52}$$

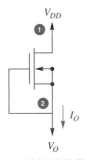

FIGURE 7.24 Depletion-mode MOSFET current source.

EXAMPLE 7.15

Use SPICE to plot the I_O versus V_O characteristic of the current source of Fig. 7.24. Let the FET parameters be $k = 0.1$ mA/V^2, $V_t = -2$ V, and $\lambda = 0.02$ V^{-1}; $V_{DD} = 6$ V.

SOLUTION   A possible program follows:

```
X7.15
VDD 1 0 DC 6
M1 1 2 2 2 M1
VO 2 0 DC 0
.MODEL M1 NMOS VTO=-2 KP=0.0002 LAMBDA=0.02
.DC VO 0 6 0.1
.PLOT DC I(VO)
.END
```

The plot is shown in Fig. 7.25. Note that the FET comes out of the saturation region into the ohmic region when $V_{DS} < V_{GS} - V_t = 0 + 2$ ($V_O = 4$). The curve shows a slope $I_D = 400(1 + 0.02\ V_{DS})\ \mu A \Rightarrow I_O = (448 - 8V_O)\ \mu A$; $r_d = 125\ k\Omega$.

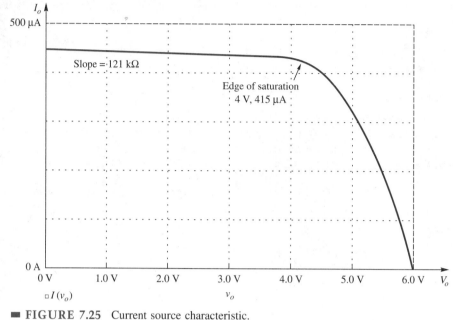

■ FIGURE 7.25 Current source characteristic.

MOSFET Current Mirror

There are many ways to realize FET current sources. Figure 7.26 shows an example of an IC current mirror using three enhancement-mode MOSFETs. Transistors M_2 and M_3 compose the current mirror, and M_1 provides the reference current instead of a resistor (as in the BJT current mirror of Fig. 7.15). Of course the FET saves considerable space on the chip. FETs M_1 and M_2 are always saturated, since $V_{DS} = V_{GS}$, and they determine

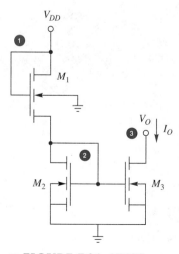

■ FIGURE 7.26 NMOS current mirror.

CHAPTER 7 BIASING AND STABILITY

the reference current. We assume that V_t is the same for all the FETs, since they are fabricated in the same process. Then the reference current, $I_{D1} = k_1(V_{GS1} - V_t)^2 = k_2(V_{GS2} - V_t)^2$ where $V_{GS1} + V_{GS2} = V_{DD}$.

This reference current is under the control of the IC designer, since the transistor parameter k is proportional to the channel W/L ratio. The output current, I_{D3}, is proportional to the reference current times the ratio of k_3 to k_2, as long as M_3 is saturated ($V_{DS3} > V_{GS3} - V_t$). The output resistance of this current source is that of the FET M_3, $r_d = 1/\lambda I_{D3}$. Consider the following example.

EXAMPLE 7.16

Assume that the current source of Fig. 7.26 has three identical MOSFETs with parameters $k = 100$ μA/V^2, $V_t = 1.5$ V, and $\lambda = 0.02$ V^{-1}; $V_{DD} = 6$ V. Find I_{D3} and R_O. Run a SPICE simulation to check the output characteristic.

SOLUTION Since the FETs are identical, $V_{GS1} = V_{GS2} = V_{DD}/2 = 3$ V. This leads to $I_{D3} = I_{D1} = 100$ μA $(3 - 1.5)^2 = 225$ μA when $V_O > 1.5$ V. Also, $R_o = r_d = 50$ V/225 μA $= 222$ kΩ. The following program yields the result of Fig. 7.27:

```
EXAMPLE7.16
VDD 1 0 DC 6
M1 1 1 2 0 M
M2 2 2 0 0 M
M3 3 2 0 0 M
VO 3 0 DC 0
.MODEL M NMOS VTO=1.5 KP=0.2M LAMBDA=0.02
.DC VO 0 6 0.1
.PLOT DC I(VO)
.END
```

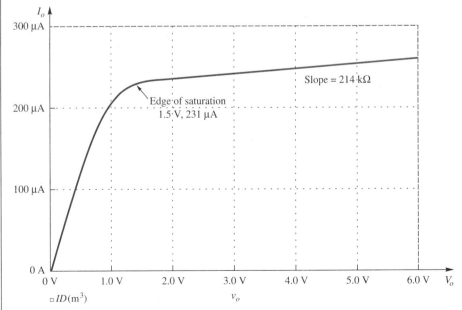

■ **FIGURE 7.27** MOSFET current mirror output characteristic (SPICE).

MOSFET Wilson Current Source

We offer one more example of a FET current source in Fig. 7.28. The reader may recognize the Wilson current source of Fig. 7.21, this time implemented with enhancement-mode MOSFETs. FET M_1, in conjunction with M_2, provides the reference current. This arrangement produces a higher output impedance for the current source, but at the expense of a restricted voltage range; M_3 and M_4 come out of the saturated region as the output voltage drops below a higher threshold voltage. If all of the FETs in Fig. 7.28(a) are in the active region, we can solve the dc situation by equating the drain currents of M_1 and M_2 and of M_3 and M_4. For a quick look we neglect λ and body effect to write

$$k_1(V_{GS1} - V_{t1})^2 = k_2(V_{GS2} - V_{t2})^2 \Rightarrow k_1(V_{DD} - V_2 - V_{t1})^2 = k_2(V_3 - V_{t2})^2, \quad (7.53)$$

$$k_3(V_{GS3} - V_{t3})^2 = k_4(V_{GS4} - V_{t4})^2 \Rightarrow k_3(V_2 - V_3 - V_{t3})^2 = k_4(V_3 - V_{t4})^2. \quad (7.54)$$

For the special case of identical FETs, these equations reduce to

$$V_{DD} - V_2 = V_3 = V_2 - V_3, \quad \text{leading to} \quad V_2 = \tfrac{2}{3}V_{DD} \quad \text{and} \quad V_3 = \tfrac{1}{3}V_{DD}.$$

The student should draw a dc model and verify these statements.

A small-signal model of the circuit is drawn in Fig. 7.28(b), with an ac test voltage v_x introduced to determine i_x, and hence $R_o = v_x/i_x$. Substituting $g_d = 1/r_d$, the current equations are

$$-g_{m1}(0 - v_2) + v_2 g_{d1} + g_{m2}v_3 + v_2 g_{d2} = 0, \quad (7.55)$$

$$i_x = g_{m3}(v_2 - v_3) + (v_x - v_3)g_{d3} = g_{m4}v_3 + v_3 g_{d4}. \quad (7.56)$$

Again, it is interesting to consider the special case of identical FETs, and also to simplify $g_m + g_d \cong g_m$, which is a good approximation. Then Eqs. (7.55) and (7.56) reduce to $R_o = v_x/i_x = 3r_d$. This illustrates the higher output impedance of the Wilson current mirror. Problems 7.47 to 7.50 relate to the Wilson circuit. The reader is encouraged to examine the output characteristic in Drill 7.11.

DRILL 7.11

Use a SPICE simulation to find the output characteristic of the Wilson circuit of Fig. 7.28. Let $V_{DD} = 8$ V, and use the MOSFET model given in Example 7.16.

ANSWER A possible program list follows, and the output is given in Fig. 7.29:

```
D7.11
M1 1 1 2 0 M1
M2 2 3 0 0 M1
M3 4 2 3 0 M1
M4 3 3 0 0 M1
.MODEL M1 NMOS
+VTO=1.5
+KP=.0002
+LAMBDA=.02
VDD 1 0 DC 8
VO 4 0 DC 0
.DC VO 0 8 .1
.PLOT DC I(VO)
.END
```

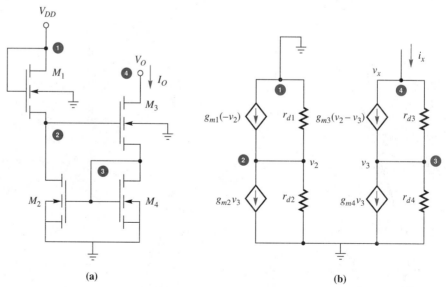

■ **FIGURE 7.28** MOSFET Wilson current source.
(a) Circuit (b) Small-signal model.

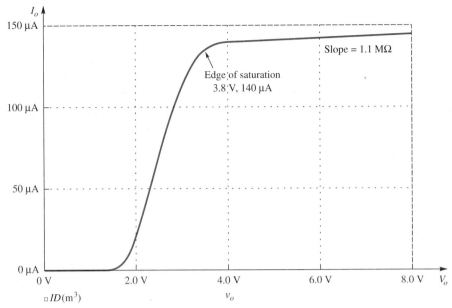

■ **FIGURE 7.29** SPICE solution of the output characteristic of a MOSFET Wilson current source, where $R_o = 1.1$ MΩ.

CHECK UP

1. **TRUE OR FALSE?** A single enhancement-mode MOSFET, operated in saturation with $V_{GS} = 0$, serves as a current source.
2. **TRUE OR FALSE?** The simple MOSFET current source (current mirror) requires enhancement devices with matched V_t and k.
3. **TRUE OR FALSE?** A MOSFET current source uses an FET rather than a resistor to obtain the reference current.
4. **TRUE OR FALSE?** A MOSFET Wilson current source has a restricted output voltage range.

SUMMARY

In this chapter we presented design criteria for establishing the Q-point for both BJT and FET single-transistor circuits. We used dc and ac load line analysis to locate the Q-point in the linear regions required to obtain amplifier as opposed to switching operation. The bias circuit design must take into account maximum power, voltage, and current ratings for the particular BJT and FET. Fabrication technology results in a wide range of device parameters. In this chapter we considered these variations by studying the Q-point bias stabilization. Bias stabilization was our introduction to negative feedback, a concept we study in considerable depth in Chapter 11.

Discrete-circuit biasing and integrated-circuit biasing design techniques are different because of the design requirements for IC fabrication technologies. Transistors are much more economical to use than large-value resistors, and capacitors are all but forbidden.

The simple, Widlar, and Wilson current sources accommodate these constraints. Both BJT and MOSFET current sources have widespread application in monolithic circuits. Even though these current sources are more complex, they offer advantages over resistor biasing with respect to impedance levels and stability. In Chapter 10, we use current sources as active loads to achieve amplifier performance not practically realizable with discrete component design approaches. Table 7.1 compares several current mirror circuits.

TABLE 7.1 Comparison of Current Mirrors, Assuming Matched Devices

	Figure	Output Resistance, R_o	Current Gain, I_O/I_{REF}
Ideal		∞	1
Simple BJT	7.15(b)	r_{o2}	$\dfrac{\beta}{\beta+2}\left(1+\dfrac{V_O+V_{EE}}{V_A}\right)$
Widlar BJT	7.18	$r_o(1+g_{m2}R_2)$	varies with R_2
Simple MOSFET	7.26	r_{d3}	$1+\lambda V_O$
Wilson MOSFET	7.28	$\approx 3r_{d3}$	$1+\dfrac{\lambda}{3}V_O$

These current ratios are also directly affected by the ratios of the device areas.

SURVEY QUESTIONS

1. What factors must be considered when establishing the Q-point for an amplifier?
2. How might the bias network load the signal source excessively?
3. Explain how to determine whether the dynamic operating point of an amplifier will remain within the linear region of the active device.
4. What constraints are imposed on R_B and R_E in a common-emitter BJT amplifier if the I_C must not change more than 10% when the β_F changes 50%?
5. Is it possible for the FET amplifier designer to guarantee an approximate I_D without knowing the parameters k and V_t? Explain.
6. Given supply voltages V_{CC} and V_{EE}, describe what is needed to obtain a simple BJT current source of I_O.
7. Given supply voltages V_{DD} and V_{SS}, describe what is needed to obtain a simple MOSFET current source of I_O.
8. A simple BJT current source of I_O becomes a Widlar current source with an output current of $I_O/2$ by the addition of an emitter resistor $R_2 = $ _____? $V_T = 26$ mV.

PROBLEMS

7.1 Refer to the amplifier circuit shown in Fig. 7.30. Let $R_1 = 30$ kΩ, $R_2 = 10$ kΩ, $R_E = 1$ kΩ, $R_C = 2$ kΩ, $R_L = 4.7$ kΩ, and $V_{CC} = 10$ V. At $T = 25°C$, $\beta_F = 150$, $V_{BE} = 0.6$ V, and $I_{CBO} = 0.01$ μA.

  (a) Find I_B, I_C, and V_{CE}.
  (b) Sketch dc and ac load lines on an I_C versus V_{CE} coordinate system. Is this an appropriate operating point for the amplifier?

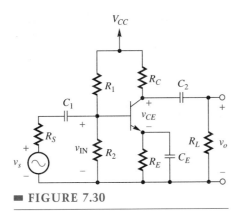

■ FIGURE 7.30

7.2 Suppose the amplifier circuit shown in Fig. 7.30 uses a *pnp* transistor having $\beta_F = 100$, $V_{BE} = -0.7$ V, and negligible I_{CO}. In addition, $V_{CC} = -12$ V, $R_E = 100$ Ω, $R_C = 1200$ Ω, and $R_L = 2200$ Ω. Find R_1 and R_2 such that $I_C = -5$ mA and the parallel combination of R_1 and R_2 is greater than 5 kΩ.

7.3 Show your results to Problem 7.2 and a load line on a set of third quadrant *pnp* $I_C - V_{CE}$ characteristics.

7.4 Verify your design of Problem 7.2 by performing a SPICE simulation. For the discrete BJT let $I_S = 10^{-14}$ A, which will yield a reasonable V_{BE}.

7.5 Consider the circuit of Fig. 7.30. For the transistor, $\beta_F = 50$, $V_{BE} = 0.7$ V, and negligible I_{CO}. Let $V_{CC} = 12$ V and $R_C = 2$ kΩ.

(a) Design a bias network for this circuit such as $I_C = 2$ mA, and the Q-point must be selected to provide operation as an amplifier in the linear active region. Find a self-consistent set of values for R_E, R_1, and R_2. The parallel combination of R_1 and R_2 must be greater than 2 kΩ. Specify your value for V_{CE}.

(b) Assume that all capacitors are large. For your design, what range of values for R_L is required to allow a minimum of a 2-V_{p-p} swing at the output, v_O, without clipping at saturation or cutoff? To demonstrate your solution, sketch and label the dc and ac load lines on a graph of I_C versus V_{CE}.

7.6 Refer to Fig. 7.30. Let the transistor be *pnp* with $\beta_F = 100$, $V_{BE} = -0.65$ V, and $I_{CO} \cong 0$. Let $V_{CC} = -10$ V, $R_C = R_L = 3$ kΩ, $R_E = 2$ kΩ, and $I_C = -1$ mA.

(a) Design a bias network with reasonable values of R_1 and R_2.

(b) Sketch and label the dc and ac load lines. Be sure to provide numerical values for key intercepts and slopes.

7.7* The common-emitter amplifier with self-bias shown in Fig. 7.30 has the dc load line sketched in Fig. 7.31. Assume $V_{CC} = 15$ V and $R_C = 5$ kΩ. For the BJT, $\beta_F = 120$, and you can neglect $V_{CE(sat)}$ and I_{CEO}.

(a) Design the bias network to achieve the given Q-point. That is, find a self-consistent set of $R_1 > 150$ kΩ, R_2, and R_E.

(b) For frequencies of operation such that $X_C \to 0$, what range of R_L values are required to obtain a symmetrical output voltage, v_O, with a minimum 4-V_{p-p} amplitude. An ac load line sketch should be used.

(c) For your design, try using another transistor whose $\beta_F = 180$. Are you still biased such that the Q-point is in the linear active region? Find I_C and V_{CE}, and show the new Q-point on the dc load line above.

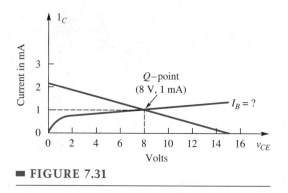

■ FIGURE 7.31

7.8 Use SPICE simulations to verify your work in Problems 7.7(a) and (c).

7.9* Refer to the circuit of Fig. 7.30. For the transistor, $\beta_F = 100$, $V_{BE} = 0.7$ V, and negligible I_{CO}. $V_{CC} = 10$ V, $R_C = 3$ kΩ.

(a) Design a bias network such that the Q-point $I_C = 1$ mA. You must specify a set of self-consistent values for R_E, R_1 and R_2 such that the circuit is biased for reasonable amplifier operation. Also, the parallel combination of R_1 and R_2 must be greater than 10 kΩ.

(b) Assume that $r_\pi = 2600$ Ω, $R_S = 600$ Ω, and $R_L = 2$ kΩ. Using the simplified small-signal model for the transistor, compute the ac voltage gain defined by $a_v = v_o/v_S$. Assume all capacitive reactances are small at the frequencies of interest.

(c) Sketch and label dc and ac load lines. Provide numerical values for the Q-point, including the quiescent power dissipation. Show the maximum ac output voltage amplitude attainable with no clipping for your design.

7.10* Solve for the Q-point values (I_{D1}, V_{DS1}, I_{C2}, V_{CE2}, I_{C3}, and V_{CE3}) in terms of the appropriate circuit elements (R_1, R_2, etc.) and device parameters for each of the four devices in the circuit of Fig. 7.32. Assume C_1, C_2, C_3, and C_4 are large.

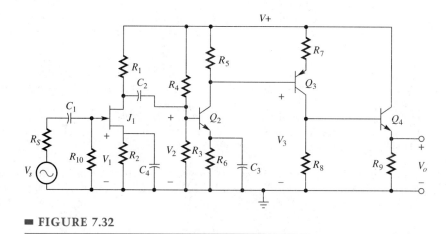

■ FIGURE 7.32

7.11 The amplifier shown in Fig. 7.30 has $R_1 = 60$ kΩ, $R_2 = 30$ kΩ, $R_E = 220$ Ω, and $R_C = 820$ Ω. Let $R_L = 2.2$ kΩ, $V_{CC} = 12$ V, and $V_{BE} = 0.7$ V. Compare the operating points when $\beta_F = 100$ and 300. Is this circuit adequately stabilized against changes in β_F? Explain your answer.

7.12 Use SPICE simulations to answer Problem 7.11.

7.13 The amplifier in Problem 7.1 is operated in an environment such that the junction temperature increases to 100°C.

(a) Use Fig. 4.31 to estimate the new β_F, and also find the new values for V_{BE} and I_{CBO}.
(b) Calculate the percentage change in I_C due to the changes in β_F, the change in V_{BE}, and the change in I_{CBO}.

7.14 The emitter-follower amplifier shown in Fig. 7.33 uses a transistor with $\beta_F = 100$ and $V_{BE} = 0.7$ V. The value of $R_1 \| R_2$ should be less than 75 kΩ.

(a) Choose R_1 and R_2 such that $V_{CE} = 2.5$ V.
(b) Find the change in I_C if β changes to 300.

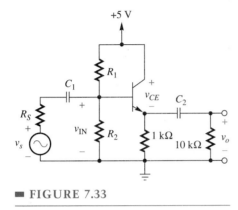

■ **FIGURE 7.33**

7.15 Consider the amplifier circuit shown in Fig. 7.34. Derive an expression for $\Delta I_C / I_{C1}$ in terms of $\Delta \beta$, assuming that V_{BE} and I_{CBO} are constant. Under what conditions may the I_{CBO} terms be neglected? [This should be analogous to Eq. (7.7)]. Compare your answer for ΔI_C with Eq. (7.9).

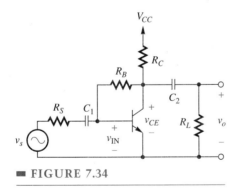

■ **FIGURE 7.34**

7.16 The amplifier shown in Fig. 7.34 has $V_{CC} = 5$ V, $R_C = 2.2$ kΩ, and $R_B = 220$ kΩ. If $V_{BE} = 0.7$ V, find I_C and V_{CE} for

(a) $\beta_F = 100$
(b) $\beta_F = 200$
(c) Is this amplifier well stabilized against changes in β_F? Explain your reasoning.

7.17 The amplifier shown in Fig. 7.34 is constructed with the resistor R_B divided into two parts, each $R_B/2$, and with the midpoint node bypassed to ground by capacitor C_3. Draw a midfrequency ac model for this amplifier, assuming C_1, C_2, and C_3 all have negligible reactance at the frequencies of interest.

7.18 The amplifier shown in Fig. 7.35(a) uses an enhancement-mode MOSFET with $i_D = 2(v_{GS} - 2)^2$ mA. Let $R_1 = R_2 = 270$ kΩ, $R_\sigma = 1$ kΩ, and $V_{DD} = 10$ V.

(a) Find I_D, and specify an R_D to give a reasonable operating point.
(b) Find the I_D that results when an FET with $i_D = 2(v_{GS} - 3)^2$ mA is substituted.

7.19 The amplifier shown in Fig. 7.35(b) uses a JFET with $I_{DSS} = 10$ mA and $V_P = -5$ V. Let $R_1 = \infty$, $R_2 = 1$ MΩ, and $V_{DD} = 10$ V.

(a) Find the values of R_σ and R_D required to yield the operating point $I_D = 0.9$ mA and $V_{DS} = 3$ V. Does this appear to be a reasonable operating point for this amplifier?
(b) Another FET, having $I_{DSS} = 20$ mA and $V_P = -8$ V, is used in the circuit designed in part (a). Find the operating point. Is the dc stabilization satisfactory?

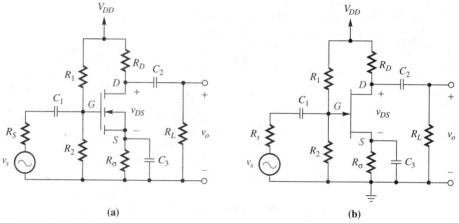

FIGURE 7.35 (a) MOSFET amplifier. (b) JFET amplifier.

7.20 A MOSFET with $i_D = 2(v_{GS} - 2)^2$ mA is used in the amplifier shown in Fig. 7.36. Let $R_G = 270$ kΩ and $V_{DD} = 10$ V.

(a) Find R_D such that $I_D = 2$ mA.
(b) Find the operating point for your design in part (a) when a FET with $i_D = 2(v_{GS} - 3)^2$ mA is substituted.

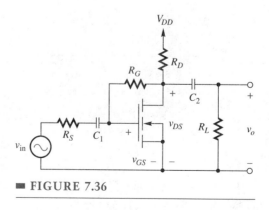

FIGURE 7.36

7.21 Given a matched pair of *pnp* transistors, and a matched pair of *npn* transistors, draw a circuit diagram showing how you could connect these transistors to achieve two simple current sources, one with +1 mA, and the other with −1 mA, using only one current reference resistor. What is its value? Use the BJT parameters of Table 18.2; let $V_{CC} = V_{EE} = 10$ V.

7.22 Refer to Fig. 7.37. Let $V_{EE} = 0$. Show that for a given I_O and V_{CC} the minimum total resistance $R_1 + R_2$ is achieved when $R_1 = \left(\dfrac{kT}{q}\right)\bigg/I_O = V_T/I_O$ and

$$R_2 \cong \frac{kT}{qI_O} \ln\left[\frac{V_{CC} - V_{BE(ON)}}{kT/q}\right]$$

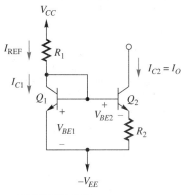

■ FIGURE 7.37

7.23 Design an *npn* Widlar current source to provide $I_O = 300$ μA for $V_{CC} = 15$ V and $V_{EE} = 10$ V. Use transistor specifications of Table 18.2 at $T = 300$ K. Select resistors R_1 and R_2, which are compatible with IC technology. Compute the output resistance, R_o.

7.24 For your design in Problem 7.23, Write a SPICE program to measure the output resistance R_o. It is suggested that you drive the output node with an ac test generator v_x, and find the ac current $i(v_x)$. Be sure that the dc voltage at the output is correct for active operation of the current source.

7.25 Repeat Problem 7.23 using the data given for a *pnp* (substrate) transistor. Let I_O be out of the collector. Sketch and label a schematic diagram.

7.26 Consider the simple multiple-current-source arrangement shown in Fig. 7.38. Let $V_{CC} = 30$ V. Specify the collector-junction area ratios and the value of R_1 necessary to obtain output currents of 200 μA, 400 μA, and 1 mA.

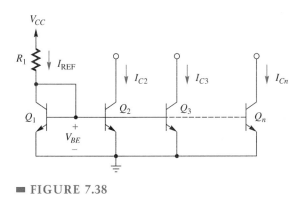

■ FIGURE 7.38

7.27 Design an *npn* Widlar current source to obtain $I_O = 10$ μA. Assume $V_{CC} = V_{EE} = 10$ V. Select values for R_1 and R_2 compatible with monolithic IC technology. Assume minimum and maximum values for V_A are 40 V and 400 V for the BJT to compute the range in R_O, the output resistance. Let $T = 300$ K.

7.28 Suppose V_{CC} increases to 20 V in Problem 7.27. For your design, what is the new value of I_O? Compare this change in I_O with the change that would occur if Problem 7.27 were realized as a simple current source.

7.29 Compute I_O for the *pnp* Widlar current source shown in Fig. 7.39. Let $T = 300$ K.

7.30 Over what range of R_L will the circuit in Problem 7.29 operate properly?

7.31 Solve Problem 7.29 by means of a SPICE simulation. Use parameters for the substrate *pnp* in Table 18.2.

7.32 Design a multiple *npn* Widlar current source to provide output currents of 20, 50, and 250 μA. Select values for a current reference resistor and the Widlar resistors, all compatible with monolithic IC technology. Assume identical transistors. Let $T = 300$ K, $V_{CC} = 15$ V and $V_{EE} = 0$.

7.33 Refer to Fig. 7.37. Let $V_{CC} = 15$ V, $V_{EE} = 0$ V, and $R_1 = 10$ kΩ. Plot the value of R_2 required for 5 μA $< I_O <$ 500 μA. Similarly, plot the value of R_{REF} required for 5 μA $< I_O <$ 500 μA in a simple current source configuration. Compare your results and suggest some design criteria useful in determining whether a simple or Widlar current source should be used.

7.34* Consider the circuit of Fig. 7.40. Assume all *pnps* are matched, and all *npns* are matched. Let $V_{CC} = V_{EE} = 10$ V, $R_1 = 15$ kΩ, $R_3 = 800$ Ω, $R_4 = 1$ kΩ, and $I_{DSS} = 8$ mA.

(a) Find the values of I_1, I_2, and I_3.
(b) What range of values for R_2 permits proper current-source operation?
(c) What value of V_P would you expect, assuming $I_D = I_3$ is as found in part (a), and that J_1 is in saturation?
(d) Suppose $V_{CC} = V_{EE}$ are increased to 12 V. Find I_1 and I_2.

7.35 Refer to Fig. 7.41, which is the dc bias circuitry for the μA741 op amp. Use the typical IC values for V_{BE} of 0.7 V for *npn* and -0.7 V for *pnp*. $V_{CC} = V_{EE} = 10$ V.

(a) Find the reference current through R_5.
(b) Assuming Q_{10} and Q_{11} are matched, find I_{C10}.
(c) Assuming Q_8 and Q_9 are matched, find the current out of the Q_8 collector.
(d) Suppose that Q_{13} has a split collector, with collector (1) having 3 times the junction area as collector (2), but the total junction area is the same as for Q_{12}. Estimate currents $I_{C13(1)}$ and $I_{C13(2)}$.

7.36 For the current sourcing system shown in Fig. 7.42, assume that the *npn* transistors Q_1 and

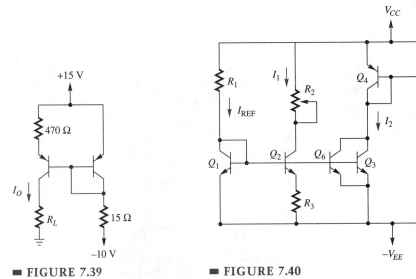

■ FIGURE 7.39   ■ FIGURE 7.40

CHAPTER 7 BIASING AND STABILITY

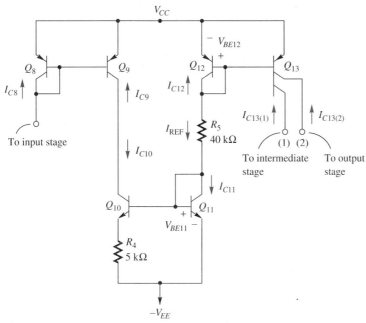

FIGURE 7.41 The dc biasing circuitry for the μA741.

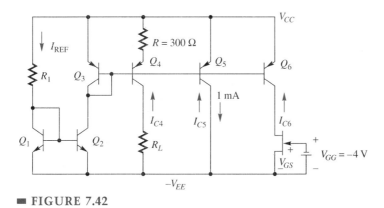

FIGURE 7.42

Q_2 are matched, and the *pnp* transistors Q_3, Q_4, Q_5, and Q_6 are matched. Assume $\beta \gg 1$ for all transistors and let $V_{CC} = V_{EE} = 10$ V. For the JFET, assume $V_P = -2$ V.

(a) Compute a value for R_1. Observe that $I_{C5} = -1$ mA.
(b) Compute a value for I_{C4}. Observe that $R_2 = 300$ Ω.
(c) Estimate a value for I_{C6}. Let $V_{GG} = -4$ V. Explain your answer.
(d) Using your value for I_{C4}, estimate the allowable range for R_L.

7.37 Complete the design for the current-source system of Fig. 7.43 (see p. 334). Let $V_{CC} = V_{EE} = 15$ V; $I_{S1} = I_{S2} = I_{S3} = I_{S4}$; $I_{S5} = 2I_{S1}$; $\beta \gg 1$ for all transistors, $I_{C1} = 1$ mA, and $I_{C2} = 50$ μA.

(a) Find R_1 and R_2.
(b) Find I_{C5}.
(c) What is the range of R_L for Q_4 to function as a current source?
(d) Suppose Q_5 is converted to a Widlar current source where the design requires $I_{C5} = 50$ μA. What value of Q_5 emitter resistor is needed?

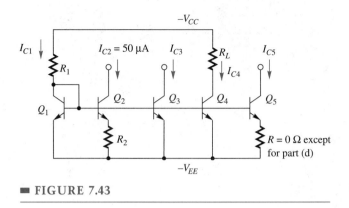

FIGURE 7.43

7.38* An equivalent schematic diagram of an LM358 op amp is shown in Fig. 7.44. Assume $V^+ = 15$ V, chip ground is 0 V, and the transistor specifications are as given in Table 18.2. You are to design a *pnp*-based current source system, using a single reference transistor, that could be used in place of the 100-μA current generator and both 6-μA current generators. That is, replace the three current-source symbols with your circuit diagram. Your resistor values must be compatible with diffused bipolar process technology.

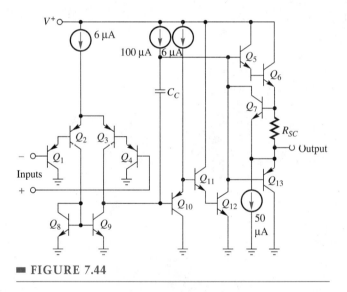

FIGURE 7.44

7.39* Refer to Problem 7.38 and Fig. 7.44. Observe that there is also a 50-μA source connected to the emitters of Q_7 and Q_{13}. Design an *npn*-based current source to provide this current. Your design could make use of the *pnp* reference transistor from Problem 7.38.

7.40 Use the circuit shown in Fig. 7.23 to stabilize the simple current source designed in Example 7.9. Assume $V_Z = 5.1$ V, $I_Z \leq 500$ μA, $V_{CC} = V_{EE} = 15$ V. Select values for R_1 and R_2.

7.41 Refer to Fig. 7.45. Assuming only that $V_D = V_{EB1}$ and that the transistor β's are large, show that I_{C2} is independent of β and V_{BE} of both transistors.

7.42 We want to obtain the three-transistor current source of the configuration of Fig. 7.46. Assume $V_{DD} = 8$ V and I_O is to be 200 μA. You must use three enhancement-mode MOSFETs, and the process being used results in $V_t = 2$ V. Specify a k that is identical for all three FETs.

7.43 For the circuit of Problem 7.42, V_{DD} increases to 10 V. What is the change in I_O?

7.44 Consider the simple current source of Fig. 7.15(b). We want to replace the reference resistor

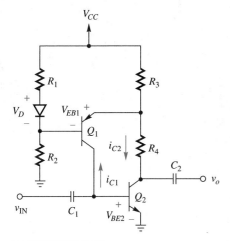

■ FIGURE 7.45

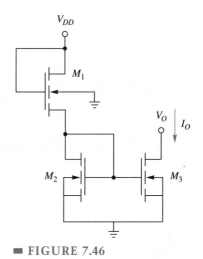

■ FIGURE 7.46

R_1 with an n-channel enhancement-mode MOSFET. Let $V_{CC} = 10$ V and $V_{EE} = 0$. Assume an FET with $V_t = 2$ V, and specify k such that $I_O = 100$ μA. Draw the circuit for your design.

7.45* Assume that the MOSFET current source of Fig. 7.46 is modified by the insertion of a 5-kΩ resistor in the source of M_3. This makes the circuit into a MOSFET Widlar current source. Assume $V_{DD} = 6$ V, and that all FETs have $V_t = 1.5$ V, $k = 100$ μA/V^2, and $\lambda = 0.02$ V^{-1}. Determine I_O.

7.46 Use a SPICE simulation to plot the output characteristic, I_O versus V_O, for the circuit of Problem 7.45. Determine the output resistance R_O.

7.47 Given a MOSFET Wilson current source as in Fig. 7.28, let $V_{DD} = 9$ V. All FETs are identical with $V_t = 2$ V, $k = 100$ μA/V^2, and λ can be neglected for the dc calculation. Calculate I_O. *Hint:* All currents are equal!

7.48 Show that for the MOSFET Wilson current source of Fig. 7.28, V_{DD} must be greater than $3V_t$, if all FETs are identical.

7.49 Show that for the MOSFET Wilson current source of Fig. 7.28, V_O must be greater than $[\frac{2}{3}V_{DD} - V_t]$ if all FETs are identical.

7.50 Use a SPICE simulation to plot the output characteristic, I_O versus V_O, for the circuit of Problem 7.47. Let $\lambda = 0.001$. Determine the output resistance R_O.

REFERENCES

1. Widlar, R. "Some Circuit Design Techniques for Linear Integrated Circuits," *IEEE Transactions on Circuit Theory,* Vol. CT-12 December 1965, pp. 586–590.

2. Wilson, G. "A Monolithic Junction FET-NPN Operational Amplifier," *IEEE Journal of Solid-State Circuits,* Vol. SC-3 December 1968, pp. 341–348.

SUGGESTED READINGS

Allen, P., and D. Holberg. *CMOS Analog Circuit Design.* New York: Holt, Rinehart, and Winston, 1987.

Colclaser, R., D. Neamen, and C. Hawkins. *Electronic Circuit Analysis.* New York: John Wiley & Sons, 1984. Chapter 10 treats biasing and current sources.

Gray, P. R., and R. G. Meyer. *Analysis and Design of Analog Integrated Circuits.* 3rd ed. New York: John Wiley & Sons, 1993. Chapter 4 deals with current sources, both BJT and MOSFET.

Grebene, A. *Bipolar and MOS Analog Integrated Circuit Design.* New York: John Wiley & Sons, 1984. Chapter 4 covers BJT biasing, and Chapter 6 applies MOS current sources.

CHAPTER 8

SMALL-SIGNAL MIDFREQUENCY AMPLIFIER ANALYSIS

8.1 The Hybrid-π Model
8.2 The Common-Emitter Amplifier: Impedances and Amplification
8.3 The Common-Base Amplifier: Impedances and Amplification
8.4 The Emitter-Follower Amplifier: Impedances and Amplification
8.5 The FET Model
8.6 MOSFET Amplifiers with Active Loads
8.7 Discrete Cascaded Amplifier Stages
8.8 Direct-Coupled and BiCMOS Amplifiers
8.9 Measurement of BJT Hybrid Parameters
Summary
Survey Questions
Problems
Reference
Suggested Readings

Having considered the establishment of an acceptable Q-point in the active region by proper bias design, we are now ready to analyze transistor circuits using a linear model. In this chapter, we define *small-signal amplifiers* as those in which the input-output transfer characteristic is approximately linear. This means that for the BJT $|v_{be}| \ll V_T$ (see Section 4.5); for the JFET $|v_{gs}| \ll |V_{GS} - V_P|$; and for the MOSFET $|v_{gs}| \ll |V_{GS} - V_t|$ (see Problem 8.18 at the end of the chapter). For *large signals*, graphical analysis and the SPICE transient analysis are more appropriate, since it is not proper to assume a linear model if the parameters change as the signal varies. In spite of this, however, a linear model is sometimes used in large-signal analysis as a simple means of obtaining very approximate results.

Further, we restrict our consideration here to *midfrequencies*.* This means that our models neglect those parasitic capacitances associated with the transistor junctions, which usually begin to affect the circuit operation above several tens of kilohertz. It also means that we neglect the capacitive reactances of the coupling and bypass capacitors, although they do have an effect below some frequency. After reviewing the small-signal ac model of Section 4.5, we consider certain refinements to better model second-order effects. We will apply this model to a comparison of various features of CE, CB, and CC amplifiers. We will also apply the small-signal field-effect transistor

---

*Most amplifiers have a *midband* range of frequencies where amplification is not attenuated by series impedances of coupling or bypass capacitors or by shunt impedances of parasitic capacitances. The concept of impedance demands a sinusoidal signal.

(FET) model to the variations in FET amplifier configurations. High-frequency models are treated in Chapter 9, as is the frequency dependence of the amplifiers.

After first considering simple discrete-transistor amplifiers, we turn to multistage integrated circuits (ICs). ICs avoid the use of capacitors by direct coupling of amplifier stages, and they often use active loads. Direct-coupled bipolar junction transistor (BJT) and metal-oxide semiconductor FET (MOSFET) amplifiers are discussed to demonstrate this technology, along with a combination of the two, called *BiMOS* or *BiCMOS*. Differential amplifiers are covered in Chapter 10.

Included is a section on the measurement of BJT common-emitter hybrid parameters, and how the hybrid-π parameters are obtained from them.

IMPORTANT CONCEPTS IN THIS CHAPTER

- Parasitic capacitances and the reactances of coupling and bypass capacitances are ignored at *midband* frequencies.
- The BJT hybrid-π model is a linear model that realistically represents the BJT at one Q-point for small signals.
- The fixed resistance from the active base region to the base contact is r_b.
- The parameter r_π models the dynamic resistance of the forward-biased base-emitter junction.
- The base-narrowing due to the Early effect is modeled by a feedback resistance r_μ, which is usually large enough to neglect, and an output resistance r_o.
- The small-signal model for a FET has no drain-to-gate feedback conductance.
- The transconductance of a BJT is higher than that of a FET of the same chip area.
- Either enhancement or depletion MOSFETs are used as active loads for CS FET IC amplifiers, because they use much less chip space than resistors.
- The voltage amplification of IC FET amplifiers is controlled by the *W/L* ratio of the devices.
- CMOS technology can produce large voltage gains, using a PMOS current mirror for the active load of an NMOS CS amplifier.
- Cascaded amplifier stages are analyzed by using the input resistance of the last stage as the load resistance of the preceding stage, etc.
- Direct-coupled circuits are used in IC design, avoiding coupling capacitances.
- The BJT cascode amplifier is a CE stage directly coupled to a CB stage.
- BiCMOS ICs combine the best features of BJT and CMOS devices on one die.

Amplifier Type	A_{vi}	R_i	R_o	A_i
CE (bypassed R_E)	High	Moderate	Moderate	Moderate
CE (unbypassed R_E)	Lower	High	Moderate	Moderate
CB amplifier	High	Low	Moderate	Low
Emitter follower	<1	Very high	Low	Moderate
CS (bypassed R_σ)	Moderate	Very high	Moderate	Very high
CS (unbypassed R_σ)	Low	Very high	Moderate	Moderate
CG amplifier	Moderate	Low	Moderate	Low
Source follower	<1	Very high	Low	Moderate

8.1 THE HYBRID-π MODEL

■ We assume $\eta = 1$ throughout. Otherwise, $g_m = \dfrac{I_C}{\eta V_T}$.

Figure 8.1(a) recalls the small-signal model for the BJT developed in Section 4.5. You are reminded that linearity requires $v_{be} \ll V_T$. At room temperature, where $V_T \approx 26$ mV, v_{be} would have to be less than about 5 mV peak. The parameters

$$g_m \equiv \left.\frac{\partial i_C}{\partial v_{BE}}\right|_{i_C = I_C} = \frac{I_C}{V_T}, \qquad (8.1)$$

$$r_\pi = \frac{V_T \beta}{|I_C|} = \frac{\beta}{g_m}, \qquad (8.2)$$

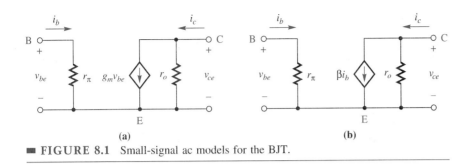

FIGURE 8.1 Small-signal ac models for the BJT.

and

$$r_o = \left.\frac{\Delta v_{CE}}{\Delta i_C}\right|_{V_{BE}} = \frac{V_A + V_{CB}}{I_C} \approx \frac{V_A + V_{CE}}{I_C} \quad (8.3)$$

all depend on the dc Q-point, so specifying the small-signal model first requires a knowledge of the operating point. Because $\beta = g_m r_\pi$, the model shown in Fig. 8.1(b) is entirely equivalent, and the two models are used interchangeably. Note that *small-signal current gain*, β, is used here, rather than dc β_F (see Section 4.9).

Base Resistance

This simple model for the BJT is somewhat deficient in that it does not distinguish the interior node at the active part of the base, as contrasted to the exterior base connection. This node is important because the parasitic junction capacitances needed in any high-frequency model must terminate at the active part of the base. The hybrid-π model, shown in Fig. 8.2, remedies this fault by separating the interior base node from the exterior contact by the ***base-spreading resistance*** r_b. The parameter r_b is determined by the geometry of the BJT structure and the base resistivity. It is known to decrease slightly with higher collector current due to *current crowding*,[1] but this is a second-order effect. This makes r_b nearly constant for a given device, on the order of tens to hundreds of ohms. By contrast, r_π varies with the operating current. Thus, whether r_b can be neglected in comparison with r_π depends largely on the Q-point. In addition, we shall see that the accuracy of the high-frequency model could be seriously impaired if r_b is neglected at high frequencies.

■ The major difference between the models of Figs. 8.1 and 8.2 is that when r_b is considered, the control voltage $v_\pi \neq v_{be}$. The difference is usually only a few percent.

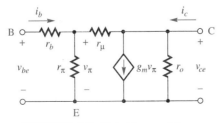

FIGURE 8.2 Hybrid-π model.

Collector-Base Resistance

As the reverse collector-base voltage increases, there is a slight decrease in base current due to base narrowing. This effect can be modeled by the passive resistance r_μ. For constant V_{BE},

$$r_\mu = \frac{\Delta V_{CE}}{\Delta I_{B1}} = \frac{\Delta V_{CE}}{\Delta I_C} \frac{\Delta I_C}{\Delta I_{B1}} = r_o \frac{\Delta I_C}{\Delta I_{B1}}, \qquad (8.4)$$

where I_{B1} is that part of the base current due to recombination. If $I_{B1} = I_B$,

$$r_\mu = r_o \beta. \qquad (8.5)$$

This represents a lower limit on r_μ. Usually, in *npn* IC transistors, the base-transport factor is better than the emitter efficiency (see Section 4.1), so that r_μ will be 2 to 10 times $r_o\beta$. Thus, often r_μ can be neglected with respect to the other resistive parameters in the circuit.

The Hybrid-π Model

Referring again to Fig. 8.2, the circuit equations are (using $g_x = 1/r_x$, where convenient)

$$v_{be} = r_b i_b + v_\pi \qquad (8.6)$$

and

$$i_c = v_{ce}(g_o + g_\mu) + v_\pi(g_m - g_\mu). \qquad (8.7)$$

DRILL 8.1

Use the BJT of Example 8.1 operating with $T = 27°C$, $I_C = 10$ mA, and $V_{CE} = 20$ V. Find the hybrid-π model for the PJT. Assume that β does not change.

ANSWER
$g_m = 390$ mS,
$r_\pi = 260$ Ω,
$r_b = 100$ Ω,
$r_o = 155$ kΩ, and
$r_\mu = 31$ MΩ

EXAMPLE 8.1

A certain 2N2222A operating at $T = 27°C$ with $I_C = 1$ mA and $V_{CE} = 10$ V has the following parameters: $\beta = 100$, $V_A = 135$ V, $r_b = 100$ Ω, and $I_{B1} = 0.5 I_B$. Find the hybrid-π model for the transistor.

SOLUTION $g_m = 1$ mA/26 mV $= 39$ mS, $r_\pi = 100/39$ mS $= 2600$ Ω, $r_b = 100$ Ω, $r_o = (135 + 10)$ V/1 mA $= 145$ kΩ, and $r_\mu = (2)(100)(145\ \mathrm{k}\Omega) = 29$ MΩ

CHECK UP

1. **TRUE OR FALSE?** BJT parameters g_m, r_π, and r_o are dependent on the Q-point.
2. **TRUE OR FALSE?** Parasitic capacitances and coupling capacitances are treated as short-circuits at *midband* frequencies.
3. **TRUE OR FALSE?** To qualify as *small signal*, the ac voltage at the base-emitter terminals, v_{be}, must be less than 5 mV.

8.2 THE COMMON-EMITTER AMPLIFIER: IMPEDANCES AND AMPLIFICATION

The typical common-emitter (CE) amplifier circuit is depicted in Fig. 8.3. For our analysis, we will consider that part of the circuit to the left of C_1 to be the signal source, that part to the right of C_2 to be the load, and that part between the coupling capacitors to be the amplifier. We assume sinusoidal voltages and currents, represented by *phasors* (as given

CHAPTER 8 SMALL-SIGNAL MIDFREQUENCY AMPLIFIER ANALYSIS

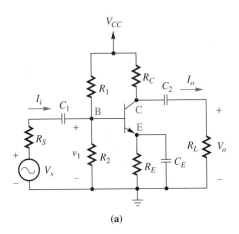

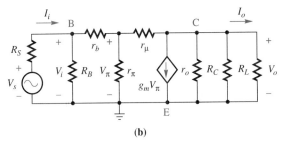

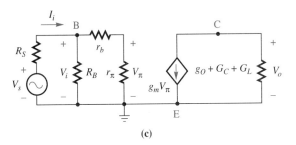

■ **FIGURE 8.3** Common-emitter amplifier. (a) Typical circuit. (b) Small-signal model for frequencies where $X_C \approx 0$. (c) Simplified model where $g_\mu = 0$.

---

in Chapter 1, our phasor notation is uppercase V or I, with lowercase subscripts). In this chapter, all impedances are real numbers, i.e., resistive. The hybrid-π model is shown in Fig. 8.3(b), assuming that all capacitors have negligible reactance at the frequencies of interest. This means that the emitter resistor R_E is effectively **bypassed** and is omitted in the ac model, having served its purpose in providing dc bias stabilization. In Chapter 9 we will examine the effect of the capacitive reactances at frequencies where they cannot be neglected. Note that $R_1 \| R_2 = R_B$.

It has been found that r_μ is so large that in most amplifiers it can be treated as an open circuit with small error. The result is shown in Fig. 8.3(c), and the simplification of having separated input and output circuits is obvious. We will examine the necessary conditions for the validity of the approximation later.

As we apply the transistor model, it is instructive to consider voltage gain, current gain, input resistance, and output resistance of the CE amplifier, and to compare it with the other configurations. Looking at the model shown in Fig. 8.3(c), we find that at the collector node (using $G_x = 1/R_x$ for convenience)

$$V_o[g_o + G_C + G_L] + g_m V_\pi = 0. \tag{8.8}$$

Substituting $V_\pi = V_i r_\pi/(r_\pi + r_b)$ into Eq. (8.8), we obtain

$$V_o(g_o + G_C + G_L) + V_i \frac{g_m r_\pi}{(r_\pi + r_b)} = 0. \tag{8.9}$$

■ This type of CE amplifier design is rarely used and obsolete, but we use this design for illustrating the analysis approach.

■ A bypass capacitor is effective only at frequencies high enough that its capacitive reactance is much lower than the impedance of the circuit it is in parallel with.

CE Voltage Amplification

The *voltage amplification* or *voltage gain* of an amplifier is defined as the ratio of the output voltage to the input voltage, which, from Eq. (8.9), is

$$A_{vi} = \frac{V_o}{V_i} = \frac{-g_m r_\pi}{(g_o + G_C + G_L)(r_\pi + r_b)} \cong \frac{-g_m}{G_C + G_L} = -g_m(R_C \| R_L). \quad (8.10)$$

The approximation assumes $g_o \ll G_C + G_L$ and $r_b \ll r_\pi$.

We can now estimate the error introduced by neglecting the current in r_μ. From Fig. 8.3(b), the r_μ current is $I_\mu = (V_\pi - V_o)g_\mu$, and substituting V_o from Eq. (8.10) gives

$$I_\mu = \left[\frac{V_i r_\pi}{r_\pi + r_b} - \frac{-V_i g_m r_\pi}{(g_o + G_C + G_L)(r_\pi + r_b)}\right] g_\mu$$

$$= \frac{V_i r_\pi}{r_\mu(r_\pi + r_b)}\left[1 + \frac{g_m}{(g_o + G_C + G_L)}\right]. \quad (8.11)$$

The current in r_π was about $I_\pi = V_i/(r_\pi + r_b)$. The ratio is

$$\frac{I_\mu}{I_\pi} = \frac{r_\pi}{r_\mu}\left[1 + \frac{g_m}{(g_o + G_C + G_L)}\right] \cong \frac{r_\pi g_m}{r_\mu(g_o + G_C + G_L)}. \quad (8.12)$$

It was certainly reasonable to neglect I_μ [as in Fig. 8.3(c)] if this ratio is less than, say, 5%, so that

$$0.05 r_\mu(g_o + G_C + G_L) > g_m r_\pi \quad \text{or} \quad r_\mu(g_o + G_C + G_L) > 20 g_m r_\pi \quad (8.13)$$

is the required condition. Substituting $r_\mu = r_o \beta$ and $\beta = g_m r_\pi$,

$$\beta r_o(g_o + G_C + G_L) > 20\beta \quad \text{or} \quad (G_C + G_L) > 19 g_o. \quad (8.14)$$

This is true for typical values of $G_L + G_C$.

CE Input Resistance

At the input node in Fig. 8.3(c), we have

$$I_i = \frac{V_i}{R_B} + \frac{V_i}{r_\pi + r_b} = V_i \frac{R_B + r_\pi + r_b}{R_B(r_\pi + r_b)}. \quad (8.15)$$

The *input resistance* of the amplifier is then

$$R_i = \frac{V_i}{I_i} = \frac{R_B(r_\pi + r_b)}{R_B + r_\pi + r_b} = R_B \| (r_\pi + r_b). \quad (8.16)$$

When the conditions of Eq. (8.14) hold, R_i is independent of R_L.

CE Current Amplification

The *current amplification* or *current gain* of an amplifier is defined as the ratio of the output (load) current to the input current:

$$A_i = \frac{I_o}{I_i} = \frac{V_o R_i}{V_i R_L} = A_{vi} \frac{R_i}{R_L} \cong \frac{-g_m(R_C \| R_L)[R_B \|(r_\pi + r_b)]}{R_L}. \quad (8.17)$$

Note that we again assume $g_o \ll G_L + G_C$.

CE Output Resistance

The *output resistance* of an amplifier is defined as the resistance seen looking back into the amplifier from the load with $V_i = 0$. To clarify the situation, we redraw Fig. 8.3(c) with $V_s = 0$ and a current into the output labeled I_x in Fig. 8.4. It is apparent that $V_i = V_\pi = 0$ and $I_x = V_o(g_o + G_C)$:

$$R_o = \frac{V_o}{I_x} = \frac{1}{g_o + G_C} = r_o \| R_C. \tag{8.18}$$

■ A knowledge of the A_{vi}, A_i, R_i and R_o of an amplifier allows us to integrate it into a larger system. The technique is illustrated in Section 8.7 and Chapter 10.

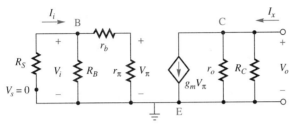

■ **FIGURE 8.4** Model for finding output resistance of the common-emitter amplifier.

■ Observe that we replace independent generators, V_s in this case, with its Thévenin equivalent resistance.

EXAMPLE 8.2

Suppose the CE amplifier shown in Fig. 8.3(a) has $V_{CC} = 6$ V, $R_1 = R_2 = 220$ kΩ, $R_E = 1$ kΩ, $R_C = 2.2$ kΩ, $R_S = 600$ Ω, and $R_L = 4.7$ kΩ. Use transistor parameters of $g_m = 42$ mS, $r_\pi = 2.4$ kΩ, $r_b = 100$ Ω, $r_\mu = 25$ MΩ, and $r_o = 125$ kΩ. Neglect capacitor reactances.

(a) Estimate the Q-point and find the values of A_{vi}, R_i, A_i, and R_o.
(b) Find $A_{vs} = V_o/V_s$.
(c) Find the *power gain*, $A_p = P_o/P_i = A_{vi} A_i$.

SOLUTION
(a) The given g_m and r_π suggest $\beta_F \cong \beta = g_m r_\pi = 100$, so the dc model gives

$$I_B = \frac{(3.0 - 0.7) \text{ V}}{[110 + 101(1)] \text{ k}\Omega} = 11 \text{ }\mu\text{A},$$

$I_C = 1.1$ mA, and $V_{CE} = 6 - 1.1$ mA $(2.2 + 1)$ kΩ $= 2.5$ V. Thus, the transistor is operating in the active region, with a possible ac signal swing of about 2 V. The condition of Eq. (8.13) is easily met.

- Equation (8.10) gives $A_{vi} = -(42 \text{ mS})(2.2 \| 4.7)$ kΩ $= -63$ (but see Drill 8.2).
- Equation (8.16) leads to $R_i = (220 \| 220 \| 2.5)$ kΩ $= 2.44$ kΩ.
- Equation (8.17) yields $A_i = -(63)\dfrac{2.44 \text{ k}\Omega}{4.7 \text{ k}\Omega} = -32.7$.
- Equation (8.18) gives $R_o = 125$ kΩ $\| 2.2$ kΩ $= 2.16$ kΩ.

(b) By voltage division,

$$\frac{V_i}{V_s} = \frac{R_i}{R_s + R_i} = \frac{2440}{600 + 2440} = 0.80.$$

DRILL 8.2

For the BJT parameters given in Example 8.2, find the error introduced in the calculation of A_{vi} by neglecting r_b and r_o.

ANSWER The error is 4% due to neglecting r_b, and 1% more to neglect r_o.

Then

$$A_{vs} = \frac{V_o}{V_s} = \frac{V_i}{V_s} A_{vi} = (0.80)(-63) = -50.$$

(c) $A_p = (-63)(-32.7) = 2060.$

Unbypassed Emitter Resistor

When the emitter resistor is not bypassed, the amplifier characteristics are modified considerably. The circuit model is as shown in Fig. 8.5(a) where r_μ is omitted as before. The further simplification of omitting r_o results in Fig. 8.5(b). At the three nodes we have

$$V_o(G_C + G_L) + g_m V_\pi = 0, \tag{8.19}$$

$$V_e = (V_\pi g_\pi + V_\pi g_m) R_E, \tag{8.20}$$

$$V_i = V_\pi g_\pi (r_b + r_\pi) + V_e = V_\pi [r_b g_\pi + 1 + (g_\pi + g_m) R_E]. \tag{8.21}$$

Combining Eqs. (8.19) and (8.21), we find the voltage gain:

$$A_{vi} = \frac{V_o}{V_i} = \frac{-g_m}{[g_\pi r_b + 1 + (g_\pi + g_m) R_E](G_C + G_L)} \cong \frac{-g_m (R_C \| R_L)}{(1 + g_m R_E)}. \tag{8.22}$$

This is the same as the A_{vi} for the bypassed R_E of Eq. (8.10), except for the division by $(1 + g_m R_E)$. When $g_m R_E \gg 1$,

$$A_{vi} \cong -\frac{R_C \| R_L}{R_E}. \tag{8.23}$$

This voltage gain is nearly independent of the transistor parameters, a very useful feature of **emitter degeneration**, which is an example of a more general class of circuits using **negative feedback**, which is covered in detail in Chapter 11. In this case, the current of the output circuit is sampled by R_E, and inserted in series with the input voltage.

We note that $I_\pi = V_\pi g_\pi$, and inserting V_π from Eq. (8.21),

$$I_\pi = \frac{V_i g_\pi}{r_b g_\pi + 1 + (g_\pi + g_m) R_E} = \frac{V_i}{r_b + r_\pi + (1 + g_m r_\pi) R_E} \tag{8.24}$$

and

$$I_i = V_i G_B + I_\pi = V_i G_B + \frac{V_i}{r_b + r_\pi + (1 + g_m r_\pi) R_E}. \tag{8.25}$$

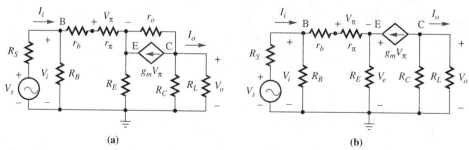

■ **FIGURE 8.5** Model of an unbypassed emitter resistor amplifier. (a) Neglecting r_μ. (b) Neglecting r_μ and r_o.

CHAPTER 8 SMALL-SIGNAL MIDFREQUENCY AMPLIFIER ANALYSIS

Recalling that $\beta = g_m r_\pi$,

$$R_i = \frac{V_i}{I_i} = \frac{R_B[r_b + r_\pi + (1 + g_m r_\pi)R_E]}{R_B + r_b + r_\pi + (1 + g_m r_\pi)R_E}$$
$$= R_B \| [r_\pi + r_b + (1 + \beta)R_E]. \quad (8.26)$$

Note that R_i increased considerably over the bypassed case due to the R_E term.

$$A_i = \frac{I_o}{I_i} = \frac{V_o}{V_i}\frac{R_i}{R_L} = A_{vi}\frac{R_i}{R_L} = \frac{-g_m(R_C\|R_L)}{(1 + g_m R_E)}\frac{R_B\|[r_\pi + r_b + (1 + \beta)R_E]}{R_L}. \quad (8.27)$$

When r_o is omitted from the model,

$$R_o \cong R_C. \quad (8.28)$$

We leave it to you (see Drill 8.3) to show that omitting r_o introduces little error when $g_o \ll (G_L + G_C)$. Otherwise, $R_o = R_C \| [r_o(1 + g_m R_E)]$.

After examining common-base and common-collector amplifiers, we will compare the circuits.

■ The effect of not bypassing the emitter resistor on the CE amplifier is to reduce the voltage amplification and increase the input resistance, and perhaps increase the output resistance.

EXAMPLE 8.3

Repeat Example 8.2, assuming that R_E is not bypassed.

SOLUTION Of course, the dc situation is the same as in Example 8.2.

(a) The ac model is as shown in Fig. 8.5(a). Using Eqs. (8.22), (8.26), (8.27) and (8.28), we obtain

$$A_{vi} = -42 \text{ mS } (2.2\|4.7) \text{ k}\Omega/(1 + 42) = -1.5,$$
$$R_i = \{220\|220\|[2.5 + (101)1]\} \text{ k}\Omega = 53 \text{ k}\Omega, \text{ and}$$
$$A_i = (-1.5)(53 \text{ k}\Omega)/(4.7 \text{ k}\Omega) = -16.9, \text{ respectively}.$$
$$R_o = R_C = 2.2 \text{ k}\Omega.$$

(b) $$A_{vs} = \frac{R_i}{R_s + R_i} A_{vi} = \frac{53}{0.6 + 53}(-1.5) = -1.5.$$

(c) $A_p = (-1.5)(-16.9) = 25.4$. We observe that compared to Example 8.2 A_{vi} is down by a factor of $(1 + g_m R_E) = 42$ because of the unbypassed emitter resistance, whereas R_i is much greater for the same reason.

DRILL 8.3

Use a SPICE solution to find R_o for the situation of Example 8.3. You will need to short V_s, and apply a test source V_x in place of R_L. Let $\beta_F = 100$.

ANSWER
$R_o = 2.198 \text{ k}\Omega$.
Apparently neglecting R_o in Eq. (8.28) was justified.

CHECK UP

1. **TRUE OR FALSE?** For a CE amplifier (R_E bypassed) if $R_B \gg r_b + r_\pi$, then $R_i \cong r_b + r_\pi$.
2. **TRUE OR FALSE?** For a CE amplifier, r_b is almost always small compared with r_π and should usually be neglected.
3. **TRUE OR FALSE?** For a CE amplifier, r_o is almost always large compared with R_C, and can usually be neglected.
4. **TRUE OR FALSE?** For a CE amplifier (R_E bypassed), the voltage amplification A_{vi} is $(1 + g_m R_E)$ times larger than the case with R_E not bypassed.

8.3 THE COMMON-BASE AMPLIFIER: IMPEDANCES AND AMPLIFICATION

■ Observe that the dc bias circuit is the same as in the CE.

A simple common-base (CB) amplifier circuit is shown in Fig. 8.6(a) and the hybrid-π transistor model is used in the ac circuit model of Fig. 8.6(b).

CB Voltage Amplification

We first neglect the current in r_o, and then show when this is valid. At the collector node,

$$V_o G_C + V_o G_L + g_m V_\pi = 0, \qquad (8.29)$$

so that

$$V_o(G_C + G_L) = g_m \left(\frac{V_i r_\pi}{r_\pi + r_b} \right), \qquad (8.30)$$

and the voltage amplification is

$$A_{vi} = \frac{V_o}{V_i} = \left[\frac{1}{(G_C + G_L)} \right] \left[\frac{g_m r_\pi}{(r_\pi + r_b)} \right] \qquad (8.31)$$

$$= \frac{\beta(R_C \| R_L)}{(r_\pi + r_b)} \cong +g_m(R_C \| R_L). \qquad (8.32)$$

The current that we neglected in r_o is seen to be

$$I_{ro} = (V_o - V_i)g_o = V_i(A_{vi} - 1)g_o \cong V_i g_m(R_C \| R_L)g_o,$$

which is small compared to the current in the controlled source when $g_o \ll (G_C + G_L)$. This is usually true. The simplified expression recognizes that g_o is much less than the other conductances, and that $r_b \ll r_\pi$. Note that for our example, this last assumption causes a 4% error. Except for the sign, equivalent to a 180° phase shift, A_{vi} is the same as for the CE amplifier.

CB Input Resistance

At the input node,

$$I_i = \frac{V_i}{R_E} + \frac{V_i}{r_\pi + r_b} + \frac{V_i - V_o}{r_o} - g_m V_\pi$$

$$= \frac{V_i}{R_E} + \frac{V_i}{r_\pi + r_b} + \frac{V_i(1 - A_{vi})}{r_o} - g_m V_\pi. \qquad (8.33)$$

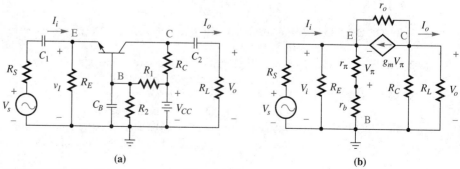

(a)   (b)

■ **FIGURE 8.6** Common-base amplifier. (a) Circuit diagram. (b) Hybrid-π model.

Inserting

$$A_{vi} = \frac{g_m r_\pi}{(r_\pi + r_b)(G_C + G_L)} \quad \text{and} \quad V_\pi = \frac{-V_i r_\pi}{r_\pi + r_b}$$

and dividing,

$$R_i = \frac{V_i}{I_i} = \frac{1}{G_E + g_o + \dfrac{1}{r_\pi + r_b} + \dfrac{\beta}{(r_\pi + r_b)}\left(1 - \dfrac{g_o}{G_C + G_L}\right)}. \tag{8.34}$$

Note that g_o always appears in a sum or factor where it is small compared to the other terms. Neglecting g_o in Eq. (8.34),

$$R_i = \frac{V_i}{I_i} = \frac{1}{G_E + \dfrac{1 + \beta}{(r_\pi + r_b)}} = R_E \left\| \frac{(r_\pi + r_b)}{1 + \beta} \right.. \tag{8.35}$$

Since often $R_E \gg r_\pi/\beta = 1/g_m$, then $R_i \cong 1/g_m$. Observe the very low R_i for the CB amplifier compared to the CE amplifier.

CB Current Amplification

$$A_i = A_{vi} \frac{R_i}{R_L} = \left[\frac{\beta(R_C \| R_L)}{(r_\pi + r_b)R_L}\right]\left[R_E \left\| \frac{(r_\pi + r_b)}{1 + \beta}\right.\right] \cong \frac{(R_C \| R_L)}{R_L}. \tag{8.36}$$

Note that A_i is less than unity for the CB.

CB Output Resistance

The output resistance is found by letting $V_s = 0$ from the model shown in Fig. 8.7. A good approximate solution is easily obtained by neglecting r_b, so that $V_\pi = -V_i$. At the emitter node,

$$V_i(G_S + G_E + g_\pi + g_o) - V_o g_o - g_m(-V_i) = 0. \tag{8.37}$$

At the collector node,

$$I_x = V_o(G_C + g_o) - V_i g_o + g_m(-V_i). \tag{8.38}$$

Combining,

$$\frac{1}{R_o} = \frac{I_x}{V_o} = G_C + \frac{g_o(G_S + G_E + g_\pi)}{G_S + G_E + g_\pi + g_o + g_m}. \tag{8.39}$$

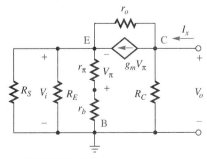

FIGURE 8.7 Model for finding output resistance of the common-base amplifier.

Since $g_o \ll G_C$, the last term is seen to be small compared to G_C, and

$$R_o \simeq R_C, \qquad (8.40)$$

or about the same as the CE amplifier.

EXAMPLE 8.4

Suppose the CB amplifier shown in Fig. 8.6 is constructed with the same components as those of Example 8.2. Find A_{vi}, R_i, A_i, R_o, and A_p.

SOLUTION This amplifier has the same Q-point as that of Example 8.2 since the dc circuit model is identical; thus, the transistor parameters are the same.

- From Eq. (8.32), $A_{vi} = (42 \text{ mS})(2.2 \| 4.7) \text{ k}\Omega = 63$. This is the same magnitude but opposite sign from Example 8.2.
- Equation (8.35) yields $R_i = 1 \text{ k}\Omega \left\| \dfrac{2500 \ \Omega}{101} \right. = 24 \ \Omega$. This low value of input resistance is typical of CB amplifiers.
- $A_i = (63)25/4700 = 0.33$ from Eq. (8.36).
- $R_o = 2.2 \text{ k}\Omega$.
- $A_p = (63)(0.33) = 21$.

DRILL 8.4

For the amplifier of Example 8.4, assume $R_s = 600 \ \Omega$. Find the source voltage gain $A_{vs} = V_o/V_s$.

ANSWER $A_{vs} = 2.5$

CHECK UP

1. **TRUE OR FALSE?** The CB amplifier is certain to have a current gain of $A_i > 1$.
2. **TRUE OR FALSE?** The CB amplifier is likely to have a low input impedance.

8.4 THE EMITTER-FOLLOWER AMPLIFIER: IMPEDANCES AND AMPLIFICATION

A typical emitter-follower, or common-collector (CC) amplifier circuit appears in Fig. 8.8. Again, our analysis makes use of the hybrid-π model for the BJT.

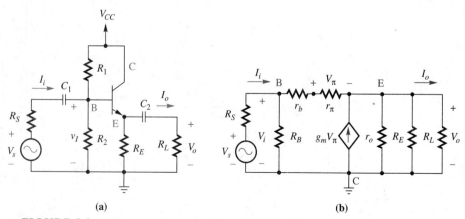

■ **FIGURE 8.8** Common-collector amplifier. (a) Typical circuit. (b) Hybrid-π model.

CHAPTER 8 SMALL-SIGNAL MIDFREQUENCY AMPLIFIER ANALYSIS

CC Voltage Amplification

At the emitter node, the current equation is

$$V_o(G_L + G_E + g_o) + \frac{(V_o - V_i)}{r_\pi + r_b} - \frac{g_m(V_i - V_o)r_\pi}{r_\pi + r_b} = 0. \quad (8.41)$$

Factoring,

$$A_{vi} = \frac{V_o}{V_i} = \frac{(1 + g_m r_\pi)/(r_\pi + r_b)}{G_L + G_E + g_o + (g_m r_\pi + 1)/(r_\pi + r_b)}$$

$$\cong \frac{\beta}{(G_L + G_E)(r_\pi + r_b) + \beta}. \quad (8.42)$$

The simplified expression assumes $\beta = g_m r_\pi \gg 1$. The value of A_{vi} is slightly less than one because $(G_L + G_E)(r_\pi + r_b)$ is usually $\ll \beta$. Because the output voltage is nearly equal to the input voltage, the amplifier is called an *emitter follower*.

■ The name *emitter follower* is more common than *common-collector amplifier* because it is descriptive of the circuit function.

CC Input Resistance

At the base node, the current equation yields

$$I_i = V_i G_B + \frac{V_i - V_o}{r_\pi + r_b} = V_i \left(G_B + \frac{1 - A_{vi}}{r_\pi + r_b} \right)$$

$$\cong V_i \left[G_B + \frac{G_L + G_E}{(G_L + G_E)(r_\pi + r_b) + \beta} \right], \quad (8.43)$$

$$R_i = \frac{V_i}{I_i} \cong \frac{1}{G_B + \dfrac{G_L + G_E}{(r_\pi + r_b)(G_L + G_E) + \beta}} = R_B \| [r_\pi + r_b + \beta(R_E \| R_L)]. \quad (8.44)$$

The assumption is that $\beta \gg 1$.

CC Current Amplification

Combining Eqs. (8.42) and (8.44),

$$A_i = \frac{I_o}{I_i} = \frac{V_o}{R_L}\frac{R_i}{V_i} = A_{vi}\frac{R_i}{R_L} = \frac{\beta R_B(R_L \| R_E)}{R_L\{[r_\pi + r_b + \beta(R_L \| R_E)] + R_B\}}. \quad (8.45)$$

CC Output Resistance

The output resistance of the emitter follower may be calculated with the aid of Fig. 8.9. We see that the base network acts as a voltage divider across V_o:

$$V_\pi = \frac{-r_\pi V_o}{r_\pi + r_b + (R_S \| R_B)}. \quad (8.46)$$

Kirchhoff's current law at the collector gives

$$I_x = V_o \left[G_E + g_o + \frac{1}{r_\pi + r_b + (R_S \| R_B)} + \frac{g_m r_\pi}{r_\pi + r_b + (R_S \| R_B)} \right], \quad (8.47)$$

$$R_o = \frac{V_o}{I_x} = \frac{1}{G_E + g_o + \dfrac{1 + g_m r_\pi}{r_\pi + r_b + (R_S \| R_B)}} \approx R_E \left\| \left[\frac{r_\pi + r_b + (R_S \| R_B)}{1 + \beta} \right] \right.. \quad (8.48)$$

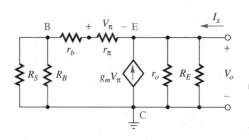

FIGURE 8.9 Model for finding output resistance of the common-collector amplifier.

EXAMPLE 8.5

Let the emitter follower shown in Fig. 8.8 be constructed with the same component values as those of Example 8.2. Find A_{vi}, R_i, A_i, R_o, and A_p.

SOLUTION   This amplifier has the same Q-point as that of Examples 8.2, 8.3, and 8.4, and the transistor model is the same.

- From Eq. (8.42), $A_{vi} = 100/[(0.21 + 1) \text{ mS } (2500) \, \Omega + 100] = 0.97$.
- From Eq. (8.44), $R_i = 110 \text{ k}\Omega \| [2.5 + 100(1\|4.7)] \text{ k}\Omega = 48 \text{ k}\Omega$, demonstrating the higher input resistance of the CC.
- From Eq. (8.45), $A_i = (0.97) \, 48 \text{ k}\Omega/4.7 \text{ k}\Omega = 9.9$.
- The output resistance of the common collector is low, as shown by

$$R_o = 1 \text{ k}\Omega \left\| \frac{[2.5 + (0.6\|110)] \text{ k}\Omega}{101} \right. = 30 \, \Omega$$

from Equation (8.48).
- $A_p = (0.97)(9.9) = 9.6$.

DRILL 8.5

For the amplifier of Example 8.5, assume $R_s = 600 \, \Omega$, and calculate A_{vs}.

ANSWER $A_{vs} = 0.96$

CHECK UP

1. **TRUE OR FALSE?** The CC amplifier is certain to have a voltage gain of $A_{vi} < 1$.
2. **TRUE OR FALSE?** The CC amplifier is likely to have a low input impedance.

BJT Amplifier Comparisons

Table 8.1 shows a comparison of the amplifications and resistances of the circuits just analyzed. The simplifications suggested in the text were made.

1. We can see that the voltage amplifications of the CE and CB amplifiers are the same, except the CE is inverting. In the case of the unbypassed emitter resistance, the gain is reduced in proportion to the factor $(1 + g_m R_E)$. The CC amplifier voltage amplification is always less than unity, typically quite close to one for reasonable values of R_L and R_E.
2. Input resistance is low for the CB, moderate for the CE (but much higher when the emitter is not bypassed), and very high for the CC amplifier.
3. Current gain is high (near $g_m r_\pi = \beta$) for the CE and CC amplifiers, but less than unity for the CB amplifier.
4. Output resistance is approximately determined by the R_C for the CE and CB amplifiers, but it is very much lower for the CC amplifier.

CHAPTER 8 SMALL-SIGNAL MIDFREQUENCY AMPLIFIER ANALYSIS

TABLE 8.1 Comparisons of CE, CB, and CC Circuits

	Common Emitter (R_E bypassed) (Fig. 8.3)	Common Emitter with R_E (Fig. 8.5)	Common Base (Fig. 8.6)	Common Collector (Fig. 8.8)
A_{vi}	$-g_m(R_C \| R_L)$	$\dfrac{-g_m(R_C \| R_L)}{(1 + g_m R_E)}$	$+g_m(R_C \| R_L)$	$\dfrac{g_m}{G_L + G_E + g_m}$
R_i	$R_B \| (r_\pi + r_b)$	$R_B \| [r_\pi + r_b + (1+\beta)R_E]$	$R_E \left\| \dfrac{1}{g_m}\right.$	$R_B \| [r_\pi + r_b + \beta(R_E \| R_L)]$
A_i	$\dfrac{-g_m(R_C \| R_L)[R_B \|(r_\pi + r_b)]}{R_L}$	$\dfrac{-g_m(R_C \| R_L)}{(1 + g_m R_E)} \dfrac{R_B \|[r_\pi + r_b + (1+\beta)R_E]}{R_L}$	$\dfrac{(R_C \| R_L)}{R_L}$	$\dfrac{\beta R_B(R_L \| R_E)}{R_L\{[r_\pi + r_b + \beta(R_L \| R_E)] + R_B\}}$
R_o	$r_o \| R_C$	R_C	R_C	$R_E \left\| \left[\dfrac{r_\pi + r_b + (R_S \| R_B)}{1 + \beta}\right]\right.$

In summary:

- *The CE amplifier* serves as a good general-purpose circuit, with the possibility of both voltage and current amplification and moderate input and output resistances. Power gain can be very good.
- *The CB amplifier* is ideal for matching a low source resistance to a high load resistance, providing power gain and possibly voltage amplification.
- *The CC (emitter-follower) amplifier* is used to match a high source resistance to a low load resistance, providing power gain and possibly current amplification.

Cascaded amplifiers are discussed in Section 8.7.

■ These amplifier circuits are seldom used alone, but are building blocks. They are usually embedded in more complex circuits, such as op amps. Other examples of amplifiers with active loads and differential amplifiers follow.

8.5 THE FET MODEL

The small-signal FET model was introduced in Chapter 5 and is redrawn in Fig. 8.10. A MOSFET and a JFET are shown, but the model will represent either of the FETs. The reverse-biased or insulated gate is assumed to be an open circuit, which makes the FET model considerably simpler than the BJT. We find g_m from Eq. (5.41) or (5.42), and r_d from (5.44):

$$g_m = 2k(V_{GS} - V_t) = 2\sqrt{kI_D}$$

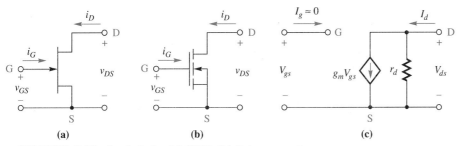

■ **FIGURE 8.10** Symbols for **(a)** JFET, **(b)** Enhancement-mode MOSFET, and **(c)** the small-signal model for either.

or

$$g_m = \frac{-2I_{DSS}}{V_P}\left(1 - \frac{V_{GS}}{V_P}\right) = -\frac{2}{V_P}\sqrt{I_{DSS}I_D},$$

$$r_d = \frac{1}{\lambda I_D}.$$

As was the case with the BJT, it is instructive to consider some of the characteristics of each of the FET amplifier configurations as we apply the model in circuit analysis. Again we consider each circuit to consist of a signal voltage source V_s with internal resistance R_S driving the amplifier through the coupling capacitor C_1, and the load resistance R_L being driven by the amplifier through the coupling capacitor C_2. Furthermore, these are again midfrequency models, and we neglect the reactances of the coupling and bypass capacitors, as well as neglecting the parasitic capacitance of the FET devices.

The Common-Source Amplifier

■ Topologically, the common-source FET amplifier is analogous to the common-emitter BJT amplifier.

Figure 8.11 presents a typical common-source (CS) amplifier (with R_σ bypassed by C_σ) and its model. Bias resistors $R_1 \| R_2$ are replaced with R_G in the ac model, and V_{gs} is seen to be equal to V_i. At the drain node [Fig. 8.11(b)],

$$V_o(g_d + G_D + G_L) = -g_m V_i, \tag{8.49}$$

so that

$$A_{vi} = \frac{V_o}{V_i} = \frac{-g_m}{g_d + G_D + G_L} \quad \text{or} \quad -g_m(R_D \| R_L \| r_d). \tag{8.50}$$

■ For this discussion, it makes little difference whether we use a JFET or MOSFET. The small-signal models are the

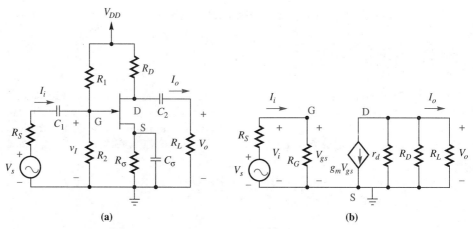

■ **FIGURE 8.11** Common-source amplifier. (a) Typical circuit. (b) ac model.

Usually, $g_d \ll G_D + G_L$, so r_d may not be a factor. At the input node, obviously,

$$R_i = \frac{V_i}{I_i} = R_G. \tag{8.51}$$

Combining,

$$A_i = \frac{I_o}{I_i} = \frac{V_o/R_L}{V_i/R_i} = A_{vi}\frac{R_i}{R_L} = \frac{-g_m R_G(R_D \| R_L \| r_d)}{R_L}. \tag{8.52}$$

We can easily find the output resistance by using the circuit shown in Fig. 8.12. When $V_s = 0$, so does V_i, which is V_{gs}. Then the current source is also zero, and $I_x = V_o(g_d + G_D)$. Dividing,

$$R_o = \frac{1}{g_d + G_D} = r_d \| R_D \cong R_D, \quad (8.53)$$

since r_d is likely to be many times larger than R_D.

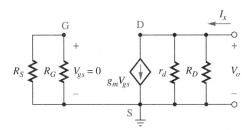

■ **FIGURE 8.12** Model for finding output resistance of the common-source amplifier.

EXAMPLE 8.6

Given a 2N5459 with $I_{DSS} = 10$ mA and $V_P = -5$ V incorporated into the CS amplifier shown in Fig. 8.11, let $V_{DD} = 15$ V, $R_1 = 1$ MΩ, $R_2 = 150$ kΩ, and $R_\sigma = R_D = R_L = 15$ kΩ. Find the Q-point of the amplifier, and calculate A_{vi}, R_i, A_i, R_o, and A_p.

SOLUTION The dc model is shown in Fig. 7.9; $I_D = 10(1 + 0.2V_{GS})^2$ mA from Eq. (7.17), and $V_{GG} = (15 \text{ V})(150)/(1000 + 150) = 1.96$ V. Using Eq. (7.21), we can calculate $I_D = 0.4$ mA and $V_{DS} = 3.0$ V. We can also use Eq. (5.40) to find $g_m = 0.8$ mS, and from the data sheet in Fig. 5.31 we can estimate $r_d = 1.25$ MΩ.

$$[g_{os} = 20 \ \mu S \text{ at } I_D = 10 \text{ mA} \Rightarrow \lambda^{-1} = 500 \text{ V}; \ r_d = (\lambda^{-1})/(I_{DQ})].$$

Then, using Eqs. (8.50) to (8.53), respectively, we calculate

$$A_{vi} = (-0.8 \text{ mS})(15 \| 15 \| 1250) \text{ k}\Omega = -6.0, \quad R_i = 1 \text{ M}\Omega \| 150 \text{ k}\Omega = 130 \text{ k}\Omega,$$

$$A_i = (-6)130/15 = -52, \quad R_o = (1250 \| 15) \text{ k}\Omega = 14.8 \text{ k}\Omega.$$

Note that the power gain is 312.

DRILL 8.6

Use SPICE to find A_{vi} and A_i for Example 8.6. Let all capacitors be large (100 μF is acceptable at 1 kHz), $R_s = 600$ Ω, and $\lambda = 0.002$ V^{-1}.

ANSWER $A_{vi} = -5.9$ and $A_i = -52$. The slight difference is in the round-off.

The Common-Source Amplifier with an Unbypassed Source Resistor

Figure 8.13 shows the circuit with R_σ not bypassed. The current source supplies two branches, r_d and $(R_D \| R_L) + R_\sigma$. As long as $r_d \gg [1/(G_D + G_L)] + R_\sigma$, we can neglect r_d. This is usually a good approximation. Then

$$V_{gs} = V_i - g_m V_{gs} R_\sigma \quad \text{or} \quad V_{gs} = V_i/(1 + g_m R_\sigma). \quad (8.54)$$

At the drain,

$$V_o(G_D + G_L) + g_m V_{gs} = 0. \quad (8.55)$$

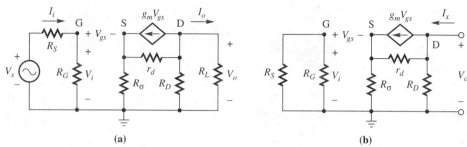

(a)   (b)

■ **FIGURE 8.13** (a) Model for the common-source amplifier with unbypassed source resistor. (b) Model for finding output resistance.

Combining,

$$A_{vi} = \frac{V_o}{V_i} = \frac{-g_m}{(1 + g_m R_\sigma)(G_D + G_L)} \cong -\frac{R_D \| R_L}{R_\sigma}. \qquad (8.56)$$

The approximation is valid if $g_m R_\sigma \gg 1$.

The input resistance is unchanged from Eq. (8.51), $R_i = R_G$.

$$A_i = \frac{I_o}{I_i} = A_{vi} \frac{R_i}{R_L} = \frac{-g_m R_G (R_D \| R_L \| r_d)}{(1 + g_m R_\sigma) R_L}. \qquad (8.57)$$

To find the output resistance, we use Fig. 8.13(b). Because $V_i = 0$, if we neglect r_d, $V_{gs} = -g_m V_{gs} R_\sigma$, the controlled source is zero, and

$$R_o = R_D. \qquad (8.58)$$

We can show that the exact solution for R_o (not neglecting r_d) is $R_o = R_D \| [r_d(1 + g_m R_\sigma)]$, which reduces to R_D in most cases.

DRILL 8.7

Find the value of R_o for the amplifier of Example 8.7 without neglecting r_d. This can be done analytically, or you may use a SPICE simulation.

ANSWER
$r_o = 14.99$ kΩ

■ **EXAMPLE 8.7**

Calculate the values of A_{vi} and A_i for the CS amplifier of Example 8.6 for the case where R_σ is not bypassed.

SOLUTION Of course, the Q-point is unchanged from the bypassed case. The gains are reduced by the factor $(1 + g_m R_\sigma) = 13$, so that

$$A_{vi} = \frac{(15 \| 15) \text{ k}\Omega (0.8 \text{ mS})}{1 + 12} = -0.46 \quad \text{and} \quad A_i = (-0.46)\frac{130}{15} = -4.0.$$

The R_i is unchanged, R_o increases slightly to 15 kΩ, and $A_p = (-4.0)(-0.46) = 1.84$.

CHECK UP

1. **TRUE OR FALSE?** The CS FET amplifier will have a moderate voltage gain, A_{vi}, if the source resistor is bypassed, but considerably less gain if it is not bypassed.
2. **TRUE OR FALSE?** The CS FET amplifier is likely to have a very high input impedance.
3. **TRUE OR FALSE?** The CS FET amplifier usually has a higher voltage gain, A_{vi}, than a BJT CE amplifier with the same Q-point current.

The Common-Gate Amplifier

A typical common-gate (CG) arrangement is shown in Fig. 8.14. Consider the ac model for the circuit in Fig. 8.14(b). Here, $V_{gs} = -V_i$, so we write for the drain node,

$$V_o(G_D + G_L + g_d) - V_i g_d + g_m(-V_i) = 0. \tag{8.59}$$

Dividing, we find that

$$A_{vi} = \frac{V_o}{V_i} = \frac{(g_m + g_d)}{G_D + G_L + g_d} \cong \frac{g_m}{G_D + G_L} \quad \text{or} \quad g_m(R_D \| R_L). \tag{8.60}$$

At the source node,

$$I_i = V_i G_\sigma + g_d(V_i + V_o) - g_m(-V_i). \tag{8.61}$$

Combining with A_{vi},

$$R_i = \frac{V_i}{I_i} = \frac{1}{G_\sigma + g_d + g_m[1 + g_d/(G_D + G_L)]} \cong \frac{1}{G_\sigma + g_m} = R_\sigma \left\| \frac{1}{g_m}, \right. \tag{8.62}$$

$$A_i = A_{vi} \frac{R_i}{R_L} = \frac{g_m\left(R_\sigma \left\| \frac{1}{g_m}\right.\right)(R_D\|R_L)}{R_L}. \tag{8.63}$$

Finally, if r_d is ignored,

$$R_o = R_D. \tag{8.64}$$

This can be seen from Fig. 8.15 since $V_{gs} = 0$ if there is no input signal. Problem 8.25 suggests finding R_D without neglecting r_d.

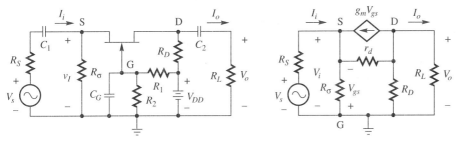

FIGURE 8.14 Common-gate amplifier. (a) Typical circuit. (b) ac model.

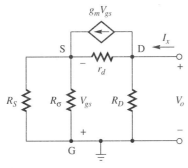

FIGURE 8.15 Model for finding output resistance of the common-gate amplifier.

EXAMPLE 8.8

Consider the CG amplifier shown in Fig. 8.14 constructed with the component values of Example 8.6. Calculate A_{vi}, R_i, A_i, R_o, and A_p.

SOLUTION Since the dc model is the same as that in the previous two examples (Fig. 7.9), the Q-point is unchanged. From Eqs. (8.60) through (8.64), we calculate

$$A_{vi} = 0.8 \text{ mS}(15\|15) \text{ k}\Omega = 6.0, \qquad R_i = \frac{1}{(0.067 + 0.8) \text{ mS}} = 1.15 \text{ k}\Omega,$$

$$A_i = 6.0 \frac{1.15}{15} = 0.46, \qquad R_o = 15 \text{ k}\Omega \| 1.25 \text{ M}\Omega = 14.8 \text{ k}\Omega.$$

Note the low input resistance and hence the low A_i compared to the CS amplifier. Also, $A_p = (6.0)(0.46) = 2.76$.

The Common-Drain Amplifier

■ The common-drain amplifier is often called a *source-follower*, analogous to the BJT emitter-follower topology.

A representative common-drain (CD) amplifier is presented in Fig. 8.16(a), with its ac model shown in Fig. 8.16(b). We find that $V_{gs} = V_i - V_o$. At the source node,

$$g_m V_{gs} = V_o(g_d + G_\sigma + G_L). \tag{8.65}$$

After combining these expressions, we obtain

$$g_m V_i = V_o(g_d + G_\sigma + G_L + g_m). \tag{8.66}$$

Rearranging results in

$$A_{vi} = \frac{V_o}{V_i} = \frac{g_m}{g_d + G_\sigma + G_L + g_m} = g_m \left(r_d \| R_\sigma \| R_L \left\| \frac{1}{g_m} \right. \right), \tag{8.67}$$

which is always less than unity.

It is readily apparent that

$$R_i = R_1 \| R_2 = R_G. \tag{8.68}$$

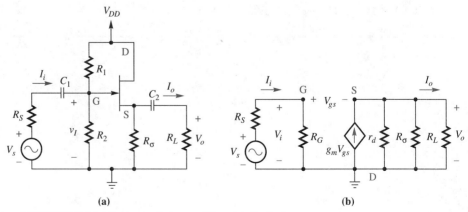

■ **FIGURE 8.16** Common-drain amplifier. (a) Typical circuit. (b) ac model.

CHAPTER 8 SMALL-SIGNAL MIDFREQUENCY AMPLIFIER ANALYSIS

The current amplification is found from the voltage amplification as follows:

$$A_i = \frac{I_o}{I_i} = A_{vi} \frac{R_i}{R_L} = \frac{g_m R_G \left(r_d \| R_\sigma \| R_L \left\| \frac{1}{g_m} \right. \right)}{R_L}. \quad (8.69)$$

Figure 8.17 aids us in finding the output resistance. The nodal equation is

$$I_x = g_m V_{gs} = V_o(g_d + G_\sigma), \quad (8.70)$$

but since $V_{gs} = -V_o$, the result is

$$I_x = V_o(g_d + G_\sigma + g_m), \quad (8.71)$$

$$R_o = \frac{V_o}{I_x} = \frac{1}{g_d + G_\sigma + g_m} = \left(R_\sigma \| r_d \left\| \frac{1}{g_m} \right. \right). \quad (8.72)$$

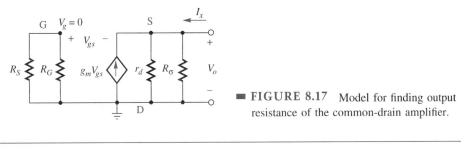

■ **FIGURE 8.17** Model for finding output resistance of the common-drain amplifier.

EXAMPLE 8.9

Using the component values of Example 8.6, draw the CD amplifier in Fig. 8.16. Find $A_{vi}, A_i, R_o,$ and A_p.

SOLUTION The $I_D = 0.4$ mA as in the three preceding examples, but this time $V_{DS} = 9$ V because $R_D = 0$. Equations (8.67) through (8.72) yield

$$A_{vi} = \frac{0.8 \text{ mS}}{(0.0008 + 0.067 + 0.067 + 0.8) \text{ mS}} = 0.86,$$

$$R_i = 1 \text{ M}\Omega \| 150 \text{ k}\Omega = 130 \text{ k}\Omega, \qquad A_i = 0.86 \frac{130 \text{ k}\Omega}{15 \text{ k}\Omega} = 7.4,$$

$$R_o = \frac{1}{(0.0008 + 0.067 + 0.8) \text{ mS}} = 1.15 \text{ k}\Omega.$$

Observe the low voltage gain and low output resistance of the common-drain amplifier. $A_p = (0.86)(7.4) = 6.4$.

CHECK UP

1. **TRUE OR FALSE?** The CG FET amplifier will have a low input impedance.
2. **TRUE OR FALSE?** The CG FET amplifier may have a moderate current gain A_i.
3. **TRUE OR FALSE?** The CD FET could serve to match a high-resistance voltage source to a low-resistance load.

FET Amplifier Comparisons

You should note that the characteristics of common-source, common-gate, and common-drain FET amplifiers are analogous to those of the CE, CB, and CC BJT amplifiers. CS amplifiers have moderate values of A_{vi}, R_i, A_i, A_p, and R_o, although when R_σ is not bypassed, A_{vi} and A_i are reduced. CG amplifiers have low input resistance and low current gain. CD (source-follower) amplifiers have low output resistance and low voltage gain. The findings in this section are tabulated in Table 8.2.

TABLE 8.2 Comparisons of FET Amplifiers

	Common Source (R_σ bypassed) (Fig. 8.11)	Common Source with R_σ (Fig. 8.13)	Common Gate (Fig. 8.14)	Common Drain (Fig. 8.16)
A_{vi}	$-g_m(R_D\|R_L\|r_d)$	$-\dfrac{R_D\|R_L}{R_\sigma}$	$g_m(R_D\|R_L\|r_d)$	$g_m\left(r_d\|R_\sigma\|R_L\|\dfrac{1}{g_m}\right)$
R_i	R_G	R_G	$R_\sigma\|\dfrac{1}{g_m}$	R_G
A_i	$\dfrac{-g_m R_G(R_D\|R_L\|r_d)}{R_L}$	$\dfrac{-g_m R_G(R_D\|R_L\|r_d)}{(1+g_m R_\sigma)R_L}$	$\dfrac{g_m\left(R_\sigma\|\dfrac{1}{g_m}\right)(R_D\|R_L)}{R_L}$	$\dfrac{g_m R_G\left(r_d\|R_\sigma\|R_L\|\dfrac{1}{g_m}\right)}{R_L}$
R_o	$r_d\|R_D$	R_D	R_D	$\left(R_\sigma\|r_d\|\dfrac{1}{g_m}\right)$

Table 8.3 is a summary of the preceding examples.

■ Examples 8.2 through 8.9 provide representative numbers for the amplifications and resistances of the eight circuits discussed. It would be instructive for you to compare them.

TABLE 8.3 Comparison of the amplifiers of Examples 8.2 to 8.9. These values are representative of amplifiers with resistive loads. Circuits with active loads could have much different numbers.

	CE	CE (w/o C_E)	CB	CC	CS	CS (w/o C_σ)	CG	CD
A_{vi}	−63	−1.5	63	.97	−6	−.46	6	0.86
R_i	2.4kΩ	53kΩ	24Ω	48kΩ	130kΩ	130kΩ	1.5kΩ	130kΩ
A_i	−33	−17	0.33	9.9	−52	−4	0.46	7.4
R_o	2.2kΩ	2.2kΩ	2.2kΩ	30Ω	14.8kΩ	15kΩ	14.8kΩ	1.15kΩ
A_p	2060	25	21	9.6	312	1.84	2.76	6.4

8.6 MOSFET AMPLIFIERS WITH ACTIVE LOADS

With integrated circuits, it is very practical to use a MOSFET as an active load in a MOSFET amplifier circuit, because the transistor effects a large resistance without using the area on the substrate that the resistor would require. We have previously considered such circuits in Section 5.4 where digital switching was our objective. Remember that NMOS is the preferred technology, other things being equal, because of the greater mobility of electrons in the *n*-channel. We briefly consider three approaches to this IC amplifier:

1. NMOS, with enhancement-mode driver, and depletion-mode load.
2. NMOS, with both driver and load enhancement-mode devices.
3. NMOS, with enhancement-mode driver and with a PMOS current source load, commonly referred to as CMOS.

These circuits represent simple common-source inverting amplifiers. In Chapter 10 we introduce the differential amplifier, which may also make use of the MOSFET active-load amplifiers.

NMOS Amplifiers with Depletion-Mode Load

The inverter circuit first considered in Fig. 5.17 is redrawn in Fig. 8.18(a) with nonzero r_d for each device, and with both substrates connected to the most negative voltage in the circuit, because there would only be one substrate in a monolithic IC.

Observe in Fig. 8.18(c) that the load, M_2, is always ON, since $v_{GS} = 0$, and the current is nearly constant at 6 mA when in the saturation region ($v_{DS2} > v_{GS2} - V_{t2}$). The characteristic in saturation has a finite slope, because the channel length is decreased when v_{DS2} increases. As has been discussed before, this relationship is described by the constant of fabrication λ, and the small-signal parameter $r_d = (\lambda I_D)^{-1}$ models this effect.

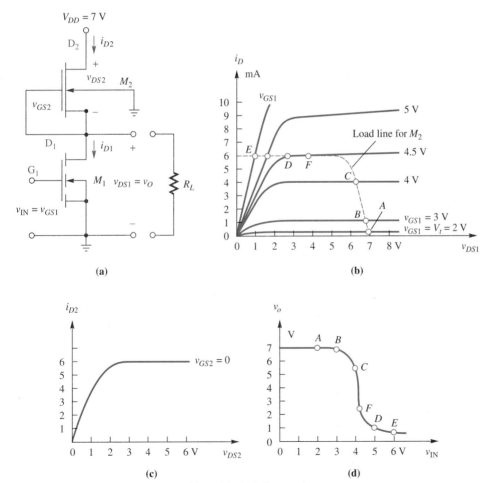

■ **FIGURE 8.18** MOSFET amplifier with depletion-mode load. (a) Circuit. (b) M_1 drain characteristic with load line. (c) M_2 drain characteristic. (d) Transfer characteristic.

Figure 8.18(b) presents the drain characteristic of FET M_1, a set of curves with several values of v_{GS1} plotted. Superimposed on this characteristic is the load line representing the locus of the equation $i_D = V_{DD} - v_{DS2}$. This can be done because both characteristics are functions of the same current $i_{DS1} = i_{DS2}$.

As v_{IN} is increased from 0, the drain currents are zero until the threshold voltage $V_{t1} = 2$ V is reached, marked point A on the M_1 drain characteristics, Fig. 8.18(b), and on the transfer characteristic, Fig. 8.18(d). At this point, and as v_{IN} reaches 3 V (point B), then 4 V (point C), M_1 is in the saturated region, but M_2 is in the ohmic region. Near $v_{IN} = 4.5$ V, both FETs are in the saturated region (point F), and it is in this vicinity that linear amplification is to be attempted. On the transfer characteristic, Fig. 8.18(d), one sees that $\Delta v_O/\Delta v_{IN}$ is greatest at F, where both devices have large output resistances. As we continue to increase v_{IN} to 5 V (point D), and to 6 V (point E), M_1 is now in the ohmic region.

The load line curve in Fig. 8.18(b) has a greater slope than the isolated characteristic of Fig. 8.18(c). This is because the threshold voltage V_t of a MOSFET increases as the source-to-substrate junction becomes reverse-biased. In our circuit, as v_{DS1} increases, the substrate-source voltage v_{BS} becomes more negative. The relationship[1] is:

$$V_t = V_{t0} + \gamma[\sqrt{2\phi_p - v_{BS}} - \sqrt{2\phi_p}], \tag{8.73}$$

where V_{t0} is the threshold voltage at $v_{BS} = 0$. The *body-effect parameter*, γ, is a fabrication constant that ranges from 0.1 to 0.5 $V^{0.5}$, and $2\phi_p$, the *surface-effect potential*, is approximately 0.6 V. You will recall that **VTO** (V_{t0}) and **GAMMA** (γ), as well as **LAMBDA** (λ), are parameters that may be specified in the SPICE model statement for MOSFETs (Section 5.5).

To find the effect of v_{BS} on i_D, we substitute Eq. (8.73) into (5.12), and take the derivative, holding v_{GS} and v_{DS} constant:

$$g_{mb} \equiv \left.\frac{\partial i_D}{\partial v_{BS}}\right|_{v_{gs}=0} = \frac{i_d}{v_{bs}} = \frac{g_m \gamma}{2\sqrt{2\phi_p - v_{BS}}} \equiv \chi g_m. \tag{8.74}$$

It is convenient to introduce the parameter

$$\chi = \frac{\gamma}{2\sqrt{2\phi_p - v_{BS}}} \tag{8.75}$$

from Eq. (8.74), which is the ratio of g_{mb} to g_m. This parameter may range from 0.1 to 0.3.

We have observed by means of the graphs of Fig. 8.18 that the circuit operates with large gain in the region where both FETs are in their saturated regions. We will assume that M_1 is biased with a dc component at v_{IN} to locate the Q-point at F. The small-signal model for this circuit is now drawn in Fig. 8.19(a).

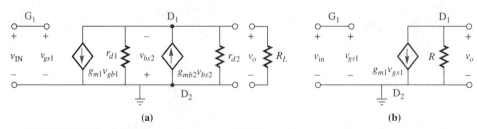

■ **FIGURE 8.19** Small-signal models for NMOS amplifier with depletion load. (a) Model. (b) Reduced model with $R = r_{d1} \| r_{d2} \| 1/g_{mb2} \| R_L$.

We can easily model M_1 with the controlled source $g_{m1}v_{gs1}$ and r_{d1}. The model for M_2 does not require the controlled source $g_{m2}v_{gs2}$, since $v_{gs2} = 0$, but it does have its own drain resistance, r_{d2}. Because the substrate of M_2 is not connected to its source, but to signal ground, i_{d2} is modulated by the substrate-to-source voltage v_{bs2}. The model shows a controlled source $g_{mb2}v_{bs2}$. The voltage v_{bs2} is from substrate (ground) to the source of M_2 (v_o). Since the current source is controlled by the voltage across it, the branch can be replaced by the conductance g_{mb2}. Figure 8.19(b) shows the three resistances combined with whatever external load R_L is connected. Then,

$$v_o = -g_{m1}v_{gs1}\left(r_{d1}\|r_{d2}\left\|\frac{1}{g_{mb2}}\right\|R_L\right). \tag{8.76}$$

Since $v_{gs1} = v_{in}$, we can find the voltage gain of the stage

$$A_{vi} = \frac{v_o}{v_{in}} = -g_{m1}\left(r_{d1}\|r_{d2}\left\|\frac{1}{g_{mb2}}\right\|R_L\right). \tag{8.77}$$

If $g_{mb2} \gg g_{d1} + g_{d2} + G_L$, as is often the case, then we have the approximate gain

$$A_{vi} \cong -\frac{g_{m1}}{g_{mb2}}. \tag{8.78}$$

It is important to note how this voltage gain is controlled by the IC designer. From Eqs. (5.12) and (5.14) we find that g_m is proportional to $\sqrt{k}$:

$$g_m = 2\sqrt{k}\sqrt{I_D}. \tag{8.79}$$

Then, because $g_{mb} = \chi g_m$ and I_D is the same in both FETs,

$$A_{vi} = -\frac{g_{m1}}{g_{mb2}} = -\frac{g_{m1}}{\chi g_{m2}} = -\frac{1}{\chi}\sqrt{\frac{k_1}{k_2}} = -\frac{1}{\chi}\sqrt{\frac{(W/L)_1}{(W/L)_2}}. \tag{8.80}$$

This demonstrates that the designer can control the voltage gain (within limits of a given technology) by choice of W/L ratios for the devices. One disadvantage of the NMOS technology that requires depletion-mode devices is an extra fabrication step to provide the n-type channel for the depletion FETs, as discussed in Chapter 18.

EXAMPLE 8.10

Consider an amplifier of the type shown in Fig. 8.18(a). Let $V_{DD} = 6$ V, and assume M_1 is biased $v_O = 3$ V so that both FETs are in saturation. Load $R_L = 1$ MΩ. Let $V_{t01} = 1.5$ V, $V_{t02} = -2$ V, $k_1 = 1$ mA/V^2, $k_2 = 0.2$ mA/V^2, $\lambda_1 = \lambda_2 = 0.01$ V^{-1}, $\gamma_1 = \gamma_2 = 0.3$ V$^{0.5}$, and $2\phi_p = 0.6$ V. Calculate A_{vi} for this amplifier.

SOLUTION For a first approximation to the drain current, we neglect λ. Then,

$$v_{T2} = v_{TO2} + \gamma_2[\sqrt{2\phi_p - v_{BS}} - \sqrt{2\phi_p}]$$
$$= -2 + 0.3[\sqrt{0.6 - (-3)} - \sqrt{0.6}] = -1.66 \text{ V}.$$
$$I_D = k_2(v_{GS2} - V_{t2})^2 = 0.2 \text{ mA/V}^2(0 - 1.66 \text{ V})^2 = 0.55 \text{ mA}.$$
$$I_D = k_1(v_{GS1} - V_{t1})^2 = 1 \text{ mA/V}^2(V_{GS1} - 1.5 \text{ V})^2 = 0.55 \text{ mA}.$$

This yields $V_{GS1} = 2.24$ V. The gate of M_1 will need a dc bias of 2.24 V to obtain the proper Q-point.

$$g_{m1} = 2k(V_{GS1} - V_{t1}) = 2(1 \text{ mA/V}^2)(2.24 - 1.5) \text{ V} = 1.48 \text{ mS}.$$

$$g_{m2} = 2k(V_{GS2}) = 2(0.2 \text{ mA/V}^2)(0 - 1.66) \text{ V} = 0.66 \text{ mS}.$$

$$\chi_2 = \frac{\gamma_2}{2\sqrt{2\phi_p - v_{BS}}} = \frac{0.3 \text{ V}^{0.5}}{2\sqrt{0.6 \text{ V} - (-3 \text{ V})}} = 0.08.$$

$$g_{mb2} = \chi g_{m2} = 0.08(0.66 \text{ mS}) = 0.053 \text{ mS}.$$

$$r_{d1} = r_{d2} = 1/(\lambda I_D) = 1/(0.01 \text{ V}^{-1})(0.55 \text{ mS}) = 182 \text{ k}\Omega.$$

$$A_{vi} = -g_{m1}\left(r_{d1} \| r_{d2} \left\| \frac{1}{g_{mb2}} \right\| R_L\right)$$

$$= -1.48 \text{ mS}\left(182 \text{ k}\Omega \| 182 \text{ k}\Omega \left\| \frac{1}{0.053 \text{ mS}} \right\| 1 \text{ M}\Omega\right) = -22.8.$$

DRILL 8.8

Estimate the range of v_o that can be achieved by the amplifier of Example 8.10 without leaving the region where both FETs are saturated.

ANSWER M_1 requires $v_o > (2.24 - 1.5)$ V, and M_2 requires $v_o < (6 - 1.66)$ V. The range is 3.6 V.

NMOS Amplifiers with Enhancement-Mode Load

An NMOS amplifier circuit that utilizes an enhancement-mode load transistor is drawn in Fig. 8.20(a). The load is connected with the gate tied to the drain. The drain characteristic of such a device is plotted in Fig. 8.20(c). Note that this FET is always in the saturated region because $v_{DS2} = v_{GS2}$.

Figure 8.20(b) presents the drain characteristic of FET M_1, a family with several values of v_{GS1} plotted. Superimposed on this characteristic is the load line representing the locus of the equation $V_{DS1} = V_{DD} - v_{DS2}$. This can be easily done because both characteristics are functions of the same current $i_{DS1} = i_{DS2}$. Note that this load line represents a much lower resistance (greater slope) than that of the load in Fig. 8.18(b). We thus expect a lower voltage gain in this circuit arrangement. The transfer characteristic is then plotted in Fig. 8.20(d).

The small-signal model for this amplifier circuit is shown in Fig. 8.21(a), where M_1 is modeled with the controlled source $g_{m1}v_{gs1}$ and r_{d1}. The model for M_2 has the controlled source $g_{m2}v_{gs2}$, its drain resistance, r_{d2}, and a controlled source $g_{mb2}v_{bs2}$ as discussed with the depletion-mode circuit. The voltages v_{bs2} and v_{gs2} are both equal to $-v_o$. Since the current sources $g_{m2}v_{gs2}$ and $g_{mb2}v_{bs2}$ are both controlled by the voltage across them, their branches may be replaced by the conductances g_{m2} and g_{mb2}, respectively. Figure 8.21(b) shows the four resistances, plus the resistance of any external load, combined. Then,

$$v_o = -g_{m1}v_{gs1}\left(r_{d1} \| r_{d2} \left\| \frac{1}{g_{m2}} \right\| \frac{1}{g_{mb2}} \| R_L\right). \tag{8.81}$$

Since $v_{gs1} = v_{in}$, we may find the voltage gain of the stage

$$A_{vi} = \frac{v_o}{v_{in}} = -g_{m1}\left(r_{d1} \| r_{d2} \left\| \frac{1}{g_{m2}} \right\| \frac{1}{g_{mb2}} \| R_L\right). \tag{8.82}$$

If $(g_{m2} + g_{mb2}) \gg (g_{d1} + g_{d2} + G_L)$, as is often the case, then we have the approximate gain

$$A_{vi} \cong -\frac{g_{m1}}{g_{m2} + g_{mb2}} = -\frac{g_{m1}}{g_{m2}(1 + \chi)} = -\frac{1}{1 + \chi}\sqrt{\frac{(W/L)_1}{(W/L)_2}}. \tag{8.83}$$

CHAPTER 8   SMALL-SIGNAL MIDFREQUENCY AMPLIFIER ANALYSIS   363

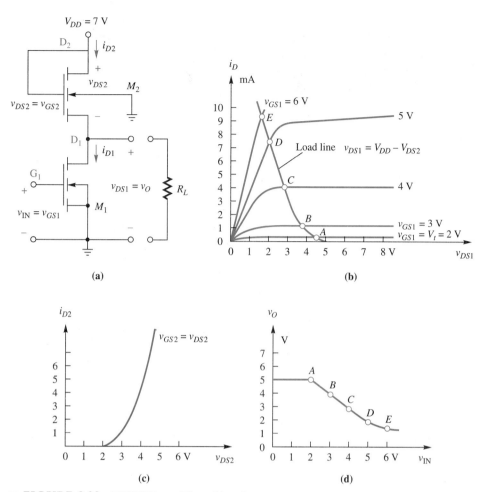

■ **FIGURE 8.20** MOSFET amplifier with enhancement-mode load. (a) Circuit. (b) M_1 drain characteristic with load line. (c) M_2 drain characteristic. (d) Transfer characteristic.

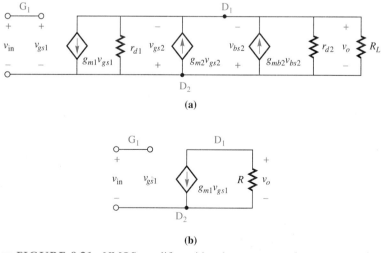

■ **FIGURE 8.21** NMOS amplifier with enhancement-mode load. (a) Small-signal model. (b) Simplified small-signal model with $R = r_{d1} \| r_{d2} \| 1/g_{m2} \| 1/g_{mb2} \| R_L$.

- Of course, the manufacturer can compensate for the lower amplification by increasing the W/L ratio in the enhancement mode case. Specific design is accomplished by varying W/L.

- Usually L is a minimum technology-driven feature size. $L \sim 0.5\ \mu m$ is now becoming common.

If we compare this result with the A_{vi} of the depletion-load amplifier [Eq. (8.80)], we find that the amplifier with the depletion load has a voltage gain larger by a factor of $(1 + \chi)/\chi$, or 5 to 10 times. So, the simpler technology of the enhancement-load amplifier comes at the expense of voltage gain (for the same geometry). Again, this voltage gain is controlled by the IC designer through the choice of W/L ratios for the devices within a given technology.

EXAMPLE 8.11

Assume an NMOS enhancement-load amplifier such as that shown in Fig. 8.20(a). Let $V_{DD} = 6$ V, $V_{t01} = V_{t02} = 1.5$ V, $\gamma_1 = \gamma_2 = 0.3$ V$^{0.5}$, $\mu_n C_{ox}/2 = 100\ \mu A/V^2$, $R_L = 1$ MΩ, $W_1 = 100\ \mu m$, $L_1 = 5\ \mu m$, $W_2 = 2.5\ \mu m$, $L_2 = 5\ \mu m$, and $\lambda_1 = \lambda_2 = 0.01$ V^{-1}. Find A_{vi} for this amplifier, assuming proper dc bias for a Q-point of $V_O = 3$ V.

SOLUTION

$$\chi_2 = \frac{\gamma_2}{2\sqrt{2\phi_p - v_{BS}}} = \frac{0.3\ V^{0.5}}{2\sqrt{0.6\ V - (-3\ V)}} = 0.08.$$

$$A_{vi} \cong -\frac{1}{1+\chi}\sqrt{\frac{(W/L)_1}{(W/L)_2}} = -\frac{1}{1+0.08}\sqrt{\frac{100/5}{2.5/5}} = -5.86$$

if the r_d terms and R_L can be neglected. To find out, we must find I_D.

$$k_2 = \mu C_{ox} W_2/2L_2 = (100\ \mu A/V^2)(2.5\ \mu m/5\ \mu m) = 50\ \mu A/V^2.$$
$$V_{GS2} = V_{DD} - V_O = 6 - 3 = 3\ V.$$
$$V_{t2} = V_{t02} + \gamma_2[\sqrt{2\phi_p - v_{BS}} - \sqrt{2\phi_p}]$$
$$= 1.5 + 0.3\ V^{0.5}[\sqrt{0.6\ V - (-3)\ V} - \sqrt{0.6\ V}] = 1.837\ V.$$
$$I_D = k_2(V_{GS2} - V_{t2})^2 = 50\ \mu A/V^2(3\ V - 1.837\ V)^2 = 68\ \mu A.$$
$$r_d = 1/\lambda I_D = 100\ V/68\ \mu A = 1.48\ M\Omega. \quad \text{Thus,}\ g_{d1} = g_{d2} = 0.68\ \mu S.$$
$$g_{m2} = 2k_2(V_{GS2} - V_{t2}) = 2(50\ \mu A/V^2)(3\ V - 1.837\ V) = 116\ \mu S.$$
$$g_{mb2} = \chi g_{m2} = 0.08(134\ \mu S) = 9.3\ \mu S.$$

Finally, it is apparent that $(g_{d1} + g_{d2} + G_D) << (g_{m2} + g_{mb2})$, and could properly be neglected. Equation (8.82) would have yielded $A_{vi} = -5.8$.

DRILL 8.9

Using the work in Example 8.11, find the value of dc voltage at v_{IN} required to bias the amplifier as specified in the example. Show that M_1 is safely in its saturated region.

ANSWER

$V_{GS1} = V_{IN} = 1.7$ V.
$V_O = V_{DS1} = 3$ V.
OK!

CMOS Amplifiers

CMOS technology makes available to the circuit designer PMOS amd NMOS devices. Consider the amplifier circuit shown in Fig. 8.22. Here we have an NMOS driver with a PMOS current mirror (Section 7.5) for a load. The advantage of this arrangement is that the load transistor offers the high output resistance of the current source, $R_o = r_{d2}$, without the additional loading caused by the body effect of M_2, since the substrate of M_2 is connected to its source. This was a serious loading problem in the NMOS loads we discussed earlier. The reference current, I_{REF}, may be provided by a resistance or another MOSFET.

The small-signal model for this circuit, drawn in Fig. 8.23, is straightforward to analyze. The model assumes that M_1 will have a dc bias at its gate such that it will operate in the

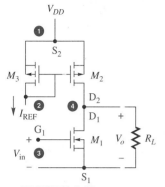

FIGURE 8.22 CMOS amplifier circuit.

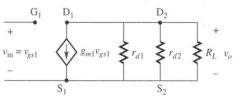

FIGURE 8.23 Small-signal model for CMOS amplifier.

saturated region at the current established by the load circuit. The ac output voltage is $v_o = -g_{m1}(r_{d1}\|r_{d2}\|R_L)$, which leads directly to

$$A_{vi} = \frac{v_o}{v_{in}} = -g_{m1}(r_{d1}\|r_{d2}\|R_L). \tag{8.84}$$

Because the output resistance of the amplifier is large, $R_o = r_{d1}\|r_{d2}$, it is important to note that R_L could load this circuit greatly. If R_L represents the input resistance of another MOS amplifier, it will be very large, but if it is the input resistance of a BJT amplifier, or a physical resistance, then it may be an important consideration. This expression predicts large voltage gains for this circuit, at the expense of the slightly more complex, and less dense, technology of CMOS as compared to NMOS. For this reason, and others, CMOS technology is more widely used than NMOS in linear circuits.

Because g_m varies as the $\sqrt{I_D}$ [Eq. (8.79)], and r_d varies as $(\lambda I_D)^{-1}$, the voltage gain will vary approximately as $(I_D)^{-0.5}$. Thus the circuit designer may determine the value of A_{vi} through the choice of the Q-point current.

■ Indeed, CMOS has become the technology of choice in both linear and digital ICs, as well as in mixed-mode ICs.

EXAMPLE 8.12

Given the CMOS amplifier circuit of Fig. 8.22, with $V_{DD} = 8$ V, $I_{REF} = 0.2$ mA, $R_L = 1$ MΩ, $k_n = 0.5$ mA/V², $|k_p| = 0.25$ mA/V², $\lambda_n = \lambda_p = 0.01$ V^{-1}, $V_{tn} = 1$ V, and $V_{tp} = -1$ V. Find the voltage amplification, A_{vi}, and the maximum range of output voltage for linear operation. As a check, use SPICE to plot v_O versus v_{IN}.

SOLUTION

$g_{m1} = 2\sqrt{k_n}\sqrt{I_D} = 2\sqrt{0.5 \text{ mA/V}^2}\sqrt{0.2 \text{ mA}} = 0.63$ mS.

$r_{d1} = r_{d2} = (\lambda I_D)^{-1} = (0.01 \times 0.2 \text{ mA})^{-1} = 0.5$ MΩ.

$A_{vi} = -g_{m1}(r_{d1}\|r_{d2}\|R_L) = -(0.63 \text{ mS})(0.5 \text{ MΩ}\|0.5 \text{ MΩ}\|1 \text{ MΩ}) = -126$.

Neglecting the effect of λ,

$$|I_D| = |k_p|(V_{GSp} - V_{tp})^2 \Rightarrow |V_{GSp} - V_{tp}| = \sqrt{0.2/0.25} = 0.89 \text{ V}.$$

Thus, to maintain M_2 in the saturation region, $|v_{DSp}| > 0.89$ V, and $v_O < V_D - 0.89 = 7.11$ V. Similarly,

$$I_D = k_n(V_{GSn} - V_{tn})^2 \Rightarrow (V_{GSn} - 1) = \sqrt{0.2/0.5} = 0.63 \text{ V}.$$

Thus, to maintain M_1 in its saturation region, $v_{DS1} = v_O > 0.63$ V. The range is from 0.63 to 7.11 V.

DRILL 8.10

Using the information given in Example 8.12, find a value of resistance such that when connected to ground from the current reference terminal in Fig. 8.22, the proper current will be provided. Suppose the designer wishes to use an NMOS FET, diode-connected (gate is tied to the drain), to replace this resistor. Let $V_{tn} = 1$ V and $\mu_n C_{ox}/2 = 50$ μA/V². Find W/L.

ANSWER
$R_{REF} = 30.5$ kΩ, $W/L = 0.15$

A listing of a SPICE program to simulate this circuit is shown:

```
EX8.12
VDD 1 0 DC 8
I1 2 0 DC 0.2M
M1 4 3 0 0 M1
M2 4 2 1 1 M2
M3 2 2 1 1 M2
RL 4 0 1E6
.MODEL M1 NMOS KP=1M VTO=1 LAMBDA=0.01 GAMMA=0.2
.MODEL M2 PMOS KP=0.5M VTO=-1 LAMBDA=0.01 GAMMA=0.2
VIN 3 0 DC
.DC VIN 1.5 2 0.01
.PLOT DC V(4)
.END
```

Figure 8.24 shows the plot of v_O versus v_{IN} for Example 8.12.

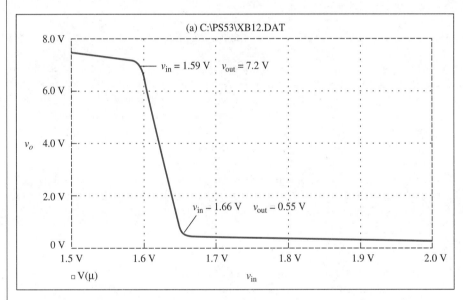

■ **FIGURE 8.24** Plot of v_O versus v_{IN} for Example 8.12.

CHECK UP

1. **TRUE OR FALSE?** MOS amplifiers use active loads because they provide a higher load resistance than it is practical to achieve with diffused resistors in an IC.
2. **TRUE OR FALSE?** The A_{vi} of a MOS amplifier with an active load is proportional to the W/L ratio of the amplifier FET divided by the W/L ratio of the load FET.
3. **TRUE OR FALSE?** The effect of the substrate-source voltage v_{bs} on the i_D of a MOSFET is about the same magnitude as the effect of the gate-source voltage v_{gs}.
4. **TRUE OR FALSE?** An NMOS amplifier has greater A_{vi} when used with an enhancement-mode load than with a depletion-mode load for equal dimensions.

CHAPTER 8 SMALL-SIGNAL MIDFREQUENCY AMPLIFIER ANALYSIS

5. **TRUE OR FALSE?** A CMOS amplifier, using an NMOS driver and a PMOS current mirror load, has high A_{vi} because there is no body effect on the load FET.

8.7 DISCRETE CASCADED AMPLIFIER STAGES

Amplifiers are often cascaded; that is, the output of one transistor amplifier is connected as the input to another. This is done to achieve more amplification than is possible in a single stage or to provide an impedance match by means of an active device. Discrete amplifiers are used as examples in this section and are treated first because the capacitance-coupled stages are independent as far as dc bias is concerned. Then examples of modern direct-coupled IC amplifiers are presented in the next section.

Voltage gain is normally provided by one or more CE or CS amplifiers. Often it is desirable to match the amplifier input resistance to the resistance of the signal source. The overall amplifier circuit may be given a large input resistance by the use of a CC or CD first stage, or it may be given a low input resistance by the use of a CB or CG input stage. The output resistance of the cascade can be made low by the use of a CC or CD output stage.

Consider the block diagram of a cascaded amplifier depicted in Fig. 8.25. Since the input resistance of the second stage serves as the output resistance of the first stage, amplification of the cascaded stages may be found as the product of the amplifications of the individual stages:

$$A_{vi} = \frac{V_{o2}}{V_{i1}} = \frac{V_{o2}}{V_{i2}} \times \frac{V_{i2}}{V_{i1}} = A_{vi1} \times A_{vi2}, \quad (8.85)$$

$$A_i = \frac{I_{o2}}{I_{i1}} = \frac{I_{o2}}{I_{i2}} \times \frac{I_{i2}}{I_{i1}} = A_{i1} \times A_{i2}. \quad (8.86)$$

Note that the overall gain calculations have been expedited by carefully defining the individual stage gains such that the output voltage and current of one stage are the input voltage and current of the next stage. Analysis of the cascade is illustrated by means of an example.

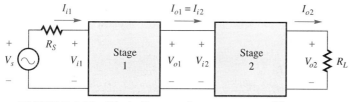

FIGURE 8.25 Block diagram of a two-stage amplifier.

EXAMPLE 8.13

The amplifier cascade shown in Fig. 8.26(a) uses a CD input FET, J_1. This presents a high input resistance to the 10-kΩ source. High voltage gain is obtained from the CE amplifier, Q_2. The CC output transistor, Q_3, produces a low output resistance to drive the 100-Ω load. The FET has a drain characteristic $i_D = 10(1 + 0.2v_{GS})^2$ mA, and $r_d = 100$ kΩ. Each BJT has $r_o = 139$ kΩ, $r_b = 100$ Ω, $r_\pi = 2600$ Ω, and $g_m = 0.039$ S at the given Q-point. Neglect coupling and bypass capacitor reactances. Find A_{vi}, A_i, and A_p for each stage and for the overall amplifier.

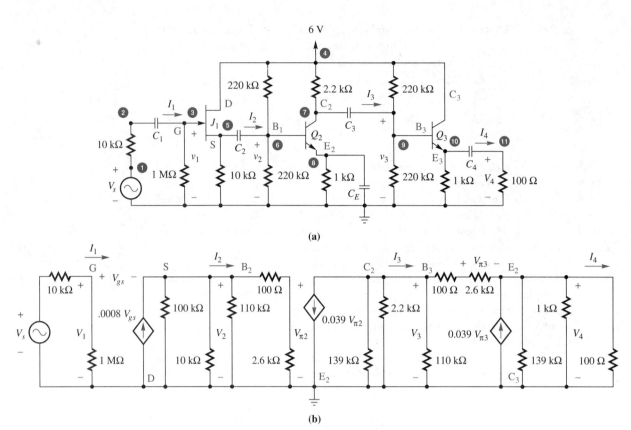

■ **FIGURE 8.26** Three-stage amplifier of Example 8.13. (a) Circuit. (b) ac model.

SOLUTION The Q-point for J_1 is found from Eq. (7.21) to be $I_D = 0.4$ mA and $V_{GS} = -4$ V. Then, from Eq. (7.19), $g_m = 0.8$ mS. The ac model can now be constructed as shown in Fig. 8.26(b). Consider the output stage first:

$$A_{vi3} = \frac{V_4}{V_3} = \frac{100}{(0.01 + 0.001)2700 + 100} = 0.77 \quad \text{from Eq. (8.42)}.$$

$$R_{i3} = \frac{V_3}{I_3} = 220\text{ k}\|220\text{ k}\|[2700 + 100(1\text{ k}\|100)] = 10.6\text{ k}\Omega \quad \text{from Eq. (8.44)}.$$

$$A_{i3} = \frac{I_4}{I_3} = 0.77\frac{10.6}{0.1} = 82 \quad \text{from Eq. (8.45)}.$$

$$A_{p3} = A_{vi3}A_{i3} = (0.77)(82) = 63.$$

We can now analyze the second stage, recognizing that the input resistance of the third stage is the load resistance of the second stage, $R_{L2} = R_{i3} = 10.6$ kΩ.

$$A_{vi2} = \frac{V_3}{V_2} = -0.039 \text{ mS } (2.2\|10.6)\text{ k}\Omega = -71 \quad \text{from Eq. (8.10)}.$$

$$R_{i2} = \frac{V_2}{I_2} = 220\text{ k}\Omega\|220\text{ k}\Omega\|2700\text{ }\Omega = 2635\text{ }\Omega \quad \text{from Eq. (8.16)}.$$

$$A_{i2} = \frac{I_3}{I_2} = -71\frac{2.63}{10.6} = -18 \quad \text{from Eq. (8.17)}.$$

$$A_{p2} = A_{vi2}A_{i2} = (-71)(-18) = 1278.$$

CHAPTER 8 SMALL-SIGNAL MIDFREQUENCY AMPLIFIER ANALYSIS

We have worked our way systematically back to the first stage, having found the input resistance of the second stage, which is the load resistance for stage one.

$$R_{L1} = R_{i2} = 2635 \; \Omega.$$

$$A_{vi1} = \frac{V_2}{V_1} = 0.8 \text{ mS } (100 \text{ k}\Omega \| 10 \text{ k}\Omega \| 2.63 \text{ k}\Omega \| 1/0.8 \text{ mS}) = 0.62 \quad \text{from Eq. (8.67).}$$

$$R_{i1} = \frac{V_1}{I_1} = R_G = 1 \text{ M}\Omega.$$

$$A_{i1} = \frac{I_2}{I_1} = 0.62 \frac{1 \text{ M}\Omega}{2635 \; \Omega} = 235 \quad \text{from Eq. (8.69).}$$

$$A_{p1} = A_{vi1} A_{i1} = (0.62)(235) = 146.$$

The overall cascaded amplifier characteristics may now be found:

$$A_{vi} = A_{vi1} A_{vi2} A_{vi3} = (0.62)(-71)(0.77) = -34.$$
$$A_i = A_{i1} A_{i2} A_{i3} = (235)(-18)(82) = -3.4 \times 10^5.$$
$$A_p = A_{p1} A_{p2} A_{p3} = (63)(1287)(146) = 1.2 \times 10^7.$$
$$A_{vs} = \frac{V_4}{V_s} = \frac{V_1}{V_s} A_{vi} = \frac{R_{i1}}{R_s + R_{i1}} A_{vi} = \frac{1 \text{ M}\Omega}{1 \text{ M}\Omega + 10 \text{ k}\Omega}(-34) = -33.7.$$

■ Example 8.13 points out the advantage of breaking up a large circuit into smaller circuits which are easier to analyze, and which allow better insight as to which components are most critical.

We observe in the preceding example that the voltage gain is obtained in the intermediate stage. The first stage serves to drive the amplifier without unduly loading the source, and the last stage drives the relatively low resistance load without excessive loading of the voltage amplifier. Thus the input and output stages provide power gain to the cascade by impedance matching. This is fairly typical in amplifier design. In Chapter 9 we examine the frequency response of cascaded amplifiers and the importance of impedance matching.

EXAMPLE 8.14

Use SPICE to find the voltage amplification of each stage of the amplifier shown in Fig. 8.26 for 1 kHz. Use the device parameters $\beta_F = 100$ and $V_A = 139$ V for the BJTs and $I_{DSS} = 10$ mA, $V_P = -5$ V, and $\lambda = 0.025$ for the JFET. Let $v_s = 1$ mV. Use large values for all capacitors (1000 μF), so that capacitive reactance is not a factor.

■ For ac analysis, SPICE assumes a linear model. Therefore, the selection of $|V_s|$ is arbitrary. A. TRAN (transient analysis) does not assume a linear model.

SOLUTION We will find the voltage magnitude at nodes 3, 6, 9, and 11. Then the stage gains may be found by division.

```
THREE-STAGE AMPLIFIER
V10 1 0 AC .001
V40 4 0 DC 6
Q2 7 6 8 Q1
RS 1 2 10K
R1 3 0 1E6
R2 5 0 10K
R3 4 7 2.2K
R4 8 0 1K
R5 4 6 220K
R6 6 0 220K
R7 4 9 220K
```

```
R8  9  0  220K
R9  10 0  1K
R10 11 0  100
C1  2  3  1E-3
C2  5  6  1E-3
C3  7  9  1E-3
CE  8  0  1E-3
C4  10 11 1E-3
Q3  4  9  10 Q1
J1  4  3  5  J1
.MODEL Q1 NPN BF=100 RB=100 VAF=139
.MODEL J1 NJF VTO=-5 BETA=.4M LAMBDA=0.025.
.AC DEC 1 1K 1K
.PRINT AC V(3) V(6) V(9) V(11)
.OP
.END
```

The program will find the device operating points, small-signal models, and the following node voltages:

$V(3) = 0.99$ mV, $V(6) = 0.61$ mV, $V(9) = 43.5$ mV, and $V(11) = 34.2$ mV.

Thus, $A_{vi1} = 0.62$, $A_{vi2} = -70$, $A_{vi3} = 0.79$, and $A_{vs} = -34.2$. These numbers compare very well with the results of Example 8.13.

CHECK UP

1. **TRUE OR FALSE?** In cascaded capacitance-coupled amplifiers, we must begin analysis at the first stage, and proceed stage by stage, multiplying the A_vs together.
2. **TRUE OR FALSE?** In cascaded capacitance-coupled amplifiers, the input resistance of the second stage is treated as the load resistance of the first stage.
3. **TRUE OR FALSE?** The voltage gains of each stage in cascaded amplifiers are independent.

8.8 DIRECT-COUPLED AND BiCMOS AMPLIFIERS

Capacitive coupling between amplifier stages allows for independence in bias design of individual stages and is a good pedagogical approach to the subject of discrete multistage amplifiers. But capacitor coupling is not appropriate for monolithic IC design. This is true because fabrication of the relatively large capacitors required for good low-frequency operation requires large areas on the substrate, and is not economically feasible.

Similarly, resistive biasing techniques, which often require large resistances, are not favored in IC design for the same reason. At the same time, biasing schemes using active devices in current sources, such as we have studied in Chapter 7, have become the choice of circuit designers. Innovative use of active devices for dc level shifting between amplifier stages not only saves space on the die, but also allows the low-frequency response to extend down to dc.

Direct-Coupled Transistor Circuits

We first look at several examples of *hybrid* amplifier circuits, in which the active devices are direct coupled and monolithic in construction, but there are some large resistors and/or capacitors external to the IC.

The amplifier drawn in Fig. 8.27 consists of an emitter follower directly coupled to the base of a second BJT. This circuit has very high input resistance, limited mostly by the bias resistors. Problem 8.46 suggests an analysis of the circuit to demonstrate this statement. See also the *Darlington* amplifier of Problem 8.14.

The circuit of Fig. 8.28 illustrates a common-emitter amplifier directly coupled to an emitter follower. The properties of this arrangement are high input impedance, low output impedance, and good voltage gain. Problem 8.47 investigates this circuit.

Figure 8.29 presents a simplified version of a circuit known as a *cascode* amplifier. The common-emitter input stage is direct coupled to a common-base output stage. This circuit has considerable advantage at high frequencies, as we shall see in Chapter 9. Refer to Problem 8.48, which provides an excellent opportunity to apply your modeling skills to this topology.

■ A *Darlington* BJT is a two-transistor array connected as shown in Fig. 8.39, with the emitter of Q_1 tied to the base of Q_2 and the two collectors tied together. The current gain of this composite arrangement is $\beta = \beta_1 \beta_2$, and it may be realized with two discrete BJTs or one 3-terminal monolithic structure.

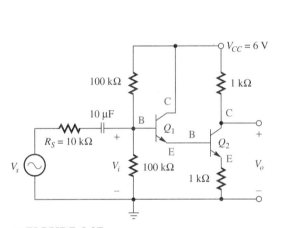

■ FIGURE 8.27

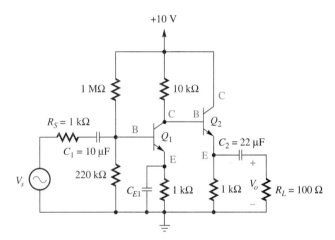

■ FIGURE 8.28

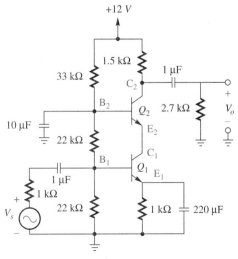

■ FIGURE 8.29 BJT cascode amplifier.

■ In the circuits of Figs. 8.29 and 8.30, very high voltage gain is obtained by the common-base BJT. Since the CB amplifier has a very low input resistance, it is buffered from the input terminals by a CE amplifier in Fig. 8.29, and by a CS amplifier in Fig. 8.30. Both arrangements have been given the name *cascode* amplifier.

BiMOS Integrated Circuits

Semiconductor technology has progressed to the point where excellent bipolar devices and MOSFET devices can be fabricated in the same monolithic structure. Such integrated circuits are referred to as **BiMOS**, or **BiCMOS**. Bipolar devices possess a much higher mutual conductance than FETs at a given current, and BJTs also have better performance at high frequencies. On the other hand, FETs offer the distinct advantage of extremely high input resistance at the gate, and lower voltage drop ($V_{DS} < V_{CE}$) when used as switches. Thus the combination of the two technologies offers the circuit designer an opportunity to optimize a design depending on the application.

You are now prepared to draw dc models, find small-signal device parameters, draw small-signal models, and then analyze the circuit for both technologies. As an example, consider the BiCMOS cascode circuit, as discussed in Example 8.15.

EXAMPLE 8.15

Refer to the BiCMOS cascode circuit of Fig. 8.30. Suppose that Q_1 has $\beta_F = 100$ and $V_A = 150$ V; M_1 has $V_t = -1$ V, $k = 50$ μA/V^2, and $\lambda = 0.02$ V^{-1}; and that M_2 is similar to M_1 except $V_t = +1$ V.

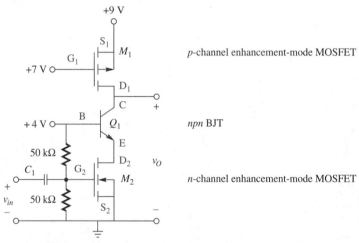

■ **FIGURE 8.30** BiCMOS cascode amplifier.

(a) Solve for the dc operating current (from M_1). Assume that the gate bias voltage has been chosen such that the FETs are operating in the pinch-off region, and the BJT is in the active region.
(b) Determine the small-signal parameters for the devices.
(c) Find A_{vim}.

SOLUTION

(a) $$I_{D1} = 50 \ \mu\text{A/V}^2 \ [-1 \ \text{V} - (-2 \ \text{V})]^2 = 50 \ \mu\text{A}.$$

(b) $$r_{d1} = r_{d2} = 1/\lambda I_D = 1/(0.02 \ \text{V}^{-1})(50 \ \mu\text{A/V}^2) = 1 \ \text{M}\Omega.$$
$$g_{mM1} = g_{mM2} = 2k(V_{GS} - V_t) = 100 \ \mu\text{A}(2 - 1) \ \text{V} = 100 \ \mu\text{S}.$$
$$g_{mQ1} = I_C/V_T = 50 \ \mu\text{A}/26 \ \text{mV} = 1.92 \ \text{mS}.$$
$$r_\pi = \beta/g_{mQ1} = 100/1.92 \ \text{mS} = 52 \ \text{k}\Omega.$$
$$r_o = V_A/I_C = 150 \ \text{V}/50 \ \mu\text{A} = 3 \ \text{M}\Omega.$$

(c) The small-signal model is drawn in Fig. 8.31. The nodal equations are

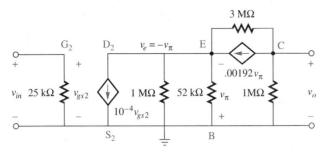

FIGURE 8.31 Small-signal model for BiCMOS cascode amplifier of Example 8.15.

$v_o(1\ \mu S + 0.33\ \mu S) = v_e(0.33\ \mu S + 1.92\ mS).$

$v_e(1\ \mu S + 0.33\ \mu S + 19\ \mu S + 1.92\ mS) - v_o(0.33\ \mu S) + v_{in}(100\ \mu S) = 0.$

$A_{vim} = (v_o/v_e)(v_e/v_{in}) = (1440)(-0.068) = -98.$

Note the high gain of the second stage!

DRILL 8.11

Solve the BiCMOS cascode circuit of Example 8.15 using a SPICE simulation. Check the values of V_{DS} and V_{CE} to verify that the devices were operating in the pinch-off and active regions, respectively, as assumed.

ANSWER
$A_{vim} = -105$,
$V_{DS1} = -2.8$ V (saturation region),
$V_{CE} = 2.8$ V (active),
$V_{DS2} = 3.4$ V (saturation region)

Some of the problems at the end of this chapter ask you to apply both types of models in the same circuit. There is a BJT emitter follower with an NMOS active load, a BJT CE amplifier with a PMOS active load, and others.

More direct-coupled circuits are discussed in detail in Chapter 10, where differential amplifiers and operational amplifiers are treated.

CHECK UP

1. **TRUE OR FALSE?** Direct coupling of IC devices offers advantages in low-frequency response and savings in the space required for large capacitors.
2. **TRUE OR FALSE?** BiCMOS combines the higher mutual conductance of FET devices and the higher input resistance of BJT devices.
3. **TRUE OR FALSE?** A CE amplifier first stage directly coupled to a CB amplifier second stage is called a cascade amplifier.

8.9 MEASUREMENT OF BJT HYBRID PARAMETERS

We now give brief consideration to the BJT arrangement and a proposed model shown in Fig. 8.32. This model is of interest because its parameters are easily measured in the laboratory, and they are usually given on manufacturers' data sheets (see Fig. 4.32). In Fig. 8.32(b) we observe an input terminal pair, with voltage v_{be} and current i_b defined, and an output terminal pair with voltage v_{ce} and current i_c. You will recognize this as the common-emitter configuration because the emitter terminal is common to both input

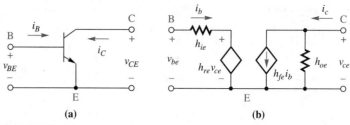

FIGURE 8.32 (a) Common-emitter configuration. (b) Hybrid-parameter small-signal model.

and output terminal pairs. Other configurations could also be used in developing a model, but the common emitter will be presented here because of its greater application. Note that the ac model for a *pnp* transistor would be no different.

Using the model shown in Fig. 8.32(b), we can write the describing equations:

$$v_{be} = h_{ie}i_b + h_{re}v_{ce}, \tag{8.87}$$

$$i_c = h_{fe}i_b + h_{oe}v_{ce}. \tag{8.88}$$

These are hybrid equations because one is a voltage equation and the other is a current equation. The constants are called *hybrid parameters* (or *h parameters*), taking their name from the equations. Their definitions follow:

$$h_{ie} = \frac{\partial v_{BE}}{\partial i_B} = \left.\frac{v_{be}}{i_b}\right|_{v_{ce}=0} = \frac{V_T\beta_F}{|I_C|} + r_b = r_\pi + r_b. \tag{8.89}$$

Thus, h_{ie} is the **CE input resistance** with collector voltage held constant at the Q-point. This partial differentiation was performed in Eq. (4.20). The Ebers-Moll equation did not take into account the bulk base-spreading resistance, r_b, so this term should be added. You can show that this expression will produce 7836 Ω at $T = 25°C$, $|I_C| = 1$ mA, $r_b = 100$ Ω, and $\beta_F = 300$. The 2N2222A data sheet lists 8 kΩ at the maximum β.

$$h_{re} = \frac{\partial v_{BE}}{\partial v_{CE}} = \left.\frac{v_{be}}{v_{ce}}\right|_{i_b=0}. \tag{8.90}$$

The parameter h_{re} is the **CE voltage feedback ratio**. This dimensionless term accounts for the slight effect on input voltage of the change in output voltage and is due to the base-narrowing (Early) effect. This term is small (the data sheet lists 8×10^{-4} maximum, at given conditions) and is often omitted from a simplified model.

$$h_{fe} = \frac{\partial i_C}{\partial i_B} = \left.\frac{i_c}{i_b}\right|_{v_{ce}=0}. \tag{8.91}$$

This is, of course, the **CE small-signal current gain**, which was first defined in Eq. (4.43), and is used interchangeably with the symbol β by many authors. Recall that $h_{fe} = \beta$ is slightly greater than $h_{FE} = \beta_F$, and both vary considerably with Q-point and temperature. Compare the data sheet values.

$$h_{oe} = \frac{\partial i_C}{\partial v_{CE}} = \left.\frac{i_c}{v_{ce}}\right|_{i_b=0}. \tag{8.92}$$

The *CE output admittance*, h_{oe}, reflects the fact that the collector current increases slightly at higher collector-emitter voltage because of the Early effect (Section 4.4). The 2N2222A data sheet gives $5 < h_{oe} < 35$ μS. This term is often neglected in comparison with larger conductances in discrete circuits, but usually needs to be considered in monolithic circuits. The manufacturer's data in Fig. 8.33 illustrates how each of the h parameters is affected by the Q point.

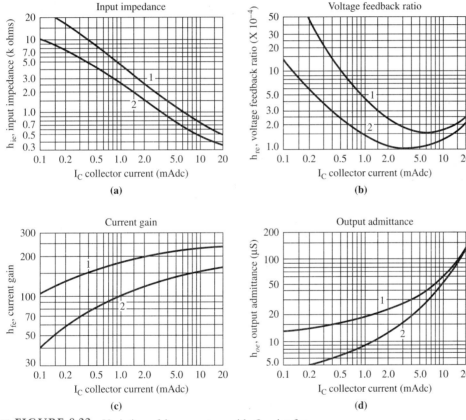

■ **FIGURE 8.33** Variation of h parameters with Q-point for the 2N2222A (courtesy of Motorola, Inc.).

Measurement of *h* Parameters

We can obtain some insight into the meaning of h parameters by considering their experimental determination. Figure 8.34(a) suggests a setup to measure h_{ie} and h_{fe}. First, it is necessary to obtain the dc operating point desired by setting V_{CC} and V_{BB} to appropriate values. The ac voltage, v_{in}, may be 0 for this part. As an example, suppose that we wish to measure parameters at $V_{CE} = 10$ V and $I_C = 1$ mA. We set $V_{CC} = 10$ V and adjust

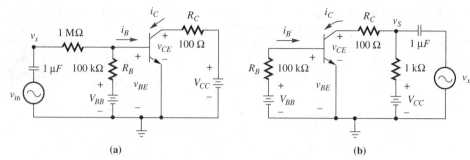

FIGURE 8.34 (a) Circuit for measuring h_{ie} and h_{fe}.
(b) Circuit for measuring h_{re} and h_{oe}.

V_{BB} until the dc current through the 100-Ω collector resistor is 1 mA. Of course, this represents a 0.1-V drop in the 100-Ω resistor, and actual V_{CE} is 9.9 V.

An ac signal of about 1 V_{rms} at about 1 kHz is used. This frequency should be low enough that the parasitic capacitances of the transistor will not have an appreciable effect on our measurements, and yet the 1-μF dc blocking capacitor has reasonably small reactance. Our assumptions are that the 100-kΩ R_B is >> than h_{ie}, so R_B can be ignored, and that the 100-Ω R_C is small enough that $v_{ce} \cong 0$. Note that both of these assumptions are easily verified.

We now measure v_s and v_{be}, calculate $i_b \cong (v_s - v_{be})/1$ MΩ and find $h_{ie} = v_{be}/i_b$. At this point, if h_{ie} is not <<100 kΩ, we could correct for this, knowing that h_{ie} is in parallel with 100 kΩ. This i_b will be about 1 μA rms, and v_{be} will be of the order of a few millivolts. It is assumed that the noise voltage at v_{be} can be kept much lower than this (measure it with $v_{in} = 0$).

For h_{fe}, we measure v_{ce}, calculate $i_c = v_{ce}/100$, and find $h_{fe} = i_c/i_b$. Since v_{ce} will be only about 10 mV, v_{CE} is usually considered to have been held constant.

In Fig. 8.34(b), the 1-V, 1-kHz signal is applied at v_x. Because of the 1-kΩ resistor, V_{CC} should be increased to 11 V to maintain the same dc V_{CE}.

Both v_s and v_{ce} are measured, and we calculate $i_c = (v_s - v_{ce})/100$. Then $h_{oe} = i_c/v_{ce}$.

Finally, we measure v_{be} and find $h_{re} = v_{be}/v_{ce}$. Since v_{be} will be less than 1 mV, $i_b = v_{be}/100$ kΩ $\cong 0$, as required.

It is possible to find the hybrid-π parameters given the h parameters. The terminal voltages and currents of the models of Figs. 8.2 and 8.32 are equated. Then, knowing $g_m = I_C/V_T$ from Eq. (8.1),

$$r_\pi = \frac{h_{fe}}{g_m}, \tag{8.93}$$

$$r_b = h_{ie} - r_\pi, \tag{8.94}$$

$$r_\mu = \frac{r_\pi}{h_{re}}, \tag{8.95}$$

$$g_o = h_{oe} - g_m h_{re}. \tag{8.96}$$

■ Of course the h-parameter model can be used in a circuit of any topology, just as we have used the hybrid-π model. It is just designed with the emitter terminal as the reference.

Equations (8.94) and (8.96) must be positive in actual transistors, forcing $h_{ie} > r_\pi$ and $h_{oe} > g_m h_{re}$. If h_{re} is neglected in Eq. (8.95), it should also be neglected in Eq. (8.96).

In summary, the h parameters determine a usable, verifiable model for the BJT. The second subscript serves to remind us that these parameters relate directly to the common-emitter configuration.

CHAPTER 8 SMALL-SIGNAL MIDFREQUENCY AMPLIFIER ANALYSIS

EXAMPLE 8.16

A certain 2N2222A operating at $T = 28°C$ with $I_C = 1$ mA and $V_{CE} = 10$ V has the following parameters: $h_{fe} = 100$, $h_{ie} = 2700 \, \Omega$, $h_{re} = 0.0002$, and $h_{oe} = 15 \, \mu S$. Find the hybrid-π model for the transistor.

SOLUTION $g_m = 1 \text{ mA}/26 \text{ mV} = 39 \text{ mS}$, $r_\pi = 100/39 \text{ mS} = 2600 \, \Omega$, $r_b = 2700 \, \Omega - 2600 \, \Omega = 100 \, \Omega$, $r_\mu = 2600 \, \Omega/0.0002 = 13 \text{ M}\Omega$, and $g_o = 15 \, \mu S - (0.039 \text{ S})(0.0002) = 7.2 \, \mu S$, or $r_o = 139 \text{ k}\Omega$.

DRILL 8.12

Compare the data given in the statement of Example 8.15 with the manufacturer's data sheet of Fig. 4.32.

ANSWER All parameters are within the range specified.

SUMMARY

Basic amplifier operation was introduced in earlier chapters. The material in this chapter built on the basic BJT operation described in Chapter 4, the basic FET operation outlined in Chapter 5, and the biasing discussed in Chapter 7 to develop a small-signal model that was used to complete a midfrequency amplifier analysis. Both the common-emitter hybrid and the hybrid-π models were defined with typical model element values obtained from actual device data sheets.

The BJT and FET are modeled as two-port devices that can be connected in three basic configurations. We analyzed each of the three configurations—common emitter or source, common base or gate, and common collector or drain—to determine the properties of amplification, input resistance, and output resistance. Tables 8.1, 8.2, and 8.3 summarize these results.

MOS integrated circuits use active loads to realize a larger effective resistance without the sacrifice of chip space required by diffused or implanted resistors. Bipolar ICs also use active loads. We considered several examples of such design, but have reserved differential amplifiers for Chapter 10.

Cascaded amplifiers are used for achieving higher amplification than is possible in a single stage and for input-output matching. We have considered an amplifier design using capacitance-coupled discrete transistors, and also direct-coupled transistors as used in monolithic design, where coupling capacitors are avoided. Included were examples of combination bipolar and MOSFET devices such as are found in BiMOS circuits. In Chapters 10 through 12, we study multistage amplifiers as they are implemented in the design of operational amplifier ICs.

SURVEY QUESTIONS

1. Why must the dc operating point of an amplifier be found before proceeding with the ac analysis?
2. Explain the relationship between the g_m, r_π, and β of a BJT.
3. Which of the following amplifiers has high R_i and low R_o: CE, CB, CC?
4. Which of the following amplifiers has low R_i and high R_o: CE, CB, CC?
5. Compare a CE amplifier with the emitter resistance bypassed to one where the emitter resistance is not bypassed.
6. Which of the following amplifiers has high R_i and low R_o: CS, CG, CD?
7. Which of the following amplifiers has low R_i and high R_o: CS, CG, CD?
8. Compare a CS amplifier with the source resistance bypassed to one where the source resistance is not bypassed.

9. What advantage does an amplifier with an active-load resistance have over one with a passive-load resistor?
10. What are the relative advantages of multistage amplifiers that are direct coupled compared to those that are capacitance coupled?
11. Describe a *cascode* amplifier.
12. What is a BiCMOS IC, and why is it important?

PROBLEMS

Problems marked with an asterisk are more challenging.

8.1 The *small-signal condition* is developed in Eqs. (4.21) to (4.26). Estimate the percentage of nonlinearity introduced in the collector current of a BJT operating in the "linear" region if $|v_{be}| = 10$ mV peak.

8.2 The amplifier shown in Fig. 8.35 uses $R_S = 1$ kΩ, $R_1 = 100$ kΩ, $R_2 = 47$ kΩ, $R_E = 1$ kΩ, $R_C = 2.2$ kΩ, $R_L = 4.7$ kΩ, and $V_{CC} = 10$ V. Neglect the reactance of C_E, and let $\beta_F = \beta = 100$. Parameter r_b is small, and r_μ and r_o are very large; these parameters can be neglected in your model.

(a) Find the operating point, and estimate g_m and r_π at 27°C.
(b) Find A_{vi}, R_i, A_i, R_o, and $A_{vs} = V_o/V_s$.

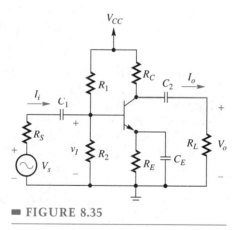

■ FIGURE 8.35

8.3* Repeat Problem 8.2, part (b), assuming $r_b = 100$ Ω, $V_A = 100$ V, and $r_\mu = r_o\beta$. Compare answers with those of Problem 8.2(b).

8.4 Use a SPICE simulation to work Problem 8.2. Show your program list. Note that for $f = 1$ kHz, $X_c \Rightarrow 0$ for $C \geq 1000$ μF, so just make them large! Also note that SPICE assumes a linear model for the ac analysis, so that even if you make $V_s = 1$ V (much too large for small-signal conditions), the program solution will tell you that $V_o = A_{vs}$ volts. Find A_{vi} by dividing: V_o/V_i.

8.5 The amplifier of Fig. 8.35 uses $R_E = 1$ kΩ, $R_C = 2.2$ kΩ, $R_L = 4.7$ kΩ, and $V_{CC} = 12$ V. Let $\beta_F = \beta = 100$, and $V_A = 100$ V. Neglect capacitive reactances. Design this circuit to have $A_{vi} = -150$.

(a) Find the required g_m and the required collector current.
(b) Choose $R_2 = 100$ kΩ, and find the R_1 required to achieve the proper Q-point.
(c) Determine A_{vs} for your circuit if $R_S = 1$ kΩ.

8.6* Work Problem 8.2 assuming that the emitter resistance is not bypassed.

8.7 Figure 8.36 uses a transistor with $\beta_F = \beta = 100$. Neglect r_μ and r_o, and neglect reactances of C_1 and C_2.

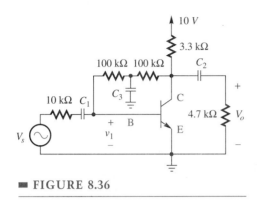

■ FIGURE 8.36

(a) Find the operating point, and estimate g_m and r_π at 25°C.
(b) Draw an ac model, and find R_i and A_{vi} if C_3 is omitted.
(c) Draw an ac model, and find R_i and A_{vi} if C_3 is large; its reactance may be neglected.

8.8 The CB amplifier shown in Fig. 8.37 has $R_S = 1$ kΩ, $R_1 = 100$ kΩ, $R_2 = 47$ kΩ, $R_E = 1$ kΩ, $R_C = 2.2$ kΩ, $R_L = 4.7$ kΩ, and $V_{CC} = 10$ V. Let $\beta_F = \beta = 100$, and neglect r_b, g_μ, and g_o. Assume C_B is large enough so that its reactance is negligible.

(a) Find the Q-point, and estimate g_m and r_π at 25°C.
(b) Find A_{vi}, R_i, A_i, R_o, and $A_{vs} = V_o/V_s$.

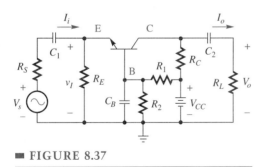

■ FIGURE 8.37

8.9* In Problem 8.8, what percentage error in A_{vi} resulted from neglecting $r_b = 100$ Ω and $g_o = 10$ μS?

8.10 Use a SPICE simulation to work Problem 8.8, only use $V_A = 100$ V and $r_b = 100$ Ω.

8.11* The CB amplifier shown in Fig. 8.37 is to be used to match a 50-Ω source to a 1-kΩ load. Select the resistor values so that $R_S = R_i$ and $R_o = R_L$. Assume $V_{CC} = 6$ V and $\beta_F = \beta = 100$.

8.12* The emitter follower shown in Fig. 8.38 is to be used to drive a 50-Ω load from a 1000-Ω source. Select resistor values such that $R_o = 50$ Ω and $R_i = 1000$ Ω. Assume $\beta_F = \beta = 100$ and $V_{CC} = 5$ V. Note that in practice, there may be no good reason to make R_i as low as R_s.

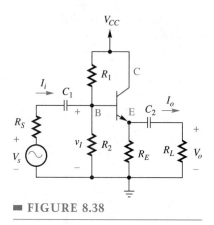

■ FIGURE 8.38

8.13* An emitter follower circuit such as that shown in Fig. 8.38 is used to drive an 8-Ω speaker as a load. Design the circuit, assuming that you need a voltage amplification A_{vi} of at least 0.9 and an output voltage as large as 1 V_{peak}. Assume $\beta = 100$ and $V_{CC} = 8$ V.

(a) Choose an $R_E \gg R_L$. What value of g_m is required? What value of collector current?
(b) Let R_2 be an open circuit, and determine an acceptable value for R_1.
(c) What is the input resistance R_i of your design?
(d) Find the power gain of the amplifier.
(e) The reactance of C_2 must be small compared to 8 Ω! What are the practical implications of this?

8.14 Consider the circuit shown in Fig. 8.39. Transistors Q_1 and Q_2, connected together as shown, are called a *Darlington circuit*. Assume transistors with $\beta_F = \beta = 100$ at 27°C.

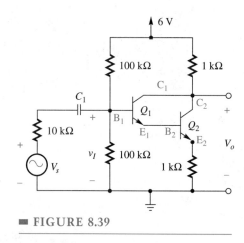

■ FIGURE 8.39

(a) Find the Q-point, and estimate g_m and r_π for each transistor.
(b) Draw an ac model, and find R_i and A_{vi}.
(c) Find the resistance at the emitter of Q_2, looking back from the 1-kΩ resistor.

8.15 Refer to the Darlington circuit of Fig. 8.39. Suppose the output voltage V_o is taken across the emitter resistor, instead of the collector. Let the collector resistor $R_C = 0$. Assume transistors with $\beta_F = \beta = 100$ at 27°C.

(a) Find the Q-point, and estimate g_m and r_π for each transistor.
(b) Draw an ac model, and find R_i and A_{vi}.

8.16* Refer to the Darlington circuit of Fig. 8.39. Suppose, as in Problem 8.15, that the collector resistor R_C is replaced by a short circuit. The emitter resistor is reduced to 10 Ω, and the emitter is coupled through a very large capacitor to the load for this amplifier, an 8-Ω speaker.

(a) Find the Q-point, and estimate g_m and r_π for each transistor.
(b) How large must the capacitor in series with the speaker be so that its reactance is small compared to 8 Ω at frequencies down to 100 Hz?
(c) Draw an ac model, and find R_i, A_i, and A_{vi}.

8.17* The circuit shown in Fig. 8.40 uses complementary transistors, one *pnp* and one *npn*. This is not a circuit where small-signal models are useful. Assume $|v_I|$ is large but $< V_{CC}$. Plot v_o vs. v_I.

8.18 Starting with Eq. (7.18), substitute $v_{GS} = V_{GS} + v_{gs}$, and then collect terms in v_{gs} and v_{gs}^2. Determine the constraint on the magnitude of v_{gs} that will cause the square term to be less than 10% of the linear term. This is a sufficient condition to justify use of the linear FET model.

8.19 The common-source amplifier shown in Fig. 8.41 uses a JFET with $I_{DSS} = 4$ mA, $V_P = -4$ V, and $\lambda^{-1} = 100$ V. Let $R_1 = \infty$, $R_2 = 1$ MΩ, $R_\sigma = 2$ kΩ, $R_D = 4.7$ kΩ, $R_L = 10$ kΩ, $R_S = 1$ kΩ, and $V_{DD} = 10$ V.

(a) Estimate g_m and r_d at the Q-point.
(b) Find A_{vi}, R_i, A_i, R_o, and A_{vs} if reactance of C_σ is negligible.
(c) Repeat part (b) if C_σ is omitted.

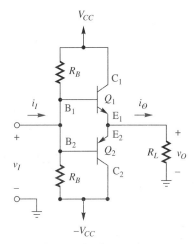

■ FIGURE 8.40

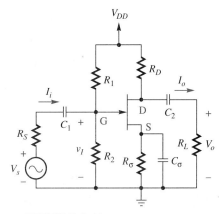

■ FIGURE 8.41

8.20 Use a SPICE simulation of the amplifier of Problem 8.19 in order to compare the results of parts (a) and (b) for JFETs with different I_{DSS}. Try the following: $I_{DSS} = 2$, 4, and 8 mA. Remember that for the SPICE JFET model, **BETA** $= I_{DSS}/V_P^2$.

8.21 Use a SPICE simulation of the amplifier of Problem 8.19 in order to determine how the amplification varies with V_{DD}. Try $V_{DD} = 8$, 9, 10, 12, and 15 V. Observe that the JFET becomes ohmic when V_{DS} becomes too small.

8.22 A 2N5458 is used in the circuit shown in Fig. 8.41. Use the average values given in the data sheet of Fig. 5.31 ($r_d = 1/g_{os}$). Let $R_1 = 1$ MΩ, $R_2 = 100$ kΩ, $R_D = 2.2$ kΩ, $R_\sigma = R_S = 1$ kΩ, $R_L = 10$ kΩ, and $V_{DD} = 11$ V.

(a) Find A_{vi} and A_{vs} if C_σ is assumed very large.
(b) Find A_{vi} and A_{vs} if C_σ is omitted from the circuit.

8.23* Derive the output resistance of the CS amplifier with unbypassed emitter resistor [Fig. 8.13(b)] without neglecting r_d.

8.24 The CG amplifier shown in Fig. 8.42 uses $R_1 = \infty$, $R_2 = 0$, $R_\sigma = 2$ kΩ, $R_D = 4.7$ kΩ, $R_L = 10$ kΩ, $R_s = 1$ kΩ, and $V_{DD} = 10$ V. The JFET has $I_{DSS} = 4$ mA and $V_p = -4$ V. Find A_{vi}, R_i, A_i, R_o, and A_{vs}.

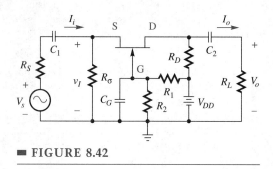

■ FIGURE 8.42

8.25* Derive the output resistance of the CG amplifier (Fig. 8.15) without neglecting r_d.

8.26 The CD amplifier as shown in Fig. 8.43 uses a JFET with $I_{DSS} = 6$ mA, $V_p = -3.5$ V, and $\lambda^{-1} = 100$ V. Assume $V_{DD} = 10$ V, $R_1 = 1$ MΩ, $R_2 = 220$ kΩ, $R_\sigma = 1$ kΩ, $R_L = 1$ kΩ, and $R_s = 10$ kΩ. Neglect the reactances of C_1 and C_2. Find A_{vi} and A_{vs}.

8.27 Design a CD amplifier similar to the one of Fig. 8.43, except the load is only 150 Ω and it is direct coupled (no need for C_2 and R_σ). Use the JFET whose parameters are given in Problem 8.26, and $V_{DD} = 6$ V. Find A_{vi}.

8.28 The enhancement-mode MOSFET used in the circuit shown in Fig. 8.44 has $i_D = 0.16(v_{GS} - 2)^2$ mA for $v_{DS} > v_{GS} - 2$, and $r_d = 50$ kΩ. Neglect the capacitive reactances.

(a) Find the Q-point and calculate g_m.
(b) Find A_{vi} and R_i.

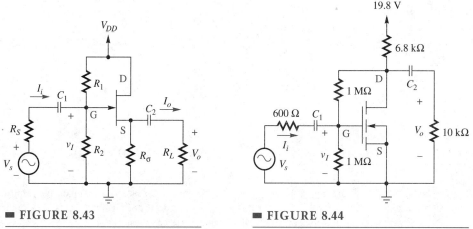

■ FIGURE 8.43   ■ FIGURE 8.44

8.29 Refer to Fig. 8.44. Suppose that the upper 1-MΩ gate resistor is divided into two 500-kΩ resistors, with the center connection bypassed to ground by means of a large capacitor. This means that the dc model for the circuit would be identical to that of Problem 8.28, but the small-signal model will have no direct connection between the drain and gate. Find A_{vi} and R_i.

8.30 A depletion-load MOSFET amplifier [Fig. 8.18(a)] has $k_1 = 9k_2$, $\chi = 0.08$.

(a) Neglecting r_{d1}, r_{d2}, and R_L, estimate the voltage amplification A_{vi}.
(b) What else would you need to know to be sure that these resistances could be neglected? What else are you assuming in part (a)?

8.31* In the depletion-load MOSFET amplifier of Example 8.10, r_{d1}, r_{d2}, and R_L were not neglected. What percentage error in A_{vi} would have resulted from ignoring these resistances?

8.32 An NMOS enhancement-load amplifier uses two identical MOSFETs with $V_t = 1.5$ V and $k = 100$ μA/V^2. The supply voltage $V_{DD} = 5$ V. What can you say about the range of output voltage v_O?

8.33* Consider an amplifier of the type shown in Fig. 8.18(a). Let $V_{DD} = 5$ V, and assume M_1 is biased with $V_{GS1} = 2.1$ V$_{dc}$ such that both FETs are in their saturation regions. You may then assume $V_O = -V_{BS} \cong 2.5$ V$_{dc}$. The load, $R_L = 10$ MΩ. Let $v_{t01} = 1.5$ V, $V_{t02} = -1.5$ V, $k_1 = 1$ mA/V^2, $k_2 = 0.25$ mA/V^2, $\lambda_1 = \lambda_2 = 0.01$ V^{-1}, $\gamma_1 = \gamma_2 = 0.3$, and $2\phi_p = 0.6$ V.

(a) Design a voltage divider R_1 and R_2 that will give the proper dc bias to the gate of M_1. What is the input resistance of your amplifier?

(b) Find the voltage amplification A_{vi} of the amplifier. This is a good problem for a SPICE simulation. If you want to solve it manually, how do you find χ, given γ?

8.34* Assume an NMOS enhancement-load amplifier such as that shown in Fig. 8.20(a). Let $V_{DD} = 5$ V, $R_L = 5$ MΩ, $V_{t01} = V_{t02} = 1.5$ V, $\gamma_1 = \gamma_2 = 0.3$ V$^{.5}$, $k_1 = 1.25$ mA/V^2, $k_2 = 0.05$ mA/V^2, and $\lambda_1 = \lambda_2 = 0.01$ V^{-1}. Find A_{vi} for this amplifier, assuming proper dc bias for a Q-point of $V_O = 2$ V.

8.35 An enhancement-load NMOS amplifier, as designed, has $(W/L)_1 = 20$, $(W/L)_2 = 0.5$, and achieves a voltage amplification of -5. How should the design be changed to realize an amplification of -10?

8.36* The CMOS amplifier depicted in Fig. 8.45 uses a single PMOS current source for its load. Let $k_n = k_p = 100$ μA/V^2, $V_{tn} = 1.5$ V, $V_{tp} = -1.5$ V, $\lambda_n = \lambda_p = 0.01$ V^{-1}, and $V_{DD} = 3$ V. A dc bias $V_{GS1} = 3$ V forces both FETs to operate in their saturated regions (the dc value of V_O will be near 1.5 V).

(a) What will be the quiescent power dissipation of this circuit?

(b) Draw a small-signal model, and calculate A_{vi} for this amplifier.

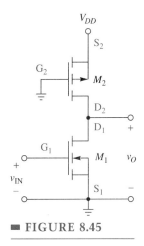

■ **FIGURE 8.45**

8.37* The same circuit with identical MOSFETs as given in Problem 8.36 does not operate well with $V_{DD} = 5$ V. Why is this? The dc bias on v_{IN} should be 5 V for proper comparison.

8.38 Use SPICE to simulate the amplifier of Problem 8.36.

(a) Plot v_O versus v_{IN} for V_{IN} from 0 to 5 V.

(b) Examine the operating point, and estimate peak ac output voltage for linear operation.

8.39 The CMOS amplifier shown in Fig. 8.46 is similar to the digital inverter circuit presented in Chapter 5, Fig. 5.19. But the resistor from the output (drains) back to the input (gates) biases the circuit in the linear range, where both FETs are in their saturation regions. Let $k_n = k_p = 50\ \mu\text{A/V}^2$, $V_{tn} = 1.5$ V, $V_{tp} = -1.5$ V, $\lambda_n = \lambda_p = 0.01$ V^{-1}, $R_G = 1$ MΩ, $R_L = 1$ MΩ, and $V_{DD} = 5$ V.

(a) Draw a dc model, and find the drain current (neglect r_d for this part).
(b) Draw a small-signal model, and find $A_{vi} = v_o/v_{in}$ for this amplifier.

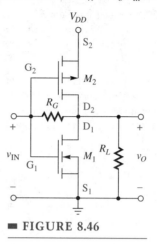

■ FIGURE 8.46

8.40 Use SPICE to simulate the circuit of Problem 8.39.

(a) Find the drain currents, and the quiescent power dissipation.
(b) Find the voltage amplification A_{vi}, input resistance R_i, and output resistance R_o of the circuit. This could be done by driving the gates with a proper dc voltage source **VIN**, and using the transfer function command **.TF VOUT VIN**, where **VOUT** is the node voltage of v_o.
(c) Find the range of output voltage where the linear model applies. This is best done with dc plot of v_O versus v_{IN} as v_{IN} varies from 0 to 5 V.

8.41 The circuit shown in Fig. 8.47 is a common-source CMOS amplifier, where M_4 is the amplifier, and M_1, M_2, and M_3 constitute a simple current source, which serves as the active load for the drain of M_4. Let $V_{DD} = 5$ V, and assume all transistors have $V_t = 1.5$ V (negative for M_4 which is PMOS), $\lambda = 0.01$ V^{-1}, $k = 0.1$ mA/V^2, and you may neglect body effect. Let v_{IN} provide 2.5 V dc bias as well as signal v_{IN}. Find $A_{vi} = v_o/v_{in}$.

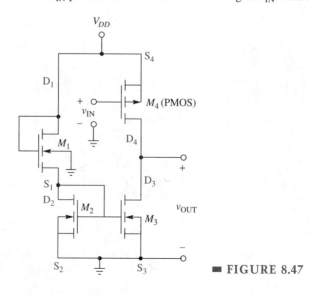

■ FIGURE 8.47

8.42 Work Problem 8.41 by means of a SPICE simulation. Use $\gamma = 0.3$ V$^{.5}$ in the model statement to account for body effect.

8.43* Refer to Fig. 8.26(a). Omit the emitter-bypass capacitor C_E, and then repeat Example 8.13. Compare your results with those of Example 8.13. Note that stage one is affected by the change of input resistance of stage two.

8.44 For the two-stage amplifier shown in Fig. 8.48, assume all capacitances are large enough to have negligible reactance for the signal frequency. Let $\beta_F = \beta = 100$ at 25°C.

(a) Find the Q-point, g_m, and r_π for each transistor.
(b) Find $A_{vi} = V_3/V_1$ and $A_i = I_3/I_1$.
(c) Remove the emitter-bypass capacitor from Q_2 and repeat part (b).

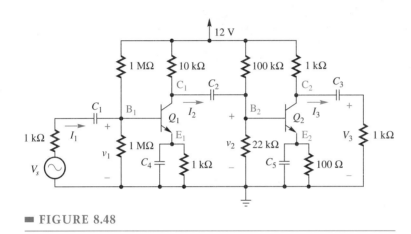

■ FIGURE 8.48

8.45* In the two-stage amplifier shown in Fig. 8.49, assume all coupling and bypass capacitors are large enough to have negligible reactance for the signal frequencies. Let $\beta_F = \beta = 80$ at 25°C.

(a) Find the Q-point, g_m, and r_π for each transistor.
(b) Find $A_{vi} = V_3/V_1$ and $A_i = I_3/I_1$.

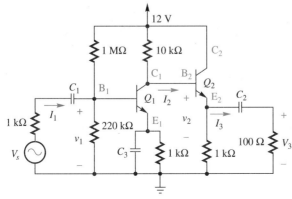

■ FIGURE 8.49

8.46 Refer to the direct-coupled amplifier of Fig. 8.27. The parameters of the BJTs are $\beta = 100$ and $V_A = 150$ V. Assume $V_{BE} = 0.7$ V.

(a) Draw a dc model, and find the Q-points of the BJTs.
(b) Draw the midband small-signal model of the amplifier.
(c) Determine the midband input impedance of this amplifier.
(d) Determine the A_{vsm}.

8.47 Draw dc and ac models, and determine the approximate A_{vsm} for the circuit shown in Fig. 8.28. The transistor parameters are $\beta = 100$ and $V_A = 150$ V. Let C_{E1} be very large.

8.48 The circuit presented in Fig. 8.29 is a simplified version of a *cascode* amplifier. Assume $\beta_F = \beta = 100$ and $V_A = 150$ V.

(a) Draw a dc model for the circuit and find the operating point.
(b) Find the hybrid-π models for the transistors and draw a small-signal model for the amplifier.
(c) Find the voltage amplification $A_{vi} = v_o/v_s$ and the input resistance R_i.

8.49* Figure 8.50 shows a CMOS version of the cascode amplifier. The active load M_1 is a PMOS current source with output resistance r_d. Assume that the common-gate transistor M_2 and the common-source transistor M_3 also have output resistance r_d and transconductance g_m. Draw a small-signal model of the amplifier and find the voltage gain v_{out}/v_{in}.

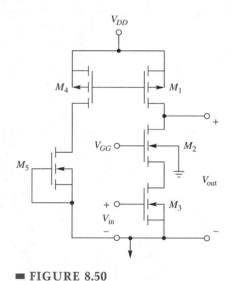

■ **FIGURE 8.50**

8.50 A 2N2222A is operating at $I_C = 1$ mA, $V_{CE} = 10$ V, and $T = 25°C$. Using the average of the two values from Fig. 8.33, find the common-emitter h parameters. Construct the hybrid-parameter model for this Q-point.

8.51 Using the h parameters found in Problem 8.50, determine the hybrid-π model for the 2N2222A at the given operating point.

8.52 Consider the transistor of Problem 8.50, which is used in a bias situation where I_C increases to 10 mA but V_{CE} remains 10 V. Find the change in each hybrid parameter.

REFERENCE

1. Antognetti, A., and G. Massobrio. *Semiconductor Device Modeling with SPICE*. New York: McGraw-Hill, 1987.

SUGGESTED READINGS

Colclaser, R., D. Neamen, and C. Hawkins. *Electronic Circuit Analysis Basic Principles.* New York: John Wiley & Sons, 1984. Chapter 4 contains a brief analysis of BJT amplifier models.

Sedra, A., and K. Smith. *Microelectronic Circuits*, 3rd ed. New York: Holt, Rinehart and Winston, 1991. Chapters 4 and 5 contain sections on classical single-stage amplifiers.

CHAPTER 9

FREQUENCY EFFECTS IN SMALL-SIGNAL AMPLIFIERS

9.1 High-frequency Small-signal BJT and FET Models
9.2 The Effect of C_μ and C_{gd}: The Miller Effect
9.3 Small-signal Amplifier Low-frequency Analysis
9.4 Direct-Coupled and BiCMOS Amplifier Examples
Summary
Problems
Survey Questions
References

In Chapter 8, we developed and used simplified small-signal models to compute key characteristics of a variety of BJT and FET amplifier circuits. All of these models and circuits were assumed to be frequency independent. Topics in Chapter 9 include the addition of parasitic capacitances in the models that will affect the high-frequency performance and the consideration of the effects of coupling and bypass capacitors on the overall circuit. As in Chapter 8, we discuss amplification of BJT and FET circuits, but now with frequency dependence added.

To tackle the analysis and design of frequency-dependent circuits, we need to introduce simplifying techniques for looking at complex circuitry. Without some of these techniques, it would be easy to lose insight into the overall operation because of algebraic complexity. The Bode plot tools of Chapter 6 are used here extensively.

The parasitic capacitance between the input and output nodes of an inverting amplifier is effectively increased by the voltage amplification of the amplifier. This result, known as the *Miller effect*, is analyzed in this chapter, and a strategy is presented to simplify the analysis of such circuits.

After a brief familiarization with the effects of the parasitic capacitances on the active devices in typical amplifier circuits, we demonstrate the power of a SPICE simulation in an analysis or design situation.

IMPORTANT CONCEPTS IN THIS CHAPTER

- The basic *active* IC BJT has three junction capacitances: the base-emitter C_π, the base-collector C_μ, and the collector-substrate C_{cs}, of which C_μ and C_{cs} are depletion-region capacitances across reverse-biased junctions.
- The *active* base-emitter junction is forward-biased, and effects of C_π include a depletion-layer capacitance and a base *diffusion capacitance*.
- All FET structures have gate-source C_{gs}, gate-drain C_{gd}, and junction capacitances to the substrate.
- Both BJT and FET high-frequency performance can be characterized by *unity gain frequencies*.
- The Miller theorem allows capacitance between the input and output of an inverting amplifier to be replaced by a larger capacitance at the input. This allows us to compute a quick estimate of the overall bandwidth of the circuit.
- Bode plots facilitate the combination of the frequency responses of midband, high-frequency, and low-frequency models of amplifier circuits.
- Each parasitic capacitance in the high-frequency model of an amplifier circuit introduces a high-frequency corner in the frequency response of the circuit.
- Each coupling and bypass capacitor in an amplifier introduces a low-frequency corner in the frequency response of the circuit.
- Direct coupling between stages in an IC is used to avoid the need for coupling capacitors, which are not economically produced in monolithic technology.
- The devices in a cascode amplifier are arranged to minimize the effect of the parasitic capacitances; this is a great advantage in high-frequency circuitry.

9.1 HIGH-FREQUENCY SMALL-SIGNAL BJT AND FET MODELS

Up to this point, we have not incorporated any elements into the small-signal models that would result in frequency-dependent operation. Here, let us start by considering a typical junction-isolated *npn* bipolar transistor whose cross section is given in Fig. 9.1.

The frequency-dependent operation arises from a combination of technological factors in the fabrication and from inherent device and junction physics. These additional, often undesirable, features are modeled as ***parasitic elements***. It is easy to observe that because these effects are distributed nonuniformly over the entire junction area or throughout the bulk semiconductor material, the voltage and current distribution is nonuniform and the resultant distributed parasitic effect is also nonuniform. To simplify matters, these distributed parasitic effects are almost always modeled as discrete lumped-parameter circuit elements at frequencies below 100 to 200 MHz.

Whether these parasitic elements must be used in the analysis or design of a circuit depends on the judgment and experience of the engineer. Of course, a good computer simulation makes it much easier to incorporate these additional circuit elements.

- Distributed models used for very high frequencies and microwave models are beyond the scope of this text.

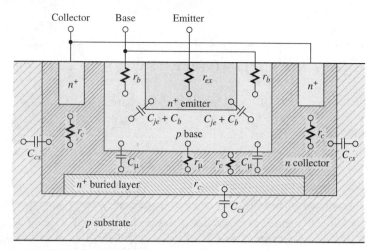

■ **FIGURE 9.1** Idealized cross section of an *npn* bipolar junction transistor showing parasitic elements (r_π, r_o, g_m not included as parasitic elements). Similar physical concepts are valid for a *pnp* transistor.

Bipolar Transistor Model

All *pn* junctions have associated voltage-dependent depletion-region capacitances. The basic *active* region IC transistor structure has three junctions:

1. The base-emitter junction, operated in forward bias
2. The collector-base junction, operated in reverse bias
3. The collector-substrate junction, always in reverse bias; the collector-substrate junction would not exist in a discrete transistor.

The most important effect occurs at the collector-base junction. This junction may behave as an abrupt junction or as a graded junction, depending on the doping profile and the reverse-bias voltage. Assuming a graded junction, from Eq. (3.49), this distributed depletion-region capacitance is modeled by

$$C_\mu = \frac{C_{\mu 0}}{\left(1 + \frac{V_{CB}}{\phi_J}\right)^{1/3}}, \tag{9.1}$$

where $C_{\mu 0}$ is the value at zero bias and ϕ_J is on the order of 0.5 to 0.7 V. Base-collector capacitance C_μ is also called C_c, C_{ob}, or C_{cb}. The value for C_μ is given in most data sheets. For instance, Fig. 4.32 shows that the 2N2222A has a C_{ob} of 8 pF at $V_{CB} = 10$ V. From a practical viewpoint, the voltage variation is rarely included in the small-signal model analytical calculation. As we illustrate later, the SPICE BJT model incorporates these voltage variations. Typically, some value from 0.1 pF to several picofarads is appropriate for small devices.

Similarly, there is an effective abrupt-junction capacitance at the collector-substrate depletion region. It can be described by

$$C_{cs} = \frac{C_{cs0}}{\left(1 + \frac{V_{CS}}{\phi_J}\right)^{1/2}}, \tag{9.2}$$

where C_{cs0} is the value at zero bias and ϕ_J is on the order of 0.5 to 0.7 V. Collector-substrate C_{cs} is about 0.5 pF to several picofarads and is considered constant since the reverse-bias voltage is relatively large and assumed constant. If the substrate-collector junction becomes forward-biased, isolation between transistors does not exist and devices are then effectively short-circuited. Other nonjunction isolation techniques are discussed in Chapter 18.

Two simultaneous effects are really occurring at the forward-biased base-emitter junction. There is a small depletion-layer capacitance, 0.1 to 0.2 pF, designated as C_{je}. The second capacitance is called the *base-charging* or **diffusion capacitance**, C_b. It is a small-signal phenomenon related to the time required for a charge carrier to traverse the base region. This time is called τ_F, the **base transit time**. Consider this effect in an *npn* transistor. Qualitatively, an input signal $\Delta v_{BE} = v_{be} = v_i$ injects a $\Delta Q_e = q_e$, which consists of minority charge carriers, into the region. To preserve charge neutrality, a $\Delta Q_h = q_h$ of majority charge carriers is supplied by the base lead. Using the basic definition of capacitance $C = Q/V$,

$$C_b = \frac{q_h}{v_i}. \quad (9.3)$$

The base transit time is defined by

$$\tau_F = \frac{\Delta Q_e}{\Delta I_c} = \frac{q_e}{i_c} = \frac{\Delta Q_h}{\Delta I_c} = \frac{q_h}{i_c} \quad (9.4)$$

so that by combining Eqs. (9.3) and (9.4), we have

$$C_b = \frac{q_h}{v_i} = \frac{\tau_F i_c}{v_i} = \tau_F g_m = \tau_F \frac{I_C}{V_T}. \quad (9.5)$$

■ $C_b = \tau_F \frac{I_c}{V_T}$ also holds for a pnp BJT.

The total **small-signal input capacitance** is given by

$$C_\pi = C_b + C_{je} \approx C_b. \quad (9.6)$$

Occasionally, $C_i = C_\pi + C_\mu$ is given in data sheets. The value of C_π is on the order of a few picofarads.

These three parasitic capacitances, along with other ohmic and junction effects, have been included in a comprehensive small-signal model; see Fig. 9.2(a). Recall that r_{ex}, r_b,

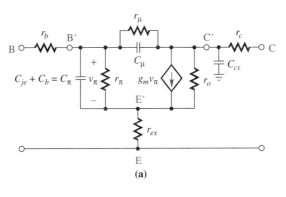

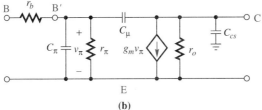

■ **FIGURE 9.2** Bipolar junction transistor small-signal model. **(a)** Complete small-signal model. **(b)** Simplified small-signal model.

■ The model of Fig. 9.2(a) is appropriately called the *high-frequency hybrid-π* model, being topologically the same as Fig. 8.2.

and r_c are the emitter, base, and collector spreading (ohmic) resistances, respectively, of the semiconductor material itself. We rarely include r_{ex} because it is on the order of an ohm or two. Sometimes r_c is included because it may be tens to hundreds of ohms, but it is usually considered insignificant compared with other resistances present in the collector circuit. Usually r_b is included because its value of tens to hundreds of ohms does affect the analysis of the transistor frequency response. Consequently, the model shown in Fig. 9.2(b) is used in most cases. For a discrete device, C_{cs} would not be included. Including r_o is usually a simple matter when it is parallel to other resistances in the circuit.

EXAMPLE 9.1

Compute all of the small-signal parameters for the *npn* transistor whose parameters and Q-point conditions at 300 K are given in Fig. 9.3.

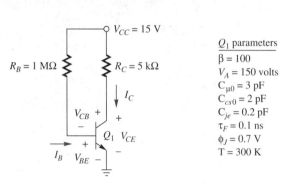

■ **FIGURE 9.3** Circuit for Example 9.1.

SOLUTION Using the loop equations

$$V_{CC} - I_C R_C + V_{CE} \quad \text{and} \quad V_{CC} = I_B R_B + V_{BE} \quad \text{results in}$$

$$15 = I_C 5\,\text{k}\Omega + V_{CE} \quad \text{and} \quad 15 - 0.7 = \frac{I_C}{(100)}(1.0\,\text{M}\Omega).$$

From these equations, $I_C = 1.43$ mA and $V_{CE} = 7.85$ V. Then $V_{CB} = V_{CE} - V_{BE} = 7.85 - 0.7 = 7.15$ V.

The small-signal parameters are computed as follows:

$$g_m = \frac{qI_C}{kT} = \frac{1.43\,\text{mA}}{26\,\text{mV}} = 55\,\text{mS} \quad \text{and} \quad r_\pi = \frac{\beta}{g_m} = \frac{100}{55\,\text{mS}} = 1818\,\Omega,$$

$$r_o = \frac{V_A}{I_C} = \frac{150\,\text{V}}{1.43\,\text{mA}} = 105\,\text{k}\Omega,$$

$$r_\mu \approx \text{several}\,\beta r_o = (100)(105\,\text{k}\Omega) \approx \text{several} \times 10.5\,\text{M}\Omega.$$

From Eqs. (9.5) and (9.6),

$$C_\pi = C_{je} + C_b = 0.2\,\text{pF} + \tau_F g_m = 0.2\,\text{pF} + (0.1 \times 10^{-9})(0.055)$$
$$= 0.2 + 5.5 = 5.7\,\text{pF}.$$

From Eq. (9.1),

$$C_\mu = \frac{C_{\mu 0}}{\left(1 + \dfrac{V_{CB}}{\phi_J}\right)^{1/3}} = \frac{3.0\,\text{pF}}{\left(1 + \dfrac{7.15}{0.7}\right)^{1/3}} = 1.34\,\text{pF},$$

$V_{CS} = V_{CE}$ since the substrate is usually implicitly grounded (connected to the emitter) in this type of circuit. Then, from Eq. (9.2)

$$C_{cs} = \frac{C_{cs0}}{\left(1 + \frac{V_{CS}}{\phi_J}\right)^{1/2}} = \frac{2 \text{ pF}}{\left(1 + \frac{7.85}{0.7}\right)^{1/2}} = 0.57 \text{ pF}.$$

DRILL 9.1

The data sheet for the 2N2222A (Fig. 4.32) gives $C_\mu = C_{ob} = 8$ pF (max) at $V_{CB} = 10$ V. Calculate $C_{\mu 0}$.

ANSWER

$C_{\mu 0} = 19.8$ pF. This is the SPICE model parameter `CJC`.

Unity Gain Frequency, f_T

Even though most data sheets specify at least the values for C_π and C_μ at some typical Q-point, the frequency capability is more often given in terms of f_T, the **unity gain** or **transition frequency**. We can relate f_T to the small-signal parameters with the aid of Fig. 9.4. The collector is *short-circuited* to the emitter in the ac model, $V_{ce} = 0$. In the dc model the Q-point must be preserved.

By definition, the **low-frequency small-signal current gain β_0** is given by I_o/I_s; see Fig. 9.4(a). The frequency at which $\beta(\omega)$ decreases to unity is f_T. This is the maximum

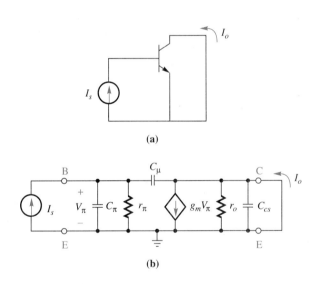

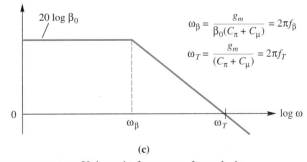

■ **FIGURE 9.4** Unity gain frequency, f_T, analysis.
(a) Circuit used to determine f_T. (b) Small-signal model.
(c) Bode analysis defining ω_β and ω_T.

frequency at which this transistor can be used as an amplifier. From Fig. 9.2(b), where r_b, r_c, and r_μ have been neglected,

$$V_\pi = I_s(r_\pi \| X_{C_\mu} \| X_{C_\pi}) = \frac{r_\pi \left[\frac{1}{j\omega(C_\pi + C_\mu)}\right]}{r_\pi + \frac{1}{j\omega(C_\pi + C_\mu)}}$$

$$= \frac{r_\pi}{1 + j\omega r_\pi(C_\pi + C_\mu)} I_s. \tag{9.7}$$

Note that r_o and C_{cs} have no effect, being shorted. If we neglect the feed-forward current in C_μ, an approximation we justify later in this chapter, then

$$I_o \approx g_m V_\pi. \tag{9.8}$$

■ Of course, the circuit shown in Fig. 9.4(a) assumes correct dc biasing.

Using this result and $\beta_0 = g_m r_\pi$ with Eq. (9.7) yields

$$\frac{I_o}{I_s}(j\omega) = \beta(\omega) = \frac{g_m r_\pi}{1 + j\omega r_\pi(C_\pi + C_\mu)} = \frac{\beta_0}{1 + j\omega \frac{\beta_0}{g_m}(C_\pi + C_\mu)}. \tag{9.9}$$

Equation (9.9) is in a form suitable for a Bode plot analysis; see Fig. 9.4(c). The low-frequency asymptote is β_0 and the 3-dB corner is at

$$\omega_C = \omega_\beta = \frac{g_m}{\beta_0(C_\pi + C_\mu)}, \qquad f_\beta = \frac{g_m}{2\pi\beta_0(C_\pi + C_\mu)}, \tag{9.10}$$

■ The term *beta cutoff frequency* implies the break point frequency in the Bode plot for β in Fig. 9.4(c). Likewise the term *unity-gain frequency*, f_T, is very descriptive.

where f_β is called the **beta cutoff frequency**. By setting $|\beta(\omega)|$ to unity in Eq. (9.9), we obtain the transition frequency.

$$\omega_T = \frac{g_m}{(C_\pi + C_\mu)}, \qquad f_T = \frac{g_m}{2\pi(C_\pi + C_\mu)}. \tag{9.11}$$

The quantity f_T is often given in data sheets. For instance, from the data sheet of the 2N2222A (see Fig. 4.32), $f_T = 300$ MHz (minimum). Thus, for $\beta_0 = 100$, $f_\beta \sim 3.0$ MHz (min).

DRILL 9.2

From the data sheet for the 2N2222A (Fig. 4.32), we have found that f_T is at least 300 MHz. Using this value, and the range of $\beta = h_{fe}$ guaranteed in the data sheet, calculate a typical range in f_β at $I_C = 10$ mA.

ANSWER
$0.8 < f_\beta < 4$ MHz

EXAMPLE 9.2

Using the model and bias conditions form Example 9.1, compute f_β. Let $g_m = 55$ mS, $r_\pi = 1818\ \Omega$, $\beta_0 = 100$, $C_\pi = 5.7$ pF, and $C_\mu = 1.34$ pF.

SOLUTION From Eq. (9.10),

$$f_\beta = \frac{g_m}{2\pi\beta_0(C_\pi + C_\mu)} = \frac{0.055}{(2\pi)(100)(5.7\ \text{pF} + 1.34\ \text{pF})} = 12.4\ \text{MHz}.$$

SPICE High-Frequency Model for the BJT

To simulate accurately the frequency-dependent effect of stored charge in the BJT junctions, we must provide additional parameters from Table 4.1 in the transistor model statement. Specifically, we will consider **CJC**, **CJE**, and **TF**.

CJC, the zero-bias base-collector junction capacitance $C_{\mu 0}$, is found from Eq. (9.1) if C_μ is known at some value of V_{BC}. The SPICE default value for the junction potential is **VJC** = 0.75 V, and for the grade coefficient is **MJC** = 0.33 (graded).

CJE is the zero-bias base-emitter transition capacitance C_{je0}. The default value for the junction potential is **VJE** = 0.75 V, and for the grade coefficient is **MJE** = 0.5 (abrupt). The variation of C_{je} with V_{BE} is similar to Eq. (9.2), but does not increase further as *forward* bias exceeds 0.375 V.

TF, the forward transit time τ_F, appears in Eq. (9.5), and with the aid of Eqs. (9.6) and (9.11), may be found to be $\tau_F \cong C_\pi/g_m$ or $\tau_F \cong (2\pi f_T)^{-1}$.

Given these parameters in the BJT model statement, SPICE can accurately determine the values of C_π and C_μ for the high-frequency model of Fig. 9.2(b).

■ A more exact relationship is given by
$$T_F = \frac{1}{\omega_T} = -\frac{C_{je} + C_\mu}{g_m}$$

MOSFET Model

Figure 9.5(a) depicts the MOSFET parasitic capacitances, assuming that the FET is biased in the saturation region, $V_{DS} > V_{GS} - V_t$. Five capacitances can be seen:

1. Capacitance C_{gs} consists of two parts: (a) capacitance across the oxide layer where the gate electrode overlaps the source contact area and (b) capacitance across the oxide layer between the gate electrode and the channel. Because the conducting

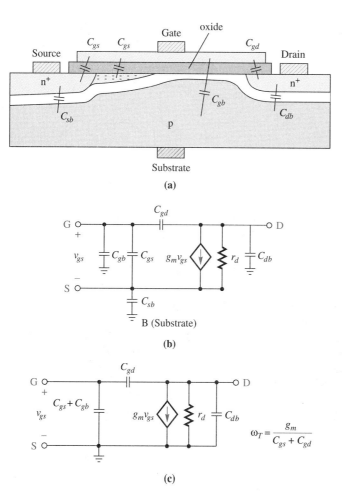

■ **FIGURE 9.5** (a) MOSFET idealized cross section showing parasitic elements (g_m and r_d are not included as parasitic elements). (b) Complete small-signal model for use at high frequencies. (c) Simplified model when source is connected to substrate.

channel is pinched off at the drain end, the capacitance is distributed nearest to the source end. Capacitance C_{gs} is usually modeled as

$$C_{gs} = \tfrac{2}{3} C_{ox} WL + C_{GSO} W,$$

where $C_{ox} = \varepsilon_{ox}/t_{ox}$ is the permittivity of SiO/oxide thickness, C_{GSO} is the gate-source overlap capacitance/unit channel width, W is channel width, and L is channel length. For $\varepsilon_{ox} = 34.5$ pF/m and $t_{ox} = 0.5$ μm, $C_{ox} = 690$ μF/m^2, and if the gate-source overlap is 0.1 μm, $C_{GSO} = 69$ pF/m. For such a MOSFET with $L = 5$ μm and $W = 100$ μm, $C_{gs} = 0.23 + 0.007 = 0.24$ pF.

2. Capacitance C_{gd} is the overlap capacitance at the drain contact, and is $C_{GDO} W$. Under the same technology as illustrated earlier, $C_{gd} \cong 0.007$ pF. With very narrow overlaps, the fringe effect may exceed the actual overlap itself.

3. Capacitance C_{gb} models the gate-substrate capacitance, where the gate electrode overlaps the substrate on both sides of the channel: $C_{gb} = C_{GBO} L$; for example,

$$C_{GBO} = C_{ox} 10 \ \mu\text{m} = 6900 \text{ pF/m} \quad \text{and} \quad C_{gb} = (6900 \text{ pF/m})(5 \ \mu\text{m}) = 0.034 \text{ pF}.$$

4. The junction capacitance between the substrate and channel is divided into two lumped capacitances C_{sb} and C_{db} at the source and drain, respectively. These capacitances are found in the abrupt-junction expressions using the reverse-junction voltages:

$$C_{sb} = \frac{C_{sb0}}{\left(1 + \dfrac{V_{SB}}{\phi_J}\right)^{1/2}} \qquad C_{db} = \frac{C_{db0}}{\left(1 + \dfrac{V_{DB}}{\phi_J}\right)^{1/2}}. \tag{9.12}$$

■ The small series resistances existing at the drain and source terminals (SPICE RD and RS) could also be used by the IC designer in CAD models for the MOSFET (see Table 5.1).

■ Compare the FET f_T with that of the BJT in Eq. (9.11).

The MOSFET high-frequency small-signal model appears as shown in Fig. 9.5(b). Often the source is connected to the substrate, so that C_{sb} would not appear, and C_{gb} would be combined with C_{gs} [Fig. 9.5(c)].

The MOSFET is described to SPICE (level 1) by providing overlap capacitances **CGDO**, **CGSO**, and **CGBO**, zero-bias junction capacitances **CBD** and **CBS**, oxide thickness **TOX**, and channel dimensions **L** and **W** in the model statement.

The high-frequency performance of MOSFETs (as well as JFETs) can be characterized by a unity gain frequency f_T, analogous to what was done for the BJT in Eq. (9.11). We urge the reader to show that

$$f_T = \frac{g_m}{2\pi(C_{gs} + C_{gd})}.$$

JFET Model

The JFET parasitic elements are easily obtained as a consequence of junction formation, although the JFET structure is more complex than a typical MOSFET. The basic JFET structure has two effective junctions [see Fig. 9.6(a)]:

1. The reverse-bias junction between the drain-source channel and gate
2. The reverse-bias junction between the gate and substrate; this is analogous to the bipolar collector-substrate junction.

The distributed depletion-region capacitance between the drain-source channel and the reverse-biased gate is divided into two lumped-element equivalents. The first lumped element is C_{gd}, the capacitance between the gate and drain region of the JFET, and the second is the lumped capacitances at the source end, C_{gd}. Typically these are diffused or

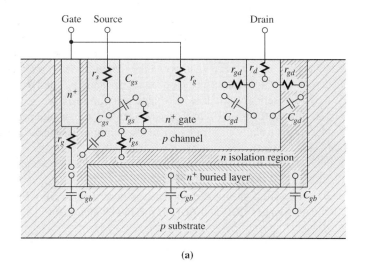

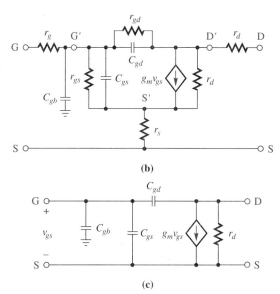

FIGURE 9.6 (a) JFET idealized cross section showing parasitic elements (g_m and r_d are not included as parasitic elements). (b) Complete small-signal model. (c) Simplified hybrid-π model.

ion-implant junctions, so the capacitance can be described by a relationship somewhere between the cube root and the square root. Assuming a square root:

$$C_{gd} = \frac{C_{gd0}}{\left(1 + \dfrac{V_{DG}}{\phi_J}\right)^{1/2}} \qquad C_{gs} = \frac{C_{gs0}}{\left(1 + \dfrac{V_{SG}}{\phi_J}\right)^{1/2}}. \tag{9.13}$$

The gate-substrate junction is abrupt, so this capacitance, C_{gb}, is given by

$$C_{gb} = \frac{C_{gb0}}{\left(1 + \dfrac{V_{BG}}{\phi_J}\right)^{1/2}}. \tag{9.14}$$

■ The junction voltages in Eq. (9.13) and Eq. (9.14) are reverse-bias voltages. Order of subscripts assumes an n-channel JFET.

The complete model is given in Fig. 9.6(b). As previously discussed, r_s, r_g, and r_d are the source, gate, and drain spreading (ohmic) resistances, respectively, of the semiconductor material itself. They are usually all less than a few tens of ohms and can be safely neglected. The neglecting of r_g is very valid when the magnitude of the almost-zero gate current is considered. The junction-effect type of resistances, r_{gd} and r_{gs}, are usually neglected since they represent very high reverse-bias junction resistances. Consequently, the model shown in Fig. 9.6(c) is used. For a discrete device, C_{gb} would not be included. The output resistance r_d is usually in parallel with other circuit resistances and can easily be combined with them.

Each of the three capacitances is on the order of 1 or 2 pF. For instance, at $V_{DS} = 15$ V and $V_{GS} = 0$ V, $C_{gd} = C_{rss} = 1.0$ pF and $C_{gs} = C_{iss} - C_{gd} = 3.5$ pF for the 2N5459 (Fig. 5.31). The JFET at high frequencies is described to SPICE by providing zero-bias junction capacitances **CGD** and **CGS** in the model statement.

■ Again, a computer simulation using this model could include these series resistances (see Table 5.2).

DRILL 9.3

Use the data for the 2N5459 given in the preceding paragraph to calculate the values for C_{gd0} and C_{gs0} for this JFET. Assume $\phi_J = 0.75$ V.

ANSWER
$C_{gd0} = 4.58$ pF and $C_{gs0} = 3.5$ pF

EXAMPLE 9.3

Compute all of the small-signal parameters for the JFET whose parameters and Q-point conditions are given in Fig. 9.7.

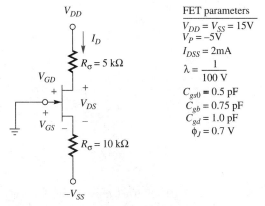

■ **FIGURE 9.7** Circuit for Example 9.3.

SOLUTION Using the loop equations,

$$V_{DD} = I_D(R_D + R_\sigma) + V_{DS} - V_{SS} \quad \text{or} \quad 30 = I_D(15 \text{ k}\Omega) + V_{DS}$$

and $0 = V_{GS} + I_D R_\sigma - V_{SS}$, which reduces to $15 = I_D(10 \text{ k}\Omega) + V_{GS}$. Neglecting the effect of λ, we can compute V_{GS} from Eq. (5.35):

$$V_{GS} = V_P\left(1 - \sqrt{\frac{I_D}{I_{DSS}}}\right) = -5\left(1 - \sqrt{\frac{I_D}{2 \text{ mA}}}\right).$$

Simultaneous solution of three equations in three unknowns yields

$$I_D \cong 1.55 \text{ mA}, \quad V_{GS} \approx -0.6 \text{ V}, \quad \text{and} \quad V_{DS} \approx 6.75 \text{ V}.$$

The small-signal parameters are computed as follows. Using Eq. (5.40),

$$g_m = \frac{-2I_{DSS}}{V_P}\left(1 - \frac{V_{GS}}{V_P}\right) = \frac{(-2)(2 \text{ mA})}{-5 \text{ V}}\left[1 - \frac{(-0.6)}{(-5.0)}\right] = 0.7 \text{ mS},$$

$$C_{gd} = \frac{C_{gd0}}{\left(1 + \frac{V_{DS}}{\phi_0}\right)^{1/2}},$$

and to obtain V_{GD}, we write $V_{GD} = I_D R_D - V_{DD} = (1.55 \text{ mA})(5 \text{ k}\Omega) - 15 = -7.25$; then

$$C_{gd} = \frac{C_{gd0}}{\left(1 + \frac{7.25}{0.7}\right)^{1/2}} = \frac{1.0 \text{ pF}}{(1 + 10.36)^{1/2}} = 0.3 \text{ pF},$$

$$C_{gs} = \frac{G_{gs0}}{\left(1 + \frac{0.6}{0.7}\right)^{1/2}} = \frac{0.5 \text{ pF}}{(1 + 0.85)^{1/2}} = 0.37 \text{ pF}.$$

$$C_{gb} = \frac{C_{gb0}}{\left(1 + \frac{V_{BG}}{\phi_J}\right)^{1/2}} = \frac{0.75 \text{ pF}}{\left(1 + \frac{15}{0.7}\right)^{1/2}} = 0.15 \text{ pF}$$

since the n-type substrate must be at the most positive point ($+V_{DD}$) in the circuit.

CHECK UP

1. **TRUE OR FALSE?** The parasitic capacitances of a BJT are junction capacitances only.
2. **TRUE OR FALSE?** The parasitic capacitances of a FET are junction capacitances only.
3. **TRUE OR FALSE?** The junction capacitances in BJTs and FETs are voltage dependent.
4. **TRUE OR FALSE?** The C_{gs} and C_{gd} capacitances in a MOSFET are voltage dependent.

9.2 THE EFFECT OF C_μ AND C_{gd}: THE MILLER EFFECT

A major objective of this chapter is to develop the ability to determine the frequency characteristics of a circuit. In pursuit of this goal, we first studied the Bode plot in Chapter 6 as a tool to obtain quickly and easily the frequency response behavior if the transfer function could be written in a standard form. Certain classes of circuit configurations lend themselves to this type of analysis. In this section, we look at circuit techniques that are useful for obtaining the frequency response effects of C_μ in a bipolar transistor common-emitter (CE) amplifier and the analogous effects due to C_{gd} in a common-source FET amplifier.

The Collector Node

A typical CE amplifier such as we investigated in Section 8.2 is redrawn in Fig. 9.8(a). The high-frequency ac model is shown in Fig. 9.8(b). This figure is identical to the ac model shown in Fig. 8.3(c) except that the transistor parasitic capacitances C_π, C_μ, and C_{cs} have been added. Note that C_w includes any stray wiring capacitance in the output circuit. As before, the coupling and bypass capacitors are considered short circuits, and r_μ is neglected. In the interest of simplifying the circuit equations, this model is further reduced in Fig. 9.8(c):

1. Rather than neglecting r_b, we incorporate it in a Thévenin equivalent of the input circuit: $R'_S = R_B \| R_S + r_b$ and $V'_S = V_S R_B/(R_B + R_S)$. The result is that we lose the B node, so that V_π represents the internal base voltage at B' rather than the external base voltage at B. This causes a small error in the calculation of A_{vi} (the same as if we had neglected r_b), which is usually on the order of 5% or less. However, this results in considerable simplification and no error in finding A_{vs}, which is usually of much greater importance.

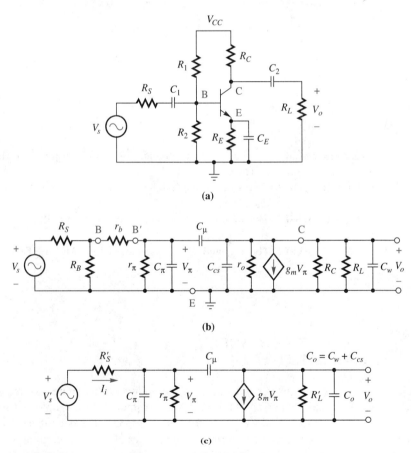

■ **FIGURE 9.8** Common-emitter amplifier high-frequency model development. (a) Common-emitter amplifier circuit. (b) High-frequency small-signal model. (c) Simplified high-frequency model.

CHAPTER 9 FREQUENCY EFFECTS IN SMALL-SIGNAL AMPLIFIERS

2. Collector node resistances are lumped together, $R'_L = R_L \| R_C \| r_o$.
3. Collector node capacitances are combined, $C_o = C_{cs} + C_w$. At the collector node,

$$V_o G'_L + V_o(j\omega C_o) + j\omega C_\mu(V_o - V_\pi) + g_m V_\pi = 0 \tag{9.15}$$

and

$$A_{vi} = \frac{V_o}{V_\pi} = \frac{-g_m + j\omega C_\mu}{G'_L + j\omega(C_\mu + C_o)} = \frac{-g_m R'_L \left(1 - \dfrac{j\omega}{g_m/C_\mu}\right)}{1 + \dfrac{j\omega}{1/[R'_L(C_\mu + C_o)]}}. \tag{9.16}$$

We make the following observations about the high-frequency transistor voltage amplification A_{vi}:

1. The *midband* A_{vi} [let $\omega \to 0$ in Eq. (9.16)] is $A_{vim} = -g_m R'_L = -g_m(R_L \| R_C \| r_o)$, exactly as it was in Eq. (8.10). This is as it should be, because neglecting the parasitic capacitances causes the high-frequency model to revert to the midband model.
2. The numerator of Eq. (9.16) has a corner frequency

$$\omega_3 = \frac{g_m}{C_\mu}. \tag{9.17}$$

At this frequency, direct transmittance between input and output nodes becomes a factor.

3. The denominator of Eq. (9.16) has a corner frequency

$$\omega_2 = \frac{1}{R'_L(C_\mu + C_o)}. \tag{9.18}$$

We illustrate the relationship of these corner frequencies by means of an example.

EXAMPLE 9.4

Consider the amplifier shown in Fig. 9.8 with $R_1 = R_2 = 220$ kΩ, $R_E = 1$ kΩ, $R_C = 2.2$ kΩ, $R_L = 4.7$ kΩ, and $R_S = 600$ Ω. The transistor parameters are $g_m = 42$ mS, $r_b = 100$ Ω, $r_\pi = 2400$ Ω, and $r_o = 125$ kΩ. (These are the same data used in Example 8.2.) Let $C_\mu = 2$ pF, $C_\pi = 10$ pF, and $C_o = 5$ pF. Calculate the amplification A_{vi} for this circuit, and sketch the resultant Bode magnitude and phase plots.

SOLUTION $R'_L = 1.48$ kΩ, $g_m R'_L = 62.2$, $\omega_3 = (42 \text{ mS})/(2 \text{ pF}) = 2.1 \times 10^{10}$ rps, and $\omega_2 = 1/(1.48 \text{ k}\Omega)(7 \text{ pF}) = 9.7 \times 10^7$ rps. Note that assuming $V_\pi = V_{be}$ causes a 4% error $\{V_\pi = V_{be}[r_\pi/[r_b + r_\pi)]\}$, which could be corrected. We neglect it here:

$$A_{vi} = \frac{-62.2(1 - j\omega/2.1 \times 10^{10})}{(1 + j\omega/9.7 \times 10^7)}.$$

We see that $|A_{vi}| = |A_{vi}(0)| = 62.2 = (35.9 \text{ dB})$, constant out to 9.7×10^7 rps, falling off at 6 dB/octave or 20 dB/decade to $|A_{vi}(\infty)| = 0.29 = (-10.8 \text{ dB})$ at 2.1×10^{10} rps. The phase is $\theta(0) = -180°$ to about 10^7 rps, $\theta(9.7 \times 10^7) = -225°$, $\theta(1.95 \times 10^{10}) = -315°$, and reaches $\theta(\infty) = 0°$ at about 2.1×10^{11} rps. These results are illustrated in Fig. 9.9 (see page 402).

We wish to stress the point that $A_{vi} = A_{vim} \approx$ constant for all frequencies from 0 to at least 10^7 rps (one decade below ω_2).

DRILL 9.4

Using the data given in Example 9.4, calculate the voltage amplification $A_{vs} = V_o/V_s$ for this circuit, assuming midband frequencies where C_1, C_2, and C_E are short and C_μ, C_π, and C_o are opens.

ANSWER
$R'_S = 697$ Ω,
$V'_s = 0.99\ V_s$, and
$A_{vs} = -47.9$

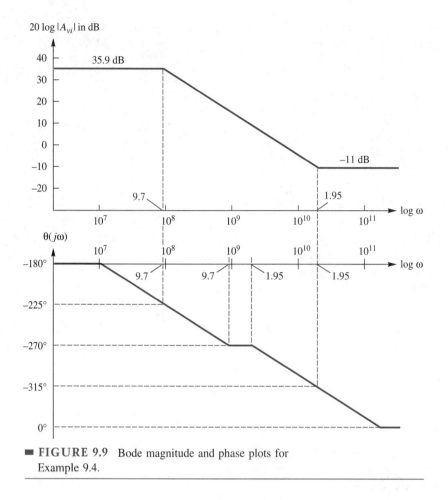

■ **FIGURE 9.9** Bode magnitude and phase plots for Example 9.4.

The Base Node and Miller Effect

The simplified CE amplifier small-signal high-frequency model is redrawn in Fig. 9.10(a). We have identified I_μ, the branch current in C_μ, with the goal of simplifying the inclusion of C_μ in the frequency-response analysis of this circuit. At the base node,

$$I_i = V_\pi g_\pi + V_\pi(j\omega C_\pi) + j\omega C_\mu(V_\pi - V_o). \tag{9.19}$$

The last term is the current that flows from the base to the collector through C_μ. This branch current is labeled I_μ in Fig. 9.10(a). Since $V_o = A_{vi}V_\pi$,

$$I_\mu = j\omega C_\mu(V_\pi - V_o) = j\omega V_\pi[C_\mu(1 - A_{vi})]. \tag{9.20}$$

If we can show that A_{vi} is real and constant, the right side of Eq. (9.20) is the current that flows from the base to the reference node through an effective capacitance

$$C_M = C_\mu(1 - A_{vi}). \tag{9.21}$$

The equivalent circuit model is shown in Fig. 9.10(b). This equivalent capacitance is called the *Miller capacitance*,[1] and it is significantly larger than C_μ. This multiplication of capacitance is a special case of a more general feedback analysis that we will study

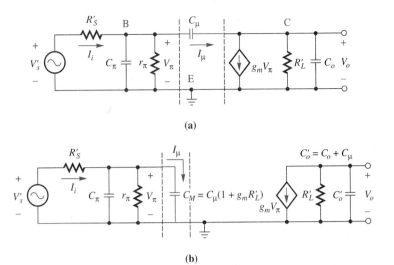

(a)

(b)

■ **FIGURE 9.10** Inclusion of Miller capacitance. (a) C_μ shown connecting base and collector. (b) Equivalent circuit, where $C_M = C_\mu(1 + g_m R'_L)$ is connected between base and reference (emitter) modes.

in Chapter 11. If valid, this approach has the obvious advantage of simplifying the base node equations because the base is effectively disconnected from the collector node.

You must be aware of the frequency dependence of A_{vi} [Eq. (9.16)], but it is enough to show that A_{vi} is constant over the range of frequencies important to our analysis. To demonstrate the implications and conditions, we proceed under the tentative assumption that $A_{vi} = A_{vim} = -g_m R'_L$. Then the Miller capacitance is

$$C_M = C_\mu(1 - A_{vi}) = C_\mu(1 + g_m R'_L). \tag{9.22}$$

Figure 9.10(b) can be redrawn as shown in Fig. 9.11, with all of the base node capacitance summed,

$$C_T = C_\pi + C_M = C_\pi + C_\mu(1 + g_m R'_L). \tag{9.23}$$

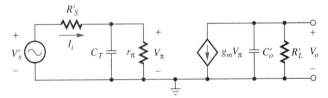

■ **FIGURE 9.11** Common-emitter amplifier high-frequency model incorporating Miller capacitance.

Rewriting the base nodal equation, we find that

$$I_i = V_\pi(g_\pi + j\omega C_T). \tag{9.24}$$

Observe that the merit of the Miller effect manipulation is that the input node may now be treated independently of the output node. This is a powerful simplification, and it works well for *inverting* amplifiers.

■ Note that at the collector node, C_μ is now in parallel with C_o, $C'_o = C_\mu + C_o$. The modified collector has a nodal equation whose denominator is identical with that of Eq. (9.16). What we have lost is the zero at $\omega_3 = g_m/C_\mu$. This is normally well beyond the frequencies of interest in the circuit.

Source-Output Voltage Amplification, $A_{vs} = V_o/V_s$

The impedance looking into the base node from R'_S is

$$z_\pi = \frac{V_\pi}{I_i} = \frac{1}{(1/r_\pi) + j\omega C_T} = \frac{r_\pi}{1 + \dfrac{j\omega}{1/r_\pi C_T}}. \quad (9.25)$$

To find the effect of the high-frequency model on the voltage amplification in a practical circuit, we now calculate the source-output amplification,

$$A_{vs} = \frac{V_o}{V_s} = \frac{V'_s}{V_s} \times \frac{V_\pi}{V'_s} \times \frac{V_o}{V_\pi} = \frac{R_B}{(R_S + R_B)} \frac{z_\pi}{R'_S + z_\pi} A_{vi} = \frac{-\dfrac{R_B}{(R_S + R_B)} \dfrac{g_m R'_L}{g_\pi + j\omega C_T}}{R'_S + \dfrac{1}{g_\pi + j\omega C_T}}$$

$$= \frac{-g_m R'_L \dfrac{R_B}{(R_S + R_B)}}{R'_S(g_\pi + j\omega C_T) + 1} = \frac{-g_m r_\pi R'_L \dfrac{R_B}{(R_S + R_B)}}{R'_S + r_\pi + j\omega r_\pi R'_S C_T} = \frac{-\dfrac{\beta R'_L}{(R'_S + r_\pi)} \dfrac{R_B}{(R_S + R_B)}}{1 + \dfrac{j\omega}{(R'_S + r_\pi)/(r_\pi R'_S C_T)}}. \quad (9.26)$$

We can make the following observations: At low to midfrequencies, where the $j\omega$ terms can be ignored,

$$A_{vsm} = -g_m R'_L \frac{r_\pi}{(R'_S + r_\pi)} \frac{R_B}{(R_S + R_B)} = -\frac{\beta(R_L \| R_C \| r_o) R_B}{(R'_S + r_\pi)(R_S + R_B)}. \quad (9.27)$$

The low-pass corner frequency is at

$$\omega_1 = \frac{R'_S + r_\pi}{r_\pi R'_S C_T} = \frac{1}{(r_\pi \| R'_S) C_T} = \frac{1}{\{[(R_B \| R_S) + r_b] \| r_\pi\}[C_\pi + C_\mu(1 + g_m R'_L)]}. \quad (9.28)$$

Note that the corner frequency is the reciprocal of the time constant formed by the product of the input C_T and the Thévenin resistance looking away from the terminals of C_T. Note also that C_M accounts for most of the high-frequency voltage-amplification degradation. The Bode magnitude and phase plots can be readily sketched (see Fig. 9.12) We now try to put this in perspective by continuing the previous example.

EXAMPLE 9.5

Find $A_{vs}(j\omega)$ for the amplifier of Example 9.4.

SOLUTION

$$R'_S = (R_B \| R_S) + r_b = (110 \text{ k}\Omega \| 600 \text{ }\Omega) + 100 = 697 \text{ }\Omega,$$

$$\omega_1 = \frac{1}{(r_\pi \| R'_S) C_T} = \frac{1}{(2400 \| 697)[10 \text{ pF} + (1 + 62.2)2 \text{ pF}]}$$

$$= \frac{1}{(540 \text{ }\Omega)(136 \text{ pF})} = 1.36 \times 10^7 \text{ rps},$$

$$A_{vsm} = -\frac{g_m r_\pi R'_L R_B}{(R'_S + r_\pi)(R_S + R_B)} = -\frac{(0.042)(2400)(1480)(110k)}{(697 + 2400)(110k + 600)} = -47.9.$$

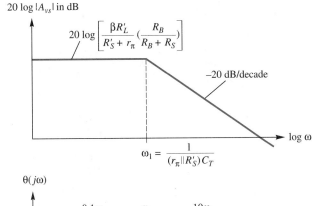

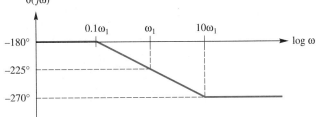

■ **FIGURE 9.12.** Bode plots for a common-emitter, Miller capacitance effect.

Thus,

$$A_{vs}(j\omega) = \frac{-47.9}{1 + j\omega/(1.36 \times 10^7)}.$$

Comparing Examples 9.4 and 9.5, we find the lowest corner frequency of A_{vi} to be nearly one decade above the corner of A_{vs}. In this situation, at least, we find it reasonable to consider the frequency dependence of A_{vi} to have little effect on A_{vs} up to its corner frequency. This is typical for CE amplifiers.

In general, use of the Miller effect gives a simple solution for the frequency response of the input circuit, but it requires $\omega \gg \omega_1$, which means that

$$\frac{1}{R'_L(C_\mu + C_o)} \gg \frac{1}{(R'_S \| r_\pi)[C_\pi + C_\mu(1 + g_m R'_L)]},$$

which is equivalent to

$$R'_L(C_\mu + C_o) \ll (r_\pi \| R'_S)[C_\pi + C_\mu(1 + g_m R'_L)]. \tag{9.29}$$

Usually $g_m R'_L \gg 1$ and $C_\mu(1 + g_m R'_L) \gg C_\pi$ so that Eq. (9.29) reduces to

$$[C_\mu + C_o] \ll (r_\pi \| R'_S) C_\mu g_m. \tag{9.30}$$

Direct Solution of A_{vs}

To gain further confidence in the Miller approach to this analysis, we compare our approximate results with the results of a direct solution of the CE amplifier shown in Fig. 9.8(c). You are encouraged to solve this two-node problem for V_o in terms of V_s.

$$\frac{V_o}{V'_s} = \frac{-G'_S(g_m - j\omega C_\mu)}{[(g_\pi + G'_S) + j\omega(C_\pi + C_\mu)][G'_L + j\omega(C_o + C_\mu)] + j\omega C_\mu(g_m - j\omega C_\mu)}, \tag{9.31}$$

DRILL 9.5

In Example 9.4 we found a corner frequency at 97 Mrps, but we treated A_{vi} as constant in finding A_{vs} in Example 9.5, where the corner then was at 13.6 Mrps. Calculate the magnitude and phase of A_{vi} at 13.6 Mrps.

ANSWER $A_{vi} = 61.8$ at $-188°$, as compared to the assumed 62.2 at $-180°$.

which can be rearranged to the form

$$\frac{V_o}{V_s} = \frac{-g_m R'_L \dfrac{r_\pi}{R'_S + r_\pi}\left(1 - \dfrac{j\omega}{g_m/C_\mu}\right)}{1 + \dfrac{j\omega}{1/[(r_\pi\|R'_S)C_T + (C_o + C_\mu)R'_L]} - \omega^2 R'_L(r_\pi\|R'_S)(C_\pi C_o + C_\pi C_\mu + C_\mu C_o)}.$$

(9.32)

We offer the following observations when comparing the exact and the approximate results:

1. Letting $\omega = 0$ in Eq. (9.32) yields identically the same A_{vsm} as found by the Miller approach in Eq. (9.27).
2. The ω^2 term will certainly have an effect at high frequencies. However, at the low-pass corner frequency it is not usually serious. For example, at the corner frequency for the amplifier of Example 9.5, the ω^2 term evaluates to 0.007.
3. Equation (9.32) has a low-pass corner frequency of

$$\omega_1 = \frac{1}{(r_\pi\|R'_S)C_T + (C_o + C_\mu)R'_L}. \qquad (9.33)$$

This corner frequency differs from the approximate value given in Eq. (9.28) by one additional term; indeed, this is the term that must be small in the inequality of Eq. (9.29). The approximate method is usually quite good, and gives a quick insight into circuit performance.

The following example is an extension of Example 9.5; in it we compare the more exact frequency response obtained by means of a SPICE solution with the Miller effect calculation.

EXAMPLE 9.6

Compare the approximate solution for A_{vs} in Example 9.5 with a SPICE solution to the problem.

SOLUTION We must write a SPICE program to plot the Bode magnitude and phase of A_{vs} from 10 Hz to 100 MHz for the amplifier of Fig. 9.8(a). Our transistor model statement is as follows: **BF** $= \beta_F = g_m r_\pi = 100$, **RB** $= r_b = 100\,\Omega$, **IS** $= I_S = 10^{-15}$ A (typical), **VAF** $= V_A = I_C r_o = (1.1\text{ mA})(125\text{ k}\Omega) = 138$ V, **CJE** $= C_{jeo} = 1$ pF (assumed),

$$\mathbf{CJC} = C_{\mu 0} = C_\mu\left(1 + \frac{V_{CB}}{\phi_J}\right)^{1/3} = 2\text{ pF}\left(1 + \frac{1.75\text{ V}}{0.75\text{ V}}\right)^{1/3} = 3\text{ pF},$$

$$\mathbf{TF} = \tau_F = \frac{C_b}{g_m} = \frac{C_\pi - C_{je}}{g_m} = \frac{10\text{ pF} - 1\text{ pF}}{42\text{ mS}} = 0.2\text{ ns}.$$

The program statements are

```
BODE PLOT OF AMPLIFIER GAIN
V10 1 0 AC 1
V40 4 0 DC 6
Q1 5 3 6 T1
RS 1 2 600
R1 4 3 220K
R2 3 0 220K
```

CHAPTER 9 FREQUENCY EFFECTS IN SMALL-SIGNAL AMPLIFIERS

```
RC 4 5 2.2K
RE 6 0 1K
RL 7 0 4.7K
C1 2 3 1E-5
C2 5 7 2.2E-6
CE 6 0 1E-4
CO 7 0 5PF
.MODEL T1 NPN BF=100 IS=IE-15 RB=100 VAF=138
+CJC=3PF CJE=1PF TF=0.2NS
.AC DEC 6 10 1E8
.PLOT AC VDB(7) VP(7) VDB(1) VP(1)
.END
```

The `.AC` statement specifies a frequency increment of six steps per decade, from 10 Hz to 100 MHz. The plot statement requests the magnitude of V_7 and V_1 in decibels ($20 \log_{10} |V_7|$) and the phase of V_7 and V_1 (in degrees). The gain in dB is the difference.

Part of the output file from this simulation is shown:

```
OPERATING POINT INFORMATION      TEMPERATURE=27.000 DEG C
BIPOLAR JUNCTION TRANSISTORS
NAME      Q1
MODEL T1
IB        1.08E-05
IC        1.09E-03
VBC      -1.79E+00
VBE       7.18E-01
VCE       2.50E+00
BETADC    1.01E+02
GM        4.21E-02
RPI       2.41E+03
RX        1.00E+02
RO        1.28E+05
CBE       1.01E-11
CBC       2.01E-12
CBX       0.00E+00
CJS       0.00E+00
BETAAC    1.01E+02
FT        5.55E+08
```

Note that the operating point agrees with the dc of Example 8.2 (the original problem), and g_m, r_π, r_b, r_o, C_π (**CBE**), and C_μ (**CBC**) all agree with the parameters given in Example 9.4. This is a check on the correctness of our model statement.

The Bode plot is shown in Fig. 9.13. Of course, to obtain the actual voltage gain directly in decibels, we let `VIN` = 1 V (0 dB) as a reference. You must be aware that in an actual circuit 1 V would not be *small signal*, but SPICE assumes a linear model for the ac analysis. Comparing these results with those of Example 9.5, SPICE finds the A_{vsm} to be -48.0 (33.63 dB), as compared to -47.9 (33.61 dB) in Example 9.5. The high-frequency corner is at 1.9 MHz according to SPICE, and at 2.16 MHz (13.6 Mrps) in the approximate solution of Example 9.5. This should give some confidence in the approximate method using the Miller effect. The difference is largely due to C_o, which the SPICE solution did not neglect.

■ Compare the SPICE results with Eq. (9.33).

DRILL 9.6

Consider the amplifier that has been analyzed in Examples 9.4, 9.5, and 9.6. Change the value of R_L from 4.7 to 1 kΩ, and find the new midband voltage gain A_{vms} and the new high-frequency break ω_1. What conclusions can be drawn from this change? The problem is an excellent application for SPICE.

ANSWER
A_{vs} = 22.2 (26.9 dB), ω_1 = 23.9 Mrps or f_1 = 3.8 MHz. Voltage amplification halved, but bandwidth doubled because the Miller capacitance is halved.

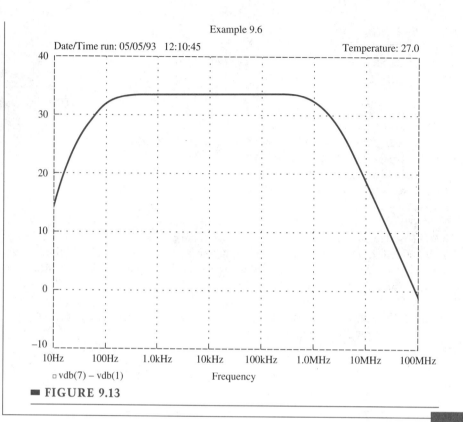

■ **FIGURE 9.13**

FET High-Frequency Analysis

The evaluation of the high-frequency performance of the FET amplifier is virtually identical to that of the bipolar transistor amplifier; the effect of C_{gd} is analogous to that of C_μ. The model used applies to either a MOSFET or a JFET.

Consider the common-source amplifier shown in Fig. 9.14(a). The high-frequency small-signal model is constructed as shown in Fig. 9.14(b). Note that for midband frequencies (and higher), the input and output coupling capacitors, and the source-bypass capacitor are considered to have negligible reactance, the parasitic gate-source and gate-drain capacitances are shown, and all of the stray capacitance-to-ground of the drain node is lumped into C_o. Let $R'_L = R_D \| R_L \| r_d$. Then, at the output node we have

$$g_m V_{gs} + V_o G'_L + j\omega C_o + j\omega C_{gd}(V_o - V_{gs}) = 0 \tag{9.34}$$

so that

$$A_v(j\omega) = \frac{V_o}{V_{gs}} = \frac{-g_m + j\omega C_{gd}}{G'_L + j\omega(C_{gd} + C_o)} = \frac{-g_m R'_L \left(1 - \dfrac{j\omega}{g_m/C_{gd}}\right)}{\left[1 + \dfrac{j\omega}{1/R'_L(C_{gd} + C_o)}\right]}. \tag{9.35}$$

A Bode plot of this frequency response would have the form of Fig. 9.9. At midband, $A_{vim} = -g_m R'_L$, as found previously in Eq. (8.50). As with the BJT, the breakpoint in

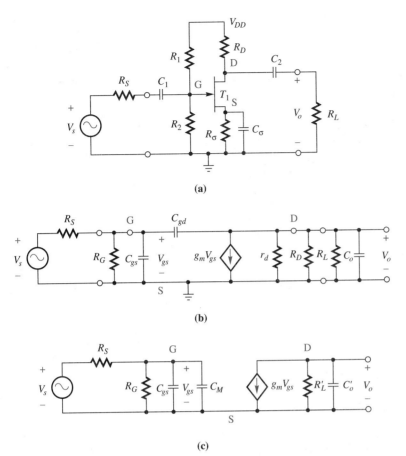

■ **FIGURE 9.14** Common-source amplifier high-frequency model development. (a) Common-source amplifier. (b) High-frequency small-signal model with C_1, C_2, C_σ neglected; C_o output stray capacitance. (c) Model incorporating Miller effect where $C_M = C_{gd}(1 - A_{vm}) = C_{gd}(1 + g_m R'_L)$ and $C'_o = C_o + C_{gd}$.

the numerator is of little consequence since it occurs at a frequency well above ω_2. This is demonstrated in Example 9.7. We observe that A_{vi} is constant to near

$$\omega_2 = 1/R'_L(C_{gd} + C_o). \tag{9.36}$$

Figure 9.14(c) shows the approximate model, with Miller capacitance, obtained from Eq. (9.21), and C_{gd} across the output node pair. The Miller approximation assumes A_{vi} constant, equal to $-g_m R'_L$. Combining, we find that

$$C_M = C_{gd}(1 - A_{vi}) = C_{gd}(1 + g_m R'_L) \tag{9.37}$$

The overall source voltage amplification is

$$A_{vs} = \frac{V_{gs}}{V_s} \times \frac{V_o}{V_{gs}} = \frac{Z_i}{R_S + Z_i} \times A_{vim} = \frac{Z_i(-g_m R'_L)}{R_S + Z_i}, \tag{9.38}$$

where

$$Z_i = R_G \left\| \frac{1}{j\omega(C_{gs} + C_M)} \right. = \frac{R_G}{1 + j\omega R_G(C_{gs} + C_M)}. \quad (9.39)$$

Combining,

$$A_{vs} = \frac{(-g_m R'_L)\dfrac{R_G}{1 + j\omega R_G(C_{gs} + C_M)}}{R_S + \dfrac{R_G}{1 + j\omega R_G(C_{gs} + C_M)}} = \frac{-g_m R'_L R_G}{R_S + R_G + j\omega R_G R_S(C_{gs} + C_M)}$$

$$= \frac{A_{vsm}}{1 + j\omega/\omega_c} = \frac{-(g_m R'_L R_G / R_S + R_G)}{1 + \dfrac{j\omega}{1/(R_G \| R_S)(C_{gs} + C_M)}}. \quad (9.40)$$

■ Note again that the modified model of Fig. 9.14(c) would lead to a nodal equation yielding an $|A_{vi}|$ and an ω_2 identical to that of Eq. (9.35). Only the zero at $\omega_3 = g_m/C_{gd}$ is lost using this approximate approach.

We see that

$$\omega_1 = \frac{1}{(R_S \| R_G)[C_{gs} + C_{gd}(1 + g_m R'_L)]}. \quad (9.41)$$

EXAMPLE 9.7

For; the CS amplifier shown in Figure 9.14, $R_S = 1$ kΩ, $R_D = 2.5$ kΩ, $R_L = 10$ kΩ, $R_\sigma = 1$ kΩ, $V_{DD} = 11$ V, $R_1 = 1$ MΩ, $R_2 = 100$ kΩ. Let $I_{DSS} = 8$ mA/V^2, $V_P = -2$ V, $C_{gs0} = 4.6$ pF, $C_{gd0} = 2.75$ pF, $\phi_J = 0.75$ V, $\lambda = 0.02$ V^{-1}. Assume C_1, C_2, and C_σ are short circuits at all frequencies of interest. Compute $A_{vs}(j\omega) = V_o/V_s$.

SOLUTION Using the dc model we find $I_D = 2$ mA, $V_{DS} = 4$ V, and $V_{GS} = -1$ V (Section 5.6). Then $g_m = (-2I_{DSS}/-V_P)(1 - V_{GS}/V_P) = 4$ mS and $r_d = (\lambda I_D)^{-1} = 50$ kΩ. Compute

$$C_{gd} = \frac{C_{gd0}}{\left(1 + \dfrac{V_{DG}}{\phi_J}\right)^{1/2}} = \frac{2.75 \text{ pF}}{\left(1 + \dfrac{5}{0.75}\right)^{1/2}} = 1 \text{ pF},$$

$$C_{gs} = \frac{C_{gs0}}{\left(1 + \dfrac{V_{SG}}{\phi_J}\right)^{1/2}} = \frac{4.6 \text{ pF}}{\left(1 + \dfrac{1}{0.75}\right)^{1/2}} = 3 \text{ pF},$$

$A_{vim} = V_o/V_{gs} = -g_m (R_L \| r_d \| R_D) = -4 \text{ (mS)}(10 \text{ k}\Omega \| 50 \text{ k}\Omega \| 2.5 \text{ k}\Omega) = -7.7$,

which is constant out to a frequency a decade below $\omega_2 = (R'_L C_{gd})^{-1} = 5.2 \times 10^8$ rps. Then the Miller capacitance is given by $C_M = C_{gd}(1 - A_{vim}) = 1$ pF $(8.7) = 8.7$ pF. Substituting this result in Eq. (9.40) yields

$$A_{vs}(j\omega) = \frac{A_{vsm}}{\left(1 + j\dfrac{\omega}{\omega_1}\right)} = \frac{(-7.7)\left[\dfrac{R_1 \| R_2}{(R_1 \| R_2) + R_S}\right]}{\left(1 + j\dfrac{\omega}{\omega_1}\right)} = \frac{-7.6}{\left(1 + j\dfrac{\omega}{\omega_1}\right)},$$

where

$$\omega_1 = \frac{1}{(R_S \| R_1 \| R_2)(C_{gs} + C_M)} = \frac{1}{(989 \text{ }\Omega)(3 \text{ pF} + 8.7 \text{ pF})} = 86.4 \text{ Mrps}.$$

Note that $\omega_1 = 86$ Mrps is not a whole decade below $\omega_2 = 520$ Mrps, so that the Miller approximation will yield a small error (see Example 9.8). Note also that the breakpoint in the numerator of Eq. (9.35), which we ignored, is at $\omega = g_m/C_{gd} = 4$ mS/1 pF $= 4$ Grps, far beyond our range of interest.

Direct Nodal Solution

Next we compare the exact representation for A_{vs} with the expression developed by using the Miller approximation. Returning to Fig. 9.14(b), we obtain the nodal equations

$$(V_s - V_{gs})G_S = V_{gs}(G_G + j\omega C_{gs}) + (V_{gs} - V_o)j\omega C_{gd} \tag{9.42}$$

and

$$(V_{gs} - V_o)j\omega C_{gd} = g_m V_{gs} + V_o(G'_L + j\omega C_o). \tag{9.43}$$

You can show that these equations can be solved for V_o in terms of V_s:

$$\frac{V_o}{V_s} = \frac{-g_m R'_L \dfrac{R_G}{R_S + R_G}\left(1 - \dfrac{j\omega}{g_m/C_{gd}}\right)}{1 + \dfrac{j\omega}{\omega'_c} - \omega^2 R'_L(R_S \| R_G)(C_{gs}C_{gd} + C_{gs}C_o + C_{gd}C_o)}, \tag{9.44}$$

where

$$\omega'_c = \frac{1}{(R_s \| R_G)[C_{gs} + C_{gd}(1 + g_m R'_L)] + (C_{gd} + C_o)R'_L}.$$

Comparing Eqs. (9.44) and (9.40), we find that

$$A_{vsm} = \frac{-g_m R'_L R_G}{R_S + R_G}$$

is identical. The corner frequency (second term in the denominator) of the equation is slightly lower than the ω_c of Eq. (9.41), but the approximation is good if

$$R'_L(C_{gd} + C_o) \ll (R_S \| R_G)[C_{gs} + C_{gd}(1 + g_m R'_L)]. \tag{9.45}$$

When, as usually holds true, $g_m R'_L \gg 1$ and $C_{gd}g_m R'_L \gg C_{gs}$, Eq. (9.45) becomes

$$(C_{gd} + C_o) \ll (R_S \| R_G)C_{gd}g_m. \tag{9.46}$$

The following example is illustrative of the differences between using the Miller effect approximation solution and the complete nodal analysis.

EXAMPLE 9.8

Using the specifications from Example 9.7, verify the validity of using the Miller effect approximation.

SOLUTION   The direct solution is given by Eq. (9.44). First observe that the corner in the numerator

$$\omega_3 = \frac{g_m}{C_{gd}} = \frac{(4 \text{ mS})}{1 \text{ pF}} = 4 \times 10^9 \text{ rps}$$

> **DRILL 9.7**
>
> Using the data from Example 9.7, perform a SPICE simulation of this amplifier. Let C_1, C_2, and C_σ be large, say 10 μF, and plot A_{vs} from 1 kHz to 100 MHz. Compare with Example 9.8.
>
> **ANSWER** The program list will have a FET model statement as follows: `.MODEL J1`
> `+NJF VTO=-2`
> `+BETA=2M`
> `+LAMBDA=0.01`
> `+CGS=4.6P`
> `+CGD=2.75P`
> `+PB=0.75`

occurs at such a high frequency that it is almost always ignored. The midband voltage gain

$$A_{vim} = -g_m(R_L\|R_D\|r_d)\left(\frac{R_1\|R_2}{R_1\|R_2 + R_S}\right) = -7.6$$

remains the same as expected. The corner is defined by

$$\omega_{c2} = \frac{1}{(R_s\|R_1\|R_2)\{C_{gs} + C_{gd}[1 + g_m(R_L\|R_D\|r_d)]\} + (R_L\|R_D\|r_d)(C_{gd} + C_o)}$$
$$= \frac{1}{(989\ \Omega)[3\ \text{pF} + 1\ \text{pF}(8.7)] + (1.92\ \text{k}\Omega)(1\ \text{pF})} = 74.1\ \text{Mrps},$$

as compared to 86.4 Mrps using the approximation technique. The percentage of error introduced by the approximations used for Eq. (9.40) is predicted by the ratio of the terms in Eq. (9.45). In this case, $(C_{gd} + C_o)R'_L = 1.92$ ns is 17% of $(R_S\|R_G)[C_{gs} + C_{gd}(1 - g_mR'_L)] = 11.6$ ns. You may note that the ω^2 term in the denominator of Eq. (9.41) is negligible at the corner frequency in this example.

CHECK UP

1. **TRUE OR FALSE?** The direct transmittance from the input node to output node through C_μ (CE amplifier) or C_{gd} (CD amplifier) has little practical effect on the output node for the frequencies of interest.
2. **TRUE OR FALSE?** The effect of C_μ (CE amplifier) or C_{gd} (CS amplifier) on the output node is totally ignored in the Miller theorem.
3. **TRUE OR FALSE?** The effect of C_μ (CE amplifier) or C_{gs} CS amplifier) on the input node is accounted for in the Miller theorem by multiplying it by $(1 - A_{vi})$ and placing it from input node to signal ground.
4. **TRUE OR FALSE?** The use of the Miller theorem is limited to CE and CS amplifiers.

9.3 SMALL-SIGNAL AMPLIFIER LOW-FREQUENCY ANALYSIS

Up to this point, we have neglected the effects of coupling and bypass capacitors in a circuit. Correctly, we have assumed that the reactance of any bypass or coupling capacitor is zero at both midband and, of course, high frequencies. Thus, we can compute the high- and low-frequency corners separately. In this section we develop techniques that, when combined with the Bode plot, allow for a quick, easy analysis or design of the complete frequency response. It will be interesting to observe that, because integrated circuit technology does not permit the use of large capacitors, the low-frequency performance calculations are often not needed. Of course, with direct coupling of signals, the circuit design must provide for appropriate biasing of cascaded states without coupling capacitors; other means of dc level shifting are used. Various approaches to this type of amplifier design are discussed later in Section 9.4 and in Chapter 10. Usually, these functions are obtained internally to the integrated circuit without using capacitors in the design; however, we may require capacitors for the system design to interface with other circuits or deliberately to obtain a specific low-frequency response.

CHAPTER 9   FREQUENCY EFFECTS IN SMALL-SIGNAL AMPLIFIERS

Low-Frequency Common-Emitter Amplifier Analysis

Consider the CE amplifier shown in Fig. 9.15(a). This circuit includes a classical self-biasing circuit, with capacitive coupling to the source and load. The next step is to draw the small-signal model. Although we have immediately included some welcome simplifications, the circuit shown in Fig. 9.15(b) is essentially complete. One quick look at this circuit and you may be tempted to throw up your hands in dismay. After all, this is only a one-transistor circuit, and already it is algebraically cumbersome. Including more devices would add considerable complexity. There are really three ways to compute the voltage gain $A_{vs}(j\omega) = V_o/V_s$ for this type of circuit:

1. A brute force algebraic effort. With six capacitors and four nodes with three or more branches, this approach quickly loses its appeal and practicality. Indeed, the final result would probably obscure key features of the solution.

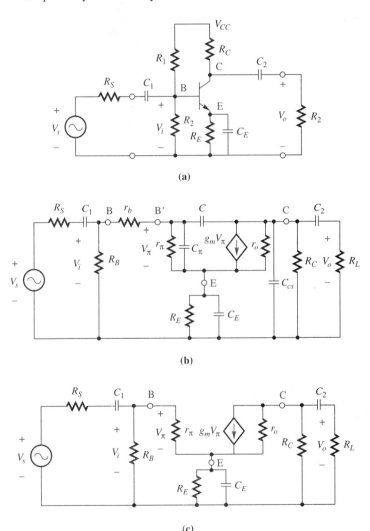

FIGURE 9.15 Common-emitter amplifier low-frequency analysis. **(a)** Common-emitter amplifier, self-biased, capacitive coupled. **(b)** Complete small-signal model. **(c)** Low-frequency small-signal model.

2. **CAD.** To obtain detailed, highly accurate results, most engineers will want to use CAD tools like SPICE. Examples will illustrate the power of this approach. Indeed, software is available for some types of circuit analysis using the more powerful programmable calculators.
3. **Approximate analysis.** The governing idea in this approach is not to obtain the most accurate result but to be able to look at a circuit and understand its key performance features by means of a series of relatively straightforward calculations, using intelligently selected approximations. These calculations then form the basis for more detailed design efforts with computer tools.

We begin by obtaining the low-frequency small-signal equivalent circuit by redrawing Fig. 9.15(b), removing all circuit elements that affect only the high-frequency performance. This circuit revision is drawn in Fig. 9.15(c) and still presents an algebraic challenge to those who persevere. The low-frequency performance of the circuit is determined by the highest corner frequency produced by C_1, C_2, or C_E. Each of the three capacitors has an effect, and those effects may be overlapping. We will find the dominant corner by considering each capacitor in turn, replacing the other two capacitors in the small-signal model with "short circuits." As you might expect, the resultant transfer function is most valid if the highest low-frequency corner is at least one decade above the next nearest corner. Essentially what we will derive is a transfer function of the form

$$A_{vs}(j\omega) = A_{vsm}[F(j\omega)], \tag{9.47}$$

where $F(j\omega)$ incorporates all frequency-dependent factors. By appropriately factoring $A_{vs}(j\omega)$ so that the terms are in a standard form, the magnitude and phase characteristics are easily sketched.

To illustrate this technique, we use the following step-by-step procedure:

1. Compute A_{vsm} by open-circuiting all capacitors affecting the high-frequency performance, such as C_π, C_μ, and C_{cs}, and short-circuiting all capacitors that affect the low-frequency performance, such C_1, C_2, and C_E.
2. Compute the low-frequency response by including either C_1, C_2, or C_E in turn, while keeping the other low-frequency capacitors short-circuited. The capacitors affecting the high-frequency response remain open-circuited.
3. Compute the high-frequency response by using the procedures outlined in Section 9.2. All capacitors affecting the low-frequency response are short-circuited.
4. Using the resultant transfer functions, sketch the Bode magnitude and phase plots, incorporating estimate of the actual response based on the asymptotic plots.

Let us apply this procedure to the CE amplifier. By means of the small-signal model shown in Fig. 9.16(a) we obtain A_{vsm}. For convenience, we will neglect r_b because usually $r_b \ll r_\pi$. Then

$$A_{vsm} = -g_m(R_L\|R_C\|r_o) \times \left(\frac{R_i}{R_i + R_S}\right), \qquad R_i = R_B\|r_\pi. \tag{9.48}$$

The Effect of the Input Coupling Capacitor C_1

To study the effect of C_1, we use the equivalent circuit given in Fig. 9.16(b). We compute $A_{vs}(j\omega)$ by using

CHAPTER 9   FREQUENCY EFFECTS IN SMALL-SIGNAL AMPLIFIERS

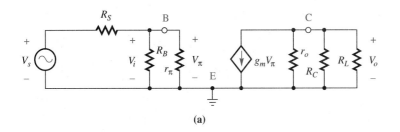

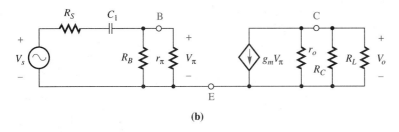

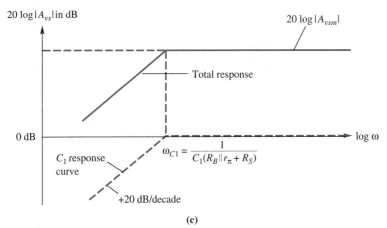

■ **FIGURE 9.16** Effect of C_1 on the frequency response.
(a) Model used to obtain midband voltage gains
$A_{vsm} = V_o/V_s$ and $A_{vim} = V_o/V_i$. (b) Model with only C_1
being considered. (c) Bode magnitude plot.

$$A_{vs}(j\omega) = \frac{V_o}{V_\pi} \times \frac{V_\pi}{V_s} = \frac{V_o}{V_s}. \tag{9.49}$$

The first term on the right side is given by $V_o/V_\pi = -g_m(R_C \| R_L \| r_o)$, which is as expected. The second term is evaluated by using voltage-divider techniques to obtain

$$\frac{V_\pi}{V_s} = \frac{R_i}{R_i + R_S + \dfrac{1}{j\omega C_1}}, \tag{9.50}$$

where $R_i = R_B \| r_\pi$ if R_E is completely bypassed by C_E. If not, R_i would be larger.

If Eq. (9.50) is incorporated into Eq. (9.49) and the result factored so that the 3-dB corner can be identified, we have

$$A_{vs}(j\omega) = \frac{[-g_m(R_L\|R_C\|r_o)] \times \left[\dfrac{R_i}{R_i + R_S}\right]}{\left[1 + \dfrac{1}{j\omega(R_S + R_i)C_1}\right]}. \tag{9.51}$$

This is easily identified as behaving as a high-pass filter of the form

$$A_{vs}(j\omega) = \frac{A_{vsm}}{\left(1 - j\dfrac{\omega_{C1}}{\omega}\right)}, \tag{9.52}$$

where

$$\omega_{C1} = \frac{1}{(R_S + R_i)C_1} = \frac{1}{(R_S + R_B\|r_\pi)C_1} \equiv \frac{1}{RC_1}. \tag{9.53}$$

By carefully studying Eq. (9.53), we can deduce an important shortcut. It is easy to observe that C_1 introduces a high-pass filter circuit behavior [see Eq. (6.28)]. Furthermore, the R in the corner frequency $\omega_{C1} = 1/(RC_1)$ is given by the Thévenin resistance across the terminals of C_1; $R = R_T = R_S + (R_B\|r_\pi)$. This result is sketched in Fig. 9.16(c). If it should turn out that R_E is not completely bypassed at frequency ω_{C1} [that is, ω_{CE2} from Eq. (9.71) is $> \omega_{C1}$], then Eq. (9.53) must be modified to include $R_E(1 + \beta)$ in series with r_π. But this would lower ω_{C1}, so Eq. (9.53) is conservative.

The Effect of the Output Coupling Capacitor C_2

To study the effect of C_2, we use the equivalent circuit shown in Fig. 9.17(a). First, the resultant $A_{vi}(j\omega)$ will be computed by means of voltage division. For the output loop,

$$V_o(j\omega) = \underbrace{-g_m(r_o\|R_C) \times V_\pi}_{\text{Thévenin's equivalent voltage source for the amplifier circuit}} \times \underbrace{\left[\frac{R_L}{(r_o\|R_C) + R_L + \dfrac{1}{j\omega C_2}}\right]}_{\text{Output voltage divider}}. \tag{9.54}$$

After some algebraic manipulation,

$$A_{vi}(j\omega) = \frac{V_o}{V_\pi} = \frac{-g_m(R_L\|R_C\|r_o)}{\left[1 - \dfrac{j}{\omega C_2(R_L + r_o\|R_C)}\right]}. \tag{9.55}$$

Since

$$A_{vs} = \frac{V_o}{V_s} = \frac{V_o}{V_\pi}\frac{V_\pi}{V_s} = A_{vi}\frac{R_i}{R_i + R_s}, \tag{9.56}$$

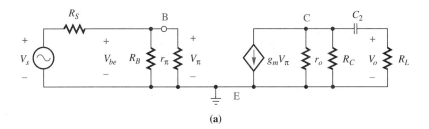

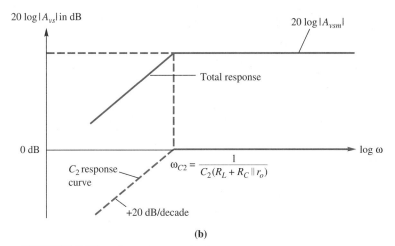

FIGURE 9.17 Effect of C_2 on the frequency response. (a) Small-signal model including C_2 only. (b) Bode magnitude plot.

we combine Eq. (9.55) with Eq. (9.56) to get the overall source voltage amplification

$$A_{vs}(j\omega) = \frac{-g_m(R_L\|R_C\|r_o)\left(\dfrac{R_i}{R_i + R_s}\right)}{\left[1 - \dfrac{j}{\omega C_2(R_L + r_o\|R_C)}\right]} = \frac{A_{vsm}}{\left(1 - j\dfrac{\omega_{C2}}{\omega}\right)}. \quad (9.57)$$

As in the case of the input coupling capacitor, we find that the output coupling capacitor modifies the midband amplification to a function having the standard form for a high-pass filter response with a corner at

$$\omega_{C2} = \frac{1}{C_2[R_L + (r_o\|R_C)]} \quad (9.58)$$

Again observe that the corner frequency could be obtained by finding the Thévenin resistance, $R_T = (R_L + r_o\|R_C)$, across C_2. The frequency response is shown in Fig. 9.17(b).

By combining Eqs. (9.52) and (9.57) to obtain

$$A_{vs}(j\omega) = \frac{A_{vsm}}{\left(1 - j\dfrac{\omega_{C1}}{\omega}\right)\left(1 - j\dfrac{\omega_{C2}}{\omega}\right)}, \quad (9.59)$$

the Bode plot for the amplifier including the effects of both C_1 and C_2 and be sketched. Assuming $\omega_{C2} = 10\omega_{C1}$, low-frequency amplitude and phase behavior is graphically presented in Figs. 9.18(a) and (b), respectively.

■ Observe the similarity here:

$\omega_{C1} = 1/C_1(R_s + R_i)$ and
$\omega_{C2} = 1/C_2(R_L + R_o)$.

The net resistance is the Thévenin equivalent resistance across C.

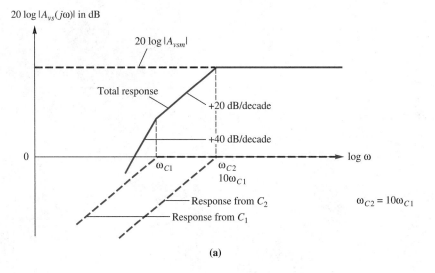

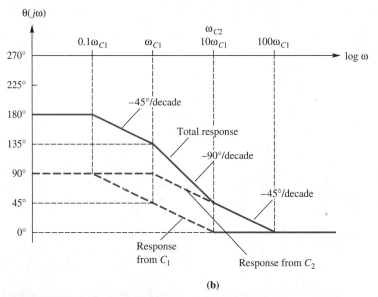

FIGURE 9.18 Bode magnitude and phase plot for

$$A_{vs}(j\omega) = \frac{A_{vsm}}{\left(1 - j\dfrac{\omega_{C1}}{\omega}\right)\left(1 - j\dfrac{\omega_{C2}}{\omega}\right)}.$$

$A_{vsm} > 0$, $\omega_{C2} = 10\omega_{C1}$.

The Effect of the Emitter Capacitor on Response

Including the effects of C_E is a bit more formidable algebraically. The equivalent circuit we will use is given in Fig. 9.19(a), where C_1 and C_2 have been replaced by short circuits. To avoid continually rewriting the same factors, we use Thévenin's theorem to reduce

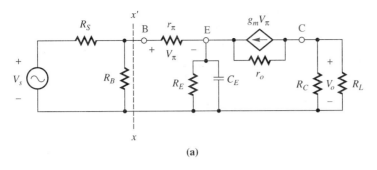

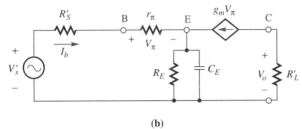

FIGURE 9.19 Effect of C_E on the frequency response. (a) Small-signal model including C_E only. (b) Simplified equivalent circuit in which $V'_s = V_s R_B/(R_B + R_S)$, $R'_S = R_s \| R_B$, $R'_L = R_C \| R_L$, $r_o \to \infty$.

the number of circuit elements to the left of x–x to that shown in Fig. 9.19(b). It is convenient to assume that $r_o \gg R_L \| R_C$ so that it can be ignored. Then

$$V'_S = \left(\frac{R_B}{R_B + R_S}\right) V_S, \quad R'_S = R_S \| R_B, \quad \text{and} \quad R'_L = R_L \| R_C. \tag{9.60}$$

The goal, as in previous steps, is to find $A_{vs}(j\omega) = V_o/V_s$ and examine the results to determine whether there is a quick, approximate technique to estimate corner frequencies. At the collector node,

$$V_o = -g_m V_\pi R'_L. \tag{9.61}$$

Summing voltages around the input loop yields

$$V'_S = I_b(R'_S + r_\pi) + (I_b + g_m V_\pi)\left(R_E \,\Big\|\, \frac{1}{j\omega C_E}\right). \tag{9.62}$$

Equation (9.62) can be expanded by using $I_b = V_\pi/r_\pi$ to obtain

$$V'_S = \frac{V_\pi}{r_\pi}(R'_S + r_\pi) + \left(\frac{V_\pi}{r_\pi} + g_m V_\pi\right)\left(\frac{R_E}{1 + j\omega R_E C_E}\right). \tag{9.63}$$

Equation (9.63) must be solved for V_π in terms of V_s and the result substituted into Eq. (9.61). Solving for V_π and collecting terms yields

$$V_\pi = \frac{V'_S}{\left(\dfrac{R'_S + r_\pi}{r_\pi}\right) + \left(\dfrac{1+\beta}{r_\pi}\right)\left(\dfrac{R_E}{1 + j\omega R_E C_E}\right)}, \tag{9.64}$$

where use has been made of $\beta = g_m r_\pi$ in the second factor. The result of combining Eq. (9.64) with Eq. (9.61) is

$$\frac{V_o}{V_s'}(j\omega) = \frac{-g_m r_\pi R_L'}{(R_S' + r_\pi) + (1 + \beta)\left(\dfrac{R_E}{1 + j\omega R_E C_E}\right)}, \tag{9.65}$$

where

$$V_s' = \frac{V_s R_B}{R_B + R_S}, \quad R_S' = R_S \| R_B, \quad R_L' = R_C \| R_L \quad \text{and} \quad r_o \to \infty.$$

Mathematically, we have accomplished our task of deriving an expression for A_{vs}. However, to use it effectively for engineering analysis or design, we would want to rewrite Eq. (9.65) in a form where the corner frequencies could be immediately identified. First, we clear fractions by multiplying by $1 + j\omega R_E C_E$ to obtain

$$\frac{V_o}{V_s'}(j\omega) = \frac{-g_m r_\pi R_L'(1 + j\omega R_E C_E)}{(1 + \beta)R_E + (1 + j\omega R_E C_E)(R_S' + r_\pi)}. \tag{9.66}$$

Expanding the denominator of Eq. (9.66),

$$\frac{V_o}{V_s'}(j\omega) = \frac{-g_m r_\pi R_L'(1 + j\omega R_E C_E)}{R_S' + r_\pi + j\omega R_E C_E(R_S' + r_\pi) + (1 + \beta)R_E}. \tag{9.67}$$

If we divide all terms by $R_S' + r_\pi + (1 + \beta)R_e$, the result is

$$\frac{V_o}{V_s'}(j\omega) = \frac{\left[\dfrac{-g_m r_\pi R_L'}{R_S' + r_\pi + (1 + \beta)R_E}\right] \times (1 + j\omega R_E C_E)}{\left[1 + \dfrac{j\omega R_e C_E(R_S' + r_\pi)}{R_S' + r_\pi + (1 + \beta)R_E}\right]}. \tag{9.68}$$

Rearranging,

■ The low-frequency voltage gain shows both a pole and a zero due to the emitter bypass capacitance CE. Fig. 9.20 shows this graphically.

$$\frac{V_o}{V_s'}(j\omega) = \frac{\left[\dfrac{-g_m r_\pi R_L'}{R_S' + r_\pi + (1 + \beta)R_E}\right] \times \left(1 + \dfrac{j\omega}{\omega_{CE1}}\right)}{\left(1 + \dfrac{j\omega}{\omega_{CE2}}\right)}. \tag{9.69}$$

There are two corner frequencies, given by

$$\omega_{CE1} = \frac{1}{R_E C_E} \tag{9.70}$$

and

$$\omega_{CE2} = \frac{R_S' + r_\pi + (1 + \beta)R_E}{C_E R_E(R_S' + r_\pi)} = \omega_{CE1}\left[\frac{R_S' + r_\pi + (1 + \beta)R_E}{R_S' + r_\pi}\right]$$

$$= \omega_{CE1}\left[1 + \frac{(1 + \beta)R_E}{R_S' + r_\pi}\right]. \tag{9.71}$$

CHAPTER 9 FREQUENCY EFFECTS IN SMALL-SIGNAL AMPLIFIERS 421

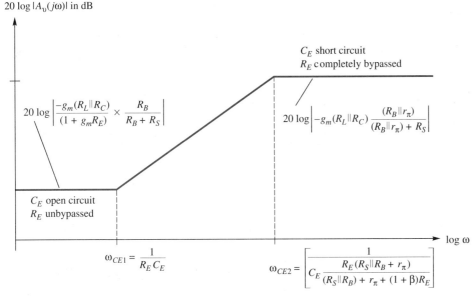

FIGURE 9.20 Effect of C_E on the voltage gain—Bode magnitude plot.

It is interesting to observe that when ω becomes large compared to ω_{CE1} and ω_{CE2}, Eq. (9.68) reduces to

$$\frac{V_o}{V'_s} = \left[\frac{-g_m r_\pi (R_L \| R_C)}{R'_S + r_\pi} \right]. \quad (9.72)$$

Upon substitution of $V'_s = V_s R_B/(R_B + R_S)$ we obtain

$$A_{vsm} = \frac{-g_m r_\pi R'_L}{(R'_S + r_\pi)} \frac{R_B}{(R_S + R_B)} = -\frac{\beta(R_L \| R_C) R_B}{(R'_S + r_\pi)(R_S + R_B)}, \quad (9.73)$$

which is the midband voltage gain previously derived in Eq. (9.27) when the emitter resistor was completely bypassed.

Similarly, if $\omega \ll \omega_{CE1}$ or ω_{CE2}, Eq. (9.69) reduces to

$$\frac{V_o}{V'_s} = \frac{[-g_m r_\pi (R'_L)]}{[R'_S + r_\pi + (1 + \beta R_E]}. \quad (9.74)$$

From Fig. 9.19, at frequencies ($\omega \ll \omega_{CE1}$) where C_E is not an effective bypass,

$$\frac{V_i}{V'_s} = \frac{[r_\pi + (1 + \beta)R_E]}{[R'_S + r_\pi + (1 + \beta)R_E]}. \quad (9.75)$$

Combining Eqs. (9.74) and (9.75), we find

$$A_{vi} = \frac{V_o}{V_i} = \frac{[-g_m r_\pi (R'_L)]}{[r_\pi + (1 + \beta)R_E]} \cong \frac{-g_m (R_C \| R_L)}{(1 + g_m R_E)}, \quad (9.76)$$

which is the result for a CE amplifier with emitter degeneration as derived in Eq. (8.22).

Using these asymptotic results, we can sketch the Bode magnitude plot as shown in Fig. 9.20. Note, from Eqs. (9.70) and (9.71), that $\omega_{CE1} \ll \omega_{CE2}$ by an amount sufficient to yield the unbypassed and completely bypassed CE voltage gains as asymptotes.

EXAMPLE 9.9

In Example 9.4 we studied the high-frequency response of a CE amplifier (Fig. 9.8). Using the same circuit and transistor parameters, analyze the low-frequency performance. Let $C_1 = 10\ \mu F$, $C_2 = 2.2\ \mu F$, and $C_E = 100\ \mu F$.

SOLUTION For the input coupling capacitor, C_1, Eq. (9.53) gives

$$\omega_{C1} = \frac{1}{(R_S + R_1\|R_2\|r_\pi)C_1} = \frac{1}{(600\ \Omega + 110\ k\Omega\|2400\ \Omega)10\ \mu F}$$
$$= 34\text{ rps or } 5.4\text{ Hz}.$$

For the output coupling capacitor, C_2, from Eq. (9.58)

$$\omega_{C2} = \frac{1}{(R_L + R_C\|r_o)C_3} = \frac{1}{(4.7\ k\Omega + 2.2\ k\Omega\|125\ k\Omega)(2.2\ \mu F)}$$
$$= 66\text{ rps or } 10.5\text{ Hz}.$$

The two corners associated with the emitter-bypass capacitor, C_E, are given by

$$\omega_{CE1} = \frac{1}{R_E C_E} = \frac{1}{(1\ k\Omega)(100\ \mu F)} = 10\text{ rps or } 1.59\text{ Hz},$$

$$\omega_{CE2} = \frac{1}{C_E\left[R_E \left\| \left(\frac{r_\pi + R_1\|R_2\|R_S}{1+\beta}\right)\right.\right]}$$

$$= \frac{1}{(100\ \mu F)\left[1\ k\Omega \left\|\left(\frac{2.4\ k\Omega + 110\ k\Omega\|600\ \Omega}{101}\right)\right.\right]} = 347\text{ rps or } 55\text{ Hz}.$$

Also, $A_{vsm} = -49$ (33.8 dB) from Eq. (9.48), which compares to -48 (33.6 dB) in Example 9.5 where r_b was not neglected. From these numbers, we can observe that the dominant low-frequency corner is established by C_E where $\omega_{CE2} = 347$ rps, and $f_{CE2} = 55$ Hz. Because C_E is not completely bypassed for frequencies below 55 Hz and based on our assumptions on the ω_1 frequency, Eq. (9.53) is incorrect, but is actually too high, an error on the conservative side (see Drill 9.8). Note that in this example, either C_E is too small (if we want good response below 55 Hz), or C_1 is larger than necessary, because its break comes below the useful bandpass of the amplifier.

DRILL 9.8

Using the data given in Example 9.9, recalculate the corner frequency for C_1, recognizing that the resistor R_E is not bypassed at the frequency of interest. This changes the input resistance R_i.

ANSWER $\omega_{C1} = 1.86$ rps or 0.3 Hz

The approximate results obtained earlier are confirmed by a detailed solution of this amplifier circuit using SPICE in Example 9.10. You are urged to refer to the complete Bode plot of Fig. 9.22, observing all four break points.

EXAMPLE 9.10

Write a SPICE program to plot the Bode magnitude of A_{vs} for the amplifier of Fig. 9.21. This is the same problem as in Example 9.6, except that now we want to observe the low-frequency response in detail. Plot from 0.01 Hz to 1 kHz. Use $\beta_F = 100$, $I_S = 10^{-15}$ A, $r_b = 100\ \Omega$, $V_A = 138$ V.

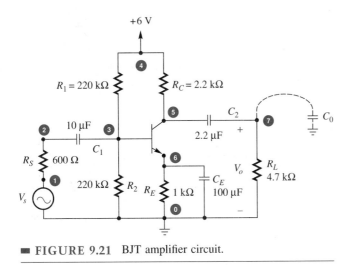

FIGURE 9.21 BJT amplifier circuit.

SOLUTION The program statements are the same as those in Example 9.6, although it is not necessary to have the parasitic capacitances in the transistor model statement, and the ac frequency analysis should range from 0.01 Hz to 1 kHz. Ten steps per decade makes an accurate plot. The plot statement requests the magnitude of V_7 in decibels ($20 \log_{10}|v_{70}|$). The Bode plot is shown in Fig. 9.22. Bode plot asymptotes are sketched in the figure, so one can see where the corners are and compare them with Example 9.9. Remember that A_{vs} in dB is V_o in dB minus V_s in dB.

■ This example was contrived to separate the break points. In practice, they probably would overlap.

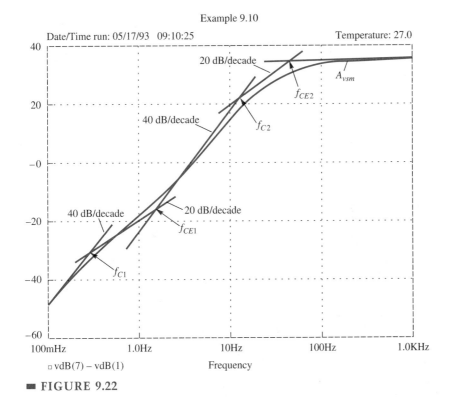

FIGURE 9.22

Complete Amplifier Frequency Response

So far we have considered the high-frequency response (Section 9.2) and the low-frequency response (Section 9.3) separately. You are now in a position to develop a transfer function for the overall frequency response by combining the various factors:

1. A midband amplification A_{vsm}
2. A low-frequency corner frequency (pole) for each coupling capacitor
3. Two low-frequency corner frequencies (pole and zero) for each BJT-emitter or FET-source bypass capacitor (Other types of *feedback* are dealt with in Chapter 11.)
4. A high-frequency corner frequency to account for the parasitic capacitances of each transistor, including the Miller effect.

For shortcuts and examples in handling *multicorner frequency response*, refer to Section 6.3.

CHECK UP

1. **TRUE OR FALSE?** There is a pair of low-frequency breakpoints for CE and CS amplifiers when their emitter (source) resistors are bypassed by a capacitor.
2. **TRUE OR FALSE?** There is a pair of low-frequency breakpoints introduced into the Bode plots of $|A_{vs}|$ for amplifiers by each coupling capacitor.
3. **TRUE OR FALSE?** When the various low-frequency breakpoints are separated by nearly a decade or more in frequency, only the highest one has practical significance.
4. **TRUE OR FALSE?** The bandwidth of an amplifier extends from the highest low-frequency breakpoint to the lowest high-frequency breakpoint.

9.4 DIRECT-COUPLED AND BiCMOS AMPLIFIER EXAMPLES

■ In the discrete transistor circuits previously discussed, capacitors were used for level shifting between sources, stages, and loads. In monolithic IC design, to avoid coupling capacitors, level shifting is accomplished by means of diodes, emitter- and source-followers, and sometimes *pnp* or *p*-channel devices.

In Section 8.8, several examples of integrated-circuit (IC) amplifiers were discussed, where the active devices were direct coupled to avoid the use of coupling capacitors. These included Darlington and cascode amplifiers, among others. The methods used to model these circuits at midband frequencies can be extended to a complete frequency analysis by including the parasitic capacitances of each device in the model. Often the signal source is capacitively coupled to the IC, and also the output of the IC may be capacitively coupled to an external load. The same observations are true with respect to BiCMOS circuits as discussed previously.

All but the simplest of these circuits can result in a rather complex model when all of the parameters are considered. Thus, a computer simulation is the best method of analysis or synthesis for many applications. But the techniques are no different than the methods

CHAPTER 9 FREQUENCY EFFECTS IN SMALL-SIGNAL AMPLIFIERS

used earlier in this chapter with discrete transistors. Two examples follow, both to give further illustration of the modeling techniques and to show how clever circuit designs, such as the cascode amplifier, can reduce the Miller effect and extend bandwidth.

The BJT Cascode Amplifier at High Frequencies

Consider the circuit shown in Fig. 9.23, consisting of a common-emitter input stage, directly coupled to a common-base amplifier. The high-frequency small-signal model is drawn in Fig. 9.24. We neglect the impedance of the coupling and bypass capacitors in this model, treating them as short circuits at the frequencies of interest; R_1 is shorted out and does not appear; and $R_B = R_2 \| R_3$. Note that there are three unknown voltages: $V_{\pi 1}$, $-V_{\pi 2}$, and V_o. It can be shown that r_{o1} and r_{o2} can be neglected. Then at node $(-V_{\pi 2})$ we write

$$g_{m1}V_{\pi 1} - g_{m2}V_{\pi 2} - g_{\pi 2}V_{\pi 2} - j\omega C_{\mu 1}(V_{\pi 2} + V_{\pi 1}) - j\omega C_{\pi 2}V_{\pi 2} = 0. \quad (9.77)$$

Grouping terms we find

$$\frac{-V_{\pi 2}}{V_{\pi 1}} = \frac{-(g_{m1} - j\omega C_{\mu 1})}{g_{m2} + g_{\pi 2} + j\omega(C_{\mu 1} + C_{\pi 2})} = \frac{\dfrac{-g_{m1}}{g_{m2} + g_{\pi 2}}\left(1 - \dfrac{j\omega C_{\mu 1}}{g_{m1}}\right)}{\left(1 + j\omega \dfrac{C_{\mu 1} + C_{\pi 2}}{g_{m2} + g_{\pi 2}}\right)} = \frac{A_2\left(1 - \dfrac{j\omega}{\omega_3}\right)}{\left(1 + \dfrac{j\omega}{\omega_2}\right)}.$$

$$(9.78)$$

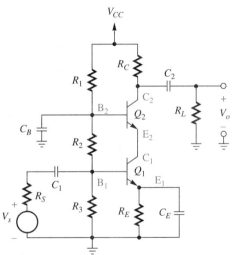

FIGURE 9.23 BJT cascode amplifier circuit.

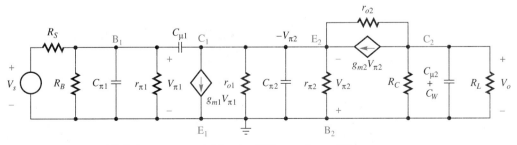

FIGURE 9.24 High-frequency model for the BJT cascode amplifier.

The midband amplification A_2 is ≈ -1, because $g_{m2} = \beta_2 g_{\pi 2}$, and $g_{m2} \cong g_{m1}$. There is a corner frequency at $\omega_3 = g_{m1}/C_{\mu 1}$, and at $\omega_2 = (g_{m2} + g_{\pi 2})/(C_{\mu 1} + C_{\pi 2})$, both at very high frequencies. At node $V_{\pi 1}$ we can write

$$(V_{\pi 1} - V_s)G_S + V_{\pi 1}g_{\pi 1} + V_{\pi 1}G_B + j\omega C_{\pi 1}V_{\pi 1} + j\omega C_{\mu 1}(V_{\pi 1} + V_{\pi 2}) = 0. \qquad (9.79)$$

The term containing $C_{\mu 1}$ is of interest. Since we have observed from Eq. (9.78) that $V_{\pi 2} \cong V_{\pi 1}$, we find that the effective Miller capacitance of the first stage is only $2C_{\mu 1}$. This is because the voltage amplification of the first stage is only -1. Rearranging Eq. (9.79), we get

$$\frac{V_{\pi 1}}{V_s} = \frac{G_S}{G_S + G_B + g_{\pi 1} + j\omega(C_{\pi 1} + 2C_{\mu 1})} = \frac{A_1}{\left(1 + \dfrac{j\omega}{\omega_1}\right)}, \qquad (9.80)$$

where

$$A_1 = \frac{G_S}{G_S + G_B + g_{\pi 1}} = \frac{R_B \| r_{\pi 1}}{R_S + R_B \| r_{\pi 1}} = \frac{R_i}{R_S + R_i},$$

and

$$\omega_1 = \frac{G_S + G_B + g_{\pi 1}}{C_{\pi 1} + 2C_{\mu 1}} \equiv \frac{1}{(R_S \| R_i)C_T}.$$

At the output node, we see

$$V_o[G_C + G_L + j\omega(C_{\mu 2} + C_w)] + g_{m2}V_{\pi 2} = 0,$$

so that

$$\frac{V_o}{-V_{\pi 2}} = \frac{g_{m2}}{[G_C + G_L + j\omega(C_{\mu 2} + C_w)]} = \frac{g_{m2}(R_C \| R_L)}{\left(1 + \dfrac{j\omega}{\omega_4}\right)} = \frac{A_3}{\left(1 + \dfrac{j\omega}{\omega_4}\right)}, \qquad (9.81)$$

where $\omega_4 = 1/(C_{\mu 2} + C_w)(R_C \| R_L)$.

The overall voltage gain of the cascode amplifier circuit is found by combining Eqs. (9.78), (9.80), and (9.81):

$$A_{vs} = \frac{V_o}{V_s} = \frac{A_1}{\left(1 + \dfrac{j\omega}{\omega_1}\right)} \frac{A_2\left(1 - \dfrac{j\omega}{\omega_3}\right)}{\left(1 + \dfrac{j\omega}{\omega_2}\right)} \frac{A_3}{\left(1 + \dfrac{j\omega}{\omega_4}\right)}$$

$$= \frac{-g_{m2}(R_C \| R_L)\left[\dfrac{R_i}{R_S + R_i}\right]\left(1 - \dfrac{j\omega}{\omega_3}\right)}{\left(1 + \dfrac{j\omega}{\omega_1}\right)\left(1 + \dfrac{j\omega}{\omega_2}\right)\left(1 + \dfrac{j\omega}{\omega_4}\right)}. \qquad (9.82)$$

The corners at ω_3 and at ω_2 are typically at least a decade higher in frequency than ω_1 and ω_4, so that the bandwidth of the amplifier is determined by ω_1 and/or ω_4. For an example consider the data of Drill 9.9.

■ The cascode connection has wide bandwidth in this circuit because the CE amplifier has $A_{vi} \simeq 1$ so Miller C_M is small, and the CB amplifier has high A_{vi} but no parasitic capacitance between the output node and the input node.

DRILL 9.9

The BJT cascode circuit of Fig. 9.24 has the following parameters: $R_S = 1$ kΩ, $R_C = 3.3$ kΩ, $R_L = 4.7$ kΩ, $R_B = 10$ kΩ, and each BJT has $I_C = 1$ mA, $\beta_F = 100$, $C_\mu = 1$ pF, and $C_\pi = 2$ pF. The output wiring capacitance $C_w = 2$ pF. Find the value of A_{vsm} and each of the high-frequency breakpoints.

ANSWER $A_{vsn} = -51$, $\omega_1 = 296$ Mrps, $\omega_2 = 9.8$ Grps, $\omega_3 = 39$ Grps, $\omega_4 = 171$ Mrps. Note that A_{vs} is down 3 dB at approximately $[1/\omega_1 + 1/\omega_4]^{-1} = 108$ Mrps.

The BiCMOS Cascode Amplifier at High Frequencies

As another example of modeling a high-fequency circuit, we revisit the BiCMOS cascode amplifier that was analyzed for midband frequencies in Example 8.15. The circuit of Fig. 8.30 is redrawn in Fig. 9.25, assuming a load resistor R_L coupled to the output node through a large capacitor. A high-frequency small-signal model is presented in Fig. 9.26, neglecting the impedance of the coupling capacitors. Note that M_1 is represented by r_d only, since $g_m v_{gs1} = 0$ because $v_{gs1} = 0$. At the output node we can write

$$V_o[G_L + g_{d1} + g_o + j\omega(C_{gd1} + C_\mu)] = -V_\pi(g_{m3} + g_o). \quad (9.83)$$

After defining $(G_L + g_{d1} + g_o) = G'_L = 1/R'_L$,

$$\frac{V_o}{-V_\pi} = \frac{g_{m3}}{G'_L + j\omega(C_{gd1} + C_\mu)} = \frac{g_{m3}R'_L}{1 + j\omega/\omega_o}, \quad \omega_o = \frac{1}{R'_L(C_{gd1} + C_\mu)}. \quad (9.84)$$

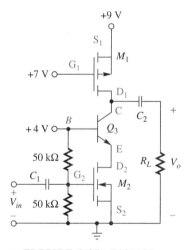

FIGURE 9.25 BiCMOS cascode amplifier circuit.

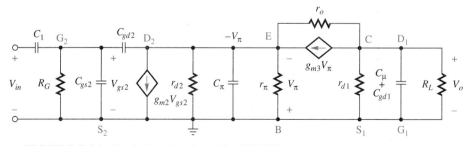

FIGURE 9.26 Small-signal model of the BiCMOS cascode amplifier circuit. Here, C_2 is assumed large.

At the emitter node,

$$-V_\pi[g_{d2} + g_o + g_{m3} + g_\pi + j\omega(C_\pi + C_{gd2})] - V_o(g_o)$$
$$+ V_{gs2}(g_{m2} - j\omega C_{gd2}) = 0. \quad (9.85)$$

Inserting $V_o = -g_{m3}R'_L V_\pi$ from Eq. (9.84), which is valid up to ω_o,

$$-V_\pi[g_{d2} + g_o + g_{m3} + g_\pi - g_{m3}(R'_L)g_o + j\omega(C_\pi + C_{gd2})]$$
$$+ V_{gs2}(g_{m2} - j\omega C_{gd2}) = 0, \quad (9.86)$$

$$\frac{-V_\pi}{V_{gs2}} = \frac{-g_{m2} + j\omega C_{gd2}}{g_{d2} + g_o + g_{m3} + g_\pi - g_{m3}g_o R'_L + j\omega(C_\pi + C_{gd2})}$$
$$= \frac{A_e[1 - j\omega(C_{gd2}/g_{m2})]}{1 + (j\omega/\omega_e)}, \quad (9.87)$$

where

$$A_e = \frac{-g_{m2}}{g_{gd2} + g_o + g_{m3} + g_\pi - g_{m3}g_o R'_L},$$

$$\omega_e = \frac{g_{gd2} + g_o + g_{m3} + g_\pi - g_{m3}g_o R'_L}{(C_\pi + C_{gd2})}.$$

The dominant high-frequency breakpoint is ω_o, and the breaks of Eq. (9.87) are of no consequence. The voltage amplification of this BiCMOS cascode circuit is

$$\frac{V_o}{V_{gs2}} = \frac{V_o}{V_e}\frac{V_e}{V_{gs2}} = \left[\frac{g_{m3}R'_L}{1 + j\omega/\omega_o}\right]\left[\frac{-g_{m2}}{g_{d2} + g_o + g_{m3} + g_\pi - g_{m3}g_o R'_L}\right], \quad (9.88)$$

where the high-frequency break is at $\omega_o = 1/R'_L(C_{gd1} + C_\mu)$.

The output resistance for this circuit $R_o \simeq r_{d1}$ (see Problem 9.43). Note also that there are low-frequency breakpoints due to C_1 and C_2:

■ Eq. (9.89) assumes $R_S = 0$.

$$\frac{V_{gs2}}{V_{in}} = \frac{R_i}{R_i + (1/j\omega C_1)} = \frac{1}{1 - (j\omega_{L1}/\omega)} \quad \text{where } \omega_{L1} = 1/R_i C_1. \quad (9.89)$$

■ EXAMPLE 9.11

Building on the midband analysis of the BiCMOS of the cascode circuit of Example 8.15, we extend the treatment to cover the frequency dependence. Recall that Q_1 has $\beta_F = 100$ and $V_A = 150$ V; M_1 has $V_t = -1$ V, $k = 50$ μA/V^2, and $\lambda = 0.02$ V^{-1}; and that M_2 is similar to M_1 except $V_t = +1$ V. In the example, we have already found $I_{DI} = 50$ μA, $r_{d1} = r_{d2} = 1$ MΩ, $g_{mM1} = g_{mM2} = 100$ μS, $g_{mQ1} = 1.92$ mS, $r_\pi = 52$ kΩ, and $r_o = 3$ MΩ.

Suppose that we now add the parasitic capacitances as follows: $C_{gs1} = C_{gs2} = 0.7$ pF. $C_{gd1} = C_{gd2} = 0.6$ pF, $C_\mu = 0.8$ pF, and $C_\pi = 1.4$ pF. Let coupling capacitor $C_1 = 1$ μF. Assume that a load resistance of 10 MΩ is added, coupled by a large capacitor. The small-signal model is drawn in Fig. 9.27. At the output node, $V_o[(1 + 0.33 + 0.1) \text{ μS} + j\omega(1.4 \text{ pF})] = -V_\pi(1.92 \text{ mS} + 0.33 \text{ μS})$ so that

$$\frac{V_o}{-V_\pi} = \frac{1.92 \text{ mS}}{1.43 \text{ μS} + j\omega\, 1.4 \text{ pF}} = \frac{1343}{1 + j\omega/1.02 \text{ Mrps}}.$$

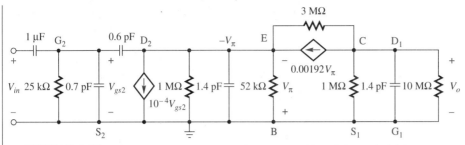

FIGURE 9.27 Small-signal model of the BiCMOS amplifier of Example 9.11.

At the emitter node,

$-V_\pi(1\ \mu S + 0.33\ \mu S + 19\ \mu S + 1.92\ mS + j\omega\ 2\ pF) - V_o(0.33\ \mu S)$
$\qquad\qquad\qquad + V_{gs2}(100\ \mu S - j\omega\ 0.6\ pF) = 0.$

Substituting $V_o = 1343$,

$-V_\pi(1\ \mu S + 0.33\ \mu S + 19\ \mu S + 1.92\ mS - 443\ \mu S + j\omega\ 2\ pF$
$\qquad\qquad\qquad + V_{gs2}(100\ \mu S - j\omega\ 0.6\ pF) = 0,$

$$\frac{-V_\pi}{V_{gs2}} = \frac{-100\ \mu S + j\omega(0.6\ pF)}{1.50\ mS + j\omega(2\ pF)} = -0.067\frac{[1 - (j\omega/167\ Mrps)]}{[1 + (j\omega/750\ Mrps)]},$$

$$\frac{V_{gs2}}{V_{in}} = \frac{25\ k\Omega}{25\ k\Omega + (1/j\omega\ 1\ \mu F)} = \frac{1}{1 - (j\ 40/\omega)},$$

$A_{vim} = (V_o/V_e)(V_e/V_{gs2})(V_{gs2}/V_{in}) = (1343)(-0.067)(1) = -90.$

There is a low-frequency breakpoint at 40 rps, and the dominant high-frequency breakpoint is at 1.02 Mrps. Note the high gain of the second stage, where the active load gives a high resistance, but also limits the bandwidth. A smaller external load connected to the output node would lower the voltage gain and affect the bandwidth. You may show that $R_o = 996\ k\Omega$.

DRILL 9.10

Example 9.11 has a low-frequency breakpoint due to the input coupling capacitor C_1 of 40 rps. What value should the output coupling capacitor C_2 have so that its breakpoint will be lower, say, 4 rps?

ANSWER
$C_2 = 0.023\ \mu F$

CHECK UP

1. **TRUE OR FALSE?** The CE first stage of the bipolar cascode amplifier has a voltage gain of -1, so the effective Miller capacitance at the input is equal to $C_{\mu1}$.
2. **TRUE OR FALSE?** The CB second stage of the bipolar cascode amplifier has no Miller capacitance.
3. **TRUE OR FALSE?** The transconductance of BJTs is much larger than that of comparable MOSFETs.

SUMMARY

In earlier chapters, BFT and FET models and circuits were assumed to be frequency independent. In this chapter, small-signal voltage-dependent capacitances inherent with junction behavior in both the BJT and FET were added to the basic frequency-independent hybrid-π mode. These capacitances, in particular, C_μ and C_{gd}, have a significant effect on the amplifier frequency dependence.

Characterization of the frequency-dependent behavior is obtained by the use of Bode magnitude and phase plots. This graphical approach is almost universally used for describing circuit behavior. We also use Bode graphical analysis in Chapter 11, where we discuss feedback amplifier frequency response and stability compensation.

Algebraic complexity is significant in frequency-response analysis; consequently, simplifying techniques, including the Miller approximation, were demonstrated in this chapter. Classical resistance-capacitance coupled common-emitter (common-source) amplifier circuits at high frequency were treated first, to emphasize the effects of parasitic capacitance in a single-stage amplifier without undue complexity. Next, low-frequency circuit behavior resulting from emitter and source bypass capacitors and coupling capacitors was demonstrated.

Modern IC design, using direct coupling to avoid capacitors, was presented. Active loads and BiCMOS designs were considered. Finally, examples incorporating both high- and low-frequency response characteristics were also given. These multicomer frequency responses are best analyzed by means of circuit analysis computer programs such as SPICE.

The subject of differential amplifiers is reserved for the next chapter. The general subject of *feedback*, where a sample of the output is inserted at the input in order to control the amplifier characteristics, is reserved for Chapter 11.

SURVEY QUESTIONS

1. Draw a high-frequency small-signal model for a BJT, for a MOSFET, and for a JFET.
2. Derive the *unity gain bandwidth* for a MOSFET with C_{gs} and C_{gd}.
3. What parameters must be given in the SPICE model statement for proper simulation of a BJT at high frequencies? A MOSFET at high frequencies? Give only the essential parameters.
4. What parameters must be known to calculate the Miller capacitance for a common-emitter amplifier?
5. In a common-emitter amplifier with large voltage gain, which capacitance (C_π or C_μ) has the most effect on high-frequency performance?
6. A signal source with resistance R_S is coupled to an amplifier with input resistance R_i by capacitance C_1. Find the resulting low-frequency corner.
7. A load resistance R_L is coupled to an amplifier with output resistance R_o by capacitance C_2. Find the resulting low-frequency corner.
8. What effect does an emitter resistance bypassed with a capacitance have on the low-frequency response of a common-emitter amplifier?
9. Compare the frequency response of a cascode amplifier with that of a common-emitter amplifier with comparable voltage gain and comparable BJTs.

PROBLEMS

The problems marked with an asterisk are more challenging.

9.1 Compute $C_{\mu O}$ for the 2N2222A if $C_{ob} = 8$ pF when $V_{CB} = 10$ V.

9.2 Find $A_{vs}(j\omega) = (V_o/V_s)(j\omega)$ for the circuit shown in Fig. 9.28. Sketch the Bode magnitude and phase plots. Consider using the Miller effect.

9.3* Proceeding with the algebraic details, show that Eq. (9.32) is the direct two-node solution for the CE amplifier circuit shown in Fig. 9.8(b).

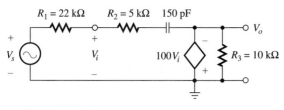

■ FIGURE 9.28

9.4 The transistor in the CE amplifier shown in Fig. 9.29 has $\beta = 100$, $V_{BE} = 0.7$ V, $C_\pi = 4$ pF, $C_\mu = 3.0$ pF, $C_W = 2$ pF, and $r_o = 100$ kΩ. You may assume $I_C = 1.13$ mA.

(a) Compute the midband voltage gain A_{vs}.
(b) Compute the high-frequency breakpoint ω_H.
  The low-frequency response of this circuit is addressed in Problem 9.20.

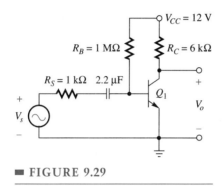

■ FIGURE 9.29

9.5 Perform a SPICE simulation of the amplifier of Fig. 9.29.

(a) Check the dc operating point and the small-signal device parameters. Show that the transistor of Problem 9.4 is accurately specified by the model statement:

**.MODEL T1   NPN   IS=2.2E-15   BF=96   VAF=109   CJE=2.5PF
+CJC=5.7PF**

(b) Use SPICE to plot $20 \log|A_{vs}(j\omega)|$.

9.6 Given the circuit shown in Fig. 9.30. Transistor parameters are $\beta = 100$, $V_A = 150$V, $C_\mu = 2$ pF, $C_\pi = 5$ pF, $r_b = 100$ Ω, and $V_{BE} = 0.7$ V.

(a) Find the dc operating point and the BJT small-signal parameters.
(b) Calculate the approximate A_{vsm} and ω_H.
(c) Ignore low frequencies (let $X_C = 0$), and plot $20 \log|A_{vs}(j\omega)|$.

The low-frequency response of this circuit is addressed in Problem 9.22.

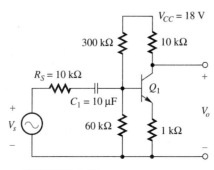

■ FIGURE 9.30

9.7 (a) A certain BJT has a base-collector parasitic capacitance $C_\mu = 2$ pF. Assuming $V_{CB} = 0.53$ V, find $C_{\mu 0} = $ CJC.
(b) Now assume $C_{je} = $ CJE ≈ 0, and $I_C = 1.52$ mA, and calculate $\tau_F = $ TF.
(c) Now use SPICE to solve Problem 9.6, using data from parts (a) and (b).
The BJT has saturation current IS $= 3 \times 10^{-15}$ A.

9.8 Calculate the value of A_{vsm} for the amplifier of Fig. 9.30, assuming that the emitter resistance is adequately bypassed at midband frequencies by means of a capacitor.

9.9* Equation (9.27) gives the midband voltage gain of the common-emitter amplifier as

$$A_{vsm} = -g_m R'_L \frac{r_\pi}{(R_S' + r_\pi)} \frac{R_B}{(R_S + R_B)},$$

where $R_S' = (R_B \| R_S) + r_b$, and $R'_L + R_C \| R_L \| r_o$. What are the design constraints and trade-offs that limit practical voltage amplification in the CE amplifier stage? Consider V_{CC}, r_b, R_L, R_S, R_B, etc.

9.10* Equation (9.28) gives the approximate bandwidth of the common-emitter amplifier as

$$\omega_1 = \frac{R_S'}{R_S' C_T},$$

where $R_S' = (R_B \| R_S) + r_b$, $C_T = C_\pi + C_\mu(1 + g_m R'_L)$, and $R'_L = R_C \| R_L \| r_o$. What must the designer do to obtain wide bandwidth in the CE amplifier?

9.11* Show that Eqs. (9.27) and (9.28) can be combined to give an expression for the gain-bandwidth product of the common-emitter amplifier stage:

$$A_{vsm} \omega_1 = \frac{-g_m R'_L r_\pi}{r_\pi R_S' C_T} \frac{R_B}{(R_S + R_B)}.$$

Now let $R_B \gg R_S$, as usually the case, and simplify. Finally for the case where $g_m R'_L \gg 1$, let

$$C_T = C_\pi + C_\mu (1 + g_m R'_L) \approx g_m R'_L C_\mu,$$

and simplify again. What are the implications of this expression to the CE amplifier designer?

9.12 Draw a small-signal high-frequency model for the common-base amplifier shown in Fig. 9.31. Neglect r_b and r_o. Show that the high-frequency response, $A_{vs}(j\omega) = V_o/V_s$, is approximately given by

$$A_{vs}(j\omega) = \frac{V_o}{V_s}(j\omega) \cong \frac{R_C/R_S}{\left(1 + j\dfrac{\omega C_\pi}{g_m}\right)(1 + j\omega R_C C_\mu)}.$$

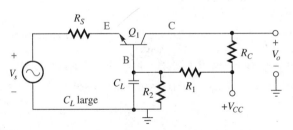

■ **FIGURE 9.31**

9.13 The emitter-follower circuit of Fig. 9.32 uses a BJT with $\beta = 100$, $V_A = 150$ V, $C_\mu = 2$ pF, $C_\pi = 5$ pF. Let $R_1 = R_2 = 100$ kΩ, $R_E = R_L = 1$ kΩ, $R_S = 4.7$ kΩ, and $V_{CC} = 9$ V.

(a) Draw a dc model, and calculate the BJT small-signal parameters.
(b) Draw a high-frequency small-signal model for the amplifier.
(c) Determine the high-frequency break point, ω_H, for the $A_{vs}(j\omega)$.

CHAPTER 9 FREQUENCY EFFECTS IN SMALL-SIGNAL AMPLIFIERS

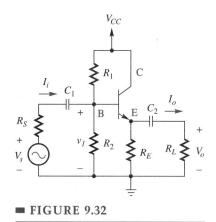

■ FIGURE 9.32

9.14 Use SPICE to work Problem 9.13. You will have to extend model parameters: CJC, CJE, IS, and TF from Problem 9.V3 results.

9.15 Determine A_{vsm} and ω_H for the circuit shown in Fig. 9.33. The transistor parameters are $\beta = 100$, $V_A = 150$ V, $C_\mu = 2$ pF, $C_\pi = 5$ pF, and $C_o = 3$ pF. Assume C_1, C_2, and C_3 are short circuits at midband.

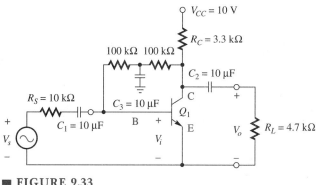

■ FIGURE 9.33

9.16 Suppose that capacitor C_3 is omitted from Fig. 9.33. Determine A_{vsm} and ω_H. The transistor parameters are $\beta = 100$, $V_A = 150$ V, $C_\mu = 2$ pF, $C_\pi = 5$ pF, and $C_o = 3$ pF.

9.17 Derive Eq. (9.44), beginning with Eqs. (9.42) and (9.43).

9.18 For the CS amplifier shown in Fig. 9.34, $R_S = 2.7$ kΩ, $R_D = 2.2$ kΩ, $R_L = 10$ kΩ, $R_\sigma = 1$ kΩ, $V_{DD} = 6$ V, $R_1 = 1.1$ MΩ, $R_2 = 100$ kΩ. Let $I_{DSS} = 4$ mA, $V_P = -2$ V, $C_{gs} = 3$ pF, $C_{gd} = 1$ pF, and $\lambda = 0.02$ V^{-1}. Assume C_1, C_2, and C_σ are short circuits at all frequencies of interest. Compute $A_{vs}(j\omega) = V_o/V_s$.

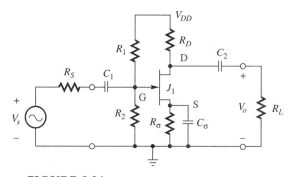

■ FIGURE 9.34

9.19 The MOSFET in the CS amplifier shown in Fig. 9.35 has $R_D = 40$ kΩ, $C_{gd} = 0.5$ pF, $C_{gs} = 1.0$ pF, $g_m = 1$ mS, $\lambda = 0$, and $R_S = 2$ kΩ. Sketch and label the Bode magnitude and phase plots.

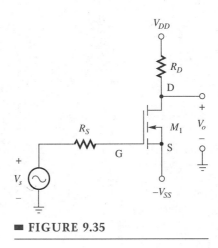

■ FIGURE 9.35

9.20 Refer to the amplifier of Fig. 9.29 and Problem 9.4.

(a) Calculate ω_L, the low-frequency break point of A_{vs} for this amplifier.
(b) Sketch the Bode amplitude and phase plots, using the data from part (a) and from Problem 9.4.

9.21 Design a circuit coupling a 10-kΩ resistive load to the collector of the amplifier of Fig. 9.29. Assume that the output resistance of the circuit of Fig. 9.29 is 6 kΩ. Your coupling capacitance must provide a low-frequency break point for the output circuit of no greater than 1 Hz.

9.22 Refer to Fig. 9.30, and assume $\beta_F = 100$ and $r_b = 100$ Ω. Calculate ω_L, the low-frequency breakpoint due to C_1.

9.23 Refer to Fig. 9.30, and assume $\beta_F = 100$ and $r_b = 100$ Ω. Modify the circuit by placing a large capacitor across the emitter resistor R_E such that $X_C \approx 0$ at the frequencies of interest. Calculate ω_L, the low-frequency break point due to C_1.

9.24 Design an amplifier using the circuit of Fig. 9.30 with a capacitor C_E to bypass the emitter resistor. Select C_E such that the low-frequency break points introduced will both be below 100 Hz. What is the minimum value of C_E?

9.25 Suppose that the transistor of Fig. 9.33 has $r_\pi = 1500$ Ω. Calculate the low-frequency break point due to C_1. Assume reactance of C_3 is negligible at this frequency.

9.26 Use SPICE to plot $20 \log|A_{vs}(j\omega)|$ from 1 Hz to 100 MHz for the amplifier of Fig. 9.33. (Use 10 steps/decade.) Use BJT parameters of Problem 9.15. What is the voltage gain and the bandwidth?

9.27 Remove C_3 from the amplifier of Fig. 9.33, and use SPICE to plot $20 \log|A_{vs}(j\omega)|$ from 1 Hz to 100 MHz for this amplifier. What is the voltage gain, and the bandwidth?

9.28* Determine the approximate A_{vsm}, ω_H, ω_L, and plot $20 \log|A_{vs}(j\omega)|$ for the circuit shown in Fig. 9.36. Assume the emitter resistors are adequately bypassed at low frequencies where C_1, C_2, and C_3 have an effect. Let $\beta = 100$, $V_A = 150$ V, $C_\mu = 2$ pF, $C_{\pi 1} = 5$ pF, $C_{\pi 2} = 35$ pF, $r_b = 100$ Ω, and $C_o = 0$.

CHAPTER 9   FREQUENCY EFFECTS IN SMALL-SIGNAL AMPLIFIERS

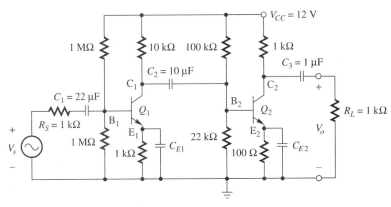

■ FIGURE 9.36

9.29 Refer to the two-stage amplifier circuit of Fig. 9.36. Assume that the coupling capacitors C_1, C_2, and C_3 are large enough that their low-frequency break points are below 10 Hz. Let $\beta = 100$ for the BJTs.

(a) How large must C_{E1} be to ensure that both of its low-frequency break points are below 100 Hz?

(b) How large must C_{E2} be to ensure that both of its low-frequency break points are below 100 Hz?

9.30 Use SPICE to plot $20 \log|A_{vs}(j\omega)|$ for the amplifier circuit of Fig. 9.36. Let both $C_E = 1000 \ \mu F$, and appropriate BJT model parameters would be **BF** $= 100$, **VAF** $= 150$ V, **IS** $= 10^{-15}$ A, **CJC** $= 3$ pF, **CJE** $= 1$ pF, and **TF** $= 0.1$ ns.

9.31* Determine the approximate A_{vsm}, ω_H, ω_L, and plot $20 \log|A_{vs}(j\omega)|$ for the circuit shown in Fig. 9.37. The transistor parameters are $\beta = 100$, $V_A = 150$ V, $C_\mu = 2$ pF, $C_\pi = 5$ pF, $r_b = 100 \ \Omega$, and $C_o = 0$.

9.32* Determine the approximate A_{vsm}, ω_H, ω_L, and plot $20 \log|A_{vs}(j\omega)|$ for the circuit shown in Fig. 9.38. The transistor parameters are $\beta = 100$, $V_A = 150$ V, $C_\mu = 2$ pF, $C_\pi = 5$ pF, $r_b = 100 \ \Omega$, and $C_o = 0$. Let C_{E1} be very large.

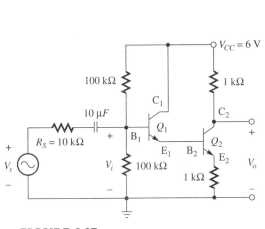

■ FIGURE 9.37

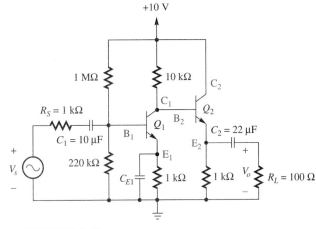

■ FIGURE 9.38

9.33 A circuit where the collector of a common-emitter amplifier is directly coupled to the emitter of a common-base amplifier is called a *cascode* circuit. Refer to Fig. 9.39.

(a) Draw a small-signal high-frequency model, making reasonable simplifications.
(b) Find the input impedance for this amplifier.
(c) Find the output impedance of this amplifier.

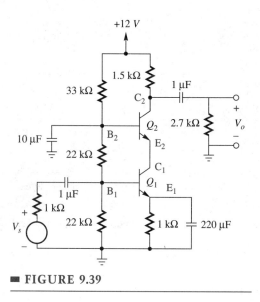

■ **FIGURE 9.39**

9.34 Both transistors in the cascode amplifier shown in Fig. 9.39 have the following characteristics: $I_C = 2.2$ mA, $\beta = 150$, $r_b = 0$, $r_o = \infty$, $C_\mu = 1$ pF, $C_\pi = 2.0$ pF, and $C_{cs} = 1$ pF.

(a) Estimate the high-frequency break point.
(b) Compute $A_{vsm} = V_o/V_s$.
(c) Sketch the Bode magnitude plot.

9.35 Problem 9.34 may be solved by SPICE using a model statement for the BJTs having `IS` $= 10^{-16}$, `BF` $= 150$, `CJC` $= 1.6$ pF, `CJE` $= 0.1$ pF, `CJS` $= 1$ pF, `TF` $= 22$ ps, and `VAF` $= 200$ V.

9.36 Derive an expression for the low-frequency $A_{vs}(j\omega) = V_o/V_s$ for the JFET common-drain (source-follower) amplifier shown in Fig. 9.40. Sketch and label the resultant Bode magnitude and phase plots.

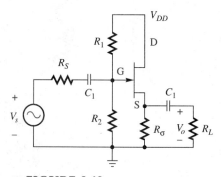

■ **FIGURE 9.40**

9.37 The transfer function for the JFET source-follower amplifier shown in Fig. 9.40 is given by

$$A_{vs}(j\omega) = \frac{V_o}{V_s}(j\omega) = \frac{A_o}{\left(1 - j\dfrac{\omega_1}{\omega}\right)\left(1 - j\dfrac{\omega_2}{\omega}\right)\left(1 + j\dfrac{\omega}{\omega_3}\right)},$$

where $\omega_1 = 5 \times 10^4$ rps, associated with C_1; $\omega_2 = 10^5$ rps, associated with C_2; and $\omega_3 = 10^7$ rps, associated with C_{gd}. Neglecting the effects of all other FET and circuit capacitances, compute values for C_1, C_2, and C_{gd}. Assume $g_m = 10$ mS, $R_S = 2$ kΩ, $R_1 = 6$ kΩ, $R_2 = 3$ kΩ, $R_L = 50$ Ω, and $R_\sigma = 100$ Ω.

9.38* Derive an expression for the complete frequency response

$$A_{vs}(j\omega) = \frac{V_o}{V_s}(j\omega)$$

for the JFET common-gate amplifier shown in Fig. 9.41. Sketch and label the resultant Bode magnitude and phase plots. Assume R_2 is completely bypassed by C_G.

9.39 Two identical, noninteracting amplifier stages are connected in cascade. The voltage gain of each stage is given by

$$A_v(j\omega) = \frac{-A_o}{\left(1 + j\dfrac{f}{f_1}\right)\left(1 - j\dfrac{f_2}{f}\right)} \qquad f_1 = 100 \text{ kHz} \quad \text{and} \quad f_2 = 20 \text{ Hz}.$$

What are the approximate high- and low-frequency 3-dB points?

9.40 The BiMOS amplifier of Fig. 9.42 consists of a BJT emitter-follower with an enhancement MOSFET active load.

(a) Draw a small-signal high-frequency model for this amplifier. Draw C_{bd} between the drain and the substrate, but of course there is no need to draw C_{bs}, since the substrate is shorted to the source in this circuit.

(b) Calculate midband voltage gain A_{vim} and the high-frequency breakpoint ω_1.

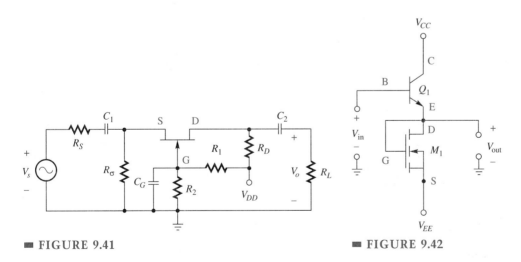

■ FIGURE 9.41   ■ FIGURE 9.42

9.41 The circuit of Fig. 9.42 has $V_{CC} = 5$ V, $V_{EE} = -5$ V, and V_{in} has no dc component. For the BJT, $\beta = 100$, $V_A = 100$ V, $C_\pi = 2$ pF, and $C_\mu = 1$ pF. For the MOSFET, $V_t = 1$ V, $k = 50$ μA/V^2, $r_d = 100$ kΩ, $C_{gs} = 1$ pF, $C_{gd} = 0.01$ pF, and $C_{bd} = 0.04$ pF.

(a) Calculate I_C, g_m, and r_o.
(b) Calculate the bandwidth. Use results of Problem 9.40 if available.

9.42★ Problem 9.41 can be simulated with SPICE. To obtain the same small-signal device models as given in Problem 9.41, you must give SPICE the appropriate model statements. For the BJT that includes `CJC` = 2 pF and `TF` = 100 ps. For the MOSFET, `KP` = $2k$, `LAMBDA` = $(I_D r_d)^{-1}$. `CBS` = 0.1 pF and `CBD` = 0.1 pF will give the information for SPICE to find C_{bs} and C_{bd}, `TOX` = $0.2 \mu m$ will provide the correct C_{gs}, and `CGDO` = 100 p (pF/m) will yield the correct C_{gd}. Try it, and experiment until SPICE model parameters are what you want. Then plot $A_{vi}(\omega)$.

9.43 Figure 9.43 shows a BiCMOS cascode amplifier.

(a) Draw a midband small-signal model for this circuit.
(b) Derive an expression for the output impedance of this amplifier.

9.44★ For the BiCMOS cascode circuit of Fig. 9.43, the device parameters are BJT has $\beta_F = 100$, $V_A = 150$ V, $C_\pi = 2$ pF, and $C_\mu = 1$ pF. MOSFET1 has $V_t = -1$ V, $k = 50\ \mu A/V^2$, $\lambda = 0.02\ V^{-1}$, $C_{gs} = 1$ pF, $C_{gd} = 0.04$ pF, and $C_{bd} = 0.1$ pF. MOSFET2 is similar to M_1 except $V_t = +1$ V.

(a) Solve for the dc operating current (from M_1) and V_{CE}, V_{DS1}, and V_{DS2}.
(b) Determine the small-signal parameters for the devices.
(c) Draw the high-frequency model for the circuit (C_1 is large).
(d) Calculate A_{vim} and high-frequency break point ω_1.

9.45★ Simulate Problem 9.44 by means of SPICE. Note suggestions given in Problem 9.42 for MOSFET model statements.

9.46★ Figure 9.44 shows a BiMOS amplifier consisting of an *npn* common-emitter with a PMOS enhancement-mode active load. C_1 is large enough to maintain constant voltage at the gate of M_1.

(a) Draw a high-frequency small-signal model for this circuit.
(b) Derive an expression for $A_{vi}(j\omega) = V_o/V_i$.

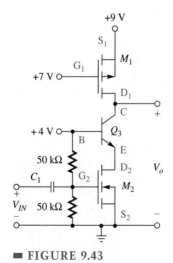

■ FIGURE 9.43

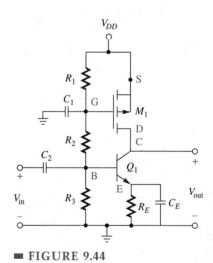

■ FIGURE 9.44

9.47 The circuit of Fig. 9.44 uses a BJT with $\beta = 100$, $V_A = 100$ V, $C_\pi = 2$ pF, $C_\mu = 1$ pF, and a MOSFET with $V_t = -1$ V, $k = 50 \mu A/V^2$, $\lambda = 0.02\ V^{-1}$, $C_{gs} = 1$ pF, $C_{gd} = 0.1$ pF, and $C_{bd} = 0.4$ pF. Let $V_{DD} = 9$ V, $R_E = 10$ kΩ, and $R_1 = R_2 = R_3 = 33$ kΩ. Assume the coupling and bypass capacitors are large. Find A_{vim} and the high-frequency break point ω_1.

9.48 How large should C_E be in the circuit of Fig. 9.44 if the amplifier is to have all low-frequency break points below 100 Hz?

9.49 How large should capacitor C_1 be in the circuit of Fig. 9.44, Problem 9.47? The amplifier must have all low-frequency breakpoints below 50 Hz.

9.50* Design a common-emitter circuit similar to Fig. 9.44, except that the active load MOSFET is a p-type depletion-mode device with the gate tied to the source. Does this arrangement have any advantages over the given circuit?

9.51 Design a bias circuit for the BiCMOS cascode amplifier of Fig. 9.43 whereby the 7- and the 4-V reference voltages are derived from the 9-V source by means of resistors and bypass capacitors. Assume that the BJT β is 100, and that these nodes must be effectively bypassed down to 100 Hz.

REFERENCES

1. Miller, J. M. "Dependence of the Input Impedance of a Three-Electrode Vacuum Tube upon the Load in the Plate Circuit," National Bureau of Standards Research Papers, Volume 15, Number 351, 1919. It is unlikely that you need to refer to this classic paper from the embryonic days of vacuum-tube circuits. However, it is interesting to observe that many of the classic circuit theorems and two-port theory developed in the 1900–1930 time frame are just as applicable to solid-state circuits as to much earlier vacuum-tube technology.

CHAPTER 10

OPERATIONAL AMPLIFIER CIRCUITRY

10.1 Differential BJT Amplifiers with Resistive Loading
10.2 Differential MOS Amplifiers with Resistive Loading
10.3 Differential JFET Amplifiers with Resistive Loading
10.4 BJT Amplifiers with Active Loading
10.5 MOS Amplifiers with Active Loading
10.6 Power Output Stages
10.7 Effects of Device Mismatch
Summary
Survey Questions
Problems
References
Suggested Readings

As discussed in Chapter 1, operational amplifiers are multistage circuit configurations especially amenable to integrated-circuit (IC) technology. They are usually characterized by high voltage gain, high input impedance, and low output impedance. However, many specialized configurations exist, and in this chapter we examine the design and application of these composite circuits. Generally speaking, an operational amplifier consists of one or more differential (difference) amplifiers, multiple current sources used for biasing and active loads, power amplifiers for output circuit drive capability, and general single-ended amplifiers for achieving voltage gain and dc level shifting. In this chapter, we examine all of these classes of circuits as they apply to the design of an integrated circuit.

In Section 10.1, we introduce the emitter-coupled differential amplifier with resistive loads. Resistively loaded metal-oxide semiconductor (MOS) amplifiers are examined in Section 10.2, followed by resistively loaded JFET amplifiers in Section 10.3. Using material related to current sources (Chapter 7), the resistive load will be replaced by a modified current source, and in that configuration we will call it an *active load*. BJT and MOS actively loaded amplifiers are discussed in Sections 10.4 and 10.5, respectively. Section 10.6 treats class B power output stages. The term *power* is used rather loosely in that for integrated circuits we rarely consider power levels over several hundred milliwatts. The discussion, however, is still applicable to audio power amplifiers using larger discrete devices such as those used in consumer audio products. We cover the effects of device mismatch in differential amplifiers in Section 10.7. SPICE simulations and related discussions are then applied to a variety of examples from earlier sections of this chapter.

Using the basic circuit configurations, definitions, and specifications in this chapter, we will proceed in Chapter 12 to the analysis and design of some typical analog ICs.

IMPORTANT CONCEPTS IN THIS CHAPTER

- Resistively loaded emitter-coupled pair circuit topology
- Definitions of input differential voltage and output differential voltage
- Emitter-coupled pair dc and ac analysis and design
- Half-circuit analysis technique
- Derivation and interpretation of the i_{C1}, i_{C2}, v_{OD} versus v_{ID} transfer functions
- Common-mode gain and common-mode rejection ratio (CMRR)
- Current source biasing
- Differential-mode and common-mode input resistance
- Resistively loaded MOS source-coupled pair circuit topology and dc and ac analyses
- Resistively loaded JFET source-coupled pair circuit topology and dc and ac analyses
- Actively loaded BJT circuit topologies and dc and ac analyses
- Actively-loaded emitter-coupled pair circuit topology and dc and ac analyses
- Actively loaded MOS amplifier topologies and load-line and dc analyses
- Definitions of class A, AB, and B power amplifiers
- Class B power amplifier circuit topology and large-signal transfer function
- Crossover distortion
- Class AB power amplifier circuit topology and large-signal transfer function
- Biasing design of class AB power amplifiers
- Back-to-back load-line analysis
- Class A power amplifier circuit topology and large-signal transfer function
- Output power and efficiency definitions
- Class A, AB, and B amplifier power and efficiency
- Short-circuit protection
- Harmonic distortion definition and computation
- Effects of input differential voltage offset voltage mismatch, collector load resistor mismatch, and input offset current mismatch

10.1 DIFFERENTIAL BJT AMPLIFIERS WITH RESISTIVE LOADING

The input stage of virtually all analog ICs uses some form of the emitter- or source-coupled differential amplifier. Figure 10.1 is a diagram of the basic emitter-coupled differential pair. We cover the MOS and JFET differential amplifiers in Sections 10.2 and 10.3. The amplifier uses both a positive power supply, V_{CC}, and a negative power supply, $-V_{EE}$. The magnitudes of the positive and negative power supplies need not be equal, although in practice, they often are.

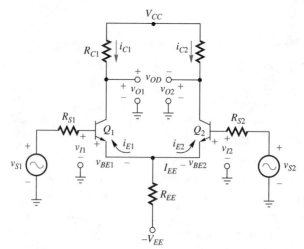

FIGURE 10.1 Emitter-coupled pair differential amplifier.

It is very important to observe that these amplifiers operate at very low frequencies, even dc, because appropriate dc level shifting is obtained without the use of coupling capacitors. No monolithic IC fabrication techniques are available for the fabrication of large capacitors (see Chapter 18).

■ Circuit topology designs and the use of junction potentials are used extensively for dc level shifting.

■ You are encouraged to use Chapter 18 as a reference whenever a fabrication or device technology issue arises.

Emitter-Coupled Pair DC Analysis

As in Chapters 4 and 5, our analysis starts with a study of the dc operation, which we will then use to set the stage for use of the small-signal model. Initially, we assume the use of matched (identical) devices, a valid assumption based on IC technology, although we will later consider the effects of small mismatches. In addition, we assume $\beta \gg 1$ so that $-i_{E1} \cong i_{C1}$ and $-i_{E2} \cong i_{C2}$. Then summing around the base loop, we have

$$v_{I1} = v_{BE1} - v_{BE2} + v_{I2}. \tag{10.1}$$

From Eq. (7.23), we get

$$v_{BE1} = V_T \ln\left(\frac{i_{C1}}{I_{S1}}\right) \quad \text{and} \quad v_{BE2} = V_T \ln\left(\frac{i_{C2}}{I_{S2}}\right). \tag{10.2}$$

Substitution of Eq. (10.2) into Eq. (10.1) yields

$$v_{I1} - v_{I2} = V_T \ln\left(\frac{i_{C1}}{I_{S1}} \times \frac{I_{S2}}{i_{C2}}\right). \tag{10.3}$$

For matched devices, the reverse-saturation currents are identical, $I_{S1} = I_{S2}$, so that Eq. (10.3) becomes

$$\frac{(v_{I1} - v_{I2})}{V_T} = \ln\left(\frac{i_{C1}}{i_{C2}}\right). \tag{10.4}$$

By defining $v_{ID} = v_{I1} - v_{I2}$ as the **input differential voltage**, Eq. (10.4) becomes

$$\frac{i_{C1}}{i_{C2}} = e^{v_{ID}/V_T}. \tag{10.5}$$

At the emitter node,

$$I_{EE} = -(i_{E1} + i_{E2}) \cong i_{C1} + i_{C2}. \quad (10.6)$$

Solving Eq. (10.6) for i_{C1} and i_{C2}, in turn, and using Eq. (10.5) results in the following two equations:

$$i_{C1} = \frac{I_{EE}}{1 + e^{-v_{ID}/V_T}}, \quad (10.7)$$

$$i_{C2} = \frac{I_{EE}}{1 + e^{v_{ID}/V_T}}. \quad (10.8)$$

These last two equations are graphed in Fig. 10.2. Let $v_{ID} = 0$. Then $i_{C1} = i_{C2} = I_{EE}/2$, a not surprising result if devices are matched. It is interesting to observe that the Q-point conditions for the collector currents have been established without the need for the self-bias R_1 and R_2 voltage dividers at the base nodes. Assuming a nominal $V_{BE(on)} = 0.7$ V, the Q-point collector currents are obtained from

$$I_{C1} = I_{C2} \cong \frac{I_{EE}}{2} = \left[\frac{-V_{BE(on)} - (-V_{EE})}{2R_{EE}}\right] = \left(\frac{-0.7 + V_{EE}}{2R_{EE}}\right). \quad (10.9)$$

We see that as v_{ID} increases from 0 V, the ratio of the collector currents changes until the point at which $v_{ID} \gg V_T$, and then one transistor is saturated and the other is cut off. In this case, $\gg$ means a factor of about 4 or so, corresponding to v_{ID} of about 100 mV at room temperature.

Rather than studying the respective collector currents, we are more interested in studying the **output differential voltage** defined by $v_{OD} = v_{O1} - v_{O2}$. The collector output voltages are given by

$$v_{O1} = V_{CC} - i_{C1}R_{C1}, \quad (10.10)$$

$$v_{O2} = V_{CC} - i_{C2}R_{C2}. \quad (10.11)$$

Assuming $R_{C1} = R_{C2} = R_C$, the output differential voltage is given by

$$v_{OD} = (V_{CC} - i_{C1}R_C) - (V_{CC} - i_{C2}R_C) = -i_{C1}R_C + i_{C2}R_C. \quad (10.12)$$

Substituting Eqs. (10.7) and (10.8) into Eq. (10.12) yields

$$v_{OD} = -I_{EE}R_C\left(\frac{1}{1 + e^{-v_{ID}/V_T}} - \frac{1}{1 + e^{v_{ID}/V_T}}\right). \quad (10.13)$$

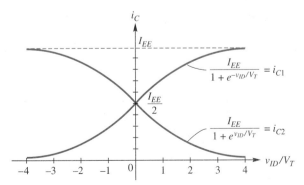

■ **FIGURE 10.2** Emitter-coupled pair collector currents as a function of the differential input voltage.

After some manipulation, Eq. (10.13) can be written

$$v_{OD} = -I_{EE}R_C\left(\frac{e^{v_{ID}/2V_T}}{e^{v_{ID}/2V_T} + e^{-v_{ID}/2V_T}} - \frac{e^{-v_{ID}/2V_T}}{e^{-v_{ID}/2V_T} + e^{v_{ID}/2V_T}}\right). \quad (10.14)$$

Combining the terms,

$$v_{OD} = -I_{EE}R_C\left(\frac{e^{v_{ID}/2V_T} - e^{-v_{ID}/2V_T}}{e^{v_{ID}/2V_T} + e^{-v_{ID}/2V_T}}\right). \quad (10.15)$$

The term in parentheses is reduced to the hyperbolic tangent:

$$v_{OD} = -I_{EE}R_C \tanh\left(\frac{v_{ID}}{2V_T}\right). \quad (10.16)$$

The differential output voltage as a function of the normalized differential input is plotted in Fig. 10.3.

The slope of this curve, v_{od}/v_{id}, is called the ***differential-mode voltage gain***, A_{dm}. We observe in Figs. 10.2 and 10.3 that for small v_{id}, the curves are quite linear. This is illustrated by computing the slope at $v_{ID} = 0$. Differentiating Eq. (10.16),

$$\frac{dv_{OD}}{dv_{ID}} = -I_{EE}R_C \frac{d}{dv_{ID}}\left[\tanh\left(\frac{v_{ID}}{2V_T}\right)\right] = -\frac{I_{EE}R_C}{2V_T}\operatorname{sech}^2\frac{v_{ID}}{2V_T},$$

$$\left.\frac{dv_{OD}}{dv_{ID}}\right|_{v_{ID}=0} = -\frac{I_{EE}R_C}{2V_T}\operatorname{sech}^2(0) = -\frac{I_{EE}R_C}{2V_T}. \quad (10.17)$$

Realizing that $I_C = I_{EE}/2$ and $g_m = I_C/V_T$, we can reduce Eq. (10.17) to the interesting result of

$$A_{dm} = -g_m R_C, \quad (10.18)$$

■ Again, small-signal is a relative term. In this case, $v_{id} \sim 10\%$ of full scale or a few mV.

■ Observe that a single-ended output would yield $A_v = \pm\frac{g_m R_C}{2}$ depending upon which collector was used.

■ There are many techniques available to expand the linear range. For example, adding an R_E in each emitter of Q_1 and Q_2. Voltage gain will be less, however. *Linearization* using *negative feedback* is discussed in Chapters 6 and 11.

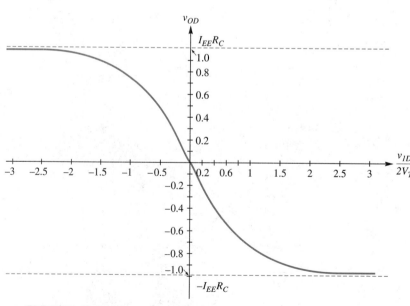

■ **FIGURE 10.3** Differential-output voltage versus normalized differential input voltage.

which is identical to the voltage gain of a single-ended common-emitter amplifier, Eq. (8.10). Furthermore, for $v_{ID} = 0$ there is a zero dc differential output voltage as input to the next stage.

Emitter-Coupled Pair AC Analysis

Equation (10.18), obtained from the slope of the transfer function, is a small-signal voltage gain expression. Intuitively, we would expect that if we incorporate the hybrid-π model into the differential amplifier illustrated in Fig. 10.1, we should be able to obtain the same result. Figure 10.4 illustrates this procedure.

A hybrid-π model, in which r_o has been neglected, has been used for both devices. Because $I_{C1} = I_{C2}$, $g_{m1} = g_{m2} = g_m = I_{C1}/V_T$. Furthermore, because we are applying a differential input voltage that can be conveniently split into two separate generators as shown in Fig. 10.4, another simplification is possible. The small-signal base-input voltages, $v_{id}/2$ and $-v_{id}/2$, are 180° out of phase, as represented by the superimposed sine-wave sketches. Therefore, these small-signal complementary voltages will exactly cancel at the coupled-emitter node. Because $v_e = 0$, this node can be considered to be at ac (small-signal) ground potential. The circuit can then be simplified as shown in Fig. 10.5. This is called the ***differential-mode half-circuit***.

■ Another way to look at this is the out-of-phase emitter currents cancel yielding $v_e = 0$.

Using Fig. 10.4, we have

$$\frac{v_{od}}{2} = \frac{-g_m R_C v_{id}}{2} \tag{10.19}$$

or

$$\frac{v_{od}}{v_{id}} = A_{dm} = -g_m R_C. \tag{10.20}$$

We can add the effect of the transistor output impedance, r_o, to Eq. (10.20) by inspection to obtain

$$A_{dm} = -g_m(R_C \| r_o). \tag{10.21}$$

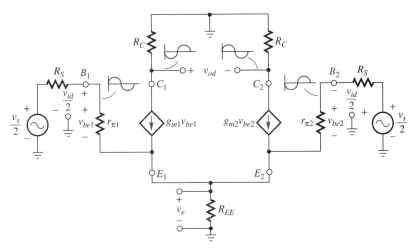

■ **FIGURE 10.4** Small-signal model with a differential input voltage.

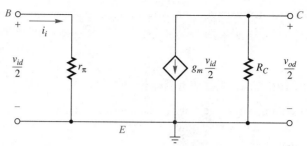

FIGURE 10.5 Differential-mode half-circuit small-signal model.

DRILL 10.1

Repeat Example 10.1 but assume $\beta = 50$ for both BJTs.

ANSWER The solution is essentially unchanged because both $\beta = 50 \gg 1$ and $\beta = 200 \gg 1$. More precisely, however, $I_E = 50$ µA and $I_C = 49$ µA, $I_B = 0.98$ µA, $g_m = 188.5$ mS, $A_{dm} = -188.5$ neglecting V_A and -179.5 including a finite V_A.

DRILL 10.2

What is the largest I_{EE} allowable for Example 10.1 with $V_{IN} = 0$ before the BJTs saturate and what value of R_{EE} is used to set this value?

ANSWER
$I_{EE} = 210$ µA,
$R_{EE} = 68$ kΩ

EXAMPLE 10.1

For the differential amplifier shown in Fig. 10.1, compute the Q-point values for I_C, I_B, I_E, and V_{CE}, and also compute A_{dm}. Assume $V_{CC} = 10$ V, $V_{EE} = 15$ V, $R_C = 100$ kΩ, $I_{EE} = 100$ µA, $R_{S1} = R_{S2}$, and $V_T = 26$ mV. For the transistors, let $V_A = 100$ V and $\beta = 200$.

SOLUTION $I_C = 50$ µA by inspection, since at the Q-point, the I_{EE} current source divides equally between Q_1 and Q_2, assuming matched transistors and matched collector resistors with $\beta \gg 1$. More precisely we have $I_E = -50$ µA and $I_C = -(I_E + I_B) = -(I_E + I_C/\beta)$ and then $I_C = -\beta I_E/(1 + \beta) = 49.75$ µA with $I_B = 0.25$ µA. We obtain V_{CE} from loop analysis:

$$V_{CE} = V_{CC} - I_C R_C + V_{BE(\text{on})}$$
$$= 10 \text{ V} - (50 \text{ µA})(100 \text{ kΩ}) + 0.7 \text{ V} = 5.7 \text{ V},$$
$$g_m = \frac{I_C q}{kT} = \frac{50 \text{ µA}}{26 \text{ mV}} = 1.92 \text{ mS},$$
$$A_{dm} = -g_m R_C = -(1.92 \text{ mS})(100 \text{ kΩ}) = -192.$$

More precisely, including the transistor output resistance $r_o = V_A/I_C = 100 \text{ V}/50 \text{ µA} = 2$ MΩ yields

$$A_{dm} = -g_m(R_C \| r_o) = -(1.92 \text{ mS})(100 \text{ kΩ} \| 2 \text{ MΩ}) = -183.$$

EXAMPLE 10.2

Rework Example 10.1 using SPICE. In addition, verify the Fig. 10.2 emitter-coupled pair collector currents as a function of the input differential voltage and the Fig. 10.3 output differential voltage as a function of the input differential voltage.

SOLUTION The node-labeled schematic diagram is given in Fig. 10.6. The resultant netlist is given next. Observe that an ac voltage source and a transfer function voltage source are in series and will yield both the ac and transfer function results using superposition principles. No frequency-dependent elements have been included. The ac solution is to be computed at 1 kHz.

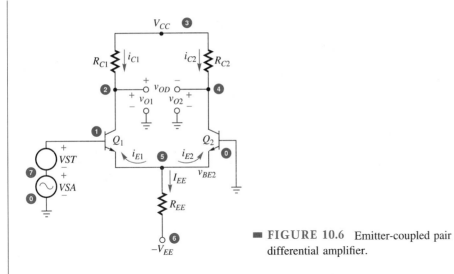

FIGURE 10.6 Emitter-coupled pair differential amplifier.

```
EMITTER-COUPLED PAIR, EXAMPLE 10.2
VCC 3 0 DC 10
VEE 6 0 DC-15
VSA 7 0 AC 1 0
VST 1 7 DC 0
RC1 3 2 100K
RC2 3 4 100K
REE 5 6 143K
Q1 2 1 5 DEV
Q2 4 0 5 DEV
.MODEL DEV NPN BF=200 VAF=100
.DC VST -200M 200M 10M
.AC DEC 1 1E3 1E3
.PRINT VM(2) VP(2) VM(4) VP(4) VM(7) VP(7)
.OP
.PROBE
.END
```

Key SPICE simulation Q-point results, with the usual approximations from Example 10.1, are given by

NAME	Q1	Q2	Ex. 10.1
MODEL	DEV	DEV	
IB	2.37E-07	2.37E-07	0.25 μA
IC	4.98E-05	4.98E-05	50 μA
VBE	6.95E-01	6.95E-01	0.7 volts
VCE	5.72E+00	5.72E+00	5.7 volts
GM	1.92E-03	1.92E-03	1.9 mS
RPI	1.09E+05	1.09E+05	104 kΩ
RO	2.11E+06	2.11E+06	2 MΩ

Figure 10.7(a) is a plot of I_{C1}, I_{C2}, and I_{EE} as a function of $VST = v_{ID}$. As expected, when $V_{ID} = 0$, $I_{C1} = I_{C2} = 50$ μA and the functions behave according to results derived in Fig. 10.2. Figure 10.7(b), the output differential voltage as a function of the input differential voltage, compares, as expected, with Fig. 10.3.

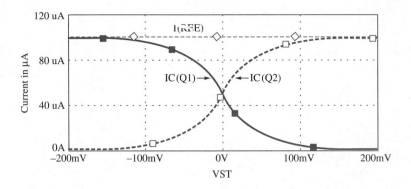

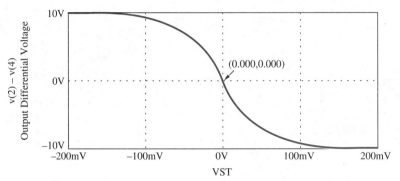

■ **FIGURE 10.7** (a) I_{C1}, I_{C2}, and I_{EE} as a function of v_{ID}.
(b) v_{OD} as a function of v_{ID}.

The ac results are summarized below. Note that the collector voltages are 180° out of phase. We then compute $A_{dm} = (V(2) - V(4))VSA = (V(2) - V(4))V(7) = -184.06$, which compares to -183 in Example 10.1.

```
   FREQ      VM(2)      VP(2)      VM(4)      VP(4)      VM(7)      VP(7)
1.000E+03  9.203E+01  1.800E+02  9.168E+01  0.000E+00  1.000E+00  0.000E+00
```

Common-Mode Rejection Ratio

To this point, we have concerned ourselves with the differential-voltage performance of an emitter-coupled amplifier. The input voltage is applied between the bases of Q_1 and Q_2, which results in the two input voltages being 180° out-of-phase. In a practical system configuration, there could be an in-phase voltage component we call the ***common-mode voltage***, v_{ic}, on each input as shown in Figure 10.8. For instance, you could be trying to amplify a weak signal (the differential component) in the presence of a large 60-Hz power-line-induced noise component, which is also present at each input as an in-phase signal. It would be extremely desirable to be able to amplify selectively the differential component much more than the common voltage component. We now demonstrate that the emitter-coupled or source-coupled amplifier does this; indeed, this ability to reject the common-mode signal provides a figure of merit for the circuit.

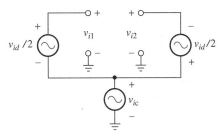

FIGURE 10.8 Linear superposition of differential- and common-mode input voltages.

As shown in Fig. 10.8, linear superposition allows us to consider the effects of v_{id} and v_{ic} independently, where

$$v_{i1} = +v_{id}/2 + v_{ic} \quad (10.22)$$

and

$$v_{i2} = -v_{id}/2 + v_{ic}. \quad (10.23)$$

A small-signal equivalent circuit, with a hybrid-π transistor model, of the emitter-coupled pair is shown in Fig. 10.9 with the same in-phase voltage, v_{ic}, applied to both inputs. We can make a few qualitative observations.

The output voltage at each collector will be in phase in contrast to the out-of-phase collector voltage relationship present for a differential input voltage. Each collector voltage will be called the ***common-mode output voltage***, v_{oc}. For the common-mode case, $v_e \neq 0$. The goal is to derive an expression for the ***common-mode voltage gain***, $A_{cm} = v_{oc}/v_{ic}$. Because $I_{C1} = I_{C2}$, $g_{m1} = g_{m2}$. It would not be too difficult to solve for A_{cm} directly using Fig. 10.9 but we can simplify the analysis by applying the half-circuit techniques in a manner very similar to that used for the differential-mode circuit.

■ Because the common-mode signals at each input are in phase, the outputs at each collector will be in phase also.

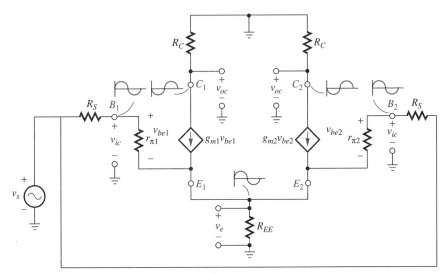

FIGURE 10.9 Small-signal model with a common-mode input voltage.

The small-signal model from Fig. 10.9 can be redrawn in the form illustrated in Fig. 10.10(a). The only difference between these circuits is that R_{EE} is modeled as two resistors, each of the value $2R_{EE}$, in parallel. Now, because the circuit is symmetrical, i_a must necessarily be zero. The circuit simplifies to that shown in Fig. 10.10(b). Common-mode circuit operation is not disturbed by opening the coupled-emitter connection since $i_a = 0$ and v_e is the same at each emitter. The resultant circuit is identical to the common-emitter amplifier with emitter resistor that we studied in Chapter 8. By comparing Figs. 8.5 and 10.10(b) and substituting Eqs. (8.22) and (8.23), the common-mode voltage gain is given by

$$A_{cm} \cong -\frac{g_m R_C}{1 + 2g_m R_{EE}}. \tag{10.24}$$

Assuming $2g_m R_{EE} \gg 1$, Eq. (10.24) reduces to

$$A_{cm} \cong \frac{-R_C}{2R_{EE}}. \tag{10.25}$$

The figure of merit comparing A_{dm} and A_{cm} is called the **common-mode rejection ratio** (CMRR) and is defined by

$$\text{CMRR} = \left|\frac{A_{dm}}{A_{cm}}\right| \tag{10.26}$$

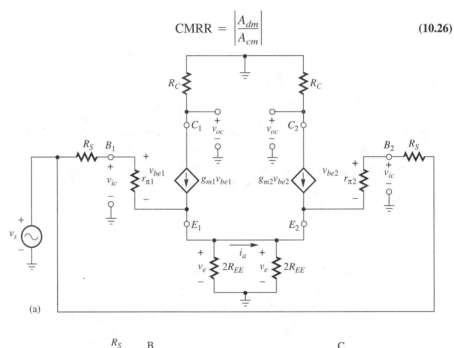

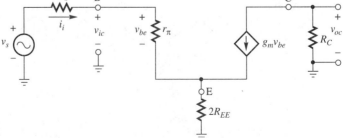

DRILL 10.3

Adjust V_{EE} in Example 10.3 to obtain CMRRs of 60 and 80 dB. Is this a reasonable technique to increase the CMRR?

ANSWERS 26.7 V (perhaps), 260.7 V (no)

■ **FIGURE 10.10** Common-mode half-circuit model development.
(a) Decomposition of R_{EE} for the common-mode small-signal model.
(b) Half-circuit model.

and is usually expressed in decibels by computing

$$\text{CMRR}_{dB} = 20 \log \left| \frac{A_{dm}}{A_{cm}} \right|. \tag{10.27}$$

Substituting Eqs. (10.18) and (10.24) into Eq. (10.26) yields

$$\text{CMRR} = \left| \frac{-g_m R_C}{\frac{-g_m R_C}{1 + 2g_m R_{EE}}} \right| = 1 + 2g_m R_{EE}. \tag{10.28}$$

This result is usually approximated by

$$\text{CMRR} \cong 2g_m R_{EE} = 2\left(\frac{I_C q}{kT}\right) R_{EE} = \frac{I_{EE} R_{EE}}{kT/q} = \frac{I_{EE} R_{EE}}{V_T}. \tag{10.29}$$

since $2g_m R_{EE} \gg 1$ and recalling that $I_{EE} = 2I_C$.

EXAMPLE 10.3

Compute the CMRR for the amplifier whose specifications are given in Example 10.1.

SOLUTION We now need to compute R_{EE} in order to apply Eq. (10.29). Realizing that the coupled emitters are at a dc potential of -0.7 V, we have

$$R_{EE} = \frac{[-0.7 - (-V_{EE})]}{I_{EE}} = \frac{(-0.7 + 15)}{100 \ \mu\text{A}} = 143 \text{ k}\Omega.$$

Substituting this result in Eq. (10.29), we have

$$\text{CMRR} = \frac{(100 \ \mu\text{A} \times 143 \text{ k}\Omega)}{0.026 \text{ V}} = 550 \text{ or CMRR}_{dB} = 20 \log 550 = 54.8 \text{ dB}.$$

(Note that $2g_m R_{EE} = 550 \gg 1$ as assumed earlier.)

An emitter-coupled pair is designed so that the CMRR is as large as possible. The CMRR can be improved by increasing I_{EE} or R_{EE}; however, this design alternative might not be acceptable because R_{EE} might be too large for monolithic IC fabrication, or the power supply voltage, V_{EE}, might become too large as illustrated in Drill 10.3. Actually, the 143-kΩ resistor required in Example 10.3 is not attractive for use in an IC either because it occupies too much die area. Refer to the design of diffused resistors discussed in Chapter 18.

■ CMRR > 90 dB can be achieved in well-designed operational amplifier input stages.

Recall from Chapter 7 that we can replace a resistor current source with a simple or Widlar current source. Let's replace R_{EE} with a Widlar current source as illustrated in Fig. 10.11. Not only have we potentially reduced the total resistance in the circuit, and thus IC die area, but we have effectively increased the CMRR, because now R_{EE} is given by Eq. (7.39), repeated here as

$$R_{EE} = R_{o4} = r_{o4}(1 + g_{m4} R_2) \tag{10.30}$$

where

$$r_{o4} = r_{onpn} = \frac{V_{Anpn}}{I_{C4}} = \frac{V_{Anpn}}{I_{EE}} \tag{10.31}$$

and V_{Anpn} is the Early voltage for Q_4.

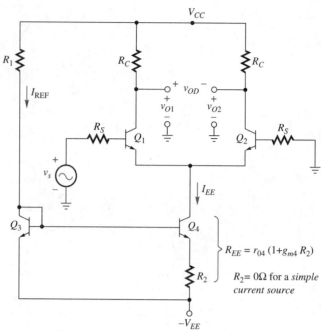

FIGURE 10.11 Emitter-coupled pair using a Widlar current source.

EXAMPLE 10.4

Replace the R_{EE} of Example 10.1 with a Widlar current source as illustrated in Fig. 10.11. Let $R_1 = 10$ kΩ. Compute the CMRR and compare the result to that obtained in Example 10.3.

SOLUTION Recall, for a Widlar current source, Eqs. (7.35) and (7.27):

$$\frac{kT}{q}\ln\left(\frac{I_{C3}}{I_{C4}}\right) = I_{C4}R_2 \quad \text{and} \quad I_{C3} = \frac{V_{CC} + V_{EE} - 0.7}{R_1}.$$

Then

$$I_{C3} = \frac{24.3}{10 \text{ k}\Omega} = 2.43 \text{ mA}$$

and solving for R_2, the result is

$$R_2 = \frac{26 \text{ mV}}{100 \text{ }\mu\text{A}}\ln\left(\frac{2.43 \text{ mA}}{100 \text{ }\mu\text{A}}\right) = 830 \text{ }\Omega,$$

$$r_o = \frac{V_{A4}}{I_{C4}} = \frac{100 \text{ V}}{100 \text{ }\mu\text{A}} = 1 \text{ M}\Omega.$$

From Eq. (10.30), the effective value of R_{EE} is given by $R_{EE} = 4.2$ MΩ, a significant increase from the 143 kΩ computed in Example 10.3. Then CMRR is computed to be

$$\text{CMRR} = \frac{I_{EE}R_{EE}}{kT/q} = \frac{(100 \text{ }\mu\text{A})(4.2 \text{ M}\Omega)}{26 \text{ mV}} = 16{,}124 \text{ or } 84 \text{ dB}.$$

This is a much higher value than is achievable by using the 143-kΩ resistor in series with $-V_{EE}$ as a current source.

DRILL 10.4

Repeat Example 10.3, keeping $I_{EE} = 100$ μA, if Widlar current source reference current $I_{C3} = 1$ mA.

ANSWERS
$R_2 = 598.7$ Ω,
$R_{EE} = 3.3$ MΩ,
CMRR = 12,702 = 82 dB

Differential and Common-Mode Input Resistance

As we have observed, the emitter-coupled amplifier exhibits different behavior depending on whether a differential-mode or common-mode signal is input. These differences carry over into the calculation of input resistances. The differential input resistance, $R_{id} = v_{id}/i_i$ is computed with the aid of Fig. 10.5:

$$\frac{v_{id}}{2} = i_i r_\pi \quad \text{or} \quad R_{id} = \frac{v_{id}}{i_i} = 2r_\pi. \quad (10.32)$$

Using Fig. 10.10(b), the common-mode input resistance, $R_{ic} = v_{ic}/i_i$, is computed from

$$v_{ic} = i_i r_\pi + (i_i + g_m v_{be})2R_{EE} = i_i[(1 + g_m r_\pi)2R_{EE} + r_\pi] \quad (10.33)$$

and, after rearranging terms,

$$R_{ic} = \frac{v_{ic}}{i_i} = r_\pi + (1 + \beta)2R_{EE} \approx 2\beta R_{EE}, \quad (10.34)$$

because usually $\beta \gg 1$. Note that $R_{ic} \gg R_{id}$.

EXAMPLE 10.5

Compute R_{id} and R_{ic} for the amplifier given in Example 10.1

SOLUTION From Eq. (10.32),

$$R_{id} = 2r_\pi = \frac{2\beta}{g_m} = \frac{400}{1.92 \text{ mS}} = 208 \text{ k}\Omega.$$

From Eq. (10.34),

$$R_{ic} = \frac{v_{ic}}{i_i} = r_\pi + (1 + \beta)2R_{EE} \approx 2\beta R_{EE} = 2(200)(143 \text{ k}\Omega) = 57.2 \text{ M}\Omega.$$

DRILL 10.5

Compute R_{id} and R_{ic} for the amplifier given in Example 10.1, but assume $\beta = 50$.

ANSWERS Recall from Drill 10.1 that the Q-point and A_{dm} are essentially unchanged, however $R_{id} = 53$ kΩ and $R_{ic} = 14.3$ MΩ.

CHECK UP

1. What conditions are placed on Q_1 and Q_2 to obtain equal Q-point collector currents?
2. **TRUE OR FALSE?** The differential output voltage is halfway between V_{CC} and V_{EE}.
3. **TRUE OR FALSE?** A_{cm} should be about the same or higher than A_{dm}.
4. Sketch a schematic diagram of an emitter-coupled pair amplifier and show the phase relationships between all voltages and currents.
5. Provide several reasons why a Widlar current source is better than a single resistor for a current source in an emitter-coupled pair amplifier.
6. What are some examples of common-mode input voltages?

10.2 DIFFERENTIAL MOS AMPLIFIERS WITH RESISTIVE LOADING

MOS technology is often used in the design of analog ICs. A principal advantage is the significant increase in the input impedance. In general, the gain is lower for MOS amplifiers because of the difference in g_m. In this section, we use our results from Section 10.1 to

> Most MOS amplifiers use active loading, not resistive loading. We cover resistive loading for MOS amplifiers to illustrate the approach and also realizing that high frequency MOS applications do use resistive loading.

analyze an MOS source-coupled differential amplifier with resistive loading. A typical NMOS circuit is shown in Figure 10.12.

Often $|V_{DD}| = |V_{SS}|$ but it is not necessary. We define the input differential voltage as

$$v_{i1} - v_{i2} = v_{ID} = v_{GS1} - v_{GS2}. \tag{10.35}$$

Typically, MOSFETs, when used as amplifiers are operated in the saturation region. Then, for each n-channel MOSFET, assuming they are matched,

$$i_{D1} = k(v_{GS1} - V_t)^2 \quad \text{and} \quad i_{D2} = k(v_{GS2} - V_t)^2 \tag{10.36}$$

and solving for v_{GS} yields

$$v_{GS1} = \sqrt{\frac{i_{D1}}{k}} + V_t \quad \text{and} \quad v_{GS2} = \sqrt{\frac{i_{D2}}{k}} + V_t, \tag{10.37}$$

with the result that

$$v_{ID} = \sqrt{\frac{i_{D1}}{k}} - \sqrt{\frac{i_{D2}}{k}}. \tag{10.38}$$

A convenient technique to analyze this circuit is through a series of half-circuit models. For example, with $v_{S1} = v_{S2} = 0$ in Figure 10.12, the gates of M_1 and M_2 are at dc ground. For matched devices, we can generate the half-circuit model shown in Figure 10.13 for obtaining the Q-point.

As expected,

$$I_{SS} = I_{D1} + I_{D2} = 2I_D. \tag{10.39}$$

Equation (10.36) is solved for I_D, V_{DS}, and V_{GS} using

$$0 = V_{GS} + I_{SS}R\sigma + (-V_{SS}) = V_{GS} + 2I_D R\sigma + (-V_{SS}) \tag{10.40}$$

around the gate loops, and

$$V_{DD} = I_D R_D + V_{DS} + (-V_{GS}) = I_D R_D + V_{DS} + 2I_D R\sigma + (-V_{SS}) \tag{10.41}$$

around the drain circuit. The simultaneous solution of Eqs. (10.37), (10.40), and (10.41) is illustrated using the load-line characteristics in Fig. 10.14. To obtain amplifier operation, the Q-point $V_{GS} > V_t$ and such that $V_{DS} > V_{GS} - V_t$.

> For simplicity, we assume discrete NMOS so that we can connect the source and substrate together. In an IC, the substrate would be connected to $-V_{SS}$.

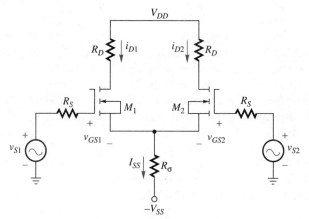

FIGURE 10.12 Source-coupled MOS pair differential amplifier.

CHAPTER 10 OPERATIONAL AMPLIFIER CIRCUITRY

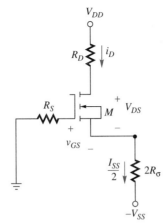

FIGURE 10.13 Half-circuit model for determining the Q-point.

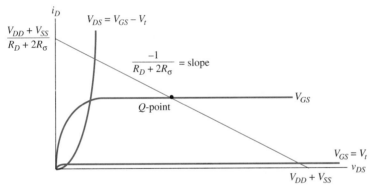

FIGURE 10.14 Q-point solution for the NMOS source-coupled pair.

To obtain the differential voltage gain defined by $A_{dm} = v_{od}/v_{id}$, we apply the half-circuit model concepts to Fig. 10.12 as shown in Fig. 10.15. From Chapter 5 and Eq. (10.36),

$$g_m = \left.\frac{\partial i_D}{\partial v_{GS}}\right|_{Q\text{-point}} = 2k(V_{GS} - V_t). \tag{10.42}$$

Similar to the derivation of Eq. (10.20), the differential-mode voltage gain is given by

$$A_{dm} = -g_m R_D. \tag{10.43}$$

The common-mode voltage gain $A_{cm} = v_{oc}/v_{ic}$ is obtained using the common-mode half-circuit model shown in Fig. 10.16. The source arrangement used in Figure 10.8 has been

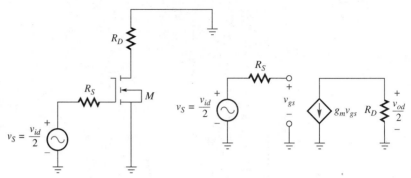

FIGURE 10.15 Half-circuit differential-mode small-signal model.

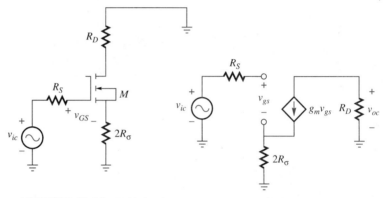

FIGURE 10.16 Half-circuit common-mode small-signal model.

DRILL 10.6

Compute g_m for an MOSFET operating at $V_{GS} = 1.5$ V, 3 V, 5 V whose specifications are $\mu_n = 600$ cm²/V-sec, $t_{ox} = 500$ Å, W = L and $V_t = 1.5$.

ANSWERS
a) $g_m|_{v_{GS}=0} = 0$
b) $g_m|_{v_{GS}=0} = 1.24 \times 10^{-4}$ S
c) $g_m|_{v_{GS}=0} = 2.9 \times 10^{-4}$ S

DRILL 10.7

If $R_\sigma = 10$ kΩ, compute the CMRR in dB for the g_m values obtained in Drill 10.6.

ANSWERS
a) 0 dB
b) 10.8 dB
c) 16.7 dB

---

applied. Observe that the common-mode half-circuit model is similar to that shown in Fig. 10.10, consequently, Eqs. (10.24) and (10.25) can be rewritten as

$$A_{cm} \cong -\frac{g_m R_D}{1 + 2g_m R_\sigma} \cong -\frac{R_D}{2R_\sigma}. \quad (10.44)$$

Similarly, comparing to Eq. (10.28)

$$\text{CMRR} = \left| \frac{-g_m R_D}{\frac{-g_m R_D}{1 + 2g_m R_\sigma}} \right| = 1 + 2g_m R_\sigma. \quad (10.45)$$

Even though there is a great deal of similarity between small-signal design equations obtained for BJT and MOS circuits, it is important to observe that the transconductance for BJTs is usually much higher than for MOSFETs at similar current levels. This is explored in the problems.

CHAPTER 10 OPERATIONAL AMPLIFIER CIRCUITRY

CHECK UP

1. **TRUE OR FALSE?** Voltage gains for MOSFET amplifiers are typically higher than for BJT amplifiers.
2. **TRUE OR FALSE?** Input resistances for MOSFET amplifiers are much higher than for BJT amplifiers.
3. Sketch what happens to the output waveform if the input voltage amplitude is so large that the MOSFETs operate in the ohmic region.
4. **TRUE OR FALSE?** Transconductance is proportional to the gate-source Q-point voltage.
5. Compare the CMRR for similarly biased BJT and MOS amplifiers.

10.3 DIFFERENTIAL JFET AMPLIFIERS WITH RESISTIVE LOADING

We approach JFET source-coupled differential amplifiers in the same way we analyzed the BJT and MOS versions. In doing so, we will observe many common results and, of course, point out key differences. Generally speaking, MOSFETs are much more widely used than JFETs.

JFET Source-Coupled Pair DC Analysis

Figure 10.17 is a diagram of a JFET source-coupled differential amplifier. As with the BJT and MOS differential amplifiers, we start with a study of the dc operation, which will set the stage for the small-signal model. We assume matched JFETs and matched drain resistors. Summing around the gate loop, we have

$$v_{I1} = v_{GS1} - v_{GS2} + v_{I2}. \tag{10.46}$$

■ JFETs are used as the input stage in some ICs. Although many ICs use total MOS technology, there are few if any ICs totally using JFETs. JFETs are available on the Tektronix (Now MAXIM) 9.5 GHz ASIC array.

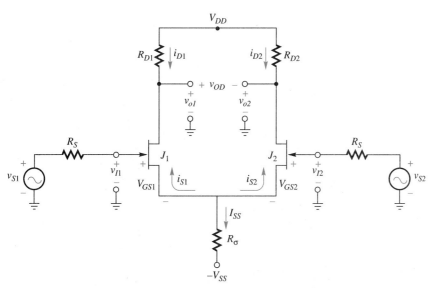

■ **FIGURE 10.17** Source-coupled pair differential amplifier.

We can rewrite Eq. (5.35) to obtain

$$V_{GS} = V_P\left(1 - \sqrt{\frac{i_D}{I_{DSS}}}\right). \tag{10.47}$$

Let $v_{ID} = v_{I1} - v_{I2}$ be the input differential voltage, $V_{P1} = V_{P2}$, and $I_{SS} = i_{D1} + i_{D2}$. We can, after some algebraic manipulation, write equations for i_{D1} and i_{D2} by substituting Eq. (10.47) into Eq. (10.46):

■ Equations 10.48 and 10.49 apply to n-channel JFETs where V_P is negative.

$$i_{D1} = \frac{I_{SS}}{2}\left[1 - \frac{v_{ID}}{V_P}\sqrt{2\left(\frac{I_{DSS}}{I_{SS}}\right) - \left(\frac{v_{ID}}{V_P}\right)^2\left(\frac{I_{DSS}}{I_{SS}}\right)^2}\right], \tag{10.48}$$

$$i_{D2} = \frac{I_{SS}}{2}\left[1 + \frac{v_{ID}}{V_P}\sqrt{2\left(\frac{I_{DSS}}{I_{SS}}\right) - \left(\frac{v_{ID}}{V_P}\right)^2\left(\frac{I_{DSS}}{I_{SS}}\right)^2}\right]. \tag{10.49}$$

Equations (10.48) and (10.49) are graphed in Fig. 10.18 for the case $I_{SS} = I_{DSS}$, or $I_D = I_{DSS}/2$. These transfer functions resemble the corresponding result for the bipolar case (Fig. 10.2) but with the linear region around $v_{ID} = 0$ much wider for the JFET. The JFET differential amplifier can handle much larger input signals than its bipolar counterpart, an important advantage.

The drain output voltages are given by

$$v_{O1} = v_{DD} - i_{D1}R_{D1} \tag{10.50}$$

and

$$v_{O2} = v_{DD} - i_{D2}R_{D2}. \tag{10.51}$$

With $R_{D1} = R_{D2} = R_D$, the output differential voltage is given by

$$\begin{aligned}v_{OD} &= (V_{DD} - i_{D1}R_D) - (V_{DD} - i_{D2}R_D) \\ &= -i_{D1}R_D + i_{D2}R_D.\end{aligned} \tag{10.52}$$

Substituting Eqs. (10.48) and (10.49) into Eq. (10.52) yields

$$v_{OD} = \frac{I_{SS}R_D v_{ID}}{V_P}\sqrt{2\left(\frac{I_{DSS}}{I_{SS}}\right) - \left(\frac{v_{ID}}{V_P}\right)^2\left(\frac{I_{DSS}}{I_{SS}}\right)^2}. \tag{10.53}$$

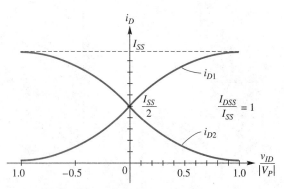

■ **FIGURE 10.18** Source-coupled pair drain current as a function of the differential input voltage, $I_{SS} = I_{DSS}$.

Again, the output differential voltage is shown to be a function of the input differential voltage, and this is plotted in Fig. 10.19. for small v_{ID}, this transfer function is approximately linear. The differential voltage gain is found by differentiating Eq. (10.53) with respect to v_{ID} and then letting $v_{ID} = 0$:

$$A_{dm} = \left.\frac{dv_{OD}}{dv_{ID}}\right|_{v_{ID}=0} = \frac{I_{SS}R_D}{V_P}\sqrt{\frac{2I_{DSS}}{I_{SS}}} = \frac{R_D}{V_P}\sqrt{2I_{SS}I_{DSS}}. \qquad (10.54)$$

At $V_{ID} = 0$, $I_{SS} = 2I_D$, and since $g_m = (2/V_P)\sqrt{I_D I_{DSS}}$, Eq. (10.54) reduces to the very familiar result that

$$A_{dm} = -g_m R_D \qquad (10.55)$$

when $v_{ID} = 0$. This is identical to the voltage gain of a single-ended common-source amplifier as shown in Chapter 5. Furthermore, for $v_{ID} = 0$, $v_{OD} = 0$, which means that there is a zero dc voltage offset as input to the next stage.

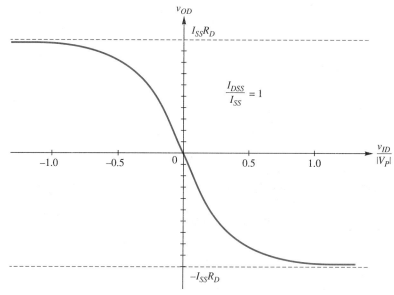

■ **FIGURE 10.19** Differential output voltage of a common-source amplifier as a function of the differential input voltage, $I_{SS} = I_{DSS}$.

JFET Common-Source Amplifier AC Analysis

In a sense, the small-signal models for both the BJT and FET amplifiers are identical in that both are modeled using voltage-controlled current generators. Of course, the transconductances and output resistances have a different physical basis and are computed differently, and the FET input resistance is essentially infinite; however, the manipulation of the model is the same. This is easily observed in Fig. 10.20(a), which is a small-signal representation of the JFET source-coupled amplifier shown in Fig. 10.15. By applying a differential input voltage, split as illustrated, we can simplify this circuit using half-circuit

■ Note that $R_i = \infty$, represented by an open circuit in the model.

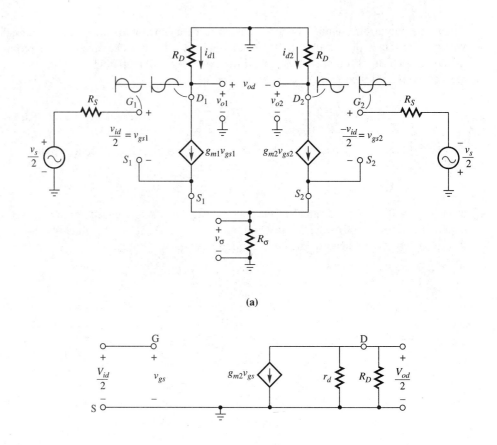

FIGURE 10.20 Common-source amplifier small-signal differential-mode models. (a) Small-signal model with differential input. (b) Half-circuit model, differential mode.

DRILL 10.8

Compute the g_m and BETA for a JFET biased at $I_D = 1$ mA. The specifications of the JFET are $I_{DSS} = 3$ mA and $V_P = -4$ V.

ANSWERS
a) $g_m = 8.7 \times 10^{-4}$ S
b) BETA = 1.875×10^{-4} A/V^2

techniques into that shown in Fig. 10.20(b). We see that the differential-mode voltage gain is given by Eq. (10.55), and by including the effect of r_d we have

$$A_{dm} = \frac{v_{od}}{v_{id}} = -g_m(R_D \| r_d). \tag{10.56}$$

EXAMPLE 10.6

For the differential amplifier shown in Fig. 10.17, compute the Q-point values for I_D, V_{GS}, and V_{DS} and also compute A_{dm}. Assume $V_{DD} = V_{SS} = 15$ V, $R_D = 50$ kΩ, and $R_\sigma = 100$ kΩ. Use the JFET parameters $I_{DSS} = 1$ mA, $V_P = -5$ V, and $r_o \to \infty$.

SOLUTION Simultaneously solving

$$I_D = I_{DSS}\left(1 - \frac{V_{GS}}{V_P}\right)^2 = 1 \text{ mA}\left(1 - \frac{V_{GS}}{-5}\right)^2$$

and

$$0 = V_{GS} + I_{SS}R_\sigma + (-V_{SS}) = V_{GS} + 2I_DR_\sigma + (-V_{SS})$$
$$= V_{GS} + 2I_D(100 \text{ k}\Omega) - 15$$

yields $I_D = 92.4$ μA and $V_{GS} = -3.48$ V. Recall from Chapter 5,

$$A_{dm} = -g_m R_D \quad \text{and} \quad g_m = -\frac{2}{V_P}\sqrt{I_D I_{DSS}}$$

$$= -\frac{2}{5}(50 \text{ k}\Omega)\sqrt{(92.4 \text{ μA})(1 \text{ mA})} = -6.08.$$

To calculate V_{DS}, $V_{DD} = I_D R_D + V_{DS} + I_{SS}R_\sigma - V_{SS}$, or $V_{DS} = V_{DD} + V_{SS} - I_D R_D - I_{SS}R_\sigma = 30 - (92.4 \text{ μA})(50 \text{ k}\Omega) - 2(92.4 \text{ μA})(100 \text{ k}\Omega) = 6.9$ V.

EXAMPLE 10.7

Rework Example 10.6 using SPICE. In addition, verify the Fig. 10.18 source-coupled pair drain currents as a function of the input differential voltage and the Fig. 10.19 output differential voltage as a function of the input differential voltage.

SOLUTION   The node-labeled schematic diagram is given in Fig. 10.21. The resultant netlist is given next. Observe that an ac voltage source and a transfer function voltage source are in series and will yield both the ac and transfer function results using superposition principles. No frequency-dependent elements have been included. The ac solution is to be computed at 1 kHz.

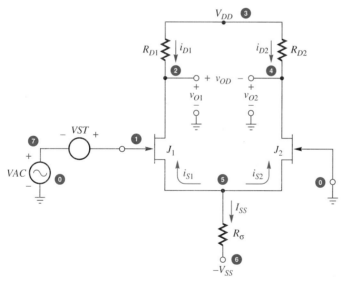

FIGURE 10.21 Source-coupled pair differential amplifier.

```
SOURCE-COUPLED PAIR, EXAMPLE 10.7
VDD  3 0 DC 15
VSS  6 0 DC -15
VST  1 7 DC 0
VAC  7 0 AC 1
RD1  3 2 50K
RD2  3 4 50K
RSIG 5 6 100K
```

```
J1 2 1 5 DEV
J2 4 0 5 DEV
.MODEL DEV NJF VTO=-5 BETA=4E-5
.DC VST -3 3 10M
.OP
.AC DEC 1 1E3 1E3
.PRINT AC VM(2) VP(2) VM(4) VP(4) VM(1) VP(1)
.PROBE
.END
```

The dc simulation results are given by

NAME	J1	J2	Example 10.6 Results for J_1 and J_2
ID	9.24E-05	9.24E-05	92.4 μA
VGS	-3.48E+00	-3.48E+00	-3.48 volts
VDS	6.90E+00	6.90E+00	6.9 volts
GM	1.22E-04	1.22E-04	1.216×10^{-4} S

Figures 10.22(a) and (b) compare with the transfer functions given in Figs. 10.18 and 10.19, respectively. The ac results, $A_{dm} = -6.08$, are in agreement with Example 10.6.

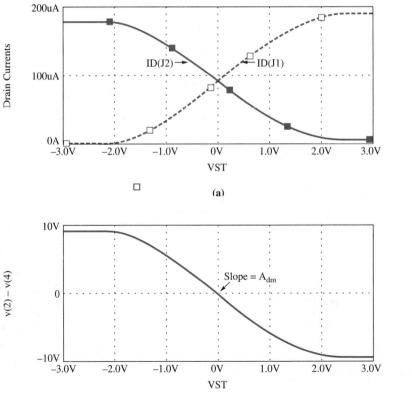

■ **FIGURE 10.22** (a) Drain currents as a function of the input differential voltage. (b) Output differential voltage as a function of the input differential voltage. Slope at $v_{ID} = 0$ yields A_{dm} as expected.

CHAPTER 10 OPERATIONAL AMPLIFIER CIRCUITRY 463

> **CHECK UP**
>
> 1. What conditions are placed on J_1 and J_2 to obtain equal Q-point drain currents?
> 2. **TRUE OR FALSE?** The differential output voltage is halfway between V_{DD} and V_{SS}.
> 3. Sketch a schematic diagram of a source-coupled pair amplifier and show the phase relationships between all voltages and currents.
> 4. What are the differences in device Q-points between JFET and MOSFET source-coupled pairs? Demonstrate using load lines.

10.4 BJT AMPLIFIERS WITH ACTIVE LOADING

The term *active load* refers to the use of current sources instead of resistors as the load in some amplifier configurations. Recall from Chapters 4 and 8 that the CE amplifier as shown in Fig. 10.23 has a voltage gain

$$A_v = \frac{v_o}{v_i} = -g_m(R_C \| r_{onpn}). \tag{10.57}$$

Usually, $R_C \ll r_{onpn}$ so that

$$A_v = -g_m R_C = \frac{-I_C}{V_T} R_C = \frac{-I_C R_C}{\left(\frac{kT}{q}\right)}. \tag{10.58}$$

To increase A_v, $I_C R_C$ must increase, and this is often not possible because it would require unacceptably large power supply voltages or resistors that are incompatible with IC fabrication technology. When we needed a large resistor to improve the CMRR in Section 10.1, we resorted to utilizing the high output impedance of a current source. Simultaneously, the current source provided the necessary circuit biasing. This same design approach is used to achieve a large, equivalent, collector load resistance.

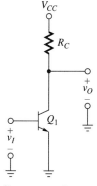

■ **FIGURE 10.23** Common-emitter amplifier with a resistive load.

Common-Emitter Amplifier with Active-Load AC Analysis

Suppose we replace R_C with a *pnp* simple current source as shown in Fig. 10.24(a). After the Q_1, Q_2, and Q_3 nodes are labeled, the small-signal model is drawn as shown in Fig. 10.24(b). Summing currents at the node where the bases of Q_2 and Q_3 and the collector of Q_2 are connected yields

$$\frac{v_3}{r_{\pi 3}} + \frac{v_3}{r_{\pi 2}} + g_{m3}v_3 + \frac{v_3}{R_1 \| r_{onpn}} = 0. \tag{10.59}$$

To satisfy this equation, v_3 must be identically zero, a very interesting result. It is not unreasonable to expect the base node to be at a very small ac level in this circuit. With $v_3 = 0$, Fig. 10.24(b) reduces to the form shown in Fig. 10.24(c). The voltage gain of this circuit is given by

$$A_v = -g_m(r_{onpn} \| r_{opnp}) = -g_m\left(\frac{r_{onpn} r_{opnp}}{r_{onpn} + r_{opnp}}\right). \tag{10.60}$$

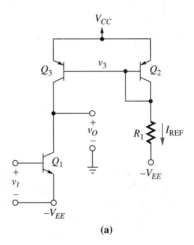

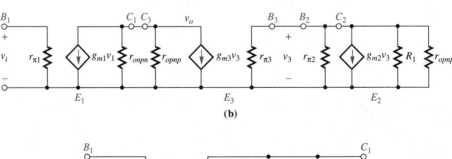

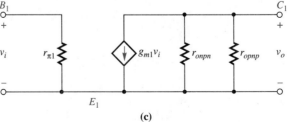

■ **FIGURE 10.24** Model development for a common-emitter amplifier with an active load.
(a) Common-emitter amplifier with an active load. (b) Complete small-signal model.
(c) Reduced small-signal model.

Since $I_{C1} = I_{C2}$, Eq. (10.60) can be written in the form

$$A_v = \frac{-\left(\dfrac{I_C}{kT/q}\right)\left(\dfrac{V_{Anpn}}{I_C}\right)\left(\dfrac{V_{Apnp}}{I_C}\right)}{\dfrac{V_{Anpn}}{I_C} + \dfrac{V_{Apnp}}{I_C}} = -\frac{V_{Anpn}V_{Apnp}}{\dfrac{kT}{q}(V_{Anpn} + V_{Apnp})}. \quad (10.61)$$

It is very interesting to observe in Eq. (10.61) that the voltage gain is independent of I_C and is a function of only the physical properties of the transistors. Using transistor specifications from Table 18.2, where $V_{Apnp} = 50$ V and $V_{Anpn} = 150$ V, the voltage gain of the amplifier is

$$A_v = \frac{-(150)(50)}{0.026(200)} = -1442.$$

Compared to the amplification available with resistive loads, this result is considerably larger, an improvement compatible with IC fabrication constraints. It is clear that any stage cascaded with this amplifier will reduce the voltage gain from the maximum theoretical value predicted by Eq. (10.60) to

$$A_v = -g_m(r_{onpn} \| r_{opnp} \| R_L), \quad (10.62)$$

where R_L is the effective load resistance presented by the input resistance of any cascaded stage. The output resistance, by inspection, is given by

$$R_o = r_{onpn} \| r_{opnp}. \quad (10.63)$$

Therefore, any design for a cascade amplifier system must take into consideration the effects of interstage loading.

Common-Emitter Active-Load DC Analysis

Figure 10.25 shows a large-signal or dc transfer characteristic for the CE amplifier with an active load. As expected, the v_O can be driven to within $V_{CE(\text{sat})}$ of each power supply. The linear region is centered about the point on the characteristic where the $V_{BE(\text{on})}$ of

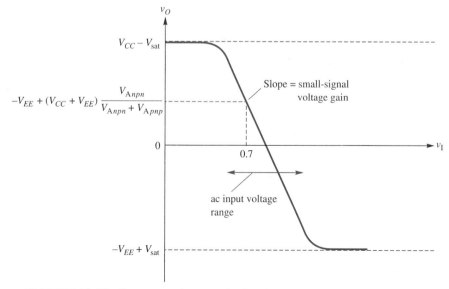

■ **FIGURE 10.25** Common-emitter transfer function.

DRILL 10.9

Explain what happens to the voltage gain and output dc voltage level as the Early voltages for both Q_1 and Q_2 become very large, i.e. V_{Anpn} and $V_{Apnp} \to \infty$.

ANSWERS The voltage gain becomes very large, Eq. (10.60). The dc output voltage is unstable in that if the transfer function shown in Fig. 10.25 becomes vertical, there is no stable location for the output voltage Q-point. Also refer to Eq. (10.64)

Q_1 is nominally 0.7 V. With no ac input signal but with Q_1 biased ON, V_O will be approximately positioned between both power supply voltages according to a voltage divider established by the Early voltages as follows:

$$V_O = -V_{EE} + (V_{CC} + V_{EE}) \frac{r_{onpn}}{r_{onpn} + r_{opnp}}$$
$$= -V_{EE} + (V_{CC} + V_{EE}) \frac{V_{Anpn}}{V_{Anpn} + V_{Apnp}}. \quad (10.64)$$

EXAMPLE 10.8

Calculate the dc Q-point for the amplifier shown in Fig. 10.24 with $V_{CC} = 15$ V, $V_{EE} = 10$ V, and $I_C = 50$ μA. Use transistor specifications from Table 18.2.

SOLUTION The dc model is given in Fig. 10.26. From Eq. (10.63),

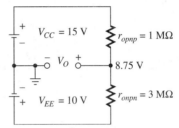

■ **FIGURE 10.26** Resultant dc model for use in Example 10.8.

$$V_O \approx -10 + 25 \frac{150}{150 + 50} = +8.75 \text{ V}.$$

EXAMPLE 10.9

Verify Example 10.8 using SPICE.

SOLUTION The node-labeled circuit, based on Fig. 10.24, is given in Fig. 10.27. The netlist (SPICE circuit description) is given by

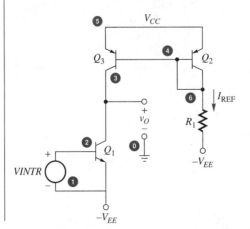

■ **FIGURE 10.27** Node-labeled active-load CE amplifier.

```
CE WITH ACTIVE LOAD
VCC 5 0 DC 15
VEE 0 1 DC 10
VINTR 2 1 DC 0
Q1 3 2 1 DEV1
.MODEL DEV1 NPN BF=200 VAF=150
Q2 4 4 5 DEV2
Q3 3 4 5 DEV2
.MODEL DEV2 PNP BF=50 VAF=50
R1 4 1 494K
.DC VINTR 0.675 0.725 0.001
.OP
.PROBE
.END
```

■ $R_1 = 494$ kΩ is very unattractive for IC designs. We address this by using a resistor in the emitter of Q_3 as well as changing bias collector current levels.

The reference resistor R_1 was set at 494 kΩ to obtain a nominal 50-μA reference current. The resultant SPICE-generated transfer function is shown in Fig. 10.28. Figure 10.28(a) shows I_{C1} as a function of the input voltage, *VINTR*, which is swept from 675 to 725 mV. Observe that when $I_{C1} = 50$ μA, nominal Q-point current $V_{BE1} = 693$ mV. As shown in Fig. 10.28(b), when $V_{BE1} = 693$ mV, the output voltage, v_O, (node 3) is about 8.75 V, as expected from Example 10.8.

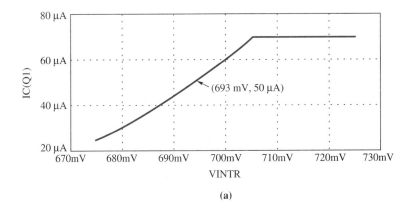

(a)

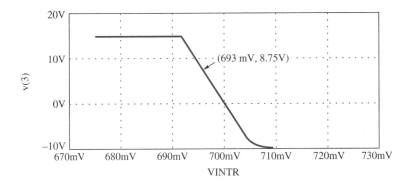

(b)

■ **FIGURE 10.28** (a) I_{C1} versus *VINTR* transfer function.
(b) v_O versus *VINTR* transfer function.

It is also interesting to note that the slope of Fig. 10.28(b) yields the voltage gain. Within graphical accuracy, we compute

$$A_v = \frac{\Delta V(3)}{\Delta VINTR} \approx -\frac{25 \text{ V}}{17 \text{ mV}} = -1470,$$

which compares well with the -1442 solution of Eq. (10.60).

Emitter-Coupled Pair with an Active Load

Several techniques are available for utilizing an active load with an emitter-coupled pair. The circuit to be described in some detail is used in one form or another as the input stage to many types of analog ICs. Consider the circuit shown in Fig. 10.29. A single-ended output, v_O, is obtained for a differential input voltage, v_{ID}. As demonstrated in Chapter 12, this approach is widely used. The current-source function is established by Q_3 and Q_4. The reference current, $I_{REF} = I_{C3}$ is established by the current source represented by the current I_{EE} in R_{EE}. Because Q_3 is the reference transistor for Q_4 in a simple current-source configuration, $I_{C3} = I_{C4}$. It is assumed that R_{IN} of the next stage is essentially an open circuit so that we can state that $I_{C1} = -I_{C3} = I_{C2} = -I_{C4}$.

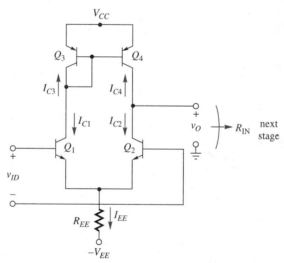

FIGURE 10.29 Emitter-coupled pair with active load.

AC Analysis

The voltage gain, $A_v = v_o/v_{id}$, is computed by drawing the small-signal model as shown in Fig. 10.30. Note that in this figure each transistor node is clearly labeled, a good practice to follow in drawing small-signal models for complicated circuits. The hybrid-π transistor models have been simplified somewhat in that r_{o3} and r_{o1} have been neglected. This circuit simplification, which will be justified shortly, is useful in reducing the algebraic complexity.

Before we apply brute algebraic force to obtain A_v, consider the effect of Q_3, the current-source reference transistor. The circuit elements within the dashed lines have been redrawn and are presented in Fig. 10.31. By applying a test source, v_T, as input and

CHAPTER 10 OPERATIONAL AMPLIFIER CIRCUITRY

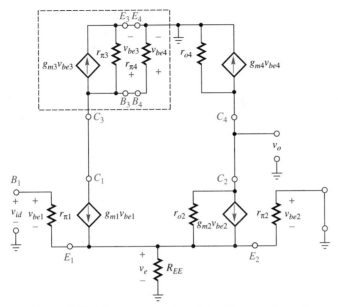

FIGURE 10.30 Small-signal model emitter-coupled pair with active load.

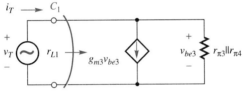

FIGURE 10.31 Circuit to support computation of r_{L1}.

solving for i_T we can obtain the net resistance, r_{L1}, exhibited by this circuit as it loads Q_1. At the C_3–C_1–B_3–B_4 node,

$$i_T = g_{m3}v_{be3} + \frac{v_{be3}}{r_{\pi 3}\|r_{\pi 4}}, \qquad (10.65)$$

and since $v_{be3} = v_T$ and $r_{\pi 3}\|r_{\pi 4} \gg 1/g_{m3}$,

$$r_{L1} = \frac{v_T}{i_T} = \frac{1}{1/(r_{\pi 3}\|r_{\pi 4}) + g_{m3}} = r_{\pi 3}\|r_{\pi 4}\left\|\frac{1}{g_{m3}} \cong \frac{1}{g_{m3}}\right.. \qquad (10.66)$$

Using the result from Eq. (10.66), Fig. 10.30 is redrawn as illustrated in Fig. 10.32. Summing currents at the output voltage node yields

$$\frac{v_o}{r_{o4}} + g_{m4}v_{be4} + g_{m2}v_{be2} + \frac{v_o + v_{be2}}{r_{o2}} = 0. \qquad (10.67)$$

At the C_1 node,

$$g_{m3}v_{be3} + g_{m1}v_{be1} = 0. \qquad (10.68)$$

■ For this model, we can show $A_{cm} = 0$. Theoretically, the CMRR $\to \infty$.

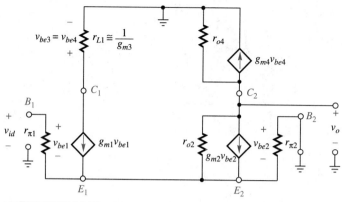

■ **FIGURE 10.32** Reduced small-signal model used to compute $A_v = v_o/v_{id}$.

Because the magnitudes of all collector currents are identical, $g_{m1} = g_{m2} = g_{m3} = g_{m4} = g_m$, Eq. (10.68) reduces to

$$-v_{be1} = v_{be3} = v_{be4}. \qquad (10.69)$$

Using this result in Eq. (10.67) yields

$$\frac{v_o}{r_{o4}} + g_m(-v_{be1}) + g_m v_{be2} + \frac{v_o + v_{be2}}{r_{o2}} = 0, \qquad (10.70)$$

■ Eqn. (10.73) can be generalized for other cases where a voltage gain is given by $a_v = g_m(r_{onpn} \| r_{opnp})$.

and realizing that

$$v_{id} = v_{be1} - v_{be2} \qquad (10.71)$$

and $r_{o2} \gg 1/g_m$, we obtain

DRILL 10.10

Compute the voltage gain for an actively-loaded emitter-coupled amplifier where $I_C = 50\ \mu A$ and $V_{Anpn} = 200$ V and $V_{Apnp} = 75$ V. What happens to the voltage gain if I_C is reduced to 25 μA? Assume $R_L \to \infty$.

$$v_o\left(\frac{1}{r_{o2}} + \frac{1}{r_{o4}}\right) = g_m\left[v_{be1} - v_{be2}\left(1 + \frac{1}{g_m r_{o2}}\right)\right] \approx g_m v_{id} \qquad (10.72)$$

or

$$A_v = \frac{v_o}{v_{id}} = g_m(r_{o2} \| r_{o4}) = \frac{q V_{Anpn} V_{Apnp}}{kT(V_{Anpn} + V_{Apnp})}. \qquad (10.73)$$

ANSWERS
a) $A_v = 2098$
b) $A_v = 2098$
   (no change)

DC Analysis

It is reasonable to expect a $\tanh(v_{ID}/2V_T)$ response, as shown in Fig. 10.28, based on the dc analysis for an emitter-coupled pair that yields the type of response as derived in Eq. (10.16). This response is illustrated in a SPICE simulation of Fig. 10.24 with BJT specifications from Table 18.2, Fig. 10.29. It can be shown[1] that between $V_{CC} - V_{CE4(\text{sat})}$ and $V_{B2} - V_{BC2(\text{sat})}$, the output voltage can be written as

$$v_O \cong V_{CC} - V_{BE(\text{on})} + \left[\frac{\dfrac{2V_{Anpn}V_{Apnp}}{V_{Anpn} + V_{Apnp}} \times \tanh\left(\dfrac{v_{ID}}{2V_T}\right)}{1 + \dfrac{V_{Anpn} - V_{Apnp}}{V_{Anpn} + V_{Apnp}} \times \tanh\left(\dfrac{v_{ID}}{2V_T}\right)} \right] \quad (10.74)$$

$$\cong V_{CC} - V_{BE(\text{on})} + \frac{2V_{Anpn}V_{Apnp}}{V_{Anpn} + V_{Apnp}} \tanh\left(\frac{v_{ID}}{2V_T}\right)$$

since $\dfrac{V_{Anpn} - V_{Apnp}}{V_{Anpn} + V_{Apnp}} \tanh\left(\dfrac{v_{ID}}{2V_T}\right)$ is often much less than 1.

■ If you include β_{pnp} in Eqn. (10.74) the argument in the tanh function becomes

$$\frac{v_{ID} - V_T \ln\left(1 + \dfrac{2}{\beta_{pnp}}\right)}{2V_T}.$$

The derivation of Eq. (10.74) is beyond the scope of this text. The large value computed for $\dfrac{V_{Anpn}V_{Apnp}}{V_{Anpn} + V_{Apnp}}$ in Eq. (10.74) means that the $\tanh(v_{ID}/2V_T)$ is reasonably linear at the shifted Q-point, and the voltage gain and slope of the transfer function can still be computed from Eq. (10.73).

The SPICE listing for the node-labeled circuit, shown in Fig. 10.33, is given by

```
EMITTER-COUPLED PAIR ACTIVE LOAD
VCC 1 0 DC 15
VEE 0 7 DC 10
VSID 4 5 AC 10
VSIDT 5 0 DC 0
REE 6 7 10K
Q1 2 4 6 MOD1
```

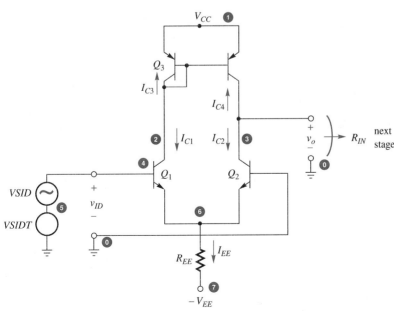

■ **FIGURE 10.33** Emitter-coupled pair with active load.

```
Q2 3 0 6 MOD1
Q3 2 2 1 MOD2
Q4 3 2 1 MOD2
.MODEL MOD1 NPN BF=200 VAF=150
.MODEL MOD2 PNP BF=50 VAF=50
.DC VSIDT -10M 10M 1M
.OP
.PROBE
.AC DEC 10 1 1E6
.END
```

The transfer function $v_{\text{out}}/v_{\text{in}}$ is plotted in Fig. 10.34. For $V_{ID} = 0$, the SPICE-generated transfer function yields a Q-point of about 12 V as shown in Fig. 10.34. Observe that the slope is steep, and relatively linear as one would expect for this topology. Indeed, if we look at the slope between -5 and 0 mV, we estimate the voltage gain by

$$A_v \approx \frac{\Delta V_{\text{out}}}{\Delta V_{\text{in}}} \approx \frac{12.5 - 5}{0.005} = 1500,$$

which compares nicely to using Eq. (10.73), which predicts

$$A_v = \frac{v_o}{v_{id}} = g_m(r_{o2} \| r_{o4}) = \frac{qV_{Anpn}V_{Apnp}}{kT(V_{Anpn} + V_{Apnp})} = \frac{(150)(50)}{(26 \text{ mV})(150 + 50)} = 1442.$$

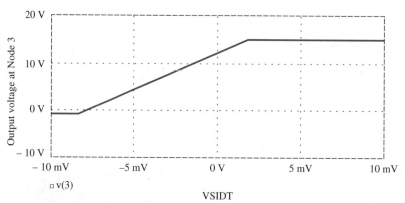

FIGURE 10.34 SPICE-generated transfer function showing a Q-point ≈ 12 V for the actively loaded emitter-coupled pair.

CHECK UP

1. **TRUE OR FALSE?** In general, actively loaded amplifiers occupy less IC die area than resistively loaded amplifiers with the same performance.
2. Discuss whether an actively loaded CE amplifier would operate if V_{Anpn} and V_{apnp} were infinite.
3. What are the output resistances for actively loaded CE and emitter-coupled pair amplifiers?

4. **TRUE OR FALSE?** The input differential resistance is higher for an actively loaded emitter-coupled pair amplifier compared to a resistively loaded emitter-coupled pair amplifier operating at the same collector currents.
5. **TRUE OR FALSE?** Inclusion of a noninfinite R_L will change the output voltage Q-point.

10.5 MOS AMPLIFIERS WITH ACTIVE LOADING

As observed with the BJT-based differential amplifier, this basic topology is one of the most useful because it utilizes the fundamental matching obtained from IC technology. Every semiconductor device technology offers advantages. For example, MOS technology provides extremely high input impedance because the SiO_2 gate dielectric is one of the best insulators known. Although MOS, in particular CMOS technology, is most often associated with digital applications, an increasing number of ICs require analog functionality. CMOS analog circuitry is widely employed to provide this functionality. As studied in Chapter 5, there are a wide variety of MOS-based devices, i.e., n- and p-channel MOSFETS operating in the enhancement or depletion-mode. These variations permit the design of a large variety of different circuit topologies. We focus on a basic CMOS (complementary n- and p-channel) analog differential amplifier.

■ A recent term is *mixed-mode* referring to analog and digital functionality on the same IC die.

Figures 10.35(a) and (b) illustrate the basic n-channel MOSFET, operating as a current sink. The current sink is operated in the saturation region, Eq. (10.36), because the current characteristics are essentially constant in saturation. The p-channel, enhancement-mode device, operating as a current source, and I-V characteristic are shown in Figs. 10.35(c)

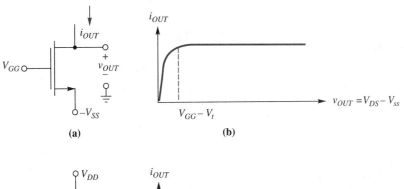

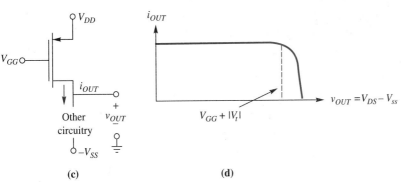

■ Observe the use of the

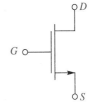

symbol in place of the

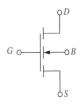

symbol. The substrate connection is assumed to connect to the appropriate potential.

■ **FIGURE 10.35** I-V Characteristics for an n-channel current sink and a p-channel current source. **(a)** n-channel current sink. **(b)** Resultant I-V characteristic. **(c)** p-channel current source. **(d)** Resultant I-V characteristic.

and (d). We observe that Eq. (10.36) can be expanded, as shown in Chapter 5, to include MOS geometric and device physics properties given by

$$i_D = k(v_{GS} - V_t)^2 = \left(\frac{\mu_n C_{ox}}{2}\right)\left(\frac{W}{L}\right)(v_{GS} - V_t)^2. \qquad (10.75)$$

The critical feature of Eq. (10.75) is the scaling provided by the (W/L) term. If we include a nonzero λ (refer to Chapter 5), the NMOS I-V characteristics described by Eq. (10.75) becomes

$$i_D = k(v_{GS} - V_t)^2(1 + \lambda V_{DS}) = \left(\frac{\mu_n C_{ox}}{2}\right)\left(\frac{W}{L}\right)(v_{GS} - V_t)^2(1 + \lambda V_{DS}). \qquad (10.76)$$

The basic current mirror, which will be used as either a current source or active load, is shown in Fig. 10.36. Because $V_{GS1} = V_{GS2} = V_{GS}$, we can compare i_O and i_1 from Eq. (10.76) as

$$\frac{i_O}{i_1} = \frac{\left(\frac{\mu_n C_{ox}}{2}\right)\left(\frac{W_2}{L_2}\right)(v_{GS2} - V_{t2})^2(1 + \lambda_2 V_{DS2})}{\left(\frac{\mu_n C_{ox}}{2}\right)\left(\frac{W_1}{L_1}\right)(v_{GS1} - V_{t1})^2(1 + \lambda_1 V_{DS1})}. \qquad (10.77)$$

Fabrication procedures yield matched devices. With matched V_t, C_{ox}, and λ Eq. (10.77) becomes

$$\frac{i_O}{i_1} = \frac{\left(\frac{W_2}{L_2}\right)(1 + \lambda V_{DS2})}{\left(\frac{W_1}{L_1}\right)(1 + \lambda V_{DS1})}. \qquad (10.78)$$

Usually λ is small and V_{DS1} and V_{DS2} are close so that to a reasonable approximation, Eq. (10.78) becomes

$$\frac{i_O}{i_1} \approx \frac{\left(\frac{W_2}{L_2}\right)}{\left(\frac{W_1}{L_1}\right)}. \qquad (10.79)$$

■ Designing by W/L ratios is one of the key advantages for MOS technology.

Equation (10.79) illustrates that we can design for a specific i_O/i_1 by lithographically defining W/L. Often, L is fixed as the minimum feature size in an IC so we are left with designing W_1 and W_2 to obtain the needed current ratio.

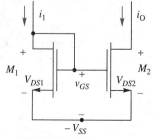

■ **FIGURE 10.36** Basic n-channel current mirror.

We are now in a position to illustrate the operation of a differential CMOS amplifier using active loads. As shown in Fig. 10.37(a), the amplifier consists of source-coupled n-channel MOSFETs, M_1 and M_2, and p-channel current mirror, M_3 and M_4. For operation as an amplifier, all MOSFETs Q-points are designed to be biased in the saturation region. There are many variations of MOS amplifiers and this example barely touches the surface of what is available. Figure 10.37(b) is a schematic diagram of the same MOS amplifier but with the I_{SS} current source replaced by a NMOS current source whose operation is described by Eq. (10.79). Our analysis proceeds in the same way in which we analyzed the BJT actively loaded amplifier. Summing voltages around the input loop and substituting Eq. (10.75), we have

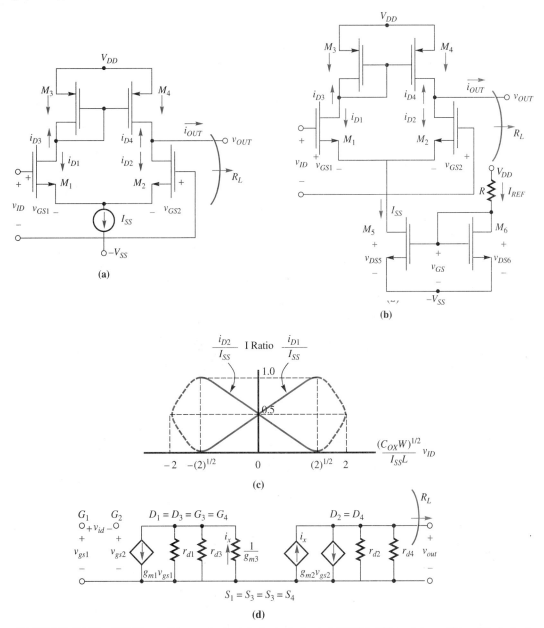

■ **FIGURE 10.37** CMOS amplifier analysis. (a) Actively loaded CMOS differential amplifier. (b) Actively loaded CMOS differential amplifier with NMOS current source. (c) Normalized transfer functions for both drain currents as a function of the input differential voltage. (d) Simplified small-signal model valid for either (a) or (b).

$$v_{ID} = v_{GS1} - v_{GS2} = \left[\frac{2i_{D1}}{\left(\frac{\mu_n C_{ox} W_1}{L_1}\right)}\right]^{\frac{1}{2}} - \left[\frac{2i_{D2}}{\left(\frac{\mu_n C_{ox} W_2}{L_2}\right)}\right]^{\frac{1}{2}}. \quad (10.80)$$

Realizing that $I_{SS} = I_{D1} + I_{D2}$ and after application of the quadratic equation, we obtain

$$i_{D1} = \frac{I_{SS}}{2} + \frac{I_{SS}}{2}\left[\frac{\left(\frac{\mu_n C_{ox} W_1}{L_1}\right) v_{ID}^2}{I_{SS}} - \frac{\left(\frac{\mu_n C_{ox} W_1}{L_1}\right)^2 v_{ID}^4}{4I_{SS}^2}\right]^{\frac{1}{2}} \quad (10.81)$$

and

$$i_{D2} = \frac{I_{SS}}{2} - \frac{I_{SS}}{2}\left[\frac{\left(\frac{\mu_n C_{ox} W_1}{L_1}\right) v_{ID}^2}{I_{SS}} - \frac{\left(\frac{\mu_n C_{ox} W_1}{L_1}\right)^2 v_{ID}^4}{4I_{SS}^2}\right]^{\frac{1}{2}}. \quad (10.82)$$

To determine the transconductance, g_m, we differentiate Eq. (10.81) or alternatively Eq. (10.82) to obtain and evaluate the result at the Q-point, $v_{ID} = 0$ to obtain

$$g_m = \left.\frac{\partial i_{D1}}{\partial v_{ID}}\right|_{v_{ID}=0} = \left[\frac{I_{SS}\mu_n C_{ox} W_1}{4L_1}\right]^{\frac{1}{2}}. \quad (10.83)$$

Implicit in this derivation is the assumption of matched FETs. That is, $W_1 = W_2$, $L_1 = L_2$, $W_3 = W_4$, and $L_3 = L_4$ and often $L_1 = L_2 = L_3 = L_4$ as a minimum dimension for the technology being used. Therefore, $-i_{D4} = -i_{D3} = I_{D1}$.

Assuming a resistive load, R_L, and summing currents at the output node, we solve for

$$v_{OUT} = (i_{D1} - i_{D2})R_L = I_{SS}\left[\frac{\mu_n C_{ox} W_1 v_{ID}^2}{I_{SS} L_1} - \left(\frac{\mu_n C_{ox} W_1 v_{ID}^2}{2I_{SS} L_1}\right)^2\right]^{\frac{1}{2}} R_L. \quad (10.84)$$

The voltage gain is the slope of Eq. (10.84) evaluated at $v_{ID} = 0$. The result is

$$A_v = \left.\frac{\partial v_{OUT}}{\partial v_{ID}}\right|_{v_{ID}=0} = \left(\frac{\mu_n C_{ox} I_{SS} W_1}{L_1}\right)^{\frac{1}{2}} R_L. \quad (10.85)$$

We can also obtain the voltage gain by constructing a small-signal model as shown in Fig. 10.37(d). This model development is analogous to the model for the BJT actively loaded amplifier shown in Figs. 10.30, 10.31, and 10.32. We observe that

$$r_{out} = r_{d2} \| r_{d4} \quad (10.86)$$

and

$$A_v = g_m[(r_{d2} \| r_{d4}) \| R_L]. \quad (10.87)$$

In terms of MOS parameters, Eq. (10.87) can be written

$$A_v = \left(\frac{\mu_n C_{ox} I_{SS} W_1}{L_1}\right)^{\frac{1}{2}} \left[\frac{1}{(\lambda_2 + \lambda_4)\left(\frac{I_{SS}}{2}\right)} \middle\| R_L\right]. \quad (10.88)$$

If $r_d = 1/\lambda I_D \gg R_L$, which might be the case if R_L represents the input of a BJT stage or external load, then A_v is given by Eq. (10.85). Otherwise, if R_L is obtained as the input resistance of a cascade CMOS stage, then A_v is given by

$$A_v = \frac{\left(\dfrac{\mu_n C_{ox} I_{SS} W_1}{L_1}\right)^{\frac{1}{2}}}{(\lambda_2 + \lambda_4)\left(\dfrac{I_{SS}}{2}\right)}. \quad (10.89)$$

CHECK UP

1. Give one major difference between a CMOS amplifier and a BJT amplifier that would favor use of the CMOS amplifier.
2. Give one major difference between a CMOS amplifier and a BJT amplifier that would favor use of the BJT amplifier.
3. **TRUE OR FALSE?** MOSFET geometries are not used to establish Q-point currents in current sources.
4. **TRUE OR FALSE?** If $V_{DS} = V_{GS}$, the MOSFET is operating in the saturation region.
5. Refer to Figure 10.37(b). What differences in circuit performance do you expect if R_L is the input to a second CMOS amplifier or a resistor?

10.6 POWER OUTPUT STAGES

The power output stage of an IC amplifier operating at the milliwatt level and the power output stage of a 200-W stereo system have a lot in common. Both must be designed to transfer power efficiently from relatively high output resistance circuits to relatively low resistance loads. This must be accomplished with minimum distortion and, when part of an IC, must be realizable in monolithic form. In general, we are interested in working with large signals, that is, signals whose peak excursions approach the maximum possible value, usually close to the power supply voltages. Load impedances will be low: in the case of operational amplifiers, several hundred to several thousand ohms, and in the case of a large audio amplifier, 4 to 16 Ω. Commensurate with these load resistance levels is the requirement to deliver large currents to the load.

DRILL 10.11

What W/L ratios are required in a current mirror to obtain currents of 10 μA, 30 μA, and 5 μA?

ANSWERS
a) $W/L = 2$
b) $W/L = 6$
c) $W/L = 1$

DRILL 10.12

Compute g_m for M_1 in Fig. 10.37. $I_{SS} = 30$ μA. The MOSFET specifications are $\mu_n = 600$ cm²/V-sec, $t_{ox} = 500$ Å, $W = 25$ μm and $L = 2$ μm.

ANSWER
$g_m = 0.62 \times 10^{-4}$ S

■ You can get ~10 watts in a single IC package. Heat sinking becomes a critical design issue. *Hybrid* packaging, more than one die in a single package is used to obtain more power dissipation capability.

Historically, power amplifiers have been divided into three classes, each class based on the *conduction angle* of the active device. The conduction angle is given in degrees, where 360° means that the current flows in the active device 100% of the time.

When the conduction angle is 360°, the amplifier is operating in *class A*. An example of a class A amplifier would be any of the amplifiers studied to this point in this text. The active device, BJT or FET, is ON during the complete sinusoidal cycle of 360°.

When the conduction angle is 180°, the amplifier is operating in *class B*. Obviously, if the transistor is conducting for only a half-cycle, the waveform is very distorted. Consequently, a class B amplifier always uses two transistors, each of which provides 180° of current conduction for a total of 360°. Occasionally, this type of operation is called *push-pull*, but this term is more of historical interest. Closely related to the class B amplifier is the *class AB* amplifier. The conduction angle is slightly more than 180° for reasons we will discuss shortly. Often, the term class B is used even when the actual circuit function is AB. The class B and class AB amplifiers are the principal focus of this section.

■ Well-designed Class C RF amplifiers have efficiencies > 75%.

When the conduction angle is less than 180° (usually much less than 180°), the amplifier is operating in *class C*. As in a class A circuit, only one device delivers power to the load, but, as you might expect, the output waveform consists of a series of sinusoidally shaped pulses that are clearly a very distorted rendition of the input sinusoid. This type of amplifier is used in radio-frequency power amplifiers, and in this application, the tuned circuit on the output stage acts as a filter for the pulse-type current waveform. The class C amplifier is beyond the scope of this text; however, the interested student is directed to Refs. 2 and 3 at the end of this chapter.

Class B Amplifier

The most common configuration for a class B amplifier is shown in Fig. 10.38(a). Transistor Q_1 is an *npn* transistor; Q_2 is *pnp*. The transistors are operated from power supplies of equal voltage magnitude, V_{CC}. The output signal, $v_O(t)$ will be obtained as a function of the input signal, $v_i(t) = V_m \sin \omega t$. To aid in this determination, the transfer characteristic,

■ For large signals, the concept of unity voltage gain is a reasonable approximation.

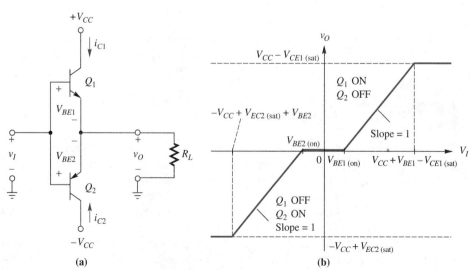

■ **FIGURE 10.38** Class B amplifier operation. (a) Class B amplifier. (b) Transfer characteristic.

$v_O(t)$ as a function of $v_i(t)$ is derived; see Fig. 10.42(b) given later. For $v_I = 0$, both Q_1 and Q_2 are off; $i_{C1} = i_{C2} = 0$, and as a result, $v_O = 0$. As v_I increases positively, Q_2 remains off, and when v_I exceeds $V_{BE(on)}$, Q_1 turns on. During this interval, Q_1 operates as an emitter follower where

$$v_I = V_{BE1} + v_O. \tag{10.90}$$

The output voltage maximum is

$$V_{O(max)} = V_{CC} - V_{CE(sat)}. \tag{10.91}$$

The input voltage necessary to drive v_O to $V_{CC} - V_{CE(sat)}$ is given by

$$V_{I(max)} = V_{BE1} + V_{CC} - V_{CE(sat)}. \tag{10.92}$$

In most circuits, because v_I cannot exceed V_{CC}, Q_1 never saturates since $V_{BE(on)} > V_{CE(sat)}$. As v_I becomes negative, Q_1 remains off, and when v_I drops below -0.7 V, the $V_{BE(on)}$ for Q_2, Q_2 turns on. Transistor Q_2 now behaves as a *pnp* emitter follower:

$$V_{O(min)} = -V_{CC} + V_{EC(sat)} \tag{10.93}$$

and occurs for an input voltage of

$$V_{I(min)} = -V_{CC} + V_{BE2} + V_{EC(sat)}. \tag{10.94}$$

The most notable feature in Fig. 10.38(b) is the 1.4-V deadband in the center of the characteristic. Obviously, the transfer characteristic leads to a distorted output waveform. This is called *crossover distortion*.

EXAMPLE 10.10

Use a SPICE simulation to demonstrate the transfer function shown in Fig. 10.38(b) for the class B amplifier shown in Fig. 10.38(a). Show the effect of crossover distortion on the output waveform. Use $V_{CC} = 12$ V with BJT specifications given in Table 18.2 and $R_L = 8$ Ω.

SOLUTION The node-labeled circuit diagram is given in Fig. 10.39. The SPICE netlist is given here:

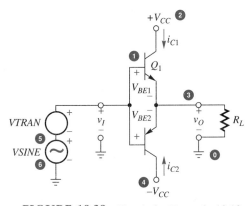

■ FIGURE 10.39 Circuit for Example 10.10.

```
CLASS B
VSINE 5 0 SIN(0 12 1000)
VTRAN 1 5 DC 0
VCC 2 0 DC 12
VEE 0 4 DC 12
Q1 2 1 3 STAN
Q2 4 1 3 BURNS
.MODEL STAN NPN BF=200 VAF=150 IS=2FA
.MODEL BURNS PNP BF=50 VAF=50 IS=2FA
RL 3 0 8
.DC VTRAN -13 13 0.1
.TRAN 10U 2M
.PROBE
.OP
.END
```

Observe that the input voltage is being swept from -13 to $+13$ V. The transfer function is plotted in Fig. 10.40 and agrees with the prediction of Fig. 10.38(b). Note the ± 0.7-V deadband, which results in crossover distortion in the output waveform as shown in Fig. 10.41.

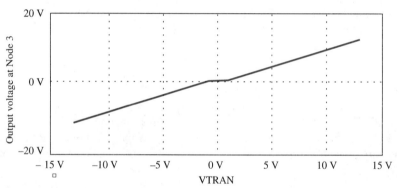

■ **FIGURE 10.40** Class B amplifier transfer characteristic illustrating the deadband region, Example 10.10.

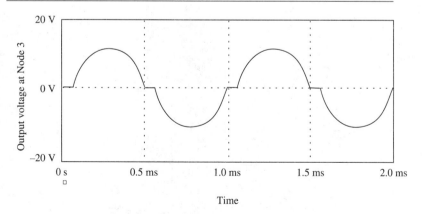

■ **FIGURE 10.41** Output waveform with crossover distortion. Clearly, if v_{in} were of lower magnitude, the resultant distortion would be more noticeable because the 0.7-V turn-on voltage would be a larger fraction of the overall output voltage magnitude.

EXAMPLE 10.11

Let $V_{CC} = 10$ V for the class B circuit shown in Fig. 10.38(a). Sketch and label key points of $v_O(t)$ for

(a) $v_i(t) = 1 \sin \omega t$ V
(b) $v_i(t) = 8 \sin \omega t$ V.

SOLUTION For both parts, $v_O(t) = v_i(t) \pm 0.7$ V. Let $\theta = \omega t$. The switch points are obtained by solving for θ in

(a) $\sin(\theta) = 0.7$, $\theta = 44.4°$, $135.60°$, $224.4°$, and $315.6°$
(b) $\sin(\theta) = 1/8(0.7)$, $\theta = 5°$, $175°$, $185°$, and $355°$.

These characteristics are plotted in Fig. 10.42. Note the symmetry about $\theta = 0°$ and $180°$ as expected. It is also important to observe that if $V_m \gg V_{BE}$, the crossover distortion is minimal.

■ In a SPICE simulation, a `.FOUR` command in a `.TRAN` analysis would yield the first 9 terms of the Fourier series of the waveform.

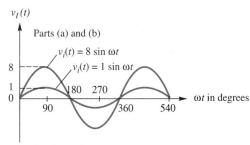

Input waveforms

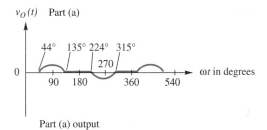

Part (a) output

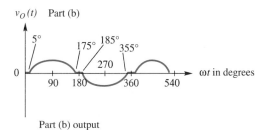

Part (b) output

■ **FIGURE 10.42** Waveforms for Example 10.11.

Class AB Amplifier

Class AB operation is used to eliminate crossover distortion. The basic idea is to have Q_1 and Q_2 both on, at least slightly, when $v_I = 0$. Consider the circuit shown in Fig. 10.43(a). Transistors Q_3 and Q_4 are connected as diodes. They need not have the drive capability of Q_1 and Q_2 since their entire function is to provide a V_{BE} drop sufficient to keep Q_1 and Q_2 barely on. Diode forward-bias current is supplied by the *pnp* current source composed of Q_5 and Q_6. This current source can either be a series resistor with V_{CC}, a simple current source, or some variation of a Widlar current source. A simple current source has been shown in Fig. 10.43(a). Quiescent operation is established by satisfying the condition

$$V_{BE3} + V_{BE4} = V_{BE1} + V_{EB2}. \tag{10.95}$$

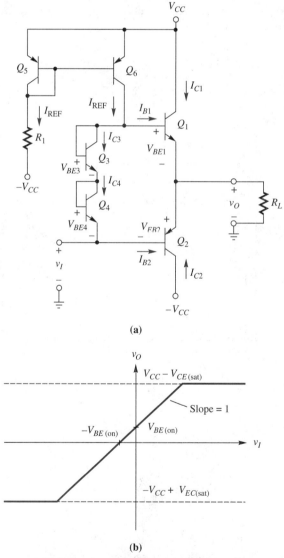

■ **FIGURE 10.43** Class AB amplifier operation.
(a) Class AB amplifier. (b) Transfer characteristic.

CHAPTER 10 OPERATIONAL AMPLIFIER CIRCUITRY

Substituting the diode equation in Eq. (10.95) yields

$$V_T \ln\left(\frac{I_{C3}}{I_{S3}}\right) + V_T \ln\left(\frac{I_{C4}}{I_{S4}}\right) = V_T \ln\left(\frac{I_{C1}}{I_{S1}}\right) + V_T \ln\left(\frac{I_{C2}}{I_{S2}}\right). \quad (10.96)$$

Since $I_{C1} = -I_{C2}$, Eq. (10.96) reduces to

$$\frac{I_{C3}I_{C4}}{I_{S3}I_{S4}} = \frac{I_{C1}^2}{I_{S1}I_{S2}} \quad (10.97)$$

or solving for I_{C1},

$$I_{C1} = -I_{C2} = \sqrt{\frac{I_{S1}I_{S2}}{I_{S3}I_{S4}}} \times \sqrt{I_{C3}I_{C4}}. \quad (10.98)$$

The reverse-saturation current ratios are established by the transistor emitter area ratios. Typically, the current capability of Q_1 and Q_2 is several times larger than Q_3 or Q_4. It is assumed that $\beta \gg 1$ for Q_1 and Q_2, so base currents can be neglected.

The transfer characteristic is shown in Fig. 10.43(b). For $v_I = 0$, $V_O = +0.7$ V since the base-emitter junction of Q_2 is slightly forward biased. The base of Q_1 is at $2 \times V_{BE(on)}$. We have emitter-follower operation; the output voltage follows the input voltage at $v_O = v_I + V_{EB2}$. The output voltage may be driven to zero by setting $v_I = -0.7$ V. The maximum output voltage is $\pm|V_{CC} - V_{CE(sat)}|$. Observe that there is an $R_{L(min)} \approx |V_{CC} - V_{CE(sat)}|/\beta_1 I_{REF}$.

DRILL 10.13

How would you add two 0.7 volt batteries to the circuit shown in Figure 10.39 to eliminate crossover distortion?

ANSWER

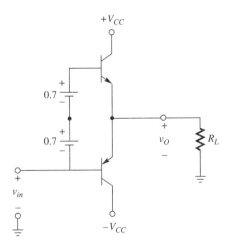

EXAMPLE 10.12

Use a SPICE simulation to demonstrate the transfer characteristic shown in Fig. 10.43(b) for the class B amplifier shown in Fig. 10.43(a). Show the effect of the crossover diodes in eliminating crossover distortion on the output waveform. Use $V_{CC} = 12$ V with BJT SPICE parameters and $R_L = 8\ \Omega$.

SOLUTION The node-labeled circuit diagram is given in Fig. 10.44. The SPICE netlist is given here:

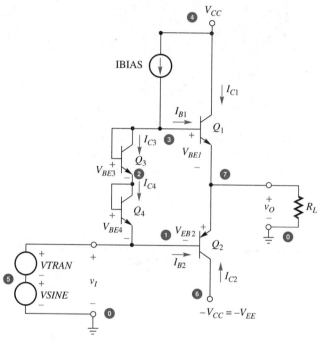

FIGURE 10.44 Circuit for Example 10.12.

```
CLASS AB EXAMPLE
VCC 4 0 DC 12
VEE 0 6 DC 12
IBIAS 4 3 DC 30M
VSINE 5 0 SIN(0 12 1000)
VTRAN 1 5 DC 0
Q1 4 3 7 STAN
Q2 6 1 7 BURNS
Q3 3 3 2 STAN
Q4 2 2 1 STAN
.MODEL STAN NPN
.MODEL BURNS PNP
RL 7 0 8
```

```
.DC VTRAN -13 13 0.1 IBIAS 30M
.TRAN 10U 2M
.PROBE
.OP
.END
```

Observe that the input voltage is being swept from -13 to $+13$ V. The transfer characteristic is plotted in Fig. 10.45 and agrees with the prediction of Fig. 10.43(b). Note the effect of the two diodes, which results in the elimination of crossover distortion in the output waveform in the class AB amplifier. There is an approximate $+0.7$-V shift in the transfer characteristic, which results in assymetrical clipping on the positive half of the cycle as shown in Fig. 10.46.

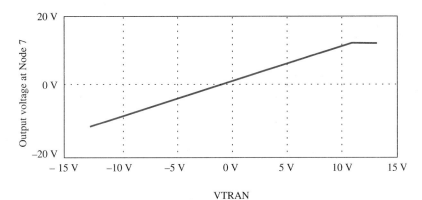

■ **FIGURE 10.45** Transfer characteristic for the class AB amplifier.

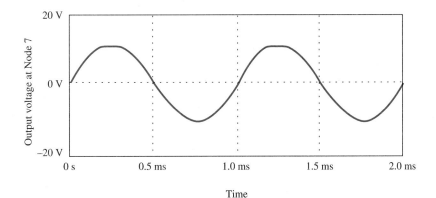

■ **FIGURE 10.46** Resultant class AB amplifier output waveform. The crossover distortion has been eliminated, but the circuit is now effectively being driven into saturation by an input voltage that is about $+0.7$ V too large.

Another way to look at the biasing of class B and AB amplifiers is to construct the back-to-back load lines as shown in Fig. 10.47 for the circuits illustrated in either Figs. 10.38 or 10.43.

The *npn*, Q_1, load line is given by

$$i_{C1} = \begin{cases} \dfrac{-v_{CE1}}{R_L} + \dfrac{V_{CC}}{R_L} & \text{for } v_{BE1} \geq 0.7 \text{ V} \\ 0 & \text{for } v_{BE1} \leq 0.7 \text{ V}. \end{cases} \qquad (10.99)$$

Also observe that the y-axis intercept is given by V_{CC}/R_L and $V_{CC} < v_{CE1} < 2V_{CC}$ when Q_1 is OFF. It is also interesting to note that the maximum collector dissipation occurs when $v_{CE1} = V_{CC}/2$ and $i_{C1} = V_{CC}/2R_L$ as illustrated by the constant power hyperbola being tangent to the load line at that point.

Similarly, the *pnp* Q_2 load line is given by

$$i_{C2} = \begin{cases} \dfrac{-v_{CE2}}{R_L} + \dfrac{-V_{CC}}{R_L} & \text{for } v_{BE2} \leq -0.7 \text{ V} \\ 0 & \text{for } v_{BE2} \geq -0.7 \text{ V}. \end{cases} \qquad (10.100)$$

The y-axis intercept is given by $-V_{CC}/R_L$ and $-2V_{CC} < v_{CE2} < -V_{CC}$ when Q_2 is OFF. The maximum collector dissipation occurs when $v_{CE2} = -V_{CC}/2$ and $i_{C2} = -V_{CC}/2R_L$ as illustrated by the constant power hyperbola being tangent to the load line at that point.

DRILL 10.14

For a class B amplifier with $V_{CC} = 30$ volts and $R_L = 8\ \Omega$, compute the required BJT specifications for $V_{CE(\text{max})}$, $I_{C(\text{max})}$, and $P_{C(\text{max})}$.

ANSWERS
a) 60 V
b) 3.75 A
c) 28.125 W

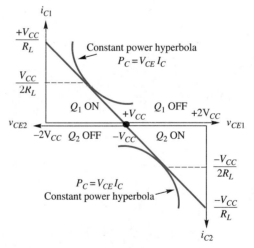

■ **FIGURE 10.47** Back-to-back load lines used for class B and AB power amplifier analysis.

EXAMPLE 10.13

For $V_{CC} = 15$ V and $R_1 = 47$ kΩ in Figure 10.43, compute all quiescent currents in the class AB amplifier if the junction areas of the output transistors are four times that of the other transistors. If $\beta_{npn} = 150$ and $\beta_{pnp} = 50$, comment on the validity of being able to neglect the base currents of Q_1 and Q_2.

CHAPTER 10   OPERATIONAL AMPLIFIER CIRCUITRY

SOLUTION   For the simple current source,

$$I_{REF} = I_{C3} = I_{C4} = \frac{(2V_{CC} - V_{BE(on)})}{R_1} = \frac{29.3 \text{ V}}{47 \text{ k}\Omega} = 623 \text{ }\mu\text{A}.$$

Then, using Eq. (10.98),

$$I_{C1} = -I_{C2} = \sqrt{\frac{4I_{S3} \times 4I_{S4}}{I_{S3}I_{S4}}} \sqrt{(623 \text{ }\mu\text{A})(623 \text{ }\mu\text{A})} = 4 \times 623 \text{ }\mu\text{A} = 2.49 \text{ mA}.$$

The quiescent base currents are

$$I_{B1} = \frac{2.49 \text{ mA}}{150} = 16.7 \text{ }\mu\text{A}$$

and

$$I_{B2} = \frac{-2.49 \text{ mA}}{50} = -49.8 \text{ }\mu\text{A}.$$

The result of $I_{B1} = 16.7 \text{ }\mu\text{A}$ would only change the value of I_{C3} and I_{C4} by about 3%; thus, neglecting base currents is a reasonable strategy.

Class A Amplifier

Another way to avoid the problem of crossover distortion is to operate an emitter follower in class A. That is, its biasing is arranged so that the transistor is conducting current during a full 360° of a cycle. Class A operation is preferred at lower relative power levels and is used extensively to buffer amplification stages in an operational amplifier. It tends to be simpler to implement because all *npn* or all *pnp* transistors are used. We will observe shortly that class B operation is preferred over class A operation in the output circuit because, for a given amount of power to the load, the power supply output power capability need not be as large.

A class A emitter-follower power amplifier is shown in Fig. 10.48. Transistors Q_2 and Q_3 are configured as a simple current source. Assuming operation between $+V_{CC}$ and $-V_{CC}$, the quiescent current is given by

$$I_Q = \frac{2V_{CC} - V_{BE(on)}}{R_1}. \tag{10.101}$$

Then at the input,

$$v_I = V_{BE1} + v_O = V_T \ln\left(\frac{i_{C1}}{I_{S1}}\right) + v_O. \tag{10.102}$$

Assuming $\beta \gg 1$ and using $i_{C1} = I_Q + \frac{v_O}{R_L}$, we have from Eqn. (10.102),

$$v_I = V_T \ln\left[\frac{I_Q + \frac{v_O}{R_L}}{I_S}\right] + v_O. \tag{10.103}$$

This equation is nonlinear and difficult to solve explicitly since it is transcendental in v_O. A transfer characteristic will be obtained by looking at case I where R_L is large, $v_O/R_L \ll I_Q$, and case II where R_L is small.

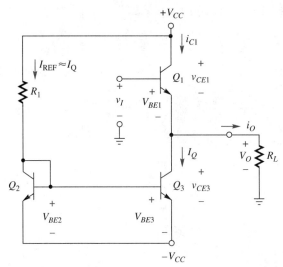

FIGURE 10.48 Class A emitter-follower output stage.

In considering case I, Eq. (10.103) becomes

$$v_I = V_T \ln\left(\frac{I_Q}{I_S}\right) + v_O = v_{BE1} + v_O, \tag{10.104}$$

where $v_{BE1} \cong V_{BE(on)} = 0.7$ V is considered essentially constant. The transfer characteristic is obtained (Fig. 10.49) by graphing

$$v_O = v_I - 0.7. \tag{10.105}$$

■ Again, unity gain is a good approximation for this large signal situation.

The maximum output voltage is

$$v_{O(\max)} = V_{CC} - V_{CE1(sat)}, \tag{10.106}$$

which is obtained when

$$v_I = V_{CC} - V_{CE(sat)} + V_{BE}. \tag{10.107}$$

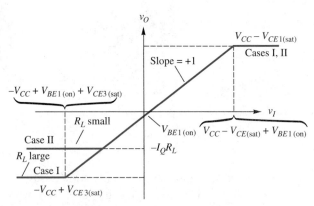

FIGURE 10.49 Class A transfer characteristic.

CHAPTER 10 OPERATIONAL AMPLIFIER CIRCUITRY

Since v_I usually does not exceed V_{CC}, Q_1 never quite saturates. The minimum output voltage is

$$V_{O(\min)} = -V_{CC} + V_{CE3(\text{sat})} \tag{10.108}$$

and occurs when

$$v_I = -V_{CC} + V_{CE3(\text{sat})} + V_{BE1}. \tag{10.109}$$

Note that Q_3 can saturate for a large negative v_I. The case I transfer function has a slope of $+1$ with an offset of 0.7 V from the origin.

For case II, the first and third quadrants of the transfer function must be considered separately. Referring to Eq. (10.103),

$$I_Q + \frac{v_O}{R_L} \gg 0$$

for all values of $v_i > 0$, so the first quadrant portion of the transfer function is virtually identical to the case I condition. This portion of the transfer function is adequately described by Eq. (10.105). For $v_I < 0$, v_O also becomes negative, and when this happens, the numerator of

$$\ln\left(\frac{I_Q + v_O/R_L}{I_S}\right)$$

could go to zero. Physically, when

$$v_O = -I_Q R_L, \tag{10.110}$$

Q_1 becomes cut off and the load current, $I_O = -I_Q$, a constant. This has the effect of yielding an output voltage saturation in the third quadrant of

$$v_O = -I_Q R_L \tag{10.111}$$

when $V_I < 0.7 - I_Q R_L$. Therefore, to obtain a full $-V_{CC}$ voltage excursion [actually $-V_{CC} + V_{CE(\text{sat})}$], the relationship between I_Q and R_L is fixed by Eq. (10.110).

DRILL 10.15

For a class A amplifier with $V_{CC} = 12$ V and $I_Q = 4$ mA, compute the range of R_L required to obtain a full rail-to-rail output voltage.

ANSWER $R_L > 3$ kΩ

EXAMPLE 10.14

Use a SPICE simulation to demonstrate the transfer function shown in Fig. 10.49 for the class A amplifier shown in Fig. 10.48. Show the effect of $R_L = 1$ kΩ being too small to obtain a full output voltage swing. Use $V_{CC} = 12$ V with BJT default SPICE specifications.

SOLUTION The node-labeled schematic diagram is given in Fig. 10.50. The netlist, with an input voltage sweep of -13 to $+13$ V, is given here:

```
CLASS A EXAMPLE
VSINE 2 0 SIN(0 12 1000)
VTRAN 1 2 DC 0
VCC 3 0 DC 12
VEE 0 5 DC 12
Q1 3 1 4 STAN
Q2 4 6 5 STAN
Q3 6 6 5 STAN
```

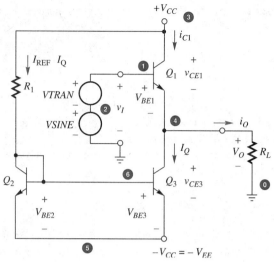

FIGURE 10.50 Class A emitter-follower output stage.

```
RREF 3 6 2400
.MODEL STAN NPN
RL 4 0 1K
.DC VTRAN -13 13 0.1
.TRAN 10U 2M
.PROBE
.OP
.END
```

The transfer function is plotted in Fig. 10.51. As expected from Eq. (10.111),

$$V_O = -I_Q R_L \cong -\left[\frac{2V_{CC} - V_{BE(on)}}{R_{REF}}\right] R_L \cong -\left(\frac{24 - 0.7}{2400\,\Omega}\right) \times (1000\,\Omega) = -9.7\,\text{V},$$

which is reasonably close to the transfer function negative saturation characteristic of Fig. 10.51. The resultant waveform is shown in Fig. 10.52. Either, the input level must be restricted to approximately ±9.7 V or R_L must increase to accommodate a larger distortion-free output voltage.

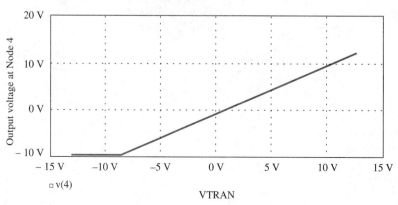

FIGURE 10.51 Transfer function for a class A amplifier with $I_Q R_L$ too small.

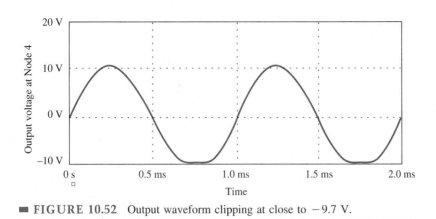

FIGURE 10.52 Output waveform clipping at close to -9.7 V.

Power and Efficiency

Efficiency or *collector efficiency* is defined by

$$\eta_C = \frac{P_L}{P_{DC}} \times 100\%, \tag{10.112}$$

where P_L is the average power dissipated in the load and P_{DC} is the average power delivered from the power supplies to the output transistors.

Although there is power dissipation in any resistor, current source, and in both the BE and CE junctions of an output transistor, usually most of the power dissipated in a circuit is the result of the relatively large $v_{CE} \times i_C$. In addition to looking at the time-averaged power relationships in a circuit, we also consider the instantaneous power values. In practice, the effect of $V_{CE(\text{sat})}$ is ignored since the 0.2 V or so is typically insignificant compared to a full voltage excursion approaching V_{CC}.

Class A Power and Efficiency Relationships

Assume $I_Q R_L$ is selected to permit $v_o(t) = V_m \sin \omega t$ where $V_m = V_{CC} - V_{CE(\text{sat})} \cong V_{CC}$ [neglecting $V_{CE(\text{sat})}$], and the output instantaneous load power is given by

$$p_L(t) = \frac{V_{CC}^2 \sin^2 \omega t}{R_L}. \tag{10.113}$$

To compute the time-averaged power in the load, we use

$$P_L = \frac{1}{T} \int_0^T p_L(t)\, dt = \frac{1}{T} \int_0^T \frac{V_{CC}^2 \sin^2 \omega t\, dt}{R_L}$$

$$= \frac{1}{T} \frac{V_{CC}^2}{R_L} \int_0^T \left(\frac{1 - \cos 2\omega t}{2} \right) dt = \frac{V_{CC}^2}{2 R_L}. \tag{10.114}$$

This result really could have been obtained by inspection, but illustrating the calculation will be useful shortly when we look at the class B case. The power supply is delivering a constant

$$P_{DC} = 2 V_{CC} I_Q \tag{10.115}$$

since both V_{CC} and I_Q are not functions of time. Substituting Eqs. (10.114) and (10.115) into Eq. (10.112) yields

$$\eta_C = \frac{V_{CC}^2/2R_L}{2V_{CC}I_Q} \times 100\% = \frac{1}{4}\frac{V_{CC}}{R_L I_Q} \times 100\%. \tag{10.116}$$

To obtain full voltage and current excursion, $V_{CC} = I_Q R_L$ so $\eta_C = 25\%$. It is interesting to observe that even if $v_I = 0$, the power supply must still deliver $2V_{CC}I_Q$ watts to the circuit. Under this situation, $\eta_C = 0\%$, a very undesirable result if overall power dissipation in a circuit is an important design consideration. In general, the class A collector efficiency is given by

$$\eta_C = \frac{V_m^2/2R_L}{2V_{CC}I_Q} \times 100\% = \frac{1}{4}\left(\frac{V_m^2}{V_{CC}^2}\right) \times 100\%. \tag{10.117}$$

The power distribution in the rest of the circuit shown in Fig. 10.48 is determined with the aid of Fig. 10.53. Assume an arbitrary input, $v_i = V_m \sin \omega t$ [part (a) of Fig. 10.53], where $V_m < V_{CC}$; then the instantaneous power in the collector junction of Q_1 is given by

$$p_C(t) = v_{CE1}(t)i_{C1}(t) = (V_{CC} - V_m \sin \omega t)(I_Q + I_m \sin \omega t), \tag{10.118}$$

where $v_{CE1}(t)$, $i_C(t)$, and the products are graphed in Fig. 10.53, parts (b), (c), and (d), respectively. The interpretation of Eq. (10.118) is that for $V_m = 0$, the instantaneous and average collector power is $V_{CC} \times I_Q$, a worst-case situation. For $V_m = V_{CC}$, Eq. (10.118) becomes

$$p_{C1}(t) = V_{CC}I_Q(1 - \sin \omega t)(1 + \sin \omega t)$$
$$= V_{CC}I_Q[1 - \sin^2 \omega t] = \frac{V_{CC}I_Q}{2}(1 + \cos 2\omega t), \tag{10.119}$$

and the time-averaged value can be obtained graphically as shown in part (d) of Fig. 10.53 or analytically by

$$P_C = \frac{1}{T}\int_0^T p_{C1}(t)\,dt = \frac{V_{CC}I_Q}{T}\int_0^T [1 - \sin^2 \omega t]\,dt$$
$$= \frac{V_{CC}I_Q}{2T}\int_0^T (1 + \cos 2\omega t)\,dt = \frac{V_{CC}I_Q}{2}. \tag{10.120}$$

The power dissipated in Q_3 is a constant given by $P_{C3} = V_{CC}I_Q$. If we do a power balance equation and neglect power dissipation in the current source, the result is

$$P_{DC} = P_{C1} + P_{C3} + P_L = \frac{V_{CC}I_Q}{2} + V_{CC}I_Q + \frac{V_{CC}I_Q}{2} = 2V_{CC}I_Q \tag{10.121}$$

as expected.

Class B and AB Power and Efficiency Relationships

For all practical cases, the class B and AB amplifiers are treated identically when computing η_C. Essentially, the crossover distortion is ignored, as is $V_{CE(\text{sat})}$ compared to the magnitude of V_{CC}. Figure 10.54 is used in this discussion. The load instantaneous power of the circuit shown in Fig. 10.38 (class B) or Fig. 10.43 (class AB) is given by

$$p_L(t) = v_O(t)i_O(t) = \frac{(V_m \sin \omega t)^2}{R_L}. \tag{10.122}$$

CHAPTER 10 OPERATIONAL AMPLIFIER CIRCUITRY

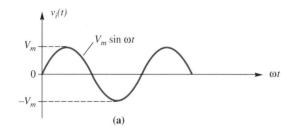

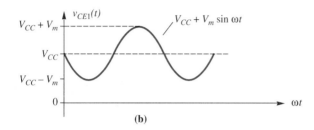

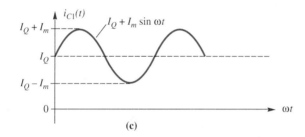

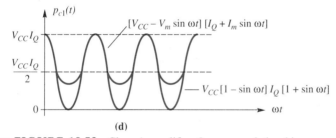

■ **FIGURE 10.53** Class A amplifier Q_1 power relationships.

and, as expected, the average load power is

$$P_L = \frac{1}{2}\frac{V_m^2}{R_L}. \tag{10.123}$$

This is no different from the class A amplifier result of Eq. (10.114) except that now $V_m < V_{CC}$.

The instantaneous power supply power over a complete cycle of operation is given by

$$p_{DC}(t) = V_{CC}i_{C1}(t) = (V_{CC})\left(\frac{V_m}{R_L}\sin \omega t\right), \qquad 0 < t < \frac{T}{2}, \tag{10.124}$$

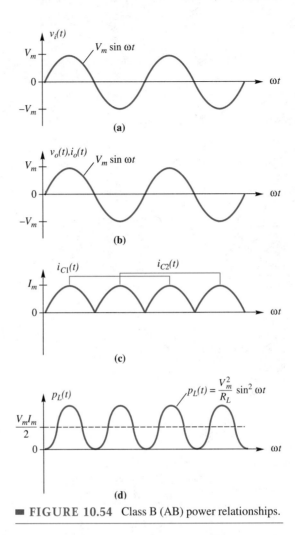

■ **FIGURE 10.54** Class B (AB) power relationships.

where i_{C1} is shown in part (b) of Fig. 10.54. For $i_{C2}(t)$, also in part (b) of Fig. 10.54, we have

$$p_{DC}(t) = -V_{CC}i_{C2}(t) = -V_{CC}\frac{V_m}{R_L}\sin\omega t, \quad \frac{T}{2} \leq t < T. \quad (10.125)$$

By computing the time-averaged value of either Eq. (10.124) or Eq. (10.125) and multiplying by 2 we will obtain P_{DC}. Then using $\omega = 2\pi f = 2\pi/T$,

$$P_{DC} = \left[\frac{1}{T}\int_0^{T/2} V_{CC}i_{C1}(t)\,dt\right] \times 2 = \frac{2V_{CC}}{T}\int_0^{T/2}\frac{V_m}{R_L}\sin\left(\frac{2\pi t}{T}\right)dt$$

$$= \frac{2V_{CC}V_m}{TR_L}\left[\frac{-T}{2\pi}\cos\left(\frac{2\pi t}{T}\right)\bigg|_0^{T/2}\right] = \frac{2V_{CC}V_m}{\pi R_L}. \quad (10.126)$$

Substituting Eqs. (10.123) and (10.126) into Eq. (10.112) yields a collector efficiency of

$$\eta_C = \frac{V_m^2/2R_L}{2V_{CC}V_m/\pi R_L} \times 100\% = \frac{\pi}{4}\frac{V_m}{V_{CC}} \times 100\%. \quad (10.127)$$

Since the maximum value of $V_m = V_{CC} - V_{CE(sat)} \cong V_{CC}$, the maximum value of the collector efficiency is

$$\eta_{C(max)} = \frac{\pi}{4} \times 100\% = 78.5\%. \quad (10.128)$$

This result is indeed much better than that obtained for a class A amplifier and is one of the most important advantages. To design a class B (AB) amplifier, the maximum collector power dissipation rating must be specified with an adequate safety margin. Worst-case transistor dissipation for a sinusoid occurs when $V_m = 2V_{CC}/\pi$ and $\eta_C = 50\%$.

Short-Circuit Protection

As previously mentioned, the output transistors probably dissipate much more power than the rest of the active and passive circuitry combined in an integrated circuit. They are designed to operate within their maximum ratings for all design values of R_L. It is easy to observe (see Fig. 10.38) that the collector currents, I_{C1} and I_{C2}, would only be limited by the power supply capabilities should $R_L = 0\,\Omega$. This relatively large value of current would quickly destroy the output transistors. To prevent this, current limiting is usually designed into the circuit. One possible design is illustrated in Fig. 10.55(a).

The protection circuitry consists of the addition of Q_2 and R_{SC}. At the base of Q_1,

$$I_I = I_{B1} + I_{C2} = I_{B1} + \beta_2 I_{B2} = I_{B1} + I_{S2} e^{qV_{BE2}/kT} \quad (10.129)$$

and since $I_{B1} \ll I_{C1}$,

$$V_{BE2} \cong I_{C1} R_{SC}. \quad (10.130)$$

DRILL 10.16

Compute the collector efficiency for a class B or AB amplifier with $V_{CC} = 12$ V and $V_m = 5$ V, 10 V, and 12 V. Compare your results with a class A amplifier with $V_{CC} = 12$ V and $V_m = 5$ V, 10 V, and 12 V. Neglect $V_{CE(sat)}$.

ANSWERS
a) Class B or AB
b) $\eta_C = 32.7\%$
c) $\eta_C = 65.5\%$
d) $\eta_C = 78.5\%$
e) Class A
f) $\eta_C = 4.34\%$
g) $\eta_C = 17.36\%$
h) $\eta_C = 25\%$

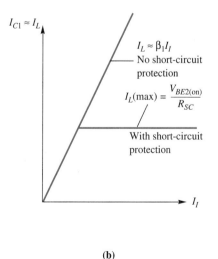

■ **FIGURE 10.55** Short-circuit protection. (a) Output stage short-circuit protection circuit. (b) Load current with and without short-circuit protection.

Then the load current is given by

$$I_L = I_{E2} + I_{C1} - I_{B2}. \qquad (10.131)$$

If R_L is large enough and R_{SC} is small enough, Q_2 is OFF, with the result that

$$I_L = I_{C1} = \beta_1 I_{B1} = \beta_1 I_I. \qquad (10.132)$$

which represents normal operation. Should R_L be short-circuited, I_{C1} would continue to increase [see Fig. 10.55(b)] toward unacceptable limits without the protection circuitry. When the voltage drop across R_{SC} approaches $V_{BE2(on)}$, nominally 0.55 to 0.7 V, Q_2 is switched on, effectively limiting I_{B1} drive current as I_{C2} increases. The load current is effectively clamped at

$$I_{L(max)} = \frac{V_{BE2(on)}}{R_{SC}}, \qquad (10.133)$$

which protects Q_1. Note that $I_{C2} \ll I_{C1}$ is obtained from the base current drive of Q_1. This clamping is illustrated in Fig. 10.55(b).

EXAMPLE 10.15

Suppose you have a stereo with a power amplifier capable of delivering 100 W to an 8-Ω load. What would be a reasonable value for R_{SC} to protect the output transistors against an output short circuit?

SOLUTION The peak load current is computed from

$$I_{Lm} = \sqrt{\frac{2P_L}{R_L}} = 5 \text{ A}.$$

Assuming that you want the output stage to clamp at $I_C = 5$ A, then $R_{SC} = 0.6$ V/5 A = 0.12 Ω. Note that the 0.6-V drop across R_{SC} is negligible compared to the peak load voltage, $V_m = 5 \text{ A} \times 8 \, \Omega = 40$ V.

Harmonic Distortion

The transfer functions have been assumed linear although they are inherently nonlinear. Even if circuit operation avoids saturation conditions, and crossover distortion has been eliminated, fundamental transistor operation is nonlinear. To this point and for ease of analysis, these nonlinearities were conveniently ignored.

For the class B or AB amplifier, referring to Figs. 10.38, 10.43, and 10.54,

$$i_{C1}(t) = I_C + A_1 \cos \omega_0 t + A_2 \cos 2\omega_0 t \\ + A_3 \cos 3\omega_0 t + \cdots + A_n \cos n\omega_0 t, \qquad (10.134)$$

$$i_{C2}(t) = -I_C - A_1 \cos(\omega_0 t + \pi) - A_2 \cos 2(\omega_0 t + \pi) \\ - A_3 \cos 3(\omega_0 t + \pi) - \cdots - A_n \cos n(\omega_0 t + \pi). \qquad (10.135)$$

The $+\pi$ represents the 180° phase shift between the collector waveforms. Then at the load,

$$i_L(t) = i_{C1} + i_{C2} = 2A_1 \cos \omega_0 t + 2A_3 \cos 3\omega_0 t + \cdots \qquad (10.136)$$

since $\cos(\omega_0 t + \pi) = -\cos \omega_0 t$ and $\cos 2(\omega_0 t + \pi) = +\cos 2\omega_0 t$. Thus the class B amplifier has the interesting property that the second harmonic and all even-order harmonics cancel. This is useful because the third harmonic would probably be less than the second harmonic. In Chapter 6, we use feedback as a key design technique in minimizing distortion in an amplifier.

Harmonic distortion (HD) is measured by assuming the output voltage across R_L:

$$v_0(t) = R_L i_L(t). \tag{10.137}$$

In this case, we will assume $i_L(t)$ has components at all harmonics. Each frequency component generates an independent contribution to the total load power according to

$$P_L = \frac{A_1^2 R_L}{2} + \frac{A_2^2 R_L}{2} + \frac{A_3^2 R_L}{2} + \cdots + \frac{A_n^2 R_L}{2}. \tag{10.138}$$

By eliminating the fundamental frequency component, $A_1 \cos \omega_0 t$, the harmonic distortion voltage is defined by

$$\text{HD} = \sqrt{A_2^2 + A_3^2 + \cdots + A_n^2}. \tag{10.139}$$

The percentage HD is defined by

$$\%\text{HD} = \frac{\sqrt{A_2^2 + A_3^2 + \cdots + A_n^2}}{A_1} \times 100\%. \tag{10.140}$$

It is not unusual for the percentage of harmonic distortion to be less than 0.1% in a well-designed amplifier, although an inspection of specifications should include the output power level since the percentage usually becomes much worse at higher signal levels in any given circuit.

■ There is a lot of "specsmanship" in providing values for %HD, especially in audio systems. One must not only look at %HD but the %HD at the power level under consideration. Obviously, if you drive an amplifier into saturation, %HD increases.

DRILL 10.17

Compute the distortion for the signal described by
$v(t) = 15 \cos(1000t) - 0.5 \cos(2000t) + 0.2 \sin(3000t) + 0.35 \cos(4000t + 30°)$

ANSWER HD = 4.28%

EXAMPLE 10.16

Suppose you measure the harmonic voltage content of the amplifier described in Example 10.10. At ω_0, $V_1 = 30$ V; at $2\omega_0$, $V_2 = 0.2$ V; at $3\omega_0$, $V_3 = 1$ V; at $4\omega_0$, $V_4 = 0.05$ V; and at $5\omega_0$, $V_5 = 0.3$ V. Compute the percentage of harmonic distortion.

SOLUTION From Eq. (10.139) or (10.140),

$$\%\text{HD} \cong \sqrt{\frac{(0.2)^2 + (1)^2 + (0.05)^2 + (0.3)^2}{(30)^2}} \times 100\% = 3.55\%.$$

CHECK UP

1. **TRUE OR FALSE?** Class A amplifiers are the most efficient.
2. **TRUE OR FALSE?** Class B amplifiers have less distortion than class AB amplifiers.
3. **TRUE OR FALSE?** The maximum power dissipation in class B amplifier output transistors occurs at maximum output signal amplitude.
4. **TRUE OR FALSE?** Maximum efficiency occurs in a class A amplifier at maximum output signal amplitude.
5. What is the voltage gain and current gain of the amplifiers discussed in this section?

10.7 EFFECTS OF DEVICE MISMATCH

One of the most important properties inherent with circuit elements fabricated by means of integrated-circuit techniques is the closeness of parameter matching (Chapter 18). Absolute values are difficult to achieve because of process variability. In this section, we examine the effects of transistor and collector resistor mismatch in an emitter-coupled pair.

■ Implicit in the following mismatch discussion is having all devices at the same temperature. IC designers do have to worry about temperature profiles across a die although it may not present a problem for less-demanding circuits.

Input Offset Voltage

Consider the emitter-coupled pair illustrated in Fig. 10.56. In this circuit, define V_{OS} as the dc *input offset voltage* that when applied between both bases yields $V_{OD} = 0$. Then, summing voltages around the input loop,

$$V_{OS} = V_{BE1} - V_{BE2} = \frac{kT}{q} \ln\left(\frac{I_{C1}}{I_{S1}}\right) - \frac{kT}{q} \ln\left(\frac{I_{C2}}{I_{S2}}\right)$$

$$= \frac{kT}{q} \ln\left(\frac{I_{C1}}{I_{S1}} \times \frac{I_{S2}}{I_{C2}}\right). \tag{10.141}$$

Because we are studying the effects of device mismatch, we will assume that $I_{S1} \neq I_{S2}$. As shown in Chapter 4, I_S is directly proportional to A_j, junction cross-sectional area, and a function of base doping density, base width, and diffusion coefficients. Thus, differences in I_S represent a composite process-controlled parameter.

To obtain $V_{OD} = 0$,

$$I_{C1}R_{C1} = I_{C2}R_{C2}. \tag{10.142}$$

This means that even if both transistors were perfectly matched so that $I_{S1} = I_{S2}$, the condition of $R_{C1} = R_{C2}$ must also hold to obtain $V_{OS} = 0$. Substituting Eq. (10.142) into Eq. (10.141) yields

■ Refer to a discussion of V_{OS} in Chapter 6.

$$V_{OS} = \frac{kT}{q} \ln\left(\frac{R_{C2}}{R_{C1}} \times \frac{I_{S2}}{I_{S1}}\right). \tag{10.143}$$

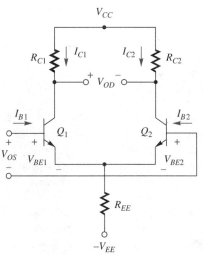

■ **FIGURE 10.56** Emitter-coupled pair with mismatched components.

CHAPTER 10 OPERATIONAL AMPLIFIER CIRCUITRY

Since we are interested in relative, as opposed to absolute, component values, we define the differences as

$$\Delta R_C = R_{C1} - R_{C2}, \qquad \Delta I_S = I_{S1} - I_{S2} \qquad (10.144)$$

and the averages as

$$R_C = \frac{R_{C1} + R_{C2}}{2}, \qquad I_S = \frac{I_{S1} + I_{S2}}{2}. \qquad (10.145)$$

Solving Eqs. (10.144) and (10.145) for R_{C1}, R_{C2}, I_{S1}, and I_{S2},

$$R_{C1} = R_C + \frac{\Delta R_C}{2} \qquad R_{C2} = R_C - \frac{\Delta R_C}{2},$$

$$I_{S1} = I_S + \frac{\Delta I_S}{2} \qquad I_{S2} = I_S - \frac{\Delta I_S}{2}. \qquad (10.146)$$

Substituting Eq. (10.146) into Eq. (10.143) yields

$$V_{OS} = \frac{kT}{q} \ln \left[\frac{\left(R_C - \frac{\Delta R_C}{2}\right)\left(I_S - \frac{\Delta I_S}{2}\right)}{\left(R_C + \frac{\Delta R_C}{2}\right)\left(I_S + \frac{\Delta I_S}{2}\right)} \right]. \qquad (10.147)$$

Factoring and cancelling Eq. (10.147) yields,

$$V_{OS} = \frac{kT}{q} \ln \left[\frac{\left(1 - \frac{\Delta R_C}{2R_C}\right)\left(1 - \frac{\Delta I_S}{2I_S}\right)}{\left(1 + \frac{\Delta R_C}{2R_C}\right)\left(1 + \frac{\Delta I_S}{2I_S}\right)} \right]. \qquad (10.148)$$

Since $R_C/2R_C \ll 1$ and $I_S/2I_S \ll 1$, and using an approximation based on the binomial expansion,

$$\frac{1}{1 \pm x} \approx 1 \mp x \qquad \text{for } x \ll 1. \qquad (10.149)$$

Equation (10.148) is rewritten as

$$V_{OS} \approx \frac{kT}{q} \ln \left[\left(1 - \frac{\Delta R_C}{2R_C}\right)\left(1 - \frac{\Delta R_C}{2R_C}\right)\left(1 - \frac{\Delta I_S}{2I_S}\right)\left(1 - \frac{\Delta I_S}{2I_S}\right) \right]. \qquad (10.150)$$

Multiplying out the terms in Eq. (10.150) and neglecting higher order terms,

$$V_{OS} \approx \frac{kT}{q} \ln \left(1 - \frac{\Delta R_C}{R_C} - \frac{\Delta I_S}{I_S} + \text{higher order terms}\right). \qquad (10.151)$$

One further simplification is possible by using the Taylor's series expansion:

$$\ln(1 + x) = x - \frac{1}{2}x^2 + \frac{1}{3}x^3 - \frac{1}{4}x^4 + \cdots, \qquad -1 < x < 1, \qquad (10.152)$$

with Eq. (10.151) to obtain

$$V_{OS} \cong \frac{kT}{q}\left(-\frac{\Delta R_C}{R_C} - \frac{\Delta I_S}{I_S}\right). \qquad (10.153)$$

DRILL 10.18

Two BJTs have matched junction areas within 1% as determined by the photolithography. Assume exactly matched collector resistors. Estimate the V_{OS}.

ANSWER
$V_{OS} = 0.26$ mV

EXAMPLE 10.17

Let the worst case $\Delta R_C/R_C = 10\%$ and the worst case $\Delta I_S/I_S = 15\%$. Compute the resultant V_{OS} at $T = 300$ K.

SOLUTION Two answers are possible using Eq. (10.153). The worst case would be $V_{OS} = 26$ mV$(-0.1 - 0.15) = 6.5$ mV. Usually, only $|V_{OS}|$ is of interest. If the individual mismatches partially compensate for each other, $V_{OS} = 26$ mV$(-0.1 + 0.15) = 1.3$ mV. Thus, Eq. (10.153) suggests a design technique for reducing V_{OS} to zero. The user has no control over the mismatch in transistor parameters; however, by deliberately changing one of the R_C values by some external circuit addition or a fabrication-process trimming procedure, V_{OS} can be adjusted to zero.

Input Offset Current

Ideally, both quiescent base currents would be identical in matched transistors. Even if $I_{C1} = I_{C2}$, it is probable that $\beta_1 \neq \beta_2$. Then the *input offset current* is defined as

■ Again, refer to Ch. 6 for additional material on I_{OS}.

$$I_{OS} = I_{B1} - I_{B2} = \frac{I_{C1}}{\beta_1} - \frac{I_{C2}}{\beta_2}. \qquad (10.154)$$

Using as an example the definitions in Eq. (10.146),

$$\Delta I_C = I_{C1} - I_{C2} \qquad \Delta \beta = \beta_1 - \beta_2,$$

$$I_C = \frac{I_{C1} + I_{C2}}{2} \qquad \beta = \frac{\beta_1 + \beta_i}{2},$$

$$I_{C1} = I_C + \frac{\Delta I_C}{2} \qquad \beta_1 = \beta + \frac{\Delta \beta}{2}, \qquad (10.155)$$

$$I_{C2} = I_C - \frac{\Delta I_C}{2} \qquad \beta_2 = \beta - \frac{\Delta \beta}{2}.$$

Problem 10.61 at the end of the chapter requires you to substitute Eq. (10.155) into (10.154) to obtain

$$I_{OS} \cong -\frac{I_C}{\beta}\left(\frac{\Delta R_C}{R_C} + \frac{\Delta \beta}{\beta}\right). \qquad (10.156)$$

It is quite common to represent V_{OS} and I_{OS} as dc sources placed external to a perfectly matched circuit as shown in Fig. 10.57.

Even though in this section we have examined the effects of mismatch only in a simple emitter-coupled pair, a single-value set of externally placed dc offset sources is often used with other idealized matched-element circuits.

CHECK UP

1. What BJT parameters are important for assuming matched BJTs?
2. What MOSFET and JFET parameters are important for assuming matched transistors?
3. **TRUE OR FALSE?** The IC fabrication process inherently yields reasonably matched devices.
4. What are some typical values for V_{OS} and I_{OS} for some operational amplifiers? (Look at Chapter 6 and preview Chapter 12.)

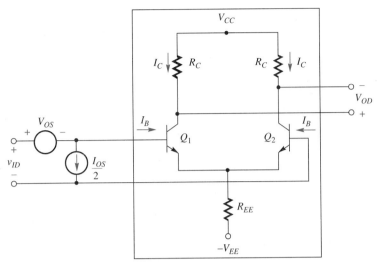

FIGURE 10.57 V_{OS} and I_{OS} external to a circuit using matched components.

SUMMARY

Virtually all analog circuits are composed of a combination of differential amplifiers with passive and active loads, single-ended amplifiers with active and passive loads, current sources, and some type of output driver. Current sources were presented in Chapter 7. The focus of this chapter was on amplifiers with a particular emphasis on the differential amplifier.

We initially studied the emitter-coupled and source-coupled pair with passive and active loads. Although the dc equivalent circuits and device models are significantly different, we found that the ac equivalent circuits were virtually identical. Of course, the computation of g_m for a BJT and FET are quite different. The common-mode and differential-mode voltage gain and input resistance calculations were simplified by use of the half-circuit model approach. Technological constraints rule out large value resistors for IC implementations; consequently, the effective large output resistance of the current source is used as an active load in both single-ended and emitter-coupled amplifier circuits. Very large voltage gains using only transistors were demonstrated.

Classes A, B, and AB power output amplifiers were presented. The large-signal analysis of the class B amplifier led to a derivation of the transfer function. Crossover distortion was eliminated by using two series diodes to force a low level of Q-point collector current in the output transistors. This is the class AB circuit, which dominates in analog ICs. The class A amplifier is also often used as an emitter-follower output stage. In general, class B and AB operations are more efficient than class A operation; however, class A operation usually exhibits less harmonic distortion than class B or AB operation. Because the power output stages are usually configured as emitter-followers, they are subject to destructive overloads if the output is short-circuited or connected to an unreasonably low load resistance. Output current-limiting circuitry to protect the output transistors was therefore discussed.

IC fabrication technology yields closely matched circuit elements. However, slight variations do occur and the effects of device mismatch were studied in the emitter-coupled pair. These device mismatches are modeled as an input offset voltage and input offset current.

Having introduced the differential amplifier and the power output amplifier, we will analyze several typical IC operational amplifiers and a related analog IC comparator in Chapter 12.

SURVEY QUESTIONS

1. Draw a circuit diagram of a resistively loaded emitter-coupled pair amplifier and define the input differential voltage and the output differential voltage.
2. Sketch and label the transfer functions for i_{C1}, i_{C2}, and v_{OD} as a function of v_{ID}.
3. What is the expression for A_{dm} for the emitter coupled pair?
4. How does A_{dm} change by including the effect of the Early voltage?
5. How do you compute R_{id}?
6. Define A_{cm} and CMRR?
7. How does the substitution of a current source for a coupled emitter resistor affect A_{dm}, A_{cm}, CMRR, and R_{id}?
8. Sketch and label the common-mode and differential-mode half-circuit models for emitter-coupled pair circuit topologies.
9. Draw a circuit diagram of a resistively loaded MOS amplifier.
10. What equations are used to relate i_D, v_{GS}, v_{ID}, and v_{OD} in a MOS amplifier?
11. Sketch and label the transfer functions for i_{D1}, i_{D2}, and v_{OD} as a function of v_{ID} in a MOS amplifier.
12. Sketch and label the common-mode and differential-mode half-circuit models for source-coupled pair MOS circuit topologies.
13. Draw a circuit diagram of a resistively loaded JFET amplifier.
14. What equations are used to relate i_D, v_{GS}, v_{ID}, and v_{OD} in a JFET amplifier?
15. Sketch and label the transfer functions for i_{D1}, i_{D2}, and v_{OD} as a function of v_{ID} in a JFET amplifier.
16. Sketch and label the common mode and differential-mode half-circuit models for source-coupled pair JFET circuit topologies.
17. Sketch and label a circuit diagram for BJT actively loaded single-ended and emitter-coupled amplifiers.
18. What equations are used for the Q-point analysis for a BJT, actively loaded, single-ended amplifier?
19. What equations are used for the Q-point analysis for a BJT, actively loaded, emitter-coupled pair amplifier?
20. What is the voltage gain, input resistance, and output resistance for actively loaded BJT amplifiers?
21. Sketch and label the I-V characteristics for an n-channel current sink and p-channel current source.
22. Explain the basic operation of an MOS current mirror.
23. What equations are used for the Q-point analysis for a CMOS actively loaded source-coupled amplifier?
24. What is the voltage gain, input resistance, and output resistance for actively loaded CMOS amplifiers?
25. Sketch and label the I-V characteristics for a CMOS amplifier.
26. What are definitions of class A, AB, B, and C amplifiers?
27. Draw schematic diagrams and transfer functions for BJT class A, B, and AB amplifiers.
28. How do you compute the efficiency of class A, AB, and B amplifiers?
29. How do you design short-circuit protection into a power amplifier?
30. How do you define and compute harmonic distortion?
31. How is input differential voltage mismatch, load resistor mismatch, and input offset current mismatch accounted for in a circuit?

PROBLEMS

Problems marked with an asterisk are more challenging.

10.1 We have already demonstrated that the slope of Eq. (10.16) at $V_i = 0$ is given by $A_{dm} = -g_m R_C$. To determine the range of validity for this small-signal model, compute the slope of Eq. (10.16) at $V_i = 0.1V_T, 1.0V_T, 2.0V_T,$ and $3.0V_T$. Then plot A_{dm} normalized to $-g_m R_C$ as a function of V_i. Recall that
$$\tanh(x) = \frac{e^x - e^{-x}}{e^x + e^{-x}}.$$

10.2 For the differential amplifier shown in Fig. 10.1, compute the Q-point values for I_C and V_{CE} and calculate $A_{dm}, A_{cm},$ and R_{id}. Let $\beta = 200$, $V_A = 100$ V, $V_{CC} = V_{EE} = 12$ V, $R_C = 100$ kΩ, and $I_{EE} = 125$ μA.

10.3 Verify your results from Problem 10.2 using a SPICE simulation.

10.4 Refer to Fig. 10.1. Let $V_{CC} = V_{EE} = 10$ V, $R_C = 50$ kΩ, $R_{EE} = 100$ kΩ with transistor $\beta = 200$ and $V_A = 150$ V. Compute $A_{dm}, A_{cm}, R_{id}, R_{ic}$, and CMRR in decibels.

10.5 Using Fig. 10.1 design an emitter-coupled pair amplifier that meets the following specifications: $R_{id} = 2$ MΩ, $A_{dm} = -500$, CMRR = 54 dB, $V_{CEQ} = 5$ V. Your design should specify values for $R_C, R_{EE}, V_{CC},$ and V_{EE}. Assume $\beta = 200$ and $V_{Anpn} = 150$ V.

10.6* Verify your results for Problem 10.4 using a SPICE simulation. Observe that you will have to include the effects of both a differential- and common-mode input signal generator.

10.7* Same as Problem 10.6 but verify your Problem 10.5 design.

10.8 Refer to Problem 10.2. If $v_{id}(t) = 2\cos(\omega_1 t)$ millivolts and $v_{ic}(t) = 5\cos(\omega_2 t)$ millivolts, write an expression for $v_{O1}(t)$, that is, the total voltage at the collector of Q_1.

10.9 Design a Widlar current source for the circuit shown in Fig. 10.11 to obtain $I_{C1} = 150$ μA. Select reasonable values R_1 and R_2. Assume $V_{CC} = V_{EE} = 15$ V. Compute the CMRR_{dB}.

10.10 The input stage, along with the appropriate current sources, of a μA725 operational amplifier is given in Fig. 10.58. For $V_{CC} = V_{EE} = 15$ V and using transistor specifications as listed in Table 18.2, compute the following:

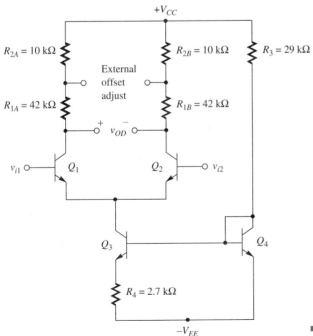

■ FIGURE 10.58

(a) A_{dm}
(b) A_{cm}
(c) CMRR in decibels
(d) R_{id}
(e) R_{ic}

10.11* Use a SPICE simulation to verify your μA725 input stage analysis of Problem 10.10. You will have to use two generators.

10.12 Compare the transconductance for a BJT biased at 1 mA and a MOSFET biased with $I_D = 1$ mA. Use with typical MOSFET specifications from Figure 5.12.

10.13 Include the effect of a nonzero λ for both MOSFETs in Fig. 10.12 and derive A_{dm}, A_{cm}, and CMRR. Use half-circuit models.

10.14 For the circuit shown in Fig. 10.12, let $R_D = 12$ kΩ, $\lambda = 0$, $V_{CC} = V_{EE} = 12$ V, $V_t = 1.0$ V, $I_{SS} = 1.1$ mA, and $k = 1.8 \times 10^{-4}$ A/V².

(a) Compute all MOSFET Q-point values.
(b) Compute A_{dm}.
(c) If $\lambda = 0.01$, recompute A_{dm}.

10.15 Refer to Problem 10.14. Design a BJT current source to bias the MOSFETs at $I_{SS} = 1.1$ mA. Use Table 18.2 for BJT data. Compute A_{cm} and CMRR for your design.

10.16* Prepare a SPICE simulation for Problem 10.14 using your BJT current source design from Problem 10.15. Compare your SPICE simulations with the analytical solution. Assume $W = L$ for the MOS models. Your solution should include a comparison of MOSFET and BJT Q-point values, A_{dm}, A_{cm}, and CMRR for both $\lambda = 0$ V^{-1} and $\lambda = 0.01$ V^{-1} cases.

10.17 Derive Eq. (10.54) by computing

$$A_{dm} = \left.\frac{dv_{OD}}{dv_{ID}}\right|_{v_{ID}=0}.$$

10.18 Refer to Fig. 10.17. Assume $R_D = 50$ kΩ, $I_{DSS} = 1$ mA, $I_{SS} = 0.5$ mA, and $V_P = -3$ V.

(a) Compute A_{dm} for this n-channel JFET amplifier.
(b) If $V_{DD} = V_{SS} = 18$ V, compute a value for R_σ.
(c) Using your results from parts (a) and (b), compute CMRR in decibels.

10.19 Verify your Problem 10.18 results using a SPICE simulation.

10.20 Using Fig. 10.17, design a source-coupled, n-channel JFET amplifier that meets the following specifications: $A_{dm} = -10$, $V_{DSQ} = 5$ V. Your design should specify a self-consistent set of values for V_{DD}, V_{SS}, and R_σ. Compute CMRR in decibels for your design. Assume $I_{DSS} = 1$ mA, $V_P = -3$ V, $R_D = 20$ kΩ, and $r_d \to \infty$.

10.21* Verify your Problem 10.20 design using SPICE.

10.22 Refer to Fig. 10.17. Let $I_{DSS} = 2$ mA, $V_p = -4$ V, $R_D = 10$ kΩ, and $\lambda = 0$. Compute A_{dm} as a function of V_{GS} for the source-coupled JFET pair.

10.23 Let $R_2 = 0$ in the CE amplifier with an active load shown in Fig. 10.59. The pnp transistors are matched. $\beta_{pnp} = 50$, $\beta_{npn} = 200$, $V_{Apnp} = 50$ V, $V_{Anpn} = 150$ V, and $T = 300$ K.

(a) What value of R_1 is required to give $I_{E3} = 1$ mA?
(b) Compute a value for $A_v = v_o/v_i$. Comment on the dependence of A_v on I_{C1}.
(c) What is the Q-point value for v_o?
(d) Repeat parts (a) through (c) but let $R_2 = 750$ Ω. Comment on the differences in the results.

10.24 If the output of the Problem 10.23 CE amplifier with an active load (Fig. 10.59) is loaded by a 1-MΩ load such as might be found as the input impedance of an oscilloscope, compute $A_v = v_o/v_i$.

10.25* Verify your results for Problems 10.23 and 10.24 using a SPICE simulation. Include transfer functions as well as the dc and ac analyses.

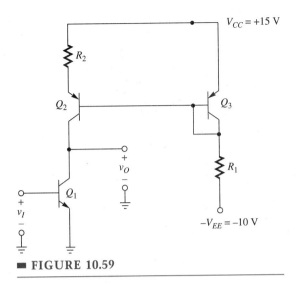

FIGURE 10.59

10.26 Assume BJT specifications as given in Table 18.2. For the circuit shown in Fig. 10.60, assume that $I_1 = 250$ μA and is synthesized from as a simple current source using an *npn* BJT (Q_3). Similarly, assume that $I_2 = 250$ μA and is synthesized from a simple current source using a *pnp* BJT (Q_4). Assume all transistors are in the linear-active region of operation.

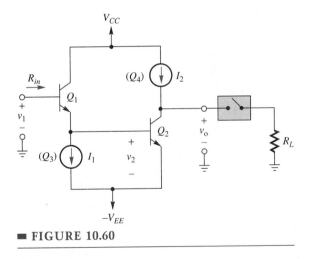

FIGURE 10.60

(a) The switch is open. Compute or estimate approximate values for R_{in} (input resistance), the voltage gain defined by $a_{v1} = v_2/v_1$, and the voltage gain defined by $a_{v2} = v_o/v_2$.
(b) The switch is now closed with $R_L = 500$ kΩ. Assume the dc conditions in the circuit are essentially unchanged. Compute an approximate value for the voltage gain $a_V = v_o/v_1$.

10.27 A Darlington differential amplifier is shown in Fig. 10.61. Observe that Q_1–Q_2 forms a CE emitter-coupled pair and Q_3–Q_4 a CC input. The v_{i1} and v_{i2} are small-signal voltage generators. Half-circuit models can be used. Compute

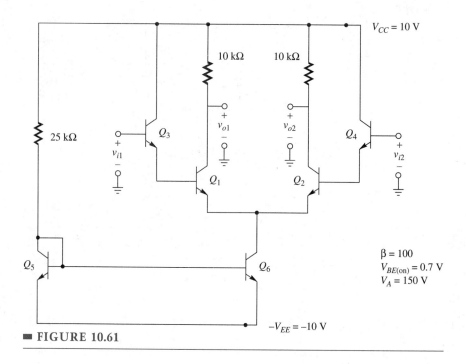
■ FIGURE 10.61

(a) dc voltage levels at the base of Q_1, collector of Q_6, collector of Q_2.
(b) Differential-mode voltage gain. Observe that you will have to make an approximation for the Q_3–Q_4 differential voltage gain.
(c) Common-mode voltage gain.
(d) CMRR in decibels.
(e) Input differential resistance, R_{id}.

10.28 Verify your results in Problem 10.27 using a SPICE simulation.

10.29 Compute values for the indicated quantities in Fig. 10.62. Assume all BJTs are matched. Use reasonable values for IC BJT properties not specified. You will have to design self-consistent values for R_1 and R_2 to establish $I_{EE} = 300$ μA.

(a) Differential-mode voltage gain, $A_{dm} = v_{od}/v_{id}$.
(b) Common-mode voltage gain, $A_{cm} = v_{oc}/v_{ic}$.
(c) CMRR in decibels.
(d) Differential input resistance, R_{id}.
(e) If $v_{id}(t) = 1.5 \cos(\omega_1 t)$ mV and $v_{ic}(t) = 2 \cos(\omega_2 t)$ mV, write an equation for $v_{o1}(t)$.
(f) Q-point values for V_{CE4}, V_{CE1}, V_{O1}, and V_{OD}.

10.30* Verify your Problem 10.29 results using a single SPICE simulation by using two sets of generators.

10.31 Use Fig. 10.62 to meet the following criteria.

(a) Design (that is, specify) a self-consistent set of values for R_C, R_1, and R_2 to obtain an approximate voltage gain $A_{dm} = v_{od}/v_{id} = -100$.
(b) Let $v_{id}(t) = 1.0 \cos(\omega_1 t)$ mV and $v_{ic}(t) = 2.0 \cos(\omega_2 t)$ mV. Write an expression for the output voltage at the collector of Q_1, $v_{o1}(t)$, based on your resistor design values. Also compute a value for the CMRR in decibels.
(c) Compute values for R_{id} and R_{ic}.

10.32 Verify your Problem 10.31 design using a SPICE simulation.

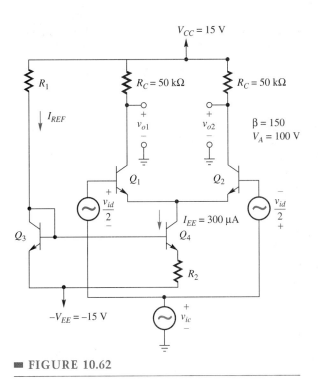

■ FIGURE 10.62

10.33 Use half-circuit models for the circuit shown in Fig. 10.63 to derive expressions for A_{dm} and R_{id}. Your results will be more meaningful if you assume that $\beta \gg 1$ and r_o can be neglected.

10.34 Using half-circuit models, derive expressions for A_{dm}, A_{cm}, R_{id}, and R_{ic} for the circuit shown in Fig. 10.64.

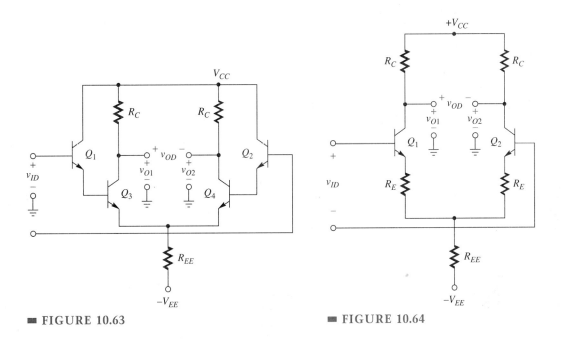

■ FIGURE 10.63

■ FIGURE 10.64

10.35 Using the same circuit information given in Problem 10.4 with the addition of $R_E = 10\ k\Omega$ in each emitter (refer to Problem 10.34 and Fig. 10.64), compute A_{dm}, A_{cm}, R_{id}, R_{ic}, and CMRR in decibels.

10.36 Refer to Fig. 10.65. Assume all transistor specifications given in Table 18.2. Let $R_{REF} = 10\ k\Omega$.

(a) Select a value for R_1 to obtain $I_{C1} = 125\ \mu A$.
(b) Compute exact Q-point values for $v_O = V_{C2}$, V_{CE6}, V_{BE5}, and V_{BE6}.
(c) What is the value of R_L if the measured value for $A_{dm} = v_o/v_{id} = 1000$?

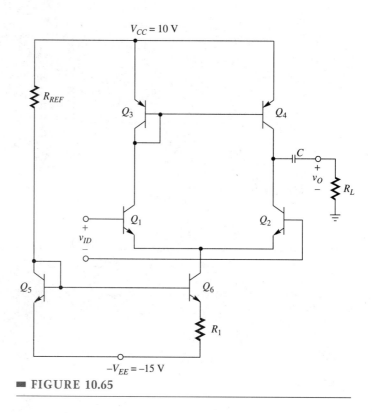

■ **FIGURE 10.65**

10.37 Refer to Fig. 10.65. Assume the following for the *npn* transistors: $\beta = 200$, $V_A = 150$ V. For the *pnp* transistors: $\beta = 50$ and $V_A = 50$ V. Also, $V_{CC} = 10$ V and $V_{EE} = 15$ V.

(a) Initially, R_L is an open circuit. Complete the design of this circuit with the requirement that differential input resistance $R_{id} = 1.0\ M\Omega$. By design, you must provide reasonable values for R_1 and R_{REF} compatible with diffused-BJT technology.
(b) What voltage gain, v_o/v_{id}, do you expect to measure with a 10-MΩ input impedance oscilloscope?
(c) Suppose a 1.0-kΩ resistor is inserted in the emitters of Q_3 and Q_4, how will that change the result computed in part (b)?

10.38 Use a SPICE simulation to verify your Problem 10.37 results.

10.39 Consider the circuit shown in Fig. 10.66. Use *npn* and lateral *pnp* specifications given in Table 18.2, and for the MOSFETs, use $V_t = 1.0$ V, $\lambda = 0$, and $k = 1.6 \times 10^{-4}$ A/V^2. Assume $V_{CC} = 6$ V and $V_{EE} = 12$ V. Compute the following quantities:

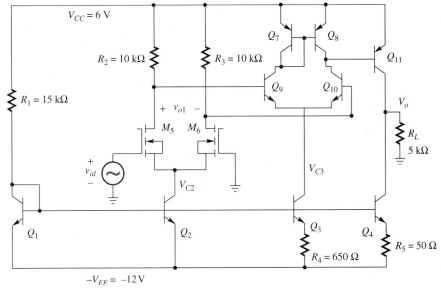

■ FIGURE 10.66

(a) Q-point values for V_{C2} and V_{C3}. Recall that the dc component of the v_{id} small-signal independent generator = 0 V.
(b) Small-signal voltage gain defined by $a_v = v_o/v_{id} = (v_o/v_{o1}) \times (v_{o1}/v_{id})$.
(c) Quiescent power dissipation for the entire circuit.

10.40* A significantly modified schematic diagram of a μA733 video amplifier is shown in Fig. 10.67. You are only going to analyze a portion of this circuit, in particular the Q_3–Q_4 emitter-coupled pair along with the Widlar current source comprised of Q_8 and Q_9. Assume BJT specifications as found in Table 18.2. Assume $V^+ = 10$ V and $V^- = -10$ V.

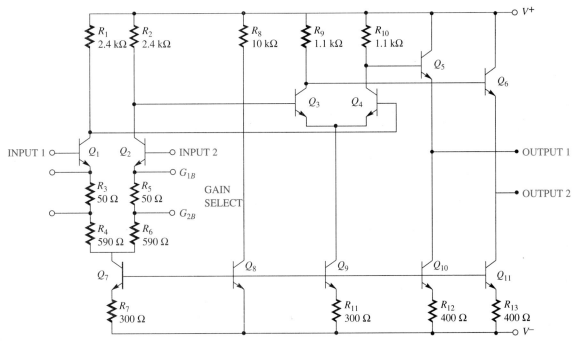

■ FIGURE 10.67

Refer to the Q_3–Q_4 emitter-coupled pair. Compute values for the following quantities. Neglect r_{o3}, r_{o4}, and the input impedance of Q_5 and Q_6.

(a) $A_{dm} = v_{od}/v_{id}$ where v_{od} is defined between the collectors of Q_3 and Q_4 and v_{id} is defined as the differential input to the bases of Q_3 and Q_4.

(b) Compute a value for the input differential resistance, R_{id}, to the Q_3–Q_4 pair.

(c) The CMRR in decibels for the Q_3–Q_4 emitter-coupled pair.

(d) Assume that the dc level at the collectors of Q_1 and Q_2 is at $+7.5$ V. Compute Q-point values for V_{CE3} and V_{CE9}.

10.41* Use a series of SPICE simulations to verify your Problem 10.40 results.

10.42 Figure 10.68 is a simplified modified diagram of a Signetics LM124 operational amplifier. Observe that the Q_1–Q_4 pair form a differential emitter-follower input stage so we estimate the v_x/v_{id} voltage gain as ≈ 1. Then, Q_2, Q_3, Q_8, and Q_9 form an emitter-coupled pair with an active load whose differential input is v_x. Assume the input impedance, R_{i10}, into Q_{10} is 10 MΩ. Use transistor data as given in Table 18.2. Compute a value for the voltage gain defined by $a_v = v_{c9}/v_{id}$.

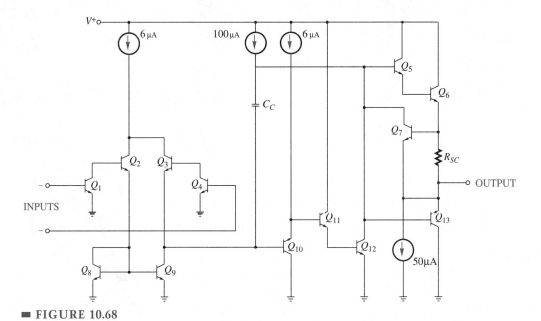

■ **FIGURE 10.68**

10.43 Complete the derivation of Eqs. (10.81) and (10.82) for the CMOS amplifier.

10.44 Design an n-channel current mirror that provides output currents of 0.5, 1.0, and 2 mA—all from a single reference. Let $V_{DD} = V_{SS} = 12$ V, $V_t = 1.0$ V, $\mu C_{ox} = 25$ μA/V^2, and $\lambda = 0$.

10.45 Draw the small-signal model for Fig. 10.37(b) and show that Fig. 10.37(d) is a good approximation. Then complete the derivation of Eq. (10.88).

10.46 Refer to Figure 10.37(a). Typical parameter values for the NMOS transistors are: $\mu_n \approx 700$ cm^2/V-sec, $t_{ox} = 300$ Å, $W_1 = 10$ L_1, and $\lambda^{-1} = 50$ volts. For the PMOS, assume $\mu_p = 300$ cm^2/V-sec and $\lambda^{-1} = 30$ volts. For the SiO$_2$ dielectric use, $\varepsilon_r = 3.9$ and $\varepsilon_o = 8.854 \times 10^{-14}$ F/cm. For $I_{SS} = 0.5$ mA, compute a value for A_v.

10.47* A differential amplifier, actively-loaded by depletion-mode NMOS transistors is shown in Figure 10.69. Using small-signal model, show that the voltage gain defined by $A_v = v_{out}/v_{in}$ is given by:

$$A_v = \frac{g_{m1}}{2}(r_{d2}\|r_{d4})$$

Then by connecting the gates of M_1 and M_2, derive an expression for A_{cm} where $A_{cm} = v_{out}/v_{in}$.

CHAPTER 10 OPERATIONAL AMPLIFIER CIRCUITRY

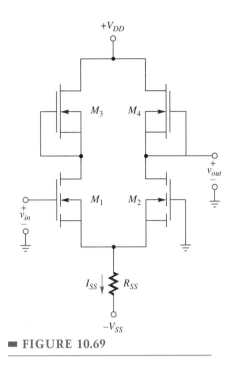

■ FIGURE 10.69

10.48 Redraw Figure 10.37 to reflect a PMOS input CMOS amplifier. Your diagram should correct labels and polarities for all Q-point voltages and currents. By analogy with the NMOS version, write an equation for the voltage gain. Provide two reasons why the NMOS input CMOS amplifier yields better performance than the PMOS input CMOS amplifier.

10.49 The circuit diagram in Fig. 10.70 consists of a resistively loaded, common-source, source-coupled pair using a simple BJT current source for biasing. For the JFET: $V_P = -4$ V,

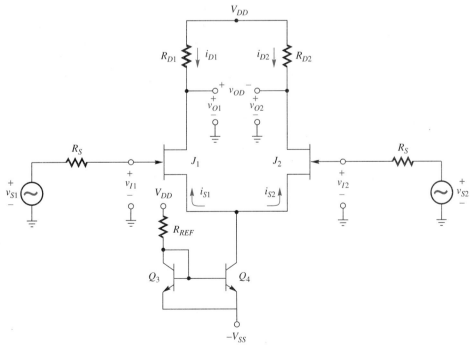

■ FIGURE 10.70

$I_{DSS} = 8$ mA, and $\lambda = 0$. For the BJT: β is large and $V_A = 100$ V. Also, $V_{DD} = 20$ V, $V_{SS} = 10$ V, and $I_{SS} = 4$ mA. $R_{D1} = R_{D2} = 6$ kΩ.

(a) Compute a value for R_1 and Q-point values for V_{DS1}, V_{CE3}, and V_{CE4}.
(b) Compute a value for the differential-mode voltage gain defined by $A_{dm} = V_{od}/V_{id}$, the common-mode voltage gain defined by $A_{cm} = V_{oc}/V_{ic}$, and the CMRR in decibels.
(c) Let $v_{id}(t) = 1.0 \cos(\omega_1 t)$ mV and $v_{ic}(t) = 2.0 \cos(\omega_2 t)$ mV. Write an expression for the output voltage at the drain of Q_1, $v_{o1}(t)$, based on your part (b) calculations.

10.50 The class B stage shown in Fig. 10.38 has an input $v_i(t) = 1.0 \sin(\omega t)$ V. Assume $V_{BE(on)} = 0.7$ V.

(a) Sketch $V_O(t)$.
(b) What is the actual conduction angle?

10.51* Use PSPICE to verify the conduction angle computed in Problem 10.50. Assume V_{BE} peak valve is 0.7 volts and conduction starts at 0.6 volts.

10.52 Refer to the class B amplifier shown in Fig. 10.38. Neglect crossover distortion. Let $v_i(t) = 10 \sin(\omega t)$ V, $V_{CC} = V_{EE} = 15$ V, and $R_L = 1$ kΩ.

(a) Compute η_C.
(b) What is the average power delivered to the load?
(c) What is the average collector dissipation in Q_1 and Q_2?

10.53 Refer to Fig. 10.71. Let $V_{CC} = V_{EE} = 15$ V, $V_{CE(sat)} = 0.2$ V, $V_{BE(on)} = 0.7$ V, $R_{REF} = 14$ kΩ, and $R_L = 5$ kΩ.

(a) Compute the collector efficiency η_C, for $v_o(t) = 5 \sin(\omega t)$ V.
(b) What is the maximum average load power and collector efficiency?

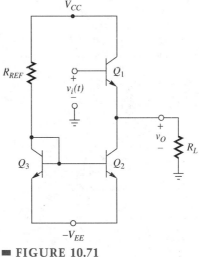

■ FIGURE 10.71

(c) What value of R_L is required to obtain the maximum average load power from this circuit?
(d) The definition of collector efficiency, of course, ignores all power lost in the rest of the circuit. Neglecting base currents but including other loss mechanisms, compute the total power required from the power supply for $v_o(t) = 5 \sin(\omega t)$ V.

10.54 Derive the collector efficiency, η_C, for the class B amplifier of Fig. 10.38 for the $v_o(t)$ triangular waveform given in Fig. 10.72. Neglect $V_{CE(sat)}$ and crossover distortion. Use symmetry in your integrations.

10.55 Derive the collector efficiency, η_C, for the class A amplifier of Fig. 10.71 for the $v_o(t)$ triangular waveform given in Fig. 10.72. Neglect $V_{CE(sat)}$ and assume $R_L I_Q$ permits a full $-V_{CC}$ to $+V_{CC}$ swing. Use symmetry in your integrations.

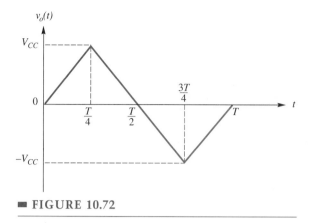

■ FIGURE 10.72

10.56* Using PWL, synthesize a 5V peak triangle waveform to the amplifier of Figure 10.38 with $V_{CC} = 10$ V and $R_L = 1$ kΩ. Using .FOUR, compare normalized odd harmonic amplitudes of $v_i(t)$ and $v_o(t)$.

10.57 Answer the following questions related to the class AB power amplifier shown in Fig. 10.73. Assume $v_i(t) = V_m \sin(\omega t)$ for parts (a) and (b).

(a) What is the maximum power that could be delivered to the 8-Ω load?

(b) Provide Q_1 and Q_2 minimum specifications for $V_{CE(max)}$, $I_{C(max)}$, and $P_{C(max)}$.

(c) Design a protection circuit for both Q_1 and Q_2 to limit current to your $I_{C(max)}$ from part (b). Your design should include a labeled circuit diagram and its interconnection with the basic class AB circuit.

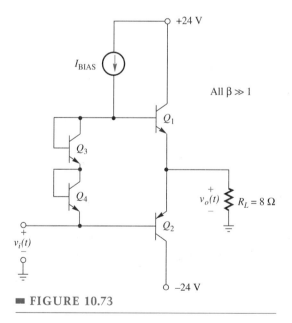

■ FIGURE 10.73

(d) Assume the cross-sectional areas of Q_1 and Q_2 are three times the cross-sectional areas of Q_3 and Q_4. What valve of I_{BIAS} is required to give output clipping close to V_{CC}? What is the quiescent valve of I_{C1} and I_{C2}?

(e) Suppose $v_i(t)$ is a symmetrical periodic square wave of appropriate amplitude such that $v_o(t) = 24\, U(t)$ for $0 < t < T/2$ and $v_o(t) = -24\, U(t)$ for $T/2 < t < T$. Compute a value for the power delivered to the 8-Ω load and the collector efficiency, η_C.

10.58 Refer to Fig. 10.74. Assume the junction areas of Q_1 and Q_2 are four times those of the Q_3 and Q_4 transistors whose specifications are given in Table 18.2.

(a) Neglecting crossover distortion effects, compute the collector efficiency, η_C when $v_i(t) = 10 \sin(\omega t)$ V.
(b) Neglecting the effects of Q_1 and Q_2 saturation, what is the maximum average power this output stage could deliver to the load?
(c) Assume the quiescent collector currents of Q_1 and Q_2 are measured at 1 mA. Neglecting base currents, design a simple current source to provide this current in place of the I_Q generator symbol.
(d) By modifying the original schematic diagram, design an overload protection circuit to remove Q_1 and Q_2 base current drive when I_{C1} or $I_{C2} > 20$ mA.

10.59 The class B output stage of a modified μA725 is given in Fig. 10.75. Let $V_{EE} = V_{CC} = 15$ V, $R_L = 2$ kΩ for parts (a) and (b). Neglect the effects of the 25-Ω protection resistor for normal operation. $R_{15} = 18$ kΩ.

(a) What is the maximum average power this circuit can deliver to the load?
(b) At the output power level in part (a), how much power is required to operate the class B output stage?
(c) What is the smallest value of R_L that can be used before the output stage protective circuitry starts to operate?

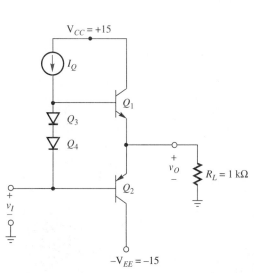

■ FIGURE 10.74

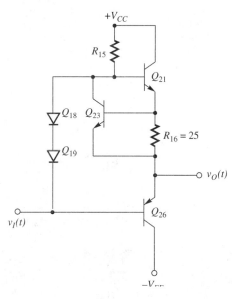

■ FIGURE 10.75

10.60 Refer to Figure 10.68. Assume $V^+ = 10$ V and $V^- = -10$ V and $R_L = 1$ kΩ is the *lowest* value of load resistor that can be safely connected to the output stage as shown. Compute values for the following:

(a) Maximum power that can be delivered to this R_L.
(b) Minimum design specifications (I_C, V_{CE}, and P_C) required for Q_6 and Q_{13}.
(c) R_{SC}.
(d) Collector efficiency for $v_o(t) = 7.0 \sin(\omega t)$ V.
(e) Show on the diagram where you would include crossover diode circuitry.

10.61 Complete the derivation of Eq. (10.156), which is used to estimate the effects of transistor mismatch on input offset current.

10.62 Compute the worst-case offset voltage if standard 5% collector resistors are used with two matched transistors in the circuit shown in Fig. 10.56.

10.63 Suppose the collector resistors are matched in Fig. 10.56. Let $50 < \beta < 300$. Estimate the worst-case input offset current at $I_C = 1$ mA for transistors selected at random.

10.64 Using Eq. (10.153), show that the input offset voltage temperature sensitivity is given by $\Delta V_{OS}/\Delta T = V_{OS}/T$.

10.65 Refer to Fig. 10.58 and Problem 10.10. An external offset potentiometer can be added to the collector circuit of the μA725 input stage as shown in Figure 10.76. What is the overall input offset voltage adjustment range possible with this circuit?

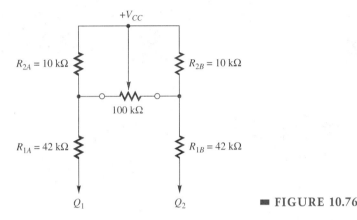

■ **FIGURE 10.76**

10.66 Suppose you measure the harmonic voltage content of the output of a class B amplifier: at ω_0, $V_1 = 10$ V; at $2\omega_0$, $V_2 = 200$ mV; at $3\omega_0$, $V_3 = 0.5$ V, and at $4\omega_0$, $V_4 = 100$ mV. Compute the percentage of harmonic distortion.

10.67* Use SPICE to obtain the harmonic distortion for Problem 10.59 if $V_i(t) = 13 \sin\omega t$ V.

10.68 Compute the percentage of harmonic distortion for a class B amplifier that does not include crossover distortion diodes. Assume that the transistors are active over the range $10° < \theta < 170°$. *Suggestion:* Review your basic circuits text to obtain the Fourier series coefficients for this type of waveform.

REFERENCES

1. Gray, P. R., and R. G. Meyer. *Analysis and Design of Analog Integrated Circuits*, 3d ed. New York: John Wiley & Sons, 1993. This is an advanced-level textbook with a good presentation of feedback, feedback stability, noise, and MOS analog circuits.

2. *RF Data Manual*. Phoenix, Ariz.: Motorola, Inc. Includes detailed specification on RF devices, some suitable for class C amplifier use, as well as a discussion of class C operation and related theory and applications; also contains additional references.

3. Chirlian, P. M. *Analysis and Design of Integrated Electronic Circuits*. New York: Harper & Row, 1981. A general textbook with some discussion of class C operation. For more applications related to higher power class C amplifiers, you can look over the suggested readings that follow.

SUGGESTED READINGS

American Radio Relay League. *The Radio Amateur's Handbook 1995* (updated annually).

Millman, J. *Microelectronics.* New York: 1979 McGraw-Hill.

Widlar, R. J. "Some Circuit Design Techniques for Linear Integrated Circuits," *IEEE Transactions on Circuit Theory*, Vol. CT-12, December 1965, pp. 586–590. R. J. Widlar is often credited with developing much of the original monolithic amplifier type of circuitry for Fairchild Semiconductors. We discuss the μA741, which is based on the μA709, the first widely used operational amplifier developed in the late 1960s.

High-Frequency ASIC Development Handbook, MAXIM Integrated Products, Sunnyvale, California, 1994. A good overview of a contemporary series of BJT ASICs operating to $f_T > 15$ GHz.

CHAPTER 11

FEEDBACK

11.1 Overview of General Feedback Concepts
11.2 Voltage-Shunt Feedback
11.3 Current-Series Feedback
11.4 Voltage-Series Feedback
11.5 Current-Shunt Feedback
11.6 Amplifier Frequency Dependence and Compensation
11.7 Oscillators
Summary
Survey Questions
Problems
Suggested Readings

As discussed in earlier chapters, the properties of an amplifier or any circuit are subject to the inherent fluctuations in active device parameters, whether due to manufacturing tolerances, power supply voltage fluctuations, temperature, or aging. In addition, the transfer characteristic of a BJT or FET is nonlinear. Although we can study the performance of a circuit using a linear small-signal model, the model is at best an approximate approach. As presented in Section 6.2, we introduced the use of feedback to improve these performance characteristics. Basically, **feedback** is the incorporation of a sample of the output signal at the input to modify the performance of a circuit.

The topic is further divided into two categories: **negative feedback** and **positive feedback**. The terms *negative* and *positive* refer to the phase relationship between the output and input signals when they are combined. As demonstrated in Section 6.2, negative feedback, often with the adjective *negative* omitted, is used to (1) stabilize the gain of an amplifier, (2) reduce distortion, (3) increase bandwidth, and (4) modify the input and output impedances.

In this chapter, we focus on the details of incorporating feedback in a circuit. Section 11.1 is a brief overview of key equations and concepts from Section 6.2. Sections 11.2 through 11.5 introduce each of the four basic feedback circuit configurations: voltage-shunt, current-series, voltage-series, and current-shunt. Each section provides a demonstration of the modification of input and output impedance and gain stabilization in general two-port configurations as well as circuit examples. One of the chief difficulties encountered when designing feedback into a circuit is the possibility that the phase relationship between the input and output signal will change with frequency, resulting in the feedback becoming positive so that the circuit will become unstable. *Unstable*, in this context, means that the circuit does not perform as an amplifier but as an *oscillator*, that is, a signal source at a given frequency. Amplifier stability is studied in Section 11.6, and an overview of oscillators is given in Section 11.7.

IMPORTANT CONCEPTS IN THIS CHAPTER

- $A = a/(1 + af) \approx 1/f$ for $af \gg 1$
- Voltage-shunt, parallel-input/parallel-output, transresistance amplifier feedback: block diagram, gain, input impedance, output impedance
- y-parameter two-port representation and unilateral network approximations
- Single-stage CE amplifier and multi-stage amplifiers with resistive, voltage-shunt feedback topology and computation of A, R_i, and R_o
- Voltage-shunt analysis of the operational amplifier and other multistage amplifiers
- Current-series, series-input/series-output, transconductance amplifier feedback: block diagram, gain, input impedance, output impedance
- z-parameter two-port representation and unilateral network approximations
- Computation of A, R_i, and R_o for the single-stage CE amplifier and multistage amplifiers with resistive, current-series feedback (emitter degeneration)
- Voltage-series, series-input/parallel-output, voltage amplifier feedback: block diagram, gain, input impedance, output impedance
- h-parameter two-port representation and unilateral network approximations
- Multistage amplifiers with resistive, voltage-series feedback topology
- Current-shunt, parallel-input/series-output, current amplifier feedback: block diagram, gain, input impedance, output impedance
- g-parameter two-port representation and unilateral network approximations
- Multistage amplifiers with resistive, current-shunt feedback topology
- Properties of feedback amplifiers where a is a function of frequency which includes positive feedback and closed-loop gain stability
- Gain margin and phase margin, compensation and dominant frequency
- Definition of an oscillator and Barkhausen criteria
- RC phase shift and Wien bridge oscillator topologies—analysis and design
- LC oscillator topologies: general circuit, Colpitts circuit, Hartley circuit
- Definition of the piezoelectric effect and the impedance characteristic of a crystal
- Properties of a crystal-controlled oscillator with examples

11.1 OVERVIEW OF GENERAL FEEDBACK CONCEPTS

Key concepts introduced in Chapter 6 are summarized in this section for convenience as we proceed with our detailed study of feedback and feedback circuit analysis. We configure all feedback circuits as shown in Figure 11.1 (same as Fig. 6.9). Any circuit or system incorporating feedback can be represented by the block diagram shown in Fig. 11.1. The use of S for signal means that it can be either a voltage or current. Repeating Eq. (6.8) we have

$$A = \frac{S_o}{S_i} = \frac{a}{1 + af}, \qquad (11.1)$$

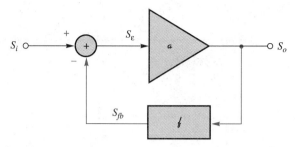

FIGURE 11.1 Basic feedback system.

- Frequency-dependent feedback elements are used in oscillator circuits and frequency-compensated amplifiers.

- Unit checking is a good approach to verifying if a and f have been correctly identified and obtained for a particular circuit.

which is the classic expression for the *closed-loop gain*, A, or the gain with feedback. A dimensional analysis of Eq. (11.1) shows that A must have the same units as the *open-loop gain*, a, and the quantity af must be dimensionless. The quantity af is called the *loop gain* and is often denoted by T. Usually a is a function of frequency, $a(j\omega)$, as expected for an amplifier. Typically, $f(j\omega)$ is not a function of frequency in simple feedback circuits, although that situation is not precluded by any of what follows.

We will apply the result that if $T = af \gg 1$, then

$$\lim_{af \to \infty} A = \lim_{af \to \infty} \frac{1/f}{(1/af) + 1} = \frac{1}{f}, \qquad (11.2)$$

where f is usually obtained from relatively precise, stable passive elements, in general resistors or capacitors so that A is now virtually independent of variations in a.

Other important concepts studied in Section 6.2 included the use of feedback for broadbanding, distortion reduction, and gain stabilization. The $(1 + af)$ term occurs throughout. For example, we have already reintroduced the closed-loop gain expression [Eq. (11.1)] and the result that A approaches $1/f$ when $af \to \infty$ [Eq. (11.2)]. If af is large but not approaching ∞, we still have significant gain stabilization as predicted by Eq. (6.12) where variations in the closed-loop gain, $\Delta A/A$, resulting from $\Delta a/a$ are reduced by $1 + af$. If the $\Delta a/a$ results in distortion because the $\Delta a/a$ yields a nonlinear transfer function, the distortion resulting from the nonlinear transfer function is also reduced by $1 + af$.

Feedback is also used for bandwidth modification as summarized by Eq. (6.34). If the original high-frequency -3-dB corner is located at frequency ω_H, the modified -3-dB corner with feedback is located at $\omega_H(1 + af)$. Of course, the trade-off is that midband gain is decreased by the same term, $1 + af$ [Eq. (11.1)].

The basic feedback system illustrated in Fig. 11.1 represents the interconnection of two two-port networks. Although the realization or recognition of these interconnected networks can be difficult in practice, it is important to present them in a simplified form so that key features of each can be extracted. Qualitatively, S_o can be either a voltage or current. Similarly, the output of the feedback network, S_{fb}, can be either a voltage or current. This suggests that there are four fundamental interconnection possibilities. Given the four basic feedback configurations, it should be increasingly obvious that there must be a systematic approach in the analysis and design of feedback circuits. A brute-force loop and node equation analysis of most practical circuits is a waste of time, especially since key properties of circuit performance are obscured by algebraic complexity. We could always work with a computer model to obtain detailed results and we will. Each of the four configurations (voltage-shunt, current-series, voltage-series, and current-shunt) is studied in the next four sections, using an approximate approach based on two-port theory with emphasis on the effect feedback has on the closed-loop gain A, the input impedance Z_i, and the output impedance Z_o. It is important to note that our definition of input and output impedance depends on the reference node under consideration subject to Thévenin and Norton source transformations required for handling the analysis. We tend to focus on voltage-shunt feedback because of its importance in operational amplifier circuit and system design as applied in Chapter 6.

■ In the balance of this chapter, we focus on circuit details. A system approach to feedback is often included in more advanced control systems or circuit theory courses.

DRILL 11.1

Assume $a = 10{,}000$. Compute the loop gain and closed-loop gain for

(a) $f = 0.0001$
(b) $f = 0.001$
(c) $f = 0.01$
(d) $f = 0.1$

Comment on the validity of the $A \approx 1/f$ for af large.

ANSWERS

	af	A	$1/f$
(a)	1	5000	10,000
(b)	10	909.1	1,000
(c)	100	99.01	100
(d)	1000	9.99001	10

The approximation works well for the last two cases.

CHECK UP

1. **TRUE OR FALSE?** Feedback amplifier calculations assume only the use of voltage gain.
2. **TRUE OR FALSE?** $A \approx 1/f$ for all values of a.
3. **TRUE OR FALSE?** There are four basic feedback circuit topologies.
4. **TRUE OR FALSE?** Loop gain is always frequency independent.

11.2 VOLTAGE-SHUNT FEEDBACK

Gain Stabilization and Impedance Modification

The basic *voltage-shunt feedback* configuration is illustrated in Fig. 11.2(a). Observe that a sample of the output voltage is connected in parallel (shunt) at the input. An input parallel connection suggests that currents are being summed. There are several other widely used names for this configuration, among them *parallel-input/parallel-output* (**PIPO**) and *shunt-shunt*. Because an output voltage is being compared to an input current, A will have units of ohms (resistance), so this configuration is referred to as a *transresistance amplifier*.

An appropriate model for the a block is given in Fig. 11.2(b), where a is a transresistance and $f = I_{fb}/V_o$. At the input summing node,

$$I_\varepsilon = I_i - I_{fb} = I_i - fV_o. \tag{11.3}$$

At the output node for $z_o \ll Z_L$,

$$V_o = aI_\varepsilon = a(I_i - fV_o). \tag{11.4}$$

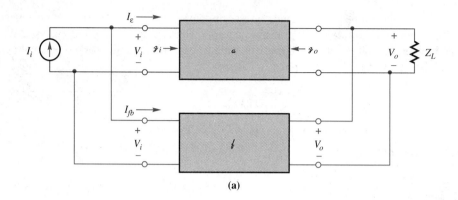

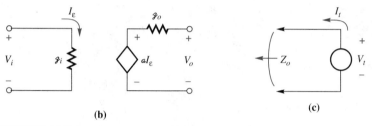

■ **FIGURE 11.2** Voltage-shunt (also known as parallel-input/parallel-output or shunt-shunt) feedback configuration. **(a)** Voltage-shunt feedback two-port interconnection. **(b)** Basic transresistance amplifier a block. **(c)** Test source in place of Z_L to determine Z_o.

Solving Eq. (11.4) for the closed-loop gain yields

$$A = \frac{V_o}{I_i} = \frac{a}{1 + af} \tag{11.5}$$

as expected.

The input impedance is defined by

$$Z_i = \frac{V_i}{I_i}. \tag{11.6}$$

Note that the uppercase Z in Z_i is used to denote input impedance for the circuit with feedback and the script z in

$$z_i = \frac{V_i}{I_\varepsilon} \tag{11.7}$$

will be used to denote the a network input impedance without feedback.

From Eqs. (11.3) and (11.4),

$$I_i = I_\varepsilon + I_{fb} = I_\varepsilon(1 + af). \tag{11.8}$$

Substituting Eq. (11.7) into Eq. (11.8) yields

$$I_i = \frac{V_i(1 + af)}{z_i}, \tag{11.9}$$

and using the Eq. (11.6) definition for input impedance,

$$Z_i = \frac{z_i}{(1 + af)}. \tag{11.10}$$

We see immediately that a shunt (parallel or current-summation) input reduces the input impedance. This is in line with the intuitive picture that a parallel connection results in an impedance reduction. When we actually use Eq. (11.10) to evaluate the input impedance, it will be important to define the input node correctly and take into account Thévenin and Norton source transformations.

To determine output impedance, a test source V_t is substituted in place of Z_L, as shown in Fig. 11.2(c), and the independent current source, I_i, is set to zero. The output impedance with feedback is defined by

$$Z_o = \frac{V_t}{I_t}. \tag{11.11}$$

Neglecting the small conductance of the f network and summing currents at the output node,

$$I_t = \frac{V_t - aI_\varepsilon}{z_o}, \tag{11.12}$$

and $I = -I_{fb}$ so that

$$I_t = \frac{V_t + aI_{fb}}{z_o} = \frac{V_t + afV_t}{z_o} \tag{11.13}$$

with the result that

$$Z_o = \frac{V_t}{I_t} = \frac{z_o}{1 + af}. \tag{11.14}$$

Again, a parallel connection results in an impedance reduction by the same $(1 + af)$ factor. As in the evaluation of the input impedance, we have to interpret the evaluation of the output impedance using Eq. (11.14) by defining the reference plane node.

Two-Port Analysis Using y Parameters

As shown in Figure 11.2, the voltage-shunt feedback configuration consists of two two-port networks whose inputs and outputs are connected in parallel. This implies that passive elements can be combined in parallel most conveniently as admittances, and sources can be combined in parallel as current generators. These are properties inherent in a ***y*-parameter two-port** model whose basic topology and definitions are provided in Fig. 11.3. The a and f networks are modeled as two parallel-connected two-port networks as shown

■ A summary of y, z, h, and g parameters is given in Appendix E.

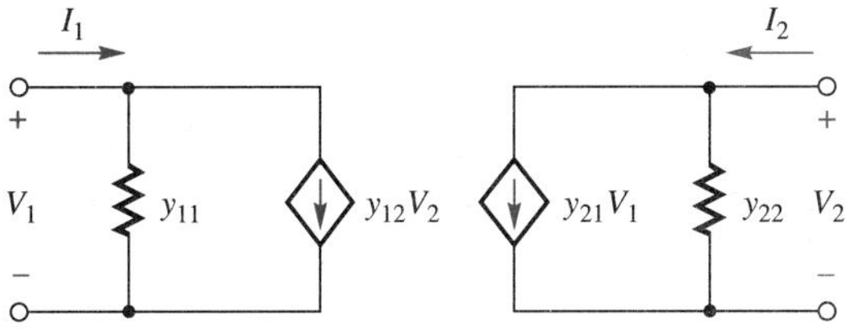

$$I_1 = y_{11}V_1 + y_{12}V_2$$
$$I_2 = y_{21}V_1 + y_{22}V_2$$

$$y_{11} = \left.\frac{I_1}{V_1}\right|_{V_2=0} \qquad y_{12} = \left.\frac{I_1}{V_2}\right|_{V_1=0}$$

$$y_{21} = \left.\frac{I_2}{V_1}\right|_{V_2=0} \qquad y_{22} = \left.\frac{I_2}{V_2}\right|_{V_1=0}$$

■ **FIGURE 11.3** The y-parameter model.

in Fig. 11.4(a). It may not be especially convenient or algebraically pleasant to obtain the y parameters of a complicated a network but in principle it can be done.

The objective is to use this circuit to solve for $A = V_o/I_s$. At the input node,

$$I_s = (Y_s + y_{11a} + y_{11f})V_i + (y_{12a} + y_{12f})V_o$$
$$= y_i V_i + (y_{12a} + y_{12f})V_o, \qquad (11.15)$$

where $y_i = Y_s + y_{11a} + y_{11f}$.

At the output node,

$$0 = (y_{21a} + y_{21f})V_i + (Y_L + y_{22a} + y_{22f})V_o$$
$$= (y_{21a} + y_{21f})V_i + y_o V_o, \qquad (11.16)$$

where $y_o = Y_L + y_{22a} + y_{22f}$. Observe, that we have included $Z_L(Y_L)$ in the y_o term. Depending on the actual circuit, there are cases where Z_L will be separated from the calculation.

Solving Eq. (11.6) for V_i and substituting in Eq. (11.15), we have

$$A = \frac{V_o}{I_s} = \frac{-(y_{21a} + y_{21f})}{y_i y_o - (y_{12a} + y_{12f})(y_{21a} + y_{21f})}. \qquad (11.17)$$

Equation (11.17) is much more recognizable if every term is divided by $y_i y_o$ to yield

$$A = \frac{\dfrac{-(y_{21a} + y_{21f})}{y_i y_o}}{1 + \left[\dfrac{-(y_{21a} + y_{21f})}{y_i y_o}\right][y_{12a} + y_{12f}]}, \qquad (11.18)$$

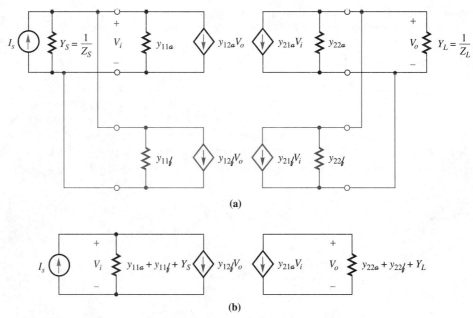

FIGURE 11.4 A y-parameter feedback model voltage-shunt circuit. (a) Complete y-parameter equivalent circuit. (b) Reduced circuit, unilateral networks.

where by comparing to Eq. (11.1), $A = a/(1 + af)$, we observe that

$$a = \frac{-(y_{21a} + y_{21f})}{y_i y_o} \quad (11.19)$$

and

$$f = y_{12a} + y_{12f}. \quad (11.20)$$

At first glance, it appears that by computing the a and f y parameters, A can be obtained by using Eq. (11.18). The difficulty is that a, in Eq. (11.19), includes a forward transmission term, y_{21f}, from the feedback network, and f, in Eq. (11.20), includes the feedback term y_{12a}, from the forward network. These are especially difficult terms to find, with y_{12a} usually being the more difficult.

Under normal circuit operation, we would expect that

$$|y_{21a}| \gg |y_{21f}|, \quad (11.21)$$

which states that all the forward gain occurs in the a network. That portion of the input signal coupled to the output through the f network will be neglected. Similarly, we would expect

$$|y_{12a}| \ll |y_{12f}|, \quad (11.22)$$

which states that reverse transmission, feedback, only occurs through the f network. Equation (11.18) then becomes

$$A \approx \frac{\dfrac{-y_{21a}}{y_i y_o}}{1 + \left(\dfrac{-y_{21a}}{y_i y_o}\right)(y_{12f})} = \frac{a}{1 + af}, \quad (11.23)$$

where

$$a \cong \frac{-y_{21a}}{y_i y_o} \quad (11.24)$$

and

$$f \approx y_{12f}. \quad (11.25)$$

Thus, the a and f networks are **unilateral**; there is only forward transmission in the a network, and there is only reverse transmission in the f network. The overall two-port y-parameter model conveniently reduces to that shown in Fig. 11.4(b). This simplified model and Eq. (11.23) suggest an analysis approach that does not require the computation of all the a and f network y parameters:

1. Compute y_{11f}, y_{22f}, and y_{12f}.
2. Set the feedback generator, $y_{12f} = 0$ temporarily.
3. Compute $a = V_o/I_s$ including the loading, y_{11f} at the input and y_{22f} at the output, provided by the feedback network. One could use y parameters or any other convenient circuit analysis techniques to compute $a = V_o/I_s$.
4. Use this value of a and the feedback ratio $f = y_{12f}$ in Eq. (11.23) to obtain A.

This method is now demonstrated in a series of examples. We will observe that one of the most difficult problems is identifying which one of the four feedback topologies can be best applied to a particular circuit.

■ We will illustrate this approximate approach for all four feedback circuit topologies. It will be useful to observe shortcuts and other simplifications.

Single-Stage CE Amplifier with Voltage-Shunt Feedback

A single-stage CE amplifier has a resistor, R_F, connected between the collector and base [Fig. 11.5(a)]. We use feedback techniques, as presented in the previous section, to obtain $A_{vs} = V_o/V_s$, R_i, and R_o. This feedback circuit is perhaps the easiest to work with, both conceptually and algebraically. In addition, it is one of the most important configurations for understanding the application of feedback to an operational amplifier.

We draw the ac model, separating the feedback network for clarity, as shown in Fig. 11.5(b). Referring to Figure 11.4, both R_L and R_S will be considered part of the amplifier, so they are included inside of the a network. Note that $Z_o' = Z_o \| R_L$. Typically, $r_o \gg R_C \| R_L$, consequently r_o has not been included. Capacitive reactances from C_1 and C_2 are assumed small at the frequencies of interest, so they have been replaced by short

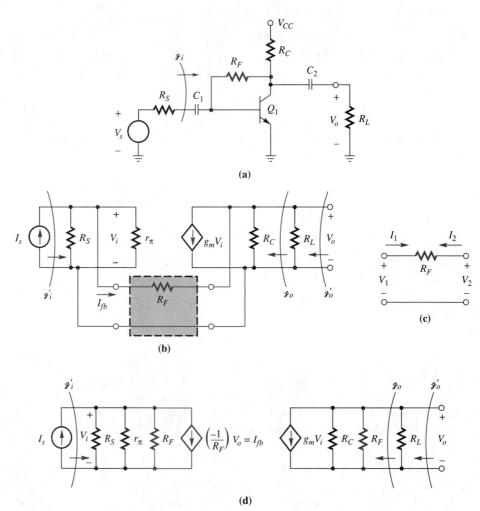

■ **FIGURE 11.5** Single-stage CE amplifier voltage-shunt feedback. **(a)** CE amplifier with R_F for biasing and feedback. **(b)** ac model with feedback network separately identified. **(c)** y-parameter definition of the feedback network. **(d)** Unilateral ac model with feedback incorporated as y_{11} and y_{22} loading and $I_{fb} = -V_o/R_F$ generator. Set $I_{fb} = 0$ to obtain $a = V_o/I_s$, z_i, and z_o.

circuits. We observe that R_F *samples the output voltage* and provides feedback in the form of a *current at the input* summing node. Therefore, this circuit utilizes **voltage-shunt feedback** and is best modeled using y parameters. Since currents are being summed, the source voltage generator and source resistance have been replaced by their Norton equivalent, $I_s = V_s/R_S$, with R_S in parallel with I_s. The isolated feedback network is given in Fig. 11.5(c). Using the y-parameter definitions given in Fig. 11.3,

$$y_{11f} = \left.\frac{I_1}{V_1}\right|_{V_2=0} = \frac{1}{R_F}, \quad y_{12f} = \left.\frac{I_1}{V_2}\right|_{V_1=0} = -\frac{1}{R_F},$$

$$y_{21f} = \left.\frac{I_2}{V_1}\right|_{V_2=0} = -\frac{1}{R_F}, \quad y_{22f} = \left.\frac{I_2}{V_2}\right|_{V_1=0} = \frac{1}{R_F}. \tag{11.26}$$

The feedback network loading, y_{11f} and y_{22f}, has been incorporated into the model as shown in Fig. 11.5(d). The transresistance gain, $a = V_o/I_s$, is computed by substituting

$$V_i = I_s(R_S \| r_\pi \| R_F) \tag{11.27}$$

into

$$V_o = -g_m V_i (R_C \| R_L \| R_F) \tag{11.28}$$

to obtain

$$a = \left.\frac{V_o}{I_s}\right|_{Y_{12f}=0} = -g_m(R_C\|R_L\|R_F)(R_S\|r_\pi\|R_F). \tag{11.29}$$

This transresistance gain derivation includes loading by the feedback network, but with the feedback generator, $I_{fb} = -V_o/R_F$, set equal to zero. Note that $f = y_{12f} = -1/R_F$, Eq. (11.25). Including the feedback generator yields

$$A = \frac{V_o}{I_s} = \frac{a}{1+af} = \frac{-g_m(R_C\|R_L\|R_F)(R_S\|r_\pi\|R_F)}{\left[1 + \dfrac{g_m(R_C\|R_L\|R_F)(R_S\|r_\pi\|R_F)}{R_F}\right]}. \tag{11.30}$$

To find the source voltage gain, A_{vs}, with feedback, use a Thévenin source transformation to obtain

$$A_{vs} = \frac{V_o}{V_s} = \frac{V_o}{I_s R_S} = \frac{1}{R_S}\left[\frac{-g_m(R_C\|R_L\|R_F)(R_S\|r_\pi\|R_F)}{1 + \dfrac{g_m(R_C\|R_L\|R_F)(R_S\|r_\pi\|R_F)}{R_F}}\right]. \tag{11.31}$$

If $af \gg 1$, Eq. (11.31) reduces to

$$A_{vs} = -\frac{R_F}{R_S}, \tag{11.32}$$

which as a resistor ratio can be made virtually independent of transistor parameter variations, temperature, and power supply changes.

Using Eq. (11.10),

$$Z_i' = \frac{z_i'}{1+af} = \frac{R_S\|r_\pi\|R_F}{\left[1 + \dfrac{g_m(R_C\|R_L\|R_F)(R_S\|r_\pi\|R_F)}{R_F}\right]}. \tag{11.33}$$

The $z_i' = R_S \| r_\pi \| R_F$ input impedance at the base input summing node is reduced by $1 + a_f$. When $a_f \gg 1$, we obtain

$$Z_i \approx \frac{R_F}{g_m(R_C \| R_L \| R_F)}. \tag{11.34}$$

The input node impedance resulting from R_F is reduced by a factor of the voltage gain, and in most situations will be very small, ideally approaching zero. To obtain the actual input impedance, R_i, at the terminals of the V_s generator we perform a Thévenin source transformation and note that the input impedance defined at the R_i plane $\approx R_S$ because the input node impedance approaches zero as shown in Eq. (11.34).

We can use the derivation of Eq. (11.34) as another technique to obtain the *Miller effect* capacitance discussed in Chapter 9. Consider the special case in which the feedback element consists of C_F only. Then, using Eq. (11.34) with $1/j\omega C_F$ substituted for R_F, we obtain a form of the familiar Miller effect capacitive reactance given by

$$Z_i' \cong \frac{1}{j\omega C_F[g_m(R_C \| R_L \| jX_{C_F})]} \approx \frac{1}{j\omega C_M} \tag{11.35}$$

■ The Miller effect was derived and discussed in Chapter 9. Note that the Miller effect is evident in any BJT or FET circuit (any two-port) where we include a C_F or a more complex feedback network.

with $C_M \approx C_F[g_m(R_C \| R_L)]$ because $R_C \| R_L$ is typically $\ll X_{C_F}$.

Using Eq. (11.14), the output impedance is reduced to

$$Z_o' \cong \frac{z_o'}{1 + a_f} \approx \frac{R_C \| R_L \| R_F}{\left[1 + \dfrac{g_m(R_C \| R_L \| R_F)(R_S \| r_\pi \| R_F)}{R_F}\right]}, \tag{11.36}$$

a much lower value, which approaches

$$Z_o' \cong \frac{R_F}{g_m(R_S \| r_\pi \| R_F)}$$

for $a_f \gg 1$. Because $R_L \gg Z_o'$, $Z_o \approx Z_o'$ as given by Eq. 11.36 where $a_f \gg 1$.

EXAMPLE 11.1

Compute A_{vs} for the circuit shown in Fig. 11.5(a). Let $R_L = 10$ kΩ, $V_{CC} = 12$ V, $R_C = 5$ kΩ, $R_S = 5$ kΩ, $\beta = 200$, $V_{BE(on)} = 0.7$ V, and $R_F = 56$ kΩ. Assume $r_o \gg R_C$.

SOLUTION The collector current is obtained from $V_{CC} = (I_C + I_B)R_C + I_B R_F + V_{BE(on)}$. Then

$$I_C = \frac{V_{CC} - V_{BE(on)}}{\left(\dfrac{1}{\beta}\right)R_F + R_C\left(1 + \dfrac{1}{\beta}\right)} = \frac{12 - 0.7}{(0.005)(56 \text{ k}\Omega) + 5 \text{ k}\Omega(1.005)} = 2.13 \text{ mA},$$

$$r_\pi = \frac{\beta}{g_m} = \frac{200}{2.13 \text{ mA}/26 \text{ mV}} = 2.44 \text{ k}\Omega.$$

From Eq. (11.32),

$$A_{vs} \approx -\frac{R_F}{R_S} = \frac{-56 \text{ k}\Omega}{5 \text{ k}\Omega} = -11.2.$$

A more exact result using Eq. (11.31) is

$$A_{vs} = \frac{1}{5\text{ k}\Omega}\left[\frac{-\left(\dfrac{2.13\text{ mA}}{26\text{ mV}}\right)(5\text{ k}\Omega\|10\text{ k}\Omega\|56\text{ k}\Omega)(5\text{ k}\Omega\|2.44\text{ k}\Omega\|56\text{ k}\Omega)}{1 + \left(\dfrac{2.13\text{ mA}}{26\text{ mV}}\right)(5\text{ k}\Omega\|10\text{ k}\Omega\|56\text{ k}\Omega)(5\text{ k}\Omega\|2.44\text{ k}\Omega\|56\text{ k}\Omega)\dfrac{1}{56\text{ k}\Omega}}\right]$$

$$= -\frac{1}{5\text{ k}\Omega}\left[\frac{411{,}000\ \Omega}{1 + \dfrac{411{,}000}{56{,}000}}\right] = -9.86.$$

In this case, $a\!f = 7.34$, which is not much greater than 1 so that the -9.8 and -11.2, exact and approximate answers, differ by $\approx 14\%$. Still, using Eq. (11.32) is an attractive alternative.

DRILL 11.2

Repeat Example 11.1 but assume $R_F = 33\text{ k}\Omega$.

ANSWERS
$I_C = 2.18\text{ mA}$
$r_\pi = 2385\ \Omega$
$A_{vs} \approx -6.6$, or more exactly, $A_{vs} = -6.087$

Multistage Amplifier with Voltage-Shunt Feedback

A three-stage, direct-coupled, cascade amplifier is shown in Fig. 11.6. The characteristics of the basic three-stage cascade amplifier are changed dramatically by connecting a feedback resistor, R_F, from the collector of Q_3 to the base of Q_1 through a large coupling capacitor, as illustrated in Fig. 11.6(a).

A sample of the ac output voltage feeds back to the Q_1 base node as a current in a parallel or shunt connection. The coupling capacitor is used so as not to disturb the original biasing on Q_1. One could eliminate the C_2 if the basic dc bias design included the effect of R_F to establish the desired Q-points. Note the location of the R_i and R_i' measurement planes.

The ac model is given in Fig. 11.6(b), where the feedback network has been identified and separated from the forward transresistance amplifier. Observe that each of the three CE stages is using an unbypassed emitter resistor whose basic topology was presented in Chapter 8. Shortly, we will identify this topology as a current-series feedback configuration and rename it single-stage *emitter degeneration feedback*.

The small-signal transistor output resistances r_o have not been included to minimize algebraic complexity.

Current summing requires that the source voltage generator and source resistance be replaced by their Norton equivalent. This voltage-shunt feedback is modeled using y parameters as defined in Fig. 11.3. The isolated feedback network is shown in Fig. 11.6(c). The y parameters for this network were computed in Eq. (11.26). The equivalent y parameters are incorporated in the small-signal model of Fig. 11.6(d) as input and output network loading, y_{11f} and y_{22f}, and a feedback current generator, y_{12f}.

■ Colored circuit elements refer to the feedback networks.

■ Same circuit topology is also used in FET circuits. It is useful to go back to Chapters 5 and 8 to note this.

Recall that, from the single-stage transresistance amplifier results of Example 11.1, the gain was in error by 14% so that $A = 1/f$ could be only marginally justifiably used. This $A = 1/f$ approximation becomes much better when applied to the three-stage amplifier. Then for this three-stage CE amplifier, we have

$$A = \frac{V_o}{I_s} \cong \frac{1}{f} = -R_F. \tag{11.37}$$

The Thévenin source transformation $V_s = I_s R_S$ is then used to obtain

$$A_{vs} = \frac{V_o}{V_s} = \frac{V_o}{I_s R_S} = -\frac{R_F}{R_S}. \tag{11.38}$$

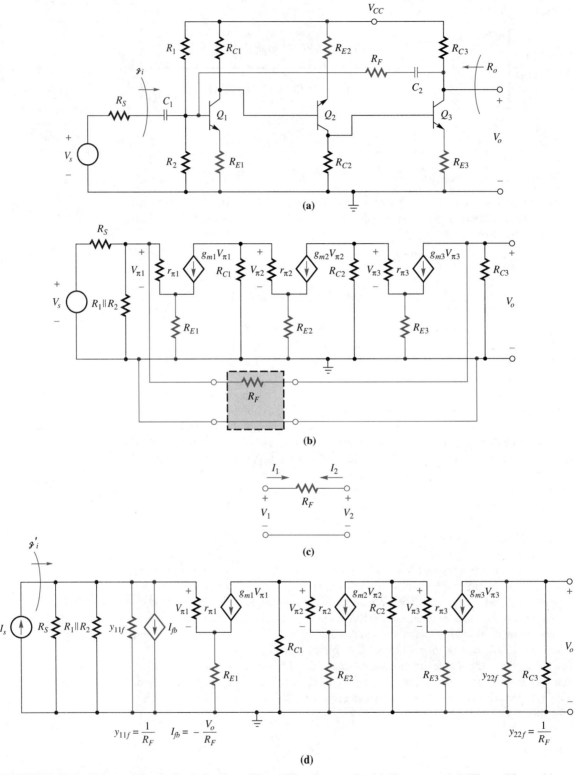

■ **FIGURE 11.6** Voltage-shunt feedback for three CE amplifiers in cascade. (a) Three cascaded CE amplifiers with feedback. (b) ac model with feedback network separately defined. (c) y-parameter definition of the feedback network. (d) ac model with feedback incorporated as y_{11f} and y_{22f} input and output loading and $I_{fb} = -(V_o/R_F)$ generator. Set $I_{fb} = 0$ to obtain $a = V_o/I_s$, z_i and z_o. Note that if there were an R_L, one could absorb it in R_{C3}.

Both the input node and output resistances, R_i and R_o, [Eqs. (11.10) and (11.14)] are low, both reduced by the term $(1 + a\beta)$, and are close to zero in practice for a multistage circuit. When we perform a Thévenin source transformation where $V_s = I_s R_S$, we observe that the input resistance at the V_s generator terminal is essentially R_S because the Q_1 base node impedance is essentially zero.

DRILL 11.3

In addition to R_F in Fig. 11.6, identify other sources of feedback explicitly shown or inherent in the devices and circuit.

ANSWERS R_{E1}, R_{E2}, R_{E3}, and C_μ for all three BJTs. More subtle and much more difficult to quantify and model are inductive and capacitive couplings between circuit elements and interconnects and the nonzero resistance in the power bus.

EXAMPLE 11.2

Use SPICE to find A_{vs}, R_i, and R_o for the voltage-shunt feedback circuit of Fig. 11.7. Assume $\beta_F = 200$, $I_S = 2 \times 10^{-15}$ A, $V_A = 150$ V, and $f = 1$ kHz. Compare results with Eqs. (11.38), (11.10), and (11.14)

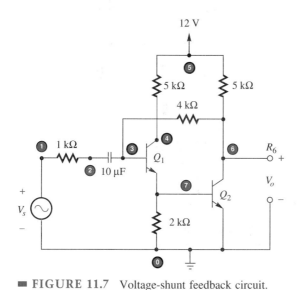

■ **FIGURE 11.7** Voltage-shunt feedback circuit.

SOLUTION The SPICE listing for this circuit is given by:

```
VOLTAGE-SHUNT FEEDBACK
VS 1 0 AC .01
VCC 5 0 DC 12
RS 1 2 1K
RC1 5 4 5K
RC2 5 6 5K
RE 7 0 2K
RF 3 6 4K
Q1 4 3 7 M1
Q2 6 7 0 M1
C1 2 3 10UF
.MODEL M1 NPN BF=200 IS=2E-15 VAF=150
.AC DEC 1 1K 1K
.PRINT AC VM(3) VM(6) VP(6) I(VS)
.END
```

Key portions of the ac output listing are

FREQ	VM(3)	VM(6)	VP(6)	I(VS)
1.000D 03	2.348D−04	3.882 D−02	−1.791D 02	9.764D−06

Thus, at 1 kHz we can calculate $A_{vs} = V(6)/VS = 0.0388\underline{/-179°}/0.01\underline{/0°} = -3.88$. The approximate $-180°$ phase shift indicates that the capacitive reactance of a 10-μF capacitor is negligible at 1 kHz. This compares to $A_{vs} = -R_F/R_S = -4$ predicted by Eq. (11.38). Resistance $R_i = $ `VM(3)/I(VS)` $= 2.35 \times 10^{-4}/9.76 \times 10^{-6} = 24\ \Omega$. Note that this is the resistance looking into node 3 from the source. To find R_o, we must remove the ac source (`VS`) and apply a test signal (`VT`) at the output through a capacitor as shown in Fig. 11.8.

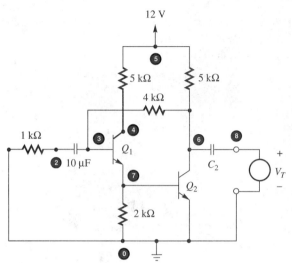

■ **FIGURE 11.8** SPICE circuit topology for obtaining the output resistance. VS set to zero and a test source, VT is capacitively coupled to Node 6.

```
OUTPUT RESISTANCE
VCC 5 0 DC 12
VT 8 0 AC .1
RS 0 2 1K
RC1 5 4 5K
RC2 5 6 5K
RE 7 0 2K
RF 3 6 4K
Q1 4 3 7 M1
Q2 6 7 0 M1
C1 2 3 10UF
C2 6 8 10UF
.MODEL M1 NPN BF=200 IS=2E-15 VAF=150
.AC DEC 1 1K 1K
.PRINT AC VM(6) I(VT)
.END
```

Then we may find $R_o = $ `V(6)/I(VT)` $= 0.1/0.00157 = 63\ \Omega$. You should note that the open-loop gain $a = $ `V(6)/IS` *cannot* be found by merely connecting R_F to ground from nodes 3 and 6, because this would change the dc condition. Rather, find the dc voltages `V(3)` $= 1.385$ V and `V(6)` $= 1.392$ V from the preceding SPICE solution. Then connect $R_F = 4$ kΩ from node 3 to a 1.392-V source, and also connect 4 kΩ from node 6 to a 1.385-V source. This effectively removes the feedback without changing the dc bias. This yields $a = -132,400$ so that $af = 33$.

Voltage-Shunt Analysis of the Operational Amplifier

The operational amplifier with feedback was first introduced in Section 1.4 with additional applications described in Chapter 6. We are now in a position to apply rigorous feedback concepts to obtain A_{vs}, R_i, and R_o and provide the justification for applying summing-point constraints. To provide continuity through this text, we will use $\mu A741$ specifications for numerical justification of any approximations in the examples.

■ We will now justify Chapter 6 operational amplifier concepts. In particular, note the development of a two-port SPICE model.

The basic model for a nonideal operational amplifier, specifically the $\mu A741$, is shown in Fig. 11.9. We now incorporate this operational amplifier in the classic inverting amplifier topology, Fig. 11.10(a). The amplifier is driven from V_s through R_S and the feedback is via R_F. A sample of the output voltage, V_o, is being summed as a current at the inverting node, the negative operational amplifier input; hence, this is a form of voltage-shunt feedback.

The small-signal model from Fig. 11.9 is incorporated in Fig. 11.10(b). As done previously for voltage-shunt feedback, the Norton source transformation

$$I_s = \frac{V_s}{R_S} \tag{11.39}$$

is required because the voltage-shunt feedback uses current summing at the input node. The feedback network has been separately identified and separated from the forward amplifier. The y parameters for this type of feedback network are given in Eq. (11.26). These equivalent y parameters are incorporated into the small-signal model as shown in Fig. 11.10(c) as input and output feedback network loading, y_{11f} and y_{22f}, and a feedback generator, y_{12f}. By setting $y_{12f} = 0$, the transresistance gain, $a = V_o/I_s$, can be computed directly using voltage division as

$$a = \frac{V_o}{I_s} = \left(\frac{-R_F \| R_L}{R_F \| R_L + r_o}\right) a_v (r_i \| R_S \| R_F). \tag{11.40}$$

■ Observe that
$$f = \frac{-1}{R_F}$$

Then, including the feedback generator,

$$A = \frac{V_o}{I_s} = \frac{a}{1 + af} \cong \frac{-\left(\frac{R_F \| R_L}{R_F \| R_L + r_o}\right) a_v (r_i \| R_S \| R_F)}{1 + \left(\frac{1}{R_F}\right)\left(\frac{R_F \| R_L}{R_F \| R_L + r_o}\right) a_v (r_i \| R_S \| R_F)}. \tag{11.41}$$

To obtain the voltage gain, substitute Eq. (11.39) into Eq. (11.41):

$$A_{vs} = \frac{V_o}{V_s} = \frac{V_o}{I_s R_S} = \frac{-1}{R_S}\left[\frac{\left(\frac{R_F \| R_L}{R_F \| R_L + r_o}\right) a_v (r_i \| R_S \| R_F)}{1 + \left(\frac{1}{R_F}\right)\left(\frac{R_F \| R_L}{R_F \| R_L + r_o}\right) a_v (r_i \| R_S \| R_F)}\right]. \tag{11.42}$$

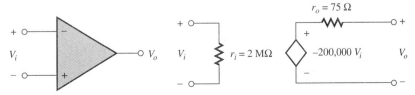

■ **FIGURE 11.9** Basic $\mu A741$ operational amplifier model.

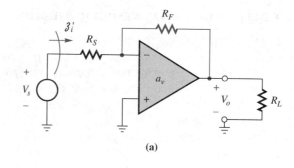

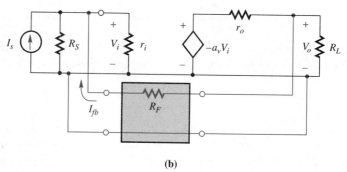

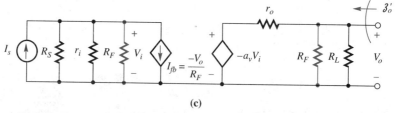

(c)

■ **FIGURE 11.10** Operational amplifier treated as a voltage-shunt feedback network. (a) Basic operational amplifier with resistive feedback. (b) Small-signal model with the feedback network identified. (c) Equivalent circuit with feedback loading.

The input impedance, where we include R_S, within the a network, is given by

$$Z'_i = \frac{\mathcal{Z}'_i}{1 + a_f} = \frac{R_S \| R_F \| r_i}{1 + \left(\dfrac{1}{R_F}\right)\left(\dfrac{R_F \| R_L}{R_F \| R_L + r_o}\right) a_v (r_i \| R_S \| R_F)}, \qquad (11.43)$$

and the output impedance, with R_L included within the a network, is given by

$$Z'_o = \left(\frac{\mathcal{Z}'_o}{1 + a_f}\right) = \frac{(R_L \| R_F \| r_o)}{\left[1 + \left(\dfrac{1}{R_F}\right)\left(\dfrac{R_F \| R_L}{R_F \| R_L + r_o}\right) a_v (r_i \| R_S \| R_F)\right]}. \qquad (11.44)$$

We now apply these results to an inverting amplifier example and compare the results to those obtained in Section 1.4.

CHAPTER 11   FEEDBACK

EXAMPLE 11.3

A $\mu A741$ is used in the inverting amplifier circuit shown in Fig. 11.11(a). Assume $R_S = 100$ kΩ, $R_L = 10$ kΩ, and $R_F = 1$ MΩ. Compute A, A_{vs}, Z_i, and Z_o.

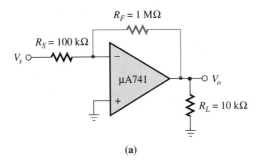

(a)

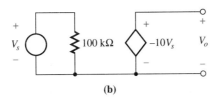

(b)

■ **FIGURE 11.11**   Circuit for Example 11.3.
(a) Inverting amplifier. (b) Equivalent circuit.

SOLUTION Substituting the $\mu A741$ specifications from Fig. 11.9 and the circuit element values from Fig. 11.10 into Eq. (11.40),

$$a = -\left(\frac{R_F\|R_L}{R_F\|R_L + r_o}\right)a_v(r_i\|R_S\|R_F)$$

$$= -\left(\frac{1\text{ M}\Omega\|10\text{ k}\Omega}{1\text{ M}\Omega\|10\text{ k}\Omega + 75\text{ }\Omega}\right)(2\times 10^5)\times(2\text{ M}\Omega\|100\text{ k}\Omega\|1\text{ M}\Omega)$$

$$= -1.726\times 10^{10}\text{ }\Omega.$$

Then,

$$a_f = 1.726\times 10^{10}\text{ }\Omega\left(\frac{1}{R_F}\right) = \frac{1.726\times 10^{10}}{10^6} = 17{,}260,$$

which is much greater than 1. The closed-loop gain is

$$A = \frac{V_o}{I_s} = \frac{a}{1 + a_f} \cong \frac{-1.726\times 10^{10}}{1 + 17260} = -999{,}942\text{ }\Omega.$$

Using Eq. (11.42),

$$A_{vs} = \frac{V_o}{V_s} = \frac{A}{R_S} = \frac{-999{,}942}{10^5} = -9.99942 \cong -10,$$

which agrees with the result [Eq. (11.20)] $A_{vs} = V_o/V_s = -R_F/R_S$, also obtained using summing point constraints.

DRILL 11.4

Study the results in Example 11.3. Would it make much difference in the overall results if a_v were only 100,000 instead of 200,000?

ANSWERS
a_f becomes 8630, $A = -999.884$, and $A_{vs} = -9.99884 \approx -10$. The summing point node input resistance increases to 9.96 Ω, still low compared to R_S. The output resistance increases to 8.62 mΩ, still low compared to R_L.

DRILL 11.5

Repeat Drill 11.4 but now assume $a_v = 1000$.

ANSWERS $a_f = 86.3$, $A = -988{,}545$, and $A_{vs} \approx -9.885$, which is not as close to the $A_{vs} \approx -10$ approximation. The summing point node input resistance increases to almost 2 kΩ, which is now starting to become significant. The output resistance increases to 1 Ω, which is still reasonably acceptable.

The input impedance, using Eq. (11.43), is

$$Z_i' = \frac{R_S \| R_F \| r_i}{(1 + a\mathit{f})} = \frac{(100 \text{ k}\Omega \| 1 \text{ M}\Omega \| 2 \text{ M}\Omega)}{17{,}261} = \frac{86 \text{ k}\Omega}{17{,}261} = 4.98 \text{ }\Omega.$$

It is very important to observe that the impedance, Z_i' at the current summing node is very low, and compared to R_S, it is effectively zero. This node is said to be at virtual ground. Consequently, at the V_S generator terminals, $R_i \approx R_S = 100$ kΩ.

From Eq. (11.44),

$$Z_o' = \frac{R_F \| R_L \| r_o}{(1 + a\mathit{f})} = \frac{1 \text{ M}\Omega \| 10 \text{ k}\Omega \| 75 \text{ }\Omega}{17261} = \frac{74.436}{17261} = 4.31 \times 10^{-3} \text{ }\Omega = 4.31 \text{ m}\Omega.$$

Similarly, the output impedance, reduced by the $(1 + a\mathit{f})$ term, is very small, effectively zero, compared to the other impedance levels in the output circuit. Removing R_L from the computation of Eq. (11.44) has a negligible effect on the results for the Z_o' of the feedback circuit.

The resultant equivalent circuit is shown in Fig. 11.11(b). Even if a_v changes significantly, $a\mathit{f}$ is still $\gg 1$ and A_{vs} is still well approximated by -10 with $R_i \approx R_S = 100$ kΩ.

■ Virtual ground is a key concept when using summing point constraints. Control systems engineers would say the error voltage is approximately zero.

CHECK UP

1. Draw and label the block diagram of a voltage-shunt feedback system.
2. Does input impedance increase or decrease with the use of voltage-shunt feedback and by what factor?
3. Does output impedance increase or decrease with the use of voltage-shunt feedback and by what factor?
4. **TRUE OR FALSE?** y parameters are used to model voltage-shunt feedback circuits.
5. **TRUE OR FALSE?** y_{12a} is used to approximate f in a unilateral approximation.
6. **TRUE OR FALSE?** $y_{12\mathit{f}}$ is used to approximate f in a unilateral approximation.
7. **TRUE OR FALSE?** All the forward gain occurs in the a network and all the reverse gain occurs in the f network when using a unilateral network approximation.
8. Sketch and label a schematic diagram for an inverting operational amplifier identifying the feedback network.
9. What are two additional names for a voltage-shunt amplifier circuit?

11.3 CURRENT-SERIES FEEDBACK

Gain Stabilization and Impedance Modification

The dual of voltage-shunt feedback is *current-series feedback*, illustrated in Fig. 11.12(a). A sample of the output current is connected in series with the input as a voltage, V_{fb}. Other widely used names for this configuration are *series-input/series-output (SISO)* and *series-series*. Since an output current is being compared to an input voltage, A will have units of siemens (admittance or conductance), so this configuration is referred to as a

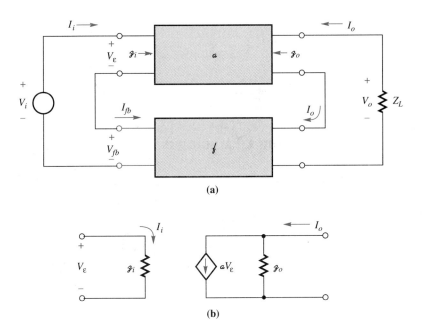

FIGURE 11.12 Current-series (also known as series-input/series-output or series-series) feedback configuration.
(a) Current-series feedback two-port interconnection.
(b) Basic transconductance amplifier a block.

transconductance amplifier. A model for the a block is given in Fig. 11.12(b), where a is a transconductance, and $f = V_{fb}/I_o$. Summing voltages around the input loop,

$$V_i = V_\varepsilon + V_{fb}. \tag{11.45}$$

At the output node for $z_o \gg Z_L$ the voltage drop of the feedback network is essentially zero,

$$I_o = aV_\varepsilon = a(V_i - V_{fb}) = a(V_i - fI_o). \tag{11.46}$$

Solving Eq. (11.46) for the closed-loop gain yields

$$A = \frac{I_o}{V_i} = \frac{a}{1 + af}. \tag{11.47}$$

The input impedance, Z_i, is computed from

$$V_i = V_\varepsilon + V_{fb} = V_\varepsilon + fI_o = V_\varepsilon + afV_\varepsilon = I_i z_i (1 + af), \tag{11.48}$$

realizing that $z_i = V_\varepsilon/I_i$. Rearranging terms,

$$Z_i = \frac{V_i}{I_i} = z_i(1 + af). \tag{11.49}$$

Thus, a series connection, as might be expected, raises the input impedance.

Using a test source at the output in place of Z_L [see Fig. 11.2(c)] and setting the independent source, V_i, to zero,

$$I_t = \frac{V_t}{z_o} + aV_\varepsilon = \frac{V_t}{z_o} - aV_{fb} = \frac{V_t}{z_o} - afI_t. \tag{11.50}$$

Equation (11.50) becomes

$$Z_o = \frac{V_t}{I_t} = z_o(1 + af). \tag{11.51}$$

The series connection results in an increase in output impedance.

Two-Port Analysis Using z Parameters

As shown in Fig. 11.12, the current-series feedback configuration consists of two two-port networks whose inputs and outputs are connected in series. This means that passive components are most conveniently combined in series as impedances and sources are combined in series as voltage generators. These are properties inherent in the **z-parameter two-port** model shown in Fig. 11.13. The a and f networks are modeled as two series-connected two-port networks as shown in Fig. 11.14(a). We could solve for $A = I_o/V_s$ directly from Fig. 11.14(a) as we did when solving for $A = V_o/I_s$ in the voltage-shunt topology. However, when we derived the final results, we observed that there were considerable simplifications possible if we assumed that the a and f networks were unilateral. By unilateral, we mean that there is forward transmission only in the a network so that

$$|z_{21a}| \gg |z_{21f}|, \tag{11.52}$$

and there is only reverse transmission in the f network such that

$$|z_{12f}| \gg |z_{12a}|. \tag{11.53}$$

Using Eqs. (11.52) and (11.53), the series-connected two-port networks then reduce to that shown in Fig. 11.14(b). From this model, compute

$$a = \frac{I_o}{V_s} = \frac{-z_{21a}}{(z_{11a} + z_{11f} + Z_S)(z_{22a} + z_{22f} + Z_L)}, \tag{11.54}$$

which includes the loading of the feedback network, z_{11f} at the input and z_{22f} at the output, but with the feedback generator $f = z_{12f} = 0$. Then

$$A = \frac{I_o}{V_s} = \frac{a}{1 + af} \cong \frac{\dfrac{-z_{21a}}{(z_{11a} + z_{11f} + Z_S)(z_{22a} + z_{22f} + Z_L)}}{1 + \dfrac{(-z_{21a})(z_{12f})}{(z_{11a} + z_{11f} + Z_S)(z_{22a} + z_{22f} + Z_L)}} \tag{11.55}$$

by including the feedback generator, z_{12f}.

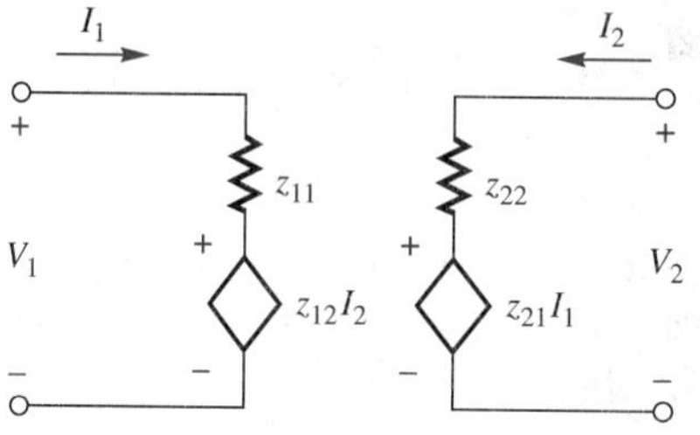

$$V_1 = z_{11}I_1 + z_{12}I_2$$
$$V_2 = z_{21}I_1 + z_{22}I_2$$

$$z_{11} = \left.\frac{V_1}{I_1}\right|_{I_2=0} \qquad z_{12} = \left.\frac{V_1}{I_2}\right|_{I_1=0}$$

$$z_{21} = \left.\frac{V_2}{I_1}\right|_{I_2=0} \qquad z_{22} = \left.\frac{V_2}{I_2}\right|_{I_1=0}$$

■ **FIGURE 11.13** The z-parameter model.

CHAPTER 11 FEEDBACK

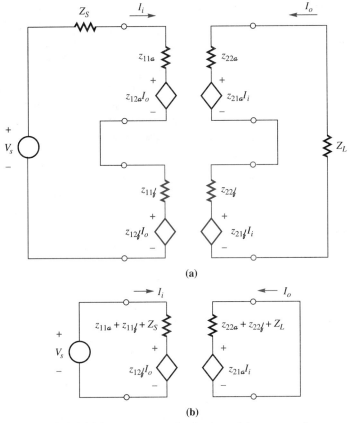

FIGURE 11.14 z-parameter feedback model current-series circuit. **(a)** Complete z-parameter equivalent circuit. **(b)** Reduced circuit, unilateral networks.

By assuming that the a and f networks are unilateral we can outline an analysis approach that does not require the computation of all of the a and f network z parameters in very much the same way we handled a voltage-shunt circuit with y parameters:

1. Compute z_{11f}, z_{22f}, and z_{12f}.
2. Set the feedback generator $z_{12f} = 0$ temporarily.
3. Compute $a = I_o/V_s$ including the loading, z_{11f} at the input and z_{22f} at the output, provided by the feedback network. One could use z parameters or any other convenient circuit analysis techniques to compute $a = I_o/V_s$.
4. Use this value of a and the feedback ratio $f = z_{12f}$ in Eq. (11.55) to obtain A.

As in Section 11.2, these techniques are now demonstrated in a series of examples.

Single-Stage CE Amplifier with Series-Series Feedback (Emitter Degeneration)

One of the most widely used feedback circuits is a CE amplifier with an unbypassed emitter resistor as shown in Fig. 11.15(a). This is called a CE amplifier with **emitter degeneration**. Although we solved this problem in Chapter 8 by using direct algebraic circuit techniques, here we use a feedback circuit approach to obtain $A_{vs} = V_o/V_s$, R_i,

■ *Emitter degeneration* is an old term but almost universally used.

and R_o. Often single-stage feedback circuits with emitter degeneration are incorporated within multistage circuits; these multistage circuits then use overall feedback, in addition to the emitter degeneration of individual stages.

The ac model is shown in Fig. 11.15(b). Again, all capacitor reactances are assumed small at the frequencies of interest. The feedback network, separated as shown in Fig. 11.15(b), consists of R_E. A sample of the output current is summed around the input loop as a voltage drop across R_E. This is a form of current-series feedback that is best modeled by z parameters. To conform to the model, the input voltage source, source resistance, and bias network have been combined in a Thévenin equivalent circuit as shown in Fig. 11.15(c), where

$$R'_S = R_1 \| R_2 \| R_S = R_B \| R_S, \qquad V'_s = \frac{V_s R_B}{R_B + R_S}. \tag{11.56}$$

Applying Fig. 11.13 to Fig. 11.15(d), the z parameters for the feedback network are computed as

$$z_{11f} = \left.\frac{V_1}{I_1}\right|_{I_2=0} = R_E, \qquad z_{12f} = \left.\frac{V_1}{I_2}\right|_{I_1=0} = R_E,$$

$$z_{21f} = \left.\frac{V_2}{I_1}\right|_{I_2=0} = R_E, \qquad z_{22f} = \left.\frac{V_2}{I_2}\right|_{I_1=0} = R_E. \tag{11.57}$$

The feedback network loading, z_{11f} and z_{22f}, has been incorporated into the model as shown in Fig. 11.15(e). The transconductance gain, $a = I_o/V'_s$, is computed by substituting the error voltage

$$V_1 = \left(\frac{r_\pi}{r_\pi + R'_S + R_E}\right) V'_s \tag{11.58}$$

into

$$I_o = g_m V_1 \tag{11.59}$$

to obtain

$$a = \left.\frac{I_o}{V'_s}\right|_{z_{12f}=0} = \frac{r_\pi g_m}{r_\pi + R'_S + R_E}. \tag{11.60}$$

This transconductance gain includes loading by the feedback network, but with the feedback generator $V_{fb} = I_o R_E$ set equal to zero. Inclusion of the feedback generator yields

$$A = \frac{I_o}{V'_s} = \frac{a}{1 + af} = \frac{\dfrac{r_\pi g_m}{r_\pi + R'_S + R_E}}{1 + \dfrac{r_\pi g_m R_E}{r_\pi + R'_S + R_E}}. \tag{11.61}$$

For algebraic convenience, R_C has been combined with R_L so that at the load,

$$V_o = -I_o(R_C \| R_L). \tag{11.62}$$

Hence, the source voltage gain with feedback is given by

$$A'_{vs} = \frac{V_o}{V'_s} = -\frac{I_o(R_C \| R_L)}{V'_s} = \frac{\dfrac{-r_\pi g_m(R_C \| R_L)}{r_\pi + R'_S + R_E}}{1 + \dfrac{r_\pi g_m R_E}{r_\pi R'_S + R_E}}. \tag{11.63}$$

CHAPTER 11 FEEDBACK

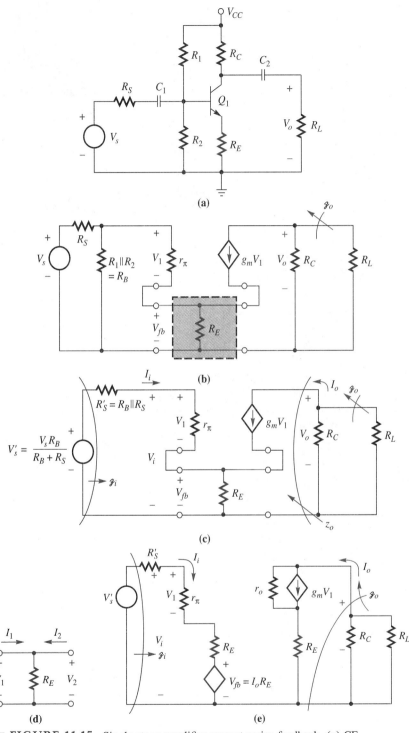

■ **FIGURE 11.15** Single-stage amplifier current-series feedback. (a) CE amplifier with an unbypassed emitter. (b) ac model with feedback network separately identified. (c) ac model with Thévenin source transformation. (d) z-parameter definition of the feedback network. (e) ac model with feedback incorporated as z_{11} and z_{22} loading and $V_{fb} = I_o R_E$ generator. Set $V_{fb} = 0$ to obtain $a = I_o/V'_s$, z_i, and z_o. Include r_o to obtain a finite R_o.

Substituting Eq. (11.56) into (11.63) yields

$$A_{vs} = \frac{V_o}{V_s} = \frac{V_o R_B}{V_s'(R_B + R_S)} = \left(\frac{R_B}{R_B + R_S}\right)\left[\frac{-r_\pi g_m(R_C \| R_L)}{r_\pi + R_S' + R_E + r_\pi g_m R_E}\right]. \quad (11.64)$$

When $a_f \gg 1$, because $r_\pi g_m R_E \gg r_\pi + R_S' + R_E$, Eqs. (11.63) and (11.64) reduce to

$$A_{vs}' = \frac{V_o}{V_s'} \approx -\frac{(R_C \| R_L)}{R_E} \quad (11.65)$$

and

$$A_{vs} \cong -\frac{(R_C \| R_L)}{R_E}\left(\frac{R_B}{R_B + R_S}\right). \quad (11.66)$$

Then A_{vs} is dependent only on passive elements. This is the same conclusion achieved by Eq. (8.23).

The input resistance is given by

$$R_i' = z_i(1 + a_f) = (r_\pi + R_S' + R_E)\left(1 + \frac{\beta R_E}{r_\pi + R_S' + R_E}\right)$$
$$= r_\pi + R_S' + R_E + \beta R_E \approx (1 + \beta)R_E, \quad (11.67)$$

where $r_\pi + R_S' \ll (1 + \beta)R_E$, which is a familiar result from Chapter 8. Usually, however, R_S is not considered as part of the amplifier input resistance. Also, self-bias is not used in IC circuits, so the gain and input impedance results do not include R_1 and R_2. Often, when self-bias is used, $R_E(1 + \beta) \gg R_B$.

The output resistance, not including R_C or R_L, for a series-connected output is given by

$$R_o = z_o(1 + a_f) = (r_o + R_E)\left(1 + \frac{g_m r_\pi R_E}{r_\pi + R_S' + R_E}\right), \quad (11.68)$$

which reduces to the familiar result obtained in Chapter 7 for a Widlar current source as well as other CE topologies (looking into the collector), introduced in Chapter 8,

$$R_o \cong r_o(1 + g_m R_E) \quad (11.69)$$

when $(R_S' + R_E)/r_\pi \ll 1$. To include R_C, we would place R_C in parallel with the R_o given in Eq. (11.69).

■ One cannot overemphasize the importance of using resistor ratios as a consequence of employing negative feedback for amplifier circuit designs. Often, this approach is called *design-by-ratios*.

EXAMPLE 11.4

Design an amplifier of the type illustrated in Fig. 11.15(a) to obtain $A_{vs} = -10$. That is, specify R_E if $V_{CC} = 12$ V, $R_C = 10$ kΩ, $R_L = 20$ kΩ, $R_1 = 1$ MΩ, $R_2 = 100$ kΩ, $\beta = 200$, $V_{BE(on)} = 0.7$ V, and $R_S = 1$ kΩ.

SOLUTION Observe that $R_B = R_1 \| R_2 = 1 \text{ M}\Omega \| 100 \text{ k}\Omega = 90.9 \text{ k}\Omega \gg R_S = 1 \text{ k}\Omega$ so that from Eq. (11.66),

$$A_{vs} = -10 \cong -\left(\frac{R_C \| R_L}{R_E}\right) = \left(\frac{10 \text{ k}\Omega \| 20 \text{ k}\Omega}{R_E}\right).$$

Thus, $R_E = 670 \ \Omega$. We would ordinarily be finished, but we should also verify that the Q-point is located in the active region. From Chapter 4,

$$V_{CC}\left(\frac{R_2}{R_1 + R_2}\right) = I_B R_B + 0.7 + (\beta + 1)I_B R_E.$$

Then

$$I_B = \frac{V_{CC}\left(\frac{R_2}{R_1 + R_2}\right) - 0.7}{R_B + (1 + \beta)R_E} = \frac{1.09 - 0.7}{90 \ \text{k}\Omega + (201)(670 \ \Omega)} = 1.7 \ \mu\text{A}.$$

Then $I_C = 200 I_B = 200 \times 1.7 \ \mu\text{A} = 340 \ \mu\text{A}$. The Q-point equation is $V_{CE} = V_{CC} - I_C(R_C + R_E) = 12 - (340 \ \mu\text{A})(10{,}670 \ \Omega) = 8.3$ V, a reasonable result. An exact result for A_{vs} from Eq. (11.64) yields $A_{vs} = -8.8$, about 12% low.

■ Note that $a_f \approx 8$ for this approximate solution.

Multistage Amplifier with Current-Series Feedback

Three CE amplifiers, all with forms of emitter degeneration, are connected in cascade as shown in Fig. 11.16(a). Direct dc coupling between stages one and two and stages two and three is made possible by using a *pnp* CE amplifier to provide the necessary dc level shifting. A sample of the output current through Q_3 feeds back as a voltage in series with Q_1. Thus, this circuit uses a form of current-series feedback. The simplified ac model is given in Fig. 11.16(b), where the feedback network has been identified and separated from the forward amplifier, and the small-signal transistor output resistances r_o have been neglected. The input network has been converted to a series circuit by using a Thévenin equivalent circuit. The ac *pnp* model is identical to the ac *npn* model. With the feedback network in series with both the input and output, z parameters (Fig. 11.13) will be used to evaluate the isolated feedback network in Fig. 11.16(c).

■ Small signal models are identical for *npn* and *pnp* BJTs. Also observe the same similarity for NMOS and PMOS transistors. Of course signs on the dc Q-point values are different.

The z parameters for this network are

$$z_{11f} = \left.\frac{V_1}{I_1}\right|_{I_2=0} = (R_F + R_{E3}) \| R_{E1}, \qquad z_{12f} = \left.\frac{V_1}{I_2}\right|_{I_1=0} = \frac{R_{E1} R_{E3}}{R_{E1} + R_{E3} + R_F} = f,$$

$$z_{21f} = \left.\frac{V_2}{I_1}\right|_{I_2=0} = \frac{R_{E1} R_{E3}}{R_{E1} + R_{E3} + R_F}, \qquad z_{22f} = \left.\frac{V_2}{I_2}\right|_{I_1=0} = R_{E3} \| (R_{E1} + R_F). \tag{11.70}$$

These equivalent circuit z parameters are incorporated into the small-signal model as shown in Fig. 11.16(d) as input and output feedback network loading, z_{11f} and z_{22f}, and a feedback voltage generator, z_{12f}. Setting $z_{12f} = f = 0$, the transconductance gain $a = I_o/V_s'$ can be computed by considering each term of

$$a \cong \frac{I_o}{V_s'} = \frac{-V_o}{R_{C3} V_3} \times \frac{V_3}{V_2} \times \frac{V_2}{V_1} \times \frac{V_1}{V_s'} \tag{11.71}$$

separately. By using

$$I_o = -\frac{V_o}{R_{C3}}, \tag{11.72}$$

■ If there were an R_L, it could be absorbed in R_{C3}.

Eq. (11.71) is evaluated as a series of three cascaded voltage gains. Furthermore, each of the three stages represents a case of single-stage emitter degeneration, so Eq. (11.65) can be used to evaluate each term. Virtually all of the output current (for $\beta \gg 1$) is

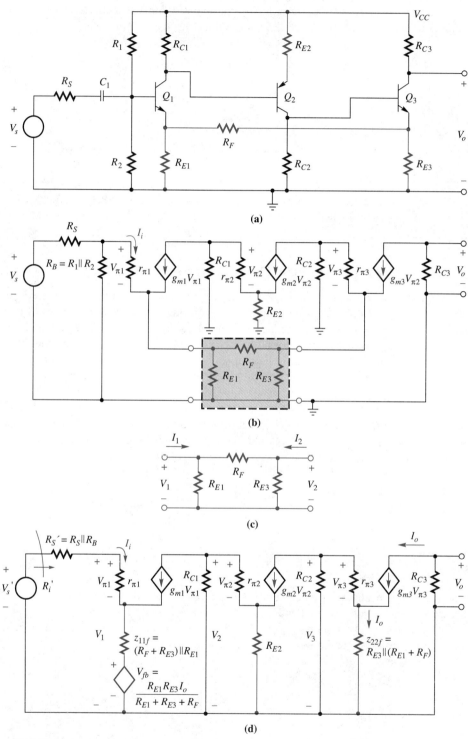

FIGURE 11.16 Current-series feedback in three CE amplifiers in cascade. (a) Three cascaded CE amplifiers with feedback. (b) ac model with feedback network separately identified. (c) z-parameter definition of the feedback network. (d) ac model with feedback incorporated. $V'_s = \dfrac{R_B}{R_B + R_S} \times V_s$.

present in this emitter circuit; thus, current sampling is the preferred feedback configuration. Then, since $-I_{E3} \approx I_o$,

$$\frac{V_o}{R_{C3}V_3} \cong -\frac{R_{C3}}{[R_{E3}\|(R_{E1} + R_F)]R_{C3}}, \tag{11.73}$$

$$\frac{V_3}{V_2} \cong -\frac{R_{C2}\|(1 + \beta_3)[R_{E3}\|(R_{E1} + R_F)]}{R_{E2}}, \tag{11.74}$$

$$\frac{V_2}{V_1} \cong -\frac{R_{C1}\|(1 + \beta_2)R_{E2}}{(R_F + R_{E3})\|R_{E1}}, \tag{11.75}$$

and using voltage division at the input,

$$\frac{V_1}{V'_s} \cong \frac{(1 + \beta_1)[(R_F + R_{E3})\|R_{E1}]}{(1 + \beta_1)[(R_F + R_{E3})\|R_{E1}] + R'_S}. \tag{11.76}$$

Often, Eq. (11.76) yields $V_1/V'_s \approx 1$. Substituting Eqs. (11.73) through (11.76) and the feedback generator term from Eq. (11.70)

$$f = z_{12f} = \left(\frac{R_{E1}R_{E3}}{R_{E1} + R_{E3} + R_F}\right) \tag{11.77}$$

into Eq. (11.71) yields

$$af \cong -\left[\frac{-1}{R_{E3}\|(R_{E1} + R_F)}\right]\left\{\frac{-R_{C2}\|(1 + \beta_3)[R_{E3}\|(R_{E1} + R_F)]}{R_{E2}}\right\}$$
$$\times \left[\frac{-R_{C1}\|(1 + \beta_2)R_{E2}}{(R_F + R_{E3})\|R_{E1}}\right]\left(\frac{R_{E1}R_{E3}}{R_{E1} + R_{E3} + R_F}\right)\left(\frac{V_1}{V'_s}\right). \tag{11.78}$$

The loop gain, af, is much greater than unity for reasonable values of circuit elements and typical values of the transistor parameters. Then

$$A = \frac{I_o}{V'_s} = \frac{1}{f} = \frac{R_{E1} + R_{E3} + R_F}{R_{E1}R_{E3}}, \tag{11.79}$$

and with a load transformation $I_o = -V_o/R_{C3}$,

$$A_{vs} = \frac{V_o}{V'_s} = -\frac{I_o R_{C3}}{V'_s} = \frac{(-R_{C3})(R_{E1} + R_{E3} + R_F)}{R_{E1}R_{E3}}. \tag{11.80}$$

Using the results from Eq. (11.78), the input resistance is given by

$$R'_i = r_i(1 + af) = \{R'_S + [r_{\pi1} + (R_F + R_{E3})\|R_{E1}]\}(1 + af) \tag{11.81}$$

and is effectively quite large.

The resistance looking into the output terminal is $\approx R_{C3}$. By not including r_{o3}, the resistance looking into the Q_3 collector is effectively infinite. Even if a finite r_o were included, the series output feedback connection would still result in a very large value looking into the collector.

DRILL 11.6

A few short questions relating to Fig. 11.16:

(a) What effect does R_{E2} have on the circuit?
(b) What would happen if you connected R_F from the emitter of Q_1 to the collector of Q_3?
(c) What would you add to the circuit design so that changing the value of R_F would not change the Q_1 and Q_3 Q-points?

ANSWERS

(a) Provides emitter degeneration and gain stabilization to just stage 2.
(b) The phase relationship would change the negative feedback to positive feedback and result in instability. That is, $af < 0$ so that $1 + af$ could be zero, resulting in $A \to \infty$.
(c) Add a capacitor in series with R_F to block dc.

CHECK UP

1. Draw and label the block diagram of a current-series feedback system.
2. Does input impedance increase or decrease with the use of current-series feedback and by what factor?

3. Does output impedance increase or decrease with the use of current-series feedback and by what factor?
4. **TRUE OR FALSE?** z parameters are used to model current-series feedback circuits.
5. **TRUE OR FALSE?** z_{12a} is used to approximate f in a unilateral approximation.
6. **TRUE OR FALSE?** z_{12f} is used to approximate f in a unilateral approximation.
7. **TRUE OR FALSE?** All the forward gain occurs in the a network and all the reverse gain occurs in the f network when using a unilateral network approximation.
8. Draw a schematic diagram of a circuit with emitter degeneration and identify the feedback circuit.
9. What are two additional names for a current-series amplifier circuit?

11.4 VOLTAGE-SERIES FEEDBACK

■ Of course, we obtain voltage gain information for the other three feedback circuit topologies using source transformations and other basic circuit analysis techniques.

The *voltage-series feedback* configuration is illustrated in Fig. 11.17(a). Here we are sampling an output voltage and summing the samples as a voltage in series at the input. This is also called *series-input/parallel-output (SIPO)* or *series-shunt* feedback. An output voltage is being directly compared to an input voltage, so A has units of volts/volt. This configuration is called a *voltage amplifier*. Using the a network model shown in Fig. 11.17(b), where f is a voltage ratio and $f = V_{fb}/V_o$,

$$V_o = aV_\varepsilon = a(V_i - V_{fb}) = a(V_i - fV_o). \tag{11.82}$$

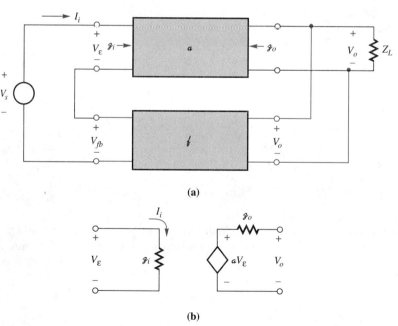

■ **FIGURE 11.17** Voltage-series (also known as series-input/parallel-output or series-shunt) feedback configuration.
(a) Voltage-series feedback two-port interconnection.
(b) Basic voltage amplifier a block.

CHAPTER 11 FEEDBACK

Rearranging Eq. (11.82), the closed-loop gain is given by

$$A = \frac{V_o}{V_i} = \frac{a}{1 + af}. \tag{11.83}$$

Since we are considering a series-input connection, intuitively from Section 11.3,

$$Z_i = z_i(1 + af). \tag{11.84}$$

Similarly, using results from Section 11.2, the output impedance for a parallel connection is given by

$$Z_o = \frac{z_o}{1 + af}. \tag{11.85}$$

Two-Port Analysis Using h Parameters

The voltage-series feedback configuration consists of two two-port networks whose inputs are in series and whose outputs are in parallel, as shown in Fig. 11.17. This configuration implies that input circuit passive components are best represented as impedances and sources as voltage generators. Similarly, output circuit passive components are best represented as admittances and sources as current generators. These are the properties of the **h-parameter two-port model** (see Fig. 11.18). The a and f networks are modeled as two hybrid-connected two-port networks as shown in Fig. 11.19(a). As done previously, this network is simplified by setting up the a and f networks as unilateral. Therefore, for forward transmission only in the a network,

$$|h_{21a}| \gg |h_{21f}| \tag{11.86}$$

and for reverse transmission only in the f network,

$$|h_{12f}| \gg |h_{12a}|. \tag{11.87}$$

The circuit model then reduces to that shown in Fig. 11.19(b). Using this model, compute

$$a = \frac{V_o}{V_s} = \frac{-h_{21a}}{(h_{11a} + h_{11f} + Z_S)(h_{22a} + h_{22f} + Y_L)}, \tag{11.88}$$

which includes the loading of the feedback network, h_{11f} at the input and h_{22f} at the output, but with the feedback generator $f = h_{12f} = 0$. Then

$$A = \frac{V_o}{V_s} = \frac{a}{1 + af} = \frac{\dfrac{-h_{21a}}{(h_{11a} + h_{11f} + Z_S)(h_{22a} + h_{22f} + Y_L)}}{1 + \dfrac{(-h_{21a})(h_{12f})}{(h_{11a} + h_{11f} + Z_S)(h_{22a} + h_{22f} Y_L)}} \tag{11.89}$$

by including the feedback generator.

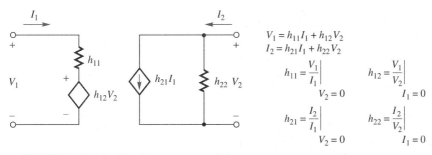

■ **FIGURE 11.18** The h-parameter model.

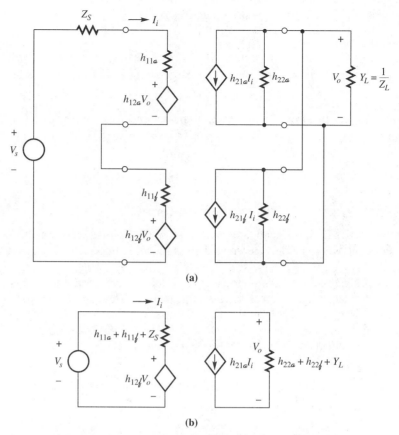

■ **FIGURE 11.19** h-parameter feedback model voltage-series circuit. (a) Complete h parameter equivalent circuit. (b) Reduced circuit, unilateral networks.

Again, by assuming the a and f networks are unilateral we can outline an analysis approach that does not require the computation of all of the a and f network h parameters in very much the same way we handled a voltage-shunt circuit with y parameters and a current-series circuit with z parameters:

1. Compute h_{11f}, h_{22f}, and h_{12f}.
2. Set the feedback generator $h_{12f} = 0$ temporarily.
3. Compute $a = V_o/V_s$ including the loading, h_{11f} at the input and h_{22f} at the output, provided by the feedback network. One could use h parameters or any other convenient circuit analysis techniques to compute $a = V_o/V_s$.
4. Use this value of a and the feedback ratio $f = h_{12f}$ in Eq. (11.89) to obtain A.

This technique is now demonstrated in an example.

Multistage Amplifier with Voltage-Series Feedback

Two CE amplifiers are connected in cascade, and an emitter follower is added as a last stage for impedance matching as illustrated in Fig. 11.20(a). In this example, a sample of the emitter-follower output voltage feeds back in series with the emitter Q_1 as a voltage. Therefore, this is a form of voltage-series feedback. It would also be possible to analyze this as a current-series feedback circuit, because a sample of the output current in R_{E3} is

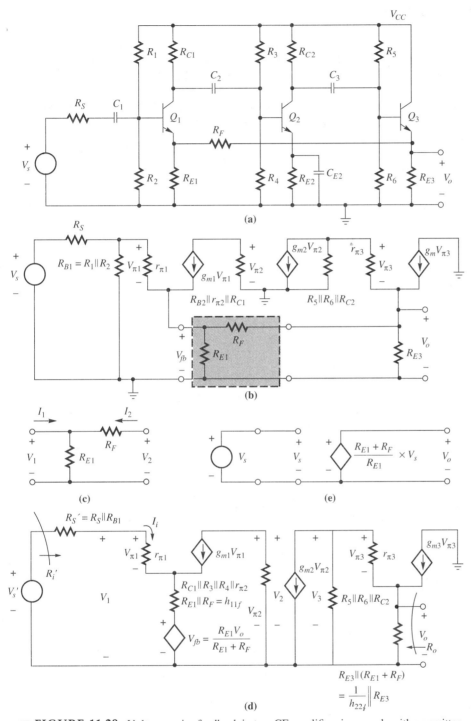

■ **FIGURE 11.20** Voltage-series feedback in two CE amplifiers in cascade with an emitter follower. (a) Two cascaded CE amplifiers with an emitter follower, all with feedback. (b) ac model with feedback network separately identified. (c) Feedback circuit definition. (d) ac model with feedback incorporated as h_{11f} and h_{22f} input and output loading and $V_{fb} = R_{E1}V_o/(R_{E1} + R_F)$ generator. Set $V_{fb} = 0$ to obtain $a = V_o/V_s$, z_i, and z_o. $V'_s = \dfrac{R_{B1}}{R_{B1} + R_S}$. (e) Equivalent circuit for $a_f \to \infty$. Note that if there were an external R_L, it would be absorbed in R_{E3}.

fed back in series with the emitter Q_1 as a voltage. However, the voltage-series analysis is more convenient and somewhat easier to treat algebraically.

A simplified ac model is given in Fig. 11.20(b), where the feedback network has been identified and separated from the forward amplifier. The small-signal transistor output impedances r_o have not been included. A Thévenin equivalent circuit has been used to combine R_S, R_1, and R_2 to R_S' and V_s' as shown in Fig. 11.20(b). Using the definition given in Fig. 11.18, the h parameters for the isolated feedback network shown in Fig. 11.20(c) are computed to be

$$h_{11f} = \left.\frac{V_1}{I_1}\right|_{V_2=0} = R_{E1}\|R_F, \qquad h_{12f} = \left.\frac{V_1}{V_2}\right|_{I_1=0} = \frac{R_{E1}}{R_{E1} + R_F} = f,$$

$$h_{21f} = \left.\frac{I_2}{I_1}\right|_{V_2=0} = \frac{-R_{E1}}{R_F + R_{E1}}, \qquad h_{22f} = \left.\frac{I_2}{V_2}\right|_{I_1=0} = \frac{1}{R_{E1} + R_F}. \tag{11.90}$$

These small-signal h parameters are incorporated into the small-signal model as shown in Fig. 11.20(d) as input and output feedback network loading, h_{11f} and h_{22f}, and a feedback voltage generator, h_{12f}. Setting $h_{12f} = 0$, $a = V_o/V_s'$ can be computed directly by using

$$a = \frac{V_o}{V_s'} = a_{v1}a_{v2}a_{v3}a_{v4} = \frac{V_1}{V_s'} \times \frac{V_2}{V_1} \times \frac{V_3}{V_2} \times \frac{V_o}{V_3}. \tag{11.91}$$

We will make use of the single-stage emitter degeneration feedback incorporated within the first stage. From Eq. (11.65),

$$a_{v2} = \frac{V_2}{V_1} \cong \frac{-(R_{C1}\|R_3\|R_4\|r_{\pi 2})}{R_{E1}\|R_F}. \tag{11.92}$$

The last stage is an emitter follower so, from inspection,

$$a_{v4} = \frac{V_o}{V_3} \cong \pm 1. \tag{11.93}$$

The input resistance of the emitter follower loads the second stage, so from Eq. (11.67), the second-stage voltage gain is given by

$$a_{v3} = \frac{V_3}{V_2} \approx -g_m\{R_5\|R_6\|R_{C2}\|(1 + \beta_3)[R_{E3}\|(R_{E1} + R_F)]\}. \tag{11.94}$$

Using Eq. (11.67), the input voltage divider ratio is

$$a_{v1} = \frac{V_1}{V_s'} \cong \frac{(1 + \beta_1)(R_{E1}\|R_F)}{(1 + \beta_1)(R_{E1}\|R_F) + R_S'}. \tag{11.95}$$

Combining Eqs. (11.92) through (11.95) and the feedback ratio as given by h_{12f}, we obtain

$$af = \left[\frac{(1 + \beta_1)(R_{E1}\|R_F)}{(1 + \beta_1)(R_{E1}\|R_F) + R_S'}\right]\left[-\left(\frac{R_{C1}\|R_3\|R_4\|r_{\pi 2}}{R_{E1}\|R_F}\right)\right]$$
$$\times (-g_{m2})\{R_5\|R_6\|R_{C2}\|(1 + \beta_3)[R_{E3}\|(R_{E1} + R_F)]\}$$
$$\times \left(\frac{R_{E1}}{R_{E1} + R_F}\right). \tag{11.96}$$

Then substitute Eqs. (11.91) and (11.96) to obtain

$$A = \frac{V_o}{V_s} = \frac{a}{1 + af}. \tag{11.97}$$

DRILL 11.7

A few short questions relating to Fig. 11.20:

(a) In addition to R_F and R_{E1} feedback in Fig. 11.20, identify other sources of feedback explicitly shown or inherent in the devices and circuit.

(b) What is the effect of removing C_{E2}? If you removed C_{E2} and moved the R_F connection from the emitter of Q_3 to the emitter of Q_2, what would happen?

ANSWERS

(a) R_{E2} provides emitter degeneration and stabilizes the Q-point. Its effect is bypassed at higher frequencies by using C_{E2}. Also C_μ for all three BJTs. As with all practical circuits, more subtle and much more difficult to quantify and model are inductive and capacitive couplings between circuit elements and interconnects and the nonzero resistance in the power buses.

(b) By moving the R_F connection, the phase relationship would change the negative feedback to positive feedback and result in instability. That is, $af < 0$ so that $1 + af$ could be zero, resulting in $A \to \infty$.

Two cascade CE amplifiers virtually assure that $a f \gg 1$, with the result that

$$A \cong \frac{1}{f} = +\left(\frac{R_{E1} + R_F}{R_{E1}}\right). \tag{11.98}$$

Using af from Eq. (11.96), the input resistance defined at the V'_s generator terminal is given by

$$R'_i \cong r_i(1 + af) = \{R'_S + [r_\pi + (R_{E1}\|R_F)(1 + \beta)]\}(1 + af) \to \infty \tag{11.99}$$

and the output resistance is given by

$$R_o \cong \frac{R_{E3}\|(R_{E1} + R_F)}{1 + af} \approx 0. \tag{11.100}$$

As $af \gg 1$, the circuit can be modeled as shown in Fig. 11.20(e), where $R_i \to \infty$, $R_o = 0$, and

$$A'_{vs} = \left(\frac{R_{E1} + R_F}{R_{E1}}\right). \tag{11.101}$$

and using the Thévenin source transformation,

$$A_{vs} = \frac{V_o}{V_s} = \left(\frac{V_o}{V'_s}\right)\left(\frac{V'_s}{V_s}\right) = \left(\frac{R_{E1} + R_F}{R_{E1}}\right)\left(\frac{R_1\|R_2}{R_1\|R_2 + R_S}\right). \tag{11.102}$$

CHECK UP

1. Draw and label the block diagram of a voltage-series feedback system.
2. Does input impedance increase or decrease with the use of voltage-series feedback and by what factor?
3. Does output impedance increase or decrease with the use of voltage-series feedback and by what factor?
4. **TRUE OR FALSE?** h parameters are used to model voltage-series feedback circuits.
5. **TRUE OR FALSE?** h_{12a} is used to approximate f in a unilateral approximation.
6. **TRUE OR FALSE?** h_{12f} is used to approximate f in a unilateral approximation.
7. **TRUE OR FALSE?** All the forward gain occurs in the a network and all the reverse gain occurs in the f network when using a unilateral network approximation.
8. What are two additional names for a voltage-series amplifier circuit?

11.5 CURRENT-SHUNT FEEDBACK

Gain Stabilization and Impedance Modification

The *current-shunt feedback* configuration is illustrated in Fig. 11.21(a). This is also called *parallel-input/series-output* or *shunt-series* feedback. An output current is being compared to an input current, so A will have units of amperes/ampere. Therefore, this is called a

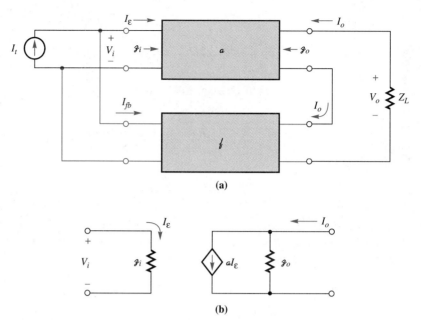

FIGURE 11.21 Current-shunt (also known as parallel-input/series-output or shunt-series) feedback configuration. (a) Current-shunt feedback two-port interconnection. (b) Basic current amplifier a block.

current amplifier. The a block model is given in Fig. 11.21(b), where a is a current gain and $f = I_{fb}/I_o$. At the output node,

$$I_o = aI_\varepsilon = a(I_i - I_{fb}) = a(I_i - fI_o) \tag{11.103}$$

or

$$A = \frac{I_o}{I_i} = \frac{a}{1 + af}. \tag{11.104}$$

As derived in Section 11.2, the input impedance for a parallel input feedback connection is given by

$$Z_i = \frac{z_i}{(1 + af)}, \tag{11.105}$$

and for a series output, as shown in Section 11.3, the output impedance is given by

$$Z_o = z_o(1 + af). \tag{11.106}$$

EXAMPLE 11.5

An amplifier without feedback has a midband current gain of 1000 (A/A), a high-frequency single corner $f_H = 100$ kHz, a low-frequency single corner $f_L = 1$ kHz, an input impedance $z_i = 10$ kΩ, and an output impedance $z_o = 50$ Ω. Feedback of $f = 0.05$ A/A is now added. Determine the values of those properties with feedback.

SOLUTION We recognize the circuit is configured as a current amplifier; i.e., a sample of the output current is being summed at the input with a current-shunt configuration.

Then

$$A = \frac{I_o}{I_s} = \frac{1000}{1 + (1000)(0.05)} = 19.6 \text{ A/A}.$$

Recall from Chapter 6 that $A = 1/f = 20$ A/A would be an acceptable answer since $af = 50 \gg 1$. The high-frequency corner $f_H = 100$ kHz $\times 51 = 5.1$ MHz, the new low-frequency corner $f_L = 1$ kHz$/51 = 19.6$ Hz, $Z_i = 10$ k$\Omega/51 = 196$ Ω, and $Z_o = 50 \times 51 = 2550$ Ω.

As can be observed from Fig. 11.21, there are two two-port networks whose inputs are in parallel and whose outputs are in series. Therefore, input circuit passive components are best represented as admittances and input circuit sources as current generators. Output circuit passive components are best represented as impedances and output circuit sources as voltage generators. These are properties of the **g-parameter two-port model**, the dual of the *h*-parameter model (see Fig. 11.22). In practice, the g-parameter model is rarely used, probably because it does not model any active device conveniently. Nevertheless, it is shown to complete the presentation of all four circuit configurations and models. The a and f networks are modeled as two hybrid-connected two-port networks as shown in Fig. 11.23(a). As with the previous cases, the network will be simplified by setting up the a and f networks as unilateral. Therefore, for forward transmission only through the a network,

$$|g_{21a}| \gg |g_{21f}|, \qquad (11.107)$$

and for reverse transmission only in the f network,

$$|g_{12f}| \gg |g_{12a}|. \qquad (11.108)$$

The circuit then reduces to that shown in Fig. 11.23(b). Using this reduced model, compute

$$a = \frac{I_o}{I_s} = \frac{-g_{21a}}{(g_{11a} + g_{11f} + Y_S)(g_{22a} + g_{22f} + Z_L)}, \qquad (11.109)$$

which includes the loading of the feedback network, g_{11f} at the input and g_{22f} at the output, but with the feedback generator $f = g_{12f} = 0$. Then

$$A = \frac{I_o}{I_s} \cong \frac{\dfrac{-g_{21a}}{(g_{11} + g_{11f} + Y_S)(g_{22a} + g_{22f} + Z_L)}}{1 + \dfrac{(-g_{21a})(g_{12f})}{(g_{11a} + g_{11f} + Y_S)(g_{22a} + g_{22f} + Z_L)}} \qquad (11.110)$$

by including the feedback generator.

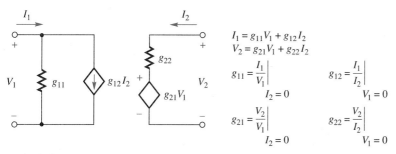

FIGURE 11.22 The g-parameter model.

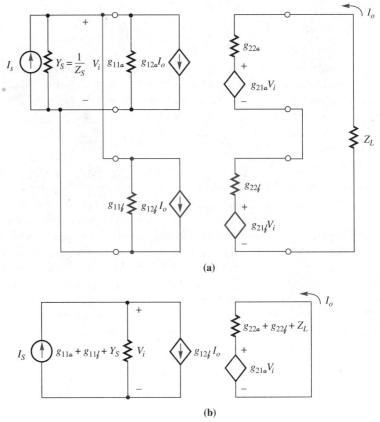

FIGURE 11.23 The g-parameter feedback model current-shunt circuit. (a) Complete g-parameter equivalent circuit. (b) Reduced circuit, unilateral networks.

Again, by assuming that the a and f networks are unilateral we can outline an analysis approach that does not require the computation of all of the a and f network g parameters in very much the same way we handled a voltage-shunt circuit with y parameters, a current-series circuit with z parameters, and a voltage-series circuit with h parameters:

1. Compute g_{11f}, g_{22f}, and g_{12f}.
2. Set the feedback generator $g_{12f} = 0$ temporarily.
3. Compute $a = I_o/I_s$ including the loading, g_{11f} at the input and g_{22f} at the output, provided by the feedback network. One could use g parameters or any other convenient circuit analysis techniques to compute $a = I_o/I_s$.
4. Use this value of a and the feedback ratio $f = g_{12f}$ in Eq. (11.110) to obtain A.

This technique is now demonstrated in an example.

Multistage Amplifier with Current-Shunt Feedback

Two CE amplifiers are connected in cascade as shown in Fig. 11.24(a). A sample of the output current through Q_2 feeds back to the Q_1 base node as a current sum, i.e., a parallel or shunt input connection. An argument might be made that a portion of the output voltage is being sampled. However, observe that the output is at the collector of Q_2, and even though fractionally lower, the voltage at the emitter of Q_2 is proportional to this collector

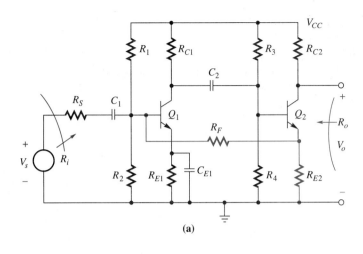

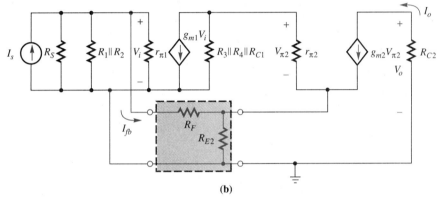

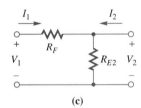

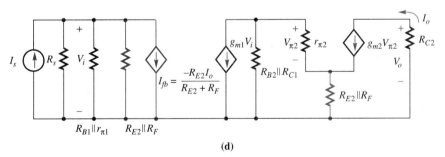

■ **FIGURE 11.24** Current-shunt feedback in two cascaded CE amplifiers. (a) Two cascaded CE amplifiers with feedback. (b) ac model with feedback network separately identified. (c) g-parameter definition of the feedback network. (d) ac model with feedback incorporated as g_{11f} and g_{22f} input and output loading and $I_{fb} = -R_{E2}I_o/(R_{E2} + R_F)$ generator. Set $I_{fb} = 0$ to obtain $a = I_o/I_s$, $\mathscr{z}_i$, and $\mathscr{z}_o$. Include r_{o2} to obtain a finite R_o and if there were an external R_L, it would be absorbed in R_{C2}.

output voltage, and virtually all of the output current (for $\beta \gg 1$) is present in this emitter circuit; thus, current sampling is the correct feedback configuration. The ac model is given in Fig. 11.24(b), where the feedback network has been identified and separated from the forward amplifier. The small-signal transistor output resistances r_o have not been included. Current summing at the input requires that the source voltage generator and source resistance be replaced by their Norton equivalent. This current-shunt feedback is best modeled by using the hybrid-g parameters. Using the g-parameter definitions provided in Fig. 11.22 and the isolated feedback network given in Fig. 11.24(c),

$$g_{11f} = \left.\frac{I_1}{V_1}\right|_{I_2=0} = \frac{1}{R_{E2} + R_F}, \quad g_{12f} = \left.\frac{I_1}{I_2}\right|_{V_1=0} = \frac{-R_{E2}}{R_{E2} + R_F} = f,$$

$$g_{21f} = \left.\frac{V_2}{V_1}\right|_{I_2=0} = \frac{R_{E2}}{R_{E2} + R_F}, \quad g_{22f} = \left.\frac{V_2}{I_2}\right|_{V_1=0} = R_{E2} \| R_F. \quad (11.111)$$

These equivalent g parameters are incorporated into the small-signal model as shown in Fig. 11.24(d) as input and output feedback network loading, g_{11f} and g_{22f}, and a feedback generator, g_{12f}. By setting $g_{12f} = 0$, $a = I_o/I_s$ can be computed directly. However, rather than computing a, observe that if $af \gg 1$, then $A = I_o/I_s = 1/f$ directly. The feedback ratio f is <1. The two cascaded CE amplifiers have a current gain

$$a = -K\beta_1\beta_2, \quad (11.112)$$

where K represents a composite current division term of <1. This result could be quite large since β_1 and $\beta_2 \gg 1$. We then have

$$af = -K\beta_1\beta_2\left(\frac{-R_{E2}}{R_{E2} + R_F}\right) \gg 1, \quad (11.113)$$

and immediately we have

$$A = \frac{I_o}{I_s} \approx \frac{1}{f} = -\left(\frac{R_{E2} + R_F}{R_{E2}}\right). \quad (11.114)$$

Realizing that $V_o = -I_o R_{C2}$ and $V_s = I_s R_S$, the source voltage gain with feedback is given by

$$A_{vs} = \frac{V_o}{V_s} = \frac{-I_o R_{C2}}{I_s R_S} = \left(\frac{R_{E2} + R_F}{R_{E2}}\right)\left(\frac{R_{C2}}{R_S}\right). \quad (11.115)$$

Although computing af exactly would permit an exact determination of the input and output resistances, they can be estimated by using the results of Eqs. (11.105) and (11.106). The parallel input current summing node of the feedback current, I_{fb}, reduces the resistance to a very low value. With this $af \gg 1$ approximation, V_s effectively looks into $R_i \approx R_S$. Similarly, at the output the resistance into the collector is very large, actually approaching ∞ for $r_o \to \infty$ so that $R_o \approx R_{C2}$ as shown.

EXAMPLE 11.6

The amplifier circuit of Fig. 11.25 uses current-shunt feedback. Find $A_{vs} = V_8/V_s$ at midband frequencies using SPICE. Compare results with Eq. (11.115). For the transistor model, use $\beta = 200$ and $V_A = 150$ V.

CHAPTER 11 FEEDBACK

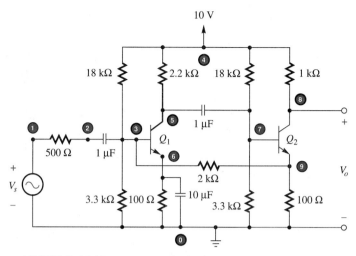

FIGURE 11.25 Current-shunt feedback circuit.

SOLUTION One such SPICE program is listed below. The frequency sweep will be extended to 100 kHz so that the effects of both the 1- and 10-μF capacitors can be neglected.

```
CURRENT SHUNT FEEDBACK
VS 1 0 AC .001
VCC 4 0 DC 10
RS 1 2 500
R1 4 3 18K
R2 3 0 3.3K
RC1 4 5 2.2K
RE1 6 0 100
RF 3 9 2K
R3 4 7 18K
R4 7 0 3.3K
RC2 4 8 1K
RE2 9 0 100
C1 2 3 1UF
C2 5 7 1UF
CE1 6 0 10UF
Q1 5 3 6 M1
Q2 8 7 9 M1
.MODEL M1 NPN BF=200 IS=2E-15 VAF=150
.AC DEC 1 10 100K
.PRINT AC VM(8) VP(8) VM(3) VP(3) I(VS)
.END
```

The printout for the ac analysis is as follows:

FREQ	VM(8)	VP(8)	VM(3)	VP(3)	I(VS)
1.000D 01	9.874D-04	1.146D 02	3.306D-05	3.824D 01	6.118D-08
1.000D 02	1.040D-03	7.584D 01	8.737D-05	2.766D 01	5.536D-07
1.000D 03	3.696D-03	2.116D 01	6.072D-05	-3.581D 01	1.813D-06
1.000D 04	3.968D-02	2.225D 00	2.945D-05	-8.682D 00	1.941D-06
1.000D 05	3.971D-02	2.226D-01	2.888D-05	-8.841D-01	1.942D-06

DRILL 11.8

A few short questions relating to Fig. 11.24:

(a) What circuit element would you change to change A_{vs} without disturbing the dc Q-points?
(b) What would you add to the circuit so you could adjust R_F and not disturb the Q-points?
(c) What would you have to do to replace C_2 with a short circuit?

ANSWERS

(a) R_S.
(b) A capacitor in series with R_F.
(c) Design the Q-points so that the dc voltage levels at the collector of Q_1 and base of Q_2 are the same. The direct-coupled circuit Q-points will have to be designed as a composite circuit rather than treating each stage individually.

■ Advanced circuit theory and control system courses discuss stability in the context of Eqn. (11.116).

At 100 kHz, $A_{vs} = V_8/V_s = 39.71$ with essentially zero phase shift as would be expected for midfrequency operation. Eq. (11.115) can be written as

$$A_{vs} = \frac{V_o}{V_s} = \frac{-I_o R_{C2}}{I_s R_S} = \left(\frac{R_{E2} + R_F}{R_{E2}}\right)\left(\frac{R_{C2}}{R_S}\right) = \left(\frac{100 \text{ k}\Omega + 2 \text{ k}\Omega}{100 \text{ }\Omega}\right)\left(\frac{1 \text{ k}\Omega}{500 \text{ }\Omega}\right) = 42.$$

This compares favorably with the SPICE solution of 39.71. At node 3, SPICE predicts $R_i = \text{VM(3)/I(VS)} = 14.87 \text{ }\Omega$, which agrees well with the R_i expected for a parallel input. Performing a source transformation at node 1, the source node, SPICE predicts $R_i' = V_s/I(VS) = 515 \text{ }\Omega \approx R_S = 500 \text{ }\Omega$.

CHECK UP

1. Draw and label the block diagram of a current-shunt feedback system.
2. Does input impedance increase or decrease with the use of current-shunt feedback and by what factor?
3. Does output increase or decrease with the use of current-shunt feedback and by what factor?
4. **TRUE OR FALSE?** g parameters are used to model current-shunt feedback circuits.
5. **TRUE OR FALSE?** g_{12a} is used to approximate f in a unilateral approximation.
6. **TRUE OR FALSE?** g_{12f} is used to approximate f in a unilateral approximation.
7. **TRUE OR FALSE?** All the forward gain occurs in the a network and all the reverse gain occurs in the f network when using a unilateral network approximation.
8. What are two additional names for a current-shunt amplifier circuit?

11.6 AMPLIFIER FREQUENCY DEPENDENCE AND COMPENSATION

So far in our study of feedback circuits, we have not included the frequency dependence of either the forward amplifier or feedback network. In general, Eq. (11.1) can be written as

$$A(j\omega) = \frac{a(j\omega)}{1 + a(j\omega)f(j\omega)} = \frac{a(j\omega)}{1 + T(j\omega)}. \quad (11.116)$$

If $T(j\omega) = -1$, $A(j\omega)$ is no longer defined, and operationally this means that $V_o(j\omega)$ exists for $V_s = 0$. The circuit will oscillate, that is, generate a signal at a frequency where $T(j\omega) = -1$. No source generator is required. This circuit behavior can be the result of a deliberate design for a signal source at some frequency; hence, we have an *oscillator*, or inherent phase and amplitude conditions in a feedback amplifier can result in oscillatory behavior that is undesirable.

The phase relationships leading to $T(j\omega) = -1$ are illustrated in Fig. 11.26. Suppose S_i has a 0° reference as shown. With the switch open, S_ε will have the same 0° phase

CHAPTER 11 FEEDBACK

reference. At some frequency, $\omega = \omega_o$, S_o could have a 180° phase reference and its amplitude will be larger than S_ε. Assume the feedback network, f, adds no additional phase shift; then S_{fb} will be in phase with S_o and at somewhat lower amplitude. If the switch is closed, S_{fb} acquires an additional 180° of phase shift as it passes through the summing (subtracting) node. It is now in phase with S_ε and reinforces S_ε. This process is self-sustaining if the gain of the forward amplifier is large enough at the frequency where the overall loop phase yields $T(j\omega) = -1$. The input generator, S_i, is no longer needed. Oscillators will be covered in Section 11.7; meanwhile, in this section, we want to develop design techniques to assure amplifier stability.

Quantitatively determining $a(j\omega)$ is a difficult algebraic problem best left for a computer model simulation. For example, to be rigorous for a μA741 analysis, C_μ, C_π, and C_{cs} would have to be considered for 26 transistors, an unattractive "by-hand" computation. Clearly, when such numbers of capacitors are included in a feedback circuit, the chances for unstable operation must be considered.

■ CAD tools like SPICE are the only practical way to get accurate results.

Let's simplify the problem by assuming there are three identical amplifier stages whose responses are each given by

$$a_1(j\omega) = \left[\frac{a_o}{1 + j(\omega/\omega_c)}\right]. \tag{11.117}$$

The transfer function of the cascade of these three amplifiers is

$$a(j\omega) = \frac{(a_o)^3}{[1 + j(\omega/\omega_c)]^3}. \tag{11.118}$$

Assume this amplifier is incorporated in a feedback circuit where the feedback network consists of resistors so that $f(j\omega) = f_o$. Then using $T = af$, we have

$$T(j\omega) = \frac{a_o^3 f_o}{\left(1 + j\dfrac{\omega}{\omega_c}\right)^3} = \frac{a_o^3 f_o}{\left(\sqrt{1 + \left(\dfrac{\omega}{\omega_c}\right)^2}\right)^3} e^{-j3\tan^{-1}(\omega/\omega_c)}. \tag{11.119}$$

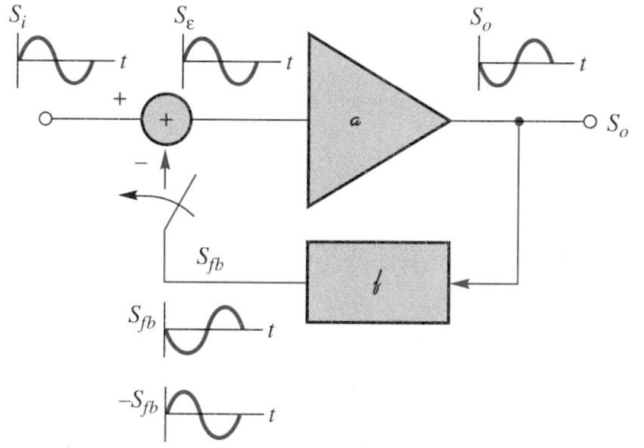

■ **FIGURE 11.26** Signal phase relationships in a feedback network.

The Bode magnitude and phase plots for $T(j\omega)$ are shown in Fig. 11.27. Observe that a_o and f_o were selected so that when $\theta(j\omega) = -180°$, $|T(j\omega)| > 1$ (>0 dB). Therefore, for this selection of $a_o f_o$, the amplifier is unstable. The amplitude of $T(j\omega)$ would have tried to pass through unity ($T = -1$) when the circuit was powered up, and as it did, it would have exhibited instability. Clearly, three alternatives to obtain stability present themselves:

1. Reduce a_o. This is often not acceptable, since the purpose of an amplifier is to provide gain despite all of the variations possible with temperature, manufacturing tolerances, and power supply fluctuations.
2. Reduce f_o. Of course, reducing the amount of feedback might not be compatible with other circuit and system design constraints, but it is an alternative.
3. Alter the phase characteristic with the addition of additional reactive passive components. This is called **compensation**.

■ Compensation is the most widely used approach of the three. Some circuits like the μA741 are *internally compensated*; some circuits require external compensation networks.

■ Another approach for stability analysis is to use the *Nyquist plot* or *pole-zero analysis*. We choose to focus on Bode plots in this text. Nyquist and pole-zero plots are widely used in control system stability analysis. Nyquist plots are also used in advanced circuit theory discussions.

Consider reducing $a_o f_o$ so that the Bode plots in Fig. 11.27 become those shown in Fig. 11.28. This feedback amplifier is stable because $|T(j\omega)| < 1$ when its phase = $-180°$, point A in the figure. The additional loop gain that could be tolerated before the amplifier became unstable, i.e., $|T(j\omega)| = 1$ when $\theta(j\omega) = -180°$, is defined as the **gain margin** (*GM*). The GM is usually expressed in decibels. See Figure 11.28 for a stable case. At point B in the figure, where $|T(j\omega)| = 1$ or 0 dB, the loop phase shift

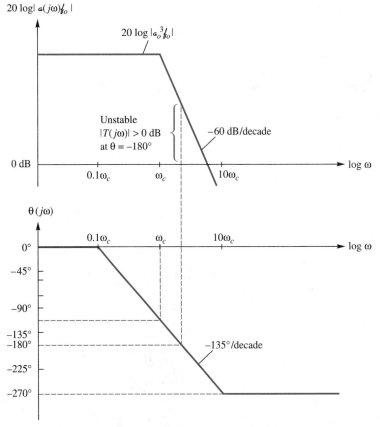

■ **FIGURE 11.27** Bode magnitude and phase plots for $T(j\omega) = a_o^3 f_o / [1 + j(\omega/\omega_c)]^3$.

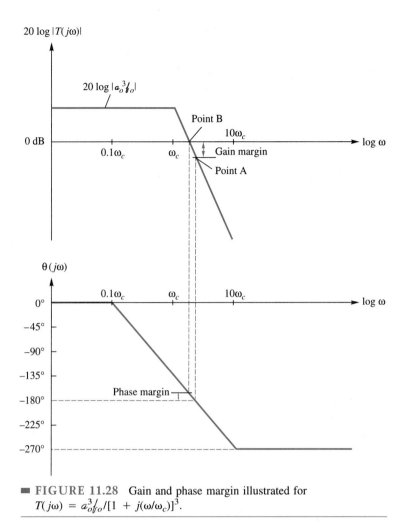

■ FIGURE 11.28 Gain and phase margin illustrated for $T(j\omega) = a_o^3 f_o / [1 + j(\omega/\omega_c)]^3$.

has not yet reached $-180°$. The additional phase shift that could be tolerated before the amplifier became unstable is defined as the **phase margin** (PM) and is computed from

$$\text{PM} = +180° + \text{phase} |T(j\omega)|, \quad (11.120)$$

where $|T(j\omega)| = 1$ and PM $> 0°$ for stability.

EXAMPLE 11.7

Compute the largest feedback factor f_o possible so that the PM $= 30°$ for a loop gain given by

$$a(j\omega) = \frac{a_o}{\left(1 + j\dfrac{f}{f_{c1}}\right)\left(1 + j\dfrac{f}{f_{c2}}\right)\left(1 + j\dfrac{f}{f_{c3}}\right)},$$

where $f_{c1} = f_{c2} = 1$ MHz, $f_{c3} = 3$ MHz, and $a_o = 200$.

SOLUTION The loop gain can be written as

$$T(j\omega) = |T(j\omega)|e^{j\theta(j\omega)} = \frac{|a_o f_o|}{\sqrt{1 + \left(\frac{f}{f_{c1}}\right)^2}\sqrt{1 + \left(\frac{f}{f_{c2}}\right)^2}\sqrt{1 + \left(\frac{f}{f_{c3}}\right)^2}} e^{j\theta},$$

where

$$\theta = -\tan^{-1}(f/f_{c1}) - \tan^{-1}(f/f_{c2}) - \tan^{-1}(f/f_{c3}).$$

From Eq. (11.120), $\theta(j\omega) = \text{PM} - 180° = 30° - 180° = -150°$. Then

$$\theta(j\omega) = -2\tan^{-1}(f/1\text{ MHz}) - \tan^{-1}(f/3\text{ MHz}) = -150°.$$

The frequency $f = 1.732$ MHz is solved for by iteration. Substituting this into the magnitude portion of $|T(j\omega)| = 1$ yields

$$\frac{200 f_o}{2\sqrt{1 + \left(\frac{1.732}{1}\right)^2}\sqrt{1 + \left(\frac{1.732}{3}\right)^2}} = \frac{200 f_o}{4.62} = 1 \quad \text{or} \quad f_o = 0.0231.$$

This means the largest f_o possible to obtain a 30° PM is $f_o = 0.0231$ and

$$A_o = \frac{a_o}{1 + a_o f_o} = \frac{200}{1 + (200)(0.0231)} = 35.6.$$

Compensation

Another approach, suggested by the plots in Figs. 11.27 and 11.28, is to introduce an additional frequency-dependent term as part of $T(j\omega)$ so that the resultant new $|T(j\omega)|$ becomes <1 when the resultant new phase becomes $-180°$ or when the desired PM is obtained. A very common approach to providing this additional open-loop gain rolloff is via a single-stage RC low-pass filter whose transfer function is

$$H(j\omega) = \left(\frac{1}{1 + j\frac{\omega}{\omega_D}}\right). \tag{11.121}$$

The corner frequency ω_D will be called the *dominant corner frequency*. The loop-gain transfer function in general now becomes

$$T(j\omega) = \frac{a_o f_o}{\left(1 + j\frac{\omega}{\omega_{c1}}\right)\left(1 + j\frac{\omega}{\omega_{c2}}\right)\left(1 + j\frac{\omega}{\omega_{c3}}\right)\left(1 + j\frac{\omega}{\omega_D}\right)}. \tag{11.122}$$

For graphical clarity, we will use the Bode magnitude and phase plots for the case of loop gain with a triple corner frequency and a dominant frequency defined by

$$T(j\omega) = \frac{a_o f_o}{\left(1 + j\frac{\omega}{\omega_c}\right)^3 \left(1 + j\frac{\omega}{\omega_D}\right)} \tag{11.123}$$

as illustrated in Fig. 11.29. Observe that for $\omega_D \ll \omega_C$ ($\ll$ means by at least one decade), an effective, constant phase shift of $-90°$ is contributed from the single-section low-pass filter characteristic. We can locate ω_D at some point so that $|T(j\omega)| < 0$ dB is assured. Of course, the design trade-off is that the lower ω_D becomes, the less bandwidth there is. Analytically, ω_D is specified from Eq. (11.123) by simultaneously solving

$$\text{phase } T(j\omega) = -\tan^{-1}\left(\frac{\omega}{\omega_{c1}}\right) - \tan^{-1}\left(\frac{\omega}{\omega_{c2}}\right) - \tan^{-1}\left(\frac{\omega}{\omega_{c3}}\right) - \tan^{-1}\left(\frac{\omega}{\omega_D}\right)$$

$$= -180°$$

(11.124)

or if a PM is specified

$$= -180° + \text{PM},$$

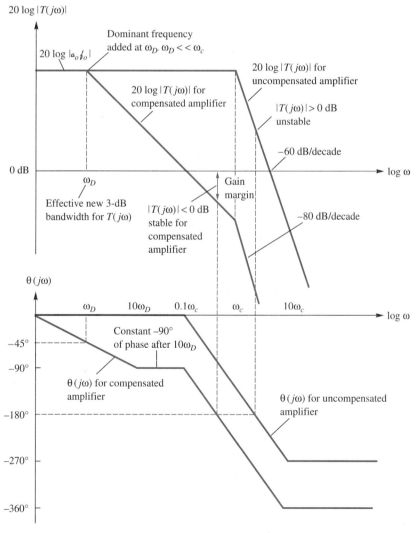

■ **FIGURE 11.29** Compensation of form $1/[1 + j(\omega/\omega_D)]$ added to $T(j\omega) = a_o f_o /\{[1 + j(\omega/\omega_c)]^3\}$.

and

$$|T(j\omega)| = 1 = \frac{a_o f_o}{\sqrt{1 + \left(\frac{\omega}{\omega_{c1}}\right)^2}\sqrt{1 + \left(\frac{\omega}{\omega_{c2}}\right)^2}\sqrt{1 + \left(\frac{\omega}{\omega_{c3}}\right)^2}\sqrt{1 + \left(\frac{\omega}{\omega_D}\right)^2}}. \tag{11.125}$$

Analogously, for a loop gain with a triple corner frequency,

$$\text{phase } T(j\omega) = -3\tan^{-1}\left(\frac{\omega}{\omega_c}\right) - \tan^{-1}\left(\frac{\omega}{\omega_D}\right) = -180° \tag{11.126}$$

or

$$= -180° + \text{PM}$$

and

$$|T(j\omega)| = 1 = \frac{a_o f_o}{\left[\sqrt{1 + \left(\frac{\omega}{\omega_c}\right)^2}\right]^3 \sqrt{1 + \left(\frac{\omega}{\omega_D}\right)^2}}. \tag{11.127}$$

These equations are transcendental in ω_D, which means they must be solved iteratively. However, this process is greatly simplified by realizing that for $\omega_D \gg$ any other corner frequency, its phase-shift contribution is an approximate $-90°$. Then

$$\text{phase } T(j\omega) = -\tan^{-1}\left(\frac{\omega}{\omega_{c1}}\right) - \tan^{-1}\left(\frac{\omega}{\omega_{c2}}\right) - \tan^{-1}\left(\frac{\omega}{\omega_{c3}}\right) - 90°$$

$$= -180° + \text{PM} \tag{11.128}$$

or

$$= -3\tan^{-1}\left(\frac{\omega}{\omega_c}\right) - 90° = -180° + \text{PM} \tag{11.129}$$

is solved simultaneously with Eq. (11.125) or (11.127).

Compensation is illustrated using the following example and, in Chapter 12, these techniques are applied to the μA741 operational amplifier internal circuitry.

EXAMPLE 11.8

Using the transfer function and corner frequencies specified in Example 11.7 and assuming $f_o = 0.2$, specify the largest f_D such that the system is stable with a 30° PM.

SOLUTION Substituting into Eq. (11.128) to solve for f yields

$$-60° = -2\tan^{-1}(f/1 \text{ MHz}) - \tan^{-1}(f/3 \text{ MHz}),$$

or $f = 477$ kHz assuming the f_D corner contributes a $-90°$ phase shift at this point. Substituting this frequency into Eq. (11.125) and solving for f_D, we have

$$1 = \frac{a_o f_o}{\left[\sqrt{1+\left(\frac{f}{f_{c1}}\right)^2}\right]^2 \left[\sqrt{1+\left(\frac{f}{f_{c2}}\right)^2}\right]\left[\sqrt{1+\left(\frac{f}{f_D}\right)^2}\right]}$$

$$= \frac{(200)(0.2)}{\left[\sqrt{1+\left(\frac{0.477}{1}\right)^2}\right]^2 \left[\sqrt{1+\left(\frac{0.477}{3}\right)^2}\right]\left[\sqrt{1+\left(\frac{0.477}{f_D}\right)^2}\right]}$$

$$= \frac{32.18}{\left[\sqrt{1+\left(\frac{0.477}{f_D}\right)^2}\right]}$$

or $f_D = 14.8$ kHz. This means that if we design with a dominant frequency of $f_D = 14.8$ kHz, the circuit would be stable with a 30° PM. In this case, f_o is ten times that computed in Example 11.7, so we are able to include much more feedback. The circuit $T(j\omega)$ bandwidth has dropped to $\approx f_D$ and $A_o \approx 1/f \approx 5$ and virtually independent of changes in a_o.

DRILL 11.9

Refer to Example 11.8. What f_D would be required if the design required a $PM = 45°$.

ANSWERS

$f = 348$ kHz, which results in $f_D = 9.82$ kHz, a significant bandwidth reduction.

CHECK UP

1. **TRUE OR FALSE?** If the signal fed back and input signal are in phase, the circuit could become unstable.
2. **TRUE OR FALSE?** If the output signal and input signal are in phase, the circuit will become unstable.
3. **TRUE OR FALSE?** The feedback network is always frequency independent.
4. Define the terms *gain margin* and *phase margin* using Bode plots.
5. Explain how a dominant frequency is used to compensate an amplifier.

11.7 OSCILLATORS

An oscillator circuit consists of a forward amplifier and a frequency-dependent feedback network. As studied in Section 11.6, at some given frequency the circuit can be designed to obtain $af = -1$ in Eq. (11.15). When applied to oscillators, the condition $af = -1$ is called the **Barkhausen criterion**. This condition of positive feedback will result in the circuit converting the dc power supply voltage to an ac signal. Theoretically, the condition $af = -1$ is satisfied at one, and only one frequency, and therefore you might expect that the output would be sinusoidal. However, this is not quite true. Equation (11.116) is based on linear circuit network theory and as soon as $af = -1$, the output voltage does not, of course, become infinite, but saturates. This distorted waveform does include a frequency component to satisfy the condition $af = -1$, and harmonics of this frequency are generated as a result of the nonlinear behavior. This nonlinear circuit analysis becomes very difficult and is beyond the scope of this book. The SPICE transient analysis can provide a good

■ The inherent nonlinear operation of oscillators results in real oscillators not having an ideal single frequency operation but a spread of frequencies around the dominant frequency output.

insight into oscillator design. However, the Barkhausen criterion $af = -1$ does predict nicely the frequency at the onset of oscillation, and as we proceed with several examples, we present techniques used to obtain a sinusoidal output.

The most important performance feature required from an oscillator is *frequency stability*, the ability to maintain a constant operating frequency under all possible circuit element and temperature variations.

RC Phase-Shift Oscillator

Assume that the common-source FET amplifier and the operational amplifier circuit shown in Fig. 11.30 have a frequency-independent gain. For the CS amplifier,

$$a(j\omega) = -g_m R_D, \tag{11.130}$$

and for the operational amplifier with negative feedback,

$$a(j\omega) = -\frac{R_2}{R_1}. \tag{11.131}$$

Both exhibit 180° of phase shift in the midband. This feature is illustrated with the sinusoidal waveforms sketched at the input and output of each circuit. Assume the three-section *RC* filter is now connected between the output and input as shown. At some unique frequency, the phase shift through the *RC* filter will be exactly 180°, and when this occurs, the output of the *RC* network will reinforce the input of the amplifier. If the gain of the amplifier is high enough to overcome losses in the *RC* network, the circuit will oscillate. Analytically, this means that R and C must be selected so that

$$\frac{V_a}{V_o} = -\frac{1}{|a|} = \frac{1}{|a|} e^{-j180°}. \tag{11.132}$$

- As evidenced by the number of technical journal articles, there is considerable interest in predicting oscillator performance based upon non-linear models.
- Note the exact 180° phase shift through this RC phase shift network at $\omega = \omega_o$.
- Also refer to Ch. 17 for digital frequency synthesis techniques for waveform generation.

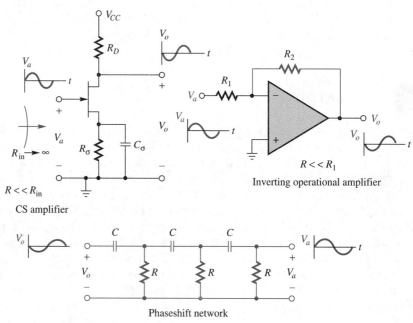

FIGURE 11.30 *RC* phase-shift oscillators.

Using loop analysis and assuming $R_{in} \gg R$ for the CS amplifier and $R_1 \gg R$ for the operational inverting amplifier,

$$\frac{V_a}{V_o} = \frac{1}{1 - 5\left(\frac{1}{\omega RC}\right)^2 - j\left[\frac{6}{\omega RC} - \left(\frac{1}{\omega RC}\right)^3\right]}. \quad (11.133)$$

The algebraic details are left for an end-of-the chapter problem. The only circuit condition yielding a 180° phase shift is

$$\frac{6}{\omega_o RC} = \left(\frac{1}{\omega_o RC}\right)^3 \quad (11.134)$$

in Eq. (11.133). This means that the frequency of oscillation is given by solving Eq. (11.134) to obtain

$$\omega_o = \frac{1}{RC\sqrt{6}} \quad \text{or} \quad f_o = \frac{1}{2\pi RC\sqrt{6}}. \quad (11.135)$$

Substituting Eq. (11.135) into Eq. (11.133), we observe that

$$\frac{V_a}{V_o} = \frac{1}{1 - 5\left(\frac{1}{\omega_o RC}\right)^2} = \frac{1}{1 - (5)(6)} = -\frac{1}{29}. \quad (11.136)$$

This means that the forward amplifiers whose voltage gains are given in Eqs. (11.130) and (11.131) must provide $|a| = 29$. In practice, $|a| > 29$ is recommended to ensure the circuit will oscillate despite variations in component values, temperature, and aging. This is called designing with **excess gain**.

■ This circuit *will not* oscillate if $|a|$ is even infinitesimally less than 29. Excess gain is important as a design criteria.

EXAMPLE 11.9

Design a 3-kHz *RC* oscillator using an inverting operational amplifier. Assume $C = 0.01 \; \mu F$.

SOLUTION Using Fig. 11.30 and from Eq. (11.135),

$$R = \frac{1}{2\pi f_o C\sqrt{6}} = \frac{1}{2\pi(3 \text{ kHz})(0.01 \; \mu F)\sqrt{6}} = 2.166 \text{ k}\Omega.$$

A standard value close to this is 2.2 kΩ yielding $f_o = 2953$ Hz. Select $R_1 = 47$ kΩ, which is $\gg 2.2$ kΩ. Then $R_2 = 29$ and $R_1 = 1.363$ MΩ. Actually, a better choice would be $R_2 = 1.5$ MΩ to allow some margin for component variation and aging.

The inherent nonlinear behavior of this oscillator yields a somewhat distorted signal at either the drain of the CS amplifier or the operational amplifier output. One way to reduce distortion is to add another operational amplifier configured as a low-pass filter, as shown in Fig. 11.31. The corner frequency is set just above the frequency of oscillation.

Generally speaking, *RC* oscillators are most useful up to several hundred kilohertz. By arranging to have all three phase-shift network resistors or capacitors simultaneously adjustable, variable frequency operation can be obtained.

■ This distorted output implies the generation of harmonics and a spectrum of signals centered about the fundamental frequency output. %HD, as discussed in Chapter 10, could be used to quantify the output signal quality (purity of spectral response).

DRILL 11.10

What range of feedback network resistor values would you use to tune the oscillator designed in Example 11.9 from 1 to 10 kHz? A frequency adjustable oscillator is sometimes called a *VFO* (*variable frequency oscillator*).

ANSWER Adjust R from 6500 to 650 Ω. This could be achieved by using a standard three-section, 10-kΩ potentiometer.

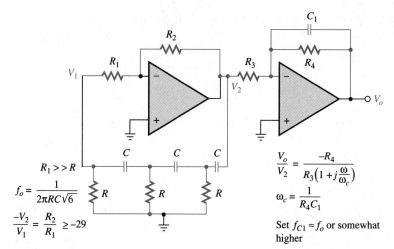

$R_1 \gg R$

$f_o = \dfrac{1}{2\pi RC\sqrt{6}}$

$\dfrac{-V_2}{V_1} = \dfrac{R_2}{R_1} \geq -29$

$\dfrac{V_o}{V_2} = \dfrac{-R_4}{R_3\left(1 + j\dfrac{\omega}{\omega_c}\right)}$

$\omega_c = \dfrac{1}{R_4 C_1}$

Set $f_{C1} \approx f_o$ or somewhat higher

■ **FIGURE 11.31** *RC* oscillator followed by an active low-pass filter.

RC Wien Bridge Oscillator

Another type of *RC* oscillator is shown in Fig. 11.32(a). A noninverting amplifier using an operational amplifier has a voltage gain given by

$$a = \dfrac{V_o}{V_b} = 1 + \dfrac{R_2}{R_1}. \tag{11.137}$$

Observe that the noninverting amplifier phase shift is 0°. The feedback network consists of a voltage divider whose transfer function is

$$f = -\dfrac{V_{fb}}{V_o} = -\dfrac{Z_2}{Z_1 + Z_2}, \tag{11.138}$$

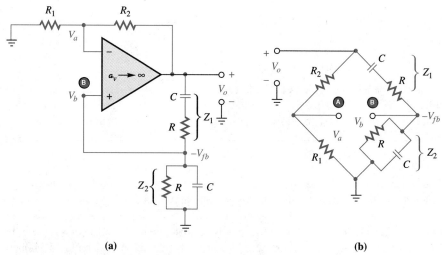

■ **FIGURE 11.32** Wien bridge oscillator. **(a)** Circuit diagram. **(b)** Bridge equivalent circuit.

CHAPTER 11 FEEDBACK

where the polarity of V_{fb} was defined in Fig. 11.15:

$$Z_2 = R \left\| \left(\frac{1}{j\omega C} \right) \right. = \frac{R}{1 + j\omega RC} \tag{11.139}$$

and

$$Z_1 = R + \frac{1}{j\omega C} = R\left(1 + \frac{1}{j\omega RC}\right). \tag{11.140}$$

If we redraw the oscillator circuit, realizing that the input impedance between points A and B is very large, we obtain the bridge configuration shown in Fig. 11.32(b); hence, the name Wien bridge oscillator. The bridge balances when the voltage at point B equals the voltage at point A. Another way of looking at this is by realizing that at one frequency, the voltage at point B is exactly in phase with the voltage at point A. The Barkhausen criterion requires that

$$a f = -1 = \left(1 + \frac{R_2}{R_1}\right) \left[\frac{\left(\frac{-R}{1 + j\omega RC}\right)}{R\left(1 + \frac{1}{j\omega RC}\right) + \frac{R}{1 + j\omega RC}} \right], \tag{11.141}$$

which, after a bit of algebraic manipulation, reduces to

$$af = \left(1 + \frac{R_2}{R_1}\right) \times \left[\frac{-1}{3 + j\left(\omega RC - \frac{1}{\omega RC}\right)} \right]. \tag{11.142}$$

The only circuit condition for which phase of $af = -180°$ is

$$\omega_o RC = \frac{1}{\omega_o RC}. \tag{11.143}$$

Solving for the frequency of oscillation,

$$\omega_o = \frac{1}{RC} \quad \text{or} \quad f_o = \frac{1}{2\pi RC}. \tag{11.144}$$

Setting $|af| = 1$ in Eq. (11.141) yields the minimum noninverting amplifier gain requirement that

$$\left| \left(1 + \frac{R_2}{R_1}\right)\left(-\frac{1}{3}\right) \right| = 1, \tag{11.145}$$

which leads to

$$\frac{R_2}{R_1} = 2, \quad a \geq 3. \tag{11.146}$$

■ Of course, this example assumes no bandwidth restrictions from the operational amplifier. This is not true, of course. The operational amplifier $a(j\omega)$ characteristic must be included.

DRILL 11.11

What range of feedback network resistor values would you use to tune the oscillator designed in Example 11.10 from 1 to 10 kHz?

ANSWER Adjust R from 15,923 to 1592 Ω. This could be achieved by using a standard two-section, 15-kΩ potentiometer in series with 1-kΩ fixed resistors.

EXAMPLE 11.10

Design a 3-kHz Wien bridge oscillator with $C = 0.01$ μF.

SOLUTION From Eq. (11.144), $R = 1/2\pi f_o C = 1/[2\pi(3 \text{ kHz})(0.01 \text{ μF})] = 5305$ Ω. A reasonably close standard value resistor of $R = 5100$ Ω results in $f_o = 3121$ Hz. Using a μA741, or any operational amplifier capable of operating at 3 kHz, select $R_2/R_1 = 2$ (plus a small amount of excess gain to compensate for component and device variations). For example, $R_2 = 22$ kΩ and $R_1 = 10$ kΩ would work well.

LC Oscillators

At frequencies from several hundred kilohertz to several hundred megahertz, it is more practical to use LC networks to provide the necessary phase-shift conditions. In general, the analysis of oscillators using inductors and capacitors is complicated by the inherent parasitic losses modeled as part of the L and C. For algebraic tractability, the following analysis assumes ideal (lossless) inductors and capacitors. The general form of the circuit is shown in Fig. 11.33(a). The forward amplifier is modeled as a frequency-independent voltage amplifier with an infinite input resistance and a finite output resistance, R_o. The pure reactive elements are modeled as Z_1, Z_2, and Z_3 in Fig. 11.33(b). The forward gain of the amplifier loaded by the feedback network is given by

$$a = \frac{V_o}{V_i} = \frac{-a_v Z_L}{Z_L + R_o}, \tag{11.147}$$

where

$$Z_L = Z_2 \| (Z_1 + Z_3) = \frac{Z_2(Z_1 + Z_3)}{Z_1 + Z_2 + Z_3}. \tag{11.148}$$

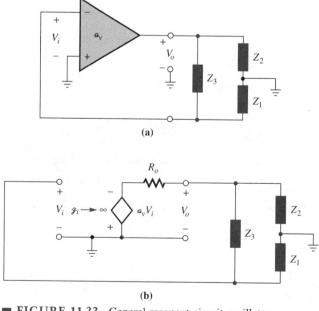

■ **FIGURE 11.33** General resonant circuit oscillator.
(a) General circuit. (b) Simplified equivalent circuit.

CHAPTER 11 FEEDBACK

The feedback ratio, using voltage division, is seen to be

$$f = \frac{-Z_1}{Z_1 + Z_3}. \tag{11.149}$$

Combining Eqs. (11.147) through (11.149), the loop gain becomes

$$T = af = \left(\frac{a_v Z_L}{Z_L + R_o}\right)\left(\frac{Z_1}{Z_1 + Z_3}\right) = \frac{a_v\left[\dfrac{Z_2(Z_1 + Z_3)}{Z_1 + Z_2 + Z_3}\right]\left(\dfrac{Z_1}{Z_1 + Z_3}\right)}{\dfrac{Z_2(Z_1 + Z_3)}{Z_1 + Z_2 + Z_3} + R_o}. \tag{11.150}$$

Clearing fractions and simplifying,

$$T = \frac{a_v Z_1 Z_2}{R_o(Z_1 + Z_2 + Z_3) + Z_2(Z_1 + Z_3)}. \tag{11.151}$$

Equation (11.151) represents a general result for any Z_1, Z_2, and Z_3. If ideal lossless reactances are assumed, that is, if

$$Z_1 = jX_1, \qquad Z_2 = jX_2, \qquad Z_3 = jX_3, \tag{11.152}$$

where $X > 0$ implies an inductive reactance and $X < 0$ implies a capacitive reactance, Eq. (11.151) becomes

$$T = \frac{a_v(jX_1)(jX_2)}{jR_o(X_1 + X_2 + X_3) + jX_2(jX_1 + jX_3)}$$
$$= \frac{-a_v X_1 X_2}{jR_o(X_1 + X_2 + X_3) - X_2(X_1 + X_3)}. \tag{11.153}$$

The phase condition can only be met if

$$X_1 + X_2 + X_3 = 0. \tag{11.154}$$

This is called *resonance* of the reactive elements. Substituting $X_1 + X_3 = -X_2$ into Eq. (11.153) yields

$$T = -1 = \frac{a_v X_1 X_2}{X_2(-X_2)} = \frac{a_v X_1}{-X_2} \tag{11.155}$$

so that

$$a_v = \frac{X_2}{X_1}. \tag{11.156}$$

The interpretation of Eqs. (11.154) and (11.156) is that for oscillation to occur

1. X_1 and X_2 must have the same sign; i.e., both are either inductive or capacitive.
2. X_3 must be of the opposite sign, satisfying Eq. (11.154).
3. $a_v = X_2/X_1$. Practically, $a_v > X_2/X_1$ to overcome circuit losses.

■ These 3 criteria are easy-to-apply good, general design rules.

We will present two principal classes of oscillator circuits (see Fig. 11.34). If X_1 and X_2 are capacitive while X_3 is inductive, the circuit is called a **Colpitts oscillator**, which is illustrated by connecting A to A and B to B in Figs. 11.34(a) and (b). The frequency of oscillation, ω_o, is found by solving Eq. (11.154)

$$-\frac{1}{\omega_o C_1} - \frac{1}{\omega_o C_2} + \omega_o L = 0 \tag{11.157}$$

for ω_o. The forward-gain stage is realized by using the CE amplifier, although any gain stage can be fabricated using operational amplifiers, discrete BJTs, or FETs. Large coupling capacitors, C_C, are required for dc isolation of the network, and their reactance is assumed to be negligible at ω_o.

■ There are literally hundreds of different oscillator configurations, each with many variations. We present only a few to illustrate the design approach.

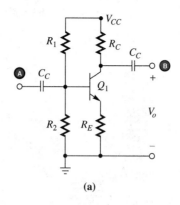

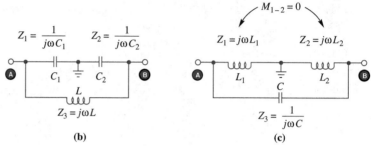

■ **FIGURE 11.34** Two forms of a resonant circuit oscillator. (a) General gain stage. (b) Colpitts feedback network. (c) Hartley feedback network.

If X_1 and X_2 are inductive and X_3 is capacitive, the circuit is called a ***Hartley oscillator***, which is illustrated by connecting A to A and B to B in Figs. 11.34(a) and (c). The frequency of oscillation, ω_o, is again found by solving Eq. (11.153)

$$\omega_o L_1 + \omega_o L_2 - \frac{1}{\omega_o C} = 0 \qquad (11.158)$$

for ω_o. It is important that the mutual inductance, $M_{1\text{-}2}$, between L_1 and L_2, be zero for Eq. (11.158) to hold.

EXAMPLE 11.11

Design a 1-MHz Colpitts oscillator using a 10-μH inductor.

SOLUTION Substituting into Eq. (11.157),

$$\frac{1}{2\pi(1\text{ MHz})C_1} + \frac{1}{2\pi(1\text{ MHz})C_2} = 2\pi(1\text{ MHz})(10\ \mu\text{H}).$$

Then

$$\left(\frac{1}{C_1} + \frac{1}{C_2}\right) = (2\pi \times 10^6)^2 (10\ \mu\text{H}) = 3.95 \times 10^8.$$

Select $C_1 = 0.01\ \mu\text{F}$; then $C_2 = 0.0034\ \mu\text{F}$. A standard value $C_2 = 0.0033\ \mu\text{F}$ would be suggested. Then, from Eq. (11.156),

$$|a_v| = \frac{X_2}{X_1} = \frac{C_1}{C_2} = \frac{0.01\ \mu\text{F}}{0.0033\ \mu\text{F}} = 3.$$

The circuit in Fig. 11.34 is a case of current-series negative feedback. Assuming $\beta \gg 1$, we can use Eq. (11.65) to obtain

$$a = \frac{V_o}{V_i} = -a_v = \frac{-R_C}{R_E} = -3.$$

The design of a CE amplifier with a gain of 3 would proceed as noted in earlier chapters, with some V_{CC} and a specific transistor having been selected.

Crystal Oscillators

In general, the *LC* oscillator suffers from frequency instability due to temperature effects and losses within the supposedly ideal *LC* elements. A less obvious difficulty is that to obtain relatively low-frequency operation, down to hundreds of kilohertz, requires physically large reactive elements. A good inductor is especially difficult to obtain. In addition, inductors have external fields whereas crystals do not. It is also difficult to obtain stable frequencies greater than tens of megahertz because of the nonideality of *L* and *C* elements. What is required is a small, physically stable, low-loss equivalent inductive reactance. This is achieved by using the **piezoelectric effect** that occurs in certain crystal structures. If a piezoelectric material is mechanically deformed, it will generate a small voltage. Conversely, application of an electric field across the crystal will produce a mechanical deformation. If a sinusoidal voltage is impressed across the crystal, it will mechanically deform in synchronism with the voltage. By selecting the material and arranging the dimensions and crystal orientation, the material will exhibit a mechanical resonance at the frequency of interest. What we have, then, is a mechanical analog of an *RLC* circuit, which we will call a **crystal**. A more advanced treatment of this topic would call this a **bulk acoustic wave resonance**. Typically, synthetic quartz is used for crystals. Other piezoelectric materials such as lithium niobate ($LiNbO_3$), zinc oxide (ZnO), and aluminum nitride (AlN), are used in specialized electronic circuits and systems.

A quartz crystal consists of a precisely machined, round or rectangular slab of quartz sandwiched between two metal plates under pressure by a spring. The entire assembly is contained in a hermetically sealed metal can. Typically, the quartz is several mils to several tens of mils thick, and on the order of 1 or 2 cm (or smaller) on a side. This sandwich structure is schematically represented by the circuit symbols shown in Fig. 11.35(a).

The crystal, as a resonant mechanical structure, is modeled as a series *RLC* circuit. It is beyond the scope of this text to derive the *RLC* element values from the underlying physics. Including the two metal plates on either side of the quartz slab, the assembly is modeled as a capacitor, C_2, in parallel with the series *RLC* circuit; see Fig. 11.35(b). Typically, the equivalent series *R* is neglected, since the quartz crystal is relatively lossless, and $C_2 \gg C_1$. The simplified equivalent circuit is shown in Fig. 11.35(c).

The quartz crystal is used in place of one of the three impedances in the general *LC* oscillator shown in Fig. 11.33. The terminal impedance is defined by

$$Z = \left(j\omega L + \frac{1}{j\omega C_1}\right) \bigg\| \frac{1}{j\omega C_2} = \frac{\left(j\omega L + \frac{1}{j\omega C_1}\right)\left(\frac{1}{j\omega C_2}\right)}{j\omega L + \frac{1}{j\omega C_1} + \frac{1}{j\omega C_2}}. \quad (11.159)$$

After algebraic manipulation, Eq. (11.159) becomes

$$Z = \frac{-j}{\omega C_2}\left(\frac{\omega^2 - \omega_s^2}{\omega^2 - \omega_p^2}\right), \quad (11.160)$$

DRILL 11.12

What parasitic circuit elements will affect the results obtained in Example 11.11?

ANSWERS A real 10-μH inductor has series resistance as well as some capacitance between the individual turns. The internal active device capacitances, especially C_μ, should be considered. Capacitors C_1 and C_2 also have some shunt resistance, but in general capacitors can be manufactured to better approximate an ideal circuit element than inductors. Even at frequencies as low as 1 MHz, care should be taken to minimize inherent coupling between components and interconnects.

■ There are a few suggested readings for those interested in AlN devices used in communications systems and sensor applications. An evolving field!

■ We won't have the opportunity to discuss *surface-acoustic wave* (*SAW*) devices in this text. They are widely used for filtering and oscillator frequency control, especially in communications systems.

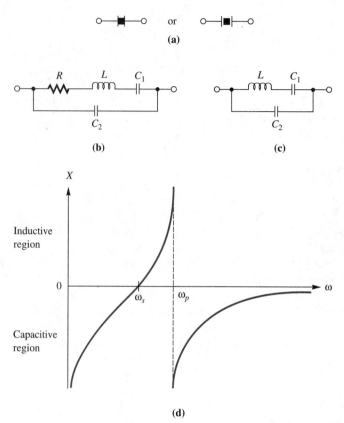

■ **FIGURE 11.35** Bulk acoustic wave (crystal) circuit diagram and equivalent circuit. **(a)** Circuit diagram symbol. **(b)** Equivalent circuit. **(c)** Simplified equivalent circuit—lossless. **(d)** Terminal reactance of the crystal.

where the crystal series resonance is defined by

$$\omega_s = \frac{1}{\sqrt{LC_1}}, \tag{11.161}$$

and the crystal parallel resonance is defined by

$$\omega_p = \frac{1}{\sqrt{L\left(\dfrac{C_1 C_2}{C_1 + C_2}\right)}}. \tag{11.162}$$

This crystal terminal impedance is graphed in Fig. 11.35(d). Observe that since $C_2 \gg C_1$, $\omega_p > \omega_s$, but by a very small amount, typically less than 0.01%. For $\omega < \omega_s$ and $\omega > \omega_p$, the terminal impedance is capacitive. For frequencies between ω_s and ω_p, the crystal looks inductive, and it is in this region where the crystal is used to synthesize a large value of low-loss inductance.

Three crystal-controlled oscillator circuits are presented in Fig. 11.36. Two circuits use a FET common-source amplifier for the gain stage. The relationship between Z_1, Z_2, and Z_3 is established by Eq. (11.154). In the circuit shown in Fig. 11.36(a), $|X_3|$ is the capacitive reactance of the parasitic C_{dg} of the FET. To use the crystal, represented by X_1, in its inductive region, the drain circuit reactance, X_2, must be tuned to be inductive.

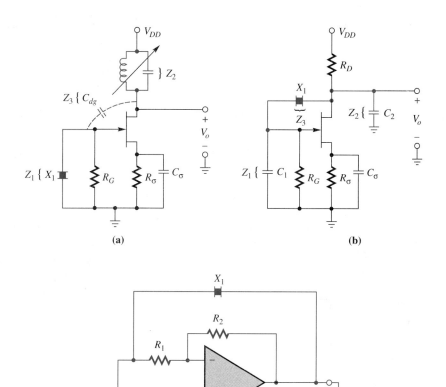

FIGURE 11.36 Three types of crystal oscillators. (a) Using C_{dg} in series resonance. (b) Pierce oscillator. (c) Operational amplifier for gain block.

The circuit shown in Figs. 11.36(b) and (c), widely used because of its simplicity, is called a *Pierce oscillator*. In the FET Pierce oscillator, the parallel combination of the crystal and C_{dg} will operate as a net inductive reactance in resonance with the capacitive reactances established by C_1 and C_2. In this case, from Eq. (11.155),

$$a = -a_v = -g_m R_D = -\frac{X_2}{X_1} = -\frac{C_1}{C_2} \quad (11.163)$$

must be satisfied, along with $X_1 + X_2 + X_3 = 0$.

The Pierce oscillator using an operational amplifier [Fig. 11.36(c)] requires the crystal to operate in its inductive region with resonance established by C_1 and C_2. In addition, from Eq. (11.156),

$$a = -a_v = -\frac{R_2}{R_1} = -\frac{X_2}{X_1} = -\frac{C_1}{C_2} \quad (11.164)$$

and $X_1 + X_2 + X_3 = 0$ must also be satisfied.

DRILL 11.13
Analysis and computations for a crystal-controlled oscillator have assumed a lossless crystal as shown in Fig. 11.35(c). Qualitatively describe what would have to happen to the results shown in Eqs. (11.163) and (1.64) if we used the crystal model in Fig. 11.35(a).

ANSWERS The series resistance of the crystal is nonzero at series resonance. The loop gain would have to increase to overcome the loss implicit in including the R of the crystal. Additional study of oscillators, which is beyond the scope of this text, would also show a lower Q (quality factor) and a concomitant broadening of the output spectrum.

■ This feedback amplifier solution algorithm is worth practicing and remembering.

CHECK UP

1. Write the Barkhausen criteria.
2. **TRUE OR FALSE?** If the output signal and input signal are in phase, the circuit could oscillate.
3. **TRUE OR FALSE?** If the output signal and input signal are in phase, the circuit will oscillate.
4. **TRUE OR FALSE?** Oscillators are linear systems.
5. Which oscillator circuits inherently operate better at audio frequencies? RF frequencies?
6. Crystal oscillators are designed to operate with the crystal representing a capacitive reactance.
7. **TRUE OR FALSE?** Crystal oscillators offer inherently better frequency stability than LC oscillators.

SUMMARY

The use of negative feedback in an individual circuit, around an operational amplifier, or in a complex electronic system is one of the most important techniques for stabilizing the performance against the inherent variations in active device parameters. We first observed that negative feedback was used to stabilize gain, reduce distortion, and increase bandwidth in Chapter 6, and then used it to modify input and output impedances in this chapter.

By treating the feedback system as the interconnection of two two-port networks, we obtained four fundamental interconnections. They were denoted as voltage-shunt (transresistance), current-series (transconductance), current-shunt (current), and voltage-series (voltage). Table 11.1 summarizes the overall two-port results.

After completing general gain and impedance derivations using basic two-port circuit theory, single and multistage circuits incorporating negative feedback were analyzed. To minimize algebraic complexity, model and circuit simplifications were used. We observed that, in the limit, the gain is dependent only on the feedback network, which is often composed only of a simple set of passive components, usually resistors.

In general, to solve for the key features of a feedback amplifier:

1. Identify the type of feedback being used.
2. Draw the small-signal model with the feedback network separately identified.
3. Compute the appropriate two-port y, z, h, or g parameters for the feedback network.
4. Include the feedback network two-port parameters in the small-signal model as input and output circuit loading and as a feedback generator. Do not include the feedforward generator term.
5. Solve for the appropriate a of the resultant small-signal model. Use all suitable approximations. Include the feedback network loading and set the feedback generator term to zero. Observe that the forward network two-port parameters are not required for obtaining a.
6. Substitute the result for a and f into $A = a/(1 + af)$.
7. Use Thévenin and Norton transformations to solve for voltage gain if required.

One of the most important applications of negative feedback is its use with an operational amplifier. This application as a voltage-shunt two-port configuration was presented. This

TABLE 11.1 Two-Port Feedback Summary

Voltage	Shunt
Sample output voltage	Sum currents in parallel at input
Parallel-input	Parallel-output
Shunt (input)	Shunt (output)
Input impedance decreases	Output impedance decreases
$A = V/I$ (Transresistance Amplifier) Use y Parameters	

Current	Series
Sample output current	Sum voltages in series at input
Series-input	Series-output
Series (input)	Series (output)
Input impedance increases	Output impedance increases
$A = I/V$ (Transconductance Amplifier) Use z Parameters	

Voltage	Series
Sample output voltage	Sum voltages in series at input
Series-input	Parallel-output
Series (input)	Shunt (output)
Input impedance increases	Output impedance decreases
$A = V/V$ (Voltage Amplifier) Use h Parameters	

Current	Shunt
Sample output current	Sum currents in parallel at input
Parallel-input	Series-output
Shunt (input)	Series (output)
Input impedance decreases	Output impedance increases
$A = I/I$ (Current Amplifier) Use g Parameters	

reinforced our work in Chapters 1 and 6 where we used the ideal operational amplifier characteristics along with the summing-point constraint analysis technique to introduce key configurations such as the inverting amplifier, noninverting amplifier, buffer, and active low-pass filter.

Because any real circuit has a frequency-dependent gain, we inadvertently introduce additional phase shift into the circuit, which could result in a positive feedback condition. This phase shift could lead to instability, which means the circuit would function as an oscillator rather than an amplifier. These effects of frequency response were presented along with a methodology for providing compensation, thereby stabilizing the circuit.

Oscillatory behavior is a desirable function in its own right. By adding additional predictable phase shift to a circuit, we obtain positive feedback and oscillator operation. Several configurations including the RC phase shift, Wien bridge, general LC, Colpitts, Hartley, and crystal-controlled oscillator configurations were presented. Table 11.2 summarizes the key properties of general sine-wave generation techniques.

Although the general feedback concept is straightforward, its application to real circuits and systems is algebraically complex. Indeed for oscillators, the problem is nonlinear. However, the two-port approach presented in this chapter offers an algebraically tractable technique for working with these circuits if you can identify the feedback network. As demonstrated, SPICE is a useful tool in analyzing and designing these complex circuits.

TABLE 11.2 Sine-Wave Oscillator Overview

Type	≈Frequency	General Comments
RC phase shift	10 Hz to low MHz	Simple. Resistivity tunable over a >2:1 range. Moderate stability, temperature sensitivity, and accuracy.
Wien bridge	<1 Hz to low MHz	Low distortion. Requires two tracking resistors for tuning. Good for precision audio and instrumentation applications.
LC	>100 kHz to low tens of MHz	Used for noncritical RF communications applications. Requires care to obtain good temperature sensitivity, accuracy, and low drift. Difficult to tune over large ranges.
Crystal	~100 kHz to 100 MHz+	Highest frequency stability and accuracy. Only tunable over ~0.001% at best. Temperature performance can be excellent. Somewhat fragile mechanically. Best choice for clocks and channelized communications systems.
Digital techniques	<1 Hz to few MHz	Digital generation of a waveform using digital-to-analog (DAC) conversion with some wave shaping (Chapter 16). Read-only-memory (ROM) look-up tables used. Widely used in signal processing and digital filtering. Usually with MOS technology.

■ Of course, we assume the active device which probably could be an operational amplifier has sufficient bandwidth. A μA741 would be useful to only a few kHz depending upon the overall design.

SURVEY QUESTIONS

1. What is the equation for closed-loop gain?
2. Draw a block diagram of a voltage-shunt feedback system. Label all voltages and currents and explain what happens to the gain and input and output resistances.
3. What two-port parameters are used to model each of the four feedback network systems?
4. What are the properties of unilateral a and f networks?
5. Why do we use the unilateral network approximations to analyze a voltage-shunt feedback circuit?
6. What happens to the gain, input resistance, and output resistance of a single-stage CE amplifier and multistage amplifiers with voltage-shunt feedback?
7. How would you model an operational amplifier system as a voltage-shunt configuration?
8. Draw a block diagram of a current-series feedback system. Label all voltages and currents and explain what happens to the gain and input and output resistances of a system incorporating current-series feedback.
9. Why do we use the unilateral network approximations to analyze a current-series feedback circuit?
10. What happens to the gain, input resistance, and output resistance of a single-stage CE amplifier and a multistage amplifier with current-series feedback?
11. Draw a block diagram of a voltage-series feedback system. Label all voltages and currents and explain what happens to the gain and input and output resistances of a system incorporating voltage-series feedback.

12. Why do we use the unilateral network approximations to analyze a voltage-series feedback circuit?
13. What happens to the gain, input resistance, and output resistance of a single-stage CE amplifier and multistage amplifiers with voltage-series feedback?
14. Draw a block diagram of a current-shunt feedback system. Label all voltages and currents and explain what happens to the gain, input resistance, and output resistance of a system incorporating current-shunt feedback.
15. Why do we use the unilateral network approximations to analyze a current-shunt feedback circuit?
16. What happens to the gain, input resistance, and output resistance of a single-stage CE amplifier and multistage amplifiers with current-shunt feedback?
17. What is the definition of positive feedback using $T(j\omega)$?
18. Define gain margin and phase margin.
19. What is meant by the term *dominant frequency* in a compensated amplifier?
20. What is the definition of an oscillator?
21. What are the Barkhausen criteria?
22. Draw schematic diagrams of *RC* phase-shift and Wien oscillators using discrete devices or operational amplifiers.
23. Demonstrate the design of a single-stage low-pass filter used to filter an oscillator output signal.
24. What are the design criteria for a general *LC* oscillator?
25. Draw schematic diagrams of Hartley and Colpitts oscillators using discrete devices and operational amplifiers.
26. What is meant by the term *piezoelectric effect*?
27. Draw the model and impedance characteristic for a crystal.
28. Draw schematic diagrams of crystal-controlled oscillators including the Pierce oscillator using discrete devices and operational amplifiers and explain their operation.

PROBLEMS

Problems marked with an asterisk are more challenging.

11.1 An amplifier without feedback has a midband *voltage gain*, $a = 1000$ V/V, a single high-frequency corner $f_H = 100$ kHz, a single low-frequency corner $f_L = 1$ kHz, an input impedance $z_i = 10$ kΩ, and an output impedance $z_o = 50$ Ω. Feedback of $f = 0.05$ V/V is now added. Determine the values of these properties with feedback.

11.2 Repeat Problem 11.1 if the midband *transresistance gain* $a = 1000$ Ω and $f = 0.05$ S.

11.3 Fill in the table assuming $a = 5000$, $f = 0.01$, $z_i = 20$ kΩ, $z_o = 100$ Ω, $f_H = 200$ kHz, and $f_L = 1$ kHz.

System Properties	Type of Amplifier Configuration			
	Current	Voltage	Transconductance	Transresistance
Closed-loop gain with units				
Input impedance, Z_i				
Output impedance, Z_o				
High-frequency corner				
Low-frequency corner				
Units of forward gain block, "a"				
Units of feedback ratio, "f"				
Output in series or parallel?				
Input in series or parallel?				

11.4 The following require one-sentence short answers:

(a) Which two of the four basic feedback configurations would best serve as a match for a low-impedance load?
(b) Which two of the four basic feedback configurations would best serve as a match to a high-impedance source?
(c) Feedback is a good approach to reduce crossover distortion. True or false? Explain.
(d) Recall from basic BJT characteristics that β is not uniform but a function of I_C. Feedback is a good approach to minimize distortion from this mechanism. True or false? Explain.
(e) Briefly explain why it is desirable for the feedback ratio, f, not to include active devices.

11.5 Compute the y, z, h, and g parameters for the resistive circuits shown in Fig. 11.37.

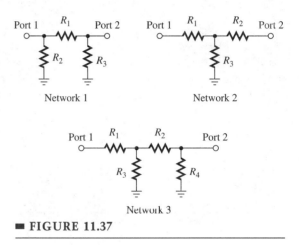

■ **FIGURE 11.37**

11.6 Compute the y, z, h, and g parameters for the RC circuits shown in Fig. 11.38.

11.7 The single-stage CE amplifier with voltage-shunt feedback shown in Fig. 11.5 has $R_C = R_L = 10$ kΩ, $V_{CC} = 12$ V, $R_F = 300$ kΩ, and $R_S = 1$ kΩ. The transistor parameters are $\beta = 200$, $V_A \to \infty$, and C_1 and C_2 are large. Use feedback principles to compute A_{vs}, R_i, and R_o.

11.8 Use SPICE to verify your dc and ac solutions to Problem 11.7. Observe that you will have to select values for C_1 and C_2 such that their capacitive reactances are low at the frequency you use for your simulation.

11.9 The CE amplifier with emitter degeneration is biased as shown in Fig. 11.39 by operating between two power supplies. Let $V_{CC} = 25$ V, $V_{EE} = 5$ V, $R_S = 600$ Ω, $R_C = 1$ kΩ, and $R_E = 200$ Ω. The transistor parameters are $\beta = 200$ and $V_A \to \infty$. Use feedback principles to compute A_{vs}, R_i, and R_o.

11.10 Use SPICE to verify your dc and ac solutions to Problem 11.9.

11.11 The direct-coupled two-stage amplifier shown in Fig. 11.40 uses a form of voltage-shunt feedback.

(a) Derive an expression for A_{vs} for this circuit.
(b) If $R_S = 1$ kΩ, $V_{CC} = 12$ V, $R_{C1} = R_{C2} = 5$ kΩ, $R_E = 2$ kΩ, $R_F = 4$ kΩ, $\beta_1 = \beta_2 = 200$, and $V_{A1} = V_{A2} \to \infty$, compute A_{vs}, R_i, and R_o.

CHAPTER 11 FEEDBACK

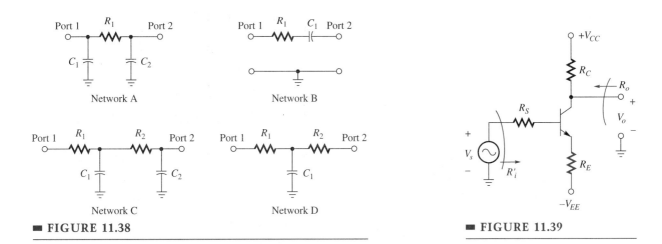

■ FIGURE 11.38

■ FIGURE 11.39

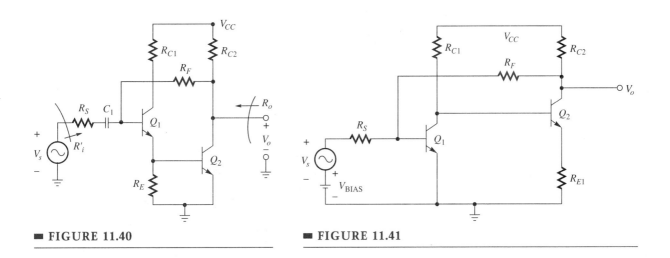

■ FIGURE 11.40

■ FIGURE 11.41

11.12 Use SPICE to verify your dc and ac solutions to Problem 11.11. Observe that you will have to select a value for C_1 such that its capacitive reactance is low at the frequency you use for your simulation.

11.13 Why doesn't the design shown in Fig. 11.41 for a two-stage amplifier with voltage-shunt feedback work as intended? You can assume that all biasing is still correct and the circuit worked as intended before the inclusion of R_F.

11.14 The ac model for the cascade amplifier shown in Fig. 11.42 has $a = -10^8 \, \Omega$. All biasing circuitry has been neglected. Assume $I_{C1} = I_{C2} = I_{C3} = 1$ mA, $R_S = 500$, and $R_{L3} = 5$ kΩ. Use transistor specifications of Table 18.2.

(a) Design a feedback amplifier whose *approximate* $A_{vs} = -50$; that is, specify a value for R_F.

(b) Compute a value for R_o and R_i.

11.15* Using the small-signal model in Fig. 11.42 and your results from Problem 11.14, design the rest of the circuit including biasing. Use BJT specifications as given in Table 18.2. Use SPICE to verify your design and obtain A_{vs}.

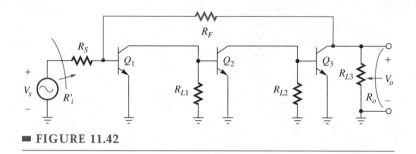

■ FIGURE 11.42

11.16 Again refer to Fig. 11.42 and your solution to Problem 11.14. Include a capacitor, C_F, in parallel with R_F to obtain a 3-dB point at 100 kHz. Sketch a Bode magnitude plot of $A_{vs}(jf)$.

11.17 *Estimate* a value for $A = I_o/I_s$ for the circuit shown in Fig. 11.43 if $R_{C2} = 5$ kΩ, $R_{E2} = 200$ Ω, and $R_F = 3$ kΩ. Compute a value for A_{vs} using appropriate source and load transformations. Let $R_S = 1$ kΩ.

11.18* Using Fig. 11.43 and your results from Problem 11.17, design the rest of the circuit including biasing. Use BJT specifications as given in Table 18.2.

11.19 Refer to your results in Problem 11.17. Suppose the R_F resistor is connected from the base of Q_1 to the collector of Q_1 instead of the emitter of Q_2 in Fig. 11.43. Now estimate A_{vs} for this circuit.

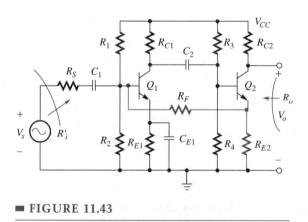

■ FIGURE 11.43

11.20 An equivalent circuit for an NE/SE592 video amplifier IC is given in Fig. 11.44. Answer the following questions. Observe that no numerical calculations are required.

(a) What type of feedback do the 7-kΩ resistors provide? How does this feedback modify the output impedance of the amplifier?
(b) Connect G_{2A} and G_{2B}. What type of feedback do the 50-Ω resistors provide?
(c) Clearly you can change the voltage gain of this IC by appropriate connections of G_{1A}, G_{2A}, G_{1B}, and G_{2B}. Describe how you would connect these terminals to achieve the following IC specifications.
   1. Maximum bandwidth.
   2. Maximum gain.
   3. Highest differential input impedance.
(d) Is feedback used anywhere else in this circuit? If so, explain qualitatively how it is used and what it achieves in the circuit performance.

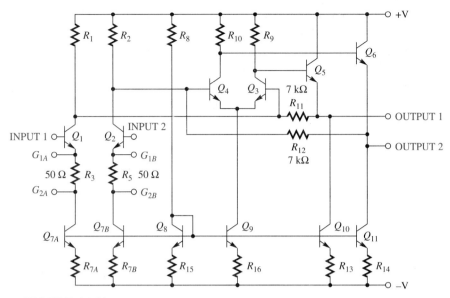

■ FIGURE 11.44

11.21 Refer to Fig. 11.43. Estimate a value for A_{vs} if $R_{E2} = 100\ \Omega$, $R_{C2} = 1\ k\Omega$, $R_S = 500\ \Omega$, and $R_F = 2\ k\Omega$. State your assumptions relating to the rest of the circuit.

11.22 Refer to Fig. 11.20. Estimate a value for A_{vs} if $R_{E1} = 1\ k\Omega$ and $R_F = 10\ k\Omega$. State your assumptions relating to the rest of the circuit.

11.23 Again refer to Fig. 11.20. Connect R_F from the emitter of Q_1 to the collector of Q_2. Estimate a value for A_{vs}. How did this change the results for R_i and R_o?

11.24 For the circuit shown in Fig. 11.16(a), assume $R_C = 10 R_E$ for each stage. Let $R_F \to \infty$, and using single stage feedback approximations, find the open loop gain.

11.25 For the circuit shown in Fig. 11.16(a), let $R_F = 5R_E$ and find $A_{vs} \approx -R_{C3}/f$. Is this a good approximation?

11.26 Refer to Fig. 11.43. Let $R_S = 1\ k\Omega$, $R_{C2} = 5\ k\Omega$, $R_{E2} = 100\ \Omega$, and $R_F = 1\ k\Omega$. Assume the rest of the circuit element values and operation permit the use of feedback analysis approximations. Estimate $A = I_o/I_s$ and $A_{vs} = V_o/V_s$.

11.27 Using feedback principles, derive an expression for A_{vs}, R_i, and R_o for the circuit shown in Fig. 11.45. Neglect capacitive reactances at the frequencies of interest.

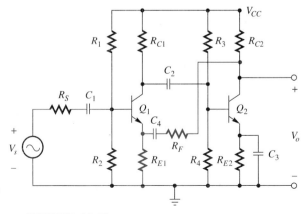

■ FIGURE 11.45

11.28 Refer to Fig. 11.45. Qualitatively describe what will happen if C_3 is removed and the R_F connection is changed from the collector of Q_2 to the emitter of Q_2.

11.29 Figure 11.46 is the single-pole, open-loop response, $a(jf)$ for a μA741 extracted from the data sheet in Chapter 6.

(a) What voltage feedback ratio, f, is required to obtain a 100-kHz bandwidth in the closed-loop response.

(b) What is the resultant closed-loop gain, $A(jf)$?

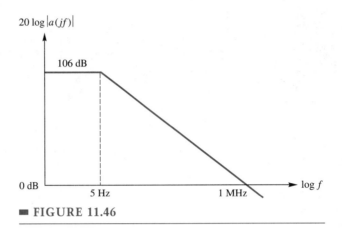

■ FIGURE 11.46

11.30 Design a three stage operational amplifier cascade to obtain a transfer function cascade

$$a(jf) = \frac{-500}{[1 + j(f/f_o)]^3}.$$

Assume ideal operational amplifiers. All resistor values should be between 2 and 100 kΩ. Each operational amplifier will require a feedback capacitor to achieve a low-pass filter 3-dB response at $f_o = 100$ kHz.

11.31* Verify your Problem 11.30 design using SPICE, with $R_{in} = 2$ MΩ, $a_o = 200{,}000$, and $R_o = 75$ Ω as a guide to a reasonable operational amplifier model.

11.32 An amplifier has a forward transfer function given by

$$a(jf) = \frac{-500}{[1 + j(f/f_o)]^3},$$

where $f_o = 100$ kHz. Add enough feedback so that the closed-loop low-frequency gain is $A_o = -20$. Compute the dominant corner frequency required to obtain stable amplifier operation with a PM of 30°.

11.33* Combine your results from Problems 11.30 and 11.32 and verify the design with compensation using SPICE. Your results should include Bode magnitude and phase plots for the closed-loop response.

11.34 An amplifier has a forward gain given by

$$a(j\omega) = \frac{-100}{\left(1 + j\dfrac{\omega}{4\pi \times 10^6}\right)\left(1 + j\dfrac{\omega}{8\pi \times 10^6}\right)}$$

and

$$f(j\omega) = \frac{-f_o}{\left(1 + j\dfrac{\omega}{12\pi \times 10^6}\right)}.$$

(a) At what frequency would this circuit exhibit instability?
(b) Specify the maximum value of f_o and the subsequent value of A_o if a phase margin of 30° is required.

11.35 An amplifier has a forward gain given by

$$a(j\omega) = \frac{-200}{[1 + j(f/f_{c1})][1 + j(f/f_{c2})]},$$

where $f_{c1} = 1$ MHz and $f_{c2} = 3$ MHz. The feedback factor is given by

$$f = \frac{-f_o}{[1 + j(f/f_{c3})]}, \text{ where } f_{c3} = 5 \text{ MHz}.$$

(a) Assuming $|f_o|$ is large enough, at what frequency would this circuit become unstable (oscillate)?
(b) These transfer functions are part of a current shunt feedback circuit. Assume $f_o = 0.016$, and the circuit is stable. Let the output resistance of the forward amplifier be $r_{oa} = 5\Omega$. Sketch $|Z_o|$ for $0 < f < 10$ MHz.

11.36 Refer to the a and f transfer functions given in Problem 11.35. Specify the largest value of f_o for a phase margin of 45°.

11.37 Consider the three-stage amplifier shown in Fig. 11.47. The operational amplifiers are μA741s, which have a GBW of 1 MHz.

(a) What is the minimum value of $|A_o| = V_o/V_s$ for a PM of 30°?
(b) Assume $|A_o| = 10$, with a PM of 30°, is required. Specify the smallest capacitor that can be used to compensate the circuit. The compensation capacitor is to be added to the feedback circuit of the second stage.

11.38* Use SPICE to list or plot open-loop gain and phase for the oscillator designed in Example 11.9 using $R = 2166 \, \Omega$.

(a) What is f_{osc} with an ideal operational amplifier?
(b) What happens if a μA741 is used instead?

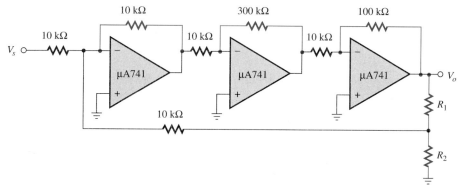

■ **FIGURE 11.47**

11.39 Derive Eq. (11.133) for the phase-shift network shown in Fig. 11.30.

11.40 Determine the frequency of operation and amplifier voltage gain required for the phase-shift network shown in Fig. 11.48. Assume an ideal voltage amplifier.

11.41 Design a 5-kHz RC phase-shift oscillator using an inverting μA741 operational amplifier. Select standard value components, and compute the actual frequency of operation with these components.

11.42 Repeat Problem 11.41, but use a Wien bridge oscillator circuit.

11.43 Design a Wien bridge oscillator to operate at a nominal frequency of 1.0 kHz. Your design should include:

(a) Standard component values appropriate for use with a μA741-type of operational amplifier
(b) All relevant design equations
(c) A single-stage, first-order active low-pass filter using an operational amplifier whose component values are selected to attenuate harmonics of the 1.0-kHz fundamental frequency.

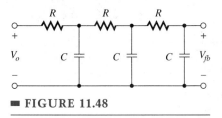

■ **FIGURE 11.48**

11.44 Design a 100-kHz Hartley oscillator using a 100-pF capacitor. By design, specify reasonable values for L_1 and L_2, as well as the R_C/R_E ratio required if the CE circuit shown in Fig. 11.34(a) is used as the amplifier.

11.45 Repeat Problem 11.44 but use an ideal operational amplifier. In this case you need to specify reasonable values for R_2 and R_1 (refer to Fig. 11.33).

11.46 Design a 100-kHz Colpitts oscillator using a 1-mH inductor. By design, specify reasonable values for C_1 and C_2, as well as the R_C/R_E ratio required if the CE circuit shown in Fig. 11.34(a) is used as the amplifier.

11.47 Repeat Problem 11.46 but use an ideal operational amplifier. In this case you need to specify reasonable values for R_2 and R_1 in the operational amplifier.

11.48 For the JFET in Fig. 11.49, let $VTO = -4$ V, $BETA = 2.5 \times 10^{-4}$ A/V^2, and $\lambda = 0$ V^{-1} for JFET parameters. All JFET internal capacitances can be ignored. Set $V_{DD} = 20$ V. Design a 1.0-MHz Colpitts oscillator using this JFET circuit and refer to Fig. 11.32. Let $C_1 = 20$ pF and $C_2 = 4$ pF. Your design should include

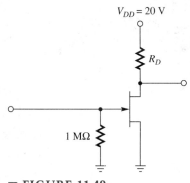

■ **FIGURE 11.49**

(1) appropriate additions to the diagram of Fig. 11.49,
(2) reasonably standard component values for all circuit components,
(3) demonstration that the Q-point is adequate and
(4) the Barkhausen criteria are being satisfied.

11.49* Verify your Problem 11.48 design with a SPICE simulation.

11.50 Show that Eq. (11.160) is derived from Eq. (11.159). Graph Eq. (11.160) for the range $0.1\omega_s < \omega < 1.1\,\omega_s$. Let $\omega_p = 1.01\omega_s$.

11.51 Assume that the inductance, L, used in your Colpitts design in Problem 11.48 is replaced by a lossless crystal with $f_s = 0.995$ MHz and $f_p = 1.005$ MHz and the rest of the design remains the same with f_o still at 1.0 MHz. Using Fig. 11.35(c), compute a value for the crystal's parallel capacitance, C_2.

11.52 Refer to Figs. 11.35 and 11.36(a). Assume a quartz crystal with equivalent circuit parameters of $L = 0.33$ H, $C_1 = 0.065$ pF, and $C_2 = 1.0$ pF. Let the FET $C_{dg} = 2.0$ pF. Compute f_p and f_s. What is the inductive frequency range? Over what range should Z_2 be tunable for oscillator operation?

11.53 Using the crystal and FET specifications of Problem 11.52, specify reasonable values for C_1, C_2, and gain for the Pierce oscillator shown in Fig. 11.36(b). Let $f_o = 1.09$ MHz.

11.54 A Pierce crystal oscillator diagram is shown in Fig. 11.50. For the lossless crystal, the series resonant frequency is $f_s = 4.998$ MHz. The parallel resonant frequency is $f_p = 5.002$ MHz. $C_2 = 0.5$ pF for the crystal.

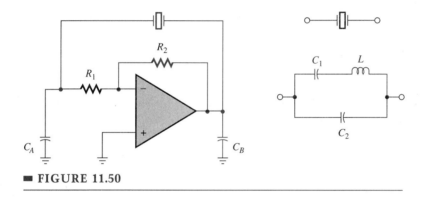

■ **FIGURE 11.50**

(a) Compute values for the crystal equivalent circuit elements C_1 and L.
(b) Consider the design of a 5.000-MHz oscillator. Compute the effective inductive reactance of the crystal at 5.000 MHz.
(c) Assume that you wanted to replace the crystal with a discrete inductor. Using your answer from part (b), compute the value of that inductor.
(d) What is the relationship that must exist between R_1, R_2, C_A, and C_B? Assume an ideal operational amplifier.
(e) Suppose you want to use a μA741 for the operational amplifier. Would the circuit work? Explain your answer.

SUGGESTED READINGS

Burr-Brown Corporation. *Burr-Brown Product Data Book*, Tucson, Arizona. 1994. A well-documented data and applications manual.

Frerking, M. E. *Crystal Oscillator Design and Temperature Compensation*. New York: Van Nostrand Reinhold, 1978. Considered a classic reference on crystal-controlled oscillators.

Gray, P. R., and R. G. Meyer. *Analysis and Design of Analog Integrated Circuits*, 3rd ed. New York: John Wiley & Sons, 1993. A good presentation of negative feedback and feedback stability.

Hakim, S. S. *Feedback Circuit Analysis*. Andover, Hants, England: Chapel River Press, Ltd., 1966. One of the earlier reference texts that form the basis for current feedback analysis techniques.

Parzen, P. *Design of Crystal and Other Harmonic Oscillators.* New York: John Wiley & Sons, 1983.

These three listings, although specialized in scope, illustrate some uses of AlN piezoelectric devices at 1 GHz.

Weber, R. J., S. G. Burns, C. F. Campbell, and R. O'Toole*. "Applications of AlN Thin-Film Resonator Topologies As Antennas and Sensors," Proc. *IEEE MTT Int. Symp.*, June 1–5, 1992, pp. 161–164.

O'Toole, R., S. G., Burns, G. Bastiaans, and M. D. Porter. "Thin Film Aluminum Nitride Resonators: Miniaturized High Sensitivity Mass Sensors," *Journal of Analytical Chemistry*, June 1992.

Burns, S. G. and P. H. Thompson. "Design, Analysis, and Performance of UHF Oscillators Using Thin-Film Resonator-Based Devices as the Feedback Element," *Proc. of 32nd Midwest Symp. on Circuits and Systems*, August 1989, p. 1005.

In addition to texts covering feedback theory and applications in electronics, there are a number of texts and papers in the control systems area that offer in-depth feedback analytical techniques. For example, we have not covered the use of Nyquist concepts in feedback analysis. Although Nyquist criteria and techniques are widely used for analyzing feedback systems and stability, most of the applications are focused in the controls area.

CHAPTER 12

OPERATIONAL AMPLIFIER EXAMPLES

12.1 Design of an Operational Amplifier Using a CA3096 *npn/pnp* Transistor Array
12.2 Analysis of a μA741 Operational Amplifier
12.3 Analysis of a CA3140 Operational Amplifier
12.4 The LM111 Comparator
12.5 Trends in Operational Amplifier Performance
Summary
Survey Questions
Problems
References
Suggested Readings

Using the fundamental circuits derived in Chapter 10, we are now in a position to design and analyze the internal circuitry of complete operational amplifiers and related analog integrated circuits (ICs). To avoid getting lost algebraically in multiple-transistor circuit analysis, we will make a deliberate effort to look for simplifications, such as neglecting base currents, and then follow a systematic approach. The approximate approach will be supported by a SPICE computer simulation. This approximate approach* is divided into three segments:

1. *Qualitative analysis:* The idea here is to identify key circuit functions such as amplification, voltage-level shifting, current sourcing, and active loading associated with a given transistor.
2. *DC circuit analysis:* The goal is to compute approximate values for all important quiescent currents and voltages.
3. *AC circuit analysis:* Using results from the dc analysis, the small-signal voltage gain, input resistance, and output resistance are computed. Phasor notation is used.

To demonstrate these procedures, we design a simple operational amplifier using a CA3096 *npn/pnp* transistor array. The design is verified using a SPICE simulation. We then apply this analysis approach to several analog circuits. Although the BJT-based μA741 general-purpose operational amplifier, the CA3140 BiMOS operational amplifier, and the LM111/211/311 BJT-based comparator have been superceded by ICs with much higher performance, it is useful pedagogically to use these circuits for an initial exposure to the basics of the internal circuitry of analog integrated circuits. Another motivation was that circuit diagrams and supporting material are readily available.

---

* The analysis approach is independent of the technology or type of circuit.

We compare our approximate and SPICE analyses of a μA741 general-purpose operational amplifier with specifications given by the manufacturer's data sheet first introduced in Chapter 6. In Section 12.3 we present an analysis of a CA3140 MOSFET-input, μA741 pin-compatible operational amplifier. Closely related to the IC amplifier is the comparator, so we devote Section 12.4 to the analysis of the LM111/211/311 family of general-purpose comparators. Abbreviated versions of the data sheets for all of these ICs are included. Section 12.5 includes an overview of trends in operational amplifier performance and how this performance compares to the μA741 and CA3140.

References are also given to a variety of manufacturers of analog ICs. Many of these manufacturers not only provide computer-based data books but complete SPICE models for use in incorporating their products' system design. These data are also often available on diskette, on CD-ROM, or through home pages on the Internet's World Wide Web node.

IMPORTANT CONCEPTS IN THIS CHAPTER

- Design of a five-transistor operational amplifier, including biasing and ac operation
- Frequency-independent and frequency-dependent SPICE modeling
- μA741 dc and ac analysis (bipolar)
- μA741 frequency response analysis, including gain and phase margin
- CA3140 dc and ac analysis (BiMOS)
- Characteristics of comparators and LM111 circuit description (bipolar)
- Definition of a precision operational amplifier and a wideband or video operational amplifier
- Trends in operational amplifier performance

12.1 DESIGN OF AN OPERATIONAL AMPLIFIER USING A CA3096 *npn/pnp* TRANSISTOR ARRAY

In general, an operational amplifier can be divided into three sections (see Fig. 12.1). The input section is used to obtain a differential input with high input resistance, appropriate level shifting for dc coupling to following circuitry, and sometimes differential-to-single-ended signal conversion, all while providing a significant portion of the overall voltage gain.

The intermediate section is designed to provide the rest of the voltage gain while not significantly loading the input circuitry and to provide the appropriate direct-coupled dc level to the output circuitry. Frequency compensation, as discussed in Chapter 11, is usually implemented in this intermediate section.

The output section usually consists of a class B or AB amplifier with unity voltage gain of sufficient power capability to drive relatively low resistance loads. This means the output resistance of this stage should be low.

CA3096 Functional Description

We will design an operational amplifier using a CA3096 as a prelude to analyzing more complex circuits available commercially. The CA3096,[1] whose data sheet is given in Fig. 12.2, is an *npn/pnp* transistor array consisting of three independent, but matched, *npn* transistors and two independent, but matched, *pnp* transistors. This is an interesting, versatile IC in that all emitters, bases, and collectors are separately accessible. The 15

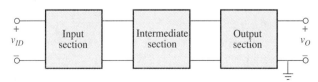

■ **FIGURE 12.1** General operational amplifier functional circuit division.

16 Pin Configuration

Q	E	B	C
1	2	1	3
2	4	5	6
3	7	8	9
4	10	11	12
5	13	14	15

Pin 16 to the most negative voltage in the circuit.

STATIC ELECTRICAL CHARACTERISTICS at $T_A = 25°C$ (For Equipment Design)

Characteristics	Symbol	Test Conditions	CA3096AE, CA3096E Limits			Units
			Min.	Typ.	Max.	
For Each n-p-n Transistor						
Collector-Cutoff Current (CA3096AE)	I_{CBO}	$V_{CB} = 10V, I_E = 0$	–	0.0013	40	nA
Collector-Cutoff Current (CA3096AE)	I_{CEO}	$V_{CE} = 10V, I_B = 0$	–	0.0055	100	nA
Collector-Cutoff Current (CA3096E)	I_{CBO}	$V_{CB} = 10V, I_E = 0$	–	0.0013	100	nA
Collector-Cutoff Current (CA3096E)	I_{CEO}	$V_{CE} = 10V, I_B = 0$	–	0.0055	1	µA
Collector-to-Emitter Breakdown Voltage	$V_{(BR)CEO}$	$I_C = 1mA, I_B = 0$	35	50	–	V
Collector-to-Base Breakdown Voltage	$V_{(BR)CBO}$	$I_C = 10µA, I_E = 0$	45	100	–	V
Collector-to-Substrate Breakdown Voltage	$V_{(BR)CIO}$	$I_{CI} = 10µA, I_B = I_E = 0$	45	100	–	V
Emitter-to-Base Breakdown Voltage	$V_{(BR)EBO}$	$I_E = 10µA, I_C = 0$	6	8	–	V
Emitter-to-Base Zener Voltage	V_Z	$I_Z = 10µA$	6	7.9	9.8	V
Collector-to-Emitter Saturation Voltage (CA3096AE)	$V_{CE(SAT)}$	$I_C = 10mA, I_B = 1 mA$	–	0.24	0.5	V
Collector-to-Emitter Saturation Voltage (CA3096E)	$V_{CE(SAT)}$	$I_C = 10mA, I_B = 1 mA$	–	0.24	0.7	V
Base-to-Emitter Voltage	V_{BE}	$I_C = 1mA, V_{CE} = 5V$	0.6	0.69	0.78	V
DC Forward-Current Transfer Ratio	h_{FE}		150	390	500	
Magnitude of Temperature Coefficient: V_{BE} (for each transistor)	$|\Delta V_{BE}/\Delta T|$	$I_C = 1mA, V_{CE} = 5V$	–	-1.9	–	mV/°C
For Each p-n-p Transistor						
Collector-Cutoff Current (CA3096AE)	I_{CBO}	$V_{CB} = -10V, I_E = 0$	–	-0.055	40	nA
Collector-Cutoff Current (CA3096AE)	I_{CEO}	$V_{CE} = -10V, I_B = 0$	–	-0.12	100	nA
Collector-Cutoff Current (CA3096E)	I_{CEO}	$V_{CE} = -10V, I_B = 0$	–	-0.12	1	µA
Collector-Cutoff Current (CA3096E)	I_{CBO}	$V_{CB} = -10V, I_E = 0$	–	-0.055	100	nA
Collector-to-Emitter Breakdown Voltage	$V_{(BR)CEO}$	$I_C = -100 µA, I_B = 0$	-40	-75	–	V
Collector-to-Base Breakdown Voltage	$V_{(BR)CBO}$	$I_C = -10 µA, I_E = 0$	-40	-80	–	V
Emitter-to-Base Breakdown Voltage	$V_{(BR)EBO}$	$I_E = -10 µA, I_C = 0$	-40	-100	–	V
Emitter-to-Base Zener Voltage	V_Z	$I_Z = 10 µA$	10	16	–	V
Emitter-to-Substrate Breakdown Voltage	$V_{(BR)ECO}$	$I_{EI} = 10 µA, I_B = I_C = 0$	40	100	–	V
Collector-to-Emitter Saturation Voltage	$V_{CE(SAT)}$	$I_C = -1 mA, I_B = -100 µA$	–	-0.16	-0.4	V
Base-to-Emitter Voltage	V_{BE}	$I_C = -100 µA, V_{CE} = -5V$	-0.5	-0.6	-0.7	V
DC Forward-Current Transfer Ratio	h_{FE}	$I_C = -100 µA, V_{CE} = -5V$	40	85	200	
		$I_C = -1mA, V_{CE} = -5V$	20	47	150	
Magnitude of Temperature Coefficient: V_{BE} (for each transistor)	$|\Delta V_{BE}/\Delta T|$	$I_C = -100 µA, V_{CE} = -5V$	–	-2.2	–	mV/°C
For Transistors Q1 and Q2 (As a Differential Amplifier): CA3096AE ONLY						
Absolute Input Offset Voltage	$|v_{IO}|$		–	0.3	5	mV
Absolute Input Offset Current	$|I_{IO}|$	$V_{CE} = 5V, I_C = 1mA$	–	0.07	0.6	µA
Absolute Input Offset Voltage Temperature Coefficient	$\left|\frac{\Delta v_{IO}}{\Delta T}\right|$		–	1.1		µV/°C
For Transistors Q4 and Q5 (As a Differential Amplifier): CA3096AE ONLY						
Absolute Input Offset Voltage	$|v_{IO}|$	$V_{CE} = -5V$	–	0.15	5	mV
Absolute Input Offset Current	$|I_{IO}|$	$I_C = -100mA$	–	2	250	nA
Absolute Input Offset Voltage Temperature Coefficient	$\left|\frac{\Delta v_{IO}}{\Delta T}\right|$	$R_S = 0\Omega$	–	0.54	–	µV/°C

■ **FIGURE 12.2** CA3096E, CA3096AE *npn/pnp* transistor array IC data sheet (courtesy of GE/RCA Solid State).[1]

nodes and the substrate connection are available in a 16-pin *dual-in-line package* (DIP). Most of the transistor parameters are specified with a minimum, typical, and maximum value. For simplicity, we will generally use the typical values, although you are encouraged to study the effects on the overall performance of using the minimum and maximum values of these parameters.

DYNAMIC ELECTRICAL CHARACTERISTICS at $T_A = 25°C$
Typical Values Intended Only for Design Guidance

Characteristics	Symbol		Test Conditions	Typical Values	Units
For Each n-p-n Transistor					
Noise Figure (low frequency)	NF		$f = 1$ kHz, $V_{CE} = 5$ V, $I_C = 1$mA, $R_S = 1$kΩ	2.2	dB
Low-Frequency, Input Resistance	R_i		$f = 1.0$ kHz, $V_{CE} = 5$ V, $I_C = 1$mA	10	kΩ
Low-Frequency, Output Resistance	R_o			80	kΩ
Admittance Characteristics:					
Forward Transfer Admittance	Y_{fe}	g_{fe} b_{fe}	$f = 1$ MHz, $V_{CE} = 5$ V, $I_C = 1$mA	7.5 $-j13$	mmho
Input Admittance	Y_{ie}	g_{ie} b_{ie}		2.2 $j3.1$	mmho
Output Admittance	Y_{oe}	g_{oe} b_{oe}		0.76 $j2.4$	mmho
Gain-Bandwidth Product	f_T		$V_{CE} = 5$V, $I_C = 1.0$ mA	280	MHz
			$V_{CE} = 5$V, $I_C = 5$ mA	335	
Emitter-to-Base Capacitance	C_{EB}		$V_{EB} = 3$V	0.75	pF
Collector-to-Base Capacitance	C_{CB}		$V_{CB} = 3$V	0.46	pF
Collector-to-Substrate Capacitance	C_{CI}		$V_{CI} = 3$V	3.2	pF
For Each p-n-p Transistor					
Noise Figure (low frequency)	NF		$f = 1$ kHz, $I_C = 100$μA, $R_S = 1$kΩ	3	dB
Low-Frequency Input Resistance	R_i		$f = 1$ kHz, $V_{CE} = 5$V, $I_C = 100$μA	27	kΩ
Low-Frequency Output Resistance	R_o			680	kΩ
Gain-Bandwidth Product	f_T		$V_{CE} = 5$V, $I_C = 100$μA	6.8	MHz
Emitter-to-Base Capacitance	C_{EB}		$V_{EB} = -3$V	0.85	pF
Collector-to-Base Capacitance	C_{CB}		$V_{CB} = -3$V	2.25	pF
Base-to-Substrate Capacitance	C_{BI}		$V_{BI} = 3$V	3.05	pF

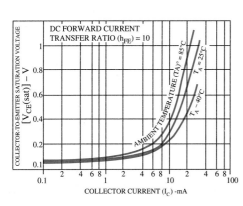

$V_{CE(SAT)}^{(n\text{-}p\text{-}n)}$ as a function of collector current.

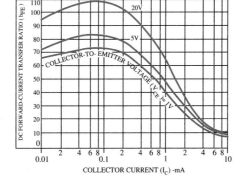

Transistor $(p\text{-}n\text{-}p)\, h_{FE}$ as a function of collector current.

■ **FIGURE 12.2** Continued

Five-Transistor Operational Amplifier: DC Analysis

All three sections of a typical operational amplifier can be implemented with the five transistors in a single CA3096, as illustrated in Fig. 12.3. A single-ended output of an emitter-coupled pair, Q_1 and Q_2, is used to drive Q_4, a *pnp* common-emitter amplifier. The remaining *npn/pnp* pair, Q_3 and Q_5, constitutes a class B amplifier driven from the collector of Q_4. A resistor is used to current-source Q_1 and Q_2. No active loads are used, and there are no crossover diodes. Since the typical h_{FE} values for the *npn* and *pnp* transistors are 390 and 85, respectively, all base currents will be neglected. It is also reasonable to assume that for $V_{ID} = 0$, $V_O = 0$.

The quiescent currents and voltages can be obtained assuming both bases are grounded. The emitter-coupled pair current source is established by

$$I_{EE} = \frac{V_{EE} - V_{BE1(on)}}{R_3} = I_{C1} + I_{C2} = 2I_{C2}. \qquad (12.1)$$

■ Observe that the bases of Q_1 and Q_2 are at zero volts. The emitters of Q_1 and Q_2 are at -0.7 volts.

Because $v_O = 0$ when $v_{ID} = 0$, V_{C4} must be zero, which means that

$$V_{R5} = -I_{C4}R_5 = V_{EE}. \qquad (12.2)$$

Biasing for Q_4 is established by arranging for V_{C2} to be sufficiently below V_{CC} that $V_{BE4} = -0.6$ V. To accomplish this,

$$V_{R2} = I_{C2}R_2 = -V_{BE4(on)} - I_{C4}R_4 \qquad (12.3)$$

or, after rearranging terms,

$$I_{C4} = \frac{-I_{C2}R_2 - V_{BE4(on)}}{R_4}. \qquad (12.4)$$

Because Q_1 and Q_2 are matched and $I_{B4} \ll I_{C2}$,

$$V_{C2} = V_{C1} = V_{CC} - I_{C2}R_2. \qquad (12.5)$$

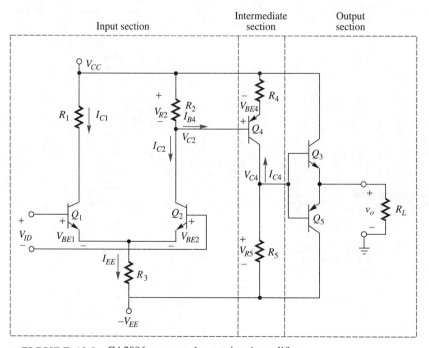

■ **FIGURE 12.3** CA3096 connected operational amplifier

Five-Transistor Operational Amplifier: AC Analysis

The small-signal model is given in Fig. 12.4, where use has been made of half-circuit techniques in the treatment of Q_1 and Q_2. Note that either Q_3 or Q_5 is active and their small-signal parameters vary over a wide range as instantaneous collector current varies. The indicated model and calculations for input resistances of Q_3 and Q_5 are merely meant to provide a lower limit loading at some large signal.

The overall voltage gain of this circuit using phasor notation is given by

$$A_v = \frac{V_o}{V_{id}} = A_{v1} \times A_{v2} \times A_{v3} = \frac{V_{o2}}{V_{id}} \times \frac{V_{o4}}{V_{o2}} \times \frac{V_o}{V_{o4}}. \tag{12.6}$$

We will proceed to estimate each of the factors in Eq. (12.6).

The input-stage voltage gain is given by

$$A_{v1} = \frac{V_{o2}}{V_{id}} = \frac{g_{m2}(r_{o2} \| R_2 \| R_{i4})}{2}. \tag{12.7}$$

The input resistance of the intermediate stage, R_{i4}, is approximately given by

$$R_{i4} \cong r_{\pi 4} + (1 + \beta_4) R_4 \tag{12.8}$$

and for $\beta_4 \gg 1$ will probably result in $R_{i4} \gg R_2$. In addition, $R_2 \ll r_{o2}$ so Eq. (12.7) can be simplified to

$$A_{v1} = \frac{g_{m2} R_2}{2}. \tag{12.9}$$

The intermediate-stage voltage gain is given by using the results from Eq. (8.23) for a CE amplifier with emitter degeneration. By neglecting the effects of R_{i5} or R_{i3} and r_{o4} when compared to R_5, we have

$$A_{v2} = \frac{V_{o4}}{V_{i4}} \cong \frac{-g_{m4} R_5}{1 + g_{m4} R_4}. \tag{12.10}$$

By inspection, for either Q_3 or Q_5 active, the voltage gain of the emitter-follower output stage is

$$A_{v3} = \frac{V_o}{V_{o4}} = +1. \tag{12.11}$$

Combining Eqs. (12.9), (12.10), and (12.11) yields

$$A_v = \left(\frac{+g_{m2} R_2}{2}\right)\left(\frac{-g_{m4} R_5}{1 + g_{m4} R_4}\right)(1). \tag{12.12}$$

From Eq. (10.32), the differential input resistance is

$$R_{id} = 2 r_{\pi 1}, \tag{12.13}$$

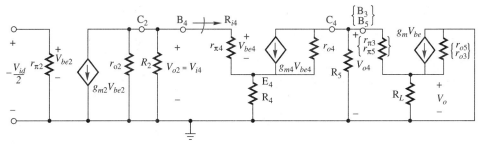

FIGURE 12.4 Small-signal model.

and using Eq. (10.34), the common-mode input resistance is

$$R_{ic} = r_{\pi 1} + (1 + \beta_1)2R_3. \tag{12.14}$$

The output resistance, R_o, is that for an emitter follower, with an interesting variation. Depending on whether Q_3 or Q_5 is active, R_o will be different on each half-cycle of operation. For instance, when Q_3 is active,

$$R_{o3} \cong \left(\frac{r_{\pi 3} + R_5}{1 + \beta_3}\right), \tag{12.15}$$

and when Q_5 is active,

$$R_{o5} \cong \left(\frac{r_{\pi 5} + R_5}{1 + \beta_5}\right). \tag{12.16}$$

In both cases, the effects of the output resistance of Q_4 have been neglected. Usually, the fact that $R_{o3} \neq R_{o5}$ is not important since R_L is much larger than either of them.

■ R_O is an average. Recall Q_3 and Q_5 are operating in a large-signal regime and we are using a small-signal approximation.

DRILL 12.1

List the values for $\beta_{\min}$ and $\beta_{\max}$ and $f_{T(\text{typical})}$ for both the *npn* and *pnp* BJTs in the CA3096. List the associated test conditions.

ANSWERS

npn $\beta_{\min} = 150$ $\beta_{\max} = 500$ $f_T = 280$ MHz $I_C = 1$ mA $V_{CE} = 5$ V

pnp $\beta_{\min} = 20$ $\beta_{\max} = 150$ $I_C = -1$ mA $V_{CE} = -5$ V
     $\beta_{\min} = 40$ $\beta_{\max} = 200$ $f_T = 6.8$ MHz $I_C = -100$ μA $V_{CE} = -5$ V

■ We find that the CA3096-based 5 transistor operational amplifier makes a very effective laboratory experiment.

EXAMPLE 12.1

Design an operational amplifier as shown in Fig. 12.3 that will obtain an approximate voltage gain of -400 using the typical specifications of the CA3096. Verify all approximations, and compute the resultant R_{id} and R_{ic}. Assume $V_{CC} = V_{EE} = 15$ V, and $T = 300$ K. Values of resistors should be compatible with monolithic IC fabrication technology. Let $R_L = 10$ kΩ.

SOLUTION These specifications are sufficiently general to allow for a large number of possible circuit designs. Our approach will be to propose a design alternative and follow it through, verifying key approximations and assumptions in the process.

Let us assume that the input-stage voltage gain should be 40. Then from Eq. (12.9),

$$A_{v1} = 40 = \frac{g_{m2}R_2}{2} = \frac{I_{C2}R_2}{2(kT/q)} = \frac{I_{C2}R_2}{(2)(26 \text{ mV})},$$

which means the $I_{C2}R_2$ product is constrained to 2.08 V. If we select $R_2 = 10$ kΩ, and $I_{C2} = 208$ μA, then $I_{EE} = 2I_{C2} = 416$ μA. This fixes the current-source resistor to be

$$R_3 = \frac{V_{EE} - V_{BE1}}{I_{EE}} = \frac{15 - 0.6}{416 \text{ μA}} = 34.6 \text{ kΩ}.$$

CHAPTER 12 OPERATIONAL AMPLIFIER EXAMPLES

Now $I_{C2}R_2 + V_{BE4} = 2.08 - 0.6 = -I_{C4}R_4$ or $I_{C4}R_4 = -1.48$ V. Also, from Eq. (12.2), $V_{EE} = -I_{C4}R_5 = 15$ V. These values for I_{C4} and V_{EE} constrain R_5/R_4 to be 10.14. Therefore, the intermediate-stage voltage gain is also constrained because when $\beta_4 \gg 1$, Eq. (12.10) reduces to

$$\frac{V_{o4}}{V_{i4}} \cong -\frac{R_5}{R_4}.$$

A reasonable value for R_5 would be 30 kΩ, which means $R_4 = R_5/10.14 = 2.96$ kΩ. Then $I_{C4} = 500$ μA, a value well within the ratings of the CA3096. The actual voltage gain of the intermediate stage is computed from

$$\frac{V_{o4}}{V_{i4}} \cong \frac{-g_{m4}R_5}{1 + g_{m4}R_4} = \frac{\left(-\dfrac{500\ \mu A}{26\ mV}\right)(30\ k\Omega)}{1 + \left(\dfrac{500\ \mu A}{26\ mV}\right)(2.96\ k\Omega)} = -9.96.$$

To verify that Q_4 is operating in the active region, compute $V_{CE4} = 30 - 1.48 - 15 = 13.52$ V, a safe, reasonable value. Using the curves for the CA3096 data sheet in Fig. 12.2, $\beta_4 = 75$. From Eq. (12.8),

$$R_{i4} = r_{\pi 4} + (1 + \beta_4)R_4 = \frac{75}{\left(\dfrac{500\ \mu A}{26\ mV}\right)} + (76)(2.96\ k\Omega) \cong 229\ k\Omega,$$

so $R_{i4} \gg R_2$ has been verified. In addition, $r_{onpn} = 80$ kΩ at 1 mA; then at $I_{C2} = 208$ μA, $r_{o2} = r_{onpn}(1\ mA/208\ \mu A) = 384$ kΩ. Therefore, since $r_{o2} \| R_{i4} \gg R_2$, Eq. (12.9) is a reasonable approximation to Eq. (12.7). The overall voltage gain is $(40)(-9.96)(1) = -398$, quite close to the voltage-gain design goal. The input resistance of the emitter-follower output stage is essentially either $\beta_3 R_L$ or $\beta_5 R_L$, and in either case it will be much larger than R_5, so its effect can be neglected. Also, r_{o4} is sufficiently large that it will not degrade the intermediate stage results very much.

From Eq. (12.13),

$$R_{id} = 2r_{\pi 2} = 2\left(\frac{\beta_2}{g_{m2}}\right) = \frac{(2)(390)}{\left(\dfrac{208\ \mu A}{26\ mV}\right)} = 97.5\ k\Omega.$$

From Eq. (12.14),

$$R_{ic} = r_{\pi 1} + (1 + \beta_1)2R_3 = \frac{390}{\left(\dfrac{208\ \mu A}{26\ mV}\right)} + (391)(2)(34.6\ k\Omega) \cong 27.1\ M\Omega.$$

The output resistances are computed from Eqs. (12.15) and (12.16). To compute R_{o3} and R_{o5}, we need to estimate some value for I_{C3} and I_{C5}. In a class B stage, these will range from zero to about V_{CC}/R_L. Assume $I_{C3} = -I_{C5} = V_{CC}/2R_L = 750$ μA. Then

■ Again, R_O is an average.

$$R_{o3} \cong \frac{r_{\pi 3} + R_5}{1 + \beta_3} = \frac{\dfrac{390}{(750\ \mu A/26\ mV)} + 30\ k\Omega}{1 + 390} = 111\ \Omega$$

and

$$R_{o5} \cong \frac{r_{\pi 5} + R_5}{1 + \beta_5} = \frac{\dfrac{55}{(750\ \mu A/26\ mV)} + 30\ k\Omega}{1 + 55} = 570\ \Omega,$$

where $\beta_{pnp} = 55$ is obtained from the *pnp* transistor h_{FE} curve in Fig. 12.2.

It is important to observe that different power supply voltages, temperature variations, manufacturing tolerances, and aging could have a significant impact on our calculations. Feedback could be used to desensitize the circuit to these parameter variations.

EXAMPLE 12.2

Verify the design in Example 12.1 using a SPICE simulation.

SOLUTION Figure 12.5 is a node-labeled circuit diagram for the circuit shown in Fig. 12.3 with all generators shown.

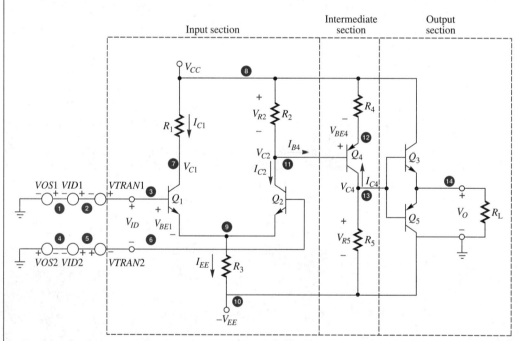

■ **FIGURE 12.5** CA3096 connected operational amplifier.

The netlist for this circuit is given by

```
CA3096 OP AMP EXAMPLE
VOS1 1 0 DC 0
VOS2 4 0 DC 0
VID1 2 1 AC 1
VID2 5 4 AC 1-180
VTRAN1 3 2 SIN(0 10M 1000)
VTRAN2 5 6 SIN(0 10M 1000)
```

CHAPTER 12 OPERATIONAL AMPLIFIER EXAMPLES

```
VCC 8 0 DC 15
VEE 10 0 DC-15
Q1 7 3 9 DEV1
Q2 11 6 9 DEV1
Q3 8 13 14 DEV1
.MODEL DEV1 NPN BF=390 VAF=80 IS=1E-14
Q4 13 11 12 DEV2
Q5 10 13 14 DEV2
.MODEL DEV2 PNP BF=75 VAF=68 IS=8E-13
R1 8 7 10K
R2 8 11 10K
R3 9 10 34.6K
R4 8 12 2.96K
R5 13 10 30K
RL 14 0 10K
.AC DEC 1 1K 1K
.TRAN 20U 2000U
.OP
.PROBE
.PRINT AC VM(14) VM(11) VM(7) VM(13)
.END
```

Three generators have been included in each input. The VOS1 and VOS2 generators are required to provide some offset voltage to turn on either Q_3 or Q_5 as necessary to obtain small-signal results. Recall that Q_3 and Q_5 are normally OFF because no crossover diodes have been included in the circuit.

Key portions of the Q-point (dc solution) are given by

NAME	Q1	Q2	Q3	Q4	Q5
MODEL	DEV1	DEV1	DEV1	DEV2	DEV2
IB	4.58E-07	4.58E-07	-1.50E-11	-5.61E-06	1.58E-11
IC	2.07E-04	2.08E-04	3.28E-11	-5.01E-04	-3.44E-11
VBE	6.11E-01	6.11E-01	3.35E-02	-5.19E-01	3.35E-02
VBC	-1.29E+01	-1.30E+01	-1.50E+01	1.29E+01	1.50E+01
VCE	1.35E+01	1.36E+01	1.50E+01	-1.35E+01	-1.50E+01
GM	8.02E-03	8.02E-03	1.49E-12	1.94E-02	1.01E-11
RPI	5.65E+04	5.65E+04	1.62E+14	4.61E+03	7.93E+12
RO	4.48E+05	4.48E+05	7.27E+11	1.62E+05	6.92E+11

The dc simulation compares quite favorably with the approximate solution. For example, we compute that $I_{C1} = I_{C2} = 208$ μA, which compares to the SPICE 207 μA and 208 μA, respectively. The value of I_{C4} is computed to be -500 μA; SPICE simulates at 501 μA. The Q_3 and Q_5 are OFF, with $|IC|$ on the order of 10^{-11} A.

The transient analysis first-stage collector of Q_2 voltage is shown in Fig. 12.6 (see page 598) for a ±10-mV input. The $\Delta V = 13.702 - 12.270 = 1.432$ V verifies the first-stage gain calculation within 10%.

■ $A_v \cong \dfrac{1.432 \text{ V}}{40 \text{ mV}} = 35.8$

The output stage voltage is shown in Fig. 12.7(a) (see page 598). Observe that for a ±10-mV input voltage, the output voltage, centered at $v_O = 0$ V, verifies the 400 voltage gain within 10%. Crossover distortion is also evident from the class B output stage. The crossover distortion is much more evident if we reduce the input voltage to ±2 mV [Fig. 12.7(b)].

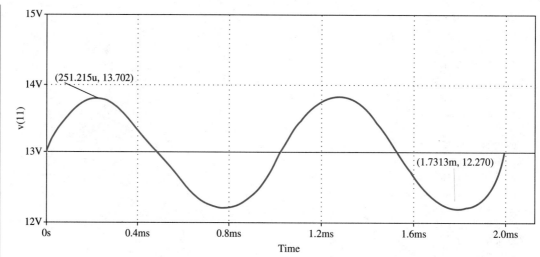

■ FIGURE 12.6 First-stage output voltage.

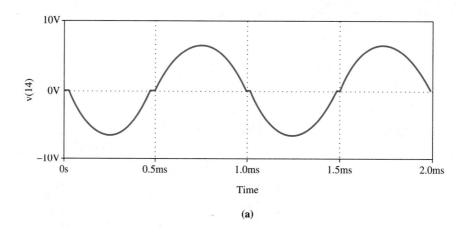

(a)

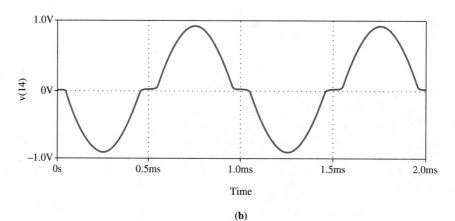

(b)

■ FIGURE 12.7 (a) Output voltage, node 14, for a ±10-mV input voltage.
(b) Output voltage for a ±2 mV input voltage to illustrate class B output stage crossover distortion. Note that parts (a) and (b) have different y-axis voltage scaling.

CHAPTER 12 OPERATIONAL AMPLIFIER EXAMPLES

> **CHECK UP**
>
> 1. Draw a schematic diagram of a five-transistor operational amplifier.
> 2. Write a SPICE netlist for your circuit.
> 3. Add crossover diodes to your schematic.
> 4. Write a SPICE netlist for your circuit that includes the crossover diodes. Perform a simulation and compare with Fig. 12.7.
> 5. Qualitatively describe what happens to the voltage gain, input resistance, and output resistance as the β_{npn} increases, as β_{pnp} increases, and as the power supply voltages increase.

12.2 ANALYSIS OF A μA741 OPERATIONAL AMPLIFIER

Early versions of the μA741 operational amplifier were introduced in 1966, and in the ensuing years it became one of the most widely used integrated circuits. Of course, literally thousands of operational amplifiers have been introduced since then, many with special features and far superior performance; however, it is interesting to observe that many of the newest designs often still advertise their pin-for-pin compatibility with a μA741. Many circuit designers will design their own application-specific integrated circuit (ASIC) operational amplifiers, using whatever technology is appropriate, for direct incorporation into an electronic system. Pedagogically, it is worthwhile to analyze the μA741 because we can exercise many of the design and analysis tools and concepts from earlier chapters to help us understand a relatively complex multistage circuit.

■ Many design engineers believe that ASICs are the best approach for both analog, digital, and mixed-mode systems. Many companies offer ASIC capability—especially in digital and mixed-mode technologies.

As in the CA3096 design exercise, we will first look at each transistor function qualitatively before proceeding with detailed dc and ac analyses. For consistency, we will use transistor parameters as given in Table 18.2, although it is almost certain every manufacturer has different individual device specifications. Indeed, we observed in the Chapter 6 data sheets that there were often four-to-one ratios on key specifications such as gain, input impedance, etc.

Qualitative Circuit Description

Several similar circuit diagrams and data sheets for the μA741 are presented by different manufacturers. This analysis will use the μA741 schematic diagram from Fairchild Semiconductor Corporation as shown in Fig. 12.8 (see page 600). The first group of transistor functions will relate to current-source biasing. Figure 12.9 (see page 600) illustrates this function.

The current-source reference resistor, R_5, provides the reference current for all Widlar and simple current sources. The reference current in Q_{11} also serves as the reference current in Q_{12}, a *pnp* source. Connected to Q_{12} is Q_{13}, which operates as a *pnp* simple current source. Note that Q_{13} has two collectors, enabling it to serve as a current-source to two separate circuits. Transistor Q_{11} also serves as a reference for Q_{10}, connected as a Widlar current source with R_4. The collector current in Q_{10} is transferred to Q_8, which biases the input stage via the simple *pnp* current source Q_9.

The input stage consists of Q_1 through Q_7 as illustrated in Fig. 12.10 (see page 601). The input consists of Q_1 and Q_2 connected as a differential emitter-follower pair. This is one design approach for obtaining a large R_{id} and low input current. The outputs of Q_1 and Q_2 drive the emitters Q_3 and Q_4 connected as a differential common-base amplifier. This provides a relatively large voltage gain. To preserve this large voltage gain, Q_5 and

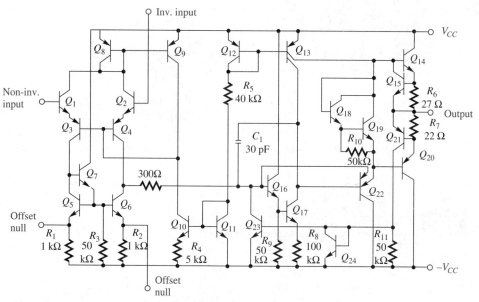

FIGURE 12.8 μA741 op amp circuit diagram (courtesy of Fairchild Semiconductor Corporation).

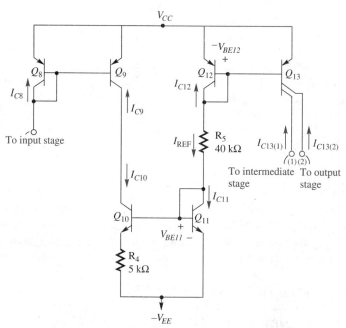

FIGURE 12.9 The dc biasing circuitry for the μA741.

Q_6 are connected as an active load. Instead of a direct connection between the base and the collector of Q_5 for a current reference, Q_7 is used. By using the inherent current gain of Q_7, the effect of I_{B5} and I_{B6} on I_{C5} is reduced by a factor of β from what it would have been, ensuring that the approximation of $I_{C5} = I_{C6}$ is quite accurate. The output signal from the input stage is derived from the collectors of Q_4 and Q_6.

The intermediate gain stage is given in Fig. 12.11 (see page 602). To minimize loading

CHAPTER 12 OPERATIONAL AMPLIFIER EXAMPLES

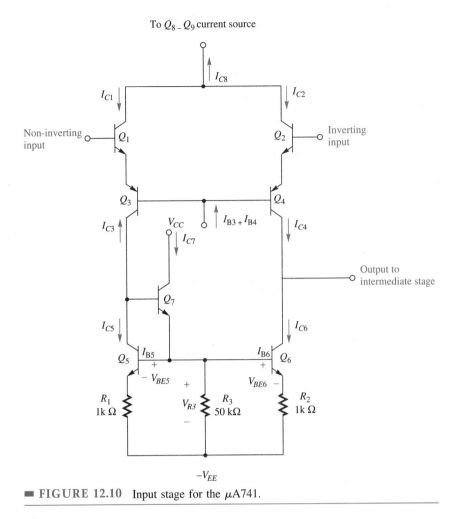

FIGURE 12.10 Input stage for the μA741.

on the input stage, an emitter follower, Q_{16}, is used to drive Q_{17} connected as a common-emitter amplifier. Some gain stabilization and voltage level shifting are provided by R_8. The current source, $Q_{13(1)}$, also serves an active load for Q_{17}, resulting in a high voltage gain for this stage. The output from the collector of Q_{17} is then input to an emitter follower, $Q_{22(1)}$, so that there is minimum loading by the output stage on the common-emitter intermediate gain stage Q_{17}. Although Q_{22} could be considered as part of the intermediate stage, it will be grouped with the output-stage circuitry.

Although the output stage looks complex (see Fig. 12.12, page 404), this apparent complexity is the result of including the internal protection circuitry for the operational amplifier. As mentioned previously, $Q_{22(1)}$ is an emitter follower used to buffer Q_{17}. This emitter follower drives the class AB output stage, Q_{14} and Q_{20}, through the crossover circuit consisting of the diode-connected transistors Q_{18} and Q_{19}. Biasing for $Q_{22(1)}$ is provided by $Q_{13(2)}$. The effective emitter load on $Q_{22(1)}$ consists of Q_{18} and Q_{19} in series with $Q_{13(2)}$—this in parallel with either Q_{14} or Q_{20}, depending on which one is conducting. Output short-circuit protection for Q_{14} is provided by Q_{15} in conjunction with R_6. Similarly, output short-circuit protection for Q_{20} is provided by Q_{21} in conjunction with R_7, Q_{24}, and Q_{23}. Input overdrive protection, which could result in excessive power dissipation in Q_{17}, is obtained using $Q_{22(2)}$.

The internal capacitor, C_1, is used to tailor the frequency response. This capacitor

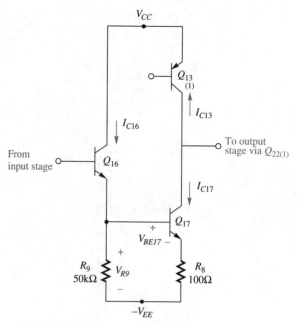

FIGURE 12.11 Intermediate stage for the μA741.

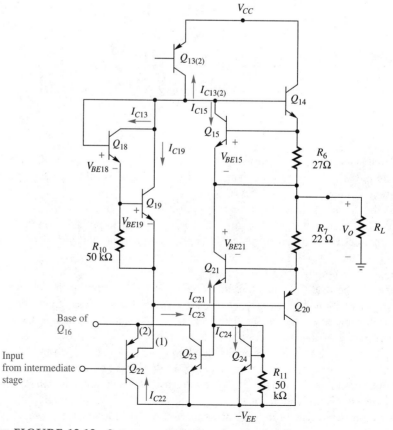

FIGURE 12.12 Output stage and protection circuitry for the μA741.

provides feedback frequency compensation. The following dc and ac analyses of the µA741 will be compared with the SPICE computer solution of this circuit.

µA741 DC Circuit Analysis

Because the data sheet in Fig. 6.2 assumes $V_{CC} = V_{EE} = 15$ V and $T = 25°C$, the analysis uses the same values. The transistor parameters used are found in Table 18.2. Clearly, variations in these transistor parameters, power supply voltages, and temperature will change the results. These variations are given by the manufacturer and should be expected by the user. Functional dependencies of key typical parameters are given graphically as part of the data sheet. We refer to these as we proceed.

Again, it is important to note that base currents are neglected in most cases. Figure 12.9 is used to compute the quiescent currents. Starting with Q_{11}, the current-source reference current is

$$I_{C11} = -I_{C12} = \frac{V_{CC} + V_{EE} - V_{BE11} + V_{BE12}}{R_5}$$
$$= \frac{30 - 0.7 - 0.6}{40 \text{ k}\Omega} = 718 \text{ } \mu\text{A}. \quad (12.17)$$

The current in Q_{10} is computed by realizing it is a Widlar current source and solving iteratively to obtain

$$I_{C10} = -I_{C9} = -I_{C8} = \frac{kT}{qR_4} \ln\left(\frac{I_{C11}}{I_{C10}}\right)$$
$$= \left(\frac{26 \text{ mV}}{5 \text{ k}\Omega}\right) \ln\left(\frac{718 \text{ } \mu\text{A}}{I_{C10}}\right) = 18.9 \text{ } \mu\text{A}. \quad (12.18)$$

Transistor Q_{13} has a split collector, with collector (1) having three times the junction area of collector (2) (see Fig. 12.13), but the total collector area is the same as for Q_{12}. Then, using the results for Q_{12},

$$I_{C13(1)} + I_{C13(2)} = I_{C13} = I_{C12} = -718 \text{ } \mu\text{A} \quad (12.19)$$

and

$$\frac{I_{C13(1)}}{I_{C13(2)}} = 3. \quad (12.20)$$

The solution is $I_{C13(1)} = -538.5 \text{ } \mu\text{A}$ and $I_{C13(2)} = -179.5 \text{ } \mu\text{A}$.

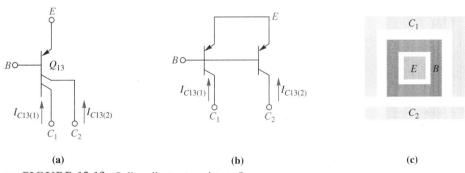

FIGURE 12.13 Split collector transistor, Q_{13}, C_1 area $= 3 \times C_2$ area. **(a)** Circuit symbol. **(b)** Alternative symbol. **(c)** Conceptual top view.

Now referring to Fig. 12.10 and assuming matched transistors,

$$I_{C1} = I_{C2} = -I_{C3} = -I_{C4} = I_{C5} = I_{C6}$$
$$= \frac{-I_{C8}}{2} = \frac{18.9 \; \mu A}{2} = 9.45 \; \mu A. \tag{12.21}$$

$$I_{C7} \cong \frac{V_{R3}}{R_3} = \frac{I_{C5}R_1 + V_{BE5}}{R_3} = \frac{I_{C5}R_1 + V_T \ln\left(\frac{I_{C5}}{I_{S5}}\right)}{R_3}$$

$$= \frac{(9.45 \; \mu A)(1 \; k\Omega) + (26 \; mV) \ln\left(\frac{9.45 \; \mu A}{2 \times 10^{-15} \; A}\right)}{50 \; k\Omega} = 11.8 \; \mu A, \tag{12.22}$$

where a more accurate result has been obtained by actually computing V_{BE5} rather than assuming $V_{BE(\text{on})} = 0.7$ V. The approximate value would then have been

$$I_{C7} \cong \frac{9.45 \; mV + 700 \; mV}{50 \; k\Omega} = 14.2 \; \mu A \tag{12.23}$$

as compared to 11.8 μA. In either case, considering $\beta_7 = 200$, the approximation that $I_{C5} = I_{C3}$ becomes quite good since $I_{B7} = 55$ nA.

Referring to Fig. 12.11, the collector current in Q_{16} is computed from

$$I_{C16} = I_{B17} + \frac{V_{R9}}{R_9} = \frac{I_{C17}}{\beta_{17}} + \frac{V_{BE17} + I_{C17}R_8}{R_9}$$

$$= \frac{I_{C17}}{\beta_{17}} + \frac{V_T \ln\left(\frac{I_{C17}}{I_{S17}}\right) + I_{C17}R_8}{R_9}.$$

Inserting $I_{C17} = -I_{C13(1)} = 538 \; \mu A$ and $\beta_{17} = 200$,

$$I_{C16} = \frac{538 \; \mu A}{200} + \frac{(26 \; mV) \ln\left(\frac{538 \; \mu A}{2 \times 10^{-15}}\right) + (538 \; \mu A)(100 \; \Omega)}{50 \; k\Omega}$$
$$= 17.4 \; \mu A. \tag{12.24}$$

The output-stage currents are computed by using Fig. 12.12. For Q_{18},

$$I_{C18} = \frac{V_{BE19}}{R_{10}} \cong \frac{0.7}{50 \; k\Omega} = 14 \; \mu A. \tag{12.25}$$

Then, at the collector of Q_{18},

$$I_{C19} = -I_{C13(2)} - I_{C18} = 180 \; \mu A - 14 \; \mu A = 166 \; \mu A. \tag{12.26}$$

To check the initial assumption of neglecting base currents, I_{C18} can be computed iteratively by using the result from Eq. (12.26) in

$$I_{C18} = \frac{I_{C19}}{\beta_{19}} + \frac{V_T \ln\left(\frac{I_{C18}}{I_{S18}}\right)}{R_{10}}$$

$$= \left(\frac{166 \; \mu A}{200}\right) + \frac{(26 \; mV)}{50 \; k\Omega} \ln\left(\frac{166 \; \mu A}{2 \times 10^{-15} \; A}\right) = 13.8 \; \mu A. \tag{12.27}$$

This result agrees quite well with the 14 μA obtained from Eq. (12.25). No further iteration is necessary. From our work with the class AB amplifier in Chapter 10, and assuming that the junction areas of Q_{14} and Q_{20} are three times that of Q_{18} and Q_{19}, we derive

$$I_{C14} \cong -I_{C20} = \sqrt{I_{C18}I_{C19}}\sqrt{\frac{I_{S14}I_{S20}}{I_{S18}I_{S19}}}$$
$$= \sqrt{(14\ \mu A)(166\ \mu A)}\sqrt{(3)(3)} = 144\ \mu A \qquad (12.28)$$

from

$$V_T \ln\left(\frac{I_{C18}}{I_{S18}}\right) + V_T \ln\left(\frac{I_{C19}}{I_{S19}}\right) = V_T \ln\left(\frac{I_{C14}}{I_{S14}}\right) + V_T \ln\left(-\frac{I_{C20}}{I_{S20}}\right).$$

Protection Circuitry

Under normal load and signal input conditions, Q_{15}, Q_{21}, Q_{24}, Q_{23}, and $Q_{22(2)}$ are OFF. Should the output become short-circuited during the positive half-cycle, Q_{15} will turn ON and effectively remove base-current drive from Q_{14}. Again from class B and AB amplifier studies in Chapter 10, this will occur when

$$I_{Lm}^+ = \frac{V_{BE15}}{R_6} = \frac{0.7}{27\ \Omega} = 26\ \text{mA}. \qquad (12.29)$$

During the negative half-cycle, Q_{21} will turn ON when

$$I_{Lm}^- = \frac{V_{BE21}}{R_7} = \frac{0.6}{22\ \Omega} = 27\ \text{mA}. \qquad (12.30)$$

When Q_{21} starts to conduct, Q_{24} will also conduct, and it then operates as a current-source reference for Q_{23}. For Q_{23} to operate, it sinks current from the base of Q_{16}, which effectively reduces the drive signal to the base of Q_{20}.

The μA741 is also protected against excessively large input signals. If a large enough signal is applied between the bases of Q_1 and Q_2, Q_1 will turn OFF. If I_{C1} is forced to zero, then $I_{C5} = I_{C6} = 0$. All of $I_{C8} = 18.9\ \mu A$ will be an input to Q_{16}. The resultant large value of $I_{C16} = \beta_{16} \times I_{C8}$ would effectively saturate Q_{17} and result in excessive power dissipation in Q_{16}. However, when V_{CB17} approaches zero volts, $Q_{22(2)}$ starts to turn ON, shunting a portion of the 18.9 μA away from the base of Q_{16}.

μA741 AC Circuit Analysis

Despite the apparent complexity of the μA741, its ac analysis proceeds in the same way that cascaded stages were treated in Chapter 8. Referring to Fig. 12.8, we will compute the overall voltage gain by considering separately each of the factors in

$$A_v = \frac{V_o}{V_{id}} = A_{v1} \times A_{v2} \times A_{v3} = \frac{V_{c6}}{V_{id}} \times \frac{V_{c17}}{V_{c6}} \times \frac{V_o}{V_{c17}}. \qquad (12.31)$$

Of course, it is crucial in this approximate analysis that simplifications be used.

To provide a relatively high input resistance, Q_1 and Q_2 are connected as a differential emitter-follower pair. The remaining elements of the input stage form a differential common-base amplifier, Q_3 and Q_4, with active loads Q_5 and Q_6. From Fig. 12.7, we can construct a model of the input stage. First, we recognize that the input is balanced and symmetrical, so that the ac currents I_{b3} and I_{b4} are equal and opposite. This forces $V_{b3} = 0$ and $I_{b4} = 0$. Likewise, I_{c1} and I_{c2} are equal and opposite, so that $V_{c1} = V_{c2} = 0$. Thus we draw the simplified ac model of Fig. 12.14. Q_5 and Q_6 are drawn to remind us that they represent a current mirror.

One way to determine V_{c6}/V_{id} is to treat the circuit shown enclosed in a box in

---

DRILL 12.2

If V_{CC} is reduced from 15 V to 12 V, compute a value for the current source reference current, I_{C11} and the collector current magnitudes in Q_1 through Q_6.

ANSWERS $I_{C11} = 568\ \mu A$ Collector current magnitudes in Q_1 through $Q_6 = 8.98\ \mu A$.

Fig. 12.14 as a two-port network [see Fig. 12.15(a)]. The first-stage voltage gain is then computed by using

$$A_{v1} = \frac{V_{c6}}{V_{id}} = -G_m(R_{o1}\|R_{i16}). \tag{12.32}$$

The output of the two-port network is short-circuited to compute

$$G_m = \frac{I_{sc}}{V_{id}}. \tag{12.33}$$

When the output is short-circuited, the circuit shown in Fig. 12.14 reduces to that shown in Fig. 12.15(b). From the symmetry, the half-circuit model approach is used [see Fig. 12.16(a)]. Since $r_o \gg r_\pi$, we neglect it to simplify this diagram. Note also that the transistors all have identical operating points, Eq. (12.21), so that $g_{m1} = g_{m2} = g_{m3} = g_{m4} = g_{m5} = g_{m6} \equiv g_m$. Using this half-circuit model, and summing currents at the emitters,

$$\left(\frac{V_{id}}{2} - V_{e3}\right)\left(\frac{1}{r_{\pi 1}} + g_m\right) = (V_{e3})\left(\frac{1}{r_{\pi 3}} + g_m\right)$$

and since $\beta \gg 1$ for both transistors, then $1/r_\pi \ll g_m$, so that

$$\left(\frac{V_{id}}{2} - V_{e3}\right)g_m = (V_{e3})g_m.$$

Thus, $V_{e3} = V_{id}/4$. Then, $I_{c3} = -g_m V_{e3} = -g_m V_{id}/4$ and, by symmetry, $I_{c4} = -g_m V_{e4} = +g_m V_{id}/4$.

At the output node

$$I_{sc} = -I_{c3} + I_{c4} = \frac{g_m V_{id}}{4} + \frac{g_m V_{id}}{4} = \frac{g_m V_{id}}{2} = G_m V_{id} \tag{12.34}$$

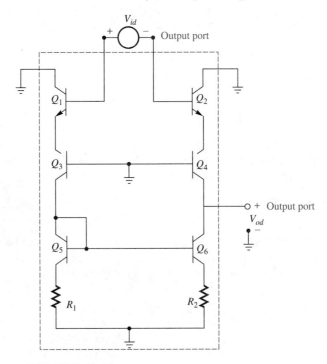

FIGURE 12.14 Simplified small-signal input-stage model.

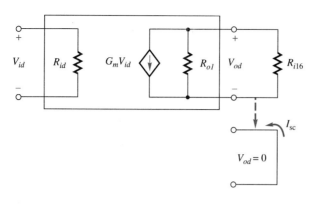

(a)

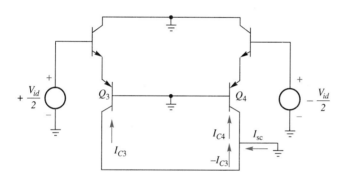

(b)

FIGURE 12.15 Input circuit analysis approach (a) Two-port representation. (b) Short-circuit reduction of the input stage.

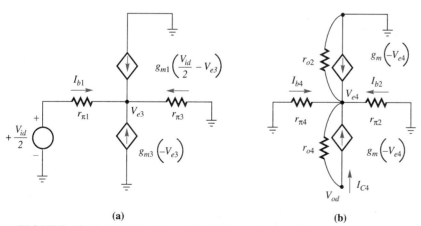

FIGURE 12.16 Half-circuit models. (a) Determination of $I_{c3}V_{id}$. (b) Determination of R_{o4}.

with the result that

$$G_m = \frac{g_{m4}}{2} = \frac{I_{C4}}{2(kT/q)}. \tag{12.35}$$

The output resistance, R_{o1}, is given by

$$R_{o1} = R_{o4} \| R_{o6}. \tag{12.36}$$

Looking at Fig. 12.16(b), we see that

$$I_{c4} = V_{e4}\left(\frac{1}{r_{o2}} + \frac{1}{r_{\pi 2}} + \frac{1}{r_{\pi 4}} + g_m\right) \approx g_m V_{e4},$$

indicating that Q_2 represents a load of $R = 1/g_m$ to the emitter of Q_4. Then, from Eq. (7.39),

$$R_{o4} = r_{o4}(1 + g_m R) \approx r_{o4}\left[1 + g_m\left(\frac{1}{g_m}\right)\right] = 2r_{o4}$$

$$= \frac{(2)(V_{Apnp})}{I_C} = \frac{(2)(50)}{9.4\ \mu A} = 10.6\ M\Omega. \tag{12.37}$$

Similarly,

$$R_{o6} = r_{o6}(1 + g_m R_2) = \frac{V_{Anpn}}{I_C} \times \left(1 + \frac{I_C}{kT/q} \times R_2\right) \tag{12.38}$$

$$= \left(\frac{150\ V}{9.45\ \mu A}\right)\left[1 + \left(\frac{9.45\ \mu A}{26\ mV}\right)(1\ k\Omega)\right] = 21.64\ M\Omega.$$

Combining Eqs. (12.37) and (12.38) with Eq. (12.36) yields

$$R_{o1} = 10.6\ M\Omega \| 21.64\ M\Omega = 7.1\ M\Omega. \tag{12.39}$$

The effective load resistance on the first stage is given by

$$R_{i16} = r_{\pi 16} + (1 + \beta_{16})(R_9 \| R_{i17}) = \frac{\beta_{16}}{g_{m16}} + (1 + \beta_{16})(R_9 \| R_{i17})$$

$$= \frac{200}{(17.4\ \mu A/26\ mV)} + (201)(50\ k\Omega \| 29.8\ k\Omega) = 4.05\ M\Omega, \tag{12.40}$$

where

$$R_{i17} = r_{\pi 17} + (1 + \beta_{17})R_8 = \frac{\beta_{17}}{g_{m17}} + (1 + \beta_{17})R_8$$

$$= \frac{200}{(538\ \mu A/26\ mV)} + (201)(100\ \Omega) = 29.8\ k\Omega. \tag{12.41}$$

Substituting the results from Eqs. (12.39) and (12.40) into Eq. (12.32) yields the voltage gain of the first stage:

$$A_{v1} = \frac{V_{c6}}{V_{id}} = \frac{-g_m}{2}(R_{o1} \| R_{i16})$$

$$= \left(\frac{-9.45\ \mu A}{2 \times 26\ mV}\right)(7.1\ M\Omega \| 4.05\ M\Omega) = -469. \tag{12.42}$$

The second-stage gain, $A_{v2} = V_{c17}/V_{c6}$, is computed from

$$A_{v2} = \left(\frac{-g_{m17}}{1 + g_{m17}R_8}\right) \times [R_{o2} \| R_{i22(1)}]$$

$$= \left(\frac{-g_{m17}}{1 + g_{m17}R_8}\right) \times [R_{o17} \| R_{o13(1)} \| R_{i22(1)}]. \tag{12.43}$$

CHAPTER 12 OPERATIONAL AMPLIFIER EXAMPLES   609

Although we could compute $R_{i22(1)}$ without too much trouble, we should first observe that it will be on the order of $\beta_{22}\beta_{14} \times R_L$ or $\beta_{22}\beta_{20} \times R_L$, which for $R_L > 2$ kΩ or so, results in $R_{i22(1)}$ in the tens of megohms. Rather than pursuing an exact answer for $R_{i22(1)}$, R_{o2} should be computed first and then compared with tens of megohms. Then,

$$R_{o2} = r_{o17}(1 + g_{m17}R_8) \| r_{o13(1)} = \left(\frac{V_{Anpn}}{I_{C17}}\right)\left[1 + \frac{I_{C17}}{(kT/q)}R_8\right] \left\| \frac{V_{Apnp}}{I_{C13(1)}} \right.$$

$$= \left(\frac{150 \text{ V}}{538 \text{ }\mu\text{A}}\right)\left(1 + \frac{538 \text{ }\mu\text{A}}{26 \text{ mV}} \times 100 \text{ }\Omega\right) \left\| \left(\frac{50}{538 \text{ }\mu\text{A}}\right) = 83.8 \text{ k}\Omega. \right. \quad (12.44)$$

Correctly neglecting $R_{i22(1)}$, the voltage gain of the second stage is

$$A_{v2} \cong \left(\frac{-g_{m17}}{1 + g_{m17}R_8}\right)(R_{o2})$$

$$= \frac{(-538 \text{ }\mu\text{A}/26 \text{ mV})}{1 + (538 \text{ }\mu\text{A}/26 \text{ mV})100 \text{ }\Omega} \times 83.8 \text{ k}\Omega = -565. \quad (12.45)$$

The 300-Ω resistor between the collector of Q_6 and the base of Q_{16} has a negligible effect on A_{v2} because $R_{i16} \gg 300$ Ω. This resistor does serve to limit I_{B16} under excessively large input signal conditions.

The third-stage voltage gain is obtained by inspection. The class AB output stage is an emitter follower whose voltage gain is $+1$. This output stage is driven by the $Q_{22(1)}$ connected as an emitter follower. The result is

$$A_{v3} = \frac{V_o}{V_{e22(1)}} \times \frac{V_{e22(1)}}{V_{c17}} \cong +1. \quad (12.46)$$

Combining Eqs. (12.42), (12.45), and (12.46), the overall μA741 computed voltage gain is

$$A_v = A_{v1} \times A_{v2} \times A_{v3} = (-469)(-565)(1) = 264{,}985. \quad (12.47)$$

The input resistance, R_{id}, is computed with the aid of Fig. 12.16(a). We see that $V_{id}/2 = I_{b1}r_{\pi1} + V_{e3} = I_{b1}r_{\pi1} + (V_{id}/4)$, which leads to $V_{id}/4 = I_{b1}r_{\pi1}$. Therefore,

$$R_{id} = \frac{V_{id}}{I_{b1}} = 4r_{\pi1}$$

$$= \frac{4\beta}{g_m} = \frac{(4)(200)}{(9.45 \text{ }\mu\text{A}/26 \text{ mV})} = 2.2 \text{ M}\Omega. \quad (12.48)$$

Determining the output resistance is somewhat more complicated in that the value will be different depending on instantaneous current and whether Q_{14} or Q_{20} is conducting. Suppose we consider the case when Q_{14} is conducting. As shown in Fig. 12.17, R'_o will not include the effect of R_6, which can be added later. Then, using results for the output resistance of an emitter follower, Eq. (8.48),

$$R'_o = \frac{r_{\pi14} + R'_s}{1 + \beta_{14}}. \quad (12.49)$$

■ Data sheets usually provide an average value for the output resistance at some given large-signal voltage.

The effective source resistance, R'_s, is given by

$$R'_s = r_{o13(2)} \left\| \left(r'_{d19} + \frac{R_{o2} + r_{\pi22}}{1 + \beta_{22}}\right). \right. \quad (12.50)$$

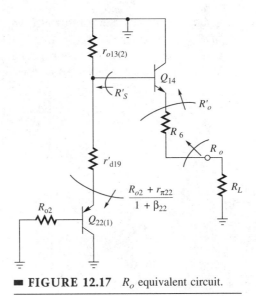

■ **FIGURE 12.17** R_o equivalent circuit.

The resistance r'_{d19} of the Q_{18}–Q_{19} circuit is found, using the small-signal model shown in Fig. 12.18, by computing

$$-I_{E18} = I_{B18} + I_{C18} = V_{\pi 18}\left(\frac{1}{r_{\pi 18}} + g_{m18}\right)$$

$$= V_{\pi 18}\frac{g_{m18}}{\beta_{18}}(1+\beta) \cong V_{\pi 18}g_{m18}. \quad (12.51)$$

Then,

$$r_{d18} = \frac{V_{\pi 18}}{-I_{E18}} = \left(\frac{\beta_{18}}{1+\beta_{18}}\right)\left(\frac{1}{g_{m18}}\right) = \left(\frac{200}{201}\right)\left(\frac{26\text{ mV}}{14\text{ }\mu\text{A}}\right) = 1848\text{ }\Omega. \quad (12.52)$$

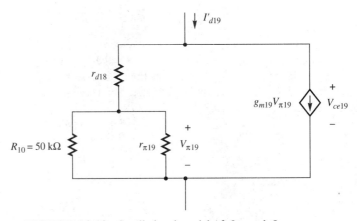

■ **FIGURE 12.18** Small-signal model of Q_{18} and Q_{19}.

CHAPTER 12 OPERATIONAL AMPLIFIER EXAMPLES 611

To obtain r'_{d19}, we compute

$$r_{\pi 19} = \frac{\beta_{19}}{g_{m19}} = \beta\left(\frac{kT/q}{I_{C19}}\right) = 200\left(\frac{26\text{ mV}}{166\ \mu\text{A}}\right) = 31.6\text{ k}\Omega \qquad (12.53)$$

and

$$V_{\pi 19} = V_{ce19}\frac{r_{\pi 19}\|50\text{ k}\Omega}{r_{\pi 19}\|50\text{ k}\Omega + r_{d18}} = 0.91 V_{ce19}. \qquad (12.54)$$

Using these results by summing currents at the top node in Fig. 12.18,

$$r'_{d19} = \frac{V_{CE19}}{I'_{d19}} = \frac{V_{CE19}}{I_{d18} + g_{m19}v_{\pi 19}}$$

$$= \frac{V_{CE19}}{0.91 V_{CE19}\left(\dfrac{1}{r_{\pi 19}\|50\text{ k}\Omega} + \dfrac{166\ \mu\text{A}}{26\text{ mV}}\right)} = 170\ \Omega. \qquad (12.55)$$

Using these results in Eq. (12.50),

$$R'_s = \left(\frac{50\text{ V}}{180\ \mu\text{A}}\right) \left\| \left[170\ \Omega + \frac{83.8\text{ k}\Omega + \dfrac{50(26\text{ mV})}{180\ \mu\text{A}}}{51}\right]\right.$$

$$= 1.94\text{ k}\Omega. \qquad (12.56)$$

This value for $R'_s = 1.94$ kΩ is substituted into Eq. (12.49) to obtain

$$R'_o = \frac{\left(200\dfrac{26\text{ mV}}{5\text{ mA}}\right) + 1.91\text{ k}\Omega}{201} = 15\ \Omega, \qquad (12.57)$$

where an estimate for an average value $I_{C14} = 5$ mA has been used. Including $R_6 = 27\ \Omega$ yields $R_o = 42\ \Omega$.

Frequency Response

Refer to the circuit diagram shown in Fig. 12.8. A 30-pF capacitor (C_1) has been incorporated into the circuit. Its purpose, as we will observe, is to introduce a corner frequency that will decrease the forward gain at a frequency well below any of the other corner frequencies inherent with the internal transistor capacitances. Because the $C_1 = 30$-pF capacitor is connected between the base of Q_{16} and the collector of Q_{17}, it can be modeled by means of a Miller effect type of analysis. The equivalent circuit based on Fig. 12.11 is given in Fig. 12.19(a).

From Eq. (12.39), the output resistance of the first stage is $R_{o1} = R_{o4}\|R_{o7} = 7.1$ MΩ. From Eq. (12.40), the input resistance of the second stage is $R_{i16} = 4.05$ MΩ. The voltage gain of the Q_{16}–Q_{17} Darlington stage is computed in Eq. (12.45) to be -565. Using the Miller effect,

$$C_M = C_1(1 - A_{v2}) = 30\text{ pF}(566) = 16.98\text{ nF}. \qquad (12.58)$$

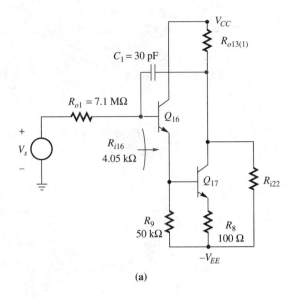

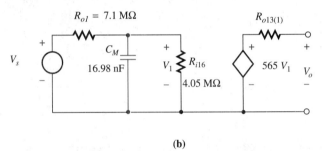

■ **FIGURE 12.19** Miller effect analysis of the μA741 compensation capacitor. **(a)** Partial circuit and functional description. **(b)** Small-signal model using Miller effect.

Using the simplified circuit shown in Fig. 12.19(b), the corner frequency is computed to be

$$f_D = \frac{\omega_D}{2\pi} = \frac{1}{2\pi C_M(R_{o1}\|R_{i16})}$$

$$= \frac{1}{2\pi(16.98 \times 10^{-9}\text{ nF})(7.1\text{ M}\Omega\|4.05\text{ M}\Omega)} = 3.6\text{ Hz}. \quad (12.59)$$

The effect of C_1 is illustrated in graph 4 of Fig. 6.2. This graph, along with a Bode phase plot, is shown in Fig. 12.20.

An important point to observe is that there are no other corners less than 1 MHz. The 30-pF capacitor is not only much larger than any other internal transistor capacitances, but its inclusion as a voltage-shunt feedback element increases its circuit effect by the voltage gain of the second stage. The forward gain decreases from 200,000 to 1 at -20 dB/decade. If this circuit were incorporated into a feedback circuit where the largest possible feedback factor (worst case for stability) of $f = 1$ were used, Fig. 12.20 could

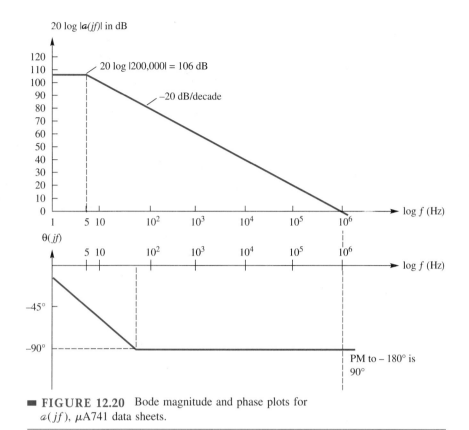

FIGURE 12.20 Bode magnitude and phase plots for $a(jf)$, μA741 data sheets.

be labeled as the Bode magnitude and phase plots for $a(jf)$. At the point where $|T(jf)| = 1$, at 1 MHz, the accumulated phase is only $-90°$. This results in a phase margin, PM = $90°$.

Qualitatively, the μA741 has sacrificed bandwidth for a very large PM. Many operational amplifiers include compensation with smaller phase margins or provide the user with circuit access for adding external compensation. Usually the data sheet provides design curves for the compensation network component values. Generally, designing for PM < $30°$ must be done very carefully because small changes in active device performance could result in instability.

After we complete a SPICE analysis, we will compare the data sheet parameters, with both the approximate calculations, and the SPICE simulations for A_v, R_{id}, and R_o.

DRILL 12.3

Tabulate typical data sheet and analytical values for a_v, R_{id}, R_o, and GBW for the μA741.

ANSWERS

	a_v	R_{id}	R_o	GBW
Data sheets	200,000	2 MΩ	75	1 MHz
Analytical calculations	264,985	2.2 MΩ	42	954 kHz

EXAMPLE 12.3

Verify the μA741 analysis using a SPICE simulation. Use the transistor models where for the *npn*: $\beta = 200$, $I_S = 2 \times 10^{-15}$ A, $V_A = 150$ V, $\beta_R = 2$, and $r_b = 200\ \Omega$. For the *pnp*: $\beta = 50$, $I_S = 10^{-14}$ A, $\beta_R = 3$, $V_A = 50$ V, and $r_b = 100\ \Omega$ with zero-voltage junction capacitances of 1 pF. The collector areas of transistors Q_{14} and Q_{20} are three times as large as the others, so use an area factor of 3. In addition, $Q_{13(1)}$ has an area factor of 0.75 and $Q_{13(2)}$ has an area factor of 0.25. To obtain a directly normalized output, use $V_{id} = 1$ V for the *a* analysis with one simulation at each decade from 1.0 Hz to 1 MHz.

SOLUTION The node-labeled circuit diagram is shown in Fig. 12.21. The netlist, based on Fig. 12.21 is given by:

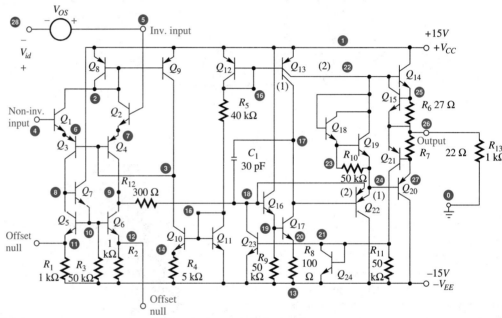

FIGURE 12.21 μA741 operational amplifier, node-labeled for SPICE.

```
SMALL SIGNAL GAIN OF UA741 OP AMP
VCC 1 0 DC 15
VEE 0 13 DC 15
VID 4 28 AC 1UV
VOS 5 28 DC .000321
VIC 5 0 DC 0
Q1 2 4 6 M1
Q2 2 5 7 M1
Q3 8 3 6 M2
Q4 9 3 7 M2
Q5 8 10 11 M1
Q6 9 10 12 M1
```

```
Q7   1   8   10  M1
Q8   2   2   1   M2
Q9   3   2   1   M2
Q10  3   15  14  M1
Q11  15  15  13  M1
Q12  16  16  1   M2
Q131 17  16  1   M2  AREA=.75
Q132 22  16  1   M2  AREA=.25
Q14  1   22  25  M1  AREA=3
Q15  22  25  26  M1
Q16  1   18  19  M1
Q17  17  19  20  M1
Q18  22  22  23  M1
Q19  22  23  24  M1
Q20  13  24  27  M2  AREA=3
Q21  21  27  26  M2
Q221 13  17  24  M2
Q222 13  17  18  M2
Q23  18  21  13  M1
Q24  21  21  13  M1
R1   11  13  1K
R2   12  13  1K
R3   10  13  50K
R4   14  13  5K
R5   16  15  40K
R12  9   18  300
R6   25  26  27
R7   26  27  22
R8   20  13  100
R9   19  13  50K
R10  23  24  50K
R11  21  13  50K
R13  26  0   1K
C1   17  18  30PF
.MODEL M1 NPN BF=200 IS=2E-15 VAF=150 RB=200 BR=2 CJE=1PF
+CJC=1PF
.MODEL M2 PNP BF=50 IS=IE-14 VAF=50 RB=100 BR=3 CJE=1PF
+CJC=1PF
.TF V(26) VIC
.AC DEC 3 .1 1MEG
.PRINT AC VM(18), VP(18), VM(17), VP(17), VM(26), VP(26)
.PLOT AC VDB(26)
.END
```

■ This is one of the longer SPICE programs in this text. At the time this text was written, the student version of PSPICE would not accept circuits with this many nodes (or more than 10 active devices). Updates on the student version are subject to change. More complete commercial versions of PSPICE or equivalent may be available at your school. SPICE 2G.6 would present no problems.

We used a common-mode voltage $V_{IC} = 0$, and found it necessary (by trial and error) to provide an offset voltage $V_{OS} = 321$ μV to force the output voltage to be near zero. SPICE dc analysis finds the operating points and small-signal models for all of the transistors. The inclusion of the .TF statement shown provides the additional information that the common-mode amplification, $v_{26}/v_{ic} = -1.29$, the common-mode input resistance $R_{ic} = 641$ MΩ, and $R_o = 60.5$ Ω. The Bode magnitude plot is shown in Fig. 12.22, which should be compared to the data sheet in Chapter 6. The corner frequency is seen to be about 3.1 Hz. It is interesting to note that when the transistor junction capacitances are omitted, there is no significant change in these data, indicating that the 30-pF compensation capacitor dominates in determining the frequency response.

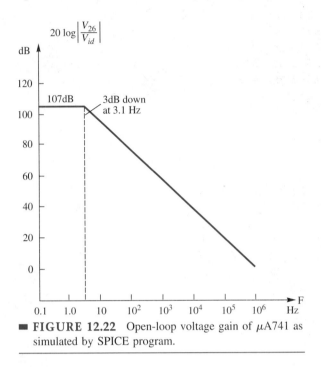

FIGURE 12.22 Open-loop voltage gain of μA741 as simulated by SPICE program.

The nodal data, shown next, show that the first-stage gain is 428 at 0.1 Hz, and the second-stage gain is 561, which verifies the calculations done earlier in this section.

	V_{18} (μV)	V_{17} (mV)	V_{26} (mV)
.1 Hz	428∠178°	240∠−2°	227∠−2°
1 Hz	408∠162°	229∠−18°	217∠−18°
10 Hz	128∠107°	71∠−73°	68∠−73°
100 Hz	13∠92°	7.5∠−88°	7.1∠−88°
1 kHz	1.3∠91°	0.75∠−89°	0.71∠−89°
10 kHz	0.14∠101°	0.075∠−90°	0.071∠−90°
100 kHz	003∠178°	0.0075∠−90°	0.007∠−90°
1 MHz	0.026∠175°	0.00075∠−94°	0.0007∠−94°

CHECK UP

1. What are the maximum, minimum, and typical values for the voltage gain, input resistance, output resistance, unadjusted input offset voltage, and −3-dB corner frequency for a μA741 operational amplifier?
2. Qualitatively describe what happens to the parameters in Checkup Question 1 as either β increases or the power supply decreases.
3. Make a list of each of the transistors used and qualitatively describe its function in the circuit.

12.3 ANALYSIS OF A CA3140 OPERATIONAL AMPLIFIER

In the mid-1970s, technology advancements permitted the incorporation of MOS and bipolar technology on the same IC die. Indeed, currently BiMOS or BiCMOS technology applied to analog circuit IC design is one of the fastest growing areas in IC product development. There is also an increasing effort in BiCMOS ASIC design. To illustrate some of the unique features of an operational amplifier incorporating this technology, we study the venerable CA3140. The CA series was manufactured by RCA and is now manufactured by Harris and others. This operational amplifier, along with many others, initially capitalized on the physical and electrical compatibility with the μA741, although the architecture and internal design permit improved performance in several areas. We need not proceed in as great a depth here as we did in our analysis of the μA741.

The schematic diagram of the CA3140 and key specifications are given in Fig. 12.23.

Qualitative Circuit Description

As we can see from Fig. 12.23 the diode function is represented by a diode symbol, although a diode is usually obtained from a transistor in which either the base and emitter or base and collector are internally connected together by metallization.

The first group of active device functions relates to current-source biasing. The current reference function is established by Q_1, D_1, and R_1. The diode-connected transistor D_1 maintains a constant current through Q_6, M_8, and Q_{12}. Observe that M_8 is a p-channel MOSFET biased by D_2 so that it provides a constant current, operating as an enhancement-mode device in the saturation region. This constant current is mirrored to Q_2 and Q_5 connected as a cascode current source. Recall that the cascode circuit consists of a common-emitter amplifier, Q_2, driving a common-base amplifier, Q_5. As given by the manufacturer, this current source provides $I_{C5} = -200$ μA to the input-stage differential pair. An identical arrangement with Q_3 and Q_4 provides $I_{C4} = -200$ μA for the second-stage biasing. The constant current established in D_2 operating as a reference also biases Q_{14} and Q_{15} so that $I_{C18} = 2$ mA.

There are three basic voltage-gain stages. The input stage uses M_9 and M_{10}, p-channel MOS enhancement-mode transistors, in an actively loaded differential amplifier. The active loads are provided by Q_{11} and Q_{12}. Offset voltage adjustment capability is available across R_4 and R_5 in an arrangement very similar to that provided in the μA741. The use of MOS input transistors yields an exceptionally high input resistance and very low input bias currents.

MOS transistors are extremely susceptible to ***electrostatic damage (ESD)***. The electrostatic voltages induced during routine handling are sufficient to puncture the <500-Å SiO_2 gate dielectric, thus destroying the device. Input ESD protection is provided by D_3, D_4, and D_5, which are connected as avalanche diodes. If the gate voltage exceeds some potentially damaging value, these diodes will provide a safer current path operating in the avalanche-breakdown mode.

The output of M_{10}, essentially obtained at the drain, drives the second stage consisting of Q_{13} connected as a common-emitter amplifier with an active load of Q_3 and Q_4. We will demonstrate shortly that most of the voltage gain is provided by the second stage. Because Q_{13} significantly loads M_{10}, the voltage gain of the first stage is relatively low.

The output stage is essentially a class B amplifier, using only *npn* transistors. During the positive half-cycle of operation, Q_{18} supplies power to the load as an emitter follower; Q_{17} is an emitter-follower driver connected as a Darlington. During the negative half-cycle of operation, Q_{16} sinks current from the load. Biasing for Q_{16} is provided by Q_{20},

■ In general, it is not a good idea to totally depend on ESD protection diodes. All MOS input devices should be stored in conducting foam and static free packages. There is an entire industry based upon handling of MOS ICs to minimize ESD.

ELECTRICAL CHARACTERISTICS FOR EQUIPMENT DESIGN
At $V^+ = 15$ V, $V^- = 15$ V, $T_A = 25°C$ Unless Otherwise Specified

Characteristic	Limits CA3140B			CA3140A			CA3140			Units
	Min.	Typ.	Max.	Min.	Typ.	Max.	Min.	Typ.	Max.	
Input Offset Voltage, $|V_{IO}|$	–	0.8	2	–	2	5	–	5	15	mV
Input Offset Current, $|I_{IO}|$	–	0.5	10	–	0.5	20	–	0.5	30	pA
Input current, I_I	–	10	30	–	10	40	–	10	50	pA
Large-Signal Voltage Gain, A_{OL}• (See Figs. 4, 18)	50k	100k	–	20k	100k	–	20k	100k	–	V/V
	94	100	–	86	100	–	86	100	–	dB
Common-Mode Rejection Ratio, CMRR (See Fig. 9)	–	20	50	–	32	320	–	32	320	μV/V
	86	94	–	70	90	–	70	90	–	dB
Common-Mode Input-Voltage Range, V_{ICR} (See Fig. 20)	–15	–15.5 to +12.5	12	–15	–15.5 to +12.5	12	–15	–15.5 to +12.5	11	V
Power-Supply Rejection Ratio, PSRR $\Delta V_{IO}/\Delta V$ (See Fig. 11)	–	32	100	–	100	150	–	100	150	μV/V
	80	90	–	76	80	–	76	80	–	dB
Max. Output Voltage ■ V_{OM}^+ (See Figs. 13,20) V_{OM}^-	+12	13	–	+12	13	–	+12	13	–	V
	–14	–14.4	–	–14	–14.4	–	–14	–14.4	–	
Supply Current, I^+ (See Fig. 7)	–	4	6	–	4	6	–	4	6	mA
Device Dissipation, P_D	–	120	180	–	120	180	–	120	180	mW
Input Offset Voltage Temp. Drift, $\Delta V_{IO}/\Delta T$	–	5	–	–	6	–	–	8	–	μV/°C
Max. Output Voltage,* V_{OM}^+ V_{OM}^-	+19	+19.5	–	–	–	–	–	–	–	V
	–21	–21.4	–	–	–	–	–	–	–	
Large-Signal Voltage Gain, A_{OL} ♦*	20k	50k	–	–	–	–	–	–	–	V/V
	86	94	–	–	–	–	–	–	–	dB

• At $V_O = 26V_{p-p}$, +12V, –14V and $R_L = 2$ kΩ.   ■ At $R_L = 2$ kΩ.

♦ At $V_O = +19V$, –21 V, and $R_L = 2$ kΩ.   *At $V^+ = 22V$, $V^- = –22V$.

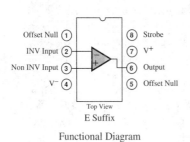

Functional Diagram

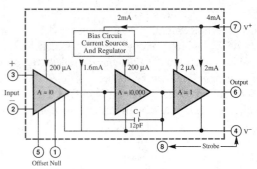

Block Diagram

■ **FIGURE 12.23** CA3140 op amp data sheet (courtesy of GE/RCA Solid State). Now manufactured by Harris Corp.

CHAPTER 12 OPERATIONAL AMPLIFIER EXAMPLES

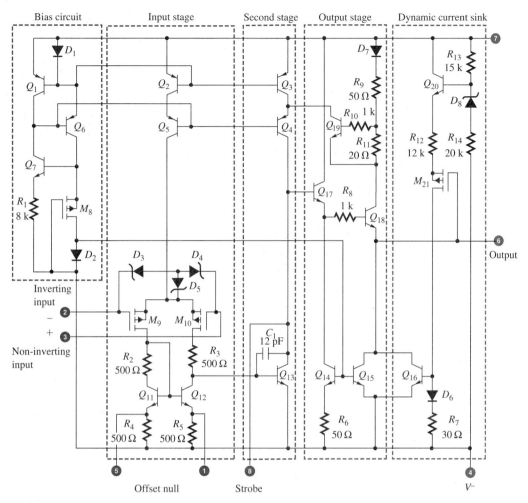

All resistance values are in ohms.

■ **FIGURE 12.23** Continued

M_{21}, and D_8. Short-circuit protection is obtained when Q_{19} turns on because of a sufficiently large voltage drop across R_{11}.

CA3140 AC Circuit Analysis

Referring to Fig. 12.23, the overall voltage gain is computed by considering separately each of the factors in

$$A_v = A_{v1} A_{v2} A_{v3} = \frac{V_o}{V_{id}} = \frac{V_{b13}}{V_{id}} \times \frac{V_{c13}}{V_{b13}} \times \frac{V_o}{V_{c13}}. \qquad (12.60)$$

Specifications for bipolar transistors are given in Table 18.2; for the *p*-channel enhancement-mode MOSFETs in the input stage, assume a threshold voltage of

$V_t = -1.0$ V and an output resistance of $r_d = 200$ kΩ. The output load $R_L = 2$ kΩ.

Recall that the input stage is configured as a source-coupled differential pair with an active load. The voltage gain of this circuit is given essentially by

$$A_{v1} = g_{m10}(r_{d10} \| R_{o12} \| R_{i13}). \tag{12.61}$$

The resistors R_2, R_3, R_4, and R_5, all equal to 500 Ω, are used both for minimizing the input offset voltage and because their values are much less than any of the resistances considered in Eq. (12.61). They are neglected in the voltage-gain calculation.

The g_m of M_{10} is computed by referring to Eqs. (5.12) and (5.14), which are repeated here for convenience:

$$i_D = k(v_{GS} - V_t)^2 \qquad v_{DS} > v_{GS} - V_t$$

and

$$g_m|_{Q\text{-point}} = 2k(v_{GS} - V_t) = 2\sqrt{k i_D}.$$

Combining these two equations to eliminate k yields

$$g_{m10} = \frac{2I_D}{V_{GS} - V_t}. \tag{12.62}$$

It is interesting to compare the MOSFET transconductance with

$$g_m = \frac{I_C}{kT/q} = \frac{I_C}{26 \text{ mV}}$$

at $T = 300$ K for the BJT. Operating in the saturation region of the FET implies that $|V_{GS}| > |V_t|$ by at least several hundred millivolts. Therefore, the MOSFET g_m is usually quite a bit less than the BJT g_m for a given current level.

To use Eq. (12.62), we must determine V_{GS}. Assume both differential inputs are grounded at zero volts; then $V_O = 0$ V, and neglecting the $I_{B18} \times R_8$ voltage drop, the base of Q_{17} is at 1.4 V. The manufacturer states that $I_{C5} = I_{C4} = -200$ μA, so the cascode current sources composed of Q_2–Q_5 and Q_3–Q_4 are operating essentially at the same Q-point V_{CE}. If $V_{B17} = V_{C4} = 1.4$ V, then $V_{C5} \cong 1.4$ V. Therefore, the coupled sources of M_9 and M_{10} are also biased at 1.4 V, with the result that $V_{GS} = -1.4$ V.

Substituting in Eq. (12.62) and recalling that the 200 μA divides equally,

$$g_{m10} = \frac{2I_{D10}}{(V_{GS} - V_t)} = \frac{2(-100 \text{ μA})}{[-1.4 - (-1.0)] \text{ V}} = 500 \text{ μS}. \tag{12.63}$$

The second-stage input resistance is given by

$$R_{i13} = r_{\pi 13} = \frac{\beta_{13}}{g_{m13}} = \frac{\beta_{13}}{\left(\dfrac{I_{C13}}{kT/q}\right)} = \frac{200}{\left(\dfrac{200 \text{ μA}}{26 \text{ mV}}\right)} = 26 \text{ kΩ}. \tag{12.64}$$

The output resistance of the active load Q_{12} is computed from

$$R_{o12} \cong r_{o12}(1 + g_{m12}R_5) = \frac{V_{Anpn}}{I_{C12}}\left(1 + \frac{I_{C12}}{kT/q} \times R_5\right)$$

$$= \frac{150 \text{ V}}{100 \text{ μA}}\left(1 + \frac{100 \text{ μA}}{26 \text{ mV}} \times 500 \text{ Ω}\right) = 4.38 \text{ MΩ}, \tag{12.65}$$

which is so much greater than r_{d10} or R_{i13} that its contribution will be ignored in Eq. (12.61). Using Eq. (12.58) the input-stage voltage gain is computed to be

$$A_{v1} = \frac{V_{b13}}{V_{id}} \approx g_{m13}(r_{d10}\|R_{i13}) = 500\ \mu S(200\ k\Omega\|26\ k\Omega) = 11.5. \quad (12.66)$$

The second-stage voltage gain is obtained from

$$A_{v2} = \frac{V_{c13}}{V_{b13}} = -g_{m13}(r_{o13}\|R_{i17}\|R_{o4}). \quad (12.67)$$

Without any hesitation, we can assume that $R_{i17}\|R_{o4} \gg r_{o13}$. Transistor Q_{17} is connected as an emitter follower whose emitter load resistance is the output resistance of Q_{14} in parallel with the input resistance of another emitter follower, Q_{18}. The composite R_{i17} is large. The output resistance of the cascode current source formed by Q_4 and Q_3 is on the order of $\beta_{pnp}r_{o4}$, a large value. The voltage gain of the second stage, using Eq. (12.67), is

$$A_{v2} \cong -g_{m13}r_{o13} = \frac{-I_{C13}}{V_T} \times \frac{V_{Anpn}}{I_{C13}} = -\frac{150}{26\ mV} = -5770. \quad (12.68)$$

The output-stage voltage gain is $+1$. Substituting the results from Eqs. (12.66) and (12.68) into Eq. (12.60), we get

$$A_v = \frac{V_o}{V_{id}} = (11.5)(-5770)(+1) = -66{,}355. \quad (12.69)$$

This result compares favorably with the typical $A = 100{,}000$ and the minimum value $A = 20{,}000$ given in the data sheet as part of Fig. 12.23.

The differential input resistance cannot be computed from the information available. It is essentially determined by the quality of the SiO_2 used to form the gate and the leakage currents of D_3, D_4, and D_5. The typical $R_{in} = 1.5\ T\Omega$ ($1.5 \times 10^{12}\ \Omega$). This is six orders of magnitude larger than that of a BJT input stage such as in the $\mu A741$. Along with this, the input bias current is specified to be about 10 pA (10×10^{-12} A), which is three orders of magnitude less than for the BJT input. Most of the other CA3140 specifications are comparable to those of a $\mu A741$ except those related to the high-frequency performance. The high-frequency characteristics are determined by the $C_1 = 12$-pF capacitor.

CHECK UP

1. What are the maximum, minimum, and typical values for the voltage gain, input resistance, output resistance, unadjusted input offset voltage, and -3-dB corner frequency for a CA3140 operational amplifier?
2. Qualitatively describe what happens to the parameters in Checkup Question 1 as either β increases or the power supply decreases.
3. Make a list of each of the transistors used and qualitatively describe its function in the circuit.

DRILL 12.4

What is the effective Miller capacitance resulting from the 12 pF compensation capacitor?

ANSWER For the computed second stage gain of -5770, $C_M = 69.3$ nF and for the data sheet nominal second stage gain of 10,000, $C_M = 120$ nF.

12.4 THE LM111 COMPARATOR

■ The comparator and operational amplifier are similar in that the voltage gain is very high.

A *comparator* is a high gain amplifier designed to convert a time-varying analog signal into a binary output. When the signal [applied to the input (+) terminal] is below a given reference voltage [applied to the (−) input terminal], the output has a LOW output state, and when the input rises above the reference, the output assumes the HIGH output state. The LM111 transfer characteristic shown in Fig. 12.24 is typical of comparator characteristics. Depending on the application, the following characteristics may be important features of a comparator design:

1. Often the input signal is produced by a low-voltage or low-current transducer, such as a photodiode sensing light intensity. Thus the comparator is required to have very large amplification to produce the relatively large output transitions from small input changes.
2. Because the input source may be high impedance, the comparator must have a very high input resistance so that the input signal is not attenuated.
3. Often switching time is of the essence, so it is necessary not only for the amplifier to have a short response time for small signals but also for the amplifier not to suffer excessive delay from storage time of saturated transistors when the inputs are overdriven. This means that the slew rate must be satisfactory when the comparator is operated in open-loop fashion.
4. Output levels must be compatible with the logic circuits to be driven by the comparator, regardless of the supply voltages to be used by the comparator circuit itself.

■ Slew rate refers to how fast a comparator or operational amplifier can switch. The units are volts/time. For a μA741, SR = 0.5 V/μsec.

At this point, you might recall the Schmitt trigger circuit of Problem 4.42. This circuit is a special case of the comparator, having hysteresis so that the positive-going signal encounters a threshold different from a negative-going signal. Also, note that these thresholds are internal to the circuit.

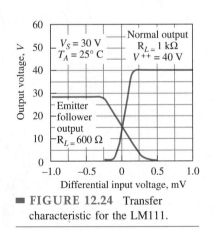

■ **FIGURE 12.24** Transfer characteristic for the LM111.

Qualitative Circuit Description

The complete circuit diagram of the LM111 is shown in Fig. 12.25. To facilitate an explanation of the circuit operation, the amplifier circuits have been redrawn in the simplified schematic shown in Fig. 12.26. Very low input currents are obtained by using the *pnp* transistors Q_1 and Q_2 as emitter-follower buffers. They drive the first standard *npn* differential stage, Q_3 and Q_4. This stage directly drives a second differential stage,

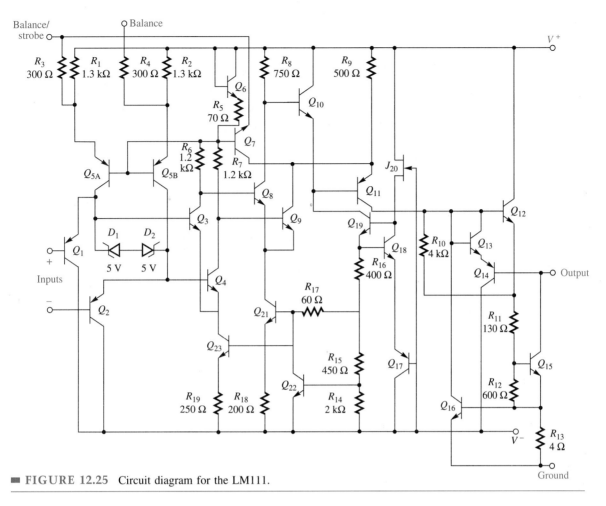

FIGURE 12.25 Circuit diagram for the LM111.

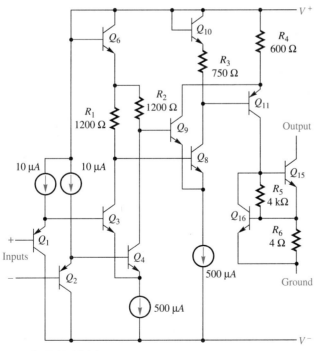

FIGURE 12.26 Simplified schematic of the LM111.

623

Q_8 and Q_9. The second differential stage drives the single-ended *pnp* amplifier Q_{11}, which in turn drives the output amplifier Q_{15}.

Comparison of Figs. 12.25 and 12.26 shows that the double BJT Q_5 (Q_{5A} and Q_{5B}) represent 10-μA current sources so that bias of Q_1 and Q_2 is not dependent on supply voltage or input common-mode voltage. Similarly, Q_{21} and Q_{23} represent 500-μA current sources in the emitter circuits of the differential amplifiers. Transistors Q_{17} through Q_{20} and Q_{22} maintain these currents constant and independent of supply voltage over the range of 5 to 15 V.

Emitter follower Q_{10} is seen to accomplish two functions. It provides the necessary level shifting between Q_8 and Q_{11} to compensate for V_{BE11}. It also provides a high input resistance for the third stage. Forcing the "strobe" terminal LOW (<0.3 V) sinks 3 mA from the terminal and disables the output because Q_7 conducts, turning OFF Q_{11}.

■ R_L connects to V^{++}, which may exceed V_{cc}.

This circuit offers two output modes of operation. In one mode, when the emitter of Q_{16} is grounded, Q_{15} operates as an open-collector output stage, pulling the output LOW (essentially zero volts) when it is ON. When Q_{15} is OFF, an external pull-up resistor, R_L, of about 1 kΩ, pulls the output HIGH to the appropriate logic voltage. In the other mode of operation, when the "ground" terminal is connected to the load, with the "output" terminal connected to the appropriate positive logic voltage, Q_{15} operates as an emitter-follower output stage. The voltage gain is much reduced in this mode. The two output modes are illustrated in the transfer function shown in Fig. 12.24.

The following measures all serve to increase the switching speed: The 5-V avalanche diodes D_1 and D_2 help to prevent large input signals from overdriving successive stages. The output transistor Q_{15} is protected from excessive current by R_{13} and Q_{16}, which diverts the base current from Q_{15} when the drop across the 4-Ω sampling resistor reaches cut-in for Q_{16}. Q_{13} and Q_{14} prevent Q_{15} from saturating.

DRILL 12.5

How does the typical voltage gain of the LM111 comparator compare with the μA741 typical voltage gain?

ANSWER The same, 200,000.

Electrical Characteristics

Figure 12.27 shows a data sheet for the LM111 comparator family.[3] You should be able to interpret these specifications in the same manner as in previous sections. Calculations and various stage gains and other circuit parameters are suggested in the problems.

> ### CHECK UP
>
> 1. What are the typical values for the voltage gain, unadjusted input offset voltage, and response time for the LM111 comparator?
> 2. Qualitatively describe what happens to the parameters in Checkup Question 1 as either β increases or the power supply decreases.
> 3. Make a list of each of the transistors used and qualitatively describe its function in the circuit.

CHAPTER 12 OPERATIONAL AMPLIFIER EXAMPLES

Voltage Comparator
DC Electrical Characteristics

LM111/211/311

Parameter	Test Conditions	LM111/LM211			LM311			Unit
		Min.	Typ.	Max.	Min.	Typ.	Max.	
Input offset voltage[4]	$T_A = 25°C$, $R_S \leq 50\ k\Omega$		0.7	3.0		2.0	7.5	mV
Input offset current[4]	$T_A = 25°C$		4.0	10		6.0	50	nA
Input bias current	$T_A = 25°C$		60	100		100	250	nA
Voltage gain	$T_A = 25°C$		200			200		V/mV
Response time[5]	$T_A = 25°C$		200			200		ns
Saturation voltage	$V_{IN} \leq -5\ mV$, $I_{OUT} = 50\ mA$ $T_A = 25°C$		0.75	1.5		0.75	1.5	V
Strobe on current	$T_A = 25°C$		3.0			3.0		mA
Output leakage current	$V_{IN} \geq 5\ mV$, $V_{OUT} = 35\ V$ $T_A = 25°C$, $I_{STROBE} = 3\ mA$		0.2	10		0.2	50	nA
Input offset voltage[4]	$R_S \leq 50\ k\Omega$			4.0			10	mV
Input offset current[4]				20			70	nA
Input bias current				150			300	nA
Input voltage range	$V = \pm 15\ V$ (Pin 7 may go to 5 V)	−14.5	13.8,−14.7	13.0	−14.5	13.8,−14.7	13.0	V
Saturation voltage	$V+ \geq 4.5\ V$, $V- = 0$ $V_{IN} \leq -6\ mV$, $I_{SINK} \leq 8\ mA$		0.23	0.4		0.23	0.4	V
Output leakage current	$V_{IN} \geq 5\ mV$, $V_{OUT} = 35\ V$		0.1	0.5				μA
Positive supply current	$T_A = 25°C$		5.1	6.0		5.1	7.5	mA
Negative supply current	$T_A = 25°C$		4.1	5.0		4.1	5.0	mA

NOTES
1. This rating applies for ± 15V supplies. The positive input voltage limit is 30V above the negative supply. The negative input voltage limit is equal to the negative supply voltage or 30V below the positive supply, whichever is less.
2. The maximum junction temperature of the LM311 is 110°C. For operating at elevated temperatures, devices in the TO-5 package must be derated based on a thermal resistance of 150°C/W, junction to ambient, in the N package, a thermal resistance of 162° C/W, or °C/W for the Ceramic package. The maximum junction temperature of the LM111 is 150°C, while that of the LM211 is 110°C. For operating at elevated temperatures, devices in the TO-5 package must be derated based on a thermal resistance of 150°C/W, junction to ambient. The thermal resistance of the Cerdip package is 110°C/W, junction to ambient.
3. These specifications apply for $V_S = \pm 15\ V$ and $0°C < T_A < 70°C$ unless otherwise specified. With the LM211, however, all temperature specifications are limited to $-25°C \leq T_A \leq 85°C$ and for the LM111 is limited to $-55°C < T_A < 125°C$. The offset voltage, offset current and bias current specifications apply for any supply voltage from a single 5V supply up to ± 15 V supplies.
4. The offset voltages and offset currents given are the maximum values required to drive the output within a volt of either supply with 1 mA load. Thus, these parameters define an error band and take into account the worst case effects of voltage gain and input impedance.
5. The response time specified is for a 100 mV input step with 5mV overdrive.
6. Do not short the strobe pin to ground; it should be current driven at 3 mA to 5 mA.

Absolute Maximum Ratings

Parameter	Rating	Unit
Total supply voltage	36	V
Output to negative suply voltage:		
LM111/LM211	50	V
LM311	40	V
Ground to negative supply voltage	30	V
Differential input voltage	±30	V
Input voltage[1]	±15	V
Power dissipation[2]	500	mW
Output short circuit duration	10	sec
Operating temperature range		
LM111	−55 to +125	°C
LM211	−25 to +85	°C
LM311	0 to +70	°C
Storage temperature range	−65 to +150	°C
Lead temperature (soldering, 10sec)	300	°C

■ **FIGURE 12.27** LM111 data sheet (courtesy of Signetics Corporation).

12.5 TRENDS IN OPERATIONAL AMPLIFIER PERFORMANCE

The μA741 was selected for analysis primarily because of its past widespread use and its illustrative classic BJT design. Similarly, the CA3140 was selected for study because it illustrates the integration of MOS and BJT technology on a single substrate. Their performance by no means represents what is available for both general and special-purpose applications. There are literally thousands of choices for general-purpose op amps whose unity gain frequency is greater than 5 MHz, slew rate 5 to 10 V/μs, a_v on the order of 10^5 to 10^6, and $V_{OS} < 1$ mV. We now discuss the trends in performance for several recent analog IC designs. It is also important to observe that the user usually pays a premium cost to obtain op amps with special performance capabilities. Often, high-performance analog circuit designs, including operational amplifiers, are incorporated in ASIC systems using a variety of technologies. For example, Tektronix, Inc., offered customers a BJT ASIC array, suitable for high-frequency amplifier design, with $f_T > 9.5$ GHz (QuickChip™).[4,11]

■ QuickChip™ and its production line was sold to MAXIM Semiconductor in mid-1994. The newest process will exceed an $f_T \sim 15$ GHz.

Precision Op Amps

A *precision operational amplifier* is defined as one having a very low V_{OS} along with minimum V_{OS} drift. A low I_B is also desirable. These precision op amps are used for low-frequency, close to dc, and low-signal-level applications. For example, biomedical sensors, thermocouples, and strain gauges are representative signal sources. There are three basic approaches to obtaining these characteristics.[5] BJT op amps, with well-matched inputs, such as the Precision Monolithics OP77AJ, offer $V_{OS} = 0.025$ mV at 25°C with $V_{OS}/T = 0.2$ μV/°C. Compare this with the $V_{OS} = 1$ mV for a μA741. Note that $a_v = 5 \times 10^5$ is quite large also.

The second approach is to use *dielectrically isolated* FET input. The input FETs are fabricated using an SOI technology (Fig. 5.2 and Chapter 18). Although dielectric isolation of the input devices does not guarantee a low V_{OS}, it does make possible very low values of I_B. The Burr-Brown OPA128 offers $I_B = 75 \times 10^{-6}$ nA, which is two to three orders of magnitude less than the I_B specified for the CA3140.

The third approach is to use a *chopper-stabilized* CMOS input stage. As we will study in Chapter 16, we can shift a low-frequency, even dc, signal higher in frequency by multiplying (chopping) the low-frequency signal by a square wave whose fundamental frequency is in a regime in which it is easy to work. The original signal can then be recovered by another multiplying (chopping) operation. The Teledyne TSC911 with $V_{OS} = 0.015$ mV for ±5-V supplies and the Maxim MAX430 with $V_{OS} = 0.005$ mV for ±15-V supplies both use a CMOS chopper-stabilized input stage. The V_{OS} drifts are on the order of 0.15 μV/°C for the TSC911 and 0.05 μV/°C for the MAX430.

It is interesting to observe that the MAX430 is designed to be pin-for-pin compatible with the μA741 pin arrangement. In a sense, this eight-pin arrangement has almost become an industry standard.

Wideband Op Amps

Video and complex high-frequency waveform processing require *wideband* or *video op amps*. These are op amps whose gain and bandwidth performance are key specifications to be maximized and then exploited. Several examples include the RCA (now Harris) CA3450 with a unity gain frequency of 170 MHz, a closed-loop gain of $A_v = 5$ with a full-power bandwidth of 10 MHz. The Comlinear CLC221 has a 170-MHz bandwidth with a 6500 V/μs slew rate. Compare this slew rate to the μA741 slew rate of 0.5 V/μs. It is interesting to observe that the CLC221's $I_B = $ 20,000 nA compared to the μA741's value of 80 nA, and V_{OS} is comparable. Most wideband op amps have a relatively low a_v on the order of a few hundred to a few thousand and require higher biasing currents.

In general, incorporating wideband op amps in a system requires that special attention be given to packaging and power supply decoupling. In addition, most radio-frequency applications and designs have standardized 50- or 75-Ω input and output impedances. For example, Signetics packages their NE5205 op amp in a shielded case with appropriate connectors and power supply decoupling components to obtain $A_v = 20$ dB from dc to 450 MHz.[6]

In this text, we have focused on silicon-based ICs. We are able to use the best features of MOS, JFET, and BJT technology on a single substrate. Applications, especially those involving fiber optic communications, are demanding improved high-frequency performance beyond what may be possible with silicon technology. As mentioned in Chapter 2, GaAs offers five times the electron mobility available in silicon; consequently, we would expect a significant improvement in high-frequency performance from GaAs ICs. Unfortunately, GaAs processing and materials technology problems have been formidable. Only recently[7] has Anadigics introduced commercially the AOP1510, a GaAs monolithic op amp with a GBW = 150 MHz, $a_v = 3000$, and a 400 V/μs slew rate. The ADA25001, also from Anadigics, is fabricated using GaAs MESFETs with 1-μm gate lengths. It offers $A_v = 22$ dB with $f_H = 2.5$ GHz. Recently, the MOSIS (MOS Implementation Service)[8] has made available GaAs foundry service for customers requiring that level of performance.

High-Power Op Amps

Typical power outputs in most op amps are on the order of a few hundred milliwatts. The μA741 is able to provide ± 13 V from a ± 15-V supply to a 2-kΩ load (Fig. 6.15), which corresponds to a load power of

$$P_L = V^2/(2R_L) = (13)^2/(2 \times 2 \text{ k}\Omega) = 42 \text{ mW}. \quad (12.70)$$

Several watts of power output delivered to an 8-Ω load is readily available by fabricating larger output devices and incorporating heat sinking in the package. The CA2004, capable of delivering 12 W at audio frequencies to an 8-Ω load, is representative of this fabrication technology. By providing adequate heat sinking for their power-output transistors in a monolithic design, National Semiconductor is able to obtain 150 W of output power from their LM12 IC.[9] These high-power op amps can be used to drive small motor loads directly or as the combination amplifier and output device in a series-pass voltage regulator (Chapter 17). Often, high-power operational amplifiers are fabricated as a hybrid to take advantage of the best heat dissipating package technology and heat sinking. VMOS and other MOS technologies are often used for high-power applications.

In general, high-power op amps will not perform well at high frequencies. The increased junction capacitance associated with large-area devices restricts their use, in general, to the audio frequency range. Usually, there are no stringent requirements on V_{OS} for these power applications.

Comparators

As with operational amplifiers, a wide variety of comparator circuits is available depending on the application: response times, sensitivity, power supply voltages, and number of devices in a package. The Texas Instruments LM306[10] comparator offers a 25-ns response time compared to a 115-ns response time for the LM111. Both utilize BJT technology. The TL714C is considered to be a very high-speed comparator using Schottky BJTs and its response time is 7 ns. Many linear CMOS comparators, such as the TLC371C, are available with response times of 200 ns. Response time assumes a 100-mV step with a 5-mV overshoot condition. The trend in comparators is to provide a TTL- or CMOS-compatible output interface for a variety of analog input signals. In addition, comparators are being directly integrated as part of A/D and D/A ICs. We will discuss A/D and D/A conversion in Chapter 16.

SUMMARY

A general operational amplifier consists of an input section, usually a differential amplifier with either a passive or active load; an intermediate section, usually a single-ended CE amplifier with either a passive or an active load; and an output section, usually a class B or AB amplifier. This system concept was demonstrated by designing a simple five-transistor operational amplifier using a CA3096 *npn/pnp* transistor array with results verified using a SPICE simulation.

A systematic application of a qualitative device-function analysis, dc circuit analysis, and ac circuit analysis was then applied to the μA741 operational amplifier. The μA741, although one of the earliest commercially available operational amplifiers, is still used as a reference for the design and operation of many newer IC amplifiers. Using an approximate approach to avoid unnecessary algebraic complexity, we obtained values for internal dc currents and the resultant ac values of small-signal voltage gain, differential input resistance, and small-signal output resistance. These computed values were compared to the manufacturer's data sheet and a SPICE simulation.

A similar but less detailed analysis was applied to a CA3140 operational amplifier. The CA3140 was selected because it represents the integration of MOS and bipolar technology on the same die. Key differences, especially the considerably higher input resistance, were noted between the CA3140 and μA741.

Many electronic systems require the conversion of an analog signal to a digital signal and vice versa. A comparator can be used for this analog signal to binary output conversion. We will explore this concept and implementation in more depth in Chapter 16. As an extension of our analysis of typical operational amplifiers, the qualitative circuit description of the LM111 comparator was presented. Additional circuit complexity is the result of incorporating strobe operation and an increase in switching speed.

The circuit analysis approach used to analyze the CA3096 transistor array operational amplifier design, as well as the μA741, CA3140, and LM111, serve as a foundation for understanding a wide variety of analog ICs. In addition we observed, by examining different IC specifications, that there have been considerable improvements in dc performance, bandwidth, and power handling capability.

SURVEY QUESTIONS

1. What are the three principal sections in an operational amplifier?
2. What are the functions and characteristics of the three principal sections of an operational amplifier?
3. Draw a schematic diagram of a basic five-transistor operational amplifier and write a corresponding SPICE netlist.
4. Summarize the dc and ac analysis steps used for the μA741 operational amplifier.
5. Sketch and label the frequency response Bode plots for the μA741.
6. Summarize the dc and ac analysis steps used for the CA 3140 operational amplifier.
7. What are the key properties of a comparator?
8. Qualitatively describe the operation of the LM111 stage by stage.
9. What are the key properties of a precision operational amplifier?
10. What are some of the current technology operational amplifier specifications?
11. What do you expect in terms of operational amplifier specifications in the future?

PROBLEMS

Assume $T = 25°C$ (approximately 300 K) for all problems. Unless otherwise stated, use the transistor parameters from Table 18.2.

Problems marked with an asterisk are more challenging.

12.1 Using the typical values on the CA3096 data sheet, compute values for the following:

(a) f_β
(b) V_A for both the *npn* and *pnp* transistors.

12.2 Draw a small-signal model for the *npn* and *pnp* transistors in a CA3096. Using whatever typical parameter information is available, label as many of the model elements as possible. Assume $I_C = 100\ \mu A$ and $V_{CE} = 5$ V.

12.3 Using *minimum* values for the *npn* and *pnp* transistors, compute A_{vi}, R_{id}, R_{ic}, and R_o for the circuit design in Example 12.1. (Example 12.1 used *typical* transistor parameter values.)

12.4 Verify your Problem 12.3 results using a SPICE simulation.

12.5 Do the same computations as in Problem 12.3, but this time use *maximum* values for both the *npn* and *pnp* transistors.

12.6 Verify your Problem 12.5 results using a SPICE simulation.

12.7 Refer to Fig. 10.24, a CE amplifier with an active load. Assume $V_{CC} = V_{EE} = 12$ V and $I_{C1} = 0.5$ mA. Compute the voltage gain if the typical specifications for CA3096 transistors were used in this circuit. What value of R_1 is required?

12.8 ★ Refer to Fig. 10.29, an emitter-coupled pair with active load. Assume $V_{CC} = V_{EE} = 12$ V and $I_{EE} = 200\ \mu A$. Compute the voltage gain, R_{id} and R_{ic}. Use the typical specifications for the CA3096 transistors.

12.9 ★ Use a SPICE simulation to verify your Problem 12.8 results. Your simulation should include a transfer function to illustrate v_O when $v_{ID} = 0$.

12.10 Design an operational amplifier of the form shown in Fig. 12.3 to obtain an approximate differential voltage gain of -500 using the minimum specifications of the CA3096. Verify all approximations, and compute the resultant R_{id}, R_{ic}, and R_o. Assume $V_{CC} = V_{EE} = 15$ V. Values of resistors should be compatible with monolithic IC fabrication technology. Let $R_L = 10$ kΩ.

12.11 Verify your Problem 12.10 design using a SPICE simulation.

12.12 Assume BJT specifications as given in Table 18.2. For the circuit shown in Fig. 12.28, assume that $I_1 = 350$ μA and is synthesized from a simple current source using an *npn* BJT (Q_3). Similarly, assume that $I_2 = 350$ μA and is synthesized from a simple current source using a *pnp* BJT (Q_4). Assume all transistors are in the linear-active region of operation.

(a) The switch is open. Compute or estimate approximate values for R_{in} (input resistance), the voltage gain defined by $a_{v1} = v_2/v_1$, and the voltage gain defined by $a_{v2} = v_o/v_2$.

(b) The switch is now closed with $R_L = 1$ MΩ. Assume the dc conditions in the circuit are essentially unchanged. Compute an approximate value for the voltage gain $a_V = v_o/v_1$.

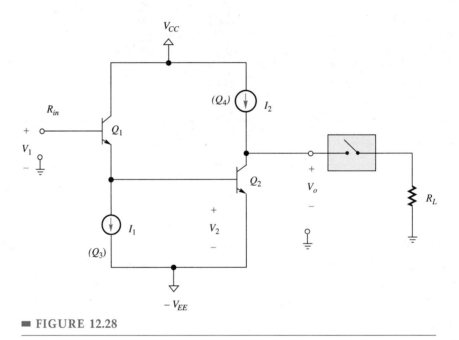

■ **FIGURE 12.28**

12.13 Estimate the dc voltage levels at each of the following points in the μA741 by using the computational results from Section 12.2. Assume $V_{CC} = V_{EE} = 15$ V, both inputs are grounded, and all devices are sufficiently well matched that the output is at 0 V. After estimating these dc levels, verify that all transistors that are supposed to be operating in the active region are actually doing so.

(a) Collector of Q_{10}
(b) Collector of Q_{11}
(c) Collector of Q_{17}
(d) Collector of Q_{12}
(e) Collector of Q_3
(f) Base of Q_{16}
(g) Base of Q_{14}

12.14 Use a dc SPICE analysis to verify your Problem 12.13 calculations.

12.15 Recompute the μA741 voltage gain if R_8 were deleted from the circuit.

12.16* The minimum recommended V_{CC} and V_{EE} for a μA741 is 5 V. Following the development in Section 12.2, compute A_v and R_{id} at this lower power supply voltage. Assume also

$\beta_{npn} = 50$ at these lower levels. Compare your results with those provided on the data sheets at that voltage level.

12.17 R_1 and R_2 are used for minimizing the effects of mismatch in the input stage of the µA741. If they were removed, what would be the new overall voltage gain? Assume $V_{CC} = V_{EE} = 15$ V.

12.18 Compute the µA741 voltage gain and differential input resistance if the bias current to the input stage is increased by 50%, i.e., from 18.9 to 28.4 µA.

12.19 Refer to Fig. 12.29, a schematic diagram of an LM124 operational amplifier. Assume $V_{CC} = 10$ V.

(a) Four current sources are explicitly shown—two 6 µA, one 100 µA, and one 50 µA. Using a single-reference transistor, design a current-source system to provide the indicated currents. Select resistor values compatible with IC technology. Observe that you will have to use both *npn* and *pnp* transistors. You should be able to substitute your design directly in place of the four current-source symbols.

(b) Estimate the small-signal voltage gain defined by V_{C9}/V_{id}. To do this easily, justify the assumption that R_{i10} can be ignored.

(c) Estimate a value for R_{id}.

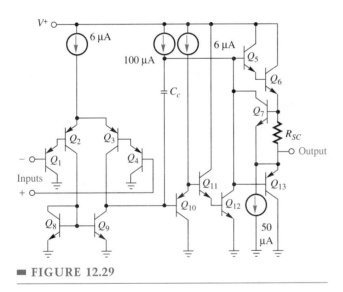

■ **FIGURE 12.29**

12.20 Fig. 12.30 includes a simplified schematic diagram for a National Semiconductor LF351 operational amplifier. Let $V_{CC} = V_{EE} = 12$ V.

(a) Briefly describe the circuit function and topology of each of the named devices.
J_1
J_2
Q_3
Q_4
Q_5
Q_6
D_1
D_2
D_3

(b) How much power could this circuit deliver to a 10-kΩ R_L?

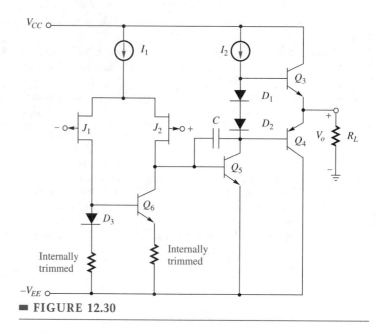

■ FIGURE 12.30

(c) Design an output stage protection circuit. The amplifier can deliver 20 V_{p-p} into a 500Ω load. Show where your design would connect in the circuit.
(d) What key specifications are enhanced by using JFETs?
(e) Assume I_1 and I_2 are about the same. Which stage contributes the most to the operational amplifier voltage gain? Briefly explain your answer assuming reasonable JFET and BJT specifications.

12.21 Figure 12.31 includes a partial schematic diagram of a Fujitsu MB47358 operational amplifier. Assume that the *npn* and *pnp* specifications are as shown in Table 18.2. Let $V_{CC} = 15$ V.

(a) Briefly, describe the circuit function of each of the named devices:
Q_1
Q_2
Q_3
Q_4
Q_5
Q_6
Q_7
Q_8
Q_9
Q_{10}
Q_{11}
Q_{12}
Q_{13}
Q_{14}

(b) Design a current-source system to provide the indicated currents I_1, I_2, and I_3. Use reasonable component values consistent with the double-diffused BJT IC process. By design, draw a labeled schematic diagram with component values. Show how your diagram connects to the circuit and replaces the current source symbols.

(c) Compute a value for the voltage gain defined by $a_v = v_{b8}/v_{id}$. Assume and justify a reasonable value for R_{i7} so that you can perform a simplified a_v calculation.

CHAPTER 12 OPERATIONAL AMPLIFIER EXAMPLES

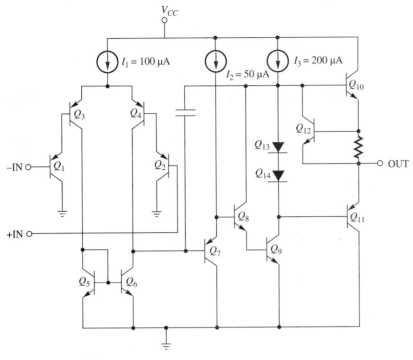

FIGURE 12.31

12.22 Compute the voltage gain of the CA3140 if the bias currents increase 50%, i.e., from 200 to 300 μA.

12.23 Suppose 5 kΩ of emitter degeneration were added to Q_{13} in the CA3140. How would this change the overall voltage gain? Would dc bias conditions and voltage levels still permit reasonable amplifier operation?

12.24* Suppose the threshold voltage V_t for the CA3140 PMOS input-stage transistors was -1.2 V instead of -1.0 V. How would this affect the voltage gain of the overall operational amplifier circuit?

12.25* Observe that the voltage gain for the CA3140 ranges from a minimum of 20,000 to a typical gain of 100,000. Suppose $V_t = -1.2$ V for typical gain and that all gain variations occurs in the PMOS input stage. What V_t will cause the gain to drop by a factor of 5? By what factor will $\mu_p \operatorname{Cox}\left(\dfrac{W}{L}\right)$ change to keep I_D constant.

12.26 Replace the M_9–M_{10} PMOS source-coupled amplifier in the CA3140 with an emitter-coupled *pnp* BJT pair. Verify that the transistors are operating in the active region. Compute the new overall operational amplifier voltage gain and differential input resistance. Compare your results with both the μA741 and CA3140.

12.27* Suppose Q_{13} in the CA3140 were replaced by an NMOS enhancement-mode transistor with $V_t = 0.5$ V, $V_{GS} = 1.0$ V, and $r_d = 200$ kΩ. Verify that the MOSFET would be operating in the saturation region. Compute new values for A_{v1} and A_{v2}, and compare the results with the original circuit.

12.28 Compute the input common-mode voltage range for the CA3140, and compare the results with the values on the data sheet shown in Fig. 12.23.

12.29 Assume for the BJTs in the direct-coupled BiMOS circuit shown in Fig. 12.32 that $r_b = 0$ Ω, $V_A \to \infty$, and for the NMOS, $1/\lambda \to \infty$. Assume $\beta_{npn} = 200$, $\beta_{pnp} = 100$, $1V_{BE(on)} = 0.7$ V. Assume for NMOS, $V_t = 1$ V and $k = 0.5$ mA/V². Neglect r_d, V_A, and λ. Assume all devices are biased to obtain amplifier operation.

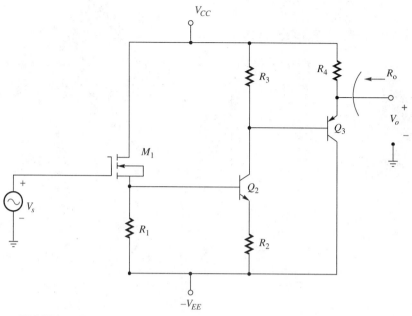

FIGURE 12.32

(a) Estimate the small-signal voltage gain, $A_v = V_o/V_s$.
(b) Suppose a C_{dg} is added to M_1, a $C_{\mu 2}$ to Q_2, and a $C_{\mu 3}$ to Q_3. Furthermore, assume $C_{dg} = C_{\mu 2} = C_{\mu 3}$. Which capacitor will degrade the bandwidth the most? Briefly explain your answer.
(c) Suppose $V_{CC} = V_{EE} = 10$ V and $I_{C3} = 2$ mA. Specify a value for R_4 that yields quiescent $V_o = $ OV (dc).
(d) Let $R_3 = 10$ kΩ and $R_2 = 2$ kΩ and $I_{D1} = 0.5$ mA. Find V_{CE2}, R_1, and the dc value of V_S.

12.30* Refer to Fig. 12.32 and the circuit results given in Problem 12.29(a). Calculate the voltage gain of each stage and compare the overall gain with your Problem 12.29(a) estimate.
Compute the following quantities:

(a) Q-point values for V_{C2} and V_{C3}. Recall that the dc component of the v_{id} small-signal independent generator = 0 V.
(b) Use SPICE to plot the dc transfer curve for V_S ranging from 2V to 6.5V. What is the approximate linear range of V_o? What dc valve of V_S gives maximum sinusoidal output swing?

12.31* Verify your Problem 12.30 analysis using SPICE.
12.32 Refer to the LM111 circuit shown in Fig. 12.25. Assume that the current through R_5 is that of the 0.5-mA current source Q_{23}. Estimate the current provided by each half of Q_5.

Let $I_{snpn} = 5$ fA and $I_{spnp} = 0.1$ fA. Use $V_{BE} = V_T \ln\left(\dfrac{I_C}{I_S}\right)$.

12.33 Repeat Problem 12.32 if the two balance terminals are connected to V^+.
12.34 Using the simplified diagram shown in Fig. 12.26, estimate the input bias currents I_{B1} and I_{B2}. Assume $\beta_{npn} = 200$ and $\beta_{pnp} = 100$.
12.35 Refer to Fig. 12.25. Let the V^- terminal be the voltage reference. Q_{20} maintains the base voltage of Q_{19} at 2.1 V and the base voltage of Q_{18} at 1.4 V, as long as V^+ is at least 5 V.

(a) Neglecting I_{B21} and I_{B23} and assuming $V_{BE22} = 0.7$ V, estimate the voltage at the collector of Q_{22}.

(b) Assuming $V_{BE21} = V_{BE23} = 0.7$ V, estimate the currents I_{C21} and I_{C23}. Compare these with the current sources indicated in Fig. 12.26.

12.36 Refer to the statement of Problem 12.35.

(a) Find the collector current of Q_{19}.
(b) What is the effective load resistance that Q_{19} presents to the emitter follower Q_{10}?

12.37 Refer to Fig. 12.25.

(a) Find the input resistance of the differential pair Q_8–Q_9.
(b) Estimate the differential voltage gain A_v for pair Q_3–Q_4.

12.38 Figure 12.26 gives $I_{EE} = 0.5$ mA for differential amplifier Q_8–Q_9. Refer to Fig. 12.25, and estimate the differential voltage gain A_v for Q_8–Q_9. Do not neglect the loading of the next stage. Assume $I_{C11} = 150$ μA.

12.39 Refer to Fig. 12.25. Assume Q_{11} is biased at 0.15 mA and its collector load is approximated at 50 kΩ. Estimate voltage gain of Q_{11} under these conditions.

12.40 Refer to Fig. 12.25. Assume that Q_{15} is conducting with $I_{C15} = 10$ mA, but it is not saturated; thus Q_{13} and Q_{14} are OFF, and Q_{16} is obviously OFF. For simplicity, let all $V_{BE(act)} = 0.7$ V. Work back through the circuit to find I_{C12}, I_{C11}, I_{C8}, and I_{C9}.

12.41 Repeat Problem 12.40 assuming that Q_{15} is at cut-in, with $V_{BE15} = 0.5$ V and $I_{C15} = 0$. Note the small difference from the results of Problem 12.40.

REFERENCES

1. Harris Corp., Melbourne, Florida. This company prints a series of data books, each covering a different product line. They are updated periodically, often annually. Harris markets the RCA CA-Series ICs.

2. *Linear Division Products Data Book 1990.* Fairchild Semiconductor Corp. (A Schlumberger Company), 333 Western Avenue, South Portland, Maine. Fairchild prints a series of data books, each covering a different product line. They are updated periodically, often annually.

3. *General-Purpose/Linear ICs Data Handbook.* Signetics, Sunnyvale, California, 1992. Signetics is a subsidiary of North American Philips.

4. *QUICKCHIP™ Design Guide, QC7.* Tektronic, Inc., Beaverton, Oregon, 1993. The QUICKCHIP™ line was sold to MAXIM[11] Semiconductor effective mid-1994.

5. Spadaro, J. "In Pursuit of the Ideal Op Amp," *Electronic Products.* Garden City, N.Y.: Hearst Business Communications. August 1, 1986, pp. 55–58.

6. Delurio, T. "UHF Amplifier Delivers Stable Gain," *Electronic Products.* Garden City, N.Y.: Hearst Business Communications. April 15, 1986, pp. 52–55.

7. Spadaro, J. "High-Performance Chips Drive Optical Fibers to Their Limit," *Electronic Products.* Garden City, N.Y.: Hearst Business Communications. June 2, 1986, pp. 63–67.

8. MOSIS MOS Implementation Service is a fabrication service, available to the academic community, subsidized by the National Science Foundation and Advanced Research Projects Agency.

9. Widlar, R., and M. Yamatake, "High-Power Op Amp Provides Diverse Circuit Functions," *EDN.* Newton, Mass.: Cahners Publishing Company. May 29, 1986, pp. 185–192.

10. *Linear Circuits Data Book, Vol 3.* Dallas, Tex.: Texas Instruments, They print a series of data books, each covering a different product line. They are updated periodically, often annually.

11. High-Frequency ASIC Development Handbook, MAXIM Integrated Products, Sunnyvale, California, 1994.

References 5, 6, 7, 9, and 10 represent but a small sample of the information available to the engineer on the technology and application of analog ICs and especially operational amplifiers. You are encouraged to review these technical and trade periodicals continually to remain abreast of the latest ICs and applications. Virtually all manufacturers and vendors of operational amplifier ICs provide a wealth of application, design, and analysis literature for the user. You should avail yourself of this information as well as the information provided in the numerous trade magazines available to the engineering community. Also, many of the IC manufacturers offer printed data books as well as software versions on diskette, CD-ROM, or through the Interest on the World Wide Web. Many of the manufacturers also prepare device SPICE model netlists for their customers in the engineering community.

SUGGESTED READINGS

"Technology '95," *IEEE Spectrum*, Vol. 32, No. 1, Jan. 1995. The Institute of Electrical and Electronics Engineers, New York. The IEEE publishes an annual technology update that provides a broad perspective on new technology, ICs, and applications in addition to other areas of current interest. In addition, you should become familiar with the *Proceedings of the IEEE, IEEE Transactions on Electron Devices*, and the *IEEE Journal of Solid-State Circuits*. These technical journals often provide an advance look at research and development areas leading to new circuits and their applications.

PART III

DIGITAL CIRCUITS

Having studied the operation and modeling of BJTs and FETs in Chapters 4 and 5, we are prepared to look in depth at typical switching circuits. In switching circuits, these active devices operate in discrete states, like ON and OFF, rather than in the active region as linear circuits do. The reader should have a basic understanding of logic functions, and our emphasis will be on the devices. Digital logic circuits are exclusively in the domain of integrated circuits (ICs). Our approach to these topics will be to progress from the elemental building blocks of digital circuits (*gates*) to larger functional circuits to subsystems to systems.

It is helpful to classify digital ICs by degree of integration, such as:

Degree of Integration	Number of Gates	Examples
Small-scale (*SSI*)	Fewer than 10	Uncommitted gates
Medium-scale (*MSI*)	From 10 to 100	Counters, shift registers
Large-scale (*LSI*)	From 100 to 1000	8-bit arithmetic logic unit
Very-large-scale (*VLSI*)	More than 1000	Microprocessor
Ultra-large-scale (*ULSI*)	More than 10^5	System on a chip

In Chapter 13, we focus on the circuits and operation of gates, which will provide familiarization with concepts that are common to all digital circuits and allow us to readily compare the various implementations of these gates. Digital ICs are realized with both FETs and BJTs, in several different technologies. Each of these approaches has certain advantages in speed, power consumption, circuit density on the die, versatility in interconnection, and cost (ease of manufacture). Most of these concepts are best understood at the gate level. Commercially available SSI examples of gate-level logic are compared (CMOS, TTL, ECL), as well as some technologies not available in SSI (NMOS, BiCMOS, MESFET, IIL).

In Chapter 14, a number of examples of digital logic functions that have been implemented in SSI and MSI are examined. These are basic building blocks for all digital circuits, and include both combinational logic (encoders, decoders, multiplexers, demultiplexers, arithmetic logic units, etc.) and sequential logic (flip-flops, shift registers, counters, etc.). Designers of systems where the production volume does not warrant custom IC design would incorporate these functions into a system design by interconnecting MSI packages on a printed-circuit board. Designers of systems of reasonable production volume would incorporate predesigned functional blocks into *application-specific integrated circuits* (ASICs).

Memories are outstanding examples of VLSI circuits. This came about for two reasons. First, memories are large and well-ordered arrays of devices, which made them easy to arrange into a very dense IC (there can be up to several million cells). Second, an intense design effort was justified because of the great potential production volume. In Chapter 15, we consider the ICs that have evolved from this application. NMOS and CMOS are the dominant technologies here, and for programming by the user, there are two NMOS devices called *FAMOS* and *FLOTOX*. *Random-access, read-only,* and *charge-coupled* memories are discussed. Because the same technology is used, we consider *programmable logic arrays* (PLAs and PALs) here.

Analog-to-digital (A/D) and *digital-to-analog* (D/A) converter ICs and systems are the subjects of Chapter 16. These provide examples of the interconnection and/or integration of the functional blocks discussed previously into subsystems that have wide application. Various system approaches to A/D conversion are evaluated.

Chapter 17 provides additional examples of IC system applications, each an important topic in its own right. Included are switched-mode power supplies and regulators, four-quadrant multipliers, and phase-locked loop circuits.

Part III may be undertaken immediately following Part I because it is independent of Part II.

CHAPTER 13

INTEGRATED-CIRCUIT LOGIC GATES

13.1 Digital Operation of Circuits
13.2 Basic Gate Terminology
13.3 Early Integrated-Circuit Logic Families
13.4 Transistor-Transistor Logic
13.5 Emitter-Coupled Logic
13.6 NMOS Logic
13.7 Complementary MOS Logic
13.8 Integrated-Injection Logic
13.9 Comparison of Logic Families
Summary
Survey Questions
Problems
References
Suggested Readings

Digital (binary) systems ranging from very complex computers down to elementary switching circuits are made up of simple logic building blocks called *gates*. These logic circuits provide both the simplest and most widespread transistor applications. At the same time, logic gates lead very naturally to the fabrication of many transistors on the same silicon chip, or integrated circuits (ICs).

This chapter assumes that the reader is familiar with the **NAND** and **NOR** functions of a binary variable (reviewed in Appendix C). If you have had no prior exposure to the subject of digital logic theory, you can benefit from some additional reading.[1] We now proceed to analyze circuit realizations of representative gates, with the goal of familiarizing the student with fundamental digital electronic circuits.

Short discussions of the now-obsolete resistor-transistor logic (RTL) and diode-transistor logic (DTL) are included to show the evolution of IC logic families. RTL is also a good pedagogical vehicle to define basic gate and transfer characteristic terminology. A *logic family* is a set of similarly configured logic circuits that can be interconnected according to simple rules without fear that driving-gate output voltages are inappropriate for driven-gate inputs. Although analysis of the internal circuit details is not necessary to enable the logic designer to use integrated logic packages in a system design, it does provide great insight into some of the design trade-offs made by the IC designer. Many engineers work at the device and gate levels of design in addition to the system level.

The dominant logic families as of this writing, transistor-transistor logic (TTL), emitter-coupled logic (ECL), and complementary MOS logic (CMOS) are treated in some depth in this chapter. Other digital technologies not associated with SSI families, such as NMOS, BiCMOS, and GaAs MESFETs, are also discussed. As an optional topic, integrated-injection logic (IIL, or I^2L) is also discussed, because many of the concepts, such as merged transistors and current-mode switching, are very much involved in present-day circuit development.

In new systems, dominance is passing from TTL to CMOS, and IC design is moving from interconnecting SSI and MSI families to *application-specific integrated circuits* (ASICs). We will refer to ASIC design in Chapters 14 and 15.

IMPORTANT CONCEPTS IN THIS CHAPTER

- Noise margin is a measure of the ability of a logic family to perform properly in the presence of temperature and supply voltage variations and other noise.
- Fan-out is the number of inputs that a given gate is capable of driving properly.
- BJT saturating logic, like RTL, DTL, and standard TTL, experiences delays due to storage time in saturated devices.
- TTL logic using Schottky BJTs avoids the storage-time delay of standard TTL.
- Tristate output circuits allow buses to be driven without conflict.
- ECL uses a current switching arrangement to avoid saturation of the BJTs.
- Subnanosecond rise times require careful circuit layout and line terminations.
- MOS logic offers circuit density and fabrication cost advantages over BJT logic.
- NMOS logic with enhancement-mode drivers and depletion-mode loads constitutes a compact technology, viable for VLSI such as memories.
- Both power dissipation and delay time in MOS circuits are proportional to C_T.
- GaAs MESFET gates are extremely fast because of the high mobility of electrons in GaAs, and the saturated velocity is higher.
- CMOS offers reduced rise time over NMOS, and negligible standby power.
- Power dissipation in CMOS is proportional to switching rate.
- BiCMOS combines the low-power, compact, and fast CMOS gate with the current capacity of the BJT output stage.
- IIL is a very compact technology, but is at a disadvantage because its logic levels are incompatible with other digital technologies.

13.1 DIGITAL OPERATION OF CIRCUITS

Ideally, the electrical voltages in *digital* circuits assume only discrete levels in contrast with *analog* circuits in which the signals vary over a continuum of levels (as long as operation is confined to the linear region of the devices). Real digital circuits only approach this ideal. Further, in most digital circuits, the voltages take on only two values. These *binary* states are variously referred to as true-false, HIGH-LOW, ON-OFF, etc. We shall also refer to them as the 1 state and the 0 state.

The actual circuit voltages used to represent these two states vary from system to system. When the voltage used to represent the 1 state is more positive than the voltage that represents the 0 state, the circuit is said to be ***positive logic***. We will use positive logic throughout this text, in accord with most designers. To accommodate normal variations in component values, temperature, and noise, the voltage levels assigned to the binary states have a range, separated by a region of unpermitted voltages. As an example, a certain logic family may define the 1 state as any voltage between 2 and 5 V, with the 0 state being defined as any voltage between 0 and 0.8 V. Voltages ranging between 0.8 and 2 V must not be used because the circuit may not be able to define the state unambiguously as a 1 or a 0. This 1.2-V difference between the highest voltage that will always be recognized as a 0 and the lowest voltage that will always be recognized as a 1 is called the *transition region* of the circuit.

The reader should keep in mind that all inputs of each gate circuit to be examined are assumed to be driven by gates that are identical in output characteristics to the one under discussion. The number of gate inputs M is referred to as the *fan-in*. Likewise, the output of the gate of interest is assumed to be connected to the inputs of N gates that have input characteristics just like the one being studied. This means that the output terminal is loaded by N input currents. The number of gates loading an output, N, is referred to as the *fan-out*.

13.2 BASIC GATE TERMINOLOGY

Diode circuit realizations of AND and OR gates were discussed in Section 3.10 and shown in Figs. 3.37 and 3.38. We found that these circuits have difficulty when they are interconnected because the input of one gate severely loads the output of another gate (Problems 3.49 and 3.50). Circuit designers have turned to active devices to provide input-output isolation and the current gain necessary to overcome these loading problems. An early attempt in this direction consisted of discrete resistors and transistors and was called *resistor-transistor logic* (RTL). RTL serves as a vehicle for basic definitions valid for all logic families.

Transfer Characteristic of an RTL Inverter

We analyzed the simple inverter shown in Fig. 13.1 in Section 4.7. The transfer characteristic reveals that the output is HIGH when the input is LOW, $<V_{BE(\text{cut-in})}$. The collector resistor is called a *pull-up* resistor because the output is pulled up toward the positive supply voltage when the transistor is OFF. The output is LOW when the input voltage is high enough to cause a base current sufficient to saturate the transistor. We now define the following general terms, which we will apply to all gate circuits:

V_{OH} = gate output voltage when the output signal is supposed to be logic 1. This varies with the loading and temperature, so manufacturers guarantee that the output will exceed a specified *minimum* value of V_{OH} for all conditions that have been specified on the data sheet.

V_{OL} = gate output voltage when the output signal is supposed to be logic 0. Data sheets will specify a *maximum* value V_{OL}, and guarantee that the output voltage will not exceed this value at all specified operating conditions.

V_{IL} = *maximum* input voltage that the gate will always properly interpret as logic 0. Allowance for variations in operating conditions within the data sheet specifications has been made.

V_{IH} = *minimum* input voltage that the gate will always properly interpret as logic 1. Allowance for variations in operating conditions within the data sheet specifications has been made.

DRILL 13.1

Refer to Example 4.15 in Section 4.12. This is a SPICE program to plot the transfer characteristic of the RTL inverter of Fig. 13.1. Compare the results, shown in Fig. 4.41, with our sketch in Fig. 13.1(b). Comments?

ANSWER The 27°C plot of Fig. 4.41 is nearly identical with Fig. 13.1(b). This suggests that our somewhat approximate assumptions were reasonable.

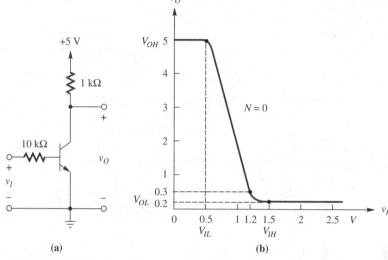

■ **FIGURE 13.1** Transistor inverter. (a) Circuit. (b) Transfer characteristic.

For the example in Fig. 13.1(b),

$V_{IL} = 0.5$ V (The transistor is OFF until $V_{BE} > V_{BE(\text{cut-in})}$.)

$V_{OH} = 5$ V (The transistor is cut off, $I_C = 0$, and $V_{OH} = V_{CC}$ if $N = 0$, $I_L = 0$.)
Note that a gate with $N = 0$ would not be very useful, so $V_{OH} = 5$ V is unlikely.

$V_{OL} = 0.2$ V (When ON, the BJT is assumed saturated, $V_{CE(\text{sat})} = 0.2$ V.)

For $V_{CE(\text{sat})} = 0.2$ V, $I_C = (5.0 - 0.2)$ V/1 kΩ = 4.8 mA for the special case of $N = 0$, $I_L = 0$. The value of I_B is greater than 4.8 mA/β_F because of saturation. Figure 4.25 predicts $\sigma = 0.7$ for $V_{CE} = 0.2$ V, thus $I_B = I_C/\sigma\beta_F$, and with $\beta_F = 100$ we obtain overdrive $I_B = 70$ μA. Now we can determine

$V_{IH} = 1.5$ V ($V_{IH} = I_B R_B + V_{BE(\text{sat})} = 70$ μA × 10 kΩ + 0.8 V = 1.5 V.)

Noise Margin

When the output of the gate is connected to the inputs of other gates, the collector voltage with the BJT cutoff, V_{OH}, is decreased because of load currents in the collector resistor. Consider the situation in Fig. 13.2, where N other inverters are connected to the output of Q_1. When the input V_I is LOW, Q_1 is off. The N base currents are each

$$I_B = \frac{V_{OH} - V_{BE(\text{sat})}}{R_B} \tag{13.1}$$

and

$$V_{OH} = V_{CC} - N I_B R_C. \tag{13.2}$$

Combining,

$$V_{OH} = \frac{V_{CC} + \dfrac{NR_C V_{BE(\text{sat})}}{R_B}}{1 + \dfrac{NR_C}{R_B}} = \frac{R_B V_{CC} + NR_C V_{BE(\text{sat})}}{R_B + NR_C} \tag{13.3}$$

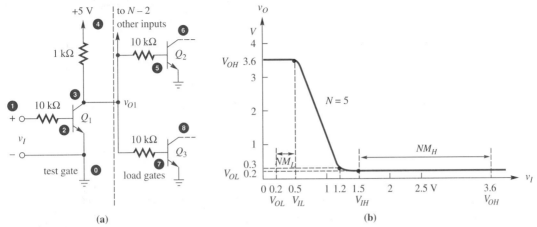

FIGURE 13.2 Inverter driving N other gates. (a) Circuit. (b) Transfer characteristics for $N = 5$. Node numbers are for Ex. 13.2.

so that for this circuit with $N = 5$, $V_{OH} = 3.6$ V. This value cannot be less than the desired minimum V_{OH}, and thus N is limited.

The transfer characteristic for $N = 5$ would appear exactly as in Fig. 13.2(b), except that V_{OH} is 3.6 V. We find that each of the base currents is 0.28 mA, much more than the 0.07 mA required to saturate the driven gates. The excess of V_{OH} over the minimum value of V_{OH} required to barely saturate the driven load gates ($V_{OH} - V_{IH} = 3.6 - 1.5 = 2.1$ V in this example) is called the **HIGH-level** *noise margin*, NM_H, of the test circuit, in this case, for $N = 5$ [see Fig. 13.2(b)].

In general, the HIGH-level output of the driving gate must serve as the HIGH-level input for the load gates, and NM_H is found as follows:

$$NM_H = V_{OH} - V_{IH}. \tag{13.4}$$

This is the safety factor that allows the driven gates to remain saturated even with the inevitable changes in supply voltages, temperatures, and manufacturing tolerances.

The difference in voltage between the maximum permitted input voltage for the LOW state and the actual driver output voltage in the LOW state ($V_{IL} - V_{OL}$) is called the **LOW-level noise margin, NM_L**. Again, refer to Fig. 13.2(b).

$$NM_L = V_{IL} - V_{OL}. \tag{13.5}$$

In this example, $NM_L = 0.5$ V $- 0.2$ V $= 0.3$ V.

■ Noise margins are a measure of how immune a logic gate is to unpredictable circuit disturbances, such as transients on the power lines. Note that noise margins will vary with fan-out.

Fan-out

Our example in Fig. 13.2 had a fan-out of 5 ($N = 5$). Observe that with input LOW, the number of gates that the collector resistor of the driving gate can "pull" HIGH limits the fan-out with RTL. The absolute maximum value of N at which this circuit could still function ($NM_H \Rightarrow 0$) is 50. With input HIGH, Q_1 saturates, and any number of load gates could be connected to its output; and since this output is below cut-in voltage for the load gate BJTs, they are all OFF.

> ### EXAMPLE 13.1
>
> Verify the maximum fan-out for the gate in Fig. 13.1(a). Assume $\beta_F = 100$, $V_{CE(\text{sat})} = 0.2$ V, $V_{BE(\text{sat})} = 0.8$ V.
>
> **SOLUTION** Use Eq. (13.3) with $V_{OH} = V_{IH} = 1.5$ V. Then $1.5[1 + (1/10)N] = 5 + (1/10)N(0.8)$ leads to $N = 50$. This is the number of load gates that this gate can pull HIGH. Fanout$_{(\text{LOW})} \to \infty$.

In addition to the dc fan-out considerations, each input contributes some parasitic capacitance to the driver output node, so large fan-out may be limited by switching-speed considerations. When Q_1 turns OFF, all of the load gate input capacitance plus the stray wiring capacitance must charge to V_{OH} through the collector pull-up resistor.

Switching Speed

Consider the waveforms shown in Fig. 13.3 that relate to an inverting logic circuit such as our example in Fig. 13.2. The numbers used assume a fan-out of 5. Suppose that the waveform v_{IN} is applied at v_{IL}. When v_{IN} makes the HIGH-to-LOW transition, v_O makes a LOW-to-HIGH transition after a delay t_{pLH}. Similarly, when v_{IN} goes HIGH, v_O switches LOW after a delay t_{pHL}. The average of these two delays is called the ***propagation delay*** of the gate, t_{pd}. These times are measured by means of the voltage levels on the waveforms midway between the two logic levels. In this case, 1.9 V is the average of 0.2 and 3.6 V.

■ Note that the identifying subscripts on these delays refer to the *output* waveform transitions.

Note that t_{pLH} consists of two consecutive delays, referred to as storage time and fall time in Section 4.11. First, Q_1 does not come out of saturation until the charge stored in the base region (in excess of that required just to reach the saturation region) is depleted by recombination and negative base current. Second, the charge represented by the excess of minority-carrier distribution over the equilibrium concentration must also be removed before the collector current falls toward cutoff.

■ $t_{pd} = \dfrac{t_{pLH} + t_{pHL}}{2}$

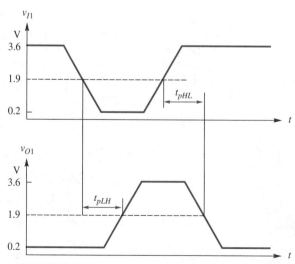

■ **FIGURE 13.3** Input and output gate voltage waveforms showing propagation delay.

Figure 13.4 shows the situation after Q_1 turns OFF. Each of the five driven inputs has a base circuit that we will attempt to model with a capacitance C_B. This, of course, is quite crude, since the charge-control model (Section 4.11) is nonlinear, but the effect will be demonstrated. As far as the output node is concerned, the circuit is equivalent if we parallel the five identical branches. The time constant is seen to be

$$\tau_1 = 5C_B\left(1\text{ k}\Omega + \frac{10\text{ k}\Omega}{5}\right).$$

Thus we see that the response time of the driven gates is affected by the fan-out, since all of the bases must charge through the one pull-up resistor.

Delay t_{pHL} involves the delay time t_d and the rise time t_r of Fig. 4.35. The minority-carrier distribution within the base must first reach the level of the active region and then the saturation level. The times required for these redistributions of charge are predictable if the appropriate time constants are known.

After Q_1 saturates, the time constant at the base of each driven transistor is $\tau_2 = C_B\ 10\text{ k}\Omega$, which in this case would be less than τ_1.

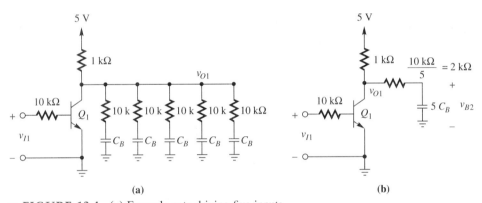

■ **FIGURE 13.4** (a) Example gate driving five inputs.
(b) Equivalent circuit.

EXAMPLE 13.2

Use SPICE to plot the response at the base of the driven transistors in the RTL circuit of Fig. 13.2. Let $N = 2$, and let all transistors have the following model statement:
.MODEL T1 NPN TF=.1N TR=10N CJC=1P CJE=2P IS=1E-14.

SOLUTION Here is a netlist:

```
RTL TRANSIENT RESPONSE
VIN 1 0 PULSE (0 5 4N 4N 4N 200N 300N)
VCC 4 0 DC 5
RB1 1 2 10K
RC1 4 3 1K
```

DRILL 13.2

Refer to Fig. 13.5, and determine $V_{BE(\text{cut-in})}$ and $V_{BE(\text{sat})}$ for the driven BJT.

ANSWER The plot V(5) indicates that $V_{BE(\text{cut-in})} = 0.6$ V, and $V_{BE(\text{sat})} = 0.7$ V.

```
Q1 3 2 0 T1
.MODEL T1 NPN TF=.1N TR=10N CJC=1P CJE=2P IS=1E-14
RB2 3 5 10K
Q2 6 5 0 T1
RC2 4 6 1K
RB3 3 7 10K
Q3 8 7 0 T1
RC3 4 8 1K
.TRAN 4N 300N
.PLOT TRAN (VIN) V(3) V(5) V(6)
.END
```

From this we obtain the plots of Fig. 13.5, where the response of the driven bases is shown by curve V(5). Our conclusions in the preceding paragraphs seem to be verified.

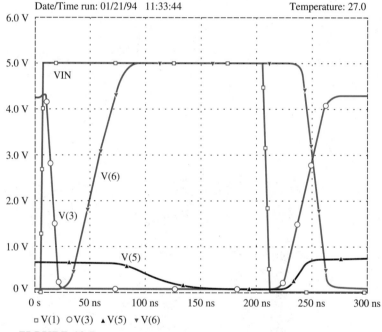

FIGURE 13.5 SPICE transient response for the RTL circuit to input pulse v_{IN}.

CHECK UP

1. **TRUE OR FALSE?** The gate output voltages V_{OH} and V_{OL} depend on loading.
2. **TRUE OR FALSE?** The noise margins NM_H and NM_L of a logic family are independent of loading.
3. **TRUE OR FALSE?** Desired switching speed limits the practical fan-out in a logic circuit.

CHAPTER 13 INTEGRATED-CIRCUIT LOGIC GATES

13.3 EARLY INTEGRATED-CIRCUIT LOGIC FAMILIES

Resistor-Transistor Logic

After initial development of RTL circuits with discrete resistors and transistors, manufacturers quickly noted the great savings in space and cost they could achieve by fabricating the whole gate, or several gates, on a single silicon chip. These circuits were realized by relatively primitive techniques compared to some of today's technology. The integrated-circuit revolution was under way by about 1961.

RTL families are made up of inverters and NOR gates. The MC914 circuit shown in Fig. 13.6 is a two-input gate from the first mass-produced IC logic family. Observe that the gate does indeed produce the NOR function: The output is LOW when either A OR B is HIGH. Circuits with more inputs are fabricated with additional transistors in parallel with Q_1 and Q_2. Note that the NOR gate is functionally complete; all other operations can be obtained by combination of NOR gates. The two-input gate is said to have a fan-in of $M = 2$, meaning that two logical variables are combined in the gate.

■ While RTL is obsolete, these simple circuits provide opportunity to define some terms applicable to all logic gates, and to observe the evolution of digital technology.

Fan-out. We now analyze the commercial RTL gate presented in Fig. 13.6(a), and compare our findings with the manufacturer's guarantees of performance. The piecewise linear transfer characteristic is very helpful in studying the operation of any logic gate, and we will develop it for each family that we investigate in this chapter. Let $v_{IN} = v_A$ or v_B, whichever is more positive.

The value of V_{IH} must be high enough to saturate the gate transistor. If we decide that the maximum value of $V_{OL} = V_{CE} = 0.2$ V, the collector current would be

$$I_C = \frac{(3.0 - 0.2) \text{ V}}{640 \text{ }\Omega} = 4.37 \text{ mA}.$$

If we assume a β_F of 50, Fig. 4.25 indicates that the required base overdrive factor is $\sigma = I_C/\beta I_B = 0.85$, and thus the required base current is

$$I_B = \frac{I_C}{0.85\beta_F} = \frac{4.37 \text{ mA}}{42.5} = 103 \text{ }\mu\text{A}.$$

At $T = 25°C$, the minimum value of V_{IH} would be

$$V_{IH} = V_{BE(\text{sat})} + (103 \text{ }\mu\text{A})(450 \text{ }\Omega) = 0.80 + 0.05 = 0.85 \text{ V}.$$

A value for $V_{BE(\text{sat})}$ of 0.80 V is fairly typical of IC transistors, subject to the modifications to be discussed. Fan-out is limited by the current that the pull-up resistor can supply to keep V_{OH} above the minimum value of $V_{IH} = 0.85$ V.

DRILL 13.3

The RTL inverter of Fig. 13.4(a) has R_B much larger than R_C, whereas the RTL inverter of Fig. 13.6(a) has R_B much smaller, even smaller than R_C. Can you see the advantage of making R_B smaller than the first case?

ANSWER As we have seen, a smaller R_B would reduce the time constant in the base circuit of the driven transistor, both for turn-on and for turn-off, assuming that C_B is not increased.

EXAMPLE 13.3

Find the maximum fan-out for the RTL gate shown in Fig. 13.6(a). Let $\beta_F = 50$, $V_{BE(\text{sat})} = 0.80$ V, and $V_{CE(\text{sat})} = 0.2$ V.

SOLUTION Let N_{max} be the maximum number of load gates that may be driven by the output. Since each load gate requires 103 μA, at 25°C

$$I_{640} = NI_B = N \times 103 \text{ }\mu\text{A} = \frac{(3.0 - 0.85)}{640} = 3.36 \text{ mA},$$

■ I_{640} is the current through the 640Ω resistor.

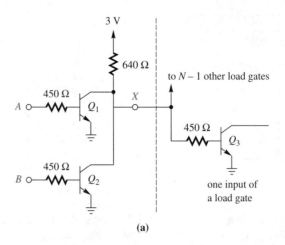

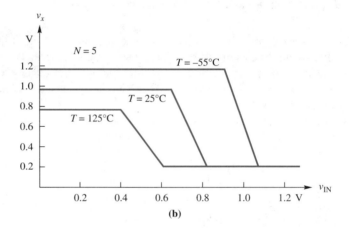

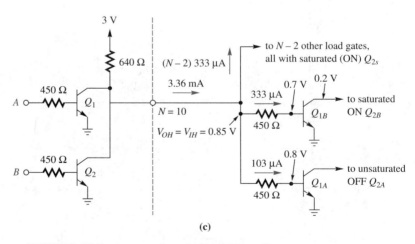

■ **FIGURE 13.6** One gate from MC914 RTL NOR.
(a) Circuit of a gate with load.
(b) Transfer characteristic for $N = 5$.
(c) Circuit illustrating current hogging.

CHAPTER 13 INTEGRATED-CIRCUIT LOGIC GATES

leading to $N_{max} = 32$. This would be unrealistic in practice for at least two reasons. (1) At lower temperatures, V_{BE} increases and β_F drops. For example, at $-55°C$, $V_{BE(sat)}$ would increase 2.2 mV/°C to $0.80 + [25 - (-55)](0.0022) = 0.98$ V, and β_F would halve to 25 (see Fig. 4.31). The same calculation we used above would now lead to $I_B = 206$ μA ($\sigma \approx 0.85$ at $T = -55°C$),

$$V_{IH} = 0.98 + (206 \text{ μA})(450 \text{ Ω}) = 1.07 \text{ V},$$

and

$$N_{max} = \frac{(3.0 - 1.07) \text{ V}}{(640 \text{ Ω})(206 \text{ μA})} = 15.$$

(2) While the $V_{BE(sat)}$ of transistor Q_1 has been estimated at 0.80 V when Q_2 is OFF, it drops to about 0.1 V less when Q_2 is saturated. This follows from Eq. (4.8), using the two different values of I_{C1} (see Problem 13.9). This results in what is called **current hogging**. As one possible example, suppose the A inputs of N gates were connected to the output of interest. Refer to Fig. 13.6(c). Suppose further that $N - 1$ of these gates had their Q_2s already saturated because their B inputs were HIGH. Thus, at 25°C, one gate requires that $I_B = 103$ μA and $V_{IH} = 0.85$ V postulated above, but all $N - 1$ other gates have a base voltage of only 0.70 V and hence a base current of

$$I_B = \frac{(0.85 - 0.70)}{450} = 333 \text{ μA}!$$

This time

$$I_{640} = (N - 1)(0.333) + 0.103 = 3.36 \text{ mA}$$

leads to $N_{max} = 10$. Note that the inputs that do not need to be turned ON are demanding the most current. At $T = -55°C$, the combined effects reduce N_{max} to about 7:

$$I_{640} = \frac{(N - 1)(1.09 - 0.90) \text{ V}}{450 \text{ Ω}} + 0.206 \text{ mA} = \frac{(3 - 1.09) \text{ V}}{640 \text{ Ω}}.$$

You can now understand why the manufacturer suggested a fan-out of only 5! The smaller N also reduces t_{pLH}, as shown previously.

EXAMPLE 13.4

Refer to Fig. 13.6. Use SPICE to demonstrate that V_{BE} of Q_1 is higher when Q_2 is OFF than when it is ON, i.e., current hogging is taking place. Use the default model for the BJTs.

SOLUTION Shown is a netlist for the situation with both BJTs ON (VA = VB = 1). the other netlist has Q_1 ON and Q_2 OFF (VA = 1 and VB = 0).

```
EXAMPLE 13.4(a)        EXAMPLE 13.4(b)
VDD  1 0 DC 3          VDD  1 0 DC 3
VA   2 0 DC 1          VA   2 0 DC 1
VB   3 0 DC 1          VB   3 0 DC 0
RA   2 4 450           RA   2 4 450
```

```
RB   3 5 450              RB   3 5 450
Q1   6 4 0 T1             Q1   6 4 0 T1
Q2   6 5 0 T1             Q2   6 5 0 T1
RC   1 6 640              RC   1 6 640
.MODEL T1 NPN             .MODEL T1 NPN
.OP                       .OP
.END                      .END
```

We find that with Q_2 ON, $V_{BE1} = 0.80$ V, but with Q_2 OFF, $V_{BE1} = 0.82$ V.

RTL Transfer Characteristic. Using a fan-out of 5, let us construct the input-output characteristic shown in Fig. 13.6(b). We have shown V_{IH} to be 0.85 V at 25°C and 1.09 V at −55°C under the assumptions made for β_F and V_{BE}. Using a worst case value of $V_{BE(sat)} = 0.90$ V at $T = -55°C$ (assumes current hogging) in Eq. (13.3) yields $V_{OH} = 1.16$ V. At 25°C, using $V_{BE(sat)} = 0.70$ V produces $V_{OH} = 0.98$ V. The V_{IL} term, or $V_{BE(cut-in)}$, for the small IC transistors is best estimated at 0.6 V for 25°C, with a -2.2 mV/°C temperature correction. Problem 13.8 verifies the plot given for $T = 125°C$.

RTL Noise Margin. As discussed in the previous section, noise margins are easily found from the transfer function. Figure 13.6(b) reveals that our example has the following values:

- $T = -55°C$ $NM_L = 0.8 - 0.2 = 0.6$ V $NM_H = 1.16 - 1.09 = 0.07$ V
- $T = 25°C$ $NM_L = 0.6 - 0.2 = 0.4$ V $NM_H = 0.98 - 0.85 = 0.13$ V
- $T = 125°C$ $NM_L = 0.35 - 0.2 = 0.15$ V $NM_H = 0.74 - 0.58 = 0.16$ V

These values are admittedly pessimistic, using worst case assumptions, but they indicate what the constraints are. In this discussion we have assumed a supply voltage of 3 V. You may show that increasing this voltage does not have an appreciable effect on fan-out; it helps the noise margin slightly, but increases the power dissipation linearly. We can summarize RTL specifications at 25°C as

- Fan-out: 5 (nominal)
- Noise margin: 0.13 V
- Propagation delay: 12 ns
- Power dissipation/gate: 11 mW (see Problem 13.10).

Because of the relatively poor fan-out and noise margin, RTL had given way to other logic families by the late 1960s.

Direct-Coupled Transistor Logic (DCTL)

As resistor-transistor logic was developing, a version of the circuit shown in Fig. 13.6(a) was attempted without the base resistors. The idea was to save the considerable chip space occupied by these resistors. Realize that the value of V_{OH} would now be $V_{BE(sat)} = V_{IH}$, and the total difference between V_{OH} and V_{OL} is only about $0.80 - .20 = 0.60$ V. The most serious fault of this design was the current-hogging problem we discussed in the previous section. Without the base resistors to aid in equalizing the base currents, it was very difficult to saturate all of the gate inputs when the fan-out was any larger than 2 or 3. For this reason, the arrangement did not become widely used. We will consider this idea again, however, in integrated-injection logic (Section 13.8), which is a technology of current interest.

Diode-Transistor Logic

We now consider a slightly more complex gate that uses diodes in addition to transistors and resistors. Diode-transistor logic (DTL) achieved greater fan-out and superior noise margins than RTL by reducing load currents, but suffered from greater propagation delay. DTL enjoyed several years of popularity, but it, in turn, was superceded by faster, more advanced circuits. We refer to DTL here as a step in the evolution of bipolar digital logic families and to further illustrate the basic requirements of digital circuits.

The circuit shown in Fig. 13.7(a) is a two-input DTL NAND gate. To begin our analysis, assume that when inputs A and B are both HIGH (5 V), diodes D_A and D_B are both OFF. Base current flows through the resistors and transistor Q_1. For simplicity, we assume that the conducting diode has a voltage drop of 0.7 V, as does the base-emitter of active transistor Q_1. We thus can estimate the voltage at point X to be

$$V_X = V_{BE2(\text{sat})} + V_{D2} + V_{BE1(\text{act})} = 0.8 + 0.7 + 0.7 = 2.2 \text{ V}.$$

Then I_{B1} is found by writing $5 - V_X = 2.8 = 1.6 \text{ k}(1 + \beta_1)I_{B1} + 2.15 \text{ k}(I_{B1})$. Assuming $\beta_1 = 30$, $I_{B1} = 54 \text{ μA}$. Also, $I_{D2} = (1 + \beta_1)I_{B1} = 1.68 \text{ mA}$, $I_{5k} = 0.8/5 = 0.16 \text{ mA}$, and $I_{B2} = I_{D2} - I_{5k} = 1.52 \text{ mA}$. Thus Q_2 is ON, and will be saturated provided that the collector current I_{C2} is limited to less than $\beta_F I_{B2} = (30)(1.52) = 45 \text{ mA}$.

DTL Transfer Characteristic and Fan-out. We see from this analysis that the output of this gate is LOW when the node marked X in Fig. 13.7(a) is 2.2 V. This requires that inputs A and B both be at least $2.2 - V_{D(\text{cut-in})} = 1.65$ V, as shown on the transfer characteristic of Fig. 13.7(b). The other breakpoint on the characteristic occurs when Q_2 is at cut-in: $V_{BE2} = 0.5$ V. Then the voltage $V_X = V_{BE2} + V_{D2} + V_{BE1} = 0.5 + 0.7 + 0.7 = 1.9$ V, and this would require the lower of the voltages at either A or B to be $V_X - V_D = 1.9 - 0.7 = 1.2$ V. Any voltage lower than this at A or B causes Q_2 to be OFF.

Noise margins can be easily determined by inspecting the transfer characteristic of Fig. 13.7(b): $NM_L = V_{IL} - V_{OL} = 1.2 - 0.2 = 1.0$ V, and $NM_H = V_{OH} - V_{IH} = 5.0 - 1.65 = 3.35$ V. These noise margins would appear to be much better than in the RTL example.

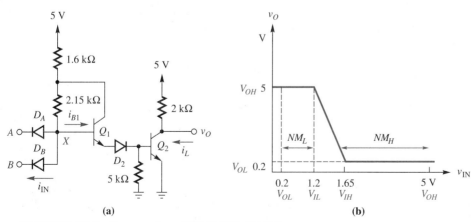

■ **FIGURE 13.7** One gate from MC949 DTL NAND.
(a) Circuit. (b) Transfer characteristic.

DRILL 13.4

Refer to the DTL circuit of Fig. 13.7(a). What is the purpose of the 5-kΩ resistor?

ANSWER The 5-kΩ resistor provides a quick discharge of Q_2 base when D_2 turns OFF. This occurs when A or B goes LOW.

EXAMPLE 13.5

Find the fan-out for the gate shown in Fig. 13.7(a); let $\beta_F = 30$.

SOLUTION When one or more of the inputs to a DTL gate are LOW, dc fan-out is no problem, since all of the inputs that are connected to this output will be pulled HIGH, where they draw no current. It is when both inputs are HIGH, and the gate output is supposed to be LOW, that current from load gates limits the fan-out. If we want $V_{CE(sat)}$ to be 0.2 V or less, a base overdrive factor of $\sigma < 0.9$ (Fig. 4.25) is required. Therefore, I_{C2} must be less than $\sigma \beta_F I_{B2} = (0.9)(30)(1.52) = 41$ mA.

The circuit can sink $I_L = 41$ mA $- (5.0 - 0.2)$ V/2 kΩ = 38.6 mA from the loads. Each load gate with a 0.2-V input will source $|I_{IN}| = (5 - 0.7 - 0.2)$ V/$(1.6 + 2.15)$ kΩ = 1.1 mA. Thus when A and B are both HIGH, the maximum dc fan-out is 38.6/1.1 = 35. This is not realistic, since the input of each gate has 5 pF of parasitic capacitance, and all inputs must charge through the 2-kΩ pull-up resistor. The time delay imposed by 35 $\times$ 5 pF charging through 2 kΩ ($\tau = 350$ ns) is unreasonable. The manufacturer suggests a maximum fan-out of 8.

We can summarize DTL specifications at 25°C as

- Fan-out: 8 (nominal)
- Noise margin: 1.0 V
- Propagation delay: 30 ns
- Power dissipation/gate: 13 mW.

We can see that the DTL gate performs better on dc fan-out and much better on noise margin than the RTL gate. The chief disadvantage is the increased propagation delay, which results from the greater capacitive loading on the output when fan-out is increased. Note the conflict between speed and fan-out.

■ High-threshold logic is still used around large industrial machinery, where large current transients are likely to produce much electrical noise.

High-Threshold Logic

The topology shown in Fig. 13.7(a) has been used with a Zener diode replacing D_2 and with higher supply voltages. This allows the realization of much higher noise margins for applications in very noisy environments. For further consideration of this subject, see Problems 13.14 and 13.15.

CHECK UP

1. **TRUE OR FALSE?** In our examples, DTL has lower propagation delay than RTL.
2. **TRUE OR FALSE?** In the logic families shown, DTL has higher noise margin than RTL.
3. **TRUE OR FALSE?** Reliable fan-out in early logic ICs is in the range of 5 to 8.
4. **TRUE OR FALSE?** A BJT logic designer can make a trade-off between delay and power.

13.4 TRANSISTOR-TRANSISTOR LOGIC

As IC technology advanced, it became economically feasible to obtain higher performance by using more transistors per gate. The *transistor-transistor logic* (**TTL**) family was developed specifically to increase switching speed without sacrificing fan-out or noise margin, and without increasing power dissipation. Introduced in 1965, TTL was leading DTL in sales by 1970 and dominated the digital IC market for more than 10 years. Indeed, TTL has spanned several technological "generations." We will illustrate some of the evolutionary changes as we discuss the families. Currently, virtually no new systems are being designed with standard TTL, and present TTL application focuses on the *low-power Schottky* (LS) and the *advanced low-power Schottky* (ALS) families.

Standard TTL Gate

Shown in Fig. 13.8 is a circuit diagram of one gate in the 7400 quad two-input NAND gate IC. *Quad* means there are four of these gates on one die. All TTL BJTs, being smaller than those of the older DTL devices, switch faster because of less parasitic capacitance. Comparing this circuit with that of the DTL gate in Fig. 13.7(a), we note the following differences:

- The input diodes D_A and D_B have become two emitters of Q_1. In other members of the TTL family, Q_1 may have from 1 to 13 emitters, depending on the number of inputs desired.
- D_2 has been replaced by Q_2 for current gain, to drive both Q_3 and Q_4.
- The output pull-up resistor has been changed to an active pull-up transistor Q_4 to source more current to load capacitances. This output circuit, with both pull-up and pull-down transistors, is called a *totem pole*.
- R_4 is smaller to aid in bringing Q_3 out of saturation faster.

■ The fan-in of 13 is the limit imposed by available pins in a 16-pin package. The other pins are required for V_{CC}, ground, and output.

We will show that when the input, A or B, is LOW, Q_1 will saturate, Q_2 and Q_3 will be OFF, and Q_4 will pull the output HIGH. When both inputs are HIGH, Q_1 is in the inverse mode, providing base current to Q_2 so that Q_2 and Q_3 will saturate. Q_4 will turn OFF, and the output will be LOW. We studied a circuit similar to that of Q_1 in Section 4.8, and we suggest that you review that section and Problem 4.41 at this time.

TTL Transfer Characteristic. We analyze detailed operation of the circuit by considering each segment of the transfer characteristic. We assume the following transistor parameters at $T = 25°C$:

$$V_{CE(\text{sat})} = 0.2 \text{ V} \qquad V_{BE(\text{sat})} = 0.8 \text{ V}$$
$$\beta_F = 30 \qquad V_{BE(\text{act})} = 0.7 \text{ V} \qquad (13.6)$$
$$\beta_R = 0.02 \qquad V_{BE(\text{cut-in})} = 0.6 \text{ V}$$

Input LOW: Refer to Fig. 13.8(c). Assume one or both of the inputs are LOW, say, $v_{\text{IN}} = 0.2$ V, because the input is being driven by a saturated transistor in a previous stage.

- $V_{B1} = V_{\text{IN}} + V_{BE1} = 1$ V and $I_{B1} = (5 - 1)$ V/4 kΩ = 1 mA. The input current is $I_{IL} = -I_{E1} = -1$ mA. To pull this input LOW (to 0.2 V), a driver must sink 1 mA.
- In the steady state, Q_1 will be very saturated because I_{C1} will be limited to the negligible base current $I_{B2} \approx 0$ of Q_2, which we will show is OFF. Figure 4.25 reveals that if Q_1 has $I_C \approx 0$ and $I_{B1} = 1$ mA, it would have $V_{CE1} \approx 0.06$ V.
- Then, $V_{B2} = V_{\text{IN}} + V_{CE1(\text{sat})} = 0.2 + 0.06 = 0.26$ V. This guarantees that V_{BE2} is below cut-in for Q_2, and Q_2 is cut off as assumed above.

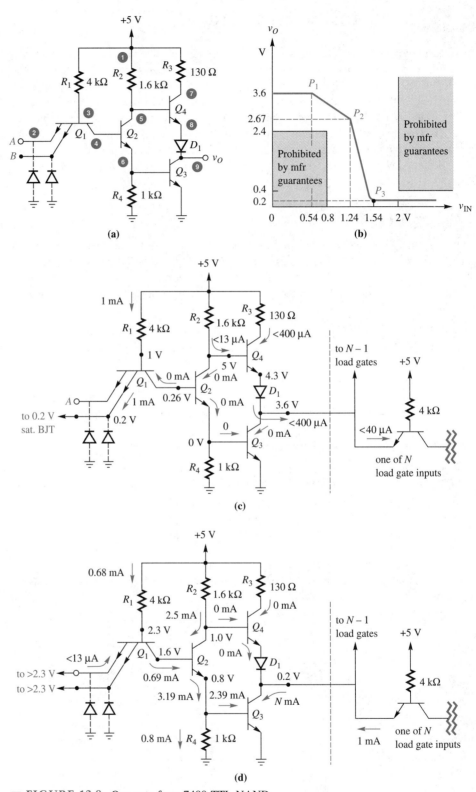

■ **FIGURE 13.8** One gate from 7400 TTL NAND.
(a) Circuit. Node numbers are for Example 13.7. (b) Transfer characteristic. (c) Static model with input LOW. (d) Static model with inputs HIGH.

CHAPTER 13   INTEGRATED-CIRCUIT LOGIC GATES   655

- Because Q_2 furnishes no current to the base of Q_3, Q_3 is also cut off.
- Q_4 will source current to any load that is connected. We will show that whether Q_4 is active or saturated will depend on this load current. When the load current $I_L = 0.4$ mA, which is the maximum for 10 standard TTL inputs, then that would be the current in D_1. In this case, $I_{B4} < I_L/\beta_F = 400/30 = 13$ μA. With so little current in R_2, $V_{B4} \approx 5$ V. Assuming the nominal voltage drop of 0.7 V for V_{BE4} and V_{D1}, $V_O = 5 - 0.7 - 0.7 = 3.6$ V. This means that $V_{OH} = 3.6$ V, and this is valid for $v_{IN} = 0.2$ V, less than V_{IL}.

As v_{IN} is increased, V_{B2} rises $V_{CE1(sat)} = 0.06$ V above v_{IN}. When V_{B2} reaches about 0.6 V, Q_2 begins to cut in. Note that V_{BE3} is still zero just before Q_2 conducts. At the point when Q_2 begins to conduct, i_{C2} causes a voltage drop in R_2, which causes an identical drop at v_O. This is P_1 in Fig. 13.8(b). The input voltage corresponding to this breakpoint in the transfer characteristic is 0.06 V less than $V_{BE2(cut-in)}$, which is $V_{IN} = 0.6 - 0.06 = 0.54$ V.

Q_2 is active from P_1 to P_2, where Q_3 comes ON. At this point, $V_{BE3} = 0.6$ V (cut-in), requiring $I_{E2} = 0.6$ V/1 kΩ = 0.6 mA. Because Q_2 is active, the collector current is

$$I_{C2} = \frac{\beta}{\beta + 1} I_{E2} = \frac{30}{31} 0.6 \text{ mA} = 0.58 \text{ mA}.$$

This causes a voltage drop in R_2, and $V_{B4} = V_{C2} = 5 - (0.58 \text{ mA})(1.6 \text{ k}\Omega) = 4.07$ V. Also, V_O will remain two diode drops ($V_{BE4} + V_{D1}$) below V_{B4} at 2.67 V. The base current of Q_2 is $I_{B2} = I_{E2}/(\beta + 1) = 0.6$ mA/31 = 19 μA, and this current must come through the base-collector diode of Q_1. $V_{B1} = V_{BE3} + V_{BE2} + V_{BC1} = 0.6 + 0.7 + 0.7 = 2$ V, and $I_{B1} = (5 - 2)$ V/4 kΩ = 0.75 mA. Then $-I_{E1} = I_{B1} + I_{C1} = 0.75 - 0.019 = 0.731$ mA. We observe that the base current divides, and most of it flows out of the input terminal, but a small fraction becomes a negative collector current driving the base of Q_2. With $I_{B1} = 0.75$ mA and $I_{C1} = -19$ μA, Eq. (4.36) predicts that $V_{CE1} = 0.06$ V (still saturated!). Finally, $V_{IN} = V_{BE3} + V_{BE2} - V_{CE1} = 0.6 + 0.7 - 0.06 = 1.24$ V. This input voltage is our V_{IL}, and it locates breakpoint P_2.

Between P_2 and P_3, as v_{IN} continues to increase, first Q_3 saturates (for normal connected load), and then Q_2 saturates, which cuts Q_4 OFF. Consider point P_3, where we assume Q_3 is saturated, with $V_O = V_{CE(sat)} = 0.2$ V. This is V_{OL}. If both Q_3 and Q_2 are saturated, $V_{C1} = V_{BE3} + V_{BE2} = 0.8 + 0.8 = 1.6$ V, and V_{B1} is one diode drop (V_{BC1}) more positive, about 2.3 V. Note that Q_1 is still saturated, so $v_{IN} = V_{C1} - V_{CE1} = 1.6 - 0.06 = 1.54$ V. This is V_{IH}.

Above P_3, further increase in v_{IN} results in no corresponding increase in V_{C1} since it is clamped at 1.6 V [$V_{BE3(sat)} + V_{BE2(sat)}$] and V_{B1} remains 2.3 V. The emitter current I_{E1} decreases, and I_{B1} is diverted through the collector junction to the base of Q_2. When v_{IN} exceeds $V_{B1} = 2.3$ V, Q_1 is in the inverse active mode.

Refer to Fig. 13.8(d) for the static situation when v_{IN} is HIGH:

- $I_{B1} = (5 - 2.3)$ V/4 kΩ = 0.68 mA, and if β_R of the inverted Q_1 is 0.02, the input current $I_{IH} = \beta_R I_{B1} = (0.02)(0.68) = 13$ μA into the emitter. Manufacturers purposely keep β_R for Q_1 small to minimize this input current, and guarantee it to be less than 40 μA. If both inputs are HIGH, I_{IH} divides between them.
- Q_2 is saturated for $v_{IN} \geq 2.3$ V, making $V_{C2} = V_{BE3(sat)} + V_{CE2(sat)} = 0.8 + 0.2 = 1$ V. This will make $I_{C2} = (5 - 1)$ V/1.6 kΩ = 2.5 mA, if $I_{B4} = 0$, as we will show. The base current $I_{B2} = I_{B1}(1 + \beta_R) = (1.02)(0.68) = 0.69$ mA. Thus we have demonstrated the saturation of Q_2 because $\beta_F I_{B2} = 20$ mA $\gg I_{C2} = 2.5$ mA.
- Summing I_{C2} and I_{B2} yields $I_{E2} = 3.19$ mA. R_4 has $I_4 = 0.8$ V/1 kΩ = 0.8 mA, leaving $I_{B3} = 2.39$ mA. Q_3 should be able to sink $(0.90)\beta_F I_{B3} = (0.90)(30)(2.39) = 64$ mA without exceeding $V_{OL} = 0.2$ V (using the overdrive factor from Fig. 4.25).

- To see that Q_4 is OFF when Q_2 is saturated, we have only to observe that $V_{B4} = 1$ V, while V_O is $V_{CE3(\text{sat})}$. This means that there is only about 0.8 V remaining for $V_{BE4} + V_{D1}$. This is not enough to bring them to conduction; they are OFF. This demonstrates the purpose of the diode.

Fan-out and Noise Margin. We have shown that $I_{IL} = -1$ mA at $v_{IN} = 0.2$ V. We have also calculated that the output should be able to sink 64 mA. Thus, this gate should be able to drive 64 similar inputs to the LOW state.

Our analysis has also shown that each HIGH input may require as much as 13 μA from the driver. Let us determine how much current Q_4 is capable of sourcing into an output load with $v_O = 2.5$ V, which is slightly above V_{IH}. We try Q_4 saturated,

$$I_{C4} = \frac{[5 - V_{CE(\text{sat})} - V_D - v_O]}{R_3} = \frac{(5 - 0.2 - 0.7 - 2.5) \text{ V}}{130 \; \Omega} = 12.3 \text{ mA},$$

$$I_{B4} = \frac{[5 - V_{BE(\text{sat})} - V_D - v_O]}{R_2} = \frac{(5 - 0.8 - 0.7 - 2.5) \text{ V}}{1600 \; \Omega} = 0.6 \text{ mA}.$$

Thus $\beta_F I_{B3} > I_C$, confirming our assumption of saturation, and Q_4 can source 12.3 + 0.6 = 12.9 mA. This current would drive 12.92/0.013 = 990 inputs!

Allowing for worst case component and temperature variation, the manufacturers conservatively suggest a fan-out of 10, which will guarantee that when the output is supposed to be LOW, Q_3 stays well saturated; $V_{OL} < 0.2$ V. Similarly, the manufacturers guarantee $V_{OH} > 2.4$ V, $V_{IL} > 0.8$ V, $V_{IH} < 2.0$ V, and $V_{OL} < 0.4$ V. These guarantees are shown in Fig. 13.8(b) by means of prohibited regions. When $v_{IN} < 0.8$ V, V_{OH} is guaranteed to be greater than 2.4 V. When $v_{IN} > 2.0$ V, V_{OL} is guaranteed to be less than 0.4 V.

Switching Speed. We are now in a position to examine the transient behavior of the gate. When an input makes a HIGH-LOW transition, v_{B2} is initially at 1.6 V, since Q_2 and Q_3 are both saturated. Thus, when an emitter of Q_1 goes LOW, Q_1 is active, with a collector current of $\beta_F I_{B1} = 30$ mA, which quickly removes the charge stored in the base of Q_2. Only after Q_2 is OFF does Q_1 saturate. When Q_2 is OFF, the base of Q_3 discharges through the relatively small (compared to DTL) value of R_4; Q_4 saturates and charges whatever parasitic capacitance is associated with the load through an effective resistance of only 130 Ω (less if Q_4 becomes active). The propagation delay, especially t_{pLH}, is thus dramatically reduced over that of DTL.

We should also mention two other items associated with switching transients. First, as Q_2 switches through the active region, there is a very short period of time when both Q_3 and Q_4 are ON (see Problem 13.16). Because the saturated Q_3 turns OFF more slowly than Q_4 turns ON, this 30-mA current pulse persists for several nanoseconds. Consequently, TTL gates are noted for introducing current "glitches" onto the power supply line each time they switch. To prevent the resulting voltage *spikes* from interacting with other circuits, it is common to **decouple** TTL packages by paralleling their power supply pins with a large capacitor (0.1 μF); that is, the capacitors provide the current spike demanded without a voltage spike developing. The novice should be warned that electrolytic capacitors are not suitable for this application because they have too much parasitic inductance!

A second problem associated with circuits capable of high-speed switching is that even short interconnections possess enough inductance to resonate with the shunt capacitance and give rise to "overshoot" or *ringing* at the steplike transitions of the waveform. This is why the diodes shown in Fig. 13.8(a) shunt the A and B inputs. They clamp the inputs to prevent the emitter(s) from becoming more negative than a diode drop. This, in turn, prevents the collector of Q_1 from going negative with respect to the IC substrate (0 V).

■ Note a 30 mA pulse with a rise time of about 1 ns would create a large voltage spike even if the power supply inductance were very small! Bypass capacitors alleviate this problem.

CHAPTER 13   INTEGRATED-CIRCUIT LOGIC GATES

The *standard* series 74xx (0 to 70°C) and 54xx (−55 to 125°C) contains a very large set of logic functions, small- and medium-scale integration.

EXAMPLE 13.6

With inputs LOW, find the output current for the gate shown in Fig. 13.8(a) if v_O were grounded. Let $\beta_F = 30$.

SOLUTION

$$I_{CE4} = \frac{[5 - V_{CE(\text{sat})} - V_{D1} - v_O]}{R_3} = \frac{(5 - 0.2 - 0.7 - 0)\text{ V}}{130\ \Omega} = 31.5\text{ mA},$$

$$I_{B4} = \frac{[5 - V_{BE(\text{sat})} - V_{D1} - v_O]}{R_2} = \frac{(5 - 0.8 - 0.7 - 0)\text{ V}}{1600\ \Omega} = 2.2\text{ mA}.$$

Because $I_C/I_B < \beta_F$, it is saturated. Output short-circuit current is
$I_{SC} = I_C + I_B = 31.5 + 2.2 = 33.7$ mA.

EXAMPLE 13.7

Use SPICE to plot the transfer characteristic of the TTL gate of Fig. 13.8. Sweep the input voltage from 0 to 2 V. For simplicity, use a single input on Q_1, and let all BJTs have a common model statement: `.MODEL T1 NPN IS=1F BF=50 BR=.1 TR=.1U TF=1N CJC=1P CJE=.2P CJS=1P RB=100`.

SOLUTION   The desired netlist is shown. An output plot is given in Fig. 13.9.

```
TTL TRANSFER CHAR
VCC 1 0 DC 5
VIN 2 0 DC 0
Q1 4 3 2 T1
Q2 5 4 6 T1
Q3 9 6 0 T1
Q4 7 5 8 T1
D1 8 9 D
R1 1 3 4K
R2 1 5 1.6K
R3 6 0 1K
R4 1 7 130
.MODEL D D IS=1F
.MODEL T1 NPN IS=1F BF=50 BR=.1 TR=.1U TF=1N CJC=1P
+CJE=.2P CJS=1P RB=100
.DC VIN 0 2 .01
.PLOT DC V(3) V(4) V(5) V(6) V(7) V(9)
.END
```

Variations on the circuits of Fig. 13.7 with different resistor sizes have been tried. Resistors that were 10 times larger reduced the power to 1 mW/gate, but unfortunately the propagation delay increased to 33 ns. Alternatively, the resistor sizes were decreased to increase the speed. Propagation delay was decreased to 6 ns, but at the high cost of 22 mW/gate dissipation. The second generation of TTLs (Schottky) has replaced these variants.

DRILL 13.5

Refer to the plots of Fig. 13.9. Using the plot for V_{c4}, estimate the peak magnitude of the pulse of current that flows thru Q_3 and Q_4 on this static plot.

ANSWER
$I_{130} = 4\text{ V}/130\ \Omega =$ 31 mA. Of course, in the dynamic case, it would depend on the rise time of V_{IN}.

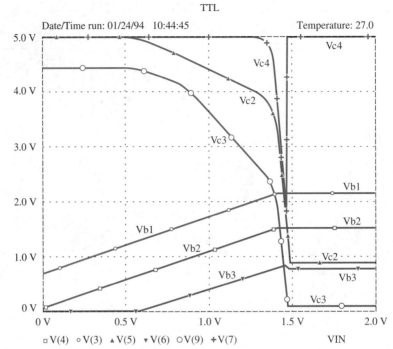

FIGURE 13.9 Results of TTL simulation. Various node voltages as a function of V_{IN}.

DRILL 13.6

According to the data of Fig. 13.9, at what input voltage does Q_2 come ON?

ANSWER Slightly above 0.5 V.

Schottky TTL

Much of the propagation delay in the TTL gate is due to the storage time of saturated transistors. As discussed in Section 4.11, the Schottky transistor is prevented from saturating by fabricating a Schottky diode in parallel with the collector-base junction, which "clamps" $V_{CE} \geq V_{BE} - V_{SD}$. In a typical **Schottky-clamped** situation, $V_{BE(SC)} = 0.75$ V, $V_{SD} = 0.40$ V, and V_{CE} is constrained to go no lower than $V_{CE(SC)} = 0.35$ V, or the edge of the active region.

The pictorial diagram in Fig. 13.10 shows that the Schottky transistor presents no

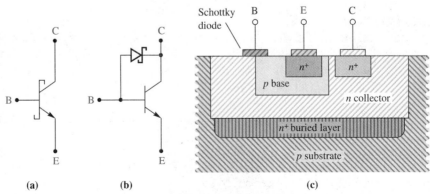

FIGURE 13.10 Schottky transistor. **(a)** Circuit symbol. **(b)** Circuit equivalent. **(c)** Fabrication pictorial.

fabrication difficulty; the aluminum, which makes an ohmic contact with the *p* base, has merely been extended to overlap the *n*-collector region, providing the necessary rectifying contact. The result of using Schottky transistors in a TTL gate is to reduce transistor turnoff time by eliminating the storage-time delay. At the same time, all diodes would be fabricated as Schottky diodes also, to prevent storage delay in diode turnoff.

A Schottky TTL gate is shown in Fig. 13.11. Note that transistors Q_1 through Q_4 serve the same functions as in Fig. 13.8. In addition to the introduction of Schottky transistors, we find two innovations, Q_5 and Q_6. The combination of Q_4 and Q_5 is known as a **Darlington circuit**, an amplifier having very low output resistance. The Darlington amplifier was discussed in Chapter 8. The combination has a very high current gain when active, and it allows the output node to charge rapidly when switched ON. The two base-emitter drops remove the need for the diode D_1 in Fig. 13.8.

■ Recall that a Darlington circuit is a compound BJT with an equivalent $\beta_D = \beta_1 \beta_2$ and input voltage drop of $2V_{BE}$.

Because of the 3.5-kΩ resistor, Q_5 is always ON. We see that when Q_2 is ON, $V_{B5} = V_{BE3} + V_{CE2(SC)} = 0.75 + 0.35 = 1.10$ V and $V_{B4} = V_{B5} - V_{BE5(act)} = 1.10 - 0.7 = 0.40$ V, so Q_4 is OFF. When Q_2 is OFF, we can find the unloaded V_{OH} by using $5 = I_{B5}(900) + V_{BE5} + (1 + \beta_F)I_{B5}(3500)$, obtaining $I_{B5} = 39$ μA for a β_F of 30.

The resulting V_{B5} is 4.96 V, leading to $V_{OH} = 3.56$ V, $2V_{BE(act)}$ lower. Note that Q_4 is not a Schottky transistor because it can never saturate in the Darlington arrangement.

Transistor Q_6 serves as an active base resistor for Q_3. Note that Q_6 does not conduct until V_{BE3} reaches cut-in. Thus, in contrast to the gate in Fig. 13.8, Q_2 does not come ON before Q_3; rather, both Q_2 and Q_3 come ON together, when $V_{B2} = V_{BE3(cut-in)} + V_{BE2(cut-in)} = 0.65 + 0.65 = 1.3$ V. This occurs when $v_{IN} = V_{B2} - V_{CE1(SC)} = 1.3 - 0.35 = 0.95$ V. We label this value V_{IL} on the transfer characteristic shown in Fig. 13.11(b). As v_{IN} continues to increase, Q_2 reaches the ON (clamped) state with $V_{B2} = V_{BE3(SC)} + V_{BE2(SC)} = 0.75 + 0.75 = 1.5$ V, corresponding to $v_{IN} = V_{IH} = 1.15$ V. At this point, Q_3 is ON and clamped with $V_{CE(SC)} = V_{OL} = 0.35$ V. Note that this transfer characteristic does not have the sloping region seen between P_1 and P_2 in Fig. 13.8(b).

The result of the Schottky design used in the 74Sxx/54Sxx series is a reduction of propagation delay to 3 ns. While Q_3 is ON (clamped), Q_6 is also ON, conducting about $(0.75 - 0.35)/250 = 1.6$ mA. This makes Q_6 equivalent to a 470-Ω resistor. When Q_2

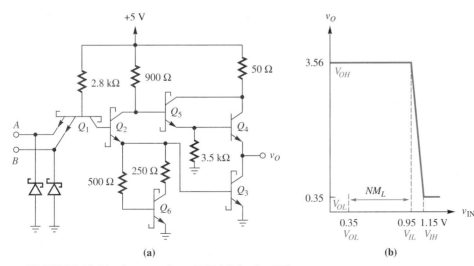

■ **FIGURE 13.11** One gate from 74S00 Schottky TTL NAND. (a) Circuit. (b) Transfer characteristic.

turns OFF, Q_6 continues to conduct just long enough to discharge the base of Q_3. Because the resistor sizes used are smaller than those of standard TTL, the power dissipation is 20 mW/gate. Figure 13.11(b) reveals $NM_L = 0.6$ V. The guaranteed value is 0.4 V.

DRILL 13.7

Refer to Fig. 13.11. Explain the statement in the text, "Q_4 is not a Schottky transistor since it can never saturate in the Darlington arrangement."

ANSWER Q_4 only has significant base current when Q_5 is ON, thus V_{CE5} is positive, requiring V_{CB4} to be positive, and Q_4 can never saturate.

EXAMPLE 13.8

Calculate the power dissipation of the 74S00 gate shown in Fig. 13.11. Assume a 50% duty cycle, no load.

SOLUTION (a) With input HIGH,

$$I_{2800} = \frac{(5 - 0.75 - 0.75 - 0.40)}{2800} = 1.11 \text{ mA}; \quad I_{900} = \frac{(5 - 0.75 - 0.35)}{900} = 4.33 \text{ mA}.$$

$I_{50} = 0; \quad$ Total $I_H = 5.44$ mA; $\quad P_H = (5 \text{ V})(5.44 \text{ mA}) = 27.2$ mW.

(b) With input LOW,

$$I_{2800} = \frac{(5 - 0.75 - 0.35)}{2800} = 1.39 \text{ mA}; \quad I_{900} = I_{B5} = 39 \text{ μA (from above)}.$$

$I_{50} = \beta_F I_{B5} = 30(39 \text{ μA}) = 1.7$ mA (no load); $\quad$ Total $I_L = 2.60$ mA.

$P_L = (5 \text{ V})(2.60 \text{ mA}) = 13.0$ mW; $\quad$ Average $P = \dfrac{(27.2 + 13.0)}{2} = 20.1$ mW.

Low-Power Schottky TTL

The third-generation TTL gate shown in Fig. 13.12 utilizes the Schottky technology with resistors about 10 times larger than those in Fig. 13.11(a). The resulting trade-off is a gate with a 10-ns propagation delay but only a 2 mW/gate power dissipation. Since

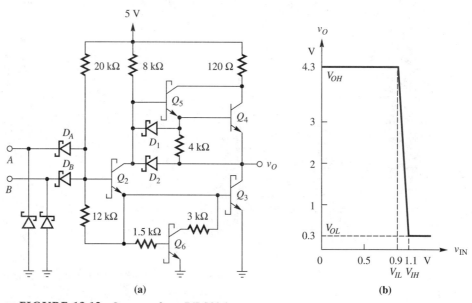

FIGURE 13.12 One gate from 74LS00 low-power Schottky TTL NAND. (a) Circuit. (b) Transfer characteristic.

CHAPTER 13 INTEGRATED-CIRCUIT LOGIC GATES

recommended operating conditions

		SN54LS00			SN74LS00			UNIT
		MIN	NOM	MAX	MIN	NOM	MAX	
V_{CC}	Supply voltage	4.5	5	5.5	4.75	5	5.25	V
V_{IH}	High-level input voltage	2			2			V
V_{IL}	Low-level input voltage			0.7			0.8	V
I_{OH}	High-level output current			−0.4			−0.4	mA
I_{OL}	Low-level output current			4			8	mA
T_A	Operating free-air temperature	−55		125	0		70	°C

electrical characteristics over recommended operating free-air temperature range (unless otherwise noted)

PARAMETER	TEST CONDITIONS†	SN54LS00			SN74LS00			UNIT
		MIN	TYP‡	MAX	MIN	TYP‡	MAX	
V_{IK}	V_{CC} = MIN, I_I = −18 mA			−1.5			−1.5	V
V_{OH}	V_{CC} = MIN, V_{IL} = MAX, I_{OH} = −0.4 mA	2.5	3.4		2.7	3.4		V
V_{OL}	V_{CC} = MIN, V_{IH} = 2 V, I_{OL} = 4 mA		0.25	0.4		0.25	0.4	V
	V_{CC} = MIN, V_{IH} = 2 V, I_{OL} = 8 mA					0.35	0.5	
I_I	V_{CC} = MAX, V_I = = 7 V			0.1			0.1	mA
I_{IH}	V_{CC} = MAX, V_I = 2.7 V			20			20	μA
I_{IL}	V_{CC} = MAX, V_I = 0.4 V			−0.4			−0.4	mA
I_{OS}§	V_{CC} = MAX	−20		−100	−20		−100	mA
I_{CCH}	V_{CC} = MAX, V_I = 0 V		0.8	1.6		0.8	1.6	mA
I_{CCL}	V_{CC} = MAX, V_I = 4.5 V		2.4	4.4		2.4	4.4	mA

† For conditions shown as MIN or MAX, use the appropriate value specified under recommended operating conditions.
‡ All typical values are at V_{CC} = 5 V, T_A = 25°C.
§ Not more than one output should be shorted at a time, and the duration of the short-circuit should not exceed one second.

switching characteristics, V_{CC} = 5 V, T_A = 25°C (see note 2)

PARAMETER	FROM (INPUT)	TO (OUTPUT)	TEST CONDITIONS		MIN	TYP	MAX	UNIT
t_{PLH}	A or B	Y	R_L = 2 kΩ,	C_L = 15 pF		9	15	ns
t_{PHL}						10	15	ns

NOTE 2: See General Information Section for load circuits and voltage waveforms.

■ **FIGURE 13.13** Data sheet for 54/74LS00 TTL NAND gates (reprinted by permission of Texas Instruments, Incorporated © copyright 1985).

Q_2 is Schottky and does not saturate, Q_1 is no longer needed to speed up the turnoff of Q_2 and has been replaced by diodes D_A and D_B. We find that $V_{IL} = V_{BE3(\text{cut-in})} + V_{BE2(\text{cut-in})} - V_{SD} = 0.65 + 0.65 - 0.4 = 1.1$ V. Diode D_1 has been included to speed the turnoff of Q_4, providing a discharge path for the base when Q_2 comes ON. The emitter resistor of Q_5 can now connect to v_O, which saves considerable power when compared to the grounded resistor of the 74S00. Additionally, it pulls the output voltage to 4.3 V for small load current. Diode D_2 speeds the pull-down of v_O. Some manufacturers include the 12-kΩ resistor. A typical data sheet for the 54/74LS00 is presented in Fig. 13.13, and typical output curves are plotted in Fig. 13.14.

Advanced Schottky TTL

Continued advances in the state of IC technology led to the introduction, in 1980, of the fourth generation of TTLs. The transistors are smaller and have oxide isolation, resulting in less parasitic capacitance. Oxide isolation between transistors is accomplished as shown in Fig. 18.39(e) in Chapter 18, where it is shown in an MOS structure.

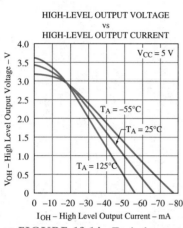 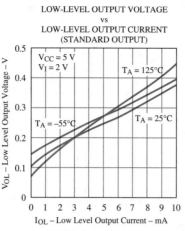 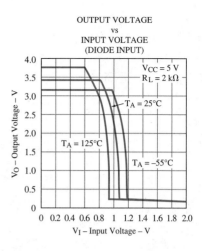

■ **FIGURE 13.14** Typical output characteristics for the 74LS00 (reprinted by permission of Texas Instruments, Incorporated © copyright 1985).

Figure 13.15 reveals that the 74ASxx/54ASxx series has the same features as the Schottky circuit in Fig. 13.11 except that the input transistor has given way to diode inputs as in the LS series, and D_1 has been added to speed the pull-down of v_O. Propagation delay is 1.5 ns, and power dissipation is 20 mW. Thus, compared to the earlier Schottky, speed is doubled without an increase in power.

An advanced low-power Schottky TTL circuit is shown in Fig. 13.16. From Q_2 to the right, the circuit appears as the low-power Schottky gate shown in Fig. 13.12. Diodes D_A and D_B serve to turn Q_2 OFF when inputs A or B go LOW. Note that Q_1 is added to speed the turning ON of Q_2 when inputs A and B go HIGH. To correct for the level shift created by V_{BE1}, the *pnp* emitter followers Q_A and Q_B are added. The result of these modifications is a gate with 1 mW of dissipation and a propagation delay of only 4 ns. Each generation of logic achieves a lower delay-power product. For a comparison, see Fig. 13.17 and Table 13.1.

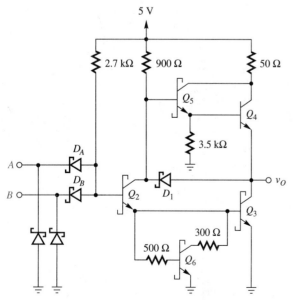 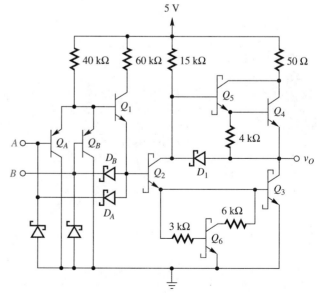

■ **FIGURE 13.15** One gate from 74AS00 advanced Schottky TTL NAND.

■ **FIGURE 13.16** One gate from 74ALS00 advanced low-power Schottky TTL NAND.

TABLE 13.1 Comparison of Characteristics of Several TTL Series

	7400	74S00	74LS00	74AS00	74ALS00
Fan-out (nominal)	10	10	10	10	10
V_{IL}-V_{OL} (worst case)	0.8–0.4 V	0.8–0.5 V	0.8–0.5 V	0.8–0.5 V	0.8–0.5 V
V_{OH}-V_{IH} (worst case)	2.4–2 V	2.7–2 V	2.7–2 V	2.7–2 V	2.7–2 V
Propagation delay	10 ns	3 ns	10 ns	1.5 ns	4 ns
Power-dissipation/gate	10 mW	19 mW	2 mW	20 mW	1 mW
Power-delay product	100 pJ	57 pJ	20 pJ	30 pJ	4 pJ

■ Newly developed TTL standards have V_{CC} = 3.3V compared to 5V, resulting in only 40% of the 5V-standard power consumption.

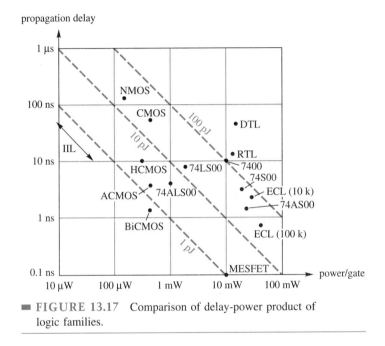

■ **FIGURE 13.17** Comparison of delay-power product of logic families.

■ Shown on Fig. 13.17 are 3 generations of CMOS. The first generation is labeled *CMOS*, followed by HCMOS and ACMOS. Often the term *CMOS* refers to any complementary MOS technology.

DRILL 13.8

From the data in Table 13.1, why does it appear that the LS series of TTL should supersede the standard TTL?

ANSWER The LS achieves the same delay with less power dissipation. The only advantage that standard TTL has is a higher output drive current capability.

A large selection of logic functions is available for each of these series. NAND gates alone come with 2, 3, 4, 8, and 13 inputs. There are AND, OR, NOR, and EOR gates, inverters, and buffers. These are all *small-scale* ICs (<10 gates/chip). In *medium-scale* ICs (10 to 100 gates/chip), we find combinational logic circuits such as *decoders* and *multiplexers* and sequential logic circuits such as *counters* and *shift-registers*. We will examine the function of these circuits in Chapter 14, but the internal circuitry consists of combinations of the gates we have discussed here.

Open-Collector and Three-State Outputs

Often it is desirable to have the outputs of several logic circuits combined at one node, the so-called **wired-AND**. The normal totem pole output has a potential conflict in this mode because one gate's pull-down transistor (Q_3) may be sinking current, while another gate's pull-up transistor (Q_4) may be sourcing current. The result is not well defined, and dissipation would be excessive. Two solutions are available.

Many logic circuits come with an **open-collector output**. This merely means that there is no internal pull-up transistor. Using Fig. 13.16 as an example, transistors Q_4 and Q_5 would be omitted. Any number of such open-collector outputs can be wired together with

■ The choice of a pull-up resistor presents design tradeoffs. Too large R would cause delay in turning on the next stage, while too small R would cause greater power dissipation. Note that this situation is what the active load (*totem pole* circuit) avoids.

a single (external) pull-up resistor, and the output state will be HIGH only when all pull-down transistors are OFF. This arrangement often saves the circuit designer an AND gate. Of course, an open-collector output cannot reach a HIGH state unless an external pull-up resistor is provided.

Another method of combining the outputs of several logic circuits at one node without conflict is known as a ***three-state*** or ***tristate bus***. The idea is to give the various output circuits access to the bus one at a time by "selecting," or *enabling*, only one output at a time. Presumably one or more inputs are also connected to the bus, but there is no conflict there unless the maximum fan-out is exceeded. Figure 13.18 shows one gate of a 74LS366 three-state hex inverter. When E (enable) is HIGH, diodes D_3 and D_4 are OFF and the gate operates normally. When E is LOW, D_4 pulls V_{B2} down to 0.75 V, holding Q_2 and Q_3 OFF, while D_3 pulls V_{B5} to 0.75 V, turning OFF Q_5 and Q_4. With both Q_4 and Q_3 OFF, the output is in the high-impedance state, and this gate does not have any effect on the bus. The E input is buffered by a four-transistor circuit (not shown) because it drives six inverters identical to the one in the figure.

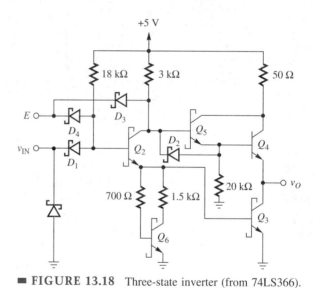

■ **FIGURE 13.18** Three-state inverter (from 74LS366).

CHECK UP

1. **TRUE OR FALSE?** For a number of years, TTL was the dominant digital technology because it was a reasonable compromise in speed, power, and cost.
2. **TRUE OR FALSE?** All other parameters being equal, TTL ICs using Schottky BJTs will be faster, but dissipate more power than standard TTL.
3. **TRUE OR FALSE?** Advanced TTL circuits achieve lower power dissipation by using oxide isolation between transistors on the substrate.
4. **TRUE OR FALSE?** Both transistors in the TTL totem-pole output circuit can be turned OFF in a tristate device.

13.5 EMITTER-COUPLED LOGIC

In our discussion of the saturating logic families, we have noted a drawback of saturation, namely storage time delay. The Schottky clamp is a means of combating this difficulty. Since about 1962, a family of nonsaturating logic has been developing concurrently with TTL and the other families. *Emitter-coupled logic* (ECL) operates in the active region, and is currently available with a propagation delay on the order of 200 ps, making it the fastest commercially available logic family implemented in silicon.

The Emitter-Coupled Pair

Consider the emitter-coupled pair (also known as a differential amplifier, discussed as an analog circuit in Chapter 10) shown in Fig. 13.19. Assuming operation in the active region, we may use Eq. (4.11) for emitter currents,

$$I_E = I_{EO} e^{V_{BE}/V_T}.$$

■ BJTs operate such that $\eta = 1$. Then $\frac{qV_{BE}}{\eta kT} \cong \frac{qV_{BE}}{kT} = \frac{V_{BE}}{V_T}.$

The ratio of these two emitter currents is

$$\frac{I_{E1}}{I_{E2}} = e^{(V_{BE1} - V_{BE2})/V_T} = e^{(V_{B1} - V_{B2})/V_T}. \quad (13.7)$$

If we arbitrarily define the edges of the switching region as those points at which

$$\frac{I_{E1}}{I_{E2}} = \frac{50}{1} \quad \text{and} \quad \frac{1}{50},$$

then these ratios occur when

$$V_{B1} - V_{B2} = \pm V_T \ln 50 = \pm 100 \text{ mV at } 25°C.$$

Thus Q_1 carries 98% of the current when V_{B1} is 100 mV greater than V_{B2}, and Q_2 carries 98% of the current when V_{B1} is 100 mV less than V_{B2}.

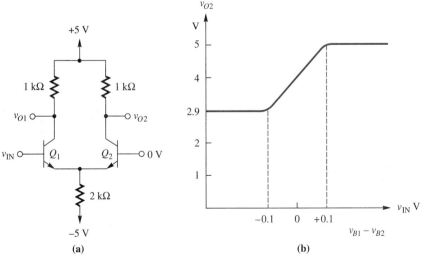

■ **FIGURE 13.19** Emitter-coupled differential amplifier.
(a) Circuit. (b) Transfer characteristic.

To analyze this emitter-coupled pair, we assume $V_{BE(\text{act})} = 0.7$ V and $\beta_F = 50$. When Q_1 is OFF and $V_{B2} = 0$, the emitters are clamped at $-V_{BE2} = -0.7$ V. The total emitter current is

$$\frac{-0.7 - (-5)}{2\,\text{k}} = 2.15\ \text{mA}.$$

When V_{B1} is -0.1 V, we have shown that Q_2 carries 98% of this current. When V_{B1} is less than -0.15 V, Q_1 is essentially cut off. You should verify this.

$$I_{C2} = \frac{\beta_F}{(\beta_F + 1)} I_{E2} = \frac{50}{51}(2.15) = 2.11\ \text{mA}.$$

We can now find $V_{C2} = 5 - 1\ \text{k}\Omega(2.11\ \text{mA}) = 2.89$ V; Q_2 is ON but not saturated. Since Q_1 is OFF, V_{C1} is 5 V.

When V_{B1} is $+0.1$ V, Q_1 conducts 98% of the current. The emitter voltage is $V_{B1} - V_{BE1} = 0.1 - 0.7 = -0.6$ V, and the total emitter current is now

$$\frac{-0.6 - (-5)}{2\,\text{k}} = 2.2\ \text{mA}.$$

■ This circuit is also known as a *current-mode switch*; we are obviously switching currents.

The total emitter current has changed only 2%, but it has switched from Q_2 to Q_1. Now V_{C1} is 2.84 V. Proper choice of resistor size prevents either Q_1 or Q_2 from saturating, and the power supply experiences no current surges. We should point out that if V_{B1} becomes more positive than 2.2 V, Q_1 will saturate; normally the input signal should not reach this level.

A Typical ECL Gate

One gate of the 10102 ECL quad two-input OR/NOR gate is shown in Fig. 13.20. This is one member of the third-generation ECL 10,000 (10K) series. Transistors Q_3 and Q_1

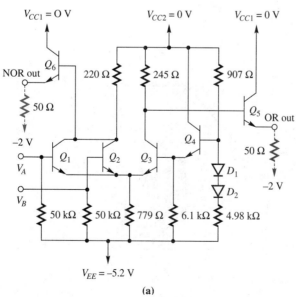
(a)

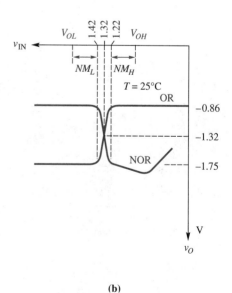
(b)

■ **FIGURE 13.20** One gate from 10102 ECL OR/NOR.
(a) Circuit. (b) Transfer characteristic.

or Q_2 constitute the emitter-coupled pair. If either V_A or V_B is HIGH, $V_{C1} = V_{C2}$ is LOW and Q_3 is OFF. A temperature-compensated reference voltage for V_{B3} is provided by Q_4 and the resistor-diode branch. The current switch outputs V_{C1} and V_{C3} are connected to the gate outputs by emitter followers Q_5 and Q_6, which provide a -0.75-V level shift and give a current amplification to drive loads that may be as low as 50 Ω.

We shall see that digital circuit designers must address many analog and high-frequency issues when operating at tens of megahertz clock rates. Figure 13.20 shows the positive supply terminal for the output transistors brought out separately. This is done because the current in the load capacitances may have large, rapid variations, which impose transients on the power supply filter. The relatively constant-current parts of the system can be somewhat isolated from this problem by means of a separate distribution bus. We also note that the positive supply voltages V_{CC1} and V_{CC2} are grounded (0 V), whereas V_{EE} is connected to a negative voltage (-5.2 V). This is done because the internal impedance from the output terminal to the positive supply terminal is much less than the impedance to the negative supply terminal. This means that any power supply noise will induce less noise in the output if the output reference is to the positive terminal.

The reference voltage V_{B3} is designed to vary with temperature. This idea did not arise with respect to the saturating logic discussed previously because the output logic levels in saturated logic are fairly well isolated from the internal parameter changes of the devices. However, with ECL, in which all transistors are operating in the active region, the variation of V_{BE} with temperature translates directly into logic level changes. By properly designing the reference voltage supply, we are able to maintain the reference midway between V_{IL} and V_{IH}.

Finally, the output of one chip is normally connected to inputs of other logic by means of a 50-Ω transmission line. With waveforms approaching 2-ns rise and fall times, care must be taken to preserve these wave shapes. Note that the propagation velocity of these signals is less than 20 cm/ns in a typical circuit layout. With a fan-out of more than one gate or an interconnection distance of more than 10 cm, the only practical means of avoiding reflections is to properly terminate a line of known characteristic impedance.[2] A printed-circuit-board layout often uses a ***microstrip***[3] distribution, which may be thought of as 50-Ω transmission lines. As suggested in Fig. 13.20, the 50-Ω termination is returned to -2 V to avoid excessive dc power dissipation. Some applications use $V_{CC1} = V_{CC2} = +2$ V, $V_{EE} = -3.2$ V, and return the 50-Ω load to 0 V (ground).

The following analysis of the circuit shown in Fig. 13.20 uses $V_{BE(\text{act})} = V_D = 0.75$ V and $\beta_F = 50$.

■ Packaging parasitics are a major consideration at these higher switching speeds.

■ Microstrip design is often studied in microwave circuits courses.

The Reference Voltage

To find V_{B3}, we write a node voltage equation at the base of Q_3, neglecting I_{B3}:

$$\frac{V_{B3} - V_{EE}}{6.1\,\text{k}\Omega} = (1 + \beta_F)(I_{907} - I_{4.98\text{k}})$$

$$= (1 + \beta_F)\left\{\left[\frac{V_{CC2} - (V_{B3} + V_{BE4})}{907\,\Omega}\right] - \left(\frac{V_{B3} + V_{BE4} - V_D - V_D - V_{EE}}{4.98\,\text{k}\Omega}\right)\right\}$$

$$\frac{V_{B3} + 5.2}{6.1\,\text{k}\Omega} = 51\left\{\left(\frac{0 - 0.75 - V_{B3}}{907\,\Omega}\right) - \left[\frac{V_{B3} + 0.75 - 0.75 - 0.75 - (-5.2)}{4.98\,\text{k}\Omega}\right]\right\}$$

$$V_{B3} = -1.32\,\text{V}$$

The effective source impedance of the reference voltage is only 15 Ω, so I_{B3} has a negligible effect on this calculation.

ECL Transfer Characteristic

When Q_1 and Q_2 are OFF ($V_A = V_B < -1.32 - 0.1$),

$$V_{E3} = -1.32 - 0.75 = -2.07 \text{ V},$$

$$-I_{E3} = \frac{[-2.07 - (-5.2)] \text{ V}}{779 \, \Omega} = 4.02 \text{ mA},$$

$$I_{C3} = 4.02 \left(\frac{50}{51}\right) = 3.94 \text{ mA}.$$

Writing a current equation at the emitter of Q_5,

$$(\beta_F + 1)(I_{245} - I_{C3}) = \frac{[V_{E5} - (-2)]}{50 \, \Omega}$$

$$(\beta_F + 1)\left\{\left[\frac{V_{CC2} - (V_{E5} + V_{BE5})}{245 \, \Omega}\right] - I_{C3}\right\} = \frac{[V_{E5} - (-2)]}{50 \, \Omega}$$

$$51\left[\frac{0 - (V_{E5} + 0.75)}{245 \, \Omega} - 3.94 \text{ mA}\right] = \frac{V_{E5} + 2}{50 \, \Omega}$$

$$V_{E5} = -1.75 \text{ V}.$$

This means that $V_{OL} = V_{E5} = -1.75$ V.

When Q_1 or Q_2 is ON, Q_3 is OFF, and the equation at the emitter of Q_5 is

$$(\beta_F + 1)(I_{245} - I_{C3}) = \frac{[V_{E5} - (-2)]}{50 \, \Omega}$$

$$(\beta_F + 1)\left\{\left[\frac{V_{CC2} - (V_{E5} + V_{BE5})}{245 \, \Omega}\right] - I_{C3}\right\} = \frac{[V_{E5} - (-2)]}{50 \, \Omega}$$

$$51\left[\frac{0 - (V_{E5} + 0.75)}{245 \, \Omega} - 0\right] = \frac{V_{E5} + 2}{50 \, \Omega}$$

$$V_{E5} = -0.86 \text{ V},$$

and $V_{OH} = V_{E5} = -0.86$ V. We find that the reference voltage V_{B3} is centered between V_{OH} and V_{OL}. In Fig. 13.20(b), we plot V_{E5} versus v_{IN} (other input held LOW).

We continue to assume 50-Ω loads as shown; if such is the case, actual dc fan-out has little effect on these numbers. Note that each HIGH input would add

$$\left[I_{50\text{k}} = \frac{[-0.86 - (-5.2)] \text{ V}}{50 \text{ k}\Omega} = 87 \, \mu\text{A}\right]$$

$$+ \left[I_{B1,2} = \frac{[-0.86 - 0.75 - (-5.2)] \text{ V}}{779 \, \Omega \, (\beta_F + 1)} = 90 \, \mu\text{A}\right] = 177 \, \mu\text{A}$$

to the load current, a very small change. With high-speed ECL, it is the capacitive load that limits the fan-out. The 50-Ω load is required to dissipate the charge stored on the line when the emitter follower turns OFF. You are reminded that the emitter follower has a very low source resistance when it is ON, but if its base goes negative faster than the emitter load capacitance can follow, then the transistor is OFF, and only the load resistance serves to discharge the line.

The noise margins are $NM_L = -1.42 - (-1.75) = 0.33$ V,

$$NM_H = -0.86 - (-1.22) = 0.36 \text{ V}.$$

CHAPTER 13   INTEGRATED-CIRCUIT LOGIC GATES   669

We may show that the NOR output at V_{E6} is complementary. When an input is pulled HIGH, say, $V_A = -0.86$ V, V_{E1} follows to -1.61 V, and I_{E1} becomes 4.6 mA, $I_{C1} = 4.5$ mA, and $V_{C1} = -0.99$ V. Thus V_{E6} will be -1.75 V when LOW, the same level as the OR output.

EXAMPLE 13.9

Calculate the power dissipation for the ECL gate shown in Fig. 13.20. The voltage reference circuit (Q_4, two diodes, and three resistors) is shared with another gate.

SOLUTION   Using the foregoing expressions, $I_{6.1\,k} = 0.64$ mA, and $I_{4.98\,k} = 0.63$ mA. Thus the total reference circuit current is 1.27 mA, and the power is $(1.27\text{ mA})(5.2\text{ V}) = 6.6$ mW. Since this reference is shared with another gate, we assign 3.3 mW/gate. We have calculated I_{779} to be 4.02 mA when Q_3 is ON and 4.6 mA when Q_1 or Q_2 is ON. Taking the average, the power of the emitter-coupled pair is $(4.3\text{ mA})(5.2\text{ V}) = 22.4$ mW. This sums to 25.7 mW/gate unloaded. If Q_5 and Q_6 are both driving 50-Ω loads returned to -2 V, they dissipate another $(23\text{ mA})(0.86\text{ V}) + (5\text{ mA})(1.75\text{ V}) = 28.5$ mW/gate internally.

DRILL 13.9

What problems might arise if different ECL packages in the same system were at different temperatures?

ANSWER   In this case, logic levels would be different, and noise margins would be unequal. For example, the gate above at 25°C has $V_{OH} = -0.86$ V, $V_{OL} = -1.75$ V, but at 125°C, $V_{IL} = -1.26$ V and $V_{IH} = -1.06$ V. When the first case drives the second case, $NM_H = 0.49$ V and $NM_L = 0.20$ V. More on this later when we discuss the 100K ECL series.

EXAMPLE 13.10

Sketch the input-output characteristic of the ECL gate for $T = 125$°C (β_F increases by 50%). Show that the temperature compensation of the voltage reference circuit maintains about equal noise margins.

SOLUTION   Using a -2.5 mV/°C temperature coefficient for V_{BE} and V_D, diode drops become 0.5 V. Repeating the earlier calculations (with $\beta_F = 75$) results in $V_{B3} = -1.16$ V, $V_{IL} = -1.26$ V, $V_{IH} = -1.06$ V, $V_{OH} = -0.60$ V, and $V_{OL} = -1.62$ V. Then $NM_H = 0.46$ V and $NM_L = 0.36$ V.

We can summarize ECL 10K series specifications at 25°C as

- Fan-out: limited by distribution capacitance
- Noise margin: 0.33 V
- Propagation delay: 2 ns
- Power dissipation/gate: 24 mW (unloaded), 52 mW (loads = 50 Ω).

■ The heat generated by ECL produces a serious cooling problem for large systems.

This delay-power product is also plotted in Fig. 13.17. A data sheet from this family is shown in Fig. 13.21.

The fourth-generation ECL is the 100K series. This series utilizes the improved technology we referred to when discussing advanced TTLs. The transistors have oxide isolation and walled-base and emitter regions, and hence they are smaller, with less parasitic capacitance. This allows propagation delays of 0.75 ns (at 40 mW/gate). In addition, a more complex compensation circuit prevents the shift in levels with temperature seen in the 10K series. This can be important in a large system where not all of the chips are at the same temperature.

ELECTRICAL CHARACTERISTICS

Each MECL 10,000 series circuit has been designed to meet the dc specifications shown in the test table, after thermal equilibrium has been established. The circuit is in a test socket or mounted on a printed circuit board and transverse air flow greater than 500 linear fpm is maintained. Outputs are terminated through a 50-ohm resistor to −2.0 volts. Test procedures are shown for only one gate. The other gates are tested in the same manner.

		@Test Temperature	TEST VOLTAGE VALUES (Volts)				
			V_{IH} max	V_{IL} min	V_{IHA} min	V_{ILA} max	V_{EE}
		−30°C	−0.890	−1.890	−1.205	−1.500	−5.2
		+25°C	−0.810	−1.850	−1.105	−1.475	−5.2
		+85°C	−0.700	−1.825	−1.035	−1.440	−5.2

Characteristic	Symbol	Pin Under Test	MC10101 Test Limits							Unit	TEST VOLTAGE APPLIED TO PINS LISTED BELOW:					(V_{CC}) Gnd
			−30°C		+25°C			+85°C			V_{IH} max	V_{IL} min	V_{IHA} min	V_{ILA} max	V_{EE}	
			Min	Max	Min	Typ	Max	Min	Max							
Power Supply Drain Current	I_E	8	−	29	−	20	26	−	29	mAdc	−	−	−	−	8	1, 16
Input Current	I_{inH}	4	−	425	−	−	265	−	265	µAdc	4	−	−	−	8	1, 16
		12	−	850	−	−	535	−	535	µAdc	12	−	−	−	8	1, 16
	I_{inL}	4	0.5	−	0.5	−	−	0.3	−	µAdc	−	4	−	−	8	1, 16
		12	0.5	−	0.5	−	−	0.3	−	µAdc	−	12	−	−	8	1, 16
Logic "1" Output Voltage	V_{OH}	5	−1.060	−0.890	−0.960	−	−0.810	−0.890	−0.700	Vdc	12	−	−	−	8	1, 16
		5	−1.060	−0.890	−0.960	−	−0.810	−0.890	−0.700		4	−	−	−		
		2	−1.060	−0.890	−0.960	−	−0.810	−0.890	−0.700		−	−	−	−		
		2	−1.060	−0.890	−0.960	−	−0.810	−0.890	−0.700		−	−	−	−		
Logic "0" Output Voltage	V_{OL}	5	−1.890	−1.675	−1.850	−	−1.650	−1.825	−1.615	Vdc	−	−	−	−	8	1, 16
		5	−1.890	−1.675	−1.850	−	−1.650	−1.825	−1.615		−	−	−	−		
		2	−1.890	−1.675	−1.850	−	−1.650	−1.825	−1.615		12	−	−	−		
		2	−1.890	−1.675	−1.850	−	−1.650	−1.825	−1.615		4	−	−	−		
Logic "1" Threshold Voltage	V_{OHA}	5	−1.080	−	−0.980	−	−	−0.910	−	Vdc	−	−	12	−	8	1, 16
		5	−1.080	−	−0.980	−	−	−0.910	−		−	−	4	−		
		2	−1.080	−	−0.980	−	−	−0.910	−		−	−	−	12		
		2	−1.080	−	−0.980	−	−	−0.910	−		−	−	−	4		
Logic "0" Threshold Voltage	V_{OLA}	5	−	−1.655	−	−	−1.630	−	−1.595	Vdc	−	−	−	12	8	1, 16
		5	−	−1.655	−	−	−1.630	−	−1.595		−	−	−	4		
		2	−	−1.655	−	−	−1.630	−	−1.595		−	−	12	−		
		2	−	−1.655	−	−	−1.630	−	−1.595		−	−	4	−		
Switching Times (50-ohm load)												Pulse In	Pulse Out	−3.2 V	+2.0 V	
Propagation Delay	$t_{4+2−}$	2	1.0	3.1	1.0	2.0	2.9	1.0	3.3	ns	−	−	4	2	8	1, 16
	$t_{4−2+}$	2									−	−		2		
	t_{4+5+}	5									−	−		5		
	$t_{4−5−}$	5									−	−		5		
Rise Time (20 to 80%)	t_{2+}	2	1.1	3.6	1.1	−	3.3	1.1	3.7		−	−		2		
	t_{5+}	5									−	−		5		
Fall Time (20 to 80%)	$t_{2−}$	2									−	−		2		
	$t_{5−}$	5									−	−		5		

■ **FIGURE 13.21** Data sheet for MC10103 ECL OR/NOR gate (courtesy Motorola, Inc.).

CHECK UP

1. **TRUE OR FALSE?** The active BJTs in ECL circuits dissipate less power than saturated transistors would.
2. **TRUE OR FALSE?** The active BJTs in ECL circuits can switch faster than saturated transistors would.
3. **TRUE OR FALSE?** The active BJTs in ECL circuits are more sensitive to temperature changes than saturated transistors would be.
4. **TRUE OR FALSE?** Interconnections between ECL chips need to be properly terminated because the logic voltages are negative.

13.6 NMOS LOGIC

MOSFET gates are the most widely used logic circuits, and have several advantages. First, MOS offers very high circuit density, typically an order of magnitude greater than bipolar. Second, the very high input resistance of the insulated gate means that drivers may be of very low current and power. Unfortunately, we run into the trade-off between speed and power once again. The FET possesses parasitic capacitance between channel and substrate and between gate and channel. The device current, then, must be high enough to charge these capacitances within the desired delay time.

VLSI circuits, such as microprocessors and memory arrays, are likely to use NMOS or CMOS. The CMOS Intel 486DX2 µP has more than 10^6 transistors with the newer INTEL P5 and P6 microprocessors approaching 200 MHz clock rates, and NMOS memories with more than 16,777,216 (16M) cells are in production. We will not find broad NMOS logic families because at the small- and medium-scale levels of integration, CMOS is the popular choice in FET logic. We discuss CMOS in Section 13.7.

The NMOS Inverter

The NMOS inverter was introduced in Section 5.4. We rejected the simple inverter shown in Fig. 5.14, which uses a resistive load, because an active FET load requires much less chip area than a diffused resistor. It is interesting to note that for a typical FET a satisfactory load resistor would be 100 kΩ. For a sheet resistance $R_\square = 100\ \Omega/\square$, a resistor geometry of $L = 1000\ W$, and a minimum feature size of 1.25 µm, the folded resistor area would be 5 mils^2, or 50 times the FET area.

■ Sheet resistance ($R_\square$), which quantifies doping levels and junction depth, is discussed in Chapter 18.

Three NMOS inverters are shown in Fig. 13.22. Although PMOS was widely used in early FETs because of fabrication difficulties with *p*-type silicon, these problems have been overcome. The *n* channel is preferred because the greater mobility of electrons offers lower channel resistance for the same dimensions. In addition, the positive supply voltage, V_{DD}, is more widely compatible with positive logic.

The enhancement-mode driver requires that when v_{IN} is LOW ($<V_{tD}$), the transistor will be OFF, which is in accord with the inverter concept. Gate and drain bias sources are of the same polarity, permitting direct coupling between stages. Because the two devices may have different thresholds, we designate the threshold of the driver as V_{tD} and that of the load as V_{tL}.

Several arrangements for the load FET are used. We will take up the case of a PMOS load, called CMOS, in Section 13.7. With NMOS, consider the three alternatives shown

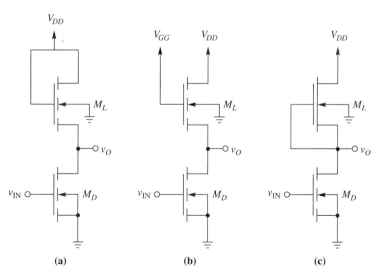

■ **FIGURE 13.22** NMOS inverters. **(a)** and **(b)** use enhancement loads. **(c)** uses depletion load.

in Fig. 13.22. In part (a) the load is an enhancement NMOS with the gate tied to the drain. The difficulty with this circuit is that the output voltage, v_O, cannot pull up to the supply voltage V_{DD}. For the channel to remain enhanced, v_{GS} must be greater than V_{tL}. If $V_{DD} = 5$ V and $V_{tL} = 1$ V, then $V_{OH} = 4$ V. The load would always be in saturation since $v_{GS} = v_{DS}$.

In part (b) the load is an enhancement NMOS with the gate connected to a voltage, V_{GG}, at least V_{tL} more positive than V_{DD}. This allows the output to pull up to V_{DD}, but at the expense of another supply and the space cost of distributing the voltage around the chip. The load would always be in the ohmic region,

$$v_{GS} > V_{tL} + v_{DS}.$$

Part (c) shows the depletion NMOS load, which is the dominant alternative because it allows v_O to pull up to V_{DD} and does not require the extra power supply. The cost is the extra fabrication step of the ion implantation of the load channel. A production CMOS process might exceed 15 mask steps.

In each of the applications of NMOS loads, the substrate is connected to ground, not to the source, because all FETs on the chip will share the common substrate. The effect of this is to cause the threshold voltage of the load, V_{tL}, to vary with the source voltage, v_O. From Section 8.6, we repeat Eq. (8.73):

$$V_t = V_{t0} + \gamma\left(\sqrt{2\phi_p - v_{BS}} - \sqrt{2\phi_p}\right), \tag{13.8}$$

where V_{t0} is the threshold voltage at $v_{BS} = 0$, the *body-effect parameter*, γ, is a fabrication constant that ranges from 0.1 to 0.5 $V^{0.5}$, and $2\phi_p$, the *surface potential*, is approximately 0.6 V. For the depletion NMOS load, V_{tL} will become less negative as V_O increases, perhaps as much as $0.3 V_O^{0.5}$ V. Thus, if $V_{DD} = 5$ V, V_t might decrease in magnitude about 0.6 V. Nevertheless, it is common to neglect this effect for a first-order solution to digital circuits. To take this effect into account with SPICE, it is only necessary to specify a value for the body-effect parameter γ in the model statement for the load FET.

- No external load resistance is shown for any of these inverter configurations, since normally the external load would consist only of more MOS gate inputs.

- The surface potential (SPICE PHI) is twice ϕ_p of the unbiased p material.

Depletion-Load Transfer Characteristic

Refer to the inverter of Fig. 13.23. We know that $v_{DSL} = V_{DD} - v_O$, $V_{GSL} = 0$, and $V_{tL} < 0$. For depletion NMOS, V_{tL} is in actuality a pinch-off voltage, being negative. Recalling Eq. (5.10), we find that FET drain current in the ohmic region is

$$i_D = k[2(V_{GS} - V_t)v_{DS} - v_{DS}^2] \qquad 0 < v_{DS} < (v_{GS} - V_t). \tag{13.9}$$

From Eq. (5.12), drain current in the saturation region is

$$i_D = k(v_{GS} - V_t)^2 \qquad 0 < v_{GS} - V_t < v_{DS}. \tag{13.10}$$

We can make some generalizations from the transfer characteristic of Fig. 13.23(b). You should verify these statements before proceeding:

- We see that M_L is always ON because $V_{GS} = 0$.
- M_L is in the ohmic region when v_O is above the line marked $v_O = V_{DD} + V_{tL}$. [Ohmic means $v_{DSL} < (v_{GSL} - V_{tL}) \Rightarrow (V_{DD} - v_O) < (0 - V_{tL}) \Rightarrow (v_O > V_{DD} + V_{tL}.$]
- M_L is in saturation when v_O is below the line marked $v_O = V_{DD} + V_{tL}$. [Saturation means $v_{DSL} > (v_{GSL} - V_{tL}) \Rightarrow (V_{DD} - v_O) > (0 - V_{tL}) \Rightarrow v_O < V_{DD} + V_{tL}.$]
- M_D is OFF when $v_{IN} < V_{tD}$.

CHAPTER 13 INTEGRATED-CIRCUIT LOGIC GATES

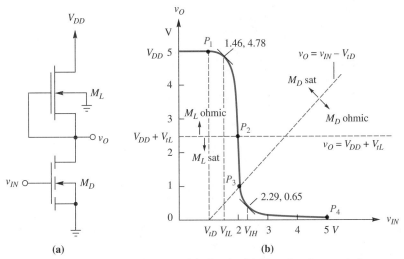

FIGURE 13.23 MOS inverter. (a) Circuit. (b) Transfer characteristic.

- M_D is in saturation when v_O is above the line marked $v_O = v_{IN} - V_{tD}$.
  [Saturation means $v_{DSD} > (v_{GSD} - V_{tD}) \Rightarrow v_O > v_{IN} - V_{tD}$.]
- M_D is in the ohmic region when v_O is below the line marked $v_O = v_{IN} - V_{tD}$.
  [Ohmic means $v_{DSD} < (v_{GSD} - V_{tD}) \Rightarrow v_O < v_{IN} - V_{tD}$.]

As v_{IN} increases from 0 to V_{tD}, $v_O = V_{DD}$, because M_D is OFF and M_L is in the ohmic region. At P_1, M_D comes ON (saturated), M_L is still in the ohmic region. From P_2, where $v_O = V_{DD} + V_{tL}$, down to P_3, where $v_O = v_{IN} - V_{tD}$, both devices are saturated. Below P_3, M_D enters the ohmic region, and M_L remains saturated. In the static situation, where $I_{DL} = I_{DD}$, the equations are easy to plot, being subject only to the small error introduced by the variation of V_{tL} with v_O.

■ Note that λ has been neglected in this analysis because for the static no-load situation it would have minimal effect.

EXAMPLE 13.11

Assume a load FET with $k_L = 0.02$ mA/V^2 and $V_{tL} = -2.5$ V (neglect body effect), and a driver with $k_D = 0.1$ mA/V^2 and $V_{tD} = 1$ V. Let $V_{DD} = 5$ V, and calculate the designated points on the transfer characteristic of Fig. 13.23(b).

SOLUTION P_1 occurs at $v_O = V_{DD}$ as v_{IN} reaches $V_{tD} = 1$ V. From P_1 to P_2, ohmic M_L has $i_{DD} = 0.02[2(2.5)(5 - v_O) - (5 - v_O)^2] = 0.02v_O(5 - v_O)$, equal to $i_{DD} = 0.1(v_{IN} - 1)^2$ mA (M_D is saturated). The result is

$$v_{IN} = 1 + (v_O - 0.2v_O^2)^{0.5}. \quad (13.11)$$

We define V_{IL} to be that value of v_{IN} where the slope of v_O versus v_{IN} is -1. Differentiating Eq. (13.11) and setting $dv_O/dv_{IN} = -1$ gives $V_{IL} = 1.46$ V.

P_2 occurs at $v_O = V_{DD} + V_{tL} = 5 - 2.5 = 2.5$ V. Here M_D and M_L are both saturated, acting as current sources; $i_{DL} = 0.02(2.5)^2 = 0.125$ mA, and $i_{DD} = 0.1(v_{IN} - 1)^2$ mA. Equating the currents yields $v_{IN} = 2.12$ V.

P_3 is at the same saturation current; thus v_{IN} is still 2.12 V, and $v_O = v_{IN} - 1 = 1.12$ V. Below P_3, M_L remains a 0.125-mA current source, and M_D is in the ohmic region. Equating $0.125 = 0.2[2(v_{IN} - 1)v_O - v_O^2]$ yields

$$v_{IN} = 1 + 0.625 v_O^{-1} + 0.5 v_O. \quad (13.12)$$

DRILL 13.10

In Eq. (5.12) we defined $k = 0.5\mu C_o W/L$. Let $\mu C_o = 20$ μA/V^2, and find the W/L ratio for the FETs in Example 13.11. If the minimum feature size (W or L) is 3 μm, what is the area of this inverter?

ANSWER
$(W/L)_L = 2$,
$(W/L)_D = 10$;
$6 \times 3 + 30 \times 3$ μm $= 108$ μm^2.

> Forcing the slope of Eq. (13.12) to be -1, we find $V_{IH} = 2.29$ V.
> Then the noise margins would be
>
> $$NM_L = V_{IL} - V_{OL} = 1.46 - 0.16 = 1.3 \text{ V}$$
> $$NM_H = V_{OH} - V_{IH} = 5 - 2.29 = 2.71 \text{ V}.$$
>
> This inverter characteristic is generated by means of a SPICE computer model of these FETs in Example 5.6, with similar results. Note that the computer model recognizes that the substrate of M_L is grounded.

■ You should locate these noise margins on the transfer characteristic of Fig. 13.23(b).

The MOS designer has many variables to consider. In Example 13.11, we note that $k_L < k_D$; in fact, $k_L/k_D = 0.2$. This results from making L/W of the load FET larger than that of the driver, which is typical. This causes the transition of the transfer characteristic to be steeper than it would be if the two FETs had identical geometry.

Switching Speed

A reasonably accurate model for the transient situation is obtained by lumping all of the parasitic capacitance of the two channels plus the external capacitance of the wiring and connected load at the output node. This is especially realistic when the node is an output terminal for the chip, where the external capacitance can be expected to dominate. The model is shown in Fig. 13.24(a).

The rise time can be found by considering M_D to turn OFF at $t = 0$. For $v_O < V_{DD} + V_{tL}$, M_L is saturated, and charges C with the constant current

$$i_C = k_L V_{tL}^2 = C_T \frac{dv_O}{dt} \tag{13.13}$$

from Eq. (13.10). Integrating Eq. (13.13), we obtain

$$t = \frac{C_T(v_O - V_{OL})}{k_L V_{tL}^2}, \tag{13.14}$$

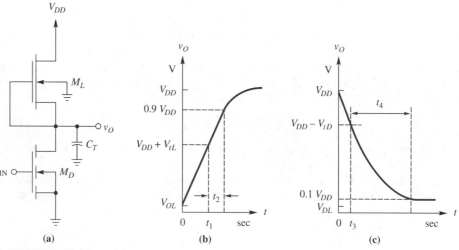

■ **FIGURE 13.24** MOS inverter timing. **(a)** Model. **(b)** Turn-OFF. **(c)** Turn-ON.

and v_O moves to $V_{DD} + V_{tL}$ from V_{OL} in

$$t_1 = \frac{C_T(V_{DD} + V_{tL} - V_{OL})}{k_L V_{tL}^2}. \tag{13.15}$$

Figure 13.24(b) shows the rising output voltage waveform, and we note that it is linear. Above $v_O = V_{DD} + V_{tL}$ [$V_{DSL} < (V_{GSL} - V_{tL})$], where the load is in the ohmic region,

$$i_C = k_L[-2V_{tL}(V_{DD} - v_O) - (V_{DD} - v_O)^2] = \frac{C_T dv_O}{dt}. \tag{13.16}$$

Integration gives

$$t = -\frac{C_T}{2k_L V_{tL}} \ln \frac{(v_O - V_{DD} - 2V_{tL})}{(V_{DD} - v_O)}. \tag{13.17}$$

To find the value of t_2, Eq. (13.17) would be evaluated at the limits $v_O = V_{DD} + V_{tL}$ and $v_O = 0.9 V_{DD}$. What we learn from these equations is that the rise time is a linear function of C_T, and varies inversely with k_L. Thus, to minimize rise time, the circuit designer obviously tries to minimize all parasitic capacitances and then increase k_L, which means more current and more power. The following example, using a C_T typical of on-chip capacitances, points out the delay problems that would be encountered by the greater parasitic capacitance of long interconnections.

EXAMPLE 13.12

Find rise time equal to $t_1 + t_2$ for Example 13.11 if $C_T = 0.5$ pF.

SOLUTION From Eq. (13.15), $t_1 = 9$ ns, and, evaluating Eq. (13.17) between $v_O = 2.5$ and 4.5 V, $t_2 = 11$ ns. Thus, $t_r = 20$ ns.

Fall time is found by setting $v_{IN} = V_{DD}$ at $t = 0$ and letting the capacitance discharge with a current $i_{DD} - i_{DL}$. The computation is similar to that for rise time, but the result is usually much less because i_{DD} is much greater than i_{DL}. If i_{DL} is small enough to be neglected, the differential equation simplifies greatly, and we find (Problem 13.34)

$$t_3 = \frac{C_T V_{tD}}{k_D(V_{DD} - V_{tD})^2}. \tag{13.18}$$

Returning to Example 13.11, C_T discharges with a current of $k_D(V_{DD} - V_{tD})^2 = 0.1(4)^2 = 1.6$ mA, or about 13 times faster than it charged.

From Fig. 13.3, using Eq. (13.13), with v_O limits 0 to $V_{DD}/2$, we are able to estimate

$$t_{pLH} = \frac{C_T V_{DD}}{2k_L V_{tL}^2}.$$

Similarly, we estimate

$$t_{pHL} = \frac{C_T V_{DD}}{2k_D(V_{DD} - V_{tD})^2}$$

(see Problem 13.36). If t_{pLH} dominates, the propagation delay is

$$t_{pd} = \frac{C_T V_{DD}}{2k_L V_{tL}^2}. \tag{13.19}$$

DRILL 13.11

The body effect was neglected in Example 13.12, and it would not be expected to have much effect on t_1, where v_O is LOW. Estimate what effect it might have on t_2 if V_{tL} changes to about -2 V in this range.

ANSWER

t_2 increases to about 12 ns, a relatively minor error. Using SPICE, with $\gamma = 0.3$ V$^{0.5}$, $t_1 = 11$ ns, $t_2 = 12$ ns, and the total rise time $t_r = 23$ ns.

For a 50% duty cycle, the power dissipation (neglecting the frequency-dependent energy lost charging and discharging the capacitor) is

$$P_D = \frac{V_{DD} I_{DL(\text{sat})}}{2} = \frac{V_{DD} k_L V_{tL}^2}{2}. \tag{13.20}$$

This would be the standby power dissipation for a system where one-half of the gates were HIGH, and half were LOW (a reasonable assumption). The resultant delay-power product is

$$P_D t_{pd} = \frac{C_T V_{DD}^2}{4}. \tag{13.21}$$

The implications of this are apparent: For a low delay-power product, we must minimize C_T and keep V_{DD} as low as practical (noise margins must be maintained).

In each switching cycle, capacitance C_T is charged to voltage V_{DD}, and then this charge, $Q = V_{DD} C_T$ is discharged to 0 V. This requires an energy from the power supply of $V_{DD}^2 C_T$ joules each cycle, and means that there is a power loss when the gate is switching equal to

$$P = V_{DD}^2 C_T f, \tag{13.22}$$

where f is the switching rate. For high-speed switching, this term could exceed that of Eq. (13.20)

DRILL 13.12

In the gate of Examples 13.11 and 13.12, assume that we are switching at a rate of 5 MHz. Find the dynamic power loss due to the capacitance.

ANSWER 62.5 μW

EXAMPLE 13.13

Calculate the t_{pd} and P_D for the gate in Examples 13.11 and 13.12.

SOLUTION Using Eqs. (13.19) and (13.20) we find $t_{pd} = 10$ ns, $P_D = 0.312$ mW, and the product is 3.1 pJ.

NMOS Gates

The circuit for a three-input NOR gate is shown in Fig. 13.25(a). If we assume that each driver transistor has the same characteristics as our inverter driver, then any input should give a transfer characteristic identical to that of the inverter in Fig. 13.23. If more than one driver is ON, then V_{OL} will be somewhat lower than that of the inverter. In estimating transient behavior, we note that each driver channel contributes some capacitance to the output node.

The NAND arrangement is shown in Fig. 13.25(b). Because the driver FETs are in series, they are wider so that the combined channels will maintain the same ON channel resistance. Because this will require more area per FET, the NOR logic is preferred.

NMOS logic is very compact, and very versatile, since design changes have mostly to do with placement of the sources and drains and the shape of the gates. Processes are standardized. NMOS readily lends itself to LSI and VLSI, where most of the interconnections are on the chip, hence low parasitic capacitance. There are no SSI NMOS logic families, because the advantages are lost if every gate has to have large area so that it can quickly charge the large off-chip capacitances that inevitably attend interconnections between IC packages.

CHAPTER 13 INTEGRATED-CIRCUIT LOGIC GATES

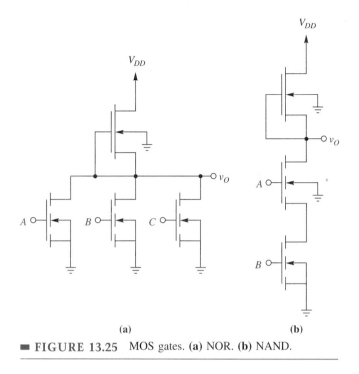

FIGURE 13.25 MOS gates. (a) NOR. (b) NAND.

EXAMPLE 13.14

Sketch the FET arrangement for a MOS gate that would satisfy the logic statement $X = \overline{(A \cdot B) + C}$.

SOLUTION Drivers A and B are in series as illustrated in Fig. 13.25(b). Driver C is in parallel with the $A \cdot B$ branch in the manner of Fig. 13.25(a), and there is one load FET for the combination.

MESFET (Gallium-Arsenide) Gates

The very high speed obtainable with GaAs (the electron mobility of n-GaAs is five to six times that of n-silicon) has prompted the development of this specialized digital technology. Gate delays as low as 20 to 100 ps have been reported; as with the other technologies, there is a cost in power dissipation. Power is on the order of 10 mW/gate, which forces low circuit density. As we shall see, noise margins are low because of small voltage swings.

The circuit topology of the NOR gate of Fig. 13.26 illustrates one approach to this technology, called ***direct-coupled FET logic***. Note the standard MESFET symbols. Recall the discussion of the MESFET in Section 5.11. Suppose that driver transistors B_{D1} and B_{D2} are enhancement MESFETs, $V_{tD} \approx 0.2$ V, and the load transistor B_L is a depletion MESFET, $V_{tL} \approx -1$ V; then V_{DD} is likely to be near 1.5 V. The output is shown connected to the inputs of other similar circuits.

When the input voltages at B_{D1} and B_{D2} are both LOW, ($<V_{tD}$), the drivers are OFF, and the load MESFET will pull the output node HIGH. But this output is connected to the gates of other MESFETs, and when those gates are more positive than about $+0.7$ V, they conduct, acting as Schottky-barrier diodes. This clamps V_{OH} at 0.7 V. It

■ Although the MESFET is not a MOSFET, being in structure more like a JFET, it is included in this section because the topology of the MESFET gates is similar to that of NMOS gates.

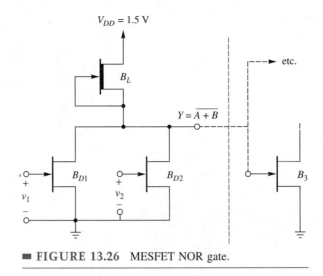

FIGURE 13.26 MESFET NOR gate.

can be shown that V_{IH} for these driven MESFETs, B_3, etc., will be the same as for B_{D1} and B_{D2} of this gate, approximately 0.6 V, and $V_{IL} \approx 0.5$ V. This would have to be found from the transfer characteristic. Thus $NM_H = 0.1$ V.

When the input voltage at either B_{D1} or B_{D2} is greater than V_{IH}, that driver MESFET will be ON, in the ohmic region, and $V_{OL} \approx 0.1$ V. This makes $NM_L = 0.4$ V. To summarize, this circuit has $t_{pd} < 0.1$ ns, $P_D \approx 10$ mW/gate, $NM_H = 0.1$ V, and $NM_L = 0.4$ V.

As one can readily see, there are many difficulties with this implementation, not the least of which is the manufacturability of enhancement MESFETs with precise V_{tD}. Some of the fabrication issues are discussed in Chapter 18. There are many other approaches to GaAs digital logic, and there is little standardization among the various circuits. No logic families exist; each application is specialized. At present, production is limited, and yield is low, but that is likely to improve as the technology matures and applications demand that level of performance.

CHECK UP

1. **TRUE OR FALSE?** NMOS is the most dense of the technologies we have studied.
2. **TRUE OR FALSE?** NMOS is preferred over PMOS because of the relative mobilities of the majority carriers.
3. **TRUE OR FALSE?** Most NMOS circuits use only enhancement FETs to save fabrication costs.
4. **TRUE OR FALSE?** The faster NMOS circuits are switched, the more power is dissipated.

13.7 COMPLEMENTARY MOS LOGIC

The **CMOS** inverter was introduced in Section 5.4. We have redrawn the circuit and its transfer characteristic as shown in Fig. 13.27 so that we can examine some of the details of the gate performance. We will show that essentially no current flows in the circuit

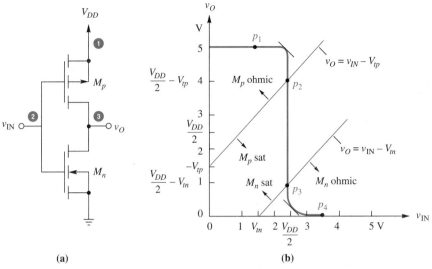

FIGURE 13.27 CMOS inverter. (a) Circuit. (b) Transfer characteristic. Nodes are numbered for Ex. 13.16.

except when it is switching. We will also find that rise time is greatly reduced from that of NMOS technology.

Transfer Characteristic

The drain current of Eqs. (13.9) and (13.10) continues to apply. The magnitude signs are a reminder that these quantities are all negative for PMOS devices.

- Ohmic: $i_{DS} = k[2(v_{GS} - V_t)v_{DS} - v_{DS}^2]$ $0 < |v_{DS}| < |v_{GS} - V_t|$
- Saturation: $i_{DS} = k(v_{GS} - V_t)^2$ $0 < |v_{GS} - V_t| < |v_{DS}|$

Assume $V_{tp} = -V_{tn}$ and $k_p = -k_n$ for symmetry, although that certainly is not required, and is actually difficult to obtain reproducibly in a manufacturing facility. Since $k \propto \mu W/L$, W/L of the p channel will have to be three times that of the n channel if the k terms are to be equal, because of the lower mobility of holes. Current flows from the positive supply, through the PMOS M_p from source to drain, and through the NMOS M_n from drain to source: The drains are connected together.

Since

- $v_{GSn} = v_{IN}$,
- $v_{DSn} = v_O$,
- $v_{DSp} = v_O - V_{DD}$,
- $v_{GSp} = v_{IN} - V_{DD}$,

we find the following:

- M_n is OFF when $v_{IN} < V_{tn}$ because $v_{GSn} < V_{tn}$.
- M_n is saturated when $v_{DSn} > (v_{GSn} - V_{tn})$, which is equivalent to $v_O > (v_{IN} - V_{tn})$. Note that this region is **above** the line marked $v_O = (v_{IN} - V_{tn})$ in Fig. 13.27(b).
- M_n is in the ohmic region when v_O is **below** the line marked $v_O = (v_{IN} - V_{tn})$.
- M_p is OFF when $v_{IN} > V_{DD} + V_{tp}$ since $(v_{IN} - V_{DD}) = v_{GSp} > V_{tp} \Rightarrow |v_{GSp}| < |V_{tp}|$. Since both v_{GSp} and V_{tp} are negative for PMOS, $v_{GSp} > V_{tp}$ is the OFF condition.

- M_p is saturated when $|v_{DSp}| > |v_{GSp} - V_{tp}|$, or for negative values, $v_{DSp} < (v_{GSp} - V_{tp})$. This is equivalent to $(v_O - V_{DD}) < (v_{IN} - V_{DD} - V_{tp})$, or v_O is less than $(v_{IN} - V_{tp})$. Note that this region is **below** the line marked $v_O = v_{IN} - V_{tp}$ in Fig. 13.27(b).
- M_p is in the ohmic region when v_O is **above** the line marked $v_O = (v_{IN} - V_{tp})$.

You should verify that these statements meet the conditions set forth in Eqs. (13.9) and (13.10).

As v_{IN} ranges from 0 to V_{tn}, $v_O = V_{DD}$, because M_n is OFF and M_p is in the ohmic region. At P_1, M_n comes ON (saturated) and M_p remains ohmic. The negative I_{DS} of M_p is equal to I_{DS} of M_n in the dc case.

$$-k_p[2(v_{IN} - V_{DD} - V_{tp})(v_O - V_{DD}) - (v_O - V_{DD})^2] = k_n(v_{IN} - V_{tn})^2. \quad (13.23)$$

From P_2, where $v_O = v_{IN} - V_{tp}$, to P_3, where $v_O = v_{IN} - V_{tn}$, both FETs are saturated. Using Eq. (13.10), and equating the currents,

$$-k_p(v_{IN} - V_{DD} - V_{tp})^2 = k_n(v_{IN} - V_{tn})^2. \quad (13.24)$$

Solving for the general case,

$$V_{IN} = \frac{\sqrt{-k_p/k_n}(V_{DD} + V_{tp}) + V_{tn}}{1 + \sqrt{-k_p/k_n}}. \quad (13.25)$$

For the symmetric case of $V_{tp} = -V_{tn}$ and $k_p = -k_n$, this reduces to

$$V_{IN} = \frac{V_{DD}}{2}. \quad (13.26)$$

- Again, the channel-length modulation term λ has been neglected as having minimal effect in this static unloaded situation.

From P_3 to P_4, M_n is ohmic and M_p is saturated. Equating the currents,

$$-k_n[2(v_{IN} - V_{tn})v_O - v_O^2] = k_p[v_{IN} - (V_{DD} + V_{tp})]^2. \quad (13.27)$$

At P_4, M_p turns OFF because $V_{IN} = V_{DD} + V_{tp}$. An example illustrates the situation in the static case.

EXAMPLE 13.15

Assume a CMOS inverter with $-k_p = k_n = 0.05$ mA/V^2, $V_{tn} = 1.5$ V, and $V_{tp} = -1.5$ V. Let $V_{DD} = 5$ V, and calculate the designated points on the transfer characteristic in Fig. 13.27(b).

SOLUTION P_1 occurs at $v_O = V_{DD}$ as $v_{IN} = V_{tn} = 1.5$ V. From P_1 to P_2, Eq. (13.23) applies. For $k_p = -k_n$ and $V_{tn} = -V_{tp} = V_t$,

$$k[2(v_{IN} - V_{DD} - V_t)(v_O - V_{DD}) - (v_O - V_{DD})^2] = k(v_{IN} - V_t)^2.$$

After considerable manipulation of the quadratic equation, it can be shown that

$$v_{IN} = v_O + V_t - V_{DD} + [2(V_{DD} - v_O)(V_{DD} - 2V_t)]^{0.5}. \quad (13.28)$$

When $dv_O/dv_{IN} = -1$, we find $v_O = 0.875V_{DD} + 0.25V_t = 4.75$ V and $V_{IL} = 2.25$ V.

From P_2 to P_3, both FETs are saturated. From Eq. (13.26), we find $v_{IN} = 2.5$ V.

At P_2, $v_O = v_{IN} - V_{tp} = 2.5 + 1.5 = 4$ V.
At P_3, $v_O = v_{IN} - V_{tn} = 2.5 - 1.5 = 1$ V.

Between P_3 and P_4, M_p is saturated and M_n is ohmic. Now Eq. (13.27) applies, and with the symmetric constants given, $k[2(v_{IN} - V_t)v_O - v_O^2] = k[v_{IN} - (V_{DD} + V_t)]^2$.

Again, after much manipulation, we may derive

$$v_{IN} = v_O + V_{DD} - V_t - [2v_O(V_{DD} - 2V_t)]^{0.5}. \quad (13.29)$$

When $dv_O/dv_{IN} = -1$, $v_O = 0.125V_{DD} - 0.25V_t = 0.25$ V and $V_{IH} = 2.75$ V. At P_4, v_{IN} reaches $V_{DD} + V_{tp}$, turning M_p OFF, and $v_O = 0$.

$$NM_L = 2.25 - 0 = 2.25 \text{ V}; \quad NM_H = 5 - 2.75 = 2.25 \text{ V}.$$

Having $NM_L = NM_H$ gives us the best noise immunity, so this is a desirable result.

EXAMPLE 13.16

Use SPICE to draw the transfer characteristic for the CMOS inverter of Example 13.15.

SOLUTION The plot of Fig. 13.28 was obtained using the following netlist:

```
EXAMPLE 13.16
VDD 1 0 DC 5
MP 3 2 1 1 MP
MN 3 2 0 0 MN
.MODEL MP PMOS VTO=-1.5 KP=0.1M
.MODEL MN NMOS VTO=1.5 KP=0.1M
VIN 2 0 DC 0
.DC VIN 0 5 .01
.PLOT DC V(3)
.END
```

■ Note the required substrate connection in the MP and MN element statements.

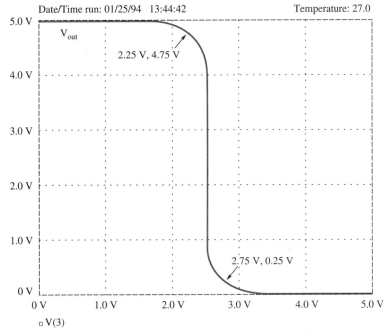

■ FIGURE 13.28 SPICE-generated transfer characteristic of Example 13.16.

DRILL 13.13

Given a CMOS inverter with FET parameters as in Example 13.15, suppose the input is LOW ($v_{IN} = 0$). How much current can the output of the gate source into a short circuit?

ANSWER 0.61 mA

We find that CMOS offers some distinct advantages over other types of logic that we have considered. When the input is LOW ($v_{IN} < V_{tn}$), M_n is OFF, the capacitance at the output node quickly charges to V_{DD}, and there is no further current flow through M_p and no steady-state power dissipation. When the input is HIGH ($v_{IN} > V_{DD} + V_{tp}$), M_p is OFF, the capacitance discharges through M_n, v_{DSn} goes to 0, and again there is no steady-state power. Note also that the substrate of each FET is connected to its source; the threshold voltage of M_p is unaffected by v_O.

Switching Speed

As with the MOS switch, we will model the transient situation by lumping all of the parasitic capacitance, C_T, at the output node. We can find the rise time by assuming that v_{IN} jumps from V_{DD} to 0 at $t = 0$, so that M_n turns OFF and M_p saturates. From Eq. (13.8),

$$I_C = -I_{DSP} = -k_p(-V_{DD} - V_{tp})^2 = C_T \frac{dv_O}{dt}. \tag{13.30}$$

Integrating Eq. (13.30) from $v_O = 0$ to $-V_{tp}$, where M_p becomes ohmic, we find that

$$t_1 = \frac{C_T(V_{tp})}{k_p(V_{DD} + V_{tp})^2}. \tag{13.31}$$

In the ohmic region, Eq. (13.9) predicts current

$$i_C = -k_p[2(-V_{DD} - V_{tp})(v_O - V_{DD})^2] = C_T \frac{dv_O}{dt}. \tag{13.32}$$

Integration gives

$$t = -\frac{C_T}{k_p(2V_{DD} + 2V_{tp})} \ln \frac{(v_O + V_{DD} + 2V_{tp})}{(V_{DD} - v_O)}. \tag{13.33}$$

To find t_2, the remainder of the rise time, we evaluate Eq. (13.33) between limits $v_O = -V_{tp}$ to $0.9V_{DD}$. Then the rise time $t_r = t_1 + t_2$.

Because of the topology of the CMOS circuit, fall time equations have the same form, and if the devices are symmetric, $t_f = t_r$. But symmetric device circuits do not occupy minimum chip area, so not all CMOS designs will be symmetric.

DRILL 13.14

A certain CMOS logic family uses an output circuit configured like Fig. 13.27(a). In addition, $k_p = k_n = 0.15$ mA/V^2 and $V_{tN} = -V_{tP} = 2$ V. Let $V_{DD} = 5$ V. Total capacitance (0.8 pF internal + 1.5 pF external load) = 2.3 pF. Estimate rise and fall times.

ANSWER 9.6 ms

EXAMPLE 13.17

Find the rise time for Example 13.15, using $C_T = 0.5$ pF.

SOLUTION Using Eq. (13.31),

$$t_1 = \frac{0.5(+1.5)}{0.05(5 - 1.5)^2} = 1.2 \text{ ns}.$$

Evaluating Eq. (13.33) from $v_O = 1.5$ to 4.5 V gives $t_2 = 3.7$ ns. Rise time is $1.2 + 3.7 = 4.9$ ns. Fall time will be the same.

If we compare this time with that required by the NMOS inverter in Example 13.12 (20 ns), we note the speed advantage of the CMOS arrangement. Note that we are comparing devices that would not be suitable for interfacing with large capacitance. The current-drive capability of a MOSFET device is not as large as for a BJT with the same die area, nor is the g_m. This subject is discussed in the section on analog BiCMOS amplifiers in Chapter 8, and in the BiCMOS logic section to follow.

Propagation delay is estimated by integrating Eq. (13.30) from $V_O = 0$ to $V_{DD}/2$.

$$t_{pd} = -\frac{C_T V_{DD}}{2k_p(V_{DD} + V_{Tp})^2}. \tag{13.34}$$

As noted previously, the power requirements of CMOS depend on the switching rate, with negligible static power dissipation. Each cycle dissipates $W = C_T V_{DD}^2$ joules. This is a marked advantage for portable equipment where standby operation should not discharge the battery. Power dissipation is $C_T V_{DD}^2$ times the switching rate.

$$P_D = C_T V_{DD}^2 \times f. \tag{13.35}$$

CMOS Gates

The 4000 series CMOS family dates from the late 1960s. An example of a two-input NOR gate from this family is shown in Fig. 13.29(a). The output is HIGH only when both inputs are LOW, so that both p-channel pull-up FETS are ON and both n-channel devices are OFF. When either or both inputs are HIGH, output will be LOW because at least one pull-down FET is ON and at least one pull-up FET is OFF.

The two-input NAND is shown in Fig. 13.29(b). Here the output will be LOW only when both inputs are HIGH, so that both n-channel FETs are ON. When either input is LOW, the output will be pulled HIGH by one or more p-channel devices and there will be no pull-down. The series FETs in these examples will be fabricated with greater W/L ratio, so that the transfer characteristic maintains reasonable symmetry.

The MOS gate inputs have extremely high resistance and therefore are susceptible to damage to the thin gate-oxide layer because of high static voltages. Static voltages gener-

DRILL 13.15

Using the device and the loading capacitance of Drill 13.14, find the (a) propagation delay and (b) the power dissipation for a switching rate of 1 MHz.

ANSWER (a) 42 ns, (b) 0.575 mW

■ Sub-threshold operation is used for very low power equipment, i.e., the digital watch.

■ t_{ox} is below 100Å in some aggressive technologies.

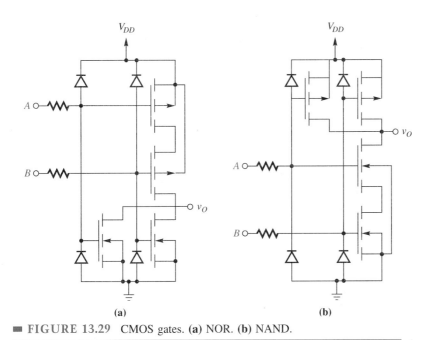

■ **FIGURE 13.29** CMOS gates. (a) NOR. (b) NAND.

ated by a person walking across a waxed floor can exceed 10 kV! The diodes built into the CMOS structure are designed to limit input voltages that fall outside the range from V_{SS} to V_{DD}. Each distributed input resistance shown is about 1.5 kΩ; a distributed *pn* junction will be between it and its *n* isolation region.

The 4000B and 74Cxx logic families are second-generation CMOS circuits. The technology is improved; polysilicon gates are used instead of metal, and the devices are smaller and faster. In addition, outputs are double-buffered. Figure 13.30 shows the incorporation of two cascaded inverters, which serve to isolate the logic from the output. The result of the high voltage gain obtained is a very steep transfer characteristic. CMOS plots in Fig. 13.17 assume $V_{DD} = 10$ V, $C_L = 50$ pF, and a switching rate of 100 kHz.

The third-generation CMOS family, the 74HCxx series, continues the trend toward smaller feature size and challenges the 74LSxx TTL series for speed, at considerably less power. In Fig. 13.17, HCMOS is plotted for $V_{DD} = 6$ V; $t_{pd} = 10$ ns and $P_D = 500$ μW at 1 MHz. A data sheet for a typical 74HCxx device is presented in Fig. 13.31.

The fourth-generation device, 74ACxx, is plotted for $V_{DD} = 5$ V; $t_{pd} = 4$ ns and same P_D.

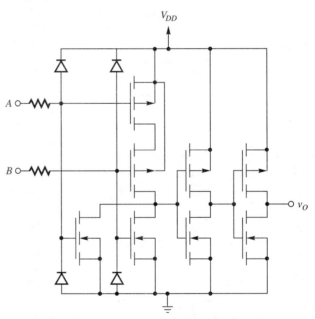

■ **FIGURE 13.30** Buffered CMOS NOR gate.

DRILL 13.16

For the same situation as Example 13.18, find the propagation delay.

ANSWER 32 ns

EXAMPLE 13.18

Referring to the data sheet in Fig. 13.31, estimate the power dissipation of a 74HCxx series gate operating at 6 V with a switching rate of 1 MHz into a 25-pF load.

SOLUTION The data sheet indicates $C_{Pd} = 20$ pF. Then $C_T = C_{Pd} + C_L = 20 + 25 = 45$ pF. From Eq. (13.35) the dynamic power dissipation is

$$P_D = C_T V_{DD}^2 f = 45 \text{ pF}(6 \text{ V})^2 \, 1 \text{ MHz} = 1.62 \text{ mW}.$$

The static power dissipation is $V_{DD}I_{CC} = 6$ V $\times$ 20 μA = 120 μW.

HIGH-SPEED CMOS LOGIC

TABLE I
SPECIFICATIONS FOR HC SSI CIRCUITS
D2804, DECEMBER 1982 – REVISED MARCH 1984

absolute maximum ratings over operating free-air temperature range†

Supply voltage range, V_{CC}	−0.5 V to 7 V
Input diode current, I_{IK} ($V_I < 0$ or $V_I > V_{CC}$)	±20 mA
Output diode current, I_{OK} ($V_O < 0$ or $V_O > V_{CC}$)	±20 mA
Continuous output current, I_O ($V_O = 0$ to V_{CC})	±25 mA
Continuous current through V_{CC} or GND pins	±50 mA
Lead temperature 1,6 mm (1/16 inch) from case for 60 seconds: FH, FK, or J package	300°C
Lead temperature 1,6 mm (1/16 inch) from case for 10 seconds: FN or N package	260°C
Storage temperature range	−65°C to 150°C

recommended operating conditions

			SN54HC' MIN	NOM	MAX	SN74HC' MIN	NOM	MAX	UNIT
V_{CC}	Supply voltage		2	5	6	2	5	6	V
V_{IH}	High-level input voltage	$V_{CC} = 2$ V	1.5			1.5			V
		$V_{CC} = 4.5$ V	3.15			3.15			
		$V_{CC} = 6$ V	4.2			4.2			
V_{IL}	Low-level input voltage	$V_{CC} = 2$ V	0		0.3	0		0.3	V
		$V_{CC} = 4.5$ V	0		0.9	0		0.9	
		$V_{CC} = 6$ V	0		1.2	0		1.2	
V_I	Input voltage		0		V_{CC}	0		V_{CC}	V
V_O	Output voltage		0		V_{CC}	0		V_{CC}	V
t_t	Input transition (rise and fall) times (except Schmitt-trigger inputs)	$V_{CC} = 2$ V	0		1000	0		1000	ns
		$V_{CC} = 4.5$ V	0		500	0		500	
		$V_{CC} = 6$ V	0		400	0		400	
T_A	Operating free-air temperature		−55		125	−40		85	°C

electrical characteristics over recommended operating free-air temperature range (unless otherwise noted)

PARAMETER	TEST CONDITIONS	V_{CC}	$T_A = 25°C$ MIN	TYP	MAX	SN54HC' MIN	MAX	SN74HC' MIN	MAX	UNIT
V_{OH} (Totem-pole outputs)	$V_I = V_{IH}$ or V_{IL}, $I_{OH} = -20$ μA	2 V	1.9	1.998		1.9		1.9		V
		4.5 V	4.4	4.499		4.4		4.4		
		6 V	5.9	5.999		5.9		5.9		
	$V_I = V_{IH}$ or V_{IL}, $I_{OH} = -4$ mA	4.5 V	3.98	4.30		3.7		3.84		
	$V_I = V_{IH}$ or V_{IL}, $I_{OH} = -5.2$ mA	6 V	5.48	5.80		5.2		5.34		
I_{OH} (Open-drain outputs)	$V_I = V_{IH}$ or V_{IL}, $V_O = V_{CC}$	6 V		0.01	0.5		10		5	μA
V_{OL}	$V_I = V_{IH}$ or V_{IL}, $I_{OH} = 20$ μA	2 V		0.002	0.1		0.1		0.1	V
		4.5 V		0.001	0.1		0.1		0.1	
		6 V		0.001	0.1		0.1		0.1	
	$V_I = V_{IH}$ or V_{IL}, $I_{OL} = 4$ mA	4.5 V		0.17	0.26		0.4		0.33	
	$V_I = V_{IH}$ or V_{IL}, $I_{OL} = 5.2$ mA	6 V		0.15	0.26		0.4		0.33	
V_{T+}†		2 V	0.8	1.2	1.5					V
		4.5 V	2	2.5	3.15					
		6 V	2.5	3.3	4.2					
V_{T-}†		2 V	0.3	0.6	0.8					V
		4.5 V	0.9	1.6	2					
		6 V	1.2	2	2.5					
$V_{T+} - V_{T-}$†		2 V	0.2	0.6	1					V
		4.5 V	0.4	0.9	1.4					
		6 V	0.5	1.3	1.7					
I_I	$V_I = 0$ to V_{CC}	6 V		±0.1	±100		±1000		±1000	nA
I_{CC}	$V_I = V_{CC}$ or 0, $I_O = 0$	6 V			2		40		20	μA
C_I		2 to 6 V		3	10		10		10	pF

† This parameter applies only for Schmitt-trigger inputs.

switching characteristics over recommended operating free-air temperature range (unless otherwise noted), $C_L = 50$ pF (see Note 1)

PARAMETER	FROM (INPUT)	TO (OUTPUT)	V_{CC}	$T_A = 25°C$ MIN	TYP	MAX	SN54HC00 MIN	MAX	SN74HC00 MIN	MAX	UNIT
t_{pd}	A or B	Y	2 V		45	90		135		115	ns
			4.5 V		9	18		27		23	
			6 V		8	15		23		20	
t_t		Y	2 V		38	75		110		95	ns
			4.5 V		8	15		22		19	
			6 V		6	13		19		16	
C_{pd}	Power dissipation capacitance per gate						No load, $T_A = 25°C$			20 pF typ	

FIGURE 13.31 Data sheet for 74HC00 CMOS NAND gate (courtesy of Texas Instruments Inc.).

BiCMOS Logic

We have found that when the load capacitance is small, CMOS is very fast, low power, compact, and has a high input resistance. The circuit of Fig. 13.32 shows this technology coupled with a BJT totem-pole output stage, which provides the high-current capability to charge the load capacitance. Essentially, BiCMOS combines CMOS logic with a BJT output buffer, and thus incorporates the best features of both technologies.

Suppose the input voltage v_{IN} is LOW (0 V). The NMOS enhancement FET M_N will be OFF, and Q_1 will be OFF because its base is effectively grounded through R_2. The PMOS enhancement FET M_P is ON, with $V_{GSP} = -V_{DD}$. Transistor M_P provides base current to Q_2, so that load capacitance C_L is quickly charged up by active Q_2 to approximately $V_{DD} - V_{BE2}$ (assuming no dc load). Load capacitance C_L then continues to charge toward V_{DD} through M_P and R_1, and V_{BE2} decreases to approximately 0 V; $V_{OH} \approx V_{DD}$.

When v_{IN} goes HIGH, M_P turns OFF, ensuring that Q_2 is OFF. Transistor M_N is ON, and the charge on C_L provides current through R_1 and M_N to the base of Q_1, and the collector current of Q_1 rapidly discharges C_L until V_{BE1} drops below cut-in. After that, C_L continues to discharge through R_1, M_N, and R_2. The value of V_{OL} approaches zero volts. The base of Q_2 also discharges through R_1. In practice, R_1 and R_2 are polysilicon resistors or MOS resistors.

An example pulse response for this circuit, with $C_L = 2$ pF, is shown by the SPICE plot of Fig. 13.33. Observe the very steep rise and fall times, which occur while the BJTs are conducting. We find that the output signal is nearly V_{DD} volts p-p. From the transfer characteristic plotted in Fig. 13.34, you can see that the noise margins for this circuit are high, both being more than 2 V.

BiCMOS logic has achieved the low propagation delays of bipolar circuits while maintaining the low-power advantage of CMOS. Note that the quiescent power of this circuit is near zero, as with CMOS. There are no families of BiCMOS logic gates. Current

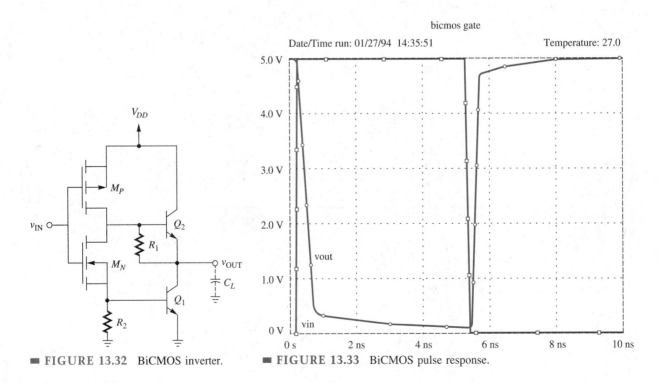

■ **FIGURE 13.32** BiCMOS inverter. ■ **FIGURE 13.33** BiCMOS pulse response.

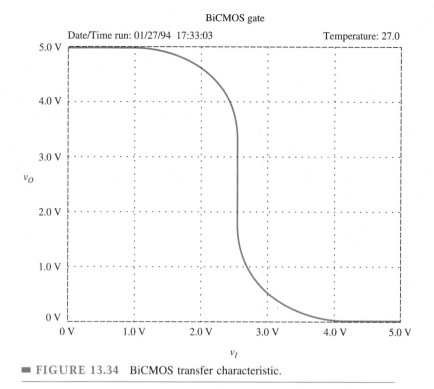

FIGURE 13.34 BiCMOS transfer characteristic.

application of this technology is largely with VLSI, in which the designer desires the density and low power of CMOS on the chip, but must not compromise speed when driving the capacitive load that will be present at any output pin of the package. Thus, a BiCMOS circuit will serve as a buffer at each output. As an example, consider a large memory array, or a microprocessor, with a very large number of CMOS circuits in a package, and a relatively few outputs that must interface with the higher capacitance of circuits outside the package.

■ A high-speed PowerPC™ microprocessor with bipolar logic in the core and BiCMOS caches is being developed.[4]

Merging the two technologies onto one chip increases the complexity of fabrication, and hence the cost, by perhaps 50 to 60%. It is interesting to note that there is a competitive alternative approach to the objectives just mentioned. The manufacturer can fabricate the VLSI CMOS chip, along with another chip that has the bipolar buffer drivers, in one package. The interconnections between the two chips are carefully designed by the fabricator, and hence the capacitance is minimized. This is an example of what is referred to as the *multichip module* or **hybrid IC.**

CHECK UP

1. **TRUE OR FALSE?** CMOS is a very dense technology, similar to NMOS in that regard.
2. **TRUE OR FALSE?** CMOS requires near zero standby power.
3. **TRUE OR FALSE?** It is possible to increase CMOS switching speed at the expense of increased power dissipation by increasing V_{DD}.
4. **TRUE OR FALSE?** BiCMOS technology permits increased output drive capability for high-speed CMOS logic.

13.8 INTEGRATED-INJECTION LOGIC

In Section 13.3 we considered and rejected direct-coupled transistor logic because of current hogging. DTL and TTL, which avoid this problem, were developed instead. As we get to large-scale integration (more than 100 gates), we find that the die area required and the power dissipated become excessive with TTL. Such circuits would require a 25-mm^2 area and dissipate 1 W. We now turn to a special type of DCTL, called *merged-transistor logic* or *integrated-injection logic* (IIL or I^{2}L), which leads to an order of magnitude reduction in area/gate and also in power. We will find that this logic is not level compatible with the other logic and hence is not easily interfaced. However, for large-scale integration, where many gates are required on a die but relatively few nodes have to be interfaced, this logic has some application. By the same token, IIL will not be used in small-scale chips, and we do not expect that gate-level logic "families" will be developed. Rather, IIL will be used in large complex functions such as microprocessors, memories, computers on a chip, and so forth. But MOS technology competes in circuit density, power economy, and often in speed, so as of this writing, NMOS and CMOS dominate this market.

To illustrate the concepts involved, consider the DCTL two-line *decoder* shown in Fig. 13.35. The circuit has four outputs, one of which is HIGH at a time, selected by a particular 2-bit input combination. This circuit actually is small scale, using only six gates (ten transistors and eight resistors), but you should imagine that the circuit is only a part of a much larger integrated circuit. Think of the outputs shown as inputs to other gates on the same chip, and imagine that the inputs connect to collectors of other gates not shown.

Suppose that the circuit is redrawn as illustrated in Fig. 13.36. The devices shown are transistors having one emitter and one base but several collectors. Theoretically, such a device would operate identically to the several transistors having their bases and emitters tied together externally so that the two figures would be functionally equivalent.

The simplified diagram presented in Fig. 13.37 attempts to show how such a device is implemented. Except for the fact that the substrate is n^+, the structure appears as a normal epitaxial-diffused transistor. However, the n region (normally used as a collector)

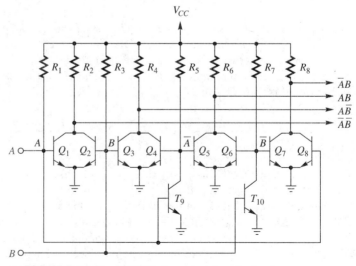

■ **FIGURE 13.35** A DCTL two-input decoder having outputs $\overline{A}B$, AB, $A\overline{B}$, $\overline{AB}$.

CHAPTER 13 INTEGRATED-CIRCUIT LOGIC GATES

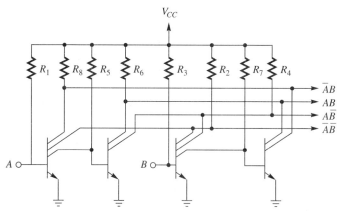

FIGURE 13.36 The DCTL two-input decoder redrawn with multiple-collector transistors.

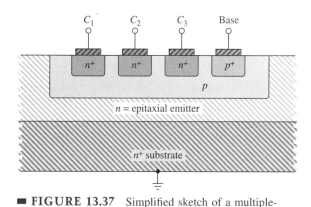

FIGURE 13.37 Simplified sketch of a multiple-collector transistor.

is grounded through the low-resistivity substrate and acts as the emitter. The p base serves its normal function, but the n^+ diffusions at the surface are the collectors. We would expect the operation of this transistor to resemble the inverse mode of a transistor connected in the normal manner.

We would expect the usual drawbacks of the inverse mode to apply:

1. The emitter efficiency is lower than normal because of the lightly doped emitter.
2. The base-transport factor is poor because the emitter will inject some electrons into the base in regions where they will have difficulty diffusing to a collector before recombining in the base region.
3. The heavily doped collectors will have a low breakdown voltage, <8 V.

By means of a proper layout, however, the β of the "upside-down" transistors can be made about 5, which is quite usable.[5] We will use voltage levels that are safely below the breakdown voltage of the device.

The important feature to note in the diagram in Fig. 13.37 is the consolidation of what was three transistors in Fig. 13.35 into the area normally occupied by one transistor. This brings us to the motivation for this arrangement:

1. Several transistors sharing a common emitter and a common base form a more compact device, which saves die area.
2. The new devices, having only one base-emitter junction, do not suffer from the current hogging that plagues ordinary DCTL.[6]
3. Free to use DCTL, the manufacturer is able to eliminate the 10 base resistors that would otherwise be required in RTL. This in itself represents a saving in die area of about 30%.

Each node in conventional DCTL has a pull-up resistor that provides base current unless that current is shunted through a saturated collector. IIL provides this current by means of a current source, known as a *current injector.* Our decoder example is redrawn in Fig. 13.38 to show *pnp* transistors acting as the current sources. These *pnp* injectors can be >80% efficient ($\alpha_{pnp} = 0.8$).[7]

You may want to refer to general fabrication details of IC technology as presented in Chapter 18 if you encounter unfamiliar terms in this paragraph. The diagram in Fig. 13.39

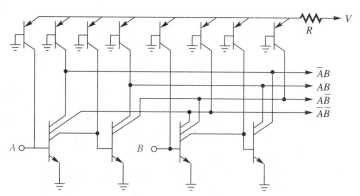

■ **FIGURE 13.38** The two-input decoder redrawn with *pnp*-current injectors replacing resistors.

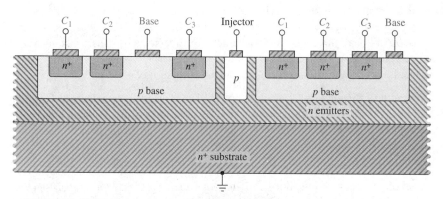

■ **FIGURE 13.39** Diagram of two IIL multiple-collector vertical *npn* transistors, with current injected into base regions by lateral *pnp* action.

shows how the circuit is implemented in IIL technology. It is apparent that this approach conserves chip area, because one *pnp*-type *injector rail* acts as the emitter for two *pnp* lateral (horizontal) current sources. The *n* region serves both as base regions for the current sources and as emitters for the gate transistors. The *p*-base regions of the gate transistors are also the collectors of the injector transistors. This system is referred to as ***merged-transistor logic*** because the *pnp* and the *npn* transistors consist of only four distinct regions. Note that fabrication requires only two diffusions and four masking steps. The latest technique forms the base in two steps, however: a p^- ion implant for the active part and a p^+ diffusion for the base-spreading resistance parts.

To complete our illustration, a possible layout for the decoder example is sketched in Fig. 13.40. Viewing the surface of the chip, we can see how compact the circuit has become. IIL allows 200 gates/mm^2, as compared to about 20 gates/mm^2 for TTL. This circuit density exceeds that of any other technology using the same design rules. The simple fabrication process also results in high yield for the manufacturer, making VLSI circuits feasible.

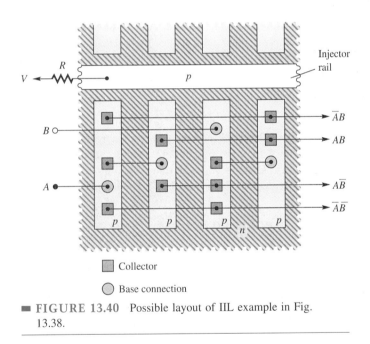

■ **FIGURE 13.40** Possible layout of IIL example in Fig. 13.38.

Logic Levels and Noise Margin

The voltage transfer characteristic for IIL is defined for one stage of a cascade of transistors. Consider Fig. 13.41, where Q_1 is an inverter driving Q_2. As long as the current at its base node is diverted by the input driver (not shown), Q_1 is OFF. As long as v_{IN} is held lower than $V_{BE(cut-in)} = 0.65$ V, $V_{CE1} = V_{BE2(sat)} = 0.75$ V. As v_{IN} exceeds 0.65 V, Q_1 begins to divert current from the base of Q_2. As v_{IN} increases slightly, in the vicinity of 0.7 V, Q_2 comes out of saturation, and v_O falls sharply to the edge of saturation, and Q_2 is OFF. As the driving current is withdrawn, v_{IN} clamps at $V_{BE1(sat)} = 0.75$ V, and $v_O = V_{CE1(sat)} = 0.05$ V for this device. In addition, $NM_L = 0.65 - 0.05 = 0.6$ V,

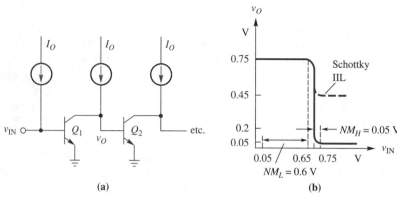

FIGURE 13.41 One stage of IIL. (a) Model. (b) voltage transfer characteristic.

and $NM_H = 0.75 - 0.70 = 0.05$ V. Ordinarily, this would be an unacceptably low value, but in IIL each driving collector is tied through a short metallization on the chip to the driven base, and external noise is of little consequence. Any input or output that must come off of the chip must be buffered, and the levels must be translated to interface with other devices. Hence these noise margins give excellent results on the chip.

Fan-out

Any collector is only required to sink current from one source I_o. The base current of each ON transistor will be I_o, and the maximum collector current(s) in a transistor will be NI_o if it has N collectors. Thus the β of the transistor must exceed $NI_o/I_o = N$. Most IIL circuits are designed with $N \le 5$, so that $\beta > 5$ is adequate.

Propagation Delay Versus Power

The voltage from the p injector rails to ground is that of a forward-biased junction, about 0.85 V in this case. This voltage is well matched at all injectors across the entire chip. Careful design avoids current hogging, and each base receives equal current. This is important because it means that all injectors can be tied together and the total current to the chip can be programmed externally by one resistor and the voltage source. We have noted in previous logic families that the manufacturer can make trade-offs between delay and power dissipation by changing the size of the on-chip resistors. Larger resistors reduce power consumption at the expense of slower charging of circuit capacitances. In the present case, *this choice may be made by the user!* The user merely determines the chip current by choice of resistor R and supply voltage (see Fig. 13.38).

The relationship between delay and power for IIL is shown in Fig. 13.42. At low currents, the delay is caused by junction and parasitic capacitances. A constant delay-power product of 0.3 pJ is realistic over the considerable range of 1 nW to 0.1 mW/gate.[7] At the bottom of the curve, the delay is determined by stored charge in the base region, and increases in injected current do not reduce the delay proportionately. Too large a current is counterproductive because of high-level injection effects. For comparison with other logic, these data are also plotted on Fig. 13.17.

We can see a striking comparison of IIL technology with TTL in LSI circuits in the following example.

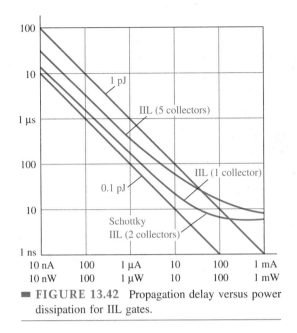

■ **FIGURE 13.42** Propagation delay versus power dissipation for IIL gates.

EXAMPLE 13.19

Assume that an LSI circuit requires 4000 gates and the propagation delay/gate must not exceed 10 ns. Compare IIL with TTL.

SOLUTION At 200 gates/mm^2, IIL uses an area of 20 mm^2 (0.18 × 0.18 in.); but with only 20 gates/mm^2, TTL would need 200 mm^2 (prohibitive!). With a power-delay product of 0.3 pJ, each IIL gate requires 30 μW, and total chip power would be 120 mW. This is quite reasonable for most IC packages, even with no heat sinks or forced convection. By contrast, at 2 mW/gate, 4000 TTL(LS) gates would dissipate 8 W!

EXAMPLE 13.20

Find the value of resistance required to properly bias the IC of Example 13.19 from a 5-V power supply.

SOLUTION Using an injector input voltage of 0.85 V, input current would be

$$\frac{200 \text{ mW}}{0.85 \text{ V}} = 235 \text{ mA}; \qquad R = \frac{(5 - 0.85) \text{ V}}{0.235 \text{ A}} = 18 \text{ }\Omega.$$

DRILL 13.17

Define the term *merged transistor*.

ANSWER When several transistors share a common region in the IC, as in Fig. 13.36 where the collector of the injector is the base of the inverter, and where the base of the injector is the emitter of the inverter, they are merged transistors.

■ The 18Ω resistor must be 1 watt.

Schottky IIL

The delay of IIL is somewhat greater than TTL, even at the bottom of the delay curve. One technique that is being used to reduce this delay while maintaining the high circuit density of IIL is diagrammed in Fig. 13.43. A single collector region is used, and the multiple collector contacts are metal to *n*-silicon Schottky diodes. Ion implantation, rather

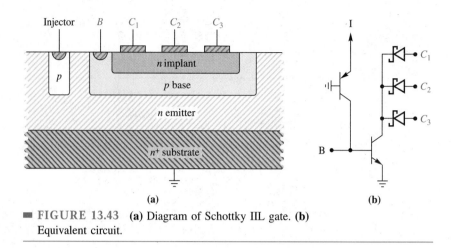

FIGURE 13.43 (a) Diagram of Schottky IIL gate. (b) Equivalent circuit.

than diffusion, is used to form a collector with a lower doping level than the base. This allows the Schottky diodes to be designed with a forward-bias voltage of about 0.4 V.

This structure causes V_{OL} to become $V_{CE(\text{sat})} + V_{SD} = 0.05 + 0.4 = 0.45$ V. The resulting reduction in logic swing from 0.7 to 0.3 V (see Fig. 13.41) also reduces the charge times of the junction capacitances by about 50%. The Schottky diodes provide isolation between the collector contacts, permitting a more compact collector, which increases β and allows an even greater circuit density. The delay-power product is typically 0.15 pJ; this value is plotted in Fig. 13.42.

Two other approaches to increasing the switching speed of IIL, integrated Schottky logic and Schottky transistor logic, are compared in the literature.[8]

13.9 COMPARISON OF LOGIC FAMILIES

Table 13.2 compares three of the most popular SSI logic families. The examples shown all represent the same state of technology to point up the differences in basic approach to logic circuits. Refer to Fig. 13.17 for a perspective on the comparative speed-power product.

In summary, it appears that:

- TTL (LS) is a practical compromise in terms of speed and power, very mature technology (hence, low cost), compatible with vast imbedded usage.
- ECL meets the need for very high speed, but at great cost in power, with logic levels different from the other major logic families, and requiring a negative supply.

TABLE 13.2 Comparison of Popular Logic Families

	TTL 74LSxx	ECL 10,000	CMOS 4000B (5 V)
V_{OH}-V_{OL}	3.4–0.25 V	−0.9–1.7 V	5–0 V
NM_H-NM_L	1.4–0.6 V	0.36–0.33 V	2.25–2.25 V
Power/gate	2 mW	24 mW (unloaded)	1.5 μW × freq (kHz)
Propagation delay	9.5 ns (15 pF)	2 ns	30 ns + 1.7 ns/pF

- CMOS offers almost zero power requirements in standby, and very low power at moderate switching rates. Speed is now competitive with TTL. This is the technology of choice for most new designs.

SUMMARY

A wide variety of digital technology is available to the logic designer. The basic binary variable functions NOT, AND, OR, NAND, and NOR are common to all digital design. Combinational logic is given a cursory review in Appendix C. We assume that most readers of this text will study digital logic theory in more depth in other courses.

Our approach to this subject has been first to consider gates as building blocks in their own right and then, in later chapters, to consider larger scale design. Thus we have given an emphasis here to those technologies for which a logic family exists, even though modern design is moving toward a more integrated approach. These logic families are divided according to IC fabrication technology and basic logic gate configuration. In most cases, fan-out for all of these technologies is limited by speed considerations rather than dc loading.

TTL is a very important BJT logic family because of its widespread maturity and application, and our emphasis on basic IC internal circuit realizations provides a good understanding of the terminal characteristics. The evolution of circuit variations gives added meaning to the general terms noise margin, propagation delay, and fan-out. New design is moving away from TTL, in favor of CMOS.

Among the other BJT logic families, ECL is very fast because of its nonsaturating logic, but at the cost of relatively high power dissipation, with its attendant problems. There is also the factor of power supply and logic level incompatibility with the other major technologies, so that its major use is in system-wide design where speed is the overriding consideration. Digital-to-analog conversion is another example.

CMOS is the technology of choice where very low standby power consumption is desired. This is a rapidly maturing technology, with a large variety of circuit functions in the families, which makes it very versatile in new design. The low cost and lower power supply requirements are making this the dominant technology as of this writing, where new design is moving toward application-specific ICs. For VLSI, it is preferred over TTL because of low power, and where there is no external capacitive load, it is just as fast. To interface with capacitive loads, BiCMOS can be used.

■ New CMOS and TTL designs specify 3.3V supply voltages to reduce power dissipation.

NMOS logic is very widely used because of its relatively simple fabrication technology and high circuit density. In general, NMOS logic is slower than BJT logic, but the former requires less power. NMOS has its greatest application in VLSI circuits where its circuit density is exploited, such as IC memories and microprocessors, but even there it is losing ground to CMOS. It is not used in small-scale integration.

GaAs MESFET technology promises extreme speeds at low voltage, but at present, the expense of fabrication limits its use to the special cases that demand that level of performance.

BiCMOS offers the best of both bipolar and CMOS technologies, at increased cost because of the more complex fabrication involved.

■ There is a class of NMOS transistors fabricated using amorphous or polysilicon as the active channel material. These are used extensively in flat-panel display pixel drivers. These specialty NMOS transistors are deposited directly on the glass display substrates.

The very compact structure of IIL illustrates its advantages in large-scale integration, but its incompatibility with the logic levels of other logic families limits its application to special-purpose circuits. It has the attractive feature of allowing the user to make the final trade-off of speed versus power in each application.

Chapter 14 provides a more systems-oriented approach to the design of more complex logic functions using basic building blocks made up of combinations of the gates discussed in this chapter. Chapter 15 considers memory circuits as very large scale integration of these gates.

SURVEY QUESTIONS

1. Why were logic "families" developed?
2. TTL replaced RTL and DTL because of what great advantage?
3. How do Schottky transistors prevent BJT logic, such as TTL, from saturating?
4. How does ECL obtain its relatively high switching speed?
5. What is the chief disadvantage of ECL?
6. NMOS logic usually has an enhancement driver and a depletion load. Compare this design to the case in which both FETs are the enhancement type.
7. CMOS is replacing NMOS as the dominant digital technology. Give the relative advantages of each.
8. What is the purpose of adding BJTs to CMOS logic in BiCMOS technology?

PROBLEMS

Problems marked with an asterisk are more challenging.

13.1 Show how to realize two-input OR, NOR, and AND functions with two-input NAND gates only.

13.2 Show how to realize two-input OR, AND, and NAND functions with two-input NOR gates only.

13.3 Use a truth table to show that $(A + B)(A + C) = A + BC$.

13.4 What Boolean logic functions are implemented by each of the circuits shown in Fig. 13.44?

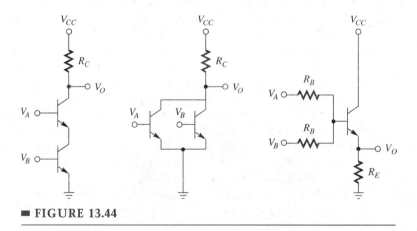

■ **FIGURE 13.44**

13.5 Calculate the noise margins for the inverter shown in Fig. 13.2(a), driving 10 identical inputs.

13.6 Consider the circuit in Fig. 13.4, where fan-out is 5. Each base $C_B = 5$ pF. Assume Q_1 has just turned OFF, and estimate the time required for the bases to reach cut-in ($V_{BE} = 0.6$ V).

13.7 Refer to your solution to Problem 13.6. Obviously, the fastest turn-on time will be when $N = 1$ (fan-out of 1). Compute this best case value for the turn-on time. What will be the turn-on time for the maximum fan-out $N = 50$ case? Refer to Example 13.1. Be sure you clearly state your equation for $v_{BE}(t)$, the voltage across each C_B.

13.8 Calculate the breakpoints for the RTL transfer characteristic in Fig. 13.6(b) for $N = 5$ and $T = 125°C$. Let $V_{BE(\text{cut-in})} = 0.6$ V, $V_{BE(\text{sat})} = 0.75$ V, and $\beta_F = 50$ at $T = 25°C$. Estimate β_F at 125°C from Fig. 4.31.

13.9* Current hogging results because, for the same base current, V_{BE} is less when I_C is less. Show this by using Eq. (4.8) with $I_E = -(\beta + 1)I_B$ for one case and $I_E = -I_B$ for another. In both cases neglect the unity term and the α_R for the typical forward-biased transistor.

13.10 Calculate the average power dissipation of the RTL gate in Fig. 13.6(a). Assume a fan-out of 3 and a gate ON time of 50%.

13.11 The circuit in Fig. 13.6(a) is constructed as a DCTL gate by omitting the base resistors.

(a) Sketch the voltage transfer characteristic.
(b) Find the noise margin.

13.12 Write a SPICE program to plot the transfer characteristic of the DTL gate of Fig. 13.7(a) at $T = 25°C$. Neglect one input, so that it is just an inverter. You may use the default models for the diodes and BJTs.

13.13 Refer to Fig. 13.7(a). Determine the logic expression for X as a function of the input variables A and B.

13.14 Suppose that the DTL circuit of Fig. 13.7(a) has two diodes where D_2 is drawn. Find the new noise margins.

13.15 Consider the HTL circuit in Fig. 13.45, using a 5.6-V Zener diode.

(a) Draw the voltage transfer characteristic, calculating the breakpoints.
(b) Calculate the noise margins.

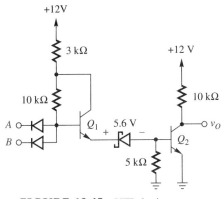

■ **FIGURE 13.45** HTL logic gate.

13.16 As the TTL gate in Fig. 13.8(a) switches, both Q_3 and Q_4 may be saturated at the same time.

(a) Find the current that flows in R_3 during this time.
(b) Find V_{CE2}, and estimate v_{IN} at this time.

13.17 The output of the 7400 gate in Fig. 13.8(a) is connected to the anode of a light-emitting diode, and its cathode is grounded. The LED has a voltage drop of 1.6 V. Find the LED current when the gate inputs are both 0.2 V. Use the transistor parameters of Eq. (13.6).

13.18 Assume the inputs of the 7400 gate in Fig. 13.8(a) are LOW. Plot the output voltage V_{OH} versus the load current I_{OH}. Use the transistor parameters of Eq. (13.6).

13.19* Repeat Problem 13.18 using a SPICE simulation. You may use default models for the devices, except $\beta_F = 50$. To simulate a two-emitter BJT, use two transistors in parallel, with bases and collectors tied together.

13.20 Assume the inputs of the 7400 gate in Fig. 13.8(a) are both HIGH (5 V). Let $\beta_F = 50$ for the BJTs. There is a load resistor of 22 Ω connected between +5 V and the output terminal of the gate. Find the load current.

13.21* Assume the A input of the 7400 gate in Fig. 13.8(a) is HIGH (5 V), and plot the voltage of the B input versus the current out of this input. Assume typical junction voltage drops. Let the output terminal of the gate be open.

13.22 Use a SPICE simulation to work Problem 13.21. For device models, see the statement of Problem 13.19.

13.23* Repeat Problem 13.21 or 13.22 with gate input terminal A grounded.

13.24* Use a SPICE simulation of the TTL gate of Fig. 13.8(a) to plot the transient response of the gate to a 25-MHz square wave, that is, let V_{IN} be stated `PULSE(0 5 .1N .1N .1N 20N 40N)`. Let your BJT model statement be `.MODEL T NPN BF=50 BR=.1 TR=100N TF=1N MJS=.5 CJS=1P CJE=1P CJC=.5P`. Estimate gate propagation delay from your plot. The element statement for the BJTs should have a 4th node (substrate) set to zero.

13.25 A partial schematic diagram of a 74L03 NAND gate with an open collector is shown in Fig. 13.46. An external 10-kΩ pull-up resistor has been added. Assume $\beta_F = 30$ and $\beta_R = 0.1$ and neglect leakage currents. Inputs A and B are connected.

(a) For $V_{IN} = 0$ V, find numerical values for I_{B1}, I_{B2}, V_O, and V_{B2}.

(b) Estimate a numerical value for V_{IN} where V_O switches to logic ZERO.

(c) For $V_{IN} = 4.5$ V, determine the region of operation for Q_1, Q_2, and Q_3 and justify your results numerically.

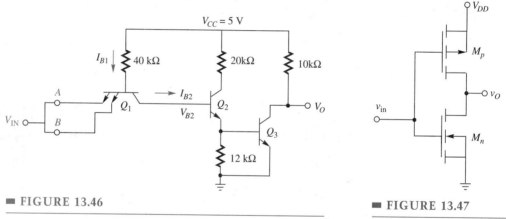

■ **FIGURE 13.46**   ■ **FIGURE 13.47**

13.26 Calculate the power dissipation of the 74LS00 gate in Fig. 13.12, assuming a 50% duty cycle.

13.27 (a) Sketch two two-input open-collector NOR gates sharing a common pull-up resistor. Find the logical output in terms of the four inputs.

(b) Repeat part (a) for two two-input NAND gates.

13.28* Verify your Problem 13.25 results using a SPICE simulation.

13.29 If the input (say, V_A) of the ECL gate in Fig. 13.20(a) is too high, Q_1 may saturate.

(a) At what V_A does this occur?

(b) Find V_O (NOR) for this value of V_A.

13.30 For the circuit shown in Fig. 13.47 (see above), let $V_{DD} = 5$ V and assume $V_{tn} = 1.0$ V, $V_{tp} = -1.5$ V, and $|k_p| = |k_n| = 0.1$ mA/V². Find a numerical value for v_O and the power being supplied from the V_{DD} supply for each of the following values of v_{IN}.

(a) $v_{IN} = 0$ V

(b) $v_{IN} = 5$ V

(c) $v_{IN} = 3.5$ V

(d) $v_{IN} = 2.0$ V

13.31 Use a SPICE simulation to obtain a transfer function for v_O versus v_{IN} and verify the results obtained in Problem 13.30.

13.32* Calculate the breakpoints and plot the transfer characteristic for the circuit in Fig. 13.23(a). Let $V_{DD} = 5$ V, $k_L = k_D = 0.1$ mA/V^2, $V_{tL} = -1$ V, and $V_{tD} = 1$ V. Compare your result with Example 13.11, where $k_D > k_L$.

13.33 Estimate the rise time of the inverter in Problem 13.32, assuming total output capacitance $C_T = 2$ pF. Compare with Example 13.12.

13.34 (a) Carry out the derivation of Eq. (13.18).
(b) Find t_3 for Example 13.12.

13.35 (a) Derive an expression for the fall time, t_4, in Fig. 13.24. This is the time for v_O to drop from $V_{DD} - V_{tD}$ to $0.1 V_{DD}$.
(b) Find t_4 for Example 13.12.

13.36 Derive the expressions given for t_{pHL} and t_{pLH} for the NMOS inverter.

13.37 Assume that a given CMOS inverter has $k_n = k_p = 0.1$ mA/V^2. Find $(W/L)_P$ and $(W/L)_N$ if $C_{ox}\mu_n = 25$ μA/V^2 and $\mu_n = 2\mu_p$.

13.38* Repeat Example 13.15 with $V_{DD} = 10$ V. Compare with the 5-V result.

13.39* Repeat Example 13.17 with $V_{DD} = 10$ V. Compare with the 5-V result.

13.40 Calculate the power dissipation of the gate in Example 13.17 as a function of switching rate.

13.41 Repeat Problem 13.40 for $V_{DD} = 10$ V and compare the results.

13.42 Figure 13.48 consists of a 4000 series CMOS schematic diagram.

(a) What logic function is implemented at Pin 6 where Pins 3, 4, and 5 are the inputs? Sketch the appropriate logic/schematic diagram symbol.

(b) What logic function is implemented at Pin 9 where Pin 8 is an input? Sketch the appropriate logic/schematic diagram symbol.

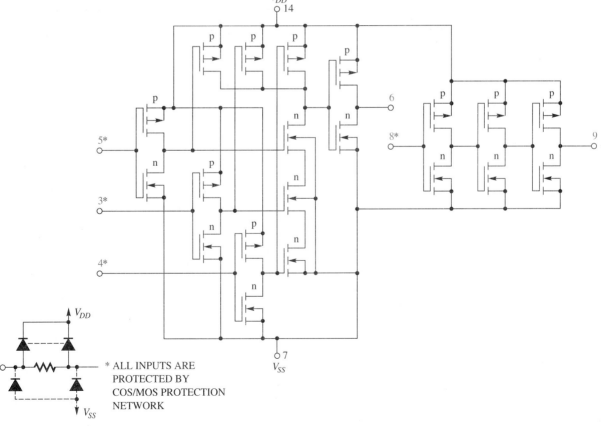

■ **FIGURE 13.48**

13.43 Refer to Fig. 13.49. Use the transfer function curves with $V_{DD} = 15$ V. Estimate values for V_{OH}, V_{OL}, V_{IL}, V_{IH}, NM_H, and NM_L and identify these points on the curves.

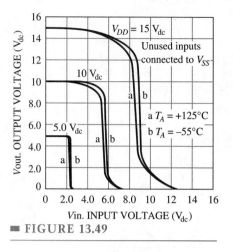

■ **FIGURE 13.49**

13.44 Refer to Fig. 13.50. What logic function is implemented at pin 6, where pins 3, 4, and 5 are the inputs? Sketch the appropriate logic/schematic diagram symbol.

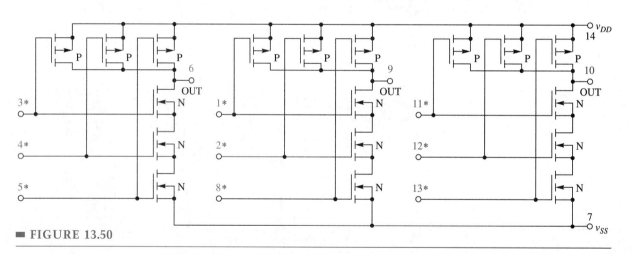

■ **FIGURE 13.50**

REFERENCES

1. Wakerly, J. F. *Digital Design Principles and Practices,* 2nd ed. Englewood Cliffs, N.J.: Prentice-Hall, 1994.
2. Cheng, D. *Field and Wave Electromagnetics,* 2d ed. New York: Addison-Wesley, 1989, Chap. 9.
3. Kaupp, H. R. "Characteristics of Microstrip Transmission Lines," *IEEE Transactions on Electronic Computers,* April 1967.
4. Fuller, Brian, and Ron Wilson. "Startup to Add Bipolar PowerPC to Road Map," *Electronic Engineering Times,* December 11, 1995, pp. 1, 130.
5. Hart, K., and A. Slob. "Integrated Injection Logic: A New Approach to LSI," *IEEE Journal of Solid-State Circuits,* October 1972, pp. 346–351.

6. Berger, H., and S. Wiedmann. "Merged-Transistor Logic (MTL): A Low-Cost Bipolar Logic Concept," *IEEE Journal of Solid-State Circuits*, October 1972, pp. 340–346.

6. Herman, J., S. Evans, and B. Sloan. "2nd Generation ILL/MTL: 20-ns Process Structures," *IEEE Journal of Solid-State Circuits*, April 1977, pp. 93–100.

7. Lohstroh, J., and R. Pluta. "Temperature Behavior of the Voltage Swings and Static Noise Margins of Integrated Schottky Logic and Schottky Transistor Logic," *IEEE Journal of Solid-State Circuits*, August 1982, pp. 677–686.

SUGGESTED READINGS

Alvarez, A., Ed. *BiCMOS Technology and Applications*. Boston: Kluwer Academic Publishers, 1989.

You should refer to current digital IC data books, such as the following:

The TTL Logic Data Book. Texas Instruments, Inc., Dallas, 1988. This volume covers standard TTL, Schottky, and low-power Schottky circuits.

The ALS/AS Logic Data Book. National Semiconductor, Santa Clara, California, 1990. This volume covers advanced low-power Schottky and advanced Schottky circuits.

MECL Device Data. Motorola, Inc., Phoenix, 1988. This data book covers several generations of ECL circuits.

F100K ECL Logic Databook and Design Guide. National Semiconductor, Santa Clara, California, 1990. This data book includes ECL logic, ECL BICMOS memories, and ECL PALs.

CMOS Logic Databook. National Semiconductor, Santa Clara, California, 1988. This data book covers the HC, HCT, C, and 4000 series of CMOS circuits.

Fact Advanced CMOS Logic Data. National Semiconductor, Santa Clara, California, 1990. This data book covers the advanced, quiet, FCT, FCTA, and FCTB series of CMOS circuits.

High-Speed CMOS Data Manual. Signetics, Sunnyvale, California, 1988. This data book covers the HC/HCT series in high-speed CMOS circuits.

Updates for most of these data books are available on diskettes, on CD-ROM, or from the World Wide Web on the Internet.

CHAPTER 14

SMALL DIGITAL SUBSYSTEMS

14.1 Combinational Logic Circuits
14.2 Sequential Logic Circuits
14.3 Monostable and Astable Timing Circuits
14.4 Data Systems Examples
Summary
Survey Questions
Problems
References
Suggested Readings

The manufacturers of logic families (CMOS, TTL, ECL) combine logic gates into arrays that accomplish specific functions. As previously noted, medium-scale integration (MSI) implies about 12 to 100 gates on the chip. A representative listing of such circuits is presented in Table 14.1. With CMOS, the variety is greater than with the bipolar families.[1-3] This chapter identifies some typical logic functions that are frequently used in logic design, and it is convenient to take examples that have been implemented in MSI circuits in the past.

TABLE 14.1 Representative List of MSI Functions Implemented in the Logic Families*

Schmitt triggers, one or two inputs	2-bit 4-line decoder/demux
BCD-to-decimal decoder	3-bit 8-line decoder/demux
BCD-to-seven-segment decoder, active high or active low	4-bit 16-line decoder/demux
D flip-flop	4-input multiplexer
JK flip-flop with + or − edge clock	8-input multiplexer
JK flip-flop with clear and/or preset	16-input multiplexer
4-bit full adder	8-line priority encoder
4-bit magnitude comparator	4 × 4 register file
Decade counters, up or up/down	4-bit arithmetic-logic unit
Decade up/down counters, with parallel load	Look-ahead carry generator
Divide-by-12 counter	Octal bus/line drivers
4-bit binary counters, up or up/down	Octal bidirectional line drivers
4-bit binary up/down counters, parallel load	8-bit addressable latch
8-bit counters, etc.	4 × 4 binary multiplier
Frequency dividers	Quad or octal transparent latch
4-bit shift-registers, shift right or bidirectional	9-bit parity generator/checker
4-bit bidirectional shift-registers, parallel load	64-bit RAM
8-bit shift-registers, etc.	Voltage-controlled oscillator
Monostables, with or without CLEAR	Frequency dividers
Retriggerable monostable	Rate multipliers

*MSI die may have multiple independent circuits on one chip, depending on the number of pins in the package. When appropriate, outputs are tristate. No attempt has been made to be exhaustive. Most of these functions are also identifiable as portions of VLSI and ASIC designs.

Having defined the functions and characteristics of several examples of these circuits, you will be able to appreciate how the circuit designer can combine these standard building blocks into larger arrays: logic systems or subsystems. Larger arrays can be in the form of MSI chips interconnected by copper traces on a printed-circuit board. Or, they may be fabricated into *large-scale integrated circuits* (LSI), having 100 to 1000 gates/chip, or even *very large scale integrated circuits* (VLSI) of 1000 to 10,000 (or more) gates/chip. Recent ULSI, ultra-large scale integrated circuits, are on the order of 10^6 gates/chip. When the IC designer integrates many of these functions specific to one application into one die, it becomes an ***application-specific integrated circuit*** (ASIC). This is justified when the specific industrial demand allows amortization of the additional design cost. ***Gate arrays*** are also available, for which only the metallization-layer steps need be modified to provide custom interconnection. In general, MSI logic families are decreasing in importance, being replaced by VLSI gate arrays or ASICs.

One class of logic functions is called ***combinational logic***, meaning that the outputs are dependent only on the present inputs. As examples, we consider multiplexers and demultiplexers, decoders and encoders, and a simple arithmetic-logic unit.

Another class of logic operations is referred to as ***sequential logic***. This means that the output depends not only on the present inputs but also on the past history of those inputs. Examples include flip-flops, registers, and counters. We present the concept of synchronous and asynchronous timing.

No matter how complex an IC may be, the basic building blocks we are going to discuss are the same, of course, replicated many times, and appropriately interconnected. Design examples are used to illustrate the application of these logical building blocks into small systems. The subject of transmission line drivers and receivers is discussed in the context of practical transmission between circuits.

---

* Although many designers associate the term *ASIC* with digital applications, ASICs are available for analog as well as mixed-signal applications.

IMPORTANT CONCEPTS IN THIS CHAPTER

- The outputs of *combinational logic* circuits depend only on the present inputs.
- Digital multiplexing allows conversion of parallel data to serial data and permits several circuits to communicate on one channel.
- A *decoder* allows an *n*-bit code word (address) to choose one of 2^n locations and can translate an *n*-bit code word into one of the 2^n functions.
- The active input of an *encoder* generates the address of that input.
- An *arithmetic logic unit* has an output that may be any one of a number of functions of two input variables, under the direction of a control word.
- The outputs of *sequential logic* circuits depend not only on the values of present inputs, but also on the past values of those inputs.
- An edge-triggered latch is called a *flip-flop*. These may also have direct set and reset.
- The *edge-triggered flip-flop* avoids race conditions and the "ones-catching" problem.
- A *register* is an array of flip-flops. This array can be arranged to shift data between stages in the register.
- A *counter* is an array of flip-flops that sequences through its states when clocked.
- A *monostable multivibrator* is a regenerative circuit with one stable state and one quasi-stable state. It is used to create a pulse of predetermined width.
- An *astable multivibrator* is a regenerative circuit that alternates continuously between its quasi-stable states. It provides a free-running digital waveform.
- *Line drivers* and *receivers* match data signals to long transmission lines.

14.1 COMBINATIONAL LOGIC CIRCUITS

Multiplexers

In general, electronic multiplexing is analogous to selecting one of several inputs by means of a multiposition switch and then distributing the resulting signal to one of several outputs by means of another multiposition switch. A digital **multiplexer** is a circuit that selects one of several inputs on the basis of another set of inputs known as the ***address***. The data at the selected input are then routed to the output. Such circuits are also called *data selectors*. Applications include data routing and time-division multiplexing (TDM).

Consider the eight-input multiplexer shown in Fig. 14.1. Note that a 3-bit address is required to select one of the eight inputs. In general, an *n*-bit address can select any one of the 2^n inputs. This multiplexer also has a ***strobe*** ($\overline{\text{Enable}}$) input. The function table in Fig. 14.1(b) reveals that the output is equal to the selected input when the strobe input is LOW, but the output is LOW when the strobe is not LOW. This arrangement allows the designer to select the input desired and then enable the output so that any switching transients generated during setup will not appear in the output. Figure 14.1(c) presents the logical realization of this function. Eight inverters, eight four-input AND gates, and an eight-input NOR gate were used. The data books[1-7] show that multiplexers are available with 2, 4, 8, and 16 inputs, with direct and/or complemented outputs and with or without strobes.

■ A *function table* or *truth table* for a combinational logic function lists the possible outputs for every combination of input variables.

CHAPTER 14 SMALL DIGITAL SUBSYSTEMS

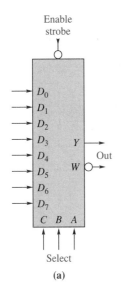

(a)

Function table

Inputs			Strobe	Outputs	
Select					
C	B	A	S	Y	W
X	X	X	H	L	H
L	L	L	L	D0	$\overline{D0}$
L	L	H	L	D1	$\overline{D1}$
L	H	L	L	D2	$\overline{D2}$
L	H	H	L	D3	$\overline{D3}$
H	L	L	L	D4	$\overline{D4}$
H	L	H	L	D5	$\overline{D5}$
H	H	L	L	D6	$\overline{D6}$
H	H	H	L	D7	$\overline{D7}$

H = High level, L = Low level, X = Irrelevant

(b)

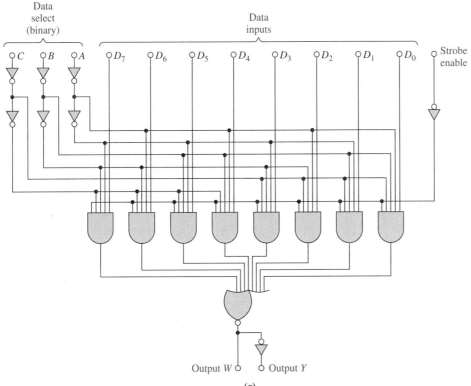

(c)

■ **FIGURE 14.1** Eight-input multiplexer. (a) Symbol. (b) Function table. (c) Logic diagram.

Demultiplexers

Just as the multiplexer selects data from one of several sources and routes it to a single output, a *demultiplexer* accepts data from a single source and distributes it to one of several outputs chosen by an address. Figure 14.2 depicts a one-line-to-eight-line demultiplexer. This is a fairly typical example; all outputs are HIGH except the one selected, which is LOW when the enable input $G = G_1 \overline{G_{2A}} \overline{G_{2B}}$ is true, so input data can be either direct or inverted.

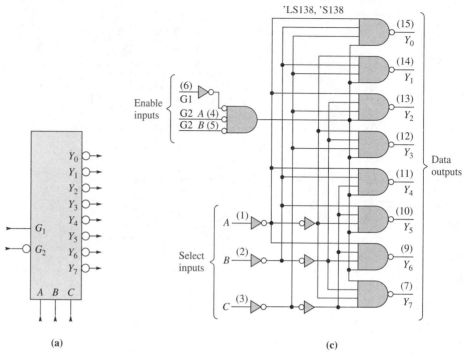

- The equation in Fig. 14.2(b) indicates that the enable input G2 is actually a Boolean function of 2 variables: G2A OR G2B. This is to simplify the function table.

'LS138, 'S138
Function table

Inputs					Outputs							
Enable		Select										
G1	G2*	C	B	A	Y0	Y1	Y2	Y3	Y4	Y5	Y6	Y7
X	H	X	X	X	H	H	H	H	H	H	H	H
L	X	X	X	X	H	H	H	H	H	H	H	H
H	L	L	L	L	L	H	H	H	H	H	H	H
H	L	L	L	H	H	L	H	H	H	H	H	H
H	L	L	H	L	H	H	L	H	H	H	H	H
H	L	L	H	H	H	H	H	L	H	H	H	H
H	L	H	L	L	H	H	H	H	L	H	H	H
H	L	H	L	H	H	H	H	H	H	L	H	H
H	L	H	H	L	H	H	H	H	H	H	L	H
H	L	H	H	H	H	H	H	H	H	H	H	L

*G2 = G2A + G2B
H = High level, L = Low level, X = Irrelevant

(b)

■ **FIGURE 14.2** One-line-to-eight-line demultiplexer.
(a) Symbol. (b) Function table. (c) Logic diagram.

Four-Channel Time-Division Multiplexing Circuit

To illustrate an application of the digital multiplexer (MUX) and demultiplexer consider the system shown in Fig. 14.3. A 2-bit control word on the select lines, A and B, connects one of the four input lines, I_0–I_3, to the connecting line. At the demultiplexing end, the same select lines choose the output line, O_0–O_3, corresponding to the input line in use. Note that this multiplexing scheme allows the four terminating lines to share a common

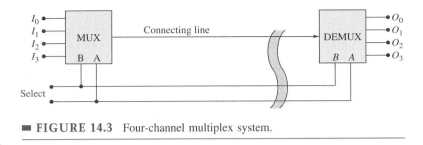

FIGURE 14.3 Four-channel multiplex system.

DRILL 14.1

Given a 16-line multiplexer and a 16-line demultiplexer, extend the idea of the system depicted in Fig. 14.3 and just described into a 16-channel system. How many connecting lines are required?

ANSWER Six: four address, one data, and one common.

connecting line. The price paid for this savings in connecting lines is that only one channel can use the connection at a time. The fact that the two ends must be synchronized is also of great importance in such systems. In the present situation, whatever determines the 2-bit control word on the select lines determines which channel can be active.

Decoders

A *decoder* is a logic circuit in which one of the outputs is specified by each input code word. Two major categories are **address decoders** and display **decoder–drivers**.

Address Decoders

By way of example, consider the address decoder in Fig. 14.4. Binary-coded-decimal (BCD) words applied as inputs activate one of ten outputs. The BCD words are written

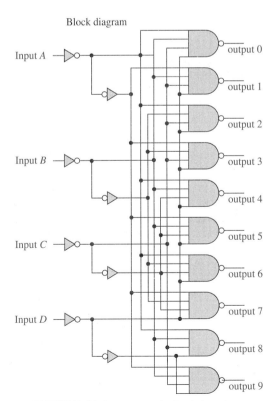

Function table

No.	Inputs				Outputs									
	D	C	B	A	0	1	2	3	4	5	6	7	8	9
0	L	L	L	L	L	H	H	H	H	H	H	H	H	H
1	L	L	L	H	H	L	H	H	H	H	H	H	H	H
2	L	L	H	L	H	H	L	H	H	H	H	H	H	H
3	L	L	H	H	H	H	H	L	H	H	H	H	H	H
4	L	H	L	L	H	H	H	H	L	H	H	H	H	H
5	L	H	L	H	H	H	H	H	H	L	H	H	H	H
6	L	H	H	L	H	H	H	H	H	H	L	H	H	H
7	L	H	H	H	H	H	H	H	H	H	H	L	H	H
8	H	L	L	L	H	H	H	H	H	H	H	H	L	H
9	H	L	L	H	H	H	H	H	H	H	H	H	H	L
	H	L	H	L	H	H	H	H	H	H	H	H	H	H
	H	L	H	H	H	H	H	H	H	H	H	H	H	H
	H	H	L	L	H	H	H	H	H	H	H	H	H	H
	H	H	L	H	H	H	H	H	H	H	H	H	H	H
	H	H	H	L	H	H	H	H	H	H	H	H	H	H
	H	H	H	H	H	H	H	H	H	H	H	H	H	H

H = High level (OFF), L = Low level (ON)

FIGURE 14.4 BCD-to-decimal decoder.

with the order of the bits increasing to the left (most significant bit to least significant), $DCBA = 0111$ representing decimal 7. The outputs are active LOW; the selected output goes LOW, with all other outputs remaining HIGH. The function of the circuit is to select some device or circuit chosen by the input address. The application may be to use high-order address bits to enable some particular memory chips or to turn on some device.

Decoder–Drivers

It is often desirable to activate a display in accordance with some input bit code. The decoder will not only need to activate the proper outputs for the particular display arrangement, but the outputs may be required to sink current from the display or source current to it. As a very popular example, consider the seven-segment display. Segment identification is shown in Fig. 14.5. Such displays are available in various forms, such as gas discharge tube, light-emitting diode, incandescent, and liquid crystal.

To introduce the fundamentals of display addressing and driving, we consider the light-emitting diode (LED), because it presents a load current to the driver. Each segment could be an array of LEDs, but it is usually a light-conducting "pipe" illuminated by one LED. The seven LEDs, or eight if there is a decimal point, may all have one terminal in common to minimize the number of connections and are known as *common-anode* or *common-cathode displays*.

Table 14.2 is the function table for a TTL 74LS47 decoder–driver, which is suitable for a common-anode seven-segment display. The circuit connections are illustrated in Fig. 14.6. The decoder is active LOW; that is, outputs that are meant to activate segments sink current from the LED. This requires that the display must be a common-anode display, with the common terminal connected to the positive supply, 5 V in this case.

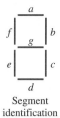
Segment identification

Numerical designations and resultant displays

■ **FIGURE 14.5** Seven-segment display. Segment identification and display patterns for the 74LS47 decoder–driver.

■ Many displays are easier to interpret if the leading zeros (to the left of the decimal point) are blanked. Compare 00110 V with 110 V.

Confirm the display patterns of Fig. 14.5 by referring to the function table. Leading and trailing zeros may be blanked by entering a LOW on the ripple-blanking input, $\overline{RBI}$, which causes a zero in this digit to be blanked and propagates a LOW for the next digit at the ripple-blanking output $\overline{RBO}$. This same pin ($\overline{RBO/BI}$), when forced LOW, acts as a blanking input, $\overline{BI}$, overriding all other inputs to turn OFF all segments. The light-test input, $\overline{LT}$, forces all segments ON when it is forced LOW. Again, refer to Table 14.2.

EXAMPLE 14.1

Calculate the values of resistance needed in the display of Fig. 14.6. Assume that a light-emitting current of 10 mA produces the desired light intensity and that the forward voltage drop is 1.5 V.

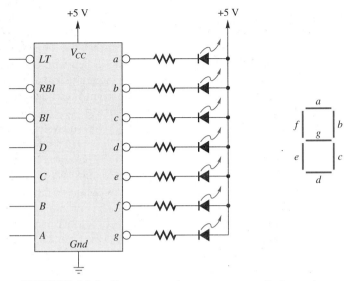

FIGURE 14.6 Common-anode seven-segment display and decoder–driver.

SOLUTION

We assume that V_{OL} for the active outputs of the 74LS47[2] is 0.2 V. To obtain the segment current of 10 mA, a current-limiting resistor of $(5 - 1.5 - 0.2)$ V/10 mA = 330 Ω is needed.

TABLE 14.2 74LS47 Decoder–Driver Function Table

Decimal or Function	Inputs						Outputs						
	$\overline{LT}$	$\overline{RBI}$	D	C	B	A	$\overline{BI/RBO}$	a	b	c	d	e	f	g
0	H	H	L	L	L	L	H	ON	ON	ON	ON	ON	ON	OFF
1	H	X	L	L	L	H	H	OFF	ON	ON	OFF	OFF	OFF	OFF
2	H	X	L	L	H	L	H	ON	ON	OFF	ON	ON	OFF	ON
3	H	X	L	L	H	H	H	ON	ON	ON	ON	OFF	OFF	ON
4	H	X	L	H	L	L	H	OFF	ON	ON	OFF	OFF	ON	ON
5	H	X	L	H	L	H	H	ON	OFF	ON	ON	OFF	ON	ON
6	H	X	L	H	H	L	H	OFF	OFF	ON	ON	ON	ON	ON
7	H	X	L	H	H	H	H	ON	ON	ON	OFF	OFF	OFF	OFF
8	H	X	H	L	L	L	H	ON	ON	ON	ON	ON	ON	ON
9	H	X	H	L	L	H	H	ON	ON	ON	OFF	OFF	ON	ON
10	H	X	H	L	H	L	H	OFF	OFF	OFF	ON	ON	OFF	ON
11	H	X	H	L	H	H	H	OFF	OFF	ON	ON	OFF	OFF	ON
12	H	X	H	H	L	L	H	OFF	ON	OFF	OFF	OFF	ON	ON
13	H	X	H	H	L	H	H	ON	OFF	OFF	ON	OFF	ON	ON
14	H	X	H	H	H	L	H	OFF	OFF	OFF	ON	ON	ON	ON
15	H	X	H	H	H	H	H	OFF	OFF	OFF	OFF	OFF	OFF	OFF
BI	X	X	X	X	X	X	L	OFF	OFF	OFF	OFF	OFF	OFF	OFF
RBI	H	L	L	L	L	L	L	OFF	OFF	OFF	OFF	OFF	OFF	OFF
LT	L	X	X	X	X	X	H	ON	ON	ON	ON	ON	ON	ON

DRILL 14.2

The display circuit of Fig. 14.6 is to be implemented using a CMOS decoder–driver, where $V_{OL} = 0.3$ V at 5 mA. Assume $V_{LED} = 1.45$ V at 5 mA. Calculate the value of the current-limiting resistors.

ANSWER 650 Ω. The display will be less bright than that of Example 14.1.

- A fluorescent lamp operating on 60 Hz pulses ON and OFF 120 times per second, but the illumination appears constant to the eye.

Because of the persistence of images on the retina, the display can be made to appear just as bright by pulsing greater current to the LEDs for a fractional duty cycle. A repetition rate of no less than approximately 50 Hz is sufficient to refresh the retinal images so that no flicker is observed. For example, our display could be driven with a current pulse of 30 mA for about 15% of the time. Because the display is more efficient at higher currents, this arrangement would appear nearly as bright as a 10-mA continuous current. As a result of this, one decoder–driver is often time-shared with several displays. CMOS chips are available with a complex six-digit display–driver subsystem in one 28-pin package.

Many MOSFET decoder–drivers are designed to drive *liquid crystal displays* (LCDs). These units have become very popular with low-power systems; because they are field-effect devices, they require very little power. For a discussion of liquid crystals, see Ref. 1. A liquid crystal, as the name implies, is a liquid having molecules that tend to align with each other. A thin layer, when sandwiched between two panes of glass, will align with fine scratches etched on the inside of the glass surface. This alignment may be chosen so that light passes through a polarizing filter without interference.

- A whole new industry has developed around the liquid crystal display. Pixel drivers are being integrated directly on the glass display surface. Ten-inch diagonal LCD color displays are widely used in lap-top computers. Thirty-inch flat-panel displays have been demonstrated. Perhaps the bulky CRT in TV will be replaced by a flat-panel LCD display.

In a common type of LCD, such as that used in digital watches, light passes through a glass polarizing filter, through the liquid crystal, reflects at a mirror surface, travels back through the liquid crystal and the polarizer. So if the surface is illuminated, it will appear bright to the eye. Transparent conducting electrodes on the two glass surfaces allow an electric field to be applied across the liquid crystal to change the molecular alignment and cause the area of the electrodes (segments) to appear dark. The disadvantages are that ambient light or backlighting is required, and the response time is slower than for LEDs. Various forms of LCDs are dominant in the display industry because of their low power, low cost, and large display area.

Encoders

Encoders are multiple-input circuits that generate the address of the active input. Applications include keyboard encoders, where the key depressed generates a binary word that is used by a computer to identify the active key. Many processors are programmed to work on routine tasks until a device needs attention. Perhaps we have a situation in which many different devices will occasionally request service. Each device can be connected to one input of an *interrupt encoder*, so that the processor, when ready to service the device(s) requesting attention, can locate the active device using the address provided by the encoder. Note that more than one input can be activated at once, and that the encoder must output the valid address of only one of them.

DRILL 14.3

Suppose switches 1, 3, and 6 are closed in the circuit of Fig. 14.7. Give the states of the encoder outputs.

ANSWER
$\overline{A_0} = H, \overline{A_1} = L,$
$\overline{A_2} = L, \overline{EO} = H,$
$\overline{GS} = L.$

A *priority encoder* generates the address of the input with the highest priority, where priority is assigned according to the position of the input. The function table for an eight-line-to-three-line priority encoder is shown in Table 14.3. This particular encoder may be cascaded with identical units by using enable input $\overline{EI}$ and enable output $\overline{EO}$. All inputs and outputs are active LOW. Observe that activating any input generates a HIGH-to-LOW transition on output $\overline{GS}$. This output can be used to signal a processor to read the encoder address. Figure 14.7 shows the application of an encoder to an array of switches. The inputs are normally held HIGH by the pull-up resistors. Suppose switch 5 closes, causing input 5 to go LOW. When line 5 is active, the output is LHL (complemented binary 5) unless the higher priority input 6 or 7 is active. Note that if the switches are of the momentary-contact type, the outputs must be read before the switch is released because the circuit shown is combinational logic, having no memory.

TABLE 14.3 Eight-Line-to-Three-Line Priority Encoder Function Table

Inputs									Outputs				
$\overline{EI}$	$\overline{I_0}$	$\overline{I_1}$	$\overline{I_2}$	$\overline{I_3}$	$\overline{I_4}$	$\overline{I_5}$	$\overline{I_6}$	$\overline{I_7}$	$\overline{GS}$	$\overline{A_0}$	$\overline{A_1}$	$\overline{A_2}$	$\overline{EO}$
H	X	X	X	X	X	X	X	X	H	H	H	H	H
L	H	H	H	H	H	H	H	H	H	H	H	H	L
L	X	X	X	X	X	X	X	L	L	L	L	L	H
L	X	X	X	X	X	X	L	H	L	H	L	L	H
L	X	X	X	X	X	L	H	H	L	L	H	L	H
L	X	X	X	X	L	H	H	H	L	H	H	L	H
L	X	X	X	L	H	H	H	H	L	L	L	H	H
L	X	X	L	H	H	H	H	H	L	H	L	H	H
L	X	L	H	H	H	H	H	H	L	L	H	H	H
L	L	H	H	H	H	H	H	H	L	H	H	H	H

H = HIGH voltage level, L = LOW voltage level, X = Don't care

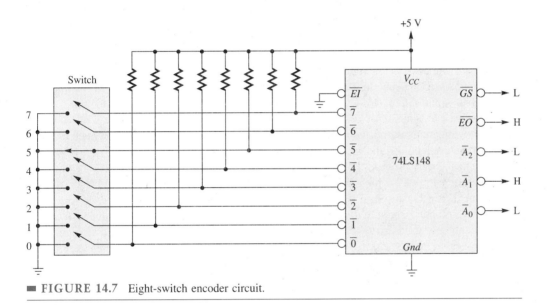

■ **FIGURE 14.7** Eight-switch encoder circuit.

Arithmetic Logic Units

Many of the MSI digital circuits have applications in arithmetic operations. We briefly consider here the arithmetic-logic unit (ALU) since this unit is representative of a basic subsystem used in microprocessors. Microprocessors, in turn, are a dominant force in the electronics industry today. They are used as controllers in systems ranging from a simple clock to a very large printing press or rolling mill, in addition to being the control components of digital computers. Modern automobiles make extensive use of microprocessors. To give some perspective to ALU application, we first examine the general organization of a computer.

A basic block diagram of a computer is shown in Fig. 14.8. This applies to the simplest microprocessor-centered microcomputer or to a mainframe computer. An embedded controller would have a similar structure. Of course, the block diagram would have to be expanded considerably to account for the many types of communications between proces-

■ Some automobile manufacturers use separate microprocessors for engine control, transmission control, antilock braking, body load compensation, instrument panel, and trip computer.

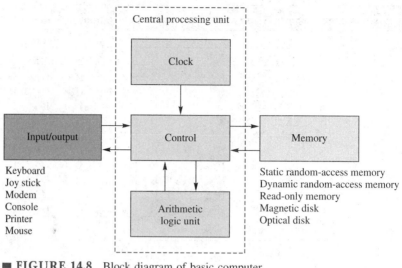

FIGURE 14.8 Block diagram of basic computer.

■ What is shown is the classic Van Neumann computer architecture. Different architectures are used for parallel processing techniques.

sors that are possible, including networks of all types, workstations, and parallel processors. We look briefly at an elementary structure that contains the basic features of a computer.

The input/output (I/O) unit is the computer's interface to the outside world. This interface can consist of devices such as keyboards, a mouse, joysticks, CRT screens, communications ports (including modems), and printers, or of transducers such as limit switches, solenoids, and pressure switches.

The memory provides both permanent and temporary storage of data for the computer. Operational codes, which are the program instructions that tell the control unit what to do next, can be stored in permanent memory. The data words that are being processed by the computer are stored in temporary memory. Relatively fast semiconductor memory is commonly used as the internal storage for small computers. Mechanical storage, such as magnetic *floppy disks*, *hard drives*, and optical *compact disks* (CD), can store larger amounts of data, but data from this relatively slow memory source are moved into faster internal memory as needed. We present the details of semiconductor memories in Chapter 15.

The control unit, driven by a clock that sequences all operations, gets a binary code word (instruction) from a permanent part of memory, executes that instruction, and then proceeds to the next instruction. The instructions can direct the control unit to move data words in either direction between I/O, memory, or registers associated with the ALU. Instructions can also direct the way in which the ALU operates on the data in the registers.

The clock, control unit, and ALU, taken together, form the ***central processing unit*** (CPU). In smaller systems, the control unit and the ALU are often fabricated on one LSI or VLSI chip called a ***microprocessor***. Microprocessors now range from 4 to 64 bits in data and addressing capacity.

Let us now consider the capabilities of an example ALU. The unit to be reviewed is available in TTL (74LS181), in ECL (MC10181), and in CMOS (MC14581). Figure 14.9 depicts the input and output terminals used.

The unit is seen to accept as inputs two 4-bit data words, A_0–A_3 and B_0–B_3. When the mode input (M) is HIGH, internal carries are inhibited, and 16 logical operations may be performed on the individual bits of A and B, as determined by the 4-bit control word, S_0–S_3. The results of these operations appear at the output terminals F_1–F_3. Table 14.4 lists the available logic functions. When the mode control input (M) is LOW, all carries

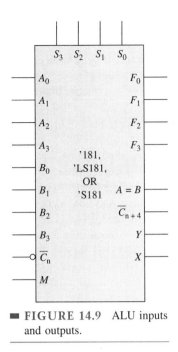

■ **FIGURE 14.9** ALU inputs and outputs.

TABLE 14.4 ALU Function Table

Mode Select Inputs				Active High Inputs and Outputs	
S_3	S_2	S_1	S_0	Logic $(M = H)$	Arithmetic* $(M = L)(\overline{C_n} = H)$
L	L	L	L	$\overline{A}$	A
L	L	L	H	$\overline{A+B}$	$A + B$
L	L	H	L	$\overline{A}B$	$A + \overline{B}$
L	L	H	H	Logical 0	Minus 1
L	H	L	L	$\overline{AB}$	A plus $A\overline{B}$
L	H	L	H	$\overline{B}$	$(A + B)$ plus $A\overline{B}$
L	H	H	L	$A \oplus B$	A minus B minus 1
L	H	H	H	$A\overline{B}$	AB minus 1
H	L	L	L	$\overline{A} + B$	A plus AB
H	L	L	H	$\overline{A \oplus B}$	A plus B
H	L	H	L	B	$(A + \overline{B})$ plus AB
H	L	H	H	AB	AB minus 1
H	H	L	L	Logical 1	A plus A†
H	H	L	H	$A + \overline{B}$	$(A + B)$ plus A
H	H	H	L	$A + B$	$(A + \overline{B})$ plus A
H	H	H	H	A	A minus 1

L = LOW voltage level, H = HIGH voltage level
*Arithmetic operations expressed in 2s complement notation.
†Each bit is shifted to the next more significant position.

are enabled, and the unit performs arithmetic operations on the two 4-bit words. These units can be cascaded so that carries (and borrows) can propagate between devices by means of carry input ($\overline{C_n}$) and carry output ($\overline{C_{n+4}}$). The function table lists the arithmetic operations that are performed without an incoming carry. An incoming carry ($\overline{C_n} = $ L) adds one to each operation. For instance, select code HLLH generates A plus B plus 1 with carry input, and select code LHHL generates A minus B with carry input.

This circuit accomplishes subtraction by complementing B (one's complement), and adding. No carry out ($\overline{C}_{n+4} = $ H) means borrow. A carry is generated when there is no underflow, so that the default -1 in the subtract operation is cancelled, and no carry is generated when there is underflow. The $A = B$ output goes HIGH when all four outputs are HIGH, so it can be used to indicate that all bits in A are equal to all bits in B when the unit is in the subtract mode with $\overline{C}_n = $ H.

The preceding discussion should convince you that this integrated circuit has considerable logic capability. To provide some perspective with regard to the complexity of this chip, the logic diagram is shown in Fig. 14.10. About 65 gates are involved.

We are now in a position to understand how a computer could be realized using ALU modules similar to the one just considered. A set of instructions (program) is stored in memory. The control unit examines these instructions sequentially. These operational codes direct control to move selected data into registers connected to the inputs of the ALU, and apply the proper codes to the *select* inputs of the ALU. The result of the operation will reside in another register, usually called the *accumulator*. Next, the output data may be moved to its destination. The control must continue to *fetch* instructions and *execute* them until the end of the stored program is reached.

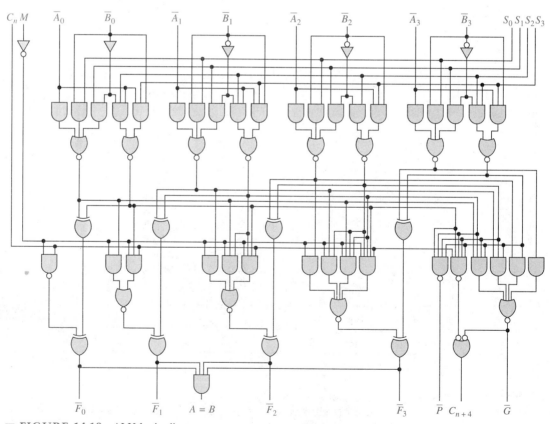

■ **FIGURE 14.10** ALU logic diagram.

EXAMPLE 14.2

Determine the state of the ALU terminals for the arithmetic operation 6 − 7 (subtraction).

SOLUTION $M = L$ gives the *arithmetic* mode. $\overline{C}_n = L$ and $S_3S_2S_1S_0 = LHHL$ puts the ALU in the operation *A minus B* (carry in defeats the default −1). $A = 6$ is binary $A_3A_2A_1A_0 = LHHL$, and $B = 7$ is binary $B_3B_2B_1B_0 = LHHH$. B is complemented and added to A with carry: $0110 + 1000 + 1$ yields $F_3F_2F_1F_0 = HHHH$ with no underflow, $\overline{C_{n+4}} = H$. In two's-complement notation, $1111 \Leftrightarrow -1$.

DRILL 14.4

See if you can find the operation "shift left" in Table 14.4.

ANSWER $S_3S_2S_1S_0 = HHLL$ shifts each bit in A to the next most significant position (A *plus* A). Note that this operation is equivalent to multiplying by two.

CHECK UP

1. **TRUE OR FALSE?** Digital multiplexing allows the use of one higher speed channel to replace several lower speed data channels.
2. **TRUE OR FALSE?** An encoder allows an *n*-bit code word to choose one of 2^n locations.
3. **TRUE OR FALSE?** An ALU can perform any arithmetic operation on two input variables.
4. **TRUE OR FALSE?** A combinational logic circuit will never have a clock input.

14.2 SEQUENTIAL LOGIC CIRCUITS

In contrast to combinational logic circuits, the pair of cross-coupled NAND gates shown in Fig. 14.11, called a *latch*, has memory. The present state of the circuit might depend on the previous states of the inputs.

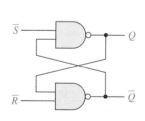

$\overline{S}$	$\overline{R}$	Q_{n+1}	$\overline{Q}_{n+1}$
0	0	1	1
0	1	1	0
1	0	0	1
1	1	Q_n	$\overline{Q}_n$

■ **FIGURE 14.11** $\overline{SET}$-$\overline{RESET}$ latch and truth table.

The SET-RESET Latch

You may recall the two-transistor latch shown in Fig. 4.36. Here we use active LOW inputs called $\overline{SET}$ and $\overline{RESET}$. Suppose both $\overline{SET}$ ($\overline{S}$) and $\overline{RESET}$ ($\overline{R}$) are HIGH, $\overline{S} = \overline{R} = 1$. Then letting $\overline{S}$ go LOW, $\overline{S} = 0$, causes $Q = 1$, and then $\overline{Q} = 0$. This is shown in line 2 in the truth table of Fig. 14.11. Now let $\overline{S} = 1$ again. The value of $\overline{Q}$ remains 0 since both inputs to the bottom NAND are 1, and Q remains 1 because $\overline{Q} = 0$ is one of the inputs to the top NAND. This is one stable *state* (one of the possible logical outcomes) for this latch. When $\overline{R} = 0$, $\overline{Q}$ is forced to 1, and Q becomes 0 (line 3 in the truth table). When $\overline{R}$ returns to 1, $\overline{Q}$ remains 1, and Q remains 0. This is the second stable state, and explains why the circuit is also called a ***bistable latch***. Thus the bottom line of the table represents the situation where Q and $\overline{Q}$ retain the $\overline{\text{SET}}$ or $\overline{\text{RESET}}$ state, whichever was more recent. The terminology $Q_{n+1} = Q_n$ means that the present value of Q equals the preceding value. The $\overline{S}$ and $\overline{R}$ should not be LOW simultaneously because this causes both outputs to be HIGH at the same time. This circuit is a 1-bit memory.

The Switch-Debounce Circuit

As an example of the use of the $\overline{S}\,\overline{R}$ latch, consider the following application. Mechanical switch contacts are notoriously noisy, often making and breaking contact several times in the first few milliseconds after being operated. This condition generates oscillating waveforms that might be misinterpreted by logic circuits. Refer to the circuit shown in Fig. 14.12, where the switch is isolated from the system input by the $\overline{S}\,\overline{R}$ latch. Note that the waveform $\overline{S}$ makes several transitions after the switch is operated, but the output at Q makes only a single transition.

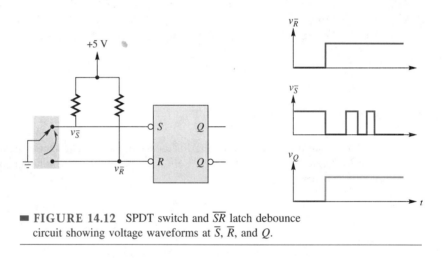

■ **FIGURE 14.12** SPDT switch and $\overline{SR}$ latch debounce circuit showing voltage waveforms at $\overline{S}$, $\overline{R}$, and Q.

Clocking the Latch

The inputs to sequential circuits such as a latch come from various parts of the system, so the relative timing of the different inputs will vary. Although this can be a problem even in combinational logic, it would be disastrous if latches were permanently left in the wrong state because of uncertainty in the timing of the input signals. For this reason, many sequential circuits are arranged so that the final state of the circuit is determined by the inputs, but the exact time at which the inputs become effective is controlled by an enabling signal called the ***clock*** (*CK*). In Fig. 14.13, we see an *SR* latch where the inputs are not effective until a HIGH clock pulse is applied.

Note that the inputs can be changed without affecting the latch if the clock is LOW. The latch is triggered in accord with the truth table when the clock goes HIGH. The clock pulse can be made narrow, so that the inputs do not have access to the latch any longer than the system requires. The minimum width t_W of the clock pulse is a parameter of the latch, as is the minimum ***setup time***, t_{su}, which is the interval that the inputs must be true before the clock pulse. These inputs are called ***synchronous*** inputs, because they are

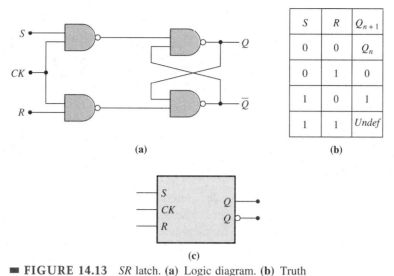

FIGURE 14.13 *SR* latch. (a) Logic diagram. (b) Truth table. (c) Logic symbol.

active only at the times the clock is HIGH. The designer of a synchronous system is able to trigger all of the synchronous devices in the circuit at the same instant by connecting them all to the same system clock.

Note that the $S = R = 1$ condition must not be used because the resulting state cannot be predicted. When the clock is active, both inputs to the latch would be LOW and both outputs would be HIGH. When the clock returns LOW, which output goes LOW depends on uncertain internal time constants and transistor speeds.

Direct Inputs

The *SR* latch can also be provided with PRESET and CLEAR inputs, as illustrated in Fig. 14.14. In the circuit shown, $\overline{S}_d$ is called a *direct set* or *PRESET* input, and it is asynchronous since it operates independently of the clock. Likewise, $\overline{R}_d$ is a *direct reset* or *CLEAR* input. These inputs can be used to put the latch into a desired state independently of the system clock.

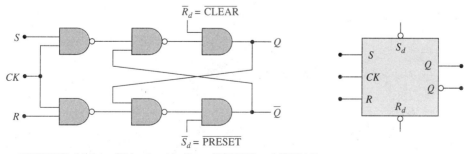

FIGURE 14.14 *SR* latch with direct PRESET and CLEAR.

The *JK* Latch

The simplified *JK* latch as drawn in Fig. 14.15 is of interest because of the useful characteristics of the truth table. Feedback from the outputs to the input gates removes the uncertainty in the truth table shown in Fig. 14.13. Suppose both direct inputs are inactive (HIGH). The first line of the truth table shows that the ouput will be unchanged when the clock becomes active since $J = 0$ disables the *J* gate, and $K = 0$ disables the *K* gate. Considering the second line of the truth table, we note that $J = 0$ prevents the clock from SETting the latch. But if the latch is already SET ($Q_n = 1$), the clock will enable the *K* gate and RESET the latch. Thus, $J = 0$ and $K = 1$ always result in $Q_{n+1} = 0$. A similar line of reasoning will enable the reader to verify the third line of the truth table.

The unique aspect of this circuit relates to the bottom line of the truth table. Suppose $Q_n = 1$ and $J = K = 1$. When the clock is activated the *J* gate is disabled by $\overline{Q}_n = 0$, but the *K* gate RESETs the latch and Q_{n+1} goes to 0. If $Q_n = 0$, the clock will SET the latch through the *J* gate, causing $Q_{n+1} = 1$. Thus the bottom line of the truth table represents the situation where the latch acts as a ***toggle***, or alternates states with each clock pulse. Output *Q* will generate a square wave with half the clock frequency, so this condition is useful as a *frequency divider*. More on this in the discussion of counters.

Race Condition. There is a problem with the operation of the *JK* latch as just described. Suppose all inputs are HIGH and $Q_n = 0$. When the clock goes HIGH, the *J* gate will SET the latch as indicated above. But if the clock persists HIGH, the *K* gate will now be enabled, and the latch will immediately RESET. The latch will oscillate between states as long as the clock remains HIGH. This situation, in which the changing output is fed back and affects the present state of the circuit, is known as a ***race*** condition. It can be eliminated by making the clock pulse narrow enough so that it returns to 0 before the

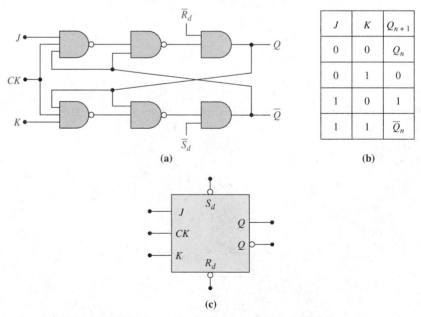

■ **FIGURE 14.15** *JK* latch. **(a)** Logic diagram. **(b)** Truth table. **(c)** Logic symbol.

change in latch output reaches the input gate. But this is an unsatisfactory solution since it might put an unreasonable constraint on the clock pulse width as far as the rest of the system is concerned. Also, it would lead to marginal operation of the latch at best. Two solutions are in common use: the *master-slave flip-flop* and the **edge-triggered flip-flop**. The term *flip-flop* is defined as a latch that is either edge triggered, or, in the case of master/slave devices, pulse triggered.

The JK Master–Slave Flip-Flop. The JK master–slave circuit shown in Fig. 14.16 is simply a cascade of two latches. The first is the master, and the JK inputs are enabled by the clock. The second is the slave, and its $\overline{SR}$ inputs are enabled by the clock inverted ($\overline{CK}$). The operation can be understood by referring to the clock waveform shown in Fig. 14.16(b). As the clock (CK) begins to rise (point 1), $\overline{CK}$ drops and disables the gates between the master and slave. As CK reaches point 2, the J and K gates of the master are enabled, including the feedback from the slave outputs Q_s and $\overline{Q}_s$. Outputs Q_m and $\overline{Q}_m$ assume their values according to the truth table shown in Fig. 14.15, but the slave remains inactive. When CK begins to fall (point 3), the J and K gates are disabled, fixing the state of the master. Finally, at point 4, the gates between the master and slave are enabled, transferring the new state of the flip-flop to the slave. The race condition has been eliminated since the JK inputs have been disabled before the output changes state.

Note that the clock signal must remain active long enough for the inputs to propagate through the master, but there is no maximum clock pulse width. This circuit does exhibit another shortcoming. Should the JK inputs change while the clock is active in such a way as to toggle the master, there is no way for the original state to be recovered. Thus,

■ The term *flip-flop* is often used synonymously with latch, but that is incorrect. It is a special type of clocked latch, not subject to *race* conditions.

■ The ⅂ in the symbol of the M/S flip-flop of Fig. 14.16 indicates a pulse-triggered device where the outputs are transferred on the *falling* edge of the clock.

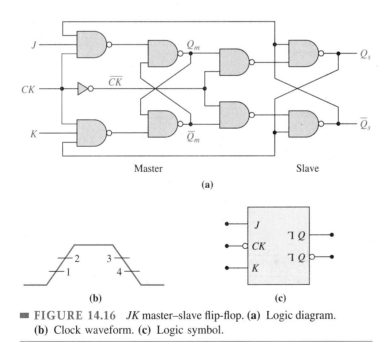

■ **FIGURE 14.16** *JK* master–slave flip-flop. **(a)** Logic diagram. **(b)** Clock waveform. **(c)** Logic symbol.

if a switching transient should switch the master while the clock is active, this "opposite state" is trapped, and it will inevitably be transferred to the slave at the falling edge of the clock (see Problem 14.7). This difficulty is relieved by using a clock pulse no wider than t_W. The edge-triggered JK flip-flop, discussed next, does not have this difficulty.

The logic symbol for the circuit in Fig. 14.16(c) has the *active LOW* symbol at the clock input to indicate that the outputs change states on the *falling edge* of the clock. JK master–slave flip-flops are also available with the opposite clock polarity.

The JK Edge-triggered Flip-Flop. In an edge-triggered flip-flop, the inputs can be changed when the clock is either LOW or HIGH. If the IC is *negative-edge triggered*, the outputs only change states at the trailing edge of the clock pulse. The inputs are only required to observe a minimum *setup time* before the trailing edge of the clock, typically a few nanoseconds. A logic diagram of a *JK* edge-triggered flip-flop is shown in Fig. 14.17(a). This example uses a negative clock pulse, but the circuit is triggered on the positive transition. The logic symbol in Fig. 14.17(b) indicates this fact by showing no inverting circle on the clock terminal and shows that this is an edge-triggered device by the wedge-shaped marking at the clock terminal.

To understand the operation of this circuit, refer to the gate output waveforms shown in Fig. 14.17(c), and assume $J = K = 1$ and $Q_n = 1$. The output of G_2 is LOW, and outputs G_4 and G_5 are HIGH. The negative-going clock is inverted at output G_1 after a propagation delay t_{1r}. Output G_2 does not respond, being disabled by $\overline{Q}_n = 0$. Output G_3 goes HIGH after another delay t_{3r}. Nothing further happens since G_3 has been disabled

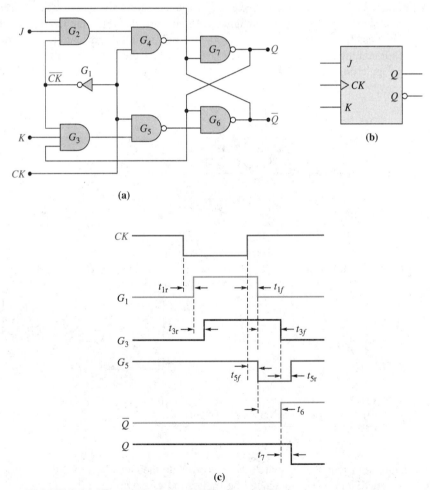

■ **FIGURE 14.17** *JK* edge-triggered flip-flop. (a) Logic diagram. (b) Logic symbol. (c) Clock waveforms.

by the LOW clock. When the clock rises, two sequences of events begin: G_1 falls after delay t_{1f}, and after a further delay t_{3f}, G_3 falls. During this time, G_5 falls after t_{5f}, causing $\overline{Q}_{n+1}$ to rise after delay t_6, which in turn causes Q_{n+1} to switch LOW after delay t_7.

Note that G_5 will switch back HIGH t_{5r} after G_3 returns LOW, but as long as G_5 remains LOW, until Q_{n+1} switches LOW, the latch is safely toggled. From the waveforms, one can observe that this requires $t_{1f} + t_{3f} + t_{5r} > t_{5f} + t_7 + t_6$. The manufacturer usually ensures that this is true by deliberately constructing G_2 and G_3 with more propagation delay than the other gates. Edge triggering requires only that the inputs be stable at the time of the trailing edge of the clock pulse and is the popular choice in IC design. Of course, asynchronous PRESET and CLEAR inputs can be added to this circuit as before.

The D Flip-Flop

A circuit in which the (single) input state is transferred to the output Q by the clock is called a D (delay) flip-flop. The circuit can be realized with a JK flip-flop by forcing $J = D$ and $K = \overline{D}$ as shown in Fig. 14.18(a). The truth table in Fig. 14.18(b) consists of the second and third lines of Fig. 14.15(b). The logic symbol for the D flip-flop is shown in Fig. 14.18(c), assuming that the rising edge of the clock is the active edge. Alternatively, the circuit shown in Fig. 14.19 could be used. The truth table and logic symbol are unchanged. The D flip-flop is popular as a 1-bit memory and is often arranged with several units on one chip, with a common clock terminal and perhaps common RESET and/or SET terminals.

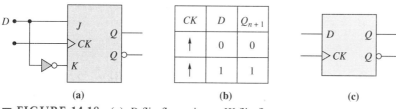

■ **FIGURE 14.18** (a) D flip-flop using a JK flip-flop. (b) Truth table. (c) Logic symbol.

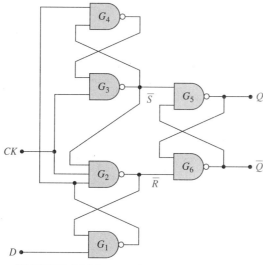

■ **FIGURE 14.19** A D flip-flop.

EXAMPLE 14.3

Verify that the D flip-flop in Fig. 14.19 obeys the logic statements of the truth table shown in Fig. 14.18(b).

SOLUTION (a) Assume $D = 0$ and $CK = 0$. Outputs G_1, $G_2 = \overline{R}$, and $G_3 = \overline{S}$ are all HIGH, and $G_4 = 0$. When CK goes HIGH, $\overline{S}$ remains HIGH, but $\overline{R}$ goes LOW, RESETting the latch, $Q_{n+1} = 0$. When CK returns LOW, $\overline{S} = \overline{R} = 1$, and Q_{n+1} remains LOW. (b) Let $D = 1$ and $CK = 0$. Again $\overline{S} = \overline{R} = 1$, but this time $G_1 = 0$ and $G_4 = 1$. When CK goes HIGH, $\overline{R}$ remains HIGH, but $\overline{S}$ goes LOW, SETting $Q_n = 1$. When CK returns LOW, $\overline{S} = \overline{R} = 1$, and the latch is unchanged.

The Transparent Latch

A relatively simple D latch whose output follows its input when its clock is enabled and holds the previous input when the clock is disabled is called a ***transparent latch***. This device is neither an edge-triggered nor a master–slave flip-flop so it is also less versatile, but it is less costly and serves well some functions of temporary storage. An example of such a circuit is shown in Fig. 14.20.

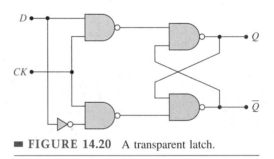

■ **FIGURE 14.20** A transparent latch.

Specifications for Flip-Flops

A check of manufacturers' data sheets reveals the typical *JK* edge-triggered flip-flop information shown in Table 14.5, from which we can compare several logic families. Speed and power were chosen in this comparison because these characteristics sum up the most important differences in technologies. Compare this table to Fig. 13.17. The high-speed limitation in any system design is the maximum toggle rate of the slowest flip-flop in a counter. In other designs, heat dissipation or power-supply requirements may be the dominant consideration.

Because a flip-flop can function as a 1-bit memory, an array of n flip-flops can store an n-bit data word. Such an array is called an ***n-bit register***. When n bits of data are simultaneously entered into an n-bit register from n data lines, we say that the data are *parallel* data, or the register is parallel-loaded. Likewise, parallel data can be transferred out of a register in one clock interval.

When the array of flip-flops is arranged to permit shifting data from one to another, it is referred to as a ***shift register***. *Serial* data are entered into a shift-register, or accepted from a shift-register, one bit at a time. Thus an n-bit data word would require n clock pulses to be entered into the register.

TABLE 14.5 Comparison of Representative *JK* Flip-Flops

Family	Example	Max Toggle Rate (MHz)	Dissipation/FF (No Load) (mW)	Reference
CMOS (5V)	74C109	4	45	2
CMOS (10V)	74C109	11	20	2
HCMOS (4.5V)	74HC109	25	12	3
ACMOS (5V)	74AC109	175	150	4
TTL	74109	33	45	5
TTL (LS)	74LS109	33	20	5
TTL (ALS)	74ALS109	34	12	5
TTL (AS)	74AS109	105	60	6
ECL (10K)	10135	125	135	8
ECL (100K)	100135	650	220	8

Serial-in Registers

Consider the array of four *D* flip-flops in Fig. 14.21. The flip-flops could as easily be *JK* or *SR*, but in any case they must be edge-triggered or master–slave flip-flops to prevent a race condition. Suppose all flip-flops are RESET or CLEARed initially by applying a positive pulse to a common CLEAR lead. Serial data consisting of the sequence of bits b_1, b_2, b_3, and b_4 are presented to the input at the clock rate and the state of the register changes as follows:

	Q_A	Q_B	Q_C	Q_D
Initially	0	0	0	0
After clock 1	b_1	0	0	0
After clock 2	b_2	b_1	0	0
After clock 3	b_3	b_2	b_1	0
After clock 4	b_4	b_3	b_2	b_1

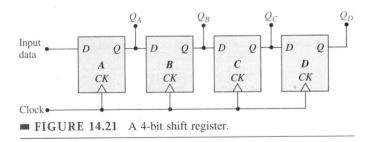

FIGURE 14.21 A 4-bit shift register.

This type of shift-register can be used in two ways. First, parallel data are available at terminals Q_A to Q_D at any time between clock pulses. This usage, called *serial-in/parallel-out*, is used to convert serial data into parallel data. Second, serial data are available at Q_D, appearing sequentially after four clock pulses. This application, termed *serial-in/serial-out*, or *first-in/first-out*, is useful as a serial memory where a data stream needs to be delayed but the order of bits is not to be changed. External terminals for Q_A, Q_B, and Q_C would not be needed in this application. Indeed, a 1024-bit serial shift register would need no more pins than does the 4-bit version.

Parallel-in Shift Registers

The logic diagram of a 4-bit parallel-load shift register is shown in Fig. 14.22. This particular circuit does not show any parallel output terminals and would be used as a *parallel-in/serial-out* shift register, or a parallel-to-serial data converter. The CLEAR input is asynchronous and overrides all other inputs to RESET all latches. CLEAR must be LOW to either load or shift. After clearing the register, the parallel enable (*PE*) is asserted HIGH to load the data at the parallel inputs into the latches by a PRESET at those terminals where the data bit is a 1. *PE* must be returned LOW before the rising edge of the clock pulse causes a shift. Again, data appear sequentially at Q_D, and each clock pulse enters whatever data are presented at the serial input into the *A* latch, shifting each bit one cell to the right.

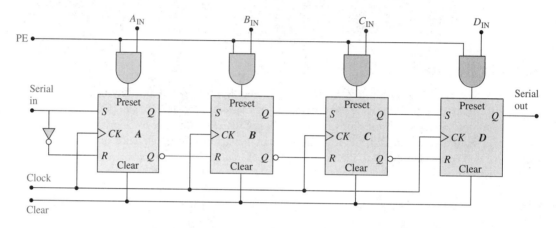

■ **FIGURE 14.22** A 4-bit parallel-load shift register (7494). PE = parallel enable.

The Universal Shift Register

■ Assuming the MSB at the right, this parallel-in/parallel-out shift register can be used to multiply a binary number by two (shift right one bit), or divide by two (shift left one bit).

For a final example of shift registers, look at Fig. 14.23, which shows a logic diagram of a parallel-access shift register that will shift right or left. We can see that this shift register is extremely versatile. The register can be cascaded by connecting the Q_D output of one stage to the shift-right serial input of the next stage, and paralleling the mode controls, clock, and clear.

Counters

We have seen that a flip-flop has two states; thus an array of *n* flip-flops has 2^n possible states. A *counter* is an array connected so that it advances from state to state when clocked by an input waveform. If the states need to be identified, then the outputs of each flip-flop in the array must be decoded. This would be the case where the purpose of the counter is to record the number of input events. In other applications, it may only be necessary to divide the input frequency by some constant. Suppose that we have a crystal-controlled clock operating at 1 MHz, and we need to derive a pulse once per second from this clock. The problem suggests a counter with 10^6 states, which would divide the input frequency down to the desired rate. A counter that sequences through *K* states, and then repeats, is called a *modulo-K-counter*. We present only a few examples of counter arrangements.

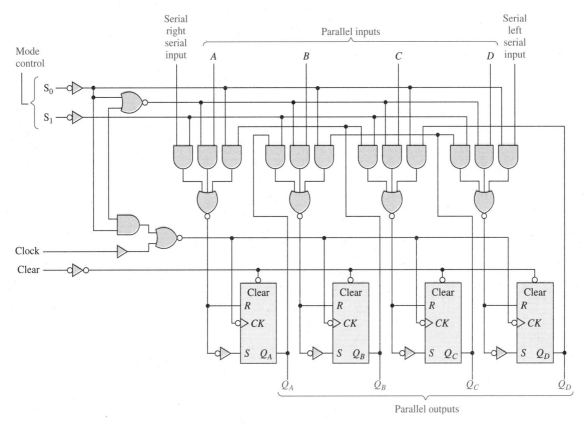

FIGURE 14.23 A parallel-access shift-right/shift-left register (74LS194).

Ripple Counters. A four-stage ripple counter is shown in Fig. 14.24(a). Each edge-triggered JK flip-flop is connected as a toggle ($J = K = 1$) and is clocked by the output of the preceding stage. Assume that all latches are initially RESET; then the input waveform is as shown in Fig. 14.24(b). Latch A will be toggled at each falling edge of the input. Similarly, latch B will be toggled each time Q_A makes the HIGH-to-LOW transition. The state table shows that the counter sequences through all 16 possible states; this is a modulo-16 counter. This is an asynchronous circuit.

Note that at the eighth count all of the flip-flops change state and that these state changes *ripple* through the counter. After the eighth input pulse falls, Q_A will fall after a propagation delay of t_A. Then Q_B will fall after a delay t_B, etc. The desired state is reached four propagation delays after the input falling edge. This ripple effect is of little consequence if our purpose is frequency division at one of the outputs, but if the code word $Q_DQ_CQ_BQ_A$ is to be decoded, there will be switching transients into the decoder when this ripple is propagating.

One possible arrangement for obtaining a modulo-5 ripple counter is shown in Fig. 14.25. After CLEAR, all outputs are LOW, and $\overline{Q}_C = J_A = 1$. Thus, at the first three input falling edges, latches A and B toggle exactly as they do in the binary counter shown in Fig. 14.24. At count 4, since Q_A and $Q_B = 1$, edge-triggered SR flip-flop C is SET at the same time that latch A is toggled, and Q_A falls to 0, toggling latch B. Now $\overline{Q}_C = J_A = 0$, so at count 5 latches A and B are inactive but C is RESET. Note that the output of stage A ripples to the input of B but that C is clocked synchronously with A.

DRILL 14.5

Given a 16-bit register of the universal type shown in Fig. 14.23, assume that the most significant bit is to the **left**. Assume that the binary data word in the register is 01100100, corresponding to decimal 100. Show that a shift to the left multiplies the number by 2, and a shift to the right divides it by 2.

ANSWER Shifting to the left gives $11001000_2 \Leftrightarrow 200_{10}$, and shifting to right gives $00110010_2 \Leftrightarrow 50_{10}$.

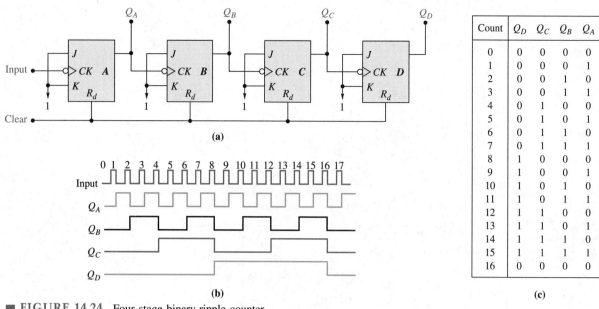

FIGURE 14.24 Four-stage binary ripple counter.
(a) Logic diagram. (b) Waveforms. (c) State table.

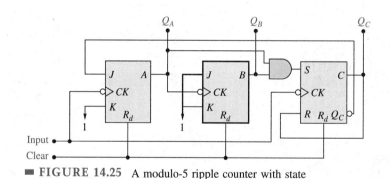

FIGURE 14.25 A modulo-5 ripple counter with state table.

A popular application for the modulo-5 counter is to cascade it with a modulo-2 counter (a flip-flop set to toggle) to constitute a decade counter. Two modulo-10 sequences are created, depending on the way in which the two counters are connected. Consider the following example:

EXAMPLE 14.4

(a) Construct a state table for the decade counter suggested in the preceding paragraph for the case with the modulo-2 driving a modulo-5 counter. (b) Repeat for the case of the modulo-5 counter driving a modulo-2 counter.

SOLUTION (a) Let the output of the modulo-2 counter be Q_A and that of the stages of the modulo-5 counter be Q_B, Q_C, Q_D, respectively. Then each negative transition of Q_A will advance the modulo-5 counter one line in the state table in Fig. 14.25. Proceeding, we show the first state output on the right to emphasize that this is the least significant bit. The solution to part (b) is obtained similarly.

(a) count	Q_D	Q_C	Q_B	Q_A
0	0	0	0	0
1	0	0	0	1
2	0	0	1	0
3	0	0	1	1
4	0	1	0	0
5	0	1	0	1
6	0	1	1	0
7	0	1	1	1
8	1	0	0	0
9	1	0	0	1
10	0	0	0	0

BCD count sequence

(b) count	Q_A	Q_D	Q_C	Q_B
0	0	0	0	0
1	0	0	0	1
2	0	0	1	0
3	0	0	1	1
4	0	1	0	0
5	1	0	0	0
6	1	0	0	1
7	1	0	1	0
8	1	0	1	1
9	1	1	0	0
10	0	0	0	0

Bi-quinary count

The result of part (a) is a counter that generates binary-coded-decimal numbers in sequence. The Q_A output of part (b) is a square wave of one-tenth the frequency of the input pulses.

DRILL 14.6

For the register of Fig. 14.23, verify the following statements:

(a) A LOW on CLEAR will RESET all latches at any time.
(b) Parallel inputs load on the positive edge of the clock when $S_0 = S_1 = 1$.
(c) The register shifts right on the positive edge of the clock when $S_0 = 1$, $S_1 = 0$.
(d) Then latch A loads from the shift-right serial input.
(e) The register shifts left on the positive edge of the clock where $S_0 = 0$, $S_1 = 1$.
(f) Then latch D loads from the shift-left serial input.
(g) The register does not respond to the clock when $S_0 = S_1 = 0$.
(h) Mode controls (S inputs) must only be changed when the clock is HIGH.

Synchronous Counters. As discussed previously, a ripple counter having n stages can have a maximum delay of n times the propagation delay of one state. This *carry delay* may exceed the time between input pulses, in which case the count state may not be read between pulses.

A *synchronous counter*, in which all flip-flops of the counter are clocked simultaneously by the input pulses, has considerably less carry delay than the ripple counter. Such a binary counter is shown in Fig. 14.26. In this four-stage counter, the maximum carry delay is the delay of one flip-flop plus one NAND gate since the carries are accomplished in parallel. We should note, however, that the output of the first NAND could also be used as an input to the second NAND, etc., so that all gates become two-input NAND circuits. Combinations of parallel and ripple carry are used. For example, six decimal counters could be cascaded to count to one million. Each decimal counter might have parallel carry within the decade, but the carries may ripple between the stages.

In this brief overview of counter circuits, we have only introduced the counter as a system design tool. The design of the optimum interconnection of flip-flops for a given count sequence is the subject of logic design texts.[9]

Commercial Counters. Four-bit counters, both synchronous and asynchronous and both binary and decade, are available in all logic families. Variations include the capability to count up or down, to be parallel loaded, and to have on-chip logic for varying the modulus. A set of synchronous examples appears in Table 14.6 for comparison.

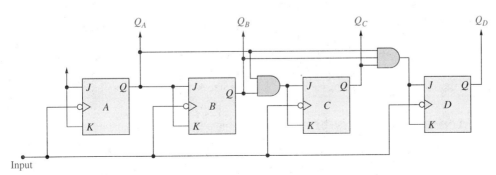

■ **FIGURE 14.26** A four-stage synchronous binary counter.

TABLE 14.6 Comparison of Representative Synchronous Parallel Load 4-Bit Counters

Family	Example Type	Count Frequency (MHz)	Reference
CMOS (5V)	74C162	8.5	2
CMOS (10V)	74C162	3	2
HCMOS (4.5V)	74HC162	25	3
ACMOS (5V)	74AC163	167	4
TTL	74162	32	5
TTL (LS)	74LS162	32	5
TTL (ALS)	74ALS162	40	5
TTL (AS)	74AS162	75	6
ECL (10K)	10136	150	8
ECL (100K)	100136	300	8

The CMOS family offers some multistage ripple counters, such as the 4040B 14-stage binary ripple counter. This type of circuit has widespread application in the derivation of related subharmonics from a single high-frequency oscillator. Examples are a music synthesizer and an electronic organ.

CHECK UP

1. **TRUE OR FALSE?** A D flip-flop is a 1-bit memory cell.
2. **TRUE OR FALSE?** An edge-triggered flip-flop is subject to race conditions.
3. **TRUE OR FALSE?** A master–slave flip-flop may be subject to "catching" unwanted states.
4. **TRUE OR FALSE?** A synchronous register or counter changes states only on the active edge of a clock signal, regardless of the states of the other inputs.

14.3 MONOSTABLE AND ASTABLE TIMING CIRCUITS

We have discussed the bistable latch and the flip-flop, recognizing that they are regenerative circuits with two stable states. The *monostable multivibrator* is a regenerative circuit with *one* stable state and one quasi-stable state. When in the stable state, it can be triggered into a quasi-stable state, where it will remain for a predetermined time and then return to the stable state. This makes this circuit extremely valuable as a source of fixed-width pulses, with obvious timing applications.

The *astable* multivibrator, as the name implies, has no stable states but two quasi-stable states. This circuit needs no external triggering but continually switches between the quasi-stable states, remaining in each state for some predictable time. The designer can use this free-running oscillator, which generates a rectangular waveform, as the source of the clocking waveform that we have found necessary in synchronous digital circuits.

Monostable Multivibrators

The two-transistor circuit shown in Fig. 14.27 was presented for analysis in Problem 4.46. The stable state is for Q_2 to be saturated, which holds Q_1 OFF, and the capacitor charges

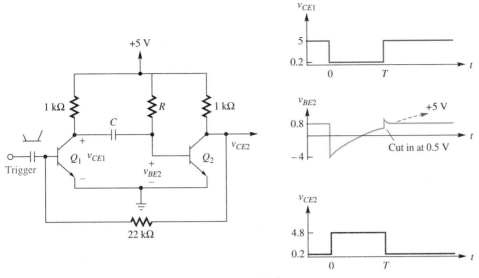

FIGURE 14.27 Two-transistor monostable multivibrator.

to $V_{CC} - V_{BE(\text{sat})} = 4.2$ V. Should Q_1 be triggered ON at time $t = 0$, say, by a narrow positive voltage pulse at its base, v_{C1} falls from 5 to 0.2 V, causing v_{B2} to fall from 0.8 to -4.0 V. The result of the trigger is that Q_1 saturates and Q_2 cuts OFF, causing the output voltage v_{CE2} to go HIGH. This, in turn, holds Q_1 saturated after the trigger pulse is removed. The base current of Q_2 remains zero while v_{B2} charges from -4 toward 5 V through resistor R. The circuit is restored to the stable state when v_{B2} reaches cut-in, 0.5 V, at time T. The equation for the base voltage of Q_2 is

$$v_{B2} = 5 - 9e^{-t/RC}. \tag{14.1}$$

Equating this expression to $v_{BE(\text{cut-in})}$ allows us to determine interval T:

$$T = -RC \ln\left(\frac{0.5 - 5}{-9}\right) = RC \ln 2 = 0.69 RC. \tag{14.2}$$

We see that this monostable circuit provides the circuit designer with a means of obtaining pulses of fixed width, determined only by the chosen R and C. The only requirements are that R must not exceed $\beta_2 R_{C2}$, so that Q_2 will saturate in the steady state, and C must have low leakage current.

■ Note that V_{CC}, $V_{BE(\text{cutin})}$, $V_{CE(\text{sat})}$, R, and C all appear in Eq. (14.2). Thus timing accuracy is dependent on all of these circuit parameters. These parameters also change with temperature.

Integrated-Circuit Monostable Multivibrators

Figure 14.28 shows a block diagram of a CMOS monostable multivibrator. The basic principle is that of Fig. 14.27 except that the BJTs have been replaced by NOR gates. The simplified waveforms shown result from assuming an ideal gate characteristic: $v_O = 5$ V when both inputs are less than $V_{TH} = 2.5$ V, and $v_O = 0$ when either input is greater than 2.5 V. The *rest state* consists of $v_3 = 5$ V, $v_4 = 0$, and $v_2 = 5$ V. A positive pulse v_1 causes v_2 to fall, and an identical drop is coupled to v_3 by the capacitor. The circuit output, v_4, must go HIGH, and this holds G_1 in the LOW state after the trigger

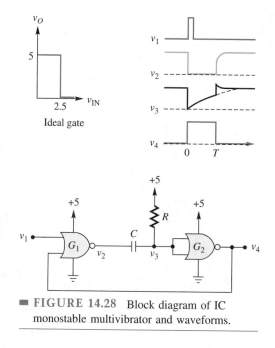

■ **FIGURE 14.28** Block diagram of IC monostable multivibrator and waveforms.

■ Again you are reminded that this expression for pulse width assumes that the gate threshold $V_{TH} = 0.5V_{DD}$, but this varies from gate to gate. Use it only as an approximation.

pulse is removed. The output pulse time is controlled by the v_3 waveform, ending when v_3 reaches 2.5 V, the threshold voltage of G_2. You should verify that if

$$v_3(t) = 5 - 5e^{-t/RC}, \qquad (14.3)$$

the gate will switch back at $T = RC \ln 2$.

Monostable multivibrators have been implemented in TTL, CMOS, and ECL technology. The period of the pulse is determined by an R and C that can be connected externally to the IC. With each circuit, the pulse width $T = kRC$, where the proportionality constant must be found from the manufacturer's specifications. The popular TTL 74121[5] and 74LS221,[5] for example, have $T = 0.69RC$ as in Eq. (14.2). The CMOS 74C221[2] has $T = RC$, and the ECL 10198[8] has $T = 1.2RC$. In each case, the relationship depends on the threshold voltages of the gates used, and is an approximation.

The IC monostable multivibrators have many variations, offering one or more of the following features. Inputs may trigger the circuit on a falling edge or on a rising edge. Some circuits have both Q (positive pulse) and $\overline{Q}$ (negative pulse) outputs. There may be an overriding CLEAR, which will terminate the pulse. A ***retriggerable*** monostable multivibrator has the capability of extending the pulse width for an additional period T each time it is retriggered. Thus, if the triggering signal occurs before the previously initiated interval has expired, the output will remain HIGH. This feature has application in the so-called ***missing-pulse detector***.

■ A missing-pulse alarm detects a break in a steady stream of pulses. The retriggerable monostable has an output pulse period greater than the time between incoming pulses, so that it is constantly being retriggered before it times out. Thus the output never goes LOW unless a pulse does not appear on time.

Astable Multivibrators

Figure 14.29 pictures an astable multivibrator consisting of two inverting gates and a timing resistor-capacitor circuit. Again assuming an ideal gate characteristic for simple analysis, the voltage waveforms are sketched in Fig. 14.30. Suppose that at time $t = 0$

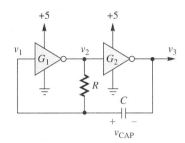

FIGURE 14.29 An astable multivibrator using CMOS gates.

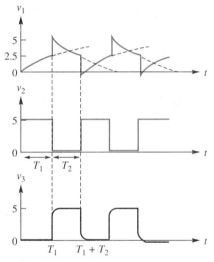

FIGURE 14.30 Voltage waveforms for the astable multivibrator.

the circuit has just been energized, C is uncharged, $v_1 = v_3 = 0$, and $v_2 = V_{CC} = 5$ V. The capacitor C charges through R toward 5 V so that

$$v_1 = 5 - 5e^{-t/RC} \tag{14.4}$$

until v_1 reaches $V_{TH} = 2.5$ V. It can be shown that this requires $T_1 = RC \ln 2 = 0.69RC$ seconds. At this point, G_1 switches and v_2 jumps to 0 V. Then G_2 switches and v_3 jumps to 5 V. Since the capacitor voltage, v_{CAP}, was 2.5 V just before the switch, v_1 could be expected to jump to 7.5 V. The gate input-protection diodes (Fig. 13.29) prevent v_1 from exceeding V_{CC} by more than a diode drop, so that there will be a rounding of the rising edge of v_3 due to the discharge of the capacitor through the resistance of the pull-up FET in G_2. The waveforms shown assume that this time constant and the propagation delays of the gates are small with respect to RC.

The timing voltage v_1 now discharges from, say, 5.5 toward 0 V, according to the exponential

$$v_1 = 5.5e^{-(t-T_1)/RC} \tag{14.5}$$

until it reaches $V_{TH} = 2.5$ V again, after which it switches back to the original state. Equation (14.5) reaches 2.5 in $T_2 = 0.79RC$. This alternate switching proceeds continu-

DRILL 14.7

Given a 5-V dc supply and two CMOS inverters having a $V_{TH} = 2.5$ V, design a circuit along the lines of Fig. 14.29 having a period of 52 μs.

ANSWER From the preceding section, RC would be 33 μs. Choose standard values, such as 0.001 μF and 33 kΩ.

ously, with a period of approximately $T = 1.58RC$. Note that the period would be a function of supply voltage, since the protective diode voltage drop would not increase proportionately with supply voltage. One method of eliminating this dependence is to insert a large resistance between v_1 and the input to gate G_1, thus preventing the clamping of v_1 by the diodes. This idea is investigated in the problems.

Gates from other logic families can be connected in a similar manner to obtain astable operation. With BJT gates, however, the input currents would affect the timing equations.

Integrated-Circuit Timers

The popular 555 IC timer[10] can be used in an astable arrangement. A block diagram of the IC and two external resistors and a capacitor are shown in Fig. 14.31. The timer itself consists of an SR latch, two comparators, a voltage divider, and two output circuits. The waveforms shown in Fig. 14.32 will facilitate an explanation of the circuit operation. Suppose that $V_{CC} = 6$ V, for example. Then each resistor in the internal voltage divider will have 2 V across it. The capacitor is assumed to be initially uncharged. Comparator 1 will be comparing 2 V on its noninverting input with v_{CAP}, and its output will be HIGH, SETting the latch. Comparator 2 will have 4 V on its inverting input and v_{CAP} on its noninverting input, and its output will be LOW. Since the latch is SET, the transistor is OFF, and the capacitor charges through R_1 and R_2 toward 6 V. At the same time, the output terminal is HIGH.

As v_{CAP} increases past 2 V, comparator 1 switches its output LOW, leaving the latch SET. When v_{CAP} reaches 4 V, comparator 2 switches its output HIGH, which causes the latch to RESET, and $\overline{Q}$ goes HIGH. The result is that the output terminal goes LOW, and R_B and β are such that the transistor saturates, pulling the node between R_1 and R_2 to 0 V. This, in turn, causes the capacitor to begin discharging toward 0 V. Comparator

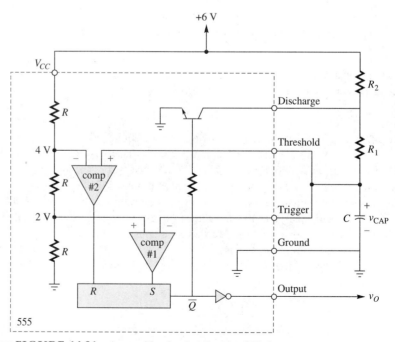

■ **FIGURE 14.31** An astable circuit using the 555 timer.

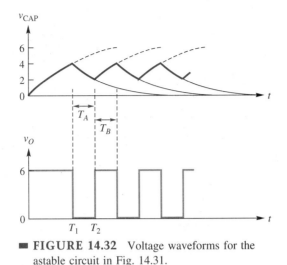
■ **FIGURE 14.32** Voltage waveforms for the astable circuit in Fig. 14.31.

2 immediately switches LOW again, but the latch remains RESET. The equation for this segment, beginning at $t = T_1$, is

$$v_{CAP} = 4e^{-(t-T_1)/R_1 C} \qquad (14.6)$$

until v_{CAP} reaches 2 V. As v_{CAP} reaches 2 V, comparator 1 SETs the latch again. At $t = T_2$, the capacitor voltage begins the trajectory

$$v_{CAP} = 6 - 4e^{-(t-T_2)/(R_1 + R_2)C}. \qquad (14.7)$$

The process continues, with the capacitor alternately charging and discharging between the comparator thresholds. The time of each interval is investigated in the problems. We can show the period of this oscillation as

$$T = 0.69(2R_1 + R_2)C. \qquad (14.8)$$

Such free-running rectangular-wave generators are used to provide periodic timing waveforms. The system clocks previously referred to in synchronous circuits are such periodic waveforms. Examples are included in the following section. In many LSI circuits, such as microprocessors, the inverters of an astable clock are incorporated into the chip design, with the R and C connected externally to determine the frequency of operation.

■ RC timing circuits have many applications, but are not considered stable enough for precision system clocks. For computer clocks, for example, quartz crystal clocks would be used.

Crystal-Controlled Clocks

Precise frequency control is realized by using a quartz crystal in the timing circuit. The circuit of Fig. 11.36(c) is used, with a MOS inverter as the gain block. Resistors R_1 and R_2 may be incorporated on the chip. Output from the gate output terminals is nearly sinusoidal and can be shaped to resemble a square wave by another inverter. Frequency stability is better than 1 part per million.

■ Clocks in present day computers are on the order of 200 MHz.

CHECK UP

1. TRUE OR FALSE? A monostable multivibrator, when triggered, generates a pulse of fixed duration.

2. **TRUE OR FALSE?** All monostable multivibrators, BJT or MOS, use single time-constant RC circuits to determine the pulse width.
3. **TRUE OR FALSE?** Astable multivibrators, when triggered, generate a rectangular wave.
4. **TRUE OR FALSE?** Astable multivibrators provide clocks for synchronous systems.

14.4 DATA SYSTEMS EXAMPLES

To illustrate the application of the IC functions described in this chapter, we next discuss a series of progressively more complex circuit designs. As a vehicle for this presentation, we assume that we wish to implement a remotely controlled display.

A Parallel Nonmultiplexed System

A block diagram for controlling a single seven-segment display by means of a 10-position single-pole switch is shown in Fig. 14.33. The encoder represents the position of the switch contact as a 4-bit data word on the lines A, B, C, and D. The decoder–driver activates the appropriate segments of the display through lines a through g.

Inputs to the encoder are normally HIGH, being connected to 5 V through pull-up resistors. One of the encoder lines is forced LOW by the switch for digits 1–9. Zero is produced when no encoder inputs are pulled LOW. The particular encoder in our example has active LOW outputs, so inverters are shown to yield a BCD representation of the selected input on lines $DCBA$, placing the least significant bit on the right. Suppose that input 7 is selected at the switch. Then the BCD word on the connecting lines is $DCBA = LHHH$.

■ Observe that segments *defg* are OFF to represent the number 7.

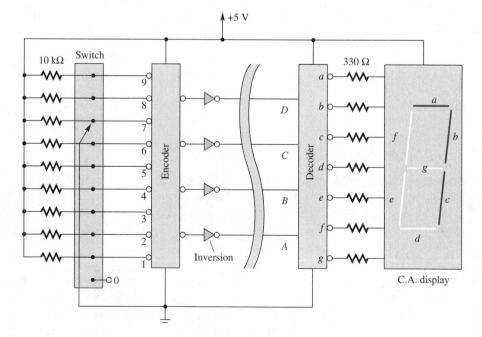

■ **FIGURE 14.33** Block diagram for digital display control.

If this were CMOS circuitry, as for example in some type of instrument system, we might use a liquid crystal display. The choice of display depends on the ambient lighting conditions, size of display, brightness desired, etc.

In this example, we have chosen a common-anode LED display. The decoder–driver must therefore provide an active LOW condition to the segments that are to be turned ON. In our illustration, segments a, b, and c will light to represent the digit 7. If we assume that our display consists of LEDs with a 1.7-V drop, then the segment current through the 330-Ω resistors will be about

$$\frac{5 - 1.7}{330} = 10 \text{ mA}.$$

To expand the dimensions of this example, you should realize that the block in our system marked "C.A. display" could just as well be a large array, where each segment is a row of incandescent lamps controlled by a solid-state relay. We will assume that the inputs to this large display system are compatible with the outputs of our decoder–driver.

Transmission Lines

As the physical distance between the display and the remote control increases, we must give some thought to the effect of long lines that connect the encoder and decoder in this case. With any of the logic families, the connections between chips must be kept short to minimize capacitance and inductance. Otherwise, the transition times are increased, and the speed of the circuit is compromised. When speed is important, as in the ECL circuits discussed in Chapter 13, a few centimeters of connection length are important. In our display system example, speed is probably not very important. After all, the switch cannot be manually changed more than once per second, and the eye cannot observe changes occurring more frequently than about 50 times per second.

■ Recall the speed of light is 30 cm/ns and nanosecond switching times are quite common in high speed systems.

More importantly in the present example, as the distance increases, the unbalanced connections shown in Fig. 14.33 become susceptible to noise. The data voltages of all of the lines in Fig. 14.33 are impressed between IC pins and ground. Noise will affect the ground lead differently from the ungrounded lead, and this connection is said to be *unbalanced*. For long leads, a balanced arrangement, or perhaps a low-impedance shielded lead, is needed.

In general, reflections are destructive to a pulse train when the line length is appreciable compared to the distance a pulse will travel in one pulse period. Transmission lines, terminated in their characteristic impedances, eliminate reflections, and thus preserve the wave shape of data signals.[11] Pulse delay and distortion for such lines are topics of some complexity and are treated in electromagnetic field theory.

Line Drivers and Receivers

The difficulties discussed above have been addressed by IC designers. In a typical **balanced line driver**, the single-ended input of the logic family concerned is converted into a balanced output signal. For example, a TTL line driver might translate an input HIGH state to a $+3$-V condition on one output lead, and a $+0.2$-V condition on the other output lead. The input LOW state would cause the output voltages to reverse. The two binary states have become balanced difference signals. The corresponding *line receiver* would accept this balanced difference signal and recover the proper HIGH or LOW single-ended output state. The key feature of this arrangement is that noise will add to both sides of the balanced line, but the differential receiver is able to discriminate against this **common-mode** noise because it is only concerned with the difference signal. This property of differential amplifiers was studied in Chapter 10. It goes without saying that both conductors of the balanced pair should be routed in close proximity so that noise signals affect both wires alike. Our example system is redrawn in Fig. 14.34 to show the four line pairs.

A typical ***unbalanced line driver*** merely buffers the logic output stage from the single-ended line by means of a high-current source and sink. Noise rejection is achieved in the unbalanced line by the shielding of coaxial lines. Typical coaxial cables have 50-Ω characteristic impedance and are terminated in a 50-Ω load. Reference has already been made in Section 13.5 to microstrip lines, which also have about 50-Ω impedance. The driver must source enough current to charge the line capacitance in the desired time and also to maintain the HIGH state at the load.

For more information on commercially available line drivers and receivers see reference 12.

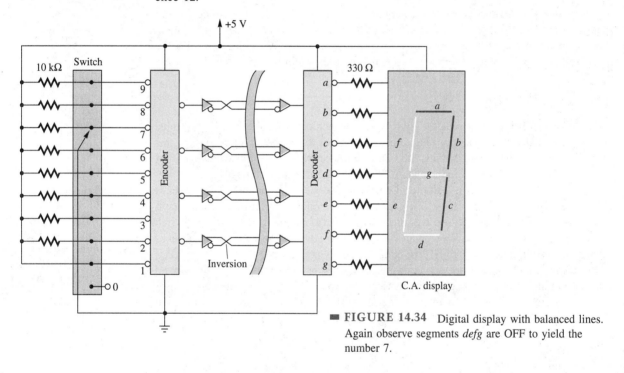

FIGURE 14.34 Digital display with balanced lines. Again observe segments *defg* are OFF to yield the number 7.

Fiber Optic Transmission Lines

Competing with the wire lines outlined above are light-conducting silicon dioxide fibers. A digital transmission system would include the same three elements: driver, channel, and receiver. The line driver would consist of a light-emitting diode or a semiconductor laser, which would produce a light wave with an intensity modulated by the digital input signal. The channel is a strand of silicon dioxide fiber, clad in an opaque sheath, and capable of light transmission for many kilometers, with an attenuation of fractions of a decibel per kilometer. The receiver is a photodiode or phototransistor. Advantages include very high *noise immunity*, very wide *bandwidth*, and very high electrical *isolation*. This technology is rapidly advancing, and no doubt will replace copper in most data circuits of more than a few meters. Telephone companies expect to replace data lines, and even the *voice-grade* lines, to the residential user with fiber in the not-too-distant future.

A Multiplexed System

For a multidigit display, such as a scoreboard, we could certainly use several of the single-digit systems discussed earlier, but the number of lines required would multiply. The cost

CHAPTER 14 SMALL DIGITAL SUBSYSTEMS

of this proliferation of conductors and associated line drivers is not justified when the digits can be multiplexed. An example of a four-digit system is shown in Fig. 14.35. Observe that the four digits require 6 connecting line pairs rather than the 16 that would be required by four setups similar to the one shown in Fig. 14.34. A practical system

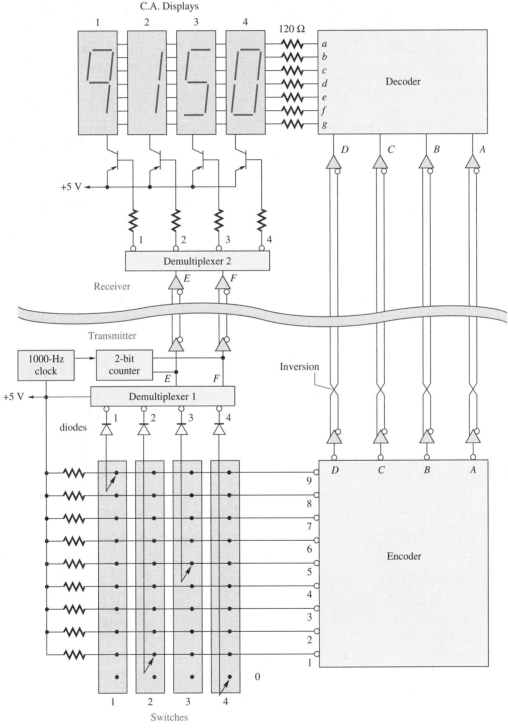

FIGURE 14.35 A four-digit multiplexed display controller.

would have many digits, but four will serve to illustrate the trade-off of some electronic complication on each end of the connecting cables, in exchange for the saving of relatively expensive conductor pairs.

A brief description of the system operation follows. The operator at the *control* end will set the four switches in accord with the digits to be displayed. Assume the number *9150*. An astable clock provides a periodic waveform to drive the multiplexing circuitry. The clock drives a four-state counter, causing lines E and F to sequence through the states 00, 01, 10, 11, 00, etc. Suppose that lines $EF = 00$. Then demultiplexer 1 will have output 1 LOW, which will pull the wiper contact of switch 1 LOW through the diode. This, in turn, causes input 9 of the encoder to be LOW, and the output on lines $DCBA = 1001$. Note that the other switches are not selected because the other outputs of the demultiplexer are HIGH and their diodes are OFF. At the receiver, $EF = 00$ also, and demultiplexer 2 will select display 1 by pulling output 1 LOW, and turning on the *pnp* transistor. All other outputs are HIGH, and this causes the common anodes of displays 2, 3, and 4 to be OFF. Thus, all segments in these displays are OFF. The 1001 at the input of the decoder thus causes a *9* to be shown on display 1.

■ The output of the encoder is actually $\overline{DCBA}$, but is inverted by the line transmission system.

As the clock causes the counter to advance, the other switches are read, and the other displays are activated in sequence. In this example, each display would be active only 25% of the time. This would be no problem for an LED display, provided that we chose our clock frequency high enough so that the eye does not perceive the flicker. We chose a clock rate of 1000 Hz, which means that each switch is scanned 250 times per second and each display digit is refreshed 250 times per second, easily fast enough to appear constant to the eye. In addition, it is necessary to increase the current in the segments while they are conducting in order to maintain the apparent brightness. The duty cycle of our example is 25%, but the LED current does not need to be quadrupled to maintain the same apparent brightness. In this case, the segment resistors are chosen to yield a current of about 25 mA. If all segments are ON (for the figure *8*), the current in the saturated *pnp* transistor would be $7 \times 25 = 175$ mA.

■ In either of these examples, the saving of 10 line pairs may easily justify the cost of the multiplexing electronics.

An alternate receiver circuit for this system is shown in Fig. 14.36. This arrangement could be used for displays that must be ON continuously. When digit 1 is transmitted, the receiver demultiplexer output 1 goes LOW, as in the other receiver. The falling edge of the demultiplexer triggers a monostable multivibrator. The LOW pulse from the monostable enables latch 1 long enough to load the BCD representation of digit 1. The monostable times out before the clock advances to digit 2, and the rising edge of the pulse latches the data. All of the digit latches are in parallel across the incoming $DCBA$ lines, but only one of them is enabled at a time. The digit stored in each latch is displayed continuously, so the multiplex clock rate can be very slow without any display flicker. Each digit has a 4-bit transparent latch, a BCD-to-seven-segment decoder, and a display driver all in a single IC.

A Serial Data Transmission System

To illustrate a simple *serial* data system, we will show the 4-digit display system example again, this time with the data being transmitted by means of a single pair. Figure 14.37 is a block diagram of such a system. We first give an overview of the system operation and then a description of the timing details.

A 1000-Hz clock and six-stage binary counter chain provide the timing at the transmitter. Each switch is scanned for 16 ms so the complete cycle requires 64 ms. During the first 8 ms of each scan, the 2-bit address of the switch, EF, and the 4-bit BCD representation of the setting, $DCBA$, are parallel loaded into a shift register. Note that bit h is always forced HIGH, and bit g is forced LOW. Then during the next 8 ms the register is shifted

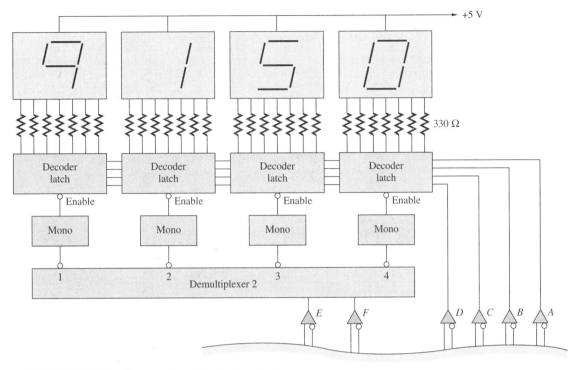

■ **FIGURE 14.36** Alternate four-digit display circuit.

right eight times, ending with the register filled with 1's since the serial input is HIGH.

The idle state on the line to the receiver is always HIGH as a result of the transmitter register bits all being HIGH after the shift, and bit h always being loaded as a HIGH. This is vital to the receiver timing because after the 8-ms idle time since the last data word, the receiver will regard the first HIGH-to-LOW transition as the beginning of a new data word. This is the reason for always loading a LOW in bit position g.

When bit g reaches the receiver, a 7-ms monostable is triggered, which enables the receiver clock. This clock operates at 16 kHz, but since it is divided by 16, the shift register is shifted exactly seven times in 7 ms. Now the data word is located in the receiver register exactly as it had been in the transmitter register before the transmission began. The address bits, EF, which tell which switch has been scanned, are connected to the demultiplexer inputs. The BCD data bits are connected to the inputs of all of the latches. The falling edge of the 7-ms monostable triggers a second monostable of 100 μs. The $\overline{Q}$ output enables the demultiplexer for 100 μs, and this latches the data into the latch corresponding to the address EF. The Q output RESETs the counter, and now the receiver waits for the next transmission. Note that the three latches that were not addressed were not enabled by the demultiplexer and do not respond to this transmission. Their displays are not changed. Of course, the one that was addressed does not change either, unless its switch had been changed since the previous scan.

This data transmission scheme is an example of *start–stop timing*. In this case, the transmitter sends one 8-bit word at the synchronous rate of 1000 bits per second and then pauses, leaving the line in the HIGH state. The receiver senses the start of a new word because a new word always begins with the LOW start bit. The receiver then operates synchronously at 1000 bits per second until it has received the word. Note that the system is independent of the delay in the line. The receiver must operate at the same data rate

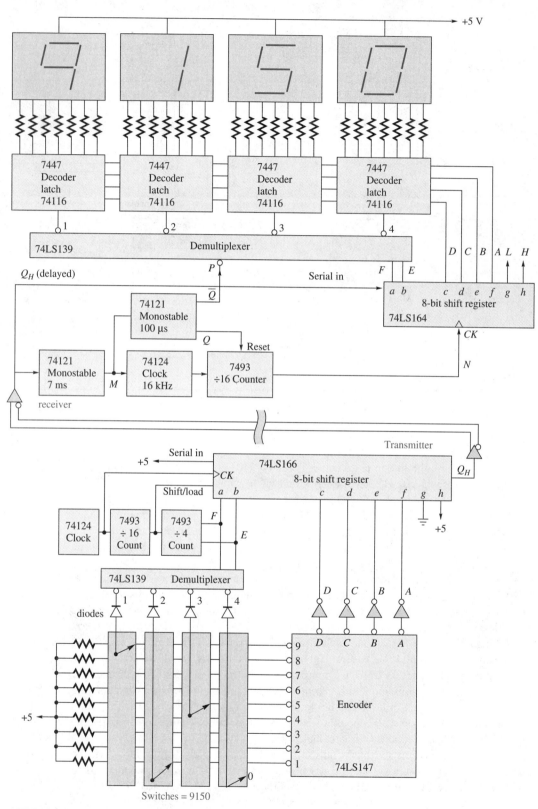

FIGURE 14.37 Diagram of the serial display control system.

as the transmitter and must know the word length. Timing is not critical if the word is not too long.

To help you understand the timing considerations in this example, a timing diagram is shown in Fig. 14.38, with waveforms at principal nodes. Clock rates are low enough that propagation delays are of no consequence. The transmitter 1-kHz clock and counter chain operate continuously. The waveforms shown assume that E and F have just switched LOW, so that switch 1 is driven by the demultiplexer. The shift/load line is LOW, so that the shift register is being parallel loaded once each clock pulse. Of course, it is not necessary to load eight successive times, but the connections are simpler. The example assumes that the BCD number 9 (1001) is loaded into register positions c, d, e, f, and switch 1 address 00 is loaded into a, b. The LOW start bit is loaded into input g, and a HIGH is entered at input h so that output Q_H does not make a premature transition. Recall that the register was left with all positions HIGH after eight right shifts because the serial input is HIGH. The line (Q_H) must not go LOW while we are loading the register, or the receiver will assume that a new transmission is beginning.

After 8 ms, the shift/load line goes HIGH, and now the rising edges of the clock cause exactly eight right shifts. The first shift pushes the LOW start bit onto the line, and when

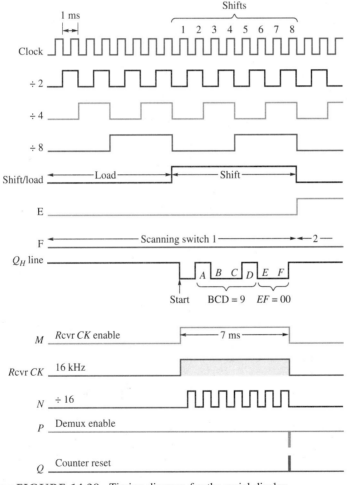

■ **FIGURE 14.38** Timing diagram for the serial display control.

it gets to the receiver, the receiver clock is enabled. After eight counts, the 4-bit counter output goes HIGH, shifting the receiver register. This is a 0.5-ms delay, ensuring that the receiver shifts occur midway between possible data transition times on the line. After seven shifts at the transmitter, the *a* bit (address bit *F*) has moved to the line, and after seven shifts at the receiver, that same bit is in position *a* in the receiver register. Now the 7-ms monostable times out, and another monostable generates the LOW pulse required to enable the demultiplexer. Output 1 pulses LOW because of the 00 address at inputs *EF*, and the data 1001 is loaded into latch 1.

These examples show applications of the various MSI circuits we have surveyed. The examples also suggest the possible trade-offs of adding some additional digital circuitry at the ends of the data channel in order to reduce the number of channels required. This is usually the proper design strategy when the distances exceed several hundred feet (until the data rates exceed the capacity of the channel).

■ The serial example saves 15 line pairs compared to the non-multiplexed case.

CHECK UP

1. **TRUE OR FALSE?** A balanced line driver is required to connect a digital circuit to a coaxial transmission line.
2. **TRUE OR FALSE?** One function of a line driver is to provide sufficient current to quickly charge the capacitance of the transmission line.
3. **TRUE OR FALSE?** A balanced line and a differential line receiver allow rejection of common-mode noise.
4. **TRUE OR FALSE?** Fiber optic lines offer high data rates and good noise immunity, but have very high loss.
5. **TRUE OR FALSE?** The cost of multiple lines in data systems motivates the use of multiplexing circuits to reduce the number of lines needed.

SUMMARY

The basic logic circuits presented in Chapter 13 are organized into a wide variety of combinational and sequential logic functions. These logic families are available in a number of technologies such as TTL, CMOS, and ECL. Here we discussed several key building blocks common to logic systems from a functional viewpoint, independent of the fabrication technology.

Today's circuit designer may never use these functional blocks in the form of MSI ICs; rather, these blocks may be implemented in a VLSI ASIC, either by custom interconnectioni of an array of logic modules at the top metallization layer, or by programming the interconnection in some type of field-programmable logic device. We pursue these options further in Chapter 15.

We presented circuit symbols, function tables, and logic diagrams for multiplexers, demultiplexers, decoders, encoders, arithmetic-logic units, latches of various types, registers, and counters. To explain the operation of these logic building blocks, we used timing diagrams as well as typical design applications in small systems.

Timing circuits, including monostable and astable multivibrators, were considered. Monostable multivibrators are used for generating pulses in digital circuits, and astable multivibrators serve to provide clocks for driving digital systems.

A very few centimeters of separation between logic circuits may be significant with rise and fall times on the order of 1 ns. However, proper termination can eliminate

■ The modern IC system designer would use a computer-aided design (CAD) program that would model the system function as opposed to the internal circuit detail. Thus our MSI design is presented only to help the reader sort out function and timing details at a relatively elemental level.

SURVEY QUESTIONS

1. Compare combinational logic circuits with sequential logic circuits.
2. What is the purpose of multiplexing digital signals? Give advantages and disadvantages.
3. Explain the function of a decoder–driver IC.
4. Sketch the connection of an active-high decoder–driver to a segment in a common cathode seven-segment LED display.
5. What is a priority encoder?
6. Give the level needed at each input of the ALU of Table 14.4 in order to realize the operation 5 plus 6.
7. Describe the operation of a JK flip-flop by means of a truth table.
8. Define a D flip-flop.
9. Using a 4-bit shift register, draw a block diagram of a circuit that will accept a 4-bit data word and transmit it serially, most significant bit first. List the control signals required.
10. Compare a ripple counter with a synchronous counter.
11. What is the principal application of a monostable multivibrator? An astable multivibrator?
12. Compare a coaxial transmission line with a balanced transmission line. What are the relative advantages of each?

PROBLEMS

Problems marked with an asterisk are more challenging.

14.1 The four-channel multiplex system in Fig. 14.3 requires three lines between multiplexer and demultiplexer. Develop an expression for the minimum number of connecting lines for an n-channel system.

14.2 Draw a diagram showing how two of the multiplexers shown in Fig. 14.1 and two of the demultiplexers shown in Fig. 14.2 could be used to multiplex 16 channels. How many SELECT lines are required?

14.3 Calculate the segment currents in the display of Fig. 14.35 when they are ON.

14.4 A large display is made up of rows of incandescent bulbs constituting the segments in a seven-segment display (similar to Fig. 14.5). The display consists of two digits: The left one has only segments b and c, and represents only a 1 when it is ON, and a *blank* when it is OFF. The right one, with all seven segments, can represent all digits from 0 to 9, and *blank*. Thus the array can represent all numbers from 0 to 19, and *blank*. Assume all switching is done by power semiconductors at the display site. How many bits are required to control this display from some remote location if no sequential logic is used?

14.5 A baseball park uses 46 of the 2 digit displays of Problem 14.4 for a scoreboard. Twenty of them are used to show runs for each team in each inning (up to a maximum of 20 innings), and up to a maximum of 19 runs per inning. The other 6 displays show total runs, hits, and errors for each team.

(a) How many lines are required to control this scoreboard if no multiplexing is used?
(b) How many lines are required if each of the 46 displays has a binary address, and the numerical data are transmitted to only one address at a time using multiplexing? Note that this requires a latch at each display, so consider how the latch will be strobed.

14.6 The display of Problem 14.4 has been redesigned, since the user is willing to set the display by incrementing a counter at each display. Explain the trade-offs that are made with this change.

14.7 The circuit shown in Fig. 14.7 uses the encoder shown in Table 14.3. What will be the output code when switches 4, 5, and 6 are all closed at the same time?

14.8 (a) Design, by drawing a functional block diagram, an electronic lock system, using three of the encoders shown in Fig. 14.7 and a few logic gates. Let the secret combination be 5–6–7. Assume that there is no problem with the user having to activate three switches at once. In practice, such a circuit will time out quickly to make each try independent.

(b) How many possible combinations are there in your lock?

(c) What would have to be done to change the combination?

14.9 Give the state of every input and output for the ALU shown in Fig. 14.9 (see Table 14.4) for each of the following cases:

(a) Adding $A = 5$ to $B = 3$ with no carry input.
(b) Subtracting $B = 3$ from $A = 5$ with no borrow input.
(c) Subtracting $B = 5$ from $A = 3$ with no borrow input.
(d) Subtracting $B = 5$ from $A = 3$ with a borrow input.

14.10 Two two-input NOR gates are cross-connected in the same manner as the NAND gates shown in Fig. 14.11. Develop the truth table.

14.11 Consider the switch-debounce circuit of Fig. 14.12. Draw a similar circuit using an SR latch instead of the $\overline{S}\,\overline{R}$ latch. Sketch the corresponding waveforms.

14.12* Investigate the "ONE's catching effect" of the master–slave flip-flop shown in Fig. 14.16. Suppose $J = K = Q_n = 0$. The clock is HIGH. Let J temporarily go HIGH while the clock is HIGH and then return LOW. Draw the waveforms at the major nodes to show that this $J = 1$ condition is trapped by the master and inevitably passed on to the slave when the clock eventually goes LOW.

14.13 Suppose that a 4-bit decoder is connected to the ripple counter shown in Fig. 14.24.

(a) If the propagation delay of each flip-flop is 20 ns, what is the maximum input frequency that allows the decoder to decode properly every state of the counter?

(b) If the decoder is fast enough, what false states will it produce as the counter ripples from state 15 to state $16 = 0$?

14.14 Suppose that a 4-bit decoder is connected to the synchronous counter shown in Fig. 14.26. If the propagation delay of each flip-flop is 20 ns and the delay of each gate is 10 ns, what is the maximum input frequency that allows the decoder to decode properly every state of the counter?

14.15 Sketch a diagram of a circuit that uses a 4-bit binary counter that will count from 0000 to 1111, and then hold until your circuit is cleared. Assume a square-wave clock signal. You will need to add some additional gates.

Problems 14.16 to 14.23 deal with the design of simple digital clocks using standard counter logic ICs.

14.16 Given a small power supply, where the transformer secondary has a voltage of $10\sin(377t)$ volts, and the dc output is $+5$ V. Draw a simple diode circuit that will provide a 60-Hz timing waveform that is compatible with logic inputs (-0.5 V $< v_o <$ 5.5 V).

14.17 Using two decade counters of the type shown in Example 14.4(a), plus a logic gate if needed, design a divide-by-60 counter circuit. The idea is to accept a 60 pulse/second pulse train, such as that generated in Problem 14.16, and convert it to a 1 pulse/second waveform.

14.18 Another divide-by-60 counter, as in Problem 14.17, will convert the 1-Hz waveform of that circuit into a 1 pulse/minute waveform. Now design a 60-second timer circuit by connecting a 2-digit display to this counter, and cascading the circuits of Problems 14.16, 14.17, and 14.18. One suggestion for a BCD display is shown in Fig. 14.6.

14.19 Continuing the problem begun in Problems 14.16–14.18, add logic to start the timer, to stop the timer, and to reset the timer. One push-button switch should quickly reset the counters, and then start the count, and another push-button switch should stop the count, holding it in the counter. Consider debouncing the switch contacts.

14.20 The circuit requested in Problem 14.18 has an output of 1 pulse/minute. Adding another stage of divide-by-60 with its display will yield a minute count, and the output will be 1 pulse/hour. Now design a divide-by-12 counter to display hours. Use components as suggested in Problem 14.17. The combination of all the stages of Problems 14.16–14.20 constitutes a complete digital clock, slaved to the 60-Hz power-line frequency.

14.21 Show how the 12-hour counter of Problem 14.20 could be changed to a 24-hour counter. Draw a diagram of a circuit that will either count to 12 or to 24, depending on the position of one single-pole double-throw switch.

14.22 Suggest a method for quickly setting the clock designed in the preceding problems. This should involve no more than adding a "set minutes" and a "set hours" switch.

14.23 There are several CMOS 14-bit binary counter chips (divide-by-2^{14}). Each stage of the counter chain has its own output after the first four stages. Show how these ICs could be used with a 33.55-MHz computer clock signal to develop a 1-Hz (approximately) output signal. How much error would result?

14.24 Sketch a diagram of a circuit that uses an 8-bit parallel-load shift register that will accept 6 bits of parallel data, $B_1B_2B_3B_4B_5B_6$, and shift it out as an 8-bit word $0B_1B_2B_3B_4B_5B_6 1$ (in the order given). Tell what control signals are needed.

14.25 Repeat Problem 14.24, using an 8-input multiplexer rather than the shift register.

14.26 The monostable circuit shown in Fig. 14.27 has transistors with $\beta_F = 50$. Choose R and C such that the quasi-stable state period will be 1 ms.

14.27 The monostable circuit of Fig. 14.39 uses CMOS NOR gates with a threshold voltage of $V_{TH} = 0.5V_{DD}$. Choose R and C such that the quasi-stable state period will be 1 ms.

14.28 Suppose the monostable circuit of Fig. 14.39 uses CMOS NOR gates with a nonideal threshold voltage $V_{TH} = 0.4V_{DD}$. Derive an expression for the output pulse time T in terms of RC.

14.29 Consider the monostable circuit of Fig. 14.39. Let the CMOS NOR gates be replaced by NAND gates, and let the resistor be connected to ground instead of to V_{DD}. If the steady-state voltage v_1 is normally HIGH, show that this circuit will be triggered to a quasi-stable state by a narrow LOW pulse at v_1. Find the period of the output pulse in terms of RC, assuming that $V_{TH} = 0.5V_{DD}$.

14.30 The two-transistor circuit shown in Fig. 14.40 is astable. Assume Q_1 comes ON and saturates at $t = 0$. Show that Q_2 will immediately turn OFF and come back ON after $0.69R_2C_2$. Then Q_1 will switch OFF for $0.69R_1C_1$ seconds. Explain the circuit operation.

14.31 Design a CMOS astable circuit of the type shown in Fig. 14.41, specifying R and C to obtain a frequency of 100 kHz. Let $V_{TH} = 2.5$ V, $V_{DD} = 5$ V, and the input-protection diode drop be 0.5 V.

14.32 As discussed, the circuit of Figure 14.41 has a period that is somewhat dependent on V_{DD}. Let $V_{TH} = 7.5$ V and $V_{DD} = 15$ V, and derive the period T in terms of RC. Assume the input-protecting diode drops are 0.5 V, as before.

14.33* Refer to the CMOS astable multivibrator of Fig. 14.41. Sometimes a large resistance is inserted in series with the input to G_1 to prevent the input-protective diodes from clamping

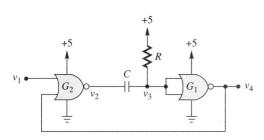

■ **FIGURE 14.39** Block diagram of IC monostable multivibrator and waveforms.

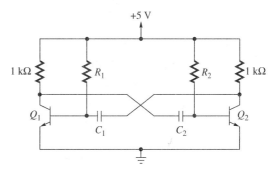

■ **FIGURE 14.40**

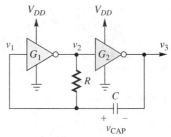

■ **FIGURE 14.41** An astable multivibrator using CMOS gates.

the voltage v_1. Assume that this resistance is 100 times the value of the R of the figure, so that the current into the gate is negligible. Now derive the new value for the astable period T. Expect about a 40% increase!

14.34 Refer to the 555 timer circuit in Fig. 14.31.

(a) Find the interval T_A, during which time C is discharging and $v_o = 0$.
(b) Find the interval T_B, during which time C is charging and $v_o = 6$ V.

14.35 Design an oscillator circuit using the 555 timer having a period of 1 second. Specify standard values for R and C, and an appropriate value of variable resistance for approximately a ±5% adjustment.

14.36 Refer to Fig. 14.31 and the results of Problem 14.34. The results show that the way to get $T_A = T_B$ (a square wave), is to make $R_2 = 0$. But this is not a good idea, because with $R_2 = 0$ there is nothing to limit the current through the transistor when it conducts, and it soon overheats, destroying itself! Show that one can obtain an approximate square wave by modifying the circuit as follows: Connect a diode across R_1, pointing downward, and then making $R_1 = R_2$.

14.37 The 555 timer can be used in the arrangement shown in Fig. 14.42 to realize a monostable circuit. The stable state is with v_o LOW and the transistor saturated. The quasi-stable state is initiated by causing the TRIGGER input to go negative with respect to 2 V, thereby SETting the latch. The output will pulse HIGH until $V_{CAP} = 4$ V and then return LOW. It is necessary that the trigger pulse be narrower than the output pulse. Derive an expression for the output pulse width.

14.38 Suppose the display system shown in Fig. 14.33 has connecting lines 10 feet long, with 20 pF/foot capacitance to ground (assume each line is twisted together with a ground lead). The inverters that drive these lines are standard TTL with 130-Ω source resistance when their output goes HIGH. Assume the decoder inputs must reach 2 V to be considered HIGH. What will be the delay in this line when the output attempts to go HIGH? Is this delay of any practical significance? (*Hint:* For a crude estimate of the time, assume all capacitance is lumped at the receiver.)

14.39 Suppose that the unbalanced system described in Problem 14.38 picks up so much induced noise that it is totally unusable. Experimentally it was demonstrated that the noise can be reduced to an acceptable level by connecting a 0.1-μF capacitor across the decoder input. What is the delay now?

14.40 Refer to the multiplexed display shown in Fig. 14.35. Suppose that the LED segments have a voltage drop of 1.7 V. The decoder and demultiplexer outputs are 0.2 V when active, and the *pnp* transistors have a β_F of 50. What maximum value of resistor should be used in the transistor base lead to ensure that the transistor saturates? Consider all possible digits.

14.41 A football scoreboard requires at least 16 BCD digits to display scores, time, quarter, down, possession, yards-to-go, and time-outs remaining.

(a) How many lines are required between controller and display for separate data and address?
(b) For serial transmission, how many bits should be sent in each packet?

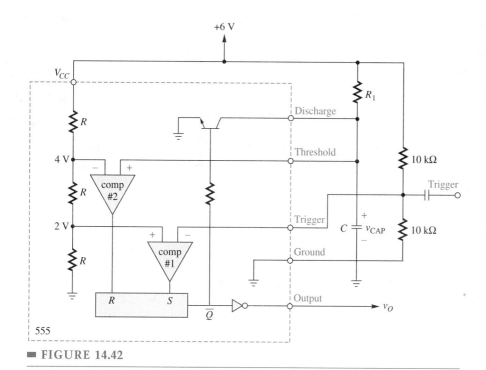

■ FIGURE 14.42

14.42* Sketch a block diagram for the serial display system shown in Fig. 14.37 using an 8-bit multiplexer instead of the 8-bit shift register at the transmitter. Show all connections to the multiplexer SELECT lines. Is it necessary to modify the receiver in any way?

REFERENCES

1. Bell, D. A. *Electronic Devices & Circuits.* Reston, Va.: Reston Publishing, 1980, pp. 401–404.
2. *CMOS Logic Databook.* National Semiconductor, Santa Clara, California, 1988.
3. *High-Speed CMOS Logic Data Book.* Texas Instruments, Inc., Dallas, 1984.
4. *FACT Advanced CMOS Logic Databook.* National Semiconductor, Santa Clara, California, 1990.
5. *The TTL Logic Data Book.* Texas Instruments, Dallas, 1988.
6. *ALS/AS Logic Data Book.* National Semiconductor, Santa Clara, California, 1990.
7. *MECL Device Data.* Motorola, Inc., Phoenix, 1988.
8. *F100K ECL Logic Databook and Design Guide.* National Semiconductor, Santa Clara, California, 1990.
9. Wakerly, J. F. *Digital Design Principles and Practices* 2nd ed. Englewood Cliffs, N.J.: Prentice-Hall, 1994, Chap. 6.
10. *Linear Applications Databook.* National Semiconductor, Santa Clara, California, 1990.
11. Cheng, D. *Field and Wave Electromagnetics*, 2d ed. New York: Addison-Wesley, 1989, Chap. 9.
12. *Semiconductor Master Selection Guide.* National Semiconductor, Santa Clara, California, 1989. Section 5.

SUGGESTED READINGS

The manufacturers listed in the preceding references, and many others, publish revised databooks periodically. These, along with their *Master Selections Guide*, which are sometimes updated annually, allow the user to stay abreast of new products and the availability of older devices. In addition, many of these vendors provide data in computer-archive format (magnetic disk or CD-ROM and on-line format).

CHAPTER 15

SEMICONDUCTOR MEMORIES

15.1 Overview of Semiconductor Memories
15.2 Introduction to Memory Organization
15.3 Read-Only Memories
15.4 Static Read/Write Random-Access Memories
15.5 Dynamic Random-Access Memory
15.6 Charge-Coupled Devices
15.7 Gate Arrays and Programmable Logic
15.8 Summary
Survey Questions
Problems
References
Suggested Reading

A means of storing information is essential to any computing system. Information can be numbers to be manipulated, the names of customers, program instructions to tell the central processing unit what to do, or a myriad of other types of data. Such data can be stored permanently outside the computer on magnetic media such as diskettes (disks), tape, or bubble memories. Compact discs, read optically, have extremely dense data storage capabilities. The relative permanence of such storage is offset by the long access time. Semiconductors are predominantly used for the high-speed memory inside the computer.

Semiconductor memories are presented in this chapter not only because of their great importance in processing systems but also as the principal application of large-scale integration (LSI)—more than 100 gates/chip—and very large-scale integration—more than 1000 gates/chip or 10,000 transistors/chip and ultra-large scale integration with millions of transistors/chip. Thus semiconductor memories are on the frontier of fabrication technology. Indeed, semiconductor memory technology has become so diverse that we will refer to the tree shown in Fig. 15.1 just to categorize the major applications. In nearly every case, the technology will be MOS.

Because of their similarity to *read-only memories* (ROMs), *programmable logic arrays* (PLAs) and *programmable array logic* (PAL) are discussed here also. Finally, we survey the development of *field-programmable gate arrays* (FPGAs). Each of these devices allows the user to achieve ASICs in the field at reasonable cost, because the many users result in high volume to the fabricator.

IMPORTANT CONCEPTS IN THIS CHAPTER

- Random-access memories include read/write and read-only.
- Dynamic R/W RAMs are very compact and low cost, but also very volatile.
- Static R/W RAMs are fast and are not volatile as long as dc power is maintained, but cost more and use more power than dynamic memories.
- Field-programmable ROMs are either permanently programmed or erasable.
- Erasable PROMs are either erased by UV radiation (FAMOS) or are electrically alterable (FLOTOX).
- FLOTOX memory cells are programmed and altered by Fowler-Nordheim tunneling.
- Nonvolatile RAM combines a static RAM with an EEPROM.
- Static R/W memory requires up to six transistors per cell.
- Dynamic R/W memory consists of one transistor and a capacitor for each cell.
- Pseudostatic or integrated RAM has refresh circuitry on the memory chip.
- Charge-coupled devices store data as charge packets on the surface of a silicon substrate, separated from a positive plate by a thin oxide layer.
- Gate arrays combine economies of scale with custom interconnections.
- PLAs and PAL are large-scale integrated circuits where certain interconnections are field programmable.

15.1 OVERVIEW OF SEMICONDUCTOR MEMORIES

Semiconductor memories are arranged in arrays of cells, where each cell stores one *bit* of information (1 or 0). Because computer operation is binary, these arrays are usually modulo-2, such as 256 bits, 1024 bits, or 65,536 bits. Many computers handle data in groups of 8 bits, which are termed a *byte*. A group of 1024 bits is called a *kilobit* (Kbit), and thus 8192 bits could be organized as a *kilobyte* (Kbyte). Thus, common usage has taken K ≡ 1024 when in a digital context. Likewise, Mbit implies 1,048,576 bits.

We can observe from Fig. 15.1 that memories are classified according to whether the information is stored serially, and can only be accessed that way, or stored in such a manner that it can be accessed in any order. Tapes, floppy disks, compact discs and bubble memories are examples of nonsemiconductor *serial memory*, in which the information is stored in sequence and the time required to access data depends on where it is in the record. By contrast, *random-access memory* (RAM) implies that the time required to retrieve data is independent of its location.

Serial Memories

Shift registers are an example of a serial memory. Data bits can be shifted sequentially into a shift register and then recovered by shifting them out the far end. This is known as a *first-in/first-out* (FIFO) memory. The most obvious application is as a data buffer, where the sequential nature of the data is preserved, but the rate is to be changed. We

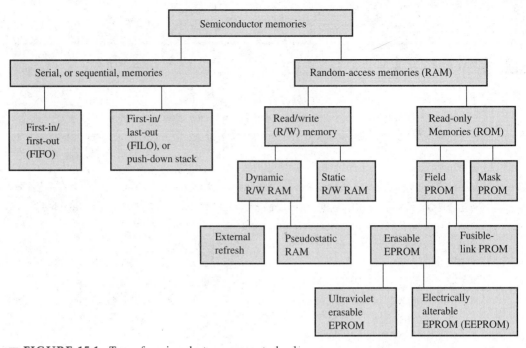

FIGURE 15.1 Tree of semiconductor memory technology.

can see an example of this when a computer sends data to a printer. The computer rapidly transmits data to the buffer and then proceeds with some other task while the relatively slow printer empties the buffer. There are many applications for this type of memory in data transmission systems for error control and encryption.

When the data are shifted into the register and then withdrawn from the input end of the register by reversing the shift direction, it is called *first-in/last-out* (FILO) memory, or a *push-down stack*. This arrangement is sometimes useful within a computer for saving data temporarily in a "nested" operation.

In applications where a serial memory can be used, it has an advantage over RAM in that the chip can be packaged very compactly, requiring very few leads. A 4Kbit RAM, for example, needs 12 address leads, whereas a 4Kbit serial memory uses only one clock lead. MOS technology is widely used for LSI registers because the devices are very compact and low in power consumption. Because MOS technology is well represented in our study of RAM, in this chapter we consider a different serial memory technology, the *charge-coupled device* (CCD).

Random-Access Memories

As stated, RAM is used for the internal memory in computers because of its low *access time*; that is, the processor can retrieve the desired data quickly from any known location in memory without sequencing through other data. RAM may be classified as either *read/write* (R/W) or *read-only memory* (ROM). Read/write would seem to be a descriptive term, implying the ability to have data written into memory by the processor and the ability to allow the data to be read from memory by the processor. However, some confusion results from the fact that R/W memory is often called RAM, which is a less specific term. Read-only memory is used to store data permanently, and indicates that the processor can only read these data, not modify them.

■ Common usage is to refer to R/W random-access memory as RAM, even though ROM is often random access also. The CD ROM, of course, is serial.

Read/Write Memories

We have alluded to the fact that R/W memory is used for the high-speed main memory of computers. Data will be moved from relatively slow external serial bulk storage into the fast internal R/W memory in blocks, where it can then be accessed randomly. A characteristic of R/W memories is *volatility*; that is, the data are lost when the power is removed from the chip. Some R/W memories are backed up with batteries to avoid loss of data due to unexpected power failure.

■ Virtually all personal computers have 5-year+ lifetime lithium batteries buried in the circuit board for this purpose.

Read/write memories are realized in all of the technologies we have previously discussed. The bipolar TTL and ECL memories are faster than MOS, but they pay the attendant price of higher power consumption. Also, NMOS has a much higher circuit density, allowing for VLSI economies. CMOS has the advantage of extremely low standby power. A technology that has a good compromise of speed and circuit density is IIL, but CMOS and BiCMOS appear to be more compatible with existing fabrication technology and embedded circuits. Here we focus our study on NMOS and CMOS, the dominant technologies as of this writing.

MOS R/W memories divide into two groups: static and dynamic. *Static* memory cells are flip-flops that require up to six transistors per cell. By contrast, *dynamic* memory cells consist of a capacitor and one transistor. Thus, dynamic memories are very compact and lend themselves to efficient VLSI and very high circuit density. The result is that dynamic MOS memories have the lowest cost per bit of the current technologies: 16 Mbit (16M $\times$ 1 bit or 4M $\times$ 4 bit) are currently available.[1] The disadvantage of dynamic memories is that since each bit is stored as a voltage on a very small capacitor, the charge tends to decay with time. This requires the system to read the data and restore the charge every few milliseconds. This process of restoring the cell voltage is termed *refresh*. Dynamic memories are usually the economic choice above 64K bytes. We will study some of the details of dynamic memories in Section 15.5.

■ 64Mbit and higher are on the way.

Static MOS R/W RAMs are very fast, and although they require more power and cost more per bit than dynamic memories for large memories, they are normally used for memories of 8K bytes or fewer because they do not require the extra circuitry for refresh. Many static memories have a standby mode in which data will be retained but cannot be accessed. This reduces power consumption considerably. Static MOS memories are considered in some detail in Section 15.4.

An alternative to dynamic R/W memory with separate refresh circuitry is the *pseudo-static RAM*, or *integrated RAM*. This is a dynamic MOS RAM with the refresh circuitry integrated onto the same substrate so that the chip appears as a static RAM at the terminals. The external circuit must know when the RAM is refreshing, however, because access is denied during refresh. The advantages of dynamic circuit density are retained at a moderate increase in cost. Application may be in the range between 8K and 256K bytes, where the choice of dynamic versus static memories is a less obvious trade-off.

Read-Only Memories

Here, the desired bit pattern can be *programmed* into the cells permanently so that the memory becomes nonvolatile. The ROM is said to be *mask programmed* when this is done by the manufacturer during fabrication. The user must provide the manufacturer with the desired bit pattern. This is not the economic choice unless the quantity of identical ROMs needed justifies the setup cost.

Another approach is to provide the user with the means to *program* the ROM. This could be done by addressing each cell where a 0 is to be stored and applying sufficient current to melt a *fusible link*. This procedure would be irreversible. When the user wants the programmable memory to be capable of being *erased* and reprogrammed, another technology is used. The MOS cells are programmed by trapping electrons on a floating gate, thus blocking the cell by removing the inversion channel connecting the source to the drain. For erasure, a window is provided in the package to allow ultraviolet radiation to reach the chip, exciting the electrons to an energy level at which they can escape from the gate. After all cells have been erased the EPROM may be reprogrammed.

Finally, devices that are read like an ordinary ROM but can be *electrically erased* are available. This means that the system can reprogram a cell without physical removal of the chip from the circuit, and without erasing the entire chip.

15.2 INTRODUCTION TO MEMORY ORGANIZATION

Before taking up the details of the various memory technologies themselves, we will consider the usual environment in which memory devices are called on to operate. First, we distinguish two levels of organization: chip interconnection within the memory system and cell layout within the integrated circuit (IC). Then we specify the general timing requirements of a memory system.

Memory System Interconnection

Figure 15.2 illustrates how semiconductor memories connect to a typical computer system. Shown is a microprocessor that uses both ROM and RAM, as well as external memory that may be accessed through the input/output (I/O) ports. The ICs are connected by *buses*, a shorthand representation of a number of parallel lines. The processor specifies the location it wishes to access by impressing address information (binary voltage levels) on the *address bus*. The figure shows a 16-bit address bus, which implies that the processor is capable of addressing $2^{16} = 65,536$ locations in the system. Some of these locations may be in ROM, some in RAM, and others in external devices accessed through the I/O ports. The processor also specifies whether it expects to READ the addressed location or WRITE to it by means of HIGH or LOW levels on the R/W line.

The *data bus*, shown in the figure as eight bits wide, is a bidirectional bus on which the processor receives a byte of data from the addressed location or sends a byte to it. We note that each block in the system that has access to the data bus must have a three-state output and be in the high-impedance state unless it is addressed. If this were not so, there would be conflicts and *contention* for the data bus. Also, it is necessary for each

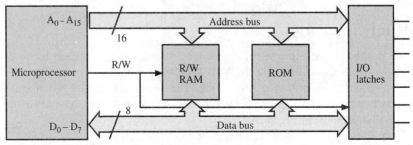

■ **FIGURE 15.2** Semiconductor memory in a typical computer system.

R/W block to interpret the R/W line and connect the memory input or output to the data bus. This is usually done on the memory IC, but sometimes it is done externally by means of a *transmission gate*.

As an aid to the proper understanding of the various control functions of memory ICs, consider the more detailed example of a memory system that is shown in Fig. 15.3. In this illustration, 4K bytes of ROM are obtained with a 4K × 8-bit EPROM. The given address decoder provides for expansion beyond what is shown. It is customary to define all possible addresses, known as *address space*, with a *memory map*. It is also customary to represent the binary addresses in the shorthand **hexadecimal** numbering system, or base 16, defined as follows:

Hex	Binary	Hex	Binary	Hex	Binary	Hex	Binary
0	0000	4	0100	8	1000	C	1100
1	0001	5	0101	9	1001	D	1101
2	0010	6	0110	A	1010	E	1111
3	0011	7	0111	B	1011	F	1111

Appendix A contains additional information on number systems. The 12 low-order address bits, A_0 through A_{11}, represent all binary numbers from 0000 0000 0000 (000_{16}) to 1111 1111 1111 (FFF_{16}). Going back to our example of Fig. 15.3, we see that when the four high-order address bits into the decoder are 0000 (0_{16}), the eight R/W chips are selected. When the high-order address bits are 0001 (1_{16}), the ROM chip is enabled. Note that the decoder is active LOW, and the *chip selects* ($\overline{CS}$) are active LOW. The complete memory map is

- 0000_{16} to $0FFF_{16}$ 4K bytes of R/W RAM
- 1000_{16} to $1FFF_{16}$ 4K bytes of ROM
- 2000_{16} to $3FFF_{16}$ Decoded for expansion (8K)
- 4000_{16} to $FFFF_{16}$ Not decoded.

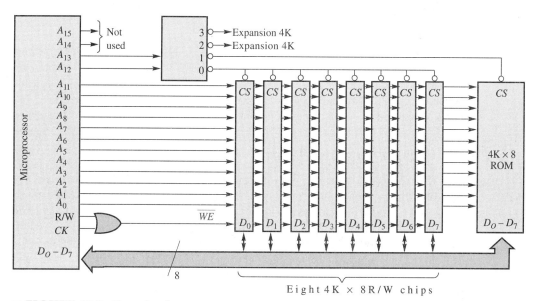

■ **FIGURE 15.3** Example of a memory system.

DRILL 15.1

Express the hex number $FB52_H$ in binary.

ANSWER
1111101101010010

EXAMPLE 15.1

Express the binary number 11011010001 in hexadecimal.

SOLUTION First, arrange the number in groups of four (from the right), adding leading zeroes if necessary. Then decode: $\underbrace{0110}_{6} \quad \underbrace{1101}_{D} \quad \underbrace{0001}_{1}$ = 6D1.

Pin Connections to Memory Chips

The CPU in our example system of Fig. 15.3 is an *8-bit* microprocessor. This means that it processes data in 8-bit groups or *words*. Since the R/W chips chosen are arranged in 4096 1-bit words, eight chips in parallel must be used to obtain 4096 words of the required 8-bit length. The ROM chosen was arranged in 4096 8-bit words, so only one chip was needed for 4096 system words. In general, each memory chip must have n address leads to allow access to 2^n words, unless the leads are multiplexed in some manner.

Note that each chip must have at least one *chip select* (or $\overline{CS}$ as shown) to allow blocks of memory to be addressed without conflicts from the other blocks. Some chips have leads with the same function called *chip enable (CE)*, and others have several **CE** or $\overline{CE}$ with on-chip decoder logic.

Every R/W RAM chip must have some method of controlling whether the memory is to output data to the data bus or to accept data from the bus. Our example has a $\overline{write\ enable}$ ($\overline{WE}$); other chips may have a lead designated **R/W**. We will have more to say about this lead when we discuss timing.

The ROM has eight *output data* leads, D_0 to D_7, in accord with the word length. We should point out that ROMs normally have the same word width as the system data bus has, since the ROM is programmed word by word, not bit by bit. R/W RAM chips have *input data* and output data leads, or else combined *input/output* leads, in accord with the chip word width. In our example, each chip has one data lead. If the input data and output data pins had been separate, they could have been connected at the data bus because the output is disabled internally by the $\overline{WE}$ lead when it is LOW. Some chips have a separate *output disable (OD)* lead.

Of course, each chip has power supply leads. The present trend is to a single power-supply voltage, usually +5 V, sometimes as low as 3.3 V. New technology is moving toward lower voltages to conserve power, reducing the heat produced. Additional leads are required for EPROMs, and we will discuss certain other functions as the need arises, but we have considered those pin connections that are common to most RAM and ROM chips.

Chip Organization and Timing

Figure 15.4 shows the block diagram of a 4096 × 1-bit static RAM, which is fairly typical of memory organization on a chip. The 4096 memory cells are arrayed in 64 columns and 64 rows. Six address bits select a row, and six address bits select a column in the matrix. The cell at the intersection of the selected row and column is connected to the I/O circuits. Only in the READ mode, when $\overline{CS}$ is LOW and $\overline{WE}$ is HIGH, does the data from the cell appear at the output. Only in the WRITE mode, when both $\overline{CS}$ and $\overline{WE}$ are LOW, will data at the input be transferred to the cell. The output lead is in the high-impedance state unless in the READ mode, and input and output pins may be tied together for a bidirectional bus. With some RAM chips $\overline{CS}$ HIGH not only disables the

■ Most ROMs are arranged in 8-bit words because programming is usually done in bytes. Data buses are in multiples of 8 bits.

■ 4096 bits is a small size, as modern memory chips range up to 16 Mbits, but this example serves our purpose without becoming pedagogically unwieldy.

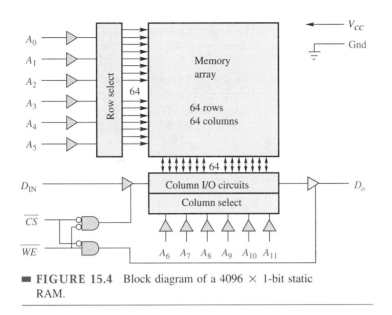

FIGURE 15.4 Block diagram of a 4096 × 1-bit static RAM.

I/O circuits but also turns OFF all decoders and amplifiers. This STANDBY mode reduces power by about 90% while preserving the data.

The same size RAM arranged as 1024 × 4 bits requires only ten address pins, but four I/O pins. In this case, input and output leads are tied together internally to conserve the number of leads for the IC package.

A mask-programmed ROM would have similar chip organization, with the obvious difference that no $\overline{WE}$ lead (or input data leads) is needed. As mentioned previously, ROMs are usually organized in 8-bit words. EPROMs need additional programming circuitry.

We will find that as the number of cells per chip gets larger, the need to hold down the number of package leads (and bus leads) suggests the multiplexing of address bits on fewer pins. This is done on larger dynamic RAMs.

Memory Timing Requirements

The following discussion of memory timing is general in nature, and the designer should always consult the manufacturer's data sheet for precise timing data for specific ICs. The waveforms shown in Fig. 15.5 relate to memory READ cycles. The upper

DRILL 15.2

A certain read-only memory is organized 65,536 × 8 bits. What is the minimum number of pins that this IC package can have?

ANSWER 27

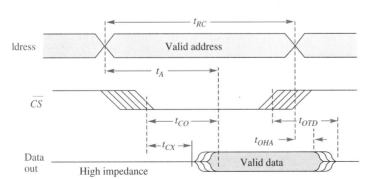

FIGURE 15.5 Timing waveforms for a READ cycle.

waveform shows a composite of the voltages on the address bus. The **READ-cycle time** t_{RC} is assumed to be the time from when the slowest address lead stabilizes until the first one changes for the next cycle. Such transitions are controlled by the processor and usually occur at fixed intervals determined by the system clock. The second waveform is that of the $\overline{CS}$ lead. The multiple transitions shown in the figure indicate the uncertainty of this timing, since $\overline{CS}$ could be just another address line, or it could be delayed by the delay of an address decoder. It is assumed that the $\overline{WE}$ lead is HIGH throughout the READ cycle.

The lower waveform shows the memory output to the data bus. The waveform midway between HIGH and LOW is the high-impedance state of the output buffer. The **access time** T_A is the time from stable address to the time valid data appear at the output, assuming that $\overline{CS}$ has enabled the output buffer by that time. If not, the time from chip selection to valid output is t_{CO}, and the time from chip select to active output is t_{CX}. Valid data remain at the output after addresses change again for time t_{OHA}, and the time from $\overline{CS}$ HIGH until the output reaches the high-impedance state is t_{OTD}.

We now examine WRITE-cycle timing with the aid of the waveforms shown in Fig. 15.6. The **WRITE-cycle time** t_{WC} is shown on the uppermost waveform, designating the valid address interval. **WRITE time** t_W is the time at which $\overline{CS}$ and $\overline{WE}$ are simultaneously LOW. The fourth waveform shows that the output goes to the high-impedance state T_{OTW} after $\overline{WE}$ goes LOW. From the lower waveform, we see that valid data must appear at the data input for a **data setup time** t_{DW} before the end of the WRITE pulse and be held for a **data hold time** t_{DH} after the WRITE pulse.

Note that the lead that rises first, $\overline{CS}$ or $\overline{WE}$, terminates the WRITE pulse and latches the data into the memory. This rising edge must occur t_{WR} before the addresses change again, which often requires the system designer to modify the R/W line from the processor, as shown in Fig. 15.3. This is because many processors change the R/W line during or after the addresses are changed. Unless the WRITE pulse is terminated before that time, the data might not be latched into memory soon enough.

Our discussion of memory timing has been quite general, but it applies broadly to many static RAMs, and the READ timing can be applied to ROMs. Dynamic RAMs will require further discussion of REFRESH timing.

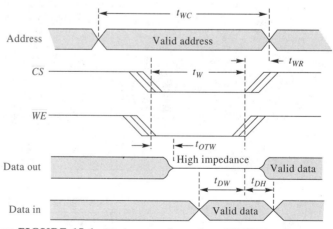

■ **FIGURE 15.6** Timing waveforms for a WRITE cycle.

CHECK UP

1. **TRUE OR FALSE?** Large serial memories must have very many pins per package.
2. **TRUE OR FALSE?** Static memories preserve their data even when the dc is removed.
3. **TRUE OR FALSE?** Read-only memories are always random access.
4. **TRUE OR FALSE?** During a write cycle, address and data must be held until after the end of the write pulse.

15.3 READ-ONLY MEMORIES

We have seen that read-only memories (ROM) are used to store fixed information and have the advantage of being nonvolatile. They are used in digital systems design to store constants, conversion tables, and program instructions. Computers have instructions stored in ROM that allow them to transfer further instructions into the R/W (or RAM) memory from external sources such as disks. ROMs are conveniently used for table look-up (for example, often-used math functions when computation time is of the essence). Another example is code conversion, such as BCD to seven-segment display. The cartridges for home video games are essentially MOS ROM chips.

Mask-Programmed ROM Cells

A simplified schematic of one MOS ROM is shown in Fig. 15.7. Address lines are decoded to select a word line. The selected word line is HIGH and enhances the *n*-channel MOSFETs that are shown with gates. These FETs, in turn, pull the corresponding bit

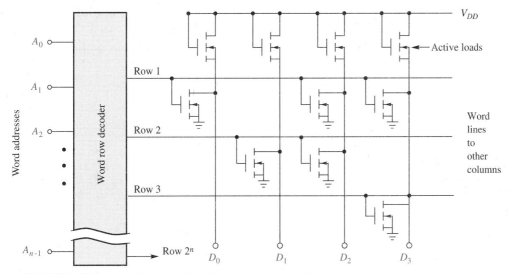

■ **FIGURE 15.7** Simplified NMOS ROM showing three words × four bits.

lines LOW, presenting a logical 0 to the output terminals. In effect, the manufacturer programs the ROM during fabrication by forming an FET in those cells where a 0 is desired, and omitting the field-effect transistor (FET) from those cells where a 1 is to be programmed. Figure 15.8 illustrates how regular and compact this technology can be. The columns are n-diffused drains connected by metal stripes. The n-diffused sources are connected to ground. In those cell positions where the manufacturer wants a FET, a thin oxide layer is formed so that the polysilicon rows act as an effective gate over the region. Another alternative is to make a p implant in those cells where no FET is desired, effectively raising the threshold voltage so that no channel can form in the cells.

■ Fabrication is discussed in more detail in Chapter 18.

The organization shown in Fig. 15.7 can also be realized with BJTs. Figure 15.9(a) depicts BJT cells with the base connected to the word row for those sites where a 1 is to reside, and the base left unconnected in those cells destined to be 0. The programming is done by the manufacturer at that metallization layer step.

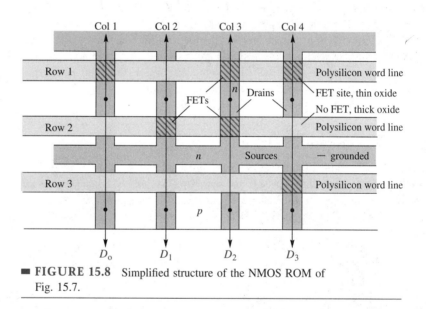

■ **FIGURE 15.8** Simplified structure of the NMOS ROM of Fig. 15.7.

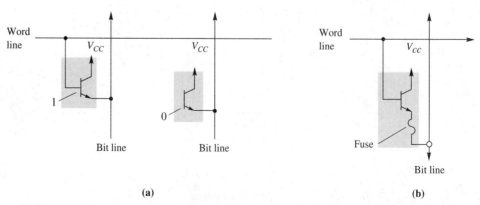

■ **FIGURE 15.9** BJT ROM cells. (a) Mask programmed. (b) Fusible link.

Programmable ROMs (PROMs)

We should note that in producing either bipolar or MOS ROMs, the manufacturer will attempt to do all of the custom programming on the chip in one of the fabrication steps, such as the metallization step. All other steps are the same for each type of ROM to keep production costs down. The ROM is said to be *mask programmed* since a custom mask is required. Because the setup charge is still very expensive, it is not economical to use a mask-programmed ROM unless at least 1000 identical units are required. There is also an inevitable delay in obtaining custom ICs.

User-programmable ROMs (PROMs) are an alternative for smaller quantities. These are also called *field-programmable* ROMs because the manufacturer does not do the programming. A typical cell is shown in Fig. 15.9(b), which reveals that each cell consists of a BJT with a fuse in the emitter lead. All cells are 1 with the fuses intact; where a 0 is desired, the fuse must be blown. The fuses are made of polysilicon, nichrome, or titanium-tungsten, and are designed to open when a large programming current is applied. Bipolar devices of this sort are available with 32×8 cells to 1024×8 cells, with access times from 35 to 75 ns, and power dissipation from 350 to 550 mW.[2]

Another type of user-programmable ROM is a device identical to the erasible PROMs that are described in the next section, except that the package has no window so that ultraviolet light cannot be used to erase the programming. Examples are CMOS "one-time programmable" ROMs available from $2K \times 8$ to $128K \times 8$ cells, with access times of 150 to 500 ns and dissipation of 50 mW (active) and 10 μW (standby).[3]

Erasable Programmable ROMs (EPROMs)

The fusible-link programming process is obviously irreversible. Often it is desirable to be able to modify a PROM that has been programmed. An example of this is likely to occur during product development, as prototype systems are being updated. The *erasable PROM* is designed to meet this need. The MOSFET cell in Fig. 15.7 is replaced by a *floating-gate transistor* or *stacked-gate cell* as shown in Fig. 15.10(a). The basic structure of this cell is diagrammed in Fig. 15.10(b). The *n*-channel MOSFET has two polysilicon gates. The first, or *floating gate*, is left unconnected. The second gate, or *select gate*, is operated to enhance the channel as in the usual MOSFET. Before being programmed, there is no charge on the floating gate, and the device displays the typical enhancement MOSFET characteristic shown by curve (a) in Fig. 15.11. The threshold voltage is low

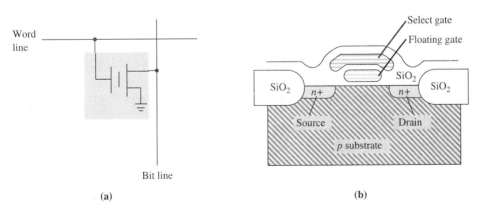

■ **FIGURE 15.10** Floating-gate transistor. **(a)** EPROM. **(b)** Basic structure.

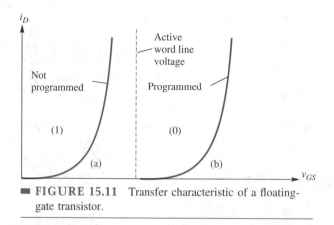

FIGURE 15.11 Transfer characteristic of a floating-gate transistor.

in this case; a HIGH condition on the word line will turn ON the cell, and the column line will go LOW. This LOW on the column line is interpreted as a 1 at the column output selector.

To program the cell, a high voltage (12 to 25 V, depending on the technology) is applied to the drain and to the select gate through the address circuitry. A channel is established, and the MOSFET operates beyond pinch-off. Electrons are accelerated in the high field between source and drain, and they acquire enough energy to enter the conduction band of the gate oxide layer. There they are attracted by the positive potential on the select gate, and many of them lodge on the floating gate. This process is self-limiting because electrons on the floating gate begin to inhibit further electron injection. The physical mechanism has led to the name *floating-avalanche MOS* (**FAMOS**) *cell.*

When the programming voltage is removed, the electrons on the floating gate are trapped, and thus the floating gate constitutes a negative charge between the select gate and the substrate. The effect is that a much higher positive voltage is required on the select gate to enhance the channel than before, moving the device characteristic to curve (b) in Fig. 15.11. The new threshold voltage is greater than the voltage applied by the row decoder, and the transistor does not come ON when addressed. The cell is now programmed to 0. Data retention time is over 10 years.

Illuminating the cell with ultraviolet (UV) light (about 2537 Å) will return it to the unprogrammed state. The EPROM package is equipped with a quartz window for this purpose. The UV light imparts photon energy to the trapped electrons, allowing them to escape through the oxide layer. All cells in the EPROM would be erased at once. Manufacturers recommend exposure of about 20 min at typical dosage. We should note that sunlight and fluorescent lamps contain some UV, and it is estimated that one week of direct sunlight or three years of room-level fluorescent lighting would erase some cells; therefore, it is suggested that the window be covered with an opaque label.

The first generation of EPROMs uses NMOS technology for the row and column drivers and I/O circuitry.[4] They operate on 5 V and use 12.5 V (or 20 V) for programming. Access time is typically 200 to 500 ns, and power is 500 mW active, 200 mW standby, with available sizes from 1K × 8 to 128 × 8 bits. Programming requires special equipment; each byte is programmed separately, requiring 50 ms. To operate in any circuit where they are to be used as a ROM, these ICs are compatible with either TTL or CMOS families.

As of this writing, CMOS is the technology of choice, yielding higher speed and lower power. CMOS EPROMs operate as detailed earlier, except that access time is typically 100 to 250 ns, and power is 100 mW active, 1 to 5 mW standby; programming one byte requires 100 μs.[1,3,4] In addition to sizes up to 128K × 8 bits, these EPROMS are available in word widths of 16 bits. Refer to Figure 15.27 at the end of this chapter for excerpts

■ The CMOS EPROM uses the same topology as the NMOS array of Fig. 15.7. The row and column drivers would be CMOS rather than NMOS. Of course, the memory cells are FAMOS for EPROMs.

from a typical EPROM, the TMS27C010A, which is a 128K × 8 CMOS UV erasable PROM. Problems 15.11 through 15.20 offer good insight into the characteristics of the EPROM in general, and the CMOS EPROM in particular.

■ Now is the best time to consult the data sheet of Fig. 15.27 and answer the short questions of Problems 15.11 through 15.20.

Electrically Erasable PROMs (EEPROMs)

For some applications, the requirement of ultraviolet light to erase the EPROM and the requirement of erasing the entire IC to change one byte are a great disadvantage. It is often desirable to be able to modify a single byte in the ROM without removing it from the circuit. The *floating-gate tunnel-oxide (FLOTOX) cell* shown in Fig. 15.12(a) has been developed to meet this application. The difference in structure from that of the FAMOS cell is an extremely thin oxide (200-Å) layer between a section of the floating gate and the FET drain. The floating gate is charged or discharged by *Fowler-Nordheim tunneling*.[5] Basically, when the electric field across the thin oxide exceeds 10^7 V/cm, electrons from the negative electrode can pass a short distance through the forbidden gap and enter the conduction band of the insulator. They can then flow freely to the positive electrode. This current flow is extremely dependent on voltage.

An early memory cell that used the FLOTOX device in an NMOS EEPROM is shown in Fig. 15.12(b). To program a word, electrons are trapped on the floating gates of the cells in the word as follows: A positive 20 V is applied to the word select line and the program line. The bit columns are grounded. Thus for each cell the top MOSFET holds the drain of the FLOTOX FET at ground potential, and electrons from this drain tunnel to the floating gate because of the concentration of the electric field at the thin oxide. After this preprogramming (erasure), the cells can be written into by connecting the column bit lines to 20 V (for a 0), or 0 V (for a 1), and then grounding the program line and activating the word select line. This will cause electrons from those floating gates where 0's are to be stored to tunnel back to the drain. The ERASE and the WRITE operations require 9 ms.

For a normal READ operation, both row select and program lines are HIGH, and those cells whose electrons are stored on the floating gate will remain OFF, and those bit lines will remain HIGH as in the FAMOS cell. READ access time is about 250 ns as in other MOS memories. Note that because two FETs are used per cell, the density of these memories is much lower than the EPROMs discussed earlier.

DRILL 15.3

Estimate the total time that would be required for an automated programming station to program an entire 128K × 8 CMOS EPROM after erasure.

ANSWER 13.1 sec

■ The floating-gate nonvolatile memory has also been demonstrated using amorphous silicon thin-film transistors.[25] These FGTFTs are fabricated on insulating substrates.

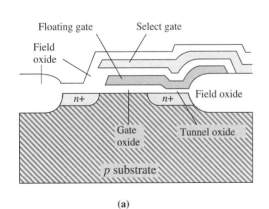

(a)

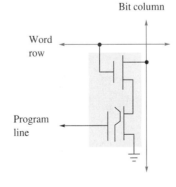

(b)

■ **FIGURE 15.12** FLOTOX cell. (a) Structure. (b) Connection in ROM.

The Intel 2816A 2K × 8-bit EEPROM[6] is an example of this technology. This chip is capable of 5-V operation, generating programming voltage and pulse shaping on the chip. It may be erased and reprogrammed on a byte basis by TTL-compatible voltages, or the entire chip may be erased in 9 ms by means of an external 10 to 15 V.

An alternative technology to achieve EEPROMs is the textured poly cell. The cross-sectional structure is shown in Fig. 15.13(a), and a single cell acts like three FETs in series, as shown in Fig. 15.13(b). Very small memory cell size is achieved by exploiting the vertical integration of the three polysilicon layers to form a merged-gate single-transistor cell. The WRITE operation proceeds as follows:

1. The poly 1 line (common to the entire array) is brought LOW, isolating the cell from ground.
2. The bit line is set up to either 0 V (for the erased state) or to 16 V (for the programmed state).
3. The poly 3 word line is ramped up to 22 V in 1 ms.

For the case where the bit line is 0 V, the channel under poly 2 is 0 V, floating-gate poly

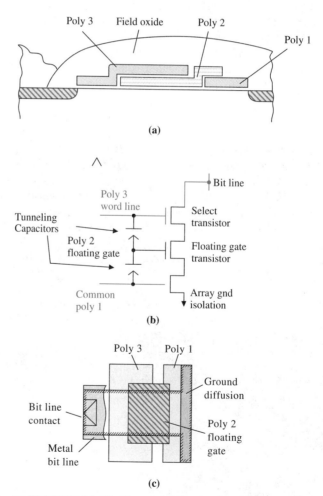

FIGURE 15.13 Textured Poly EEPROM cell.
(a) Cell cross section. (b) Cell equivalent circuit.
(c) Cell top view.

2 is capacitively pulled toward 0 V, and the field concentrates across the poly 2–3 thin oxide layer, causing any electrons that might be trapped on poly 2 to tunnel off to poly 3.

For the case where the bit line is HIGH, the channel under poly 2 capacitively pulls poly 2 positively, and the field concentrates across the poly 2–1 thin oxide layer, causing an injection of electrons onto the floating gate.

For a given field strength, the textured poly silicon allows a thicker oxide for the same tunnel current. The cell is read in the same way as the FAMOS is read.

The Xicor X2816A[7] is an example of this technology, with a read access time of 250 ns and programming time of <10 ms. Latches store the word to be written, so that the write pulse width is only 150 ns, and the IC automatically finishes the program timing. On-chip circuits derive the high voltages needed from the 5-V supply (more on this concept later).

Flash EEPROMs

The challenge to produce more compact, faster, and lower power EEPROMs has resulted in the marriage of the FAMOS and the FLOTOX devices into a single FET memory cell, developed with CMOS technology.[8] The *flash* cell programs as does the EPROM, and erases as does the EEPROM cell. Consider the cell structure in Fig. 15.14(a). This cell appears very similar to the FAMOS cell of Fig. 15.10(b), except that the oxide layer between the floating gate and the FET source is extremely thin, 100 Å. Beginning with all cells erased, programming of one word (usually a byte) is done by forcing the select gates to +12 V, and the drains of those bits that are to become 0's are connected by the column decoders to about +7 V. These high voltages are generated on the chip from the single +5-V supply. As with the FAMOS cell, the accelerating electrons in the channel produce a flood of electrons, and the field from channel to select gate causes some of them to lodge on the floating gate, where they are trapped, and thus the cell is programmed to logic 0. This is accomplished in <10 μs. A data word to be written into the flash memory is entered into a temporary latch at a normal data bus write-time, where it is held until the word has been programmed.

Erasure (removing the charge from the floating gate) is accomplished when +12 V. is applied to the FET source with the select gate grounded and the FET drain open so that the electrons trapped on the floating gate tunnel back to the FET source [Fig. 15.14(b)]. Entire arrays (or at least entire blocks of cells) are erased at once, in about 10 ms, although byte-by-byte verify will run the total erase time up to 1 to 2 sec. The speed of erasure and programming has given rise to the name *flash* memories. Many variations of this

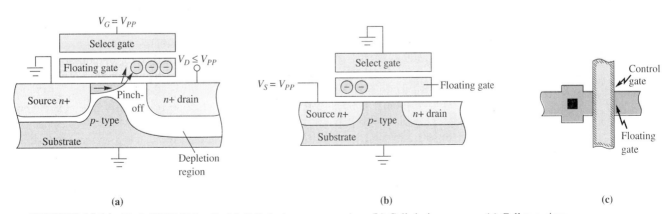

■ **FIGURE 15.14** Flash EEPROM cell. (a) Cell during programming. (b) Cell during erasure. (c) Cell top view.

■ You are reminded that the flash EEPROM is CMOS technology, so all row and column drivers would be CMOS.

DRILL 15.4

Using the information given regarding programming speeds, how long would it take the system to completely program a 256K × 8 flash ROM?

ANSWER 2.6 sec

---

technology exist among manufacturers; for a comparison see Ref. 9. Some technologies guarantee 1000 program-erase cycles, others over 100,000 cycles.

Nonvolatile RAM

The EEPROM cell has another application, in the so-called **nonvolatile RAM**. This chip operates as a R/W static RAM, with each cell backed up by an EEPROM cell. The application is for very rapid saving of large amounts of data at power failure. One example of this memory is the Xicor X2004 512 × 8 bit NOVRAM.[10] System operation is as follows: The RAM is used by the processor in the normal fashion, with read and write times typical of MOS memories (less than 300 ns). Should the need arise, a 200-ns pulse to the STORE lead will cause all of the data in RAM to be replicated in the EEPROM cells. The store operation requires 10 ms, so if the need to save the data was a failure of the primary power, it will be necessary to maintain 5 V at this chip for 10 ms after the STORE pulse is given. Upon restoration of power, all data can be recalled from EEPROM to the RAM in 10 μs by applying a 400-ns pulse to the RECALL lead.

CHECK UP

1. **TRUE OR FALSE?** ROMS using the FAMOS cell may not be reprogrammed cell by cell, but the entire array must be reprogrammed in order to change one cell.
2. **TRUE OR FALSE?** Tunneling is used both to program and to erase the FLOTOX cell.
3. **TRUE OR FALSE?** Tunneling is used to program the cells in a flash EEPROM because tunneling is faster than the avalanche process.
4. **TRUE OR FALSE?** A nonvolatile RAM consists of a static RAM and an EEPROM.

15.4 STATIC READ/WRITE RANDOM-ACCESS MEMORIES

Each memory cell in a static read/write RAM (SRAM) is a latch, and data are retained as long as power is maintained to the chip. These RAMs are realized with MOS technology such as NMOS or CMOS, or with bipolar technology such as TTL or ECL. Bipolar RAMs are relatively fast, having access times of 10 to 100 ns. Power dissipation is also high, typically 0.1 to 1.0 mW/bit. By contrast, the NMOS RAM described next, the 2147H,[11] has an access time of 35 ns, an active power dissipation of 200 μW/bit, and standby power dissipation of 25 μW/bit. A CMOS version of the static RAM has a typical access time of 100 ns, an active power that depends on switching speed, and a standby dissipation of 1 nW/bit.[12] The combination of high circuit density, low power dissipation, and reasonable access time has led to the dominance of MOS technology in the manufacture of RAM, and BiCMOS is replacing the bipolar arrays, since it encompasses the best of both technologies. In addition, SRAMs with BiCMOS I/O drivers are available that are compatible with ECL systems, with access times of 5 ns.[13] SRAMs are widely used as high-speed cache memories in personal computers.

Static RAM Chip Organization

As an example of a fairly typical NMOS SRAM, consider again Fig. 15.4, the block diagram of an Intel 2147,[11] which has 4096 1-bit words. The memory cells are arrayed in 64 rows and 64 columns. The lower order six address bits are decoded to select one of 64 rows, and the upper order six address bits are decoded to select the column. Additional logic controls the flow of data to or from the selected cell. When $\overline{CS}$ is HIGH, all of the address buffers are disconnected from power, which essentially turns OFF all transistors on the chip except the active transistor in each memory cell. This standby mode reduces the power dissipation of the chip by about 90%.

■ Again, this 4 Kbit SRAM is relatively small. SRAM chips range from 256 bits to 512 Kbits.

When $\overline{CS}$ is LOW, so that the chip has been enabled, the output data leads remain in the high-impedance state unless $\overline{WE}$ is HIGH. Thus, both input and output data leads may be tied to the data bus without conflict. We now briefly consider the type of circuits used in each of the functional blocks.

Static MOS Memory Cell

Figure 15.15 shows a standard NMOS six-transistor cell, where M_1 and M_2 are depletion-mode active loads for the cross-coupled latch, M_3 and M_4, and M_5 and M_6 serve as bidirectional transmission gates to couple the cell to the column data lines when enabled by the row selects. Note that a CMOS cell would be identical in operation, with M_1 and M_2 replaced by p-channel enhancement-mode MOSFETs. You are advised to work Problem 15.22.

■ The best way to visualize the CMOS SRAM is to sketch the CMOS version of Fig. 15.15, Problem 15.22.

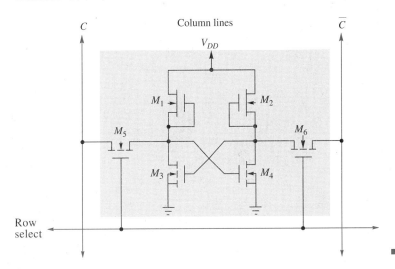

■ **FIGURE 15.15** Static MOS memory cell.

■ **EXAMPLE 15.2**

Show that M_3 OFF and M_4 ON is a stable state for the latch shown in Fig. 15.15. Assume $k_L = 0.01$ mA/V^2 and $V_{tL} = -1$ V for M_1 and M_2, and $k_D = 0.2$ mA/V^2 and $V_{tD} = 1$ V for M_3 and M_4. $V_{DD} = 5$ V.

SOLUTION If M_3 is OFF, There will be no voltage drop in M_1 and $V_{G4} = 5$ V. Assuming that V_{G3} is LOW, M_2 will be saturated with $I_D = k_L(V_{GS2} - V_{tL})^2 = 0.01(0 + 1)^2 = 10$ μA. M_4 will be in the linear region with $I_D = k_D[2(V_{GS4} - V_{tD})V_{G3} - V_{G3}^2] = 10$ μA. This quadratic yields $V_{G3} = 20$ mV, so that the original assumption that M_3 is OFF is verified. By symmetry, M_4 OFF and M_3 ON is also a stable state.

DRILL 15.5

Refer to Example 15.2. What is the static power dissipation of this memory per cell?

ANSWER 50 μW

DRILL 15.6

What is the advantage of the CMOS cell over the NMOS cell?

ANSWER The steady-state power dissipation is near zero.

■ Note that all FETs in the circuit of Fig. 15.16 have the same substrate, which is at ground potential. To simplify the schematic, the substrate-to-ground connection is not shown at every FET.

The READ operation for the cell is as follows: The row pertaining to the cell having been selected, M_5 and M_6 will be ON. Suppose that the cell contains a binary 0; that is M_3 is ON and M_4 is OFF. Then column data line C will be pulled LOW since its current will flow to ground through M_5 and M_3. Column data line $\overline{C}$ will be pulled HIGH as M_2 and M_6 source current to $\overline{C}$ from V_{CC}. At the same time, the column select decoder will have connected this column's data sense lines to the chip data sense lines, and the $\overline{WE}$ will have enabled the output data driver. We should recognize at once that the time required by the cell transistors to charge and discharge the capacitance of the column data lines C and $\overline{C}$ constitutes much of the access time of the RAM.

To picture the WRITE operation, assume that the column selectors have gated a binary 1 onto this column's data lines; that is, C is HIGH, and $\overline{C}$ is LOW. When the row selectors have enabled M_5 and M_6, M_6 will pull V_{G3} down to ground, turning OFF M_3 if it is ON, and leaving it OFF if it had been in that state. If necessary, M_5 and M_1 now pull V_{G4} HIGH, and the latch is left with M_3 OFF and M_4 ON, which represents the 1 state for the cell. Again, a minimum WRITE pulse time is required because of the necessity for the input data buffer to charge the capacitance of the data lines C and $\overline{C}$.

Column Input/Output Circuits

The method by which the flow of data to and from the column sense circuitry is controlled is best seen by referring to Fig. 15.16, which shows the diagram for one of the n columns.

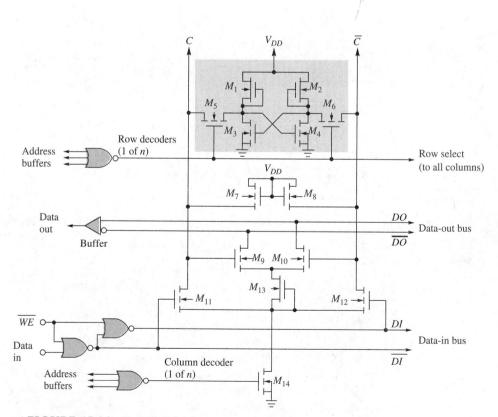

■ **FIGURE 15.16** Static RAM column sense circuitry.

M_7 and M_8 serve as active pull-ups for the column sense lines C and $\overline{C}$, charging the line to $V_{DD} - V_{tL}$, or about 3.5 V when the chip is deselected. When the chip is selected and one of the rows is enabled, C and $\overline{C}$ take on the state of the cell. For example, if the cell shown is in the 1 state, M_4 is ON and M_3 is OFF. If this row is selected, M_4 will pull $\overline{C}$ LOW through M_6. To a lesser extent, M_1 will pull C slightly higher through M_5.

When this column is selected, the source of M_{12} is grounded through M_{14}. M_{13} is a depletion-mode device that serves as the source resistor for the source-coupled pair M_9 and M_{10}. The active loads for this differential amplifier are part of the data output buffer, which serves whichever column is activated.

When $\overline{WE}$ is LOW, the input data state is applied to the gate of M_{12} and its complement to the gate of M_{11}. Thus, a 1 input would turn ON M_{12} and pull V_{G3} LOW through M_6, leaving the cell in the 1 state.

Address Buffer

Figure 15.17 shows an address buffer of the type used in the Intel 2147H. As we shall see, each address line must drive a large number of decoders, so buffering is required. In addition, the complement of each address state is also used, to minimize decoding logic. ENABLE signal E, and its complement $\overline{E}$, are derived from $\overline{CHIP\ SELECT}$ such that they respond quickly to select but are slightly delayed for deselect. This allows the amplifiers to remain active for very short deselect intervals, and it speeds access time.

M_1, M_2, and M_3 are enhancement-mode devices with very low threshold voltage and serve as switches to turn off power to these three stages of the amplifier when the chip is deselected, putting them into the standby mode. M_4, M_5, and M_6 are depletion-mode active loads for the three amplifiers M_7, M_8, and M_9. The A output driver, M_{13}, has active load M_{12}, and $\overline{A}$ output driver, M_{15}, has active load M_{14}. The $\overline{E}$ line, through M_{10} and M_{11}, deactivates the drivers in standby mode. We should recognize the nonnegligible savings in power resulting from standby operation. There are 12 address buffer drivers on this chip, and each of them drives 64 decoders, representing a rather large current requirement.

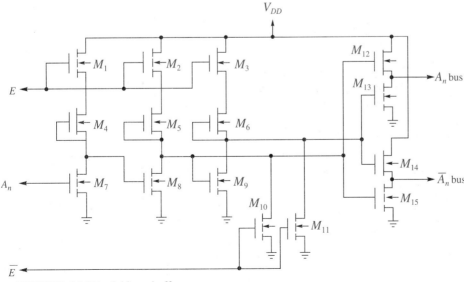

■ **FIGURE 15.17** Address buffer.

Address Decoder

■ The address buffer and decoder of Figs. 15.17 and 15.18 would be modified in the CMOS version of the SRAM. Can you suggest the necessary changes?

In the case of a 64 × 64 memory array, there are 64 row decoders and 64 column decoders. Each of these is implemented as a six-input NOR, as shown in Fig. 15.18. By proper connection of each address line or its complement to the inputs, each of the 64 decoders has a unique 6-bit address, one in which all six inputs are LOW. M_1 is a low-threshold switch that applies power to the decoder when the chip is selected, and M_2 serves as the active load for the NOR circuit. When all inputs are LOW, the active pull-up drives the row or column line associated with this decoder HIGH. Since this line may represent considerable capacitance, the output is often buffered with a two-stage driver.

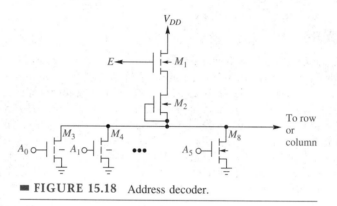

■ **FIGURE 15.18** Address decoder.

Substrate Bias

■ Note that the generation of dc voltage greater than the power supply voltage is an application of the astable circuits of Chapter 14 and the voltage doubler circuit of Chapter 3.

As we have noted, memory performance is limited by the parasitic capacitance of the row and column lines. To reduce this capacitance, the substrate may be operated at a negative potential. To allow the chip to operate on a single 5-V supply, this bias may be generated by an oscillator internal to the chip. The oscillator voltage is capacitively coupled to the substrate through diodes in the manner shown in Fig. 3.36(a) for the voltage doubler.

EXAMPLE 15.3

Estimate the number of transistors used in the 4096 × 1-bit memory described in the preceding paragraphs, including the array and address buffers and drivers.

SOLUTION Assuming the six-transistor cell, the array has 4096 × 6 = 24,576. Assuming NOR decoders of the type shown in Fig. 15.18, with six pull-down transistors and a five-transistor buffer, we have 64 × 2 × 13 = 1664. Each of the 64 columns has eight transistors in its sense circuit (see Fig. 15.16), 64 × 8 = 512. The total is 24,576 + 1664 + 512 = 26,752 transistors. In addition, a few more transistors would be used in buffering $\overline{CS}$, input data, and output data.

■ A typical high-speed cache memory is 15–25 ns access time, 256 K × 8 bits. These SRAMs could be either CMOS or NMOS, using an 8-chip module.

Refer to Figure 15.28 at the end of this chapter for excerpts from the data sheets of a typical SRAM, the NMC2147H. Problems 15.26 through 15.35 provide a good review of characteristics of SRAMs in general.

CHAPTER 15 SEMICONDUCTOR MEMORIES

> **CHECK UP**
>
> 1. **TRUE OR FALSE?** A static NMOS RAM has six transistors per cell, whereas the CMOS RAM has four transistors per cell.
> 2. **TRUE OR FALSE?** A static MOS memory cell is essentially a 1-bit latch.
> 3. **TRUE OR FALSE?** Standby mode in a static memory reduces the power usage by 50%.
> 4. **TRUE OR FALSE?** Parasitic capacitance between memory row and column lines and the substrate can be reduced by operating the substrate at a negative voltage.

15.5 DYNAMIC RANDOM-ACCESS MEMORY

To increase radically the number of cells on a single IC, the cells must be made smaller. Cells consisting of four transistors and three transistors have been used. At present, the ultimate in compactness is the cell that uses a capacitor to store a charge and one transistor to gate it to sense circuits. This cell has two drawbacks: (1) Readout is destructive and (2) charge leaks off. The result is that the circuit, although simple and compact, requires more support circuitry than static RAM. In particular, charge must be restored when the cell is read, and the charge in every cell must be *refreshed* periodically.

The description that follows is fairly general in nature, but it is in the context of the 64K × 1-bit Intel 2164A,[14] which is fairly typical of NMOS dynamic (DRAM) technology. Typical access time is 100 ns, active power is 200 mW, and standby power is 15 mW. This is a 5-V only circuit; other dc voltages, such as substrate bias, are derived on the chip from the single supply. The CMOS DRAMs also achieve access times of 100 ns, and active power under 200 mW, but standby power is less than 1 mW.[15]

■ Again we have chosen a fairly small memory IC as our example so that the organization example would not be unwieldy. Current DRAM chips are typically 4 M × 1 bit, with multiple chips in a module.

One-Transistor Memory Cell

Consider the cell shown in Fig. 15.19(a), which consists of a polysilicon-oxide-silicon capacitor and an NMOS enhancement-mode transistor. Row lines enable the transistor, connecting the capacitor to the column sense line. We assume that the logic 1 is stored

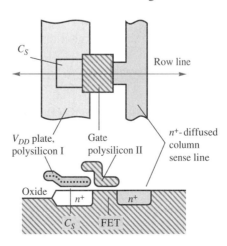

■ **FIGURE 15.19** The one-transistor cell. **(a)** Equivalent circuit. **(b)** Layout.

as the voltage $V_{DD} = 5$ V, and a logic 0 is represented by the voltage $V_{SS} = 0$ V. Note in the layout shown in Fig. 15.19(b) how compact the array can be, with the column line doubling as the transistor drains and the metal row line contacting the polysilicon gates at each cell. The diffused bottom side of the capacitor is also the transistor source, while the top side of the capacitor is a polysilicon layer connected to V_{DD}.

It has been found that when the polysilicon V_{DD} storage line overlaps the polysilicon gate, the diffused region shown in the figure to represent the bottom capacitor plate and FET source is unnecessary. A potential well is created under the storage plate, and electrons may be stored for short periods of time. There are many different arrangements of this basic cell; for a detailed comparison, refer to Ref. 16. For this introduction, we ignore such details and concentrate on the circuit operation. A typical cell is 20×10 μm, and the oxide thickness at the capacitor site is about 320 Å.

Before we consider the sense amplifier circuitry, an overview of cell operation is in order. Before each READ operation, rows are precharged LOW, to 0 V, and columns are precharged HIGH, to V_{DD}. When a cell is read, its row goes HIGH, causing the cell transistor to connect the cell capacitor to the column line. The cell capacitance is about 0.04 pF, and that of the column line is perhaps 20 times this. Thus, when the cell's charge is input to the column line, the column voltage does not change much, and the charge stored in the cell is lost. An example will illustrate the point.

EXAMPLE 15.4

Assume that the cell storage $C_S = 0.04$ pF and that of the column line circuitry is 1.2 pF. If the the column line has been precharged to 5 V, how much will the line voltage change when a cell storing a 0 (0 V) is accessed?

SOLUTION The charge stored in the line, 1.2 pF $\times$ 5 V, is distributed over 1.24 pF. The equilibrium voltage is $5 \times 1.2/1.24 = 4.84$ V, and the change is only 160 mV.

Read-Refresh Circuitry

The sense amplifier is required to detect the small change in sense line voltage, amplify it for the output line, and replenish the charge lost from the cell as it is read. As a further challenge, the small desired signal is contaminated by stray signals originating elsewhere on the chip. The simplified diagram of the sense amplifier shown in Fig. 15.20 will aid in understanding the operation of the sense circuits. We note at once that the column sense line has been split into two halves. Cells for 64 rows plus a reference, or **dummy** cell, are both above and below a latch, which serves as a balanced regenerative amplifier.

A READ operation proceeds as follows:

1. Before any cell is selected, all row lines, including both dummy lines, are forced to 0 V, and the P line is pulsed HIGH. Transistors M_5 and M_6 precharge both halves of the column line to $V_{DD} = 5$V, and transistors M_9 and M_{10} precharge the dummy cells to $V_{SS} = 0$. Precharge requires about 100 ns, and then P goes LOW, allowing the column lines to float.
2. The row corresponding to the cell to be accessed is strobed HIGH. Suppose that row 0 has been selected by the row address latch-decoder. Then M_{11} will allow the charge from its capacitor to reach equilibrium with that of the top column line. If a logic 1 was stored in this cell, its voltage originally was 5 V, but because of leakage it may have decreased to perhaps 3 V. The result of the charge redistribution is a very slight

CHAPTER 15 SEMICONDUCTOR MEMORIES

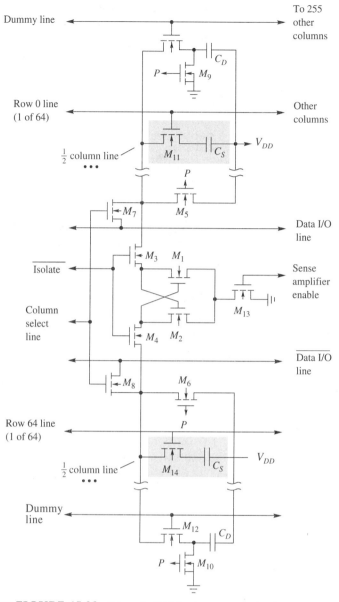

FIGURE 15.20 Dynamic RAM column sense circuitry.

decrease in voltage on the upper column sense line. If this cell contained a logic 0, its voltage was 0, and after access the upper column sense line will have dropped in voltage about 500 mV.

3. Simultaneously with the selection of row 0, the row decoder logic will also strobe the lower dummy line. The lower dummy cell is accessed in order to balance the selection of a row from the upper half. The 0 V from the dummy cell is connected to the lower column sense line by M_{12}. The dummy cell is sized such that the equilibrium voltage on the lower column line is decreased about 250 mV, or midway between the two possible states for the upper lines.

4. A sense amplifier strobe pulse now activates the latch by grounding the sources of M_1 and M_2 through M_{13}. The column line isolation transistors M_3 and M_4 now gate the small difference signal existing between the column lines to the latch. The regenerative circuit amplifies the difference signal, and the line with the lower voltage is driven to 0 V. The other line remains at a relatively high level and is then boosted back above the V_{DD} level by a column boost capacitor-transistor combination (not shown in the figure). If we assume that a logic 1 existed in our selected cell at row 0, then the upper column sense line is now HIGH, and the charge in the cell will be restored through M_{11}, which is still selected. Note that every cell on row 0 (there are 255 other columns) will have been refreshed at once.
5. If this column has been selected by the column address latch-decoders, M_7 and M_8 will now be clocked to gate the information read from this cell to the data I/O lines. Continuing our example, because the row 0 cell contains a logic 1, I/O will be HIGH, and $\overline{\text{I/O}}$ will be LOW. These lines drive a balanced data output buffer, which presents the logic 1 at the data output. Note that if the selected row had been in the bottom half (rows 64 to 127), the data on the data I/O lines would be inverted from that stored in the cell. The row select logic would insert an inversion at the output buffer.

We recognize that dividing the column sense lines into two parts not only halved the capacitance on each but also provided the sense amplifier with a balanced input, providing common-mode noise rejection.

The WRITE operation proceeds in a similar fashion in all columns except the column selected to be written into by the column address. The write enable pulse causes the input data buffer to impress the data on the data I/O line and its complement on the $\overline{\text{data I/O}}$ line. Of course, this data will be inverted when the selected row is in the bottom half of the array. Suppose a logic 1 is to be written into the cell at row 64 in this column. Then $\overline{\text{data I/O}}$ will be HIGH, and data I/O will be LOW. Row select will have turned ON M_{14}. When column select occurs, M_2 will be turned OFF through M_7 and M_3, while M_1 will be turned ON through M_8 and M_4. The bottom column sense line will be driven HIGH, and the cell at row 64 will be charged HIGH.

The Refresh Operation

We have taken note of the fact that the small charge stored in the cell capacitance tends to leak off, which is why the memory is called dynamic. Typically, the charge remains at a level where it can be safely detected for only a few milliseconds, and most DRAM must be refreshed every few ms. This time varies with the technology from 2 to 64 ms. As discussed, any operation that selects a row in the array will refresh all the cells in that row. Thus a satisfactory refresh procedure consists of the first four steps in the READ cycle.

If we assume that each cycle requires a minimum of 100 ns for precharge and 150 ns for row address strobe, each REFRESH cycle would require a minimum of 250 ns. Further, assuming the chip has 128 rows, it would require at least 128×250 ns = 32 μs to refresh the array. This is 1.6% of 2 ms, so the memory is unavailable for random access during this small fraction of the time. Of course, all DRAM chips connected to the same data bus can be refreshed at the same time.

The systematic memory refresh will either require additional circuitry operating in parallel with the main system processor or else that processor will be required to devote a fraction of its time to managing the process. We shall see that this additional circuitry can be integrated into each memory chip so that the refresh becomes transparent to the user.

Chip Organization of a 64K DRAM

The simplified block diagram shown in Fig. 15.21, adapted from the Intel 2164A, will serve as a vehicle for our discussion of typical DRAM operation and system interfacing.[14] All inputs and the output are TTL compatible. The system processor is required to present the eight low-order address bits to the memory on the address bus and then activate the row address strobe ($\overline{RAS}$) lead. After a delay of 30 to 60 ns, where the eight high-order address bits are present on the address bus, the column address strobe ($\overline{CAS}$) lead is activated. The precharge of memory internal lines and dummy cells occurs when $\overline{RAS}$ is HIGH.

■ Currently, the support circuitry (row and column address latches, decoders, and drivers) would be CMOS technology.

When $\overline{RAS}$ goes LOW, the bits on the eight address inputs A_0 through A_7 are latched into the row address buffer-latch. The seven bits A_0 through A_6 are decoded to select one of 128 rows and the appropriate dummy row. At the same time, control timing is initiated to enable the 512 sense amplifiers. Note that the 512 cells are addressed because each of the four quadrants in the array has one row and one dummy line activated.

When $\overline{CAS}$ goes LOW, the bits on the eight address inputs are latched into the column address buffer-latch. The seven bits A_8 through A_{14} are decoded to connect one of the 128 columns in each quadrant to the quadrant data I/O lines. One of the four quadrant I/O pairs is switched to the I/O gating block on the basis of the status of address bits A_7 and A_{15}.

Input data are latched into the input data buffer-latch when $\overline{RAS}$, $\overline{CAS}$, and $\overline{WE}$ are all LOW. Thus, during a WRITE cycle, when $\overline{WE}$ is low, data are routed into the I/O lines and to the selected column. When $\overline{WE}$ does not go LOW, the data read from the selected

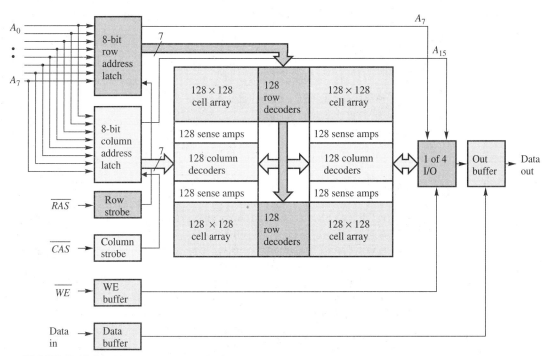

■ **FIGURE 15.21** 64K DRAM organization.

cell become available at the output buffer. As discussed previously, a REFRESH-ONLY cycle consists of addressing one of the 128 rows by means of $\overline{RAS}$, and $\overline{CAS}$ is not used.

Memory Control and Integrated DRAM

Figure 15.22 shows a simplified block diagram of some of the circuitry required to manage a DRAM system. Control directs the multiplexing of the 16 address bits from the CPU in coordination with the generation of the $\overline{RAS}$ and $\overline{CAS}$ control signals. When the refresh timer determines that it is time for the memory to be refreshed, control signals to the CPU to wait while the counter sequences through all of the row addresses.

This circuitry may be an IC DRAM controller chip or other logic ICs external to the CPU and the memory. Some or all of these functions could be integrated within the CPU or within the DRAM chip itself. When the refresh circuitry is incorporated on the memory chip, it is called a pseudostatic or *quasistatic RAM*. A device that has all of the controller functions shown in Fig. 15.22 on the memory chip has been introduced; it is called an ***integrated RAM***.[17] For a large memory system using many memory chips, the redundancy of having the controller on every chip would not be economical.

■ SIMMs are readily available for virtually all personal computers.

Large memory design utilizes blocks of memory chips affixed to a small printed-circuit board, called a *Single In-line Memory Module* (SIMM). An example would be nine 1M $\times$ 1 bit ICs in a 1 M $\times$ 9 bit SIMM. This package would then snap into a socket in the system memory board.

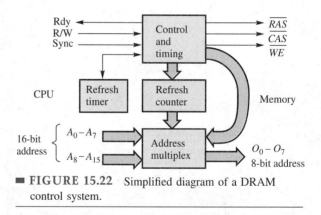

■ **FIGURE 15.22** Simplified diagram of a DRAM control system.

CHECK UP

1. **TRUE OR FALSE?** Dynamic RAMs are very compact, using only two MOSFETs per cell.
2. **TRUE OR FALSE?** Dynamic cells must be refreshed whenever they are read, because reading the cell is destructive to the data.
3. **TRUE OR FALSE?** A read operation for a dynamic RAM compares the charge stored at the addressed cell with that in a dummy cell.
4. **TRUE OR FALSE?** The column sense amplifier in a dynamic RAM must distinguish between line voltages that are only tenths of a volt different.

15.6 CHARGE-COUPLED DEVICES

We have seen that the NMOS dynamic RAM stores data as a charge packet at the surface of a *p*-substrate, separated from a positive plate by a thin oxide layer. This storage principle can be applied in a much more compact memory by giving up the random-access property of the DRAM. The charge packet may be gated to an adjacent cell, as are data in a long shift register. The data that are in storage are continually shifted through the register, regenerated, and then recirculated. Thus, the data can be accessed only at the end of the register.

To understand the operation of a typical charge-coupled device (CCD), refer to the simplified diagram shown in Fig. 15.23. The input and output diodes, D_{IN} and D_{OUT}, are normally HIGH so that electrons are not injected into the *p* region. At time T, the polysilicon gates labled ϕ_2 are driven positive by the ϕ_2 clock waveform, creating a depleted region, or an empty potential well, under each ϕ_2 gate. Gates ϕ_1 and ϕ_3 are all LOW so that these empty potential wells are isolated from the wells under gates ϕ_4, which may or may not contain charge packets.

At T_2, ϕ_1 goes HIGH, allowing the charge packets that may be at the ϕ_4 cells to expand into the ϕ_2 cells. At this time, if a logic 1 is to be introduced into the leftmost cell, D_{in} goes LOW, injecting electrons into the first cell. For storing a logic 0, no charge would be injected.

At T_3, ϕ_4 goes LOW, removing the potential well under the ϕ_4 electrodes and forcing any charge packets to continue moving to the right, out of the ϕ_4 sites toward ϕ_2 cells. The oxide layer under the odd-numbered gates is thicker than that of the others so that

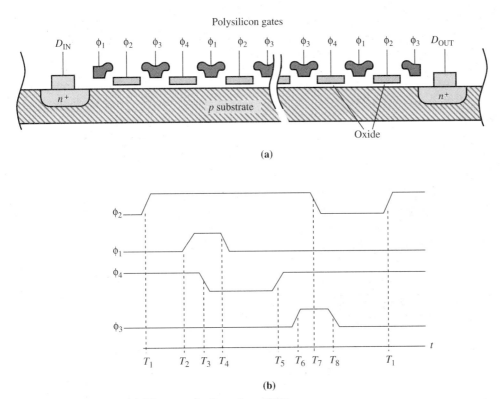

■ **FIGURE 15.23** (a) Diagram of a four-phase CCD structure. (b) Clock waveforms.

the ϕ_1 potential wells are not as deep as the ϕ_2 wells, and a majority of each charge will move into ϕ_2 locations. By this time, any injection from D_{in} into the left cell has been completed.

At T_4, ϕ_1 returns LOW, forcing any charge remaining in ϕ_1 out, and into ϕ_2 cells. One shift of data in the register has occurred, from the ϕ_4 to ϕ_2 cells, requiring about 750 ns in a typical situation. This minimum shift rate is imposed by the charge-transfer efficiency. It has been found that better than 99.99% of the charge packet can be transferred in one shift.

Similarly, ϕ_4 cells are prepared to receive charge form ϕ_2 cells at T_5. This is accomplished by pulsing ϕ_3 HIGH at T_6, forcing the charge out of ϕ_4 cells at T_7, and exhausting the ϕ_3 regions at T_8. During this part of the cycle, the charge packet that has reached the rightmost cell is delivered to the output diode.

The structure we have described is called a *four-phase CCD* because four interlocking clock waveforms are used to shift the charge packets. Three-phase systems are also used. As in NMOS dynamic memory cells, the data stored as charge packets are perishable. Some of the electrons in a charge packet leak off through the depletion layers, and some thermally generated electrons fall into empty potential wells. Thus, the data must be refreshed periodically, which requires that the shifting be a continuous process. Therefore, the shift frequency must be maintained above some mimimum, typically 100 kHz.

CCD Memories

Several CCD memory chips are available. Reference 18 describes the Intel 2416 16Kbit device in detail. A block diagram of the more recent 64Kbit chip is presented in Fig. 15.24. The chip may be thought of as an array of 256 tracks laid out in parallel, with the four-phase electrodes running across all tracks. This is a compact arrangement, with tracks separated by narrow strips of thick oxide. Actually, each track is a circuit of four 64-bit registers as far as access is concerned.

Access to any bit is achieved by shifting all registers until the desired bit is at the end of its CCD register and then latching it into a holding register. At a shift frequency of 1 MHz, the maximum wait (called *latency time*) is about 130 μs. Each holding register

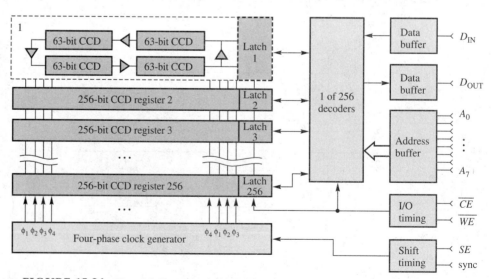

FIGURE 15.24 Pictorial diagram of a 64K bit CCD memory chip.

holds four bits from each track; thus a total of 1024 bits may be accessed before another search is performed. The proper 1-of-256 register is accessed by means of an 8-bit address. Access time to one of these holding registers is less than 285 ns so that sequential data I/O rates may be more than 3.5 Mb/s.

In summary, CCDs provide for large serial memories that are higher in speed than moving-surface magnetic memories. However, because CCDs are volatile, they will not replace the magnetic media for bulk storage. Similarly, because faster DRAMs have been developed to a point where they are not significantly more costly than CCDs, it is unlikely that CCD memories will replace DRAM for high-speed memories either. Applications at present are mainly in analog devices.

Analog Signal Processing and Imaging with CCDs

Nothing about the CCD requires that the charge packet be quantized, and thus CCDs have important application in analog signal processing and imaging.[19] The CCD register serves as a delay line, where the amount of delay is controlled by the shift frequency. Often in signal processing, the signal must be compared to itself or another signal delayed in time. The CCD register can be constructed with as many access points, or taps, as desired. Many kinds of transversal filters can be realized by appropriate weighting and summing of signals available at the taps.

An area of great interest at the present time is that of generation of video signals by means of CCD array. Small, low-cost portable video cameras and sensitive satellite transducers make use of CCD imaging. The image is focused on the substrate, and charge is introduced in each cell proportional to the average light intensity impinging on the cell.

■ One only has to note how the price and quality of camcorders has changed over the past several years to realize that CCD imaging is a rapidly maturing technology.

Periodically, the charge captured in each cell is transferred to a CCD register that is shielded from the light source and then shifted to the output. This output appears as a sampled signal representing the intensity of light at the cell locations along one particular row in the array, the same as the output for one scan of a vidicon. Such CCD arrays are also approximately the same diameter as the one-inch vidicon. For details on the construction and performance of such devices, refer to Ref. 20.

15.7 GATE ARRAYS AND PROGRAMMABLE LOGIC

In this chapter, we have concentrated on memories, a rather specific application of VLSI. The economics of producing a very large IC requires amortization of developmental costs over a large number of units. Memories are a unique combination of a specialized application with very large numbers of chips in demand, and they have well-ordered structure. It is natural that they are at the forefront of VLSI development.

In general, higher levels of integration become more specialized, leading to the lower production volume, which may not justify the developmental costs involved. Several methods are used to exploit the advantages of VLSI with limited customization costs for applications with limited production volume. These techniques will also reduce the time required to implement the desired IC.

Gate Arrays

A gate array, as the name implies, consists of a VLSI circuit fabricated with a large number of gates that are not interconnected until the last top metallization step. This allows the IC manufacturer to build a volume of identical units and then let the end user design only the interconnection mask, a relatively inexpensive procedure.

We should point out that several challenges face the logic array user:

1. Designing interconnections for all but the simplest arrays requires fairly sophisticated computer-aided design (CAD). Some semiconductor vendors provide help on this, and some independent firms offer CAD services.
2. Vendors require two to eight weeks processing time.
3. Testing a large logic array is a complex task, which is complicated by the fact that such devices have a large number of terminals and nonstandard packages. Computer-aided test programs must be developed.
4. There will probably be no second source for a part because this is a customized device.
5. Functional density on the chip is less than for a custom chip.

In spite of these problems, many users find the logic array the best solution to their logic and design at a lower cost than a completely custom design. A recent innovation is to make the gate array interconnections field programmable, and this approach is rapidly capturing the attention of many logic designers. More on this subject a little later.

Standard Cells

Users have only to specify standard circuit modules in their design in order for custom IC vendors to automate the design of a custom chip. For example, latches, comparators, and decoders are standard cells. The fabricator has all of the design work done for these cells, and it is built into the CAD programs.

Field-Programmable Logic

Logic arrays that are field-programmable are another solution to the problem of consolidating random logic circuits into fewer packages.[21] Typically, there are 8 to 16 inputs. Each input is both buffered and inverted. A large number of AND gates is available, and any of the AND inputs can be connected to any of the chip inputs or any of the inverted inputs. The outputs of the AND gates can, in turn, be connected to the inputs of any of a number of NOR gates. The NOR outputs become the array outputs.

Several such arrangements deserve special mention. Consider the schematic of a PROM in Fig. 15.25(a), where programmable-logic symbology is used to simplify the diagram. We note that a PROM is a special case of *field-programmable logic device* (FPLD), where the address decoders act as the AND functions, which are called *product lines* in FPLDs. For the PROM, each possible combination of inputs results in one product line, and these connections are fixed during fabrication. The fixed connections are represented in the figure by dots at the intersections. The outputs are formed by field-programmable OR gates, as we have seen in our study of PROM, EPROM, and EEPROM column driver circuits.

Programmable Logic Array (PLA)

Fig. 15.25(b) shows a symbolic diagram of the PLA, where both the AND matrix and the OR matrix connections are programmable. With n inputs, there will be fewer than 2^n product lines. Product line connections to the OR matrix are unrestricted. Field-programmable logic arrays (FPLAs) were first introduced in the mid-1970s by Signetics Corporation, using bipolar technology, and were programmed by fusing titanium-tungsten links. The

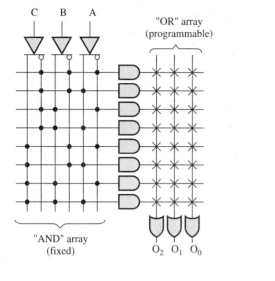

(a)

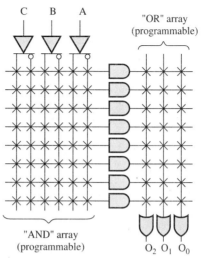

(b)

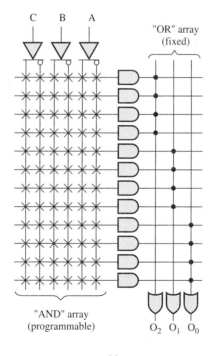

(c)

•— Fixed connection   ✳— Programmable connection

FIGURE 15.25 The basic architecture of (a) PROM. (b) PLA. (c) PAL.

circuit designer must determine which nodes in the matrices must be preserved, and then means are provided to apply fusing current to each of the other links.

Figure 15.26 depicts the structure of an 82S100 bipolar $16 \times 48 \times 8$ FPLA. The significance of the three numbers is easily seen in the figure. Sixteen inputs and 16 complemented inputs may be ANDed in any of 48 combinations by the Schottky diodes onto the vertical *product* lines. In turn, any of the product lines may be ORed onto the 8 output lines by means of the transistors. In this particular drawing, that part of the circuit that is shown with the fuses intact implements the logical expression $O_0 = A_0 A_{15} + A_0 \overline{A}_{15} + \overline{A}_0 A_{15}$.

The connections can be made by the vendor according to a connection diagram provided by the user, or they can be programmed in the field by blowing fuses in a manner similar to that used with fusible-link ROMs. Most logical expressions can be realized by the PLA, and one PLA will eliminate 10 to 20 SSI chips. The cost of the FPLA is relatively low.

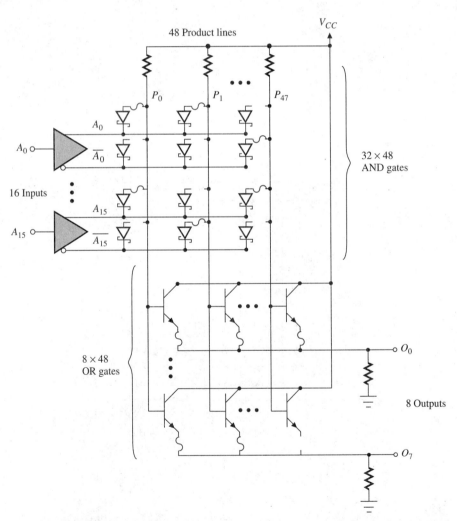

FIGURE 15.26 Structure of a bipolar $16 \times 48 \times 8$ FPLA.

Programmable Array Logic (PAL)

Also in the mid-1970s, Monolithic Memories Inc. introduced the PAL™, which is a registered trademark of MMI. Figure 15.25(c) shows that the difference between the PLA and PAL is that the latter circuit fixes the inputs to the OR matrix to certain product lines. The restriction on which product terms are combined in an OR gate is partially offset by the variety of part types with different OR array configurations offered by PAL manufacturers. The PAL has proven very popular with circuit designers, partly because they are inherently easier to use than PLAs, and partly because MMI (and others) have supported the programming with excellent software. The rapid advances in microcomputer CAD workstations have enabled the circuit designer to use these devices efficiently.

The basic PAL was first implemented with bipolar technology, using fusible links for nodal connections and the simple structure suggested by Fig. 15.25(c). The rapidly increasing demand for greater flexibility, larger arrays, and lower power, as well as keen competition in the industry, quickly resulted in a proliferation of enhancements and innovations, both in functional utility and in technological improvements.

The evolution of modern PAL devices includes the following:

- Outputs of some of the OR lines have been made available as inputs to other product lines.
- Three-state buffers have been added to these feedback outputs, so that the output pin can be programmed as either an input or an output.
- Exclusive-OR gates were inserted into the output lines, allowing programming of the output polarity.
- *D* flip-flops were added to the output circuits, allowing sequential operation of those circuits using feedback arrangements.
- JK flip-flops were added with asynchronous preset and clear inputs, so that programmed *logic sequencers* could be designed and tested.

Examples of each of these variations, and others, are detailed in Ref. 22.

The methods for realizing the connections within the arrays have diversified. We have mentioned the fusible links used in the first bipolar arrays. Because usually relatively few nodes in the array will be connected, most of the links had to be fused. It is natural to develop the concept of the *antifuse*, where only those nodes where a connection is desired will be programmed. This is accomplished in bipolar technology by placing a tiny *npn* device at each potential contact site. The base layer is unconnected, the typical operating collector-emitter voltages are below breakdown, and the unprogrammed switches are OFF. To program this particular antifuse, a higher voltage is applied to the array in such a manner as to cause the BJT to go into avalanche breakdown. The overcurrent in the device causes an alloying through the base region, and a permanent short-circuit from collector to emitter results. As with the fuse, this programming cannot be reversed.

CMOS arrays have been introduced that use EPROM or EEPROM cells to control the matrix switches. This means that the ICs can be programmed, and then erased by UV or electrical means and reprogrammed. Advantages of the CMOS arrays are lower power, ability to correct programming errors without wasting the chip, and 100% factory testing. Obviously, however, fuse technology cannot be 100% factory tested! CMOS arrays have also been created using SRAM cells to operate the matrix switches. In this case, the IC is dynamically programmed by storing the data for the contact matrix in RAM, permitting the configuration to be altered during operation. In a mature design, the user may download the switch configuration data from a PROM into the array SRAM.

Field-Programmable Gate Array (FPGA)

Further evolution and expansion of the various technologies previously discussed are loosely grouped under the broad title of FPGAs. Strictly speaking, the name implies departure from the highly structured arrays of an AND matrix followed by an OR matrix, and incorporates a large number of uncommitted gates and other logic modules that can be interconnected by user-programmed switches. There are as many approaches to this programmable logic device as there are manufacturers in this arena. For a survey of this topic, see Ref. 23. The trend of the 1990s of VLSI logic design has been to use FPGAs for low-volume production and/or quick turnaround prototyping, and to use custom interconnected gate arrays for volume production.

The greatest difficulty with this approach to the VLSI logic design problem is the complexity of the software required to do the logic synthesis for the larger logic systems. For example, different programming of the interconnections will yield different delays through the logic paths. For a discussion of this and other related issues, see Ref. 24.

CHECK UP

1. TRUE OR FALSE? Programmable logic allows the end user to obtain all of the advantages of a custom IC much sooner and at a much lower cost.

2. TRUE OR FALSE? Users of *gate arrays* must design only the interconnections on an IC that has been fabricated up to the metallization step by the manufacturer.

3. TRUE OR FALSE? PLA and PAL devices consist of an AND matrix followed by an OR matrix and have connections that can be programmed in the field.

4. TRUE OR FALSE? PAL devices use bipolar technology.

SUMMARY

Large semiconductor memories are relatively inexpensive because VLSI technology has made great advances in compacting more bits per chip. This has come about because the great demand for larger memories has fueled much research and development, and because the structure of a memory is amenable to scaling and expanding.

Table 15.1 compares representative commercial memory chips.[1,2,8] An attempt was made to compare similar speeds and technologies and to use current prices in tens quantity. Power is dependent on clock rate.

■ Vendors distinguish between "full custom" designs, and those designs where only the interconnections between standard cells or standard gates are tailored to the specific application.

The other chief applications of VLSI technology are also in high production volume, rather structured ICs, like microprocessors and logic arrays. We have avoided much discussion of microprocessors, considering them a subject in their own right. ASIC custom gate arrays are very important in designs where the volume warrants the development costs and time. A very attractive alternative for prototypes, for low production volumes or for quick development time, is one of the various programmable logic devices. CMOS PALs, controlled by an internal EEPROM switch matrix, seem appropriate for logic systems of under 1000 gates, while larger systems should consider FPGA with CMOS EEPROM technology. Table 15.2 summarizes some of the important considerations.

TABLE 15.1 IC Memory Comparisons

	Type		t_A (ns)	Power (mW)	Cost/bit (cents)
DRAM	144Mb	CMOS SIMM	60	3250	0.0006
SRAM	1Mb	NMOS	15	1600	0.006
SRAM	256Kb	CMOS	85	400	0.0037
ROM	512Kb	CMOS	120	450	0.0012
EPROM	512Kb	CMOS	120	158	0.0015
EEPROM	64Kb	CMOS	120	79	0.01
FLASH	256Kb	CMOS	120	79	0.01

TABLE 15.2 Logic Design Considerations

	Unit Price	Development Costs	Design Time	Design Aids Needed	Space Needed	Ease of Change	Manufacturing Costs
Fixed function devices	Low	Low	Short	Data books	High	Good	High
Custom gate arrays	High	High	Long	Complex CAD	Low	Poor	LOW
Programmable logic devices	Low to medium	Medium	Short to medium	Simple CAD	Medium	Good	Low to medium

SURVEY QUESTIONS

1. What advantages and disadvantages do serial memories have?
2. Name some applications for read-only memories.
3. Describe the mechanisms for programming and erasing a FAMOS cell and a FLOTOX cell.
4. Compare static and dynamic R/W memories, giving relative advantages.
5. Explain why the column lines in a dynamic memory chip are split into halves.
6. What is the purpose of having dummy cells in the dynamic R/W memory chip?
7. Name some applications for CCDs.
8. In application-specific ICs, when might a gate array be used rather than a fully custom chip?
9. What is the difference between a PAL and a PLA?

PROBLEMS

15.1 Draw a block diagram (similar to the one shown in Fig. 15.3) of a 512 × 8-bit memory using 256 × 4-bit chips. Show how the address leads are connected.

15.2 Draw a block diagram of a 16K × 8-bit memory using 4K × 8-bit chips. Show address lead connections.

15.3 A certain 2K × 8-bit ROM has a 128 × 128 cell array. How many address bits are used in (a) the row decoder? (b) the column decoder?

15.4 Draw a block diagram such as the one shown in Fig. 15.4 for a 1024 × 4-bit static RAM. Assume that each of the four input data leads is tied to the corresponding output data lead internally to conserve package leads.

15.5 A CPU has a read cycle time of 1 μs. $\overline{CS}$ is delayed from the address change by 20 ns. Assume a RAM with t_A = 350 ns, t_{CO} = 120 ns, t_{CX} = 20 ns, t_{OTD} = 100 ns, and t_{OHA} = 50 ns.

(a) How long does the CPU have to read valid data?
(b) What would be the minimum read cycle time for this RAM?

15.6 Suppose that an NMOS microprocessor can source a maximum of 400 μA to each address line when that line is HIGH. Eight memory chips are addressed in parallel, and each device has a typical input capacitance of 3 pF. The address line itself has a capacitance to ground of 6 pF. Estimate the rise time on the address line if V_{IH} = 2.4 V. Assume $V_{OL} \cong 0$ V.

15.7 Suppose the addressing scheme of Problem 15.6 is unacceptably slow. In an effort to reduce the rise times, a 5-kΩ pull-up resistor is connected between each address line and V_{DD} = 5 V. Find the new rise time. Assume the processor has no difficulty sinking the resistor current when the line is LOW.

15.8 You have 1/16-in.-thick glass-filled G-10 epoxy printed-circuit board using 1-ounce copper cladding on each side. A microstrip trace 0.1 in. wide has an inductance of 7.5 nH/in. Suppose a 4K × 1-bit memory chip receives its dc power through such a trace that is 10 in. long. When the memory chip switches from standby to normal operation, its dc operating current switches from 30 to 180 mA in 10 ns.

(a) What would be the transient voltage effect in this trace if other parasitic effects can be ignored?
(b) What could be done to reduce this power supply disturbance at the chip?

15.9 Assume 1-ounce copper-clad printed-circuit boards (1 ounce of copper per square foot) have a 1.35-mil copper thickness. What is the dc resistance of a 0.1-in.-wide trace 10 in. long?

15.10 Explain how a ROM could be used to realize the BCD-to-seven-segment decoder of Table 14.2. Consider only the BCD inputs and the seven-segment outputs. What is the minimum size ROM that would be acceptable?

Problems 15.11 through 15.20 require the use of the EPROM data sheets of Figure 15.27 (see page 788).

15.11 This particular manufacturer (Texas Instruments) markets several versions of this EPROM. What four maximum access times are guaranteed?

15.12 What power supply voltage V_{CC} is required for this EPROM? Give tolerances.

15.13 What are the requirements for erasure?

15.14 What is the purpose of the VERIFY mode?

15.15 The MODE of this EPROM is controlled by voltages on pins $\overline{E}$, $\overline{G}$, and V_{PP}.

(a) What voltages are required for READ?
(b) What voltages are required for PROGRAM?
(c) What voltages are required for OUTPUT DISABLE?
(d) What voltages are required for STANDBY?
(e) What voltages are required for VERIFY?

15.16 What are the EPROM power requirements during

(a) ACTIVE READ cycle?
(b) STANDBY?
(c) PROGRAM pulse?

15.17 (a) The HIGH outputs are capable of sourcing what current to the load?
(b) The LOW outputs are capable of sinking what current from the load?
(c) What type of load is assumed?

CHAPTER 15   SEMICONDUCTOR MEMORIES

15.18   (a) What is the maximum input current at the logic inputs?
(b) What is the maximum input capacitance at the logic inputs?

15.19   Explain why the data sheet says that the nominal programming time for this EPROM is 13 sec. Show calculations. What conclusion can be drawn from these numbers?

15.20   Describe the *Signature Mode*.

15.21   A certain nonvolatile RAM is powered by a regulated power supply such as that shown in Fig. 3.29. Assume that the power supply filter capacitor will maintain adequate dc into the regulator for 35 ms after ac power is interrupted. With sufficient warning, the system processor can signal the nonvolatile RAM with a 200-ns store pulse, which transfers the volatile memory data over into the nonvolatile cells. The save process requires 10 ms. Needed is a circuit that will sense ac power failure in time to give a HIGH-to-LOW transition to the procesor, commanding it to save the data. Draw a circuit that will monitor that ac power at the transformer secondary and generate such a warning within 15 ms. A full-wave peak rectifier with a short time constant and a comparator are suggested. Assume a comparator that will give the desired transition whenever its input falls below 1 V. Show your timing calculations. Note that the worst case is when ac fails at the peak of the ac wave.

15.22   Sketch the CMOS implementation of the six-transistor cell shown in Fig. 15.15. Assume the only change is that M_1 and M_2 are *p*-channel enhancement-mode FETs, with different gate connections.

15.23   A 4096-bit RAM dissipates 150 mW in standby at $V_{DD} = 5$ V. The memory cells are as shown in Fig. 15.15. You may assume that only the cell array remains active in standby. Let $V_{tL} = -1$ V. Estimate the value of k for the load transistors, M_1 and M_2.

15.24   Sketch the circuit of a decoder array using 3-bit decoders of the type shown in Fig. 15.18. Show how the three address lines are connected to each of the eight decoders.

15.25   A certain CMOS RAM is powered by a 5-V dc supply. The RAM data are to be protected from unexpected ac power failure by a battery backup. Assume that the system processor is able to sense ac power failure (Problem 15.21) in time to set a latch to switch the RAM into standby mode. This prevents bad data from entering the RAM as the system crashes, and also reduces the power requirement for the RAM. Assume further that the RAM, whose normal supply voltage is $5 \pm 10\%$ V, will retain data in the standby mode with a supply voltage as low as 2 V. Draw a circuit to show how the backup battery could be connected to support only the low power RAM and the latch.

Problems 15.26 through 15.35 refer to the SRAM data sheet excerpts in Fig. 15.28 (see page 796).

15.26   This particular manufacturer (National Semiconductor) sorts these SRAMs by access times. What four maximum access times are guaranteed?

15.27   Compare the active power with the standby power for this SRAM.

15.28   This SRAM has separate pins for *Data in* and *Data out*. How can this memory be used with a bidirectional data bus?

15.29   Under what conditions can *read-cycle time*, t_{RC}, be equal to *address access time*, t_{AA}, for this SRAM?

15.30   This chip can be read two ways for successive read cycles:

(a) $\overline{WE}$ remains HIGH, and $\overline{CS}$ remains LOW, and only the address changes between cycles,
(b) $\overline{WE}$ remains HIGH, and $\overline{CS}$ goes LOW when the address changes.

Which method presents valid data sooner?

15.31   For this SRAM, answer the following:

(a) The HIGH outputs are capable of sourcing what current to the load?
(b) The LOW outputs are capable of sinking what current from the load?
(c) What type of load is assumed?

15.32 The *write pulse* is the time when both $\overline{WE}$ and $\overline{CS}$ are LOW. Thus, the write pulse is ended whenever either $\overline{WE}$ or $\overline{CS}$ goes HIGH. Usually $\overline{CS}$ changes at about the same time as the address changes. Suppose a processor has a cycle time of 100 ns, and valid data are presented to the data bus during the last half of a write cycle. What must be the $\overline{WE}$ timing to ensure that valid data are latched into the addressed cell?

15.33 $\overline{CS}$ controls the active-standby condition.

(a) How long after $\overline{CS}$ goes LOW until power-up is accomplished?
(b) How long after $\overline{CS}$ goes HIGH until standby is accomplished?

15.34 Under what conditions can the *write-cycle time*, t_{WC}, be equal to *address valid to end of write time*, t_{AW}, for this SRAM?

15.35 When the output of this SRAM is switching at a 10-MHz rate, how much power is lost in charging and discharging the output circuit capacitances? Assume the data bus adds 5 pF to C_{OUT}.

15.36 A DRAM storage cell has a capacitance of 0.04 pF and is charged to 5 V. If the cell voltage has decreased to 4 V in 2 ms, estimate the leakage current and the equivalent resistance of the leakage path.

15.37 Draw a block diagram of a refresh circuit that would generate 128 row addresses incrementing them every 250 ns. The circuit should cycle through the sequence once when triggered by a 2-ms timer.

15.38 Make a list of the control signals that must be generated on the dynamic RAM chip to accomplish the READ operation.

15.39 A certain SIMM (memory module) contains nine 4Mbit $\times$ 1 DRAM chips. Each DRAM has 11 address pins, $\overline{RAS}$, $\overline{CAS}$, $\overline{W}$, data in, data out, V_{DD}, and V_{SS}. What contacts are needed on the card edge? Assume a 9-bit bidirectional data bus.

15.40 A serial memory consists of circular 256-bit registers. If the shift clock is 1 MHz, what is the maximum wait required to access a particular cell?

15.41 The position of the data in the registers of Problem 15.40 can be monitored by means of an 8-bit counter. Show by means of a simple block diagram how one could access a particular register position by means of an 8-bit address.

15.42 Write the logical expression for output O_7 of the PLA in Fig. 15.26.

15.43 A PLA such as suggested in Fig. 15.26 is to be used to implement a BCD-to-seven-segment decoder.

(a) Using only the *A, B, C,* and *D* inputs from the truth table of Table 14.2, explain how the PLA could be connected.
(b) Write an expression for the *a* output in terms of the products of the inputs (and their complements). Sketch the device connections required to realize this output.

REFERENCES

1. *MOS Memory Databook.* Texas Instruments Inc., Houston, 1991, Chap. 7.
2. *Memory Databook.* National Semiconductor Corp., Santa Clara, California, 1990, Chap. 3.
3. Ref. 2, Chap. 1.
4. *Memory Components Handbook.* Intel Corp., Santa Clara, California, 1988, Chap. 4.
5. Johnson, W., *et al.* "16K EEPROM Relies on Tunneling for Byte-Erasable Program Storage," *Electronics*, February 28, 1980, pp. 113–117.
6. *Memory Components Handbook.* Intel Corp., Santa Clara, California, 1984, Chap. 5.
7. *Xicor Data Book.* Xicor, Inc., Milpitas, California, 1988, Chap. 3.
8. *Memory Products.* Intel Corp., Santa Clara, California, 1993, Chap. 3.
9. Ref. 8, pp. 3-684–3-687.
10. Ref. 7, Chap. 1.

11. Ref. 4, Chap. 3.
12. Ref. 1, Chap. 5.
13. Ref. 1, Chap. 4.
14. Ref. 6, pp. 3-23–3-39
15. Ref. 6, pp. 3-251–3-252.
16. Rideout, V. "One-Device Cells for Dynamic Random-Access Memories: A Tutorial," *IEEE Transactions on Electron Devices*, June 1979, pp. 839–852.
17. Ref. 6, pp. 3-41–3-69.
18. Chou, S. "Design of a 16384-bit Serial Charge-Controlled Memory Device," *IEEE Transactions on Electron Devices*, February 1976, pp. 78–86.
19. Melen, R., and D. Buss. *Charge-Coupled Devices: Technology and Applications*. New York: IEEE Press, 1977.
20. Tompsett, W., et al. "Charge Coupling Improves Its Image, Challenging Video Camera Tubes," *Electronics*, January 18, 1973, pp. 162–169.
21. Kazami, S. "Design Prototypes Quickly with Programmable Arrays," *Electronic Design*, February 19, 1981, pp. 121–124.
22. Haznedar, H. *Digital Microelectronics*. Redwood City, Calif: Benjamin Cummings, 1991, pp. 491–523.
23. Pellerin, D., and M. McClure. "Not All FPGAs Are Created Equal," *Electronic Products*, July 1991, pp. 33–40.
24. Goering, R. "Promise Yet To Be Fulfilled, *Electronic Engineering Times,* April 11, 1994, pp. 1, 37–38.
25. Burns, S. G., H. R. Shanks, A. Constant, C. Gruber, D. Schmidt, C. Thielen, F. Olympie, T. Schumacher, *Design and Fabrication of A-Si: H-Based EEPROM Cells,* Proceedings of the Electrochemical Society Meeting, October 1994, Miami, Florida.

SUGGESTED READING

Haznedar, H. *Digital Microelectronics*. Redwood City, Calif: Benjamin Cummings, 1991. Chapter 10.

TMS27C010A 1 048 576-BIT UV ERASABLE PROGRAMMABLE READ-ONLY MEMORY
TMS27PC010A 1 048 576-BIT PROGRAMMABLE READ-ONLY MEMORY

SMLS110 – NOVEMBER 1990

- Organization ... 128K x 8
- Single 5-V Power Supply
- Operationally Compatible With Existing Megabit EPROMs
- Industry Standard 32-Pin Dual-in-line Package and 32-Lead Plastic Chip Carrier
- All Inputs/Outputs Fully TTL Compatible
- Max Access/Min Cycle Time

$V_{CC} \pm 5\%$	$V_{CC} \pm 10\%$	
'27C010A-100		100 ns
'27PC010A-100		100 ns
'27C010A-120	'27C010A-12	120 ns
'27PC010A-120	'27PC010A-12	120 ns
'27C010A-150	'27C010A-15	150 ns
'27PC010A-150	'27PC010A-15	150 ns
'27C010A-200	'27C010A-20	200 ns
'27PC010A-200	'27PC010A-20	200 ns

- 8-Bit Output For Use in Microprocessor-Based Systems
- Very High-Speed SNAP! Pulse Programming
- Power-Saving CMOS Technology
- 3-State Output Buffers
- 400-mV Minimum DC Noise Immunity With Standard TTL Loads
- Latchup Immunity of 250 mA on All Input and Output Pins
- No Pullup Resistors Required
- Low Power Dissipation (V_{CC} = 5.5 V)
  — Active ... 165 mW Worst Case
  — Standby ... 0.55 mW Worst Case
    (CMOS-Input Levels)
- PEP4 Version Available With 168 Hour Burn-In and Choices of Operating Temperature Ranges

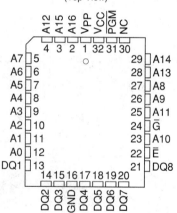

description

The TMS27C010A series are 1 048 576-bit, ultraviolet-light erasable, electrically programmable read-only memories.

PIN NOMENCLATURE	
A0–A16	Address Inputs
$\overline{E}$	Chip Enable
$\overline{G}$	Output Enable
GND	Ground
NC	No Internal Connection
$\overline{PGM}$	Program
DQ1–DQ8	Inputs (programming)/Outputs
V_{CC}	5-V Supply
V_{PP}	13-V Power Supply †

† Only in program mode.

■ **FIGURE 15.27** TMS 27C010A EPROM data sheet (courtesy Texas Instruments).

CHAPTER 15   SEMICONDUCTOR MEMORIES

TMS27C010A 1 048 576-BIT UV ERASABLE PROGRAMMABLE READ-ONLY MEMORY
TMS27PC010A 1 048 576-BIT PROGRAMMABLE READ-ONLY MEMORY

SMLS110 – NOVEMBER 1990

The TMS27PC010A series are 1 048 576-bit, one-time, electrically programmable read-only memories.

These devices are fabracated using power-saving CMOS technology for high speed and simple interface with MOS and bipolar circuits. All inputs (including program data inputs) can be driven by Series 74 TTL circuits without the use of external pullup resistors. Each output can drive one Series 74 TTL circuit without external resistors.

The TMS27C010A EPROM is offered in a dual-in-line ceramic package (J suffix) designed for insertion in mounting hole rows on 15,2-mm (600-mil) centers. The TMS27C010A is also offered with two choices of temperature ranges of 0°C to 70°C and −40°C to 85°C (TMS27C010A-__JL and TMS27C010A-__JE, respectively). The TMS27C010A is also offered with 168 hour burn-in on both temperature ranges (TMS27C010A-__JL4 and TMS27C010A-__JE4, respectively). (See table below).

The TMS27PC010A OTP PROM is offered in a 32-lead plastic leaded chip carrier package using 1,25-mm (50-mil) lead spacing (FM suffix). The TMS27PC010A is offered with two choices of temperature ranges of 0°C to 70°C and −40°C to 85°C (TMS27PC010A-__FML and TMS27PC010A-__FME, respectively). (See table below).

EPROM AND OTP PROM	SUFFIX FOR OPERATING TEMPERATURE RANGES WITHOUT PEP4 BURN-IN		SUFFIX FOR PEP4 168 HOUR BURN-IN VS TEMPERATURE RANGES	
	0°C to 70°C	−40° to 85°C	0°C to 70°C	−40° to 85°C
TMS27C010A-xxx	JL	JE	JL4	JE4
TMS27PC010A-xxx	FML	FME		

These EPROMs and OTP PROMs operate from a single 5-V supply (in the read mode), thus are ideal for use in microprocessor-based systems. One other 13-V supply is needed for programming. All programming signals are TTL level. These devices are programmable using the SNAP! Pulse programming algorithm. The SNAP! Pulse programming algorithm uses a V_{PP} of 13 V and a V_{CC} of 6.5 V for a nominal programming time of thirteen seconds. For programming outside the system, existing EPROM programmers can be used. Locations may be programmed singly, in blocks, or at random.

operation

There are seven modes of operation listed in the following table. The read mode requires a single 5-V supply. All inputs are TTL level except for V_{PP} during programming (13 V for SNAP! Pulse) and 12 V on A9 for signature mode.

FUNCTION	MODE							
	READ	OUTPUT DISABLE	STANDBY	PROGRAMMING	VERIFY	PROGRAM INHIBIT	SIGNATURE MODE	
$\overline{E}$	V_{IL}	V_{IL}	V_{IH}	V_{IL}	V_{IL}	V_{IH}	V_{IL}	
$\overline{G}$	V_{IL}	V_{IH}	X†	V_{IH}	V_{IL}	X	V_{IL}	
$\overline{PGM}$	X	X	X	V_{IL}	V_{IH}	X	X	
V_{PP}	V_{CC}	V_{CC}	V_{CC}	V_{PP}	V_{PP}	V_{PP}	V_{CC}	
V_{CC}	V_{CC}	V_{CC}	V_{CC}	V_{CC}	V_{CC}	V_{CC}	V_{CC}	
A9	X	X	X	X	X	X	VH‡	VH‡
A0	X	X	X	X	X	X	V_{IL}	V_{IH}
DQ1-DQ8	Data Out	HI-Z	HI-Z	Data In	Data Out	HI-Z	CODE	
							MFG	DEVICE
							97	D6

† X can be V_{IL} or V_{IH}.
‡ V_H = 12 V ± 0.5 V.

■ **FIGURE 15.27**   (cont)

TMS27C010A 1 048 576-BIT UV ERASABLE PROGRAMMABLE READ-ONLY MEMORY
TMS27PC010A 1 048 576-BIT PROGRAMMABLE READ-ONLY MEMORY

SMLS110 – NOVEMBER 1990

read/output disable

When the outputs of two or more TMS27C010As or TMS27PC010As are connected in parallel on the same bus, the output of any particular device in the circuit can be read with no interference from competing outputs of the other devices. To read the output of a single device, a low-level signal is applied to the $\overline{E}$ and $\overline{G}$ pins. All other devices in the circuit should have their outputs disabled by applying a high level signal to one of these pins.

latchup immunity

Latchup immunity on the TMS27C010A and TMS27PC010A is a minimum of 250 mA on all inputs and outputs. This feature provides latchup immunity beyond any potential transients at the P.C. board level when the devices are interfaced to industry standard TTL or MOS logic devices. The input/output layout approach controls latchup without compromising performance or packing density.

For more information see application report SMLA001, *"Design Considerations; Latchup Immunity of the HVCMOS EPROM Family"*, available through TI Sales Offices.

power down

Active I_{CC} supply current can be reduced from 30 mA to 500 µA for a high TTL input on $\overline{E}$ and to 100 µA for a high CMOS input on $\overline{E}$. In this mode all outputs are in the high-impedance state.

erasure (TMS27C010A)

Before programming, the TMS27C010A EPROM is erased by exposing the chip through the transparent lid to a high intensity ultraviolet light (wavelength 2537 Å). The recommended minimum exposure dose (UV intensity x exposure time) is 15-W·s/cm^2. A typical 12-mW/cm^2, filterless UV lamp will erase the device in 21 minutes. The lamp should be located about 2.5 cm above the chip during erasure. After erasure, all bits are in the high state. It should be noted that normial ambient light contains the correct wavelength for erasure. Therefore, when using the TMS27C010A, the window should be covered with an opaque label. After erasure (all bits in logic high state), logic lows are programmed into the desired locations. A programmed low can be erased only by ultraviolet light.

initializing (TMS27PC010A)

The one-time programmable TMS27PC010A PROM is provided with all bits in the logic high state, then logic lows are programmed into the desired locations. Logic lows programmed into an OTP PROM cannot be erased.

SNAP! Pulse programming

The TMS27C010A and TMS27PC010A are programmed using the TI SNAP! Pulse programming algorithm illustrated by the flowchart in Figure 1, which programs in a nominal time of thirteen seconds. Actual programming time will vary as a function of the programmer used.

The SNAP! Pulse programming algorithm uses an initial pulse of 100 microseconds (µs) followed by a byte verification to determine when the addressed byte has been successfully programmed. Up to 10 (ten) 100-µs pulses per byte are provided before a failure is recognized.

The programming mode is achieved when V_{PP} = 13 V, V_{CC} = 6.5 V, $\overline{E} = V_{IL}$, $\overline{G} = V_{IH}$. Data is presented in parallel (eight bits) on pins DQ1 through DQ8. Once addresses and data are stable, PGM is pulsed low.

More than one device can be programmed when the devices are connected in parallel. Locations can be programmed in any order. When the SNAP! Pulse programming routine is complete, all bits are verified with $V_{CC} = V_{PP}$ = 5 V ± 10%.

program inhibit

Programming may be inhibited by maintaining a high level input on the $\overline{E}$ or $\overline{PGM}$ pins.

program verify

Programmed bits may be verified with V_{PP} = 13 V when $\overline{G} = V_{IL}$, $\overline{E} = V_{IL}$, and $\overline{PGM} = V_{IH}$.

■ **FIGURE 15.27** (cont)

TMS27C010A 1 048 576-BIT UV ERASABLE PROGRAMMABLE READ ONLY MEMORY
TMS27PC010A 1 048 576-BIT PROGRAMMABLE READ-ONLY MEMORY

SMLS110 – NOVEMBER 1990

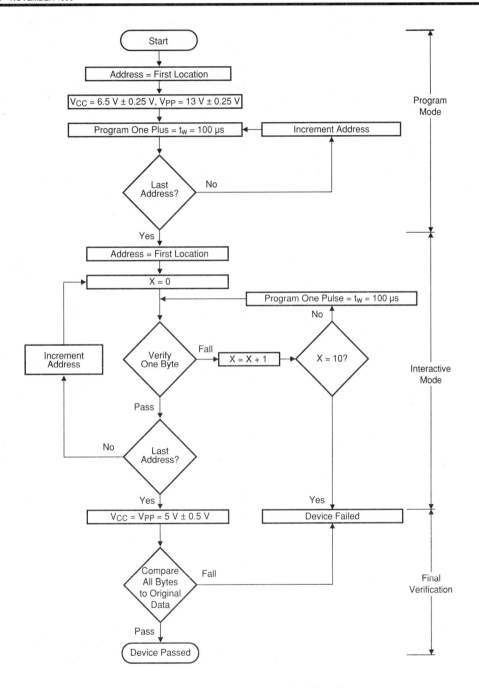

Figure 1. Snap! Pulse Programming Flow Chart

■ FIGURE 15.27 (cont)

TMS27C010A 1 048 576-BIT UV ERASABLE PROGRAMMABLE READ-ONLY MEMORY
TMS27PC010A 1 048 576-BIT PROGRAMMABLE READ-ONLY MEMORY

SMLS110 – NOVEMBER 1990

absolute maximum ratings over operating free-air temperature range (unless otherwise noted)†

Supply voltage range, V_{CC} (see Note 1)	–0.6 V to 7 V
Supply voltage range, V_{PP}	–0.6 V to 14 V
Input voltage range, All inputs except A9	–0.6 V to V_{CC} +1 V
A9	–0.6 V to 13.5 V
Output voltage range, with respect to V_{SS} (see Note 1)	–0.6 V to V_{CC} +1 V
Operating free-air temperature range ('27C010A-__JL and JL4, '27PC010A-__FML)	0°C to 70°C
Operating free-air temperature range ('27C010A-__JE and JE4, '27PC010A-__FME)	–40° to 85°C
Storage temperature range	–65° to 150°C

†Stresses beyond those listed under "Absolute Maximum Ratings" may cause permanent damage to the device. This is a stress rating only, and functional operation of the device at these or any other conditions beyond those indicated in the "Recommended Operating Conditions" section of this specification is not implied. Exposure to absolute-maximum-rated conditions for extended periods may affect device reliability.
NOTE 1: All voltage values are with respect to GND.

recommended operating conditions

			'27C010A/PC010A-100 '27C010A/PC010A-120 '27C010A/PC010A-150 '27C010A/PC010A-200			'27C010A/PC010A-12 '27C010A/PC010A-15 '27C010A/PC010A-20			UNIT
			MIN	TYP	MAX	MIN	TYP	MAX	
V_{CC} Supply voltage	Read mode (see Note 2)		4.75	5	5.25	4.75	5	5.25	V
	SNAP! Pulse programming algorithm		6.25	6.5	6.75	6.25	6.5	6.75	V
V_{PP} Supply voltage	Read mode (see Note 3)		V_{CC} – 0.6	V_{CC}	V_{CC} + 0.6	V_{CC} – 0.6	V_{CC}	V_{CC} + 0.6	V
	SNAP! Pulse programming algorithm		12.75	13	13.25	12.75	13	13.25	V
V_{IH} High-level input voltage		TTL	2.0		V_{CC} + 0.5	2.0		V_{CC} + 0.5	V
		CMOS	V_{CC} – 0.2		V_{CC} + 0.5	V_{CC} – 0.2		V_{CC} + 0.5	
V_{IL} Low-level input voltage		TTL	–0.5		0.8	–0.5		0.8	V
		CMOS	–0.5		GND + 0.2	–0.5		GND + 0.2	
T_A Operating free-air temperature	'27C010A-__JL,JL4 '27PC010A-__FML		0		70	0		70	°C
T_A Operating free-air temperature	'27C010A-__JE,JE4 '27PC010A-__FME		–40		85	–40		85	°C

NOTES: 2. V_{CC} must be applied before or at the same time as V_{PP} and removed after or at the same time as V_{PP}. The device must not be inserted into or removed from the board when V_{PP} or V_{CC} is applied.
3. V_{PP} can be connected to V_{CC} directly (except in the program mode). V_{CC} supply current in this case would be $I_{CC} + I_{PP}$. During programming, V_{PP} must be maintained at 13 V ± 0.25 V.

■ **FIGURE 15.27** (cont)

CHAPTER 15   SEMICONDUCTOR MEMORIES

TMS27C010A 1 048 576-BIT UV ERASABLE PROGRAMMABLE READ-ONLY MEMORY
TMS27PC010A 1 048 576-BIT PROGRAMMABLE READ-ONLY MEMORY

SMLS110 – NOVEMBER 1990

electrical characteristics over full range of operating conditions

PARAMETER		TEST CONDITIONS	MIN	MAX	UNIT
V_{OH} High-level output voltage		$I_{OH} = -20\ \mu A$	$V_{CC} - 0.2$		V
		$I_{OH} = -2.5\ mA$	3.5		
V_{OL} Low-level output voltage		$I_{OL} = 2.1\ mA$		0.4	V
		$I_{OL} = 20\ \mu A$		0.1	
I_I Input current (leakage)		$V_I = 0\ to\ 5.5\ V$		±1	μA
I_O Output current (leakage)		$V_O = 0\ to\ V_{CC}$		±1	μA
I_{PP1} V_{PP} supply current		$V_{PP} = V_{CC} = 5.5\ V$		10	μA
I_{PP2} V_{PP} supply current (during program pulse)		$V_{PP} = 13\ V$		50	mA
I_{CC1} V_{CC} supply current (standby)	TTL-input level	$\overline{E} = V_{IH}, V_{CC} = 5.5\ V$		500	μA
	CMOS-input level	$\overline{E} = V_{CC} \pm 0.2\ V, V_{CC} = 5.5V$		100	
I_{CC2} V_{CC} supply current (active) (output open)		$\overline{E} = V_{IL}, V_{CC} = 5.5V$, t_{cycle} = minimum cycle time†, outputs open		30	mA

† Minimum cycle time = maximum access time.

capacitance over recommended ranges of supply voltage and operating free-air temperature, f = 1 MHz‡

PARAMETER	TEST CONDITIONS	MIN	TYP§	MAX	UNIT
C_i Input capacitance	$V_I = 0, f = 1\ MHz$		4	8	pF
C_o Output capacitance	$V_O = 0, f = 1\ MHz$		6	10	pF

‡ Capacitance measurements are made on sample basis only.
§ All typical values are at $T_A = 25°C$ and nominal voltages.

switching characteristics over full ranges of recommended operating conditions (see Notes 4 and 5)

PARAMETER	TEST CONDITIONS (SEE NOTES 4 & 5)	'27C010A-100 '27PC010A-100		'27C010A-120 '27PC010A-120 '27C010A-12 '27PC010A-12		'27C010A-150 '27PC010A-150 '27C010A-15 '27PC010A-15		'27C010A-200 '27PC010A-200 '27C010A-20 '27PC010A-20		UNIT
		MIN	MAX	MIN	MAX	MIN	MAX	MIN	MAX	
$t_{a(A)}$ Access time from address	$C_L = 100\ pF$, 1 Series 74 TTL load, Input $t_r \leq 20\ ns$, Input $t_f \leq 20\ ns$		100		120		150		200	ns
$t_{a(E)}$ Access time from chip enable			100		120		150		200	ns
$t_{en(G)}$ Output enable time from $\overline{G}$			55		55		75		75	ns
t_{dis} Output disable time from $\overline{G}$ or $\overline{E}$, whichever occurs first¶		0	50	0	50	0	60	0	60	ns
$t_{v(A)}$ Output data valid time after change of address, $\overline{E}$, or $\overline{G}$, whichever occurs first		0		0		0		0		ns

¶ Value calculated from 0.5-V delta to measured output level.

NOTES: 4. For all switching characteristics the input pulse levels are 0.4 V to 2.4 V. Timing measurements are made at 2 V for logic high and 0.8 V for logic low (reference AC Testing Wave Form).
5. Common test conditions apply for t_{dis} except during programming.

■ FIGURE 15.27   (cont)

TMS27C010A 1 048 576-BIT UV ERASABLE PROGRAMMABLE READ-ONLY MEMORY
TMS27PC010A 1 048 576-BIT PROGRAMMABLE READ-ONLY MEMORY

SMLS110 – NOVEMBER 1990

switching characteristics for programming: V_{CC} = 6.5 V and V_{PP} =13 V (SNAP! Pulse), T_A = 25°C (see Note 4)

PARAMETER		MIN	NOM	MAX	UNIT
t_{disG}	Output disable time from $\overline{G}$	0		130	ns
$t_{en(G)}$	Output enable time from $\overline{G}$			150	ns

recommended timing requirements for programming: V_{CC} = 6.5 V and V_{PP} = 13 V (SNAP! Pulse), T_A = 25°C, (see Note 4)

			MIN	TYP	MAX	UNIT
$t_{w(PGM)}$	Program pulse duration	SNAP! Pulse programming algorithm	95	100	105	µs
$t_{su(A)}$	Address setup time			2		µs
$t_{su(E)}$	$\overline{E}$ setup time			2		µs
$t_{su(G)}$	$\overline{G}$ setup time			2		µs
$t_{su(D)}$	Data setup time			2		µs
$t_{su(VPP)}$	V_{PP} setup time			2		µs
$t_{su(VCC)}$	V_{CC} setup time			2		µs
$t_{h(A)}$	Address hold time			0		µs
$t_{h(D)}$	Data hold time			2		µs

NOTES: 4. For all switching characteristics the input pulse levels are 0.4 V to 2.4 V. Timing measurements are made at 2 V for logic high and 0.8 V for logic low (reference AC Testing Wave Form).

PARAMETER MEASUREMENT INFORMATION

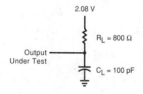

Figure 2. AC Test Output Load Circuit

AC testing input/output wave forms

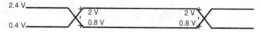

AC testing inputs are driven at 2.4 V for logic high and 0.4 V for logic low. Timing measurements are made at 2 V for logic high and 0.8 V for logic low for both inputs and outputs.

■ **FIGURE 15.27** (cont)

CHAPTER 15 SEMICONDUCTOR MEMORIES

TMS27C010A 1 048 576-BIT UV ERASABLE PROGRAMMABLE READ-ONLY MEMORY
TMS27PC010A 1 048 576-BIT PROGRAMMABLE READ-ONLY MEMORY

SMLS110 – NOVEMBER 1990

program cycle timing (SNAP! Pulse programming)

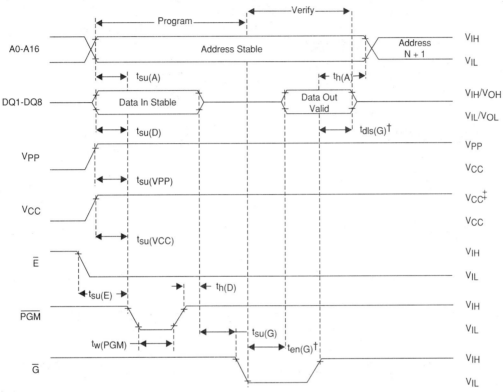

† $t_{dis(G)}$ and $t_{en(G)}$ are characteristics of the device but must be accommodated by the programer.
‡ 13-V V_{PP} and 6.5-V V_{CC} for SNAP! Pulse programming.

read cycle timing

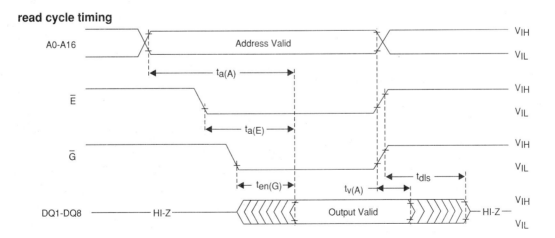

■ **FIGURE 15.27** (cont)

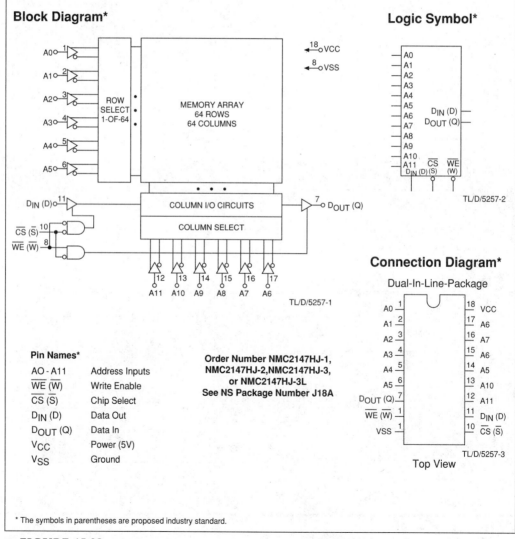

■ FIGURE 15.28 NMC2147H Data sheet (courtesy National Semiconductor).

NMC2147H

Absolute Maximum Ratings
If Military/Aerospace specified devices are required, please contact the National Semiconductor Sales Office/Distributors for availability and specifications.

Voltage on Any Pin Relative to VSS	−3.5V TO + 7V
Storage Temperature Range	−65°C TO + 150°C
Power Dissipation	1.2W
DC Output Current	20 mA
Bias Temperature Range	−65°C TO + 135°C
Lead Temperature (Soldering, 10 sec.)	300°C

Truth Table*

$\overline{CS}$ (S)	$\overline{WE}$ ($\overline{W}$)	DIN (D)	DOUT (Q)	Mode	Power
H	X	X	Hi-Z	Not Selected	Standby
L	L	H	Hi-Z	Write 1	Active
L	L	L	Hi-Z	Write 0	Active
L	H	X	DOUT	Read	Active

DC Electrical Characteristics
$T_A = 0°C$ to $+70°C$, $V_{CC} = 5V \pm 10\%$ (Notes 1 and 2)

Symbol	Parameter	Conditions	NMC2147H-3L Min	NMC2147H-3L Max	NMC2147H-1 NMC2147H-2 NMC2147H-3 Min	NMC2147H-1 NMC2147H-2 NMC2147H-3 Max	NMC2147H Min	NMC2147H Max	Units
$|I_{LI}|$	Input Load Current (All Input Pins)	VIN = 0V to 5.5V, VCC = Max		10		10		10	µA
$|I_{LO}|$	Output Leakage Current	$\overline{CS}$ = VIH, VOUT = GND to 4.5V, VCC = Max		50		50		50	µA
VIL	Input Low Voltage		−3.0	0.8	−3.0	0.8	−3.0	0.8	V
VIH	Input High Voltage		2.0	6.0	2.0	6.0	2.0	6.0	V
VOL	Output Low Voltage	IOL = 8.0 mA		0.4		0.4		0.4	V
VOH	Output High Voltage	IOH = −4.0 mA	2.4		2.4		2.4		V
ICC	Power Supply Current	VIN = 5.5V, TA = 0°C, Output Open		125		180		160	mA
ISB	Standby Current	VCC = Min to Max, $\overline{CS}$ = VIH		20		30		20	mA
IPO	Peak Power-On Current	VCC = VSS to VCC Min, $\overline{CS}$ = Lower of VCC or VIH Min		30		40		30	mA

Capacitance
$T_A = 25°C$, f = 1 MHz (Note 3)

Symbol	Parameter	Conditions	Min	Max	Units
CIN	Address/Control Capacitance	VIN = 0V		5	pF
COUT	Output Capacitance	VOUT = 0V		6	pF

Note 1: The operating ambient temperature range is guaranteed with transverse air flow exceeding 400 linear feet per minute.
Note 2: These circuits require 500 µs time delay after VCC reaches the specified minimum limit to ensure proper orientation after power-on. This allows the internally generated substrate bias to reach its functional level.
Note 3: This parameter is guaranteed by periodic testing.

AC Test Conditions

Input Test levels	GND to 3.0 V
Input Rise and Fall Times	5 ns
Input Timing Reference Level	1.5 V
Output Timing Reference Level (H-1)	1.5 V
Output Timing Reference Level (H-2, H-3, H-3L)	0.8 V and 2.0 V
Output load	See Figure 1

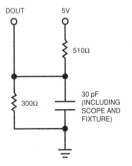

FIGURE 1. Output Load

* Symbols in parentheses are proposed industry standard.

■ **FIGURE 15.28** (cont)

Read Cycle AC Electrical Characteristics $T_A = 0°C$ to $+70°C$, $V_{CC} = 5V \pm 10\%$ (Note 1)

SYMBOL		Parameter	NMC2147H-1		NMC2147H-2		NMC2147H-3 NMC2147H-3L		NMC2147H		Units
Alternate	Standard		Min	Max	Min	Max	Min	Max	Min	Max	
t_{RC}	TAVAV	Read Cycle Time	35		45		55		70		ns
t_{AA}	TAVQV	Address Access Time		35		45		55		70	ns
t_{ACS}	TSLQV	Chip Select Access Time (Note 4)		35		45		55		70	ns
t_{LZ}	TSLQX	Chip Select to Output Active (Note 5)	5		5		10		10		ns
t_{HZ}	TSHQZ	Chip Deselect to Output TRI-STATE (Note 5)	0	30	0	30	0	30	0	30	ns
t_{OH}	TAXQX	Output Hold from Address Change	5		5		5		5		ns
t_{PU}	TSLIH	Chip Select to Power-Up	0		0		0		0		ns
t_{PD}	TSHIL	Chip Deselect to Power-Down		20		20		20		30	ns

Max Access/Current	NMC2147H-1	NMC2147H-2	NMC2147H-3	NMC2147H-3L	NMC2147H
Access (TAVQV—ns)	35	45	55	55	70
Active Current (ICC—mA)	180	180	180	125	160
Standby Current (ISB—mA)	30	30	30	20	20

Read Cycle Waveforms*

Read Cycle 1 (Continuous Selection $\overline{CS}$ = VIL, $\overline{WE}$ = VIH)

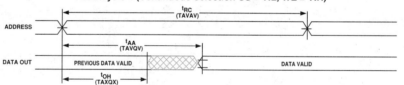

TL/D/5257-5

Read Cycle 2 (Chip Select Switched, $\overline{WE}$ = VIH) (Note 4)

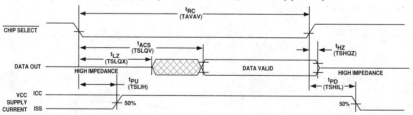

TL/D/5257-6

Note 4: Addresses must be valid coincident with or prior to the chip select transition from high to low.

Note 5: Measured ±50 mV from steady state voltage. This parameter is sampled and not 100% tested.

*The symbols in parentheses are proposed industry standard.

■ **FIGURE 15.28** (cont)

CHAPTER 15 SEMICONDUCTOR MEMORIES

NMC2147H

Write Cycle AC Electrical Characteristics TA = 0°C to 70°C, V_CC = 5V ± 10% (Note 1)

SYMBOL		Parameter	NMC2147H-1		NMC2147H-2		NMC2147H-3 NMC2147H-3L		NMC2147H		Units
Alternate	Standard		Min	Max	Min	Max	Min	Max	Min	Max	
t_{WC}	TAVAV	Write Cycle Time	35		45		55		70		ns
t_{CW}	TSLWH	Chip Select to End of Write	35		45		45		55		ns
t_{AW}	TAVWH	Address Valid to End of Write	35		45		45		55		ns
t_{AS}	TAVLS TAVWL	Address Set-Up Time	0		0		0		0		ns
t_{WP}	TWLWH	Write Pulse Width	20		25		25		40		ns
t_{WR}	TWHAX	Write Recovery Time	0		0		10		15		ns
t_{DW}	TDVWH	Data Set-Up Time	20		25		25		30		ns
t_{DH}	TWHDX	Data Hold Time	10		10		10		10		ns
t_{WZ}	TWLQZ	Write Enable to Output TRI-STATE (Note 5)	0	20	0	25	0	25	0	35	ns
t_{OW}	TWHQX	Output Active from End of Write (Note 5)	0		0		0		0		ns

Write Cycle Waveforms* (Note 6)

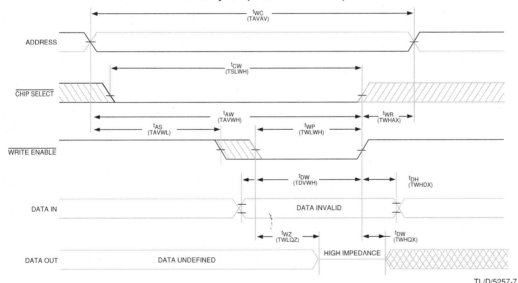

Note 6: The output remains TRI-STATE if the CS and WE go high simultaneously. WE or CS or both must be high during the address transitions to prevent an erroneous write.

* The symbols in parentheses are proposed industry standard.

■ **FIGURE 15.28** (cont)

CHAPTER 16

ANALOG-TO-DIGITAL AND DIGITAL-TO-ANALOG CONVERSION

16.1 Analog-To-Digital Conversion Process
16.2 Digital-To-Analog Converters
16.3 Analog-To-Digital Conversion
16.4 Summary
Survey Questions
Problems
References

To this point in the text, we have categorized electrical signals as analog (continuous) or digital (discrete). An *analog* signal is defined as a continuous function in time with a continuous first derivative. Fundamentally, all variables such as voltage, current, pressure, temperature, velocity, etc., are analog signals. This is because physical processes tend to oppose discontinuities in these variables by preventing infinite rates of change.

To use a digital computer or microcomputer for computation and subsequent signal processing, these analog signals are coded as a binary digital signal by means of an *analog-to-digital (A/D) converter.* This approach is used in remote-control telemetry applications and for communication systems signal processing. Other applications include digital instruments such as digital voltmeters and computer-networked instrumentation.

A complete A/D converter (ADC) system that could be used as an interface for a computer controller is shown in Fig. 16.1. Typically, a single ADC circuit is used to process a number of analog channels, $v_i(t)$, where $i = 0, 1, 2, \ldots, m$, where m is usually designed to accommodate multiples of eight input channels. The sampling order and interval is under computer control logic. To obtain the digitized value of each of the analog inputs at some given time, a sample-and-hold circuit, using a circuit with some type of capacitive storage element, is used to obtain an instantaneous time-invariant value of one of the analog inputs when it is being sampled. This voltage sample taken at time t_S, $V_S(t_S)$, is then input to the ADC for digitizing. The binary digitized signal is then stored or buffered in a parallel latch unit it is needed. The resultant n-bit digital output word $d_{-1}d_{-2}\cdots d_{-n}$ relates $V_S(t_S)$ as a binary fraction to the full-scale reference voltage V_{FS}:

$$V_O(t_S) = (d_{-1}2^{-1} + d_{-2}2^{-2} + \ldots + d_{-n}2^{-n})V_{FS}. \tag{16.1}$$

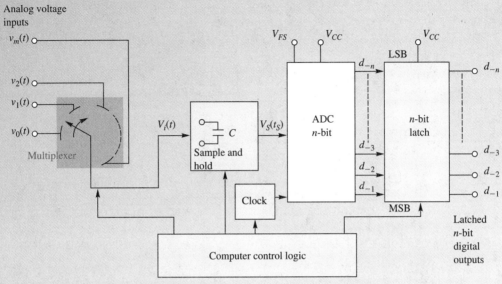

■ **FIGURE 16.1** A/D system with an *m*-channel input signal multiplexer.

Bit d_{-n} is called the ***least significant bit*** (LSB) because it represents the smallest weighting, and bit d_{-1} is called the ***most significant bit*** (MSB) because it represents the largest weighting. A representative transfer characteristic for a 3-bit ADC is given in Fig. 16.2(a). The ***resolution*** is one LSB and is given by

$$\Delta V = V_{FS} 2^{-n}. \tag{16.2}$$

The resultant ***quantizing error*** or error voltage is defined as the difference between the ideal transfer characteristic and the digitized equivalent. This error voltage, which lies between $\pm 1/2\ \Delta V$, is plotted in Fig. 16.2(b). Note that this characteristic has rounded the input analog value to the *nearest* digital code value, which minimizes the error. There are 2^n binary code values. We discuss a number of types of ADCs in Section 16.3.

Consider the A/D example of the process of reading a thermometer. One usually looks at a volume of liquid in a glass tube, assumes that the level is proportional to the temperature, and then expresses the temperature by *rounding off* this level to the nearest discrete code on a conveniently calibrated scale.

---

* Consider a Fahrenheit thermometer calibrated from $-17°F$ to $110°F$ in the decimal code. It would take 7 bit code words to cover this range in binary ($2^7 = 128$).

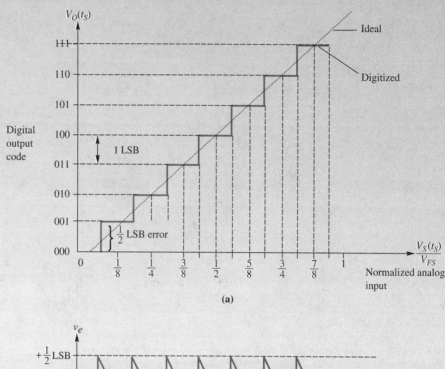

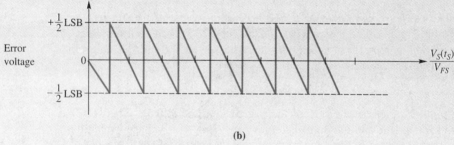

■ **FIGURE 16.2** Quantization of the linear characteristic. (a) Transfer characteristics, ideal and digitized. (b) Error voltage, ideal and digitized transfer characteristics.

On the other hand, having manipulated these variables in digital form, it is often necessary to convert them back to analog signals (D/A) in order for them to be compatible with physical machines, transducers, etc. We shall examine both of these interfaces between analog and digital signals and circuits in this chapter.

> ## IMPORTANT CONCEPTS IN THIS CHAPTER
>
> - A/D conversion includes sampling, quantizing, and coding
> - The sampling theorem and sample-and-hold circuits
> - Quantization resolution and noise
> - Binary-weighed resistor DACs and ladder DACs
> - Bipolar and CMOS examples of DAC
> - Parallel or flash ADCs and cascaded flash ADCs
> - Successive-approximation ADCs
> - Single-slope and dual-slope ADCs
> - Counting or tracking converters
> - Switched-capacitor ADCs
> - Comparisons of ADCs speed and complexity

16.1 ANALOG-TO-DIGITAL CONVERSION PROCESS

To convert a time-varying analog signal into a digital signal, it is first necessary to represent the continuous signal as a train of sample values. The samples themselves are constants, but may have any value. The next step is to quantize each sample into a discrete value. This step constitutes the actual A/D conversion. Finally, the digital sample may be recoded into a binary code word if that is desired. The ADC may accomplish the code conversion as part of the quantization process.

Sampling Theorem

It's been shown that a continuous signal may be uniquely specified by 2B samples/second, where B is the *bandwidth*, or highest frequency in a band-limited signal. This is called the **Nyquist theorem**, and it says that the signal is not compromised in any way if it is sampled at any rate higher than 2B samples/second. This analog signal can be completely restored by passing the samples through a low-pass filter having a bandwidth of B Hz. As a practical matter, the signal is usually *oversampled* (sampled at a rate somewhat higher than 2B) in order to be able to restore the signal with a less-than-ideal low-pass filter.

Sample-and-Hold Circuit

The basic idea of the **sample-and-hold** (S/H) circuit is illustrated in Fig. 16.3(a). The switch is closed at the sampling time for a short interval. The capacitor charges to the sample voltage, and then holds this voltage when the switch opens. This interval is small compared to the sampling period; thus, it is certain to be short compared to the period of the signal being sampled, and will approximate an instantaneous sample value. A more

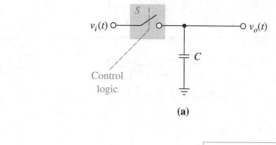

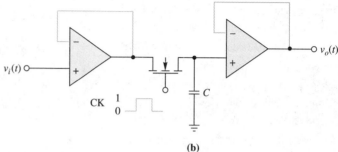

FIGURE 16.3 (a) Ideal S/H circuit. (b) Practical S/H circuit.

■ The sample value information resides in the height of the sample (voltage), not in its width. The sample time is short, to obtain a nearly instantaneous sample of the signal. The hold time may persist until the next sample time to allow the converter time to operate.

realistic S/H circuit is depicted in Fig. 16.3(b). Here the first op amp serves to buffer the input signal, so that the capacitance charges quickly from a low source resistance. The MOS switch is controlled by the sampling signal and has a low channel resistance. The second op amp has a very high input resistance and isolates the capacitance from the load.

Quantization

The sample will be held until it is time to acquire another sample. During this time, the ADC has a (nearly) constant voltage, which it must convert to a digital code word. At this time, the sample must be rounded off into one of a finite set of digital code words. This is analogous to the reading of the thermometer to the nearest degree. Suppose that the signal to be converted varies over a peak-to-peak range of 10 V, and we want our output to be accurate to within ±10 mV. **Accuracy** is the difference between the analog input voltage and the FS-weighed equivalent of the binary output code. That means that the digital samples must be no more that 20 mV apart, and it would require 10 V/20 mV = 500 code words. For a binary code, that requires 9-bit words (2^9 = 512). In general, for the ADC to obtain an accuracy of $V_e = \pm 0.5 \, \Delta V$ for an input voltage range of $\pm \overline{V}_{in} = 2 \, \overline{V}_{in}$ volts,

$$n = \log_2(\overline{V}_{in}/v_e) \text{ bits} \qquad (16.3)$$

are required. Because of the error introduced by rounding off, the sampled signal now has what is called **quantization noise**. Assuming $V_i(t)$ is sinusoidal, the mean square signal is

$$S = \overline{V}_{in}^2/2 \text{ volt}^2. \qquad (16.4)$$

The sample amplitude is approximated by the midpoint in the interval in which it lies, so that the probability density of error voltage is a constant $1/2v_e$ over the range $-v_e$ to $+v_e$. The noise power (the mean square error voltage) is

$$N = \int_{-v_e}^{v_e} \frac{q^2}{2v_e} \, dq = \left. \frac{q^3}{6v_e} \right|_{-v_e}^{v_e} = \frac{v_e^2}{3} \text{ volt}^2. \qquad (16.5)$$

Combining,

$$\frac{S}{N} = \frac{3\overline{V}_{in}^2}{2v_e^2} = 1.5 \times 2^{2n}. \tag{16.6}$$

The quantization signal-to-noise level in decibels is

$$\left(\frac{S}{N}\right)_Q = 10 \log_{10}\left(\frac{S}{N}\right) = (1.76 + 6n) \text{ dB}. \tag{16.7}$$

Observe that the more bits there are in the code word, the more resolution the sampling process has, and the greater the $(S/N)_Q$. The ADCs that we will examine in Section 16.3 all make the conversion to a binary code directly, so this can be considered one step. Many of the ADC circuits make use of a *digital-to-analog converter* (DAC) as part of their system, so we will consider the DAC circuit first.

Before we proceed, let us consider a conversion example, which will give some perspective to the system requirements.

■ Once the analog signal has been converted to digital, it has a finite S/N and the output will never be better than this $(S/N)_Q$. But in most binary communications systems it does not have to degrade further. In analog systems, S/N degrades at every point along the channel.

EXAMPLE 16.1

Suppose we have voice signal, band-limited to 3.5 kHz, having a peak voltage of 5 V (10 V$_{p-p}$).

(a) How often must it be sampled for faithful reconstruction?
(b) If each sample is encoded with 8 bits, what will be the greatest sample error, and what will be the S/N ratio at the receiver?

SOLUTION
(a) The sampling rate must exceed 7000 samples/second; 8000 is usually chosen for ease of reconstruction. The signaling rate is $8 \times 8000 = 64,000$ pulses/second, with perhaps some additional pulses for synchronization.
(b) With $n = 8$, there are $2^8 = 256$ levels, and the LSB spacing between quantization levels is 10 V/256 = 39 mV. The peak error would then be 19.5 mV. If we can assume that the peak voice signal can be maintained at 5 V, $(S/N)_Q = 1.76 + 6(8) = 50$ dB.

DRILL 16.1

Given a color TV signal band-limited to 4 MHz, what is the minimum sampling rate to satisfy the sampling theorem?

ANSWER 8,000,000 samples/second.

CHECK UP

1. **TRUE OR FALSE?** A 15-kHz sound recording would have to be sampled 15,000 times per second in order to preserve all of the signal in digital form.
2. **TRUE OR FALSE?** The analog sample voltage must be held constant while the conversion to digital format takes place.
3. **TRUE OR FALSE?** A 6-bit digital communications system could be expected to have a signal-to-noise ratio of no more than about 38 dB.
4. **TRUE OR FALSE?** An analog signal is sampled 2000 times per second. The signal out of the sampler is a digital signal before it enters the quantization circuit.

16.2 DIGITAL-TO-ANALOG CONVERSION

After processing as a digitally encoded signal, a DAC is used to obtain an equivalent analog signal. A simplified DAC is illustrated in Fig. 16.4. A binary fraction, $d_{-1} \ldots d_{-n}$, is converted to an analog voltage, V_O. A sequential presentation of digitally encoded samples, when properly LP filtered, becomes $v(t)$. There are a large number of applications in communications and audio systems and in the control of signal generation, programmable controllers, and control systems.

The output voltage of a DAC is given by

$$V_O = KV_{FS}(d_{-1}2^{-1} + d_{-2}2^{-2} + d_{-3}2^{-3} + \cdots + d_{-n}2^{-n}), \quad (16.8)$$

where K is a gain constant for the system and the digits $d_{-n}, \ldots, d_{-1}$ are binary levels (0 or 1) for an n-bit word. Note that V_O is proportional to the product of V_{FS} and the digital input. Thus, if V_{FS} can be varied, this circuit is called a *multiplying* **DAC**. Some DACs accept both polarities of V_{FS}, and some accept bipolar digital signals.

■ Bipolar signals use a 3-s state code (+V, 0, −V) encoding a binary ZERO as 0 V and a binary ONE alternately as +V and −V.

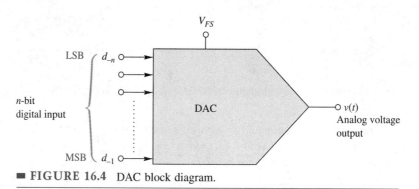

■ **FIGURE 16.4** DAC block diagram.

Binary-Weighted Resistor DAC

A DAC that uses a binary-weighted resistor array is illustrated in Fig. 16.5. The availability of each bit is established by control circuitry and the input latch. The inverting input of

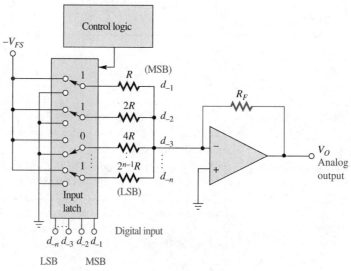

■ **FIGURE 16.5** An n-bit DAC, basic binary resistor array.

the op amp is a virtual ground due to the negative feedback resistor R_F; thus, if the MSB is a ONE, the current in the top resistor is $I_{-1} = -V_{FS}/R$, and selected currents decrease in a binary fashion as we move down in the array. Because the current into the op amp input is negligible, the sum of the resistor currents flows through R_F, and the output voltage is given by

$$V_O = R_F V_{FS} \left(\frac{d_{-1}}{2^0 R} + \frac{d_{-2}}{2^1 R} + \frac{d_{-3}}{2^3 R} + \cdots + \frac{d_{-n}}{2^{n-1} R} \right)$$

$$= \frac{2R_F}{R} V_{FS} (d_{-1} 2^{-1} + d_{-2} 2^{-2} + d_{-3} 2^{-3} + \cdots + d_{-n} 2^{-n}). \quad (16.9)$$

This connection to an op amp is known as a *current-to-voltage converter*. The scaling constant used in Eq. (16.8) is given by

$$K = 2R_F/R. \quad (16.10)$$

The LSB is d_{-n} and the MSB is d_{-1}. This technique, although simple in concept, is difficult to implement for 6-bit or larger DACs because a very large span of resistor values is required. This ratio, observed from Fig. 16.5, is

$$\frac{R_{\text{MSB}}}{R_{\text{LSB}}} = \frac{1}{2^{n-1}} \quad (16.11)$$

for an *n*-bit converter. Basic monolithic IC fabrication techniques must be expanded to include laser trimming or similar *in situ* manufacturing adjustment techniques.

R-2R Ladder DAC

To alleviate this large span of resistors, the current division technique, illustrated in Fig. 16.6, is used. The total current into the top of this **R-2R** *ladder* is

$$I_{\text{REF}} = V_{\text{REF}}/R. \quad (16.12)$$

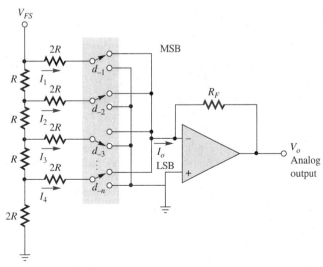

■ **FIGURE 16.6** An *n*-bit *R-2R* ladder DAC that uses current division.

An input digit of binary ONE switches a particular branch current to the summing node, whereas binary ZERO switches that branch current to ground. Note that this branch current is identical in either case, since the summing node is virtual ground. Current division for this parallel $2R$, series R network yields

$$I_1 = 2I_2 = 4I_3 = 2^{n-1}I_n \tag{16.13}$$

and each of these currents, representing a given bit from d_{-1} to d_{-n}, sums conventionally at the op amp input node:

$$I_O = -\frac{V_{FS}}{R}(d_{-1}2^{-1} + d_{-2}2^{-2} + d_{-3}2^{-3} + \cdots + d_{-n}2^{-n}). \tag{16.14}$$

The voltage out of the current-to-voltage converter is

$$V_O = \frac{V_{FS}}{R}R_F(d_{-1}2^{-1} + d_{-2}2^{-2} + d_{-3}2^{-3} + \cdots + d_{-n}2^{-n}). \tag{16.15}$$

and with

$$K = -R_F/R, \tag{16.16}$$

Eq. (16.15) also becomes Eq. (16.8). Although twice as many resistors are required compared to the binary-weighted resistor array, the ease in obtaining only a 2:1 resistor ratio makes this an attractive alternative for DACs having more than six bits.

Bipolar DAC

Figure 16.7 shows a BJT 8-bit DAC, the MC1408.[1] Note that the output current I_O flows into the switches from the top, and the ladder's reference current flows out the bottom of the ladder. The reference current is generated internally, programmed by a current into terminal 14, $I_{REF} = V_{REF}/R_{14}$ (or a negative V_{REF} could be used similarly at pin 13).

To aid in understanding the digital switches, Fig. 16.8 shows a simplified circuit for one of the eight switches. The switch, Q_1 is OFF when the input bit d_n is HIGH (>2.1 V), because the emitter of Q_1 is clamped at ~ 2.1 V by diode D_1 (assuming $V_D = 0.7$ V). Then D_2 is OFF, and the ladder current for this "rung" must flow through the unity-gain amplifier, Q_2 and Q_3, and D_3 to combine with the other output current terms. The small current sources, I_B, together with Q_4, serve to bias the amplifier to maintain the voltage into the ladder constant. When the input bit is LOW (~ 0 V), the switch, Q_1, is ON, pulling the collector of Q_2 up to 0.7 V, where it is clamped by D_2. This unhooks D_3, because the output is 0 V, and the ladder current now is supplied through Q_1.

Figure 16.9 is a schematic diagram of a typical application of this DAC. The resistance R_{15} is chosen equal to R_{14} so that both inputs to the current reference op amp will see the same source resistance.

CHAPTER 16 ANALOG-TO-DIGITAL AND DIGITAL-TO-ANALOG CONVERSION

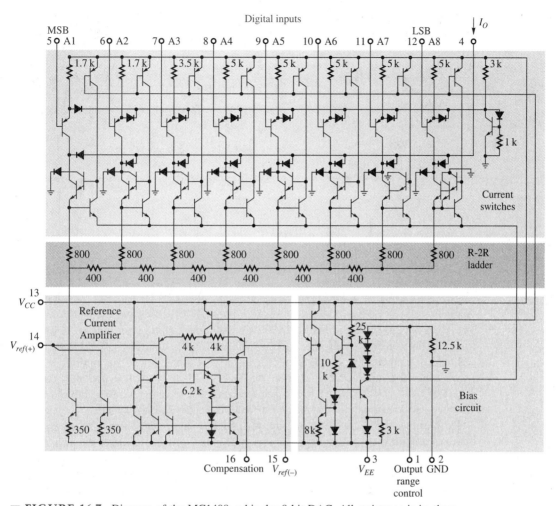

■ **FIGURE 16.7** Diagram of the MC1408, a bipolar 8-bit DAC. All resistance is in ohms.

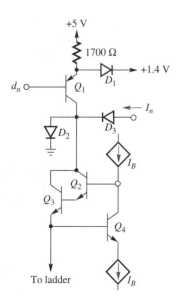

■ **FIGURE 16.8** Simplified diagram of one switch in the MC1408 DAC.

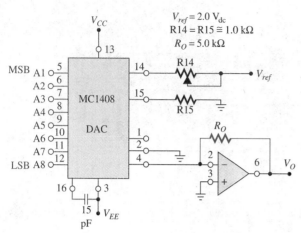

FIGURE 16.9 Typical MC1408 DAC circuit.

Theoretical V_O

$$V_O = \frac{V_{ref}}{R15}(R_O)\left[\frac{A1}{2} + \frac{A2}{4} + \frac{A3}{8} + \frac{A4}{16} + \frac{A5}{32} + \frac{A6}{64} + \frac{A7}{128} + \frac{A8}{256}\right]$$

Adjust V_{ref}, R14, or R_O so that V_O with all digital inputs at high level is equal to 9.961 V.

$$V_O = \frac{2\,V}{1\,k}(5\,k)\left[\frac{1}{2} + \frac{1}{4} + \frac{1}{8} + \frac{1}{16} + \frac{1}{32} + \frac{1}{64} + \frac{1}{128} + \frac{1}{256}\right]$$

$$= 10\,V\left[\frac{255}{256}\right] = 9.961\,V$$

CMOS DAC

The CMOS version of the ladder DAC has the same topology as Fig. 16.7, but each switch is implemented with the double-buffered CMOS circuit of Fig. 16.10. CMOS technology has a number of advantages over the BJT technology. The input impedances for the transistors are much higher, there are fewer problems with the temperature compensation of bias circuitry, and CMOS is more adaptible to integrating digital and analog circuitry on the same die. Furthermore, CMOS is nearly an ideal switch.

In the past, BJT converters had an advantage over CMOS in that speed could be obtained at the expense of greater currents, since the g_m of BJT's increases with current. But as circuit density continues to increase, forcing a trend to lower power, CMOS begins to be more attractive than BJTs. Thus, as we observe higher levels of integration, with DACs and ADCs imbedded in ASICs, CMOS is the technology of choice, together with BiCMOS.

Performance

We have seen that the resolution of a DAC circuit is determined by the number of bits used. Thus a very important measure of system performance is established when the

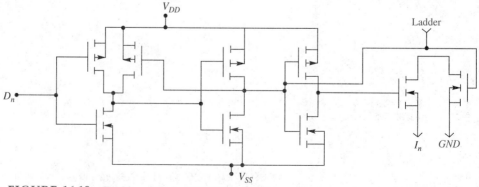

FIGURE 16.10 CMOS switch for CMOS DAC.

resolution has been specified. The accuracy, or worst case deviation from the ideal response, of the DAC is a function of the terms in Eq. (16.8). This involves the accuracy of the reference voltage, the accuracy of the op amp current-to-voltage converter, the precision of the resistor network, the matching of the current switches, and the reference-voltage-to-reference-current circuits. These last four sources of error are normally parameters of the DAC integrated circuit (IC).

The speed of the DAC is referred to as the *settling time*, the elapsed time from the application of the digital word until the output has reached to within some defined distance from its final value. Both the current switches and the op amps have some delay.

■ Integrated circuit manufacturers are able to control resistor and device matching to close tolerances, so that the burden for accuracy falls on the voltage reference.

EXAMPLE 16.2

A DAC circuit such as that shown in Fig. 16.9 has a reference voltage of -5 V.

(a) We want the reference current to be 2 mA, and the full-scale output voltage to be $+10$ V. Find proper values for R_{14}, R_{15}, and R_O, and explain how the negative supply could be used.
(b) For a digital input word of 10110001, what is the analog output voltage?

SOLUTION

(a) The negative supply would be connected to R_{15}, while R_{14} must be grounded.
$R_{14} = R_{15} = 5$ V/2 mA $= 2500$ Ω, $R_O = 10$ V/2 mA $= 5$ kΩ.
(b) $V_O = 10(128 + 32 + 16 + 1)/256 = 6.91$ V.

DRILL 16.2

What is the usable range of a current-to-voltage converter such as that found in Figs. 16.5, 16.6, and 16.9?

ANSWER $V_O = -I_{\text{IN}} \times R_F$ must not exceed the saturation voltages of the op amp, which is usually 1 or 2 V less than its power-supply voltages.

CHECK UP

1. **TRUE OR FALSE?** The output of a DAC is proportional to a binary fraction times a reference voltage, sometimes called the full-scale voltage.
2. **TRUE OR FALSE?** The weighted-resistor DAC is difficult to fabricate because of the precision required of the resistors.
3. **TRUE OR FALSE?** The ladder DAC requires only two sizes of precision resistors.
4. **TRUE OR FALSE?** CMOS DAC circuits are inherently faster than bipolar DAC circuits.

16.3 ANALOG-TO-DIGITAL CONVERSION

Conceptually, A/D conversion is usually illustrated as in Fig. 16.11, where $v(t)$ is the analog input and $d_{-1} \ldots d_{-n}$ is the digital output. The 8-bit ($n = 8$) ADC is widely used, with a 14-bit ADC technologically realizable at a reasonable cost. As has been discussed, a time-varying signal must be sampled and held while the conversion takes place, and this process must be completed in time for the next sample to be taken.

We now describe several types of A/D conversion. These involve the *parallel* or *flash converters*, the *successive-approximation converters*, the *digital-tracking* or *servo-type converters*, and the *integrating A/D converters*, including the *single-slope* and *dual-slope* versions. All use analog and digital building blocks presented previously in this text. Often, both the analog and digital circuits, as well as computer control circuitry, are included within the same integrated circuit.

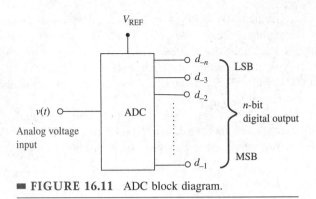

FIGURE 16.11 ADC block diagram.

Parallel or Flash ADC

The term *flash* refers to the quickness of this circuit. The parallel or flash-type converter ideally requires only one clock cycle. This is accomplished by applying the sampled analog input voltage simultaneously to $2^n - 1$ comparators, with each comparator requiring a separate reference input voltage. The digital n-bit word is defined by

$$D = d_{-1} + d_{-2} + d_{-3} + \cdots + d_{-n}. \tag{16.17}$$

A 3-bit parallel converter is shown in Fig. 16.12. Required are $2^n - 1 = 2^3 - 1 = 7$ comparators. That means that seven reference levels in addition to zero are required. To establish the $\frac{1}{2}$LSB voltage level, the first comparator input voltage starts at $\frac{1}{14}V_{REF}$, the reference voltage. Here we use only $7R$ in the reference divider, so the LSB resolution is $\frac{1}{7}V_{REF}$, and the digital output would be

$$V_s(t_s) = \frac{V_{REF}}{7}(d_{-1}4 + d_{-2}2 + d_{-3}1). \tag{16.18}$$

Digital logic is then used to decode the comparator outputs for the 3-bit digital output. Note that this particular ADC would saturate (all comparators are ON) when $V_s(t_s) > \frac{13}{14}V_{REF}$, and the error would exceed the $\frac{1}{2}$LSB voltage level for $V_s(t_s) > V_{REF}$. The system clock is used to control the latches and the decoding logic.

There are many possible variations on this basic idea. Different offsets in the comparator reference voltages are possible by merely connecting the bottom of the resistor array to a nonzero reference voltage. This would allow inputs that might be positive or negative. Different decoding schemes can be used also, for instance two's-complement representation of signed voltages.

Although this parallel technique is very fast, it can also be very complex for high-resolution A/D conversion. For example, an 8-bit converter requires $2^n - 1 = 2^8 - 1 = 255$ comparators, a 256-precision-resistor array, and a 255-to-8 decoder. Anything larger than this is considered impractical with BJT technology, but with CMOS VLSI, it is becoming economically feasible to produce very high-speed ADCs. Note that the speed is limited not only by the comparator settling time, but also by the comparator input capacitance, which imposes time constants on the resistor array. Here again the CMOS comparators have an advantage, with low input capacitance. The accuracy of the flash ADC depends on comparator sensitivity, the precision of the resistor divider, and the reference voltage.

CHAPTER 16 ANALOG-TO-DIGITAL AND DIGITAL-TO-ANALOG CONVERSION

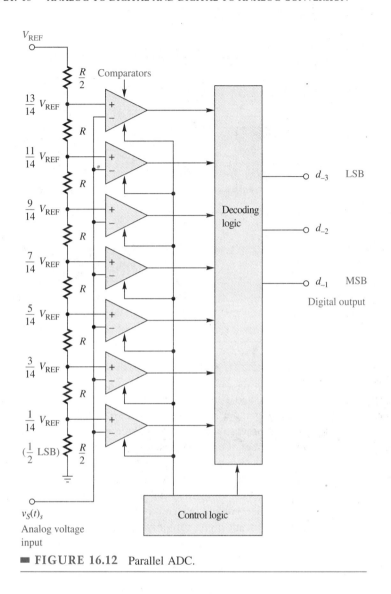

FIGURE 16.12 Parallel ADC.

■ Note that the accuracy of this ADC is dependent on the voltage reference, as it is in all these converters. The other factors are more easily controlled by the manufacturer of the IC.

DRILL 16.3

A 6-bit flash ADC is connected in a positive, monopolar manner, analogous to Fig. 16.12. We desire 0.1-V resolution.

(a) What must be the reference voltage?
(b) What is the range of this ADC?
(c) What is the worst-case quantization error, assuming perfect components?

ANSWERS

(a) 6.3 V (assuming $R/2$ for top and bottom resistors)
(b) 0 to 6.35 V
(c) 0.05 V

EXAMPLE 16.3

What modifications should be made to the ADC of Fig. 16.12 so that it could decode both positive and negative input voltage?

SOLUTION The negative end of the resistor divider must be connected to a negative V_{REF}. Then the range would be $\pm V_{REF}$. The decoder logic could be constructed to output the digits in two's-complement signed number notation.

Cascaded-Flash ADC

A block diagram of an ADC circuit that trades some speed for less complex logic is shown in Fig. 16.13. This arrangement uses a cascade of two $n/2$-bit flash ADCs to obtain a total decoding capability of n bits. The analog signal is sent to the first flash ADC,

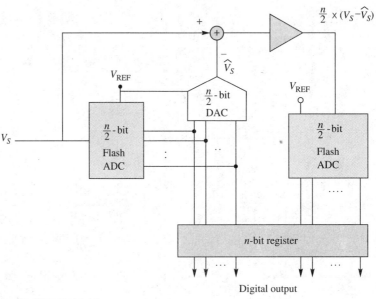

FIGURE 16.13 Block diagram of cascaded flash ADC.

which has a coarse resolution of $2^{n/2} \times$ LSB. Then its decoded output, which contains the $n/2$ most significant bits of the total conversion, is input to an $n/2$-bit DAC. The resultant DAC output is subtracted from the original analog signal, which leaves the difference voltage. Now the remainder is decoded by the second flash ADC to obtain the $n/2$ least significant bits. This process requires about three times the conversion time of a single flash ADC, but the number of precision resistors and comparators is reduced from $2^n - 1$ to $2(2^{n/2} - 1)$. For a 12-bit system, this is a reduction from 4095 to 126 comparators!

If the cascaded flash ADC system is to work properly, some precautions must be observed. The accuracy of *both* flash converters and the DAC must equal that desired of the overall system. The difference voltage presented to the second ADC must always be positive, because the least significant digits can only add to the final result. This can be assured by proper offsets in the ADCs. Once again the accuracy is dependent on the reference voltage.

Another method of accomplishing the cascaded flash ADC is to use a multiplex arrangement so that only one flash converter is used. First the flash ADC is connected to the analog input to obtain the most significant half of the output bits. These bits are latched, and the DAC uses these bits to generate the coarse signal to be subtracted from the input. The difference signal is amplified, which effectively improves the resolution of the second conversion. Finally, the ADC is switched to convert this amplified difference signal and obtain the least significant half of the output bits. This design is more hardware efficient than the first design.

EXAMPLE 16.4

A 12-bit cascaded flash ADC system is monopolar, with two 6-bit ADCs and a 6-bit DAC. For each ADC the reference voltage is 6.4 V, and the resistor divider has 64 equal resistors.

(a) What is the resolution of the first ADC?
(b) Assuming that the input analog signal remains in the range of 0 to 6.3 V, what is the maximum error voltage after the first conversion?

(c) How much must the difference signal be amplified before the second ADC to allow the system to have 12-bit resolution?
(d) What is the maximum error in the final output?

SOLUTION
(a) The step size is 6.4/64 = 0.1 V.
(b) This ADC has no offset voltage, so the error could approach 0.1 V.
(c) 64 times.
(d) 0.1/64 = 1.56 mV.

DRILL 16.4
Refer to the cascaded flash ADC of Example 16.4.

(a) For an analog input voltage of 3.19 V, what will the digital output be, assuming ideal components?
(b) What is the quantization error?

ANSWERS
(a) 011111111001
(b) 0.15 mV

Successive-Approximation ADC

The *successive-approximation* (SA) ADC has both relatively high speed and high resolution, with the result that it is the most popular of the converters. Figure 16.14 illustrates such a system. An *n*-bit *SA register* (SAR) is used to accumulate the output digits as follows: At the beginning of a conversion, the register is cleared. The MSB is set to ONE, causing the DAC to output one-half of full-scale voltage. If the DAC output is larger than the input sample voltage, the MSB is reset; if not it remains set. On the following clock cycle, the next MSB is tested; if DAC output from the new approximation exceeds the input voltage, then this bit is reset. This testing of each bit proceeds until the LSB has been tested, requiring n clock cycles. The SAR now contains the desired digital representation of the analog voltage. The accuracy of this system depends on the DAC and its reference voltage, as well as on the comparator accuracy. It should be apparent that the quantization error of this system as described could approach +1 LSB.

The successive-approximation ADC, while much less complex then the flash-type ADC, shares the very desirable characteristic of a fixed conversion time. This makes them particularly suited to microprocessor applications if the speed of the flash ADC is not required.

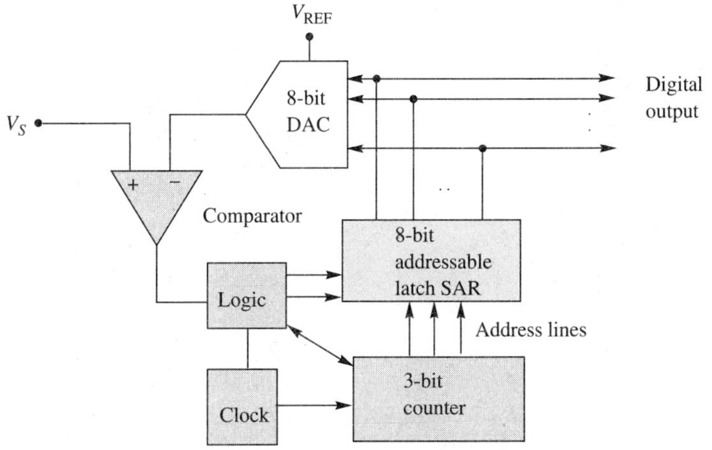

■ **FIGURE 16.14** Block diagram of an 8-bit successive-approximation ADC.

DRILL 16.5

Suppose that the reference voltage of the SA ADC in Example 16.5 has dropped 0.5% to 10.19 V. For the same 7.99-V input, what would be the output digital indication?

ANSWER 11001000, representing 8.00 V on the designed scale. The reference error is in the opposite direction from the quantization error.

EXAMPLE 16.5

Given an 8-bit SA ADC with a reference voltage of 10.24 V. The voltage to be measured is 7.99 V. Neglect any error in the reference or in the DAC. Give the results of each of the successive approximations.

SOLUTION
1. 10000000 $\Rightarrow$ 5.12 V < 7.99 V, so MSB remains set.
2. 11000000 $\Rightarrow$ 7.68 V < 7.99 V, so next MSB remains set.
3. 11100000 $\Rightarrow$ 8.96 V > 7.99 V, so next bit is reset.
4. 11010000 $\Rightarrow$ 8.32 V > 7.99 V, so next bit is reset.
5. 11001000 $\Rightarrow$ 8.00 V > 7.99 V, so next bit is reset.
6. 11000100 $\Rightarrow$ 7.84 V < 7.99 V, so next bit remains set.
7. 11000110 $\Rightarrow$ 7.92 V < 7.99 V, so next bit remains set.
8. 11000111 $\Rightarrow$ 7.96 V < 7.99 V, so LSB remains set.

The quantization error is 7.96 − 7.99 = −0.03 V, or −0.3% of full-scale. 1 LSB is 0.4% full-scale.

Single-Slope ADC

Figure 16.15 shows a block diagram of a *single-slope* ADC. The input voltage, $0 < V_S < V_{CC}$, is the sample analog input voltage and is input to the noninverting terminal of a comparator. Initially, at $t = 0$, S_1 is closed and S_2 is open, and the capacitor C is discharged. The counter is at zero. The comparator output voltage is called the gated voltage, v_G, and is plotted in Fig. 16.16(a).

At $t = 0$, the control circuitry opens S_1 and closes S_2. A constant current source charges C. The voltage on the capacitor is given by

$$v_C(t) = \frac{1}{C}\int_0^t I_O \, dt + V(0) = \frac{I_O t}{C} + V(0) \tag{16.19}$$

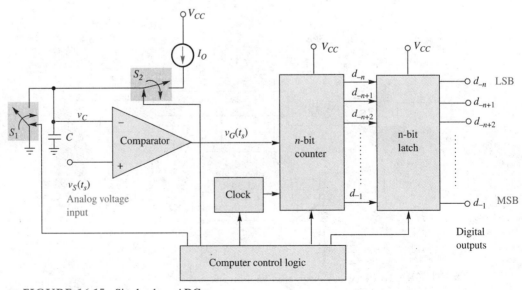

FIGURE 16.15 Single-slope ADC.

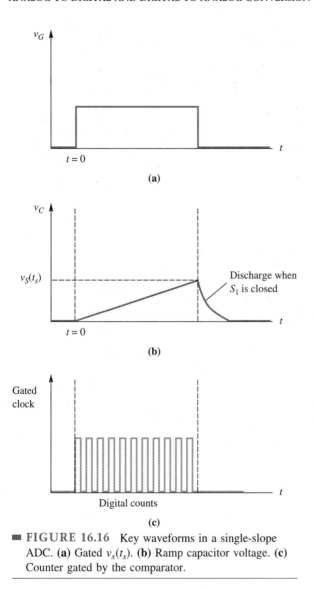

FIGURE 16.16 Key waveforms in a single-slope ADC. **(a)** Gated $v_s(t_s)$. **(b)** Ramp capacitor voltage. **(c)** Counter gated by the comparator.

and is the equation for a linear ramp; see Fig. 16.16(b). Voltage v_C is input to the inverting terminal of the comparator. As long as $v_C < V_S$, the comparator output is HIGH, and the $(n + 1)$-bit counter is enabled and digitally counts at the clock rate; see Fig. 16.16(c). When $v_C = V_S$, the comparator output, v_G, switches LOW, and the counter stops. The counter retains a digital value equivalent to V_S. The clock rate must be designed such that the full-scale input voltage does not cause the counter to overflow. The entire cycle repeats with the opening of S_2 and the closing of S_1 to discharge C. The counter is also reset to zero.

Even though the single-slope converter is easy to build, its accuracy is strongly dependent on the degree of capacitor leakage and aging, an accurate stable current source, and an accurate and stable clock. The measurement requires up to 2^n clock pulses, and the clock period must be greater than the delay time of the comparator. This makes the single-slope ADC a relatively slow ADC. Note that conversion time varies with the voltage to be converted.

■ The single-slope ADC depends not only on the accuracy of a reference source, but also on an accurate clock and a low-leakage, constant value capacitor.

DRILL 16.6

What is the useful range for the input signal $v_s(t_s)$ for the ADC of Fig. 16.15?

ANSWER The current source must not saturate, thus $v_s(t_s)$ must not exceed the maximum voltage for which the current source is linear (less than V_{CC}).

EXAMPLE 16.6

The ADC shown in Fig. 16.15 uses a current source of 100 μA and a capacitance of 0.01 μF. We want each count of the counter to represent 0.1 V. What should the clock rate be?

SOLUTION The slope of the ramp is $I/C = 10^{-4}$ A/10^{-8} F = 0.01 V/μs, therefore the period of the clock is 10 μs, or the frequency is 100 kHz. The ADC can measure 10 V in 1 ms.

Dual-Slope ADC

Many of the error problems inherent in the single-slope ADC are eliminated in the *dual-slope* ADC. A block diagram is given in Fig. 16.17. This time, let $V_S < 0$ be the sample of the analog input voltage. Initially, the capacitor C is discharged by the closed switch S_2, and the counter is at zero. At $t = 0$, S_1 is connected to V_S for a fixed time T_1 as shown in Fig. 16.18, and S_2 opens. The integrator output voltage is

$$v_I(t) = \frac{1}{C}\int_0^t I\, dt = \frac{1}{C}\int_0^t \frac{-V_S}{R}\, dt = -\frac{V_S t}{RC}. \tag{16.20}$$

This is a positive slope, since V_S was chosen negative. As soon as the ramp starts up, the comparator output goes HIGH, and the counter begins counting at the clock rate. The fixed time T_1 is determined by the time it takes the counter to reach overflow (N_1 clock pulses) and

$$v_I(T_1) = -\frac{V_S T_1}{RC}. \tag{16.21}$$

At time T_1, switch S_1 selects the reference voltage V_{REF}, and v_I integrates back down with a slope of $-V_{REF}/RC$. The counter begins a new count from zero, since it overflowed at T_1. The downward integration requires time T_2, and, since the charging voltage equals the discharging voltage,

$$-\frac{V_S T_1}{RC} = \frac{V_{REF} T_2}{RC}, \tag{16.22}$$

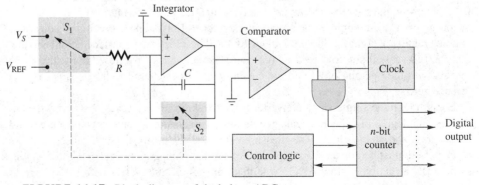

■ **FIGURE 16.17** Block diagram of dual-slope ADC.

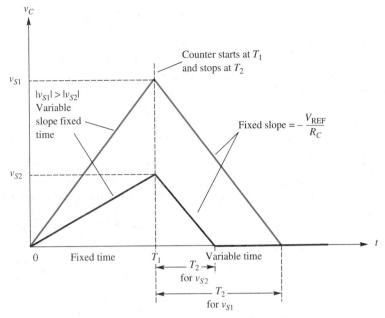

FIGURE 16.18 Graphical interpretation of the dual-slope ADC.

which results in $-V_S T_1 = V_{REF} T_2$. Now $T_1 = N_1$ clock periods, and $T_2 = N_2$ clock periods, so that $-V_S N_1 = V_{REF} N_2$. Finally,

$$-V_S = V_{REF} \frac{N_2}{N_1}. \tag{16.23}$$

What we should observe here is that as long as R, C, and the clock frequency do not change in the short term (one measurement cycle), they cancel out of the equation. We need know only V_{REF} and the two counter readings. Of course, N_1 is a constant of the design, and could be some convenient value like 1000. Note that if the ADC were to measure positive V_S, a negative V_{REF} would be required. After $T_1 + T_2$, the controller will close S_2 (the capacitor voltage will be zero already), save the counter value N_2, and then clear the counter. A new measurement may now begin.

EXAMPLE 16.7

Given an ADC system such as that diagrammed in Fig. 16.17, let the counter consist of three decades, the clock frequency be 1 MHz, $V_{REF} = -5$ V, $R = 100$ kΩ, and $C = 0.01$ μF.
(a) For an analog input voltage $V_S = 4$ V, what will the counter read?
(b) How long did the measurement take?
(c) What is the range of this ADC?

SOLUTION The fixed time $T_1 = 1000/f = 1$ ms. From Eq. (16.20) the slope of the ramp $0 < T < T_1$ is $-V_S/RC = -4000$ V/s, and the ramp reaches -4 V. For the time $T_1 < t < T_2$ the slope is $-V_{REF}/RC = 5000$ V/s, and $T_2 = 4/5000 = 0.8$ ms.

DRILL 16.7

What could be changed in the ADC of Example 16.7 so that the counter would read volts directly with no scale factor?

ANSWER If the reference voltage were -10 V, the counter would read directly in one-hundredths of a volt. Insert a decimal point after the first digit.

(a) The counter will reach 0.8 ms × 1 MHz = 800.
(b) $T_1 + T_2 = 1.8$ ms.
(c) The counter will reach 999 at $-V_{REF}(0.999) = 4.995$ V. If one kept track of the overflow, the range could be increased, but the ramp would reach a voltage lower than -5V, and the power supplies would need to be large enough to maintain linear operation on the op amp integrator.

Counting or Digital-Tracking Converter

This type of ADC uses a DAC to generate a staircase approximation to a ramp instead of a constant current source charging a capacitor to generate a ramp. As shown in Fig. 16.19(a), the sampled analog voltage, V_S, is input to the noninverting terminal of a

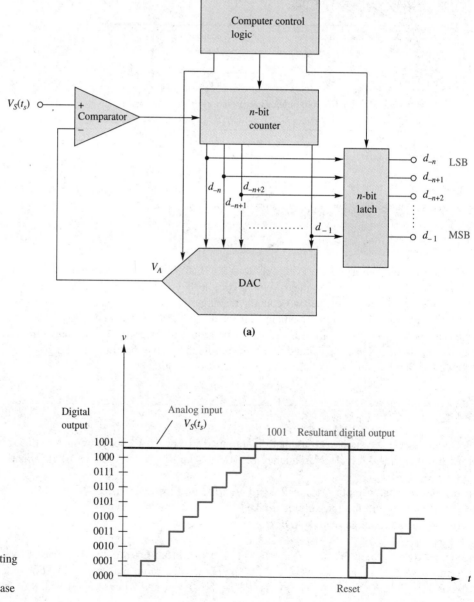

■ **FIGURE 16.19** Counting ADC. **(a)** counting ADC block diagram. **(b)** Staircase $v_A(t)$ comparator inputs.

comparator. At $t = 0$, the counter is set to zero; consequently, the DAC voltage, V_A, which is input to the inverting terminal of the comparator, is also zero. When $V_S > V_A$, the counter will increment UP at the clock rate. This is illustrated in Fig. 16.19(b). When V_A exceeds V_S, the comparator output changes sign, which stops the counter. The resultant digital word, $d_{-n} \ldots d_{-1}$, is stored in a latch, and the control circuitry resets the entire cycle of operations.

An extension of this type of ADC, useful for tracking small changes in V_S, is illustrated in Fig. 16.20(a). As in the previous case, at $t = 0$ the UP-DOWN counter is set to zero;

■ The counting ADC is subject to the reference voltage accuracy. Note also that the conversion time depends on the voltage to be converted, as in the slope converters.

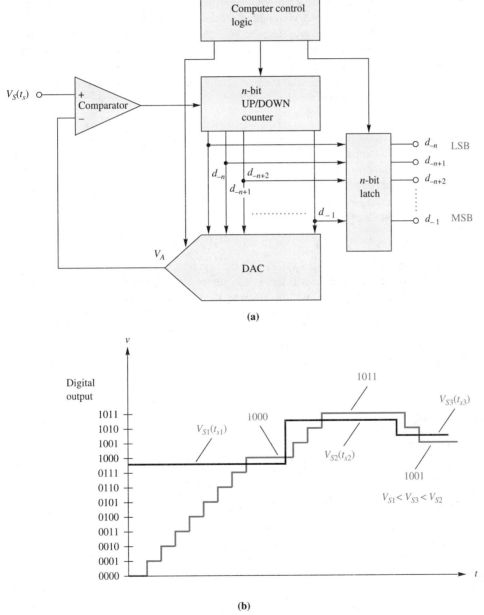

■ **FIGURE 16.20** Tracking-servo ADC. (a) Tracking-servo ADC block diagram. (b) Tracking digital output.

DRILL 16.8
What is the resolution and the maximum error of a counting ADC that uses an 8-bit counter, an 8-bit DAC, and a 5.12-V reference?
ANSWER Both the resolution and the maximum quantization error would be one LSB = 0.02 V. This assumes negligible error in the reference or the DAC.

consequently the D/A voltage, V_A, which is input to the comparator, is also zero. When $V_S > V_A$, the counter will increment UP at the clock rate. This is illustrated in Fig. 16.20(b). When V_A exceeds V_S, the comparator output changes sign, which increments the counter DOWN. The resultant digital word exhibits one LSB of ambiguity. As shown, $D = 1000$ for V_{S1}, $D = 1011$ for V_{S2}, and $D = 1001$ for V_{S3}. Small changes in V_S are reflected quickly in the digitized equivalent.

Switched-Capacitor ADC

Especially appropriate for CMOS technology is a method of ADC that uses the successive-approximation technique by means of *charge redistribution*.[2] Instead of using precision resistances, this circuit uses binary-weighted precision capacitances, which are relatively easy to obtain in MOS technology. Refer to Fig. 16.21 for a simplified block diagram of the major features of the system. Dummy capacitance C_D has the weight of the LSB capacitance C_0, and it serves to make the total capacitance of the array equal to $2C$.

The following is a presentation of the basic steps needed to take one sample of voltage V_S, and convert it to a binary word:

1. S_o is closed, and switches S_{-1} to S_D all connect their respective C's to the unknown voltage V_S. This means that the voltage on all C's is tracking the value of V_S, since the comparator input is at virtual ground.
2. S_C is opened, and now the comparator input floats, and the paralled capacitors have a voltage sample of V_S, and the combined charge is $Q_T = 2C\overline{V}_S$.
3. S_{-1} is switched to V_{REF}, and all the other switches connect their respective C's to ground. Total charge is conserved, so that

$$2CV_{REF} = C(V_{REF} - V_i) + C(0 - V_S) \quad \text{thus} \quad V_I = \tfrac{1}{2}V_{REF} - V_S.$$

This means that if $V_S > \tfrac{1}{2}V_{REF}$, V_i will be negative, the comparator output will be HIGH, and the ADC logic will leave S_{-1} set on V_{REF}, and determine that the MSB is ONE. If $V_S < \tfrac{1}{2}V_{REF}$, V_I will be positive, the comparator output will be LOW, the logic will reset S_{-1} to ground, and determine that the MSB is ZERO.

4. Now S_{-2} is switched to V_{REF}, and the charge again redistributes. This time the algorithm determines whether the second MSB should be ONE ($\tfrac{1}{2}C$ left connected to V_{REF}), or ONE ($\tfrac{1}{2}C$ reconnected to ground).

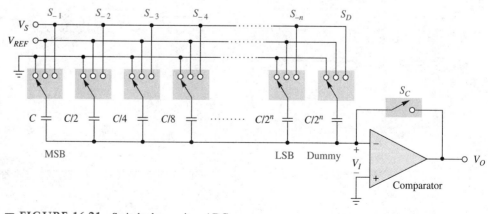

■ **FIGURE 16.21** Switched-capacitor ADC.

5. This procedure continues, until all of the bits have been tested, as in the resistive version of the SA ADC. The final position of the switches will correspond to the binary result stored in SAR, with a ONE indicating that the respective switch is still connected to V_{REF}, and a ZERO meaning that the switch has been reset to ground.

As has been indicated, this type of circuitry lends itself well to the CMOS technology, and is increasingly the ADC of choice for data conversion.

Voltage References and Clocks for Converters

We have seen that the accuracy of every conversion scheme discussed in this chapter depends on the accuracy of a reference voltage. Indeed, accurate voltage references are of great concern to most circuit designers. Reference diodes using avalanche or Zener breakdown have been considered in Chapter 3. Zener-referenced bias circuits have the drawback that to obtain zero temperature coefficient (TC), a breakdown diode would be avalanching, which would generate noise which would be objectionable in a sensitive converter.

The preferred choice for voltage references is the so-called *band-gap* reference circuit, which combines the negative TC of the BJT base-emitter V_{BE} (~ -2 mV/°C) with the positive TC of V_T ($\sim +0.086$ mV/°C). Proper choice of circuit and device parameters allow the circuit designer to fabricate a voltage (or current) reference with zero TC at say 300K. This subject is treated in detail in Ref. 4.

Each of the converters also requires a clock. Most converters are used in a larger system, and the clock is shared with other components. Typically, a crystal oscillator (Chapters 11 and 14) provides the master clock, and other timing is derived from this clock. The accuracy of the crystal clock would be sufficient for the converters, even the single-slope ADC, which is sensitive to clock rate.

CHECK UP

1. **TRUE OR FALSE?** The n-bit flash ADC requires $n + 1$ comparators.
2. **TRUE OR FALSE?** The n-bit cascaded-flash ADC requires $n/2$ clock pulses to make a conversion.
3. **TRUE OR FALSE?** The n-bit successive-approximation ADC requires n clock pulses to make a conversion.
4. **TRUE OR FALSE?** The n-bit counting ADC is by far the slowest of the converters presented here.
5. **TRUE OR FALSE?** The single-slope ADC, while slow, is very accurate.

SUMMARY

It is becoming more and more common to convert analog signals to digital form for processing or communications, and then convert them back to analog. In communication systems, it is a common method for multiplexing many signals into a common channel, called ***time-division multiplexing*** (TDM). It is also a common method for communication systems to trade bandwidth for signal-to-noise ratio. There are countless other applications.

The process of converting a time-varying analog signal to a digital signal involves sampling frequently enough to specify the signal completely, and then quantizing each sample into a digital code word. If the analog signal is band-limited with a bandwidth B, it has been determined that the signal can be represented by at least 2B samples/second without any loss of information. When the analog signal varies appreciably during the conversion time, it is necessary to use the sample-and-hold circuit. We have seen how the maximum error in the quantizing process is affected by the number of bits in the sample code word.

The need to work with both analog and digital techniques is even more explicit when studying A/D and D/A converters. Current fabrication technology permits the combination of many of these functions on the same die. This is especially true with CMOS, where each of the various ADC and DAC schemes that we have discussed has been integrated into self-contained converter systems, or into ASIC systems.

Using the basic IC building blocks and techniques, we discussed a variety of ADC and DAC circuits. Table 16.1 summarizes the key features of several ADCs.

■ Note that for a fixed comparator settling time T_s, clocking slower than the minimum rate of $1/T_s$ increases the resolution, but not the accuracy.

TABLE 16.1 Comparison of ADC Characteristics

Type of n-bit ADC	Relative Conversion Time	Complexity (# of transistors)	Typical CMOS 8-bit ADC Conv. Time[3]
Flash	Fastest	Very complex	100 ns
Cascade flash	Very fast	Complex	1 μs
Successive-approx.	Fast	Moderate	17 μs
Counting	Slow	Simple	125 μs*
Integrating (slope)	Slow	Simple	1–30 ms#

*Assumes DAC settling time of 100 ns and comparator settling time of 300 ns.
#The longer times on slope ADCs reflect 4½ decimal digits (>14 bits).

One ADC technology that is especially suited to CMOS cirucits is the *charge-redistribution* method of successive-approximation. This scheme uses binary-weighted capacitors rather than resistors in the internal DACs ladder. Another recent innovation is BiCMOS converters, where the best features of fast bipolar switches and low-power CMOS logic can be combined.

SURVEY QUESTIONS

1. Name the steps requried in converting an analog signal to a digital signal.
2. State the sampling theorem in your own words.
3. Describe the sample-and-hold function, and sketch a simple S/H circuit.
4. How many digital code words are required to represent a 25-V_{p-p} analog signal with no more than 1-mV error?
5. What is the main difficulty in implementing a weighted-resistor DAC?
6. Draw a block diagram of a ladder DAC.
7. What technologies are used to fabricate a DAC?
8. Sketch a block diagram and briefly describe the flash ADC system, the cascaded flash ADC system, the successive-approximation ADC system, the counting ADC system, and the double-slope ADC system.
9. Rank order the above five ADC systems as to speed of conversion.
10. Rank order the above five ADC systems as to circuit complexity (cost).

CHAPTER 16 ANALOG-TO-DIGITAL AND DIGITAL-TO-ANALOG CONVERSION   825

PROBLEMS

Problems marked with an asterisk are more challenging.

16.1 An audio signal band-limited to 15 kHz is sampled with a 16-bit ADC. What is the output S/N when the signal is a full-scale sine wave?

16.2 What is the minimum bit rate out of the system of Problem 16.1?

16.3 Digital telephone voice channels use 8000 samples/second, and are encoded 8 bits/sample. What would be the maximum frequency allowed in the analog voice signal?

16.4 Equation (16.7) would predict a quantization S/N of about 50 dB for the 8-bit system of Problem 16.3. What assumptions were made in developing this equation that are probably not valid for a voice signal?

16.5 Twenty-four of the voice channels described in Problem 16.3 are time-division multiplexed, with a synchronization bit added after each of the 24 channels has been sampled once. What is the total bit rate of this TDM system?

16.6 Specify the number of bits needed for an ADC requiring a resolution of less than (a) 3%, (b) 1%, and (c) 0.1%.

16.7 Sketch and label the digitized form of $v(t) = 10 \cos \omega t$ volts for one period using a 4-bit ADC.

16.8 A 1-kHz, 2-V peak sine wave is sampled without a S/H circuit. How much change would occur during conversion if the ADC process requires 10 μs?

16.9 What gate voltages would be needed to control the FET switch in the S/H circuit of Fig. 16.3 if $v_i(t)$ varies between ± 1V? Assume the NMOS FET has $V_t = +2$ V. and $k = 1$ mA/V^2. Estimate R_{ON} for your answer.

16.10 Design the binary resistor ladder for an 8-bit DAC of the form illustrated in Fig. 16.5. Assume $R_{MSB} = 1$ kΩ. Discuss the resistor precision and tolerance required. Use these results to comment on the realization with IC technology.

16.11 Refer to the ladder DAC of Fig. 16.6. Suppose the $2R$ resistor for the MSB were in error by 1%. What effect would that have on the output voltage?

16.12 Suppose that the NMOS switches in the CMOS DAC of Fig. 16.10 have $V_t = 2$ V, $k = 1$ mA/V^2, and that $V_{DD} = 5$ V. Estimate the value of R_{ON}.

16.13 Design a 2-bit flash ADC of the form shown in Fig. 16.12. This includes the design of the decoding logic using combinations of NAND and NOR gates and inverters.

16.14 Design the decoding logic for the 3-bit ADC illustrated in Fig. 16.12.

16.15* Show how a PAL could be used to provide the decoding logic for the 3-bit ADC illustrated in Fig. 16.12.

16.16 Consider the cascaded flash ADC diagram of Fig. 16.13. Why does the accuracy of the first ADC have to be so much greater than its resolution?

16.17 Consider the cascaded flash ADC diagram of Fig. 16.13. Sketch an op amp circuit that will perform the operation $V_2 = (n/2)(V_S - \hat{V}_S)$.

16.18 Give the results of each bit test in an 8-bit SA ADC that has a reference voltage of 10.00 V if the analog input voltage is 9.00 V.

16.19 Refer to the block diagram of a SA ADC in Fig. 16.14. A delay flip-flop is required in the logic between the comparator and the SAR input to prevent a race condition. Explain.

16.20* The block diagram of a SA ADC in Fig. 16.14 suggests the use of an addressable latch to implement the SAR. Only the 3 MSB of a 4-bit counter need be used to address the eight positions in the register, leaving the LSB to subdivide the time needed for each bit into two parts. Also, the clock waveform itself has two parts. Using these two inputs and the comparator output, design a simple logic circuit that will properly drive the two addressable latch inputs: D*ata in*, and $\overline{E}$*nable*. The addressed latch is transparent to data on D when $\overline{E}$ is LOW, and data on D is latched when $\overline{E}$ goes HIGH. The address should not change unless $\overline{E}$ is HIGH.

16.21 Refer to the block diagram of the single-slope ADC of Fig. 16.15. Is the switch S_2 really needed? Explain.

16.22 Refer to the block diagram of the single-slope ADC of Fig. 16.15. Show by means of a sketch how switch S_1 could be implemented by means of a BJT. Assume that the logic circuitry uses $V_{CC} = 5$ V.

16.23 Refer to the block diagram of the single-slope ADC of Fig. 16.15. What parameter(s) in this circuit could be varied to calibrate the ADC?

16.24* Figure 16.22 depicts a circuit called a *boot-strap sweep*, which can be used to implement the constant current source (and capacitor C and switches) of the single-slope ADC of Fig. 16.15. When the transistor is ON, C is discharged and the ramp voltage is zero. The capacitance C_2 charges up to V_{CC} through the diode. When the transistor is turned OFF, C begins to charge toward V_{CC} through R, the ramp voltage follows the voltage on C because the op amp has unity gain. The voltage across R remains constant because C_2 is so large that it charges C without significant discharge. The result is that the current through R (and C) is approximately constant.

(a) What is this constant current charging R?
(b) What is the range on the ramp output voltage?
(c) What size base resistor is needed for the BJT? Explain how this resistance affects the ramp recovery time.

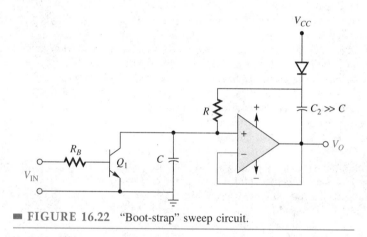

■ **FIGURE 16.22** "Boot-strap" sweep circuit.

16.25 What determines the resolution of the single-slope ADC?

16.26 Refer to the block diagram of the dual-slope ADC of Fig. 16.17. What parameter(s) in this circuit could be varied to calibrate the ADC?

16.27 What determines the resolution of the dual-slope ADC?

16.28 Show how a MOSFET could be used as the S_2 switch in the dual-slope ADC of Fig. 16.17.

16.29 Show how two MOSFETs could be used as the S_1 switch in the dual-slope ADC of Fig. 16.17.

16.30 For the dual-slope ADC, derive an expression for the clock frequency in terms of R, C, and the number of counts desired for full-scale.

REFERENCES

1. *Linear and Interface Data Book,* Motorola, Inc., Phoenix, Arizona, 1990.
2. Haznedar, H. *Digital Microelectronics,* Redwood City, Calif.: Benjamin Cummings, 1991, pp. 427–428.
3. *Linear Circuits—Data Conversion, DSP Analog Interface, and Video Interface,* Texas Instruments, Inc., Dallas, 1992.
4. Gray, P. and R. Meyer. *Analysis and Design of Analog Integrated Circuits.* 3rd ed. New York: Wiley & Sons, 1994, pp. 338–346.

CHAPTER 17

ADDITIONAL EXAMPLES OF ANALOG INTEGRATED CIRCUITS

17.1 Analog Systems Overview
17.2 Series-Pass Voltage Regulator
17.3 Switching Regulators
17.4 Analog Multipliers
17.5 Phase-Locked Loop, System, and Circuit Description

17.6 Phase-Locked Loop Applications
Summary
Survey Questions
Problems
References

So far, we have focused on the internal circuitry and function of operational amplifiers, as well as their overall systems operation utilizing feedback. A large variety of analog integrated circuits (ICs) whose function is not explicitly amplification can be studied using techniques developed in the preceding chapters. Section 17.1 presents an overview of analog IC systems with a focus on the burgeoning use of *ASICs* (application-specific integrated circuits) for embedded applications. Indeed, most electronic systems employ a mix of analog functions, digital microprocessors, and digital "glue and control logic." In addition, we observe that all electronic systems require compact, regulated, efficient power supplies.

Many systems are required to communicate with other systems via digital-based protocol networks using copper, optical, or radio-frequency links. Contemporary electronic circuit design mirrors these needs by recognizing the wide variety of digital and analog functions required. This chapter, along with the Chapter 16 A/D and D/A material, presents a brief discussion of some of the more widely used functions. In Section 17.2, we introduce the **series-pass voltage regulator** as an example of an integrated circuit that utilizes amplification and negative feedback to obtain a function other than amplification. Similarly, in Section 17.3 we present examples of *switched-mode power supplies* to illustrate the operation of negative feedback to control pulse duration rather than amplitude. After all, all electronic systems require some type of dc power.

In Section 17.4 we examine analog circuit techniques that are used in the multiplication of signals to obtain *modulation, demodulation, mixing,* and *phase-detection* functions. These are very important for communications systems-based functions. Phase detection and negative feedback are combined in the **phase-locked loop**, whose system and circuit descriptions are presented in Section 17.5, along with selected applications in Section 17.6. Again, many of the applications are based on data-link communications needs.

IMPORTANT CONCEPTS IN THIS CHAPTER

- Definition of voltage regulation and a block diagram of a voltage regulator
- Voltage-series feedback applied to a series-pass voltage regulator
- Qualitative operational description of a switched-mode voltage regulator
- Comparison between series-pass and switched-mode voltage regulators
- Three regions of operation of a Gilbert cell multiplier
- System applications for multipliers including modulator, demodulator, phase detector and phase-locked loop
- Block diagram and qualitative operational description of a phase-locked loop
- Properties of zero- and first-order phase-locked loops
- Definition of lock range and capture range
- Application of a phase-locked loop for FM demodulation and for frequency synthesis

17.1 ANALOG SYSTEMS OVERVIEW

As we have explored in earlier chapters, to meet system requirements, most electronic systems include both analog and digital circuits. We exist in an analog environment. In general, however, there is a trend to represent and manipulate these analog signals digitally because embedded microcomputer integrated circuits can process these data effectively. Indeed, *digital-signal processing* (DSP) integrated circuits and algorithms for digitally representing visual, sound, and environmental signals is a major effort within the electrical engineering and computer science communities. For example, consider the electronic thermometer shown in Fig. 17.1. Temperature is an analog quantity. The sensor could be a *thermistor*, which is a temperature-sensitive resistor, or a *thermocouple*, which is a sensor based on a temperature-dependent voltage developed at the junction between two dissimilar conductors. The sensor will probably be an input to an amplifier, which is an analog circuit. An A/D converter will then digitize the analog signal representation of temperature, and provide a digital display. A clock (oscillator) will be required. If the

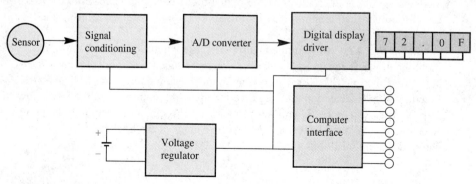

■ FIGURE 17.1 Electronic digital read-out thermometer.

display is an LCD, the digital display driver will need to generate a tens of kHz signal for polarizing the display. The computer interface will have to provide appropriate clocked signals to a particular computer bus standard. Of course, all electronic circuits need a source of dc power. In this case, a battery, followed by a voltage regulator will be used to power the rest of the system. We study two basic types of voltage regulators in the next two sections.

Communications systems provide excellent examples for systems using both analog and digital circuits. A simplified block diagram for an FM radio is shown in Fig. 17.2. The 88- to 108-MHz input signal into the antenna must be amplified in a band-pass rf amplifier. Efforts have also been aimed at performing direct analog-to-digital (A/D) conversion at these frequencies. As we will study in this chapter, a mixer (multiplier) will be used to shift the incoming frequency to some intermediate value. Additional amplification takes place before the signal is demodulated to an audio signal and amplified to drive a speaker. Tuning is accomplished by using a phase-locked loop to obtain the local oscillator signal as input to the mixer; in addition, the phase-locked loop could be configured to yield a digital readout of the frequency.

You could prepare similar block diagrams for electronic systems you run into every day. It is important to remember that the most complex system can be broken into simpler subsystems and circuits, many of which, whether primarily digital or analog in function, have been studied in previous chapters. Many of these complex systems are packaged within a single IC or hybrid. Figure 17.3 illustrates just a very few samples of IC configurations and packaging. Observe that both digital and analog functions are represented. One could also take the view that as digital operating speeds increase, one has to apply many of the analog circuit concepts to their analysis and design especially with respect to active and passive element parasitics and frequency response. In the remaining sections of this chapter, we will combine circuit ideas previously introduced to design additional circuits with wide utility.

■ One can buy almost complete AM and FM radios on a chip as well as major television sub-systems. Often a company will design and implement an ASIC that contains proprietary circuits and systems and are often labeled with a special custom part number. There are literally tens of thousands of functions of varying complexity available as an off-the-shelf IC or configured from ASIC digital and analog arrays.

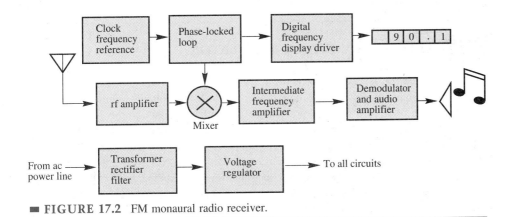

■ **FIGURE 17.2** FM monaural radio receiver.

CHECK UP

1. **TRUE OR FALSE?** Most systems contain both analog and digital circuits.
2. Draw a block diagram of an electronic system of your choice that includes both circuits with primarily analog and digital functions.

■ **FIGURE 17.3** (1) 100 mm wafer. Each die is an ASIC BJT array with $f_T = 2$ GHz npn transistors using 2 μm design rules. Includes capacitors in a digital-programmable array, diffused resistor arrays, and test structures. (2) Burr-Brown OPA 600 wide bandwidth operational amplifier for use in video, D/A, and VCO applications requiring GBW = 5 GHz, FET input stages. Hybrid internal construction with *surface-mount devices* (SMD). Gold-plated package—ceramic substrate. Lid has been removed. (3) LM320K high-power negative voltage regulator in a TO-3 package. Requires heat sinking for optimum use in a system. (4) National DAC 0800, 8-bit DAC with 100 ns settling time in a 16-pin DIP. (5) and (6) High-speed microprocessor system. Ceramic substrate with gold metalization. Lid on and off to show placement of the IC.

17.2 SERIES-PASS VOLTAGE REGULATOR

A *voltage regulator* is used to provide a constant output voltage despite reasonable variations in input supply voltage and output load. Most voltage regulators are also designed to be relatively temperature insensitive and to provide some mechanism to protect themselves and the load against overload. Recall from Section 3.7 and subsequent discussions that the equivalent source resistance, R_o, from the transformer windings, diodes, and wiring results in a decrease in output voltage as the load current increases (refer to Fig. 17.4). This output voltage degradation is illustrated in Fig. 17.4(b). The power-supply voltage regulation, defined in Eq. (3.41) and repeated here for convenience, is given by

$$VR = \frac{V_{\text{NO LOAD}} - V_{\text{LOAD}}}{V_{\text{NO LOAD}}} \times 100\%. \tag{17.1}$$

Clearly, one characteristic of a good power supply is a low output resistance, R_o, which will result in a low VR. Any practical circuit has a *maximum load current*, $I_{L\max}$, which if exceeded, usually results in damage to the power supply and load.

A voltage regulator is typically inserted between the output filter capacitor and the load, as shown in Fig. 17.5(a). Ideally, it provides an operating characteristic like that illustrated in Fig. 17.5(b). Normal operation occurs between points A and B on the characteristic where V_O remains essentially constant. Some voltage regulators will current

CHAPTER 17 ADDITIONAL EXAMPLES OF ANALOG INTEGRATED CIRCUITS

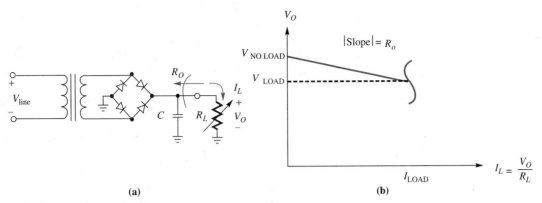

FIGURE 17.4 Unregulated dc power supply. (a) Full-wave bridge rectifier. (b) Typical load characteristic.

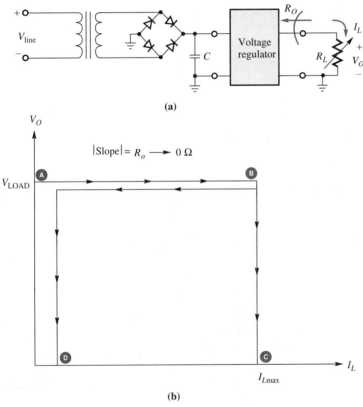

FIGURE 17.5 Regulator block diagram and ideal load characteristic. (a) Full-wave bridge with a voltage regulator. (b) Typical load characteristic illustrating foldback.

limit at point C. While operation at point C may prevent damage to the voltage regulator circuit, the load could be damaged. A more desirable approach is to *fold back* the current by traversing the characteristic from point B to point D. Power to the entire circuit must then be interrupted before normal operation can be resumed.

The basic voltage regulator circuit consists of four sections (see Fig. 17.6).

1. *Reference source.* This is illustrated using an avalanche or Zener diode (Section 3.8). It is important that the reference source provide a voltage relatively independent of input source voltage and temperature changes over some operating range.
2. *Error detector.* This consists of a differential amplifier and the R_1–R_2, output-voltage sampling network. The circuit is in a voltage-series feedback configuration where the error voltage, V_ε, is defined in Eq. (11.82). If R_1 and R_2 are included within the IC, the result is a ***fixed-output voltage regulator***. If R_1 and R_2 are added externally, the result is an ***adjustable-output voltage regulator***.
3. *Control device.* In the case of a series-pass regulator, this is a BJT operating as an emitter follower with the output load in the emitter circuit as shown. For voltage regulators with I_{Lmax} rated below 1 A, the series-pass transistor is often incorporated within the same IC. Higher currents can be controlled by using an external series-pass transistor where the regulator output current serves as the base current drive for the external series-pass transistor.
4. *Current limit.* This function is represented by $R_{SC} \ll R_L$ and operates in a similar fashion to the output short-circuit protection often included in an operational amplifier as described by Eq. (10.133).

We are now prepared to derive an expression for the regulation, VR, of this circuit. From Eq. (11.2), we have

$$A_{vs} = \frac{V_O}{V_S} = \frac{V_O}{V_R} = \frac{a}{1 + af} \approx \frac{1}{f}\bigg|_{af \gg 1}, \qquad (17.2)$$

where $a = V_o/V_\varepsilon$, and $f = h_{12f} = R_2/(R_1 + R_2)$.

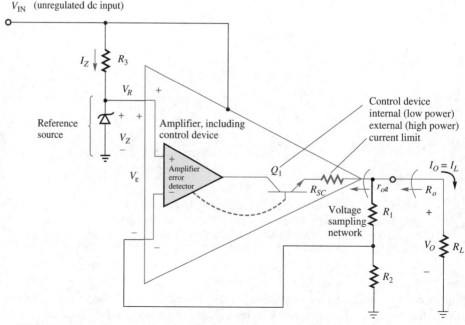

■ **FIGURE 17.6** Basic series-pass voltage regulator.

CHAPTER 17 ADDITIONAL EXAMPLES OF ANALOG INTEGRATED CIRCUITS

We immediately observe that for $a_f \gg 1$, the output voltage is given by

$$V_O = V_R \left(\frac{R_1 + R_2}{R_2} \right). \tag{17.3}$$

Therefore, over the operating range of the circuit, the output voltage is determined by a stable avalanche-diode reference voltage and the ratio of two resistors.

The output resistance, using Eq. (11.85), is given by

$$R_o = \frac{r_{oa} \left\| \frac{1}{h_{22f}} \right.}{1 + a_f} = \frac{r_{oa} \| (R_1 + R_2)}{1 + a_f}. \tag{17.4}$$

Typically, $r_{oa} \ll (R_1 + R_2)$ and $a_f \gg 1$ so that combining Eqs. (17.3) and (17.4) yields

$$R_o \cong \frac{r_{oa}}{a_f} = \frac{r_{oa}}{a} \left(\frac{R_1 + R_2}{R_2} \right) = \frac{r_{oa}}{a} \left(\frac{V_O}{V_R} \right). \tag{17.5}$$

The effective output resistance has been reduced by the gain of the amplifier, a, resulting in the desired low-output resistance characteristic of a good voltage source.

The voltage regulation in terms of circuit parameters is determined by using Fig. 17.7. Substituting Eq. (17.5) into Eq. (17.1),

$$VR = \frac{V_{\text{NO LOAD}} - V_{\text{LOAD}}}{V_{\text{NO LOAD}}} = \frac{\Delta V_O}{V_{\text{NO LOAD}}} = \frac{R_o \Delta I_O}{V_{\text{NO LOAD}}} = \frac{r_{oa} \Delta I_O}{a V_R} \times 100\%. \tag{17.6}$$

Strictly speaking, Eq. (17.6) refers to load regulation. Changes in the input dc voltage, often resulting from fluctuations in the 60-Hz power lines, will have an effect on V_R and the forward voltage gain, a. The **line regulation** is defined by

$$VR_{\text{line}} = \frac{\Delta V_O}{\Delta V_{\text{line}}} \times 100\%. \tag{17.7}$$

DRILL 17.1

A voltage-regulated power supply has $V_O = 11.5$ V at $I_L = 150$ mA and $V_O = 11.2$ V at $I_L = 475$ mA which is considered full load. Compute VR and R_o.

ANSWERS
$VR = 3.77\%$
$R_o = 0.92\ \Omega$

DRILL 17.2

Refer to Fig. 17.6. A 6.8 V, 100 mW avalanche diode with $I_{Z(\min)} = 0.5$ mA is used as a reference in a voltage regulator. What values of R_3 are possible for $V_{\text{IN}} = 9.0$ V?

ANSWERS
$R_3 = 4400\ \Omega$
to 150 Ω

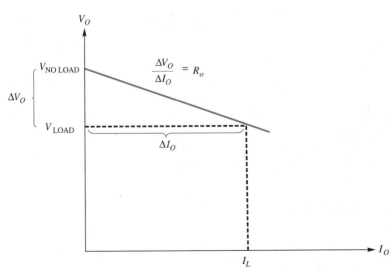

■ **FIGURE 17.7** Output characteristic.

Analysis of a Typical Series-Pass Regulator

We will study the operation of the Signetics NE550 adjustable-output voltage regulator[1,2] (see Fig. 17.8). It is similar to the μA723 except for the output protection circuitry. The analysis of the 78HVXX series of positive fixed-output voltage regulators and the 79HVXX series of fixed-output negative voltage regulators is similar. The "XX" denotes the nominal output voltage established by an internal R_1–R_2 resistor network. For example, the 79HV15 is a -15-V voltage regulator. The NE550 diagram is shown as it would be connected for basic operation with foldback current limiting as illustrated by point D in Fig. 17.5(b). The *reference source, error detector, control device,* and *current-limiting circuitry* will be identified as we proceed with the qualitative circuit description.

■ The 78 and 79 series are classics. The analysis approach would be the same for more complex or newer designs.

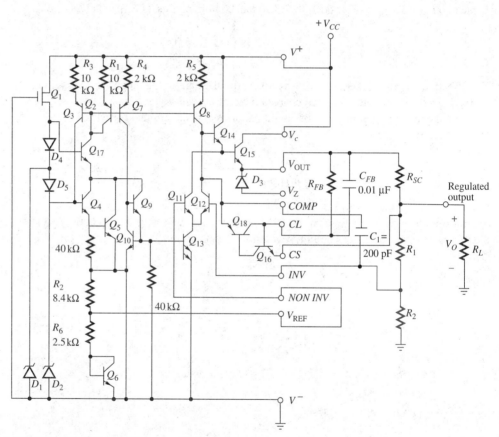

■ **FIGURE 17.8** NE550 schematic diagram configured with foldback current limiting.

Qualitative Circuit Description

The reference voltage, V_{REF}, is obtained from the avalanche diode D_2. The voltage developed across D_2 is buffered by the Q_2 through Q_5 combination. These four transistors, along with Q_6, R_2, and R_6 provide a near-zero temperature coefficient for V_{REF} (see

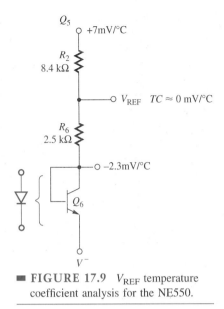

FIGURE 17.9 V_{REF} temperature coefficient analysis for the NE550.

Fig. 17.9). The temperature coefficient of the voltage at the emitter of Q_5 is about $+7$ mV/°C. Using voltage division, the TC of V_{REF} is given by

$$TC|_{V_{REF}} = \frac{R_6}{R_2 + R_6}[+7 \text{ mV/°C} - (-2.3 \text{ mV/°C})] - 2.3 \text{ mV/°C}$$

$$= \frac{2.5 \text{ k}\Omega}{8.4 \text{ k}\Omega + 2.5 \text{ k}\Omega}(9.3 \text{ mV/°C}) - 2.3 \text{ mV/°C}$$

$$= -0.17 \text{ mV/°C}. \tag{17.8}$$

■ Ideally, R_2 and R_6 should be designed for $TC \cong 0$, however these values are obtained directly from the schematic diagram.

To operate properly, D_2 needs a constant current, I_Z, over a range of power-supply voltages and temperature. In addition, biasing is required for the error detector (amplifier) and control device (output transistor). The FET, Q_1, functions as a constant current source, as shown in Fig. 17.10, by providing an appropriate value of I_D for a given V_{GS}. Initially, when power is applied to the circuit, there is no current in D_2. The I_D of Q_1 will turn ON Q_4. The base of Q_4 is two diode drops below the source of Q_1. With Q_4 ON, Q_{17} is switched ON and is able to sink current for the bases of Q_3, Q_2, Q_7, and Q_8, thus activating these devices as current sources. Observe that Q_2 is a current reference for the circuit and operates in the Widlar configuration so that the current in Q_3 keeps Q_4 operating and provides current for D_2. When this happens, the avalanche voltage drops across D_2 and D_1 are about the same, so D_5 is effectively OFF. The current in Q_7, established by the current in Q_2, also references current for Q_{10}. The current gain of Q_9 ensures that I_{B10} and I_{B13} can be neglected when estimating $I_{C13} = 2I_{C10}$. The simple current source Q_{13} is used to bias the differential amplifier composed of Q_{11} and Q_{12}. The active load is obtained by using Q_8. The output control device is a Darlington pair, Q_{14} and Q_{15}, connected as a composite emitter follower. The compensation capacitor (Section 11.5), is used to ensure internal amplifier stability.

The protection circuitry illustrates the versatility of the devices that can be fabricated in an integrated circuit. Transistors Q_{16} and Q_{18}, shown in Fig. 17.11(a), are connected as a *silicon-controlled rectifier* (SCR), also called a *thyristor* which is a form of a *silicon-controlled switch* (SCS). The symbol of this *pnpn* composite transistor latch structure is

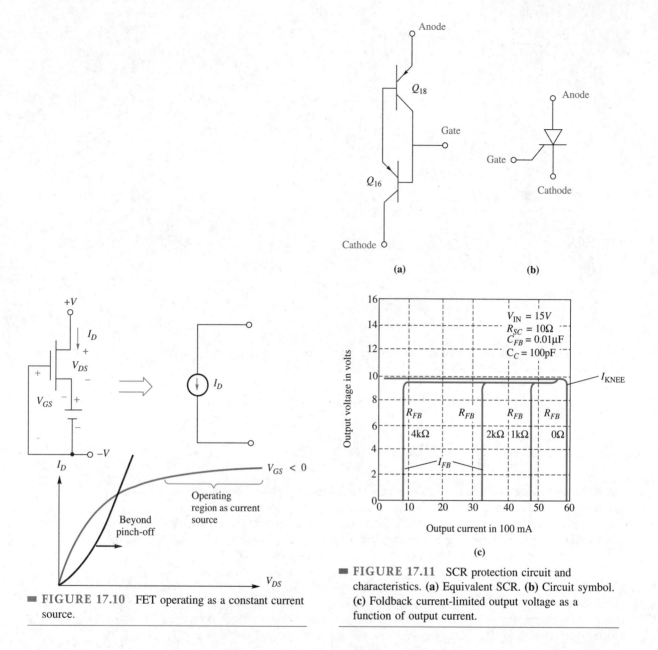

■ FIGURE 17.10 FET operating as a constant current source.

■ FIGURE 17.11 SCR protection circuit and characteristics. (a) Equivalent SCR. (b) Circuit symbol. (c) Foldback current-limited output voltage as a function of output current.

■ A detailed discussion of thyristors including *TRIACs* and *SCRs* and their applications in power switching and power systems control is beyond the scope of this text. One familiar application is that of a light-dimmer.

given in Fig. 17.11(b). If the current through the sense resistor, R_{SC}, develops a voltage in excess of about 0.6 V due to an output circuit overload, Q_{16} will switch ON. The collector current of Q_{16} will provide the base current for the *pnp*, Q_{18}, and switch it ON. With Q_{18} ON, the collector current from Q_8 that would normally drive the first stage of the Darlington output device, Q_{14}, is diverted to the emitter of Q_{18}, thus reducing the load current to some fraction of I_{C18}. Depending on the value of R_{FB}, the current folds back to a much lower value, thus protecting the load and regulator. In addition, the SCR circuit element requires that to restore normal operation, input power to the entire circuit must be removed and then reapplied, or the load must be momentarily disconnected. This is because the Q_{16}–Q_{18} is latched ON, with base current for Q_{16} being derived continuously from I_{C18}. The resultant transfer function for the SCR and its inclusion in the overall circuit is given in the foldback current load characteristic shown in Fig. 17.11(c).

EXAMPLE 17.1

Select values of R_1 and R_2 to obtain the regulated output voltage, $V_O = 7.5$ V. Specify R_{SC} so that the short-circuit current is limited to 100 mA. Assume $V_{REF} = 1.63$ V.

SOLUTION From Eq. (17.3),

$$V_O = V_R \left(\frac{R_1 + R_2}{R_2} \right) \quad \text{or} \quad \frac{R_1 + R_2}{R_2} = \frac{7.5}{1.63} = 4.60.$$

It appears we can select any set of R_1 and R_2 values to satisfy this ratio. However, matching between the noninverting (V_R) input and the inverting input ($R_1 - R_2$ divider node) is desirable for obtaining the best temperature coefficient and amplifier drift performance. The output resistance at the V_R node is given approximately by $R_6 \| R_7 \approx$ 2.5 kΩ $\|$ 8.4 kΩ $\approx$ 2 kΩ. Then in addition to satisfying $(R_1 + R_2)/R_2 = 4.60$, the design should also satisfy $R_1 R_2/(R_1 + R_2) = 2$ kΩ. Then $R_1 = 9.0$ kΩ and $R_2 = 2.5$ kΩ, resulting in $V_O = 7.5$ V.

Short-circuit protection is established by turning ON Q_{17}. Then

$$R_{SC} = \frac{V_{BE16}}{I_{SC}} = \frac{0.6 \text{ V}}{100 \text{ mA}} = 6 \, \Omega.$$

CHECK UP

1. **TRUE OR FALSE?** Ideal voltage sources have an infinite output resistance.
2. What type of feedback does a series-pass voltage regulator use?
3. What are the definitions of *load* and *line* regulation?
4. **TRUE OR FALSE?** Foldback is a desirable feature in a practical voltage regulator.
5. **TRUE OR FALSE?** The output stage of a series-pass voltage regulator is a CE amplifier.

17.3 SWITCHING REGULATORS

The principal disadvantage with series-pass regulators is the relatively low efficiency. Using Fig. 17.12, we define efficiency as

$$\eta = \frac{V_O I_O}{V_{IN} I_{IN}} \times 100\% \cong \frac{V_O}{V_{IN}} \times 100\%, \tag{17.9}$$

where $I_O \approx I_{IN}$ for $\beta \gg 1$. Typically, $\eta = 50$ to 70% so that 30 to 50% of the power is dissipated in the output transistor because it is operating in the active region. Most IC regulators can safely dissipate up to 1 W depending upon packaging. For regulators with V_O significantly lower than V_{IN} and with I_O unchanged, the increased power lost in the series-pass transistor means that additional heat sinking is required with a cost, weight, and size penalty.

The *switched-mode power supply* (SMPS) or *switching regulator* controls the load power by varying the *duty cycle* of the series-pass transistor. Essentially, the series-pass transistor is either ON or OFF, and in either state, its power dissipation is quite low. Qualitatively, a SMPS can be represented by the diagram shown in Fig. 17.13. The SMPS

■ The TO-3 package shown in Figure 17.1 is good for 5 to 10 watts depending upon heat sinking.

■ Of course, the output regulating devices must have breakdown voltages in excess of the input voltage requiring regulation.

■ This is also sometimes called a *buck pulse-width-modulation* (PWM) converter.

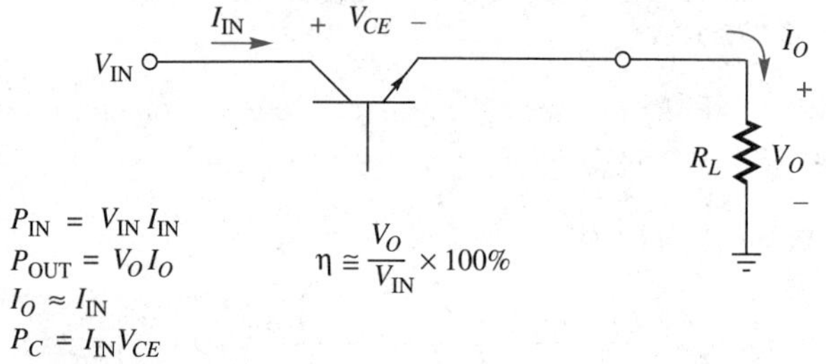

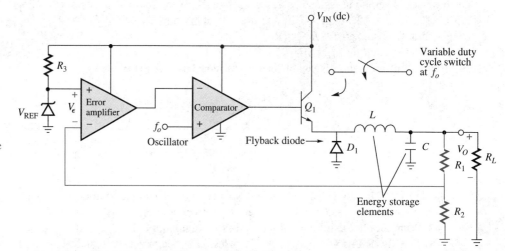

$P_{IN} = V_{IN} I_{IN}$
$P_{OUT} = V_O I_O$
$I_O \approx I_{IN}$
$P_C = I_{IN} V_{CE}$

$\eta \cong \dfrac{V_O}{V_{IN}} \times 100\%$

■ **FIGURE 17.12** Series-pass transistor efficiency.

■ Another common approach is to use a power MOSFET (VMOS) for the switching transistor.

■ **FIGURE 17.13** Basic SMPS functional diagram.

is essentially a ***dc-to-dc converter***. The dc output from the basic rectifier and capacitor filter powers an oscillator operating at frequencies from several hundred Hz to 20 kHz or so. This rectangular wave signal is used to switch a series-output transistor, Q_1, between ON and cutoff. This switched output is connected to the load, R_L. To provide for a uniform power flow to the load, L and C energy storage elements are used during the switch-OFF interval. A feedback circuit is used to control the duty cycle of the series-pass transistor. If the output voltage V_O drops, the difference between it and some reference voltage, $V_R = V_{REF}$, as shown in Fig. 17.13, is amplified and used to change the transition point of the comparator and thus increase the duty cycle of Q_1. As the output increases to the desired level, the duty cycle stabilizes at some increased value. This controlled change of the duty cycle of a rectangular wave signal is called ***pulse-width modulation***. The fundamental frequency of the rectangular wave is established by a fixed-frequency oscillator at frequency ω_o.

To illustrate this feedback-controlled SMPS, we will outline the operation of the ***flyback dc-to-dc converter*** (see Fig. 17.14).[3] This is a form of a self-oscillating switching regulator that is interesting because positive feedback from R_4 is used to obtain oscillatory behavior and negative feedback is used to obtain a regulated output voltage. The step-by-step circuit description will be keyed to developing the waveforms shown in Fig. 17.15.

CHAPTER 17 ADDITIONAL EXAMPLES OF ANALOG INTEGRATED CIRCUITS

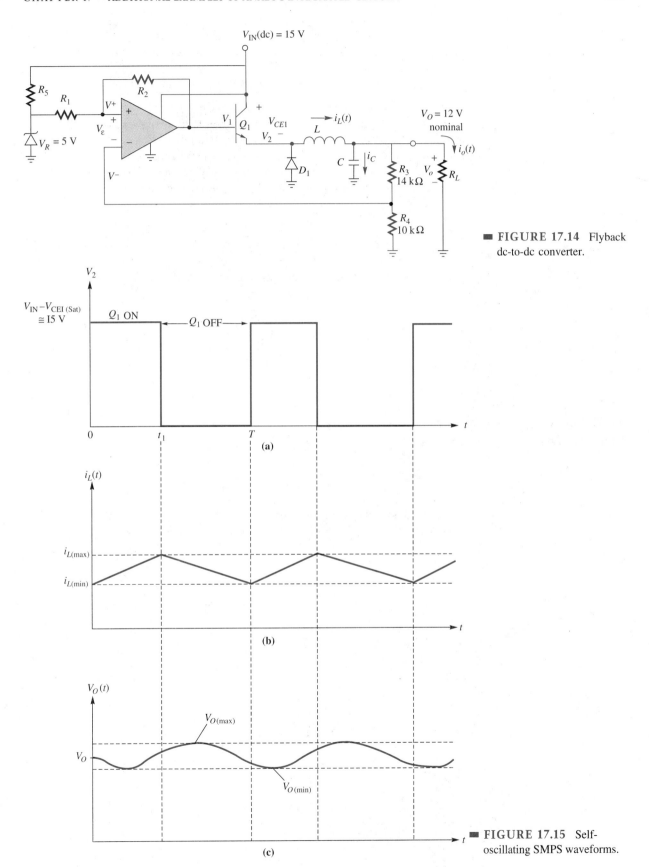

FIGURE 17.14 Flyback dc-to-dc converter.

FIGURE 17.15 Self-oscillating SMPS waveforms.

For example, assume that the circuit shown in Fig. 17.14 is designed to operate as a 12.0-V regulator and V_O has decreased to 11.95 V because of a decrease in R_L. Also assume $V_R = 5.0$ V established by an avalanche reference diode. Let $V_{IN} = 15$ V, which also supplies power for the operational amplifier.

1. Initially we will assume Q_1 is ON. Then $V_2 \cong V_{IN} - 1 - V_{BE1} \cong V_{IN} - 1.7 \cong V_{IN}$ [see Figs. 17.14 and 17.15(a)]. Observe that the collector power dissipation during this interval is given by $P_C = I_L (1.7 \text{ V})$ which is relatively low.

■ This equation assumes a 1-volt drop for V_{CB}. Q_1 cannot saturate.

2. Diode D_1 is reverse biased and behaves as an open circuit.
3. The inductor current, I_L, charges the output capacitor and simultaneously increases V_o by supplying energy to the $R_L C$ output. The inductor current [Fig. 17.15(b)] is computed from

$$i_L(t) \cong \frac{1}{L} \int_0^t (V_{IN} - V_O)\, dt + i_L(0). \qquad (17.10)$$

4. The R_3–R_4 voltage divider samples the output, in this case $V_O = 12$ V, and feeds the sample voltage,

$$V^- = \frac{R_4}{R_3 + R_4} \times V_O = \left(\frac{10\text{ k}\Omega}{10\text{ k}\Omega + 14\text{ k}\Omega}\right) \times 12\text{ V} = 5.0\text{ V} \qquad (17.11)$$

to the inverting input of the operational amplifier.

5. The operational amplifier is configured as a comparator with hysteresis, as illustrated in Fig. 17.16(a). The transfer function for this comparator is given in Fig. 17.16(b). By observing this transfer function, the voltage at which the comparator switches state is given by

■ You will recognize this configuration as a positive feedback topology.

$$V^- = V_R \pm \frac{R_1}{R_1 + R_2} \times V_1. \qquad (17.12)$$

For this example, we will set the hysteresis at ±10 mV or 20 mV by an appropriate

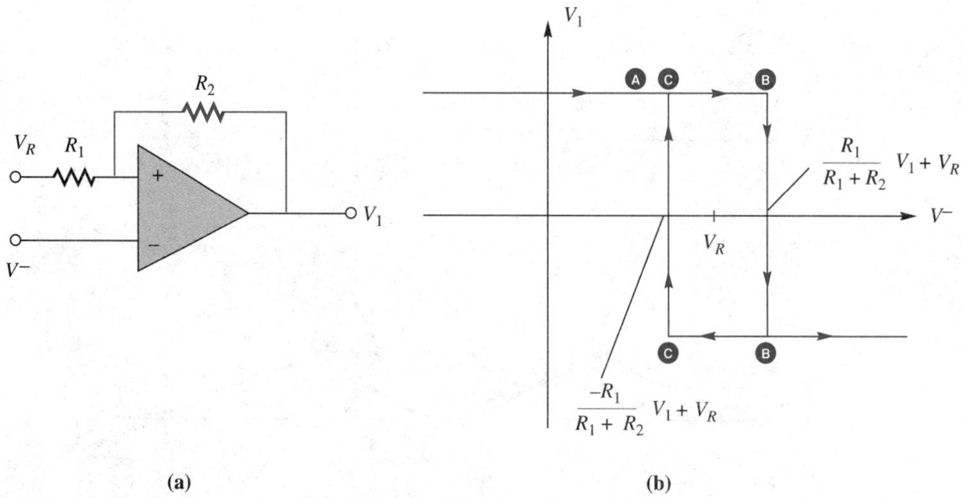

■ **FIGURE 17.16** Comparator operation with hysteresis in a positive feedback circuit. **(a)** Comparator with hysteresis circuit. **(b)** Transfer characteristic.

selection of R_1 and R_2; that is, the comparator will switch state at $V^- = 4.990$ and 5.010 V. Since $V^- = 4.979$ V when $V_O = 11.95$ V using Eq. (17.11), Q_1 is kept in saturation by the comparator output, V_1, as illustrated by point A in Fig. 17.16(b).

6. Then $i_L = i_C + i_O$, and V_O continues to increase [see Fig. 17.15(c)]. If the comparator were not in the circuit, V_O would eventually reach $V_{IN} - V_{CE1(sat)} = V_{IN} = 15$ V.

7. When V^- increases to 5.010, point B in Fig. 17.16(b), corresponding to $V_O = 12.024$ V, the comparator output V_1 switches and turns Q_1 OFF. The Q_1 power dissipation is essentially zero during this interval. Thus, the Q_1 ON time, t_1 [Fig. 17.15(a)], is defined. Inductor current i_L has reached a maximum.

8. The energy stored in the inductor and capacitor tends to keep V_O increasing slightly beyond Q_1 ON time. The RLC circuit configuration of the output yields an underdamped sinusoidal-shaped V_O.

9. With i_L and i_C now supplying energy to the load, D_1 is forward biased. The diode D_1 is sometimes called a *flyback diode*. Its inclusion is necessary to provide a return circuit for I_L.

10. As the stored energy in the inductor and capacitor decreases, V_O starts to decrease. When V^- decreases to 4.990 V, point C in Fig. 17.16(b), corresponding to $V_O = 11.976$ V, the comparator changes state, turning ON Q_1 and thus repeating the cycle of operation with step 1 again. The OFF interval is given by $T - t_1$ in Fig. 17.15(a). The frequency of operation is given by[4]

$$f_o = \sqrt{\frac{\frac{4}{\pi} \times \frac{V_R}{V_O} \times \frac{R_1 + R_2}{R_1}}{(2\pi)^2 LC}}. \quad (17.13)$$

The ripple, R, can be approximated from[3]

$$R = \frac{V_{MAX} - V_{MIN}}{V_O} \times 100\% = \frac{1}{15 f_o^2 LC} \times 100\%. \quad (17.14)$$

Since the power dissipation for Q_1 operating as a switch is essentially zero, the overall regulator efficiency ranges between 70 and 90%.

We may observe that the SMPS is especially useful as a dc-to-dc converter supplying multiple dc voltages from a single dc source. Conceptually, this is illustrated in Fig. 17.17.

■ Recall $v_L = L \dfrac{di}{dt}$ and if $\dfrac{di}{dt}$ is large, so is v_L resulting in destructive transient voltages in the circuit. Usually Q_1 will fail from exceeding $V_{CE[max]}$.

■ A SMPS is electrically noisy. That is, the 10's of kHz switching generates harmonics well into the radio spectrum. A SMPS must meet radio frequency interference regulations as specified by the Federal Communications Commission. You can qualitatively observe this switching noise by holding a portable radio next to a SMPS.

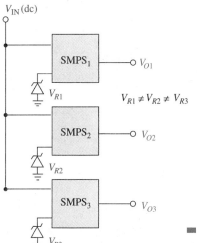

■ **FIGURE 17.17** Multiple dc regulated voltages from a single unregulated dc voltage input.

EXAMPLE 17.2

Complete the analysis and design of the $V_O = 12$-V SMPS by computing the switching frequency f_o, the ripple R, and values for R_1 and R_2 in the comparator hysteresis circuit. Assume $L = 10$ mH and $C = 10$ μF.

SOLUTION For ± 10 mV of hysteresis, we have, from Eq. (17.12),

$$\left(\frac{R_1}{R_1 + R_2}\right)V_1 = |V^- - V_R| = 10 \text{ mV} \quad \text{or} \quad \frac{R_1}{R_1 + R_2} = \left|\frac{0.01}{V_1}\right|.$$

The comparator output will be within a volt or two of $V_{IN} = 15$ V. Assume $V_1 = 13$ V, then

$$\frac{R_1}{R_1 + R_2} = \frac{0.01}{13} = 7.7 \times 10^{-4}.$$

Not much positive feedback hysteresis is required. This is because 10 mV of hysteresis will yield a very low ripple. Then $R_2 = 1300 R_1$ or let $R_1 = 1$ kΩ and $R_2 = 1.3$ MΩ. Substituting in Eq. (17.13) yields

$$f_o = \left[\frac{\frac{4}{\pi} \times \frac{V_R}{V_O} \times \left(\frac{R_1 + R_2}{R_1}\right)}{(2\pi)^2 LC}\right]^{1/2} = \left[\frac{\left(\frac{4}{\pi}\right)\left(\frac{5}{12}\right)(1300)}{(2\pi)^2(10 \text{ mH})(10 \text{ }\mu\text{F})}\right]^{1/2} = 13.2 \text{ kHz}.$$

From Eq. (17.14), the ripple is given by

$$R = \frac{1}{15 f_o^2 LC} \times 100\% = \frac{1}{(15)(13.2 \text{ kHz})^2(10 \text{ mH})(10 \mu\text{F})} = 0.38\%,$$

which is quite good in practice.

DRILL 17.3

Compute hysteresis and ripple for the circuit discussed in Example 17.2 with $R_1/(R_1 + R_2) = 5 \times 10^{-3}$. What is the resultant switching frequency?

ANSWERS
$f_o = 5.2$ kHz,
hysteresis $= \pm 65$ mV,
Ripple $= 2.48\%$

CHECK UP

1. **TRUE OR FALSE?** SMPS usually operate at frequencies above the power line frequency, 60 Hz in the United States.
2. Compare the efficiencies of SMPS and series-pass voltage regulators.
3. **TRUE OR FALSE?** Both negative feedback and positive feedback is used in SMPS voltage regulators.
4. **TRUE OR FALSE?** The higher the ripple, the better the design.
5. Why are inductors and transformers lighter in switched-mode power supplies?

17.4 ANALOG MULTIPLIERS

An *analog multiplier* accepts as input two signals, $v_1(t)$, and $v_2(t)$, and produces as an output,

$$v_o(t) = K v_1(t) v_2(t). \tag{17.15}$$

The circuit symbol is given in Fig. 17.18. In particular, we will study a *four-quadrant multiplier* that can accept positive and negative input voltages to produce the appropriate positive or negative output voltage.

CHAPTER 17 ADDITIONAL EXAMPLES OF ANALOG INTEGRATED CIRCUITS

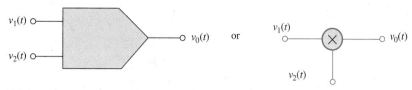

■ **FIGURE 17.18** Circuit symbol for a multiplier.

Virtually all analog multipliers use a form of the *Gilbert cell*.[5] This is diagrammed in Fig. 17.19. It is composed of three emitter-coupled pairs, Q_1–Q_2, Q_3–Q_4, and Q_5–Q_6. A Widlar current source, Q_7 and Q_8, provides biasing for the entire circuit. Matched resistor loads, R_C, are used. It is important to observe that the operation of this circuit is made possible by the ability to obtain matched transistors in the fabrication process:

$$i_{C1} = \frac{i_{C5}}{1 + e^{-v_1/V_T}}, \tag{17.16}$$

$$i_{C2} = \frac{i_{C5}}{1 + e^{v_1/V_T}}, \tag{17.17}$$

$$i_{C3} = \frac{i_{C6}}{1 + e^{v_1/V_T}}, \tag{17.18}$$

$$i_{C4} = \frac{i_{C6}}{1 + e^{-v_1/V_T}}. \tag{17.19}$$

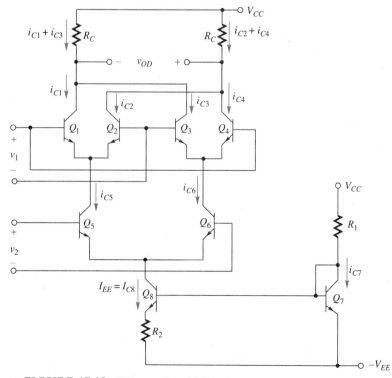

■ **FIGURE 17.19** Gilbert cell multiplier.

Recall $V_T = kT/q$. The biasing current,

$$I_{EE} = I_{C8} = \frac{V_T}{R_2} \ln\left(\frac{I_{C7}}{I_{C8}}\right) \tag{17.20}$$

is obtained by using Eq. (7.35) for a Widlar current source. The overall circuit biasing is established by

$$i_{C5} = \frac{I_{EE}}{1 + e^{-v_2/V_T}} \tag{17.21}$$

and

$$i_{C6} = \frac{I_{EE}}{1 + e^{v_2/V_T}}. \tag{17.22}$$

The differential output voltage is obtained from

$$v_{OD} = v_{C2} - v_{C1} = [(i_{C1} + i_{C3}) - (i_{C2} + i_{C4})]R_C. \tag{17.23}$$

Substituting Eqs. (17.16) through (17.22) into Eq. (17.23) and factoring yields

$$v_{OD} = \frac{I_{EE}R_C}{(1 + e^{-v_1/V_T})}\left(\frac{1}{1 + e^{-v_2/V_T}} - \frac{1}{1 + e^{v_2/V_T}}\right)$$
$$- \frac{I_{EE}R_C}{(1 + e^{v_1/V_T})}\left(\frac{1}{1 + e^{-v_2/V_T}} - \frac{1}{1 + e^{v_2/V_T}}\right). \tag{17.24}$$

The hyperbolic trigonometric substitution used to derive Eq. (10.16) is used to reduce Eq. (17.24) to

$$v_{OD} = \frac{I_{EE}R_C}{(1 + e^{-v_1/V_T})}\tanh\left(\frac{v_2}{2V_T}\right) - \left[\frac{I_{EE}R_C}{(1 + e^{v_1/V_T})}\right]\tanh\left(\frac{v_2}{2V_T}\right), \tag{17.25}$$

and again to obtain the final result,

$$v_{OD} = I_{EE}R_C \tanh\left(\frac{v_2}{2V_T}\right)\tanh\left(\frac{v_1}{2V_T}\right). \tag{17.26}$$

We can obtain the multiplication function for small signals by realizing that

$$\tanh x = x - \frac{x^3}{3} + \frac{2x^5}{15} - \frac{17x^7}{315} \approx x \tag{17.27}$$

for $x \ll 1$. For $(v/2V_T) \ll 1$, Eq. (17.26) reduces to

$$v_{OD} = \frac{I_{EE}R_C}{(2V_T)^2} v_1(t)v_2(t) = Kv_1(t)v_2(t). \tag{17.28}$$

For small input signals, on the order of a few millivolts, the circuit in Fig. 17.19 works quite well. For larger input signals, we can linearize Eq. (17.26) by predistorting each input signal by performing the $\tanh^{-1}(v/2V_T)$ operation. Then Eq. (17.26) becomes

$$v_{OD} = I_{EE}R_C \tanh\left[\tanh^{-1}\left(\frac{v_1}{2V_T}\right)\right]\tanh\left[\tanh^{-1}\left(\frac{v_2}{2V_T}\right)\right]$$
$$= I_{EE}R_C\left(\frac{v_1}{2V_T}\right)\left(\frac{v_2}{2V_T}\right). \tag{17.29}$$

CHAPTER 17 ADDITIONAL EXAMPLES OF ANALOG INTEGRATED CIRCUITS

This operation is available to us in circuit form. Consider the circuit shown in Fig. 17.20, where the differential output voltages v_1 and v_2 are now input as the original v_1 and v_2. Then

$$\Delta v_1 = v_{C12} - v_{C11} = v_{BE9} - v_{BE10}. \quad (17.30)$$

Substituting Eq. (7.23) into Eq. (17.30) yields

$$\Delta v_1 = V_T \ln\left(\frac{i_{C9}}{i_{S9}}\right) - V_T \ln\left(\frac{i_{C10}}{i_{S10}}\right) = V_T \ln\left(\frac{i_{C9}}{i_{C10}}\right) = V_T \ln\left(\frac{i_{C11}}{i_{C12}}\right). \quad (17.31)$$

Using the simplified transistor model shown in Fig. 17.21, and assuming Q_9, Q_{10}, Q_{11}, and Q_{12} are matched,

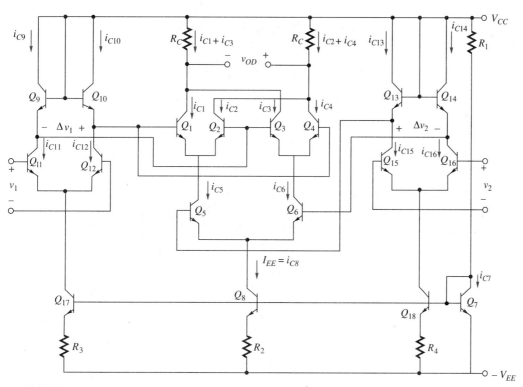

FIGURE 17.20 Gilbert cell multiplier with input $\tanh^{-1}$ predistortion circuits.

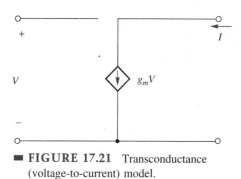

FIGURE 17.21 Transconductance (voltage-to-current) model.

$$v_{OD1} = V_T \ln\left[\frac{I_{C11} + g_m v_1}{I_{C12} - g_m v_1}\right] = V_T \ln\left[\frac{1 + \dfrac{g_m v_1}{I_{C11}}}{1 - \dfrac{g_m v_1}{I_{C11}}}\right]. \tag{17.32}$$

We will substitute the hyperbolic trigonometric identity

$$\tanh^{-1}(x) = \frac{1}{2}\ln\left(\frac{1+x}{1-x}\right) \tag{17.33}$$

into Eq. (17.32). The result is

$$v_{OD1} = V_T \tanh^{-1}\left(\frac{g_m v_1}{I_{C11}}\right). \tag{17.34}$$

When this result is applied to both inputs, Eq. (17.26) becomes

$$v_{OD} = I_{EE} R_C \left(\frac{g_m v_1}{I_{C11}}\right)\left(\frac{g_m v_2}{I_{C11}}\right) = K v_1(t) v_2(t), \tag{17.35}$$

usable as a multiplier for both large and small signals.

EXAMPLE 17.3

Use a SPICE simulation to illustrate the operation of the Gilbert cell shown in Figure 17.19. Show the linear region of operation. Use default npn BJT models. Assume $V_{CC} = V_{EE} = 15$ V and design for $I_{C1} = I_{C2} = I_{C3} = I_{C4} = 100$ μA using reasonable IC-compatible diffused resistor values. Let $R_{C1} = R_{C2} = 25$ kΩ.

SOLUTION The node-labeled circuit for Figure 17.19 is shown in Figure 17.22. Transfer functions are obtained by a VTRAN generator at each input, each incremented by 10 mV over a range from -100 mV to $+100$ mV. To establish $I_{C1} = I_{C2} = I_{C3} = I_{C4} = 100$ μA, we observe that for $\beta \gg 1$, $I_{C8} \approx 400$ μA. Assume $I_{C7} = 2$ mA, then

$$R_1 = \frac{30 - 0.7 \text{ V}}{2 \text{ mA}} = 14650 \ \Omega$$

and for a Widlar current source,

$$R_2 = \frac{V_T}{I_{C8}} \ln\left(\frac{I_{C7}}{I_{C8}}\right) = \frac{26 \text{ mV}}{400 \ \mu\text{A}} \ln\left(\frac{2 \text{ mA}}{400 \ \mu\text{A}}\right) = 105 \ \Omega.$$

The resultant netlist is written as

```
Gilbert Cell
VTRAN1 1 4 DC 0
VTRAN2 8 9 DC 0
VCC 7 0 DC 15
VEE 0 12 DC 15
Q1 2 1 3 DEVICE
Q2 6 4 3 DEVICE
```

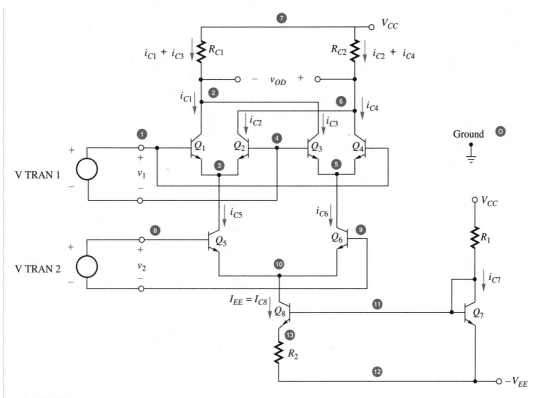

■ **FIGURE 17.22** Gilbert cell multiplier.

```
Q3 2 4 5 DEVICE
Q4 6 1 5 DEVICE
Q5 3 8 10 DEVICE
Q6 5 9 10 DEVICE
Q7 11 11 12 DEVICE
Q8 10 11 13 DEVICE
R1 7 11 15.5K
R2 13 12 100
RC1 7 2 25K
RC2 7 6 25K
.MODEL DEVICE NPN
.PROBE
.DC VTRAN1 -100M 100M 10M VTRAN2 -100M 100M 20M
.TF V(6) VTRAN1
.TF V(6) VTRAN2
.END
```

The resultant set of transfer characteristics are shown in Fig. 17.23(a). As expected, these characteristics behave according to the $\tanh(v/2V_T)$ relationship from Eqn. (17.26). If we change the input scale from ± 100 mV to ± 20 mV, we observe, as shown in Fig. 17.23(b), a linear set of characteristics as predicted by Eqn. (17.29)

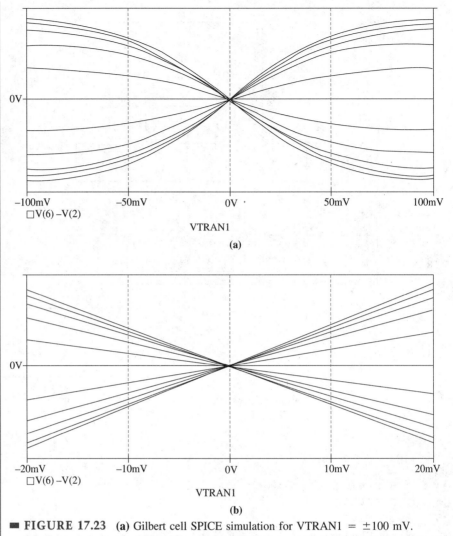

■ **FIGURE 17.23** (a) Gilbert cell SPICE simulation for VTRAN1 = ±100 mV.
(b) Expanded view for VTRAN1 = ±20 mV to show small-signal linear operating region.

Multiplier Applications

There are four main classes of application for multipliers.

Class 1: Modulation, Demodulation, and Frequency Translation. Using Eq. (17.15), if $v_1(t) = \cos(\omega_1 t)$ and $v_2(t) = \cos(\omega_2 t)$, then

$$v_o(t) = K \cos \omega_1 t \cos \omega_2 t$$
$$= \frac{K}{2}[\cos(\omega_2 + \omega_1)t + \cos(\omega_2 - \omega_1)t]. \tag{17.36}$$

■ A mathematical treatment offered in communications system textbooks would show that the frequency translation is the result of the *convolution* of two impulse functions, $\delta(\omega_1)$ and $\delta(\omega_2)$.

If ω_1 represents an audio signal and ω_2 is in the radio-frequency spectrum, then the multiplication process shifts the audio signal to the rf spectrum where it can be transmitted as an rf signal. The entire process is illustrated in Fig. 17.24. Demodulation, that is, recovery of the original audio signal in addition to other terms, $v'_o(t)$, at the receiver is accomplished by multiplying the transmitted signal by $v_3(t) = \cos(\omega_2 t)$ to obtain

CHAPTER 17 ADDITIONAL EXAMPLES OF ANALOG INTEGRATED CIRCUITS

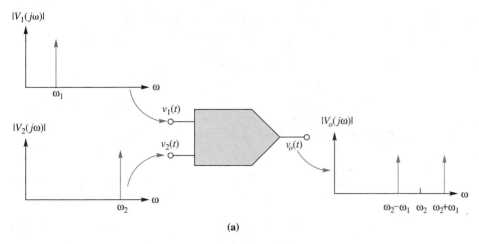

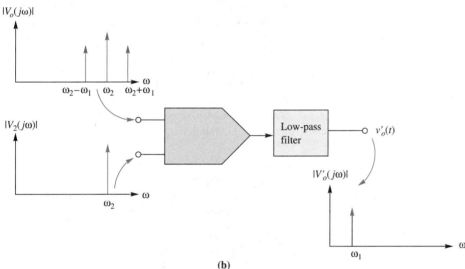

■ **FIGURE 17.24** (a) Modulation (frequency translation) multiplier operation. (b) Demodulation (frequency translation) multiplier operation.

$$v'_O(t) = \frac{K}{2} \cos \omega_2 t [\cos(\omega_1 + \omega_2)t + \cos(\omega_1 - \omega_2)t]$$
$$= A \cos \omega_1 t + \text{other terms.} \qquad (17.37)$$

The original signal, $A \cos(\omega_1 t)$ is obtained after low-pass filtering. A is a gain constant. It is beyond the scope of this text to spend very much time on this application. You are encouraged to review related communications textbooks.[6]

Class 2: Chopper and Gating. If one of the input signals is a rectangular wave or gate function of amplitude $\pm V_1$, its Fourier series is

$$v_1(t) = \sum_{n=1}^{\infty} V_1 C_n \cos n\omega_1 t, \qquad C_n = \frac{\sin\left(\dfrac{n\pi}{2}\right)}{\left(\dfrac{n\pi}{4}\right)}. \qquad (17.38)$$

■ Communication system applications of multipliers is one of the most important global uses. Whether we call it multiplication, mixing, modulation, sampling, frequency translation or demodulation, this application is the subject of many textbooks and course offerings.

If $\omega_2 \ll \omega_1$, $v_o(t)$ will be samples of the other input signal, $v_2(t) = V_2 \cos \omega_2 t$. The resultant spectrum is shifted just as in the previous case and is computed from

$$v_o(t) = v_1(t)v_2(t) = V_1 V_2 \sum_{n=1}^{\infty} C_n \cos n\omega_1 t \cos \omega_2 t. \qquad (17.39)$$

Observe that the shifted signal of ω_2 is centered around odd multiples of ω_1.

Class 3: Function Generation. If we let $v_1(t) = v_2(t) = v(t)$, then

$$v_o(t) = K[v(t)]^2, \qquad (17.40)$$

which is the equation for a parabola. Since most waveforms can be reasonably accurately constructed from the first several terms of a MacLaurin series of the form

$$f(x) = f(0) + x\frac{df(0)}{dx} + \frac{x^2}{2!}\frac{d^2f(0)}{dx^2} + \frac{x^3}{3!}\frac{d^3f(0)}{dx^3} + \cdots, \qquad (17.41)$$

virtually any function can be synthesized by using multipliers and summing amplifiers with appropriately scaled resistors. A description of commercial function generator modules and related topics is given in selected references.[7-9]

Class 4: Phase Detectors. If $v_1(\omega_o t)$ and $v_2(\omega_o t)$ are of the same frequency, but differ in phase by θ, $v_o(\omega_o t)$ will include a dc component proportional to θ. This phase detector function is a key component for a phase-locked loop, which we will study in Section 17.5. Let $v_1(\omega_o t)$ and $v_2(\omega_o t)$ be represented by the signals at frequency ω_o sketched in Fig. 17.25, parts (a) and (b), respectively.

A point-by-point multiplication of $v_1(t)$ and $v_2(t)$ results in the waveform given in Fig. 17.25(c). Although there are frequency components at $2\omega_o$, as well as higher order sum and difference frequencies present in $v_o(t)$, we are only interested in the dc or average value of $v_o(\omega t)$. This can be obtained graphically as

$$V_{\text{AVE}} = \frac{1}{2\pi}\int_0^{2\pi} v_o(t)\, d(\omega t) = \frac{1}{2\pi}(A_1 - A_2 + A_1 - A_2)$$

$$= \frac{1}{\pi}(A_1 - A_2) = \overline{v_O}(\theta), \qquad (17.42)$$

where

$$A_1 = V_O(\pi - \theta) \qquad (17.43)$$

and

$$A_2 = V_O \theta. \qquad (17.44)$$

Graphically in Fig. 17.25(c), we subtract the areas below the x-axis from the areas above the x-axis. Combining Eqs. (17.42) through (17.44) yields

$$\overline{v_O}(\theta) = \frac{1}{\pi}[V_O(\pi - \theta) - V_O\theta] = \frac{V_O}{\pi}[\pi - 2\theta] = V_O\left(1 - \frac{2\theta}{\pi}\right). \qquad (17.45)$$

Equation (17.44) is plotted in Fig. 17.26(a). We see immediately that $\overline{v_O}$ is proportional to the phase difference θ, a result previously mentioned, which we will use in our phase-locked loop study. If $v_1(\omega t)$ and $v_2(\omega t)$ are inputs to the Gilbert cell shown in Fig. 17.19, the dc average output voltage is obtained between the collectors of Q_1 and Q_4. This voltage is sketched in Fig. 17.26(b).

DRILL 17.4

The signals $v_1(t) = A\cos(1500t) + B\sin(1200t)$ and $v_2(t) = C\sin(10{,}000t)$ are input to a multiplier. List the frequency spectral components present at the output. Design a system to demodulate the $v_1(t)$ from the output signal.

ANSWERS The frequency spectrum of $v_o(t)$ includes $\omega = 8{,}500, 8{,}800, 11{,}200,$ and $11{,}500$ rad/sec. Multiply $v_o(t)$ by $v_3(t) = D\sin(10{,}000t)$ in a second multiplier. The output spectrum consists of signals with $\omega = 1{,}500, 1{,}200, 18{,}500, 18{,}800, 21{,}200,$ and $21{,}500$ rad/sec. Then use a LPF with a 3dB corner just above $1{,}500$ rad/sec but well below $18{,}500$ rad/sec.

CHAPTER 17 ADDITIONAL EXAMPLES OF ANALOG INTEGRATED CIRCUITS

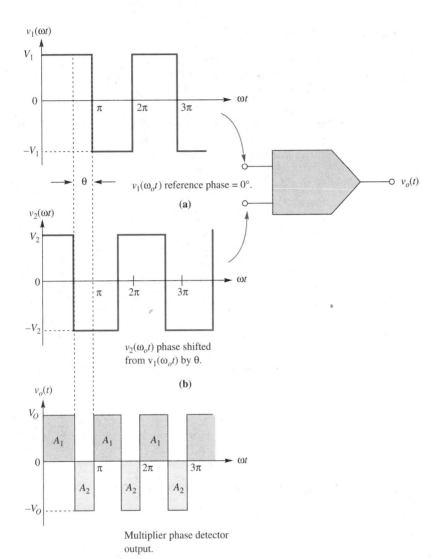

■ **FIGURE 17.25** Phase detector input and output waveforms.

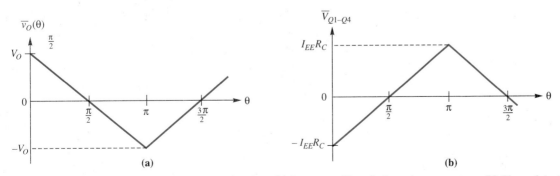

■ **FIGURE 17.26** Phase detector transfer function. (a) Low-pass filtered phase detector output. (b) Phase detector output from Fig. 17.19.

Multiplier Data Sheet

The MC 1495/1595L is a wideband monolithic four-quadrant multiplier IC. Excerpts from the data sheet and a simplified schematic diagram are given in Fig. 17.27. The Gilbert cell can be identified as Q_5, Q_6, Q_7, Q_8, Q_9, and Q_{10}. The $\tanh^{-1}$ operation for v_y is obtained from Q_1, Q_2, Q_{11}, Q_{12}, and the diode-connected transistors D_1 and D_2. For the second input signal, v_x, nonlinearity is compensated for directly. Current sourcing for the v_y input is accomplished by using the diode connected transistor D_3 as a reference Q_{13} and Q_{14} as the current sources. A similar arrangement is accomplished with D and Q_{15} and Q_{17}.

Observe that the $V_D - V_x$ characteristic in Figure 17.27 is similar to the Gilbert Cell simulation shown in Figure 17.23(b).

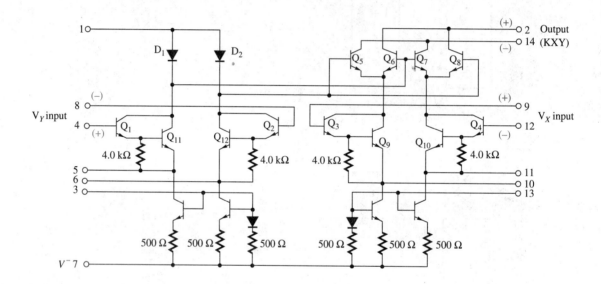

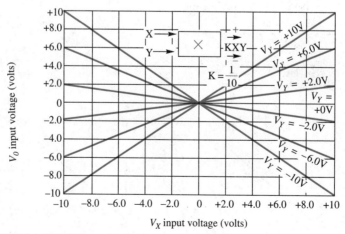

■ **FIGURE 17.27** MC 1495/1595L multiplier schematic diagram and data sheet (courtesy of Motorola, Inc.).

CHAPTER 17 ADDITIONAL EXAMPLES OF ANALOG INTEGRATED CIRCUITS

Electrical Characteristics ($V^+ = +32V$, $V^- = -15V$, $T_A + 25°C$, $I_3 = I_{13} = 1$ mA, $R_X = R_Y = 15$ kΩ $R_L = 11$ kΩ unless otherwise noted)

Characteristics		Figure	Symbol	Min	Typ	Max	Unit
Linearity Output Error in Percent of Full Scale: $T_A = +25°C$		5					%
$-10 < V_X < +10 (V_Y = \pm 10V)$	MC1495 MC1595		E_{RX}	–	±1.0 ±0.5	±2.0 ±1.0	
$-10 < V_Y < +10 (V_X = \pm 10V)$	MC1495 MC1595		E_{RY}	–	±2.0 ±1.0	±4.0 ±2.0	
$T_A = 0$ to $+70°C$	MC1495						
$-10 < V_X < +10 (V_Y = \pm 10V)$			E_{RX}	–	±1.5	–	
$-10 < V_Y < +10 (V_X = \pm 10V)$			E_{RY}	–	±3.0	–	
$T_A = -55°C$ to $+125°C$	MC1595						
$-10 < V_X < +10 (V_Y = \pm 10V)$			E_{RX}	–	+0.75	–	
$-10 < V_Y < +10 (V_X = \pm 10V)$			E_{RY}	–	±1.50	–	
Squaring Mode Error: Accuracy in Percent of Full Scale After Offset and Scale Factor Adjustment		5	E_{SQ}				%
$T_A = +25°C$	MC1495 MC1595			– –	±0.75 ±0.5	– –	
$T_A = 0$ to $+70°C$	MC1495				±1.0		
$T_A = -55°C$ to $+125°C$	MC1595				±0.75		
Scale Factor (Adjustable) $\left(K = \dfrac{2R_L}{I_3 R_X R_Y}\right)$		–	K		0.1	–	
Input Resistance (f = 20 Hz)	MC1495 MC1595 MC1495 MC1595	7	R_{INX} R_{INY}		20 35 20 35	– – – –	MegOhms
Differential Output Resistance (f = 20 Hz)		8	R_O		300		k Ohms

■ **FIGURE 17.27** *(Continued).*

CHECK UP

1. **TRUE OR FALSE?** A Gilbert cell is used for time-domain signal multiplication.
2. **TRUE OR FALSE?** Mixing of two different frequency sinusoids yields signals at the sum and different frequencies.
3. Explain how the output signal given by Eq. (17.26) can yield an essentially linear result. Provide two techniques.
4. **TRUE OR FALSE?** Frequency translation is a common property of Gilbert cell multipliers operating as a mixer, modulator, or demodulator.
5. **TRUE OR FALSE?** Phase detectors operate at lower signal levels.

17.5 PHASE-LOCKED LOOP, SYSTEM, AND CIRCUIT DESCRIPTION

The *phase-locked loop* (**PLL**) is an analog circuit that uses a negative feedback control loop to produce both an oscillator output frequency, which is synchronized with an input signal frequency, and an output voltage proportional to input signal frequency changes. The PLL feedback loop shown in Fig. 17.28 consists of a phase detector, a low-pass filter, an amplifier, and a *voltage-controlled oscillator* (**VCO**). To analyze this PLL as a linear

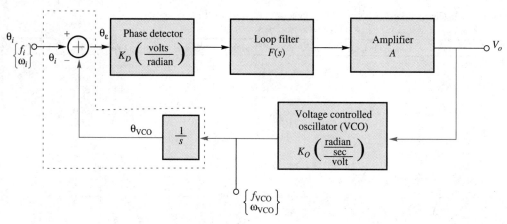

■ **FIGURE 17.28** Phase-locked loop block diagram.

circuit, we assume that the loop is *locked*, that is, that the VCO output frequency ω_{VCO} is identical to the input frequency, ω_i.

Although the PLL is used for locking to an input frequency, the fundamental transfer function for a phase detector [Eq. (17.45)] relates output voltage to input phase. The general relationship between frequency and phase for both the input signal and the VCO signal is given by

$$\theta(t) = \theta(0) + \int_0^t \omega(t)\, dt \tag{17.46}$$

and

$$\omega(t) = \frac{d\theta(t)}{dt}. \tag{17.47}$$

Typically, ω_{VCO} cannot be changed much beyond a ±25% range so that the output frequency of the VCO is given by

$$\omega_{VCO} = \omega_o + K_O V_o \tag{17.48}$$

instead of just $\omega_{VCO} = K_O V_o$. The frequency ω_o is called the **center or free-run frequency**.

Observe that the functional elements within the dashed lines in Fig. 17.28 are operationally included within the phase detector.

Although we assume operation in the sinusoidal frequency domain, we initially derive the PLL transfer function in the ***complex frequency domain*** where $s = \sigma + j\omega$. Using Fig. 17.28,

$$V_o(s) = AF(s)K_D\theta_\varepsilon = AF(s)K_D(\theta_i - \theta_{VCO}). \tag{17.49}$$

■ Refer to circuit theory, where $s = \sigma + j\omega$ in general, and the use of LaPlace transforms.

Realizing that the integration operation called for in Eq. (17.46) corresponds to multiplying by $1/s$ in the complex frequency domain,

$$\theta_{VCO} = \frac{1}{s}\omega_{VCO} = \frac{1}{s}(K_O V_o). \tag{17.50}$$

Substituting Eq. (17.50) in Eq. (17.49) and solving for the PLL transfer function yields

$$\frac{V_o}{\theta_i}(s) = \frac{AF(s)K_D}{1 + \frac{K_D F(s) A K_O}{s}}. \tag{17.51}$$

The final form of the transfer function is obtained by substituting $\theta_i(s) = [\omega_i(s)]/s$ into Eq. (17.51) to obtain

$$\frac{V_o}{\omega_i}(s) = \frac{\frac{K_D AF(s)}{s}}{1 + \frac{K_D AF(s) K_O}{s}}. \tag{17.52}$$

Equation (17.52) is in the form for the classic closed-loop feedback circuit, Eq. (15.4), where $a = K_D AF(s)/s$ and $f = K_O$.

The special PLL closed-loop response, V_o/ω_i, will be determined by the selection of the filter transfer function, $F(s)$. Although many forms are available for the synthesis of $F(s)$, forms often suggested by the manufacturer,[10] we will consider only two basic forms of $F(s)$.

First-Order Loop

The *first-order loop* uses a *zero-order filter*, $F(s) = 1$, as shown in Fig. 17.29(a). Substituting $F(s) = 1$ into Eq. (17.52) and letting $K_v = K_O K_D A$ results in

$$\frac{V_o}{\omega_i}(s) = \frac{K_v}{K_O(s + K_v)} \tag{17.53}$$

in the complex frequency domain.

The time response of the first-order loop is illustrated by letting the input frequency $\omega_i(t)$ change instantaneously from ω_1 to ω_2, $\omega_i(t) = \omega_1 + (\omega_2 - \omega_1)U(t)$, as shown in Fig. 17.30(a). The resultant output voltage from a step change in input frequency [Fig. 17.30(b)] is obtained by using the unit step time response of Eq. (17.53):

$$v_O(t) = \frac{[\omega_2 - \omega_1]U(t)}{K_O}(1 - e^{-tK_v}) + \frac{\omega_1}{K_O}. \tag{17.54}$$

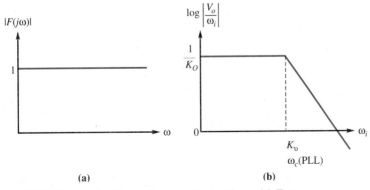

■ **FIGURE 17.29** First-order phase locked loop. **(a)** Zero-order filter transfer function. **(b)** First-order PLL response.

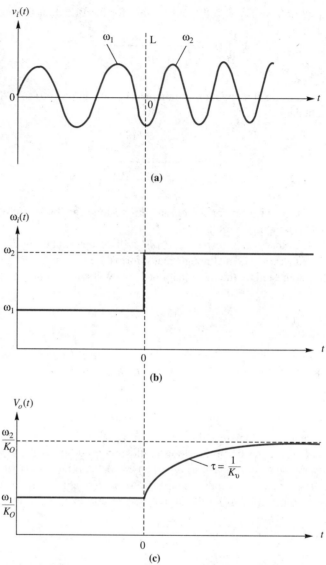

FIGURE 17.30 First-order PLL response to $\omega_i(t) = (\omega_2 - \omega_1)U(t) + \omega_1$. (a) PLL input, $v_i(t)$. (b) PLL input, $\omega_i(t)$. (c) $V_o(t)$.

The PLL time constant, $\tau = \dfrac{1}{K_v}$, is illustrated in the $v_O(t)$ response [see Fig. 17.30(c)].

Equation (17.53) is converted to the sinusoidal frequency domain by setting $s = j\omega$ and rearranging in a standard form to obtain

$$\frac{V_o}{\omega_i}(j\omega) = \frac{1}{K_O\left(1 + j\dfrac{\omega}{K_v}\right)}. \tag{17.55}$$

We recognize that $K_v = \omega_{C(PLL)}$, which is then defined as the **loop bandwidth**. The Bode magnitude plot of the first-order PLL response is shown in Fig. 17.29(b). We illustrate the application of Eq. (17.55) to the demodulation of frequency-modulated signals in Section 17.5.

Second-Order Loop

The *second-order loop* uses a *first-order filter*,

$$F(s) = \frac{1}{1 + (s/\omega_{C1})} \quad \text{or} \quad F(j\omega) = \frac{1}{1 + j(\omega/\omega_{C1})}, \quad (17.56)$$

■ Refer back to Chapter 6 where we discussed this single-pole low-pass filter and its R-C circuit realization and design.

which can be synthesized as a single-section RC low-pass filter, $\omega_{C1} = 1/RC$, as shown in Fig. 17.31(a). Substituting Eq. (17.56) into Eq. (17.52) and again letting $K_v = K_O K_D A$ results in

$$\frac{V_o}{\omega_i}(s) = \frac{1}{K_D} \frac{1}{\left(1 + \dfrac{s}{K_v} + \dfrac{s^2}{\omega_{C1} K_v}\right)}. \quad (17.57)$$

The time response of Eq. (17.57) is given by

$$v_o(t) = \frac{1}{K_O}(e^{s_1 t} + e^{s_2 t}), \quad (17.58)$$

which represents a second-order response. This characteristic equation describing the second-order time response is a staple of electric circuit analysis textbooks.[11,12]

The resultant $v_o(t)$ for the input signal, $\omega_i(t) = (\omega_2 - \omega_1)U(t) + \omega_1$ [Fig. 17.32(a)], exhibits three basic forms [Fig. 17.32(b), (c), and (d)], depending on the relationship between ω_C and K_v.

Define the *damped natural frequency* as

$$\omega_n = (\omega_{C1} K_v)^{1/2} \quad (17.59)$$

and the *damping coefficient* as

$$\zeta = \frac{1}{2}\left(\frac{\omega_{C1}}{K_v}\right)^{1/2}. \quad (17.60)$$

For $\zeta < 1$, we have an underdamped response as illustrated in Fig. 17.32(b). The degree of oscillatory behavior depends on the exact value of ζ, with smaller values of ζ producing a more pronounced, longer duration, damped oscillation. This is illustrated for the two curves where, $\zeta_1 < \zeta_2 < 1$.

For $\zeta > 1$, we have an overdamped response as illustrated in Fig. 17.32(c). For ζ large, the overdamped response becomes more pronounced. Compare Fig. 17.32(c) with Fig. 17.32(d).

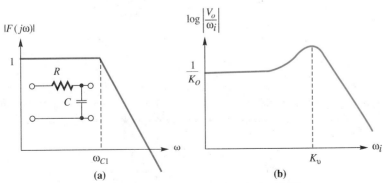

■ **FIGURE 17.31** Second-order phase locked loop. **(a)** First-order filter transfer function. **(b)** Second-order PLL response.

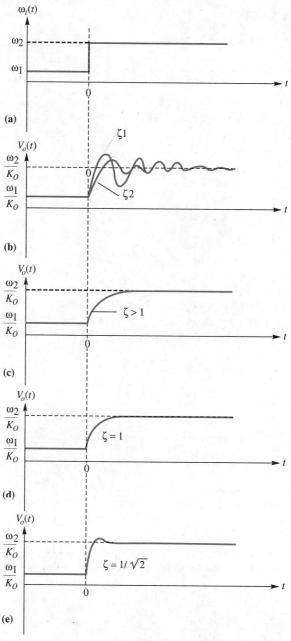

FIGURE 17.32 Second-order PLL response to $\omega_i(t) = (\omega_2 - \omega_1)U(t) + \omega_1$.

Critical damping occurs when $\zeta = 1$ and is illustrated in Fig. 17.32(d). This represents a reasonable compromise for obtaining $v_O(t)$, which closely reproduces the input $\omega_i(t)$.

The sinusoidal frequency-domain magnitude plot is obtained by substituting $s = j\omega$ in Eq. (17.57) to obtain

$$\frac{V_o}{\omega_i}(j\omega) = \frac{1}{K_O}\left[\frac{1}{1 + \dfrac{j\omega}{K_\nu} + \dfrac{(j\omega)^2}{\omega_{C1}K_\nu}}\right]. \qquad (17.61)$$

Another widely used design solution occurs when $\zeta = 1/\sqrt{2}$, which represents a slightly undamped response. In the sinusoidal frequency domain, this results in a *maximally flat response*, which is also illustrated in the family of curves in Fig. 17.32(b). The time response is illustrated in Fig. 17.32(e). Setting $\zeta = 1/\sqrt{2}$ in Eq. (17.60) establishes the $F(s)$ design criteria

$$\omega_{C1} = 2K_v. \tag{17.62}$$

Lock Range and Capture Range

All of the previous discussion required that the PLL VCO be synchronized in frequency with the $\omega_1(t)$ input. As illustrated in Fig. 17.33, the **lock** or **lock-in range** extends to $\pm \omega_L$ around ω_o. The **tracking range** extends from ω_o to either $+\omega_L$ or $-\omega_L$, which is one-half the lock range.

The process in which the PLL becomes locked is a complicated nonlinear process that is beyond the scope of this text. Until lock is established, the fundamental frequency component of $v_o(t)$ is given by the first-order multiplication product of $\omega_i(t)$ and $\omega_{VCO}(t)$, which is the difference frequency at any given time. This process is illustrated in Fig. 17.34. When the PLL becomes fully locked, $\omega_i(t)$ and $\omega_{VCO}(t)$ are operating in synchronism with some constant phase difference represented by a constant $v_o(t)$ output. The input frequency range over which this phase-locking process occurs is called the **capture range**. As illustrated in Fig. 17.34, the capture range, $\pm \omega_C$, is less than the lock range.

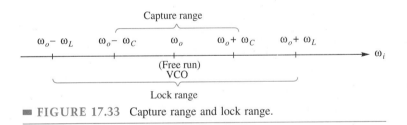

FIGURE 17.33 Capture range and lock range.

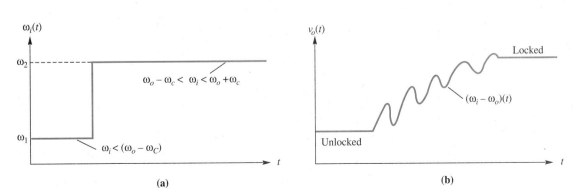

FIGURE 17.34 PLL locking process. (a) $\omega_i U(t)$ from outside the capture range to within the capture range. (b) Capture process for $v_o(t)$.

> **CHECK UP**
>
> 1. Which portion of a PLL requires positive feedback?
> 2. **TRUE OR FALSE?** The VCO and input signal are operating at the same frequency when the phase-locked loop is locked.
> 3. **TRUE OR FALSE?** The VCO phase and input signal phase are identical when the phase-locked loop is locked.
> 4. **TRUE OR FALSE?** A zero-order loop has a smaller capture range than a first-order loop.
> 5. **TRUE OR FALSE?** Lock range is less than or equal to the capture range.

17.6 PHASE-LOCKED LOOP APPLICATIONS

Two important applications for the monolithic PLL IC are in the areas of FM demodulation and frequency synthesis. To support the study of these applications, we will use the specifications and circuit description for the SE/NE 565 PLL (see Fig. 17.35).[13]

All key functions of the PLL can be qualitatively examined by using the equivalent schematic in conjunction with Fig. 17.28. The differential input signal, terminals 2 and 3, are input to a Gilbert cell operating as a phase detector. The VCO output, terminal 4, can be connected externally to the second Gilbert cell input, the phase comparator input, terminal 5, to close the feedback loop. The free-run frequency of the VCO, f_o, is set by connecting a capacitor, C_1, between terminals 1 and 9 and a resistor, R_1, between terminals 8 and 10. The free-run frequency is computed from

$$f_o \cong \frac{1.2}{4R_1 C_1} \qquad (17.63)$$

with a suggested range of 2 to 20 kΩ for R_1. The voltage output from terminal 7, the collector terminal of an emitter-coupled pair amplifier, is also input to the VCO internally. The loop filter function, $F(s)$, is implemented at terminal 7. For a zero-order PLL, no additional connections are required. For a first-order loop, a capacitor C_2 is connected from terminal 7 to terminal 10, which is effectively at small-signal ground. A single-section low-pass filter, as shown in Fig. 17.29(a), is formed with the internal resistance $R = 3.6$ kΩ and C_2. A compensation capacitor, which we studied in Chapter 15, on the order of $C_3 = 0.001$ μF is connected between terminals 7 and 8.

The capture range and lock range are computed from[13]

$$f_C \cong \pm \frac{1}{2\pi}\sqrt{\frac{2\pi f_L}{\tau}}, \qquad (17.64)$$

where $\tau = 3.6$ k$\Omega \times C_2$ and

$$f_L \cong \pm \frac{8 f_o}{2 V_{CC}}. \qquad (17.65)$$

Equation (17.65) predicts a lock range in excess of $\pm 60\% \, f_o$. Although this may be required for very wide bandwidth systems, often a much narrower lock range is desirable. To reduce the lock range, the loop gain can be decreased by decreasing the voltage gain of the emitter-coupled pair. An additional resistance is effectively placed in parallel with the 3.6-kΩ internal resistor by connecting an external resistor R_2 between terminal 7, the

■ The 565 PLL is a venerable old workhorse. The analysis techniques and applications demonstrated would apply to virtually any PLL circuit. One also finds PLL circuits as portions of larger ICs configured from ASICs.

CHAPTER 17 ADDITIONAL EXAMPLES OF ANALOG INTEGRATED CIRCUITS

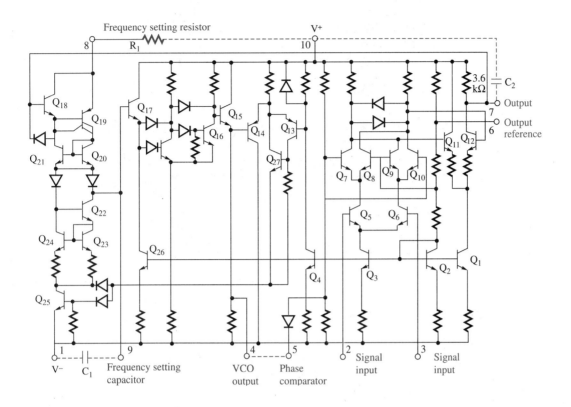

Parameter	Test Conditions	SE565 Min	SE565 Typ	SE565 Max	NE565 Min	NE565 Typ	NE565 Max	Unit
SUPPLY REQUIREMENTS								
Supply voltage		12		±12	±6		±12	V
Supply current			8	12.5		8	12.5	mA
INPUT CHARACTERISTICS								
Input impedance	f_o = 50 kHz, ±10%	7	10		5	10		kΩ
Input level required for tracking	frequency deviation	10	1		10	1		mVrms
VCO CHARACTERISTICS								
Center frequency	C_1 = 300pF		500			500		kHz
Maximum value	Distribution taken about	300						
Distribution	f_o = 50 kHz, R_1 = 5.0 kΩ, C_1 = 1200pF	−10	0	+10	−30	0	+30	%
Drift with temperature	f_o = 50 kHz		200			300		ppm/°C
Drift with supply voltage	f_o = 50 kHz, V_{CC} = ±6 to ±7 volts		0.1	1.0		0.2	1.5	%/V
Triangle wave								
Output voltage level		1.9	0		1.9	0		V
Amplitude			2.4	3		2.4	3	Vp−p
Linearity			0.2			0.5		%
Square wave								
Logical "1" ouput voltage	f_o = 50 kHz	+4.9	+5.2		+4.9	+5.2		V
Logical "0" ouput voltage	f_o = 50 kHz		−0.2	+0.2		−0.2	+0.2	V
Duty cycle	f_o = 50 kHz	45	50	55	40	50	60	%
Rise time			20	100		20		ns
Fall time			50	200		50		ns
Output current (sink)		0.6	1		0.6	1		mA
Output current (source)		5	10		5	10		mA
DEMODULATED OUTPUT CHARACTERISTICS								
Output voltage level	Measured at pin 7	4.25	4.5	4.75	4.0	4.5	5.0	V
Maximum voltage swing[3]			2			2		Vp−p
Output voltage swing	±10% frequency deviation	250	300		200	300		mVp−p
Total harmonic distortion			0.2	0.75		0.4	1.5	%
Ouput impedance			3.6			3.6		kΩ
Offset voltage (V6 − V7)			30	100		50	200	mV
Offset voltage vs temperature (drift)			50			100		μV/°C
AM rejection		30	40			40		dB

■ **FIGURE 17.35** SE/NE 565 PLL equivalent schematic and specifications (courtesy of Signetics Corporation).

output, and terminal 6, an internal node in the bias circuit whose dc potential is about the same as the dc potential of terminal 7. For $\pm V_{CC} = 6$ V, $R_2 = 2$ kΩ will reduce the lock range to $f_L = \pm 25\% f_o$. For $R_2 = 10$ kΩ, $f_L = \pm 50\% f_o$.

DRILL 17.5

Refer to Figure 17.35. Write an equation for the I_{REF} in terms of the circuit elements.

ANSWERS There are three resistors in series with the collector of Q_2 and one resistor in the emitter of Q_2. Call these R_1, R_2, R_3, and R_4 respectively. Define $I_{\text{REF}} \cong I_{C2}$ for $\beta \gg 1$; we then have

$$I_{\text{REF}} \cong \frac{V^+ - V^- - V_{BE2(\text{on})}}{R_1 + R_2 + R_3 + R_4} \quad \text{assuming } V^- < 0.$$

Observe, pin 6, the output reference is sampled between R_1 and R_2.

EXAMPLE 17.3

Design a PLL using an NE 565 so that it will track an input signal whose frequency components lie between 75 and 125 kHz. The PLL should be capable of capturing for input frequency components between 90 and 110 kHz. Assume $V_{CC} = V_{EE} = 6$ V. Include a schematic diagram with standard value components selected to meet these specifications.

SOLUTION By using the equivalent schematic diagram and specifications for the NE 565 shown in Fig. 17.35 and the circuit description, the schematic diagram shown in Fig. 17.36 is drawn. Observe that the circuit component values for R_1, R_2, C_1, C_2, and C_C must be specified.

Using Eq. (17.63) with $f_o = 100$ kHz, let $C_1 = 0.001$ μF; then

$$R_1 = \frac{1.2}{4C_1 f_o} = \frac{1.2}{(4)(0.001 \; \mu\text{F})(100 \text{ kHz})} = 3 \text{ k}\Omega,$$

which is within the recommended 2 to 20-kΩ range for this component. Both $C_1 = 0.001$ μF and $R_1 = 3$ kΩ are standard values. To achieve a lock range of ± 25 kHz, which is equivalent to $\pm 25\% f_o$, R_2 must be set to 2 kΩ. A capture range of ± 10 kHz is required. Solving Eq. (17.64) for τ yields

$$\tau = \frac{2\pi f_L}{(2\pi)^2 f_C^2} = \frac{(2\pi)(25 \text{ kHz})}{(2\pi)^2 (10 \text{ kHz})^2} = 40 \; \mu\text{s}.$$

Then $C_2 = \tau/3.6$ k$\Omega = 40$ μ/3.6 k$\Omega = 11$ nf. Loop compensation is obtained by using a 0.001-μF capacitor between terminals 7 and 8. An input coupling capacitor, C_C, must be selected so as to present a very low reactance, X_C, to input signals whose frequency is as low as 50 kHz.

From the device specifications, the typical input resistance is 10 kΩ, so if we assume that the $X_C \leq 1\%$ of 10 k$\Omega = 100$ Ω, then

$$C_C \geq \frac{1}{\omega X_C} = \frac{1}{(2\pi)(50 \text{ kHz})(100 \; \Omega)} = 0.0318 \; \mu\text{F}$$

so that the standard value of 0.033 μF can be used.

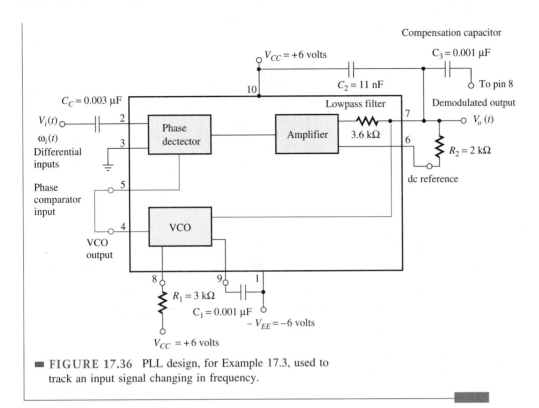

FIGURE 17.36 PLL design, for Example 17.3, used to track an input signal changing in frequency.

FM Demodulation

FM refers to the encoding of information as changes in frequency. As discussed in Section 17.4, when the PLL is locked to the incoming FM signal, the VCO will track the frequency variations. The filtered and amplified error voltage from the phase detector, used as input to the VCO, is the demodulated FM output. To assure accurate demodulation, the VCO gain, K_O, must be linear. A more specialized type of FM signal is called *frequency-shift keying* (FSK). Digital data are encoded as two separate preset frequencies, the frequencies corresponding to a logic one or a logic zero. The center frequency of the VCO is usually set midway between the two encoded data frequencies. Both FSK and frequency demodulation in general can be illustrated in the following example.

■ One of the most important applications is tone decoding, i.e., the touch-tone telephone dialing system. Consider how many data services depend upon accurate decoding of the ten digits and the * and # symbols.

EXAMPLE 17.4

The FSK signal composed of two frequencies, $f_1 = 1500$ Hz corresponding to a logic ONE and $f_2 = 1000$ Hz corresponding to a logic ZERO, is input to a zero-order PLL with $K_v = 10^4$ sec^{-1}, $f_o = 1200$ Hz, and $K_O = 2\pi \times 500$ rad/V. The FSK signal is sketched in Fig. 17.37(a). Sketch and label the output signal, $v_o(t)$.

SOLUTION Using Eq. (17.54), the time constant of the step frequency response is given by $\tau = 1/K_v = 0.1$ ms. From an $f_1 = 1500$ Hz, the steady-state value is

$$V_o = 2\pi\left(\frac{f_1 - f_o}{K_o}\right) = 2\pi\left(\frac{1500 - 1200}{2\pi \times 500}\right) = \frac{300}{500} = +0.6 \text{ V}.$$

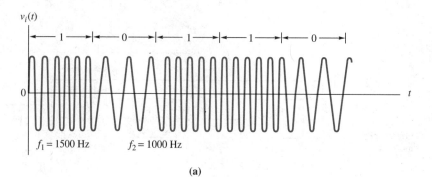

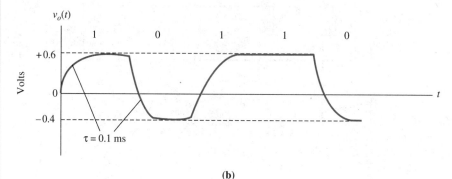

- **FIGURE 17.37** FSK demodulation example using a PLL. (a) Input FSK signal. (b) Demodulated FSK output.

For $f_2 = 1000$ Hz, the steady-state value is

$$V_O = 2\pi\left(\frac{f_2 - f_o}{K_O}\right) = 2\pi\left(\frac{1000 - 1200}{2\pi \times 500}\right) = \frac{-200}{500} = -0.4 \text{ V}.$$

The resultant decoded digital signal is sketched in Fig. 17.37(b), where $+0.6$ V corresponds to a logic 1 and -0.4 V corresponds to a logic 0.

Frequency Synthesis

PLL *frequency synthesis* is used to generate a large number of precisely spaced frequencies, all referenced to a stable frequency source. This is the approach used in most communications systems requiring precisely spaced channels. Although this could be done by using a precision quartz crystal for each output frequency, it is more cost effective to generate, that is, synthesize, each of the needed output frequencies from a single stable quartz crystal reference. The frequency stability of each output frequency is established by the frequency stability of the quartz crystal oscillator reference. Digital divider circuitry, perhaps under microprocessor control, can then establish the desired channel. Consider the PLL integer frequency synthesizer diagrammed in Fig. 17.38. A digital counter operating as a divider is inserted in the loop between the VCO and phase detector. One input to the phase comparator, a signal with frequency ω_1, is obtained from a system standard or clock usually designed using a quartz crystal oscillator. The second input signal frequency,

■ Usually the synthesizer output also controls a digital display for direct read-out of frequency. This display circuitry is discussed in Chapter 14.

CHAPTER 17 ADDITIONAL EXAMPLES OF ANALOG INTEGRATED CIRCUITS   865

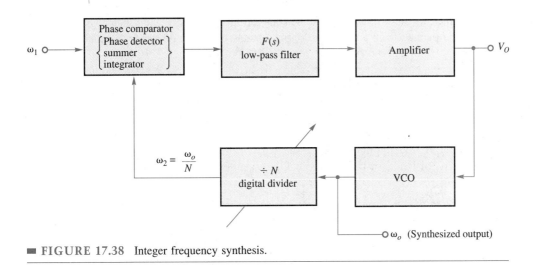

■ **FIGURE 17.38** Integer frequency synthesis.

ω_2, will be identical to ω_1 when the PLL is locked. The frequency relationships in the loop are given by

$$\omega_1 = \omega_2 = \frac{\omega_o}{N}. \quad (17.66)$$

The VCO is actually operating at an exact phase-locked multiple of ω_1,

$$\omega_o = N\omega_1. \quad (17.67)$$

The only restriction on N is that the VCO frequency must remain within the lock range. The amplified filtered dc error voltage present at V_o is often used to control some type of indicator that shows lock has been established.

A large value of N is required to generate a large number of closely spaced frequencies. This means that ω_1 may have to operate at a low frequency, which might not be practical. Greater versatility is obtained by using a prescaler, fractional-frequency synthesis technique illustrated in Fig. 17.39. The signal frequency input to the phase detector, $\omega_2 = \omega_1/M$

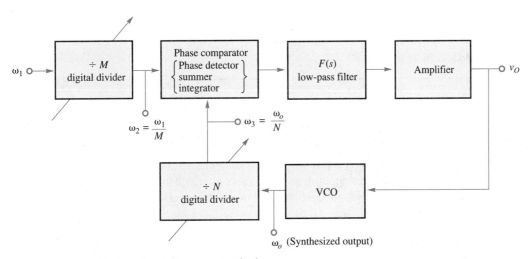

■ **FIGURE 17.39** Fractional frequency synthesis.

is obtained as the output from the selectable digital divider. The input signal operating at ω_1 is obtained from an accurate stable crystal-controlled reference oscillator. A digital divider operating as a counter is inserted in the loop between the VCO and phase detector. The second phase detector input signal is at $\omega_3 = \omega_o/N$. When the loop is locked, $\omega_3 = \omega_2$, with the result that

$$\omega_o = N\omega_3 = \frac{N\omega_1}{M}. \tag{17.68}$$

As before, the amplified filtered dc error voltage present at V_o may be used to control some type of indicator that shows lock has been established.

DRILL 17.6

Repeat Example 17.5 but the channel spacing is reduced to 20 kHz.

ANSWERS
$N/M = 51/50$
for obtaining 1020 kHz.
With $M = 50$,
$50 < N < 100$.

EXAMPLE 17.5

Specify a set of values for N and M to obtain a frequency synthesizer output between 1.0 and 2.0 MHz with 25-kHz spaced output frequencies. Assume a 1-MHz clock.

SOLUTION Select the center frequency of 1.5 MHz so that 1.0 MHz $< f_L <$ 2.0 MHz. To generate $f_o = 1025$ kHz from a $f_1 = 1000$ kHz clock requires $N/M = f_o/f_1 = 1025/1000 = 41/40$ so that by setting $M = 40$, $f_2 = 25$ kHz, $40 \leq N \leq 80$ would yield 25-kHz channels between 1.0 and 2.0 MHz. Observe that different N and M values would result if f_1, the clock, were specified differently. Often in a communications system, several PLL synthesizers are used to cover a large frequency range and a large number of precisely spaced channels.

CHECK UP

1. **TRUE OR FALSE?** A signal proportional to phase difference is the output of a PLL with FM input.
2. List several applications requiring frequency synthesis.
3. Draw and label block diagrams of an integer and a fractional-frequency synthesizer.

SUMMARY

Using basic IC building blocks and techniques, we discussed a variety of primarily analog circuit functions, including series and switching power supply regulators, multipliers, and phase-locked loops.

Voltage regulator design requires combining basic amplifier, power amplifier, and both negative and positive feedback concepts. In addition, packaging and heat dissipation are important issues for this application. Multipliers, as circuit elements or as subsystem phase detectors in a PLL, depend quite strongly on the matched device fabrication inherent in IC technology. When included as part of a PLL, negative feedback for the entire loop and positive feedback for the VCO are important. Frequency synthesis techniques illustrate the need to understand both the analog and digital aspects of any electronic system. Fabrication technology permits combining many of these functions on the same die.

CHAPTER 17 ADDITIONAL EXAMPLES OF ANALOG INTEGRATED CIRCUITS

SURVEY QUESTIONS

1. Draw some block diagrams, with as much detail as possible, of some typical electronic systems; i.e., a digital voltmeter, a home heating system control system, battery charger, television receiver, cellular radio, digital clock, FAX transmitter and receiver, motor speed control, etc.
2. What is the definition of voltage regulation?
3. How does a series-pass and switched-mode voltage regulator operate?
4. Draw a schematic diagram of a Gilbert cell multiplier and explain its operation qualitatively and its three regions of operation.
5. Explain the operation of a Gilbert cell as a mixer, modulator, and demodulator.
6. How does a phase detector operate?
7. Draw and label a block diagram of a phase-locked loop and explain its operation.
8. What are the differences in design and operation of a zero- and first-order phase-locked loop?
9. What are definitions of lock and capture range?
10. How does a phase-locked loop decode information from an FM signal?
11. How do you configure a phase-locked loop to obtain integer and fractional frequency synthesis?

PROBLEMS

Problems marked with an asterisk are more challenging.

17.1 The output voltage of a regulated power supply decreases from 5 V at no load to 4.95 V at a load current of 500 mA. Compute the effective output resistance and voltage regulation VR.

17.2 A 12-V automobile battery decreases to 11.3 V under a starter current load of 200 A. Compute the effective output resistance and voltage regulation, VR.

17.3* Using Figure 17.4 and SPICE, design a dc supply that can provide 15.0 V at no load and no less than 14.8 V with a 1.0-A load. Your design must specify values for R_L and C.

17.4 A series-pass voltage regulator has $V_R = 2.0$ V, $V_O = 8$ V, $a = 10^4$, and $R_o = 0.1$ Ω.

   (a) Compute r_{oa}.
   (b) Compute VR for $I_O = 100$ mA.

17.5 A series-pass voltage regulator $V_R = 2.0$ V, $V_O = 8$ V, $a(\omega = 0) = 10^4$, and $R_o(\omega = 0) = 0.1$ Ω. Assume the unity gain frequency of the amplifier portion is 1 MHz and $a(jf)$ is a single-pole response.

   (a) Compute and sketch $|z_{oa}(\omega)|$.
   (b) Compute and sketch $|z_o(\omega)|$.

17.6 For the circuit given in Fig. 17.6, let $V_R = 4.7$ V, $8 \leq V_{IN} \leq 12$ V, $I_{Z(min)} = 1.0$ mA; the avalanche diode maximum power is 250 mW. Select a minimum and maximum value for R_3 that will permit operation over all values of V_{IN} and satisfy the avalanche-diode operating conditions.

17.7 Select values for R_1 and R_2 to obtain the regulated output voltage $V_O = 5.0$ V using an NE 550 voltage regulator. The parallel combination of R_1 and R_2 should be 2 kΩ.

17.8 If $V_{CC} = 12$ V for the NE 550 5.0-V regulator of Problem 17.7, estimate the efficiency. If $I_{L(max)} = 100$ mA, how much power is Q_{15} dissipating?

17.9 The NE 550 is limited to $I_L < 100$ mA (Example 17.1). Using Fig. 17.8, show how you would connect an external series-pass npn transistor so that $I_{L(max)} = 2$ A, $V_O = 7.5$ V, and $V_{CE} = 12$ V. Specify key requirements for this external pass transistor such as β_{min} and $P_{C(max)}$.

17.10 The amplifier, including the control device (everything within the large operational symbol in Fig. 17.6), for a series-pass voltage regulator consists of a μA741 operational amplifier with an input resistor $R_S = 10$ kΩ and a feedback resistor $R_F = 1$ MΩ. Refer to Figs. 1.13 and 17.6. Assuming that $V_R = 4.7$ V, $R_1 = 180$ kΩ, and $R_2 = 100$ kΩ, compute the following:

(a) V_O, the regulated output voltage.
(b) R_o, output resistance.
(c) $R_{L(\min)}$.

17.11 Design a SMPS of the type illustrated in Fig. 17.14. Assume you require $V_O = 9.0$ V and $V_R = 4.7$ V. Let $L = 10$ mH and $C = 10$ μF. Select values of R_3, R_4, R_1, and R_2 to obtain < 20 mV of hysteresis. Compute the resultant switching frequency ω_o and the ripple R. Let $V_{IN} = 15$ V and the comparator output equal ± 13 V.

17.12 Design a SMPS system of the type illustrated in Fig. 17.17. Assume you require output voltages of $V_O = 9.0$ volts and 15 volts and $V_R = 4.7$ volts. Let $L = 4.7$ mH and $C = 22$ μF. Select values of R_3, R_4, R_1, and R_2 to obtain <20 mV of hysteresis for each SMPS regulator. Compute the resultant switching frequency ω_o and the ripple R. Let $V_{IN} = 18$ volts and the comparator output can approach no closer than 2 volts of the rail.

17.13* Refer to Figs. 17.14 and 17.15. Use a series of SPICE simulations to verify the operation of the SMPS shown in Fig. 17.14. Your results should include scaled waveforms of the form shown in Fig. 17.15.

17.14* Verify your results computed in Problem 17.11 using SPICE simulations.

17.15 Using analog multipliers, sketch block diagrams for circuits that will realize the following:

(a) $v_o(t) = K v_i^2(t)$, a "squarer" circuit.
(b) $v_o(t) = K v_i^3(t)$, a "cuber" circuit.

K is a different circuit constant in each case.

17.16* Use a SPICE simulation to verify your designs in Problem 17.15. Consider using the operational amplifier models presented in Chapter 1. The multiplier can be simulated by using higher order terms of controlled generators.

17.17 Show that the circuit given in Fig. 17.40 functions as an analog divider in which

$$v_o(t) \approx \frac{-v_1}{Kv_2}(t) \quad \text{for } R_1 = R_2.$$

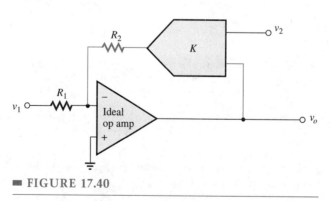

■ FIGURE 17.40

17.18 Using Problem 17.17 as a hint, design a circuit block diagram with ideal operational amplifiers and multipliers to generate the square root of a signal, i.e.,

$$v_o(t) = \left[-\frac{R_2 v_1(t)}{R_1 K} \right]^{1/2}.$$

What restrictions are required on $v_1(t)$?

17.19 Using Problems 17.15 and 17.18 as guides, design a circuit block diagram with ideal operational amplifiers and multipliers to generate the cube root of a signal; i.e.,

$$v_o \alpha [v_1(t)]^{1/3}.$$

17.20* Design a circuit block diagram to obtain the transfer function

$$v_{out} = \frac{K}{RC} \int (v_x)^2 \, dt$$

where v_x is the input analog voltage. K is a system constant and the RC product points to the use of an integrator in your design. Assume ideal operational amplifiers and multipliers. What signal processing algorithm is obtained from this system?

17.21* Refer to Fig. 17.19, the basic Gilbert cell multiplier. Assume all BJTs are matched with specifications of $\beta = 200$, $V_A = 150$ volts, and $V_{BE(on)} = 0.7$ V. Let $V_{CC} = V_{EE} = 15$ V, $I_{C7} = 1$ mA, and $R_C = 20$ kΩ. Use SPICE to graphically verify Eqns. (17.26) and (17.29) for v_1 and v_2 where v_1 and v_2 are ramped from -1.0 to $+1.0$ volts.

17.22* Use your SPICE analysis and listing from Problem 17.21. Let $v_1(t) = 1.0 \sin(2\pi 10{,}000 t)$ V and $v_2(t) = 1.0 \sin(2\pi 3{,}000 t)$ V. Use a SPICE transient analysis and the .FOUR command to show that $v_{od}(t)$ exhibits spectral components at 7.0 kHz and 13.0 kHz.

17.23 Using the curves shown in Figure 17.23, explain why the Gilbert cell multiplier is often referred to as a *variable transconductance amplifier*.

17.24 Show that the circuit shown in Figure 17.41 will function as a two-quadrant multiplier. Specify the regions of operations and voltage range constraints on v_1 and v_2.

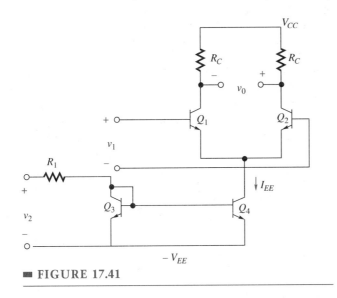

■ **FIGURE 17.41**

17.25 Verify your analysis obtained in Problem 17.24 for the two-quadrant multiplier using a SPICE simulation with a .TRAN analysis. Assume $V_{CC} = V_{EE} = 15$ V and $I_{EE} = 1$ mA. Let $R_C = 10$ kΩ. Use the default SPICE *npn* BJT model.

17.26 Refer to Fig. 17.35, the equivalent schematic diagram and specifications for an SE/NE 565 PLL. List all of the transistors and diode-connected transistors. After each device, briefly identify its key function in the IC. For example, the transistors whose bases are connected to pins 2 and 3 are the input differential emitter-coupled pair and part of the Gilbert cell.

17.27 The diode bridge circuit shown in Figure 17.42 was introduced in Chapter 3. This circuit is often used as a mixer (multiplier). Assume $v_1(t)$ is a zero-centered square wave with a fundamental frequency f_1 and $v_2(t) = A \cos(2\pi f_2 t)$ where $f_2 \ll f_1$. Assume both signals are of sufficient amplitude to completely switch the diodes.

(a) Sketch $v_1(t)$, $v_2(t)$, and $v_o(t)$ to show the chopping of the $v_1(t)$ waveform.
(b) Sketch the spectrum of $v_o(t)$.

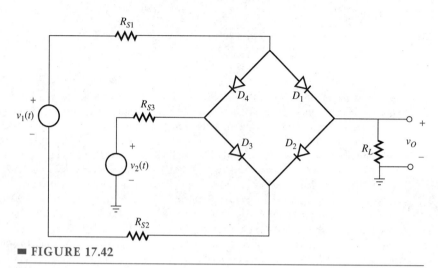

■ FIGURE 17.42

17.28* Refer to your results in Problem 17.27 and Figure 17.42. Let $R_{S1} = R_{S2} = R_{S3} = 50\ \Omega$ and $R_L = 1\ \text{k}\Omega$. Assume the $v_1(t)$ square wave frequency, f_1, is 20 kHz with a peak-to-peak amplitude of 10 V and $v_2(t) = 3\cos(2\pi\ 1000t)$. Use a SPICE simulation with a transient analysis to obtain $v_o(t)$. Use the .FOUR command to obtain the spectral components of $v_o(t)$. Assume a SPICE default diode model

17.29 An NMOS source-coupled pair is shown in Figure 17.43. Assume $\lambda = 0\ V^{-1}$ and both MOSFETS are matched. Show that

$$\Delta i = i_{D1} - i_{D2} \cong \left(\frac{\mu_n C_{ox}}{2}\right)\left(\frac{W}{L}\right) v_{id} \left[\frac{[4 i_{SS}]}{\left(\frac{\mu_n C_{ox}}{2}\right)\left(\frac{W}{L}\right)}\right]^{1/2}$$

for $v_{id}^2 \ll \left[\dfrac{[4 i_{SS}]}{\left(\frac{\mu_n C_{ox}}{2}\right)\left(\frac{W}{L}\right)}\right]$.

Therefore, we have obtained an NMOS analog multiplier circuit for the functions $v_{id}(t)$ and $\sqrt{i_{SS}(t)}$.

17.30 We now cascade two of the NMOS amplifiers shown in Fig. 17.43 to obtain the circuit shown in Fig. 17.44. Assume $\lambda = 0\ V^{-1}$, all four MOSFETS are matched, and both drain resistors are also matched. Show that

$$\Delta i = i_{D3} - i_{D4} \cong -R_D i_{SS}(t) v_{id}(t) \left[(\mu_n C_{OX})\left(\frac{W}{L}\right)\right]\left[1 - \frac{\left((\mu_n C_{OX})\left(\frac{W}{L}\right)\right)^2 R_D^2 v_{id}^2(t)}{4}\right]^{1/2}.$$

Therefore, we have now obtained an NMOS analog multiplier circuit for the functions $v_{id}^2(t)$ and $i_{SS}(t)$.

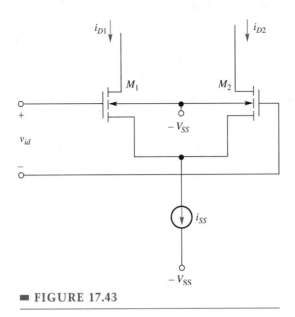

■ FIGURE 17.43

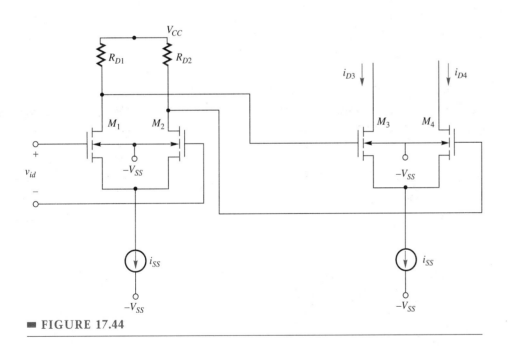

■ FIGURE 17.44

17.31* By combining the ideas in Problems 17.29 and 17.30, we can construct a two-quadrant multiplier as shown in Figure 17.45. Assume $\lambda = 0 \text{ V}^{-1}$, all six MOSFETS are matched, and both drain resistors are matched. Show that

$$\Delta i = i_{o7} - i_{o8} \cong K v_a(t) v_b(t)$$

where K is a constant and $V_G > V_t$ with I_{SS} designed so that M_3 and M_4 are operating in saturation.

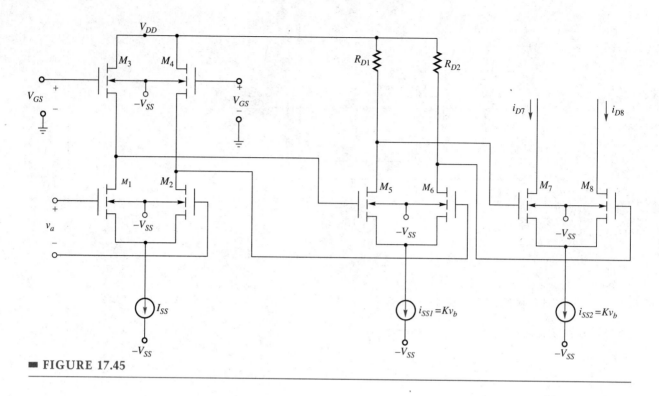

■ FIGURE 17.45

17.32 Refer to Fig. 17.28, a standard PLL block diagram. Let $K_\nu = 1000$ sec^{-1}, $K_O = 4\pi \times 10^3$ (sec-V)$^{-1}$, $F(s) = 1$, and $f_o = 5$ kHz.

(a) Compute a maximum value for $f_i(t) = f_{i(\max)}$ to ensure locked PLL operation for the input signal shown in Fig. 17.46.

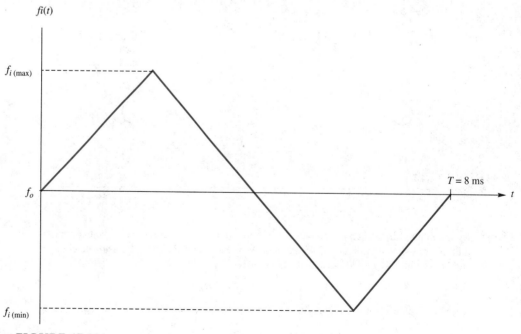

■ FIGURE 17.46

(b) Observe that $T = 8$ ms. Sketch and label the approximate output waveform $v_o(t)$.

(c) Design, that is, specify, an appropriate first-order $F(s)$ with numerical values to achieve a maximally flat response.

17.33 Refer to Fig. 17.47. Let $K_D = 0.3$ V/rad, $K_O = 0.2\omega_o$ rad/sec-V^{-1}, and $f_o = 2$ kHz.

(a) The switch is open. Then $v_1(t) = \cos(2\pi \times 2000t + 45°)$ and $v_2(t) = \cos(2\pi \times 2000t - 15°)$. Assume $F(s)$ is an ideal low-pass filter with a cutoff frequency of 3 kHz. Sketch and label $v_o(t)$.

(b) If the low-pass filter from part (a) is replaced by $F(s) = 1$, what other spectral components are present in v_o?

(c) Repeat part (a) but assume $v_1(t) = \cos(2\pi \times 2020t)$ and $v_2(t) = \cos(2\pi \times 2000t)$.

(d) Close the switch and let $F(s) = 1$. Let $v_1(t) = \cos(2\pi \times 2100t)$, $0 < t < 5$ ms, and $v_2(t) = \cos(2\pi \times 1950t)$, $5 < t < 10$ ms, and assume the loop is locked. Sketch and label $v_O(t)$.

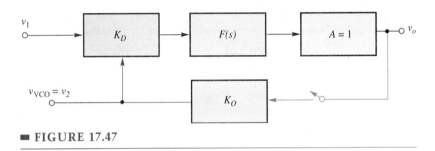

■ **FIGURE 17.47**

17.34 Refer to the PLL block diagram of Fig. 17.28. Assume for the first-order PLL that $f_o = 10$ kHz, $K_v = 1000$/sec, and $K_O = 4\pi \times 10^3$/(sec-V). Sketch and label $v_o(t)$ for the given $f_i(t)$.

17.35 Use an NE 565 to phase lock a 38-kHz signal. Assume the capture range is ± 3 kHz and $V_{CC} = V_{EE} = 6$ V. Specify values for R_1, C_1, R_2, C_2, and C_C. As an added note, this type of circuit is used as the stereo signal decoder for standard FM broadcast receivers with a standard 38-kHz subcarrier.

17.36 Commercial FM radio stations sometimes use their wide bandwidth allocation to transmit private entertainment services. They use a 67 kHz subcarrier. Use an NE 565 as a PLL to decode this 67-kHz signal. Assume the capture range is ± 5 kHz and $V_{CC} = V_{EE} = 6$ V. Specify values for R_1, C_1, R_2, C_2, and C_C.

17.37 Design a signal generator using fractional frequency synthesis techniques employing a PLL to meet the following specifications:

(a) The clock frequency shall be 1.0 MHz as derived from a crystal-controlled clock.
(b) The overall range shall be from 1.0 to 2.0 MHz.
(c) The resolution shall be 10 kHz.

Your design must provide a well-labeled fractional-frequency PLL block diagram with values for a suggested free-run VCO frequency and divider numbers.

17.38 The U.S, AM broadcast band as established by the Federal Communications Commission runs from 535 through 1605 kHz although there is some legislation being discussed to increase this frequency allocation. Each broadcast station is spaced by 10 kHz with the first channel starting at 540 kHz and continuing with a 10-kHz channel spacing to 1600 kHz.

Design a first-order, PLL-based, fractional frequency synthesizer using a 1-MHz clock reference to obtain all the AM broadcast band carrier frequencies. By design, we mean provide a well-labeled block diagram with numerical values for a suggested free-run VCO frequency and divider numbers.

17.39 The U.S. FM broadcast band as established by the Federal Communications Commission runs from 88 MHz through 108 MHz. Each broadcast station is spaced by 200 kHz with the first channel starting at 88.1 MHz.

Design a first order, PLL-based, fractional frequency synthesizer using a 10.7 MHz clock reference to obtain all the FM broadcast band carrier frequencies. The design should include a well-labeled block diagram with numerical values for a suggested free-run VCO frequency and divider numbers.

REFERENCES

It is important to note that many of the data books provided by key players in the semiconductor device and circuit industry are available on magnetic diskette or CD-ROM media as well as hardcopy. The list of manufacturers of these products is extremely large and fluid. You are encouraged to review the trade literature to keep current on the market.

1. *Signetics Analog Data Manual* and *General-Purpose/Linear ICs Data Handbook.* Signetics Corporation, Sunnyvale, California, 1992.
2. *Signetics Analog Applications Manual.* Signetics Corporation, Sunnyvale, California, 1992. Section 4 discusses voltage regulators. Several versions incorporate these data within the data manual.
3. Holt, Charles A. *Electronic Circuits, Digital and Analog.* New York: John Wiley & Sons, 1978.
4. Soclof, S. *Design and Applications of Analog Integrated Circuits.* Englewood Cliffs, N.J.: Prentice-Hall, 1991.
5. Gilbert, B. "A Precise Four-Quadrant Multiplier with Subnanosecond Response," *IEEE Journal of Solid-State Circuits*, December 1978, Vol. SC-3, pp. 365–373.
6. Stremler, Ferrel G. *Introduction to Communication Systems.* Reading, Mass.: Addison-Wesley Publishing Co., 1990.
7. *Burr-Brown Product Data Book Series.* Tucson, Ariz.: Burr Brown Corporation, 1984, 1989, 1994.
8. *Analog and Telecommunications Product Data Book.* Harris Corporation, Melbourne, Florida, 1990.
9. *Motorola Linear Interface Integrated Circuits.* Motorola Semiconductor Products, Phoenix, Arizona, 1991.
10. Ref. 2, Sec. 10.
11. Nilsson, J. W. *Electric Circuits*, 5th ed. Reading, Mass.: Addison-Wesley Publishing Co., 1996.
12. Hayt, Jr., W. H., and J. E. Kemmerly. *Engineering Circuit Analysis.* New York: McGraw-Hill Book Co., 1978.
13. Ref. 1, Sec. 16.

PART IV

SEMICONDUCTOR TECHNOLOGY

This section of this text includes only one chapter but it is a very important chapter in that it provides an overview of the IC technology fabrication justification for the specifications of all the active and passive devices used in the preceeding 17 chapters. Chapter 18 is written as a stand-alone chapter. Therefore, you could use this chapter either as a reference for this book or include it directly as part of the course. It would fit well right after Chapter 5 since by then you will have been introduced to all key devices. It is also useful to review the definitions and topics in Chapter 2 before proceeding to the study of Chapter 18. Chapter 18 only presents an overview and doesn't even come close to a comprehensive presentation; consequently, you should avail yourself of opportunities to take complete courses, perhaps with a laboratory, in semiconductor device fabrication and semiconductor device physics. As you continue your studies in electronic circuits and systems, whether digital or analog, it will become increasingly important to understand the basics of device fabrication so that one can effectively compare the large and increasing number of technologies available to the designer. Therefore, Chapter 18 discusses the fabrication technology and resultant specifications obtained for resistors, capacitors, diodes, BJTs, JFETs, and MOSFETs. One should observe that SPICE models for all these elements are directly keyed to fabrication parameters and geometries.

CHAPTER 18

BASIC FABRICATION TECHNOLOGY AND DEVICE CONSTRAINTS

18.1 Impurity Diffusion
18.2 Ion Implantation
18.3 Resistive Properties of Doped Layers
18.4 Photolithography and Masking
18.5 Resistors
18.6 Capacitors
18.7 *npn* Transistors
18.8 *pnp* Transistors
18.9 Diodes
18.10 Junction Field-Effect Transistors
18.11 Metal-Oxide Semiconductor Transistors
18.12 Summary
Survey Questions
Problems
References
Suggested Readings

Integrated-circuit technology presents the designer and user of integrated circuits (ICs) with many choices with respect to performance and functionality. Tens or even hundreds of thousands of individual transistors are fabricated on a single die using a variety of technologies, and each addressing different applications. Therefore, it is quite important that a designer become familiar with the various factors and technologies associated with the fabrication of ICs. The objectives of this chapter are twofold. The first is to provide an overview and reference of the various technologies used for IC manufacture so that the designer and end user can understand the constraints of each technology. The second objective is to provide a foundation for advanced study.

The term *monolithic* is derived from the ancient Greek *mono*, meaning "single" or "one," and *lithos*, meaning "stone." A monolithic circuit is an entire circuit fabricated within a single piece of suitably doped crystalline semiconductor, usually silicon. Fig. 18.1 shows the progression of silicon ingot sizes from 19 mm used in the 60s to the very common 100-mm ingot used in the 80s and currently. A 200-mm ingot is now being used by some of the large manufacturers with 300-mm sizes on the near horizon. These ingots are nominally single-crystal doped silicon usually grown by the *Czochralski* process. This process is diagrammed in Figure 18.2. Basically, the crystal is grown by immersing a seed crystal in molton silicon (suitably doped) and then cooling gradually to $T < 1412°C$ as the crystal is pulled from the melt. Again, we will mention certain applications in GaAs and other III-V compounds only in passing because, for the most part, such circuits are beyond the scope of this book. No integrated circuits are being fabricated in germanium, although the use of germanium emitters in high-frequency bipolar technology is being explored.

■ **FIGURE 18.1** Czochralski-grown ingots of single-crystal silicon. We observe the development of the technology as it proceeds from (1) a 19 mm diameter ingot in the late 60s to (2) a 200 mm diameter ingot currently being introduced into the production environment. (3) The 100 mm diameter ingot is widely used but many new semiconductor foundry installations are being designed for 100 mm and beyond capability. Some of the larger semiconductor manufacturers are even considering using 300 mm wafers. (4) A 125 mm wafer with 120 VLSI die is shown. Each die is 9 mm on a side.

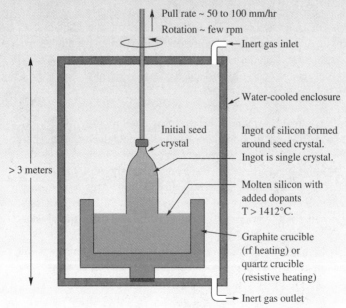

■ **FIGURE 18.2** Czochralski process for growing ingots of single-crystal silicon. Diameter and defects are controlled by the rotation and pull rates.

We first consider the fundamental steps in *planar* technology, in which all components are fabricated on a single surface of the semiconductor. These processes are oxidation, solid-state diffusion of dopants, ion implantation, photolithography, the growth of epitaxial layers, and general material deposition. Early pioneers in these basic processes were Jack St. Clair Kilby at Texas Instruments, who in 1958–1959 demonstrated an RC oscillator and flip-flop, and Robert N. Noyce at Fairchild Semiconductor, who in 1959–1961 proceeded to develop the foundation for the commercial manufacture of a logic family.[1] In addition to the basics of solid-state diffusion, we will discuss ion implantation of impurities and the use of very-short-wavelength photolithographic techniques. As a fundamental introduction to semiconductor device processing, we have elected to provide a basic overview of key steps using fundamental analytical equations. An engineer working in the semiconductor device processing industry would use some variation of the semiconductor process modeling program SUPREM[2]; however, the use of SUPREM is beyond the scope of this text.

It is important to realize that the material and processing properties graphs presented in this chapter are only approximations. Variations on the values and functional behavior of many of these parameters are often presented in the literature and most semiconductor manufacturing facilities spend considerable time and money on generating precise test data used to obtain reproducible results. With this basic caveat, we will illustrate how the basic steps are employed in the fabrication of passive components, such as resistors and capacitors, and active components, such as diodes, bipolar transistors, and FETs of various types.

IMPORTANT CONCEPTS IN THIS CHAPTER

- Qualitative description of semiconductor impurity diffusion.
- Use of the one-dimensional diffusion equation.
- Infinite (constant) and finite source diffusion equation solutions and application to the formation of doping profiles.
- Technique for growing of epitaxial layers.
- Use of ion implantation.
- Definition and computation of a dose.
- Resistive properties of doped layers.
- Definition and computation of sheet resistance for uniform and nonuniform layers.
- Growth of SiO_2 and applications of SiO_2 in semiconductor device fabrication.
- Basic steps in the photolithography/masking process.
- Fabrication and properties of diffused and pinch resistors.
- Fabrication and properties of MOS and junction capacitors.
- Fabrication and properties of *npn* transistors and of lateral and substrate *pnp* transistors.
- Fabrication and properties of JFETs.
- Fabrication and properties of MOSFETs, including CMOS, VMOS, DMOS, and SOI devices and circuits.

18.1 IMPURITY DIFFUSION

In general, IC devices require multiple layers of *n*- and *p*-type semiconductor material, suitably interconnected to obtain the desired function. One of the primary techniques used to obtain *n*- and *p*-type layers is by the diffusion of impurity materials into the semiconductor.

The one-dimension, constant diffusivity diffusion equation, derived as Eq. (2.24), is repeated here for convenience:

$$\frac{\partial N(x,t)}{\partial t} = D \frac{\partial^2 N(x,t)}{\partial x^2}. \tag{18.1}$$

Mobile charge carriers will diffuse across a concentration gradient, resulting in a current flow. In the derivation of this equation, there was no requirement for having charged particles as the diffusing particles. Equation (18.1) is just as valid for impurity atoms that will become mobile in the silicon crystal if the temperature is high enough, as it was for charged-particle diffusion in a semiconductor. This impurity diffusion property has been extensively studied and the diffusion coefficients measured. Figure 18.3 presents a graph plotting the **diffusion coefficient**, **D**, for several typical dopants in silicon as a function of temperature.[3,4] Analytically, the temperature dependence illustrated in Fig. 18.3 is given by the Arrhenius relationship

$$D = D_0 e^{-K/T}, \tag{18.2}$$

where K is a material-dependent constant and T is in kelvins.

CHAPTER 18 BASIC FABRICATION TECHNOLOGY AND DEVICE CONSTRAINTS

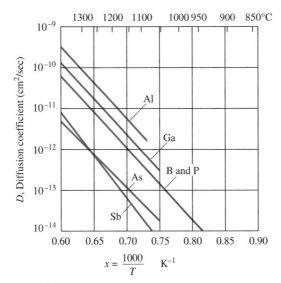

FIGURE 18.3 Diffusion coefficients of Column III and Column V elements in silicon as a function of temperature (from Refs. 3 and 4).

Impurity atoms are introduced at the surface of the silicon. During the process in which the silicon and impurity atoms are heated, usually to more than 900°C, the impurity atoms substitutionally replace the silicon atoms in the lattice; hence the name **substitutional impurities**. The silicon can be made either *n*- or *p*-type, depending on the impurity being used.

Intuitively, we expect the doping concentration to be highest at the surface and to decrease with time and distance from the surface, as predicted by Eq. (18.1). Also, as the temperature increases, the impurity concentration at a given depth into the material will also increase. The story is a bit more complex than this. Equation (18.1) is a differential equation, which implies that to obtain the solution, we must consider boundary conditions. Two basic processes are used in practice. Each creates a different diffusion profile because the boundary conditions are different.

■ We restrict our diffusion discussion to one dimension. Strictly speaking, it is three dimensional and the equation $\frac{\partial N}{\partial t} = \nabla^2 DN$ or $\frac{\partial N}{\partial t} = D\nabla^2 N$ must be solved for concentration-dependent and concentration-independent conditions respectively.

Infinite or Constant Source

The first case is that of the **infinite** or **constant source**. Suppose we expose an arbitrarily thick silicon wafer to a large volume of an impurity gas so that the impurity concentration at the surface is given by N_0. The term *infinite* or *constant source* implies that this concentration will be kept constant at the surface. Mathematically, we can state the boundary conditions as

1. $N(0, t) = N_0 \quad 0 < t < \infty$.
2. $N(x \to \infty, t) = 0 \quad 0 < t < \infty$.

Under these conditions, it can be shown that the solution is given by

$$N(x, t) = N_0 \left(1 - \text{erf} \frac{x}{2\sqrt{Dt}} \right) = N_0 \, \text{erfc}\left(\frac{x}{2\sqrt{Dt}} \right), \qquad (18.3)$$

where *erf* is referred to as the error function and *erfc* is referred to as the complementary

> ■ SUPREM (**S**tanford **U**niversity **PR**ocess **E**ngineering **M**odels) is available from commercial vendors. The original source is the Applied Electronics Labs, Stanford University. Stanford can be contacted for obtaining a version at nominal cost.

error function where erfc(z) = 1 − erf(z). These functions have been tabulated and are available from a number of sources.[5] Of more interest than the actual values is a sketch of the ***diffusion profile*** as illustrated in Fig. 18.4. The actual material-dependent physical processes and related material coefficients are sufficiently complex, and in many cases only empirically determined, such that computer simulations (SUPREM) are widely used to obtain the actual profiles. In practice, even though Eq. (18.3) offers only a very rough approximation to the real profiles, it is useful because it offers reasonably tractable closed-form solutions.

Typically in semiconductor fabrication work, the impurity concentration values span several orders of magnitude; consequently, $N(x,t)$ is plotted logarithmically. Note that the surface concentration is a constant, and the impurity concentration profile looks approximately like an exponential when plotted with a logarithmic concentration scaling. Indeed, for ease of computation, a very rough approximation to Eq. (18.3) at a given temperature and time is given by

$$N(x) = N_0 e^{-x/L}, \quad (18.4)$$

where L is some characteristic length defining the steepness of the profile; L would depend on processing temperature and time.

It is now possible to see how we can create a *pn* junction. Suppose we start out with a uniformly *n*-doped wafer of silicon of concentration N_D. We then expose the wafer to a *p* dopant with a uniform concentration of N_A. At a certain temperature and at a given time, we observe an impurity profile as sketched in Fig. 18.5. A *pn* junction is produced where $N_A(x_j,t) = N_D$. The junction occurs at $x = x_j$, the ***junction depth***. The complementary error function diffusion profile is used to create relatively abrupt junctions.

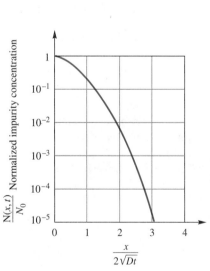

■ **FIGURE 18.4** Normalized logarithmic plot of $N(x,t)/N_0$ = erfc $[x/(2\sqrt{Dt})]$.

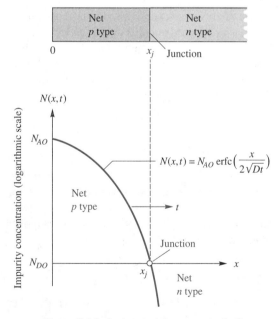

N_{DO} — Original substrate donor concentrationß

■ **FIGURE 18.5** Junction formation using "erfc" diffusion.

EXAMPLE 18.1

Consider a silicon substrate uniformly doped with 10^{15} atoms/cm^3 of arsenic. The substrate is exposed to an infinite source of boron, resulting in a surface concentration of 10^{18} atoms/cm^3 and a concentration decreasing exponentially to $1/e$ of its surface value at a depth of 1.0 μm. What is the depth of the junction x_j?

SOLUTION The substrate is n-type (arsenic is a Column V material) with a concentration of 10^{15} atoms/cm^3. The infinite-source doping profile can be approximated by Eq. (18.4), and solving for L, we obtain

$$\frac{N(1.0\ \mu m, t)}{N_0} = \frac{1}{e} = e^{-x/L},$$

so $L = 1.0\ \mu$m (1 μm = 10^{-6} m = 10^{-4} cm). The junction will occur when $N(x_j, t) = 10^{15}$ atoms/cm^3.

$$10^{18} e^{-x/1.0\ \mu m} = 10^{15}\ \text{atoms/cm}^3$$
$$e^{-x/1.0\ \mu m} = 10^{-3}$$
$$x_j = 1.0\ \mu m (\ln 10^3) = 6.9\ \mu m.$$

EXAMPLE 18.2

Suppose we diffuse boron into the substrate of Example 18.1 for 1 h at $T = 1100°$C. What will be the depth of the junction, x_j?

SOLUTION From Fig. 18.3, $D = 3.5 \times 10^{-13}$ cm^2/sec at 1100°C (1100°C = 1373 K). Substituting into Eq. (18.3), we have

$$N(x,t) = N(x_j, 3600\ \text{sec}) = 10^{15}\ \text{atoms/cm}^3$$
$$= 10^{18}\ \text{erfc}\left[\frac{x_j}{2\sqrt{(3.5 \times 10^{-13})(3600)}}\right].$$

Using Fig. 18.4,

$$10^{-3} = \text{erfc}\left[\frac{x_j(\text{cm})}{7.1 \times 10^{-5}\ \text{cm}}\right]$$
$$\left(\frac{x_j}{7.1 \times 10^{-5}}\right) = 2.33$$
$$x_j = 1.65\ \mu m.$$

DRILL 18.1

Refer to Example 18.2 and obtain the depth of the junction for a 30- and 240-min, $T = 1100°$C, constant-source diffusion.

ANSWERS
30 min, $x_j = 1.30\ \mu$m;
240 min, $x_j = 3.67\ \mu$m.

DRILL 18.2

Suppose the substrate doping from Example 18.1 is given by $N_D = 10^{16}$ donor atoms/cm^3. For $T = 1100°$C and $t = 1$ hr, compute the junction depth.

ANSWER
$x_j = 1.3\ \mu$m

Finite Source

The second set of boundary conditions we will apply to Eq. (18.1) provides for a fixed amount of dopant initially at the surface, a ***finite source***. Thus, as the diffusion process proceeds, the surface concentration decreases. Mathematically, we can state the boundary condition as

1. $Q(x,t)$ is a sheet impurity concentration given in terms of atoms/cm^2 and $Q(0,0) = Q_0$.

2. $N(x, t \to \infty) = 0$ for an arbitrarily thick silicon slab. Under these conditions, it can be shown that the solution is given by

$$N(x,t) = \frac{Q_0}{\sqrt{\pi Dt}} e^{-x^2/4Dt}. \qquad (18.5)$$

This function is known as the **Gaussian distribution**, and values are tabulated in a number of books,[5] although it may be computed directly. The Gaussian diffusion profile is sketched in Fig. 18.6. Note that since the surface concentration is continually decreasing as a function of time, one would expect the junction to be somewhat less abrupt with a

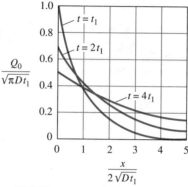

FIGURE 18.6 Normalized logarithmic plot of the Gaussian diffusion profile (with respect to $t = t_1$). Note that surface concentration decreases with time.

shallower concentration gradient than that achieved by the complementary error function diffusion gradient.

EXAMPLE 18.3

Compare the impurity concentration at 1 μm with that of the surface concentration for a 1- and a 2-hr diffusion of boron into an n-type substrate at 1100°C.

SOLUTION Using Eq. (18.5) and from Fig. 18.3 ($D = 3.5 \times 10^{-13}$ cm²/sec),

$$N(1\ \mu\text{m}, 3600\ \text{sec}) = \frac{Q_0}{\sqrt{(\pi)(3.5 \times 10^{-13})(3600)}} \times e^{-(1 \times 10^{-4}\ \text{cm})^2/(4)(3.5 \times 10^{-13})(3600)}$$
$$= Q_0 \times 2.2 \times 10^3\ \text{atoms/cm}^3,$$

and, similarly,

$$N(1\ \mu\text{m}, 7200\ \text{sec}) = Q_0 \times 4.16 \times 10^3\ \text{atoms/cm}^3.$$

However, while the concentration at 1 μm approximately doubled, at the surface ($x = 0$) the concentration dropped by a factor of

$$\frac{N(x = 0, 3600)}{N(x = 0, 7200)} = \frac{\dfrac{Q_0}{\sqrt{(\pi)(3.5 \times 10^{-13})3600}}}{\dfrac{Q_0}{\sqrt{(\pi)(3.5 \times 10^{-13})7200}}} = \sqrt{2}.$$

By contrast, the surface concentration is a constant for an erfc function.

Clearly, the actual processing is affected by wafer size and manufacturing control parameters, but the most critical factor is precision control of high oven temperatures.

Epitaxial Layers

We have been assuming that the impurities are diffused into a uniform lightly-doped layer of n- or p-type silicon. The composition and uniformity of this first layer are very important for the overall device performance and in determining the precise junction depths. This layer is formed by a process called *epitaxy* (from the Greek *epi*, meaning "upon," and *taxis*, meaning "arrangement"). Growth of this layer consists of a condensation of either intrinsic or doped silicon onto the substrate in such a way that the basic substrate crystalline structure remains continuous across the interface. For intrinsic silicon, we use silane (SiH_4) decomposition or silicon tetrachloride ($SiCl_4$) reacting with hydrogen at 1200°C according to

$$SiH_4 \xrightleftharpoons{T=1200°C} Si + 2H_2 \quad \text{or} \quad SiCl_4 + 2H_2 \xrightleftharpoons{T=1200°C} Si + 4HCl.$$

This process takes place in an *epitaxial reactor*, a specially designed chamber in which gas flows and temperatures can be carefully controlled to achieve good crystal structure.

To obtain n- or p-type epitaxial layers, phosphine (PH_3), arsine (AsH_3), or diborane (B_2H_6) is introduced into the *epitaxial reactor* along with silane or silicon tetrachloride and hydrogen. The result of these reactions is a uniformly-doped layer from 1 to 25 μm thick. Typically, the entire integrated circuit then lies within the epitaxial layer.

In comparison with diffusion, the epitaxial process offers (1) relatively abrupt junctions, (2) more precise control of layer thickness, and (3) the ability to apply a lightly doped layer on top of a more heavily doped region. Selective area epitaxy is much more difficult to do than selected area diffusion.

CHECK UP

1. **TRUE OR FALSE?** Diffusion of semiconductor impurities occurs even at room temperature.
2. Which set of boundary conditions yields the flattest impurity profile?
3. What are the two key processing parameters that can be adjusted to obtain a particular junction depth?
4. **TRUE OR FALSE?** One can obtain any value for a surface concentration by adjusting doping gas flow.
5. **TRUE OR FALSE?** Epitaxial layers can be formed on any type of substrate.

DRILL 18.3

Compare the total amount of dopant for both the 1- and 2-hr diffusions presented in Example 18.3.

ANSWER Both are the same. The areas under each of the diffusion profiles are the same. Mathematically,

$$\int_0^\infty N(x, t_1)\, dx = \int_0^\infty N(x, t_2)$$

where $t_1 = 1$ hr and $t_2 = 2$ hr.

18.2 ION IMPLANTATION

The diffusion processes discussed in Section 18.1 are used for a wide variety of devices and ICs. However, to obtain extremely thin, selectively placed, doped layers, the industry uses a process called *ion implantation*, which was developed in the mid-1970s. Electrically accelerated charged ions of the desired impurity atom are implanted into the substrate for the purpose of obtaining the desired dopant concentration profile. The result is similar to that achieved by the diffusion process—a controlled change of the dopant concentrations over a defined depth. The depth these ions implant is a complex function of an accelerating electric potential, the substrate composition, the ionic mass, and the particular characteristics of the apparatus. Typical ion energies are from a few tens of kiloelectron volts (keV) to several hundred keV. Figure 18.7 presents a graph of projected ion-implantation depth as a function of impurity and accelerating energy.[6] It is interesting to note that we can now achieve impurity-layer thickness of tenths of a micrometer or less, with reasonably precise control over the profile as long as we maintain precise control over the accelerating potential. Even though the process and mathematical description are complex, we can estimate the total number of ions required to achieve a given profile with the aid of Fig. 18.8. Although the profile can be shown to be roughly Gaussian, we will approximate this relatively flat profile as uniform. If we permit this, the total number of singly-charged ions needed is approximated by

$$N_I = \frac{I \times t}{q} \tag{18.6}$$

where I is the beam current in amperes and t is the time in seconds.

To compute the total number of ions needed, N_I, consider the geometry illustrated in Fig. 18.8. Assume that a uniform donor concentration of N_D is needed over a rectangular area $(y = A) \times (z = B)$ and an abrupt junction depth of x_j; then

$$N_I = A \times B \times x_j \times N_D \tag{18.7}$$

■ Some ion implanters now exceed one-million electron volts (1 MeV).

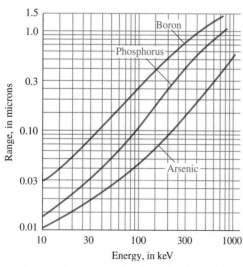

■ **FIGURE 18.7** Ion-implantation depth for B, P, As in silicon.[6]

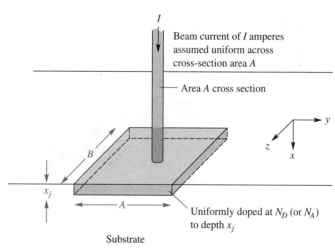

■ **FIGURE 18.8** Ion-implantation geometry.

is the total number of ions required from the ion-implantation source. Each ionic charge equals 1.6×10^{-19} C. Therefore, a current of 1 A equals 6.24×10^{18} ions per second.

EXAMPLE 18.4

How many ions are required to obtain a uniform doping profile of $N_D = 5 \times 10^{17}$ atoms/cm^3 to depth of 1 μm? How long will it take to ion implant a wafer 4 in. (100 mm) in diameter? Assume a beam current of 100 μA.

SOLUTION From Eq. (18.7), the number of ions per unit area is given by

$$\frac{N_I}{A \times B} = N_D \times x_j = \left(\frac{5 \times 10^{17}}{\text{cm}^3}\right)(1 \ \mu\text{m})\left(\frac{1 \text{ cm}}{10^4 \ \mu\text{m}}\right) = 5 \times 10^{13} \text{ ions/cm}^2.$$

The area of a 100-mm wafer is about 81 cm^2. The total implant time, t, required using Eq. (18.6) is

$$t = \frac{N_I}{(A \times B)I} = \left(\frac{5 \times 10^{13} \text{ ions}}{\text{cm}^2}\right)\left(\frac{81 \text{ cm}^2}{10^{-4} \text{ A}}\right) \times \left(\frac{1 \text{ A}}{6.24 \times 10^{18} \text{ ions/sec}}\right)$$

$$= 6.48 \text{ sec}.$$

■ Some ion implanters also generate doubly-ionized ions ($2 \times 1.6 \times 10^{-19}$ C).

We can expand on the results shown in Fig. 18.8. The theoretical doping profile is Gaussian [Fig. 18.9(a)]. The *range R* is the distance from the surface to the peak value. The standard deviation in doping concentration along the direction of the implant is called the *straggle*, ΔR_p. The doping concentration perpendicular to the implant direction is the *transverse straggle*, ΔR_T. Straggles are usually <10% of the range. A comparison between a more precise simulation and measured profiles for boron at 30, 100, 300, and 800 keV is shown in Fig. 18.9(b). They are roughly Gaussian in shape but the process simulation model does include secondary effects as well as industrial empirical data for this particular process. We can obtain the *dose*, the number of ions per square centimeter, by integrating the area under the respective curves in Fig. 18.9. Observe that a simplified approximation to the dose was computed in Example 18.4.

A simplified ion implanter is sketched in Fig. 18.10. Note the need to be able to scan the beam accurately over the intended area. Beam or spot size is usually less than 1 cm^2, with implant geometries defined by the photolithographic techniques described in Section 18.4.

It is interesting to note that the ion-implantation process is a relatively low temperature one. Usually after ion implant occurs, the wafer is **annealed** at a temperature of no higher than 900°C for a few minutes to allow the natural repair of the crystal lattice after the ion bombardment. This relatively low-temperature process assures that junction positions will not change significantly since the diffusion coefficients become quite small at these temperatures (see Fig. 18.3).

In summary, this process has the following advantages:

1. Precise control over the impurity dose
2. Control over the profile, with a peak in the impurity concentration now achievable below the surface
3. Relatively low-temperature processing
4. Pattern registration for small geometries.

DRILL 18.4

Approximately, how does the whole wafer dose change from a reference value individually if the (a) wafer diameter is doubled, (b) beam current is doubled, (c) implant energy is doubled, and (d) doubly ionized ions instead of singly ionized ions are used.

ANSWERS
(a) ×4
(b) ×2 (c) very roughly ×2; see Figs. 18.8 and 18.9. (d) ×2

■ There is considerable interest in *rapid-thermal anneal* (RTA) and synonomously *rapid-thermal processing* (RTP) to minimize junctions moving while annealing the crystal. Lasers are being used as well as high-power rf heating.

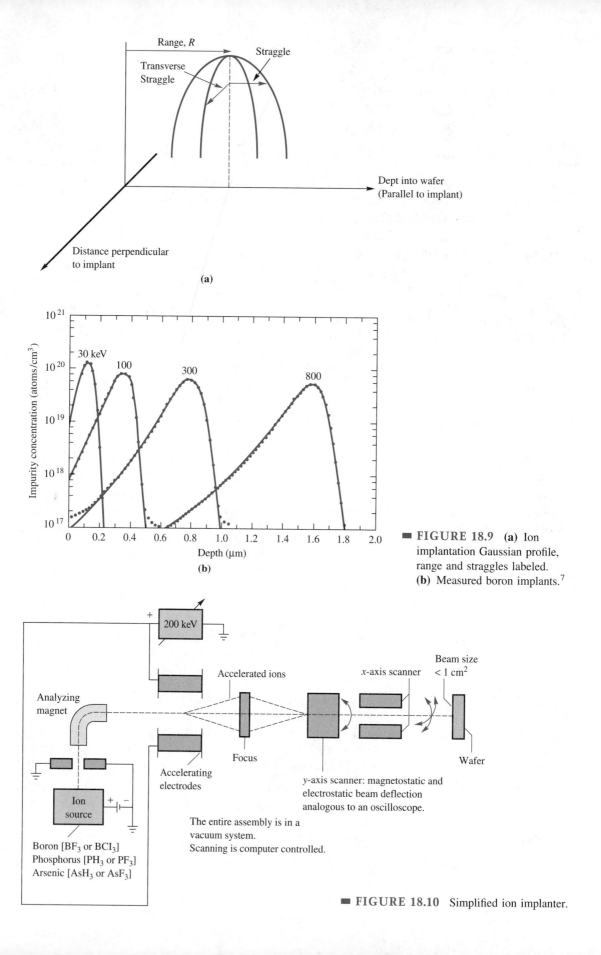

■ **FIGURE 18.9** (a) Ion implantation Gaussian profile, range and straggles labeled. (b) Measured boron implants.[7]

■ **FIGURE 18.10** Simplified ion implanter.

CHECK UP

1. **TRUE OR FALSE?** Ion implantation allows for deeper junction depths than those obtained from diffusion.
2. **TRUE OR FALSE?** Dose is more easily controlled with ion implantation compared to diffusion.
3. **TRUE OR FALSE?** All impurities are accelerated to exactly the same depth in a wafer.
4. Why is annealing required after an ion implantation?
5. Sketch a typical ion-implantation profile labeling range, straggle, and lateral straggle.

18.3 RESISTIVE PROPERTIES OF DOPED LAYERS

The diffusion or ion-implantation process results in a relatively thin surface layer whose impurity concentration is different from that of the original material. Therefore, if this layer consists of an *n*-type material, the conductivity is given by Eq. (2.34), which is repeated here:

$$\sigma_n = q\mu_n N_D. \tag{2.34}$$

If it consists of a *p*-type, the conductivity is approximately given by Eq. (2.36):

$$\sigma_p = q\mu_p N_A. \tag{2.36}$$

Suppose we fabricate this doped semiconductor in the form of a rectangular slab with a geometry as shown in Fig. 18.11. The resistance of this uniformly doped slab is given by

$$R = \frac{\rho L}{A} = \frac{\rho L}{WT} = \frac{L}{\sigma WT}. \tag{18.8}$$

Combining Eq. (18.8) with either Eq. (2.34) or (2.36), we can write the resistance for *n*-type material as

$$R = \frac{L}{WT(q\mu_n N_D)}, \tag{18.9}$$

and for *p*-type material as

$$R = \frac{L}{WT(q\mu_p N_A)}. \tag{18.10}$$

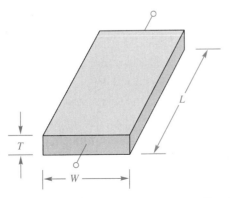

■ **FIGURE 18.11** Geometry for a uniformly doped slab used to compute sheet resistance.

It has become commonplace in the semiconductor industry to characterize a semiconductor's resistive property by its **sheet resistance**, $R_\square$, sometimes also referred to as R_S. Sheet resistance is defined for a material by combining the layer thickness along with the conductivity. We can see this by rewriting Eq. (18.9) as

$$R = \frac{L}{W} \frac{1}{(q\mu_n N_D T)}, \tag{18.11}$$

where the sheet resistance is given by

$$R_\square = \frac{1}{q\mu_n N_D T} \tag{18.12}$$

for an *n*-type layer. Similarly, for a *p*-type layer, we can define the sheet resistance as

$$R_\square = \frac{1}{q\mu_p N_A T} \tag{18.13}$$

It is important to observe that the units for sheet resistance are in ohms. However, to distinguish the unique definition of sheet resistance, the units are often given as ohms per square ($\Omega/\square$). Resistance is obtained by multiplying by the number of squares (L/W). The "square" is dimensionless.

Sheet resistance is also applicable to conducting **thin films**. Consider a conducting material deposited with uniform thickness, T, on an insulating substrate. Equation (18.8) is used with $L = W$ so that

$$R = \frac{1}{\sigma T} = \frac{\rho}{T}, \tag{18.14}$$

where ρ or σ is the resistivity or conductivity of the thin-film material.

■ Process simulation programs use numerical methods that allow incorporation of μ as a function of concentration as well as D as a function of concentration.

So far we have been discussing uniformly-doped layers. However, recall from Section 18.1 that the doping, i.e., carrier concentration, is not uniform with depth. Two possible doping profiles are the complementary error function and the Gaussian. We can modify the computation for sheet resistance by assuming the doped layer consists of a number of infinitesimally thin sheets, each with a uniform doping concentration. This is illustrated in Fig. 18.12. The resistance, ΔR, of one of these sheets is given by

$$\Delta R = \frac{L}{qW \overline{\mu}_n N_D(x) \Delta x}, \tag{18.15}$$

where $N_D(x)$ could be either the erfc or Gaussian function, and $\overline{\mu}_n$ is assumed to be the average mobility. If you look at Fig. 2.18, you will see that the mobilities are a complicated function of doping concentration. To make the calculations manageable, some average mobility is selected from the data given in Fig. 2.18. If we sum all of the ΔR contributions to the resistance between the surface and x_j and let $\Delta x \to dx$, we obtain

$$R = \frac{L}{qW \overline{\mu}_n \int_0^{x_j} N_D(x)\, dx}. \tag{18.16}$$

The quantity

$$R_\square = \frac{1}{q\overline{\mu}_n \int_0^{x_j} N_D(x)\, dx} \tag{18.17}$$

is defined as the sheet resistance. A refinement to Eq. (18.17) would be the inclusion of the original background doping: $N_A(x)$ for an *n*-type diffusion into a *p* material or $N_D(x)$ for a *p*-type diffusion into an *n*-type material. For an *n*-type layer, Eq. (18.17) becomes

$$R_\square = \frac{1}{q\left[\int_0^{x_j} \overline{\mu}_n N_D(x)\, dx - \int_0^{x_j} \overline{\mu}_n N_A(x)\, dx\right]}, \tag{18.18}$$

CHAPTER 18 BASIC FABRICATION TECHNOLOGY AND DEVICE CONSTRAINTS

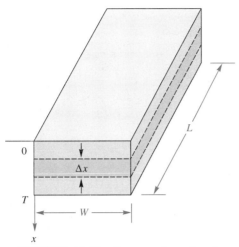

FIGURE 18.12 Geometry for a doped layer where $N_D(x)$ or $N_A(x)$ is used to compute sheet resistance.

and for a *p*-type layer this becomes

$$R_\square = \frac{1}{q\left[\int_0^{x_j} \overline{\mu}_p N_A(x)\, dx - \int_0^{x_j} \overline{\mu}_p N_D(x)\, dx\right]}. \quad (18.19)$$

EXAMPLE 18.5

What is the sheet resistance $R_\square$ of a uniformly boron-doped sample 2 μm thick, where $N_A = 10^{18}$ acceptor atoms/cm³?

SOLUTION   Boron is an acceptor material. From Fig. 2.18, $\mu_p \cong 120$ cm²/V-sec. Using Eq. (18.13),

$$R_\square = \frac{1}{q\mu_p N_A T} = \frac{1}{(1.6 \times 10^{-19})\left(\dfrac{120\ \text{cm}^2}{\text{V-sec}}\right)\dfrac{10^{18}}{\text{cm}^3} \times 2\ \mu\text{m} \times \left(\dfrac{10^{-4}\ \text{cm}}{\mu\text{m}}\right)}$$

$$= 260\ \Omega/\square.$$

EXAMPLE 18.6

Consider a silicon substrate doped with 10^{15} atoms/cm³ of boron. Assume a phosphorus diffusion profile approximated by $N_D(x) = 10^{18} e^{-x/1\ \mu\text{m}}$. What is the sheet resistance of the *n* layer?

SOLUTION   From Fig. 2.18 $\overline{\mu}_n = 350$ cm²/V-sec. If we neglect the substrate doping concentration, we have, using Eq. (18.17),

$$R_\square = \frac{1}{q\int_0^{x_j} \mu_n N_D(x)\, dx},$$

DRILL 18.5

What are the sheet resistances $R_\square$ of a uniformly boron-doped sample 2 μm thick, where $N_A = 10^{14}$ atoms/cm³, 10^{16} atoms/cm³, and 10^{20} acceptor atoms/cm³?

ANSWERS   694 kΩ/□; 8 kΩ/□; and 7.1 Ω/□ within graphical accuracies.

DRILL 18.6

What are the sheet resistances $R_\square$ of a uniformly phosphorus-doped sample 2 μm thick, where $N_D = 10^{14}$/cm³, 10^{16}/cm³, and 10^{20} donor atoms/cm³?

ANSWERS   223 kΩ/□; 2840 Ω/□; and 4 Ω/□ within graphical accuracies.

where $x_j = 6.9$ μm from Example 18.1.

$$R_\Box = \left(q\bar{\mu}_n \int_0^{x_j} 10^{18} e^{-x/1.0\,\mu m}\right)^{-1}$$

$$= [q\bar{\mu}_n 10^{18}(-1.0\,\mu m)(e^{-x/1.0\,\mu m})]^{-1} \bigg|_0^{6.9\,\mu m}$$

$$= \left[(1.6 \times 10^{-19}\,C)\left(\frac{350\,cm^2}{V\text{-sec}}\right)\frac{10^{18}}{cm^3} \times (-1\,\mu m)(10^{-4}\,cm/\mu m)(e^{-6.9} - 1)\right]^{-1}$$

$$= 179\,\Omega/\Box.$$

If we include the substrate doping and realize that $\bar{\mu}_n \approx 1200$ cm^2/V-sec, in this region and using the above result we have,

$$R_\Box = \frac{1}{0.005587 - (1.6 \times 10^{-19})(1200)(10^{15})(6.9)(10^{-4})} = 183\,\Omega/\Box.$$

which is not significantly different from the approximation of ignoring the substrate concentration. By contrast, assuming a uniformly doped layer approximately 6.9 μm deep with $N_D \approx 10^{18}$ donor atoms/cm^3, and using Eq. (18.12), we obtain

$$R_\Box = \frac{1}{q\bar{\mu}_n N_D T} = \frac{1}{(1.6 \times 10^{-19})(350)(10^{18})(6.9)(10^{-4})} = 26\,\Omega/\Box,$$

which is significantly different because $N_D(x)$ is definitely *not* uniform. This simplified type of calculation will yield a better result for ion-implanted layers because of the inherent doping uniformity achieved with ion implantation. Observe that if we would have used $N_D = 1.5 \times 10^{17}$ donor atoms/cm^3 as an estimated average uniform doping, then $R \cong 86\,\Omega/\Box$.

CHECK UP

1. **TRUE OR FALSE?** The unit for sheet resistance is Ω/cm^2.
2. **TRUE OR FALSE?** For a given doping concentration and junction depth, a donor-doped layer exhibits a lower sheet resistance than an acceptor-doped layer.
3. How does mobility affect layer resistivity and sheet resistance?
4. **TRUE OR FALSE?** In general, the thinner the layer, the higher the sheet resistance.
5. Why is it easier to compute the sheet resistance of an epitaxial layer compared to a diffused or an ion-implanted layer?

18.4 PHOTOLITHOGRAPHY AND MASKING

One of the fundamental steps in the fabrication of integrated circuits is the use of photolithographic techniques to define a particular geometry that will then become a resistor, transistor, FET, etc. We will use the term *photolithography* loosely, in that *photo* generally refers to a technique implying a visible light source of 4000 to 7000 Å. Because the resolution in defining a line edge is related to wavelength, shorter wavelengths such as ultraviolet are usually used. Typically, the *minimum feature size* (smallest lateral dimension) is at least several wavelengths long. This limitation is in addition to resolution limits

CHAPTER 18 BASIC FABRICATION TECHNOLOGY AND DEVICE CONSTRAINTS   893

imposed by lens performance and mechanical alignment tolerances of the equipment. Indeed, in the very densely-packed circuits, *electron beam* and *X-ray lithography* techniques are now used especially for the photomask generation. No matter what wavelength of the illuminating radiation is used, the process can be illustrated by studying the steps in Fig. 18.13. Note that in order to show detail in these cross sections, the vertical scaling is exaggerated in comparison to the horizontal scaling.

■ There are currently discussions taking place on a government-funded initiative on developing ion-beam lithography techniques. Basically, lithography requirements are driven by minimum feature sizes, throughput, and yield.

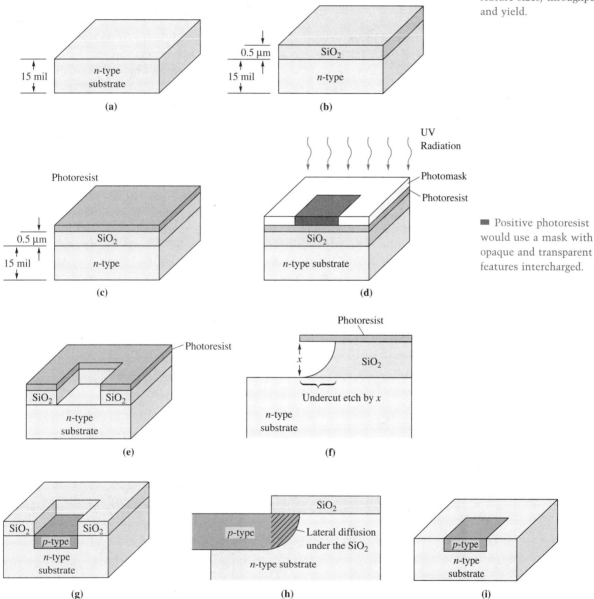

■ Positive photoresist would use a mask with opaque and transparent features interchanged.

■ **FIGURE 18.13** Photolithographic process (negative photoresist). **(a)** *n*-type substrate. **(b)** SiO_2 layer formed on silicon surface. **(c)** Thin layer of photoresist covering the oxide. **(d)** UV exposure of photoresist through the mask. **(e)** Etch through SiO_2 after removal of exposed photoresist. **(f)** Etching undercut. **(g)** Remove photoresist. Diffuse *p* dopant into substrate. **(h)** Lateral diffusion. **(i)** Remove SiO_2 using HF.

- Technical issues and costs associated with using 300 mm (12 inch) wafers are being studied by industrial consortia.

Let us assume we want to diffuse a rectangular-geometry *p*-type layer into an *n*-type substrate [see Fig. 18.13(a)]. The substrate for a 100-mm-diameter wafer is usually 20 to 22 mils thick, which is thick enough to be handled without breakage. The semiconductor industry often uses the term *mil* = 0.001 in. = 25.4 μm for dimensions in excess of 1 mil. The substrates are thicker for the 150-mm wafers that many fabrication facilities use. There is even limited production now occurring in which 200-mm (8-in.) "dinner plate" size wafers are being used. These larger wafers are being used for larger ICs where the capital equipment cost can be justified.

We will first place our *n*-type sample in an oxidizing atmosphere, usually dry oxygen or steam, at a temperature of 1000°C+. A "dry" oxide (SiO_2, essentially a very pure glass) is formed by the following chemical reaction: $Si + O_2 \rightarrow SiO_2$. An estimate of the resultant oxide thickness as a function of time and temperature can be obtained by using Fig. 18.14. Dry oxides are relatively thin but dense. Thicker oxides are grown using the "wet" oxidation process governed by $Si + 2H_2O \rightarrow SiO_2 + 2H_2$ where the H_2O is in the form of steam. An estimate of the resultant oxide thickness as a function of time and temperature can be obtained from Fig. 18.15.

- Concommitant with using larger wafer sizes is the development of large-area lithography equipment. Equipment to expose a 12-inch wafer is now available although its primary use is in liquid-crystal flat panel display manufacturing.

Typically, most oxidations are at least a two-step process; a dry oxidation followed by a wet oxidation with resultant thicknesses on the order of 0.2 to 0.5 μm. In reality, about 46% of the oxide consists of Si consumed during the oxidation process, as shown in Fig. 18.16; consequently, a 0.5-μm oxide uses 0.5 μm × 0.46 = 0.23 μm of the original silicon, and the oxide extends 0.27 μm above the original surface as shown. However, for clarity of presentation, the total 0.5 μm of SiO_2 is almost always shown to be on top of the Si surface [Fig. 18.13(b)].

The thickness required is a compromise between the desired feature resolution and the need to have a good barrier to dopants. In general for the 1.0- to 3.0-μm minimum feature size, which is characteristic of most production ICs using optical lithography,

- <100> refers to the crystallographic orientation of the wafer. <100> is the most commonly used Si orientation.

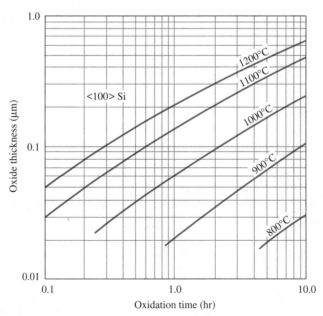

■ **FIGURE 18.14** SiO_2 growth rate for dry O_2 (from Ref. 8).

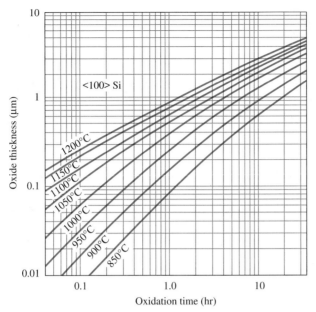

■ **FIGURE 18.15** SiO$_2$ growth rate for wet O$_2$ (from Ref. 8).

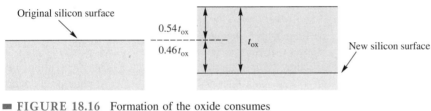

■ **FIGURE 18.16** Formation of the oxide consumes (oxidizes) silicon resulting in a new silicon surface.

the oxide thickness should be no greater than 0.5 μm. This oxide layer is useful in the following ways:

1. It serves as an effective high-temperature barrier to dopant gases.
2. Because the resistivity is very high ($\rho > 10^{16}$ Ω-cm), it behaves as an insulator.
3. The thickness is relatively easy to control by means of a combination of temperature and time exposure in an oxidizing atmosphere as shown in Figs. 18.14 and 18.15.
4. It can be removed selectively by means of a hydrofluoric acid (HF)-based etching solution.
5. The layer is chemically passive to atmospheric constituents so that the silicon surface underneath is protected. This is called **passivation**.

The oxide layer is now coated with a photosensitive emulsion called **photoresist** [see Fig. 18.13(c)]. This resist layer can be easily removed with certain solvents. We can use either a positive or negative resist, for example, Kodak KPR. The negative-resist procedure is illustrated. If we expose this photosensitive material to ultraviolet

■ *Silicon nitride*, Si$_3$N$_4$ deposited by a low-temperature CVD process is also widely used for passivation. There is some development in using conformal polymers as passivation layers and to improve surface topographical variations.

- *Lift-off* photolithography is used for metal-pattern definition in certain process designs. Basically, photoresist is deposited everywhere with vias (openings) patterned in the photoresist where needed. Metal is then deposited everywhere, including over the photoresist. The photoresist is then dissolved using acetone, and the metal only remains in the via location. The metal on top of the photoresist is lifted off the wafer.

- The exposed wafer can also have dopants added by ion implantation. Metals can also be sputtered or evaporated to form contacts.

DRILL 18.7

How long does it take to form a 0.1-μm SiO_2 layer using a dry oxidation at $T = 900°C$, $1000°C$, and $1100°C$?

ANSWERS
9 hr; 2.2 hr; 36 min.

DRILL 18.8

How long does it take to form a 0.5-μm SiO_2 layer using a wet oxidation at $T = 900°C$, $1000°C$, and $1100°C$?

ANSWERS
4.2 hr; 1.4 hr; 36 min.

---

light, it is polymerized, which makes it very resistant to a number of acid etches. For example, suppose we have a mask that transmits the ultraviolet in all regions except the rectangular area in which we want to diffuse an impurity [see Fig. 18.13(d)]. After exposure and a photolithographic development process, we can remove the resist from the area shown. In this area, bare oxide is exposed. Usually the SiO_2-etched walls are shown as vertical [see Fig. 18.13(e)]. However, the etching process is essentially *isotropic*; that is, it proceeds in all directions under the photoresist, so that the walls really etch as illustrated in Fig. 18.13(f). If we are trying to define a 3-μm line on the substrate with a 0.5-μm-thick oxide, the resultant line width at the surface of the oxide will be 4 μm. The **undercut etch** is approximately the same amount as the vertical etch. Using an HF etch, we selectively remove the oxide.

After we remove the remaining resist with an organic solvent, the oxide serves to mask the substrate. The wafer is then exposed to a *p*-type dopant in a diffusion furnace. Because the oxide is essentially impervious to the dopant gas, the dopant atoms [Fig. 18.13(g)] will only penetrate into the areas no longer covered by SiO_2. The demarcation where *p*-type material changes to *n*-type material is the location of the junction. It is also important to observe that not only is there vertical diffusion through the oxide openings, but the dopants diffuse laterally under the oxide. If we expect a 3-μm vertical diffusion to form a junction, then we can also expect about a 3-μm lateral diffusion, or perhaps a bit less. This is illustrated in Fig. 18.13(h). This lateral diffusion further complicates the layout design of circuits. Ion-implantation techniques minimize, if not eliminate, the lateral diffusion because the ion beam is essentially perpendicular to the wafer. For clarity, we will ignore this lateral diffusion in the cross sections. The SiO_2 can then be removed in preparation for subsequent steps [see Fig. 18.13(i)].

As we will soon observe, the photolithographic process is repeated as often as necessary to form passive and active devices. Each masking step must be precisely aligned with preceding steps to obtain accurate device geometry registration. Each of the circuits is replicated hundreds of times across the wafer. This replication uses a step-and-repeat projection system as illustrated in Fig. 18.17. A photographic mask, 5× or 10× final dimensions, of a single IC die is projected onto a photosensitized wafer. Each projection area or field can be as large as 14 × 14 mm. These fields are projected across the wafer in sequence. The system illustrated in Fig. 18.17 is capable of 0.8-μm resolution. As a rule, successive layer projections must be accurate to at least 10% of this 0.8-μm minimum feature size. To achieve this, the mechanical platform is positioned using a laser interferometer keyed on alignment marks that must be included in the basic mask layout.

CHECK UP

1. **TRUE OR FALSE?** SiO_2 is formed at room temperature.
2. **TRUE OR FALSE?** The growth rate for SiO_2 is higher for a dry oxidation process than for a wet oxidation process.
3. What are typical ranges for minimum photolithography feature sizes found in routine production, advanced development, and research type devices?
4. List at least five reasons why SiO_2 is important in Si semiconductor device processing. Explain.
5. **TRUE OR FALSE?** A photolithographic process is usually only done once when fabricating a device.

CHAPTER 18 BASIC FABRICATION TECHNOLOGY AND DEVICE CONSTRAINTS

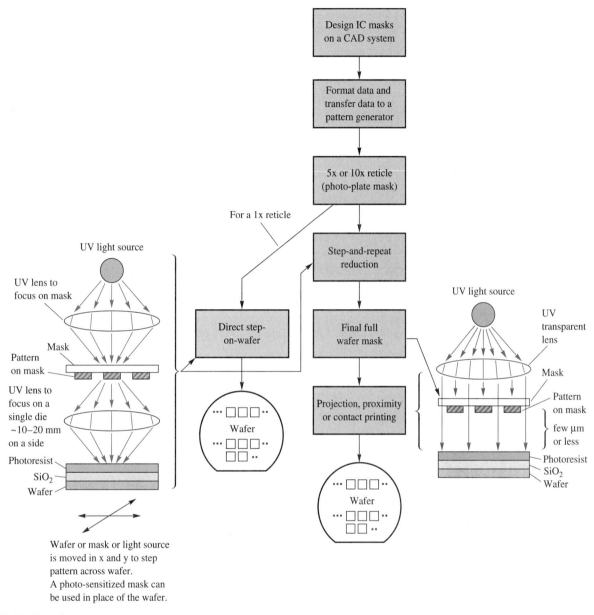

■ **FIGURE 18.17** Basic photolithography process.

18.5 RESISTORS

Typically, the beginning raw silicon wafer consists of a *p*-type substrate that is 20 to 22 mils thick for a 100-mm (4-in.) wafer. Exposing the heated substrate to a silane gas ($SiCl_4$), H_2, and a dopant gas will result in the crystal-ordered growth of a thin layer of uniformly doped silicon whose doping is relatively independent of the underlying substrate. This thin layer (10 to 20 μm), called the ***epitaxial layer***, serves as the foundation for the component fabrication. For clarity, we will simplify all cross-sectional views. Part of this simplification includes a vertical scale exaggeration. Because vertical features are measured

■ There are a series of wafer size standards employed in the industry so that the wafer handling equipment can be manufactured cost effectively.

in microns and horizontal features in mils, it is quite common to find the vertical scale expanded when looking at device cross sections. In addition, we will neglect the undercutting of the etch and rounding of edges due to imperfections in the photolithography and the isotropic characteristics of etching processes because it is difficult to observe them in the cross-sections as a result of scaling.

Resistors are formed either during the base or emitter transistor diffusions (to be discussed shortly) or as suitably biased FETs operating in the ohmic region. Logically, the first type of resistor is called a *base-diffused* or *emitter-diffused resistor*, and the second type is called a *pinch resistor*. Thin-film-deposited resistors of high resistivity metals such as nichrome are also used.

■ In MOS processes, the source and drain diffusions may be used to form resistors.

Diffused Resistors

Figure 18.18 depicts the top and cross-sectional views of a base-diffused resistor. The first masking, photolithography, and p^+ diffusion sequence is used to define the *isolation region*. By reverse-biasing the diode effectively formed between the isolation region and substrate, we can electrically isolate, for the most part, any component fabricated within this region.

The second step in the sequence is to diffuse the *p*-type material, which is then used as the resistance. A commonly used *p*-type material is boron. In a gas-diffusion system, which is quite common in the industry, the boron is in the form of diborane gas, B_2H_6. Increasing in popularity is the use of boron nitride (BN) wafers. The BN wafers are placed next to the silicon wafer, and the boron is transferred from the BN wafer to the silicon wafer by means of an O_2 and N_2 gas flow at a suitably high temperature. Spray-on dopants have also been used but have been superseded by gas and solid-source systems.

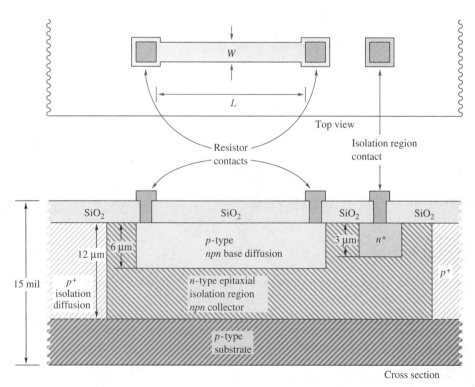

■ **FIGURE 18.18** Base-diffused resistor.

To ensure a good ohmic contact with the isolation region, an n^+ diffusion is used. Common n-type materials are phosphorus and arsenic. In a gas-diffusion system, phosphorus is supplied in a phosphine gas and arsenic is supplied in an arsine gas. Both of these gases, as well as diborane, must be handled carefully because of their toxicity. Increasing in popularity is the use of phosphorus pentoxide (P_2O_5) wafers. Their application is similar to that for BN wafers. The last masking operation is used to define the metallization pattern. Typically, 2 to 4% copper-doped aluminum or a more complex multilevel metallization system is used.[9] Using results from Section 18.3, the resistance of this structure is given by

$$R = \frac{L}{W} \times R_\square. \tag{18.20}$$

Because this structure is formed by the same p diffusion that is used to form bases for npn transistors elsewhere on the substrate, it is referred to as a *base-diffused resistor*. Its use is defined at the metal interconnection step. Typical values for $R_\square$ are on the order of 100 $\Omega/\square$ to 200 $\Omega/\square$.

Often, large L/W ratios are achieved by fabricating the resistor in a folded pattern as illustrated in Fig. 18.19. Recall that for an $R_\square = 100\ \Omega/\square$, an $L/W = 50$ squares would be required to obtain a modest-sized 5-kΩ resistor. A resistor with this length would be difficult to use because the distance between the two end contacts would be on opposite ends of a die, hardly practical in a compact circuit. It would be more reasonable to wind the resistor back and forth about four or five times to obtain the necessary L/W. Even so, diffused resistors much larger than a few tens of kΩ are uneconomical in terms of die area. As we studied in earlier chapters, innovative circuit design yields large equivalent resistances in the form of active loads—that is, suitably biased BJTs or FETs are used instead of resistors to perform the same circuit function.

To be complete, the isolation region reverse-biased junction contributes a parasitic capacitance. Even though this is a distributed circuit element, it is modeled by separating the total into two lumped portions as shown in Fig. 18.20.

Typical minimum feature sizes will run between <1 μm to perhaps 10+ μm so that the maximum resistor size is dictated not only by the economics of the silicon area being used but by the increasing parasitic capacitance associated with large resistors and their associated junction area.

■ The semiconductor industry adheres to very stringent standards in the handling and disposal/treatment of toxic chemicals and gases.

■ Silicon-doped aluminum (~2%) is also used to avoid spiking through shallow surface diffusions. Usually, multilevel metallization systems are used to improve yield and long-term reliability.

■ There are correction factors required for corners and end contacts, but we neglect these in the basic $\frac{L}{W}$ calculation.

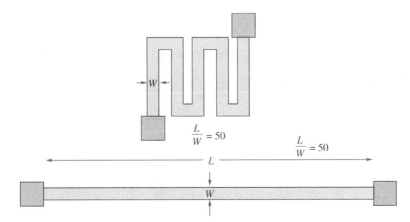

■ **FIGURE 18.19** Comparison between a folded-pattern and linear-pattern diffused resistor with $L/W = 50$ squares. The R values will be about the same.

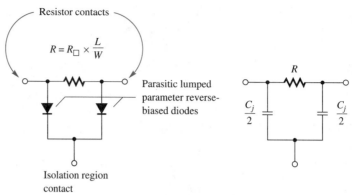

FIGURE 18.20 Base-diffused resistor model including parasitic elements.

EXAMPLE 18.7

Using the sheet resistance obtained from Example 18.5, design an IC resistor with a value of 10 kΩ.

SOLUTION From Example 18.5, $R_\square = 260\ \Omega/\square$. Using Eq. (18.20),

$$\frac{L}{W} = \frac{R}{R_\square} = \frac{10\text{ k}\Omega}{260\ \Omega} \approx 38.5 \text{ squares.}$$

If we use a minimum feature size of 10 μm, then $W = 10$ μm and $L = 385$ μm = 15.2 mils. Because 15.2 mils could represent a large fraction of a complete IC die linear dimension, it would be advisable to use a folded pattern as illustrated in Fig. 18.19.

DRILL 18.9

Refer to Drill 18.5. Which p-doping density would you choose to obtain a 40-kΩ diffused resistor? Why?

ANSWER For a reasonable L/W ratio use $R_\square = 8$ kΩ/□ corresponding to $N_A = 10^{16}/\text{cm}^3$.

■ Minimum feature size is usually defined as the minimum line width.

The emitter-diffused resistors are fabricated similarly except the n^+-type emitter diffusion is used for the resistive element (see Fig. 18.21). Typically, this material has a sheet resistance of only a few ohms to a few tens of ohms per square so this type of resistor is inherently of low value and often used as a *crossunder* interconnect.

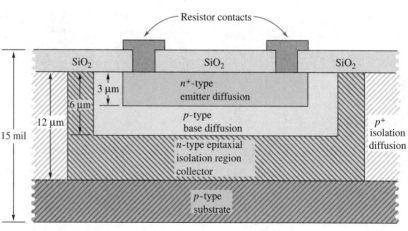

FIGURE 18.21 Emitter-diffused resistor.

EXAMPLE 18.8

Could you use an emitter diffusion of $N_D = 10^{20}$ atoms/cm^3 and depth $= 2$ μm for obtaining a 10-Ω resistor with reasonable geometries?

SOLUTION We can estimate R from Eq. (18.12) as

$$R_\square = \frac{1}{q\mu_n N_D T} = \frac{(10^4\ \mu\text{m/cm})}{(1.6 \times 10^{-19})(76\ \text{cm}^2/\text{V-sec})(10^{20}/\text{cm}^3)(2\ \mu\text{m})} = 4.1\ \Omega/\square,$$

where $\mu_n = 76$ cm^2/V-sec is obtained from Fig. 2.18. Therefore, $L/W = 2.4$, which would be easy to do. This may be useful as a circuit **crossunder** instead of trying to rearrange crossing metallization patterns. Actually, an L/W ratio of a bit less than 3 rather than 2.4 would provide a more exact construction of a 10-Ω resistor since the aluminum pad and interconnect metallization have a resistance of a few tenths of an ohm.

DRILL 18.10

Would it be reasonable to construct a 10-kΩ resistor using the emitter diffusion presented in Example 18.8? Explain.

ANSWER No. $L/W = 2500$ would use an unreasonable amount of die area even if arranged as a folded resistor.

The numerous processing steps in a diffused resistor make it difficult to obtain good absolute accuracies; ±20% is typical, but since all resistors on a wafer are subjected to the same processes, the resistors are matched usually to ±1% or less. It is important that the metal make an ohmic (nonrectifying) contact with the resistor.

Pinch Resistors

Recall from Chapter 5 that a JFET, when operated at voltages below pinch-off, and in the ohmic region, exhibits a resistive i_D versus v_{DS} characteristic (see Fig. 5.27). Furthermore, this "resistor" can be of somewhat higher value than that achievable with base diffusion because, by cleverly arranging the bias voltages, the channel thickness can be made smaller. The cross section of a base-pinch resistor, along with its equivalent circuit schematic diagram symbol, is illustrated in Fig. 18.22. The resistor contacts behave as

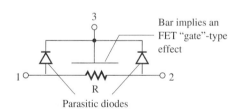

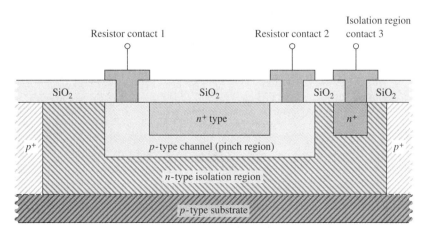

FIGURE 18.22 Base-pinch resistor cross section and equivalent circuit.

source and drain connections, respectively. The *p*-channel region formed from the *p*-diffusion is then pinched down in accordance with the depletion region resulting from the reverse-bias diode formed between the epitaxial layer isolation region contact and the resistor terminal. The n^+ diffusion is electrically in contact with the *n*-type epitaxial isolation region. Although the sheet resistances for a thin channel are high (greater than 5 kΩ/□), the resistor value is not very accurate or constant if the terminal voltages on the resistor change. The reverse-breakdown voltage of this junction is only a few volts, limiting the range.

EXAMPLE 18.9

The base doping yields a resistivity of $\rho = 1$ Ω-cm. Compare the sheet resistances between a base-diffused resistor of thickness 10 μm and a base-pinch resistor with a 0.50-μm channel.

SOLUTION

$$R_\square \text{ (base-diffused resistor)} = \frac{\rho}{T} = \frac{1 \, \Omega\text{-cm}}{10 \, \mu\text{m}} = \frac{1 \, \Omega\text{-cm}}{10 \times 10^{-4} \, \text{cm}} = 1000 \, \Omega/\square,$$

$$R_\square \text{ (base-pinch resistor)} = \frac{\rho}{T} = \frac{1 \, \Omega\text{-cm}}{0.5 \times 10^{-4} \, \text{cm}} = 20 \, \text{k}\Omega/\square.$$

This higher sheet resistivity is useful under limited circuit conditions to obtain higher resistor values.

A similar structure using the epitaxial layer as the pinch region can also be fabricated as illustrated in Fig. 18.23. The epitaxial layer, used as the isolation region,

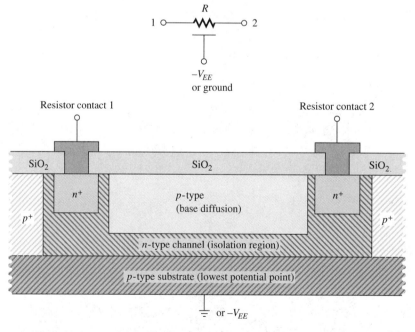

FIGURE 18.23 Epitaxial-pinch resistor cross section and equivalent circuit.

CHAPTER 18 BASIC FABRICATION TECHNOLOGY AND DEVICE CONSTRAINTS

is pinched between the *p*-type substrate and the base diffusion. Again, appropriate biasing is required for what is essentially an *n*-channel JFET operating below pinch-off. Because of lighter doping densities, the breakdown voltage is higher than for a base-pinch resistor.

Resistor Parameter Summary

Table 18.1 summarizes key resistor parameters. It is important to note that tolerances can be tightened by using component screening and testing in conjunction with computer-controlled laser trimming.

Economic compromise is required for fabricating increasingly accurate components and designing circuits tolerant to typical manufacturing or temperature variations.

TABLE 18.1 Key IC Resistor Properties (5-μm Linewidths)

Process	$R_\square$ ($\Omega/\square$)	Absolute Accuracy (%)	Matching Accuracy (%)	Comments
Base-diffused	50–300	±20	±2 ($W = 5$ μm) ±0.2 ($W = 50$ μm)	Most widely used in analog circuitry. Maximum value to few tens of kΩ.
Emitter-diffused	1–10	±20	±2 worst case	Used for $R \cong 0$ Ω crossovers or for power device matching and parallel device ballast
Ion-implant	50–1000	±3	±2 ($W = 5$ μm) ±0.1 ($W = 50$ μm)	Most accurate because of ion-implantation control and use of computer-controlled laser trimming
Base-pinch	1–10 kΩ	±50	±10	Used for current sources and as an "active load." Extremely bias-voltage dependent
Epitaxial-pinch	<10 kΩ	±50	±10	Same as base-pinch resistor

CHECK UP

1. **TRUE OR FALSE?** As a practical matter, any resistance value for a particular circuit design can be obtained by changing the *L/W*.
2. **TRUE OR FALSE?** Resistors on the same die are very well matched.
3. Why do designers implement folded resistor structures?
4. **TRUE OR FALSE?** Acceptor-doped layers offer higher value resistors than donor-doped layers of the same thickness and doping density.
5. What advantages do pinch resistors have compared to diffused resistors?

18.6 CAPACITORS

■ Recall from Chapter 5, MOS capacitors are a key component in memory and CCD elements.

Monolithic integrated-circuit capacitors are inherently limited to values under 100 pF because of the large die area required. Innovative design techniques have been developed for ICs to minimize the number and size of required capacitors. There are two basic types of IC capacitors: the junction capacitor and the metal-oxide semiconductor (MOS) capacitor. The MOS capacitor is the more widely used; however, the junction capacitor appears at every junction as a parasitic element.

MOS Capacitor

The MOS capacitor most closely resembles its discrete counterpart. Two conductors are separated by a thin dielectric, in this case, SiO_2. Figure 18.24 illustrates this construction. The capacitor is fabricated within the n-type epitaxial layer, which has been appropriately masked to obtain an isolation region. The n^+ emitter diffusion is used as one side of a parallel-plate capacitor. A thin oxide of thickness < 500 to no more then 1000 Å covers the n^+ layer. The aluminum metallization is used for the second parallel-plate conductor. From classical electrostatics, the capacitance is given by

■ Si_3N_4 (silicon nitride) is also used for gate dielectrics but it is lossy compared to SiO_2 because its resistivity is several orders of magnitude lower.

$$C = \frac{\varepsilon A}{d} = \frac{\varepsilon LW}{d}. \tag{18.21}$$

Typically, values of 0.2 to 0.3 pF/mil² are obtained, depending on oxide thickness. Here we assume that $\varepsilon = \varepsilon_0 \varepsilon_r$ where $\varepsilon_0 = 8.854 \times 10^{-14}$ F/cm² and $\varepsilon_r = 3.9$ for SiO_2. As with resistors, some parasitic elements are also generated. The isolation region junction

■ We use aluminum metallization generically in that more often, heavily-doped polysilicon or multilevel metallization techniques are used.

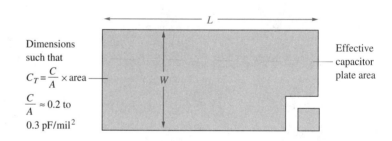

■ For Si_3N_4, we use $\varepsilon_r \sim 5.8$.

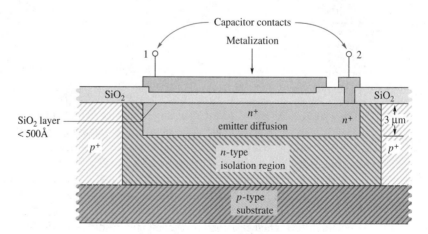

■ **FIGURE 18.24** MOS capacitor configuration.

diode, with its associated reverse-bias capacitance and the *n*-region generated series resistance, is illustrated in Fig. 18.25. This MOS capacitor has three major advantages:

1. The capacitance value is not a function of terminal voltage.
2. The voltage polarity on each plate is of no concern as long as the isolation region is reverse biased with respect to the substrate, i.e., the capacitor is nonpolar.
3. The oxide dielectric exhibits extremely low loss since it is an insulator. This, in combination with the relatively low series resistance generated by the n^+ contact, results in a high-Q (low dissipation factor) capacitor.

Fortunately, the MOS capacitor is rarely directly connected to the outside world because it, like the MOSFET gate contact, is extremely vulnerable to punch-through from stray electrostatic potentials.

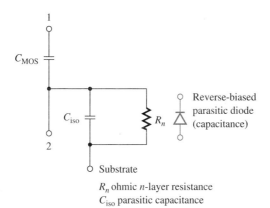

FIGURE 18.25 Equivalent circuit of an MOS capacitor.

EXAMPLE 18.10

What is the SiO_2 thickness for a MOS capacitor with a value of 0.3 pF/mil^2? What would be a safe maximum operating voltage if the breakdown electric field $E_{crit} = 10^7$ V/cm? ε_r for $SiO_2 = 3.9$.

SOLUTION $C = \varepsilon A/d$ or

$$d = \frac{\varepsilon_0 \varepsilon_r}{\dfrac{C}{A}} = \frac{(3.9)(8.854 \times 10^{-12}\,\text{F/m})\left(\dfrac{1\,\text{m}}{10^6\,\mu\text{m}}\right)\left(\dfrac{10^{12}\,\text{pF}}{\text{F}}\right)}{\left(\dfrac{0.3\,\text{pF}}{\text{mil}^2}\right) \times \left(\dfrac{1\,\text{mil}}{25.4\,\mu\text{m}}\right)^2}$$

$$= 0.0742\,\mu\text{m} = 742\,\text{Å}.$$

The breakdown voltage is given by

$$V_B = E_{crit} \times d = 10^7\,\text{V/cm} \times 742\,\text{Å} \times \frac{1\,\text{cm}}{10^8\,\text{Å}} = 74\,\text{V}.$$

More realistically, however, you would operate at 10 to 25% of V_B (7 to 20 V) since localized thin spots or defects would result in breakdown at a lower voltage.

DRILL 18.11

Compute the C/A and breakdown voltage for a 400-Å-thick SiO_2 layer.

ANSWERS
$C/A = 0.86\,\text{fF}/\mu\text{m}^2 = 0.00086\,\text{pF}/\mu\text{m}^2 = 0.56\,\text{pF/mil}^2$; $V_B = 40$ V.

- MOS and junction capacitors are comparable in values. Often, junction capacitors are considered a parasitic element that degrades high-frequency performance. There is an entire technology based upon dielectrically isolated devices whose high-frequency (speed) performance is significantly enhanced over junction-isolated devices.

Junction Capacitor

As studied in Chapter 2, a reverse-biased junction exhibits a voltage-dependent capacitance. Figure 18.26[10] is a plot of the capacitance per unit area as a function of the doping concentration on the lightly-doped side, the side on which the depletion region is predominantly located. As illustrated in Fig. 18.27, a capacitor fabricated in this way typically uses the collector-base junction. A junction capacitor is not quite as useful as the MOS capacitor because of the voltage modulation of the capacitance. Also, the capacitance per unit area is not very large, so obtaining large capacitors in this way is not economically justifiable in terms of area. This capacitance is important in that it always exists between the isolation region and the substrate and as such represents a parasitic circuit element.

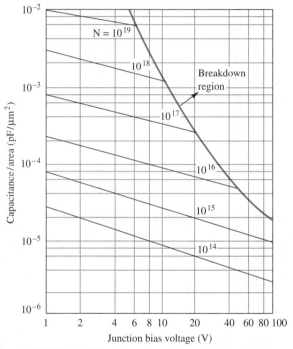

■ **FIGURE 18.26** Abrupt junction capacitance versus voltage and impurity concentration on the lightly doped side.

EXAMPLE 18.11

Compare the capacitance per unit area for the MOS capacitor and a junction capacitor with a doping of 10^{17} atoms/cm^3. Both capacitors are biased at 10 V.

SOLUTION From Fig. 18.26, $C/A \sim 3 \times 10^{-4}$ pF/μm^2 = 0.19 pF/mil^2, which is quite comparable to the 0.2 to 0.3 pF/mil^2 for the MOS device. Note, however, that the breakdown voltage of approximately 20 V for a junction capacitor is much less than for the MOS capacitor and the capacitance is a strong function of the voltage across it.

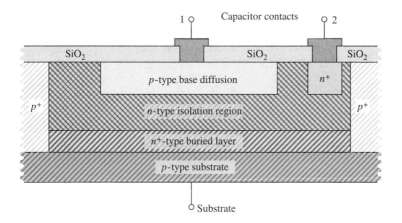

■ Refer to the discussion supporting Figure 18.28(a) for the implementation of a *buried layer*.

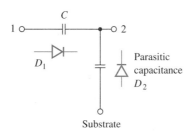

■ **FIGURE 18.27** Junction capacitor. D_1 and D_2 must be reverse biased, preferably with a constant voltage.

CHECK UP

1. **TRUE OR FALSE?** Capacitors larger than 100 pF are not practical.
2. **TRUE OR FALSE?** All diffused resistors have an associated junction capacitance.
3. **TRUE OR FALSE?** Designers prefer to use junction capacitors rather than MOS capacitors.
4. **TRUE OR FALSE?** MOS capacitors are essentially voltage independent.
5. **TRUE OR FALSE?** Metallization interconnects across a die form parasitic MOS capacitors.

18.7   *npn* TRANSISTORS

We have already alluded to many of the fabrication steps required to obtain a monolithic *npn* transistor. In this section, we illustrate the steps in a series of cross-sectional sketches. It is beyond the scope of this text to consider detailed device or processing design. Some of the references at the end of the chapter expand on these topics.[7–11] The important feature to observe is that by subjecting all devices on a single wafer to the same processing, we can achieve a high degree of matching even if absolute parameter tolerances are more difficult to control.

An n^+ layer is diffused into, or epitaxially grown on, a p-type substrate. This n^+ layer, as illustrated in Fig. 18.28(a), will have a relatively low resistance and will form the buried layer for the collector region. Its function is to provide a uniformly distributed low-resistance current path for the collector current. Sheet resistance $R_\square$ is on the order of a few tens of $\Omega/\square$. The n dopant is usually As or Sb because these dopants are least likely to rediffuse during subsequent high-temperature diffusion processing steps (see the diffusion coefficients given in Fig. 18.3).

Figure 18.28(b) illustrates the subsequent growth of an n-type epitaxial layer with $N_D = 10^{15}$ atoms/cm^3. This layer is on the order of 10 to 25 μm thick. By means of a photolithographic masking and p-type diffusion process, usually using boron, we can define n-type isolation regions as illustrated in Fig. 18.28(c). This p^+ diffusion must penetrate through to the p-type substrate, thus electrically isolating the n-type collectors. This is perhaps the longest process, on the order of several hours, because of the depth required. Sheet resistance $R_\square$ for the epitaxial isolation region, which also forms the npn collector region, is a few tens of $\Omega/\square$.

The next step is the base masking and diffusion as illustrated in Fig. 18.28(d). A p-type boron diffusion, with the associated photolithographic masking, is used and results in a base region of 2 to 5 μm in depth. The surface concentration is on the order of 10^{19}

■ In order to reduce parasitic collector substrate and capacitance between individual transistors, oxide isolation is widely used. Where the p^+ isolation diffusion is shown, trenches are etched and filled with oxide. There are a variety of approaches to obtain oxide isolation.

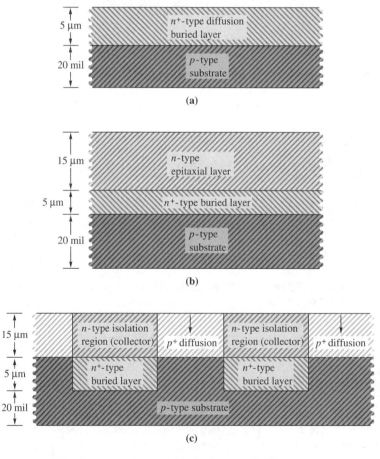

■ **FIGURE 18.28** Key steps in fabricating a monolithic npn transistor. (a) n^+ buried layer selectively diffused. (b) Formation of n-type epitaxial layer. (c) p^+ diffusion isolates n regions.

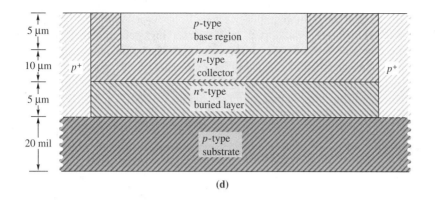

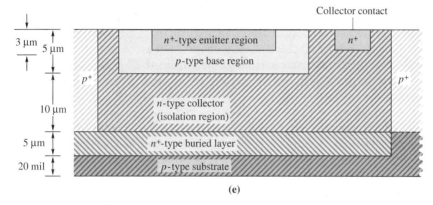

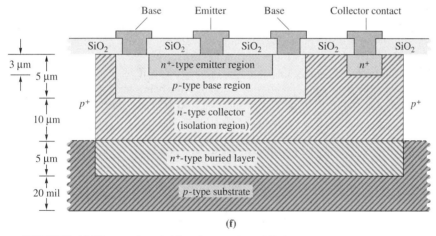

■ **FIGURE 18.28** continued. **(d)** *p*-base region diffusion. **(e)** n^+ emitter diffusion. **(f)** Cross section of completed *npn* transistor. Thickness of 20–22 mil assumes a 100 mm wafer.

atoms/cm^3. The layer concentration is usually achieved with a Gaussian profile. To obtain a fixed surface concentration, which can then diffuse according to Eq. (18.5), a ***predeposition*** step is used. The surface is exposed to an "infinite" dopant concentration for a few minutes. This results in a very thin (tens to hundreds of angstroms) layer of *p*-type material deposited on the surface. An oxide is then formed over the surface, which effectively blocks out the addition of more impurities and also prevents out-diffusion of the p-dopant during the next high-temperature step. In this second high-temperature process, called the ***base drive***, the thin surface layer diffuses into the isolation region, now following a Gaussian profile since the thin predisposition represents a finite source. The resultant layer has a sheet resistance of a few hundred $\Omega/\square$. It is important to note that this step also forms the base-diffused resistors used in the circuit (see Section 18.5).

The emitter and contacts to the collector result from an n^+-type diffusion with a surface concentration in the range of 10^{20} to 10^{21} atoms/cm^3 [see Fig. 18.28(e)]. An n^+ diffusion into the n-type collector is needed to obtain a good ohmic contact[12-14] between the copper-doped aluminum metallization and the relatively high resistance collector isolation region. The sheet resistance is a few $\Omega/\square$, and the depth is only a few tenths of a micrometer to perhaps as much as 3 μm. The result is a thin base region that is, of course, needed for a high-β device.

The entire wafer is now coated with a thin, 2% copper-doped aluminum layer (less than 1 μm) by a deposition process. The unwanted material is then etched away by another photolithographic and masking operation to define the interconnections. Figure 18.28(f) shows a cross section of the completed transistor. Figure 18.29 presents a sketch with the dimensions of the actual mask set for the 2N2222A, the BJT we discussed extensively in Chapter 4.

We have by no means considered the many process variations used in industry to achieve or optimize certain characteristics. You should review the literature including the Suggested Readings noted at the end of this chapter. Research and development is ongoing into the underlying device physics, device and circuit design, and process technology.

Figure 18.30 is a sketch of the doping profile for the *npn* transistor. Compare this profile with the cross section in Fig. 18.28(f). Because of the number of variables in the fabrication process, specifications can vary widely; indeed, data sheets often show a two or three to one variation in key parameters, or a data sheet may only state a minimum or maximum value for a parameter. For example, the 2N2222 data sheet presented in Chapter 4 shows a β variation from 50 to 300 even at the same Q-point.

The individual transistors or integrated circuits on the wafer are separated into individual ***die*** before packaging. Usually the individual transistor or IC is tested for both dc and ac performance before circuit separation and again after being packaged. Devices used for more exacting applications, such as for space or military hardware, undergo considerably

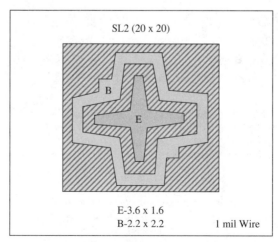

■ **FIGURE 18.29** 2N2222A mask set. Dimensions in mils (1 mil = 25.4 μ).

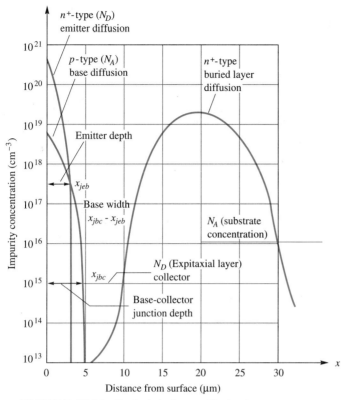

■ **FIGURE 18.30** Typical doping profile for the *npn* transistor.

more testing at all fabrication milestones, which includes temperature performance evaluation and radiographic inspection of the completed package.

CHECK UP

1. What key element of the hybrid-π model is changed by using a buried layer?
2. What key process parameters yields the highest β?
3. **TRUE OR FALSE?** Junction isolation is the only approach to isolating multiple BJTs.

18.8 *pnp* TRANSISTORS

Any *pnp* transistor structure within an IC must be compatible with the fabrication of the basic *npn* epitaxial transistor. Illustrating the basic compatibility of the most common *pnp* transistors with the *npn* transistors, Figs. 18.31 and 18.32 show both *npn* and *pnp* transistors fabricated within a common substrate using the same diffusions but different photolithographic masking to define the terminal functions. The *pnp* transistors are required for biasing, dc level shifting, and complementary-symmetry power output stages. Because the processing is usually optimized for *npn* devices, *pnp* specifications are not as good.

■ Virtually all ICs are 100% tested for all key parameters and functions at least at room temperature. Automated equipment and the increased emphasis on quality control have made IC reliability a serious competitive and economic issue. Usually, military- and space-qualified ICs must meet a variety of very stringent requirements that have a significant cost impact.

■ One, of course, can obtain high-performance *pnp* and *npn* devices on the same substrate by increasing the number of mask steps. Essentially, both diffusions and ion-implantation levels are optimized for each. This may require ~20 mask steps.

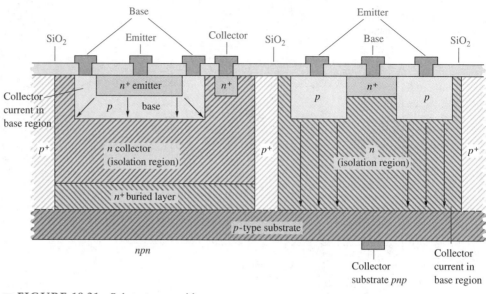

■ **FIGURE 18.31** Substrate *pnp* with *npn*.

A comparison of key *npn* and *pnp* specifications for typical analog IC devices is given in Table 18.2.

TABLE 18.2	Key Typical BJT Parameters at 300°K (Room Temperature)		
Parameter	**npn**	**pnp (Substrate)**	**pnp (Lateral)**
β_F	200	50	30
β_R	2	4	3
V_A (volts)	150	50	50
ϕ_J (volts)	0.7	0.55	0.55
I_S (amperes)	2×10^{-15}	10^{-14}	2×10^{-15}
r_b (ohms)	200	100	300

Substrate (Vertical) Transistor

Referring to Fig. 18.31, we see that the base diffusion, epitaxial layer, and the substrate used to form the *npn* transistor are now connected to form the emitter, base, and collector of a *pnp* transistor. It is easy to observe why this *pnp* transistor does not have the β_F of its *npn* counterpart. The base region, because it is formed from the *n*-type isolation region, is relatively thick. Therefore, the recombination current in the base is much larger, lowering β_F. To further illustrate this, the minority-carrier current paths are illustrated in Fig. 18.31. Unfortunately, the collectors of multiple *pnp* transistors on a single IC will be connected, thus limiting the usefulness of this geometry. This geometry is useful for fabricating one of the output transistors in a class B amplifier, a topic we studied in Chapter 10. An advantage is that the substrate, which is the *pnp* collector, can handle the heat dissipation when the substrate is in thermal contact with the package.

Lateral Transistor

As shown in Fig. 18.32, for the lateral transistor, the *npn* base diffusion is used for the *pnp* emitter and collector, and the *npn* collector diffusion is used for the *pnp* base. The *npn* n^+ emitter diffusion is used to obtain a good ohmic contact with the *n*-type base region of the *pnp*. As illustrated in Fig. 18.32, the lateral base dimension is somewhat larger than the *npn* vertical diffused base, resulting in a somewhat lower β_F because of recombination. The minority-carrier current paths qualitatively illustrate this effect. This particular geometry permits multiple *pnp* transistors on a single die. The major disadvantage of lateral *pnp* transistors is the relatively low current gain compared to *npn*. An application of the lateral transistor is presented in Section 13.7 (IIL).

It is also interesting to observe that a parasitic *pnp* transistor is formed with the substrate as illustrated in Fig. 18.33. The substrate "collects" carriers emitted into the base by surface emitter or collector.

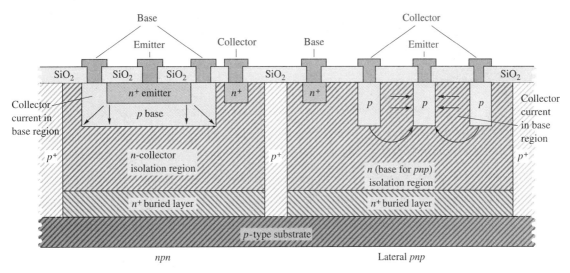

FIGURE 18.32 Lateral *pnp* transistor fabricated with an *npn* transistor.

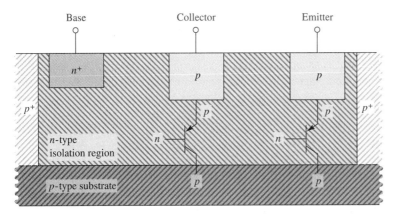

FIGURE 18.33 Parasitic *pnp* substrate transistors fabricated along with the desired lateral *pnp* device.

> **CHECK UP**
>
> 1. **TRUE OR FALSE?** The β for *npn* and *pnp* transistors fabricated with the same diffusions are the same.
> 2. **TRUE OR FALSE?** All substrate *pnp* collectors are connected.
> 3. What is one key advantage of a substrate *pnp* transistor compared to a lateral *pnp* transistor?

18.9 DIODES

Once we have the basic *npn* or *pnp* transistor, it is a straightforward matter to arrange the metallization to achieve one of the diode configurations pictured in Fig. 18.34. Usually, the collector-base junction is connected and the base-emitter junction serves as the diode. Whether a designer uses a *pnp* or an *npn* transistor or the base-emitter (BE) or base-collector (BC) junction depends on the dc bias constraints in the circuit.

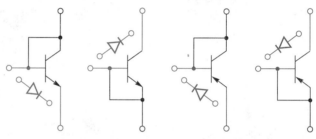

FIGURE 18.34 Diode-connected transistors.

> **CHECK UP**
>
> 1. **TRUE OR FALSE?** A diode formed from the BE junction has a higher breakdown voltage compared to a diode formed using a BC junction.
> 2. Why is it poor design practice to use the diode formed between the collector and substrate?
> 3. Why do junction-isolated IC diodes require three connections?

■ Recall pinch resistors are essentially a JFET structure. Far more BJTs and MOSFETs or CMOSFETs are fabricated on the same die as compared to BJTs and JFETs fabricated on the same die.

18.10 JUNCTION FIELD-EFFECT TRANSISTORS

It is sometimes advantageous to have both BJTs and JFETs on the same die since each contributes unique properties to a circuit. This is also the case for integrating BJTs and MOSFETs on the same die. Because the diffusion or ion-implantation depths are uniform across a die, photographic masking changes are used to obtain different devices.

Figure 18.35 shows an idealized cross section of an *npn* transistor along with a *p*-channel JFET. A typical mask geometry is given for the 2N5457-59 *n*-channel JFET in Fig. 5.32. This is called a double-diffused JFET because, to optimize the *p*-type base width at the same time as the *p*-type JFET channel, two separate n^+ diffusions are used to form the emitter and top-side gate connection. Obviously, this double-diffused JFET requires an additional masking fabrication step, necessary to obtain compatible devices, each with optimal parameters. A *p*-channel JFET cross section using ion implantation for more precise impurity profile control is illustrated in Fig. 18.36.

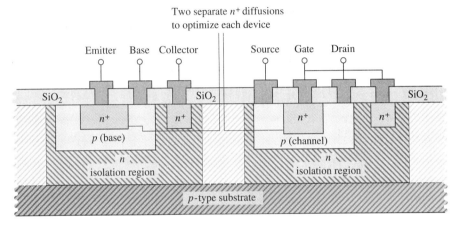

■ **FIGURE 18.35** *npn* BJT—double-diffused, *p*-channel JFET idealized cross section.

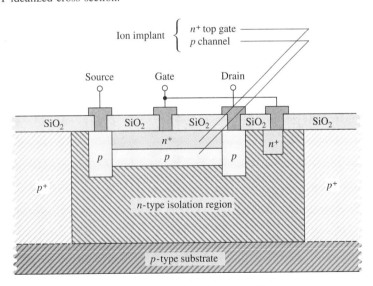

■ **FIGURE 18.36** Ion-implanted *p*-channel JFET.

CHECK UP

1. Why is it more difficult to fabricate an *npn* BJT and an *n*-channel JFET than an *npn* BJT and a *p*-channel JFET?
2. Sketch a cross section of an *n*-channel JFET compatible with a lateral *pnp* BJT.

18.11 METAL-OXIDE SEMICONDUCTOR TRANSISTORS

Most digital ICs use only MOSFETs; however, we can also fabricate MOS devices on the same die with *npn* transistors. This type of process is called *BiMOS* or *BiCMOS*. We have already considered a MOS structure for use as a capacitor. Geometries are realistically portrayed in Chapter 5. Figure 18.37 illustrates a compatible *npn* transistor with a *p*-channel MOSFET.

Because of the need for thousands or even millions of MOSFETs on a single die to obtain large-scale memory and microprocessor functions, the gate length has dropped below 1.25 μm for many standard commercial circuits. Gate lengths below 0.7 μm have been introduced into the marketplace to meet the need for higher functionality and larger memory storage. To obtain the required accuracy in photomask alignment, most MOSFETs are fabricated by means of a **self-aligned** (heavily doped) **polysilicon gate** technique. *Polysilicon* refers to polycrystalline silicon, or silicon that does not have a single-crystal structure. In a sense, the polysilicon is being treated as an ohmic conducting material. It is easy to deposit, and its use results in good ohmic contacts. Figure 18.38 illustrates this procedure for a *p*-channel enhancement device.

■ Minimum feature sizes are a moving target. There is some indication that commercial production in the 0.25 μm area is forthcoming. There are really two problems as one moves to sub -0.5 μm geometries. The first is the technology limitations and cost of photolithography equipment. The second, and potentially most serious, is the issue of short-channel effects and the physics of device operation moving to a different regime. After all, 0.25 μm is only a few hundred atomic layers thick.

We will start out in a conventional manner by diffusing the p^+ source and drain regions [see Fig. 18.38(a)]. The separation between source and drain, which normally defines the gate length, will be somewhat larger than is finally required. A thin (<500 Å and as low as 125 Å for some advanced processes) SiO$_2$ layer is thermally grown over the surface [see Fig. 18.38(b)]. The polysilicon gate is now deposited onto the oxide layer. An older technology used Al metal as the gate, which is how one obtains the term *metal* in MOS. Even though polysilicon is predominantly used, the MOS terminology remains. Note that the gate does not extend the full distance between the source and drain [see Fig. 18.38(c)].

The entire wafer is now subjected to *p*-type ion implantation, usually using boron. The energy in this beam, in excess of 50 to 80 keV, is such that the boron ions easily penetrate through the oxide, effectively doping the *n*-type substrate. However, the boron ions are stopped by the polysilicon layer as shown in Fig. 18.38(d). In effect, the gate polysilicon forms the mask. Therefore, the gate length is now defined rather well as the lateral

■ **FIGURE 18.37** *npn* BJT and *p*-channel MOSFET requires a *p* and p^+ diffusion.

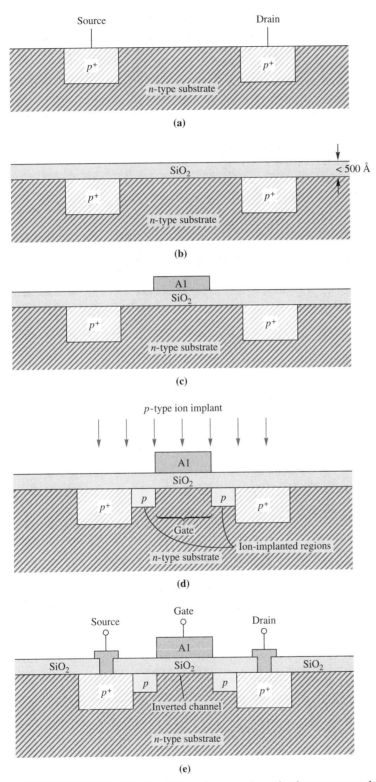

FIGURE 18.38 Metal self-aligned gate p-channel enhancement-mode MOSFET. (a) Diffusion of p^+ source and drain. (b) Thin SiO_2 layer grown on silicon surface. (c) Aluminum gate deposition and masking. (d) Gate length definition using Al gate as mask. (e) Final MOSFET with metallization.

dimension of the gate conductor. The use of ion-implantation techniques has minimized the effect of lateral diffusion and registration tolerances. The resultant p implant offers $N_A = 10^{14}$ acceptors/cm^3. Figure 18.38(e) illustrates the final result with deposition and masking of the source and drain metallization.

Building on the structure shown in Figure 18.38, we will illustrate a procedure for obtaining a pair of n-channel enhancement-mode devices. We will start out with a lightly doped p-type substrate on which we have a 500-Å thermally grown SiO$_2$ layer, a 500-Å layer of Si$_3$N$_4$ (silicon nitride), and a 1000-Å layer of SiO$_2$. These last two layers are deposited from a condensing gas. This deposition process is called *chemical vapor deposition* (CVD). This "sandwich" is illustrated in Fig. 18.39(a).

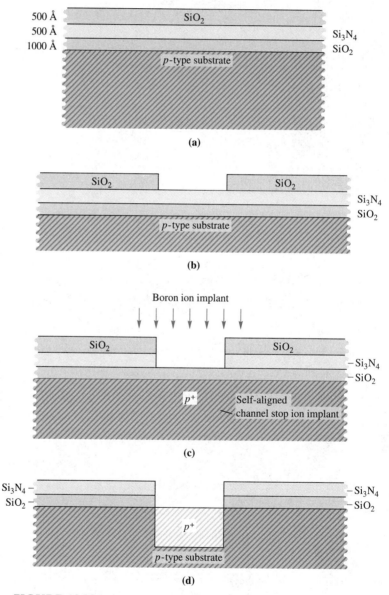

■ **FIGURE 18.39** Polysilicon gate, self-aligned structure, n-channel, enhancement mode. (a) SiO$_2$-Si$_3$N$_4$-SiO$_2$ silicon "sandwich." (b) Masking and SiO$_2$ etch to define MOSFET. (c) Channel stop ion-implantation. (d) Removal of SiO$_2$ covering layer.

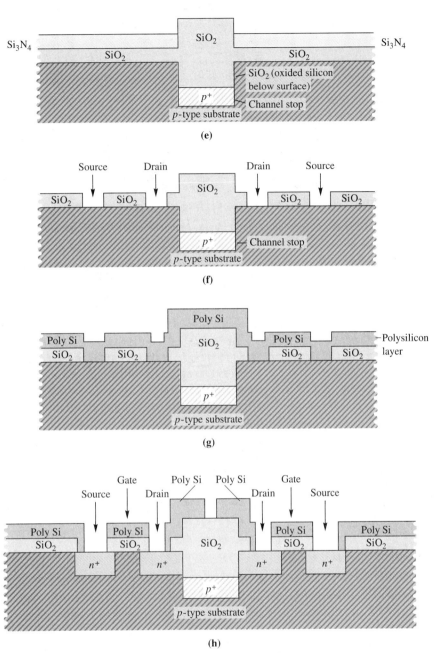

FIGURE 18.39 *continued.* **(e)** Field oxide growth. **(f)** Opening of source and drain regions. **(g)** Polysilicon deposition over wafer surface. **(h)** Source and drain diffusion using polysilicon mask.

The first mask step is used to define those areas that will contain transistors. The top layer of SiO_2 is etched away in the unmasked areas [Fig. 18.39(b)]. The wafer is now subjected to *p*-type (boron) ion implantation that forms a self-aligned p^+ region called a *channel stop* [Fig. 18.39(c)]. The self-alignment results from the effect of the Si_3N_4 serving as a mask. The exposed SiO_2 is now etched. Note that the SiO_2 under the Si_3N_4 is protected [Fig. 18.39(d)]. A new SiO_2 layer is thermally grown. This layer is about 1.5 μm thick, and because the top surface of the silicon is being oxidized, the SiO_2 is

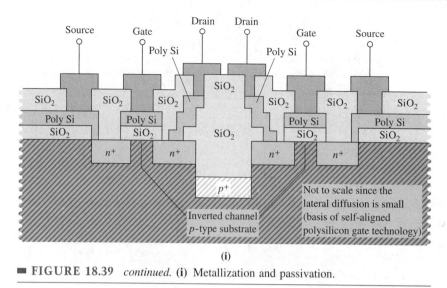

■ **FIGURE 18.39** *continued.* **(i)** Metallization and passivation.

shown as extending below the surface for about half its thickness [Fig. 18.39(e)]. The thick oxide layer is called the *field oxide,* and the thin oxide layer is called the *gate oxide*. The Si_3N_4 is then removed. Holes for the source and drain are now opened [Fig. 18.39(f)]. A polysilicon layer is deposited over the surface [Fig. 18.39(g)]. Holes are etched into the polysilicon, and the wafer is subjected to an *n*-type diffusion (usually phosphorus), creating the source and drain regions. The polysilicon acts as a mask to the diffusions [Fig. 18.39(h)]. Again, observe the short gate length. After top metallization is added, an SiO_2 passivation layer is deposited as a final passivation step to protect the metallization as well as any exposed silicon surface [Fig. 18.39(i)].

There are a number of different, though related, fabrication procedures for MOS devices, with the self-aligned gate processes we just described being but a small sample. You are encouraged to look at the Suggested Readings at the end of this chapter and to keep abreast of latest developments by making it a practice to read current technical and trade journals.

Complementary MOSs

By referring to Fig. 5.18, we can see that *p*-channel and *n*-channel MOSFETs can be fabricated on the same substrate to obtain a CMOS inverter. Initially, a *p*-region (called a *well*) is formed to serve as an isolated region for the *n*-channel MOSFET. The *n*-channel MOSFET is formed in this well by diffusing an n^+ source and drain. The *p*-channel MOSFET is formed directly in the *n*-type substrate by diffusing a p^+ source and drain.

■ MOSIS is sponsored by the U.S. Advanced Research Projects Agency (ARPA) with academic support by the National Science Foundation. Additional information is available from USC Information Sciences Institute, 4676 Admiralty Way, Marina Del Ray, California 90292-6695 (tel. 213-822-1511).

Current high-density memory and digital technology has moved in the direction of using self-aligned polysilicon CMOS and NMOS. Many details have been omitted, but this simplified description illustrates the basic compatibility of many MOS processes. One example is the CMOS devices used in the MOSIS (MOS Implementation Service) process (Figure 18.40).

Vertical MOS (VMOS) Field-Effect Transistors

Lower power dissipation is achievable if the ohmic channel resistance can be made small in the ON state. One way to achieve this is to provide for a short channel length, but current production technology yields minimum feature sizes on the order of <0.75 μm. However, this minimum feature size refers to a planar dimension. We already know that

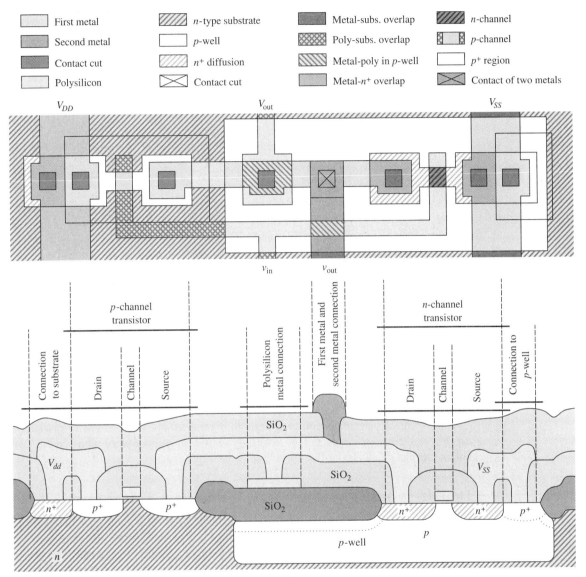

FIGURE 18.40 Basic features of the standard *p*-well MOSIS process (from Ref. 15).

it is possible to control the depth and thickness of doping layers to a fraction of a micron by means of ion implantation or even by diffusion techniques. We can take advantage of this by fabricating the MOSFET as illustrated in Fig. 18.41.

An n^+-type substrate is used as a source connection. A *p*-type thin layer, 1 μm thick or less, is formed on the substrate. A lightly doped p^--type layer, several microns or more thick, is then diffused as the top layer of the sandwich. Using photolithographic techniques, a groove is etched so that it penetrates into the n^+-type substrate. It is possible to etch the "V" selectively by using the anisotrophy of the ⟨100⟩-oriented silicon crystal. The crystal structure in combination with the appropriate etching chemical will yield a 54.7° etching angle. An n^+-type diffusion is then used to form the drain. The gate and drain metallization is then deposited. Recall that the source contact is to the substrate. The device is operated as an *n*-channel MOSFET in that, with appropriate biasing, the narrow *p*-type region will invert yielding an *n*-type short channel.

■ Potassium hydroxide (KOH) and hydrazine (N_2H_4) are typical selective etches. Safety is a critical issue when working with hydrazine, which is also used as a rocket fuel.

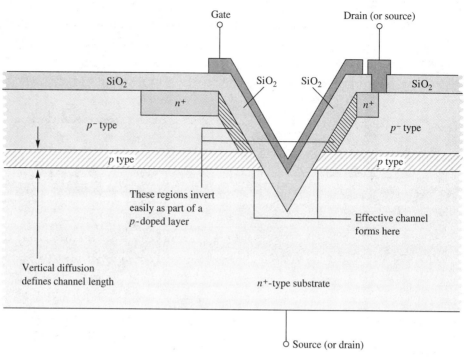

FIGURE 18.41 Vertical MOSFET (VMOS) cross section.

Because of the short channel and large area substrate connection, VMOS transistors are often used as discrete power devices. Typically, they offer $r_{DS(on)} = 1$ to $2\,\Omega$, several amperes of drain current capability, and up to several tens of watts of power dissipation capability. Operating voltages with a $V_t \sim 1$ V ensures compatibility with typical logic voltage levels.

Double-Diffused MOS (DMOS) Field-Effect Transistors

The DMOS structure, which is shown in Fig. 18.42, achieves a short channel length by using the lateral diffusion properties of an n- and a p-type region. The difference between the lateral diffusion yields the narrow channel without resorting to a difficult additional series of precision masking steps. Actually, with the rapid advances in lithography tech-

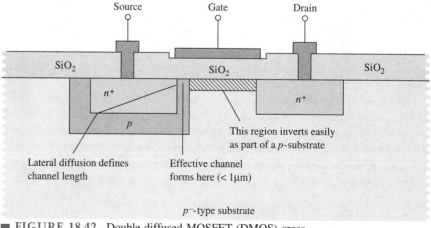

FIGURE 18.42 Double-diffused MOSFET (DMOS) cross section.

niques, the benefits of the DMOS structure are now available in simpler structures. As with VMOS, which is more widely used, this technology is often used for high-power discrete devices.

Silicon-on-Sapphire (SOS) / Silicon-on-Insulator (SOI) Technology

A major disadvantage to CMOS is that the parasitic capacitance between the common drain connection and the substrate limits the frequency of operation to about 100 MHz. Essentially, because the silicon substrate is a conductor, it forms one side of the parasitic capacitor. If the application requires higher switching speeds, an excellent, though expensive, alternative is the use of a sapphire, Al_2O_3, aluminum oxide substrate, which is a good electrical insulator. Because the sapphire has a crystal structure that closely resembles silicon, a thin silicon layer can be epitaxially grown on the surface. By means of lithographic techniques, this epitaxial layer can be partitioned into electrically isolated regions (see Fig. 18.43). Because reverse-biased junction diodes, with their concomitant junction capacitance, are no longer needed for isolation between devices, the parasitic capacitance is essentially zero. Figure 18.43 illustrates two NMOS devices fabricated on sapphire.

■ Current SOI focus is on buried silicon dioxide (SIMOX).

SOS is not widely used because of its high material cost. An Al_2O_3 75-mm wafer may cost up to 50 times as much as a 100-mm silicon wafer. In addition, the fabrication is more complex. It may be that GaAs, which offers inherently faster operation, will supersede the SOS technology in many applications. There is no clear-cut choice for high-frequency MOS operation with respect to the basic material or topology. Most approaches to achieving higher speeds have focused on finer resolution lithographies.

■ Amorphous silicon and polycrystalline-based NMOS transistors can be fabricated directly on glass substrates. This is done in the flat-panel display industry. Research is also being done on using polymide (plastic) substrates for amorphous silicon transistor fabrication[16,17].

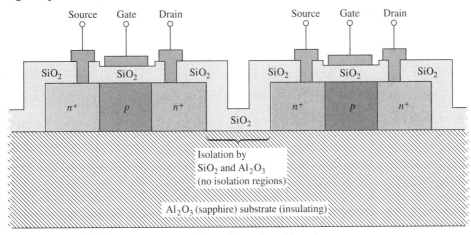

■ **FIGURE 18.43** Silicon-on-insulating substrate/sapphire (SOI/SOS) cross section, *n*-channel enhancement mode

---

CHECK UP

1. **TRUE OR FALSE?** V_t can be reduced using a thinner gate oxide.
2. Describe two ways MOSFETs are isolated electrically.
3. Provide one reason why SOI technology is more radiation resistant than MOS technology.
4. **TRUE OR FALSE?** VMOS topologies provide higher power dissipations than planar MOS topologies.
5. Why are self-aligned gates used?

SUMMARY

Each integrated-circuit fabrication technology offers both constraints and advantages for the designer and end user of electronic circuits. This chapter provided an overview of some of the major fabrication technologies as well as their applications.

Although detailed semiconductor device processing information is beyond the scope of this text, we reviewed the fundamentals of impurity diffusion, ion implantation, and photolithography as they relate to the manufacture of both passive and active components. Passive components include both diffused and pinch resistors and junction and MOS capacitors. A series of cross-sectional sketches illustrates the fabrication of the *npn* transistor. Many of the other passive and active devices are manufactured by means of the *npn* transistor fundamental process, but with different circuit functions defined by different photolithographic masking and metallization processes. Similarly, we also presented descriptions and cross-sectional sketches of several different MOSFET families, with NMOS and CMOS being of most current interest.

SURVEY QUESTIONS

1. What is the relationship between μm, mil, Å, and cm? Compare to some common objects.
2. What is the one-dimensional diffusion equation (define all terms) and what are the solutions for the diffusion equations subject to either infinite or finite source boundary conditions?
3. How is diffusion used to form a doping profile and junction?
4. How is an epitaxial layer grown?
5. What is the profile resulting from ion implantation and how does it compare to a diffusion profile?
6. How do you compute the resistivity and sheet resistance of a uniformly doped and nonuniformly doped layer?
7. Describe the processes to grow SiO_2 layers of particular thicknesses.
8. Using a series of sketches, describe the photolithography/masking process.
9. How do you fabricate diffused resistors, pinch resistors, MOS capacitors, and junction capacitors? List typical properties.
10. How do you fabricate *npn* transistors? List typical properties.
11. How do you fabricate lateral and substrate *pnp* transistors? List typical properties.
12. How do you fabricate JFETs? List typical properties.
13. How do you fabricate MOSFETs including CMOS, VMOS, DMOS, and SOI devices and circuits? List typical properties of each device.

PROBLEMS

Problems marked with an asterisk are more challenging.

18.1 Verify that Eq. (18.3) satisfies the diffusion equation, Eq. (18.1).

18.2 Verify that Eq. (18.5) satisfies the diffusion equation, Eq. (18.1).

18.3 Consider a silicon substrate uniformly doped with 10^{16} atoms/cm^3 of phosphorus. The substrate is now exposed to an infinite source of boron where $N_0 = 10^{20}$/cm^3 for 1 hr at $T = 1100°C$. What will be the depth of the junction?

18.4 Using the information in Example 18.2, plot x_j versus diffusion time at $T = 1100°C$. Consider the range of 5 min $< t <$ 8 hr. How much time is required to form the junction at 1 mil = 25.4 μm?

18.5 Using the information in Example 18.2, plot x_j versus diffusion temperature for a 1-hr diffusion time. Consider the range $900°C < T < 1250°C$.

18.6 Using the information in Example 18.3, compare the impurity concentration at 1 μm with that of the surface concentration for boron diffusion into an n-type substrate over the range $900°C < T < 1250°C$.

18.7 Assume a p-type substrate doping concentration of 10^{16} atoms/cm^3. From an infinite arsenic source concentration of 10^{19} atoms/cm^3, compute the junction depth x_j, for a diffusion process lasting 2 hr at $1100°C$.

18.8 Explain why it is important to arrange for higher temperature diffusions to be done first in a processing sequence.

18.9 Often diffusions are accomplished using a two-step process where the first step, called *predeposition*, is given by Eq. (18.3),

$$N(x,t_1) = N_0 \text{erfc}\left(\frac{x}{2\sqrt{D_1 t_1}}\right),$$

where D_1 is the diffusion coefficient for time t_1, the predeposition time. The *drive* step follows Eq. (18.5),

$$N(x,t_2) = \frac{Q_0}{\sqrt{\pi D_2 t_2}} e^{-x^2/4D_2 t_2},$$

where D_2 is the diffusion coefficient for time t_2.

(a) Show that for a thin layer predeposition,

$$Q_0 = 2N_0 \left(\frac{D_1 t_1}{\pi}\right)^{1/2}.$$

(b) Show that the final impurity concentration can then be given by

$$N(x,t_1,t_2) = \frac{2N_0}{\pi} \left(\frac{D_1 t_1}{D_2 t_2}\right)^{1/2} e^{-x^2/4D_2 t_2}.$$

18.10 You wish to grow a 0.5-μm SiO$_2$ layer on a 1×10^{17} atoms/cm^3 boron-diffused silicon substrate. Design a reasonable three-step (dry-wet-dry) oxidation sequence such that the first dry layer is 400 Å and the last dry layer is 100 Å.

18.11 How many ions/cm^2 are required to obtain a uniform doping profile of $N_D = 10^{15}$ donor atoms/cm^3 to a depth of 0.75 μm? Assume a beam current of 100 μA. How long will it take to ion implant a 100-mm wafer?

18.12 What is the sheet resistance $R_\square$ of a uniformly arsenic-doped sample 2 μm thick where $N_D = 10^{19}$ atoms/cm^3?

18.13 Let $N_D(x) = N_0 e^{-x/L}$ [Eq. (18.4)]. Derive the average value of $\bar{N}_D$ for some junction depth x_j.

18.14* Let $N_D(x) = N_0 \text{erfc}(x/2\sqrt{Dt})$ [Eq. (18.3)]. Derive the average value of $\bar{N}_D$ for some junction depth x_j. This derivation requires that you use erfc integral tables.

18.15 Compute the $R_\square$ with and without including the substrate doping for a silicon substrate doped with 10^{14} atoms/cm^3 of boron. Assume a phosphorus diffusion profile of $N_D(x) = 10^{18} e^{-x/1.5 \, \mu m}$ atoms/cm^3.

18.16 Assume that an infinite boron-doping profile will be approximated by an exponential of the form

$$N_A(x) \cong N_0 \exp[-x/2(Dt)^{1/2}] = N_0 \exp(-x/0.25 \, \mu m),$$

where $N_0 = 4 \times 10^{20}$ atoms/cm^3. Assume the wafer doping is $N_D = 10^{17}$ atoms/cm^3.

(a) Compute a value for the sheet resistance $R_\square$ for the original uniformly doped n-type layer.

(b) Compute a value for the boron-doped layer sheet resistance $R_\square$. Observe that to do this you will have to estimate a value for the μ_p from Figure 2.18. What is the average resistivity, ρ, of this layer?

18.17 By photolithography, a square 10×10 μm is masked on the surface of an SiO_2 layer 8000 Å thick. Estimate the undercutting of the opening on the SiO_2 surface.

18.18 You need a 20-kΩ resistor fabricated using a p uniform base diffusion of 10^{18} acceptor atoms/cm^3 of thickness 5 μm. Give the dimensions of the resistor, realizing that you wish to minimize die area, yet not require a technologically impossible minimum feature size. Compute σ and $R_\square$ along with the design.

18.19 Assume you start with an n-type substrate whose impurity concentration is given by $N_D = 10^{16}$ donor atoms/cm^3. Into this substrate positioned at $x = 0$, you diffuse boron having an approximate concentration of $N_A(x) = 10^{18} e^{-x/2.0\,\mu m}$/cm^3. $T = 300$ K.

(a) What are the electron and hold concentrations just inside the surface $(x = 0^+)$?
(b) What is the sheet resistance of this diffused layer, (1) taking into account the original substrate doping and (2) ignoring the original substrate doping?
(c) Assume an average doping density of $N_A(x) = 10^{17}$ atoms/cm^3. Now find the sheet resistance and compare this approximation with the results from part (b).
(d) For a minimum feature of 5 μm, how much die area would a 10-kΩ resistor require?

18.20 A base-diffused resistor of 40 kΩ is required. If the minimum feature size is 10 μm and the $R_\square = 100$ $\Omega/\square$, calculate the resistor length. If $R_\square = 10$ $\Omega/\square$ as a result of ion implantation, calculate the resistor length. What is the thickness ratio between the base-diffused resistor and the ion-implanted resistor? Assume equal average doping densities.

18.21 Provide estimates on value ranges for diffused resistors with an $L/W = 10$ for the doping levels given. The idea here is to consider that IC circuit designs must accommodate technology-driven resistor values: A base-diffused resistor is formed during the npn BJT base diffusion. Usually, this is a lightly doped layer <3 μm deep with N_A between 10^{14} and 10^{18} atoms/cm^3.

18.22 Provide estimates on value ranges for diffused resistors with an $L/W = 10$ for the doping levels given. The idea here is to consider that IC circuit designs must accommodate technology driven resistor values: An emitter-diffused resistor is formed during the BJT emitter diffusion. Usually this is a heavily doped layer <1 μm deep with N_D between 10^{19} and 2×10^{20} atoms/cm^3.

18.23 An ion-implanted "base" resistor of value 50 kΩ is manufactured with $\rho = 0.1$ Ω-cm material. Assume a minimum feature size of 10 μm. How thick is the ion-implanted layer?

18.24 For the device described in Problem 18.23, the ion-implanter has a beam current of 50 μA. How long will it take to fabricate the resistor? Assume a uniform spot size of 1 cm^2. $\rho = 0.1$ Ω-cm corresponds to $N_A = 6 \times 10^{17}$ acceptor atoms/cm^3.

18.25 If the base sheet resistance is within $\pm 5\%$ and the minimum feature size (resistor line width) is within ± 2.5, plot the percent change in resistance for 10 μm $< W <$ 100 μm.

18.26 Design a square capacitor whose value is 15 pF utilizing an abrupt-junction pn diode. Assume an impurity doping of 10^{16} atoms/cm^3 and $V_{bias} = 8$ V. Compute the dimensions in mils.

18.27 Perform the same design as in Problem 18.22 only use the MOS capacitor structure with a technologically reasonable SiO_2 dielectric thickness.

18.28 Compute an approximate value for the parasitic capacitance that exists between the collector and substrate. The dimensions of the collector isolation region are $100 \times 100 \times 10$ μm deep. The substrate is uniformly doped at 10^{14} atoms/cm^3, and the collector doping is much larger. The dc voltage between the collector and substrate is 8 V.

18.29 Compute the die area required for a MOS capacitor of 20 pF when the SiO_2 layer is 500 Å thick.

18.30 Assume the base diffusion impurity profile is given by $N_A(x) \cong 10^{18} e^{-x/1.0\,\mu m}$ atoms/cm^3 and the emitter diffusion impurity concentration is given by $N_D(x) = 10^{20} e^{-x/0.5\,\mu m}$ atoms/cm^3. Neglect the collector and substrate doping concentrations.

(a) What is the depth of the emitter-base junction?
(b) Determine the sheet resistance $R_\square$ of the emitter region. Compare your result, assuming a uniform doping concentration of $N_D = 10^{20}$ atoms/cm^3, with the $N_D(x) = 10^{20} e^{-x/0.5\,\mu m}$ atoms/cm^3 profile.

18.31 Simplified doping profiles for an *npn* transistor are as follows:

$$\text{Collector} \quad N_{DC}(x) = 10^{15} \text{ atoms/cm}^3$$
$$\text{Base} \quad N_{AB}(x) = 10^{18} e^{-x/1.0 \, \mu m} \text{ atoms/cm}^3$$
$$\text{Emitter} \quad N_{DE}(x) = 10^{20} e^{-x/0.25 \, \mu m} \text{ atoms/cm}^3$$

(a) What is the depth of the collector-base junction? Ignore the substrate doping level. There is no buried layer.

(b) What is the depth of the emitter-base junction? Ignore the collector doping level.

18.32 An *npn* transistor is built within a *p*-type substrate. There is a parasitic *pnp* transistor between the base, collector, and substrate.

(a) If the *npn* transistor is in the active region, what is the operating condition of the *pnp* transistor?

(b) Explain the operating condition of the *pnp* transistor when the *npn* is in cutoff, saturation, and reverse operation.

18.33 Sketch the cross section for a multiple-emitter transistor.

18.34 As you recall, most BJT processes are keyed to the fabrication of the *npn* transistor, whose simplified cross section is shown in Fig. 18.44(a). Quantitatively discuss why it is unreasonable to use the same double-diffused process for fabricating a high-performance *pnp* transistor as shown in Fig. 18.44(b).

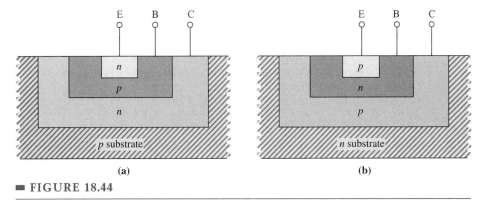

■ **FIGURE 18.44**

18.35 Suppose you diffuse a collector to a junction depth of $x_j = 20 \, \mu m$. How close in lateral dimensions could you space two collector regions?

18.36 Gold wire 1 mil in diameter is typically used to connect IC elements to the outside world. If this wire is to conduct 10 mA, what is the current density for a wire? If the wire is 100 mils long, how much power is dissipated? For gold, $\sigma = 4.25 \times 10^5 \, (\Omega\text{-cm})^{-1}$.

18.37 Compare the number of VLSI circuits, 250×250 mil, that can be fabricated on a 4-in. and a 6-in. wafer. Estimate the edge yield losses by counting the number of incomplete die.

18.38 For a given wafer size and 2-μm minimum feature size technology, we know that the overall wafer yield is 75%. Is it worthwhile to go to a much more difficult 0.8-μm minimum feature size technology whose overall wafer yield is only 35%?

18.39 Suppose we require a 30% wafer yield for some complex device process requiring 12 equally difficult mask steps. What must the yield be after each mask step? To make some dollars on the deal, a good yield might be 80%. Now compute the individual mask step yield.

18.40 Using silicon epitaxy followed by a diffusion process, outline a process flow that yields a diffusion with a junction depth of 2.0 μm into a 10-μm epitaxial layer doped at 2×10^{17} atoms/cm^3 with $N(0) = 10^{19}$ atoms/cm^3. The idea here is that you suggest reasonable values for appropriate times, temperatures, and materials.

18.41 Consider using $E_{\text{crit}} = 10^7$ V/cm. Compare the size of this MOS capacitor with that of active devices to establish a reasonable maximum for IC capacitor values. Discuss restrictions on operating voltages.

18.42 Assume you wish to use either the BE or BC junctions in reverse bias to obtain a 10-pF capacitor. Assume uniform values for $N_{DE} = 10^{20}$ atoms/cm^3, $N_{AB} = 10^{17}$ atoms/cm^3, and $N_{DC} = 10^{18}$ atoms/cm^3 and $V_R = 10$ V. Now compute the dimensions and compare to the MOS capacitors. Compare the breakdown potentials and capacitor value voltage sensitivity.

18.43 Assume a phosphorus-doped 20-μm epitaxial layer where $n(x) = 2 \times 10^{17}$ atoms/cm^3, this epitaxial layer having been deposited on a heavily boron-doped substrate. For ease in algebraic manipulation and computation, we assume that the infinite boron source base doping profile normally given by $n(x) = N_0 \operatorname{erfc}[x/2(Dt)^{1/2}]$ will be approximated by an exponential of the form

$$n(x) \cong N_0 \exp[-x/2(Dt)^{1/2}] = N_0 \exp(-x/0.25 \ \mu\text{m}),$$

where $N_0 = 4 \times 10^{20}$ atoms/cm^3.

(a) Compute a value for the sheet resistance $R_\square$ for the original epitaxial layer.
(b) Compute a value for the base-diffused region sheet resistance.
(c) Assume the process described is the predeposition step in a two-step diffusion process. The infinite boron source is now removed. Specify a reasonable set of a drive time and temperature such that the junction depth is at 5.0 μm. Also compute a resultant surface impurity dopant concentration, $N(x = 0)$.
(d) Recommend a minimum separation distance between two base-diffused resistors formed in the epitaxial layer after the two-step diffusion process. This minimum separation distance should include an allowance for a 2-μm depletion width. Support your answer with a sketch.

18.44 A 0.5-μm SiO$_2$ layer is used as the dielectric for an MOS capacitor where the underlying p diffusion forms one contact and the second contact consists of a top 4- $\times$ 4-mil metallization. Compute this capacitance value. These dimensions are similar to those one expects from the bonding pads in an IC and thus represents a parasitic capacitance.

18.45 You want to fabricate a 30-pF capacitor using either the BE or BC junctions in reverse bias. Assume uniform values for $N_{DE} = 2 \times 10^{20}$ atoms/cm^3, $N_{AB} = 5 \times 10^{16}$ atoms/cm^3, and $N_{DC} = 10^{18}$ atoms/cm^3 and $V_R = 15$ V. Compute the dimensions and compare to typical dimensions for a MOS capacitor where $t_{ox} = 3500$ Å. Compare the breakdown potentials and capacitor value voltage sensitivity.

18.46* Table 18.3 is a Typical TTL Process Summary flowchart. Address the following questions referenced to the appropriate steps in the process.

(a) Estimate the doping density, mobility, and majority- and minority-carrier concentration in the starting material wafer.
(b) Assume that the initial oxide $t_{ox} = 10{,}000$ Å is only an approximation. Using appropriate curves, provide a more exact t_{ox} for the initial oxide step. Also determine the t_{ox} that exists after the 2-hr wet O$_2$ portion of this procedure.
(c) Assume that the resistivity of the n-type epitaxial layer is given by $\rho = 0.17$ Ω-cm. Demonstrate by showing computationally that the infinite-source isolation diffusion with parameters as shown will perform the needed function.
(d) Consider the emitter deposition step. Assume that instead of a diffusion process, we use ion-implantation phosphorus. Suppose we then end up with a *uniformly doped* emitter where $x_{jBE} = 0.5$ μm and $R_\square = 10$ $\Omega/\square$. Compute the ion implant dose. Also provide a reasonable set of values for the beam current and the length of time you must keep the beam current on.

18.47* Table 18.3 is a Typical TTL Process Summary flowchart. Address the following questions referenced to the appropriate steps in the process.

(a) *Initial oxide step:* Is the t_{ox} correct for the specified growth conditions?
(b) *Initial oxide step:* Suppose this step were not done correctly in that the wet O$_2$ step was not included. Would the resultant dry O$_2$ oxide be able to mask effectively against the buried layer diffusion? Explain

CHAPTER 18 BASIC FABRICATION TECHNOLOGY AND DEVICE CONSTRAINTS   929

(c) *Buried layer diffusion step:* Assume this is a constant source diffusion where As is introduced at the solid solubility limit. Compare your results for the junction depth, x_j, and the sheet resistance, $R_\square$, with the stated values. Assume the substrate is <100> material. Note that the terminology $\rho_S = R_\square$.

(d) *Epitaxial deposition step:* What is the range of doping densities for this material? Resistors are often fabricated using epitaxial layers. What range of resistor values could you obtain if you restricted $L/W < 20$?

(e) *Epitaxial deposition step:* Design a diagnostic test pattern and procedure for evaluating the x_j, resistivity, and doping densities. Show typical scaled I-V plots that one might obtain.

(f) Suppose a 3- × 3-mil bonding pad used the initial oxide to provide electrical isolation. What value of parasitic capacitance would be expected?

(g) *Glassivation step:* We call this passivation. Could this 2% phosphorus-doped SiO_2 be thermally grown? How would you deposit the SiO_2?

TABLE 18.3 Basis TTL BJT Process

STARTING MATERIAL:	"*p*"-TYPE Cz SILICON BORON-DOPED	
	DIAMETER:	3-in.
	THICKNESS:	20 mils
	RESISTIVITY:	5–10 Ω-cm
INITIAL OXIDE:	TEMPERATURE:	1100°C ± 1°C
	TIME:	2-HR STEAM + 2-HR DRY O_2
	t_{ox}:	10,000 Å
BURIED LAYER PR		
BURIED LAYER DIFFUSION:	ARSENIC TRIOXIDE SOURCE AT 480°C	
	TEMPERATURE:	1225°C
	TIME:	16 HR
	x_jBL:	≈ 5–7 μm
	ρ_SBL:	<13 Ω/□
OXIDE REMOVAL:	48% HF	
	TIME:	1 min
BACK ETCH:	LDA TIME:	1 min
EPITAXIAL DEPOSITION:	"*n*"-TYPE ARSENIC DOPED	
	EPI THICKNESS:	5–7 μm
	RESISTIVITY:	0.14–0.20 Ω-cm
THERMAL OXIDATION:	TEMPERATURE	1100°C ± 1°C
	TIME:	1 HOUR STEAM + 30 minutes DRY O_2
	t_{ox}:	7500 Å ± 250 Å
ISOLATION PR		
ISOLATION DIFFUSION:	BORON SPIN-ON SOURCE	
	TEMPERATURE:	1175°C ± 1°C
	TIME:	2.5–3.0 HR
	x_jISO:	5–7 μm
	ρ_SISO:	7 ± 1 Ω/□
BORON GLASS REMOVAL:	48% HF FOR 60 seconds	

(continued on page 930)

TABLE 18.3 Basis TTL BJT Process *(Continued)*

REOXIDATION:		TEMPERATURE:	1000°C ± 1°C
		TIME:	15 min STEAM + 15 min DRY O_2
		t_{ox}:	3500 Å ± 200 Å
BASE PR			
BASE DEPOSITION:		BBr_3 SOURCE	
		TEMPERATURE:	980°C ± 1°C
		TIME:	3 min WARMUP + 12 min SOURCE + 12 min PURGE (TYPICAL)
		ρ_S:	40 ± 1 Ω/□
BORON GLASS REMOVAL:			1 min IN 10:1 H_2O/HF
			10 min BOILING HNO_3 (CONCENTRATED)
			10 sec in 10:1 H_2O/HF
REDISTRIBUTION AND OXIDATION:		TEMPERATURE:	1100°C ± 1°C
		TIME:	APPROXIMATELY 10 min STEAM, 2 HR DRY O_2
		x_{jb}:	1.6 μm ± 0.1 μm (5–7 FRINGES)
		ρ_S:	190 ± 5 Ω/□
		t_{ox}:	4000 Å ± 200 Å
EMITTER PR			
GOLD BACKING:		Au SPIN 4K rpm FOR 20 sec, Au ALLOY:	
		Au EMITTER TUBE VESTIBULE BLOCK FOR 15 sec	
EMITTER DEPOSITION:		SPIN ON PHOSPHORUS	
		TEMPERATURE:	1000°C ± 1°C
		TIME:	15 min (TYPICAL)
		ρ_S:	10–11 Ω/□ TEST WAFER "P"-TYPE
PHOSPHORUS GLASS REMOVAL:		5 sec 10 H_2O/HF	
REDISTRIBUTION:		TEMPERATURE:	1060°C
		TIME:	10 min STEAM (TYPICAL), 10 min DRY O_2
		ρ_S:	8–12 Ω/□
		t_{ox}:	1100 Å ± 200 Å
		x_{je}:	1.5 μm
		w_b:	0.25 μm
		β:	50–60 at 10 mA/5 V
		V_{OFFSET}:	90–120 mV
PREOHMIC PR			
BACK ETCH:		20 sec IN 48% HF	
ALUMINUM DEPOSITION:		SUBSTRATE TEMP:	300°C 30°C
		t_{Al}:	12 kÅ ± 2 kÅ
METAL PR			
ALUMINUM ALLOY:		30 sec AT 580°C	
GLASSIVATION:		2% PHOSPHORUS	
		THICKNESS:	12 kÅ ± 2 kÅ
PROBE:		β:	AT 10 mA/5 V 30–65
		OFFSET:	90–140 mV

Courtesy Texas Instruments.

CHAPTER 18 BASIC FABRICATION TECHNOLOGY AND DEVICE CONSTRAINTS

18.48 An *n*-type silicon wafer, uniformly doped at 3×10^{15} atoms/cm^3 is exposed to a 30-min boron constant source diffusion at 950°C. Compute the junction depth x_j. Sketch the profile labeling all key values.

18.49 Estimate $R_\Box$ for the diffusion in Problem 18.48.

18.50 Suppose part of a process traveler for CMOS requires an *n*-type phosphorus implant in a *p*-type wafer. Assume a 200-keV implant for 2 minutes with a 20-μA/cm^2 beam current. The *p*-type wafer is doped at $N_A = 10^{16}$ atoms/cm^3. Compute values for the dose and junction depth. Sketch the doping profile labeling the range and junction depth.

18.51 Sometimes, what are called **crossunders** are used as interconnects. Crossunders are obtained from shallow *erfc* diffusions. What would be the sheet resistance, $R_\Box$, for 0.25-μm-thick uniformly doped As and B layers where you are to use the maximum achievable electrically active doping concentrations?

18.52 Use results from Problem 18.51 to obtain the parasitic capacitance for a 500-μm-long $\times$ 5-μm-wide crossunder fabricated from As or B. Assume an abrupt junction.

18.53 Using silicon epitaxy followed by a two-step diffusion process, outline a process flow that yields a base diffusion with a junction depth of 2.0 μm into a 10-μm epitaxial layer doped at 2×10^{17} atoms/cm^3 with $N(0) = 10^{19}$ atoms/cm^3. The idea here is that you suggest reasonable values for appropriate times, temperatures, and materials. Indicate your sources for any material processing constants and parameters.

REFERENCES

1. "The Solid-State Era." *Electronics*, April 17, 1980, pp. 322–371. This is a special fiftieth-anniversary commemorative issue of *Electronics* devoted to the historical background of all fields of electronics. Very interesting and informative reading although it provides an overview only through 1979.

2. Hansen, Stephen E. *SUPREM III User's Manual*, Version 8628. Stanford University, Palo Alto, California, August 1986. This is the core document for the simulation program supported by the federal government. Numerous commercial versions are available.

3. Fuller, C. S., and J. A. Ditzenberger. "Diffusion of Donor and Acceptor Elements in Silicon," *Journal of Applied Physics*, May 1956, pp. 544–553.

4. Kendall, D. L., and D. B. DeVries. "Diffusion in Silicon." In *Semiconductor Silicon* (R. R. Haberecht and E. L. Kern, Eds.), p. 358. New York: Electrochemical Society, 1969.

    These graphical data are continually referenced and reproduced in virtually all semiconductor technology and processing texts and references, despite the age of the Refs. 3 and 4,

5. *Handbook of Mathematical Functions,* (M. Abramwitz and I. A. Stegun, Eds.). *Applied Mathematics Series 55,* 1964. Washington, D.C.: National Bureau of Standards. Everything you ever cared to know about obscure functions, including formulas, graphs, and tables. Note that the National Bureau of Standards is now the National Institute for Science and Technology. More recent versions have been printed. Computer tools such as MATHCAD™ also include these functions.

6. Pickar, K. A. "Ion Implantation in Silicon-Physics, Processing and Microelectronic Devices." In *Applied Solid-State Science,* Vol. 5 (R. Wolfe, Ed.). New York: Academic Press, 1975. An advanced treatment published during the early stages of industrial utilization of ion implantation.

7. Jaeger, Richard C., *Introduction to Microelectronic Fabrication,* Volume V Modular Series in Solid State Devices. (Georold W. Neudeck and Robert F. Pierret, Eds.), Reading, Mass: Addison-Wesley, 1988.

    This text offers a concise treatment of fundamental microelectronic processing technology steps at the late junior-senior level. Many of the earlier courses from the technical journals are reproduced in this volume.

8. Ghandhi, S. K. *VLSI Fabrication Principles—Silicon and Gallium Arsenide,* 2nd ed. New York: John Wiley & Sons, 1994. Advanced, current text. Good chapter references.

9. Sze, S. M. *Physics of Semiconductor Devices,* 2nd ed. New York: Wiley/Interscience, 1981. An all-purpose comprehensive reference written at an intermediate to advanced level. Primarily devoted to

device physics, but includes much discussion of how fabrication technology and physics interact to obtain a particular device specification.

10. Gray, P. R., and R. G. Meyer. *Analysis and Design of Analog Integrated Circuits,* 3rd ed. New York: John Wiley & Sons, 1993. Chapter 2 presents a good beginning- to intermediate-level overview of IC components construction.

11. Sze, S. M. *VLSI Technology.* New York: McGraw-Hill, 1988. Advanced-level reference, very descriptive and complete. Primarily an organized summary of the technical literature on a wide variety of current topics. Especially good on metallization.

12. van der Ziel, A. *Solid-State Physical Electronics.* Englewood Cliffs, N.J.: Prentice-Hall, 1976. See Section 4.5, Contact Problem in Metals.

13. Donovan, R. P. "Oxidation." In *Fundamentals of Silicon Integrated Device Technology* (R. M. Burger and R. P. Donovan, Eds.). Englewood Cliffs, N.J.: Prentice-Hall, 1967, pp. 41, 49.

14. Reinhard, D. K. *Introduction to Integrated Circuit Engineering.* Boston: Houghton Mifflin Company, 1987. Some basic laboratory information has been provided.

15. Maly, W. *Atlas of IC Technologies.* Menlo Park, Calif.: Benjamin/Cummings, 1987. Provides qualitative cross-sectional views of many bipolar and MOS technologies.

16. Burns, Stanley G., Howard R. Shanks, Alan P. Constant, Carl Gruber, Dave Schmidt, Allen Landin, Casey Thielen, Florence Olympic, Tracy Schumacher, and Jeromeo Cobbs. *Design and Fabrication of α-Si:H Based EEPROM Cells, Proceedings of the 1994 Electrochemical Society,* pp. 711–712, Oct. 9–14, 1994.

17. Constant, Alan P., Stanley G. Burns, Howard R. Shanks, Carl Gruber, Allen Landin, David Schmidt, Casey Thielen, Florence Olympic, Tracy Schumacher, and Jeromeo Cobbs. *Development of Thin-Film Transistor Based Circuits on Flexible Polyimide Substrates, Proceedings of the 1994 Electrochemical Society,* pp. 715–716, Oct. 9–14, 1994.

SUGGESTED READINGS

You should be aware that advances in integrated-circuit and device engineering are rapid. An easy-to-read overview of current advances is typically available in the trade literature. More specifically, you are encouraged to scan back and current issues of *Electronics, Electronic Design, EDN, IEEE Spectrum, IEEE Transactions, IEEE Proceedings,* and *Solid State Technology.* Many large national and international conferences also focus semiconductor device design and processing and you are encouraged to obtain their published abstracts and proceedings. In addition, the following is a partial list of other texts focusing on Chapter 18 topics.

Colclasser, R. A. *Microelectronics Processing and Device Design.* New York: John Wiley & Sons, 1980. An older senior-level text.

Ghandhi, S. K. *VLSI Fabrication Principles—Silicon and Gallium Arsenide,* 2nd ed. New York: John Wiley & Sons, 1994. Advanced, current text. Good chapter references.

Glaser, A. B., and G. E. Subak-Sharpe. *Integrated Circuit Engineering.* Reading, Mass.: Addison-Wesley, 1979. General reference at the senior or graduate level. Broad treatment with some circuits for illustration of the technology. Fundamentals are good but it is an older text.

Wolf, S., and R. N. Tauber. *Silicon Processing for the VLSI Era, Vol. I: Process Technology.* Lattice Press, 1986.

APPENDIX A

LIST OF SYMBOLS

Å	angstrom, 10^{-8} cm = 10^{-10} m = 10^{-4} μm
A_{cm}, A_{dm}	voltage gain: common mode, differential mode
A_i	current amplification
A_j	area of junction
A_p	power gain of amplifier
A_{vi}, A_{vs}	voltage amplification: output/input, output/source
A_{vsm}	midband output/source voltage amplification
a	forward gain of feedback amplifier
α_F, α_R	common base dc current gain: forward, reverse
B	magnetic field intensity
BV_{CBO}	breakdown voltage, collector to base
BV_{CEO}	breakdown voltage, collector to emitter
BV_{GSS}	breakdown voltage: gate to source
β	common-emitter small-signal current gain, $\beta = h_{fe}$
β_F, β_R	common-emitter dc current gain: forward, reverse
°C	Celsius, temperature unit
C_0	transition capacitance of junction at zero bias
C_D	diffusion capacitance of junction
C_M	Miller capacitance
CMRR	common-mode rejection ratio
C_T	transition capacitance of junction
C_T	total base capacitance, including Miller effect
C_b	base diffusion capacitance
C_{cs}, C_{cs0}	collector-to-substrate capacitance: with bias, with zero bias
C_{gb}, C_{gb0}	gate to substrate: with bias, with zero bias
C_{gd}, C_{gd0}	gate-to-drain capacitance: with bias, with zero bias
C_{gs}, C_{gs0}	gate-to-source capacitance: with bias, with zero bias
C_{gss}, C_{gss0}	gate-to-substrate capacitance: with bias, with zero bias
C_{ib}, C_{je}	emitter-base transition capacitance, emitter-box depletion capacitance
C_{ob}	collector-base transition capacitance
C_{sb}, C_{sb0}	source to substrate: with bias, with zero bias
C_π	emitter-base capacitance
$C_\mu, C_{\mu 0}$	collector-base capacitance: with bias, with zero bias
C_w	stray wiring capacitance
c	velocity of light in a vacuum, 3×10^8 m/sec
D, D_n, D_p	diffusion coefficient, for electrons, for holes
E	electric field intensity as a vector
E	energy
E_A, E_D	energy level of acceptor states, of donor states
E_C, E_V	energy level at bottom of conduction band, at top of valence band
E_a	ionization energy
E_g	band-gap energy
eV	electron volt = 1.602×10^{-19} J
ϵ_0	permittivity of free space, 8.854×10^{-12} F/m = 8.854×10^{-14} F/m
ϵ_r	relative dielectric constant
η	emission coefficient in diode equation
η_C	collector efficiency
f	particle flux
f	frequency, Hz
f_C, f_L, f_T	capture frequency, lock frequency, unity-gain frequency
f_o	free-run frequency or frequency of operation

APPENDIX A

f	feedback network gain
G	electron-hole generation rate
GBW	gain-bandwidth product
g_{ij}	g parameters of a two-port network
g_m, g_{m0}	transconductance: with bias, with zero bias
g_o	output conductance of a two-port network
$H(j\omega)$	voltage transfer function of a two-port network, $\lvert H(j\omega)\rvert e^{j\theta(j\omega)}$
HD	harmonic distortion
h	Planck's constant $= 6.626 \times 10^{-34}$ J-sec
h_{ij}	h parameters of a two-port network
h_{FE}	dc current gain, common emitter
h_{fe}	small-signal short-circuit gain, common emitter
h_{ie}	input resistance, common emitter
h_{oe}	output admittance, common emitter
h_{re}	voltage feedback ratio, common emitter
I_{CBO}	dc collector leakage current with emitter open
I_{CEO}	dc collector leakage current with base open
I_{CO}	collector saturation current, emitter open
I_{CS}	collector current in a saturated circuit
I_{DSS}	drain-source saturation current
I_{EO}	emitter saturation current with collector open
I_{GSS}	gate reverse-saturation current (JFET), gate leakage current (MOS)
I_{OS}	input offset current for operational amplifier
I_S	diode reverse-saturation current
i_B, i_C, i_E	total current into BJT terminal: base, collector, emitter
i_D	total current into diode anode
i_D, i_G, i_S	total current into FET terminal: drain, gate, source
i_Z	total avalanche diode breakdown current
J_n, J_p	current density: electron, hole
K	Kelvin, absolute temperature unit
K	proportionality constant for a multiplier
K_0, K_v	VCO gain, loop bandwidth
k	Boltzmann's constant $= 1.381 \times 10^{-23}$ J/K
k_D, k_L, k_n, k_p	MOSFET current proportionality constant, A/V^2: driver, load, n driver, p load
L	inductance
L_n, L_p	diffusion length for electrons, holes
λ	wavelength
λ	slope of drain characteristic/I_{DSS} at $V_{GS} = 0$, V^{-1}
M	mutual inductance
m_e	mass of electron $= 9.1 \times 10^{-31}$ kg
μ, μ_n, μ_p	mobility: charged particle, electron, hole
μm	micron, one micrometer $= 10^{-4}$ cm $= 10^{-6}$ m $= 10^4$ Å
N	number of charge carriers or carrier concentration (number/cm^3)
N_A, N_D	acceptor concentration, donor concentration (number/cm^3)
NM_H, NM_L	noise margin high, noise margin low
n	avalanche multiplication constant
n, n_o	charged-particle concentration, surface concentration
n, n_i	electron concentration, intrinsic electron concentration
n_n, n_p	electron concentration: in n region, in p region
n_{no}, n_{po}	equilibrium electron concentration: in n region, in p region

n^+	heavily doped n region
Ω	ohm
ω, ω_o	angular frequency, fundamental angular frequency
ω_β	beta cutoff angular frequency
$\omega_c, \omega_H, \omega_L$	corner frequency, high-frequency corner, low-frequency corner
ω_T	unity-gain or transition frequency
ω_i, ω_{VCO}	VCO input frequency, VCO output frequency
ω_n	natural damped frequency
P_D, P_J, P_L	power dissipation, junction power, load power
p^+	heavily doped p region
p, p_i	hole concentration, intrinsic hole concentration
p_n, p_p	hole concentration in n region, in p region
p_{no}, p_{po}	equilibrium hole concentration in n region, in p region
ϕ_J	junction potential
Q_D	stored charge of excess minority carriers
Q_n, Q_p	stored charge in n region, in p region
q	magnitude of electronic charge $= 1.6 \times 10^{-19}$ C
R	electron-hole recombination rate
R	ripple
R	electrical resistance, Ω
R_F	forward resistance in diode model
R_i	input resistance of a two-port network
R_{ic}, R_{id}	input resistance: common mode, differential mode
R_o	output resistance of a two-port network
$R_\square$	sheet resistance, $\Omega/\square$
r_{BE}	dynamic base resistance
$r_{DS(ON)}$	ON drain resistance
r_e	dynamic emitter resistance
r_b	base-spreading resistance
r_d	dynamic diode resistance
r_d	small-signal drain output resistance
r_o	collector output resistance
r_π	base-emitter resistance, hybrid-π model
r_μ	collector-base feedback resistance, hybrid-π model
ρ	electrical resistivity (Ω-cm)
ρ	charge density
$S_i, S_\epsilon, S_o, S_f$	signals: input, error, output, feedback
s	complex frequency
σ	electrical conductivity $(\Omega\text{-cm})^{-1}$
σ	base overdrive factor, $I_C/\beta_F I_B$
T	open-loop gain
T	temperature
T_A, T_C, T_J, T_S	temperature: ambient, case, junction, heat sink
TC	temperature coefficient
t_A, t_W	access time, write pulse time
t_{CO}, t_{CX}	chip select-to-valid output time, CS-to-active output time
t_{DW}, t_{DH}	data setup time, data hold time
t_{OH}, t_{OTD}	output hold time, deselect-to-high Z output time
t_{RC}, t_{WC}	read cycle time, write cycle time
t_c	average time between collisions
t_d	base turn-on delay time

APPENDIX A

t_p, t_{pd}	average gate propagation delay
t_{pHL}, t_{pLH}	propagation delay: high to low, low to high
t_r	collector current rise time
t_{rr}	diode reverse recovery time
t_s	storage time for excess carriers
$\theta(j\omega)$	angle of voltage transfer function
$\theta_{JC}, \theta_{CS}, \theta_{SA}$	thermal resistance: junction to case, case to heat sink, heat sink to ambient
τ_F	base transit time
τ_n, τ_p	electron lifetime, hole lifetime
V_0	junction contact potential
V_A	Early voltage
V_{IH}	minimum gate high-level input voltage
V_{IL}	maximum gate low-level input voltage
V_{OH}	minimum gate high-level output voltage
V_{OL}	maximum gate low-level output voltage
V_{OS}	input offset voltage for operational amplifier
V_P	peak voltage
V_P	pinch-off voltage of JFET
V_{SD}	forward voltage drop of Schottky diode
V_T	thermal voltage, kT/q (do not confuse with V_t)
V_t	threshold voltage of MOSFET (do not confuse with V_T)
V_{tD}, V_{tL}	threshold voltage: driver FET, load FET
V_{tn}, V_{tp}	threshold voltage: n FET, p FET
V_o or V_F	offset voltage of diode model
V_Z	magnitude of avalanche diode breakdown voltage
v_D	external diode voltage
v_J	junction potential, biased
v_{OD}	differential output voltage
v_R	external diode reverse voltage
v	velocity
v_d	drift velocity
v_d, v_{max}	average electron velocity, maximum electron velocity
v_n, v_p	electron velocity, hole velocity
v_s	scatter-limited velocity
w_n, w_p	width of space-charge region on n side, on p side
x_j	junction depth
y_{ij}	admittance parameters of a two-port network
y_{fs}	mutual conductance with zero bias, same as g_{m0}
y_{os}	small-signal drain output conductance
z_{ij}	impedance parameters of a two-port network
ζ	damping coefficient

APPENDIX B

PHYSICAL CONSTANTS AND CONVERSION FACTORS

TABLE B1 Physical Constants

Quantity	Symbol	Value
Boltzmann's constant	k	1.38066×10^{-23} J/K
Elementary charge	q	1.60218×10^{-19} C
Electron rest mass	m_e	9.1095×10^{-31} kg
Electron volt	eV	1 eV = 1.60218×10^{-19} J
Permittivity of free space	ϵ_0	8.85418×10^{-14} F/cm
		8.854×10^{-12} F/m
Planck's constant	h	6.62617×10^{-34} J-sec
Speed of light in vacuum	c	3.0×10^{10} cm/sec = 3×10^8 m/sec
Thermal voltage at 300 K	kT/q	0.0259 V

TABLE B2 Conversion Factors and Prefixes

Unit	Symbol	Conversion
Angstrom unit	Å	1Å = 10^{-10} m = 10^{-8} cm = 10^{-4} μm
Angular frequency	ω	$\omega = 2\pi f$ rps (radians/sec)
Frequency	f	$f = \omega/2\pi$ Hz (cycles/sec)
Micron, micrometer	μm	10^{-6} m = 10^{-4} cm = 10^{-3} mm
Mil	mil	10^{-3} in. = 25.4 μm
Temperature, Celsius	°C	°C = K − 273.15°
femto	f	$\times 10^{-15}$
pico	p	$\times 10^{-12}$
nano	n	$\times 10^{-9}$
micro	μ	$\times 10^{-6}$
milli	m	$\times 10^{-3}$
kilo	k	$\times 10^{3}$
mega	M	$\times 10^{6}$
giga	G	$\times 10^{9}$
tera	T	$\times 10^{12}$

APPENDIX C

NUMBER SYSTEMS AND BOOLEAN ALGEBRA

APPENDIX C

C.1 NUMBER SYSTEMS

The *base* or *radix* of a number system is the number of symbols used. A number system can be formed from any base greater than one. As the most familiar example, the *decimal* number system uses 10 symbols and has base 10. A number in the system is represented by the coefficients of a series expansion in powers of the base.

$$1029_{10} = (1 \times 10^3) + (0 \times 10^2) + (2 \times 10^1) + (9 \times 10^0) = 1000 + 20 + 9.$$

By convention, the one's (10^0) coefficient is positioned at the right, the ten's (10^1) coefficient is positioned next, and so on.

The *binary* number system has two symbols; hence, the base is two.

$$101110_2 = (1 \times 2^5) + (0 \times 2^4) + (1 \times 2^3) + (1 \times 2^2) + (1 \times 2^1) + (0 \times 2^0).$$

We see that the base of the number may be shown as a subscript to avoid confusion so that the binary number is not mistaken for a number in some other system that uses 0 and 1 in its set of symbols.

The binary number in the preceding example may be converted to a decimal number—for instance, by performing the series addition after representing each term with its decimal equivalent: $2^5 = 32_{10}$, $2^3 = 8_{10}$. Then

$$101110_2 = (1 \times 32_{10}) + (0 \times 16_{10}) + (1 \times 8_{10}) + (1 \times 4_{10}) + (1 \times 2_{10}) + (0 \times 1) = 46_{10}.$$

To convert from decimal to binary, it is often more convenient to perform a series of divisions of the decimal number by the binary base (2). For instance, 501_{10} may be converted to binary as follows:

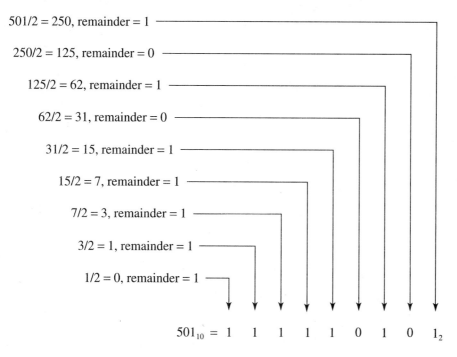

The *hexadecimal* number system (base 16) is useful as a shorthand way of expressing binary numbers. This is because a 4-bit binary number can be represented by one hexadecimal symbol. Recall that a 4-bit binary number has $2^4 = 16_{10}$ possible states, and this is

TABLE C1

Decimal	Binary	Hexadecimal	BCD
0	0000	0	0000
1	0001	1	0001
2	0010	2	0010
3	0011	3	0011
4	0100	4	0100
5	0101	5	0101
6	0110	6	0110
7	0111	7	0111
8	1000	8	1000
9	1001	9	1001
10	1010	A	
11	1011	B	
12	1100	C	
13	1101	D	
14	1110	E	
15	1111	F	

the number of symbols in the hexadecimal number system. A binary number can be expressed as a hexadecimal number by grouping the binary symbols (bits) in groups of four from the right and then replacing each group by its hexadecimal equivalent as shown in Table C1. For the previous example,

$$501_{10} = 111110101_2 = 0001\ 1111\ 0101 = 1\ F\ 5_{16} \text{ or } 1F5_H.$$

$$1F5_H = (\underline{1} \times 16^2) + (\underline{15} \times 16)^1 + (\underline{5} \times 16^0) = 256 + 240 + 5 = 501_{10}.$$

Also by division,

$$501/16 = 31, \text{ remainder} = 5$$
$$31/16 = 1, \text{ remainder} = 15$$
$$1/16 = 0, \text{ remainder} = 1$$

$$501_{10} = 1 \quad F \quad 5 = 1F5_H$$

The *binary-coded-decimal* (BCD) scheme is used to preserve the decimal digit's identity while representing it with a binary code. This is useful when decimal arithmetic is desired or decimal digits are to be displayed. The binary words 1010 to 1111 are not defined in this code. Table C1 shows equivalent numbers in these systems for comparison.

C.2 BASIC FUNCTIONS OF BINARY VARIABLES

All logic operations, no matter how complex, are composed of a very few basic functions.

The Buffer

A binary circuit that has one input and an output with the same state as the input is called a ***buffer***. Although it may not appear to be a particularly useful circuit, it does, as the name implies, provide isolation between the output and the input and is widely used. If A is the input and X is the output, the equation is

$$X = A. \tag{C.1}$$

■ By equation, we mean the Boolean equation/expression.

Figure C.1 shows the two possible results summarized in a compact form known as a ***truth table***, as well as the circuit symbol for the buffer.

A	X
0	0
1	1

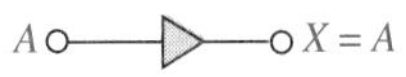

(a) $A \circ\!\!-\!\!\triangleright\!\!-\!\!\circ X = A$ (b)

■ **FIGURE C.1** Buffer. **(a)** Truth table. **(b)** Circuit symbol.

The NOT Gate, or Inverter

A binary circuit that has one input and an output with the opposite state from the input is called a ***NOT gate***, or more commonly, an ***inverter***. Let the input be A and the output be X. The equation is

$$X = \overline{A}. \tag{C.2}$$

The bar over the A signifies inversion and the expression is read "X equals NOT A" or "X equals the ***complement*** of A." Figure C.2 shows the truth table for this equation, as well as the circuit symbol for an inverter. The little circle in the logic symbol signifies inversion.

A	X
0	1
1	0

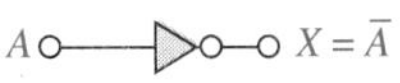

(a) $A \circ\!\!-\!\!\triangleright\!\!\circ\!\!-\!\!\circ X = \overline{A}$ (b)

■ **FIGURE C.2** Inverter. **(a)** Truth table. **(b)** Circuit symbol.

The AND Gate

A gate that has two or more inputs whose output assumes the ONE state if, and only if, *all* inputs are ONEs, is called an ***AND gate***. For two inputs, A and B, the equation is

$$X = AB = A \cdot B = A \times B \tag{C.3}$$

and is referred to as "A AND B." Usually the center dot or multiplication sign is omitted. The truth table for this two-input AND gate is shown in Fig. C.3. One may say that the output is true if A is true AND B is true. The figure also shows the logic symbol for a two-input AND.

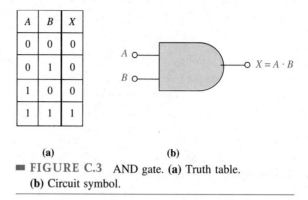

(a)   **(b)**

■ **FIGURE C.3** AND gate. (a) Truth table. (b) Circuit symbol.

The NAND Gate

A circuit that has an output that is the inverted result of the AND operation is called a **NAND gate**. The equation is

$$X = \overline{ABC} \qquad (C.4)$$

for a three-input NAND. The truth table is shown in Fig. C.4 with a corresponding circuit symbol. Note that the small circle indicates inversion of the AND function.

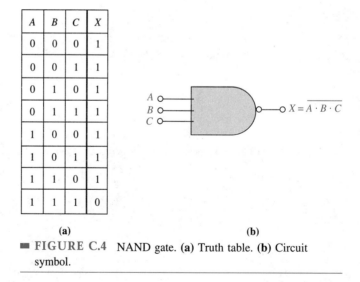

(a)   **(b)**

■ **FIGURE C.4** NAND gate. (a) Truth table. (b) Circuit symbol.

The OR Gate

A gate that has two or more inputs whose output is ONE whenever one or more of the inputs are ONEs is an **OR gate**. Such a gate with three inputs is described by

$$X = A + B + C. \qquad (C.5)$$

The truth table for this is shown in Fig. C.5. We note that the output is true if A is true OR B is true OR C is true. The figure also shows the circuit symbol for a three-input OR.

APPENDIX C

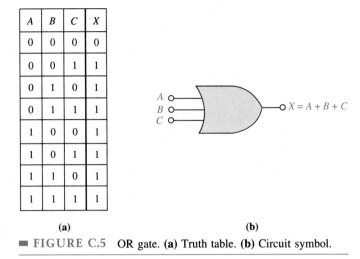

(a)   **(b)**

■ FIGURE C.5 OR gate. **(a)** Truth table. **(b)** Circuit symbol.

The NOR Gate

A circuit that has an output that is the complement of the OR operation is a *NOR gate*. The expression for a two-input NOR is

$$X = \overline{A + B}. \tag{C.6}$$

The resulting truth table is shown in Fig. C.6, along with its circuit symbol. Again we see the small circle, indicating inversion of the OR function, X is true when (A OR B) is *not* true.

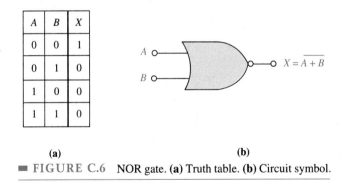

(a)   **(b)**

■ FIGURE C.6 NOR gate. **(a)** Truth table. **(b)** Circuit symbol.

The EXCLUSIVE-OR Gate

A two-input gate whose output is a ONE whenever *either* input is a ONE, but not when both inputs are ONEs, is termed an *EXCLUSIVE-OR gate*. The equation is written

$$X = A \oplus B. \tag{C.7}$$

The truth table is shown in Fig. C.7, together with the circuit symbol. This operation is related to binary addition, where

$$0 + 0 = 0,$$
$$0 + 1 = 1,$$
$$1 + 0 = 1,$$
$$1 + 1 = 10 \text{ (binary representation of decimal 2)},$$

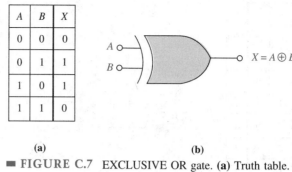

A	B	X
0	0	0
0	1	1
1	0	1
1	1	0

(a)   (b)

FIGURE C.7 EXCLUSIVE OR gate. (a) Truth table. (b) Circuit symbol.

or ZERO is the ONE's position and a carry into the 2's position. Without the carry, we have what is called *modulo 2 addition*, or the EXCLUSIVE OR.

DeMorgan's Law

The Boolean expression

$$\overline{AB} = \overline{A} + \overline{B} \tag{C.8}$$

(known as DeMorgan's Law) can be verified by comparing the relevant columns of the truth table shown in Fig. C.8. By complementing both sides of Eq. (C.8), we obtain

$$\overline{\overline{AB}} = AB = \overline{\overline{A} + \overline{B}}. \tag{C.9}$$

The importance of Eq. (C.9) is that it reveals that it is possible to achieve the AND function with inverters and an OR gate. Similarly, by substituting $\overline{C} = A$ and $\overline{D} = B$ in Eq. (C.8), we find that

$$\overline{\overline{CD}} = C + D, \tag{C.10}$$

which demonstrates that the OR operation can be accomplished by means of inverters and a NAND gate. You are encouraged to show that all of the preceding binary functions can be realized by means of either NAND gates or NOR gates.

This short description of basic logical operations will allow you to study the modern integrated-circuit realizations of these gates with an understanding of their functions. If you have had no prior exposure to the subject of digital logic theory, you can benefit from some additional reading.

A	B	$\overline{A}$	$\overline{B}$	$\overline{A} \cdot \overline{B}$	$\overline{A}+\overline{B}$
0	0	1	1	1	1
0	1	1	0	1	1
1	0	0	1	1	1
1	1	0	0	0	0

FIGURE C.8 Truth table to demonstrate $\overline{AB} = \overline{A} + \overline{B}$.

APPENDIX D

INTRODUCTION TO SPICE

This appendix is used to support the syntax statements used throughout the text where **SPICE** has been employed as an analysis or design tool. An electronic circuit does not have to be very complex or contain many elements before the manual computational effort becomes unwieldy. Indeed, for many of our earliest examples we were forced to use a number of device-model approximations and circuit simplifications. These circuit solutions are usually quite adequate for a first look; however, for a more detailed design and analysis *computer-aided design* (CAD) and *computer-aided engineering* (CAE) tools are used.

Computer-based workstations used in integrated-circuit layout and electronic circuit design provide or support a variety of proprietary or public domain software useful in CAD or CAE. There are two main categories of CAD and CAE tools. The first category contains software tools used to lay out the multiple layers of an integrated circuit for a given function using a particular set of technology-driven design rules. A discussion of these tools is beyond the scope of this text. The second category of CAD and CAE tools consists of software used to solve an array of linear algebraic or piecewise-linear differential equations generated by application of circuit loop and node analysis. These programs can accommodate more complex circuit device models than can practically be incorporated in a manual calculation.

Currently, one of the more widely used general-purpose circuit simulation programs for industrial and academic computer systems is called SPICE, <u>S</u>imulated <u>P</u>rogram, <u>I</u>ntegrated <u>C</u>ircuit <u>E</u>mphasis. SPICE was originally developed at the University of California, Berkeley, by Lawrence Nagel, with significant modifications by Ellis Cohen. SPICE has undergone many modifications over the years. SPICE 2G.6[1] from UC–Berkeley is occasionally used in this text to illustrate electronic circuit solutions of large systems. Often we use the compatible PSPICE®,[2] a very popular version developed by MicroSim Corporation[3] to run on DOS and MacIntosh personal computers. Commercial versions are also available from many vendors, with MicroSim, Silvaco, Hewlett Packard, Cadence Design Systems, Metasoftware, and Tektronix having a significant impact on the market.

This appendix includes a summary of the most common default representations for SPICE circuit models along with advanced device model listings. Throughout the text, these models have been introduced and applied in a variety of examples and problems.

Protocols and formats vary with computer systems and software packages, although the models are essentially the same. Applying the programs in this text will require a knowledge of the local computer platform and the SPICE software package resident on the platform. Each system may have different constraints with respect to the maximum number of nodes as well as output format.

Many versions of SPICE are linked to a ***schematic-capture*** program. That is, the SPICE line-by-line circuit description, called a ***netlist***, is generated by entering the schematic diagram. We chose not to use the schematic capture in this text, believing that it is important to be able to write the netlist directly.

You will have to obtain the mechanics of graphical output from the version of SPICE that you are using. For example, PSPICE® has a powerful graphical interface invoked by .PROBE®[3] that allows very versatile display of output data.

D.1 BASIC CIRCUIT MODELS FOR SPICE

SPICE can be used to simulate circuits containing resistors, capacitors, inductors, mutual inductors, independent and dependent voltage and current sources, and basic semiconductor devices. Separate SPICE models are used for the BJT, JFET, MOSFET, and diode. For example, the BJT model can include the ohmic resistances such as r_b, charge storage

APPENDIX D

effects, and a current-dependent output resistance r_o. The MOSFET models can include the effects of the scatter-limited velocity, charge-controlled capacitances, short-channel effects to the degree they are understood, and channel-length variations as a function of terminal voltages.

Initially, the user must identify each *node* of the circuit. A node is defined as the connection of two or more elements. The ground or reference node is **NODE 0**. The remaining nodes are numbered with integers >0. After labeling all nodes in a circuit, we are now in a position to provide the formalism for labeling each circuit element.

■ In the text, we have used **N** to show a node for SPICE examples.

There are two basic model format statements. The first is called the *element statement* and is used for all passive elements and generators. The second is called the *model statement* and is used for semiconductor devices along with an element statement. SPICE allows varying degrees of circuit element model complexity.

A statement may be continued by entering a "+" in column 1 of the following line; SPICE continues reading beginning with column 2. Either fixed or floating-point numbers may be used. Element statement scale factors, as illustrated in Table D.1, are used as part of the number.

TABLE D.1 Element Statement Scale Factors

Element Statement	Scale Factors
F	$10^{-15} = 1E-15$
P	$10^{-12} = 1E-12$
N	$10^{-9} = 1E-9$
U	$10^{-6} = 1E-6$
M	$10^{-3} = 1E-3$
K	$10^{3} = 1E3$
MEG	$10^{6} = 1E6$
G	$10^{9} = 1E9$

D.2 ELEMENT STATEMENTS

Element Statements for Resistors

The resistor element model and reference directions are shown in Fig. D.1. The statement format is given by

$$\text{RXXX NI NJ VALUE } [\text{TC} = \text{TC1}[\,,\text{TC2}]\,] \quad \text{(D.1)}$$

where **RXXX** is a unique resistor name or label; **NI** and **NJ** are the two element nodes; and **VALUE** is a nonzero resistance value in ohms. An element name or label can include up to eight alphanumeric characters. The **TC**, temperature coefficient, specification within the brackets is optional. If not specified, the default value is zero. When **TC1** and **TC2** are specified, the algorithm

$$\text{VALUE(T)} = \text{VALUE(TNOM)}[1 + \text{TC1}(T-\text{TNOM}) + \text{TC2}(T-\text{TNOM})^2] \quad \text{(D.2)}$$

is used. The default value for **TNOM**, the nominal temperature, is 27°C. For example, the

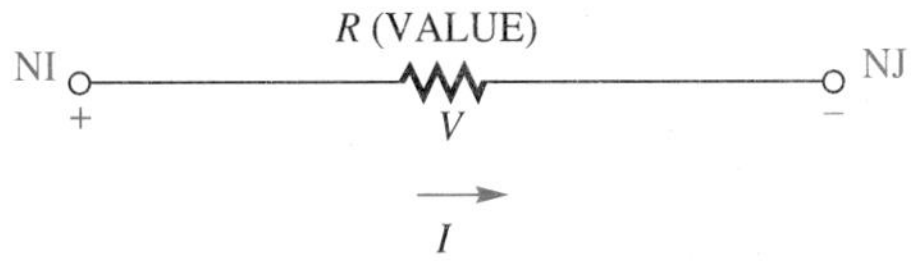

■ FIGURE D.1 Resistor element model.

element statement for a 10-kΩ resistor connected between nodes `I=3 J=6` with a zero default temperature coefficient is given by

$$\text{R1 3 6 10K} \tag{D.3}$$

and if `TC1=.003` and `TC2=.0001`, then the element statement is changed to

$$\text{R1 3 6 10K TC}=.003,.0001 \tag{D.4}$$

Element Statement for Capacitors

The capacitor element model and reference directions are shown in Fig. D.2. The element statement format for a constant value capacitor is given by

$$\text{CXXX NI NJ VALUE [IC}=\text{INCOND]} \tag{D.5}$$

where `INCOND` refers to an initial voltage condition, V_{IJ}. The default value for `IC` is zero. For example, the element statement for a 270-pF capacitor connected between nodes `I=5` and `J=0` with a 3-V initial condition is given by

$$\text{CTWO 5 0 270P IC}=3 \tag{D.6}$$

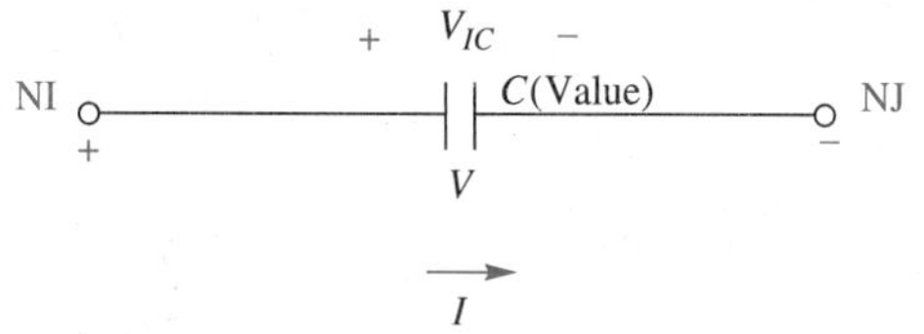

■ **FIGURE D.2** Capacitor element model.

Element Statement for Inductors

The inductor element model and reference directions are shown in Fig. D.3. The element statement format for a constant value inductor is given by

$$\text{LXXX NI NJ VALUE [IC}=\text{INCOND]} \tag{D.7}$$

where `INCOND` refers to an initial condition, I_{IJ}. The default value for `IC` is zero. For example, the element statement for a 1.5-mH inductor connected between nodes `I=2` and `J=4` with a 22.5-mA initial condition is given by

$$\text{LALPHA 2 4 1.5M IC}=.0225 \tag{D.8}$$

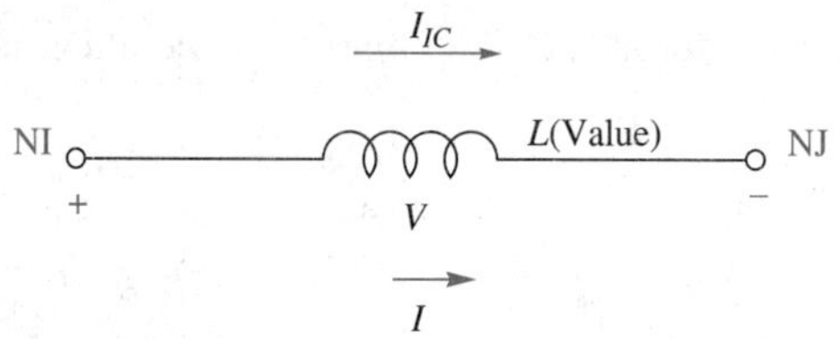

■ **FIGURE D.3** Inductor element model.

APPENDIX D

D.3 SOURCES

Independent Sources

The independent voltage and current-source elements are illustrated in Figs. D4(a) and (b), respectively. The general element statement forms are

VXXX NI NJ [DC DCVALUE][TRANSIENT VALUE] + [AC ACMAG [ACPHASE]] (D.9)

IXXX NI NJ [DC DCVALUE][TRANSIENT VALUE] + [AC ACMAG [ACPHASE]] (D.10)

For example, the element statement for a 12-V dc source connected between nodes I = 1 and J = 0 is given by

V1 1 0 DC 12V (D.11)

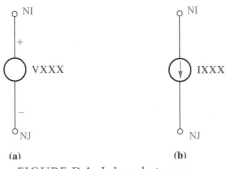

■ FIGURE D.4 Independent source element models: (a) Voltage source. (b) Current source.

The alphanumeric V label after the 12 is ignored in SPICE and is included for the convenience of the user. If ACPHASE is omitted following the keyword AC, zero is assumed. The element statement for a 15-mA peak sinusoidal current with a 30° phase reference connected between nodes I = 4 and J = 7 is given by

ITWO 4 7 AC 15M 30 (D.12)

There are five transient source functions, each beginning at $t = 0$. The element statements for each type follow:

Pulse:

VXXX N$^+$ N$^-$ PULSE(V_{INIT} V_{PULSE} T_{DELAY} T_{RISE} T_{FALL} T_{WIDTH} T_{PERIOD}) (D.13)

where $V_{INITIAL}$ and $V_{PULSE\ HEIGHT}$ must be specified in volts or amps. The default value of T_{DELAY} is 0 sec; the default values for T_{RISE} and T_{FALL} are TSTEP sec, and the default for $T_{PULSE\ WIDTH}$ and $T_{PULSE\ PERIOD}$ is TSTOP sec.

Sinusoidal:

VXXX N$^+$ N$^-$ SIN(V_{OFFSET} V_{AMPL} Freq T_{DELAY} DF PHASE) (D.14) where V_{OFFSET} and V_{AMPL} must be specified in volts or amps and `Freq` is in hertz (the default is `1/TSTOP`). The default T_{DELAY} is 0 sec; the default damping factor `DF` is 0 s^{-1}, and the default `PHASE` is 0 degrees.

Exponential pulse:

$$\text{VXXX N}^+ \text{ N}^- \text{ EXP}(V_{INIT} V_{PULSE} T_{D1} \tau_1 T_{D2} \tau_2) \tag{D.15}$$

where $V_{INITIAL}$ and V_{PULSE} must be specified in volts or amps. Rise delay time T_{D1} (when the pulse begins) has a default value of 0 sec, and rise time constant τ_1 has a default value of `TSTEP` sec. Fall delay time T_{D2} (when the pulse starts down) has a default value of $T_{D1} + $ `TSTEP` sec, and fall time constant τ_2 has a default value of `TSTEP` sec.

Piecewise-linear:

$$\text{VXXX N}^+ \text{ N}^- \text{ PWL}(T_1 V_1 T_2 V_2 \ldots) \tag{D.16}$$

Each time and voltage (or current) pair is a point on the graph of the waveform versus time. Between these coordinates, the function is piecewise-linear.

Single-frequency FM:

$$\text{VXXX N}^+ \text{ N}^- \text{ SFFM}(V_{OFFSET} V_{AMPL} F_C \text{ MDI } F_S) \tag{D.17}$$

where V_{OFFSET} and V_{AMPL} must be specified in volts or amps. Carrier frequency F_C and F_S are in hertz, with default value `1/TSTOP`. Modulation index `MDI` must be specified.

The use of transient and pulse sources is illustrated in the text examples.

Dependent Sources

From two-port analysis, we observe that there are four types of linear-dependent sources. These are illustrated in Fig. D.5. The element statements for each of these dependent sources follow:

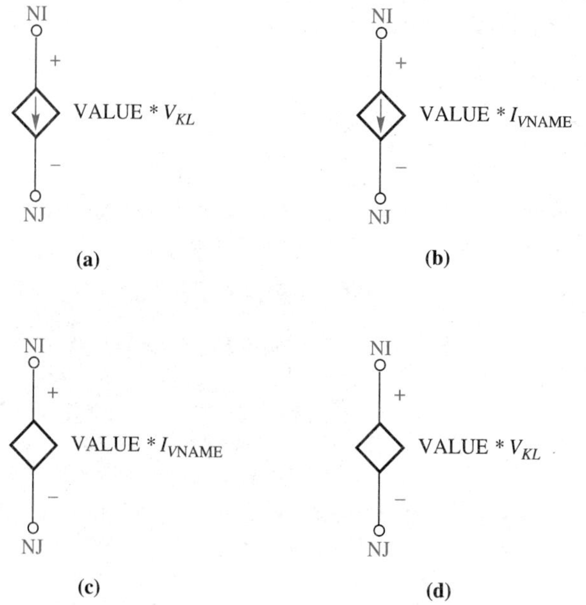

■ FIGURE D.5 Dependent source element models: (a) Voltage-controlled current source. (b) Current-controlled current source. (c) Current-controlled voltage source. (d) Voltage-controlled voltage source.

Voltage-controlled current source, Fig. D.5(a):

$$\text{GXXX NI NJ NK NL VALUE} \tag{D.18}$$

The positive and negative nodes are denoted by **NI** and **NJ**, respectively. Current direction is from **NI** to **NJ**. The nodes **NK** and **NL** are the positive and negative nodes of the controlling voltage. **VALUE** is expressed in siemens.

Current-controlled current source, Fig. D.5(b):

$$\text{FXXX NI NJ VNAME VALUE} \tag{D.19}$$

In some versions of SPICE, the reference branch with the controlling current must contain a voltage source, **VNAME**. If a voltage source does not exist, it is required to incorporate a dummy zero-volt independent source in that branch. The dummy voltage source functions as an ammeter. **VALUE** is the current gain of the current-controlled current source. For example, the element statement for the dependent source in Fig. D.6 is given by

$$\text{F1 2 0 VTEST 150} \tag{D.20}$$

■ Non-linear controlled sources can also be included where the controlling parameter is a multi-dimensional polynomial.[1] This application is beyond the scope of this text.

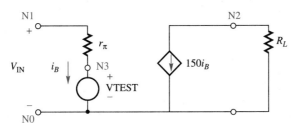

■ **FIGURE D.6** Current-controlled current source with **VTEST** as a dummy voltage generator.

Current-controlled voltage source, Fig. D.5(c):

$$\text{HXXX NI NJ VNAME VALUE} \tag{D.21}$$

As with the current-controlled current source, the reference current is obtained from the current through the **VNAME** voltage source. If **VNAME** does not exist, then a dummy zero-volt source must be added to the circuit. **VALUE** is the transresistance in ohms.

Voltage-controlled voltage source, Fig. D.5(d):

$$\text{EXXX NI NJ NK NL VALUE} \tag{D.22}$$

The positive and negative nodes of the dependent source are **NI** and **NJ**, respectively. The positive and negative nodes of the controlling voltage are given by **NK** and **NL**, respectively. **VALUE** is the voltage gain of the voltage-controlled voltage source.

D.4 ACTIVE DEVICE MODELS FOR SPICE

Device models for diodes, BJTs, and FETs are discussed in detail in the appropriate section of the text (Chapters 3, 4, and 5 respectively).

Diodes

Section 3.16 presents the SPICE diode model, with current and voltage conventions shown in Fig. 3.46, and with element and model statements provided in Eqs. (3.59) and (3.60).

Table 3.1 lists selected model parameters, which are described in the text. Table D.2 shows a more comprehensive list of diode parameters available for advanced simulation.[2]

■ Model parameters found in Tables D.2, D.3, D.4, and D.5 are specific to PSPICE™. Other syntax variables may be present in other SPICE versions. All SPICE models, however, are relatively consistent between versions.

TABLE D.2 Diode Model Parameters

Name	Parameter	Units	Default
IS	saturation current	A	$1.0E-14$
N	emission coefficient	—	1
ISR	recombination current parameter	A	0
NR	emission coefficient for ISR	—	2
IKF	high-injection knee current	A	infinite
BV	reverse breakdown voltage	V	infinite
IBV	current at breakdown voltage	A	$1.0E-3$
NBV	reverse breakdown ideality factor	—	1
IBVL	low-level reverse breakdown knee current	A	$1.0E-3$
NBVL	low-level reverse breakdown ideality factor	A	$1.0E-3$
RS	ohmic resistance	Ω	0
TT	transit-time	sec	0
CJO	zero-bias junction capacitance	F	0
VJ	junction potential	V	1
M	grading coefficient	—	0.5
FC	forward-bias depletion capacitance coefficient	—	0.5
EG	activation energy	eV	1.11
XTI	saturation current temperature exponent	—	3.0
TIKF	IKF temperature coefficient (linear)	C^{-1}	0
TBV1	BV temperature coefficient (linear)	C^{-1}	0
TBV2	BV temperature coefficient (quadratic)	C^{-2}	0
TRS1	RS temperature coefficient (linear)	C^{-1}	0
TRS2	RS temperature coefficient (quadratic)	C^{-2}	0
KF	flicker noise coefficient	—	0
AF	flicker noise exponent	—	1

Bipolar Junction Transistors

■ Observe that a fourth node—the substrate—must be included for BJTs and MOSFETs. Default is 0.

Section 4.12 presents the SPICE BJT model, with current and voltage conventions for both *npn* and *pnp* devices shown in Fig. 4.38, and with element and model statements provided in Eqs. (4.49) and (4.50). Table 4.1 lists selected model parameters, which are described in the text. Table D.3 shows a more comprehensive list of BJT parameters available for advanced simulation.[2] Section 9.1 shows how junction capacitances and transit time are modeled in SPICE.

MOSFETs

Section 5.5 presents a SPICE MOSFET model, with current and voltage conventions for both *n*- and *p*-channel devices shown in Fig. 5.22, and with element and model statements provided in Eqs. (5.20) and (5.22). Table 5.1 lists selected LEVEL 1 model parameters, which are described in the text. The MOSFET high-frequency parameters are discussed in Section 9.1. The more sophisticated higher level MOSFET models are not treated in this text, but are discussed in the references. Table D.4 shows a more comprehensive list of MOSFET parameters available for advanced simulation.[2]

TABLE D.3 BJT Parameters

Name	Parameter	Unit	Default
IS	transport saturation current	A	1.0E−16
BF	ideal maximum forward beta	—	100
NF	forward current emission coefficient	—	1.0
VAF	forward Early voltage	V	infinite
IKF	corner for forward beta high current roll-off	A	infinite
ISE	BE leakage saturation current	A	0.10E−13
NE	BE leakage emission coefficient	—	1.5
BR	ideal maximum reverse beta	—	1
NR	reverse current emission coefficient	—	1
VAR	reverse Early voltage	V	infinite
IKR	corner for reverse beta high current roll-off	A	infinite
ISC	BC leakage saturation current	A	0
ISC	BC leakage saturation current	A	0
NC	BC leakage emission coefficient	—	21.5
NK	high current roll-off coefficient	—	0.5
ISS	substrate p-n saturation current	A	0
NS	substrate p-n emission coefficient	—	1
RE	emitter resistance	Ω	0
RB	zero-bias base resistance	Ω	0
RBM	minimum base resistance	Ω	RB
IRB	current where RB falls halfway to RBM	A	infinite
RC	collector resistance	Ω	0
CJE	BE zero-bias depletion capacitance	F	0
VJE	BE built-in potential	V	0.75
MJE	BE junction exponential factor	—	0.33
CJC	BC zero-bias depletion capacitance	F	0
VJC	BC built-in potential	V	0.75
MJC	BC junction exponential factor	—	0.33
XCJC	fraction of BC capacitance connected to base	—	1
CJS	zero-bias collector-substrate capacitance	F	0
VJS	substrate junction built-in potential	V	0.75
MJS	substrate junction exponential factor	—	0
FC	coefficient, forward-bias depletion capacitance	—	0.5
TF	ideal forward transit time	sec	0
XTF	coefficient for bias dependence of TF	—	0
VTF	voltage describing VBC dependence of TF	V	infinite
ITF	TF dependency on IC	A	0
PTF	excess phase at freq = 1.0/(TF*2PI) Hz	deg	0
TR	ideal reverse transit time	sec	0
QCO	epitaxial region charge factor	C	0
RCO	epitaxial region resistance	Ω	0
VO	carrier mobility knee voltage	V	10
GAMMA	epitaxial region doping factor	—	1E−11
EG	energy gap for temperature effect on IS	eV	1.11
XTB	forward and reverse beta temperature coefficient	—	0
XTI	temperature exponent for effect on IS	—	3
TRE1	RE temperature coefficient (linear)	C^{-1}	0
TRE2	RE temperature coefficient (quadratic)	C^{-2}	0
TRB1	RB temperature coefficient (linear)	C^{-1}	0
TRB2	RB temperature coefficient (quadratic)	C^{-2}	0
TRM1	RBM temperature coefficient (linear)	C^{-1}	0
TRM2	RBM temperature coefficient (quadratic)	C^{-2}	0
TRC1	RC temperature coefficient (linear)	C^{-1}	0
TRC2	RC temperature coefficient (quadratic)	C^{-2}	0
KF	flicker noise coefficient	—	0
AF	flicker noise exponent	—	1

TABLE D.4 MOSFET Parameters

Name	Parameter	Units	Default
LEVEL	model index	—	1
VTO	zero-bias threshold voltage	V	0.0
KP	transconductance parameter	A/V^2	$2.0E-5$
GAMMA	bulk threshold parameter	$V^{0.5}$	0.0
PHI	surface potential	V	0.6
LAMBDA	channel-length modulation (MOS1 and MOS2 only)	1/V	0.0
RD	drain ohmic resistance	Ω	0.0
RS	source ohmic resistance	Ω	0.0
RG	gate ohmic resistance	Ω	0.0
RB	bulk ohmic resistance	Ω	0.0
RDS	drain-source shunt resistance	Ω	infinite
CBD	zero-bias BD junction capacitance	F	0.0
CBS	zero-bias BS junction capacitance	F	0.0
IS	bulk junction saturation current	A	$1.0E-14$
PB	bulk junction potential	V	0.8
PBSW	bulk p-n sidewall potential	V	PB
TT	bulk p-n transit time	sec	0.0
CGSO	gate-source overlap cap. per meter channel width	F/m	0.0
CGDO	gate-drain overlap cap. per meter channel width	F/m	0.0
CGBO	gate-bulk overlap cap. per meter channel length	F/m	0.0
RSH	drain and source diffusion sheet resistance	$\Omega/\square$	0.0
CJ	zero-bias bulk junction bottom cap. per square meter of junction area	F/m^2	0.0
MJ	bulk junction bottom grading coefficient	—	0.5
CJSW	zero-bias bulk junction sidewall capacitance per meter of junction perimeter	F/m	0.0
JSSW	bulk p-n saturation sidewall current/length	A/m	0.0
N	bulk p-n emission coefficient	—	1
MJSW	bulk junction sidewall grading coefficient	—	0.3
JS	bulk junction saturation current per square meter of junction area	A/m^2	$1.0E-8$
TOX	oxide thickness	m	$1.0E-17$
NSUB	substrate doping	$1/cm^3$	0.0
NSS	surface state density	$1/cm^2$	0.0
NFS	fast surface state density	$1/cm^2$	0.0
TPG	type of gate material: +1 opp. to substrate −1 same as substrate 0 Al gate	—	1.0
XJ	metallurgical junction depth	m	0.0
LD	lateral diffusion	m	0.0
UO	surface mobility	$cm^2/V\text{-}s$	600
UCRIT	critical field for mobility degradation (MOS2 only)	V/cm	$1.0E4$
UEXP	critical field exponent in mobility degradation (MOS2 only)	—	0.0
UTRA	transverse field coefficient (mobility) (deleted for MOS2)	—	0.0
VMAX	maximum drift velocity of carriers	m/s	0.0
NEFF	total channel charge (fixed and mobile) coefficient (MOS2 only)	—	1.0
XQC	thin-oxide capacitance model flag and coefficient of channel charge share attributed to drain (0–0.5)	—	1.0
KF	flicker noise coefficient	—	0.0
AF	flicker noise exponent	—	1.0
FC	coefficient for forward-bias depletion capacitance formula	—	0.5
DELTA	width effect on threshold voltage (MOS2 and MOS3)	—	0.0
THETA	mobility modulation (MOS3 only)	1/V	0.0
ETA	static feedback (MOS3 only)	—	0.0
KAPPA	saturation field factor (MOS3 only)	—	0.2

JFETs

Section 5.10 presents a SPICE JFET model, with current and voltage conventions for both *n*- and *p*-channel devices shown in Fig. 5.41, with an element statement in Eq. (5.61) and with an example model statement in Eq. (5.63). Table 5.2 lists selected JFET model parameters, which are described in the text. The JFET high-frequency parameters are also discussed in Section 9.1, showing how the junction capacitances are modeled in SPICE. Table D.5 shows a more comprehensive list of JFET parameters available for advanced simulation.[2]

TABLE D.5 JFET Parameters

Name	Parameter	Unit	Default
VTO	threshold voltage	V	−2.0
BETA	transconductance parameter	A/V^2	1.0E−4
LAMBDA	channel-length modulation parameter	1/V	0
IS	gate junction saturation current	A	1.0E−14
N	gate junction emission coefficient	—	1
ISR	gate junction recombination current parameter	A	0
NR	emission coefficient for ISR	—	2
ALPHA	ionization coefficient	1/V	0
VK	ionization knee voltage	V	0
RD	drain ohmic resistance	Ω	0
RS	source ohmic resistance	Ω	0
CGD	zero-bias GD junction capacitance	F	0
CGS	zero-bias GS junction capacitance	F	0
M	gate junction grading coefficient	—	0.5
PB	gate junction potential	V	1
FC	forward-bias depletion capacitance coefficient	—	0.5
VTOTC	VTO temperature coefficient	V/C	0
BETATCE	BETA exponential temperature coefficient	%/C	0
XTI	IS temperature coefficient	—	3
KF	flicker noise coefficient	—	0
AF	flicker noise exponent	—	1

Control Commands for Analysis

Each program begins with a `TITLE` statement, and ends with an `.END` statement. Control statements are used to prompt SPICE to perform the desired analysis, and provide the desired output. These statements are briefly explained:

dc analysis:

$$.\text{DC } S_1 \, V_{START} \, V_{STOP} \, \text{INC} \tag{D.23}$$

is a command that steps the dc voltage (or current) source named `S`$_1$ from V_{START} to V_{STOP} in steps of `INC`, all specified in volts or amps. Capacitors are treated as open circuits, inductors as short circuits, and the results may be displayed by a `.PRINT` or a `.PLOT` statement.

ac analysis:

$$.\text{AC XXX NP } F_{START} \, F_{STOP} \tag{D.24}$$

is a command that will sweep all ac sources in the circuit from frequency F_{START} to frequency F_{STOP} using a scale `XXX NP` (where `LIN NP` $\Rightarrow$ linear using `NP` points, `DEC NP` $\Rightarrow$ `NP` points/decade, and `OCT NP` $\Rightarrow$ `NP` (points/octave).

Transient analysis:

$$\texttt{.TRAN TSTEP TSTOP TSTART TMAX}$$

is a command exercising one of the transient type of sources mentioned previously. `TSTEP` specifies the interval for plotting and printing the results of this analysis. `TSTOP` specifies the time of the last analysis. The default value of `TSTART` is zero. If `TSTART` is specified greater than 0, analysis begins at zero, but the output is printed or plotted starting at time `TSTART`. `TMAX` is the largest computing time step allowed. If not specified, the default will be the smaller of `TSTEP` or `(TSTART-TSTOP)/50`.

SPICE always performs a `DC` analysis to find the operating point when an `AC` or transient analysis is done, but to print the operating point parameters of all of the nonlinear or active devices in the circuit, use this command:

Operating point analysis:

■ Some SPICE versions print the operating point (.OP) information as a default.

$$\texttt{.OP} \tag{D.25}$$

Transfer function analysis:

$$\texttt{.TF } V_{OUT} \; V_{IN} \tag{D.26}$$

commands a small-signal dc analysis between the input source V_{IN} and the output at V_{OUT}, printing V_{OUT}/V_{IN}, the input resistance at V_{IN}, and the output resistance at V_{OUT}. V_{OUT} and V_{IN} may be any combination of voltages or currents.

■ Note that the .TF analysis only functions for direct-coupled circuits.

Many more advanced features of the program have not been described here, such as noise analysis, distortion analysis, Fourier analysis, nonlinear dependent sources, subcircuit definition, and numerical solution control options. All of these features, plus details on those functions that were discussed, can be found in the references.

REFERENCES

1. Vladimirescu, A., Kaihe Zhang, A. R. Newton, D. O. Peterson, and A. Sangiovanni-Vincentelli. *SPICE Version 2G User's Guide.* Department of Electrical Engineering and Computer Science, University of California, Berkeley, 10 August 1981.

2. Banzhaf, W. *Computer-Aided Circuit Analysis Using PSPICE,*® 2nd ed. Englewood Cliffs, N.J.: Regents/Prentice-Hall, 1992.

3. MicroSim Corporation, 20 Fairbanks, Irvine, California 92718 USA. PSPICE and PROBE are registered trademarks of MicroSim Corporation.

APPENDIX E

TWO-PORT MODEL SUMMARY

Figure E.1 is a summary of the four topological variations for modeling two-port networks.

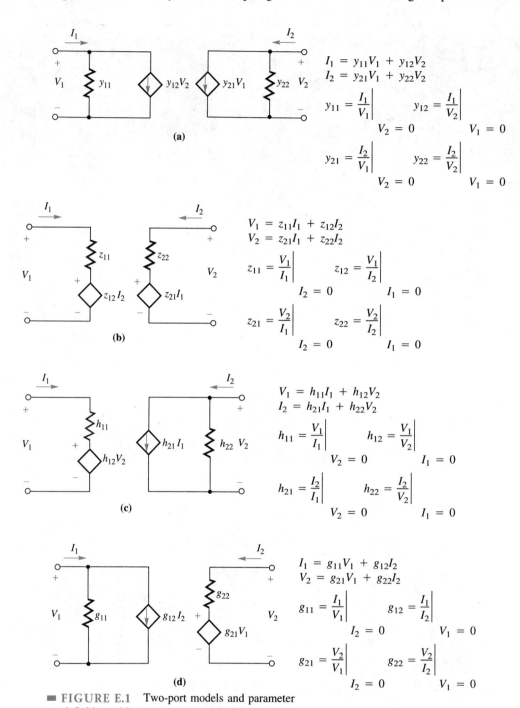

FIGURE E.1 Two-port models and parameter definitions. **(a)** y-parameter model. **(b)** z-parameter model. **(c)** h-parameter model. **(d)** g-parameter model.

APPENDIX F

ANSWERS TO SELECTED PROBLEMS

Chapter 1

1. (a) $150 \angle 30°$ V (b) $150 \angle -60°$ V (c) $279.8 \angle 30.36°$ 6. $2.718 \, I_o/V_R$ 9. 9.13 V, $60.87 \, \Omega$ 11. 4.67 V, $155.6 \, \Omega$
14. $v_o = -56.2 \cos \omega t$ mV 18. 6.32 W
20. $R_1 = 1$ kΩ and $R_2 = 31.6$ kΩ are 1% values, and $R_1 = 1.5$ kΩ and $R_2 = 47$ kΩ are 5% values; there are many others.
24. $R_3 = R_4$, $R_2 = 2R_4$, $R_1 = 4R_4$ 27. $v_o = -R_2/R_1 \, (1 + R_4/R_2 + R_4/R_3) v_s$ 30. $v_o = -R_2/KR_1 \times v_s/v_x$

Chapter 2

2. 6.24×10^{15} electrons/s 4. $\sigma = 5.8 \times 10^7 (\Omega\text{-m})^{-1}$, $\mu = 45$ cm^2/v-s, $v_d = 9.48 \times 10^{-6}$ m/s
6. (a) $1.14 \, \Omega$ (b) $0.98 \, \Omega$
11. Paschen > 10000 Å > 7200 Å, so infrared   Lyman < 1216 Å < 4000 Å, so ultraviolet
14. $\lambda = 1107$ nm = 11070 Å, infrared   17. blue 2.48 eV, red 1.77 eV   20. exponential term dominates
22. $0.0223 (\Omega\text{-cm})^{-1}$ from data of Table 2.2, $0.0212 (\Omega\text{-cm})^{-1}$ from Fig. 2.3
27. from Fig. 2.18(b), $N_D = 5 \times 10^{18}$ donors/cm^2, $N_A = 10^{19}$ acceptors/cm^2 30. $n_i \approx 0$
32. (a) $n = 8 \times 10^{17}$ electrons/cm^3, $p = 281$ holes/cm^3 (b) $\sigma = 38 (\Omega\text{-cm})^{-1}$ 35. $\sigma = 2.4 (\Omega\text{-cm})^{-1}$, $\rho = 0.42 \, \Omega\text{-cm}$
39. 10^{13} majority carriers/cm^3, 2.25×10^7 minority carriers/cm^3, $\rho_n = 430 \, \Omega\text{-cm}$, $\rho_p = 1300 \, \Omega\text{-cm}$
40. (a) 63 MΩ (b) $35 \, \Omega$ (c) $12 \, \Omega$
44. (300K): $n = 2.6 \times 10^{15}$ electrons/cm^3, $p = 8.65 \times 10^4$ holes/cm^3;
    (400K): $n = 2.6 \times 10^{15}$ electrons/cm^3, $p = 4.4 \times 10^9$ holes/cm^3
47. no   50. $n_i = 2.3 \times 10^{13}$, -somewhat less than typical doping levels
51. $n = 2.4 \times 10^{14}$ electrons/cm^3, $p = 9.4 \times 10^5$ holes/cm^3 53. $1.44 \, \mu s$

Chapter 3

1. $n_n = 10^{14}$, $p_n = 2.25 \times 10^6$, $p_p = 10^{17}$, $n_p = 2250$ 5. 296 mV   8. 25.5 mA, no   10. 10.9 A
12. Spectral density would not match film, but could use a blue filter.   15. $+1.4$ mV/°C at 9.8 V   17. 6.41 V
18. $I_{D1} = I_{D2} = -1 \, \mu A$, $V_{D1} = -99.964$ V, $V_{D2} = -36$ mV   22. $340 \, \Omega$ 25. $3.24 \, \Omega$
30. For 60 Hz: $v_s = 5.7$ V$_{peak}$, $C = 167{,}000 \, \mu F$ 31. 12 mA   34. (a) $0 \le I_L \le 95$ mA (b) 30 mA $\le I_L \le 125$ mA
36. Choose 1N5233: $R_x = 79.2 \, \Omega$; if $I_{z \, min} = 5$ mA, $I_{l \, max} = 78$ mA.   37. 10.7%
41. Use Fig. 3.34 with with $V_1 = 1.4$ V, $V_2 = -1.4$ V, $R_1 = 3.3$ kΩ
44. (a) 3 V (b) During negative half cycle, v_o changes from -8 V to -7.9 V.   46. very slight effect
53. (a) 5.78 V (b) 0 V (c) >2 V   55. 11.8 μm, 17.9 kV/cm   58. (a) 0.877 V (b) 3.38×10^{-5} cm (c) 309 pF
63. 0.49 V   64. 0.385 μF   67. 0.693 μs   68. 5 mA   69. (a) $T_J = 166$°C (b) $\theta_{CA} = 31.25$°C/W

Chapter 4

2. (a) active (b) inverse (c) saturated (d) cutoff   4. 1 fA   5. 0.802 V, 0.580 V   7. 110 Ω
8. (a) $I_C = 8.61$ mA, $V_{CB} = 1.39$ V (b) $I_C = 5.25$ mA, $V_{CB} = -0.5$ V   13. r_o 160 kΩ, $V_A = 78 \, \Omega$
16. $\beta_F = 80$, $\alpha_F = 0.9877$ 19. $R_B = 53$ kΩ, $R_C = 300 \, \Omega$
20. $I_B = -199 \, \mu A$, $I_C = -19.9$ mA, $V_{CE} = -3.44$ V
24. (a) $I_B = 20.25 \, \mu A$, $I_C = 1.01$ mA, $V_{CE} = 2.95$ V (b) $I_B = 35.8 \, \mu A$, $I_C = 1.79$ mA, $V_{CE} = 1.38$ V
25. $I_B = 11.4 \, \mu A$, $I_C = 1.14$ mA, $V_{CE} = 2.71$ V   27. 28 kΩ   28. $I_C = 2.91$ mA, $V_{CE} = 2.07$ V, active
29. $I_C = 3.71$ mA, $V_{CE} = 1.28$ V, because $V_{CE} > V_{BE}$ 31. $g_m = 192$ mS, $r_\pi = 1300 \, \Omega$, $r_o = 22$ kΩ
32. no   33. 19.2%   37. (a) off (b) $I_C/I_B = 12.3 < \beta_F$ 38. no   40. $\alpha_R = 0.333$, $\beta_R = 0.5$
41. $v_o = 5$ V for $v_{IN} < 0.438$ V, $v_o < 0.3$ V for $v_{IN} > 0.64$ V   45. (a) $n = 6$ (b) 26 V
47. (a) 2.665 V (b) 2.664 V   48. active, 51 μA, 5.1 mA, 380 μA   50. 70°C, 57.5°C   51. 100°C/W, 52.5°C

Chapter 5

1. 302 μA/V^2 3. $g_{mn}/g_{mp} = 1.67$ 4. knees at 1 V, 0.3 mA; at 2 V, 1.21 mA; at 3 V, 2.72 mA
7. knees at -1 V, $-108 \, \mu$A; at -2 V, $-432 \, \mu$A; at -3 V, 972 μA   12. $R_1/R_2 = 14$ V/6 V, $0 < R_L < 800 \, \Omega$
15. KP = 625 μA/V^2, VTO = 2 V
17. (a) 1.216 mA (b) $R_1 = 34.1$ kΩ, $R_2 = 65.9$ kΩ (c) $R_1 = 57.9$ kΩ, $R_2 = 42.1$ kΩ   19. 2 V $< v_{IN} <$ 2.6 V
21. (a) 10 V, 0 mA (b) 25 mV, 1.995 mA (c) 50 mV, 1.99 mA (d) 132 mV, 1.97 mA

APPENDIX F

25. (a) 4 V (b) −3 V **26.** ∞, 500 Ω, 250 Ω **27.** (a) +0.5 V (b) −7 V **31.** −2.4 V + ϕ_J
33. (a) −1 V (b) −3 V **34.** 1250 Ω vs. 470 Ω **36.** $V_{GS} = +1.72$ V, $r_d = 100$ kΩ, $g_m = 1.56$ mS
41. If $R_\sigma = 2$ kΩ: $I_D = 1$ mA, $V_{GS} = -2$ V, $V_{DS} = 6$ V; $V_{DS} > V_{GS} - V_P$ **43.** $I_D = 9$ mA, $V_{GS} = -1$ V, $V_{DS} = 7$ V
45. One solution: omit R_1; $R_\sigma = 293$ Ω, $V_{GS} = -1.46$ V, $R_D = 1707$ Ω
46. (a) $V_{GS} = 2.5$ V, $U_D = 1.25$ mA, $V_{DS} = 5$ V, $g_m = 2.5$ mS (b) $A_{vi} = -0.91$ (c) $A_{vi} = -10$
48. One solution: $R_1 = 630$ kΩ, $R_2 = 293$ kΩ (b) $g_m = 3.16$ mS (c) $A_{vi} = -12.6$
  (d) $C_1 > 0.16$ μF, $C_\sigma > 133$ μF to make $f_L = 5$ Hz
53. $I_{D1} = 1.028$ mA, $I_{D2} = 0.353$ mA **60.** $V_{DD} \geq 2000 R_D + 4$ **61.** $R_D \leq 1.4$ kΩ

Chapter 6

1. 94 dB **2.** 13.6 mV **3.** $I_B = 95$ nA, $I_{OS} = 10$ nA. μA741 **4.** $R_2 = R_1 \| R_F$
5. T decreases β decreases; if I_E is constant, I_B increases. **9.** $V_o = -\dfrac{R_4}{R_1}\left[V_1 + \dfrac{V_2}{2} + \dfrac{V_3}{4}\right]$
14. $f = 0.0015$ **15.** $f = 0.004$ **17.** $a_1 a_2/(1 + f_2 a_2 + f_1 a_1 a_2)$ **22.** $0.158576\,\omega_C$, $6.30611\,\omega_C$
24. $C = 1.59$ nF **26.** $R = 79.6$ kΩ **28.** $A_v = -\dfrac{R_2}{R_1}\dfrac{1}{(1 + j\omega C R_2)}$ **30.** $A_v = -\dfrac{R_2}{R_1}\dfrac{1}{\left(1 - \dfrac{j}{\omega C R_1}\right)}$

36. (a) $\dfrac{3981\left(1 + j\dfrac{\omega}{112.5}\right)^2}{\left(1 + j\dfrac{\omega}{20}\right)^2 \left(1 + j\dfrac{\omega}{1991}\right)}$ **37.** (a) $\dfrac{5\left(1 + j\dfrac{f}{2}\right)}{\left(1 + j\dfrac{f}{100}\right)\left(1 + j\dfrac{f}{50000}\right)\left(1 + j\dfrac{f}{12000000}\right)}$

40. (a) $\dfrac{-100}{\left(1 - j\dfrac{20}{f}\right)\left(1 + j\dfrac{f}{20000}\right)}$ (b) $C_1 = 7.96$ nF, $C_2 = 7.96$ μF **44.** (a) $\dfrac{562\left(1 + j\dfrac{f}{1687000}\right)}{\left(1 - j\dfrac{10}{f}\right)\left(1 + j\dfrac{f}{30000}\right)}$

47. change + feedback to − feedback **50.** dc gain is unbounded **51.** $A_o = -10$, $f_L = 106$ Hz, $f_H = 21.2$ kHz
53. inverse of Problem 51 **55.** (a) $f_L = 20 \times 3$ Hz, $f_H = 100/3$ kHz (b) These points are only 1.6 dB down.

Chapter 7

2. Let $R_B = 6$ kΩ; then $R_1 = 47.8$ kΩ, $R_2 = 6.86$ kΩ.
5. If $R_B = 2.5$ kΩ and $R_E = 500$ Ω, then (a) $R_1 = 16.5$ kΩ, $R_2 = 2.95$ kΩ, $V_{CE} = 6.98$ V; (b) $R_L \geq 667$ Ω
7. (a) $R_E = 1983$ Ω; if $R_1 = 180$ kΩ, then $R_2 = 45$ kΩ. (b) $R_L \geq 3.33$ kΩ (c) $I_C = 1.047$ mA, $V_{CE} = 8$ V, yes
9. If $R_B = 20$ kΩ and $R_E = 2$ kΩ, then (a) $R_1 = 69.5$ kΩ, $R_2 = 28.25$ kΩ (b) $A_v = -36.6$ (c) 5 V, 1 mA, 5 mW, 1.2 V_{peak}
11. 7.816 mA, 3.85 V ⇒ 11.48 mA, 0.05 V; no, saturated! **14.** (a) must let $R_2 = \infty$, $R_1 = 72.7$ kΩ (b) $I_C \Rightarrow 3.45$ mA
16. (a) 0.972 mA, 2.8 V (b) 1.299 mA, 2.13 V (c) 33.6%
18. (a) $I_D = 2$ mA. If $R_D = 2$ kΩ, then $V_{DS} = 4$ V. (b) $I_D = 1.219$ mA, $V_{DS} = 6.34$ V
19. (a) $R_\sigma = 3.89$ kΩ, $R_D = 3.89$ kΩ (b) saturated model contradicted; no!
20. (a) $R_D = 3.5$ kΩ (b) $I_D = 1.734$ mA $V_{DS} = 3.93$ V **23.** Let $R_1 = 24.1$ kΩ, then $R_2 = 104$ Ω, $R_o = 1.1$ MΩ.
25. Let $R_1 = 23.7$ kΩ, then $R_2 = 104$ Ω, $R_o = 367$ kΩ. **26.** $R_1 = 28.9$ kΩ, $A_2 = A_1$, $A_3 = 0.4 A_1$, $A_4 = 0.2 A_1$
27. If $R_1 = 19.1$ kΩ, then $R_2 = 12$ kΩ; 22 MΩ < R_L < 224 MΩ. **29.** 136 μA **30.** $R_L < 107$ kΩ
32. If $R_1 = 14.2$ kΩ, then $R_{2A} = 144$ Ω, $R_{2B} = 1.56$ kΩ, $R_{2C} = 5.09$ kΩ.
35. (a) 465 μA (b) 17.1 μA (c) 17.8 μA (d) 335 μA, 112 μA
36. (a) 19.3 kΩ (b) −159 μA (c) 0 (d) $R_L < 123$ kΩ **40.** $R_1 = 47$ kΩ, $R_2 = 10.2$ kΩ **42.** $k = 50$ μA/V^2
43. I_D changes from 200 μA to 450 μA **47.** 100 μA

Chapter 8

2. (a) 1.878 mA, 4 V, 72.6 mS, 1377 Ω (b) −108.8, 1320 Ω, −30.6, 2.2 kΩ, −61.9
5. (a) 100 mS, 2.7 mA (b) 140 kΩ (c) −72.9
6. (a) 1.878 mA, 4 V, 72.6 mS, 1377 Ω (b) −1.464, 24.4 kΩ, −7.59, 2.2 kΩ, −1.41

7. (a) 1.744 mA, 4.2 V, 67.9 mS, 1473 Ω (b) -130.4, 749 Ω (c) -129.1, 1452 Ω
8. (a) 1.876 mA, 4 V, 73.1 mS, 1368 Ω (b) 109.5, 13.36 Ω, 0.311, 2.2 kΩ, 1.44
11. One solution: $R_C = 1$ kΩ, $R_E = 2$ kΩ, $R_1 = 66$ kΩ, $R_2 = 28.7$ kΩ
13. One solution: (a) $R_E = 40$ Ω, $I_C = 155$ mA, $g_m = 5.96$ S (b) 717 Ω (c) 352 Ω (d) $A_P = 41.8$ (e) 1000 μF
14. (a) $I_{C1} = 16.6$ μA, $g_{m1} = 638$ μS, $r_{\pi 1} = 156.7$ kΩ, $I_{C2} = 1.675$ mA, $g_{m2} = 64.4$ mS, $r_{\pi 2} = 1550$ Ω
    (b) $R_i = 49.76$ kΩ, $A_{vi} = -0.97$ (c) $R_o = 31.55$ Ω
19. (a) $g_m = 1$ mS, $r_d = 100$ kΩ (b) -3.11 MΩ, -309.8, 4.7 kΩ, -3.095
    (c) -1.055, 1 mΩ, -105.5, 4.628 kΩ, -1.053
22. (a) -4.34, -4.29 (b) -1.274, -1.26 24. 3.2, 667 Ω, 0.213, 4.7 kΩ, 1.28   26. 0.55, 0.52
28. (a) 1.44 mA, 10 V, 960 μS (b) -3.58, 179 kΩ   30. (a) -37.5 (b) λ_1, λ_2, V_{DD}, V_{t1}, V_{t2}, λ_2 34. -4.48
35. increase $(W/L)_1$ to 80   39. (a) 50 μA (b) -66.7 41. -102.5
44. (a) 0.882 mA, 2.29 V, 34.3 mS, 2.914 kΩ, 5.2 mA, 6.27 V, 202.5 mS, 494 Ω (b) 1594, 4620 (c) 648, 1877
45. (a) 448 μA, 17.4 mS, 4.59 kΩ, | 6 mA, 223 mS, 343 Ω (b) -72.5, -3244
50. $h_{ie} = 3.5$ kΩ, $h_{fe} = 138$, $h_{re} = 3 \times 10^{-4}$, $h_{oe} = 14$ μS

Chapter 9

1. 20 pF   4. (a) -171 (b) 1.83 Mrps   7. $C_{\mu 0} = 2.39$ pF, $\tau_f = 85.4$ ps   8. -80.55
13. (a) $g_m = 96.8$ mS, $r_\pi = 1033$ Ω, $r_o = 62$ kΩ (c) 120 Mrps   15. $A_{vsm} = -16$, $\omega_H = 2.96$ Mrps
16. $A_{vsm} = -9$, $\omega_H = 5.16$ Mrps   18. $A_{vsm} = -4$, $f_H = 6.95$ MHz   20. (a) $\omega_L = 138$ rps
22. $\omega_L = 2.3$ rps   23. $\omega_L = 8.5$ rps   24. $C_E > 18$ μF   25. $f_L = 1.39$ Hz
28. $A_{vsm} = 1105$, $f_H = 1.55$ MHz, $f_L = 81$ Hz   29. (a) 41 μF (b) 40 μF
31. $A_{vsm} = 0.8$, $f_H = 9.5$ MHz, $f_L = 0.266$ Hz   32. $A_{vsm} = -59.5$, $\omega_H = 6.8$ Mrps, $\omega_L = 235$ rps
34. (a) $f_H = 51$ MHz (b) $A_{vsm} = -49.3$ 37. $C_1 = 5$ nF, $C_2 = 100$ nF, $C_{gd} = 100$ pF
41. (a) $I_C = 587$ μA, $g_m = 356$ μS, $r_o = 180$ kΩ, $\omega_H = 7.55$ Grps   47. $A_{vim} = -1363$, $\omega_H = 4.19$ Mrps
48. $C_E > 5.9$ μF   49. C_1 is unnecessary.

Chapter 10

2. $I_C = 62$ μA, $V_{CE} = 6.48$ V, $A_{dm} = -226$, $A_{cm} = -0.55$, $R_{id} = 167$ kΩ
4. $A_{dm} = -87.7$, $A_{cm} = -0.248$, $R_{id} = 225$ kΩ, $R_{ic} = 37$ MΩ, CMRR $= 51$ dB
5. $R_C = 2.73$ MΩ, $R_{EE} = 1.36$ MΩ, $V_{CC} = 18.5$ V, $V_{EE} = 14.9$ V
9. If $R_1 = 10$ kΩ, $R_2 = 196$ Ω, CMRR $= 85$ dB.
10. $A_{dm} = -32.6$, $A_{cm} = -0.0012$, CMRR $= 88.9$ dB, $R_{id} = 635$ kΩ, $R_{ic} = 1639$ MΩ
12. $g_{mMOS} = 1$ mS, $g_{mBJT} = 38.5$ mS   14. (a) $I_D = 0.55$ mA, $V_{DS} = 8.15$ V (b) -7.55 (c) -7.4
18. (a) -16.67 (b) 39 kΩ (c) 28.6 dB   22. $A_{dm} = -10 - 2.5V_{GS}$, $-4 < V_{GS} < 0$
23. (a) 23.8 kΩ (b) -1442 (c) 7.5 V (d) 24.3 kΩ, -3216, 7.5 V   24. -1539, -1619
26. (a) 3.93 MΩ, 0.995, -1442 (b) -1100
29. (a) -268 (b) If $R_2 = 104$ Ω, $A_{cm} = -0.029$. (c) 79.3 dB (d) 52 kΩ
    (e) 201 cos ωt mV (f) $V_{CE1} = 8.15$ V, $V_{CE4} = 14.4$ V, $V_{O1} = 7.55$ V, $V_{OD} = 0$
35. $A_{dm} = -4.7$, $A_{cm} = -0.236$, $R_{id} = 4.25$ MΩ, $R_{ic} = 39.5$ MΩ, CMRR $= 26$ dB
36. (a) 235 Ω (b) $V_O = 6.9$ V, $V_{BE6} = 0.662$ V, $V_{BE5} = 0.723$ V, $V_{CE6} = 14.29$ V (c) 613 kΩ
39. (a) $V_{C2} = -2.94$, $V_{C3} = -0.64$ (b) 4610 (c) 52.6 mW
40. (a) -4.29 (b) 103 kΩ (c) 85.8 dB (d) $V_{CE3} = 3.03$ V, $V_{CE9} = 16.8$ V   42. 641
49. (a) $V_{CE4} = 8$ V, $V_{DS1} = 10$ V, $V_{CE3} = 0.7$ V, $R_{REF} = 7.86$ kΩ
    (b) $A_{dm} = -12$, $A_{cm} = -0.111$, CMRR $= 40.7$ dB (c) -6.22 cos ωt mV
52. (a) 52.4% (b) 50 mW (c) 22.75 mW each   55. 16.7%
57. (a) 36 W (b) $V_{CE(max)} > 48$ V, $I_{C(max)} > 3$ A, $P_{C(max)} > 7.3$ W
    (c) $R_{SC} = 0.2$ Ω (e) $P_L = 72$ W, $\eta = 100\%$
59. (a) 46.8 mW (b) 77.3 mW (c) 500 Ω   62. 2.6 mV   65. ± 5.1 mV

Chapter 11

1. $A_o = 19.6$, $Z_i = 510$ kΩ, $Z_o = 0.98$ Ω, $f_H = 5.1$ MHz, $f_L = 19.6$ Hz
2. $A_o = 19.6$ Ω, $Z_i = 196$ Ω, $Z_o = 0.98$ Ω, $f_H = 5.1$ MHz, $f_L = 19.6$ Hz

7. $A_{vs} = -102.3$, $R_i = 1237\ \Omega$, $R_o = 4794\ \Omega$ 9. $A_{vs} = -4.87$, $R_i = 40.4\ k\Omega$, $R_o = 1\ k\Omega$
11. (b) $A_{vs} = -3.886$, $R_i = 23.3\ \Omega$, $R_o = 63.5\ \Omega$ 13. It's positive feedback!   16. 63.7 pF
17. $A = -16$, $A_{vs} = +80$ 19. $A_{vs} \approx 75$ (bias must be changed)   21. 42   22. 11
26. -11, 55   27. $1 + R_F/R_{E1}$
29. (a) 0.1   (b) $\dfrac{10}{1 + j\dfrac{f}{100\ kHz}}$ 32. $f_D = 1.9657$ kHz   34. (a) 6.6 Mhz   (b) $f_o = 0.045$, $A_o > -18.2$
37. (a) -36.3 (b) 104 pF
40. $\omega = \dfrac{1}{RC\sqrt{6}}$, $A_o > -\dfrac{1}{29}$ 44. $L_1 = 2$ mH, $L_2 = 23$ mH, $R_C/R_E > 11.5$
46. $C_1 = 10$ nF, $C_2 = 3.39$ nF, $|A| > 2.95$ 47. same as 11.46
52. $f_S = 1.0867$ Mhz, $f_P = 1.1215$ Mhz; from f_S to f_P 38 kHz   53. $C_2 = 4$ pF, $C_1 = 8.37$ pF, $|A| > 2.1$

Chapter 12

1. (a) npn: 720 kHz, pnp: 65 kHz   (b) npn: 75 V, pnp: 63 V   3. $A_v = 35$, $R_{id} = 37.9\ k\Omega$, $R_{ic} = 8.9\ m\Omega$, $R_o = 233\ \Omega$
5. $A_v = 38$, $R_{id} = 125\ k\Omega$, $R_{ic} = 29.5\ M\Omega$, $R_o = 94.5\ \Omega$ 7. $R_1 = 45.4\ k\Omega$, $A_v = -1593$
10. Possible: $R_2 = 6.16\ k\Omega$, $R_3 = 35.6\ k\Omega$, $R_4 = 1\ k\Omega$, $R_5 = 30\ k\Omega$
13. (a) -1.2 V   (b) -14.3 V   (c) -1.2 V   (d) 14.3 V   (e) -13.8 V   (f) -13.65 V   (g) 0.6 V
15. 335,000   17. 254,000   22. 82,000   26. 548,000 26 kΩ 28. $-15.3 < v_{IN} < 12.6$
29. (a) $-R_3/R_2$ (b) $C_{\mu 2}$ (c) 5 kΩ (d) $V_{CE2} = 7.11$ V, $R_1 = 5.8\ k\Omega$, $V_S = 4.875$ V
32. 16.7 μA   34. 87 nA   36. (a) 1.35 mA   (b) 131 kΩ 37. (a) 41.6 kΩ (b) 10.9   38. -3.6 39. -73
40. 1.036 mA, -0.18 mA, 269.5 μA, 230.5 μA

Chapter 13

4. (a) **NAND**   (b) **NOR**   (c) **OR**   5. $NM_H = 1.4$ V, $NM_L = 0.3$ V   6. 6.5 ns
8. $V_O = 0.79$ V at $V_{IN} = 0.38$ V; $V_O = 0.2$ V at $V_{IH} = 0.64$ V   10. 9.2 mW   13. X = AB
14. $NM_L = 1.7$ V, $NM_H = 2.65$ V   16. (a) 30 mA   (b) $V_{CE2} = 0.9$ V, $V_{IN} = 1.4$ V   17. 20.4 mA   20. 119 mA
25. (a) $I_{B1} = 105\ \mu$A, $I_{B2} = 0$, $V_o = 5$ V, $V_{B2} = 0.06$ V   (b) 1.4 V   (c) Q_1 reverse active, Q_2 and Q_3 saturated
26. 2.2 mW   29. (a) -0.62 V   (b) -1.77 V   30. (a) 5 V   (b) 0 V   (c) 0 V   (d) 4.6 V   33. 88.4 ns
35. (b) 1.69 ns   37. $(W/L)_n = 8$, $(W/L)_p = 16$ 40. 12.5 pW/Hz   41. 50 pW/Hz
43. $V_{OH} = 13$ V, $V_{OL} = 1.6$ V, $V_{IH} = 10$ V, $V_{IL} = 8$ V, $NM_L = 6.4$ V, $NM_H = 3$ V

Chapter 14

1. $k = \log_2 n + 2$ lines (including common)   3. 24.2 mA/segment   4. 5 bits   7. $A_2 A_1 A_0 = $ LLH
13. (a) 12.5 Mhz   (b) 1111, 1110, 1100, 1000, 0000   14. 33.3 MHz
23. Use 25 binary stages = 2^{25} = 33554432. Error is 0.01 3 % = 11 s/day   27. R = 144 kΩ, C = 10 nF
32. T = 1.452 RC   37. T = 1.10 RC   40. max R = 1.21 kΩ

Chapter 15

4. (a) 7 bits   (b) 4 bits   5. (a) 700 ns   (b) 350 ns   7. 63 ns   9. R = 0.05 Ω
13. 15 Ws/cm^2 of UV at 2537 angstroms   16. (a) 2.5 mA   (b) 2.1 mA   (c) TTL load(s)   18. (a) 1 μA   (b) 8 pF
23. 7.3 μA/V^2 26. 35, 45, 55, 70 ns   28. D_{IN} and D_{OUT} are tied together.
31. (a) 4.0 mA   (b) 8.0 mA   (c) TTL load(s)   35. 2.75 mW   39. 11 address, 9 data, RAS, CAS, W, V_{DD}, ground.
42. $O_7 = A_0 A_{15} + \overline{A_0} \overline{A_{15}}$

Chapter 16

2. 480 kb/s   5. 1.544 Mb/s   6. (a) 6 bits   (b) 7 bits   (c) 10 bits   11. 0.5% of V_{FS} 12. 167 Ω
16. The error of the first DAC is multiplied by n/2   18. 11100110
20. No, but opening S_2 slows the discharge   26. V_{REF} 30. $f = N/RC$

Chapter 17

1. $0.1\ \Omega$ and VR = 1%   **4. (a)** $250\ \Omega$ **(b)** 0.125%   **6.** $137\ \Omega < R_3 < 3.3\ k\Omega$ **8.** 700 mW and $\eta = 41.7\%$
11. If $R_1 = 1\ k\Omega$, $R_4 = 10\ k\Omega$, then $R_2 = 649\ k\Omega$, $R_3 = 9.1\ k\Omega$, $F_o = 10.5$ kHz, $R = 0.76\%$ **18.** set $v_2(t) = v_o(t)$
32. (a) 5.25 kHz   **(b)** Triangle wave $V_m = 3.14$ V   **(c)** $R = 5\ k\Omega$, $C = 0.1\ \mu F$
35. $R_1 = 10\ k\Omega$, $C_1 = 790$ pF, $C_2 = 0.124\ \mu F$, $C_C = 0.001\ \mu F$ $R_2 = 3.6\ k\Omega$

Chapter 18

3. $1.95\ \mu m$ **7.** $0.617\ \mu m$ **11.** 5.4 ms   **12.** $28.4\ \Omega/\square$ **17.** $8000\ \text{Å} = 0.8\ \mu m$ **20.** 157 mils, 1.57 mils, 100
23. $0.2\ \mu m$ for $L = 10\ W$ **24.** $0.38\ \mu s$ **26.** 15.25 mils square   **27.** 8.2 mils square   **28.** 0.14 pF
29. $170\ \mu m$ square   **35.** $40\ \mu m$ plus design margin   **36.** 2×10^6 A/cm^2, 1.18 mW   **38.** Yes   **43. (a)** $25\ \Omega/\square$
(b) $10\ \Omega/\square$ **(c)** 3.5 hours at 1150°C   **(d)** $6\ \mu m$

INDEX

Important definitions appear on pages that are printed in **boldface** type.

Abrupt junction, **66**
ac load line, 297
Acceptor atom, **47**
Acceptor energy states, **47**
Access time, **750**
Acid etch, 896
Active filter, 285
Active load, **463**–477
Active region, **130**, 131
Address, **704**
Address bus, **752**
Address decoder, **707**
Address space, **753**
Alpha (α_F, α_R), **128**, **130**
Amorphous, **40**
Amplifier, 14, *See also the device type*
   biasing, 295
   BiCMOS, 370–373, 427–429
   Cascode, **372**–373, 425–429
   class A, AB, B, and C, **748**–795
   common-base (CB), 133–138, 346–347, 351
   common-collector (CC), 348–350, 351
   common-emitter (CE), 138–154, 340–345, 351
   common-source (CS), 224–225, 352–354
   current, **550**
   differential, 441–463
   direct-coupled, 370–373, 424–429
   emitter-follower, 348–350
   MOSFET, active load, 358–356
   operational, *See* Op Amp
   output stage, 249
   small-signal, **336**–377
   stability, **556**–560
   transconductance, **535**
   transresistance, **519**
   voltage, **544**
Analog gate, 204–205
Analog multiplier, **842**–853
   data sheet, **852**–853
Analog signal, **800**
Analog system, 828–829
Analog-to-digital converter (ADC), 800, **811**–824
   comparison of characteristics, 824
   digital-tracking (servo), **811**, 820–822
   flash (parallel), **811**, 812–814
      cascaded-flash, 813–814
   integrating, **811**
      dual-slope, **818**–820
      single-slope, **816**–818
   successive-approximation, **811**, 815–816
      switched capacitor, 822–823

Analog voltage, **10**
AND circuit, 96, *See also* Digital circuit
Annealing, **887**
Anode, **68**
Arithmetic-logic-unit (ALU), **711**–715
ASIC, **639**, 703
Astable multivibrator, **728**, 730–733
Avalanche breakdown, **75**
Avalanche multiplication, **161**, 162

Background doping, 890
Balanced line, **735**–736
Band gap, **36**
Band-gap reference, **823**
Bandwidth, *See* Corner frequency
Barkhausen criterion, **563**
Base, **124**
   diffusion, **910**
   overdrive, **124**
   resistance, **141**, 339
   transit time, **391**
   width modulation, **135**
Base-diffused resistor, **898**–900
Base-emitter offset voltage, **142**
Base-spreading resistance, **141**, 339
Base-transport factor, **128**
Beta (β_F, h_{FE}), **139**
Beta cutoff frequency, **394**
Beta variation, 156–157
Bias, **10**, 67
Bias circuit, BJT, 295–299
Bias circuit, FET, 304–310
Biasing, *See* Operating point
BiCMOS, BiMOS amplifier, 370–373
Binary signal, **97**
Binary variables, 943
Binary-weighted resistor DAC, 806–807
Bipolar junction transistor, *See* BJT
Bistable multivibrator (latch), **716**
Bit, **749**
BJT, 6, **123**–167, *See also circuit types*
   active model, **142**
   active region, **130**–131
   amplifier, 145, *See also* Amplifier
   array, 589
   base resistance, **141**, 339
   base-transport factor, **128**
   biasing, 295–299
   CB, 133–138
   current source, 310–320
   cutoff region, **131**–132
   Early effect, **135**, 138
   data sheet, 158–159
   emitter efficiency, **127**
   emitter resistance, **134**

   equivalent circuit, 130
   fabrication, 907–913
   frequency effects, 390–395
   hybrid parameter measurement, 373–376
   inverted mode, **131**–132, 154–155
   lateral *pnp* transistor, 913
   model, hybrid-π, **338**–340
   operating point selection, 297–299
   operation, 130
   pnp structure, 912–913
   saturated model, 148–152
   saturation region, **131**, 142–146
   substrate pnp transistor, 912
   switching times, 162–164
   symbols, 124–125
   typical parameters, 152, 912
Bode plots, standard forms, 261–272
Body-effect parameter, **360**–364
Boltzmann's constant, 32
Breakdown diode, **75**
Breakdown-diode voltage regulator, 90
Broadbanding, 270–272
   gain bandwidth product, **272**
   high-frequency corner, 264, **269**
   low-frequency corner, 267, **269**
   midband gain, 271
Bulk-acoustic wave resonance, **571**
Bypass capacitor, **341**, 418–423
Byte, **749**

Capacitance, depletion. *See* Depletion capacitance
Capacitance-coupled amplifier, 412–418
Capacitor, IC diffused, 906–907
Capacitor, IC MOS, 904–905
Capture range, **859**
Carrier concentration, 37
Carrier multiplication, **147**
Cascaded amplifier, **15**, 367–370, 527, 541, 546, 552
Cathode, **68**
CB (common-base), **133**–138
   amplifier. *See* Amplifier.
   circuit model, **135**
   collector characteristic, **133**
   emitter characteristic, **134**
CE (common-emitter), **138**–154
   active model, **142**
   amplifier, *See* Amplifier
   base characteristic, **141**
   collector characteristic, **139**–140
   current gain, 156–157
   cutoff, 146–147
   dc current gain, 139

CE (common-emitter), (*continued*)
  input resistance, h_{ie}, **374**–375
  large-signal current gain, **156**
  output admittance, h_{oe}, **374**–375
  output resistance, **144**
  saturation model, 148–152
  small-signal current gain, h_{fe} or β, **157**, **374**–375
  voltage feedback ratio, h_{re}, **374**–375
Center frequency, **854**
Central processing unit (CPU), **712**
Channel stop, **919**
Charge storage, 100–101
Charge-control model, 132, 164
Charge-coupled device (CCD), 775–777
  imaging, 777
Charge redistribution, **822**
Chemical vapor deposition (CVD), **918**
Chopper, 849–850
Chopper-stabilized op amp, **626**
Clamp circuit, **94**
Class A, **478**, 487–495
Class AB, **478**, 482–487
Class B, **478**, 478–481
Class C, **478**
Clipper circuit, **92**
Clock, **716**, 733
Closed-loop gain, 258–259, 518
CMOS (complementary MOS), **201**–203
  amplifier, 364–366
  DAC, 810
  logic
    data sheet, 685
    high speed, 684
    input protection, 683–684
    inverter, **201**–203
    noise margin, 681
    NOR/NAND, 683
    power dissipation, 683
    propagation delay, 683
    switching time, 682–683
    transfer characteristic, 201, 679–681
  structure, 202, 920
Collector, **124**
  current, 132
    change with beta, 300–301
    change with I_{CBO}, 302
    change with V_{BE}, 301
  cutoff current, **147**
  efficiency, **491**
  reverse saturation current I_{co}, **130**, 139
Collector-base breakdown voltage, **161**
Collector-base resistance, **340**
Collector-base transition capacitance, 164, 390. *See also* Depletion capacitance.
Collector-emitter breakdown voltage, **161**
Combinational logic, **703**–715
Common-anode display, **708**–710
Common-base configuration. *See* CB.
Common-cathode display, **708**

Common-collector configuration. *See* Emitter follower.
Common-emitter configuration. *See* CE.
Common-mode
  half circuit, **449**
  input resistance, **453**
  noise, **735**
  rejection ratio (CMRR), **450**–451, 456
  voltage gain, **448**–450, 456
Compact disc (CD), **712**
Comparator, **622**–624, 628, 840
  data sheet, 625
  internal circuit, 622–624
Compensation. *See* Frequency compensation
Compensation capacitor, 611, 616
Complementary error function (erfc), 881–882
Complementary MOS logic. *See* CMOS.
Complex frequency, **854**
Computer system, 711–712
Computer-aided design (CAD), **948**
Computer-aided engineering (CAE), **948**
Concentration gradient, 32
Conduction angle, **478**
Conduction band, **36**
Conductivity, **28**, 30
Conductor, **29**
Contact potential, **65**
Corner frequency, **264**
  dominant, **560**
Counter, **724**–728
Coupling capacitor, 414–418
Covalent bond, **40**
Critical damping, 858
Crossover distortion, **479**
Crystal-controlled clock, 733
Crystal oscillator, 571–573. *See also* Oscillator.
Crystalline, **40**
Current
  amplification, **342**, 345, 347, 349, 352, 354, 355, 357, 358
  amplifier, **550**
  density, **29**, 41
  division, 807
  gain. *See* Current amplification.
  hogging, **649**
  injector, **690**
  limiting, **832**
  mirror, 322–323
  source, BJT, **310**–320
    MOSFET, 321–325
Current-series feedback, **534**–544
  analysis, 536–537
  cascaded amplifier, 541–543
  input impedance, 535
  output impedance, 536
Current-shunt feedback, **549**–556
  analysis, 551–552
  cascaded amplifier, 552

input impedance, 550
output impedance, 550
Current-to-voltage converter, **807**
Cut-in, **75**
Cut-in voltage, **134**
Cutoff region, **131**-132, 146–147

D flip-flop, 721
Damped natural frequency, **857**
Damping coefficient, **857**–859
Dark current, **74**
Darlington pair, **371**, 380, 659
Data bus, **752**
Data selector, **704**
Data word, **754**
dc load line, 297
dc restorer, **95**
dc-to-dc converter, **838**
De Morgan's Law, **946**
Decibel (dB), **14**
Decoder, 688, **707**
Decoder-driver, **707**–710
Decoupling capacitor, **656**
Delay time, BJT, **162**
Delay-power product, **663**
Demodulation, 848–849
Demultiplexer, **705**
Depletion capacitance, 99–100, 391
Depletion region, **63**, 70
Die, **910**
Dielectric breakdown, 73
Differential amplifier, **441**–463
  input resistance, **453**
  voltage gain, **444**–445, 455, 459
Differential-mode half circuit, **445**
Diffused resistor, **898**
Diffusion, **26**, 31
  capacitance, **100**–101, 391
  coefficient, **32**, 880
  constant, **32**
  current, **31**
  equation, **33**
  length, **50**
  of impurities, 880–885
  profile, **882**
Digital circuit, **10**, 640
  AND gate, 96, **943**
  buffer, **943**
  exclusive OR gate, **945**
  NAND gate, **944**
  NOR gate, **945**
  NOT gate, **943**
  OR gate, 97, **944**
Digital display systems, 734–742
Digital-to-analog converter (DAC), **805**, 806–811
  binary-weighted resistor DAC, 806–807
  bipolar DAC, 808–810
  CMOS DAC, 810
  R-2R ladder DAC, 808–808
Digital-tracking ADC converter, **811**, 820–822

INDEX

Diodes, 72
    analog switch, 97–98
    breakdown, 75
        data sheet, 76
    capacitance, 99–101
    characteristics, 73–78, 99–108
    clamp, 94
    clipper, 92
    data sheet, 77
    dynamic resistance, 80–82
    equation, 72
    fabrication, 914
    gate/switch, 97–98
    graphical analysis, 78
    heating, 107–108
    ideal, 79
    light-emitting (LED), 105
    limiter, 92
    logic circuit, 96
    model, piecewise-linear, 79–80
    protection, 617
    rectifiers, 83. *See also* Rectifier circuits.
    saturation current, 72
    Schottky, 104
    temperature effects, 74
    tunnel, 106
    varactor, 100
Diode-connected BJT, 914
Diode-transistor logic (DTL), 651–652
Direct-coupled amplifier, 370–373
Direct-coupled transistor logic (DCTL), 650, 677
Distortion,
    crossover, 479
    harmonic, 496–497
    reduction with feedback, 259
Dominant corner frequency, 560
Donor atom, 45
Donor energy states, 44–45
Doping, 40, 44
Dose, 33, 887
Double-diffused MOSFET (DMOS) structure, 922–923
Double-subscript notation, 9
Drain current change with V_P, I_{DSS}, 304
Drain resistance, 222
Drift, 26
Drift velocity, 27
DTL (diode-transistor logic), 651–652
    basic gate, 651
    fan-out, 651
    NAND gate, 651
    noise margin, 651
    speed/power, 652
    transfer characteristic, 651
Dual-slope ADC, 818–820
Dummy cell, 770–772
Dynamic RAM (DRAM), 751, 769–774
Dynamic resistance, 81, 141

Early effect, 135, 139

Early voltage, 140
Ebers-Moll equations, 130
ECL (emitter-coupled logic), 665–670
    basic gate, 666–667
    data sheet, 670
    fan-out, 668
    noise margin, 668
    OR/NOR, 666
    power dissipation, 669
    temperature compensation, 667
    transfer characteristic, 668–669
Effective mass, 37
Edge-triggered flip-flop, 719, 720–721
Efficiency, collector, 491
Einstein relation, 32
Electrically erasable PROM (EEPROM), 761–763
Electron mobility, 27, 45
Electron volt, 34
Electron-hole pair generation, 41
Emitter, 124
    degeneration, 344, 537
    diffusion, 910
    efficiency, 127
    follower, 348–350
    resistance, 134
    reverse saturation current I_{EO}, 130
Emitter-base breakdown voltage, 155
Emitter-coupled logic (ECL), 665, 670
Emitter-coupled pair, 442–453, 665. *See also* Differential amplifier.
Emitter-diffused resistor, 900
Encoder, 710–711
Energy gap, 36
Energy-band theory, 34–38
Energy-level diagram, 34, 35, 36
Epitaxial layer, 885, 897
Epitaxial reactor, 885
Equilibrium situation, 65
Erasable PROM (EPROM), 759–761
    data sheet, 788–795
Error function (erf), 881
Error signal, 258
Extrinsic, 42

Fall time, BJT, 164
Fall time, MOSFET, 675
Fan-in, 641
Fan-out, 641
Feature size, 892
Feedback,
    advantages, 259–260
    concepts, 257–260, 517–518
    current-series, 534–544
    current-shunt, 549–556
    network (ψ), 519
    network configurations, 520, 535, 544, 550
    parallel-input/parallel-output, 519
    parallel-input/series-output, 541

    series-input/parallel-output, 544
    series-input/series-output, 534
    series-series, 534
    series-shunt, 544
    shunt-series, 549
    shunt-shunt, 519
    voltage-series, 544–549
    voltage-shunt, 519–534
FET. *See* JFET, MOSFET
FET, small-signal model, 351
Fiber optic transmission line, 736
Fick's First Law, 32
Fick's Second Law, 33
Field effect transistor. *See* JFET, MOSFET.
Field oxide, 920
Field-programmable gate array, 782
Field-programmable logic device, 778
Filter capacitor, 88–90
Finite source (diffusion), 883–885
First-order filter, 857
First-order loop, 855–856
Flash ADC, 811, 812–814
Flash EEPROM, 763
Flip-flop, 719
Floating-avalanche MOS (FAMOS) cell, 760
Floating-gate transistor, 759–760
Floating-gate tunnel oxide (FLOTOX) cell, 761
Floppy disk, 712
Flyback dc-to-dc converter, 838–842
Flyback diode, 841
FM demodulation, 860, 863–864
Foldback circuit, 831
Forbidden energy band, *See* Band gap
Forward alpha (α_F), 128
Forward amplifier (a), 258
Forward beta (β_F), 139
Forward resistance of diode, 80
Forward-biased junction, 68
Four-quadrant multiplier, 842–847
Free-run frequency, 854
Frequency
    bandwidth, 272. See also Corner frequency.
    beta cutoff, 394
    broadbanding, 270
    compensation, 558, 560–563
        741 op amp, 611
    corner. *See* Corner frequency.
    divider, 718
    multicorner response, 276–279
    stability, 556–560
    synthesis, 860, 864–866
    transition, 393
    translation, 848–849
    unity-gain, 393
Frequency independent BJT model, 338
Frequency-shift keying (FSK), 863
Function generation, 850
Fusible link, 751

g-parameter model, **551**
Gain
  common-mode, **449**, 456
  differential, **444**, 459
  power, **14**
  voltage, **14**
Gain margin, **558**
Gain stabilization, 260
Gain-bandwidth product, **272**
Gallium arsenide (GaAs), 43
Gate arrays, VLSI, 775–778
Gate oxide, **920**
Gaussian diffusion profile, **884**
Germanium (Ge), 39–40
Gilbert cell, **843**–847
Graded junction, **60**

h-parameters (common-emitter), measurement, 373–376
h-parameters (2–port), **545**
Harmonic distortion, 496–**497**
Heat sink, **107**
Heterojunction, 74
Hexadecimal number, **753**, 941
High-frequency model, 389–398
High-level injection, **69**
High-pass filter, **266**–268
High-power op amp, 727
High-threshold logic, 650, 697
Hole, **41**
Hole mobility, 41, 45
Hole-electron pair generation, **41**
Hybrid IC, **687**
Hybrid- model, **338**–340
Hyperabrupt junction, **100**

I-V characteristic, **12**
IC resistor parameters, summary, 903
I_{CBO}, I_{CEO}, I_{CO}, I_{EO}, **130, 139, 147**
Ideal diode, **79**
Ideal op amp, 17, 249
IIL (integrated injection logic), 688–694
  basic operation, 688–690
  fan-out, 692
  injector, **690**
  noise margin, 691
  physical structure, 690–691
  power dissipation, 692–693
  propagation delay, 692–693
  Schottky, 693
  transfer characteristic, 691–692
Imaging with CCD, 777
Impurity concentration, 45
Impurity diffusion, 880–885
Infinite source (diffusion), **881**–883
Infrared emitter, 104, 105
Injected minority carriers, **68**
Injection junction laser, **105**–106
Injector rail (IIL), **691**
Input bias current, **249**
Input capacitance, **391**

Input offset current, **252**, 500
Input offset voltage, **249**, 498–499
Input resistance, 252, 342, 345, 346, 349, 352, 354, 355, 356, 358
Insulator, **29**
Integrated circuit, 7
Integrated RAM, **751**, 774
Integrated-injection logic. *See* IIL.
Integrating ADC, **811**
Interrupt encoder, **710**
Intrinsic, **40**, 41
Intrinsic carrier concentration, **42**
Inverse mode, **131**–132, 154–155
Inverter, 153–154, 198–203
Ion implantation, **886**–888
Ionization energy, **45**
Isolation diffusion, 898, 908
Isolation region, **898**

JFET. *See also circuit types.*
  biasing, 304
  channel shape, 209,213
  common source amplifier, 224–225, 352–354
  common-drain amplifier, 226–227, 356–357
  common-gate amplifier, 226, 355
  data sheet, 217–220
  drain characteristic, 213–215
  drain-source saturation current, **213**
  dynamic drain resistance, **222**
  fabrication, 914–915
  forward transfer admittance, **220**
  frequency effect, 396–398
  gate current, 220
  load line, 224–225
  maximum power dissipation, 219
  maximum voltage ratings, 217
  models, 221–223
  n- and p-channel, 209, 215
  ohmic region, **211**–212
  ON drain resistance, 211–212
  operating point, 304
  output admittance, 220, 222
  physical operation, 209–213
  pinch resistor, 901–903
  pinch-off, 212–213
  pinch-off voltage, **212**–220
  saturation region, 213
  source follower, 226–227
  static characteristics, 213–215
  structure, 209
  symbols, 184–185
  temperature effects, 309
  transconductance, **220**
  voltage gain, 224, 226, 227
JK flip-flop, 719–721
JK latch, **718**
Junction capacitor, **99**, 390, 396–399, 906–907
Junction depth, **882**

Junctior diode, 59, **63**
Junction FET (JFET). *See* JFET.
Junction temperature, 107, 157–160

Kilobit, **749**

Large-scale integration (LSI), **637**
Large-signal, **80**
Large-signal voltage gain, **252**
Latch, **715**–723
Latency time, **776**
Law of mass action, **42**
Least significant bit (LSB), **801**
Lifetime, minority carrier, 49, 101
Light-emitting diode (LED), **105**
Limiter circuit, **92**
Limiting diode, 92, 617
Line drivers and receivers, **735**–736
Line regulation, **833**
Liquid-crystal display (LCD), **710**
Lithography, x-ray/electron beam, **893**
Load line, **78**, 198–201, 297
Lock range, **859**
Logic
  BiCMOS, 686–687
  CMOS, 678–685
  comparison, 694–695
  ECL, **665**–670
  family, **639**
  functions, 663
  IIL, 688–694
  levels, 641
  merged-transistor, **688, 691**
  MESFET, **677**–678
  NMOS, 670–677
  noise margin, 642–**643**, 650
  nonsaturating, 658, 665
  power dissipation, 663
  propagation delay, **644**–645
  RTL, **641**–650
  speed, 656, 663
  tristate, **664**
  TTL, **650**–664
  wired-AND, **663**
Loop bandwidth, **856**
Loop gain, **259, 518**
Low-frequency amplifier analysis, 412–423
Low-level injection, **69**
Low-pass filter, **262**–265, 285

Magnitude plot, 261
Majority carrier, 47
Mask, photolithographic, **892**–896
Mask-programmed ROM, **751**, 757–759
Master-slave flip-flop, **719**
Maximally flat response, **859**
Maximum collector current, 160, 298
Maximum collector-base voltage, 161
Medium-scale integration (MSI), 637
Memory
  access time, **750**

INDEX

CCD, 775–777
chip organization, 754–755, 765
comparison, 782–783
DRAM control, 774
dynamic chip organization, 773
dynamic circuitry, 770–772
map, **753**
one-transistor cell, 679–770
RAM, **749**, 750
read-only (ROM), **750**, 751–752, 757–764
read/write (R/W), **750**, 751
read/write operation, 755–756
refresh, **769**
refresh operation, 772
serial, **749**–750
static RAM, **751**, 754, 764–768
system organization, 752–753
Merged-transistor logic (MTL), **688**, **691**
MESFET, **231**–233, 677–678
Metal-oxide-semiconductor, *See* MOSFET
Metal-semiconductor FET. *See* MESFET
Metallization, 899
Micro strip line, **667**
Microprocessor, **712**
Midfrequency (midband) model, **296**–299
Midfrequency small-signal model, 336–377
Miller capacitance, **402**–403, 409
Miller effect, 399–412
    741 op amp, 611–612
Minimum feature size, **892**
Minority carrier lifetime, **49**
Minority carrier suppression, **46**
Minority carriers, **46**
Mismatch, 498–501
Missing-pulse detector, **730**
Mobility, 27. *See also* Electron *or* Hole mobility.
Model. *See also* SPICE model.
    BJT, active CE, **142**–146
    BJT, saturated, **148**–152
    charge-control, 132, 164
    diode, piecewise-linear, 79–80
    Ebers-Moll, **130**
    ideal diode, **79**
    hybrid-, **338**–340
    JFET, 221–223
    MOSFET, 221–223
    op amp, 273
Modulation, 848–849
Modulator, 79
Modulo-2 addition, **946**
Monolithic, **877**
Monostable multivibrator, **728**–730, 746
MOS capacitor, 904–905
MOSFET. *See also* circuit types.
    amplifiers, active load, **358**–366
    analog switches, 204–205
    biasing, 304–310
    characteristics, 189–192
    circuit symbols, 185

complementary-symmetry. *See* CMOS.
current source, 321–325
data sheets, 194–196
depletion mode, **192**–193
drain-current characteristics, 189–192
dynamic model, 221–222
enhancement mode, **186**–192
fabrication, 916–924
frequency effects, 395–396
gate, 186
gate capacitance, 295–396
gate current, 194
inverter, active load, 200–201
inverter, passive load, 198–199
inverter, switching time, 674–675
inverter, transfer characteristic, 672–674
load line active load, 200–201
maximum voltage ratings, 194
n-channel, 186–188
NOR/NAND, 676–677
ohmic region, **188**–191
operation, 186–189
p-channel, **191**
saturation region, **191**
static model, 191
structure, 186
subthreshold operation, **188**
symbols, 184–185, 198
temperature dependence, 309
threshold voltage, **188**, 194, 360–364
Most significant bit (MSB), 764, **801**
Multichip module, **687**
Multicorner frequency response, **268**–270, 276–279
Multiple-collector transistor, 688–689
Multiplexed data transmission system, 736–738
Multiplexed display, 710, 738
Multiplexer, **704**
Multiplier, 842–853

n-type semiconductor, **46**
Negative feedback, **344**, 516
NMOS, 302–309. *See also* MOSFET.
    delay-power product, 674–676
    propagation delay, 674–676
    transfer characteristic for depletion load, 672–674
Noise margin, **642**
Nonvolatile RAM, **764**
Norton, **13**
Notation, 9
npn transistor, **124**, 907–911
Number systems, 941–942
Nyquist theorem, **803**

Offset voltage, **79**
Ohm's Law, 11
Ohmic contact, **103**
Op amp, **16**

ac analysis, 593–594
chopper stabilized, **626**
dc analysis, 592
design of, 589–598
examples
    active low-pass filter, 285
    difference amplifier, **19**
    inverting amplifier, **18**, 273–275
    noninverting amplifier, **19**
    unity-gain buffer, **19**
frequency response, 273
ideal, **17**, 249
input stage, 248, 592
intermediate stage, 249, 593
inverting/noninverting input, 17
offsets, 249, 252
open-loop gain, **518**
output stage, 249, 593–594
slew rate, 250
virtual ground, 534
wideband, **626**–627
Op amp, 741, 17
    ac analysis, 605–611
    circuit description, 599–602
    data sheet, 249–251
    dc analysis, 603–604
    frequency response, 273, 611–613
    input stage, 605–608
    intermediate stage, 608–609
    output stage, 609–611
Op amp, 3140
    ac analysis, 619–621
    circuit, 617–619
    data sheet, 254–255, 618–619
Open collector, TTL, **663**
Open-loop gain, **258**, 252, 518
Open-loop voltage gain, **252**
Operating point (Q point), **10**, 78, 295
Operating point stabilization, **299**–303
Operational amplifier. *See* Op amp.
Opposite-state catching, 719
Optical coupler, **105**
Optical isolator, **105**
OR circuit, 97. *See also* Digital circuit.
Oscillator, 516, 563–576
    Barkhausen criterion, **563**
    Colpitts, **569**
    crystal, **571**–573
        equivalent circuit, 571
        parallel resonance, 572
        piezoelectric effect, **571**
        series resonance, 572
    general resonant, **569**
    Hartley, **570**
    LC, 568–570
    Pierce, **573**
    RC phase-shift, 564–565
    RC Wien bridge, 566–567
Output protection, 605
Output resistance, **144**, 252, 343, 345, 347, 349, 353, 354, 355, 357, 358

972 INDEX

Output resistance, current source, 313, 316, 318
Output short-circuit current, **252**
Output short-circuit protection, 495
Output stage, 249, 593–594, 609
Overdamped response, 857
Oversampled, **803**
Oxide layer, **186**, 894–896

p-type semiconductor, **47**
Parallel ADC, **811**, 812–814
Parallel data transmission system, 734–736
Parallel-input/parallel-output feedback. *See* Voltage-shunt feedback.
Parallel-input/series-output feedback. *See* Current-shunt feedback.
Parasitic capacitance, 389–391, 395–398
Parasitic resistance, 389, 391–392, 398
Particle motion, 26–28
Passivation, **895**
Peak rectifier, **88**
Permittivity, **63**
Phase detector, 850–851
Phase margin, **559**
Phase plot, 261
Phase-locked loop (PLL), **853**–859
 data sheet, 861
 time constant, 856
Phasor, **9**, 340
Phonon, **35**, 40
Photo diode, **74**, 106
Photolithography, **892**–896
Photon, **35**
Photoresist, 893, **895**
Piecewise-linear model, 79, **90**
Piezoelectric effect, **571**
Piezoelectric materials, 571
Pinch resistor, 901–903
Planar technology, **879**
Planck's constant, **35**
pn junction, **59**, 61
pnp transistor, **124**, 911–913
Polycrystalline, **40**
Polysilicon gate, **916**
Positive logic, **640**
Power amplifier, 447–497
Power dissipation, 157–159
Power gain, **14**–15
Power supply, 88–90
Power transformer, 83
Precision op amp, **626**
Predeposition (buried layer), **909**
Prescaler, 865
Priority encoder, **710**–711
Programmable array logic (PAL), **781**
Programmable logic arrays (PLA), 778–780
Programmable ROM (PROM), 759
Propagation delay, **644**–645
Psuedostatic RAM, **751**, 774
Pull-up resistor, **641**
Pulse-width modulation, **135**, 838
Punch-through, **135**
Push-pull operation, 478

Quantization accuracy, **804**
Quantization noise, **804**
Quantizing error, **801**
Quiescent point (Q point), **10, 78,** 295

Race condition, **718**–719
RAM (random-access memory), **749**, 750
 dynamic (DRAM), **751**, 759–774
 dynamic cell, 769–770
 dynamic chip organization, 773
 refresh operation, **759**, 772
 static, **751**, 754, 764–768
  data sheet, 796–799
 static address buffer, 767
 static address decoder, 768
 static chip organization, 765
 static column, 766–767
 static MOS cell, 765–766
 substrate bias, 768
Read operation, 755–756
Read-only memory (ROM), **750**, 751–752, 757–764
Read/write memory, **750**, 751
Read/write operation, 755–756
Recombination, **42**
Rectifier circuits, 83–99
 bridge, **86**
 full-wave, **85**
 half-wave, **83**
 precision half-wave, **85**
Reference source, **832**, 834
Reference voltage, 823
Refresh, **769**
Registers, **722**–724
Regulation, *See* Voltage regulation
Relaxation time, **27**
Resistance of doped layers, 889–892
Resistivity, **29**
Resistor, diffused, **898**
Resistor parameter summary, 903
Resistor-transistor logic (RTL), **641**–650
Resolution (ADC), **801**
Retriggerable multivibrator, **730**
Reverse alpha (α_R), **130**
Reverse beta (β_R), **137**
Reverse diode current, **71**
Reverse-biased junction, **70**
Reverse-recovery time, **102**
Ripple counter, **725**
Ripple voltage, **88**
Rise time, **162**–164
Rise time, MOSFET, 674–675
ROM cells, 759–764
RTL (resistor-transistor logic), **641**–650
 basic gate, 641
 fan-out, 643, 647–649
 noise margin, 642–**643**, 650
 NOR gate, 647
 transfer characteristic, 641–642, 650

Sample-and-hold circuit, **803**
Sampling theorem, **803**

Saturated CE voltage, **149**
Saturation, **72**
Saturation in CE, 142–146
 Saturation region, BJT, **131**
  MOSFET, 191
  JFET, 213
Scatter-limited velocity, 189
Schottky diode, **104**
Schottky IIL, 693
Schottky transistor, **164, 658**
Second-order loop, **857**–859
Self-aligned gate, **916**
Semiconductor, **29**
Sequential logic, **703**, 715–728
Serial data transmission system, **738**–742
Serial memory, 723, **749**–750
Series resonance, 572
Series-input/parallel-output feedback. *See* Voltage-series feedback.
Series-input/series-output feedback. *See* Current-series feedback.
Series-pass voltage regulator, **830**–833
Series-series feedback. *See* Current-series feedback.
Series-shunt feedback. *See* Voltage-series feedback.
Servo-type ADC, **811**, 820–822
Semi-insulating, **43**
Set-reset latch, 715
Settling time, **811**
Setup time, **716**, 720
Seven-segment display, 708
Sheet resistance, **890**
Shift register, **722**–724
Shift register, CCD, 775–776
Short circuit protection, 495
Shunt-series feedback. *See* Current-shunt feedback.
Shunt-shunt feedback. *See* Voltage-shunt feedback.
Signal processing and imaging, 777
Silicon, 39–48
Silicon-controlled rectifier (SCR), **835**
Silicon-controlled switch (SCS), **835**
Silicon-on-insulating substrate (SOI), 923
Silicon-on-sapphire (SOS) structure, 923
Simple current source, **311**–314
Simulation program (SPICE). *See* SPICE.
Single-slope ADC, **816**–818
Slew rate, 250
Small-signal condition, **12**, 81, 82, 143
Solar cells, **106**
Source-coupled pair, **453**–463
Space-charge layer, **63**
Space-charge region, **63**
SPICE, **108**
 control commands
  ac analysis, 957–958
  dc analysis, 957
  operating point statement, 958
  transient analysis, 958
 default parameters

BJT, 166, 955
diode, 109, 954
JFET, 229, 927
MOSFET, 206, 956
element statement, **949**
BJT, 165
capacitor, 950
dependent source, 952–953
diode, 108–112
independent source, 951–952
piecewise-linear statement, 952
pulse definition statement, 951
sinusoidal definition statement, 951
inductor, 950
JFET, 228
MOSFET, 205
resistor, 949
end statement, 957
initial conditions, 950
model statement, **949**
BJT, 165, 394
diode, 109–112, 953
JFET, 229, 396–398
MOSFET, 207, 394
netlist, 948
node, **949**
plot statement, 957
print statement, 957
schematic-capture, **948**
sensitivity statement, 167–168
temperature analysis, 168–169
title statement, 957
transfer function command, 958
transient analysis, 958
transient statement, 958
Spikes, **656**
Stability factor, 313, 317
Standard cells, 778
Start-stop timing, **739**
Static RAM, **751**, 754, 764–768
Storage time, BJT, **164**
Storage time, diode, **102**
Straggle, **887**
Strobe input, **704**
Substitutional impurities, **881**
Summing-point constraints, **18**
Surface potential, **360**
Switch debounce, 716
Switched-mode power supply (SMIPS), **837**-842
Switching regulator, **837**–842
Switching speed, 644–646, 656
Switching time, **162**–164
Synchronous counter, **727**

Temperature compensation, 667
Temperature effects
beta, 156–157
I_{CBO}, 147
junction current, 74

V_{BE}, 153
V_P, I_{DSS}, 309
Temperature stabilization, 319–320
Thermal generation current, **74**
Thermal resistance, **107**, 157, 160
Thermal voltage, **32**
Thermistor, **828**
Thermocouple, **828**
Thévenin, **12**
Thin films, **890**
Three-dB (3–dB) point. *See* Corner frequency *and* Bode plot.
Thyristor, **835**
Time constant, 94
Time-division multiplexing (TDM), 706, **823**
Timer, 732–733
Timing diagram, 755–756
Toggle, **718**
Totem-pole output stage, **653**
Tracking range, **859**
Transconductance,
BJT, **144**
MOSFET, **196**
Transconductance amplifier, **535**
Transfer characteristic, **59**, 83
Transfer function, 258, **261**, 854–858
Transient response, 111–113
Transistor. *See* BJT, JFET, *or* MOSFET
Transistor array, 589–591
Transistor-transistor logic (TTL), **650**–664
Transition capacitance, 99–100. *See also* Depletion capacitance.
Transition frequency, **393**–394
Transition region, **11**, 641
Transition time, **102**
Transmission lines, 735
Transparent latch, **722**
Transresistance amplifier, **519**
Tristate output, **664**
Truth table, **704**
TTL (transistor-transistor logic), 650–664
advanced low-power Schottky, 662
advanced Schottky, 661–662
basic gate, 653
data sheet, 661
fan-out, 656
low-power Schottky, 660–661
NAND, 653
noise margin, 656
open collector, **663**
output characteristics, 662
Schottky, 658–660
speed, 656, 663
transfer characteristic, 653–656
tristate output, **664**
wired-AND, **663**
Tunneling, **761**
Two-port analysis
current-series feedback, 536–537
current-shunt feedback, 551–552

voltage-series feedback, 545–546
voltage-shunt feedback, 521–523

Unbalanced line, **635**-636
Unbypassed emitter resistor, 344
Unbypassed source resistor, 353
Uncovered charge, **63**
Undercut etch, **986**
Underdamped response, 857
Unilateral network, **523**
Unity-gain frequency, **393**–394, 396
Universal shift register, 724

Valence band, **36**
Varactor diode, **100**
Varicap, **100**
Vertical MOSFET (VMOS) structure, 920–922
Virtual ground, 534
VLSI, **637**
Volatile data, **751**
Voltage
amplification, **342**, 344, 346, 349, 352, 354, 355, 356, 358
amplifier, **544**
breakdown, **73**
doubler, 95
gain, **14**. *See also* Voltage amplification
reference, 823
regulation, **90**
regulator fixed/variable-output, **832**
regulator series-pass, **830**–837
regulator, IC, **832**
Voltage-controlled oscillator (VCO), **853**, 864–866
Voltage-series feedback, **544**–549
analysis, 545–546
cascaded amplifier, 546–549
input impedance, 545
output impedance, 545
Voltage-shunt feedback, **519**–534
analysis, 521–523
cascaded amplifier, 527–530
CE amplifier, 524–526
op amp, 531–534
input impedance, 520
output impedance, 521

Wideband op amp, **626**–627
Widlar current source, **314**–317
Width of space-charge layer, 66
Wilson current source, **318**, 324
Wired-AND, OR, **663**
Write operation, 756

y-parameter model, **521**

z-parameter model, **536**
Zener breakdown, **73**
Zero-order filter, **855**